SOLUTION FORMS OF A SECOND ORDER LINEAR NETWORK WITH CONSTANT INPUTS

General solution of the differential equation

$$\frac{d^2x}{dt^2} + 2\sigma\frac{dx}{dt} + \omega_n^2 x = F = \text{constant}$$

having characteristic equation $s^2 + 2\sigma s + \omega_n^2 = (s - s_1)(s - s_2) = 0$

$$s_{1,2} = -\sigma \pm \sqrt{\sigma^2 - \omega_n^2}$$

Case 1. Real distinct roots, i.e., $\sigma^2 > \omega_n^2$
(overdamped response for $\sigma > 0$)

$$x(t) = Ae^{s_1 t} + Be^{s_2 t} + X_f$$

$$s_{1,2} = -\sigma \pm \sqrt{\sigma^2 - \omega_n^2}$$

Case 2. Complex distinct roots, i.e., $\sigma^2 < \omega_n^2$
(underdamped damped response for $\sigma > 0$)

$$x(t) = e^{-\sigma t}(A\cos\omega_d t + B\sin\omega_d t) + X_f$$

$$= Ke^{-\sigma t}\cos(\omega_n t + \theta) + X_f$$

where notationally

$$s_{1,2} = -\sigma \pm j\omega_d \quad \text{and} \quad \omega_d = \sqrt{\omega_n^2 - \sigma^2}$$

Case 3. Real identical roots, i.e., $\sigma^2 = \omega_n^2$
(critically damped response for $\sigma > 0$)

$$x(t) = (A + Bt)e^{-\sigma t} + X_f$$

LINEAR CIRCUIT ANALYSIS

TIME DOMAIN, PHASOR, AND LAPLACE TRANSFORM APPROACHES

RAYMOND A. DeCARLO
Purdue University

PEN-MIN LIN
Purdue University

PRENTICE HALL, Englewood Cliffs, New Jersey 07632

Library of Congress Cataloging-In-Publication Data

DeCarlo, Raymond A.,
Linear circuit analysis : time domain, phasor, and Laplace transform
approaches / Raymond A. DeCarlo, Pen-Min Lin.
p.   cm.
Includes index.
ISBN 0-13-473869-1
1. Electric circuit analysis. 2. Electric circuits, Linear.
3. Laplace transformation. 4. Time-domain analysis. I. Lin, Pen-Min,
II. Title.
TK454.D44 1995
621.319'2—dc20   94-26845
                                              CIP

Publisher: Alan Apt
Development Editor: Sondra Chavez
Editor-in-Chief: Marcia Horton
Production Editor: Jennifer Wenzel
Marketing Manager: Gary June
Designers: Kenny Beck and Joanne Kaiser

Cover Designer: Jeannette Jacobs
Buyer: Lori Bulwin
Photo Editor: Lorinda Morris-Nantz
Photo Researchers: Chris Pullo and Melinda Reo
Supplements Editor: Alice Dworkin
Editorial Assistant: Shirley McGuire

©1995 by Prentice-Hall, Inc.
A Simon & Schuster Company
Englewood Cliffs, NJ 07632

TRADEMARK INFORMATION
Mathematica is a registered trademark of Wolfram Research, Inc.
MATLAB is a registered trademark of the MathWorks, Inc.
Theorist is a registered trademark of Allan Bandio Association
TTC is a registered trademark of Telecommunication Technology Corp.
PSpice is a registered trademark of MicroSim.

Printed in the United States of America

10 9 8 7 6 5 4 3 2 1

ISBN 0-13-473869-1

Prentice-Hall International (UK) Limited, *London*
Prentice-Hall of Australia Pty. Limited, *Sydney*
Prentice-Hall Canada Inc., *Toronto*
Prentice-Hall Hispanoamericana, S.A., *Mexico*
Prentice-Hall of India Private Limited, *New Delhi*
Prentice-Hall of Japan, Inc., *Tokyo*
Simon & Schuster Asia Pte. Ltd., *Singapore*
Editora Prentice-Hall do Brasil, Ltda., *Rio de Janeiro*

PREFACE

PRESENT AND PAST

For several decades in the United States, most electrical engineering undergraduates have taken two semesters of linear circuit analysis in their sophomore year. Those in other engineering disciplines often have taken only one semester. Standard topics include laws and techniques of resistive circuit analysis, time domain transient analysis of first and second order linear circuits, a phasor approach to sinusoidal steady state analysis, resonant and magnetically coupled circuits, transformers, elementary 3-phase circuits and some Fourier series analysis.

The late 70's and the decade of the 80's marked a time of revolutionary growth in undergraduate electrical engineering education. Discrete time system and computer engineering concepts grew into integral parts of the engineer's required graduation toolbox. Software programs such as PSpice®, MATLAB® and its toolboxes, Theorist®, Mathematica®, and a host of others appeared to streamline the computational drudgery of engineering analysis and design. Paralleling this were advances in research and technology. IC implementations of active filters and especially switched capacitor filters have become shelf items in supply houses. Matrix-based numerical algorithms for large scale circuit simulation, parameter optimization, layout, and design are widely available. Our research colleagues have dramatically deepened our ability to analyze, design, and control nonlinear circuits and systems.

APPROACHING THE YEAR 2000

In order to meet the needs of the next decade, our approach was to more efficiently package the traditional circuits courses in an up-to-date framework without sacrificing rigor. For example, we have introduced the matrix formulation of node and mesh equations for solution with software programs such as MATLAB® or its equivalent, and have added a special section on the Modified Nodal Method of circuit analysis. The ubiquitous presence of active circuits built around the op amp, prompted us to integrate op amps throughout the text. In support of this we have a more careful treatment of circuits containing controlled sources and op amps; in particular a careful restatement of the traditional Thévenin and Norton Theorems usually stated (rigorously) in many texts only for passive circuits. We have unified the treatment of RL and RC first order circuits and emphasized the use of the characteristic equation for the solution of second order circuits in contrast to the usual formula-development for each of the parallel and series cases. Classical phasor analysis is introduced in step-by-step fashion beginning with a review of its complex variable foundation. These ideas underlie our treatment of ac steady state power considerations and applications to power systems which we treat in the text.

Many texts introduce a notion of generalized phasor analysis. Our imperative was to replace this often ill-used approach with an early introduction to Laplace transforms without first covering the Fourier series and Fourier transform. This allows us to recast the usual second semester circuits topics (resonant and magnetic circuits, two ports, filters, etc.) with a systems flavoring underpinned on the foundation of the Laplace transform. For example, with resonant circuits we lead the student quickly through the ordinary material spiced with an application or two and then unify the ideas under the umbrella of a bandpass transfer function. In fact, when teaching this material we ordinarily begin with the bandpass transfer function so that the student sees the unity from the onset.

This menu allows us to expand the boundaries of traditional coverage without additional investment in time while simultaneously developing advanced skills: students are able to tackle both transient and steady state circuit analysis for any excitation with a rational Laplace transform besides the usual steps, ramps, exponentials, sinusoids and sinusoidally modulated exponentials for circuits of small to moderate order. This framework offers students an entire semester of problem-solving practice in the Laplace transform context, thereby significantly honing their analytical skills. This is not achievable in the more traditional setting for teaching the course. The approach also enables us to unify the frequency and time-domain approaches to circuit analysis in the systems context by covering continuous-time convolution. Here we introduce the basic definitions, the integration and graphical approaches, and show where it might be useful for circuit analysis.

With the Laplace transform, the concepts of impedance and transfer function are cleanly and rigorously introduced. Further, there is the potential for the incorporation of synthesis techniques into homework exercises otherwise precluded in the traditional treatment. In sum, this approach better prepares students for the rigors of the signals and systems course of the junior year and frees up time in the junior year for the exploration of discrete time systems concepts.

To better implement our approach we developed a software program entitled tfc which automates the entire approach to Laplace transform analysis. This, of course, can be complemented by MATLAB® commands which also accommodate Laplace transform analysis and convolution.

To allow flexibility in curriculum development and to introduce (nonlinear) electronic circuits at a very elementary level, we provide two optional chapters on piecewise linear analysis of diode and transistor amplifier circuits. The piecewise linear approach enables the analysis of many simple and useful electronic circuits via the techniques of linear resistive circuits. It permits us to address the growing need to expose students to nonlinear behavior in a simple understandable way.

Personal Reasons

All of the above reasons not withstanding, starting in 1986 and continuing to this day, we have had a genuine desire to significantly improve the content and complexion of the basic circuits courses. We wanted to tie circuits concepts to real world devices such as microwave ovens, stereo amplifiers, adapters for portable computers, etc. With the suggestions and help of our editor, we did this with chapter openers. Further, we wanted to present challenging homework exercises and to present the material in a way consistent with our perception of research and development over the last decade.

I. KEY CONCEPTS

1. Balanced Emphasis on Concepts and Calculation

Quality software programs numerically automate important aspects of circuit analysis and design, relieving engineers of tedious and often impossible hand calculations. However, numerical algorithms implemented as canned programs are no substitute for an understanding of the basic circuit principles and properties which govern circuit behavior nor for a firm understanding of the steps necessary for the solution of a problem, because from our perspective, there are five educational tasks of a circuits course:

1. Create an environment where the student has the opportunity to learn the basic vocabulary, principles, analysis methods and design techniques of circuit theory as gleaned from the accumulated experience of past engineers and physicists.
2. Create an environment for applying the principles of mathematics and physics to engineering problems.
3. Develop a qualitative understanding of circuit behavior.
4. Foster the development of rational thinking patterns in the context of circuit analysis and design in order to prepare students for solving a broad variety of engineering problems.
5. Help develop the student's ability to critically evaluate their chosen problem-solving technique and the accuracy of their answer.

In order to meet these needs we have included some advanced sections marked with an asterisk (*) and a variety of problems and exercises from the simple to the difficult. (The more difficult exercises are also flagged with an asterisk.) Some problems apply the ideas of the text while a few enhance or extend the ideas described in the text. Although numerical approaches and problems are included throughout the text, we have included two appendices, one for each of the traditional semesters of circuits, on numerical problems for exploration by the student. These use software packages such as MATLAB®, tfc (a supplement to this text and described earlier), and SPICE. Of course, there are many other programs that can be substituted for these.

We believe that using such programs not only allows the student to more easily calculate numbers but reinforces the delineated properties of circuit behavior covered in the text. On the other hand, the principles and properties developed throughout the text allow the students to assess the reasonableness and accuracy of answers computed using a software program. Software programs are not infallible and it is always possible to construct an example which will cause a program to fail to produce meaningful numbers.

2. Text Exercises and Homework Problems

The text contains a wide variety of analysis, design, and computer-oriented problems. We have tried to provide a balance between problems emphasizing concepts and those emphasizing the mechanics of computing solutions. Design problems and software assisted problems are marked in the text with ◆ and ◆ respectively.

3. Chapter Openers/Real World Applications

The book includes a variety of real world applications that introduce and motivate almost every chapter in the book. Students immediately preview how circuits concepts underlie a wide variety of applications. Some of these are:

- **(a)** Car Heater Fan Speed Control (Chapter 2)
- **(b)** Digital-to-Analog Converter (Chapter 3)
- **(c)** Protection Circuits Against Overvoltage and Polarity Reversal (Chapter 6)
- **(d)** Capacitive Voltage Regulator (Chapter 7)
- **(e)** Sawtooth Waveform Generation (Chapter 8)
- **(f)** Microwave Oven (Chapter 9)
- **(g)** Capacitive-bridge Pressure Sensor (Chapter 10)
- **(h)** Fluorescent Light (Chapter 14)
- **(i)** DC Motor (Chapter 15)
- **(j)** Averaging by a Finite Time Integrator Circuit (Chapter 16)
- **(k)** How a Touch Tone Phone Signals the Numbers Dialed (Chapter 17)
- **(l)** Rectifier Circuit (Chapter 22)

Each of these openers is discussed in a simplified form either within a section of the associated chapter or as a chapter problem. In addition, we have included career boxes as an added feature so students will be better informed about career opportunities in the field.

4. In-depth Coverage

The traditional topics are covered in a more modern format, utilizing available software for problem solving where necessary. As mentioned in the Preface, we introduce a matrix-based approach to nodal and mesh analysis and include a special section on the Modified Nodal approach to circuit analysis. Teamed up with MATLAB® or its equivalent, the student has a powerful analysis tool not directly available to engineering students of past decades. (A MATLAB® supplement which expands upon the numerical implementation of the course material is available through Prentice Hall.) Op amp coverage is integrated throughout the text. Rigorous treatments of Thévenin and Norton theorems for active circuits are given. The Laplace transform is introduced after phasor analysis and utilized through the remainder of the text. This allows a unifying systems approach to impedance concepts, circuit transfer functions, convolutional approaches to circuit analysis, resonance, magnetic circuits, two ports and filtering. The software program, tfc, for use on a PC is available from Prentice Hall to enhance the Laplace transform analysis required in the text. The program computes roots, Laplace transforms, partial fraction expansions, and inverse Laplace transforms; it also has a provision for the manipulation of two-port parameters. Fourier series with applications to power supplies and distortion in an amplifier ends our treatment.

5. Review and Development of Text

Extensive review of both volumes and extensive class testing of Volume II.

6. Chapter Pedagogy

- **(a)** Each chapter begins with an overview of chapter contents and ends with a summary and a glossary of terms and concepts used throughout the chapter.
- **(b)** Throughout each chapter "key concept boxes" highlight important definitions, laws, and properties of circuit analysis and design.
- **(c)** Many examples illustrate concepts and techniques. Each example is set off with its conclusion clearly marked.
- **(d)** More advanced sections and problems are marked with an asterisk for students desiring more challenging material.

7. Presentation of Piecewise Linear Analysis

Presentation of piecewise linear analysis of diode and transistor amplifier circuits as a first step in approaching nonlinear circuits: This material is optional and may be omitted without any loss of continuity.

8. Software for Laplace Transform Analysis

Free software for Laplace transform and two-port analysis for IBM PC or equivalent is available from the publisher.

9. Instructor's Manual

An Instructor's Manual is available from the publisher.

10. Matlab® Supplement

Jim Gottling of the Department of Electrical Engineering of The Ohio State University has put together a Matlab® supplement that can be used in conjunction with the text.

II. HOW TO USE THIS BOOK

This is the only circuits book available in one comprehensive book and two separate volumes. The purpose is to give instructors and students some cost effective options.

1. Volume Selection

Volume I best fits a one semester course primarily for those not taking a second course with a price tag much lower in cost than competing two semester texts. Volume II is for those taking only the second course or for those using a different text for the first course. Alternately, one can choose a combined volume for those taking a two semester sequence or equivalent.

As mentioned earlier, most electrical engineering sophomores take two semesters of linear circuit analysis. Those in other engineering disciplines often take only one semester. The two-volume book contains an abundance of material that can be used in a variety of ways. An instructor can choose the proper ingredients for the intended course. Some possible plans are given in the following table:

	Course length	Intended Course Structure	Chapters used
1.	Two semesters (3 credit hours each)	Traditional linear circuit analysis	Semester 1 vol. 1, with ch. 6 and 12 omitted Semester 2 vol. 2, ch. 13 to 20, plus ch. 21 or ch. 22
2.	Three quarters (2 credit hours each)	Traditional linear circuit analysis	quarter 1 vol. 1, ch. 1, 2, 3, 4, 5, 7, 8 quarter 2 vol. 1, ch. 9, 10, 11, and vol. 2, ch. 13, 14, 16, 17 quarter 3 vol. 2, ch. 18 to 21
3.	One-semester (4 credit hours)	Introductory linear circuit analysis and Elementary electronic circuits	vol. 1, ch. 1 through 12
4.	One-semester (3 credit hours)	Introductory linear circuit analysis	vol. 1, with ch. 6 and 12 omitted
5.	One-semester (3 credit hours)	Laplace transform analysis of linear circuits	vol. 2, ch. 13 through 21

For the first semester of option 1 or for option 4, the lecture by lecture topics can be chosen as:

1	General circuit element, charge, current
2	Voltage, sources, power
3	Resistance, Ohm's Law, power reprise
4	Kirchhoff's Laws, single loop, node circuits
5	R combinations, v & i division
6	Op amp basics
7	Nodal analysis
8	Nodal analysis (cont.)
9	Mesh analysis
10	Superposition and linearity
11	Linearity (cont.) & source transformations
12	Review
13	**TEST #1**
14	Thévenin's and Norton's Theorems
15	Thévenin's and Norton's Theorems (cont.)
16	Inductance
17	Capacitance
18	L and C combinations, duality
19	First order undriven circuits: RL case and RC case; use of R_{th}; initial conditions from past dc excitations; sequential switching
20	First order undriven circuits (continued); recovery of stored energy; differential equation, and mathematical solution for first order circuits with constant inputs
21	RC emphasis; examples/applications
22	Unit step function; DC response of first order circuits; the 3-parameter formula; exponential decay and growth curves; inductive and capacitive circuit examples; piecewise constant inputs
23	Transient analysis of first order circuits (cont.); unit step, natural and forced response; solution to real exponential input; integral solution for arbitrary inputs
24	Switching & elapsed time calculation; negative time constant and stability
25	RC op-amp circuits
26	Differential equation models of parallel and series RLC circuits; derivation of solution for the undriven case
27	Review
28	**TEST #2**
29	Solution of 2nd order differential; equation; models for undriven circuits continued: overdamped, critically damped, and underdamped cases
30	Solution of 2nd order differential; equation; models for circuits driven by constant input excitations
31	Sinusoidal forcing function
32	Complex forcing function
33	Phasors: KVL & KCL with phasors
34	Impedance & admittance
35	Sinusoidal steady state (SSS) analysis using phasors; general case requiring (matrix) mesh and node equations
36	SSS phasor diagrams; examples of applications
37	Frequency response: magnitude and phase plots; lowpass and highpass examples
38	Instantaneous power absorbed by a general 2-terminal element and special cases of R, L, and C; average power over a specified time interval; average power when $p(t)$ is periodic; sinusoidal steady state average power
39	Maximum power transfer in SSS; adjustable impedance/resistive load; application in communication circuits

40	Review
41	**TEST #3**
42	Effective (rms) value; definition and examples of sine and triangular waves; counter-example to superposition of power; RMS value of response when all inputs are at different frequencies

43	Complex power, VA; reactive power, var conservation of Power
44	Balanced 3-phase circuits; economic consideration in power transmission
45	**FINAL EXAM**

For the second semester of option 1 or for option 5, the lecture-by-lecture topics can be chosen as:

1	Motivation for studying Laplace transform
2	Laplace transform, transforms of basic signals
3	Inverse transform, partial fraction expansion

4	Basic properties
5	Solution of linear DE's with initial conditions
6	Z(s),Y(s), series-parallel & other manipulations

7	Transfer function
8	Equivalent circuits for L and C with I.C.
9	Nodal and mesh analyses in s-domain

10	Switching in linear circuits
11	Switched capacitor circuits
12	H(s),poles, zeros, s-plane plot and stability

13	**Test #1**
14	Decomposition of the complete response
15	Sinusoidal steady state analysis

16	Frequency response from pole-zero plot
17	Impulse response, h(t); relation to H(s); additional Laplace properties and review
18	Convolution, definition, and integral evaluation; relationship to L.T.; time domain derivation

19	Circuit response calculation using convolution; graphical convolution; additional properties
20	Additional properties; convolution algebra;
21	SPICE, and tfc

22	Parallel resonance, calculation of Ωr
23	Parallel resonance, exact analysis
24	Other resonance forms; approximate analysis

25	Pole-zero approach to resonant circuit analysis
26	Frequency and amplitude scaling
27	Mutual inductance, the dot convention

28	Mutual inductance: s-domain and phasor equations
29	**Test #2**
30	Coefficient of coupling stored energy

31	Ideal transformer
32	Coupled inductors modeled with ideal transformer
33	One-port networks

34	y parameters
35	Equivalent circuit & analysis of terminated 2-port
36	z parameters

37	h parameters
38	t parameters; reciprocity; conversion table
39	2-port interconnections

40	Test #3
41	Indefinite admittance parameters
42	Lowpass Butterworth response; loss in dB; from $|H(j\omega)|$ to $H(s)$
43	Calculation of filter order and cut-off frequency; basic passive realization 2nd order filters
44	Basic active realization of 2nd order filters

For non-electrical engineering majors, spending three semester credit hours in linear circuit analysis, use Volume I with Chapters 6 and 12 omitted. For non-electrical engineering majors, spending four semester credit hours in linear circuit analysis and basic electronic circuits, use Volume I in its entirety.

Of course, many other arrangements are possible. The ingredients are there. An instructor can choose a proper combination for the intended course. For example, if Fourier Series analysis needs to be covered, then other topics (such as two-port interconnections) may be omitted.

2. Special Notes on Terminology

Although the Institute of Electrical and Electronic Engineers (IEEE) has a standard dictionary for electrical terms, the engineering literature does not always adhere to the standards. In one sense, students need to know the standard terminology and also the variations which is what they will encounter when they work in the real world. This variation also shows up in the book in some places. For example, we have used the following terms synonymously:

node	terminal
circuit	network
dependent source	controlled source

Although the Standard International unit for conductance is Siemens, the term "mho" (ohm spelled backward) is still widely used in engineering literature and manufacturers' data sheets. This book used the term mho and the symbol inverted Ω as the unit for conductance.

ACKNOWLEDGMENTS

A work of this magnitude has input from many people over many years. We are particularly indebted to our colleagues Mark Wicks, Gerry Heydt, LeRoy Silva, Chin-Lin Chen, Jim Krogmeier, George Wodicka, Barrett Robinson, Phil Peleties, and Tan-Li Chou who have added in various ways to the quality of the text, but in no way are responsible for any of its shortcomings. These, of course, rest solely with the authors.

We would also like to acknowledge the many teaching assistants of EE-201 and EE-202 at Purdue University who have contributed corrections and suggestions to the many chapter exercises and homework problems throughout. Our secretaries, Cathy Tanner and Nicky Danaher, who have helped with the mechanics and proofing of the text deserve a special note of thanks. They freed us from the many mailings and correspondences that allowed us to focus on the writing. Also Jean Jackson, who helped set up guidelines for color usage in the figures.

We would like to thank all of the reviewers listed below, especially Irwin Sandberg and Peter Aronhime, whose critiques have helped us improve the text in many different areas.

Peter Aronhime - University of Louisville
John A. Fleming - Texas A & M University
Edwin W. Greeneich - Arizona State University
James V. Krogmeier - Purdue University
Derek Lile - Colorado State University
Charles P. Neuman - Carnegie Mellon University
Burks Oakley II - University of Illinois at Urbana-Champaign
Edgar A. O'Hair - Texas Tech University
Irwin W. Sandberg - University of Texas at Austin

We also must acknowledge our editor Alan Apt for his many ideas and suggestions, our development editor Sondra Chavez for her coaxing and coaching, our production editor Jennifer Wenzel for her patience and skill in dealing with the many tribulations of production, and the other PH staff who have either quietly or indirectly helped to shape the face (quite literally) of this text.

Our friends from industry who have supplied technical information on products and photos for use as chapter openers, Bill Lorenz of AlliedSignal and Howard Babbitt III of The Mitre Corporation must also be mentioned. And Mr. Thackery, W3IU, who showed one of the authors the magic of ham radio.

We would also like to thank all those of the School of Electrical Engineering at Purdue University who encouraged us to revise and update the content of our basic circuits courses. The update led to a set of notes and homework exercises that in turn led to the volumes.

More importantly, we must thank our wives, Chris and Louise, who have supported us throughout this eight-year project. Last but not least we thank our attorney, Scott Sullivan, who has guided us through several skirmishes and one of the authors apologizes for all the lawyer jokes he has told over the last eight years.

Dedication

To our wives
Chris and Louise,
and to
Frank and Adele DeCarlo

CONTENTS

Note: This book is also available in individual volumes entitled, *Linear Circuit Analysis: Time Domain and Phasor Approach Volume One (Chapter 1–12 and Appendices A1–A3)* and *Linear Circuit Analysis: A Laplace Transform Approach Volume Two (Chapters 13–22 and Appendices B1–B3)*

Chapter 11 SINUSOIDAL STEADY-STATE POWER CALCULATIONS   **326**

Chapter 12 PIECEWISE LINEAR ANALYSIS OF TRANSISTOR   **368**
AMPLIFIER CIRCUITS

Chapter 13 LAPLACE TRANSFORM ANALYSIS, 1: BASICS   **400**

Chapter 14 LAPLACE TRANSFORM ANALYSIS, 2:   **442**
CIRCUIT APPLICATIONS

CHAPTER 1

CHAPTER OUTLINE

Introduction and Basic Concepts

L OOK around your home or apartment. Electrical energy provides light. An electric motor runs the compressor in the refrigerator, air conditioner, and the pump in the dishwasher. Electricity supplies energy to a special resonator in the microwave oven to heat food quickly. The radio, TV, and stereo operate by channeling electrical energy through special circuits so that we can be entertained. The computer contains millions of small circuits that allow us to solve complex engineering problems and write books, such as this one, with interesting examples. The battery in a car drives an electric motor that starts the engine. These devices operate according to the basic principles of circuit theory and this text is about circuit theory. ◆

◆ CHAPTER OBJECTIVES

- Introduce and investigate three basic electrical quantities—charge, current, and voltage—and the conventions for their reference directions.
- Investigate energy and power conversion in electric circuits and demonstrate that these quantities are conserved.
- Define independent and dependent voltage and current sources that act as energy or signal generators in a circuit.
- Classify memoryless circuit elements by their terminal voltage-current relationships.
- Explain Ohm's law for a resistor.
- Investigate power dissipation as heat in a resistor.
- Explain the difference between a device and its circuit model.

1. INTRODUCTION AND BASIC CONCEPTS

What Are Circuits?

Circuits are energy or signal/information processors consisting of interconnections of "simple" elements or devices. The processing of energy or information takes place through various time signals called *voltages* and *currents*. An element called a *source* will generate a voltage or a current representing some type of information. An interconnection of circuit elements will process this signal into a new voltage or current.

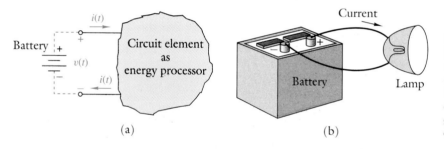

Careers in Electrical Engineering

It is sometimes difficult for students to see how introductory circuits leads to careers in the field. The Career Boxes that will be included in some of the chapters will let you know about related career opportunities in electrical engineering that are exciting and challenging.

When choosing an area of specialization, remember that salary considerations can be of secondary importance when compared to: doing a job you enjoy, being in a geo-

graphic location you like, being in a comfortable corporate culture, and working for a firm that produces products with which you are in ethical agreement.

Courses like introductory circuits are really entrees into an exciting world of almost limitless opportunities. We hope these Career Boxes will give you a sense of how this course is a means to a fascinating end.

Five common circuit elements are the *battery*, the *resistor*, the *capacitor*, the *inductor*, and the *operational amplifier*. Using these and a number of other elements, this text develops an understanding of basic circuits and simultaneously constructs a set of tools essential for analyzing and designing circuits. Because many engineering disciplines, such as bioengineering and mechanical systems, have circuit analogies, the tools we develop will have a very general applicability.

Our trek begins with the introduction of an arbitrary two-terminal circuit element, illustrated in figure 1.1. As stated, a circuit element can be thought of as an energy processor at its most general level and as an information or signal processor at a practical level. In processing energy, the circuit element either absorbs energy, stores energy, delivers energy, or converts energy from one form to another. It processes the energy by modifying the current waveform or signal $i(t)$ and the voltage signal $v(t)$.

Figure 1.1 (a) General circuit element (connected to a battery) as an energy or signal processor; $v(t)$ is the voltage developed across the element, and $i(t)$ is the current flowing through the element. (b) A practical example of a general circuit element (a car headlight) connected to a car battery.

Figure 1.1 shows a standard voltage-current labeling of a two-terminal general circuit element. As will be seen, the voltage causes the current $i(t)$ to flow through the circuit element or device. Voltage is analogous to water pressure, while current is analogous to water flowing through a pipe. A device specifies a relationship between the voltage $v(t)$ and the current $i(t)$. For example, if the device is an ideal resistor, then $v(t) = Ri(t)$, where R is a constant of resistance. Such relationships specify whether power, and thus energy, is absorbed, stored, delivered, or converted into another form. For example, one can think of the heating element in an electric oven as a circuit element called a *resistor*. The resistor absorbs the electrical energy and converts it into heat energy, which cooks, among other things, turkey dinners.

2. FIELDS, CHARGE, AND CURRENT

Fields

Imagine a ball dropping from your hand to the ground. Falling is the effect of letting go of the ball. The ball falls due to the force of gravitational attraction between the ball and the earth. Because the mass of the ball is minuscule compared to the mass of the earth, it is convenient to think of the origin of the force in another manner: The earth creates a *gravitational field*, and the field exerts a force on any object placed within its boundaries. The force on our ball is proportional to the mass of the ball and to the strength of the field. The ball moves with the direction of the field. This everyday phenomenon, as we will see, parallels the behavior of current and voltage in a circuit.

The concept of a *field* is common to all science and engineering. Of interest to electrical engineering and circuit theory are **electric fields**, **magnetic fields**, and **electromagnetic fields**. All fields have two important properties: **magnitude** (field strength) and **direction**. Everyday experience points this out for the gravitational field. For a magnetic field, the classic experiment of a bar magnet held beneath a thin piece of rigid paper sprinkled with iron filings demonstrates the direction and strength of the field: The filings move into patterns called *lines of force* or *flux lines*.

Some fields change direction and magnitude with time. A changing electric field produces a changing magnetic field and conversely. Electric motors and electric generators exploit this interrelationship. A changing electric field at a sufficiently high frequency will radiate energy into the atmosphere from an antenna and is the basis of radio communication.

Charge

Charge is an electrical property of matter. Matter consists of atoms. Naively speaking, an atom is a nucleus of positively charged protons and uncharged (i.e., neutral) neutrons surrounded by a cloud of negatively charged electrons. The basic unit of charge is the **coulomb**, denoted by uppercase C. The accumulated charge on 6.2415×10^{18} electrons equals -1 coulomb. The charge on an electron is -1.6019×10^{-19} C. Two equally charged particles 1 meter apart in free space have charges of 1 coulomb each if they repel each other with a force of $10^{-7}c^2$ newtons, where $c = 3 \times 10^8$ m/sec. The force is attractive if one particle has a positive charge and the other has a negative charge. The force of attraction or repulsion between two charged bodies is inversely proportional to the square of the distance between them (assuming that the dimensions of the bodies are very small compared with the distance). Notationally, Q will denote a fixed charge and q or $q(t)$ a time-dependent or time-varying charge.

> *Exercise.* **1.** How many electrons have a combined charge of 53.406×10^{-12} C?
> **ANSWER:** 333,333,333.
> **2.** Sketch the charge profile $q(t) = 3(1 - e^{-2t})$, $t \geq 0$, that is present on a plate.

A **conductor** is a material, usually a metal, in which electrons can move to neighboring atoms with relative ease. By contrast, **insulators** oppose easy electron movement. An **ideal conductor** offers zero resistance to electron movement, while an **ideal insulator** offers infinite opposition to electron movement. Wires are assumed to be ideal conductors of current, unless otherwise indicated. Metals, carbon, and acids are common conductors. Common insulators include dry air, dry wood, porcelain, glass, and rubber. Copper wire is probably the most common conductor.

Current

When there is a net flow of charge across any area, we say that there is a **current** across the area. Quantitatively, the movement of 1 coulomb of charge through a cross section of a conductor in 1 sec produces an electric current of 1 ampere, denoted by A. In other words, the ampere is the basic unit of electric current and equals 1 coulomb/sec. The direction of current flow, however, is taken by convention as opposite to the direction of electron flow, as illustrated in figure 1.2. This is because our knowledge of electric current preceded our understanding of the atomic structure of matter: Early in the history of electricity, the assumption that current was the movement of positive charges, as illustrated in figure 1.3, was widespread.

As our understanding of device physics advanced, we learned that in ionized gases, in electrolytic solutions, and in some semiconductor materials, the movement of positive charges constitutes part or all of the current, while in metallic conductors, the current consists solely of the movement of electrons. Figures 1.2 and 1.3 both depict a current of 1 A flowing from the left to the right. In circuit analysis, we do not distinguish between these two cases: Each is represented symbolically by figure 1.4a. The arrowhead serves as

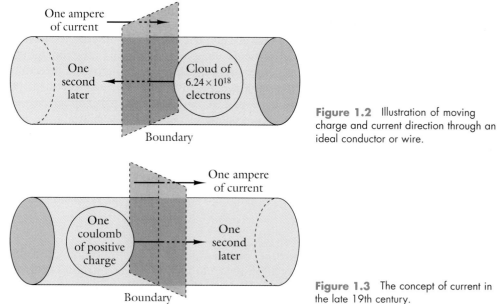

Figure 1.2 Illustration of moving charge and current direction through an ideal conductor or wire.

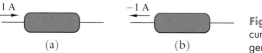

Figure 1.3 The concept of current in the late 19th century.

a reference for determining the true direction of the current. The current is in the same direction as the arrow if its value is positive and in the opposite direction of the arrow if its value is negative. In other words, in each of figures 1.4a and 1.4b, the current is 1 A flowing from left to right.

Figure 1.4 In both (a) and (b), 1 A of current flows from left to right through the general circuit element.

Two situations dictate the need to allow a negative value for the current:

1. The current actually reverses its direction from time to time, as in the case of ac current, to be described shortly;
2. Even though the current does not change direction, its direction is not known *a priori*.

One arbitrarily assigns a reference direction, indicated by an arrow, for the current. Subsequent analysis of the circuit leads to a positive or negative value for the current which determines its true direction. The following example will further clarify these conventions.

◆ EXAMPLE 1.1

Figure 1.5 shows a slab of semiconductor material in which

1. positive charge carriers move from left to right at the rate of 0.2 C/sec and
2. negative charge carriers move from right to left at the rate of 0.48 C/sec.

(a) Find I_a and I_b.
(b) Describe the movement of the charge carriers on the wire at the boundaries A and B.

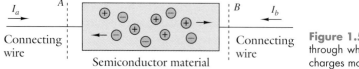

Figure 1.5 A semiconductor material through which positive and negative charges move.

SOLUTION

Part (a). The current from left to right due to the movement of the positive charges is 0.2 A. The current from left to right due to the movement of the negative charges is

0.48 A. Therefore, I_a, the *total* current from left to right, is $0.2 + 0.48 = 0.68$ A. Since I_b is the current from right to left, its value is then -0.68 A.

Part (b). The wire is a metallic conductor in which only electrons move. Therefore, at the boundaries A and B, negative charges (carried by electrons) move from right to left at the rate of 0.68 C/sec.

If a charge (more precisely, *net* charge) Δq crosses a boundary in a short time frame of Δt seconds, then the approximate current flow is

$$I = \frac{\Delta q}{\Delta t}, \tag{1.1}$$

where I in this case is a constant. The instantaneous current flow as a function of time is the limiting case of equation 1.1, i.e.,

$$i(t) = \frac{dq(t)}{dt}, \tag{1.2}$$

where $q(t)$ is the amount of charge that has crossed the boundary in the time interval $[t_0, t]$.

Of course, the integral counterpart of equation 1.2 is

$$q(t) = \int_{-\infty}^{t} i(\tau)d\tau. \tag{1.3}$$

 EXAMPLE 1.2

The charge crossing a boundary in a wire varies as $q(t) = [1 - \cos(120\pi t)]/120\pi$, for $t \geq 0$. Determine the current flow, and then verify equation 1.3.

SOLUTION

Since current flow is simply the time derivative of $q(t)$, for $t \geq 0$,

$$i(t) = \frac{dq(t)}{dt} = \sin(120\pi t).$$

For this current, equation 1.3 implies that

$$q(t) = \int_{0}^{t} \sin(120\pi\tau)d\tau = \left. -\frac{\cos(120\pi\tau)}{120\pi} \right|_{0}^{t} = \frac{[1 - \cos(120\pi t)]}{120\pi},$$

as expected.

 EXAMPLE 1.3

Determine the total charge flowing through a cross section of a conductor if the current through the conductor is given by the waveform of figure 1.6.

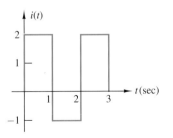

Figure 1.6 Square wave current signal.

SOLUTION

From equation 1.3, the total charge, say, Q, is given by the integral of the current over the time interval $[0, \infty]$. By simply summing the area under each part of the current waveform, by inspection, the total charge is

$$Q = \int_0^\infty i(\tau)d\tau = 2 - 1 + 2 = 3 \text{ C}.$$

> **Exercise.** If the current flow through a cross section of conductor is $i(t) = \cos(120\pi t)$ for $t \geq 0$ and 0 otherwise, determine $q(t)$ for $t \geq 0$.

If the force that moves charge along a wire is constant, then the rate of charge transferred is constant; i.e., $dq/dt = I$ is constant. Such a current is called **direct current** (abbreviated **dc**). Figure 1.7a shows a graphical representation of a direct current. Figure 1.7b shows an **alternating current** (abbreviated **ac**), which means sinusoidal, i.e., of the form $A \sin(\omega t + \phi)$, where A is the peak magnitude, ω is the angular frequency, and ϕ is the phase angle of the sine wave. With alternating current, the instantaneous value of the waveform changes regularly through negative as well as positive values. Thus, the direction of the current flow changes regularly, as indicated by the $+$ and $-$ values of the current. Lastly, figure 1.7c shows a current that is neither dc nor ac, but that nevertheless will appear in later circuit analyses.

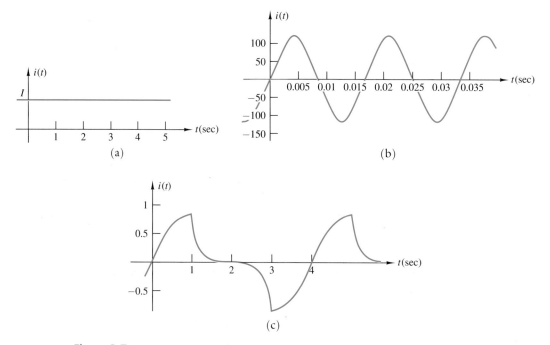

Figure 1.7 (a) Direct current or dc: $i(t) = I_0$. (b) Alternating current or ac: $i(t) = 120 \sin(120\pi t)$. (c) Neither ac nor dc current.

Because the value of an ac waveform changes with time, ac is measured in different ways. Suppose the instantaneous value of the current at time t is $K \sin(\omega t + \phi)$ evaluated at the proper time. The term **peak value** refers to K in $K \sin(\omega t + \phi)$. The **peak-to-peak value** is $2K$. Another measure of ac current, indicative of its heating effect, is called the **root mean square** (rms), or **effective value**, of the current, which is related to the peak value by the formula

$$\text{rms} = \frac{\sqrt{2}}{2}\text{peak value} = 0.7071K. \tag{1.4}$$

A derivation of equation 1.4 will be given in chapter 11.

A special instrument called an *ammeter* measures current. Some ammeters read the peak value, while others read the rms value. One type of ammeter, based on the interaction between the current and a permanent magnet, reads the **average value** of the current. From calculus, the average value of any function $f(t)$ over the time interval $[0, T]$ is

$$F_{\text{ave}} = \frac{1}{T} \int_0^T f(t)dt. \tag{1.5}$$

For the absolute value of an ac waveform, the average value is 0.636 times the peak value. The average value is computed according to equation 1.5 as follows:

$$\text{Average value} = \frac{K}{T} \int_0^T |\sin(\omega\tau)| \, d\tau = \frac{2K}{T} \int_0^{\frac{T}{2}} \sin(\omega\tau)d\tau$$

$$= \frac{2K}{T} \left[\frac{-\cos(\omega\tau)}{\omega} \right]_0^{\frac{T}{2}} = \frac{2}{\pi}K = 0.636K. \tag{1.6}$$

3. VOLTAGE

Another important physical quantity used in circuit analysis is **voltage.** To facilitate an understanding of this new concept, we draw an analogy from a familiar experience illustrated in figure 1.8, in which a ball rests on a shelf above the floor before falling.

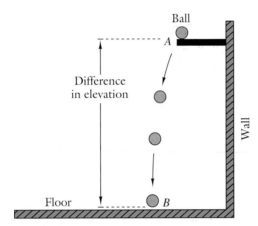

Figure 1.8 Conversion of potential energy into the kinetic energy of a falling ball.

Nudging the ball off the shelf starts its fall to the floor due to the earth's gravitational field. A ball always falls from a higher elevation to a lower elevation. As the ball falls, it picks up velocity. In terms of energy, the ball loses gravitational potential energy and gains kinetic energy. Energy is converted from one form to the other, but the total energy (kinetic plus potential) remains constant because of the principle of conservation of energy. At point B, just before the ball touches the floor, the kinetic energy gained by the ball can be determined from the principle of conservation of energy as follows:

energy converted = kinetic energy gained

= gravitational potential energy lost

= gravitational force on the ball × difference in elevation

This relationship has the alternative form,

$$\text{difference in elevation} = \frac{\text{energy converted}}{\text{weight of body}}. \tag{1.7}$$

 EXAMPLE 1.4

In figure 1.8, if the mass of the ball is 5 kg and the shelf is 2 m above the floor, find the velocity of the ball just before it hits the floor.

SOLUTION

The kinetic energy gained (in joules, denoted J) is

$$0.5(\text{mass})(\text{velocity})^2 = 0.5 \times 5(\text{velocity})^2 \text{ J}.$$

The gravitational potential energy lost is

$$(\text{weight of ball}) \times (\text{elevation difference}) = (5 \times 9.8) \times 2 = 98 \text{ J},$$

where the gravitational acceleration constant equals 9.8 m/sec^2. Equating these two expressions leads to

$$\text{velocity of ball} = 1.895 \text{ m/sec}.$$

With the above background, we are now ready to describe an analogous phenomenon in an electric field. Figure 1.9 shows two conducting plates with opposite but equal amounts of charge distributed on their surfaces. Assume that the charges are static (i.e., they do not move) and that the space in between is a vacuum (free space).

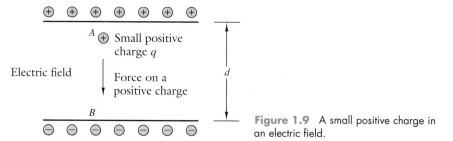

Figure 1.9 A small positive charge in an electric field.

Any *positive* charge placed between the plates is subject to a downward-pulling force, part of which is due to repulsion by the positive charges on the top plate and part of which is due to the attraction of the negative charges on the bottom plate. This attraction and repulsion indicates the presence of an electric field produced by the charges on the two plates. (Other means of producing an electric field will be discussed later.) The positive charge q experiences a force proportional to both the field strength at the point of placement and the magnitude of q. As long as q is very small compared to the charges on the plates, its presence will not alter their distribution. Hence, the electric field strength is approximately constant, except near the edges of the plates. Thus, the force acting on the charge q by the electric field is constant. This is analogous to the constant force acting on a ball by the gravitational field.

Since the positive charge q at A tends to move toward B, we say, *qualitatively*, that the point A in the electric field is at a *higher potential* than point B or, equivalently, that point B is at a *lower potential* than point A. An analogy to mechanics is now evident: A positive charge in an electric field "falls" from a point of higher potential to a point of lower potential, just as a ball falls from a higher elevation to a lower elevation in a gravitational field.

While this gravitational analogy facilitates an understanding of a force on a charge in an electric field, it has its limitations. For example, when a positive charge "falls" in an electrical field, the action does not refer to any *spatial* orientation: If we turn the whole setup of figure 1.9 upside down, the positive charge q still moves from point A to point B, an upward movement in space. Furthermore, the analogy used a positive test charge q. What about a negative charge $-q$ placed at B, as in figure 1.10? Here, the negative charge experiences an upward-pulling force and moves from a point B of lower potential to a point A of higher potential. In a gravitational field, we know very well that there is no such thing as a negative mass that falls from a lower elevation to a higher elevation.

Despite these limitations, a very strong *quantitative* analogy between figures 1.8 and 1.9 is possible. In figure 1.9, as the charge q moves from A toward B, it picks up velocity and gains kinetic energy. Just before q hits the bottom plate, the kinetic energy gained equals the force (assumed constant here) acting on q multiplied by the distance traveled *in*

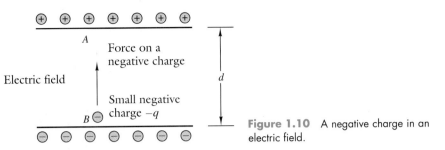

Figure 1.10 A negative charge in an electric field.

the direction of the force. The kinetic energy is thus proportional to q and to the "distance traveled," which equals d; therefore,

$$\text{energy converted} = \text{kinetic energy gained} \propto q.$$

The proportionality constant in this relationship is defined as the **potential difference or voltage between A and B.** The term "voltage" is synonymous with "potential difference." Mathematically,

$$\text{potential difference} = \frac{\text{energy converted}}{\text{magnitude of charge}}. \tag{1.8}$$

Notice the similarity between equations 1.7 and 1.8, where "potential difference" replaces "difference in elevation" and "magnitude of charge" replaces "weight of body."

The standard unit for measuring differences in elevation is the meter abbreviated m. The standard unit for measuring potential difference or voltage is the volt, abbreviated V. *According to equation 1.8 (with standard units of V, J, and C), if 1 joule of energy is converted from one form to another when moving 1 coulomb of charge from point A to point B, then the potential difference or voltage between A and B is 1 V.* In equation form with units, we have

$$1 \text{ V} = \frac{1 \text{ J}}{1 \text{ C}}. \tag{1.9}$$

The use of terms such as "difference in elevation," "energy converted," "potential difference," and "voltage" implies that they all have positive values. If the word "difference" is changed to "drop" (or to "rise"), then potential drop and drop in elevation have either positive or negative values, as the case may be. To illustrate this point in the context of figure 1.8, we make the following four statements:

The difference in elevation *between A and B* is 2 m.
The difference in elevation *between B and A* is 2 m.
The drop in elevation *from A to B* is 2 m.
The drop in elevation *from B to A* is −2 m.

Similarly, for figure 1.9, we can make the following four statements:

The voltage *between* (or across) A and B is 2 V.
The voltage *between* (or across) B and A is 2 V.
The voltage drop *from A to B* is 2 V.
The voltage drop *from B to A* is −2 V.

 EXAMPLE 1.5

In figure 1.9, if the voltage between the two plates is 100 V and a small body carrying a positive charge of 10^{-12} C moves from A to B, find the kinetic energy of the body just before it hits the bottom plate.

SOLUTION

From equation 1.8, using standard units,

$$\text{Kinetic energy of body} = (100) \times (10^{-12}) = 10^{-10} \text{ J}.$$

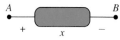

EXAMPLE 1.6

In a cathode ray tube, such as in a computer monitor or an oscilloscope, the electrons emitted from the cathode are accelerated toward the anode (the monitor screen), which is positively charged. If the voltage between the anode and cathode is 5,000 V, find the velocity of each electron when it reaches the anode. The mass of an electron is 9.1066×10^{-31} kg.

SOLUTION

The calculation is analogous to that in example 1.4. The charge on each electron is -1.60219×10^{-19}C. From equation 1.8, the amount of energy converted after the electron accelerates through a 5,000 V potential difference is

$$1.6019 \times 10^{-19} \times 5,000 = 8.01 \times 10^{-16} \text{ J.}$$

This must equal the kinetic energy of the electron, which is given by

$$0.5(\text{mass})(\text{velocity})^2 = 0.5 \times 9.1066 \times 10^{-31}(\text{velocity})^2.$$

Equating these two expressions yields

$$\text{Velocity of electron} = 41.9 \times 10^6 \text{m/sec.}$$

Example 1.6 shows that, because of its very small mass, the electron can reach an extremely high speed after accelerating through a high voltage. In the present case, the speed reaches about 14% of the speed of light. These high-speed electrons impinge on a fluorescent screen, which then emits visible light.

Exercise. The mass of a proton is 1,832 times that of an electron. If a proton accelerates through a potential difference of 5,000 V, what is the speed of the proton?
ANSWER: 9.79×10^5 m/sec.

In writing equations for circuit analysis, it is often important to know which of two points is at a higher potential. To this end, we speak of the *voltage drop* from point A to point B, conveniently denoted by V_{AB}. If the value of V_{AB} is positive, then point A is at a higher potential than point B. On the other hand, if V_{AB} is negative, then point B is at a higher potential than point A. Since V_{BA} stands for the voltage drop from point B to point A, it is obvious that $V_{AB} = -V_{BA}$.

The preceding double subscript convention is one of three methods commonly used to specify a voltage drop unambiguously. Such a method requires labeling all points of interest with letters or integers so that V_{AB}, V_{AC}, V_{12}, V_{13}, etc., make sense. A second, more common, convention uses + and − markings at two points, together with a variable or constant labeling the voltage drop *from the point marked + to the point marked −*. Figure 1.11 illustrates this convention; observe that x denotes the voltage drop from A (marked +) to B (marked −). If x is positive, then A is at a higher potential than B; if x is negative, then B is at a higher potential than A. The markings + and − by themselves do not indicate which point is at a higher potential. On any general circuit element, the markings may be *arbitrarily* assigned. They serve as a reference direction for unambiguously specifying the voltage drop. A third method for specifying a voltage drop, using a single subscript, will be discussed in chapter 4.

Figure 1.11 The + and − markings establish a reference direction for a voltage drop. For accuracy, we always place the (+,−) markings reasonably close to the circuit element, so that no ambiguity arises in interpreting which two points the variable x is associated with.

The following example illustrates the use of the double subscript and (+,−) markings for specifying voltage drops.

EXAMPLE 1.7

Figure 1.12 shows a circuit consisting of four general circuit elements with indicated voltage drops. Suppose it is known that $V_{AB} = 4$ V and $V_{AD} = 5$ V. Find the values of x, y, z, V_{BC}, and V_{CD}.

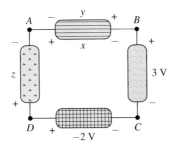

Figure 1.12 Four arbitrary circuit elements for illustrating the use of (+,−) markings for specifying voltage drops.

SOLUTION

The meaning of the double subscript notation and the (+,−) markings for a voltage imply that

$$x = 4 \text{ V}, y = -4 \text{ V}, z = -5 \text{ V}, V_{BC} = 3 \text{ V}, \text{ and } V_{CD} = 2 \text{ V}.$$

Exercise. In figure 1.12, find V_{CB} and V_{DC}.

ANSWERS: −3 V; −2 V.

Exercise. The convention for (+,−) markings is used in many electrical apparatus. Figure 1.13 shows an old 12-V automobile battery whose (+,−) marking cannot be seen due to corrosion of the terminals. A digital voltmeter (DVM) is connected as shown. The display reads −12 V. Determine the (+,−) marking of the battery terminals.

ANSWER: Left terminal, −; right terminal, +.

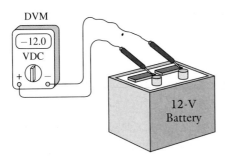

Figure 1.13 Determination of the (+,−) marking on a battery.

Our discussion of potential difference (voltage) in this section has centered around a charge moving through an electric field generated by charges present on two parallel plates separated by a vacuum. It is not necessary that a charge move through the field for the existence of a potential difference. In figure 1.9, there is a potential difference between points A and B, whether or not the charge q is placed in the field. This is similar to figure 1.8, where a difference in elevation exists between the shelf and the floor, whether or not a ball falls from the shelf.

Finally, just like current, there are different types of voltages: dc voltage, ac voltage, and general voltage waveforms. Figure 1.7, with the vertical axis relabeled as $v(t)$, illustrates the different voltage types.

4. ENERGY CONVERSION IN AN ELECTRIC CIRCUIT

In this section, we explore the energy relationships present when charge moves continuously in an electric field. Imagine again a ball falling from a shelf in the presence of a gravitational field, as shown in figure 1.14. Just before the ball hits the floor, its original gravitational potential energy is converted into kinetic energy, as was investigated in example 1.4. After hitting the floor and bouncing a few times, the ball comes to rest, all its kinetic energy completely spent as heat.

Now suppose a person standing nearby picks up the ball and returns it to the shelf. In doing this, the person works against the gravitational force. The work expended by

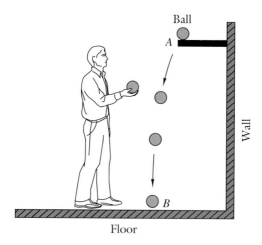

Figure 1.14 A continuous conversion between different forms of energy.

Ball

A

Wall

B

Floor

the person and transferred to the ball equals the weight of the ball times the difference in elevation. The energy supplied by the person is stored as gravitational potential energy associated with the ball on the shelf. Suppose next that the person nudges the ball off the shelf again and then returns the ball to the shelf after it has stopped bouncing. Suppose further that this process is repeated indefinitely 10 times per minute. Then continuous conversion of energy into different forms takes place: When the ball falls, its gravitational potential energy is first converted into kinetic energy and then into heat; as the person lifts the ball back to the shelf, the ball gains energy from an outside agent (the person) and stores the energy as gravitational potential energy. The principle of conservation of energy dictates that the potential energy gained must equal the energy lost to heat. The total change in energy sums to zero during each complete cycle of the process. Note that if the energy is converted at the rate of P_0 J/sec, then there is a power conversion of P_0 watts.

A phenomenon similar to that shown in figure 1.14 occurs in electric circuits. Figure 1.15a shows two general circuit elements connected together at points A and B. Suppose there exists a *constant* voltage drop from A to B, denoted by V_{AB}, and assume that $V_{AB} > 0$, i.e., A is at a higher electric potential than B. What causes the potential difference to appear between A and B will be explained in the next section. For now, we might think that one of the two elements is a battery, which maintains approximately a constant voltage between its terminals. Also, assume that a *constant* current $I_0 > 0$ flows in the circuit as shown. For the sake of simplicity in deriving an energy relationship, we will assume the current to consist of a movement of positive charges, although in reality, this need not be the case. (See example 1.1.)

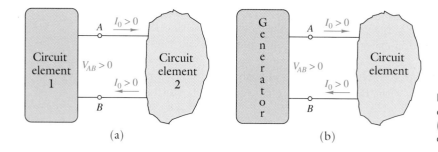

Figure 1.15 (a) A simple two-element circuit with dc current and voltage. (b) Connection of a generator and a circuit element.

During a time interval of T sec, $(I_0 \times T)$ C of charge move through circuit element 2 from A to B, analogously to the successive fallings of the ball in figure 1.14. In "falling" from a higher potential point A to a lower potential point B, the charge loses electric potential energy. The lost potential energy is converted into some other form of energy, heat being only one of several possibilities. This conversion occurs in circuit element 2, and the element is said to *absorb energy*. According to equation 1.8, the amount of energy absorbed is $V_{AB} \times (I_0 \times T)$. The **power** *absorbed* by element 2 is by definition the rate at which element 2 absorbs energy. This rate equals $[V_{AB} \times (I_0 \times T)]/T = V_{AB}I_0$.

Next, consider what happens within circuit element 1 in figure 1.15a. Here, the charge is moved from point B at a lower potential to point A at a higher potential, analogously to the person's lifting the ball in figure 1.14. Some external agent, analogy the person, must do work against the force on the charge due to the electric field in order to "lift" the charge to a higher potential. This can happen in a couple of ways. On the one hand, if element 1 is a battery, then the energy required to move the charge is derived from chemical action among the compounds inside the battery. On the other hand, if element 1 is a dc voltage generator, then the energy required to move the charge is derived from a conversion of mechanical energy to electrical energy by a process that is beyond the scope of this text. For our present objectives, it is sufficient to recognize that as charges move from a lower potential to a higher potential in element 1, some form of energy is converted into electric potential energy. Element 1 is said to *generate* electrical energy. By definition, the electric power *generated* by element 1 is the rate at which that element generates electric energy. According to equation 1.8, this rate equals $[V_{AB} \times (I_0 \times T)]/T = V_{AB}I_0$. Thus, this simple connection of two circuit elements demonstrates the principle of conservation of power: The power generated by one element equals the power absorbed by the other element. To picture this principle more vividly, call the generic element 1 in figure 1.15a a "generator" or "source," as shown in figure 1.15b.

The analogy of a gravitational field has proven useful in explaining the notions of voltage and energy conversion in an electric field. An alternative mechanical analogy based on fluid flow may provide some additional intuitive explanation of the electrical phenomena. Since an electric current is a flow of charge, it is similar in many respects to the flow of water in a pipe. Figure 1.16a shows the familiar cooling system seen in an automobile. The pump is mechanically driven by the running engine through a belt. The pump builds up pressure, which forces the cooling water to circulate through the radiator, where the heat is dissipated into the ambient air with the aid of a fan. The water returns to the pump and recirculates through the system. When we compare figure 1.16a to the simple circuit of figure 1.16b, it is reasonable, at least *qualitatively*, to liken the electric current to the water flow and the battery to the pump. The pump builds up a pressure to force the current to flow. In like manner, the battery builds up a voltage to force the current to flow. From this point of view, voltage (electrical potential difference) in electrical systems is analogous to pressure in fluid systems. A stronger, *quantitative* analogy to fluid systems can even be given. However, unlike the gravitational analogy, which is an everyday experience, the quantitative analogy to fluid flow requires a considerable background in hydrodynamics.

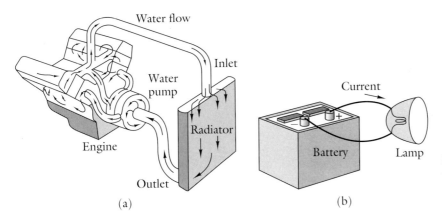

Figure 1.16 Analogy between (a) a fluid flow cooling system of a car engine and (b) a simple electric (battery lamp) circuit.

5. RELATIONSHIPS AMONG VOLTAGE, CURRENT, POWER, AND ENERGY

Light bulbs come in all sorts of shapes, sizes, and wattages. **Wattage** is a measure of the power consumed by a bulb. Some typical wattages are 15, 25, 40, 60, 75, and 100 watts. These power consumptions differ because the current required to light each bulb is different. Generally, a higher wattage bulb requires more work to convert electrical energy into light energy: The higher the wattage, the brighter the light.

Power is the rate of change of work per unit time. The ability to determine the power absorbed by each circuit element is of utter importance to every electrical engineering student. Using a circuit element beyond its power-handling capability could lead to a disaster.

Our discussion of these topics will begin by considering the simplest case of dc voltage and current. We then extend the results to the case of a time-varying voltage and current.

Power and Energy for dc Voltages and Currents

Consider a circuit element with a dc voltage V across it and a dc current I through it, with reference directions as shown in figure 1.17. During a time interval of T seconds, the charge passing through the circuit element is $I \times T$ C. Since the voltage drop from A to B is V volts, according to equation 1.8, the energy absorbed by the elements equals $V \times I \times T$. The rate of energy conversion is constant and equals $(V \times I \times T)/T = VI$. Therefore, for dc voltages and currents, we reach the very simple relationship

$$P = VI, \tag{1.10}$$

where P is the power in watts absorbed by the circuit element. Consequently, the energy in joules absorbed by the element during the time interval T is

$$W = PT. \tag{1.11}$$

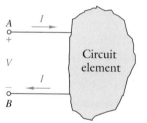

Figure 1.17 A circuit element excited by a dc voltage and current. With the passive sign convention, the absorbed power is $P = VI$.

◆ EXAMPLE 1.8

An electroplating apparatus uses 220 V dc voltage and draws 10 A dc current.

(a) What is power consumed by the apparatus?
(b) If electric energy costs 20 cents per kwh (kilowatt-hour), what will it cost to operate the apparatus for a single 10-hour day?

SOLUTION

(a) From equation 1.10, the power consumed is

$$P = 220 \times 10 = 2,200 \, \text{W}, \quad \text{or} \quad 2.2 \, \text{kilowatts}.$$

(b) According to equation 1.11, the energy consumed per 10-hour period is

$$W = 2.2 \times 10 = 22 \, \text{kwh}.$$

Therefore, the cost to operate the apparatus is $22 \times 0.2 = 4.4$ dollars per day.

It is important to remember that in using equations 1.10 and 1.11 to calculate the absorbed power and energy, the voltage and current reference directions, indicated by $(+,-)$ and an arrow, must conform with those shown in figure 1.17. This is usually called the **passive sign convention** (or the associated voltage-current references, in some texts). If either $(+,-)$ or the arrow *alone* is reversed, then the equation for absorbed power becomes $P = -VI$. For arbitrary assignments of $(+,-)$ and the arrow into a circuit element, the rule for calculating the absorbed power P in the dc case may be stated in words as follows:

> **Rule for calculating absorbed power** The power absorbed by any circuit element with terminals labeled A and B is equal to the voltage drop from A to B multiplied by the current through the element from A to B.

The calculated value of the absorbed power P may be negative. If it is, then the circuit element actually generates power or, equivalently, delivers power to the remainder of the circuit.

 EXAMPLE 1.9

For the circuit of figure 1.18, determine the power absorbed by each circuit element. Which elements are delivering power? Verify the principle of conservation of power for this circuit.

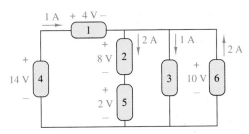

Figure 1.18 A circuit containing several general circuit elements.

SOLUTION

Using either equation 1.10 or the power consumption rule, we find that the power absorbed by each element is as follows:

1. For element 1, $P_1 = 4 \times 1 = 4$ watts.
2. For element 2, $P_2 = 8 \times 2 = 16$ watts.
3. For element 3, $P_3 = 10 \times 1 = 10$ watts.
4. For element 4, $P_4 = 14 \times (-1) = -14$ watts.
5. For element 5, $P_5 = 2 \times 2 = 4$ watts.
6. For element 6, $P_6 = 10 \times (-2) = -20$ watts.

Since P_4 and P_6 are negative, element 4 delivers 14 watts and element 6 delivers 20 watts of power, while the remaining four elements absorb power. Since the total power generated is $(14 + 20) = 34$ watts and the total power absorbed is $(4 + 16 + 10 + 4) = 34$ watts, conservation of power is verified for this example.

If the power absorbed by a circuit element is positive, the exact nature of the element determines the type of energy conversion that takes place. For example, a circuit element called a *resistor* (to be discussed shortly in detail) converts electrical energy into heat. If the circuit element is a battery that is being charged, then electrical energy is converted into chemical energy. If the circuit element is a dc motor turning a fan, then electrical energy is converted into mechanical energy. Finally, a circuit element called an *inductor* (to be discussed in chapter 7) stores electrical energy in a magnetic field which may be reclaimed later.

Non-dc Power and Energy Calculations

Consider figure 1.19, in which $i(t)$ is an arbitrary time-dependent current entering a general circuit element and $v(t)$ is the time-dependent voltage across the element. Because the voltage and current are functions of time, the power is also a function of time and is therefore denoted $p(t)$. For any specific value of $t = t_1$, the value $p(t_1)$ indicates the power absorbed by the element at that particular time—hence the terminology **instantaneous power** for $p(t)$.

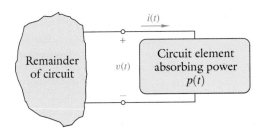

Figure 1.19 Calculation of absorbed power for time-varying voltages and currents in which $p(t) = v(t)i(t)$.

The relationship

$$p(t) = v(t)i(t) \tag{1.12}$$

extends equation 1.10 in the obvious way; i.e., the **instantaneous power** $p(t)$ in watts is the product of the voltage $v(t)$ in volts and the current $i(t)$ in amps. This product also makes sense from a dimensional point of view:

$$\text{volts} \times \text{amps} = \frac{\text{joules}}{\text{coulomb}} \times \frac{\text{coulombs}}{\text{second}} = \frac{\text{joules}}{\text{second}}.$$

Knowing the power $p(t)$ absorbed by a circuit element as a function of t allows one to compute the energy $W(t_0, t_1)$ absorbed by the element during the time interval $[t_0, t_1 > t_0]$. $W(t_0, t_1)$ is the integral of $p(t)$ with respect to t over $[t_0, t_1]$; i.e.,

$$W(t_0, t_1) = \int_{t_0}^{t_1} p(t)dt, \tag{1.13}$$

where t_0, the lower limit of integration, could be $-\infty$. Note that for the dc case, $p(t) = P = \text{constant}$, and equation 1.13 reduces to $W = P \times (t_1 - t_0) = PT$, which is equation 1.11.

In equation 1.13 , if $t_0 = -\infty$ and $t_1 = t$ is variable, then $W(-\infty, t)$ becomes a function only of t, which, for brevity, is denoted by

$$W(t) = \int_{-\infty}^{t} p(t)dt. \tag{1.14}$$

$W(t) = W(-\infty, t)$ represents the total energy absorbed by the circuit element from the beginning of time to the present time t. It is often miscalled the "instantaneous work done by the circuit element at t."

Since energy is the integral of power, power is the rate of change (derivative) of energy. Hence, differentiating both sides of equation 1.14 yields

$$p(t) = \frac{dW(t)}{dt} = v(t)i(t) \tag{1.15}$$

where $p(t)$ is the power in watts and $W(t)$ the energy in joules.

◆ EXAMPLE 1.10.

In the circuit of figure 1.19, the current $i(t)$ and voltage $v(t)$ have waveforms graphed in figures 1.20a and b, respectively. Sketch $p(t)$, the instantaneous power absorbed by the circuit element.

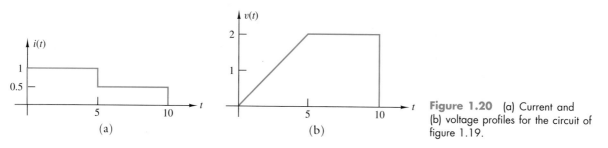

Figure 1.20 (a) Current and (b) voltage profiles for the circuit of figure 1.19.

SOLUTION

A simple graphical multiplication of figures 1.20 (a) and (b) yields a sketch of the instantaneous power, shown in figure 1.21.

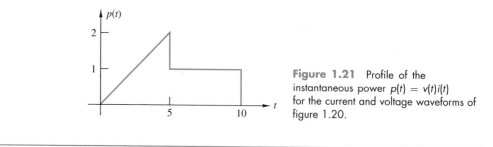

Figure 1.21 Profile of the instantaneous power $p(t) = v(t)i(t)$ for the current and voltage waveforms of figure 1.20.

Charge, Voltage, Current, and Energy Considerations

To conclude this section, we present some relationships involving a time-dependent charge $q(t)$, which is the integral of the current $i(t)$ as per equations 1.2 and 1.3. Substituting equation 1.2 into the rightmost side of equation 1.15 produces

$$\frac{dW(t)}{dt} = v(t)\,\frac{dq(t)}{dt},\qquad(1.16a)$$

or equivalently,

$$v(t) = \frac{\dfrac{dW(t)}{dt}}{\dfrac{dq(t)}{dt}}.\qquad(1.16b)$$

From calculus, if W is expressed as a function of q, then

$$\frac{dW(t)}{dt} = \frac{dW}{dq}\,\frac{dq}{dt},$$

in which case equation 1.16b reduces to

$$v(t) = \frac{dW}{dq};\qquad(1.16c)$$

i.e., the voltage drop is the derivative of the instantaneous work (in joules) delivered to the circuit element with respect to the instantaneous charge (in coulombs) passing through the element.

Equations 1.15 and 1.16 are differential equations. Hence, we can integrate both sides of equation 1.16a with respect to t, from t_0 to t, to produce the dual integral relationship

$$\int_{t_0}^{t} \frac{dW}{d\tau}\,d\tau = W(t) - W(t_0) = \int_{t_0}^{t} v(\tau)\,\frac{dq}{d\tau}\,d\tau.$$

When v is expressed as a function of q, the variable of integration may be changed from t to q, following the rules of calculus. The result is

$$W(t_0, t) = W(t) - W(t_0) = \int_{q(t_0)}^{q(t)} v(q)\,dq.\qquad(1.17)$$

Equation 1.17 suggests a very interesting property: If v is a function of q, then the energy absorbed by the circuit element during the time interval $[t_0, t]$ depends only on the initial value of the charge $q(t_0)$ and the final value of the charge $q(t)$. Thus, no knowledge of the complete waveform of $q(t)$, $i(t)$, or $v(t)$ is ever needed! In chapter 7, equation 1.17 will be used to calculate the energy stored in a circuit element called a *capacitor*. The next two examples illustrate the application of equation 1.16.

◆ EXAMPLE 1.11

If 12 kJ of energy are required of an external generator to move 3 kC of charge uniformly through a circuit element in 1 sec (from node A to node B), determine the voltage across the element, as depicted in figure 1.22.

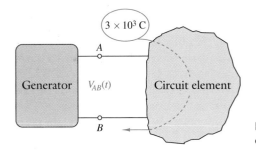

Figure 1.22 Charge moving through an arbitrary circuit element.

SOLUTION

From equation 1.16c,

$$v_{AB}(t) = \frac{\Delta W}{\Delta q} = \frac{12 \times 10^3 \text{ J}}{3 \times 10^3 \text{ C}} = 4 \text{ V}.$$

◆ **EXAMPLE 1.12**

A generator, as illustrated in figure 1.23, delivers a current to the attached circuit element so that the charge crossing the top node during the time interval $[0, t]$ is

$$q(t) = \frac{1 - \cos(2\pi t)}{2\pi}.$$

If the work required of the generator over $[0, t]$ to deliver this charge is

$$W(t) = q(t) - \pi q^2(t),$$

determine
(a) the current $i(t)$, for $t \geq 0$, through the circuit element, and
(b) the voltage $v(t)$, for $t \geq 0$, across the circuit element.

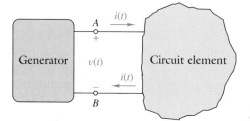

Figure 1.23 A generator attached to a circuit element.

SOLUTION

From equation 1.2,

$$i(t) = \frac{dq(t)}{dt} = \sin(2\pi t),$$

and from equation 1.16c,

$$v(t) = 1 - 2\pi q(t) = \cos(2\pi t).$$

In the previous two examples, we have assumed that some voltage exists between the two terminals A and B of a circuit element. But how is this voltage generated? The next section addresses that question.

6. IDEAL VOLTAGE AND CURRENT SOURCES

Equations 1.11 through 1.17 are valid for a general circuit element, which may be a piece of wire, a lamp, a battery, or even a piece of rock. Using their terminal voltage-current relationships, we may classify circuit elements into different categories. The goal of this section is to specify ideal voltage and current sources by their terminal voltage-current relationships.

The wall socket of a typical home represents a voltage source. One plugs appliances into these sockets. Those appliances then consume energy. This causes a current flow through the appliance. For modest amounts of current draw (those below the fuse setting), the voltage remains nearly a 115-V sinusoid; i.e., 115 V is the effective value of the sinusoid. The peak value of the sinusoid is $115\sqrt{2}$ V $= 163$ V. It is necessary to idealize such real-world situations with a symbol for circuit analysis. Figure 1.24a presents such a symbol (a circle with the $\pm$ sign inside). The idealization is called an **ideal voltage source** or, more commonly, an **independent voltage source.**

The waveform or signal $v(t)$ in the figure represents the voltage produced by the generator or source at each time t. The plus and minus signs on the source indicate polarity, which

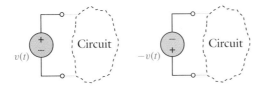

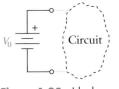

Figure 1.24 Equivalent ideal voltage source representations with attached hypothetical circuit.

is a labeling or reference frame that designates how one measures the voltage across the source. For example, the voltage V_{AB} discussed in the previous section could just as well be denoted V_{+-}. The $+-$ does not mean that the $v(t)$ is positive. Indeed, figure 1.24b shows an equivalent source representation with opposite polarities. Finally, the voltage source is **ideal** because it maintains the given voltage regardless of the current that may flow from the source to an attached circuit or circuit element.

Batteries are a special kind of voltage source. In a battery, chemical energy and electrical energy are interconverted. From equation 1.12, if the power absorbed by the battery is positive, and the battery is of the rechargeable type, then electrical energy is converted into chemical energy, i.e., the battery is charging. On the other hand, for typical battery applications, such as in flashlights and for starting a car engine, the absorbed power is negative. This means that the battery delivers power by converting chemical energy into electrical energy.

Figure 1.25 shows the symbol for an **ideal battery**. The voltage drop from the side with the long dash to the side with the short dash is V_0 V, with $V_0 > 0$. In commercial products, the terminal marked with a + sign corresponds to the side with the long dash. An ideal battery produces a constant voltage under all operating conditions, i.e., regardless of any current drawn from an attached circuit or circuit element. Real batteries are not ideal, but approximate the ideal case over a manufacturer-specified range of current requirements. Other devices, that approximate ideal dc and ac generators, convert mechanical energy to electrical energy.

Figure 1.25 Ideal battery representation of ideal dc voltage source.

In addition to devices that maintain fixed voltage waveforms, there are devices that maintain fixed current waveforms. Such an idealization is given in figure 1.26, which presents the symbol (a circle with an arrow inside) for an **ideal current source,** or an **independent current source.** An ideal current source produces and maintains a current $i(t)$ under all operating conditions, i.e., regardless of the voltage requirements imposed by an attached device. Of course, the current appearing across each source can be a constant (dc), sinusoidal (ac) or any other time-varying function.

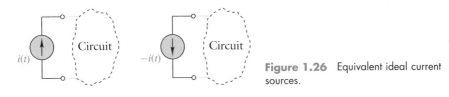

Figure 1.26 Equivalent ideal current sources.

An example of an approximately ideal current source found in nature is lightning. When lightning strikes a lightning rod, the path to ground is almost a short circuit, and very little voltage is developed between the top of the rod and ground. However, if lightning strikes a tree, the path of the current to ground is impeded, and a large voltage (pressure) appears across the tree, often causing it to split.

Independent and dependent sources have a labeling, shown in figure 1.27, that is different from the passive sign convention of figure 1.17. An independent source is delivering power if $p(t) = v(t)i(t) > 0$ and absorbing power if $p(t) = v(t)i(t) < 0$. Note that sources can absorb power. Physically, this is the situation in a battery charger, wherein a battery (one source) is absorbing power from another source, the charger.

Another type of ideal source is a **dependent,** or **controlled, source.** Such a source produces a current or voltage that depends on the voltage or current of some other device in the circuit. In the text, the symbol for a dependent source is a diamond shape—a **dependent voltage source** if $\pm$ appears inside and a **dependent current source** if an arrow appears inside. Figure 1.28 depicts the situation for dependent voltage sources. Here, the voltage across the diamond-shaped source depends either on a current, labeled i_x, through

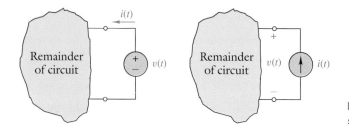

Figure 1.27 Common voltage-current source labeling.

some other circuit device or on the voltage, v_x, across it. If the voltage across the source depends on the voltage v_x, i.e., if $v(t) = Av_x$, then the source is called a **voltage-controlled voltage source**. If the voltage across the source depends on the current—i.e., $v(t) = Bi_x$, the source is called a **current-controlled voltage source**.

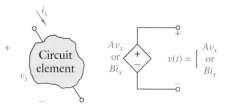

Figure 1.28 The right element is a voltage-controlled voltage source if $v(t) = Av_x$ and a current-controlled voltage source if $v(t) = Bi_x$.

A similar terminology exists for the current source case. In figure 1.29, the symbols illustrate the **voltage-controlled current source**, i.e., $i(t) = Av_x$, and the **current-controlled current source**, for which $i(t) = Bi_x$.

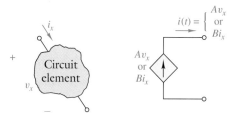

Figure 1.29 The right element is a voltage-controlled current source if $i(t) = Av_x$ and a current-controlled current source if $i(t) = Bi_x$.

Later in the text, with the introduction of a circuit device known as an *operational amplifier*, dependent sources will receive a fuller discussion. Our purpose here is to introduce the concept and use it in some simple situations.

Source voltages or currents are called **excitations**, inputs, or input signals. A constant voltage will be denoted by an uppercase V, V_0, V_1, V_A, etc. A constant current will typically be denoted by I, I_0, I_1, I_A, etc. Often, the units of volts or amps prove inconvenient, and we shall substitute smaller or larger quantities as required. Some convenient notation, given in table 1.1, aids the expression of these larger and smaller units.

TABLE 1.1 SHORTHAND NOTATION FOR LARGE AND SMALL VALUES OF VOLTAGE, CURRENT, ETC.		
Name	Abbreviation	Value
Pico	p	10^{-12}
Nano	n	10^{-9}
Micro	μ	10^{-6}
Milli	m	10^{-3}
Kilo	k	10^{3}
Mega	M	10^{6}

7. RESISTANCE

Different materials allow electrons to move from atom to atom with different levels of ease. Suppose the same dc voltage is applied to two conductors, one carbon and one copper, of the same size and shape. Then two different currents would flow. The current flow depends on a property of the conductor called **resistance**: The smaller the resistance, the larger the current flow. The ohm, denoted Ω, is the basic unit of resistance. A material has a 1-Ω resistance if a one-volt excitation causes one ampere of current to flow through it.

One could imagine water flowing through pipes of different diameters as analogous to current flow. For a given pressure, a pipe with a larger diameter allows a larger volume of water to flow and therefore has a smaller resistance than a pipe with half its diameter.

Ohm's Law

A **resistor** is a device that impedes the flow of current. Just as dams impede the flow of water and provide electric power, recreation, and flood control for rivers, resistors are a means of controlling current flow in a circuit. In addition, resistors are a good approximate model to a wide assortment of devices, such as light bulbs and heating elements in ovens. Figure 1.30a shows the standard symbol for a resistor; the voltage and current reference directions have been marked in accordance with the passive convention. Figure 1.30b pictures a resistor connected to an ideal battery that is a source of energy.

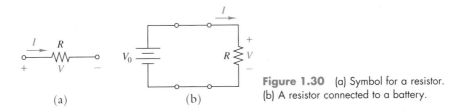

(a) (b)

Figure 1.30 (a) Symbol for a resistor. (b) A resistor connected to a battery.

The resistance R in ohms measures the resistance of a device to current flow. In 1827, Ohm observed that the dc current through a conductor is proportional to the voltage across the conductor, i.e.,

$$I \propto V.$$

Thus, one can write

$$I = \frac{V}{R}, \tag{1.18a}$$

which has the equivalent formulation

$$V = RI, \tag{1.18b}$$

where the proportionality constant R is the **resistance** of the conductor in ohms. In equation 1.18a, the proportionality constant is the reciprocal of R, i.e., $G = 1/R$, which is called the **conductance** of the device. The International System of Units (SI) for conductance is **siemens**, abbreviated S. However, in the United States, the **mho**, the older name for the unit of conductance, is still widely used. In this text, we will use the mho (*ohm* spelled backward) as the unit of conductance. If a device or wire has zero resistance, i.e., if $R = 0$, it is said to be a **short circuit**. A short circuit has infinite conductance. On the other hand, if a device or wire has infinite resistance (and hence zero conductance), it is said to be an **open circuit**. In precise language, *resistor* denotes the actual physical device and *resistance* is a property of the device. However, in most of the literature on electronic circuits, the two words are used synonymously, and we will continue this convention.

OHM'S LAW

Ohm's law, as observed for constant voltages and currents, is given in equation 1.18b. However, it is true for all time-dependent waveforms that excite a linear resistor. Thus, we can generalize equation 1.18b to

$$v(t) = Ri(t). \tag{1.19}$$

It is very important to remember that in using equations 1.18 and 1.19, the voltage and current reference directions must conform to the passive sign convention shown in figure 1.30a. If either, but not both of, the voltage or the current direction is reversed, then Ohm's law becomes $v = -Ri$. As an aid in writing the correct v-i relationship for a resistor, Ohm's law is stated in words as follows:

For a resistor connected between terminals A and B, the voltage drop from A to B is equal to the resistance multiplied by the current flowing from A to B through the resistor.

Once the voltage and current associated with a resistor are known, the power absorbed by the resistor is easily calculated. Assuming the passive convention, we combine equation 1.12 for power with Ohm's law, equation 1.19, to obtain

$$p(t) = v(t)i(t) = i^2(t)R = \frac{v^2(t)}{R},$$ (1.20a)

which reduces to

$$P = VI = I^2R = \frac{V^2}{R}$$ (1.20b)

for the dc case.

Equation 1.20 brings forth a very important property of a resistor. Unlike other types of circuit elements, which may absorb power at one moment and generate power at another, a resistor *always* absorbs power. The power is dissipated in heat. An elementary explanation of the generation of heat in a wire conductor is as follows. In the absence of an applied field, the free electrons move *randomly* within a conductor, and hence, no net flow of charge is observed, making the current zero. When an electric field is established within the conductor by an externally applied source, there is an average drift of electrons in the direction opposing the field (recall that electrons have negative charges), producing an electric current. The energy an electron gains through accelerating to a certain velocity will be partially or wholly surrendered to the particle with which it collides. The energy given up will appear in the form of heat. After the collision, the electron will again be accelerated and continue to move along the conductor, colliding with other particles in the process. This may be compared to what happens in a pinball machine. As a ball rolls down from a higher elevation to a lower elevation, it successively collides with other pegs. Each collision causes part of its kinetic energy to be converted into heat.

Resistance depends on the geometrical structure and the material of the conductor. For a specific temperature, R is proportional to the length L of the conductor and inversely proportional to its cross-sectional area A; i.e.,

$$R \propto \frac{L}{A}.$$

Writing this proportion as an equation yields

$$R = \rho\frac{L}{A},$$ (1.21)

where the proportionality constant ρ is called the **resistivity** in ohm-meters. The resistivity of copper at $20°C$ is 1.7×10^{-8} ohm-m. Table 1.2 lists the relative resistivity of various materials with respect to copper.

TABLE 1.2 RESISTIVITY OF VARIOUS MATERIALS RELATIVE TO COPPER. THE RESISTIVITY OF COPPER AT $20°C$ IS 1.7×10^{-8} OHM-M					
Silver	0.94	Chromium	1.8	Iron	5.68
Copper	1.00	Zinc	3.4	Tin	6.7
Gold	1.4	Cadmium	4.4	Carbon	2.4×10^3
Aluminum	1.6	Nickel	5.1	Silicon	135×10^9

EXAMPLE 1.13

Sixteen-gauge copper wire has a resistance of 4.094 Ω for every 1,000 feet of wire. Determine the resistance of 1,000 feet of aluminum wire and 1,000 feet of nickel wire. Determine the power absorbed by the conductor and given off as heat if a 10-A dc current flows through 100 feet of wire.

SOLUTION

The resistivity of aluminum with respect to copper is 1.6. Hence, 100 feet of aluminum wire has a resistance of $1.6 \times 0.4094 = 0.655$ Ω. According to equation 1.20b, if a 10-A current flows through 100 feet of aluminum wire, the absorbed power given off as heat is

$$P = VI = [RI]I = RI^2 = 65.5 \text{ watts.}$$

Notice that every 100 feet of 16-gauge aluminum wire would absorb $(65.5 - 40.9)$ watts = 24.6 watts more power than copper would absorb.

The resistivity of nickel wire is 5.1 times greater than copper; 100 feet of nickel wire would therefore have an approximate resistance of $5.1 \times 0.4094 = 2.088$ Ω. Hence, 16-gauge nickel wire excited by a 10-A dc current absorbs 208.8 watts of power per 100 feet, or 8.5 times more power than copper absorbs per unit length.

Because of the form of equations 1.20, the resistive power loss is usually called the *i*-squared-*r* loss. The interested reader may consider Table 1.3, which lists the resistances in ohms of different gauges of copper wire.

TABLE 1.3 RESISTANCE OF WIRE PER 1,000 FEET. (NOTE: 1 MIL = 0.001 INCH. THE MIL IS NOT AN SI UNIT, BUT IT IS WIDELY USED BY THE WIRE INDUSTRY IN THE UNITED STATES)		
Gauge	$\Omega/10^3$ft	Diameter (mils)
12	1.619	80.8
14	2.575	64.1
16	4.094	50.8
18	6.510	40.3
20	10.35	32

In addition to depending on the type of material of which a conductor is made, resistance depends on temperature. For most metallic conductors, resistance increases with increasing temperature. Carbon is contrary, its resistance decreasing as the temperature rises. However, a resistance absorbs power, and most of the power is given off as heat, causing the temperature of the resistor to rise. Usually, resistors that must absorb a great deal of power are made physically larger so they can better dissipate the heat that is generated. Clearly, an awareness of the temperature characteristics of a material is important when designing circuits. Bulbs, for example, have a *hot resistance* and a *cold resistance*. Of course, since bulbs are of use only when lit, the hot resistance is the more important characteristic.

 EXAMPLE 1.14

The hot resistance of a light bulb is 120 Ω. Find the power absorbed by the bulb if it is connected across a constant 90-V source, as illustrated in figure 1.31.

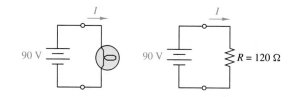

Figure 1.31 Light bulb and equivalent resistive circuit model.

SOLUTION

From Ohm's law (equation 1.18a), the current is

$$I = \frac{V}{R} = \frac{90}{120} = 0.75 \text{ A.}$$

From equation 1.20b, the power absorbed by the lamp is

$$P = 0.75^2 \times 120 = 67.5 \text{ watts.}$$

Observe that the power delivered by the source is $90 \times 0.75 = 67.5$ watts. Therefore, the power delivered by the source equals the power absorbed by the resistor. This illustrates the principle of conservation of power.

◆ **EXAMPLE 1.15**

Two bulbs are connected in series across a constant 120-V battery, as illustrated in figure 1.32. The first bulb uses 18.5 watts of power, and the second bulb uses 24.5 watts of power.

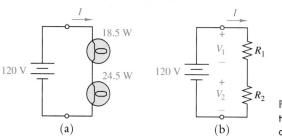

Figure 1.32 (a) Series connection of two bulbs and (b) equivalent resistive circuit model.

(a) Determine the current $i(t)$ through each bulb.
(b) Determine the voltages $v_1(t)$ and $v_2(t)$ across each bulb.
(c) Determine the hot resistances, R_1 and R_2, of each bulb.
(d) Determine the power used by each bulb when it is connected directly to the 120-V source.

SOLUTION

(a) The total power delivered by the source is 43 watts, which is the sum of the power consumed by each bulb. The power delivered by the source must satisfy

$$P = VI = 120I = 43 \text{ watts.}$$

Hence, $I = 0.3583$ A.

(b) By the definition of a two-terminal circuit element and because charge through a cross section of an ideal conductor must be conserved, the current into each resistor equals the current leaving the resistor. Therefore, the current through each resistor is the same. Again, using the equation $P = VI$, the voltage across each resistor is easily calculated to be

$$V_1 = \frac{P_1}{I} = \frac{24.5}{0.3583} = 68.37 \text{ V}$$

and

$$V_2 = \frac{P_2}{I} = \frac{18.5}{0.3583} = 51.63 \text{ V.}$$

(c) Finally, from Ohm's law for the dc case,

$$R_1 = \frac{V_1}{I} = 190.8 \ \Omega$$

and

$$R_2 = \frac{V_2}{I} = 144 \ \Omega.$$

As a check, the power absorbed by each resistor must satisfy the formula

$$P = VI = RI^2.$$

For the first resistor, $P_1 = 190.8(0.3583)^2 = 24.5$ watts, and for the second resistor, $P_2 = 144(0.3583)^2 = 18.5$ watts, as expected.

(d) When each bulb is connected directly to the source, $P_1 = (120)^2/190.8 = 75.5$ watts and $P_2 = 100$ watts. The curious aspect of this result is that the light bulb using the most power (24.5 watts) in the series connection uses the least power (75.5 watts) in the parallel connection and conversely.

Exercise. Using the formula $P = V^2/R$ given in equation 1.20b, compute the power absorbed by each resistor in example 1.8.

One of the problems with series connections of bulbs is circuit failure. If one bulb burns out—i.e., if the filament in the bulb open-circuits—then all the lights go out. By contrast, parallel circuits continue to operate in the presence of such failures and are easier to fix, as only the bulb that has burnt out (open circuited) is unlit. The following example illustrates power consumption for a parallel connection of light bulbs.

◆ EXAMPLE 1.16

Consider the parallel circuit of figure 1.33. Determine (1) the effective hot resistance of each bulb and (2) the total power delivered by the source. (3) If at $t = 10$ hours, the current supplied by the source is 0.6471 A, determine which bulb has burnt out.

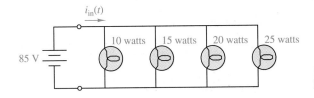

Figure 1.33 Parallel connection of light bulbs.

SOLUTION

(1) Rearranging the formula $P = V^2/R$, we obtain

$$R_{10W} = \frac{85^2}{10} = 722.5 \ \Omega, \qquad R_{15W} = \frac{85^2}{15} = 481.667 \ \Omega,$$

$$R_{20W} = \frac{85^2}{20} = 361.25 \ \Omega, \qquad R_{25W} = \frac{85^2}{25} = 289 \ \Omega.$$

(2) The total power delivered by the source equals the sum of the powers consumed by each bulb, which is 70 watts.

(3) If at $t = 10$ hours, the current supplied by the source is 0.6471 A, then the power delivered by the source at $t = 10$ hours is $85 \times 0.6471 = 55$ watts. The bulb that has burnt out was using $70 - 55 = 15$ watts of power. Hence, the 15-watt bulb has burnt out.

8. NONIDEAL SOURCES

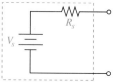

Figure 1.34 A battery model that accounts for an internal resistance.

Real batteries have an internal resistance due to the chemicals and electrodes inside them. Therefore, a model of a real battery must include an internal resistance in series with an ideal battery. Figure 1.34 illustrates the idea.

◆ EXAMPLE 1.17

A nickel-cadmium battery has an open-circuit voltage of 6 V. When connected across a 1-Ω resistor, the voltage drops to 5.97 V. Determine the internal resistance of the battery.

SOLUTION

Figure 1.35a, which shows a model of a battery that includes an internal resistance R_s, illustrates the situation. Since the battery is not connected to a circuit element, no current flows through the internal resistance, in which case the so-called open-circuit voltage of the battery is 6 V.

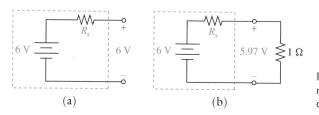

(a)   (b)

Figure 1.35 Battery with internal resistance (a) open-circuited and (b) connected to a 1-Ω load.

Figure 1.35b shows the battery connected to a 1-Ω resistive load across which appears 5.97 V. The voltage that appears across the internal resistance, R_s, is 0.03 V. From the definition of a two-terminal circuit element, the current that flows through R_s is the same current that flows through the 1-Ω resistor. This current is $5.97/1 = 5.97$ A. By Ohm's law, the ratio of the voltage across R_s to the current through R_s equals R_s. Hence,

$$R_s = \frac{0.03}{5.97} = 0.005 \ \Omega.$$

Ideal voltage sources have zero internal resistance. Real voltage sources have an internal resistance. The value of this resistance may change with the current draw. There may also be other effects. However, for our purposes, a more realistic model of an independent voltage source contains a constant (positive) internal resistance, as illustrated in figure 1.36a.

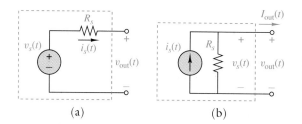

(a)   (b)

Figure 1.36 (a) A nonideal voltage source as an ideal voltage source with an internal series resistance. (b) A nonideal current source as an ideal current source with a parallel internal resistance.

Ideal current sources have infinite resistance. Real current sources have a finite internal resistance. Figure 1.36b depicts a model of a current source that accounts for internal resistance. Notice that the resistor is in parallel with the current source.

Voltage sources and current sources have a graphical interpretation in terms of their input and output currents. In the ideal case, the so-called v-i (voltage-current) characteristic of an **ideal constant voltage source** is a horizontal straight line. This means that the voltage supplied by the source is fixed for all possible current draws. An **ideal constant current source** has a vertical straight line for its v-i characteristic, which means that the current is fixed for all possible voltages across the source. These ideas are illustrated in figure 1.37.

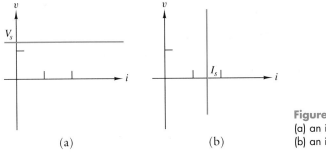

Figure 1.37 v-i characteristics of (a) an ideal constant voltage source and (b) an ideal constant current source.

The nonideal case is quite different. Because of the internal resistance, a **nonideal constant voltage source** v-i characteristic satisfies the linear relationship

$$v_{\text{out}} = -R_s i_s + V_s, \tag{1.22}$$

and for a **nonideal constant current source**, in which $G_s = 1/R_s$,

$$i_{\text{out}} = -G_s v_s + I_s. \tag{1.23}$$

Equations 1.22 and 1.23 are illustrated by the graphs in figure 1.38 when $v_s(t) = V_s$ for the nonideal voltage source and $i_s(t) = I_s$ for the nonideal current source. If the value of R_s is very small in comparison with potential load resistances, as ordinarily expected, then the line in figure 1.38a approximates a horizontal line, the ideal case. On the other hand, the line in figure 1.38b approximates a vertical line whenever R_s is much much larger than a potential load resistance. This would then approximate the ideal case.

In a similar way, a **nonideal dependent voltage source** is a connection of an ideal dependent source with a series resistance, and a **nonideal dependent current source** is a connection of an ideal dependent current source with a parallel resistance.

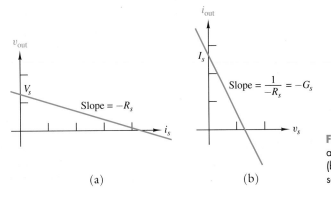

Figure 1.38 (a) v-i characteristic of a nonideal constant voltage source and (b) i-v characteristic of a nonideal current source.

9. ADDITIONAL REMARKS

The Notion of a Device

The ideal resistor is a device that satisfies Ohm's law. Ohm's law specifies a relationship between the current through the resistor and the voltage across it. A graph of this relationship is known as the **v-i characteristic** of the resistor. Figures 1.37 and 1.38 show the v-i characteristics of ideal and nonideal voltage and current sources. This might suggest characterizing all devices by a v-i characteristic, which would simplify matters immensely if it were possible. However, only memoryless single-input, single-output devices can be specified by a v-i characteristic. A single-input, single-output device is **memoryless** if there is a fixed functional relationship between its input current at time t and its output voltage at time t or between its input voltage at time t and its output current at time t, for each t, i.e., if $v = f(i)$ or $i = f(v)$. In general, the relationship can also depend on time, but this generality is beyond our present needs. The resistor, the dependent and independent ideal source, the dependent and independent nonideal source, and a whole variety of other devices are all memoryless. Although the ideal resistor studied before has a

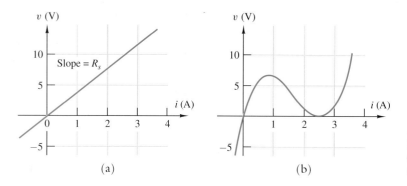

Figure 1.39 (a) Linear resistor characteristic. (b) A possible nonlinear resistor characteristic.

straight-line characteristic as illustrated in figure 1.39a, there are commercially available nonlinear resistors with characteristics typified by figure 1.39b.

Devices that have memory have voltage and current relationships which satisfy a differential equation or whose output voltage or current at time t depends on past values of the input current or voltage. A device whose output voltage satisfies $v(t) = i(t-1)$ is not memoryless, because the value of the output voltage, $v(t)$, at time t depends on the value of the input current at time $(t-1)$. A device whose input and output voltages satisfy a differential equation is the *capacitor*, studied in chapter 7. The voltage and current of a capacitor satisfy the differential equation

$$i_C(t) = C\frac{dv_C(t)}{dt}.$$

As we will see, this relationship allows the capacitor to store energy. Storing energy is possible only if the device has memory.

> **Exercise.** Which of the following relationships are memoryless: (a) $v(t) = -3i(t)$; (b) $i(t) = v^3(t) + 2v(t)$; (c) $i(t) = v(t) + dv(t)/dt$; (d) $v(t) = i(t) - 16$.

The Notion of a Model

The ideal resistor satisfies Ohm's law for all currents and voltages. Real-world resistors have a behavior that approximates Ohm's law over a reasonable range of voltages and currents. The interconnections between resistors and ideal sources allow us to approximate the behavior of real-world devices in a systematic way. These interconnections are called **circuit models**. The models become increasingly sophisticated as one introduces other types of circuit elements (such as capacitors, inductors, diodes, etc.) for interconnection. It is precisely through idealized circuit models that engineers are able to design appliances such as televisions, videocassette recorders, stereos, etc., and carry out systematic analyses of real-world processes.

Frequency, Wavelength, and the Notion of a Lumped Circuit Element

As already mentioned, sinusoidal waveforms have a frequency measured in hertz (Hz) or radians/second (rad/sec). Frequencies from 5 Hz to 20 kHz are typically called **audio frequencies**. Frequencies above 20 kHz are typically called **radio frequencies**. The letter f ordinarily denotes frequency in Hz. The Greek letter ω ordinarily denotes frequency in rad/sec. The relationship between the two variables is $\omega = 2\pi f$.

Associated with a sinusoidal frequency is a **wavelength**, denoted by λ. In general,

$$\lambda \text{ (wavelength in m)} = \frac{c}{f \text{ (frequency in Hz)}},$$

where $c = 3 \times 10^8$ m/sec is the speed of light. The definition of a device used in this chapter has some hidden assumptions that depend on the notion of a wavelength. The definition of a **circuit element** presupposes that the current entering the element equals the current leaving the element. For this to be true, the physical dimensions of the element must be small relative to the wavelength of the sinusoidal frequencies expected to drive the device. For example, an audio frequency of 20 kHz has a wavelength

$$\lambda = \frac{3 \times 10^8 \text{ m/sec}}{20 \text{ kHz}} = 15 \times 10^3 \text{ m}.$$

Suppose a device is 1,500 meters long. Then the current leaving the device would differ from the current entering the device by about 1%. Fortunately, an ordinary circuit element is not 1,500 meters long. On the other hand, suppose one injects a current $i = \cos(2\pi 37.5 \times 10^6 t)$ A (i.e., at 37.5 MHz) into a 2-meter antenna. Then the wavelength is $\lambda = 8$ meters. The antenna thus is a quarter wavelength long, which means that if the current at the base of the antenna is 1 A, then the current at the tip of the antenna is zero.

This text deals only with circuit elements whose dimensions are small with respect to the wavelength of signals that excite the elements. Such circuit elements are termed **lumped**, whereas the antenna described in the preceding paragraph would be called a *distributed* circuit element. Distributed devices are typically covered in a course in electromagnetic fields.

The Skin Effect

When flowing through a conductor, sinusoidal currents exhibit an interesting behavior called the **skin effect**. Suppose the conductor has a circular cross section. Then, according to the skin effect, as the frequency of the sinusoidal current increases, the current flow (the transport of charge) along the conductor becomes concentrated on its outer rim. This reduces the effective cross-sectional area through which charge is transported and thereby increases the effective resistance of the conductor. One concludes that radio frequency resistance is much greater than dc resistance and that a hollow tube would be as effective (and cheaper) than a solid conductor of the same diameter at high frequencies.

Power Reprise and Efficiency

Recall that power is the rate of doing work. For a resistor, there are three formulas for computing power:

$$p(t) = v(t)i(t) = Ri^2(t) = \frac{v^2(t)}{R}.$$

In a resistor, absorbed power appears as heat energy. Not all devices convert all absorbed electric energy to heat, although almost all devices convert some electrical power to heat. Electric energy that is converted to heat or that is used to overcome friction or some other obstacle to a desired behavior is typically called a **loss**. On the other hand, the heating element of a stove has the sole purpose of converting as much electrical energy to heat as possible.

The heating effect in a resistor is the basis for the operation of fuses. A **fuse** is a short piece of conductor with a very low resistance and a predetermined current-carrying capacity. It is inserted in a circuit and carries the current of equipment or appliances that it is designed to protect. When the current goes beyond the rating of the fuse, the heat that is generated burns up the fuse and opens the circuit. In this way, the fuse protects a more expensive piece of equipment from damage. A burnt-out fuse should never be replaced by a larger fuse or by a heavy piece of conductor. Oversize fuses or solid wire jumpers defeat the purpose of fuses and permit the operation of a device at overload capacity, with consequent damage to equipment.

A device that converts mechanical energy into electrical energy is the *generator*. The power sent into a radio transmitter is used to convert information into electromagnetic energy, which radiates into the atmosphere from an antenna. A television antenna or dish picks up a weak state of this energy, sends it to a television set or special amplifier powered by electrical energy, and converts the electromagnetic radiation to a picture with sound. The electrical energy from a stereo is converted to mechanical energy in the speaker, which produces sound by vibrating a cone. The analysis and design of these types of devices requires a much deeper understanding of circuit principles than that given in this first chapter.

One might wonder how one pays for the electricity used in the home. Basically, one pays for the *work* the electricity does, not the rate at which it is consumed. Electrical work equals power multiplied by time. Therefore, one pays the utility company for kilowatt-hours used.

Unfortunately, not all devices utilize all the electrical energy they consume for a useful purpose. The ratio of the output power to the input power times 100% is called the percent **efficiency** of a device. Because energy is expensive, it is important to design energy-efficient devices.

SUMMARY

This chapter contains many topics that are often introduced for the first time in basic physics courses. Since electrical engineers have to work constantly with charge, current, voltage, energy, and power, we have reintroduced these basic notions in an electrical engineering context for the purpose of analyzing and designing circuits. To help clarify the fundamental concepts of an electric field, voltage, charge transport, current, and energy, we utilized the familiar concept of the gravitational field. We illustrated these concepts and the principles of conservation of power and energy with numerous examples.

At the introductory level, it is very fruitful to classify and study circuit elements from their measured terminal voltage-current characteristics. This approach avoids all involvement with field theory and device physics, which would have unduly complicated the presentation of introductory circuit analysis. Ideal voltage and current sources were defined in this manner. Then, after introducing the resistor, we showed how a practical voltage or current source can be modeled by an ideal source and a resistor.

We carefully pointed out the distinction between an actual circuit and the mathematical model on which we perform analysis. Only some extremely simple circuits were discussed in this chapter. Subsequent chapters will develop methods by which we can analyze practical circuits of a much more sophisticated variety.

TERMS AND CONCEPTS

Alternating current: a sinusoidally time-varying current signal having the form $K \sin(\omega t + \phi)$.

Audio frequencies: frequencies below 20 kHz.

Battery: a device that converts chemical energy into electrical energy and maintains approximately a constant voltage between its terminals.

Charge: an electric property of matter, measured in coulombs. Like charges repel and unlike charges attract each other. Each electron carries an amount of charge equal to -1.6×10^{-19} C.

Circuit model: an approximation to a real circuit that is amenable to mathematical analysis. One can think of a circuit model as an interconnection of ideal circuit elements (circuit diagram) or as some mathematical equation.

Conductance: reciprocal of resistance, with mhos (or siemens) as its unit.

Conductor: a material, usually a metal, in which electrons can move to neighboring atoms with relative ease.

Conservation of power (energy): the sum of the powers (energies) generated by a group of circuit elements is equal to the sum of the powers (energies) absorbed by the remaining circuit elements.

Current: movement of charges. Current is measured in amperes. One ampere is the movement of charges through a surface at the rate of 1 C/sec.

Dependent (controlled) current source: a current source whose strength depends on the voltage or current of some other element in the circuit.

Dependent (controlled) voltage source: a voltage source whose strength depends on the voltage or current of some other element in the circuit.

Direct current: a current that remains constant with time.

Efficiency of a device: the ratio of the output power to the input power times 100%.

Electric field: property such that if a charge is placed at a point in space, then the charge experiences a force. If this is true, we say that an electric field exists at that point. A rigorous definition of an electric field is given in a subsequent course in fields, but is not needed in circuit analysis.

Ideal conductor: material that offers zero resistance to the movement of electrons.

Ideal insulator: material that offers infinite opposition to the movement of electrons.

Independent (ideal) current source: an ideal device that delivers current as a prescribed function of time (e.g., $\{2\cos t + 12\}$ A), no matter what circuit element is connected across its terminals.

Independent (ideal) voltage source: an ideal device whose terminal voltage is a prescribed function of time (e.g., $\{2\cos t +12\}$ V), no matter what current goes through the device.

Instantaneous power: the value of $p(t) = v(t)i(t)$ at a particular instant of time.

Insulator: a material that opposes the easy movement of electrons.

Lumped circuit element: circuit element whose dimensions are small compared to the wavelength of frequencies that excite the element.

Mho: unit of conductance; equal to inverse ohm. (See siemens.)

Ohm: unit of resistance. One ohm equals the ratio of 1 V to 1 A.

Ohm's law: for any metallic conductor, the current i through the conductor at any time t is proportional to the voltage v across the conductor at that time.

Open circuit: a connection of infinite resistance or zero conductance.

Passive sign convention: convention by means of which voltage and current reference directions are indicated. (See figure 1.17.)

Peak-to-peak value: $2K$ in $K\sin(\omega t + \phi)$ of an ac waveform.

Peak value: K in $K\sin(\omega t + \phi)$ of an ac waveform.

Power: the rate of change of work per unit time.

Resistance: for a resistor, the proportionality constant R in the formula $v(t) = Ri(t)$. Resistance is measured in ohms. One ohm means that the voltage is 1 V when the current is 1 ampere.

Resistivity: the resistance of a conductor is proportional to its length and inversely proportional to its cross section area. The proportionality constant, usually denoted by ρ, is called the resistivity of the material. The resistivity of copper at $20°C$ is 1.7×10^{-8} ohm-m.

Resistor (linear): a physical device that obeys Ohm's law. (There are commercially available nonlinear resistors that do not obey Ohm's law.)

Root mean square (rms) or effective value: a measure of ac current; related to the peak value by the formula rms $= 0.7071K$, where $K\sin(\omega t + \phi)$ is the ac waveform.

Short circuit: a connection of zero resistance or infinite conductance.

Siemens: international unit for mhos, or inverse ohms.

Skin effect: as the frequency of a sinusoidal current through a conductor increases, the current flow (the transport of charge) along the conductor becomes concentrated on the outer rim of the conductor.

Source: a device that approximates either an ideal voltage source or an ideal current source.

v-i characteristic: a graphical or functional representation of a memoryless circuit element.

Voltage (synonymous with potential difference): measurement obtained when a positive charge moves from a point of higher potential to a point of lower potential, accompanied by a conversion of energy. Voltage is measured in volts. The existence of 1 V between two points A and B means that the energy converted in moving one coulomb of charge between A and B is one joule.

Wattage: a measure of the consumption of power.

Wavelength of a sinusoid: denoted by λ; equals the speed of light divided by the frequency in Hz.

PROBLEMS

1. Consider the diagram of figure P1.1a.
   (a) Determine the charge on 7.573×10^{37} electrons.
   (b) If this number of electrons moves uniformly from the left end of a wire to the right in 1 ms (millisecond), what current flows through the wire?
   (c) How many electrons must pass a given point in 1 minute to produce a current of 10 A?
   (d) If the charge profile across the cross section of a conductor from left to right is given by $q(t) = t + 0.2e^{-5t}$ C for $t \geq 0$ and zero for $t < 0$, plot the profile of the current that flows across the boundary. In what direction would the current flow?
   (e) Repeat part (d) for the charge profile (in C) sketched in figure P1.1b.

2. (a) The current in an ideal conductor is given by $i(t) = 1 - 2e^{-2t} + e^{-3t}$ A for $t \geq 0$. Determine the charge transferred, $q(t)$, as a function of time for $t \geq 0$.
   (b) Repeat part (a) for the current profile sketched in figure P1.2.

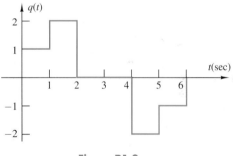

Figure P1.2

3. A current $i(t)$ is as shown in the graph of figure P1.3. Determine the charge that is transported by $i(t)$ through the surface S_0 of the conductor, also illustrated in the figure, during the time interval $0 \leq t \leq 3$ sec and over $0 \leq t \leq 6$ sec.

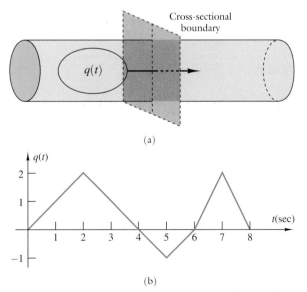

(a)

(b)

Figure P1.1 (a) Charge crossing a boundary. (b) Hypothetical charge profile.

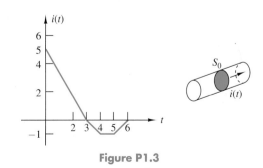

Figure P1.3

4. It is possible to show that voltage is the derivative of work $W(t)$ (change in energy) with respect to charge (change in charge); i.e., $v = dW/dq$, where $W(t)$ is the work being expended at time t to move the charge $q(t)$. It is also known that

$$W(t) = \int_{-\infty}^{t} p(t)dt,$$

where $p(t) = v(t)i(t)$ is the power absorbed by a circuit element, illustrated in figure P1.4a.

(a) For the device shown, suppose that

$$p(t) = \begin{cases} P_0 \text{ for } t \geq 0 \\ 0 \text{ for } t < 0 \end{cases}$$

and that the charge delivered to the device is $q(t) = t$ coulombs for $t \geq 0$ and zero otherwise.

(1) Compute $W(t)$.
(2) Express $W(t)$ as a function of $q(t)$.
(3) Compute $dW(t)/dq(t)$.
(4) Differentiate $q(t)$ to find $i(t)$, and solve for $v(t)$ in $p(t) = v(t)i(t)$.
(5) Do the answers in parts 3 and 4 coincide? Should they? Explain.

(b) Suppose the voltage across and the current through the device are as sketched in figure P1.4b and c respectively.

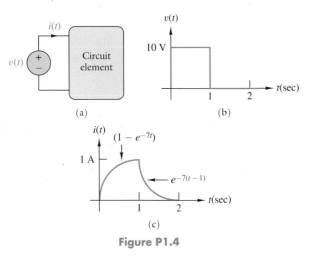

(a)　　　　　　(b)

(c)

Figure P1.4

(1) Sketch the power $p(t)$ absorbed by the device.
(2) Sketch the work $W(t)$ expended by the source.

5. Determine the value of the current I_x in the dc resistive circuit of figure P1.5, given that the unknown devices absorb the powers indicated.

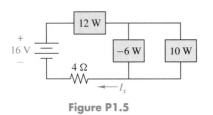

Figure P1.5

6. (a) Determine the power absorbed by each of the circuit elements in figures P1.6a and b.

(b) Show that the algebraic sum of the absorbed powers is zero. Be careful of signs.

(*Note*: In figure P1.6b, $v_1(t) = 3(1 - e^{-t})$ V for $t \geq 0$, $i_1(t) = 3e^{-t}$ A for $t \geq 0$, $v_2(t) = 3e^{-t} - 1$ V for $t \geq 0$, and $i_2(t) = 3e^{-t} - 1$ A for $t \geq 0$.)

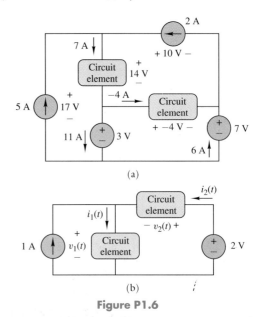

(a)

(b)

Figure P1.6

7. (a) Consider the circuit of figure P1.7a, which shows three lamps, AA, BB, and CC, in a parallel circuit. This is a simplified example of a light circuit on a car, in your house, or possibly on a Christmas tree. The AA bulb uses 36 watts when lit, the BB bulb uses 24 watts when lit, and the CC bulb uses 14.4 watts when lit.

(1) Determine the current through each bulb.
(2) Determine the current I_{in} delivered by the battery.
(3) Determine the total power delivered by the source, and show that the powers absorbed by each circuit element sum to zero.

(b) Determine the number of AA bulbs in parallel that would be required to blow the 15-A fuse in the circuit of figure P1.7b.

(c) Repeat part (b) for CC bulbs.

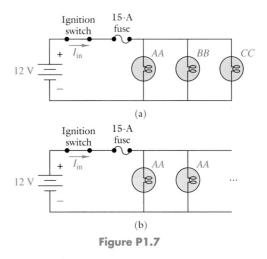

(a)

(b)

Figure P1.7

8. The constant voltage V_R across a resistor R is scaled to a new value αV_0, where α is some real scalar. By what factor

does the old power change? That is, determine the new absorbed power as a function of the old absorbed power.

9. In the circuit of figure P1.9, there are three independent sources and five ordinary resistors.

    (a) Determine which of the circuit elements are sources and which are resistors.

    (b) Determine the value of the resistance for each of the resistors.

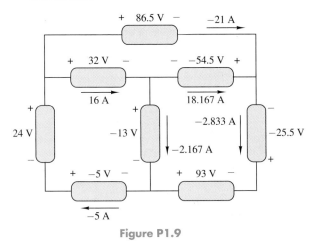

Figure P1.9

10. If 500 feet of 20-gauge copper wire is soldered end-to-end to 55 feet of nickel wire, determine the total resistance of the wire.

11. For the circuit of figure P1.11:

    (a) Determine the output voltage and output current.

    (b) Determine the voltage gain $|V_{out}/V_{in}|$.

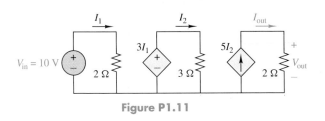

Figure P1.11

12. For the circuit of figure P1.12:

    (a) Determine the output voltage and output current.

    (b) Determine the voltage gain $|V_{out}/V_{in}|$.

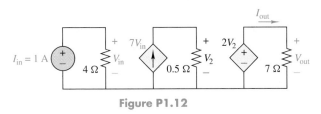

Figure P1.12

13. Repeat problem 12 for the circuit of figure P1.13.

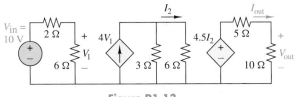

Figure P1.13

14. A nonideal constant voltage source, an ordinary resistor, and a nonideal constant current source have v-i characteristics given in figures P1.14(a), (b), and (c), respectively. Determine the values of the source voltage or current, the value of the source internal resistance, and, finally, the value of the resistance for figure P1.14c.

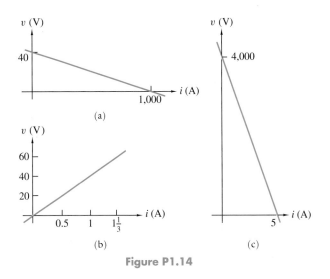

Figure P1.14

15. Consider the circuit of figure P1.15. The power consumed by each resistor is as follows: $p_2(t) = 98$ W, $p_3(t) = 12$ W, $p_4(t) = 16$ W, $p_5(t) = 768.8$ W, and $p_6(t) = 486$ W. The subscript corresponds to the resistor value.

    (a) For each resistor, determine the indicated voltage or current.

SCRAMBLED ANSWERS: 3, 14, 9, 2, 8.

    (b) Determine the total power delivered by the two sources.

    (c) Determine (if you can) the value of V_{in} and the value of I_{in}. Explain your reasoning.

SCRAMBLED ANSWERS: 12.4, 68.

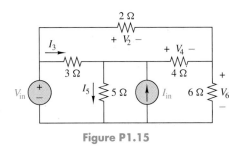

Figure P1.15

16. Consider the circuit of figure P1.16.

    (a) Determine an expression for V_{out} in terms of R_1, R_2, α, and V_{in}.

    (b) If $R_2 = 10$ Ω and $\alpha = 100$, determine the value of R_1 so that the voltage gain $|V_{out}/V_{in}|$ is 5.

    (c) Given your answer to (b), determine the power gain, which is the ratio of the power delivered to R_2 divided by the power delivered by the source.

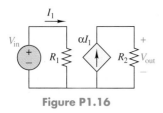

Figure P1.16

17. The voltage V_L in figure P1.17 is related to the current I_L according to

$$V_L = \begin{cases} 16 - I_L^2 & \text{for } 0 \le I_L \le 4 \\ 0 & \text{for } I_L \ge 4 \end{cases}$$

(a) Find the powers absorbed by the load when $I_L = 2$ A and $I_L = 3$ A.
(b) Find the value of I_L that maximizes the power absorbed by the load.

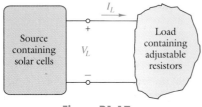

Figure P1.17

18. The power absorbed by the resistor in the circuit of figure P1.18 is 25 watts. Determine the value of R.

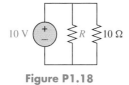

Figure P1.18

19. The power delivered by the source in the circuit of figure P1.19 is 325 watts. Determine the value of R.

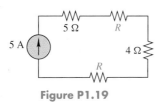

Figure P1.19

20. What is the speed in m/sec of an electron that has been accelerated through 20 kV? The mass of an electron is 9.1066×10^{-31} kg.

21. (a) How much energy in joules is required to heat a 30-gallon water heater from 40°F to 140°F? (It may be assumed that 2.5 watt-hours are required to raise the temperature of a gallon of water 1°F.)
(b) If the water heater of part (a) is an electric type operating at 120 V dc and drawing 10 A of current, how long does it take to reach the desired temperature of 140°F?

22. A water heater supplies water at 80°C at a rate of 40 gallons per hour. The water at the inlet is 25°C. The heating element used in the water heater has an efficiency of 90%.

If the energy cost is 8 cents per kwh and the heater is used 6 hours each day, determine the monthly bill for a 30-day month.

23. (a) What is the current-carrying capacity of a 0.5-w, 1-MΩ resistor used in a radio receiver?
(b) What is the current-carrying capacity of a 5-kw, 2-Ω resistor used in an electric power station?

24. Determine the resistance of 120 m of copper wire that has a 0.5 by 2 cm cross section.

25. Determine the resistance of a copper tube having the following dimensions:

Length: 12 m
Outside diameter: 0.6 cm
Wall thickness: 0.12 cm

26. Determine the resistance of a nickel ribbon having the following dimensions:

Length: 20 m
Width: 1.5 cm
Thickness: 0.1 cm

27. The resistance R of a conductor is a function of the temperature T (in °C). Over a range of temperature that is not too great, the relationship between $R(T)$ and T is linear and can be expressed as

$$R(T) = R(20)[1 + \alpha(T - 20)],$$

where α is called the *temperature coefficient* of the conducting material. For copper $\alpha = 0.0039$ per degree celsius. If the resistance of a coil of wire is 21 Ω at -10°C, what is the resistance when the wire is operating at 10°C?

28. A resistance coil of tungsten wire is used for temperature measurement. Its resistance is 200 Ω at 20°C. What is the resistance at 150°C? What is the change in resistance in ohms per degree celsius? The temperature coefficient for tungsten is 0.0045 per degree celsius.

29. The resistance of the copper winding in a certain machine, measured at 20°C, was 0.002 Ω. After the machine was run for 2 hours, the resistance was measured again and found to be 0.0022 Ω. What is the hot temperature of the copper wire?

30. If energy costs 8 cents per kwh, how much does it cost to use a 7-w night light from 8 P.M. to 6 A.M.? What is the monthly (30 day) cost?

31. If energy costs 8 cents per kwh, how much does it cost to run a television set 6 hours per day? The typical set consumes 120 watts of power.

32. A resistor is used to load a 2-kw, 120-V dc generator. What is the full-load current rating of the resistor?

33. A 50-cell lead storage battery has an open-circuit voltage of 102 V and a total internal resistance of 0.05 Ω.
(a) If the battery delivers 80 A to a load resistor, what is the terminal voltage?
(b) What is the terminal voltage when the battery is being charged at a 50-A rate?
(c) What is the power delivered by the charger in part (b)? How much of the power is lost in the battery as heat?

34. In the circuit of figure P1.34, the voltages, currents, and powers of some elements have been measured and indicated in the diagram.

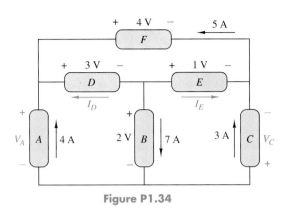

Figure P1.34

(a) If element A generates 20 W of power, find V_A.
(b) Find the power absorbed by element B.
(c) If element C generates 3 W of power, find V_C.
(d) If element D absorbs 27 W of power, find I_D.
(e) If element E absorbs 2 W of power, find I_E.
(f) Find the power absorbed by element F.

35. The switch S in figure P1.35 is assumed to be ideal; i.e., it behaves as a short circuit when closed, and as an open circuit when open. If the switch repeatedly closes for 1 ms and opens for 1 ms, what is the average value of $i(t)$?

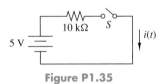

Figure P1.35

36. In figure P1.36, the switch S alternately stays at position A for 1 ms and at position B for 4 ms. Find the average value of $i(t)$.

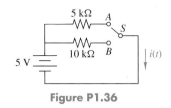

Figure P1.36

37. The input current to the circuit of figure P1.37 is $i_s(t) = 20\cos(2\pi t)$ A for $t \geq 0$ and zero for $t < 0$.
(a) Compute the instantaneous power delivered by the source. Using a graphing program, graph the power delivered as a function of time for $0 \leq t \leq 3$.
(b) Now compute and graph an expression for the energy dissipated in the resistor as a function of t for $0 \leq t \leq 3$.

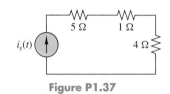

Figure P1.37

38. A nonideal voltage source, as sketched in figure P1.38, has internal resistance $R_s = 2\ \Omega$. If the voltage source is a constant 12 V, plot the power delivered by the source as a function of R for $0 \leq R \leq 5\ \Omega$. For what value of R is the power absorbed by R a maximum? Can you derive this result analytically? (*Hint*: In calculus, one learns that for differentiable functions, the values of the independent variable leading to a minimum or maximum are found by setting the derivative of the function to zero.)

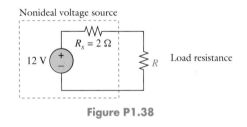

Figure P1.38

CHAPTER 2

CHAPTER OUTLINE

Kirchhoff's Current and Voltage Laws and Series-Parallel Resistive Circuits

A CAR HEATER FAN SPEED-CONTROL APPLICATION

RESISTORS have many uses in electronic circuits. One such use is to control the current flow just as dams control the flow of water along rivers. The ability of a resistor to control current flow comes from Ohm's law, $V = RI$. For a fixed voltage, high values of resistance lead to small currents, whereas low values of resistance lead to higher currents. This property is put to work in controlling the speed of the heater motor in a typical car. The following diagram illustrates the idea:

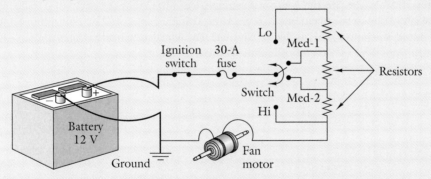

In the diagram, there are three resistors in series. As we will learn in this chapter, the resistance of a series connection of resistors is the sum of the resistances of each resistor. So, with the switch in the low position, the 12-V car battery "sees" three resistors in series with a motor. The three resistors in series, which represent a relatively large resistance, severely restrict the current through the motor. With less current, there is less power and the motor speed is slow. When the switch moves to the Med-1 position, there is less resistance in the circuit, allowing more current to flow and thereby increasing the motor speed. Each successive position removes resistance from the circuit, and the motor speed increases accordingly.

The material in this chapter will allow us to quickly analyze practical circuits such as the one just described. In problem 33, the student is asked to do the analysis.

CHAPTER OBJECTIVES

- Define and utilize the two fundamental laws of circuit theory, Kirchhoff's current law and Kirchhoff's voltage law, which govern the distribution of currents and voltages in a circuit.
- Introduce series and parallel resistive circuits.
- Develop a voltage division formula that specifies how voltages distribute across series connections of resistors.
- Develop a current division formula that specifies how currents distribute through a parallel connection of resistors.
- Show that a series connection of resistors has an equivalent resistance equal to the sum of the resistances in the series connection.
- Show that a parallel connection of resistors has an equivalent conductance equal to the sum of the conductances in the parallel connection.
- Explore the calculation of the equivalent resistance or conductance of a series-parallel connection of resistances, i.e., a circuit having a mixed connection of series and parallel connections of resistors.
- Explore the calculation of voltages, currents, and power in a series-parallel connection of resistances.
- Illustrate how the above laws and techniques can be applied to the design of instruments called multimeters, which measure voltage, current, and resistance.

---

1. INTRODUCTION

This chapter explores the very basic laws and properties of circuits. Any interconnection of circuit elements is a circuit. For example, figure 2.1a shows a **series circuit** consisting of a sequential connection of two-terminal circuit elements (resistors), end to end. An important property of a series connection is that all two-terminal elements carry the same current. Figure 2.1b shows a **parallel circuit** in which the top terminals of each resistor are wired together, as are the bottom terminals. An important property of parallel circuits is that each of the circuit elements has the same voltage across it. The common connection point, noted in the parallel circuit of figure 2.1b, is called a **node**. Formally, a node is a connection point of two or more circuit elements.

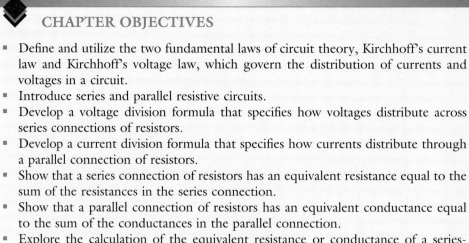

Figure 2.1 (a) Series connection of resistors. (b) Parallel connection of resistors.

Sources interconnected with circuit elements produce currents through the elements and voltages across the elements. For example, a voltage source connected across figure 2.1a would generate voltages v_1 through v_n. Similarly, a current source connected across the circuit of figure 2.1b would produce currents i_1 through i_n. A law called **Kirchhoff's current law (KCL)** governs the flow of currents into and out of a connection point or node, as in the top connection of figure 2.1b. A law called **Kirchhoff's voltage law (KVL)** governs the distribution of voltages around loops of circuit elements, as in figure 2.1a. These two laws are stated and illustrated in this chapter.

For practical reasons, this text explores only interconnections of *lumped* circuit elements, i.e., circuits whose physical dimensions are far smaller than the wavelengths of the signals that excite them. Hence, charge is transported more or less instantaneously through the conductor. Analyzing devices excited by signals whose wavelength is comparable to the devices' dimensions requires the theory of electromagnetic fields built around Maxwell's equations. These equations govern the behavior of electric and magnetic fields, which cause charge to be transported and current to flow. KVL and KCL are approximations to Maxwell's equations for lumped circuit elements when the voltage and current excitations have sufficiently large wavelengths. Hence, KVL and KCL apply only to lumped circuits. Fortunately, such circuits are found in home appliances, televisions, video cassette recorders, stereos, and a host of other electronic mechanisms. This suggests that the material covered in this chapter has broad applicability.

A proper statement of KVL and KCL requires the notions of a branch and a node. A **branch** of a circuit is simply a two-terminal circuit element. The notation for a branch is a line segment. The endpoints of the branch (the terminals of the circuit element) are sometimes called **nodes**. However, in general, a **node** is a point of connection of two or more circuit elements (branches), as described previously. Figure 2.2 shows two illustrations of branch and node concepts.

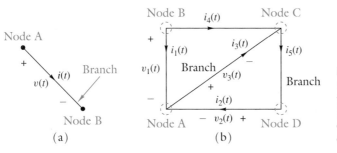

Figure 2.2 (a) Single branch representing a circuit element with terminals labeled as nodes A and B. (b) Interconnection of branches (circuit elements) with connection points labeled as nodes A through D.

The labeling in figure 2.2 shows each branch with the conventional reference directions. The arrowhead on a branch denotes the reference current direction, which typically is from plus to minus. Reference directions can be arbitrarily assigned. For a voltage source, it is often convenient to have the current arrowhead point from − to +. Similarly, for a current source, the reference voltage polarity is often taken as + to − from the arrow's head to the arrow's tail.

Recall that reference directions constitute a reference frame for consistent measurements. The + to − direction does not mean that the voltage is always positive if measured this way.

2. KIRCHHOFF'S CURRENT LAW

Before stating Kirchhoff's current law, imagine a number of branches connected at a node, as in node A of figure 2.2b. The current through each branch has a reference direction indicated by an arrow. If the arrow points toward the node, the current is entering; if the arrow points away from the node, the current is leaving. If a current is referenced as leaving a node, then the negative of the current enters the node and conversely.

KIRCHHOFF'S CURRENT LAW (KCL)
Kirchhoff's current law states that the algebraic sum of the currents entering a node of a circuit consisting of lumped elements is zero for every instant of time. Equivalently, the algebraic sum of the currents leaving a node is zero for every instant of time.

From physics, we know that charge is neither created nor destroyed. Hence, the charge transported into a node must equal the charge leaving the node, since charge cannot accumulate at a node. Kirchhoff's current law expresses this conservation law in terms of branch currents.

KCL applied at node A in figure 2.2b implies that $i_1(t) + i_2(t) - i_3(t) = 0$ for all t. Applying KCL to node B requires that $i_4(t) = -i_1(t)$. Finally, applying KCL to node D implies that $i_5(t) = i_2(t)$. Similarly, in figure 2.3, by KCL, the current $i_R(t) = 6\cos(2t) - 3\cos(2t) - \cos(2t) - 2\cos(2t)$.

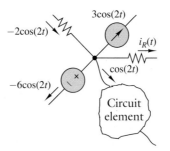

Figure 2.3 A connection of five circuit elements at a single node. Their currents are constrained by KCL.

Exercise. Suppose the current through the voltage source in the direction of the arrow in figure 2.3 is changed to $-2\cos(2t)$. Find $i_R(t)$.

ANSWER: $-4\cos(2t)$.

Two implications of KCL are of immediate interest. First, as a general rule, KCL implies that current sources cannot be connected in series. For example, figure 2.4 shows an invalid connection of two arbitrary current sources $i_1(t)$ and $i_2(t)$, where $i_1(t) \neq i_2(t)$. The connection is invalid because KCL requires that $i_1(t) = i_2(t)$.

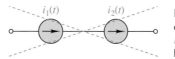

Figure 2.4 Invalid connection of two arbitrary current sources, given that $i_1(t) \neq i_2(t)$. Such a connection violates KCL.

A second immediate consequence of KCL is that a current source supplying zero current is equivalent to an open circuit, as illustrated in figure 2.5. This would suggest that a **current source** has **infinite internal resistance**. (See the model of figure 1.36b.) To grasp this from another perspective, note that a current source is represented by a vertical line in the i-v plane. (See figure 1.37b.) The slope of the vertical line, which is infinite, determines the internal resistance of the source.

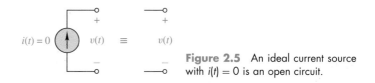

Figure 2.5 An ideal current source with $i(t) = 0$ is an open circuit.

One of the more usual applications of KCL is given in the following example.

EXAMPLE 2.1

In the parallel resistive circuit of figure 2.6, the voltage across each resistor is $6\cos(t)$ V. Determine the current through each resistor and the current $i_{\text{in}}(t)$ supplied by the voltage source.

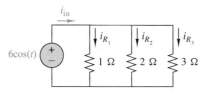

Figure 2.6 Parallel resistive circuit for example 2.1.

SOLUTION

By Ohm's law,

$$i_{R_1}(t) = 6\cos(t) \text{ A},$$

$$i_{R_2}(t) = 3\cos(t) \text{ A},$$

and

$$i_{R_3}(t) = 2\cos(t) \text{ A}.$$

By KCL,

$$i_{\text{in}}(t) = 6\cos(t) + 3\cos(t) + 2\cos(t) = 11\cos(t) \text{ A}.$$

Exercise. Suppose the source voltage in the circuit of figure 2.6 were changed to $-12\cos(2t)$. Find $i_{\text{in}}(t)$.

ANSWER: $-22\cos(2t)$ A.

Kirchhoff's current law works for special closed surfaces, called Gaussian surfaces, as well as for single nodes as just illustrated. A **Gaussian surface** is a closed curve in the plane or a closed surface in three dimensions *with a well-defined inside and outside*. Making the last phrase more mathematically rigorous would only serve to confuse matters; it is easier to lean upon a more intuitive understanding. Figure 2.7 illustrates the idea of a Gaussian surface for three situations. Observe that such a surface can enclose more than one node, as shown in figure 2.7a.

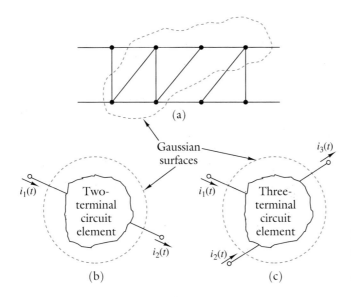

Figure 2.7 Illustrations of a Gaussian surface. (a) Arbitrary network. (b) Two-terminal device. (c) Three-terminal device.

KCL for Gaussian surfaces reads as follows: *For lumped circuits, the algebraic sum of the currents entering (leaving) a Gaussian surface is zero at every instant of time.* For the two-terminal device of figure 2.7b, this implies that $i_1(t) = i_2(t)$, which is precisely the definition of a lumped **two-terminal device**. For the three-terminal device of figure 2.7c, $i_3(t) = i_1(t) + i_2(t)$. Use of a Gaussian surface in circuit analysis sometimes offers the advantage of quickly determining branch currents.

 EXAMPLE 2.2

Figure 2.8 shows a circuit in which each element has been replaced by a branch. Such a figure is called a *graph* of the network. The objective of this example is to determine as many of the currents, I_1 through I_8, as possible, using only KCL and the concept of a Gaussian surface.

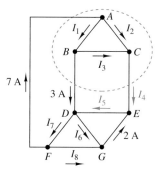

Figure 2.8 Graph of a network for example 2.2 showing a Gaussian surface to compute I_4 directly.

SOLUTION

The Gaussian surface indicated in figure 2.8 allows us to compute I_4 directly. We have $I_4 = 7 - 3 = 4$ A. It is not possible to compute I_4 directly by applying KCL only to nodes. Knowing I_4 allows us to compute $I_5 = I_4 + 2 = 6$ A.

Exercises.

1. Draw a Gaussian surface in figure 2.8 that allows one to determine I_5 directly.
2. Suppose $I_6 = 1$ A. Apply KCL to nodes D and F. Solve the resulting equations to obtain I_7 and I_8.

ANSWERS: in random order: 1 A, 8 A.

As a final comment on KCL observe that it holds regardless of the type of lumped two-terminal device represented by each branch of the circuit. This is because KCL specifies how branch currents interact at nodes or through Gaussian surfaces, regardless of the type of lumped element connected to the node or within the Gaussian surface.

3. KIRCHHOFF'S VOLTAGE LAW

Kirchhoff's voltage law (KVL) specifies how voltages distribute across the elements of a circuit. Before conveying four equivalent versions of KVL, we set forth several basic concepts needed for those statements. The first is the notion of a closed path. A **closed path** in a circuit is a connection of two-terminal elements that ends on the node where it began and that traverses each node in the connection only once. Figure 2.9 illustrates several closed paths. One such path is A-B-C-D-E-A, i.e., it begins at node A, moves to node B, drops to node C, moves through element 4 to node D, drops down through element 6 to reference node E, and finally, moves back through the voltage source to A. A second closed path is A-B-C-E-A, and a third is B-D-C-B.

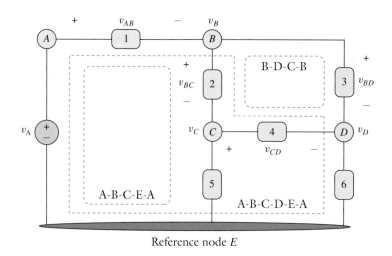

Reference node E

Figure 2.9 A circuit diagram illustrating three closed paths (e.g., *A-B-C-D-E-A*), the concept of node voltages (e.g., v_A and v_B), and the concept of branch voltages (e.g., v_{AB} and v_{BC}).

A second concept that will underlie our statement of KVL is that of a node voltage. A **node voltage** of a circuit is the voltage drop from a given node to a reference node. The circuit of figure 2.9 is labeled with nodes A through E, with node E taken as the reference node, branches 1 through 6, and node voltages $v_A, \ldots, v_D$. As E is the reference node, v_E is taken as zero. v_A denotes the voltage drop from node A to node E, v_D denotes the voltage drop from node D to node E, and similarly for the remaining node voltages.

A third important concept is that of a closed node sequence, which generalizes the notion of a closed path. A **closed node sequence** is a finite sequence of nodes that begins and ends at the same node. Finally, we define the notion of a connected circuit. A **connected circuit** is a circuit for which any node can be reached from any other node by some path through the circuit elements. Figures 2.9 and 2.10 depict connected circuits. Note that the closed node sequence A-B-C-D-E-A in figure 2.10 is not a closed path, because there is no circuit element between nodes B and C.

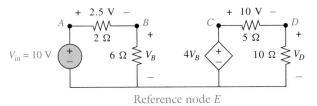

Figure 2.10 Simple dependent source circuit for illustrating the concepts of a connected circuit and a closed node sequence.

This brings us to our statements of KVL.

KIRCHHOFF'S VOLTAGE LAW (KVL)

Kirchhoff's voltage law can be stated in different ways. Following are four equivalent statements of the law.

Statement 1: For lumped circuits, the algebraic sum of the voltage drops around any closed path is zero at every instant of time.

Statement 2: For lumped connected circuits and any node sequence, say, A-D-B- $\ldots$ -G-P, the voltage drop

$$v_{AP} = v_{AD} + v_{DB} + \ldots + v_{GP}$$

at every instant of time.

Statement 3: For lumped circuits and all pairs of nodes j and k, the voltage drop v_{jk} from node j to node k is

$$v_{jk} = v_j - v_k$$

at every instant of time. Here, v_j is the voltage at node j with respect to the reference node, and v_k is the voltage at node k with respect to the reference node. Note that j and k stand for arbitrary indices and could be any of the nodes in figure 2.9, e.g., $j, k \in \{A, B, C, \ldots, E\}$.

Statement 4: For lumped connected circuits, the algebraic sum of the node-to-node voltages for any closed node sequence is zero for every instant of time.

Referring back to figure 2.9 and the closed path A-B-C-D-E-A, we see that KVL implies that $v_{AB} + v_{BC} + v_{CD} + v_{DE} + v_{EA} = 0$. Note that $v_{EA} = -v_{AE} = -v_A$. Further, from statement 2 of KVL, $v_A = v_{AB} + v_{BC} + v_{CD} + v_{DE} = v_{AB} + v_{BC} + v_C$. Finally, $v_{AB} = v_A - v_B$ and $v_{CD} = v_C - v_D$, as asserted by statement 3 of KVL.

As a second illustration of KVL, refer to the circuit of figure 2.10. Consider the node sequence E-A-B-E. Here, $V_{in} = 10 = V_{AB} + V_B = 2.5 + V_B$. Hence, $V_B = 7.5$ V. Now consider the node sequence E-C-D-E. For this sequence, $4V_B = V_{CD} + V_D$, or equivalently, $30 = 10 + V_D$. This implies that $V_D = 20$ V. Finally, consider the node sequence E-B-C-E. This node sequence is not a closed path, because there is no element between nodes B and C. Nevertheless, it follows that $-V_B + V_{BC} + V_C = 0$, or equivalently, $V_{BC} = V_B - V_C = 7.5 - 30 = -22.5$ V.

All of the preceding four statements, as well as their equivalence, can be justified using the definition of and the notation for "voltage" drop presented in section 3 of chapter 1. The justification is particularly easy to comprehend if one uses the gravitational field analogy, as was done in that section.

> **Exercises.**
>
> 1. In the circuit of figure 2.9, suppose $v_C = 10$ V and $v_D = -3$ V. Find v_{DC}.
> 2. Suppose $v_B = 120\cos(120\pi t)$, $v_{BD} = 18\cos(120\pi t)$, and $v_C = 32\cos(120\pi t)$. Find v_{CD} at $t = 0.5$ sec.
> 3. Find v_{AD} when $v_A = 100$ V, $v_{DC} = -10$ V and $v_C = 25$ V.
> 4. True or false? Two voltage sources of different voltages can be connected in parallel.
>
> **SCRAMBLED ANSWERS:** 85 V, -13 V, -70 V, false.

Two further implications of KVL are of immediate interest. First, as a general rule, KVL implies that two voltage sources of different voltages, say, $v_1(t) \neq v_2(t)$, cannot be connected in parallel. Such a connection would violate KVL, which requires that $v_1(t) = v_2(t)$ for a parallel connection. Second, a voltage source supplying zero V for all possible values of current through the source is equivalent to a short circuit. This suggests that the **internal resistance** of a **voltage source** is zero. One can see this by referring to the fact that in the i-v plane an ideal dc voltage source is represented by a horizontal line. The slope of the line is zero and represents the resistance of the source. These ideas are dual to those expressed for current sources earlier.

As a final comment, observe that KVL holds for all closed node sequences, independently of the lumped device represented by each branch of the connected circuit. The distribution of voltages around closed paths can be viewed as a special case of this general statement.

4. APPLICATION OF KVL: VOLTAGE DIVISION AND SERIES RESISTANCE

During holidays one often sees strings of lights hanging between poles or trees. Many times, these strings contain a series connection of light bulbs. Inside each light bulb is a coil of wire called a *filament*, which gives off an intense light when hot. From a circuit-theoretical perspective, the filament has an equivalent hot resistance. Accordingly, one can model the series connection of bulbs by a series connection of resistors with each resistor paired with a specific bulb. Determining the voltage across each of the lights would then be equivalent to determining the voltage across each of the resistors in the model. It is quite common to model electrical loads by resistors. Also, it is important to calculate the voltage across each resistor, which represents an electrical load such as a light bulb. We can illustrate such a calculation by studying the case of three resistors in series across a voltage source $v_{in}(t)$, as illustrated in figure 2.11.

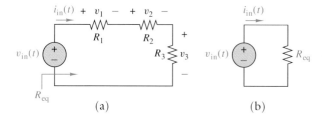

Figure 2.11 (a) Three series resistors connected across a voltage source. By KCL, the current through each resistor is $i_{in}(t)$. (b) The equivalent resistance, $R_{eq} = R_1 + R_2 + R_3$, "seen" by the source, i.e., $v_{in}(t) = R_{eq} i_{in}(t)$.

In the circuit of figure 2.11, the current through each resistor is $i_{in}(t)$, by KCL. From Ohm's law, the voltage across each resistor is

$$v_j(t) = R_j i_{in}(t),$$

for $i = 1, 2, 3$. These two facts suggest that if a resistance R_i is small relative to the other resistances, then only a small portion of the source voltage will develop across it. On the other hand, if R_i is large relative to the other resistances, then a larger portion of the source voltage will develop across it. These two notions in turn suggest that voltage distributes around a loop of resistors in proportion to each of the resistances.

To see this, apply KVL to the circuit. The source voltage must equal the sum of the resistor voltages, i.e.,

$$v_{in}(t) = v_1(t) + v_2(t) + v_3(t) = (R_1 + R_2 + R_3)i_{in}(t), \tag{2.1}$$

where we have substituted $R_j i_{in}(t) = v_j(t)$. Dividing equation 2.1 by $(R_1 + R_2 + R_3)$ yields

$$i_{in}(t) = \frac{v_{in}(t)}{R_1 + R_2 + R_3}.$$

Since $v_j(t) = R_j i_{in}(t)$ for $j = 1, 2,$ and 3,

$$v_j(t) = R_j i_{in}(t) = \frac{R_j}{R_1 + R_2 + R_3} v_{in}(t). \tag{2.2a}$$

This is a **voltage division formula** for a three-resistor series circuit. The formula expresses each resistor voltage in the series connection as a proportion of the input voltage. The proportion is simply the ratio of the branch resistance to the total series resistance. For a circuit composed of n resistors in series, as illustrated in figure 2.12, the general voltage division formula is

$$v_j(t) = \frac{R_j}{R_1 + R_2 + \cdots + R_n} v_{in}(t), \tag{2.2b}$$

for $j = 1, \ldots, n$.

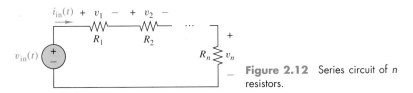

Figure 2.12 Series circuit of n resistors.

There are a few other interesting aspects to the preceding analysis. The total resistance in equation 2.1 is $R_{eq} = R_1 + R_2 + R_3$, which satisfies Ohm's law for the input voltage $v_{in}(t)$ and input current $i_{in}(t)$; i.e., $v_{in}(t) = R_{eq} i_{in}(t)$. Therefore, one concludes that $R_{eq} = R_1 + R_2 + R_3$ is the equivalent resistance "seen" by the source. It also follows that *resistances in series add*; i.e., resistors in series can be combined into a single resistor whose resistance is the sum of the resistances. This is expressed in figure 2.11b.

Exercise. In figure 2.12, suppose each resistor is 1 Ω. Determine the equivalent resistance.

ANSWER: n Ω.

◆ EXAMPLE 2.3

Find the equivalent resistance "seen" by the source for the circuit of figure 2.13.

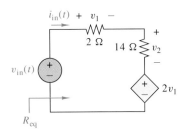

Figure 2.13 Series circuit containing a dependent voltage source.

SOLUTION

To determine the equivalent resistance requires that one deduce the relationship between the input voltage and the input current.

Step 1. The first step is to write KVL for the circuit, i.e.,

$$v_{in} = v_1 + v_2 + 2v_1 = 3v_1 + v_2. \tag{2.3}$$

Step 2. The second step is to express v_1 and v_2 in terms of i_{in} using Ohm's law: $v_1 = 2i_{in}$ and $v_2 = 14i_{in}$. Substituting into equation 2.3 yields

$$v_{in} = 20i_{in} = R_{eq}i_{in}.$$

Therefore, $R_{eq} = 20\ \Omega$. Notice that the dependent source increases the resistance of the two series resistors by $4\ \Omega$. Dependent sources can increase or decrease the resistance of a circuit. They may even make the resistance negative.

Exercise. Suppose the dependent source in the circuit of figure 2.13 has its value changed to $2(v_1 + v_2)$. Find R_{eq}.
ANSWER: $48\ \Omega$.

5. APPLICATION OF KCL: CURRENT DIVISION AND PARALLEL RESISTANCE

The outlets in the average home are connected in parallel. As numerous people have experienced, too many appliances connected to the same outlet or set of outlets on the same fused circuit cause the fuse to blow or breaker to open. Although each appliance may use only a portion of the maximum allowable current for the circuit, together the total current exceeds the allowable limit.

Fundamental to the analysis that follows is the question of how current distributes through a parallel connection of resistors. To keep the analysis simple, consider a set of three parallel resistors driven by a current source, as illustrated in the circuit of figure 2.14. Our goal is to derive a formula that relates the current through each resistor, $i_j(t)$, to the input current $i_{in}(t)$. The variable that links $i_i(t)$ to $i_{in}(t)$ is the voltage $v_{in}(t)$, which, by KVL, appears across each resistor.

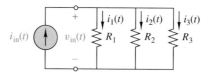

Figure 2.14 Three parallel resistors driven by a current source.

Since $v_{in}(t)$ appears across each of the resistors, Ohm's law implies that each resistor current is

$$i_j(t) = \frac{v_{in}(t)}{R_j} = G_j v_{in}(t), \tag{2.4}$$

where $G_j = 1/R_j$ is the conductance in mhos and $j = 1, 2,$ and 3. Applying KCL to the top node of the circuit yields

$$i_{in}(t) = i_1(t) + i_2(t) + i_3(t).$$

Using equation 2.4 to substitute for each $i_j(t)$ yields the relationship between $v_{in}(t)$ and $i_{in}(t)$:

$$v_{in}(t) = \frac{1}{\dfrac{1}{R_1} + \dfrac{1}{R_2} + \dfrac{1}{R_3}} i_{in}(t) = \frac{1}{G_1 + G_2 + G_3} i_{in}(t). \tag{2.5}$$

This formula is a form of Ohm's law for the parallel circuit, where $G_{eq} = G_1 + G_2 + G_3$ is the equivalent conductance of the parallel circuit and $R_{eq} = 1/G_{eq}$ is the equivalent resistance. We shall have more to say on this shortly.

To obtain a relationship between $i_{in}(t)$ and $i_j(t)$, we substitute equation 2.5 into equation 2.4 to obtain

$$i_j(t) = \frac{\dfrac{1}{R_j}}{\dfrac{1}{R_1} + \dfrac{1}{R_2} + \dfrac{1}{R_3}} i_{in}(t) = \frac{G_j}{G_1 + G_2 + G_3} i_{in}(t). \tag{2.6}$$

This is called a **current division formula.** It says that currents distribute through the branches of a parallel resistive circuit in proportion to the conductance of the particular branch, G_j, relative to the total conductance of the circuit, $G_{eq} = G_1 + G_2 + G_3$.

Exercises.

1. In figure 2.14, suppose $R_1 = 1\ \Omega$, $R_2 = 0.5\ \Omega$, and $R_3 = 0.5\ \Omega$. Find the current through R_1 if $i_{in}(t) = 10e^{-t}$ A.

   ANSWER: $i_1(t) = 2e^{-t}$ A.

2. Refer again to figure 2.14, but suppose there are only two resistors in parallel. Show that

$$i_1(t) = \frac{R_2}{R_1 + R_2} i_{in}(t) \quad \text{and} \quad i_2(t) = \frac{R_1}{R_1 + R_2} i_{in}(t).$$

There is another very important aspect to the preceding derivation. Recall that equation 2.5 relates the input current $i_{in}(t)$ to the input voltage $v_{in}(t)$, in the form of Ohm's law. One concludes, then, that $G_{eq} = G_1 + G_2 + G_3$ is the equivalent conductance "seen" by the source of the parallel circuit and $R_{eq} = 1/G_{eq}$ is the equivalent resistance. We can further interpret this to mean that *conductances in parallel add to form equivalent conductances.* This is analogous to the fact that resistors in series add to form equivalent resistances. On the other hand, resistances in parallel do not add, and conductances in series do not add. We can therefore conclude that from the perspective of the source, the parallel circuit of figure 2.14 has the equivalent representations given in figure 2.15.

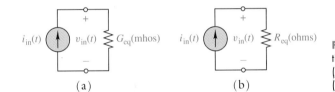

(a) (b)

Figure 2.15 Equivalent representations of the parallel circuit of figure 2.14. (a) Conductance specified in mhos. (b) Resistance specified in ohms.

Exercises.

1. Show that the equivalent resistance of two resistors in parallel (see figure 2.16) is given by the formula

$$R_{eq} = \frac{R_1 R_2}{R_1 + R_2}.$$

2. In figure 2.16, suppose $i_{in}(t) = 12$ A and $R_1 = 10\ \Omega$. Find R_2 so that $i_2(t) = 4$ A.

   ANSWER: $R_2 = 20\ \Omega$.

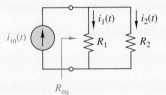

Figure 2.16 Two resistors in parallel driven by a current source.

The general case of n resistors in parallel is shown in figure 2.17. The equivalent resistance R_{eq} of the parallel set of resistors in figure 2.17 is

$$R_{eq} = \cfrac{1}{\cfrac{1}{R_1} + \cfrac{1}{R_2} + \cdots + \cfrac{1}{R_n}} = \frac{1}{G_1 + G_2 + \cdots + G_n} = \frac{1}{G_{eq}} \qquad (2.7)$$

i.e., $G_{eq} = G_1 + G_2 + \cdots + G_n$. Further, the current through each branch satisfies the current division formula

$$i_j(t) = \cfrac{\cfrac{1}{R_j}}{\cfrac{1}{R_1} + \cfrac{1}{R_2} + \cdots + \cfrac{1}{R_n}} i_{in}(t) = \frac{G_j}{G_1 + G_2 + \cdots + G_n} i_{in}(t). \qquad (2.8)$$

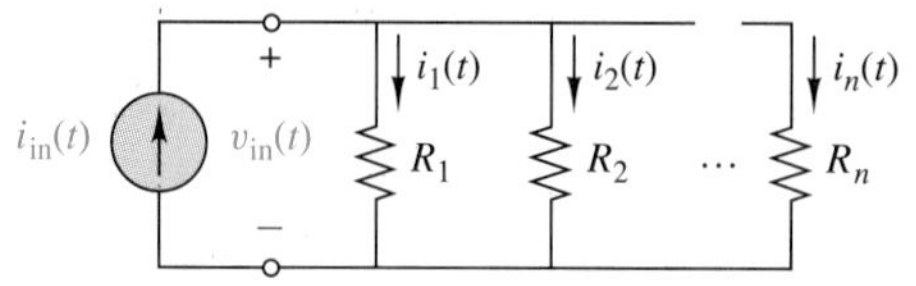

Figure 2.17 Parallel connection of n resistors driven by a current source.

◆ EXAMPLE 2.4

Find the input voltage and the current through R_2 in the circuit of figure 2.18 when

$$i_{in}(t) = \begin{cases} 5e^{-t} \text{ A} & \text{for } t \geq 0 \\ 0 & \text{for } t < 0 \end{cases}.$$

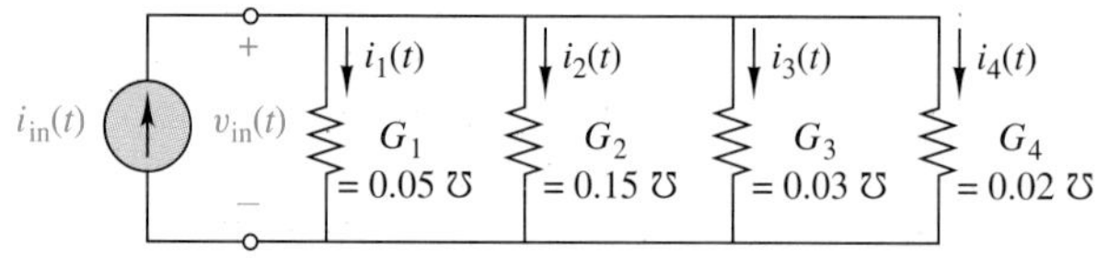

Figure 2.18 Parallel connection of four resistors.

SOLUTION

Step 1. Compute the equivalent conductance and equivalent resistance of the circuit:

$$G_{eq} = G_1 + G_2 + G_3 + G_4 = 0.25 \text{ mho},$$

and

$$R_{eq} = \frac{1}{G_{eq}} = 4 \ \Omega.$$

Step 2. From Ohm's law,

$$v_{in}(t) = R_{eq} i_{in}(t) = \begin{cases} 20e^{-t} \text{ V} & \text{for } t \geq 0 \\ 0 & \text{for } t < 0 \end{cases}.$$

Step 3. Using the current division formula of equation 2.8 to find $i_2(t)$ yields

$$i_2(t) = \frac{G_2}{G_{eq}} i_{in}(t) = \frac{0.15}{0.25} i_{in}(t) = \begin{cases} 3e^{-t} \text{ A} & \text{for } t \geq 0 \\ 0 & \text{for } t < 0 \end{cases}.$$

◆ EXAMPLE 2.5

A nonideal source in figure 2.19 supplies power to a load R_L. Determine (1) the load current I_L, (2) the voltage across the load, V_L, (3) the power absorbed by the load, and (4) the power delivered by the ideal current source.

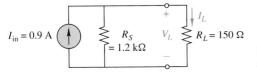

Figure 2.19 Parallel resistive circuit with nonideal source.

SOLUTION

Step 1. By the current division formula, the load current is

$$I_L = \frac{\dfrac{1}{R_L}}{\dfrac{1}{R_S} + \dfrac{1}{R_L}} I_{in} = \frac{\dfrac{1}{150}}{\dfrac{1}{150} + \dfrac{1}{1.2 \times 10^3}} 0.9 = 0.8 \text{ A}.$$

Step 2. By Ohm's law, the load voltage is

$$V_L = 150 I_L = 120 \text{ V}.$$

Step 3. The power absorbed by the load is

$$P_L = (0.8) \times (120) = 96 \text{ W}.$$

where W indicates watts.

Step 4. The power absorbed by the 1.2-kΩ resistor is

$$P_{abs} = \frac{V_L^2}{1.2 \times 10^3} = 12 \text{ W}.$$

Step 5. The power delivered by the ideal current source is

$$P_{del} = 12 + 96 = 108 \text{ W}.$$

This can be checked as $P_{del} = I_{in} \times V_L = 108$ W.

6. SERIES-PARALLEL INTERCONNECTIONS

Finding the equivalent resistance often simplifies computations of input and output currents, as well as calculations of voltages. The examples that follow illustrate the two principles developed in the last section, namely, that (1) resistances in series add to form an equivalent resistance and (2) conductances in parallel add to form an equivalent conductance.

 EXAMPLE 2.6

Find the equivalent resistance and the voltage across the source for the circuit of figure 2.20.

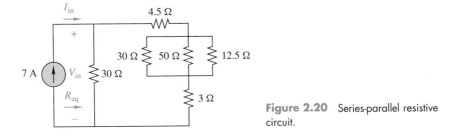

Figure 2.20 Series-parallel resistive circuit.

SOLUTION

To compute the equivalent resistance, we first find the equivalent resistance of the parallel combination of the 30-Ω, 50-Ω, and 12.5-Ω resistors on the right side of the circuit. We use the property that *conductances in parallel add* and that the equivalent resistance is the reciprocal of the conductance. In particular, the equivalent resistance of these parallel branches is given

$$R_{eq1} = \frac{1}{G_{eq1}} = \frac{1}{\dfrac{1}{30} + \dfrac{1}{50} + \dfrac{1}{12.5}} = 7.5 \ \Omega.$$

With this resistance, the rightmost side of the circuit is equivalent to three series resistances whose equivalent resistance is their sum; i.e.,

$$R_{eq2} = 4.5 + 7.5 + 3 = 15 \ \Omega.$$

Finally, the desired R_{eq} is the parallel combination of the 30-Ω resistor with R_{eq2}. The result is

$$R_{eq} = \frac{1}{G_{eq}} = \frac{1}{\dfrac{1}{30} + \dfrac{1}{15}} = 10 \ \Omega.$$

The voltage across the source is simply

$$V_{in} = R_{eq}I_{in} = 10 \times 7 = 70 \ V.$$

◆ EXAMPLE 2.7

In this second example of finding equivalent resistances, consider the circuit of figure 2.21. This circuit is a simplified approximation of an amplifier circuit.

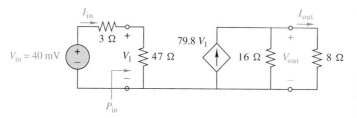

Figure 2.21 Crude approximation of an amplifier circuit. The dependent source acts as a voltage amplifier.

(a) Compute I_{out}.
(b) Compute V_{out}.
(c) Compute P_{out}, the power absorbed by the 8-Ω resistor.
(d) Compute I_{in}.
(e) Compute the input power to the amplifier, P_{in}.
(f) Compute the power gain P_{out}/P_{in}.

SOLUTION

(a) To compute I_{out}, one must first compute V_1 by voltage division. Here,

$$V_1 = \frac{47}{50} 40 \times 10^{-3} = 37.6 \times 10^{-3} \ V.$$

Using this value of V_1 and current division on the right half of the circuit yields

$$I_{out} = \frac{0.125}{0.125 + 0.0625} \left(79.8 \times 37.6 \times 10^{-3}\right) = 2 \ A.$$

(b) V_{out} follows by Ohm's law, i.e.,

$$V_{out} = 2 \times 8 = 16 \ V.$$

(c) P_{out} is simply the product of the voltage and current delivered to the load; i.e.,

$$P_{out} = 2 \times 16 = 32 \ W.$$

(d) Clearly,

$$I_{in} = \frac{40 \times 10^{-3}}{50} = 0.8 \text{ mA}.$$

(e) The power delivered to the amplifier is

$$P_{in} = (37.6 \times 10^{-3})(0.8 \times 10^{-3}) = 30.08 \times 10^{-6} \text{ W}.$$

(f) The resulting power gain is the ratio of the power absorbed by the 8-Ω load to the power delivered by the independent source. Thus,

$$\text{Power gain} = \frac{P_{out}}{P_{in}} = \frac{32}{30.08} \times 10^6 = 1.064 \times 10^6.$$

Example 2.6 points out a very interesting fact: The equivalent resistance of a series-parallel connection of resistors requires only two types of arithmetic operations—adding two numbers and taking the reciprocal of a number—no matter how complex the network. A hand calculator easily executes both operations. Such is not the case with a non-series-parallel network. To find the equivalent resistance of a non-series-parallel network, one usually must write simultaneous equations and evaluate determinants, a topic to be studied in detail in chapter 4.

It is important, then, to recognize when a problem belongs to the series-parallel category, in order to take advantage of the series-parallel operations. In the previous series-parallel examples, one and only one independent source has been explicitly indicated on the circuit diagram. This is part of the definition of a series-parallel network: The independent source must be indicated, or equivalently, the pair of terminals, called *input terminals*, at which the source is to be connected must be specified. The specification of the input terminals determines whether or not a network is series parallel. The following example illustrates this point.

EXAMPLE 2.8

Consider the circuit of figure 2.22a. Determine whether or not the network is series parallel for each of the following pairs of terminals:

1. Case 1: (A, B);
2. Case 2: (A, C);
3. Case 3: (C, D).

If the answer is affirmative, give an expression as well as the numerical values for the equivalent resistance, using the notation $//$ (double slash) for combining resistances in parallel; i.e., $R_1 // R_2$ means that R_1 and R_2 are in parallel, and $R_1 // R_2 // R_3$ means that R_1 is in parallel with R_2 which is in parallel with R_3.

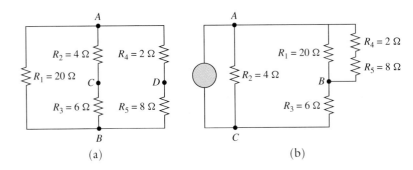

Figure 2.22 (a) The concept of a series-parallel and a non-series-parallel network. (b) Series-parallel network seen at terminals (A, C) of (a).

SOLUTION

Case 1. With an independent source connected to nodes A and B, the source "sees" a series-parallel network. By inspection of figure 2.22a, the equivalent resistance is

$$R_{eq} = R_1 // [(R_2 + R_3) // (R_4 + R_5)] = 20 // [(4 + 6) // (2 + 8)] = 4 \text{ Ω}.$$

Case 2. With (A, C) as the input terminal pair, the network is series-parallel. This is made apparent by redrawing the network as shown in figure 2.22b, from which it follows that

$$R_{eq} = R_2 // \{ R_3 + [(R_4 + R_5)//R_1] \} = 4//\{6 + [(2 + 8)//20]\} = 3.04 \ \Omega.$$

Case 3. With (C, D) as the input terminal pair, the network is not series-parallel. The calculation of R_{eq} for this case requires methods to be discussed in chapter 4.

7. CIRCUIT APPLICATIONS: DESIGN OF ANALOG MULTIMETERS

A **multimeter** (also called a *volt-ohm meter*, abbreviated VOM) is an instrument capable of measuring voltage, current, and resistance. There are two varieties: the analog multimeter and the digital multimeter. An analog meter displays a measurement as the deflection of a needle pointer on a calibrated scale. A digital meter displays a measurement value on an LED bank or on an LCD readout as in a digital watch. Each meter runs on an entirely different principle of operation. Although analog multimeters are less common than in the past, their principles of operation show a remarkably useful application of simple resistive circuits to practical design, which we shall explore in this section.

The heart of an analog multimeter is the meter movement illustrated in figure 2.23a. A movable coil, consisting of many turns of fine wire, is mounted on a pivot between the north and south poles of a permanent magnet. When a current passes through the coil, it interacts with the magnetic field to produce a torque, which then rotates the coil, deflecting the attached needle to the right. A spiral spring restrains the rotation of the coil. The needle comes to rest when the electromagnetic force and the spring force balance each other out. When the current is removed, the spring returns the needle to its initial position on the extreme left, marked as 0 on the scale. The spring, coil, and needle are designed so that the total assembly produces a deflection proportional to the coil current.

A manufactured meter movement has two important specifications:

1. R_m, the meter's internal resistance or the coil's dc resistance.
2. I_m, the full-scale current, i.e., the current through the coil when the needle has a full-scale deflection to the right.

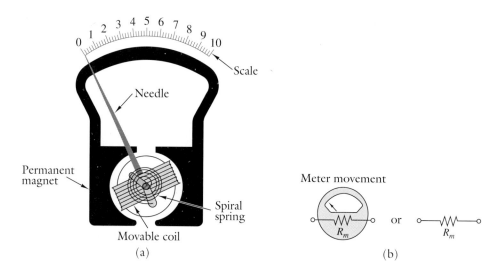

Figure 2.23 (a) A meter movement and (b) its circuit representation.

With this information, the full-scale voltage $V_m = R_m I_m$, by Ohm's law. Since the rotation of the needle is designed to be proportional to the current through R_m, the entire movement is represented by R_m, as in figure 2.23b. The drawing of the scale and needle there is not essential, but helps to identify R_m as a measuring device.

In commercial products, the rotating coil is usually designed so that I_m has standard values, such as 1 ma, 50 μA, 20 μA, etc. Generally speaking, the smaller I_m, the more sensitive the meter is, with an accompanying higher price for the unit. To achieve a precise value of I_m, one adjusts the number of turns of the coil, assuming a fixed permanent magnet. The coil's resistance depends on the number of turns and size of the wire. Table 2.1 lists some typical values of I_m, R_m, and V_m.

TABLE 2.1 DATA FOR TYPICAL ANALOG METER MOVEMENTS				
Item	I_m	R_m	V_m	Sensitivity
1	50 μA	1,140 Ω	57 mV	20 kΩ/V
2	1 mA	105 Ω	105 mV	1 kΩ/V
3	50 μA	2,000 Ω	100 mV	20 kΩ/V
4	1 mA	46 Ω	46 mV	1 kΩ/V

Directly using meter movement 1 in table 2.1 has two drawbacks: the meter can measure only very low voltages, up to 57 mV; and construction of its scale is difficult, since the full-scale voltage of 57 mV is a poor value for uniform scale labeling. The other meter movements have similar problems. To remedy these drawbacks, designers add series *multiplier* resistors to construct a voltmeter and parallel *shunt* resistors to construct an ammeter.

Voltmeters

Several examples will illustrate the basic ideas behind the design of a voltmeter.

EXAMPLE 2.9

Using meter movement 1 in table 2.1, construct a voltmeter having a range of 0–5 V.

SOLUTION

Figure 2.24 illustrates the voltmeter setup. In connecting the two meter probes across 5 V, one desires a full-scale deflection, at which time the current should be 50 μA. This implies that the total resistance seen when looking into the probes should be (5 V)/(50 μA) = 10^5 Ω. The coil has a dc resistance of 1,140 Ω. Therefore, the necessary series multiplier resistor should be $R_1 = 100,000 - 1,140 = 98,860$ Ω.

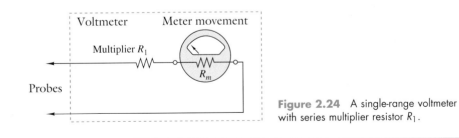

Figure 2.24   A single-range voltmeter with series multiplier resistor R_1.

EXAMPLE 2.10

Redesign the voltmeter of example 2.9 for the following multiple ranges: 0–1 V, 0–5 V, 0–150 V.

SOLUTION

Following the reasoning of example 2.9, a formula for the multiplier resistance is

$$R_x = \frac{V_{\text{fs}} - V_m}{I_m},$$

where V_{fs} denotes the desired full-scale voltage. Substituting for the known values of I_m, V_m, and the three values of V_{fs} yields

$$R_1 = 18,860 \ \Omega, \quad R_2 = 98,860 \ \Omega, \quad R_3 = 2,998,860 \ \Omega.$$

The complete schematic diagram for the multirange voltmeter is shown in figure 2.25.

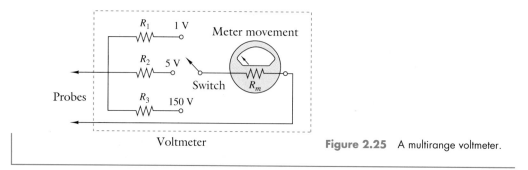

Voltmeter

Figure 2.25   A multirange voltmeter.

The equivalent resistance R_{eq} "seen" looking into the two probes, or test leads, of the voltmeter depends on the selected voltage range. Table 2.2 summarizes the design parameters for the voltmeter of figure 2.25.

TABLE 2.2 DATA ASSOCIATED WITH THE VOLTMETER OF FIGURE 2.25			
Full-scale voltage	1 V	5 V	150 V
Multiplier resistance R_x	18,860 Ω	98,860 Ω	2,998,860 Ω
Total resistance R_{eq}	20 kΩ	100 kΩ	3 MΩ

From the table, the ratio

$$\frac{\text{Total resistance}}{\text{Full-scale voltage}},$$

in ohms per volt, is constant and equal to the reciprocal of the full-scale current I_m. This ohm-per-volt value is called the **sensitivity** of the voltmeter. The voltmeter of example 2.10 has a full-scale current of 50 μA and therefore a sensitivity of 20 kΩ per volt. The larger this value, the better the voltmeter is, in the sense of having reduced loading effects on the circuit. The next example illustrates the effect of loading.

 **EXAMPLE 2.11**

Consider the circuit of figure 2.26a.

(a) Calculate the voltage V_o.
(b) A voltmeter with a 1-kΩ-per-V sensitivity is used to measure V_o over a range of 0–10 V. Determine the meter reading.
(c) If a voltmeter with a 20-kΩ-per-V sensitivity is used to measure V_o over a range of 0–10 V, determine the meter reading.

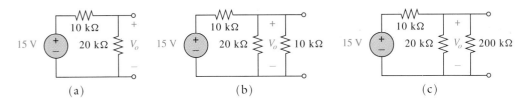

Figure 2.26 The loading effect due to a voltmeter. (a) Circuit whose output V_o is to be measured. (b) Loading effect of low-sensitivity voltmeter. (c) Loading effect of high-sensitivity voltmeter.

SOLUTION

(a) Voltage division on figure 2.26a indicates that

$$V_o = \frac{20}{10 + 20} 15 = 10 \text{ V}.$$

(b) Over the range of 0–10 V, the resistance between the probes of the voltmeter is 10 V × (1 kΩ/V) = 10 kΩ. This represents a 10-kΩ load connected in parallel with a 20-kΩ resistance, as shown in figure 2.26b. Voltage division then yields $V_o = 6$ V. This is a 40% deviation from the true answer of $V_o = 10$ V.

(c) Over the range of 0–10 V, the resistance between the probes of the voltmeter is 10 V × (20 kΩ/V) = 200 kΩ. This represents a 200-kΩ load connected in parallel with a 20-kΩ resistance, as shown in figure 2.26c. Again, voltage division yields $V_o = 9.677$ V. This is within a reasonable tolerance of the precise answer.

The foregoing example shows very clearly the effect of loading due to the measuring instrument and thus emphasizes the importance of choosing a voltmeter with adequate sensitivity. Although modern-day voltmeters typically have a sensitivity better than 20 kΩ/V, a meter with a sensitivity of 1 kΩ/V is used in the example to dramatize the effect of loading.

Exercise. Repeat the design of example 2.10, using meter movement 2 in table 2.1.

Ammeters

A major difference between voltage and current measurement is how the probes are connected to the circuit. For voltage measurement, the two test leads are connected *across* the nodes of interest. For current measurement, the proper circuit branch must be broken at a proper point in order to insert the meter probes.

The meter movement may be used directly as an ammeter. However, the useful range is limited to very small currents. To extend the range, one places a *shunt* resistor in parallel with the meter movement, i.e., in parallel with R_m. Thus, only a very small fraction of the total current flows through the meter coil. A few examples will illustrate the basic ideas behind the design of an ammeter.

 EXAMPLE 2.12

Using meter movement 2 in table 2.1, construct an ammeter having a range of 0–10 mA.

SOLUTION

Figure 2.27 illustrates the structure of the problem. When 10 mA flows into one probe and out of the other, a full-scale deflection will require a coil current of 1 mA. Therefore, the shunt resistor, R_{sh}, must carry $(10 - 1) = 9$ mA. Use of the current division formula, equation 2.8, yields

$$9 \text{ mA} = \frac{105}{105 + R_{sh}} 10 \text{ mA}.$$

Thus, $R_{sh} = 105/9 = 11.67$ Ω.

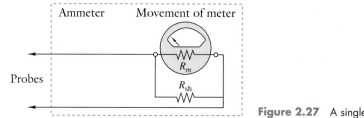

Figure 2.27 A single-range ammeter.

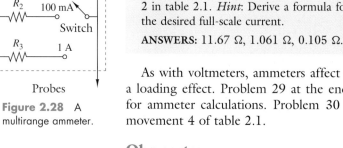

Figure 2.28 A multirange ammeter.

Exercise. Find the resistances R_1, R_2, and R_3 in figure 2.28 necessary to achieve the following multiple measuring ranges: (1) 0–10 mA, (2) 0–100 mA, (3) 0–1 A. Use meter movement 2 in table 2.1. *Hint*: Derive a formula for R_{sh} in terms of V_m, I_{fs}, and I_m, where I_{fs} denotes the desired full-scale current.

ANSWERS: 11.67 Ω, 1.061 Ω, 0.105 Ω.

As with voltmeters, ammeters affect the circuit current to be measured. Again, this is a loading effect. Problem 29 at the end of the chapter investigates the effect of loading for ammeter calculations. Problem 30 asks for the design of an ammeter, using meter movement 4 of table 2.1.

Ohmmeter

An ohmmeter is an instrument that measures the dc resistance of a two-terminal device. In its simplest form, an ohmmeter can be constructed using a basic meter movement, a battery (typically 1.5 V or 3 V), and an adjustable resistor, as illustrated in figure 2.29.

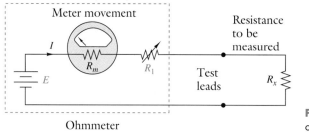

Figure 2.29 A basic single-range ohmmeter.

The resistance R_1 of figure 2.29 is adjusted so that when the test leads are shorted, the meter has a full-scale deflection; i.e., $I = I_m$ when $R_x = 0$. This requires that

$$R_1 = \frac{E - R_m I_m}{I_m}.$$

Because the battery voltage E may change with time, R_1 must be variable to compensate for the change. Usually, R_1 is a potentiometer, i.e., a variable resistor that can be adjusted by the user. The knob of the potentiometer serves as a fine-tuning adjustment, so that the needle comes to rest on the rightmost position, marked 0 ohms on the scale, when the leads are shorted. With the test leads open, no current flows through the meter coil. The needle returns to the leftmost position, marked ∞ ohms on the scale, by the action of the spiral spring. When the test leads are connected to a resistance of $R_1 + R_m$ ohms, the current through the meter is reduced to half of the full-scale value. Therefore, the middle point of the scale should be marked with a resistance value of $R_1 + R_m$ ohms, which equals E/I_m. Figure 2.30 shows a typical scale for an ohmmeter.

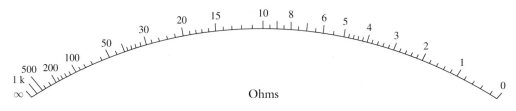

Figure 2.30 A typical ohmmeter scale.

The simple ohmmeter of figure 2.29 has some very serious drawbacks. First, the resistance reading on the meter is very sensitive to the battery voltage. Second, if the unknown resistance falls into a range close to the left end of the scale, it will be hard to read the value accurately because of the crowded markers. (See figure 2.30.) A multirange ohmmeter alleviates these difficulties. To prepare for the design of such an ohmmeter, consider first the design of a single-range ohmmeter that uses two resistors inside the meter, as shown in figure 2.31. The additional resistor, R_a, in parallel with R_m, provides additional flexibility in the design and allows the mid-scale resistance to be any value among a wide range of possibilities.

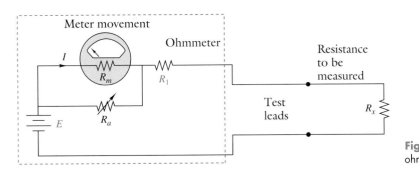

Figure 2.31   An improved single-range ohmmeter.

Analysis of the ohmmeter circuit of figure 2.31 requires no more than series-parallel circuit calculations. Unfortunately, the determination of R_1 and R_a is messy if done without foresight. A brief explanation of the design procedure follows.

As before, when the test leads are shorted, the meter should show a full-scale deflection. The circuit is shown in this condition in figure 2.32a.

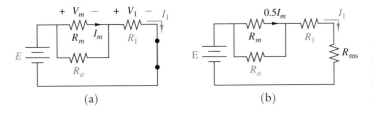

Figure 2.32   Analysis of the series-parallel circuit of an ohmmeter. (a) Circuit with test leads shorted for full-scale deflection. (b) Circuit measuring R_{ms} for mid-scale deflection.

We denote the parallel combination of R_m and R_a by R_p, i.e.,

$$R_p = \frac{R_m R_a}{R_m + R_a}.$$

From figure 2.32a, R_p and R_1 are series connected. Therefore,

$$\frac{R_1}{R_p} = \frac{I_1 R_1}{I_1 R_p} = \frac{V_1}{V_m} = \frac{E - V_m}{V_m}. \tag{2.9}$$

Suppose the midpoint of the scale is to be marked with a resistance value R_{ms}. (The subscript ms stands for <u>m</u>id<u>s</u>cale.) When the test leads are connected to R_{ms}, the current through the meter coil must be $0.5I_m$, i.e., one-half of the full-scale current. Figure 2.32b shows this condition. Comparing figure 2.32b with figure 2.32a, we find that the currents through R_a and R_1 must also be half of their values shown in figure 2.32a. This implies that R_{ms} is equal to the series combination of R_p and R_1:

$$R_{ms} = R_1 + R_p = R_1 + \frac{R_m R_a}{R_m + R_a}. \tag{2.10}$$

Equations 2.9 and 2.10 may now be used to calculate R_1 and R_a as follows. Look up the data, I_m, R_m, for the desired meter movement. Choose the battery voltage E to be used. Choose a desired midscale resistance value R_{ms}. Then equation 2.9 gives the ratio R_1/R_p, and equation 2.10 gives the sum $R_1 + R_p$. From these two equations, calculate R_1 and R_p. Finally, from $R_p = R_m R_a/(R_m + R_a)$, calculate R_a. The following example illustrates the complete design procedure.

EXAMPLE 2.13

Using a meter movement that has $R_m = 10 \text{ k}\Omega$ and $I_m = 50 \text{ μA}$ (and hence, $V_m = 0.5 \text{ V}$), design a dual-range ohmmeter. A 1.5-V battery (assumed ideal) is used. The first range has a midscale resistance value of $1,000 \ \Omega$, and $10,000 \ \Omega$ for the second range.

SOLUTION

Using the procedure just described, we tabulate the results in table 2.3, with all resistances in ohms. The complete circuit diagram is shown in figure 2.33.

TABLE 2.3 DESIGN VALUES FOR A DUAL-RANGE OHMMETER OF EXAMPLE 2.13					
Range	$R_1 + R_p$	R_1/R_p	R_p	R_1	R_a
1	1,000	2	333.33	666.66	344.83
2	10,000	2	3,333.30	6,666.60	5,000.00

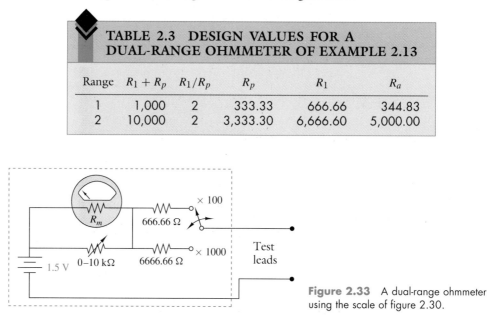

Figure 2.33 A dual-range ohmmeter using the scale of figure 2.30.

Note that fixed resistors are used for R_1, whereas a 0–10-kΩ potentiometer is used for R_a. With the test leads shorted, the potentiometer is first adjusted so that the needle indicates the zero-ohm condition. If the scale shown in figure 2.30 is used for both ranges, then the reading from the scale must be multiplied by 100 for range 1 and by 1,000 for range 2. These scaling constants are marked on the range selector switch.

SUMMARY

This chapter has presented the essential building blocks of linear lumped circuit theory, beginning with the two fundamental laws for interconnecting circuit elements: KVL and KCL. KVL states that for lumped circuits, the algebraic sum of the voltages around any closed node sequence of a circuit is zero. Similarly, KCL says that for lumped circuits, the algebraic sum of the currents entering a node is zero.

In conjunction with Ohm's law, these laws allowed us to develop voltage division and current division formulas. The voltage division formula applies to series resistive circuits driven by a voltage source. The voltage developed across each resistor was found to be proportional to the resistance of the resistor relative to the equivalent resistance "seen" by the source. For example, in a two-resistor circuit, R_1 in series with R_2, we found that

$$v_1 = \frac{R_1}{R_1 + R_2} v_{\text{in}}.$$

The current division formula applies to parallel resistive circuits driven by a current source. Here, the current through

each resistor with conductance G_i was found to be proportional to G_i divided by the equivalent conductance "seen" by the source. Since conductance is the reciprocal of resistance, the idea can also be expressed in terms of the resistances of the circuit. For example, in a two-resistor parallel circuit, R_1 in parallel with R_2,

$$i_1 = \frac{G_1}{G_1 + G_2} i_{\text{in}} = \frac{\dfrac{1}{R_1}}{\dfrac{1}{R_1} + \dfrac{1}{R_2}} i_{\text{in}} = \frac{R_2}{R_1 + R_2} i_{\text{in}}.$$

In deriving the voltage division formula, we learned that the resistances of a series connection of resistors may be added together to obtain an equivalent resistance, prompting the statement that resistors in series add. Analogously, the derivation of the current division formula for parallel circuits led us to conclude that a parallel connection of resistors has an equivalent conductance equal to the sum of conductances. This is sometimes expressed in terms of resistances as the inverse of the sum of the reciprocals; i.e., for n resistors in parallel,

$$R_{eq} = \cfrac{1}{\cfrac{1}{R_1} + \cdots + \cfrac{1}{R_n}},$$

which leads to the very special formula for two resistors in parallel,

$$R_{eq} = \frac{R_1 R_2}{R_1 + R_2},$$

often referred to as the product-over-sum rule.

All of these ideas were applied to the analysis of series-parallel networks, which are interconnections of series and parallel groupings of resistors. Our analysis showed us how to compute the equivalent resistance of such circuits.

The ideas were then applied to the design of analog voltmeters, ammeters, and ohmmeters. Although not as common now as in the past, analog voltmeters were the mainstay of voltage, current, and resistance measurements for many years. What is truly important, however, is that their utility is derived not from sophisticated theory, but from clever applications of the very basic laws of circuit theory.

Finally, the problems at the end of the chapter apply the ideas described throughout the chapter to the design of meters and to the analysis of the resistive speed control of a dc heater blower-motor found in a car. Thus, with the basic knowledge of circuits presented in this chapter, it is possible to understand practical applications.

TERMS AND CONCEPTS

Branch: a two-terminal circuit element, denoted by a line segment.

Closed node sequence: a finite sequence of nodes that begins and ends with the same node.

Closed path: a connection of devices or branches through a sequence of nodes that ends on the node where it began and that traverses each node only once.

Connected circuit: a circuit for which any node can be reached from any other node by some path through the circuit elements.

Current division: formula showing that the distribution of current in a branch of a parallel resistive circuit is proportional to the conductance of the particular resistor, G_j, divided by the total parallel conductance of the circuit, $G_{eq} = G_1 + G_2 + G_3$.

Gaussian surface: a closed curve in the plane or a closed surface in three dimensions with a well-defined inside and outside.

Kirchhoff's current law (KCL): law asserting that the algebraic sum of the currents entering a node of a circuit consisting of lumped elements is zero for every instant of time. In general, for lumped circuits, the algebraic sum of the currents entering (leaving) a Gaussian surface is zero at every instant of time.

Kirchhoff's voltage law (KVL): law asserting that for lumped circuits, the algebraic sum of the voltage drops around any closed path in a network is zero at every instant of time. In general, for lumped connected circuits, the algebraic sum of the node-to-node voltages for any closed node sequence is zero for every instant of time.

Multimeter or volt-ohmmeter (VOM): an instrument capable of measuring voltage, current, and resistance.

Node: the end point of a branch that represents the terminals of a circuit element. In general, a node is a point of connection between two or more circuit elements (branches).

Node voltage: the voltage drop from a given node to the reference node.

Parallel circuit: a side-by-side connection of two-terminal circuit elements whose top terminals are wired together and whose bottom terminals are wired together.

Series circuit: a sequential connection of two-terminal circuit elements, end to end.

Voltage division: formula showing that each resistor voltage in a series connection is a proportion of the input voltage that is equal to the ratio of the branch resistance to the total series resistance.

// (double slash): notation for combining resistances in parallel. $R_1//R_2$ means that R_1 and R_2 are in parallel, and $R_1//R_2//R_3$ means that R_1 is in parallel with R_2, which is in parallel with R_3.

PROBLEMS

1. Use KVL to determine the voltage V_x in the circuit of figure P2.1.

2. Consider the circuit of figure P2.2, in which each shaded box is a general circuit element. What is the current I?

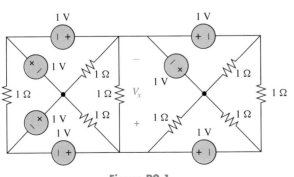

Figure P2.1

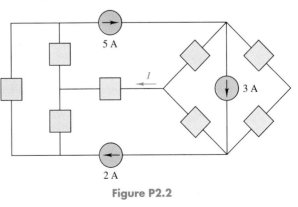

Figure P2.2

3. If $I_{in} = 7$ A, determine the current I_{out} for the circuit of figure P2.3.

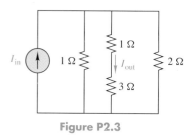

Figure P2.3

4. This is a conceptual problem and requires no calculations for the answer. Consider circuits 1 and 2 of figure P2.4. All resistors are 1 Ω, except the one labeled R Ω. The difference between R_{eq1} and R_{eq2} is the presence of the positive nonzero R-Ω resistor between points a and b. The equivalent resistances seen between the inputs of circuits 1 and 2, R_{eq1} and R_{eq2}, respectively, satisfy which of the following:

(a) $R_{eq1} > R_{eq2}$ for any R.
(b) $R_{eq1} < R_{eq2}$ for any R.
(c) $R_{eq1} = R_{eq2}$ for any R.
(d) There is no general relationship between R_{eq1} and R_{eq2}. Any relationship depends on the value of R.

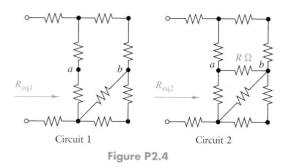

Circuit 1   Circuit 2

Figure P2.4

5. For the circuit of figure P2.5, determine (a) the voltage drop V_1 and (b) the voltage drop V_2 across the independent current source.

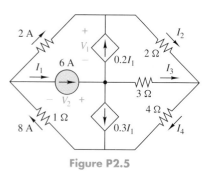

Figure P2.5

6. In figure P2.6, switch S_1 closes at $t = 0$ and switch S_2 closes at $t = 5$ sec. At $t = 10$ sec, both switches open. Sketch the voltage $v_R(t)$ for $0 \leq t \leq 15$ sec.

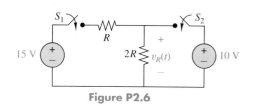

Figure P2.6

7. (a) For the circuit of figure P2.7a, determine the value of the current I.
(b) Given the circuit of figure P2.7b and the indicated currents and voltages, determine (1) the voltages V_1 through V_4 and (2) the currents I_1 through I_4.

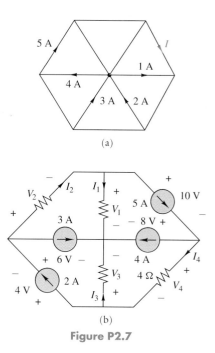

(a)

(b)

Figure P2.7

8. (a) Find the current I_R and the voltage V_{out} for the circuit of figure P2.8.
(b) If a resistor of R Ω is placed across the output terminals, determine the current I_R and the voltage V_{out}.

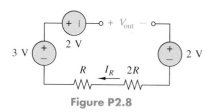

Figure P2.8

9. For the circuit of figure P2.9, use KCL and KVL to find the voltage across each current source and the current through each voltage source.

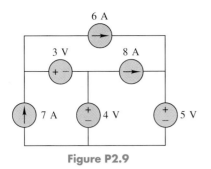

Figure P2.9

10. For the circuit of figure P2.10, use a *single* application of KCL to find I_1. Each shaded box is a general circuit element. The value of I_1 is independent of the nature of these elements.

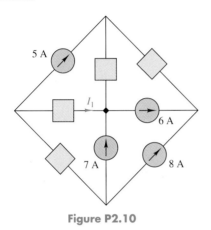

Figure P2.10

11. By the use of KCL, KVL, Ohm's law, and the equation for power, find the power *delivered* by each independent source and the power *absorbed* by each resistor for the circuit of figure P2.11. (*Check*: Total of delivered power = total of absorbed power.)

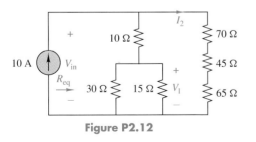

Figure P2.11

12. For the circuit of figure P2.12, determine (1) R_{eq}, (2) V_{in}, (3) V_1, and (4) I_2.

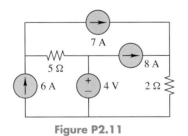

Figure P2.12

13. (a) In figure P2.13a, $V_2 = 24$ V. Determine I_{in}, V_1, and R.

(b) In figure P2.13b, a dependent voltage source has been added to the circuit of figure P2.13a. Determine V_1 in terms of a and R. If $I_{in} = 0.25$ A and $a = 4$, determine R.

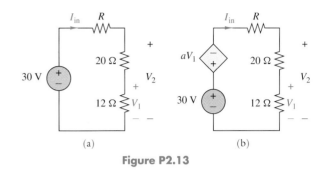

Figure P2.13

14. (a) For the circuit of figure 2.14a, it is known that $I_2 = 3$ A. Determine (1) V_{in}, (2) R_L, (3) I_3, and (4) the power delivered to the load R_L.

(b) For the circuit of figure 2.14b, determine I_1 in terms of a and R_L. Now suppose that $I_1 = 2$ A, determine R_L and the power delivered to R_L.

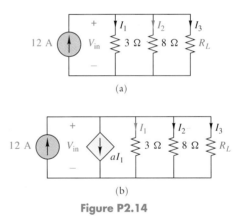

Figure P2.14

15. For the circuit of figure P2.15, suppose $i_s(t) = 6$ A for $t \geq 0$ and zero otherwise.

(a) Determine $i_1(t)$ and $v_1(t)$.

(b) Determine $v_o(t)$.

(c) Calculate the power supplied by the current source $i_s(t)$ and the power consumed by the 27-Ω resistor. Where did the additional power come from?

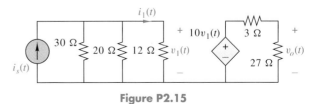

Figure P2.15

16. For the circuit of figure P2.16, write a single equation that expresses I_{in} in terms of R_1, R_2, μ, and V_{in}.

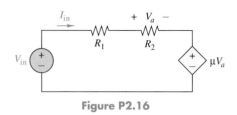

Figure P2.16

17. For the circuit of figure P2.17, write a single node equation that allows you to find I_{in}/V_{in}.

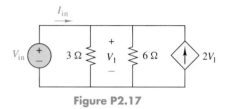

Figure P2.17

18. Consider the "ladder network" of figure P2.18.

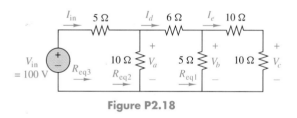

Figure P2.18

(a) Find successively the equivalent resistances R_{eq1}, R_{eq2}, and R_{eq3}.
(b) Using the voltage division formula, find successively V_a, V_b, and V_c.
(c) Using the current division formula, find I_{in} and then I_d and I_e.

19. In the circuit of figure P2.19, determine A so that the power delivered to the 2-Ω load resistor is $10P_{in}$, where P_{in} is the instantaneous power consumed by the 4-Ω and 18-Ω resistors. Equivalently, P_{in} is the power delivered by the nonideal voltage source.

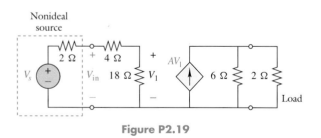

Figure P2.19

20. Some physical problems have models that are infinite ladders of resistors, as illustrated in figure P2.20.
(a) Determine the equivalent resistance R_{eq} at the terminals a-b in figure P2.20a. (*Hint*: Since the resistive network is infinite, the resistance seen at the terminals a-b is the same as the resistance seen at the terminals c-d.)

(b) Determine R_{eq} at the terminals a-b for the ladder network of figure P2.20b.

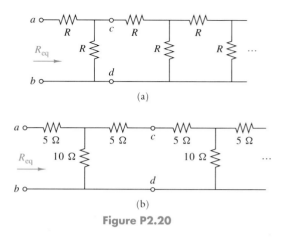

Figure P2.20

21. For the circuit of figure P2.21, determine the voltages $v_1, \ldots, v_4$.

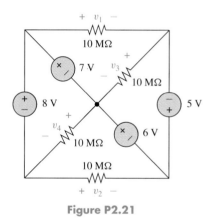

Figure P2.21

22. For the circuit of figure P2.22:
(a) Calculate R_{AC}, the equivalent resistance seen at terminals A and C, using the formulas for series and parallel combinations of resistors.
(b) Can the equivalent resistance R_{AB} be calculated by the same method as in part (a)? How about R_{BC}? State your reason without performing any actual computation.

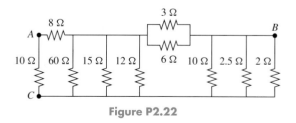

Figure P2.22

23. For the circuit of figure P2.23, determine the value of R that makes $R_{eq} = 6\ \Omega$.

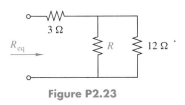

Figure P2.23

24. Determine the value of the voltage v_x for the circuit of figure P2.24.

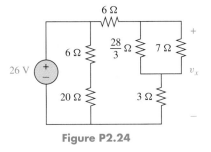

Figure P2.24

◆ 25. For the circuit of figure P2.25, $V_1 = 70$ V and $V_2 = 20$ V. Determine the value of R_1 that is necessary to achieve these voltages.

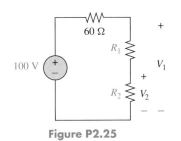

Figure P2.25

26. Find the value of I_x in the circuit of figure P2.26.

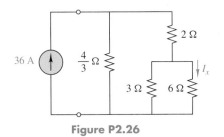

Figure P2.26

◆ 27. In the circuit of figure P2.27, find the value of R (in Ω) for which $I_1 = 2$ A.

Figure P2.27

28. Find the equivalent resistance R_{eq} "seen" by the current source I_s in the circuit of figure P2.28.

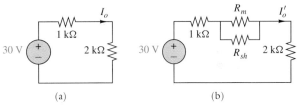

Figure P2.28

29. Consider the circuits of figure P2.29. These circuits illustrate the effect of loading on current measurements by an ammeter.
   (a) Calculate the current I_o in figure 2.29a.
   (b) If the ammeter of example 2.12 is used to measure the current, and the meter is set at a range of 0–10 mA, determine the meter reading, I_o'.

ANSWERS: 10 mA, 9.965 mA.

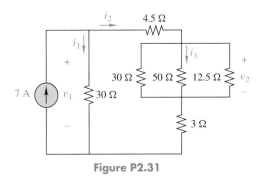

Figure P2.29 The loading effect due to an ammeter. (a) Current before the insertion of the ammeter. (b) Current after insertion of the ammeter.

◆ 30. Using meter movement 4 of table 2.1, repeat example 2.12 for the following ranges: (1) 0–10 mA, (2) 0–100 mA, and (3) 0–1 A.

31. For the circuit in figure P2.31, use current division successively to find i_1, i_2, and i_3. Then find v_1, and apply voltage division to find v_2. Check your answer by computing v_2 using Ohm's law and i_3.

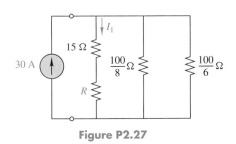

Figure P2.31

◆ 32. Design a dual-range ohmmeter using a meter movement that has $R_m = 100$ Ω and $I_m = 1$ mA. A 1.5-V battery (assumed ideal) is used. The first range has a midscale resistance value of 120 Ω, and the second stage has one of 1,200 Ω.

33. The circuit shown in figure P2.33 is a blower motor control for a typical car heater. In this circuit, resistors are used to control the current through a motor, thereby controlling the fan speed.

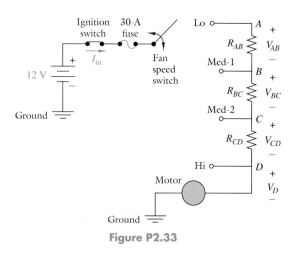

Figure P2.33

(a) With the switch in the Lo position, the current supplied by the battery is 2.5 A. The voltage drops across the resistors and motor are $V_{AB} = 6.75$ V, $V_{BC} = 1.5$ V, $V_{CD} = 0.625$ V, and $V_D = 3.125$ V. Consider the motor, as represented by a load resistance.
   (1) Determine the value of each resistance and the value of the equivalent resistance representing the motor.
   (2) Determine the power dissipated in each resistor and the power used by the motor.
   (3) Determine the relative efficiency of the circuit, which is the ratio of the power used by the motor to the power delivered by the battery.
(b) With the switch in the Med-1 position, determine:
   (1) the voltage drops across each resistor.
   (2) the current delivered by the battery.
   (3) the relative efficiency of the circuit.
(c) Repeat (b) with the switch in position Med-2.
(d) The switch is in the high position. A winding in the motor shorts out. The fuse blows. What is the largest equivalent resistance of the motor that will cause the fuse to blow?

34. (a) For the circuit of figure P2.34, find the approximate value of R_{L2} that will cause the fuse to blow.
(b) Repeat part (a) for the case where $R_{L1} = 40$ Ω.
(c) Repeat part (a) for the case where $R_{L1} = 15$ Ω.

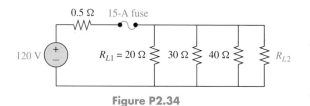

Figure P2.34

35. In the circuit of figure P2.35, switch S_1 closes at $t = 5$ sec, S_2 closes at $t = 10$ sec, and S_3 closes at $t = 15$ sec. Plot $v_{\text{out}}(t)$ for $0 \leq t \leq 20$ sec.

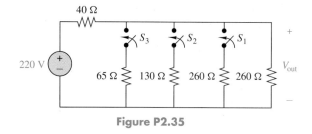

Figure P2.35

36. Find v_d for the circuit of figure P2.36.

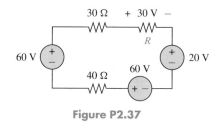

Figure P2.36

37. For the circuit of figure P2.37, find the value of R.

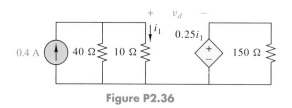

Figure P2.37

38. For the circuit of figure P2.38, find $i_s(t)$.

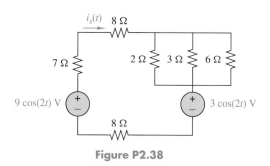

Figure P2.38

39. Construct a series voltage divider circuit whose total resistance is $1,400$ Ω, with the requirement that one can pick off three-fourths and one-fourth of the source voltage.

40. An electric energy transmission line has a rated power capacity of 50 megawatts. A commercial electrical code requires that the line carry at most 80% of its rated capacity. The source voltage driving the line is 750 kV. Determine the minimum resistive load that this energy transmission system can handle.

41. The Wheatstone bridge circuit is shown in figure P2.41. The bridge circuit is said to be balanced if $R_a R_d = R_b R_c$. In this case, the voltage $V_{\text{out}} = 0$ for any voltage V_{in}. Develop a voltage division argument to show that $V_{\text{out}} = 0$ if and only if $R_a R_d = R_b R_c$.

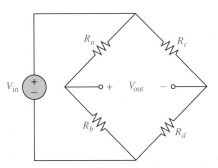

Figure P2.41   Wheatstone bridge circuit.

CHAPTER 3

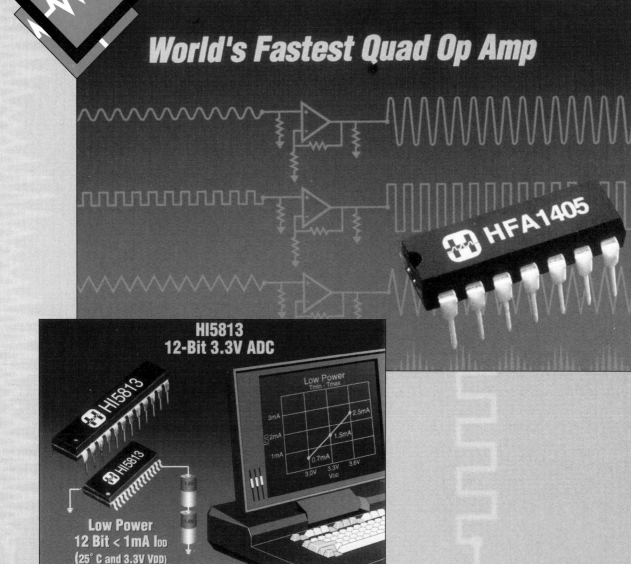

World's Fastest Quad Op Amp

HFA1405

HI5813
12-Bit 3.3V ADC

Low Power
12 Bit < 1mA I_{DD}
(25° C and 3.3V V_{DD})

Dependent Sources and the Operational Amplifier

CHAPTER OPENER

SIGNAL processing and the control of a physical process have become staples in the technological side of today's world. The workhorse of practical process control and signal processing is the microprocessor, a miniature computer. This device does everything from processing information signals for television and radio to controlling oil refineries. Microprocessors can be found in car and jet engine controllers, anti-lock brake systems, robots, electric drills, table saws, coffee brewers, etc.

The microprocessor is widely used because of its relatively low cost, its adaptability to a host of diverse applications, and the ease with which it can be reprogrammed to meet changing demands. For example, the microprocessor that controls a car's engine can be reprogrammed to meet future demands for increased engine performance and stricter emission constraints. This reprogrammability provides a cost-effective means for updating existing controllers to meet future needs.

Many of the controllers and signal processors that contain microprocessors interact with analog devices and processes; i.e., they produce continuous output signals and, often, continuous input signals for processing and control objectives. For example, temperature sensors, voltage and current sensors, air pressure sensors, and the like are inherently analog devices that produce continuous outputs for signal processing and control. On the other hand, the microprocessor processes numbers—in particular, binary numbers. Therefore, it is necessary to have an interface called an analog-to-digital converter (A/D converter), which converts analog signals to binary representations, before the microprocessor can work its "magic."

At the other end, the microprocessor produces a string or sequence of binary numbers in real time as its output. These binary numbers must in turn be converted into useful analog signals to activate, for example, readouts on your car's dashboard, to control the speed of your electric drill, or to activate the mixing of chemicals in a vat. The digital-to-analog converter (D/A converter) is a device that accomplishes this task.

The current chapter is not about microprocessors, A/D converters, or D/A converters. Rather, it is about a new device called an operational amplifier. Armed with the principles of KVL, KCL, and Ohm's law, as well as two devices—the resistor and the operational amplifier—one can build practical A/D and D/A converters. Section 5 describes how to build a simple commercially available D/A converter using these principles and basic circuit elements. It is such principles that allow us to harness the power of the microprocessor.

1. INTRODUCTION

Chapter 1 defined and discussed independent voltage and current sources. An independent source has the property that its voltage (or current) is independent of the value of any other network variable. The voltage (or current) of a **dependent** or **controlled source**, on the other hand, depends on some other voltage or current in the network. Controlled sources are **active** elements, in contrast to resistors, which are **passive** elements. In electronic circuits such as amplifiers and oscillators, controlled sources are instrumental in providing a mathematical model for the circuit. This mathematical model allows us to analyze and predict the performance of the circuit.

The main focus of this chapter is the operational amplifier (op amp), which is a special type of controlled source. It is now a basic circuit element in the design of many practical electronic circuits. After describing the basic operation and basic model of the op amp, we introduce some simple applications whose understanding requires no more than KCL, KVL, and the voltage divider formula. Later chapters will present other applications, after introducing additional circuit elements and discussing more advanced analysis techniques. These applications hint at the importance of the op amp and furnish a motivation for the study of electronic circuits.

2. DEPENDENT VOLTAGE AND CURRENT SOURCES

There are four types of controlled sources:

1. voltage-controlled voltage source (VCVS)
2. voltage-controlled current source (VCCS)
3. current-controlled voltage source (CCVS)
4. current-controlled current source (CCCS)

Each of these depends on whether the source is a voltage or a current source and whether the controlling variable is a voltage or a current. The various symbols for controlled sources appear in figure 3.1.

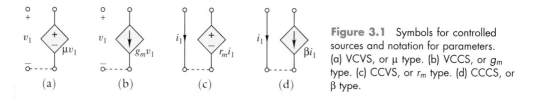

Figure 3.1 Symbols for controlled sources and notation for parameters. (a) VCVS, or μ type. (b) VCCS, or g_m type. (c) CCVS, or r_m type. (d) CCCS, or β type.

In practical controlled sources, the controlling voltage or current is usually associated with a circuit element, as illustrated in figure 1.28. However, it is often convenient, especially in computer-aided circuit analysis, to view the controlling current as that flowing through a short-circuit branch, i.e., a resistor with zero resistance or an independent voltage source with zero voltage, as is done in the well-known circuit simulation program SPICE. When the controlling variable is a voltage, the voltage may be across a network element or even across two nodes between which no element is connected. In either case, we can always view the controlling voltage as that across an open-circuit branch. This convention is adopted in figure 3.1.

In the figure, the diamond shape denotes a controlled source. An arrow appearing inside indicates a controlled current source, with the reference direction of the current given by the arrow. When "±" appears inside, the source is a controlled voltage source, with the voltage reference direction given by the ± signs.

A parameter value completes the specification of a *linear* controlled source. In figure 3.1, the parameters are denoted by μ, g_m, r_m, and β, common notations in electronic circuit texts. To maintain consistency, a g_m type of controlled source means a VCCS, a μ type of source means a VCVS, etc.

In figure 3.1b, once the controlling voltage v_1 is known, the source on the right-hand side behaves as a current source of magnitude $g_m v_1$ ampere. Because the unit for $g_m v_1$ is ampere and the unit for v_1 is volt, it follows that the unit for g_m is ampere per volt, or mho. Since g_m has the same unit as a conductance, and the controlling and controlled variables belong to two different network branches, g_m is called a **transfer conductance** (or transconductance). Other controlled sources have a similar interpretation: The parameter r_m is called a **transfer resistance** and has the ohm as its unit, the parameter μ is a ratio of two voltages and therefore is dimensionless, and the parameter β, a ratio of two currents, is also dimensionless.

From figure 3.1, it is seen that each controlled source, in its broadest sense, is a **four-terminal device**. In practice, the great majority of controlled sources have one terminal common to the two branches, making them **three-terminal devices.** The dashed lines joining the two bottom nodes in the figure illustrate this point.

The controlled sources of figure 3.1, in which μ, g_m, r_m, and β are constants, have linear v-i relationships. Controlled sources may also have **nonlinear** v-i relationships. In such a case, the element will be called a **nonlinear controlled source**. Only linear controlled sources are considered in this text. Thus, the term "controlled source" will mean one of the four types in figure 3.1.

Analyses of some simple circuits containing controlled sources have been given in chapters 1 and 2. The next few examples emphasize some special properties of controlled sources that are not shared by resistors.

◆ EXAMPLE 3.1

Find the power dissipated or generated in each branch of the network of figure 3.2.

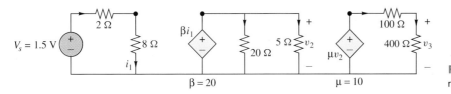

Figure 3.2 A circuit showing power relationships.

SOLUTION

Step 1. *First, calculate i_1.* From Ohm's law,

$$i_1 = \frac{1.5}{2 + 8} = 0.15 \text{ A}.$$

Step 2. *Calculate βi_1, $i_{5\Omega}$ and $i_{20\Omega}$.* The controlled source behaves as a current source of $\beta i_1 = 20 \times 0.15 = 3$ A. Using the current divider formula,

$$i_{5\Omega} = (-3)\frac{20}{5 + 20} = -2.4 \text{ A}$$

and

$$i_{20\Omega} = (-3)\frac{5}{5 + 20} = -0.6 \text{ A}.$$

Step 3. *Calculate* μv_2, $i_{100\Omega}$ *and* $i_{400\Omega}$. The controlled source behaves as a voltage source of $\mu v_2 = 10 \times (-2.4 \times 5) = -120$ V. Therefore, $i_{10\Omega} = i_{40\Omega} = -120/(100 + 400) = -0.24$ A.

Step 4. *Compute the power dissipated in each resistor.* The resistive power dissipation is given by $P_R = i^2 R$. Hence,

$$P_{2\Omega} = (0.15)^2 \times 2 = 0.045 \text{ W},$$

$$P_{8\Omega} = (0.15)^2 \times 8 = 0.18 \text{ W},$$

$$P_{5\Omega} = (2.4)^2 \times 5 = 28.8 \text{ W},$$

$$P_{20\Omega} = (0.6)^2 \times 20 = 7.2 \text{ W},$$

$$P_{100\Omega} = (0.24)^2 \times 10 = 5.76 \text{ W},$$

$$P_{400\Omega} = (0.24)^2 \times 40 = 23.04 \text{ W}.$$

Step 5. *Compute the power generated by each source.* Here, we need to know the voltage across and the currents through each branch. Note that the voltage across the βi_1 source is $v_2 = 5i_{5\Omega} = 5(-2.4) = -12$ V, and the current through the μv_2 source is $-i_{100\Omega} = 0.24$ A. Therefore, the powers *delivered by* various sources are:

$$P_s = v_s(i_1) = 1.5(0.15) = 0.225 \text{ W},$$

$$P_\beta = v_2(-\beta i_1) = (-12)(-3) = 36 \text{ W},$$

$$P_\mu = \mu v_2(i_{100\Omega}) = (-120)(-0.24) = 28.8 \text{ W}.$$

It is easy to verify that the circuit conserves power, since

$$\text{total power dissipated in the 6 resistors} = 65.025 \text{ W}$$

and

$$\text{total power delivered by one independent source} \\ \text{and two controlled sources} = 65.025 \text{ W}.$$

Unlike a resistor, which always dissipates power, a controlled source may generate power in some cases and dissipate power in others. Since a controlled source has the potential to generate power, it is called an **active element**.

An examination of the answers in example 3.1 may arouse some wishful thinking. The fact that the 1.5-V independent source delivers 0.225 W of power is easy to accept, because the source could have been a battery. On the other hand, the fact that one controlled source is delivering 36 W and the other 28.8 W of power seems a little puzzling. Why not purchase such a device and use it to power, say, a lamp? Where does the power come from? At this point, it is important to understand that a controlled source is not a component that can be picked off the shelf like a resistor. It is usually constructed from some semiconductor device and requires a dc power supply for proper operation. (See example 3.4 in the next section.) The power delivered by the controlled source actually comes from the dc power supply.

Chapter 2 described how to find the equivalent resistance of any number of resistors connected in the series-parallel fashion. With such networks, the equivalent resistance is always a positive number. When controlled sources are present, a strange result may happen, as illustrated in the next example.

◆ EXAMPLE 3.2

Find the equivalent resistance, $R_{eq} \triangleq v_s/i_s$, for the circuit of figure 3.3.

SOLUTION

Applying KVL to the single loop, and noting that $v_1 = v_s$, we obtain

$$-v_s + Ri_s + 2v_1 = -v_s + Ri_s + 2v_s = v_s + Ri_s = 0.$$

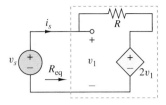

Figure 3.3 Calculation of R_{eq} for a circuit with a controlled source.

Consequently,

$$R_{eq} = \frac{v_s}{i_s} = -R. \tag{3.1}$$

Thus, the circuit enclosed in dashed lines has a negative equivalent resistance!

An important conclusion can be drawn from this example: In the study of linear circuit analysis, controlled sources admit the possibility of having negative resistances. It is easy to show that a negative resistance always delivers power and therefore behaves as an active element. Like a controlled source, it is not a component to be picked off the shelf. A negative resistor is typically constructed from a semiconductor device, which requires a dc power supply for proper operation.

The analysis of the controlled sources in the preceding examples required only KCL, KVL, and simple voltage divider or current divider formulas. More complicated linear circuits necessitate a more systematic approach. To see this need, add a resistor across any two nodes shown in solid circles in figure 3.2. The method of solution used in example 3.2 immediately breaks down. The study of more systematic methods will take place in chapter 4.

3. THE OPERATIONAL AMPLIFIER

Introduction

Resistors and independent sources are two-terminal devices. By contrast, the dependent or controlled sources described in section 3.2 are either four-terminal or three-terminal devices. A slight variation of the dependent source is the **operational amplifier**, abbreviated op amp. A real op amp is a semiconductor device, consisting of nearly two dozen transistors and a dozen resistors. However, advances in integrated circuit manufacturing technology have made the op amp only slightly more expensive than a single transistor. Because of its versatility and low cost, the op amp has become a basic building block in communication, control, and instrumentation circuits. This section sketches the basic properties of an op amp—just enough to understand some of its interesting applications. More applications will appear in later chapters.

Circuit Design Engineer

Circuit design engineers, in conjunction with solid state device and computer engineers, develop the chips that lead to new and faster computer systems. This process is called VLSI design. Besides a thorough understanding of circuits, this field requires knowledge of computer architecture, switching theory, computer design, numerical methods, operating systems, artificial intelligence, data communications, and voice communications.

Circuit design engineers design everything from the electronic or digital circuits that control computers to the printed circuit boards on which a computer's components are mounted. These printed circuit board designers etch the engineer's circuit design in copper on a laminated board, which becomes a wireless framework forming the computer.

Circuit engineers also design and improve the switching circuitry that continually monitors the quality and use of communications traffic, rerouting traffic when a problem is detected. Switching circuitry makes it possible for us to direct-dial long distance telephone numbers. Our excellent telephone system is proof of how effective today's circuit design engineers have been as they work in conjunction with communications and computer engineers.

The actual op amp circuit comes in a sealed package. Only a few terminals protrude from the package for connection to other circuit components. Figure 3.4a shows two common op amp packages. Figure 3.4b shows a typical arrangement of terminals for the dual-in-line package. The terminal markings and the symbol of figure 3.4b do not appear on the actual device, but may be found in the device data sheet or in reference handbooks.

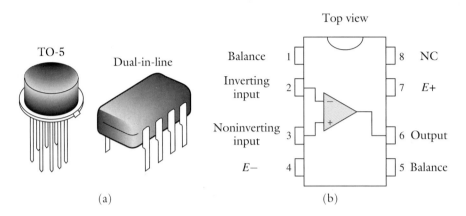

Figure 3.4 (a) Typical packages and (b) typical arrangement of terminals of an op amp.

Of the eight terminals in figure 3.4b, the one labeled "NC," for no connection, is unused. The two terminals labeled "balance" (or "offset") have importance only in implementing the circuit. When setting up an actual circuit, resistors of appropriate values are connected to these terminals, so that the output voltage is zero with no applied input. This is a balancing adjustment that is best discussed in a laboratory session. For the time being, we shall ignore the "balance" terminals. This leaves only five essential terminals. Of these, the $E+$ and $E-$ terminals are connected to a dual power supply, shown in figure 3.5a. Typically, V_{DC} is in the range of 5 to 15 V. Although the power supply connection is essential for the op amp to behave properly, it does not directly affect the analysis of the device's input-output properties. Only the remaining three terminals interact with the input and output signals. Two of these are input terminals and the other is an output terminal.

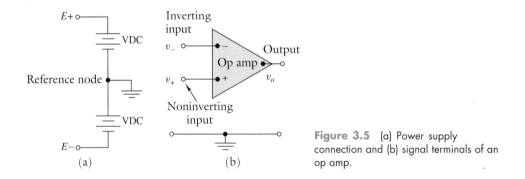

Figure 3.5 (a) Power supply connection and (b) signal terminals of an op amp.

Operation of op amp

Recall that V_{ab} typically signifies the voltage drop from node a to node b. If only one node is mentioned—for example, V_c—then the voltage drop is from node c to a reference node. The ground symbol in a circuit diagram usually denotes the reference node, although the node may or may not be "grounded," i.e., connected to earth. For an op amp circuit using a dual power supply, the reference node is not one of the pins on the op amp package. Instead, it is the node where the two power supplies join, as shown in figure 3.5a. The two power supply voltages are set equal in magnitude, but opposite in polarity. In figure 3.5b, v_o denotes the voltage drop from the output terminal to ground; v_- denotes the voltage drop from the inverting input terminal to ground; and v_+ is similarly defined.

It is customary in op amp circuits to show only the essential connections of the op amp to other circuit components. These connections are shown by the triangle symbol of figure 3.5b. The power supply connection is understood and omitted from the circuit diagram. Often, even the ground symbol is omitted. Among the many characteristics of an op amp, the most important is the curve that specifies how the open circuit output voltage v_o varies with the input voltages v_+ and v_-. Since two independent variables v_+ and v_- are present, one might expect that a three-dimensional model would be needed to display this relationship. Fortunately, the output v_o of a typical op amp depends only on the *difference*, $v_d \triangleq v_+ - v_-$, between the two input voltages. This greatly simplifies the display of the input-output relationship.

Figure 3.6a shows a typical curve defined by the function $v_o = f(v_d)$, where v_o is the open-circuit (no-load) output voltage. If the $-$ terminal is grounded, i.e., if $v_- = 0$ and $v_d = v_+$, then v_o and v_+ are of the same sign. This suggests the adjective "noninverting" for the $+$ input terminal. On the other hand, the figure shows that if the $+$ terminal is grounded, i.e., if $v_+ = 0$ and $v_d = -v_-$, and an input voltage is applied to the $-$ terminal, v_o and v_- are of opposite signs. Hence, we have the terminology "inverting" for the $-$ input terminal.

An elementary analysis of op amp circuits usually builds on segmented straight-line approximation to the smooth curve, as shown in figure 3.6b. Such piecewise linear approximation will underlie the op amp circuits in this text. The essential features may be roughly described as follows. When the differential input voltage v_d is small in magnitude, the output voltage is proportional to the input. When the input voltage magnitude exceeds some critical value, denoted V_{sat}/A, the output voltage ceases to increase and remains constant, a condition called **saturation**.

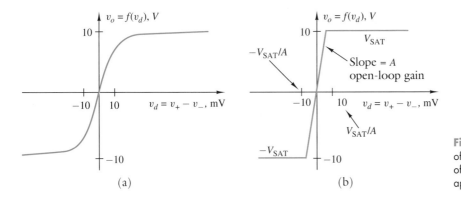

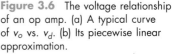

Figure 3.6 The voltage relationship of an op amp. (a) A typical curve of v_o vs. v_d. (b) Its piecewise linear approximation.

More accurately, with regard to figure 3.6b, the output voltage saturates at two levels, $\pm V_{sat}$. The two corresponding horizontal straight-line segments are called the **saturation regions.** Actual measurements on the op amp show that V_{sat} is slightly smaller than V_{DC}, the power supply voltage. Furthermore, the positive and negative saturation voltages are usually not equal. For the sake of simplicity, we shall take V_{sat} equal to the power supply voltage V_{DC} and consider $\pm V_{sat}$ as the saturation voltages.

The straight-line segment through the origin in figure 3.6b is called the **linear active region.** In this region, the op amp provides a very high voltage gain. The slope of the segment, denoted by A, is called the **open-loop gain** of the op amp. It is the voltage gain when the output terminals are open circuited—hence the descriptive term open-loop gain. Typical values of A range as high as 10^5 to 10^6. The critical input voltages at which saturation occurs are seen to be $\pm V_{sat}/A$. If $V_{sat} = 15$ V and $A = 10^5$, the critical input voltages are ± 0.15 mV. This is why the axes in figure 3.6 have different scales—volts for the v_o axis and mV for the v_d axis. In linear circuits, the active region is the normal region of operation. The saturation voltages limit the range of output voltage obtainable when the op amp acts as a high-gain device.

Mathematically, the curve of figure 3.6b has the equation form:

$$v_o = Av_d = A(v_+ - v_-) \text{ for } |v_o| < V_{\text{sat}} \text{ or } |v_d| < \frac{V_{\text{sat}}}{A}, \tag{3.2a}$$

$$v_o = V_{\text{sat}} \text{ for } v_d > \frac{V_{\text{sat}}}{A}, \tag{3.2b}$$

$$v_o = -V_{\text{sat}} \text{ for } v_d < -\frac{V_{\text{sat}}}{A}. \tag{3.2c}$$

More equations are necessary to completely describe the operation of the op amp. Fortunately, we can construct a circuit model, which embodies all the properties measurable at the terminals. The theoretical derivation of such a model and the experimental verification of its validity are beyond the scope of this text. However, since we are using the op amp as a basic circuit component, it suffices to know its terminal properties without really delving into the internal mechanisms that make it work. This approach should be of no surprise to us, as we have treated the resistor in a similar manner: A resistor is a two-terminal device whose voltage and current satisfy $v = Ri$; how the electrons actually move about inside the resistive material is unspecified.

Model of the Op Amp

A circuit model for dc applications of the op amp is shown in figure 3.7a. The functional relationship $f(*)$ is given by the curve of figure 3.6b. If the op amp is operating in its linear active region, then the nonlinear controlled source may be replaced by a linear controlled source. The circuit model for linear operation reduces to that shown in figure 3.7b. Typical values of R_i are on the order of megohms, with R_o less than 100 ohms. If the input terminals of an op amp are connected to a source having an internal resistance $R_s << R_i$, then the effect of R_i may be neglected. Neglecting R_i means replacing R_i by an open circuit. Further, if the output terminal of an op amp is connected to a load having $R_L >> R_o$, then the effect of R_o may be neglected. Neglecting R_o means replacing R_o by a short circuit. After neglecting these two resistors, the circuit model for an op amp in the linear region reduces to figure 3.7c. Using the model of figure 3.7c sacrifices some accuracy of results, but greatly simplifies the analysis.

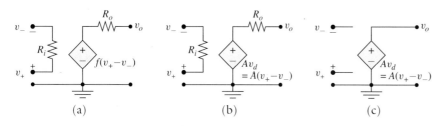

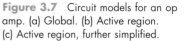

Figure 3.7 Circuit models for an op amp. (a) Global. (b) Active region. (c) Active region, further simplified.

Comparing figure 3.7c with figure 3.1a, we see that an op amp, operating in its active region, acts like a VCVS. Interestingly enough, we may proceed in the opposite direction; i.e., it is possible to construct an ordinary VCVS by the use of an op amp. Nevertheless, what distinguishes an op amp from an ordinary VCVS is the op amp's very large gain, $\mu = A > 10^4$. Such a high-gain voltage amplification device permits the construction of many useful circuits that can perform the mathematical operations of addition, subtraction, multiplication, differentiation, integration, and other operations discussed in later chapters. This is why the device is called an *operational* amplifier. Op amps are the key elements in an analog computer and were even built with vacuum tubes in the 1940s. Nowadays, practically all commercially available op amps are built with transistors.

4. ANALYSIS OF BASIC OP AMP CIRCUITS BY VIRTUAL SHORT CIRCUIT

One method for analyzing op amp circuits is to replace each op amp by one of the models shown in figure 3.7. The choice depends on whether the operating point is known a priori and on the level of accuracy desired. If an op amp is operating in its active region and extreme accuracy of the circuit response is not required, the simplest model shown in figure 3.7c is satisfactory. The problem then reduces to that of analyzing a circuit containing VCVSs. There is, however, another method of analysis, based on the concept of a virtual short circuit. This method will allow us to analyze many simple op amp circuits by inspection.

Consider the global model for an op amp shown in figure 3.7a, with the function $f(*)$ given by figure 3.6b. An op amp is said to be **ideal**, if the following three properties hold:

$$A \to \infty \text{ (infinite gain)}, \tag{3.3a}$$

$$R_i \to \infty \text{ (infinite input resistance)}, \tag{3.3b}$$

$$R_o = 0 \text{ (zero output resistance)}. \tag{3.3c}$$

The infinite gain property leads to the voltage transfer characteristic shown in figure 3.8a. To distinguish an ideal op amp from a nonideal op amp, we may add the word *ideal* to the circuit symbol, as shown in figure 3.8b. An ideal op amp has three circuit models, each for a particular region of the $f(v_d)$ curve. Figures 3.8c, d, and e illustrate these models.

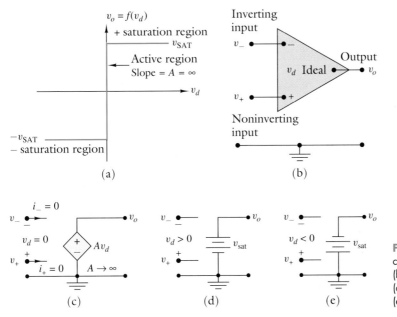

Figure 3.8 Ideal op amp and its circuit models. (a) v_o vs. v_d curve. (b) Circuit symbol. (c) Active region. (d) Region of positive saturation. (e) Region of negative saturation.

Although the active region model for an ideal op amp has the same components and connections as those shown in figure 3.7c for the finite gain case, the fact that the op amp is operating on the vertical line segment with $v_d = 0$ (see figure 3.8a) has a very important implication. There is no wire connected between the two input terminals, i.e., they are not *physically* short circuited. Yet the voltage difference between the two input terminals is zero, meaning that they behave *virtually* as if they are short circuited. This phenomenon is called a **virtual short circuit.**

If one of the input terminals is grounded, the op amp is said to operate in the *single-ended* mode. For a single-ended op amp, the virtual circuit becomes what is commonly called a **virtual ground.** For an ideal op amp in its active region (see figure 3.8c); the properties of a virtual short circuit and virtual ground may be described by the equations

$$i_+ = 0, \tag{3.4a}$$

$$i_- = 0, \tag{3.4b}$$

$$v_d = 0, \text{ or } v_+ = v_-. \tag{3.4c}$$

The recommended way to solve problems is to replace an ideal op amp by the model of figure 3.8c and make use of the special properties prevailing at the input terminals, as summarized in equation 3.4. It must be emphasized, though, that in using the circuit model of either figure 3.7c or figure 3.8c, we tacitly assume that the op amp operates

---

in its active region. The validity of this assumption must be verified after obtaining the solution. For example, if the input has a specific value, then we must verify that the output magnitude is less than V_{sat}. On the other hand, if the input is left as a variable, then it is understood that the solution is valid only if the input magnitude is small enough to keep the output below saturation. When saturation occurs, a different model, such as figure 3.8d or 3.8e, must be used to analyze the circuit.

An alternative way of analyzing a circuit containing *ideal* op amps is to replace the op amp by the finite gain model of figure 3.7c and obtain the answer containing A as a variable. By letting A approach infinity, we obtain the answer for the ideal op amp case. This, of course, is a rather roundabout way to solve the problem.

The next two examples will illustrate the first approach. To obtain the solutions by inspection, it is important to remember that V_{ab}, the voltage drop from node a to node b, satisfies $V_{ab} = V_{ac} - V_{bc}$, where V_{ac} and V_{bc} are the voltage drops from nodes a and b, respectively, to an arbitrary third node c. As shown in chapter 2, this relationship is a simple consequence of KVL.

◆ EXAMPLE 3.3

The op amp circuit of figure 3.9a is called an *inverting amplifier*. Assume that the output magnitude is below saturation. Find the voltage gain V_o/V_s.

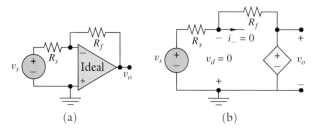

Figure 3.9 Analysis of an inverting amplifier. (a) Circuit configuration. (b) Linear circuit model for analysis.

(a) (b)

SOLUTION

We replace the ideal op amp in figure 3.9a by the model of figure 3.8c. The result is shown in figure 3.9b. Applying KCL to the inverting input terminal, and utilizing the properties $i_- = 0$ and $v_- = v_+ = 0$, one may write

$$\frac{v_s - v_-}{R_s} + \frac{v_o - v_-}{R_f} = \frac{v_s}{R_s} + \frac{v_o}{R_f} = 0.$$

This immediately leads to the answer

$$\frac{v_o}{v_s} = -\frac{R_f}{R_s}. \qquad (3.5)$$

This inverting amplifier circuit is used in a wide range of commercial circuits. Note that:

1. The input and output voltages are always of opposite polarity; hence the name *inverting amplifier*.
2. In theory, by choosing proper values for R_f and R_s, a voltage gain of any magnitude may be realized. In practice, however, other factors limit the range of gain that is attainable.

It is also possible to obtain the solution by the following reasoning, which provides additional insight into the circuit's operation. Observe that in figures 3.9a and 3.9b, the noninverting input terminal is grounded. Because of the virtual short-circuit property, the inverting input terminal is also at ground (i.e., zero) potential. The full input voltage then appears across R_s. Accordingly, the current through R_s (flowing to the right) is v_s/R_s. Since no current can enter the op amp through the input terminals ($i_- = 0$), the current v_s/R_s must flow through the resistor R_f, causing a voltage drop across R_f of $(v_s/R_s)R_f$

volts. Because the negative terminal is at virtual ground, this voltage drop has the same magnitude as the output voltage. From the direction of the current flow, we see that $(v_s/R_s)R_f$ is the negative of the output voltage. Hence, $v_o/v_s = -R_f/R_s$.

> **Exercise.** In example 3.3, it is further given that $R_s = 10$ kΩ, $R_f = 50$ kΩ, and $V_{\text{sat}} = 8$ V. Find v_o for $v_s = \pm 1$ V, ± 1.5 V, and ± 2 V.
>
> **ANSWERS:** $-5, 5, -7.5, 7.5, -8, 8$ V.

 EXAMPLE 3.4

Figure 3.10a shows a noninverting amplifier. Assume that the output is below saturation. Find the voltage gain v_o/v_s.

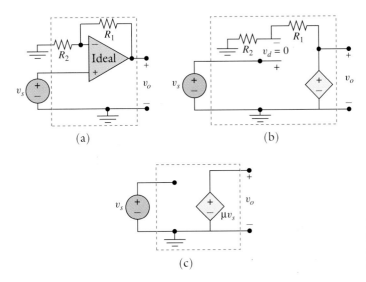

(a)

(b)

(c)

Figure 3.10 A noninverting amplifier as a VCVS. (a) The circuit configuration. (b) Linear circuit model for analysis. (c) Equivalence to a VCVS.

SOLUTION

Replace the ideal op amp in figure 3.10a by the model of figure 3.8c. The result is shown in figure 3.10b. In this case, neither input terminal is grounded.

After applying the voltage divider formula to find v_-, one obtains

$$v_- = \frac{R_2}{R_1 + R_2}v_o. \tag{3.6}$$

From figure 3.10b, $v_+ = v_s$. From the virtual short-circuit property of the ideal op amp, $v_+ = v_-$. Substitution of these relationships into equation 3.6 yields

$$v_s = \frac{R_2}{R_1 + R_2}v_o.$$

The desired voltage gain is

$$\frac{v_o}{v_s} = 1 + \frac{R_1}{R_2}. \tag{3.7}$$

From equation 3.7, we see that the voltage gain has a magnitude greater than or equal to 1. Because v_o and v_s are always of the same polarity, the circuit is called a *noninverting amplifier*.

The circuits in figure 3.10 may be used as three-terminal devices. Solid circles denote the three external terminals. Dashed boxes delineate the three-terminal device. When viewed as a three-terminal device, the circuit of figure 3.10b has exactly the same properties as the three-terminal VCVS shown in figure 3.10c, with $\mu = (1 + R_1/R_2) \geq 1$. This shows that

---

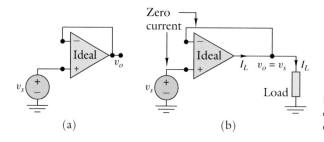

Figure 3.11 A voltage follower and its use as a buffer. (a) The circuit configuration. (b) Action as a buffer.

when a VCVS with $\mu \geq 1$ is needed, we can actually build it with an op amp as shown in figure 3.10a. Again, a power supply is needed for the active element to behave properly.

A special case of the noninverting amplifier that is of great interest is obtained from figure 3.10a by choosing $R_2 = \infty$ and $R_1 = 0$. The resulting circuit is shown in figure 3.11a and has a gain $v_o/v_s = 1$. The circuit is called a **voltage follower** because $v_o = v_s$. It is used in practical electronic circuits to prevent the load from drawing current directly from a source. When used in this fashion, it is often called a **buffer**. The load voltage is the same as the source voltage when the source is not loaded. The current to the load is actually supplied by the op amp. Figure 3.11b illustrates this point. It is important that the current drawn by the load does not exceed the maximum output current rating of the op amp.

> **Exercise.** In example 3.4, if $R_1 = 4$ kΩ, $R_2 = 1$ kΩ, and $V_{\text{sat}} = 8$ V, find v_o for $v_s = \pm 1$ V, ± 5 V, and ± 10 V.
>
> **ANSWERS:** ± 1, ± 5, ± 8 V.

In all of the preceding examples, we assumed that the op amp operates in its active region, as per the model of figure 3.8c. Our final example will illustrate the need for and inadequacy of all three models in figure 3.8. The example shows that these models, which represent only *algebraic* properties, cannot *uniquely* predict the correct output voltage; additional types of circuit elements are necessary to explain certain op amp circuit behaviors observed in the laboratory.

◆ EXAMPLE 3.5

In trying to build an inverting amplifier with voltage gain equal to -4, a student inadvertently reverses the connection of the two input terminals, resulting in the circuit of figure 3.12a. The op amp is assumed to be ideal (see figure 3.8) with $V_{\text{sat}} = 9$ V. The student observes a 9-V output in the laboratory.

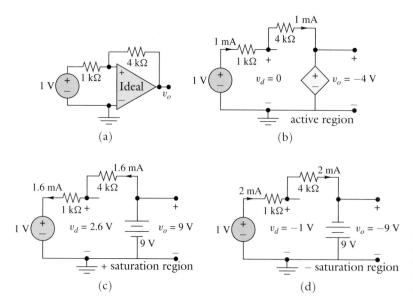

Figure 3.12 Analysis of an incorrectly connected inverting amplifier. (a) The circuit configuration. (b) Solution using active region model. (c) Solution using + saturation model. (d) Solution using − saturation model.

1. Can we explain this phenomenon in terms of our model of figure 3.8?
2. Does the circuit of figure 3.12a behave as an inverting amplifier?

SOLUTION

(1) As per figure 3.12a, $R_s = 1$ kΩ, $R_f = 4$ kΩ, and the op amp is assumed ideal with $V_{\text{sat}} = 9$ V. The observed behavior depends on the region of operation, i.e., (1) the active region (figure 3.12b), (2) the positive saturation region (figure 3.12c), or (3) the negative saturation region (figure 3.12d). A valid mathematical solution exists in each region and is summarized in the following table:

Case	Circuit model	Op amp region	v_o, volts	v_d, volts
1	figure 3.12b	active	−4	0
2	figure 3.12c	+ saturation	9	2.6
3	figure 3.12d	− saturation	−9	−1

All three solutions satisfy the *algebraic* equations for the op amp, KCL, KVL, and Ohm's law. The current in each resistor has been indicated in figures 3.12b through 3.12d to help verify KCL and KVL. Each of these figures is a valid mathematical solution based on the idealized models of figure 3.8. However, how can the observed output of 9 V, instead of −4 V, be explained?

Unfortunately, we cannot give a satisfactory explanation with the algebraic models of figure 3.8. We need the notion of dynamics to explain the observed behavior. In particular, in chapters 7 and 8, after introducing the concept of a capacitance and first-order circuits, we will return to this example and explain in a mathematical way why 9 V is observed. Nevertheless, the reader should be aware that the algebraic op amp model is inadequate for certain applications and for explaining certain behaviors.

(2) Since a 9-V output is observed for an input of 1-V, it is clear that when the connections of the two input terminals are reversed, resulting in figure 3.12a, the op amp circuit ceases to be an inverting amplifier.

5. AN APPLICATION TO ANALOG-TO-DIGITAL CONVERTERS

Elements of A/D and D/A Conversion

As mentioned in the chapter opener, many real-world control and signal-processing applications require the conversion of analog signals to digital signals. Analog-to-digital conversion, or **A/D conversion**, maps the level or magnitude of the analog signal at each instant of time to a binary number that approximates its actual value. This binary number representation is said to be a *quantized* version of the signal. The input-output curve of figure 3.13 illustrates the idea behind the conversion process.

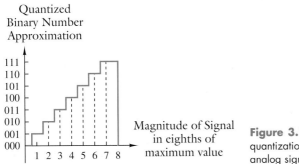

Figure 3.13 Three-bit binary quantization of the magnitude of an analog signal.

In the figure, a level between 0 and 0.5 maps to 000, a level between 0.5 and 1.5 maps to 001, etc. The binary representation is said to be a *three-bit* quantization because only three places of the binary number are used. Commercially available devices can have 10-bit or more quantization levels and can convert millions of samples per second into binary numbers.

As mentioned in the chapter opener, the binary numbers are processed by a microprocessor to produce new binary numbers. These new binary numbers represent the levels of a new waveform at successive time instants. This information must be converted to a usable analog signal through the use of a digital-to-analog converter, or **D/A converter**. For the sake of illustration, suppose that at $t = 0, 1, 2, 3, 4, 5, \ldots$ seconds a microprocessor produces the binary numbers 011, 100, 110, 001, 010, 101, ... respectively. Then the D/A converter would produce the piecewise constant analog signal illustrated in figure 3.14.

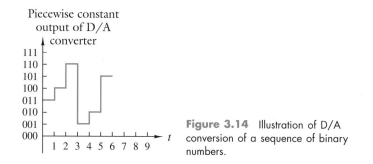

Figure 3.14 Illustration of D/A conversion of a sequence of binary numbers.

There are many types of circuits that convert binary numbers to piecewise constant waveforms. In the following subsection, we will describe one such circuit that is commercially available.

A Binary-Weighted Summing Circuit

Figure 3.15 shows a circuit with n inputs, $v_{n-1}, \ldots, v_2, v_1, v_0$, and one output, v_{out}. Suppose that each input voltage can have only one of two values: 0 or E V. With this assumption, the states of the inputs at any time can be described by a row vector

$$[v_{n-1}, \ldots, v_2, v_1, v_0] = [b_{n-1}, \ldots, b_2, b_1, b_0]E,$$

where $b_k = 1$ if v_k is E V and $b_k = 0$ if v_k is zero. Our goal here is to design a D/A converter—a circuit—whose input, $[b_{n-1}, \ldots, b_2, b_1, b_0]E$, produces an output voltage v_{out} proportional to the integer represented by the binary number $[b_{n-1}, \ldots, b_2, b_1, b_0]$, i.e.,

$$v_{\text{out}} = E_0[b_{n-1}2^{n-1} + \ldots + b_2 2^2 + b_1 2^1 + b_0 2^0] \text{ V}. \tag{3.8}$$

For example, the input $[0\ 0\ 0\ 1]E$ yields $v_{\text{out}} = 1 \times E_o$, and the input $[1\ 1\ 1\ 1]E$ yields $v_{\text{out}} = 15 \times E_o$.

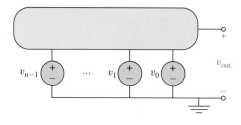

Figure 3.15 A circuit with n inputs and one output.

The circuit design and analysis aspects of this problem require no more background than an understanding of series-parallel circuits, as covered in chapter 2, and the properties of an ideal op amp described in section 3 of the current chapter. To maintain reasonable simplicity in the figures and equations, we will consider only the case $n = 4$. Extensions to higher values of n are straightforward in principle and are left as exercises. In particular, problem 11 at the end of the chapter shows a basic weighted summing circuit using an op amp, and problem 12 shows how the design of this circuit can produce the binary weighted sum given in equation 3.8. One serious drawback of the circuit is the wide spread of the resistor values. For example, if $n = 10$, then the ratio of the largest resistance to the smallest resistance in the circuit is $2^{10} = 1,024$.

There are several op amp circuits that avoid this wide range of element values. Figure 3.16 is one such circuit that utilizes an R–$2R$ resistor **ladder network**, i.e., a network that has successive series and parallel resistors, each of the same value.

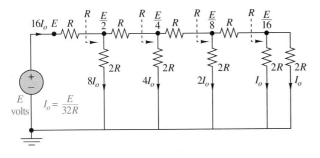

Figure 3.16 Current and voltage distributions in an R–$2R$ ladder. The dashed lines at the top of the figure indicate that the resistance seen looking into the remaining part of the circuit is always R.

Using the formulas in sections 3 and 4 of chapter 2 for series- and parallel-connected resistors, we can compute the equivalent resistances seen toward the right at different points of the ladder. For the R–$2R$ ladder, starting from the rightmost point of figure 3.16 and moving successively to the left, one encounters a particularly easy set of calculations: $2R//2R = R$ and $R + R = 2R$. The dashed arrows in the figure indicate four places where the equivalent resistances (looking toward the right) equal R.

Using the voltage division formula, equation 2.2, the node voltages (with respect to the reference node) decrease successively from left to right in a binary-weighted fashion, i.e., $E/2$, $E/4$, $E/8$, and $E/16$. By Ohm's law, then, the currents in the shunt $2R$ resistors are also binary weighted, according to $8I_0$, $4I_0$, $2I_0$, and I_0. Figure 3.16 identifies each of these binary-weighted voltages and currents. Since the total current delivered by the source is $16I_0$, and the source "sees" an equivalent resistance of $2R$, Ohm's law yields $16I_0 \times 2R = E$, or equivalently,

$$I_o = \frac{E}{32R}.$$

To obtain the output voltage specified by equation 3.8, we still have to devise a means for summing up the currents from some selected shunt $2R$ resistors and diverting them through a resistor to produce the desired voltage. The summing is done with an op amp; the selective diverting of current is done with the aid of electronically operated switches.

Figure 3.17a shows a single-pole, double-throw (SPDT) switch. A connection is made between either the contacts (A, B) or the contacts (A, C). The switching may be done manually, as in switches for selecting the front speakers or rear speakers in a car stereo system. However, if rapid switching is necessary, an electronic device is typically used. Although the details of an electronic switch are beyond the scope of this text, suffice it to say that the connection between the contacts is governed by a controlling voltage, $V_{control}$. When $V_{control}$ is zero, a connection is made between one set of contacts, say (A, B). When $V_{control}$ has a sufficiently large magnitude, a connection is made between the other set of contacts, say, (A, C). This action is shown schematically in figure 3.17b, where the dashed lines indicate that the switch is controlled by the voltage source.

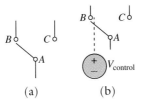

(a)   (b)

Figure 3.17 (a) A single-pole, double-throw (SPDT) switch. (b) An electronically controlled SPDT switch.

By connecting an (ideal) op amp, four electronically controlled switches, and the R–$2R$ ladder as shown in figure 3.18, we achieve the desired binary-weighted summing specified by equation 3.8. To see this, consider the specific inputs shown in figure 3.18, i.e., $[v_3, v_2, v_1, v_0] = [1001]E$. The switch positions are then (S_0, S_3) up and (S_1, S_2) down. All of the $2R$ shunt resistors in figure 3.16 have their lower terminals connected to the ground, resulting in the binary-weighted voltage and current distributions. In figure 3.18, the lower terminal of some $2R$ resistors are connected to ground through the switches being in the downward position. Other $2R$ shunt resistors are connected to the inverting input terminal of the op amp, through switches in the upward position. Since the op amp is assumed ideal, a virtual ground exists at the input terminals. Therefore, regardless of the position of the switch, the lower terminal of each shunt $2R$ resistor is held at the ground potential. Consequently, the same voltage and current values exist in the ladders of figure 3.16 and 3.18. However, an ideal op amp has an infinite input impedance, meaning that no current flows into the input terminals. Therefore, the currents I_o and $8I_o$ are forced to go through the feedback resistor connected between the op amp output and inverting input terminals. The output voltage is thus proportional to the sum of the diverted currents, which, in turn, is proportional to the integer represented by the binary number $[b_3, b_2, b_1, b_0]$. It follows that v_{out} given by equation 3.8 has been implemented.

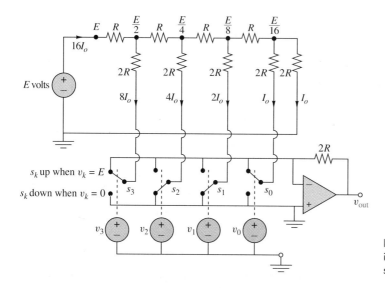

Figure 3.18 An op amp circuit implementing the binary-weighted summing specified by equation 3.8.

Another op amp circuit can be constructed to implement equation 3.8 without the use of electronically controlled switches. The analysis of that circuit, however, requires some techniques to be discussed in chapters 4 and 5. For this reason, the discussion of the circuit is postponed until after chapter 5.

SUMMARY

This chapter introduced five new active circuit elements: four dependent source elements and the operational amplifier. The four dependent source devices are the voltage-controlled voltage source (VCVS), the voltage-controlled current source (VCCS), the current-controlled voltage source (CCVS), and the current-controlled current source (CCCS). Each of these devices is either a four-terminal or a three-terminal device. If one of the input terminals and one of the output terminals are tied together as a common reference, then each dependent source is a three-terminal device. Otherwise it is a four-terminal circuit element. Further, each device is an active element, because each has the potential to generate power for the network. (Practically speaking, such devices require an external power supply.) They are used to model the behavior of real-world circuits in which the voltage or current into a

node of the circuit depends on a voltage or current at some other point of the circuit. Such a configuration is common to almost all electronic devices.

A slight variation of the dependent source is the operational amplifier, abbreviated op amp. A real op amp is a semiconductor device consisting of nearly two dozen transistors and a dozen resistors. There are two input terminals to the op amp. The output voltage of the op amp is proportional to the voltage across the input terminals, provided that the input voltage is small in magnitude. When the input voltage exceeds some critical value, the output voltage ceases to increase and remains constant, a condition called saturation.

There are several levels of op amp modeling, as illustrated in figure 3.7. For much engineering analysis, the op amp is assumed to be ideal. An ideal op amp has infinite input impedance and thus perfectly measures the voltage appearing across its input terminals, with no loading to the rest of the circuit. Practically speaking, the input impedance is very large in relation to the impedances of the circuit elements connected to it. The output voltage is modeled as the voltage across a dependent voltage source proportional to the measured input voltage. The gain or proportionality constant is typically very large, often in the range of 10^5, which is a good approximation to infinity for many engineering applications. These assumptions lead to some interesting properties of op amps when interconnected with other circuit elements. In particular, when the op amp is connected in a feedback configuration—i.e., when there is some circuit element providing a current path between the output and input terminals—the voltage across the input terminals is driven to zero. In this case, there is said to be a virtual short across the input terminals. Further, because the input impedance is assumed to be infinite, there is no current entering the input terminals of the op amp. Hence, if one of the input terminals is grounded, then the other terminal is said to be at virtual ground. The word *ground* is used because there is no voltage between the grounded terminal and the ungrounded terminal. The word *virtual* is used because no current can flow between the two terminals, and thus no ground in the ordinary sense of the term is present.

Once the concept of the op amp was firmly established, the chapter went on to describe some basic resistive circuits typical of op amp applications. The chapter ended with a very important application of the op amp to a D/A converter that is available commercially.

TERMS AND CONCEPTS

Active element: a circuit element that requires an outside power supply for proper operation and that has the potential to generate power on an average basis.

A/D converter: a circuit that converts a continuously varying signal (voltage or current) into a digital output. The digital output is a binary code representing the continuous signal amplitude at uniformly spaced time instants called sampling points.

Buffer: a circuit designed to prevent the loading effect in a multistage amplifier. A buffer isolates two successive amplifier stages. An ideal buffer has infinite input impedance, zero output impedance, and constant voltage gain.

D/A converter: a circuit that converts a sequence of numbers (usually binary) into a waveform piecewise constant over each sampling period. The level of the piecewise constant waveform at successive points in time corresponds to the level represented by the binary code produced at the beginning of the sampling period.

Dependent or controlled source: a voltage source or current source whose value depends on some other voltage or current in the network.

Four-terminal device: a circuit element having four accessible terminals for connection to other circuit elements.

Ideal op amp: an operational amplifier with infinite input resistance and infinite open-loop gain.

Inverting amplifier: an operational amplifier connected to provide a negative voltage gain at dc.

Linear active region: region where the curve of an op amp's output vs. input transfer characteristic is essentially a straight line through the origin.

Noninverting amplifier: an operational amplifier connected to provide a positive voltage gain at dc.

Open-loop gain: the ratio of the output voltage (loaded, but without any feedback connection) to the voltage across the two input terminals of an op amp. The slope μ of the straight line in the active region of an op amp is the open-loop gain under the no-load condition. When a load R_L is present, the open-loop gain is reduced to $\mu R_L/(R_L + R_o)$, where R_o is the output resistance of the op amp.

Operational amplifier (abbreviated op amp): a multistage amplifier with very high voltage gain (exceeding 10^4) used as a single circuit element. The name stems from the fact such amplifiers, with proper feedback connections, are used for performing mathematical operations, such as addition, subtraction, differentiation, integration, etc.

Passive element: a circuit element that cannot generate power on an average basis.

Quantization: the process whereby the amplitude of a signal is divided into a finite number (usually a power of 2) of levels, each represented by an integer.

Saturation regions: region where the curve of an op amp's output vs. input transfer characteristic is essentially a horizontal line. There are two such regions, one for positive input voltage and the other for negative input voltage.

Three-terminal device: a circuit element having three accessible terminals for connection to other circuit elements. Examples include bipolar junction transistors and two-resistor voltage dividers.

Transfer conductance (or transconductance): the parameter g_m in the equation $i_2 = g_m v_1$ describing a voltage-controlled current source. It has the mho as its unit.

Transfer resistance (or transresistance): the parameter r_m in the equation $v_2 = r_m i_1$ describing a current-controlled voltage source. It has the ohm as its unit.

Virtual ground: condition wherein, when an ideal op amp has one of its input terminals grounded and is operating in the active region, the other input terminal is also held at the ground potential because of the virtual short effect. (See next.) A *virtual* ground is in contrast to a *physical* ground.

Virtual short circuit: condition wherein, when an ideal op amp is operating in the active region, the voltage across the two input terminals is zero, even though the terminals are not hard-wired together. A *virtual* short circuit is in contrast to a *physical* short circuit.

Voltage follower: a voltage-controlled voltage source with gain equal to 1. An op amp can be easily connected to become a voltage follower. (See figure 3.11a.)

PROBLEMS

1. In the active circuits shown in figure P3.1, $g_m = 0.0125$ mho and $r_m = 10^5$ Ω. Find v_o and the power dissipated in each of the six branches.

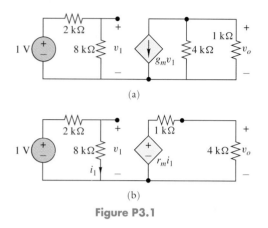

(a)

(b)

Figure P3.1

ANSWERS: (a) -8 V ...; (b) 8 V ....

2. The circuit shown in figure P3.2 contains a CCCS (current-controlled current source).
   (a) Find v_o/i_s. (*Hint*: Apply KCL to the top node.)
   (b) Repeat part (a) with the direction of the CCCS reversed.

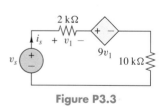

Figure P3.2

ANSWERS: 62.5 Ω; -250 Ω.

3. The circuit shown in figure P3.3 contains a VCVS.
   (a) Find the equivalent resistance "seen" by the source v_s (i.e., the ratio v_s/i_s). (*Hint*: Apply KVL to the loop.)
   (b) Repeat part (a) with the direction of the VCVS reversed.

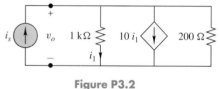

Figure P3.3

ANSWERS: 30 kΩ; -6 kΩ.

4. One of the two circuits shown in figure P3.4 contains a negative resistance and the other a VCCS. Find v_o/v_s, and verify that both circuits provide the same voltage amplification.

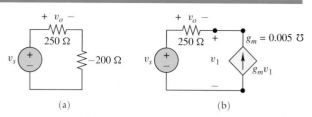

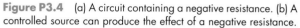

(a)   (b)

Figure P3.4 (a) A circuit containing a negative resistance. (b) A controlled source can produce the effect of a negative resistance.

5. In figure P3.5, determine the value of R for which the power delivered to the 10-Ω resistor is 10 watts.

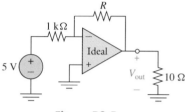

Figure P3.5

ANSWER: 2 kΩ.

6. The op amp shown in figure P3.6a is assumed to be ideal and has $V_{sat} = 15$ V.
   (a) Plot the v_o-vs.-v_s transfer characteristic for $-5 \le v_s < 5$ V.
   (b) If $v_s(t)$ is as shown in figure P3.6b, plot $v_o(t)$ for $0 \le t \le 6$ sec.

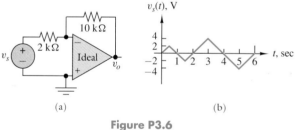

(a)   (b)

Figure P3.6

   (c) If the op amp is not ideal, but has a finite gain of $A = 1,000$, what is the value of v_o when v_s is 2 V? What is the percent error caused by the assumption that the op amp is ideal ($A = \infty$)?

ANSWER: $-1,000/100.6 = -9.94$ V, 0.6%.

7. For each of the circuits shown in figure P3.7, find the voltage ratio v_o/v_s.

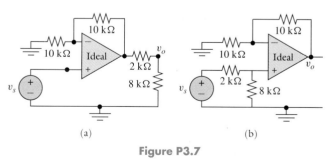

(a)   (b)

Figure P3.7

ANSWER: 1.6 for both cases.

8. A signal source is represented by a 1-V voltage source in series with a 10-kΩ resistance. A load is represented by a 10-kΩ resistance.

(a) If the load is connected directly to the source, as shown in figure P3.8a, find the load voltage, load current, and source current.

(b) If a buffer is inserted between the source and the load, as shown in figure P3.8b, find the load voltage, load current, source current and current supplied by the op amp.

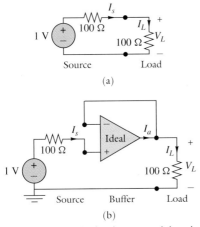

(a)

(b)

Figure P3.8 (a) A load connected directly to source. (b) The effect of a buffer.

ANSWERS: (a) 0.5 V, 5 mA, 0.5 mA; (b) 1 V, 10 mA, 0 A, 10 mA.

9. If the op amp of the previous problem is not ideal, but has a finite gain $A = 1,000$, what is value of V_L? What is the percent error caused by the assumption that the op amp is ideal ($A = \infty$)?

ANSWER: $1,000/1,001 = 0.999$ V, 0.1%.

10. For the circuit shown in figure P3.10, find the voltage ratio v_o/v_s.

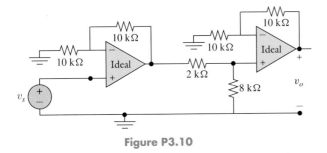

Figure P3.10

ANSWER: 3.2.

11. The circuit shown in figure P3.11 is called a summing amplifier. Assume that the op amp is ideal. Prove that

$$V_o = -\left[\frac{R_f}{R_1}V_1 + \frac{R_f}{R_2}V_2 + \frac{R_f}{R_3}V_3\right].$$

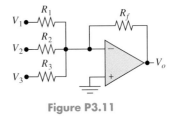

Figure P3.11

(*Hint*: Apply KCL to the noninverting input terminal, and make use of the virtual ground property of an ideal op amp.)

12. In the op amp circuit shown in figure P3.12, $R_f/R_0 = 1$, $R_f/R_1 = 2$, $R_f/R_3 = 4$, and $R_f/R_4 = 8$. Each applied input V_k is either zero or E (positive) V.

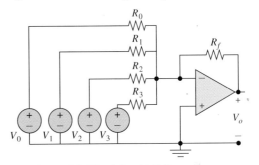

Figure P3.12 A four-bit D/A converter.

(a) Find $|V_o|$ if $[V_3V_2V_1V_0] = [1001]E$ V.

(b) Find $|V_o|$ if $[V_3V_2V_1V_0] = [0110]E$ V.

(c) If $|V_o| = 13E$ V is desired, what should $[V_3V_2V_1V_0]$ be?

(d) If $|V_o| = 7E$ V is desired, what should $[V_3V_2V_1V_0]$ be?

Remark: The circuit is a four-bit digital-to-analog converter. Another circuit that performs the same function, but has a smaller spread of resistance values, is described in chapter 5.

ANSWERS: (a) $9E$, (b) $6E$, (c) $[1\,1\,0\,1]E$, (d) $[0\,1\,1\,1]E$.

13. Making use of the principle demonstrated in the previous problem, design a five-bit digital-to-analog converter. Let $E = 1$ V.

(a) Find $|V_o|$ if $[V_4V_3V_2V_1V_0] = [11001]E$ V.

(b) If $|V_o| = 27$ V is desired, what should $[V_4V_3V_2V_1V_0]$ be?

14. For the circuit shown in figure P3.14, if the input voltage $V_{s1} = 10$ V and the input voltage $V_{s2} = 5$ V, determine V_{out}. Assume ideal op amps.

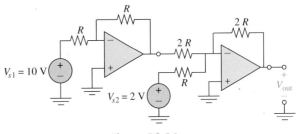

Figure P3.14

ANSWER: 6 V.

15. For the circuit shown in figure P3.15, determine the gain V_o/V_i.

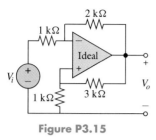

Figure P3.15

ANSWER: −8.

16. For the circuit shown in figure P3.16, prove that

$$V_o = -\frac{R_f}{R_1}V_1 + \left(\frac{R_f}{R_1}+1\right)\left(\frac{R_3}{R_2+R_3}\right)V_2.$$

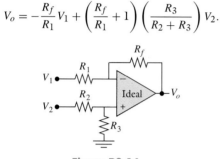

Figure P3.16

(*Hint*: Apply KCL to the inverting input terminals, and solve for V_-. Use the voltage divider formula to express V_+ in terms of V_2. Next, make use of the virtual short-circuit property of an ideal op amp to equate V_+ to V_-. Finally, solve for V_o in terms of V_1 and V_2. An alternative, simpler solution to the problem based on the superposition theorem is described in chapter 5.)

17. A special case of the previous circuit is shown in figure P3.17. Show that $V_o = k(V_2 - V_1)$.

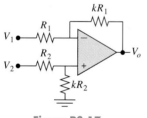

Figure P3.17

Remark: Since the output is k times the difference between the two inputs, the circuit is called a **difference amplifier** or a **differential amplifier**.

18. (a) In the circuit of figure P3.18, $V_{sat} = 18$ V and the input voltage $V_s = 1$ V. Determine the output voltage V_{out}, using virtual short circuit.

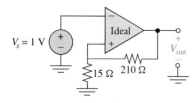

Figure P3.18

(b) If the circuit is built in the laboratory with a practical op amp, do you expect to observe V_{out} as calculated in part (a)? (Read example 3.5 before answering this question.)

19. Consider the circuit of example 3.5. Suppose the op amp is not ideal and is represented by the model of figure 3.7c with gain $A = 1{,}000$. The resistor values remain unchanged. Recalculate the values of V_o and V_d. As before, there are three possible mathematical solutions. Which solution would you predict to be the one observed in the actual circuit?

20. The op amp shown in figure P3.20, assumed to operate in the active region, is represented by the model of figure 3.7c with $A = 1{,}000$. Find V_o and I_s. What is the equivalent resistance "seen" by the voltage source? Can you expect the circuit to remain in the active region?

ANSWERS: 1 V, −248.75 µA, −4.02 Ω, no.

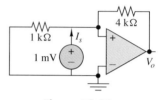

Figure P3.20

21. For the circuit of figure P3.21, show that the output voltage V_o is equal to $I_s R_f$. This op amp circuit converts a current source into a voltage source.

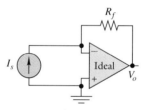

Figure P3.21 Converting a current source into a voltage source.

22. For the circuit of figure P3.22, show that the load current I_L is equal to V_i/R_a. This op amp circuit converts a voltage source into a current source.

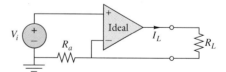

Figure P3.22 Converting a voltage source into a current source.

23. In the previous problem, since I_L depends on V_i and R_a only, the load need not be a resistor. For example, R_L may be replaced by a light-emitting diode (LED), as shown in figure P3.23. Then, by turning the knob of the 10-kΩ potentiometer, one can control the brightness of the LED. Note that the current through the load is not supplied by the voltage source V_i. Determine the magnitude of the LED

current if the potentiometer is set at (1) $R_1 = 5$ kΩ and (2) $R_1 = 8$ kΩ.

ANSWER: 1.32 mA, 2.1 mA.

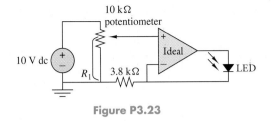

Figure P3.23

24. *Measurement of the open-loop gain* A *for a specified load* R_L. In theory, A can be measured with the simple circuit shown in figure P3.24a. In practice, however, the measurement is inaccurate because of the very high gain of the op amp. The problem is that the noise at the input terminals greatly affects the meter readings. Two practical circuits for measuring the open-loop gain are that of figure P3.24b, for the case where A is below 10,000, and that of figure 3.24c, for the case $A \gg 10{,}000$. In both circuits, the load resistance has been adjusted to account for the presence of feedback resistor R_f. The equivalent load connected to the output terminal is $R_L = R_L'//R_f$. Therefore, the value of the adjusted load resistance R_L' in figures P3.24b and c is $R_L' = R_L R_f/(R_f - R_L)$.

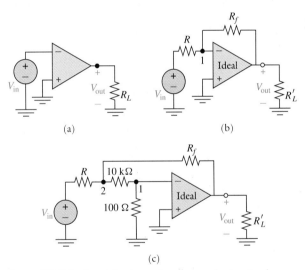

Figure P3.24 Circuit for measuring the open-loop gain of an op amp. (a) Measured results are inaccurate for the simple circuit. (b) A practical circuit for measuring the open-loop gain. (c) A variation of (b).

(a) Show that from the voltage readings of figure P3.24b, the open-loop gain of the op amp in figure P3.24a, with load R_L, is given approximately by V_{out}/V_1.

(b) Show that from the voltage readings of figure P3.24c, the open-loop gain of the op amp in figure P3.24a, with load R_L, is given approximately by $101V_{\text{out}}/V_2$.

25. Prove that in the op amp circuit of figure P3.25, v_o is related to the four inputs by

$$v_o = -4v_1 - 5v_2 + 7v_3 + 6v_3.$$

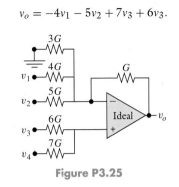

Figure P3.25

Observe that the sum of the conductances connected to the inverting input terminal is equal to the sum of the conductances connected to the noninverting input terminal.

26. Prove that in the op amp circuit of figure P3.26, v_o is related to the four inputs by

$$v_o = 4v_1 + 8v_2 - 6v_3 - 7v_3.$$

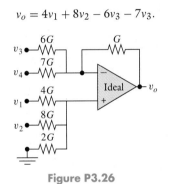

Figure P3.26

Observe that the sum of the conductances connected to the inverting input terminal is equal to the sum of the conductances connected to the noninverting input terminal.

27. (a) Using the experience gained in problems 25 and 26, design an op amp circuit having five inputs such that

$$v_o = 4v_1 + 8v_2 - 6v_3 - 7v_4 - 3v_5.$$

(b) Repeat part (a) with

$$v_o = -4v_1 - 8v_2 + 6v_3 + 7v_4 + 3v_5.$$

28. Extend the A/D converter circuit to six bits. Select resistor values so that the total power dissipated in the R–$2R$ ladder is less than 20 mW when the applied voltage is $E = 10$ V.

CHAPTER 4

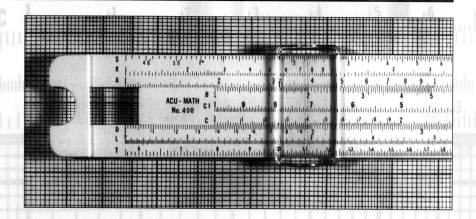

Node and Loop Analysis

HISTORICAL NOTE

F OR a network consisting of resistors and independent voltage sources, one can apply KCL, KVL, and Ohm's law to construct a set of simultaneous equations whose solution yields all currents and voltages in the circuit. In theory, the basic analysis problem is completely solved. In practice, this approach is impractical because it requires a very large number of equations, even for a small network. For example, a six-branch, four-node network with each node connected to the other nodes through a single element leads to a set of 12 equations in 12 unknowns: 3 equations from KCL, 3 equations from KVL, and 6 equations from v-i relationships among elements. The 12 unknowns are the 6 branch currents and 6 branch voltages.

Before the advent of digital computers, simultaneous equations were solved manually, possibly with the aid of a slide rule or some primitive mechanical calculating machines. Any method that reduced the number of equations was highly treasured. It was in such an environment that Maxwell's mesh analysis technique (1881) received much acclaim and credit. Through the use of a fictitious circulating current, called a mesh current, Maxwell was able to reduce the number of equations greatly. For example, for the network mentioned in the preceding paragraph, the number of equations drops from 12 to 3. These 3 equations are called *mesh equations*, in which the mesh currents are the only unknowns.

An alternative technique using nodal equations appeared in the literature as early as 1901. However, the method did not gain momentum until the late 1940s, because most problems in the early days of electrical engineering could be solved efficiently using mesh equations in conjunction with some network theorems. However, with the invention of multiterminal vacuum tubes having interelectrode capacitances, there appeared compelling reasons to use the node method—primarily because it accounts for the presence of capacitances without introducing more equations. In addition, some types of vacuum tubes behave very much like current sources and are more easily accommodated with node equations. By the late 1950s, almost all texts on electrical circuits presented both the mesh and node methods.

Since the 1960s, many digital computer software programs have been developed for the simulation of electronic circuits that would otherwise be impossible to analyze by hand. SPICE is one of the more well known of these. All such simulation packages use the node equation method over mesh equations. One reason for this is that a person can easily identify a mesh by the eye, whereas it is very difficult for a computer to recognize a mesh.

For networks consisting of only resistors and current sources, writing node equations is straightforward. Because of certain difficulties in writing node equations when independent and dependent voltage sources are present, a modification of the conventional node method by a research group at IBM resulted in the modified nodal analysis (MNA) method in the middle 1970s. With the MNA method, the formulation of network equations, even in the presence of voltage sources and all types of dependent sources, becomes very systematic.

This chapter discusses the writing of equations for linear resistive networks. Its coverage ranges from the classic mesh equations to the present-day computationally straightforward modified nodal analysis method. ◆

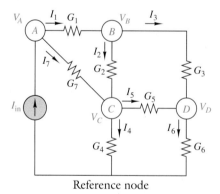

CHAPTER OBJECTIVES

- Describe and illustrate the method of node analysis for the computation of node voltages in a circuit. Knowledge of the node voltages of a circuit allows one to compute all the branch voltages and, with knowledge of the element values, all the branch currents.
- Define the notion of a loop current and describe and illustrate the method of mesh, or more generally, loop analysis for the computation of loop currents in a circuit. Knowledge of all the loop currents of a circuit allows one to compute all the branch currents. In conjunction with knowledge of the branch element values, this enables one to compute all the branch voltages.
- Formulate the node analysis and loop analysis equations as matrix equations and use matrix methods in their solution, emphasizing the use of existing software for the general solution.
- Describe and illustrate the modified nodal approach to circuit analysis. This method underlies the general software algorithms that are available for computer simulation of circuits.
- Explain and illustrate some of the theoretical questions that underlie the validity and use of both the nodal and loop analysis methods.

1. INTRODUCTION

Recall that a **node voltage** in a circuit is the voltage drop from a given node to a reference node. Figure 4.1 shows a circuit labeled with nodes A through D with associated node voltages $V_A, \ldots, V_D$ and seven branches, one for the current source and one for each of the seven conductances $G_1, \ldots, G_7$. (Throughout this chapter we will deal almost exclusively with dc; hence, the uppercase notation for voltages and currents will be most common.)

Figure 4.1 Circuit with labeled node voltages $V_A, \ldots, V_D$. Node voltages are the potential difference between the given nodes and the reference node.

Given every node voltage in a circuit, KVL tells us that every branch voltage is the difference of the node voltages present at the terminals of the branch. Mathematically, for lumped circuits and all pairs of nodes j and k, the voltage drop V_{jk} from node j to node k is

$$V_{jk} = V_j - V_k$$

at every instant of time, where V_j is the voltage at node j with respect to the reference node and V_k is the voltage at node k with respect to the reference node. Here, j and k stand for arbitrary indices and could be any of the nodes A, B, C, or D in figure 4.1. For the purposes of circuit analysis, this means that knowledge of all node voltages, in conjunction with device information paints a rather complete picture of the circuit behavior.

Nodal analysis is an organized means of computing all node voltages of a circuit. Nodal analysis builds around KCL, i.e., at each node of the circuit, the sum of the currents leaving (entering) the node is zero. Each current in the sum is associated with a branch incident on the node. The current in each branch generally depends on a subset of the circuit node voltages, the branch conductance, and possibly source values. After substituting this branch-related information for each current in a node's KCL equation, one obtains a nodal equation. For example, the nodal equation at node A in figure 4.1 is $I_{in} = I_1 + I_7 = G_1(V_A - V_B) + G_7(V_A - V_C)$. The nodal equation at node C is $-I_2 + I_4 + I_5 - I_7 = G_2(V_C - V_B) + G_4V_C + G_5(V_C - V_D) + G_7(V_C - V_A) = 0$. By writing such an equation at each node of the circuit (except for the reference node), one obtains a set of independent equations. Of course, one could substitute a KCL equation at the reference node for any of the other equations and still obtain an independent set. The solution of the resulting set of nodal equations specifies all the node voltages of the circuit. Once all the node voltages are known, all branch voltages and branch currents can be found. Note that the reference node may be chosen arbitrarily and can sometimes be chosen to greatly simplify the analysis.

The set of nodal equations is amenable to solution by computer. This permits an engineer to analyze very large and complex circuits that would defy hand calculation. The special form of nodal analysis used in general-purpose software packages for circuit analysis is called *modified nodal analysis*, a topic taken up in section 5. All such techniques build on a matrix formulation of the circuit equations. In view of this and the widespread use of matrices in circuits, systems, and control, we will stress a matrix formulation of equations throughout the chapter.

The counterpart to nodal analysis is **loop analysis**. In loop analysis, the counterpart of a node voltage is a **loop current**, which circulates around closed paths of the circuit. A **loop** or **closed path** in a circuit is a contiguous sequence of branches that begins and ends on the same node and touches no other node more than once. For each loop in the circuit, one defines a **loop current**, as illustrated in figure 4.2, which shows three loops or closed paths denoting loop currents I_1, I_2, and I_3. Of course, one can draw other loops for this circuit.

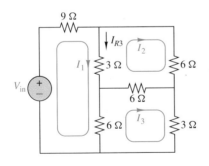

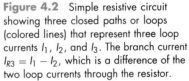

Figure 4.2 Simple resistive circuit showing three closed paths or loops (colored lines) that represent three loop currents I_1, I_2, and I_3. The branch current $I_{R3} = I_1 - I_2$, which is a difference of the two loop currents through the resistor.

Using a fluid flow analogy, one can think of loop currents as fluid circulating through closed sections of pipe. The fluid in different closed paths may share a segment of pipe. This segment is analogous to a branch of a circuit on which two or more loop currents are incident. The net current in the branch is analogous to the net fluid flow. Thus, we note that each branch current can be expressed as a sum or difference of loop currents. For example, the branch current $I_{R3} = I_1 - I_2$. Using loop currents, element resistance values, and source values, we can express the sum of the voltages around each loop in terms of the loop currents. For example, the first loop, labeled I_1 in figure 4.2, has the loop equation

$$V_{in} = 9I_1 + 3[I_1 - I_2] + 6[I_1 - I_3].$$

We will explore this concept more thoroughly in section 4. Here, we see that loop analysis builds on KVL, whereas node analysis builds on KCL.

This section describes nodal analysis for circuits containing dependent and independent current sources, resistances, and independent voltage sources that are grounded to the reference node. Floating independent or dependent voltage sources are covered in section 3 and require the use of a Gaussian surface for systematic construction. As mentioned earlier, nodal analysis is a technique for finding all node voltages in a circuit. With knowledge of all the node voltages and all the element values, one can compute all branch voltages and currents.

For the class of circuits discussed in this section, it is possible to write a nodal (KCL) equation at each node that is not connected to a voltage source. If a node is connected to a voltage source grounded to the reference node, then the voltage of the node is specified by the voltage source. For the other nodes, each nodal equation will sum the currents leaving the node. Each current in the sum will be expressed in terms of dependent or independent current sources or branch conductances and node voltages. The set of these equations will have a solution that yields all the pertinent node voltages of the circuit. Example 4.1 is a simple introduction to the method.

◆ EXAMPLE 4.1

Consider the circuit of figure 4.3, which contains an independent voltage source, an independent current source, and four resistances whose conductances in mhos are given in place of their resistances. The objective of this example is to determine the node voltages V_a and V_b and then $V_x = V_a - V_b$.

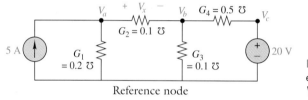

Figure 4.3 Resistive circuit for example 4.1. Note that the node voltage V_c is specified by the voltage source.

SOLUTION

Step 1. *Consider node* a *labeled with node voltage* V_a. Using KCL to sum the currents leaving this node yields the equation

$$G_1 V_a + G_2(V_a - V_b) - 5 = 0.$$

Grouping coefficients of V_a and V_b and moving the source current of 5 amps to the right yields the equivalent equation,

$$(G_1 + G_2)V_a - G_2 V_b = 5. \tag{4.1}$$

Step 2. Summing the currents leaving node b yields the equation

$$G_2(V_b - V_a) + G_3 V_b + G_4(V_b - 20) = 0,$$

or equivalently,

$$-G_2 V_a + (G_2 + G_3 + G_4)V_b = 20G_4. \tag{4.2}$$

Step 3. *Consider node* c. Because this node is connected to the top terminal of the grounded independent voltage source, the node voltage is specified as 20 V. This fact was utilized in writing the node equation at node b. Application of KCL to this node is necessary only if we want to find the current through the voltage source.

Step 4. *Plug in numbers and write the node equations in matrix form.* With numbers, equations 4.1 and 4.2 reduce to

$$0.3V_a - 0.1V_b = 5$$

and

$$-0.1V_a + 0.7V_b = 10.$$

It is possible to solve these two equations simultaneously by adding and subtracting equations. However, because many calculators now do matrix arithmetic, because of the widespread use of matrix software packages such as MATLAB, and because equation solution techniques in circuits, systems, and control heavily utilize matrix methods, we consider the matrix form of the equations, namely,

$$\begin{bmatrix} 0.3 & -0.1 \\ -0.1 & 0.7 \end{bmatrix} \begin{bmatrix} V_a \\ V_b \end{bmatrix} = \begin{bmatrix} 5 \\ 10 \end{bmatrix}. \tag{4.3}$$

Step 4. Solving equation 4.3 by the inversion method (or equivalently, by Cramer's rule), one obtains the solution

$$\begin{bmatrix} V_a \\ V_b \end{bmatrix} = \begin{bmatrix} 0.3 & -0.1 \\ -0.1 & 0.7 \end{bmatrix}^{-1} \begin{bmatrix} 5 \\ 10 \end{bmatrix} = 5 \begin{bmatrix} 0.7 & 0.1 \\ 0.1 & 0.3 \end{bmatrix} \begin{bmatrix} 5 \\ 10 \end{bmatrix} = \begin{bmatrix} 22.5 \\ 17.5 \end{bmatrix} \text{ V.}$$

The branch voltage $V_x = V_a - V_b = 22.5 - 17.5 = 5$ V.

The matrix in equation 4.3 is **symmetric**, i.e., its transpose is itself. (The entries above and below the diagonal are mirror images of each other.) Whenever there are no dependent sources present, the coefficient matrix of the node equations is symmetric. Moreover, the 1–1 entry of the matrix is the sum of the conductances at node a, the 2–2 entry is the sum of the conductances at node b, etc. The 1–2 entry of the matrix is the negative of the sum of the conductances between nodes a and b. If there were a 1–3 entry, it would be the negative of the sum of the conductances between nodes a and c. Thus, whenever dependent sources are absent, it is easy to write the nodal matrix by inspection. Further, if independent voltage sources are absent, then the right-hand side of the nodal matrix equation can also be written by inspection: The entry is simply the sum of the independent source currents injected into the node at which KCL is applied. When voltage-controlled current sources are present, the steps for writing nodal equations are the same as illustrated in example 4.1. Generally, in this case, the resultant matrix is not symmetric.

The next example illustrates how the equations are written when a current-controlled current source is present in the circuit.

◆ **EXAMPLE 4.2**

Consider the circuit of figure 4.4, which contains a dependent current source. The objective here is to write a set of nodal equations that accounts for the dependent source.

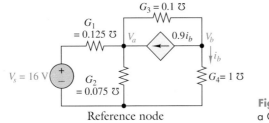

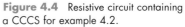

Reference node

Figure 4.4 Resistive circuit containing a CCCS for example 4.2.

SOLUTION

The steps in this example proceed in the same sequence as the steps in example 4.1. The only difference is that here there will be a third equation, which specifies the current in the dependent current source in terms of the node voltages.

Step 1. *Write a node equation at node* a. Summing the currents leaving node a yields the equation

$$(G_1 + G_2 + G_3)V_a - G_3V_b - 0.9i_b = G_1V_s. \tag{4.4}$$

In step 3, we will substitute for i_b.

Step 2. *Write a node equation at node* b. Again, summing the currents leaving node *b* yields

$$-G_3 V_a + (G_3 + G_4)V_b + 0.9i_b = 0. \tag{4.5}$$

Step 3. *Determine* i_b, *and substitute into equations 4.4 and 4.5.* Clearly,

$$i_b = G_4 V_b,$$

or equivalently, $G_4 V_b - i_b = 0$. Substituting into equations 4.4 and 4.5 yields

$$(G_1 + G_2 + G_3)V_a - (G_3 + 0.9G_4)V_b = G_1 V_s,$$
$$-G_3 V_a + (G_3 + G_4 + 0.9G_4)V_b = 0. \tag{4.6}$$

Step 4. *Insert numbers and write the node equations in matrix form.* Plugging numbers in for each G_i and writing the nodal equations of equation 4.6 in matrix form yields

$$\begin{bmatrix} 0.3 & -1 \\ -0.1 & 2 \end{bmatrix} \begin{bmatrix} V_a \\ V_b \end{bmatrix} = \begin{bmatrix} 2 \\ 0 \end{bmatrix}.$$

Step 5. Solving by the matrix inverse method or by a canned software program produces

$$\begin{bmatrix} V_a \\ V_b \end{bmatrix} = \begin{bmatrix} 0.3 & -1 \\ -0.1 & 2 \end{bmatrix}^{-1} \begin{bmatrix} 2 \\ 0 \end{bmatrix} = 2 \begin{bmatrix} 2 & 1 \\ 0.1 & 0.3 \end{bmatrix} \begin{bmatrix} 2 \\ 0 \end{bmatrix} = \begin{bmatrix} 8 \\ 0.4 \end{bmatrix} \text{V}.$$

An alternative method of carrying out the simple substitution $i_b = G_4 V_b$ is to do it right on the circuit diagram, even before writing the node equations. Thus, the controlled current source, $0.9i_b$, becomes $0.9G_4 V_b = 0.9V_b$. This amounts to transforming a current-controlled current source into a voltage-controlled current source, to which the simpler procedure of example 4.1 applies.

With matrix methods in conjunction with either a calculator or a software package that easily executes matrix manipulations, it would have been better to avoid substituting for i_b. It would have been easier to write three equations (equations 4.4, 4.5, and $G_4 V_b - i_b = 0$) in matrix form as

$$\begin{bmatrix} G_1 + G_2 + G_3 & -G_3 & -0.9 \\ -G_3 & G_3 + G_4 & 0.9 \\ 0 & G_4 & -1 \end{bmatrix} = \begin{bmatrix} V_a \\ V_b \\ i_b \end{bmatrix}$$

$$\begin{bmatrix} 0.3 & -0.1 & -0.9 \\ -0.1 & 1.1 & 0.9 \\ 0 & 1 & -1 \end{bmatrix} \begin{bmatrix} V_a \\ V_b \\ i_b \end{bmatrix} = \begin{bmatrix} 2 \\ 0 \\ 0 \end{bmatrix}.$$

Then, direct matrix inversion achieved through a software program such as MATLAB, or a trifold application of Cramer's rule yields the same answers as example 4.2:

$$\begin{bmatrix} V_a \\ V_b \\ i_b \end{bmatrix} = \begin{bmatrix} 4 & 2 & -1.8 \\ 0.2 & 0.6 & 0.36 \\ 0.2 & 0.6 & -0.64 \end{bmatrix} \begin{bmatrix} 2 \\ 0 \\ 0 \end{bmatrix} = \begin{bmatrix} 8 \text{ V} \\ 0.4 \text{ V} \\ 0.4 \text{ A} \end{bmatrix}.$$

A calculator or a software program that allows for matrix computation permits a straightforward solution process. It is also less prone to error, as substitution of one equation into another with the consequent regrouping of terms becomes unnecessary.

The next example illustrates the nodal equation method when all independent and dependent voltage sources have one terminal at the reference node, i.e., when all such sources are grounded or nonfloating. (**Floating** means that neither node of the source is connected to the reference node.)

EXAMPLE 4.3

Consider the circuit in figure 4.5, which contains three voltage sources and one independent current source. Find the node voltages V_a and V_b.

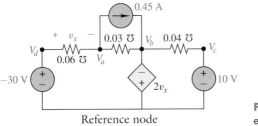

Reference node

Figure 4.5 Resistive circuit for example 4.3.

SOLUTION

There are four nodes, labeled a, b, c, and d. However, V_d is specified by the -30-V voltage source. V_c is specified by the 10-V independent voltage source, and V_b is specified by a dependent voltage source, i.e., $V_b = -2v_x$. Therefore, there is only one unknown node voltage, V_a.

Step 1. *Write a node equation at node* a. Clearly,

$$(0.06 + 0.03)V_a - 0.06(-30) - 0.03V_b + 0.45 = 0,$$

or equivalently,

$$0.09V_a - 0.03V_b = -2.25. \tag{4.7}$$

Step 2. *Determine an equation for* V_b. Since $v_x = V_d - V_a$, it follows that

$$V_b = -2v_x = -2(-30 - V_a) = 60 + 2V_a. \tag{4.8}$$

Step 3. Substitute equation 4.8 into equation 4.7 to obtain

$$0.09V_a - 0.06V_a = -2.25 + 1.8 = -0.45$$

in which case

$$V_a = -15 \text{ V and } V_b = 30 \text{ V}.$$

This completes our discussion of writing nodal equations for circuits that do not contain floating voltage sources. The difficulties encountered in the presence of floating voltage sources is taken up in the next section.

3. NODAL ANALYSIS II: FLOATING VOLTAGE SOURCES

When a floating dependent or independent voltage source is present in a circuit with respect to a given reference node, a direct application of KCL to either terminal node of the voltage source is not fruitful. There are several ways to handle this situation. One way is to enclose both of the terminal nodes at each end of the floating voltage source by a **Gaussian surface**, i.e., a closed curve. Figure 4.6 illustrates such an enclosure for the circuit of example 4.4. The nodes contained in this enclosure define what many texts call a **supernode**. We shall follow that terminology.

Since the surface enclosing the two nodes is a Gaussian surface, KCL requires that the sum of the currents leaving the supernode be zero. Nodal analysis would then proceed by writing KCL equations at each supernode, plus all remaining circuit nodes, in terms of the node voltages. Not unexpectedly, this will not produce a sufficient number of equations to determine the unknowns. It is necessary to write additional equations that specify the relationship between the terminal node voltages contained within each supernode. At this point, further explanation is best provided by an example.

◆ EXAMPLE 4.4

Determine the node voltages V_a, V_b, and V_c in the circuit of figure 4.6, in which the bottom node is taken to be the reference node.

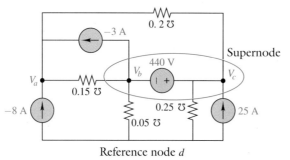

Figure 4.6 Resistive circuit containing a floating voltage source for the given reference. Generally, the reference node may be chosen arbitrarily.

SOLUTION

Step 1. *Write a node equation at node* a. Summing the currents leaving node a yields

$$(0.15 + 0.2)V_a - 0.15V_b - 0.2V_c + 8 + 3 = 0,$$

or equivalently,

$$0.35V_a - 0.15V_b - 0.2V_c = -11. \tag{4.9}$$

This provides one equation in three unknowns.

Step 2. *Write a nodal equation at the indicated supernode.* Since the surface enclosing the supernode is a Gaussian surface, the sum of the currents leaving the surface is zero. Hence,

$$-(0.15 + 0.2)V_a - 3 + (0.05 + 0.15)V_b + (0.25 + 0.2)V_c - 25 = 0.$$

Simplifying this expression leads to

$$-0.35V_a + 0.2V_b + 0.45V_c = 28. \tag{4.10}$$

Step 3. *Write an equation relating the voltages* V_b *and* V_c *inside the supernode.* Clearly, the voltages V_b and V_c are constrained by the voltage source between their respective nodes. Hence,

$$-V_b + V_c = 440. \tag{4.11}$$

Step 4a. At this point, one alternative is simply to write the preceding three equations in matrix form and solve. In matrix form, equations 4.9, 4.10, and 4.11 are

$$\begin{bmatrix} 0.35 & -0.15 & -0.2 \\ -0.35 & 0.2 & 0.45 \\ 0 & -1 & 1 \end{bmatrix} \begin{bmatrix} V_a \\ V_b \\ V_c \end{bmatrix} = \begin{bmatrix} -11 \\ 28 \\ 440 \end{bmatrix}.$$

Solving this matrix-vector equation by the matrix inversion method yields

$$\begin{bmatrix} V_a \\ V_b \\ V_c \end{bmatrix} = \begin{bmatrix} 0.35 & -0.15 & -0.2 \\ -0.35 & 0.2 & 0.45 \\ 0 & -1 & 1 \end{bmatrix}^{-1} \begin{bmatrix} -11 \\ 28 \\ 440 \end{bmatrix} = \begin{bmatrix} -90 \\ -310 \\ 130 \end{bmatrix} \text{ V.} \tag{4.12}$$

The solution of equation 4.12 can proceed by any number of methods, but is easily achieved directly using the equation solver found in MATLAB or on any number of advanced scientific calculators.

Step 4b. If hand calculation is desired, one could eliminate V_c from equations 4.9 and 4.10. An alternative method of carrying out the simple substitution $V_c = V_b + 440$ is to do it right on the circuit diagram, even before one starts to write the node equations. Thus, the voltage at node c will be marked as $(V_b + 440)$. This way, the variable V_c will never appear when writing the node equations. Application of KCL at node a and the supernode leads directly to two equations in two unknowns, V_a and V_b. The details of this method are left as an exercise.

Exercise. Substitute equation 4.11 into equations 4.9 and 4.10. Solve, and verify the answers obtained in equation 4.12.

Exercise. Compute the voltages V_{db}, V_{ab}, and V_{cb} for the circuit of figure 4.6.
ANSWERS in random order: 220 V, 310 V, 440 V.

In the foregoing example, node d was taken as the reference node. However, node b could just as easily be chosen, in which case the voltage source would not have been floating. Problem 4 investigates this choice of reference node.

The next example deals with a circuit having a floating dependent voltage source. By convention, the reference node of this circuit, shown in figure 4.7, and all subsequent circuits will be the bottom node of the circuit unless stated otherwise.

◆ EXAMPLE 4.5

For the circuit of figure 4.7, in which $I_s = 2$ A, determine the node voltages V_a, V_b, and V_c. Also, determine i_x.

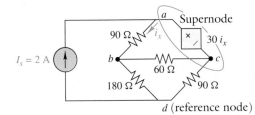

Figure 4.7 Resistive circuit containing a floating dependent voltage source when node d is chosen as the reference node.

SOLUTION

Step 1. **Sum the currents leaving the supernode.** At the indicated supernode,

$$I_s = \frac{1}{90}V_a - \left(\frac{1}{90} + \frac{1}{60}\right)V_b + \left(\frac{1}{90} + \frac{1}{60}\right)V_c.$$

Multiplying both sides of this equation by 180 yields the simplified equation

$$2V_a - 5V_b + 5V_c = 180I_s. \qquad (4.13)$$

Step 2. **Sum the currents leaving node** b. Summing the currents leaving node b yields

$$-\frac{1}{90}V_a + \left(\frac{1}{90} + \frac{1}{60} + \frac{1}{180}\right)V_b - \frac{1}{60}V_c = 0.$$

Multiplying both sides by 180 produces the simplified equation

$$-2V_a + 6V_b - 3V_c = 0. \qquad (4.14)$$

Step 3. **Write an equation relating the terminal node voltages of the supernode.** From figure 4.7,

$$V_a - V_c = 30i_x = \frac{30}{90}(V_a - V_b),$$

which reduces to

$$2V_a + V_b - 3V_c = 0. \tag{4.15}$$

Writing equations 4.13 through 4.15 in matrix form yields

$$\begin{bmatrix} 2 & -5 & 5 \\ -2 & 6 & -3 \\ 2 & 1 & -3 \end{bmatrix} \begin{bmatrix} V_a \\ V_b \\ V_c \end{bmatrix} = \begin{bmatrix} 180I_s \\ 0 \\ 0 \end{bmatrix} = \begin{bmatrix} 360 \\ 0 \\ 0 \end{bmatrix}.$$

Of course, one can solve for V_a, V_b, and V_c by Cramer's rule. For example, to find V_a by Cramer's rule, one has

$$V_a = \frac{\det \begin{bmatrix} 360 & -5 & 5 \\ 0 & 6 & -3 \\ 0 & 1 & -3 \end{bmatrix}}{\det \begin{bmatrix} 2 & -5 & 5 \\ -2 & 6 & -3 \\ 2 & 1 & -3 \end{bmatrix}} = \frac{-5{,}400}{-40} = 135 \text{ V}.$$

Most often, it is better to use a matrix equation solver found in such programs as MATLAB, on a calculator, or on a computer, rather than employing Cramer's rule. Also, one may compute the inverse of the coefficient matrix via a software package or with a calculator and then solve for the node voltages as follows:

$$\begin{bmatrix} V_a \\ V_b \\ V_c \end{bmatrix} = \begin{bmatrix} 2 & -5 & 5 \\ -2 & 6 & -3 \\ 2 & 1 & -3 \end{bmatrix}^{-1} \begin{bmatrix} 360 \\ 0 \\ 0 \end{bmatrix} = \begin{bmatrix} 0.375 & 0.25 & 0.375 \\ 0.3 & 0.4 & 0.1 \\ 0.35 & 0.3 & -0.05 \end{bmatrix} \begin{bmatrix} 360 \\ 0 \\ 0 \end{bmatrix} = \begin{bmatrix} 135 \\ 108 \\ 126 \end{bmatrix} \text{ V}.$$

With these node voltages, one computes

$$i_x = \frac{1}{90} (V_a - V_b) = 0.3 \text{ A}.$$

Exercise. Compute the voltages V_{dc}, V_{ac}, and V_{bc}.
ANSWERS in random order: -18 V, -126 V, 9 V.

The concept of a supernode can be extended to the case where the Gaussian surface encloses more than two nodes connected by independent and/or controlled voltage sources.

This completes our discussion of the standard nodal equation method of circuit analysis. We will take up the discussion of nodal analysis again in section 5 on the modified nodal analysis technique, which is the basis of many computer software packages for general circuit analysis.

4. LOOP ANALYSIS

Loop analysis is a second general analysis technique for computing the voltages and currents in a circuit. **Mesh analysis** is a special case of loop analysis for planar circuits (i.e., circuits that can be drawn without branch crossings) in which the loops are chosen to be the meshes, as illustrated in figure 4.2. In contrast to KCL, which underlies nodal analysis, KVL underlies loop analysis. In loop analysis, one sums the voltages around a loop in terms of loop currents, as illustrated in figures 4.2 and 4.8. **Loop currents** circulate around closed paths in the circuit. Similarly, in the mesh analysis of planar circuits, the term **mesh current** is used for loop current (more about this in section 6). Certain branches of the circuit may be common to two or more loop currents. The **branch current**, then, equals the net flow of the loop currents incident on the branch. Hence, once loop currents are defined, one can express each branch current in terms of the loop currents and the sum of the voltages around each loop in terms of the source values and the loop currents multiplied by appropriate element values. This produces a set of equations called **loop equations.**

One solves the loop equations for the loop currents. Once the loop currents are known, we reverse the process and compute desired branch voltages and branch currents in the circuit.

 EXAMPLE 4.6

Consider the circuit of figure 4.8 with the three specified loops. (Other loops are possible—see problem 10.) The loop currents for each loop are I_1, I_2, and I_3. The objective is to find each of these currents.

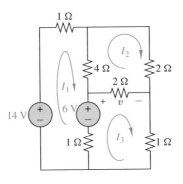

Figure 4.8 Resistive circuit for the loop analysis of example 4.6.

SOLUTION

Step 1. *Write a loop equation based on I_1.* Summing the voltages around loop 1 in terms of the loop currents produces

$$14 \text{ V} = I_1 + 4(I_1 - I_2) + 6 + (I_1 - I_3) = 6I_1 - 4I_2 - I_3 + 6.$$

Hence,

$$6I_1 - 4I_2 - I_3 = 8 \text{ V}.$$

Step 2. *Write a loop equation based on I_2.* Writing a loop equation for loop 2 yields

$$0 = 4(I_2 - I_1) + 2I_2 + 2(I_2 - I_3) = -4I_1 + 8I_2 - 2I_3.$$

Step 3. *Consider loop 3.* Summing the voltages around loop 3 yields

$$6 \text{ V} = -I_1 - 2I_2 + 4I_3.$$

Step 4. Writing the preceding three loop equations in matrix form results in

$$\begin{bmatrix} 6 & -4 & -1 \\ -4 & 8 & -2 \\ -1 & -2 & 4 \end{bmatrix} \begin{bmatrix} I_1 \\ I_2 \\ I_3 \end{bmatrix} = \begin{bmatrix} 8 \\ 0 \\ 6 \end{bmatrix}.$$

Solving by the matrix inverse method or by Cramer's rule yields the loop currents in A:

$$\begin{bmatrix} I_1 \\ I_2 \\ I_3 \end{bmatrix} = \begin{bmatrix} 6 & -4 & -1 \\ -4 & 8 & -2 \\ -1 & -2 & 4 \end{bmatrix}^{-1} \begin{bmatrix} 8 \\ 0 \\ 6 \end{bmatrix} = \begin{bmatrix} 0.35 & 0.225 & 0.20 \\ 0.225 & 0.2875 & 0.20 \\ 0.20 & 0.20 & 0.40 \end{bmatrix} \begin{bmatrix} 8 \\ 0 \\ 6 \end{bmatrix} = \begin{bmatrix} 4 \\ 3 \\ 4 \end{bmatrix} \text{ A}.$$

With these loop currents, it is possible to determine all voltages and currents in the circuit. For example, the voltage

$$v = 2(I_3 - I_2) = 2 \text{ V}.$$

The matrix in the equation

$$\begin{bmatrix} 6 & -4 & -1 \\ -4 & 8 & -2 \\ -1 & -2 & 4 \end{bmatrix} \begin{bmatrix} I_1 \\ I_2 \\ I_3 \end{bmatrix} = \begin{bmatrix} 8 \\ 0 \\ 6 \end{bmatrix}$$

from step 4 of the foregoing example is symmetric. Whenever there are no dependent sources present, the coefficient matrix of the loop equations is symmetric. Moreover, the 1–1 entry of the matrix is the sum of the resistances in loop 1, the 2–2 entry is the sum of the resistances in loop 2, etc. The 1–2 entry of the matrix is $\sum(\pm R_k)$, where each R_k is a resistance common to both loops 1 and 2; we use the $+$ sign when both loop currents circulate through R_k in the same direction and use the $-$ sign otherwise. Thus, whenever dependent sources are absent, it is easy to write the loop matrix by inspection. Further, if independent current sources are absent, then the right-hand side of the loop equations can also be written by inspection: The entry is simply the net voltage of the sources in the loop that tends to deliver a current in the direction of the loop current.

If an independent current source coincides with a single loop current, then the analysis becomes simpler because that loop current is no longer an unknown: The loop current has the same value as the source current if their directions coincide; it has the same value with opposite sign if their directions are opposing. It is not necessary to apply KVL to that loop, unless one wishes to compute the voltage across the independent current source. For example, the circuit of problem 5 can be solved with two loop equations in two unknowns.

This next example illustrates a phenomenon of loop analysis analogous to the use of supernodes in nodal analysis. Many texts define something called a supermesh and write a special loop equation for this supermesh. This is often confusing for the beginner. There is an easier way, which the following example illustrates.

◆ EXAMPLE 4.7

Consider the circuit of figure 4.9, which contains a 7-A current source common to loops 1 and 2. Using the indicated loops, find values for the loop currents I_1, I_2, and I_3, and the voltage v.

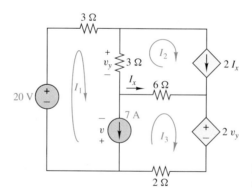

Figure 4.9 Circuit containing a current source between loops.

SOLUTION

To begin the solution, define a voltage v as indicated, across the current source. This is an extra variable that can be eliminated during the solution process. There are three specified loops. Observe that the current I_2 is specified by the dependent current source $2I_x$. Also, the currents I_1 and I_3 are constrained by the 7-A independent current source.

Step 1. *Sum the voltages around loop 1.* In a straightforward manner,

$$20 \text{ V} = 3I_1 + 3(I_1 - I_2) - v = 6I_1 - 3I_2 - v. \tag{4.16}$$

Step 2. *Consider loop 2. The goal of summing the voltage around a loop is to constrain the loop current and its relationship to adjacent loop currents by an independent equation.* In contrast to loop 1, loop 2 contains a dependent current source that constrains I_2 as follows:

$$I_2 = 2I_x = 2(I_3 - I_2).$$

Equivalently,

$$0 = 3I_2 - 2I_3. \tag{4.17}$$

Note that this constraint is not obtained through the application of KVL; it is necessary to apply KVL around loop 2 only if one wishes to determine the voltage across the dependent current source $2I_x$.

Equation 4.17 is a second equation for a problem with a total of four unknowns: I_1, I_2, I_3, and v. Two further equations are needed.

Step 3. *Consider the effect of the 7-A independent current source common to loops 1 and 2.* The 7-A source constrains the difference of I_1 and I_3 as follows:

$$7 = I_1 - I_3. \tag{4.18}$$

Step 4. At this point, we need a fourth equation that independently relates v to the other loop currents. To get this equation, we sum the voltages around loop 3:

$$0 = 6(I_3 - I_2) + 2v_y + 2I_3 + v.$$

But $v_y = 3I_1 - 3I_2$. Hence,

$$0 = 6I_1 - 12I_2 + 8I_3 + v. \tag{4.19}$$

Step 5. Putting equations 4.16 through 4.19 in matrix form yields

$$\begin{bmatrix} 6 & -3 & 0 & -1 \\ 0 & 3 & -2 & 0 \\ 1 & 0 & -1 & 0 \\ 6 & -12 & 8 & 1 \end{bmatrix} \begin{bmatrix} I_1 \\ I_2 \\ I_3 \\ v \end{bmatrix} = \begin{bmatrix} 20 \\ 0 \\ 7 \\ 0 \end{bmatrix}. \tag{4.20}$$

Solving equation 4.20 by the matrix inverse method or by an available software package produces the solution given by

$$\begin{bmatrix} I_1 \\ I_2 \\ I_3 \\ v \end{bmatrix} = \begin{bmatrix} 0.10 & 0.50 & -0.20 & 0.10 \\ 0.0667 & 0.667 & -0.80 & 0.0667 \\ 0.10 & 0.50 & -1.20 & 0.10 \\ -0.60 & 1.00 & 1.20 & 0.40 \end{bmatrix} \begin{bmatrix} 20 \\ 0 \\ 7 \\ 0 \end{bmatrix} = \begin{bmatrix} 0.60 \\ -4.2667 \\ -6.40 \\ -3.60 \end{bmatrix} \tag{4.21}$$

with currents in amps and v in volts.

In terms of hand calculation, the 4×4 matrix of equation 4.20 is unwieldy. Using equation 4.17, one can eliminate I_3 (or I_2) from all remaining equations, reducing the problem to three equations in three unknowns. Alternatively, one can denote this relationship directly on the circuit diagram, so that I_3 (or I_2) never appears in any of the equations. Further, one can solve equation 4.16 (or 4.19) for v and substitute into equation 4.19 (or 4.16) to reduce the problem to two equations in two unknowns. These can then be solved by a variety of hand calculation methods. Other ways of simplification for hand calculation are also possible and are left as exercises.

Exercise. Reduce the number of equations and unknowns in equation 4.20 by using equation 4.17 to eliminate I_3 and by using equation 4.16 to eliminate v in equation 4.19. Your answers for I_1 and I_2 should be the same as in example 4.7.

Exercise. Reduce the number of equations and unknowns in equation 4.20 by using equation 4.17 to eliminate I_2 and by using equation 4.18 to eliminate I_1. Again, your answers for I_3 and v should be the same as in example 4.7.

In the preceding example, there was an independent current source between two loops. The independent current source presents an obstacle to a straightforward writing of the loop equations. To deal with this difficulty, a voltage v was defined across the independent current source. The normal loop equations were then written with the extra variable v included. Whenever there is an independent or dependent current source between two loops, the standard procedure is to define a voltage across the source and write the usual loop equations. The added variable is eliminated in the solution process or can be eliminated earlier by a hand substitution.

As mentioned before, current sources constrain the relationship among loop currents common to the source, as illustrated in step 5 of example 4.7. To obtain enough independent equations, it is necessary to write down the constraining equation explicitly. The idea is similar to the constraints imposed between node voltages separated by an independent or dependent voltage source.

We conclude this section with an example of a one-transistor amplifier equivalent circuit.

◆ EXAMPLE 4.8

The circuit of figure 4.10 is a simplified small-signal equivalent circuit for a one-transistor amplifier. Find

(a) the voltage gain, v_o/v_s;
(b) the input resistance seen by the source, i.e., $R_{in} = v_s/i_b$.

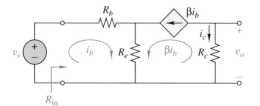

Figure 4.10 Small-signal equivalent circuit for a one-transistor amplifier. The signals here are usually time dependent, so we have adopted the lowercase notation for voltages and currents.

SOLUTION

Step 1. *Determine* v_s *in terms of* i_b. A loop equation around the first loop yields

$$v_s = R_b i_b + R_e(i_b + \beta i_b) = [R_b + (1+\beta)R_e]i_b. \tag{4.22}$$

Step 2. *Determine* v_o *in terms of* i_b. Since $i_c = -\beta i_b$, it follows that

$$v_o = R_c i_c = -R_c \beta i_b. \tag{4.23}$$

Step 3. To determine the voltage gain, one takes the ratio of equation 4.23 to equation 4.22, to obtain

$$\text{Voltage gain} = \frac{v_o}{v_s} = -\frac{R_c \beta i_b}{[R_b + (1+\beta)R_e]i_b} = -\frac{R_c \beta}{R_b + (1+\beta)R_e}.$$

Step 4. *Determine the input resistance* R_{in}. The input resistance is given by the ratio v_s/i_b. From equation 4.22,

$$R_{in} = [R_b + (1+\beta)R_e].$$

There is one final point to make before closing our discussion of loop analysis: Loops can be chosen in different ways. Cleverly choosing loops can sometimes simplify the solution of the loop equations. For example, by choosing a loop that passes through a current source so that no other loop passes through that source, we automatically specify the loop current by the current source.

5. MODIFIED NODAL ANALYSIS

To this point, the text has discussed two general circuit analysis methods: node analysis and loop analysis. In many simple circuits, the computational effort needed by either method is about the same. The choice is mainly a personal preference. However, as the size of an arbitrary circuit grows larger, there are two obvious reasons for choosing the nodal method over the loop method: (1) the number of nodal equations is usually smaller than the number of loop equations (more on this in section 6) and (2) formulating nodal equations for computer solution is easier than methods based on loop equations. This second assertion follows because in a manual analysis the loops are easily identified by *inspection*, whereas in an automated formulation of loop equations some algorithm must construct a set of independent loop equations.

On the other hand, writing nodal equations is particularly easy if the circuit contains only resistances, independent current sources, and voltage-controlled current sources—for short, an $R-I_s-g_m$ network. For such a network, one simply applies KVL to every node (except the reference node) and obtains a set of node equations directly, without any finesse.

The treatment of grounded independent and dependent voltage sources is illustrated in section 2 by examples 4.1 and 4.3. The treatment of floating independent or dependent voltage sources is discussed in section 3, where the special concept of a supernode is utilized. Example 4.2 illustrated a case where, besides the node voltages V_a and V_b, an additional unknown i_b has been included. This special case sparks an extension to what are called **modified nodal analysis** (MNA) equations. Our goal is to show that the MNA method can easily accommodate independent voltage sources and all types of dependent sources. The method retains the simplicity of the nodal method, while removing its limitations. The MNA method is the most commonly used method in present-day computer-aided circuit analysis programs.

In the MNA method, the unknowns are the usual nodal voltages, plus some naturally occurring *auxiliary currents*. These unknown auxiliary currents include the following:

1. currents through independent voltage sources;
2. currents through dependent voltage sources;
3. currents through short-circuit elements;
4. the controlling currents of appropriate dependent sources;
5. currents declared as output quantities.

Note that item 3 is actually a special case of item 1 because a short-circuit element can be viewed as an independent voltage source of 0 V.

Suppose one has identified the auxiliary currents in a network N. Then it is possible to write the MNA equations for N, step by step, as follows:

Step 1. For every element x whose current I_x has been chosen as an auxiliary current, replace that element *temporarily* by an independent current source having the value I_x. Also, replace every current-controlled current source βI_x by an independent current source having value βI_x. The resulting network, denoted N^*, is an $R-I_s-g_m$ network. Write the conventional nodal equation for N^*.

Step 2. For every element x whose current I_x has been chosen as an auxiliary current, write an equation describing its constitutive relationship in the original network N.

Step 3. Combine the equations of steps 1 and 2, and write them in the form of a single **linear matrix equation**:

$$[\text{coefficient matrix}] \times [\text{unknown vector}] = [\text{known vector}].$$

The next example will illustrate the mechanics of the MNA method and the rationale behind the procedure.

◆ **EXAMPLE 4.9**

Write the modified nodal equations for the circuit of figure 4.11a.

SOLUTION

Step 1. In this circuit, only one auxiliary current, I_x, the current through the 440-V voltage source, is necessary. Replacing the 440-V source by an independent current source,

---

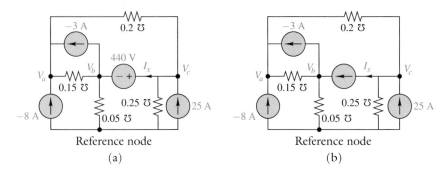

Figure 4.11 (a) Original network N. (b) Modified network N^*, which replaces N in the MNA method.

I_x, results in the network N^* shown in figure 4.11b. N^* contains only resistors and independent current sources; its nodal equations are very easy to write. Following the procedure of section 4.2, one obtains

$$0.35V_a - 0.15V_b - 0.2V_c = -11, \tag{4.24a}$$

$$-0.15V_a + 0.2V_b = I_x + 3, \tag{4.24b}$$

$$-0.2V_a + 0.45V_c = 25 - I_x. \tag{4.24c}$$

Step 2. The constitutive relationship for the 440-V voltage source whose current I_x has been chosen as an auxiliary current is

$$V_c - V_b = 440. \tag{4.24d}$$

Step 3. Writing equations 4.24a through 4.24d as a single matrix equation gives the desired modified nodal equation

$$\begin{matrix} \textit{nodal equation} \\ \textit{for } N^* \\ \\ \textit{auxiliary branch} \\ \textit{equation} \end{matrix} \left\{ \begin{matrix} \\ \\ \\ \\ \left\{ \end{matrix} \right. \begin{bmatrix} 0.35 & -0.15 & -0.20 & 0 \\ -0.15 & 0.2 & 0 & -1 \\ -0.2 & 0 & 0.45 & 1 \\ \hdashline 0 & -1 & 1 & 0 \end{bmatrix} \begin{bmatrix} V_a \\ V_b \\ V_c \\ \hdashline I_x \end{bmatrix} = \begin{bmatrix} -11 \\ 3 \\ 25 \\ \hdashline 440 \end{bmatrix}. \tag{4.25}$$

Solving equation 4.25 by the matrix method yields

$$V_a = -90 \text{ V}, \qquad V_b = -310 \text{ V}, \qquad V_c = 130 \text{ V}, \qquad I_x = -51.5 \text{ A}.$$

Of course, the answers for the node voltages agree with those obtained in example 4.4 using the supernode concept.

Since the matrix in equation 4.25 is larger than that in equation 4.12 for the same example, one may wonder whether there is any advantage of the MNA method over the supernode method. The answer is yes, because the supernode method, as described earlier, relies on inspection to delineate the supernode. On the other hand, the MNA method is systematic and easily automated for implementation on a computer.

It should become clear now that the purpose of constructing the $R-I_s-g_m$ network N^* is to make the writing of the nodal equation for N^* very straightforward and systematic. Once this construction is understood and familiarity with the procedure gained, it is really not necessary to draw the network N^* explicitly.

At this point, an inquisitive student will ask: How can we justify replacing the 440-V source in example 4.9 by an independent current source (step 1 of the procedure)? Doesn't that yield a different network from the one we intend to solve? It is true that the network N^* is different from the original network N in that the 440-V voltage source is no longer there. However, this is only an intermediate step in writing the set of equations for the original network. Referring again to example 4.9, we see that in the last scalar equation of 4.24 (step 2 of the procedure), we added the constraint equation, $V_c - V_b = 440$. This addition restores the true property of the associated branch. The constraint equation $V_c - V_b = 440$ says that it is, in fact, an independent voltage source of 440 V. Thus, every scalar equation in equation 4.24 corresponds to the original network N. The network N^* is helpful, but not *indispensable,* in writing the MNA equations.

Example 4.9 illustrates the basic idea of the MNA method. However, it is probably too simple a circuit to convince anyone of the usefulness of the method. The next example should demonstrate the power of MNA. A person who is skeptical should try to solve the problem by some other method and compare. The main concern here is not the number of equations, but the simplicity and generality of the procedure for writing the network equations.

 EXAMPLE 4.10

The circuit of figure 4.12a contains all types of network elements considered so far, except for ideal op amps. Find the node voltages by writing and solving the modified nodal equations.

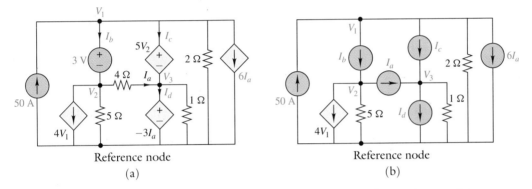

Figure 4.12 (a) Network N containing all types of sources for example 4.10. (b) Network N^* associated with network N for writing MNA equations.

SOLUTION

It is necessary to choose four auxiliary currents: I_a (controlling current of a dependent source), I_b (current through an independent voltage source), I_c (current through a dependent voltage source), and I_d (another current through a dependent voltage source). Replacing these four elements by independent current sources yields the network N^* shown in figure 4.12b, which is an R–I_s–g_m network.

The nodal equations for N^* are obtained by applying KCL to nodes 1, 2, and 3. The results are:

node 1: $$0.5V_1 + 6I_a + I_b + I_c = 50; \qquad (4.26a)$$

node 2: $$4V_1 + 0.2V_2 + I_a - I_b = 0; \qquad (4.26b)$$

node 3: $$-I_a - I_c + I_d + V_3 = 0. \qquad (4.26c)$$

The four auxiliary branch constituent relationships are obtained by referring to the original network N:

For the 4-Ω resistance: $$V_2 - V_3 = 4I_a; \qquad (4.27a)$$

For the 3-V voltage source: $$V_1 - V_2 = 3; \qquad (4.27b)$$

For the $5V_2$ dependent source: $$V_1 - V_3 = 5V_2; \qquad (4.27c)$$

For the $-3I_a$ dependent source: $$V_3 = -3I_a. \qquad (4.27d)$$

Writing equations 4.26 and 4.27 as a single matrix equation yields the desired modified nodal equation:

$$
\begin{matrix}
\textit{nodal equations} \\ \textit{for N*} \\ \\ \textit{constituent} \\ \textit{relationships for} \\ \textit{branches with} \\ \textit{auxiliary currents}
\end{matrix}
\left\{
\begin{bmatrix}
0.5 & 0 & 0 & 6 & 1 & 1 & 0 \\
4 & 0.2 & 0 & 1 & -1 & 0 & 0 \\
0 & 0 & 1 & -1 & 0 & -1 & 1 \\
0 & 1 & -1 & -4 & 0 & 0 & 0 \\
1 & -1 & 0 & 0 & 0 & 0 & 0 \\
1 & -5 & -1 & 0 & 0 & 0 & 0 \\
0 & 0 & 1 & 3 & 0 & 0 & 0
\end{bmatrix}
\begin{bmatrix}
V_1 \\ V_2 \\ V_3 \\ I_a \\ I_b \\ I_c \\ I_d
\end{bmatrix}
=
\begin{bmatrix}
50 \\ 0 \\ 0 \\ 0 \\ 3 \\ 0 \\ 0
\end{bmatrix}
\right.
. \tag{4.28}
$$

Solving the modified nodal equations by the matrix method (with a software package such as MATLAB) yields the nodal voltages

$$V_1 = 6 \text{ V}, \qquad V_2 = 3 \text{ V}, \qquad V_3 = -9 \text{ V}$$

and the auxiliary currents

$$I_a = 3 \text{ A}, \qquad I_b = 27.6 \text{ A}, \qquad I_c = 1.4 \text{ A}, \qquad I_d = 13.4 \text{ A}.$$

We conclude the section with some remarks on the MNA method. First, it would appear that the procedure described for formulating the MNA equations still needs some inspection to construct the network N^*. The $R-I_s-g_m$ network N^* is constructed to emphasize the underlying principles and to facilitate the writing of the nodal equations. In an automated formulation of the MNA equations using a computer-implemented algorithm, N^* is not constructed. Instead, one uses what are called "element stamps" to assemble the MNA equations directly from each branch description. Some simple element stamps for an $R-I_s-g_m$ network are described in the problems at the end of the chapter. A complete set of element stamps for all linear network elements may be found in advanced books on computer-aided circuit simulation.

Our second remark is that in an automated formulation of the MNA equations, it is customary to require that the controlling current of any dependent source be associated with an independent voltage source or a short-circuit element. This reduces the number of element stamps in the complete catalog.

Third, the MNA equations have a larger dimension than equations using the supernode concept. However, special methods, called "sparse matrix techniques," make the computational effort depend on the number of nonzero entries in the nodal matrix, not its row or column dimension. Since each node in a circuit has only a small number of branch connections to other nodes, the MNA matrix, although of larger dimension, has mostly zero entries for large circuits. Because of this, sparse matrix techniques allow very efficient and accurate solution of the MNA equations. Details are reserved for more advanced courses on circuit theory.

6. THEORETICAL CONSIDERATIONS

Selection of Loop Currents for a General Network Configuration

The circuits studied in section 4 are **planar**, i.e., are simple enough to be drawn in a plane, *without branches having to cross each other*. Figure 4.13a illustrates a planar circuit. Our goal is to investigate writing KCL and KVL for planar and nonplanar circuits.

Recall that KCL and KVL equations do not depend on the true nature of each branch (element), as long as the element is lumped. For KVL and KCL, the important information is "to which two nodes each branch is connected." For this reason, it is convenient to replace each network branch by a line segment. This simplified representation of the connectedness of the circuit elements is called the **graph** associated with the circuit. The graph for the circuit of figure 4.13a appears in figure 4.13b, where one branch crossing is present. However, it is easy to redraw the graph to eliminate the branch crossing, as shown in figure 4.13c. Hence, the circuit of figure 4.13a is planar.

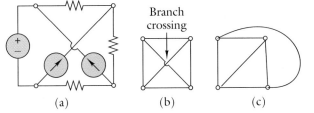

(a)   (b)   (c)

Figure 4.13 (a) Original circuit. (b) Its graph. (c) A redrawing of (b) to eliminate branch crossing.

Although the great majority of practical circuits are planar, the application of loop analysis to nonplanar circuits is very important. Our analysis methods would be inadequate otherwise. A circuit is **nonplanar** whenever it cannot be drawn in a plane without branch crossings. Two simple nonplanar circuits appear in figure 4.14. As drawn, each graph has only one branch crossing. Encouraged by the success of redrawing figure 4.13b to eliminate branch crossings, one may try different ways of redrawing the graphs of figure 4.14. Unfortunately, all attempts to eliminate the branch crossing fail. In graph theory, these two graphs are called *Kuratowski's basic nonplanar graphs.*

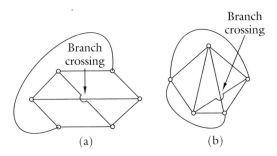
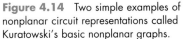

(a)   (b)

Figure 4.14 Two simple examples of nonplanar circuit representations called Kuratowski's basic nonplanar graphs.

In the case of a planar circuit, after drawing the circuit without branch crossings, a set of "window panes" or meshes can be precisely identified. Mathematically, a *mesh* is the boundary of a region with a finite area. Intuitively, the word *mesh* suggests a resemblance to the openings of a fish net. When the loop currents are chosen to be these fictitious currents circulating around the meshes, they are usually called **mesh currents**. Figures 4.2, 4.8, and 4.9 illustrate the idea. *Mesh equations* are the result of applying KVL to meshes.

For nonplanar circuits, such as those shown in figure 4.14, the concept of a mesh is ill defined because of the branch crossings. How, then, does one write loop equations for these circuits? In courses in advanced circuit analysis, this is tackled in a very systematic manner by using a concept called the *tree* of a graph or circuit. Such an approach introduces a great deal of nomenclature that is not very useful for students whose future technical interest is outside of analog circuit analysis. For that reason, this text takes a new approach.

The basic idea is to utilize the simple scheme of mesh analysis as much as possible and to handle the troublemaking branches in a special manner. In fact, example 4.7 serves to introduce the basic idea of this strategy. There, the troublemaking 7-A independent current source was implicitly and temporarily replaced by an independent voltage source to facilitate the writing of the mesh equations. Later, its true nature was restored by the constraint equation 4.18.

Our strategy for writing loop equations for nonplanar circuits is as follows. First, given a nonplanar circuit N, make a preliminary effort to redraw the circuit, with the aim of reducing the number of branch crossings. There is no need to minimize the number of crossings, although a redrawing with a minimum number of crossings will be the best. After this preliminary effort, suppose there are still m crossings. By removing a sufficient number of branches from the circuit, we can eliminate all of the crossings. Denote the resultant planar circuit by N^*. A satisfactory set of loop currents for the original circuit N may then be chosen to consist of

1. mesh currents in the planar circuit N^* and
2. circulating currents in loops, each of which is formed by one of the removed branches and some set of branches in N^*.

One then applies KVL to each of the loops to obtain the loop equations. The procedure is best illustrated with some examples.

EXAMPLE 4.11

Figure 4.15 shows a nonplanar circuit N. All resistors have a value of 1 ohm. Write the loop equations and find the currents in all the resistors.

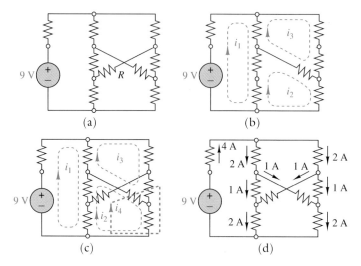

(a)   (b)

(c)   (d)

Figure 4.15 Loop analysis of a nonplanar circuit. All resistances have a value of 1 Ω. (a) Nonplanar network N. (b) Derived planar network N^*. (c) Loop currents for N. (d) Branch current values.

SOLUTION

Removing the resistor labeled R in figure 4.15a to eliminate the branch crossing yields the planar circuit N^* of figure 4.15b. N^* contains three easily identified mesh currents, labeled i_1, i_2, and i_3. Restoring the resistor R creates an additional loop current labeled i_4. This loop current circulates in R and three other resistances, as shown by the heavy lines of figure 14.15c. Applying KVL to these four loops and using the "inspection rule" described in section 4 yields the loop equations

$$\begin{bmatrix} 4 & -2 & -1 & -1 \\ -2 & 4 & -1 & 2 \\ -1 & -1 & 4 & 1 \\ -1 & 2 & 1 & 4 \end{bmatrix} \begin{bmatrix} i_1 \\ i_2 \\ i_3 \\ i_4 \end{bmatrix} = \begin{bmatrix} 9 \\ 0 \\ 0 \\ 0 \end{bmatrix}. \tag{4.29}$$

Solving equation 4.29 results in $i_1 = 4$ A, $i_2 = 3$ A, $i_3 = 2$ A, and $i_4 = -1$ A. Using figure 4.15c, we obtain the branch currents (i.e., the currents in each of the resistors) as simple linear combinations of loop currents. Figure 4.15d displays the results.

In some situations, it is advantageous to apply the foregoing strategy for handling nonplanar circuits to a planar circuit. This is particularly true if any independent current sources are first removed to form N^*. It is easy to identify the mesh currents in N^*. Then, upon restoring the removed independent sources one by one, the corresponding loop currents become known quantities. It is unnecessary to apply KVL to these loops, unless one wants to know the voltages across the independent current sources. The next example illustrates this point.

EXAMPLE 4.12

Consider the planar circuit N shown in figure 4.16a. To focus our attention on the topological aspect of the problem, assume that all resistances are 1 ohm. Determine the currents delivered by the two voltage sources.

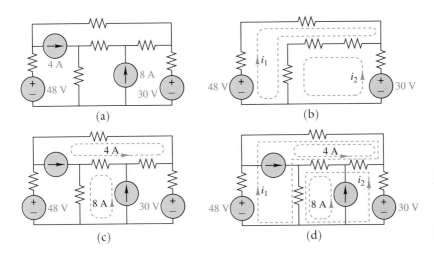

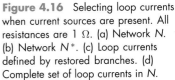

Figure 4.16 Selecting loop currents when current sources are present. All resistances are 1 Ω. (a) Network N. (b) Network N^*. (c) Loop currents defined by restored branches. (d) Complete set of loop currents in N.

SOLUTION

Removing the two independent sources results in network N^*, shown in figure 4.16b, for which two mesh currents i_1 and i_2 are easily identified. Restoring the 4-A current source defines a new loop current, as shown in figure 4.16c. Two facts about this new loop current must be pointed out: since it is a known quantity, 4 A, there is no need to introduce another current variable; and the 4-A loop current could also be chosen to go through the 48-V source. We choose not to do this, because in this way, the current delivered by the 48-V source is simply i_1. Similarly, the restoration of the 8-A source introduces another known loop current, shown in figure 4.16c. The complete set of four loop currents is shown in figure 4.16d.

Since there are only two unknown loop currents, only two equations in these two unknowns are necessary. Applying KVL to each of the indicated loops results in the following two loop equations: For loop 1,

$$i_1 + (i_1 - 4) + (i_1 + i_2 - 4) + (i_1 + i_2 - 4 + 8) + (i_1 + i_2 + 8) = 48,$$

or equivalently,

$$5i_1 + 3i_2 = 44; \qquad (4.30a)$$

and for loop 2,

$$i_2 + (i_1 + i_2 - 4) + (i_1 + i_2 - 4 + 8) + (i_1 + i_2 + 8) = 30,$$

or equivalently,

$$3i_1 + 4i_2 = 22. \qquad (4.30b)$$

Solving equations 4.30a and 4.30b yields $i_1 = 10$ A and $i_2 = -2$ A.

In the case of planar circuits, this solution method is similar to the "supermesh" method discussed in some texts. However, as illustrated in example 4.15, the present method is more general because it is also applicable to nonplanar circuits.

Numbers of Independent KCL and KVL Equations

Previous sections described systematic methods for writing node and loop equations. In nodal analysis, one applies KCL to all but the reference nodes. In loop analysis, one applies KVL to the meshes (usually, but not always) of a planar circuit or to a set of properly selected loops of a nonplanar circuit. Then, for the equations to be uniquely solvable they must be independent; i.e., each equation must contain information that is not contained in the set of remaining equations. It is now time to address questions on the numbers of independent KCL and KVL equations that are required to yield a unique solution. These questions are of theoretical interest in that they provide a justification for the widespread and reliable use of the methods described.

Before stating the classical fundamental results, it is necessary to mention some introductory concepts. First is the notion of a "connected circuit." Intuitively speaking, a circuit is **connected** if every pair of nodes in the circuit is joined by some set of branches. All the examples in this chapter contain connected circuits. If a circuit is not connected, one can analyze each connected subcircuit separately. Thus, it is sufficient to restrict our attention to only connected circuits. (We need not provide a formal definition of a connected circuit here.)

A second pertinent concept, "linearly independent equations," comes from the mathematics of linear algebra: A set of equations, denoted as $\{F_1(x_1, x_2, \ldots, x_n) = 0, \ldots, F_n(x_1, x_2, \ldots, x_n) = 0\}$ is said to be **linearly independent** if no one equation can be expressed as the scaled sum of the remaining equations. If the set is not linearly independent, then it is said to be **linearly dependent**. From this definition, a set of equations is linearly dependent if and only if one of the equations is a scaled sum of the remaining equations, e.g., $F_n = a_1 F_1 + \ldots + a_{n-1} F_{n-1}$. The constants $a_1, \ldots, a_{n-1}$, that scale the equations $F_1, \ldots, F_{n-1}$ are real numbers.

The following two theorems provide a justification for using node and loop analyses as discussed in this chapter.

> **Node Analysis Theorem:** For a connected circuit with n nodes and b branches, KCL applied to any $n - 1$ nodes yields a set of $n - 1$ linearly independent equations. Further application of KCL to the remaining node does not increase the number of independent equations.

(This theorem explains why, in writing node equations, we apply KCL to all but the reference node of the circuit.)

> **Loop Analysis Theorem:** For a connected circuit with n nodes and b branches, the maximum number of linearly independent equations based on KVL applied to the loops of the circuit is $b - n + 1$.

Most rigorous proofs of these theorems employ graph-theoretic concepts and results that, unfortunately, will be of little benefit to most students in their future work. Alternative proofs of these theorems employ the more elementary techniques of contradiction and mathematical induction that are ordinarily taught in early college mathematics courses. However, such a digression is not warranted in a first course on circuits, so we end our discussion of nodal and loop analysis at this point.

SUMMARY

This chapter introduced nodal analysis, a technique for writing a set of equations whose solution yields all node voltages in a circuit. Knowing all the node voltages and all the element values, one can compute all branch voltages and currents. Whenever there are no dependent sources present, the coefficient matrix of the node equations is symmetric. Hence, whenever dependent sources are absent, it is possible to write the nodal matrix by inspection. Further, if independent voltage sources are absent, then the right-hand side of the nodal matrix equation can also be written by inspection;

the entry is simply the sum of the independent source currents injected into the node at which KCL is applied. When voltage-controlled current sources are present, the steps for writing nodal equations are the same as illustrated in example 4.1. Generally, in this case, the resultant matrix is not symmetric.

When a floating dependent or independent voltage source is present in a circuit with respect to a given reference node, it is necessary to enclose both of the terminal nodes of the floating voltage source by a **Gaussian surface**, i.e., a closed curve. Figure 4.6 illustrated such an enclosure for the circuit of example 4.4. The nodes contained in the enclosure define what is called a **supernode**. Since the sum of the currents leaving a Gaussian surface is zero, the sum of the currents leaving the supernode is zero. For each supernode containing one voltage source, one additional equation is necessary. Together, all the additional equations must specify the relationship between the terminal node voltages contained within each supernode. If a supernode contains m voltage sources, then m additional equations are necessary.

Also introduced in this chapter was the technique of loop analysis. Mesh analysis is a special case of loop analysis for planar circuits when the loops are chosen to be the meshes. In loop analysis, one sums the voltages around a loop in terms of (fictitious) loop currents (or equivalently, mesh currents for planar circuits) that circulate around closed paths in the circuit. The branch current of a circuit is equal to the net flow of the loop currents incident on a particular branch of the circuit. Hence, once loop currents are defined, one can express each branch current in terms of the loop currents. One can then sum the voltages around each loop in terms of the source values, loop currents, and element values. This produces a set of equations called loop equations, which one then solves for the loop currents. Once the loop currents are known, one reverses the process and computes desired branch voltages and branch currents in the circuit.

Whenever there are no dependent sources present, the coefficient matrix of the loop equations is symmetric, and it is easy to write the loop matrix by inspection.

As the size of an arbitrary circuit grows larger, there are two good reasons for choosing the nodal method over the loop method: The number of nodal equations is usually smaller than the number of loop equations, and the formulation of nodal equations for computer solution is easier than methods based on loop equations. Writing nodal equations is particularly easy if the circuit contains only resistances, independent current sources, and voltage-controlled current sources—in short, if the circuit is an R–I_s–g_m network. For such a network, one simply applies KVL to every node except the reference node and obtains a set of node equations directly. For floating independent or dependent voltage sources, the task is more complex. The **modified nodal analysis** (MNA) method adds additional auxiliary variables to the formulation of the nodal equations. The method retains the simplicity of the nodal method while removing its limitations and is the most commonly used method in present-day computer-aided circuit analysis programs. The unknowns of the MNA method are the usual nodal voltages, plus some naturally occurring auxiliary currents. These auxiliary currents include currents through independent voltage sources, currents through dependent voltage sources, currents through short-circuit elements, the controlling currents of appropriate dependent sources, and currents declared as output quantities.

The chapter also focused on several theoretical considerations. First, it examined the difficulty with and approach to writing loop equations for nonplanar circuits, i.e., circuits whose graphs cannot be drawn without branch crossings. Then it looked at the numbers of equations needed to write node or loop equations. Finally, it presented two theorems, the nodal and loop analysis theorems. The nodal analysis theorem says that for a connected circuit with n nodes and b branches, KCL applied to any $n-1$ nodes yields a set of $n-1$ linearly independent equations, so that further application of KCL to the remaining node does not increase the number of independent equations. The loop analysis theorem says that for a connected circuit with n nodes and b branches, the maximum number of linearly independent equations based on KVL applied to the loops of the circuit is $b - n + 1$.

TERMS AND CONCEPTS

Connected circuit: circuit in which every pair of nodes is joined by some set of branches.

Cramer's rule: a method for solving a linear matrix equation for its unknowns, one by one, through the use of determinants; the method has serious numerical problems when implemented on a computer, but is often convenient for small, 2×2 or 3×3, hand calculations.

Floating source: source, neither of whose nodes is connected to the reference node.

Gaussian surface: a closed curve or closed surface surrounding two or more nodes.

Graph of a circuit: diagram in which each branch (element) of a network is replaced by a line segment; thus, a graph represents information about the connections between the circuit elements.

Linear matrix equation: an equation of the form $Ax = b$, where A is an $n \times n$ matrix, x is an n-vector of unknowns,

and b is an n-vector of constants.

Linearly dependent: a set of equations is linearly dependent if and only if one of the equations is a scaled sum of the remaining equations, e.g., $F_n = a_1 F_1 + \ldots + a_{n-1} F_{n-1}$. The constants $a_1, \ldots, a_{n-1}$ that scale the equations $F_1, \ldots, F_{n-1}$ are real numbers.

Linearly independent set of equations: the set of equations $\{F_1(x_1, x_2, \ldots, x_n) = 0, \ldots, F_n(x_1, x_2, \ldots, x_n) = 0\}$ is linearly independent if no one equation can be expressed as the scaled sum of the remaining equations.

Loop (closed path): a contiguous sequence of branches that begins and ends on the same node and touches no other node more than once.

Loop analysis: an organized method of circuit analysis for computing loop currents in a circuit. Knowledge of the loop currents allows one to compute the individual element currents and, consequently, the element voltages.

Loop current: a (fictitious) current circulating around a closed path in a circuit.

Matrix inverse: the inverse, if it exists, of an $n \times n$ matrix A, denoted by A^{-1}, satisfies the equation $AA^{-1} = A^{-1}A = I$, where I is the $n \times n$ identity matrix. The solution of the linear matrix equation $Ax = b$ is given by $x = A^{-1}b$.

Mesh: boundary of any region with a finite area obtained after drawing a planar graph in a plane without branch crossing. Intuitively, meshes resemble the openings of a fish net.

Mesh analysis: the special case of loop analysis for planar circuits in which the loops are chosen to be the meshes.

Mesh current: a fictitious current circulating around a mesh in a planar circuit.

Modified nodal analysis: a modification of the basic nodal analysis method in which the unknowns are the usual nodal voltages, plus some naturally occurring auxiliary currents. These unknown auxiliary currents include currents through independent voltage sources, currents through dependent

voltage sources, currents through short-circuit elements, the controlling currents of appropriate dependent sources, and currents declared as output quantities.

Nodal analysis: an organized method of circuit analysis built around KCL for computing all node voltages of a circuit.

Node voltage: the voltage drop from a given node to a reference node.

Nonplanar circuit: a circuit that cannot be drawn in a plane without branch crossings.

Planar circuit: a circuit that is simple enough to be drawn in a plane, *without branches having to cross each other*.

Supernode: a situation in which a Gaussian surface encloses the nodes associated with one or more floating independent and/or dependent voltage sources. These voltage sources must form a *connected* subcircuit by themselves within the Gaussian surface.

Symmetric matrix: a matrix whose transpose is itself.

PROBLEMS

1. Consider the circuit of figure P4.1.
   (a) Write an appropriate number of node equations.
   (b) Put the node equations in matrix form.
   (c) Solve the node equations for the voltages V_a and V_b.
   (d) Determine V_x, V_{da}, and V_{db}.
   (e) Determine the power delivered by each source.
   (f) Compute the currents I_1, I_2, and I_3.

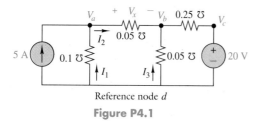

Reference node d

Figure P4.1

2. Consider the circuit of figure P4.2.
   (a) Write an appropriate number of node equations.
   (b) Put the node equations in matrix form.
   (c) Solve the node equations for the voltages V_A and V_B.
   (d) Determine V_o.
   (e) Determine the power delivered by each source. (Be careful of signs.)

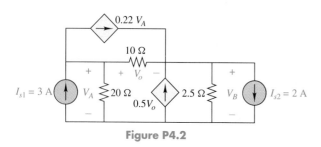

Figure P4.2

3. Consider the circuit of figure P4.3.

   (a) Identify the appropriate supernode.
   (b) Write appropriate nodal equations.
   (c) Solve the nodal equations for the node voltages V_B and V_C; find the current i_x.
   (d) Determine the power delivered by each of the sources.

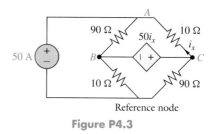

Reference node

Figure P4.3

4. Consider the circuit of figure P4.4. Choose node b as the reference node. This choice eliminates the floating voltage source, and hence, the nodal equations can be written without the need for a supernode. Write and solve a set of nodal equations for the voltages V_{ab}, V_{cb}, and V_{db}. Compare your answers with the results of the second exercise following example 4.4. It is important to note that in some cases, the choice of reference node can simplify the construction of the nodal equations.

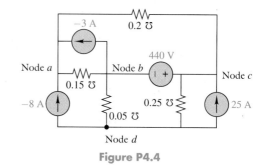

Node d

Figure P4.4

5. Consider the circuit of figure P4.5. Choose node c as the reference node, and write an appropriate set of nodal equations. Solve the nodal equations for the voltages V_{ac}, V_{bc}, and V_{dc}. Compare your answers with those of the exercise following example 4.5.

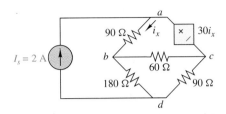

Figure P4.5 By choosing node c as the reference node, it is possible to simplify the construction of the node equations.

6. Consider the circuit of figure P4.6.
   (a) Identify the appropriate supernode.
   (b) Write an appropriate number of node equations.
   (c) Put the node equations in matrix form.
   (d) Solve the node equations.
   (e) Determine the power delivered by the dependent voltage source. (*Check*: $19 \leq |V_B| \leq 23$, and all node voltages are whole numbers.)

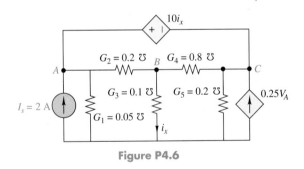

Figure P4.6

7. Use nodal analysis to find V_o/I_{in} for the circuit of figure P4.7.

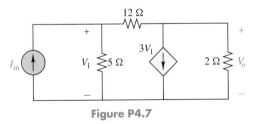

Figure P4.7

8. The circuit shown in figure P4.8 contains two grounded independent voltage sources. Use nodal analysis to find V_1 and V_2.

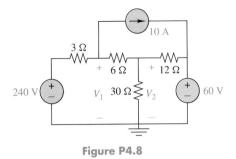

Figure P4.8

9. Consider the circuit of figure P4.9.
   (a) Choose a reference node, and solve for all remaining node voltages.
   (b) Determine the power delivered by each source.
   (c) Determine the power absorbed by each resistor.
   (d) Repeat parts (a) through (c) when the dependent source labeled $2i_y$ is changed to $2i_z$.

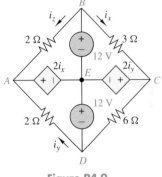

Figure P4.9

10. Consider the loop shown in figure P4.10, which is a modification of the circuit of example 4.6. Write loop equations and solve for the loop currents. Compute the voltage v. Your answer should be the same as that given in example 4.6.

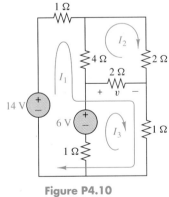

Figure P4.10

11. It is sometimes possible to simplify the equations of loop analysis by cleverly choosing loops. Consider the circuit of figure P4.11, which coincides with that of example 4.7, except for loop 1. This choice allows for a direct specification of I_3. Write loop equations and solve for the loop currents. Then determine v_y, I_x, and v, and compare your answers with those of example 4.7.

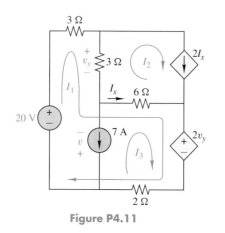

Figure P4.11

12. For the circuit of figure P4.12:
(a) Determine a set of three independent loop equations and one equation for an auxiliary variable associated with the dependent current source.
(b) Solve for the loop currents and the auxiliary variable.
(c) Determine the equivalent input resistance seen at the source.

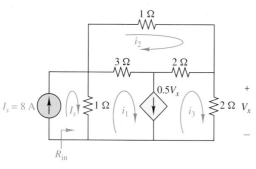

Figure P4.12

◆ 13. The mesh equations for the circuit of figure P4.13 are

$$\begin{bmatrix} 2 & -4 & -0.5 \\ -1.5 & 6.5 & -2.5 \\ -0.5 & -2.5 & 3.5 \end{bmatrix} \begin{bmatrix} i_1 \\ i_2 \\ i_3 \end{bmatrix} = \begin{bmatrix} v_1 \\ -v_2 \\ 0 \end{bmatrix}.$$

Determine R_1, R_2, R_3, and R_4.

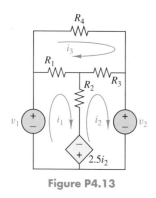

Figure P4.13

14. Determine the node voltages V_A and V_B and also the voltage V_x in the circuit of figure P4.14. It is suggested that

you write your equations in matrix form and solve using the formula

$$\begin{bmatrix} a & b \\ c & d \end{bmatrix}^{-1} = \frac{1}{ad - bc} \begin{bmatrix} d & -b \\ -c & a \end{bmatrix}$$

for a 2×2 inverse.

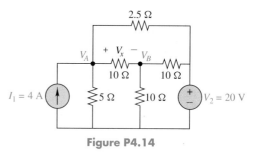

Figure P4.14

15. Once again, determine the voltages V_A, V_B, and V_x, this time in the circuit of figure P4.15. This problem is a slight variation on problem 14.

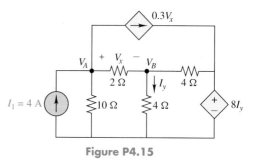

Figure P4.15

16. Consider the circuit of figure P4.16.
(a) Identify the appropriate supernode.
(b) Write a set of nodal equations for the circuit.
(c) Reduce the nodal equations to a set of two equations in the variables V_A and V_C only.
(d) Solve the nodal equations of part (c) by any means you choose, but preferably by matrix inverse. Specify V_A, V_B, and V_C.

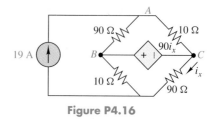

Figure P4.16

◆ 17. The nodal equations for the circuit of figure P4.17 are

$$11V_A - 3V_B - 2V_C = 0,$$

$$-5V_A + 3V_B + 32V_C = 6,$$

$$3V_A - V_B - 2V_C = 0.$$

Determine all four resistor values and β. (*Hint*: Find all the conductances first and then convert to resistances.)

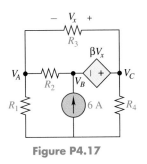

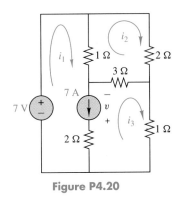

Figure P4.17

Figure P4.20

18. Write a set of mesh equations for the circuit of figure P4.18. Find I_1, V_x, and V_y. (*Hint*: You should be able to reduce the mesh equations to a single equation in I_1.)

21. Write the modified nodal equation for the circuit of problem 2, using the procedure outlined in section 5.

22. Repeat problem 21 for the circuit of problem 3.

23. Repeat problem 21 for the circuit of problem 6.

24. Derive the "element stamps" shown in figure P4.24 for modified nodal analysis; i.e., show that the parameters G, J, and E will each appear in the coefficient matrix and the right-hand side of the MNA equation exemplified by equation 4.25.

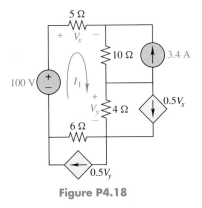

Figure P4.18

19. Write a set of mesh equations for the circuit of figure P4.19. Determine V_A, V_B, and I_x.

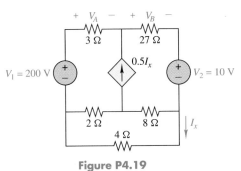

Figure P4.19

20. For the circuit of figure P4.20:
   (a) Show that the loop currents satisfy

$$\begin{bmatrix} -1 & 6 & -3 \\ 1 & 0 & -1 \\ 1 & -4 & 4 \end{bmatrix} \begin{bmatrix} i_1 \\ i_2 \\ i_3 \end{bmatrix} = \begin{bmatrix} 0 \\ 7 \\ 7 \end{bmatrix}.$$

   (b) Solve the matrix loop equation for i_1, i_2, and i_3.

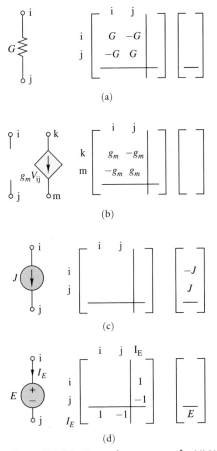

Figure P4.24 Some element stamps for MNA. (a) Conductance. (b) Voltage controlled current source. (c) Independent current source. (d) Independent voltage source.

25. Using the element stamps of problem 24, write the modified nodal equation for the circuit of figure 4.11a. Does the result agree with equation 4.25?

26. Consider the circuit of problem 19. First number the nodes, and then, using the element stamps derived in problem 24, write the modified nodal equations.

27. Figure P4.27a shows an electric locomotive propelled by a dc motor. The locomotive pulls a train of 12 cars back and forth between two stations. The motor behaves like a 590-V battery in series with a 1.296-Ω resistor. Suppose the train is midway between stations on the west side and the east side, where 660-V dc sources provide electricity. The resistance of the rails affects the current received by the locomotive. The equivalent circuit diagram is given by figure P4.27b, in which $R = 0.15$ Ω. Using mesh analysis:
  (a) Find the currents I_1 and I_2.
  (b) Find the current in the locomotive motor.
  (c) Repeat parts (a) and (b) when the locomotive is one-third distant from either station.

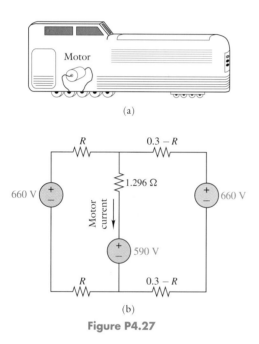

Figure P4.27

28. Reconsider problem 27. This time, suppose there are two locomotives on the track. One is one-third distant from the east side station, and the other is one-third distant from the west side station.
  (a) Draw the equivalent circuit diagram.
  (b) Write a set of three mesh equations, and solve for the mesh currents.
  (c) Determine the two motor currents.
  (d) Determine the power delivered by each of the 660-V sources.

29. For the circuit of figure P4.29:
  (a) Use nodal analysis to determine the node voltages. (*Hint*: Solve the nodal equations using MATLAB or some other software program.)
  (b) Compare the use of nodal analysis with that of loop analysis for this circuit. For example, how many equations are needed for a loop analysis?

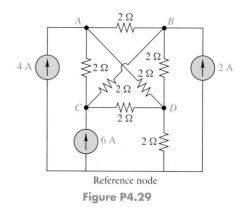

Figure P4.29

30. Consider the circuit of figure P4.30. This circuit represents a Wheatstone bridge circuit (see problem 41 of chapter 2) used in conjunction with a thermistor, denoted by R_T, to measure temperature at a remote site. A thermistor is a sensor whose resistance changes as a function of temperature. The thermistor resistance R_T has a value of 250 Ω at 0°C and a value of 80 Ω at 50°C. As a function of temperature, the resistance can be approximated by a straight line between these points. The nominal value of the environment in which the thermistor is placed is 25°C. The resistance $R_m = 10$ kΩ denotes the internal resistance of a voltmeter. The value of R is 250 Ω. The resistance $R_L = 2.5$ Ω is the resistance of the wires to the thermistor. The resistance R_x is variable and is adjusted so that the bridge circuit is balanced at the nominal value of the thermistor resistance; i.e., R_x is chosen so that the meter reads 0 V at the nominal value of the environment temperature.
  (a) Determine the equation of the straight line that expresses the thermistor resistance as a function of temperature.
  (b) Determine the nominal value of the thermistor resistance.
  (c) Determine the value of R_x.
  (d) Use nodal analysis to determine the output voltage at 0°C and at 50°C.
  (e) Keep R_T as a variable, and derive a formula for the meter voltage reading in terms of R_T. Use this formula to develop a chart that shows the meter voltage reading for 5° changes in temperature beginning at 0°C.

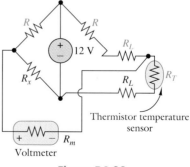

Figure P4.30

31. For the circuit of figure P4.31, find all node voltages by node analysis. Use the indicated reference node.

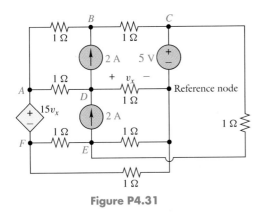

Figure P4.31

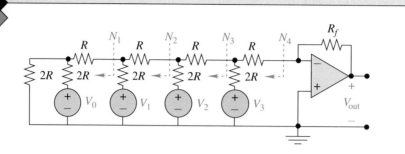

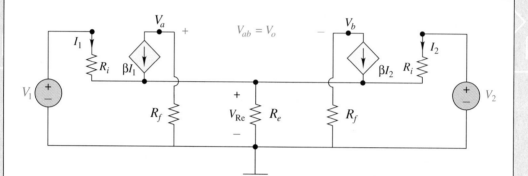

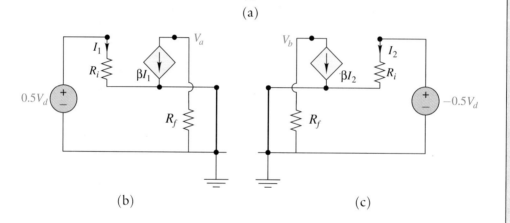

(a)

(b)   (c)

CHAPTER OUTLINE

Network Theorems

HISTORICAL NOTE AND TWO INTERESTING APPLICATIONS

IN the late 19th century, before the introduction of the alternating current (ac), electricity was available mainly in the form of direct current (dc). Basic laws—Kirchhoff's current law, Kirchhoff's voltage law, and Ohm's law—for the analysis of electrical circuits having dc voltage sources and resistors were developed in the mid-19th century. A direct application of these laws to a linear circuit usually results in a large number of simultaneous equations. Efforts to reduce the number of these equations have led to the development of the nodal and mesh analysis methods. In the precomputer era, manual solution of a large system of equations was a very difficult task. Any technique that alleviated this difficulty was enthusiastically received by engineers. As a result, many network theorems were developed that not only simplified manual analysis, but also provided a better understanding of circuit behavior. One such theorem that can be accurately dated and credited to its originator is Thévenin's theorem (1883).

Because of the insight they provide, network theorems are still useful today, despite the availability of high-power computer-aided circuit simulation software. Section 6 describes the use of network theorems to simplify the analyses of two very useful circuits: a digital-to-analog converter and a difference amplifier. The analysis of the former will utilize two new techniques: the superposition principle and the Thévenin equivalent circuit replacement technique. Their utilization shows a tremendous advantage over the method of solving simultaneous equations. The analysis of the difference amplifier exploits a symmetry that is present in the circuit and the principle of superposition. The advantage it affords is a deeper insight into circuit operation than a pure simultaneous equation approach. ◆

◆ CHAPTER OBJECTIVES

- Understand the meanings of the following basic theorems for linear resistive networks: proportionality, superposition, linearity, source transformation, Thévenin and Norton equivalent circuits, and maximum power transfer.
- Use of the theorems to simplify manual analysis and to gain a better insight into circuit behavior.
- Learn a variety of techniques for finding the Thévenin equivalent of a given circuit.

1. INTRODUCTION

Chapter 4 discussed several general methods of linear circuit analysis, including mesh analysis, nodal analysis, and modified nodal analysis. In theory, these tools are sufficient for the analysis of any linear resistive circuit. The problem is no more than to formulate network equilibrium equations, and solve them by any method studied in algebra. Indeed, computer programs can be written to carry out all the tedious computational tasks.

It might appear that little is left to learn in the area of linear resistive circuit analysis. This is far from true: Even though digital computers can relieve us from tedious numerical calculations, *manual* analysis of basic circuits remains a very important part of the training of any electrical engineering student. The reason is that an automatic computer solution of a circuit usually obscures the properties of various components making up a complex network. On the other hand, a manual analysis, when feasible, often leads to a better understanding of the circuit properties. Only when one understands the procedures for manual analysis, thereby developing a sense of the forms solutions take, can one feel confident in the use of computers for circuit analysis.

Very few students have ever tried to solve a set of four equations in four unknowns manually. This again points to a very serious drawback of manual analysis: the small size of the circuit that one can comfortably handle. Fortunately, with the use of theorems studied in this chapter (and more in later chapters), many seemingly complex circuit problems can be reduced to simple ones amenable to manual analysis. The objectives for introducing some network theorems in this chapter are twofold: to better understand the fundamental properties of linear circuits and to reduce the manual computational effort.

2. PROPORTIONALITY, SUPERPOSITION, AND LINEARITY

Linear and Nonlinear Resistive Elements

Chapter 1 introduced the v-i curves for a linear resistor, for dc independent voltage sources, and for dc independent current sources. Figure 5.1 resketches these curves with specific values and some alternative arrangements for the v-i axes. Figures 5.1a and 5.1b show the curves for dc voltage and current sources, respectively. Figure 5.1c shows the curve for a linear resistor, a straight line passing through the origin. Figure 5.1d portrays a nonlinear resistor that models a tungsten lamp. A **nonlinear resistor** has a v-i curve that is not a straight line, but that still passes through the origin. All practical nonlinear resistors absorb instantaneous power and therefore have their v-i curves restricted to the first and third quadrants of the v-i plane. Observe that the tungsten lamp behaves as a linear resistor for very small values of current and displays nonlinear effects when the current is large.

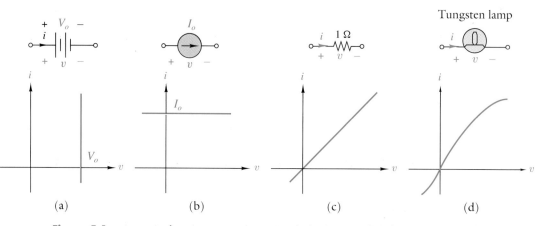

Figure 5.1 v-i curves of some common two-terminal elements. (a) dc voltage source. (b) dc current source. (c) Linear resistor. (d) Tungsten lamp.

For a linear resistor, because the v-i curve is a straight line through the origin, or $v = Ri$, the following two properties must hold:

1. The current i is proportional to the voltage v (**proportionality property**).
2. If $v = v_1$ produces $i = i_1$ and $v = v_2$ produces $i = i_2$, then, when the applied voltage is the sum of v_1 and v_2, the resultant current will be the sum of i_1 and i_2 (**superposition property**). To see this, simply note that if $i_1 = v_1/R$, $i_2 = v_2/R$, and $v = v_1 + v_2$, then

$$i = \frac{v}{R} = \frac{v_1 + v_2}{R} = \frac{v_1}{R} + \frac{v_2}{R} = i_1 + i_2.$$

The foregoing two properties also hold for circuits much more complex than a single linear resistor. To see this extension, it is necessary to define some new terms.

First, an n-terminal ($n \geq 2$) circuit element is a **linear resistive element** if the relationships among its terminal voltages and currents can be described by a set of *linear, homogeneous, algebraic* equations (rather than differential or integral equations). The linear resistors, linear controlled sources, and ideal op amps studied in the previous chapters are all linear resistive elements. This is easily seen from equation 1.18b, figure 3.1, and figure 3.8c.

A **linear resistive circuit** is a circuit consisting of only linear resistive elements and independent voltage and current sources. As mentioned in section 9 of chapter 1, a linear resistive circuit is memoryless in the sense that the output voltage or current at time t depends only on inputs at time t.

Proportionality and Superposition Properties

In a circuit containing several independent sources, one often wants to know the effect of a particular independent source *acting alone*. "Acting alone" means that the independent source (either a voltage source or a current source) has a nonzero value, while all other independent sources have zero values. The term "deactivated," or more colloquially, "killed" or "dead," refers to a source that is set to zero. Since a "killed" independent voltage source has zero V across its terminals no matter what current passes through, it behaves like a resistor of zero ohms, or a short circuit. Similarly, since a "killed" independent current source has zero amperes passing through it no matter what voltage appears across its terminals, it behaves like a resistor of infinite ohms, i.e., an open circuit.

Before stating some important properties of a linear circuit, it is necessary to clarify some terminology common to advanced courses in systems theory, but also applicable to elementary circuit analysis. When we speak of the *output* space or the *response* of a circuit, we usually mean a certain voltage or current declared to be of interest. The term *output* is not associated with power, unless the word *power* is explicitly used. In a general electromechanical system, the inputs may be any of a variety of physical quantities, e.g., voltages, currents, forces, torques, etc. However, for electric circuit analysis, *inputs* or *excitations* to a circuit always refer to the independent voltage and current sources applied to the circuit.

Using this terminology, we can extend the two properties for a linear resistor to general linear resistive circuits.

THE PROPORTIONALITY PROPERTY.

In a linear resistive circuit, when any one of the independent sources is acting alone, the output is proportional to that single input. Further, if the single input is multiplied by a constant K, the output is also multiplied by K.

THE SUPERPOSITION PROPERTY.

In any linear resistive circuit containing more than one independent source, any output (voltage or current) in the circuit may be calculated by adding together the contributions

---

due to each independent source acting alone, with the remaining independent sources deactivated.

A simple example, rather than a formal proof, best illustrates the meaning of these properties. Consider the circuit shown in figure 5.2. Let i_a be the desired response.

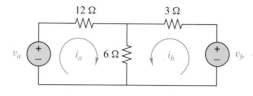

Figure 5.2 Demonstration of the proportionality and superposition properties.

Following the procedure of section 4.4, the mesh equations are

$$\begin{bmatrix} 18 & 6 \\ 6 & 9 \end{bmatrix} \begin{bmatrix} i_a \\ i_b \end{bmatrix} = \begin{bmatrix} v_a \\ v_b \end{bmatrix}.$$

Applying Cramer's rule to find i_a yields

$$i_a = \frac{1}{14} v_a - \frac{1}{21} v_b. \tag{5.1}$$

Now, if v_a is acting alone, then $v_b = 0$, and $i_a = v_a/14$, indicating that i_a is proportional to v_a. Similarly, if v_b is acting alone, then $v_a = 0$, and $i_a = -v_b/21$, indicating that i_a is proportional to v_b. Thus, the proportionality property follows from equation 5.1.

Next, the overall response i_a is the sum of two terms: $v_a/14$ and $-v_b/21$. But these are just the responses when v_a and v_b are respectively acting alone. Hence, the superposition property also follows from equation 5.1.

A very interesting and significant application of the proportionality property occurs in the analysis of a ladder network. A ladder network is a network that has the patterned structure shown in figure 5.3, where each box represents a two-terminal circuit element.

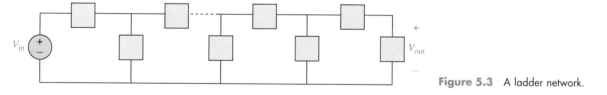

Figure 5.3 A ladder network.

A typical analysis problem is: Given V_{in} and all resistances in figure 5.3, find the output voltage V_{out}. One can, of course, solve the problem by writing a set of mesh equations or node equations; but then one must solve these equations simultaneously. It is possible, on the other hand, to solve the ladder network (no matter how many resistive elements are present) without simultaneous equations by the following method: Assume that $V_{out} = 1$ V, and work backward toward the source end to find the corresponding source voltage. Divide this voltage into V_{in} to obtain a constant K. Then, by the proportionality property, the actual output voltage corresponding to the applied source voltage is K volts.

◆ **EXAMPLE 5.1**

Find V_1 in the resistance ladder network of figure 5.4.

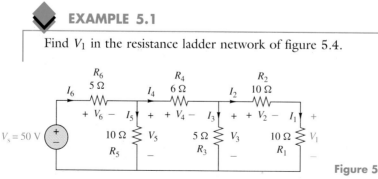

Figure 5.4 A resistive ladder network.

SOLUTION

Assume that $V_1 = 1$ V. By repeatedly applying Ohm's law and Kirchhoff's current and voltage laws, we generate the following sequence of simple equations (standard units of Ω, V, and A are used):

$$I_1 = \frac{V_1}{R_1} = \frac{1}{10} = 0.1, \qquad \text{(Ohm's law)}$$

$$I_2 = I_1 = 0.1. \qquad \text{(KCL)}$$

$$V_2 = R_2 I_2 = 10 \times 0.1 = 1, \qquad \text{(Ohm's law)}$$

$$V_3 = V_1 + V_2 = 1 + 1 = 2, \qquad \text{(KVL)}$$

$$I_3 = \frac{V_3}{R_3} = \frac{2}{5} = 0.4, \qquad \text{(Ohm's law)}$$

$$I_4 = I_2 + I_3 = 0.1 + 0.4 = 0.5. \qquad \text{(KCL)}$$

$$V_4 = R_4 I_4 = 6 \times 0.5 = 3, \qquad \text{(Ohm's law)}$$

$$V_5 = V_3 + V_4 = 2 + 3 = 5, \qquad \text{(KVL)}$$

$$I_5 = \frac{V_5}{R_5} = \frac{5}{10} = 0.5, \qquad \text{(Ohm's law)}$$

$$I_6 = I_4 + I_5 = 0.5 + 0.5 = 1. \qquad \text{(KCL)}$$

$$V_6 = R_6 I_6 = 5 \times 1 = 5, \qquad \text{(Ohm's law)}$$

$$V_s = V_5 + V_6 = 5 + 5 = 10. \qquad \text{(KVL)}$$

The preceding analysis shows that for $V_1 = 1$ V, the corresponding input voltage is $V_s = 10$ V. Therefore, $K = 50/10 = 5$. By the proportionality property, when the input $V_s = 50$ V, the corresponding value for V_1 is 5 V.

In the solution to example 5.1, we have separated the expressions into blocks to emphasize the repetitive pattern. For example, the expressions in the third block are obtained from the second block simply by increasing all subscripts by 2. When the ladder network has more elements, the sequence of expressions contains more blocks, each of which entails two additions and two multiplications. It is obvious that with this method of ladder network analysis, the computational effort increases only linearly with the size of the network. The advantage of this special technique is more impressive when the size of the network is large.

Exercise. In example 5.1, change all resistances to 2 Ω and find V_1.
ANSWER: $V_1 = 100/13$ V.

Although we have written a system of circuit equations to demonstrate the superposition property, in the actual application of superposition we can often avoid writing simultaneous equations. In fact, this is one of the reasons for using the method of superposition. The following example illustrates this point.

 EXAMPLE 5.2

For the circuit of figure 5.5a, find V_o across the 3-Ω resistor by superposition.

SOLUTION

Step 1. When V_a is acting alone, the circuit reduces to that of Figure 5.5b. The voltage divider formula gives $V_o = 18 \times 3/(3 + 6) = 6$ V.

Step 2. When I_b is acting alone, the circuit reduces to that of figure 5.5c. The 6-Ω and 3-Ω resistors are connected in parallel, yielding an equivalent resistance of 2 Ω. The voltage across the parallel combination is $V_o = 2 \times 2 = 4$ V.

Step 3. When both sources are present, by superposition, $V_o = 6 + 4 = 10$ V.

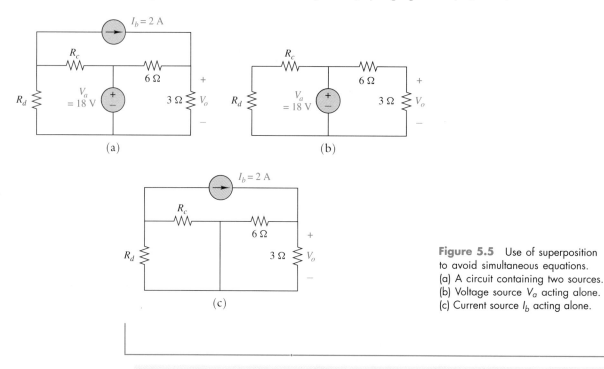

(a)

(b)

(c)

Figure 5.5 Use of superposition to avoid simultaneous equations.
(a) A circuit containing two sources.
(b) Voltage source V_a acting alone.
(c) Current source I_b acting alone.

Exercise. Switch the positions of V_a and I_b in figure 5.5a, and set $R_c = 3$ Ω, and $R_d = 6$ Ω. Find the voltage V_o by superposition.

ANSWER: Depending on the polarities, V_o is 10 or -10 and 2 or -2 V.

Exercise. In the circuit of figure 5.2, let $V_a = 84$ V and $V_b = 42$ V. Find the voltage across the 6-Ω resistor using superposition and the voltage divider formula.

ANSWER: 12 + 24 = 36 V.

When a circuit containing controlled sources has multiple inputs, the use of superposition reduces voltage and current response calculations to a series of single-input problems. It is important to keep in mind that each of the reduced single-input circuits has all the original controlled sources intact. The next two examples illustrate this point.

EXAMPLE 5.3

For the circuit of figure 5.6a, find V_o by superposition.

SOLUTION

When each independent voltage is acting alone, the circuit reduces to the single-input circuits of figure 5.6b and 5.6c, respectively.

Step 1. The voltage divider formula applied to figure 5.5b yields

$$V_o = \frac{R_L}{R_o + R_L}(-\mu V_g) = \frac{R_L}{R_o + R_L}(-\mu)\frac{R_2}{R_2 + R_1}V_1.$$

Step 2. Similarly, applying voltage division to figure 5.5c yields

$$V_o = \frac{R_L}{R_o + R_L}(-\mu V_g) = \frac{R_L}{R_o + R_L}(-\mu)\frac{R_1}{R_2 + R_1}V_2.$$

Step 3. Using superposition, the sum of the preceding two expressions gives the desired output voltage:

$$V_o = \frac{-\mu R_L}{R_o + R_L} \left(\frac{R_2}{R_1 + R_2} V_1 + \frac{R_1}{R_1 + R_2} V_2 \right).$$

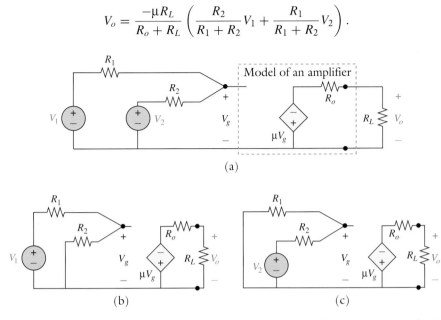

(a)

(b)   (c)

Figure 5.6 Application of the superposition property. (a) An amplifier with two inputs. (b) V_1 acting alone. (c) V_2 acting alone.

Exercise. A resistor R_3 is connected across the input terminals of the amplifier of figure 5.6a. If $R_1 = R_2 = R_3 = R$, $R_o = 0$, and $\mu = 3$, find V_o.
ANSWER: $V_o = -(V_1 + V_2)$.

◆ EXAMPLE 5.4

Figure 5.7a shows an inverting summer circuit. Find the output voltage V_o in terms of the two input voltages by superposition.

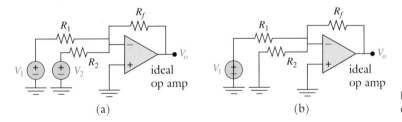

(a)   (b)

Figure 5.7 (a) An inverting summer circuit. (b) V_1 acting alone.

SOLUTION

When the independent source V_1 is acting alone, the circuit reduces to that of figure 5.7b. Because of the "virtual ground" property of an ideal op amp, R_2 has no voltage across it and thus has no effect on V_o. When $V_2 = 0$, equation 3.5 implies that

$$V_o = -\frac{R_f}{R_1} V_1.$$

Similarly, when $V_1 = 0$,

$$V_o = -\frac{R_f}{R_2} V_2.$$

By superposition, with both sources active,

$$V_o = -\left(\frac{R_f}{R_1} V_1 + \frac{R_f}{R_2} V_2 \right).$$

If all resistors have the same value, then $V_o = -(V_1 + V_2)$, leading naturally to the name of the circuit as an **inverting summer.**

Both circuits of examples 5.3 and 5.4 yield a *scaled sum* of two inputs, i.e.,

$$V_o = -(\alpha_1 V_1 + \alpha_2 V_2).$$

There are two advantages of using the op amp circuit of figure 5.7: A change in the source resistance R_k affects the coefficient α_k only; and extending the method to produce a scaled sum of more than two inputs is easy.

> **Exercise.** Design an op amp circuit having three inputs (V_1, V_2, V_3) and one output V_o, such that $V_o = -(2V_1 + 3V_2 + 7V_3)$. Use resistors in the kΩ range.

Although there is a controlled source in the single-input circuits of figures 5.6b and 5.6c, each circuit configuration is simple enough for one to write down the output expression directly using the voltage divider formula. For more complex circuit configurations, the appearance of controlled sources usually makes the analysis tricky, especially for beginning students. However, using the superposition property, one can reduce the analysis of a single-input circuit with a single controlled source to the analyses of *two* single-input circuits *without controlled sources*. The basic idea of the method is described with the aid of figure 5.8.

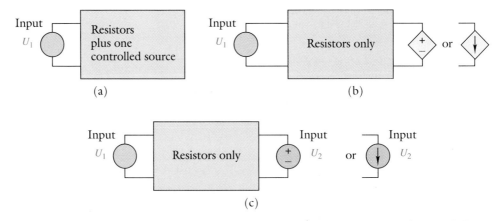

Figure 5.8 (a) A single-input circuit with a single controlled source. (b) Extracting the controlled source. (c) The derived two-input circuit without the controlled source.

Figure 5.8a shows the given circuit, wherein the single input U_1 may be a voltage source V_1 or a current source I_1. To accommodate both possibilities, no ± sign or arrow has been put inside the circle symbol. Extracting the single controlled source yields the configuration of figure 5.8b, in which the box contains resistors only. The controlled source may be any one of the four types defined in section 2 of chapter 3. Figure 5.8c is derived from figure 5.8b by replacing the controlled source by an *independent* source. A controlled voltage source is replaced by an independent voltage source carrying the same ± sign polarities. A controlled current source is replaced by an independent current source carrying the same reference arrow direction. The analysis method now proceeds as follows:

Step 1. In the derived two-input passive circuit of figure 5.8c, find expressions for the desired output and the *controlling variable* (for the controlled source) by superposition. The result may be written as

$$\text{output} = aU_1 + bU_2, \tag{5.2}$$

$$\text{controlling variable} = cU_1 + dU_2. \tag{5.3}$$

Step 2. Write the equation characterizing the controlled source:

$$U_2 = \alpha \, (\text{controlling variable}). \tag{5.4}$$

Step 3. Eliminate the controlling variable from equations 5.3 and 5.4:

$$U_2 = \frac{\alpha c}{1 - \alpha d} U_1. \tag{5.5}$$

Step 4. Substitute equation 5.4 into equation 5.2 to obtain the answer:

$$\text{output} = \left(a + \frac{\alpha bc}{1 - \alpha d} \right) (\text{input}). \tag{5.6}$$

In applying equation 5.6, we need only calculate the four coefficients $\{a, b, c, d\}$. These are obtained by solving the passive two-input circuit of figure 5.8c. Such an approach actually forms the basis of a very important method, called the "return difference method" for the analysis of feedback amplifiers, a topic discussed in courses in advanced circuit and controls. Since the derivation of equation 5.6 from equations 5.2 through 5.4 is straightforward, it is hardly necessary to memorize equation 5.6. On the other hand, if one has to solve the same type of problems many times, repeated derivations of the same formula are not warranted; we may as well use the formula directly. The next example illustrates the use of equation 5.6.

EXAMPLE 5.5

In the amplifier circuit of figure 5.9a, $\mu = 90$. Find the voltage gain V_o/V_i by the use of superposition.

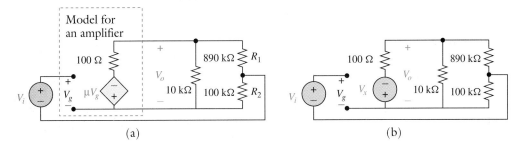

(a)   (b)

Figure 5.9 (a) An amplifier circuit. (b) The associated two-input passive circuit.

SOLUTION

As per the preceding algorithm, replace the controlled voltage μV_g by an independent voltage source V_x to obtain the two-input passive circuit of figure 5.9b. Superposition and voltage division are the essential solution techniques.

Step 1. Determine the load resistance seen by the amplifier. This resistance, a series-parallel combination of three resistances, has the value

$$10 \text{ k}\Omega // (890 \text{ k}\Omega + 100 \text{ k}\Omega) = 9.9 \text{ k}\Omega.$$

Step 2. When the source V_x is acting alone, voltage division implies

$$V_o = \frac{9,900}{9,900 + 100} (-V_x) = -0.99 V_x$$

and

$$V_g = \frac{100}{100 + 890} V_o = \frac{100}{990} (-0.99 V_x) = -0.1 V_x.$$

Step 3. When the source V_i is acting alone, the solutions of figure 5.9b are, obviously, $V_o = 0$ and $V_g = V_i$.

Step 4. Apply superposition. According to the superposition property, the solutions of figure 5.9b are the sum of the results of steps 2 and 3, i.e.,

$$V_o = -0.99 V_x \tag{5.7a}$$

and

$$V_g = V_i - 0.1 V_x. \tag{5.7b}$$

---

2. PROPORTIONALITY, SUPERPOSITION, AND LINEARITY

Further, the controlled source is characterized by

$$V_x = 90V_g. \tag{5.7c}$$

Step 5. *Eliminate* V_g. The foregoing three equations correspond to equations 5.2 through 5.4. After eliminating V_g from equations 5.7b and 5.7c, we obtain

$$V_x = 90V_g = 90(V_i - 0.1V_x) = 90V_i - 9V_x,$$

from which it follows that $V_x = 9V_i$. Substituting this result into equation 5.7a yields the desired voltage gain:

$$\frac{V_o}{V_i} = -0.99 \times 9 = -8.91.$$

Alternatively, we can use equation 5.6 to find the answer. Comparing equations 5.7 with equations 5.2 through 5.4, we identify

$$a = 0, \quad b = -0.99, \quad c = 1, \quad d = 0.1, \quad \alpha = 90.$$

Substituting these numbers into equation 5.6 yields the same answer as above.

Exercise. Repeat example 5.5 if the controlled source parameter μ is changed from 90 to 108 (a 20% change). What is the percent change in V_o/V_i?

Linearity

Nonlinear resistors violate both the proportionality and the superposition properties. To see this, consider the circuit of figure 5.10a, which contains one nonlinear resistor characterized by $i = v|v|$, as depicted in figure 5.10b.

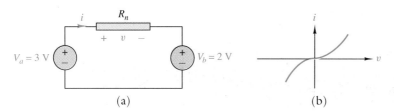

(a)   (b)

Figure 5.10 Superposition does not hold for a nonlinear resistor. (a) A circuit containing one nonlinear resistor. (b) *v-i* curve of the nonlinear resistor.

From the curve in figure 5.10b, it is obvious that the proportionality property fails, as only a straight line through the origin would satisfy a proportional relationship. To check the superposition property, we calculate the current i. The voltage across R_n is $v = 3 - 2 = 1$ V, resulting in a current $i = 1$ A. When $V_a = 3$ V is acting alone, the current is $i = 3 \times 3 = 9$ A. When $V_b = 2$ V is acting alone, the current is $i = (-2) \times |-2| = -4$ A. When both sources are present, the current is $i = 1$ A, which is not the sum of 9 and -4. Therefore, the superposition property does not hold for a nonlinear resistor.

It is possible to construct special circuits containing nonlinear resistors that possess the superposition property but not the proportionality property, and vice versa. (See problem 2.) The two properties are independent of each other. The fact that linear resistive circuits always satisfy both the properties is significant. For this reason, it is usually stated as a theorem.

LINEARITY THEOREM

For any linear resistive circuit, the outputs are related linearly to the inputs in the sense that both proportionality and superposition hold.

A rigorous proof of this theorem involves the formulation of equilibrium equations for networks containing all four types of controlled sources—for example, by the modified nodal analysis method of section 5 of chapter 4. Such discussions are hardly fruitful for beginners in circuit analysis, whose primary aim of learning the theorem is to make problem solving easier. For purely resistive circuits having no dependent sources, the proof follows the same line of reasoning as the example illustrating figure 5.2. Let the m inputs be denoted $(u_1, u_2, \ldots, u_m)$ and the output be denoted by y. The nodal or loop analysis methods of chapter 4 lead to the solution form

$$y = \alpha_1 u_1 + \alpha_2 u_2 + \ldots + \alpha_m u_m, \tag{5.8}$$

where $\alpha_1, \alpha_2, \ldots, \alpha_m$ are constants independent of the inputs.

We can easily demonstrate that equation 5.8 implies both the proportionality and the superposition properties. Let u_k be an independent source acting alone. By equation 5.8, the output is simply $y = \alpha_k u_k$, indicating very clearly the proportionality property. Equation 5.8 also shows that y is the sum of m terms $(\alpha_1 u_1), (\alpha_2 u_2), \ldots, (\alpha_m u_m)$. But these are just the outputs when each of the inputs $(u_1, u_2, \ldots, u_m)$ is acting alone. This establishes the superposition property. Thus, *equation 5.8 is a mathematical statement of the linearity theorem.*

The linearity theorem has some other implications. Suppose a network has only one input $u(t)$, which is the sum of m terms; i.e.,

$$u(t) = u_1(t) + u_2(t) + \ldots + u_m(t).$$

If $u(t)$ is an independent voltage source, one can view this equation as representing m independent voltage sources $u_1(t), u_2(t), \ldots, u_m(t)$ connected in series. If $u(t)$ is an independent current source, we may view this equation as representing m independent current sources $u_1(t), u_2(t), \ldots, u_m(t)$ connected in parallel. In either case, the superposition property applies; i.e., the total response $y(t)$ is the sum of the individual responses due to $u_1(t), u_2(t), \ldots, u_m(t)$. Such a use of the superposition property provides the basis for the Fourier series method of steady-state analysis of dynamic networks with a periodic but nonsinusoidal input.

Further Examples and Remarks

Suitable use of the linearity theorem often simplifies circuit analysis and provides better insights into circuit behavior. Some examples illustrate these points.

 EXAMPLE 5.6

A linear resistive circuit has two inputs v_a and i_b. Table 5.1 lists the results of two sets of measurements. Find the output voltage when the inputs are $v_a = 10$ V and $i_b = 8$ A.

TABLE 5.1
TWO SETS OF MEASUREMENTS OF A LINEAR CIRCUIT

v_a (V)	i_b (A)	v_{out} (V)
5	0	3
0	2	6

SOLUTION

By the proportionality property, the responses due to 10 V and 8 A, each acting alone, are 6 V and 24 A, respectively. By superposition, the output is $6 + 24 = 30$ V.

◆ EXAMPLE 5.7

A linear resistive circuit has two inputs, v_a and i_b. Results from two sets of measurements are listed in table 5.2. Find the output voltage when the inputs are $v_a = 10$ V and $i_b = 8$ A.

◆ TABLE 5.2   DATA FOR EXAMPLE 5.7

v_a (V)	i_b (A)	v_{out} (V)
4	14	50
2	8	28

SOLUTION

According to the linearity theorem, we may write

$$v_{out} = Av_a + Bi_b,$$

where A and B are two constants. To determine these constants, we substitute the measured data into the equation and obtain the two equations,

$$50 = 4A + 14B,$$

$$28 = 2A + 8B,$$

from which it follows that $A = 2$ and $B = 3$. Finally, the output is $v_{out} = 2 \times 10 + 3 \times 8 = 44$ V.

Exercise.   For example 5.6, determine the coefficients α_1 and α_2 for the linear relationship $v_{out} = \alpha_1 v_a + \alpha_2 v_b$.

ANSWER: 0.6 and 0.3.

Several remarks about the concepts of linearity and superposition are in order. First, for linear resistive circuits, the principle of superposition holds only for the calculation of voltages and currents; it does not hold for the calculation of power. To see this, consider a simple linear circuit obtained from figure 5.10a by replacing R_n with a 1-Ω resistor. When both sources are present, the power absorbed by the resistor is $P = (3 - 2)^2/1 = 1$ W. When V_a is acting alone, $P = 3^2/1 = 9$ W. When V_b is acting alone, $P = 2^2/1 = 4$ W. Obviously, $1 \neq 9 + 4$, indicating that superposition is not valid for the calculation of power.

Second, the proportionality property is sometimes referred to as the *homogeneity* property. Likewise, some texts call the superposition property the *additive* property. And the linearity theorem is called the superposition theorem in some texts. The nomenclature is not standardized and may be a source of confusion. However, the basic facts remain unchanged, regardless of what they are called. Our use of the term "superposition" is consistent with the intuitive meaning of the word, and our use of the term "linearity" is consistent with the usage of the term *linear transformation* in linear algebra. As pointed out in chapter 1, a circuit processes the signal, or transforms a set of input signals into a set of output signals. In this sense, the transformation achieved by a linear resistive circuit satisfies the linear transformation property in algebra.

Last, our demonstration of the proportionality and superposition properties made use of mesh equations. Other simple circuits can be constructed that use node equations. To establish the properties for a general linear resistive circuit, one can use modified nodal equations, which allow all four types of dependent sources. A detailed and rigorous proof of the superposition property for active circuits, however, is beyond the scope of this introductory text.

3. SOURCE TRANSFORMATIONS

As a motivation for studying the theorems of this section, consider the problem of finding the current I_{AB} in the circuit of figure 5.11.

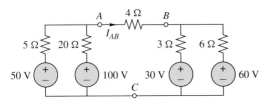

Figure 5.11 A circuit illustrating the use of a source transformation.

The mesh analysis of chapter 4 requires three mesh equations. Node analysis requires two node equations: Choose node C as the reference node, and apply KCL to nodes A and B. In both cases, the solution of a system of simultaneous equations is necessary.

A network theorem known as the *source transformation theorem* makes the problem much easier. Before presenting the theorem, however, we must define some new terms. First, a **two-terminal network** is an interconnection of circuit elements inside a box having only two accessible terminals for connection to other networks. If the box contains any dependent source, the controlling voltage or current must also be *within* the box. Figure 5.12 pictures four two-terminal networks identified by the dashed-line boxes. Later, to avoid crowding the figures, we often omit the outlines of the box and simply highlight the two external terminals. Nonetheless, it is important to be aware of the existence of such a box, at least conceptually, and to be ready to draw the box when necessary.

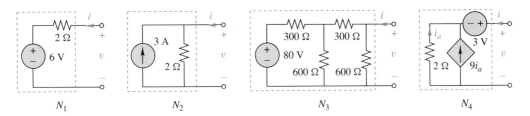

Figure 5.12 Examples of two-terminal networks, i.e., networks in which only two terminals are available for connection to other networks.

Second, **equivalent two-terminal networks** are two-terminal networks with the same terminal voltage-current relationship. For example, N_1 and N_2 in figure 5.12 are equivalent networks. To see this, write the v-i relationship for each network. By inspection, N_1 has the v-i relationship

$$v = 2i + 6,$$

and N_2 satisfies

$$v = 2(i + 3) = 2i + 6.$$

These two equations are identical. Therefore, N_1 and N_2 are equivalent.

Because equivalent two-terminal networks have the same v-i relationship, by definition, it should be obvious that if one network is interchanged with its equivalent network, all currents and voltages *outside the box* should remain unaffected. The network outside the box may or may not be linear.

> **Exercise.** Change the parameters in N_1 and N_2 so that (1) N_1 is equivalent to N_3 and (2) N_2 is equivalent to N_4.

We now generalize the equivalence between N_1 and N_2 of figure 5.12, stating it as a theorem.

SOURCE TRANSFORMATION THEOREM

A two-terminal network consisting of a series connection of an independent voltage source V_s and a nonzero resistance R is equivalent to a two-terminal network consisting of a parallel connection of an independent current source $I_s = V_s/R$ and a resistance R.

Conversely, a two-terminal network consisting of a parallel connection of an independent current source I_s and a finite resistance R is equivalent to a two-terminal network consisting of a series connection of an independent voltage source $V_s = RI_s$ and a resistance R. The reference directions for voltages and currents are as shown in figure 5.13a.

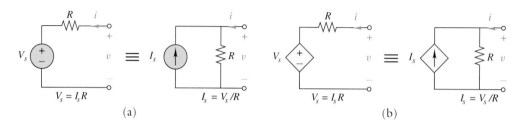

Figure 5.13 Source transformation theorem. (a) Independent sources. (b) Controlled sources.

Verification of the source transformation theorem is very simple. Both two-terminal networks have the same v-i relationship, namely, $v = Ri + V_s$. Therefore, they are equivalent. Note that the boxes are not explicitly shown in figure 5.13a.

Although the source transformation theorem is usually stated for the case of independent sources, as shown in figure 5.13a, it is easy to show that the same transformation applies to the case of controlled sources. Figure 5.13b illustrates the transformation involving controlled sources.

The following example illustrates how the source transformation simplifies solving problems.

◆ EXAMPLE 5.8

Find I_{AB} in figure 5.11 by repeated applications of the source transformation theorem.

SOLUTION

The first step is to substitute all series V_s-R combinations by their parallel I_s-R equivalents. This results in figure 5.14a. The next step is to combine the parallel resistances and the

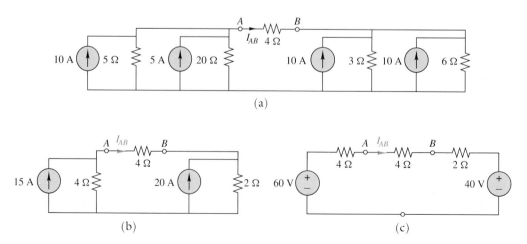

Figure 5.14 Repeated applications of the source transformation theorem. (a) Result of replacing V_s-R by I_s-R. (b) Result of combining parallel elements. (c) Result of replacing I_s-R by V_s-R.

parallel current sources. This step results in figure 5.14b. Finally, we interchange the parallel I_s-R combinations in figure 5.14b with their series V_s-R equivalents. Figure 5.14c shows the result.

Throughout the preceding operations, i.e., replacing two-terminal networks by their equivalents, the 4-Ω resistor remains *outside* of the replaced boxes. Therefore, the current I_{AB} is not obscured or hidden by the substitutions. Figure 5.14c allows us to calculate I_{AB} easily as

$$I_{AB} = \frac{60 - 40}{4 + 4 + 2} = 2 \text{ A}.$$

In passing, we note that the node equation method of chapter 4 required solving two simultaneous equations to obtain the current. Sometimes the present approach is preferable.

Exercise. A 600-Ω resistor R_L is connected across the external terminals of the network N_3 of figure 5.12. Find the voltage across R_L by repeated applications of the source transformation theorem.

ANSWER: Magnitude of voltage = 20 V.

EXAMPLE 5.9

For the circuit of figure 5.15a, find V_1 and V_2.

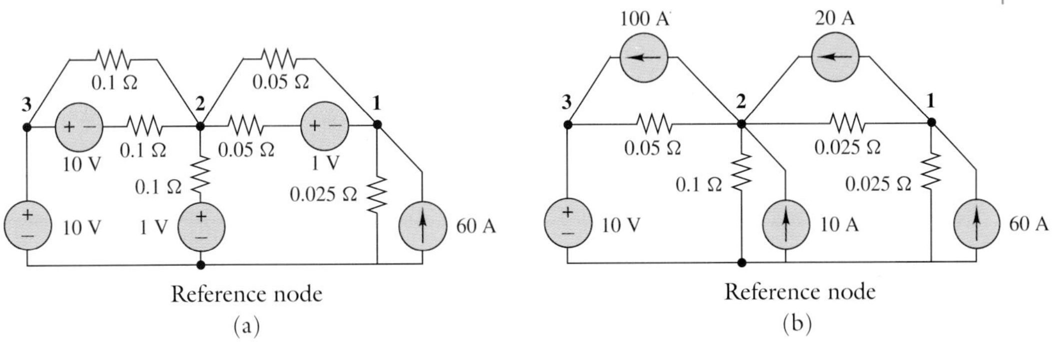

Figure 5.15 (a) The given circuit. (b) The circuit after source transformations.

SOLUTION

The circuit contains both grounded and floating independent voltage sources. Instead of using the supernode concept of chapter 4, we apply source transformations and combine all parallel resistances. The result is figure 5.15b. Applying KCL to nodes 1 and 2 results in the following nodal equations: At node 1,

$$40(V_1 - V_2) + 40V_1 = 60 - 20,$$

and at node 2,

$$20(V_2 - 10) + 10V_2 + 40(V_2 - V_1) = 20 + 10 - 100.$$

In matrix form,

$$\begin{bmatrix} 80 & -40 \\ -40 & 70 \end{bmatrix} \begin{bmatrix} V_1 \\ V_2 \end{bmatrix} = \begin{bmatrix} 40 \\ 130 \end{bmatrix}.$$

Solving the equations using MATLAB, the inverse matrix method, or Cramer's rule yields the answers $V_1 = 2$ V and $V_2 = 3$ V. In particular, for the inverse matrix method,

$$\begin{bmatrix} V_1 \\ V_2 \end{bmatrix} = \begin{bmatrix} 80 & -40 \\ -40 & 70 \end{bmatrix}^{-1} \begin{bmatrix} 40 \\ 130 \end{bmatrix} = \begin{bmatrix} 0.0175 & 0.01 \\ 0.01 & 0.02 \end{bmatrix} \begin{bmatrix} 40 \\ 130 \end{bmatrix} = \begin{bmatrix} 2 \\ 3 \end{bmatrix}.$$

Introduction

In the analysis of an electrical network, very often one is interested only in the voltage, current, or power associated with a particular *load*. The load may be a single resistor or a general two-terminal network. The effect on the load voltage or absorbed power when the load is varied may also be of importance. In such cases, the analysis can be greatly simplified by replacing the linear network, exclusive of the load, by a simple equivalent circuit. The equivalent circuit will consist of a single resistance and one independent source. Figure 5.16 illustrates this substitution.

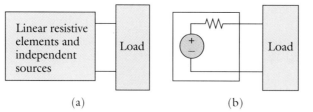

(a)   (b)

Figure 5.16 Substitution of a complex network (a) by a simple equivalent (b).

A French telegraph engineer M. L. Thévenin stated the equivalent circuit idea in 1883. He said that the equivalent circuit would consist of an independent voltage source in series with a resistor. At that time, no one had any concept of a dependent source or an independent current source. Thévenin's idea applied only to networks containing resistances and independent voltage sources.

This section will restate the essence of Thévenin's idea using modern terminology. A simple proof of his basic theorem is presented using superposition. An extended version of the theorem valid for dependent sources will be stated without proof. The emphasis for this latter case of linear active networks will be on the calculation of the parameters in the equivalent circuit. Several examples will illustrate the applications.

Electrical Engineering Professor

Teaching electrical engineering can be a rewarding profession because of the challenge of interacting with students, the excitement of working on the cutting edge of technology, the stimulating professional contacts, and having opportunities and the facilities for basic research, technological development, and writing. Electrical engineering professors can choose to emphasize undergraduate education, graduate education, or maintain a balance between the two.

When focusing on undergraduate education, a professor must learn to lead students from an understanding of basic principles to the application and design of real-world devices. At the same time, the professor must constantly update curriculums by incorporating new theories and trends into courses. Graduate education requires the professor to be thoroughly knowledgeable in cutting edge research and technological development to be able to teach meaningful graduate courses and guide graduate students through research projects and writing theses. The professor must also teach the skills and tools necessary for the completion of these projects.

In addition to needing a Ph.D. to teach at most universities, electrical engineering professors must be able to communicate well with students. In order to retain good students, professors in introductory classes are challenged to find innovative methods for facilitating the learning process and practical applications to excite student interest. Watching students successfully complete investigative projects, publish research papers, graduate with their B.S., M.S., or Ph.D., and contribute to the betterment of society through their careers, can provide a sense of fulfillment to electrical engineering professors.

Many schools require electrical engineering professors to be actively involved in educational research, i.e., research needed for the completion of an M.S. or Ph.D. Unlike industrial careers, electrical engineering professors usually have the freedom to choose their research topics. In addition to working with graduate students on research projects, professors collaborate with other professors and professionals in industry, which creates a very exciting learning environment. This collaboration often leads to worthwhile inventions and discoveries that contribute to the field of electrical engineering, and more importantly to society at large.

Thévenin's Theorem for Passive Networks

Figure 5.17a denotes a network consisting only of resistors and independent sources. Figure 5.17b shows the so-called Thévenin equivalent consisting of a voltage source $v_{oc}(t)$ in series with a resistor R_{th}. The voltage $v_{oc}(t)$ is simply the voltage appearing across the terminals A and B assuming that no load is attached. Figure 5.17c emphasizes this point. Finally, figure 5.17d indicates that R_{th} is the equivalent resistance seen at the terminals A and B when all independent sources are deactivated, i.e., when independent voltage sources are short circuits and independent current sources are open circuits.

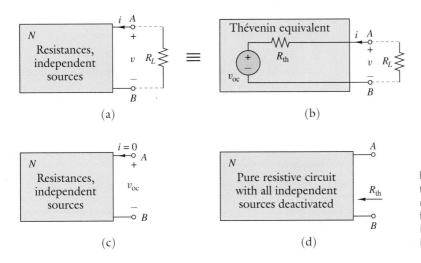

Figure 5.17 (a) Network N attached to a load R_L. (b) Thévenin equivalent network attached to load. (c) Circuit for computing v_{oc} in which $i = 0$. (d) Circuit for computing R_{th} in which all independent sources are deactivated.

The celebrated Thévenin theorem for passive networks capsulizes this discussion.

THÉVENIN'S THEOREM FOR PASSIVE NETWORKS

Any two-terminal linear resistive network N consisting of resistances and independent sources is equivalent to a two-terminal network consisting of a resistance R_{th} in series with an independent voltage source $v_{oc}(t)$ where $v_{oc}(t)$ is the open-circuit voltage appearing across the terminals of N and R_{th}, called the **Thévenin equivalent resistance**, is the equivalent resistance of N when all independent sources are deactivated. Figure 5.17 shows the appropriate polarity for $v_{oc}(t)$.

A simple example illustrates Thévenin's theorem and how its use can significantly reduce one's computational effort.

 EXAMPLE 5.10

For the circuit shown in figure 5.18a, find the voltage V and power P_L for 11 values of R_L, ranging from 200 Ω to 400 Ω in 20-Ω steps.

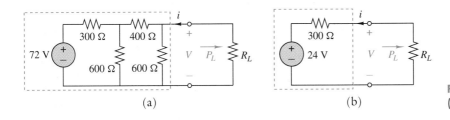

Figure 5.18 (a) Network with load. (b) Thévenin equivalent with load.

SOLUTION

It is possible, but not advisable, to use the general analysis methods of chapter 4. With such an approach, it would be necessary to solve a set of simultaneous equations for each value of R_L. This in turn would require an elevenfold repetition of the process to complete the solution—a very time-consuming task.

An alternative approach utilizes Thévenin's theorem. Using repeated source transformations on the dashed box of figure 5.18a yields the Thévenin equivalent circuit shown in figure 5.18b. The result is a 24-V source in series with a 300-Ω resistor. The use of voltage divider and power formulas immediately yields the desired answers, tabulated as follows:

R_L (Ω)	200	220	240	260	280	300	320	340	360	380	400
V (V)	9.6	10.1	10.7	11.4	11.6	12	12.4	12.7	13.1	13.4	13.7
P_L (mW)	461	469	474	477	479	480	479	478	476	473	470

Of singular interest is the column in which $R_L = 300 = R_s$. Here, the absorbed power, 480 mW, is the largest among the values listed for P_L. In section 5, we will prove that for fixed R_s, $R_L = R_s$ is indeed the condition under which the maximum power is transferred from the source to the load.

◆ EXAMPLE 5.11

Find the voltage V_{AB} across the 8-Ω resistance in figure 5.19a by the use of Thévenin's equivalent circuit.

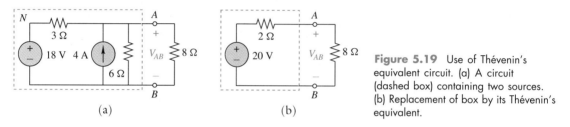

(a)   (b)

Figure 5.19 Use of Thévenin's equivalent circuit. (a) A circuit (dashed box) containing two sources. (b) Replacement of box by its Thévenin's equivalent.

SOLUTION

Figure 5.19a shows the two-terminal network N, enclosed in a dashed-lined box. To find R_{th}, deactivate the independent sources in N. In other words, replace the 18-V voltage source by a short circuit and the 4-A current source by an open circuit. The resultant network is a parallel connection of the 3-Ω and 6-Ω resistances. Therefore, $R_{\text{th}} = 3 \times 6/(3+6) = 2$ Ω.

The open-circuit voltage v_{oc} is easily determined by the principle of superposition and the voltage and current divider formulas as follows:

$$v_{\text{oc}}^1 \text{ (due to 18-V source)} = \frac{6}{6+3} \times 18 = 12 \text{ V},$$

$$v_{\text{oc}}^2 \text{ (due to 4-A source)} = 4 \times 2 = 8 \text{ V}.$$

Consequently, $v_{\text{oc}} = v_{\text{oc}}^1 + v_{\text{oc}}^2 = 12 + 8 = 20$ V.

Figure 5.19b shows the Thévenin equivalent circuit of N. From this circuit,

$$V_{AB} = \frac{20 \times 8}{2+8} = 16 \text{ V}.$$

Exercise. In figure 5.19a, suppose the 3-Ω resistor is changed to 5 Ω and the 6-Ω resistor to 20 Ω. Find V_{AB} by the use of Thévenin's theorem.
ANSWER: $304/15 \simeq 20.3$ V.

A Proof of Thévenin's Theorem for Passive Circuits

The verification of the passive form of Thévenin's theorem entails a simple application of superposition to show that the network of figure 5.17a is equivalent to that of figure 5.17b.

The two networks are equivalent if they have the same v-i relationship at the terminals (A,B).

To prove this, apply an independent current source $i(t)$ to the terminals (A,B) of the circuit in figure 5.17a. The resultant network is shown in figure 5.20. By the principle of superposition (section 2), the terminal voltage v must have the form

$$v(t) = \text{(sum of responses due to internal independent sources)}$$
$$+ \text{ (response due to external current source } i)$$
$$= \alpha(t) + \beta i(t) \tag{5.9}$$

for an appropriate voltage $\alpha(t)$ and an appropriate constant β.

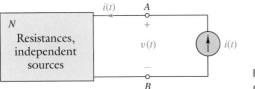

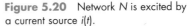

Figure 5.20 Network N is excited by a current source $i(t)$.

Let us now examine the physical meanings of the quantities $\alpha(t)$ and β in equation 5.9. The quantity $\alpha(t)$ represents the terminal voltage due to all *internal sources*. In other words, it is the terminal voltage $v(t)$ when $i(t) = 0$, i.e., when the terminals (A,B) are left open circuited as in figure 5.17c. This suggests the more descriptive notation $v_{oc}(t)$ for $\alpha(t)$. From equation 5.9, the quantity β is simply the ratio $v(t)/i(t)$ when $\alpha(t) = 0$, i.e., when all *internal independent sources* are deactivated. Hence, β is the equivalent resistance of the "dead" network shown in figure 5.17d. Although chapter 2 used the notation R_{eq} for this equivalent resistance, the notation R_{th} emphasizes its association with Thévenin's theorem.

Equation 5.9 for figure 5.17a has the equivalent form

$$v(t) = R_{th}i(t) + v_{oc}(t). \tag{5.10}$$

This is precisely the v-i relationship for the two-terminal network of figure 5.17b. This then establishes the equivalence between figures 5.17a and b, and proves Thévenin's theorem.

As a final point, suppose one connects a load resistance R_L to the terminals of N in figure 5.17a, with the objective of determining I_L, the current through R_L from A to B. Replacing N by its Thévenin equivalent, figure 5.17b, leads to the obvious result

$$I_L = \frac{v_{oc}}{R_{th} + R_L}. \tag{5.11}$$

In some literature (including Thévenin's original work and some dictionaries of electronics), equation 5.11 is stated as Thévenin's theorem. We have, however, considered the equation to be a simple consequence of equation 5.10 and set forth Thévenin's theorem in terms of equivalent two-terminal networks without reference to an attached load. This treatment of Thévenin's theorem is more general because, as stated in section 2, the equivalence of the two-terminal networks remains valid even if the load is a very complex linear or—worse yet—a nonlinear network. In other words, the equivalence in terms of the v-i relationship of two-terminal networks is more basic.

The Norton Equivalent Circuit

In equation 5.10, if R_{th} is not zero, then

$$i(t) = \frac{v(t)}{R_{th}} - \frac{v_{oc}(t)}{R_{th}}. \tag{5.12}$$

Equation 5.12 may be interpreted as representing a circuit in which a current source of $v_{oc}(t)/R_{th}$ amperes is in parallel with a resistance R_{th}. Figure 5.21b illustrates this circuit configuration.

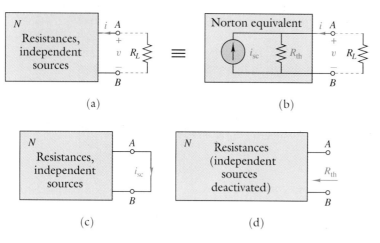

Figure 5.21 (a) N as a two-terminal network. (b) Norton equivalent of N. (c) Meaning of $i_{sc}(t)$, where $v(t) = 0$. (d) Meaning of R_{th}.

Further, if $v(t) = 0$ in equation 5.12, i.e., if the terminal pair (A, B) is shorted, then

$$\frac{v_{oc}(t)}{R_{th}} = -i(t) \equiv i_{sc}(t).$$

This follows because if $v(t) = 0$, $i_{sc}(t)$ denotes a short-circuit current, as illustrated in figure 5.21c. In the context of this terminology, if $R_{th} \neq 0$, then

$$i_{sc}(t) = \frac{v_{oc}(t)}{R_{th}}, \tag{5.13a}$$

$$v_{oc}(t) = i_{sc}(t) R_{th}, \tag{5.13b}$$

and

$$R_{th} = \frac{v_{oc}(t)}{i_{sc}(t)}. \tag{5.13c}$$

The associated equivalent circuit, shown in figure 5.21b, is called the **Norton equivalent circuit** and has an i-v relationship dual to equation 5.10; i.e.,

$$i(t) = \frac{1}{R_{th}} v(t) + i_{sc}(t). \tag{5.14}$$

The appearance of the Norton equivalent circuit was a natural outcome of advances in technology. When E. L. Norton was a scientist with Bell Laboratories, the invention of vacuum tubes made independent current sources a realistic possibility. Many electronic circuits were modeled with independent and dependent *current sources* which lead to a modification of Thévenin's equivalent to the Norton equivalent. In fact, the Thévenin equivalent circuit of figure 5.17b and Norton equivalent circuit of figure 5.21b are inter-related by the source transformation theorem presented in section 3. Given equations 5.13, a knowledge of any two of the three quantities (v_{oc}, i_{sc}, R_{th}) is sufficient for constructing the Thévenin and Norton equivalent circuits.

We have described two simple representations of the two-terminal network of figure 5.17a: the Thévenin equivalent circuit of figure 5.17b and the Norton equivalent circuit of figure 5.21b. When all of the independent sources inside the two-terminal network are dc, there is yet a third method of characterizing the network that is often quite useful for the graphical solution of electronic circuit problems.

In figure 5.22a, since all independent sources inside the network are dc, v_{oc} in equation 5.10 is a constant. Thus, the i-v curve represented by the linear equation is a straight line. The third method of characterizing a network simply represents the two-terminal dc network by its straight-line i-v or v-i curve. There are different ways of choosing the axes and the variables for constructing the curve. For example, either i or $i_L \equiv -i$ may be used, with either v or i_L as the independent variable. The quantities R_{th}, v_{oc}, and i_{sc} all have very clear meanings, as illustrated in the graphs of figures 5.22b to 5.22d.

In certain situations, it is easier to determine the Thévenin equivalent for a two-terminal network from measured data and i-v plots than from writing and solving network equations. The next example illustrates such a situation.

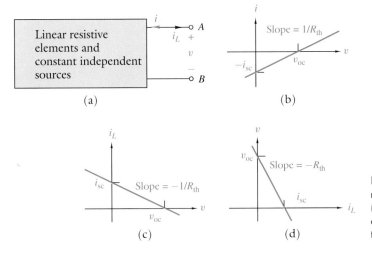

Figure 5.22 (a) A two-terminal resistive network with dc sources. (b),(c),(d): Three different representations of the v-i (i-v) curve that clearly identify the meanings of v_{oc}, i_{sc}, and R_{th}.

◆ EXAMPLE 5.12

A certain linear resistive two-terminal network has a very complex schematic diagram. It is decided to determine the parameters of the Thévenin equivalent circuit experimentally.

(a) If $v_{oc} = 10$ V and $i_{sc} = 100$ A are measured, determine the parameters of the Thévenin equivalent circuit.
(b) If two data points ($v = 10$ V, $i_L = 0$) and ($v = 9$ V, $i_L = 10$ A) are measured, determine the parameters of the Thévenin equivalent circuit.
(c) Which of the above two schemes is preferable from a measurement point of view?

SOLUTION

(a) By direct measurement, $v_{oc} = 10$ V; hence, by equation 5.13c, $R_{th} = v_{oc}/i_{sc} = 10/100 = 0.1\ \Omega$.

(b) In figure 5.22d, $v = 10 = v_{oc}$. R_{th} is the negative of the slope of the curve of v vs. i_L. Therefore,

$$R_{th} = -\frac{9 - 10}{10 - 0} = 0.1\ \Omega.$$

Finally, $i_{sc} = v_{oc}/R_{th} = 100$ A, as in part (a).

(c) In this particular example, the measurement of (b) is preferable because it avoids handling the large short-circuit current.

Thévenin's Theorem for Linear Active Circuits

Up to now, we have excluded dependent sources in our statements of Thévenin and Norton equivalent circuits for linear two-terminal networks. Extension of the theorems to include dependent sources requires the attachment of a condition under which these theorems remain valid. The need for this additional condition arises from the fact that when dependent sources are included, the v-i curve of a linear resistive two-terminal network may or may not be a straight line such as those shown in figure 5.22. Two other possibilities exist: There may be only one possible v-i point, or the possible v-i points may fill the entire plane. The simple circuits in figure 5.23 illustrate these pathological cases, which exist in theory, but have zero probability of occurring in practical circuits. For this reason, we will pay little attention to such pathological cases in this chapter.

> **Exercise.** Verify that ($v = 2$ V, $i = 3$ A) are the only possible values for the terminal voltage and current of the network of figure 5.23a.

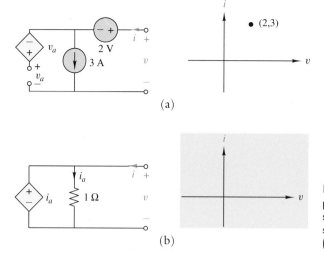

(a)

(b)

Figure 5.23 Pathological cases made possible by the presence of dependent sources for which the v-i curve is not a straight line. (a) Single point in v-i plane. (b) Entire v-i plane filled.

Exercise. Verify that $v = E$ and $i = J$ are possible values of the terminal voltage and current in figure 5.23b, for arbitrary values of E and J.

For a general linear resistive two-terminal network, Thévenin's theorem is as follows:

THÉVENIN'S THEOREM FOR ACTIVE LINEAR NETWORKS

Almost any two-terminal linear resistive network is equivalent to either the Thévenin equivalent shown in figure 5.17b, or the Norton equivalent shown in figure 5.21b, or both. The meanings of R_{th}, v_{oc}, and i_{sc} are the same as before, and equations 5.10–5.14 continue to hold. "Almost any" means "for all two-terminal networks, except the rare circuits exemplified in figure 5.23."

One may still have the following concern: Given a linear two-terminal resistive network, how do we know whether it is a pathological case or not? A practical way of tackling this problem is to assume initially that the circuit is not pathological and proceed to find R_{th}. Any pathology that exists will then reveal itself when neither R_{th} nor $1/R_{\text{th}}$ can be uniquely determined.

Computation of Thévenin and Norton Equivalent Circuits

Later chapters will explore many important applications of the Thévenin and Norton equivalent circuits. The present subsection focuses on finding the parameters (v_{oc}, i_{sc}, R_{th}). For circuits without dependent sources, such computations are not difficult. There are even *explicit* formulas for the parameters of the Thévenin equivalent. (See chapter 19 of volume 2.) When dependent sources are included, the general problem becomes much more difficult—and sometimes even intriguing, as illustrated by the simple circuits of figure 5.23. Except for some simple circuits or some special circuit configurations, the general problem will inevitably boil down to the solution of a general linear resistive circuit containing dependent sources. That is why solving the general problem is difficult: The general analysis methods of chapter 4 require the formulation and solution of sometimes numerous simultaneous equations. Fortunately, in many practical cases, it is possible to avoid simultaneous equations.

We next illustrate a variety of methods with examples. As always, each problem can be solved by several methods. The given solution does not imply that the best one has been chosen; rather, the aim is to illustrate the method itself. A guide for selecting a good method will be presented later.

In general, it is a good strategy to compute R_{th} first and then v_{oc} or i_{sc}, for several reasons: (1) With all internal independent sources deactivated, one can tell easily whether the "dead" network is of the series-parallel type, which can be solved without writing simultaneous equations. (2) The nonexistence of the Thévenin equivalent or Norton equivalent will become apparent upon computation of R_{th}: If $R_{th} = 0$, then the given network has no Norton equivalent; on the other hand, if $R_{th} \to \infty$, then the given network has no Thévenin equivalent. (3) For some types of problems, such as the design of a load to extract maximum power from a fixed linear resistive network containing independent sources, the only information needed is R_{th}. This problem will be discussed in section 5.

◆ EXAMPLE 5.13

Find the Thévenin equivalent of the two-terminal network sketched in figure 5.24a.

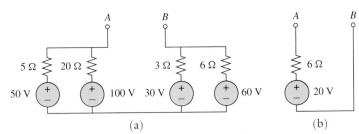

Figure 5.24   (a) A series-parallel circuit. (b) The Thévenin equivalent.

(a)   (b)

SOLUTION

Following the aforesaid strategy, we first calculate R_{th}. With all independent sources deactivated, the resultant network is a series-parallel connection of four resistances. Hence, by direct calculation,

$$R_{th} = (5//20) + (3//6) = 4 + 2 = 6 \ \Omega.$$

Using superposition and voltage division, the open-circuit voltage is

$$v_{oc} = v_{AB} = \frac{20}{20+5}50 + \frac{5}{20+5}100 - \frac{6}{3+6}30 - \frac{3}{3+6}60$$
$$= 40 + 20 - 20 - 20 = 20 \text{ V}.$$

Figure 5.24b shows the Thévenin equivalent of the circuit of figure 5.24a. Since the "dead" network is a series-parallel connection of resistances, we could also have found the Thévenin equivalent by repeated application of source transformations, as was done in example 5.8. Figure 5.14c shows an intermediate step when using source transformations. Combining the series-connected voltage sources and resistances in figure 5.14c leads to the circuit of figure 5.24b.

Exercise. Repeat example 5.13 with all resistances changed to 40 Ω.

ANSWERS: $R_{th} = 40 \ \Omega$, $v_{oc} = 30$ V.

◆ EXAMPLE 5.14

Find the Thévenin equivalent of the network of figure 5.25a.

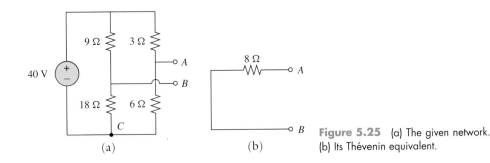

(a)   (b)

Figure 5.25   (a) The given network. (b) Its Thévenin equivalent.

SOLUTION

Again, we first calculate R_{th} and then v_{oc}.

Step 1. *Determine R$_{th}$.* With the 40-V source deactivated, the resistances are series-parallel connected. By inspection,

$$R_{th} = (9//18) + (3//6) = 6 + 2 = 8 \ \Omega.$$

Step 2. *Determine v$_{oc}$.* Using the voltage divider formula,

$$v_{oc} = v_{AB} = v_{AC} - v_{BC} = 40 \left(\frac{18}{18 + 9} - \frac{6}{6 + 3} \right) = 0.$$

Figure 5.25b shows the Thévenin equivalent of the network of figure 5.25a. In spite of the presence of an independent voltage source in the original network, its Thévenin equivalent is a "dead" network. Note that in this example R_{th} cannot be computed from equation 5.13c, because both v_{oc} and i_{sc} are zero.

◆ EXAMPLE 5.15

Find the Thévenin equivalent of the two-terminal network (enclosed by dashed lines) in figure 5.26a for $\mu = -79$ and $\mu = 101$. Think of the dependent source as a voltage amplifier.

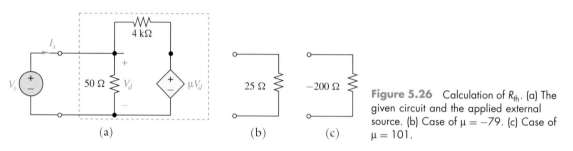

(a) (b) (c)

Figure 5.26 Calculation of R_{th}. (a) The given circuit and the applied external source. (b) Case of $\mu = -79$. (c) Case of $\mu = 101$.

SOLUTION

Step 1. *Determine R$_{th}$ as a function of μ.* Because of the presence of the dependent source, R_{th} cannot be calculated by series-parallel combinations of resistances. Further, since the two-terminal network contains no internal independent sources, $v_{oc} = 0$ and $i_{sc} = 0$. Hence, equation 5.13c cannot be used either. The only recourse left is to apply an external independent voltage source and find the current delivered by the source, or to apply an external independent current source and find the voltage across the source. Figure 5.26a depicts the former approach. Since $V_d = V_s$, applying KCL to the top terminal yields

$$I_s = \frac{V_s}{50} + \frac{V_s - \mu V_s}{4,000} = [0.02 + 0.00025(1 - \mu)]V_s.$$

Therefore, according to Thévenin's theorem, $R_{th} = R_{eq}$ of the "dead" network $= V_s/I_s$, or

$$R_{th} = \frac{V_s}{[0.02 + 0.00025(1 - \mu)]V_s} = \frac{1}{0.02 + 0.00025(1 - \mu)}.$$

Step 2. *Consider R$_{th}$ for two given values of μ:*

1. For $\mu = -79$, we have $R_{th} = 25 \ \Omega$ (figure 5.26b).
2. For $\mu = 101$, we have $R_{th} = -200 \ \Omega$ (figure 5.26c).

Step 3. *Determine v$_{oc}$.* As mentioned earlier, v_{oc} is zero because the given network contains no internal independent sources.

Three remarks about this example are in order:

1. For circuits containing controlled sources, the existence of a negative resistance is definitely a possibility. This was first pointed out in chapter 3.
2. In calculating R_{th}, the applied voltage V_s may have any value, since it cancels out in the final answer for R_{th}. In particular, we may let $V_s = 1$ V, as is done in many textbooks, to simplify the appearance of the equations.
3. The example suggests that for networks without internal independent sources, one might artificially attach a source to compute R_{th}. In fact, this is often done.

Exercise. Solve example 5.15 by the second suggested approach, i.e., applying an external current source I_s and finding V_s, the voltage across the source. R_{th} is then equal to V_s/I_s.

◆ **EXAMPLE 5.16**

Find either the Thévenin or the Norton equivalent circuit of the two-terminal network of figure 5.27a.

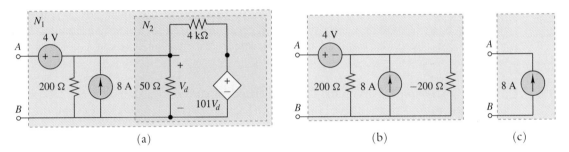

(a)   (b)   (c)

Figure 5.27 (a) Two-terminal network N_1 having only a Norton equivalent circuit. (b) Thévenin equivalent of example 5.15 is substituted for the inner box N_2 in N_1. (c) Norton equivalent of N_1.

SOLUTION

One useful strategy for tackling complex circuits is to replace portions of the circuits by their simpler equivalents. Here, we recognize that the inner box, N_2, in the given circuit is the same circuit as was analyzed in example 5.15. Therefore, N_2 is equivalent to a -200-Ω resistance. Replacing N_2 by this resistance results in the circuit of figure 5.27b. We can now analyze this simpler circuit.

Step 1. Determine R_{th}. With the 4-V and 8-A independent sources in figure 5.27b deactivated, the equivalent resistance "looking into" terminals (A, B) is a parallel combination of 200 Ω and -200 Ω (from the result of example 5.15). This yields

$$\frac{1}{R_{th}} = \frac{1}{200} - \frac{1}{200} = 0.$$

Then $R_{th} \to \infty$, and the Thévenin equivalent circuit does not exist.

Step 2. Determine i_{sc}. By inspection of figure 5.27b, the current from the 8-A source simply passes through terminals (A,B) when these terminals are short-circuited. Therefore, $i_{sc} = 8$ A. Figure 5.27c shows the Norton equivalent circuit of the network N_1.

It may be a little surprising to find that figure 5.27a has no Thévenin equivalent circuit, despite the prominent appearance of the 4-V independent source. Furthermore, trying to find R_{th} in figure 5.27b by injecting a 1-A current source into the "dead" network will lead to a violation of KCL, because the "dead" network actually behaves like an open circuit.

EXAMPLE 5.17

Find either the Thévenin or Norton equivalent circuit of the two-terminal network of figure 5.28a.

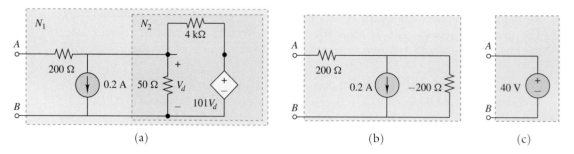

Figure 5.28 (a) Two-terminal network N_1 having only a Thévenin equivalent circuit. (b) The Thévenin equivalent of example 5.15 is substituted for the inner box N_2 in N_1. (c) Thévenin equivalent of N_1.

SOLUTION

Using the same strategy as in example 5.16, we first replace the inner box, N_2, in figure 5.28a by its equivalent, which is a -200-Ω resistance (from the result of example 5.15). Figure 5.28b shows the result of this replacement. We then analyze the simpler circuit therein.

Step 1. *Determine* R_{th}. With the 0.2-A independent current source in figure 5.28b deactivated, the equivalent resistance "looking into" the terminals (A,B) is a series combination of 200 Ω and -200 Ω. This results in

$$R_{\text{th}} = 200 - 200 = 0.$$

Step 2. *Determine* v_{oc}. From figure 5.28b, the open-circuit voltage is

$$v_{\text{oc}} = (-0.2)(-200) = 40 \ V.$$

Figure 5.28c shows the resultant Thévenin equivalent circuit. Because $R_{\text{th}} = 0$, the given network has no Norton equivalent circuit.

The preceding two examples reinforce the statement that finding the Thévenin or Norton equivalent circuit for a network containing dependent sources is not always straightforward. In both examples, had we not replaced the inner box N_2 by a resistance, we would have had to write two simultaneous equations to solve the problem. Recall the third remark following example 5.15. We attach an external source to the dead network to find R_{th}. In most cases, the applied source may be either a current source or a voltage source, as shown in example 5.15 and the exercise immediately after. However, in the case of example 5.16, applying a current source and writing two nodal equations will result in an inconsistent set of equations. Similarly, in the case of example 5.17, applying a voltage source and writing loop equations will also result in an inconsistent set of equations. The inconsistency manifests itself in the form of a zero-valued determinant for the system of equations. The experience gained from these two examples suggests that at the first attempt at analysis, one may apply an external voltage source or current source, whichever is more convenient. If the resultant equations are inconsistent, then one changes the type of applied external source and solves the problem again. If in both cases the systems of equations are inconsistent, then the network belongs to the rare pathological cases illustrated in figure 5.23.

In the next three examples, we find the Thévenin equivalent of a given circuit by three different methods.

Find the Thévenin equivalent circuit parameters for the two-terminal network of figure 5.29a by computing v_{oc} and i_{sc}.

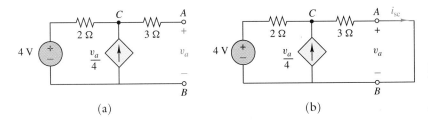

(a)   (b)

Figure 5.29 Derivation of the Thévenin equivalent by finding v_{oc} and i_{sc}. (a) The given circuit. (b) Terminals A and B shorted for calculating i_{sc}.

SOLUTION

Step 1. *Determine* v_{oc}. In figure 5.29a, the open-circuit voltage is $v_{oc} = v_{AB} = v_a$. Applying KCL to node C yields

$$\frac{v_a}{4} = \frac{v_{CB} - 4}{2}.$$

Now, $v_{CB} = v_{AB} = v_a$, because there is no current passing through the 3-Ω resistor. Hence, the foregoing equation becomes

$$\frac{v_a}{4} = \frac{v_a - 4}{2},$$

from which it follows that $v_{oc} = v_a = 8$ V.

Step 2. *Determine* i_{sc}. In figure 5.29b, because of the short-circuited terminals (A, B), $v_a = 0$. As a result, the controlled current source $v_a/4$ is also zero. The short-circuit current is then easily calculated to be

$$i_{sc} = \frac{4}{2 + 3} = 0.8 \text{ A}.$$

Step 3. *Determine* R_{th}. From equation 5.13c, $R_{th} = v_{oc}/i_{sc} = 8/0.8 = 10$ Ω.

■ **EXAMPLE 5.19**

Find the Thévenin equivalent circuit parameters for the two-terminal network of figure 5.29a by computing R_{th} and v_{oc}.

SOLUTION

Step 1. *Determine* R_{th}. Following the method used in example 5.15, we deactivate the 4-V voltage source in figure 5.29a and apply an external voltage source of 1 V (or any other value). Figure 5.30 shows the resultant circuit.

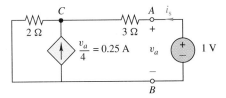

Figure 5.30 Finding R_{th} for the circuit of figure 5.29a.

The current i_s delivered by the 1-V source can be computed by a number of methods. For example, after noting the relationship $v_a = v_s$, we can write down the result for i_s directly by superposition as follows:

Due to the 1-V source,

$$i_s^1 = \frac{1}{2 + 3} = 0.2 \text{ A}.$$

Due to the 0.25-A source, from the current division formula,

$$i_s^2 = -0.25 \frac{2}{2+3} = -0.1 \text{ A}.$$

By superposition, the total current delivered by the 1-V source is

$$i_s = i_s^1 + i_s^2 = 0.2 - 0.1 = 0.1 \text{ A}.$$

Therefore, $R_{\text{th}} = v_s/i_s = 1/0.1 = 10 \ \Omega$.

Step 2. *Determine* v_{oc}. This was done in example 5.18. The result is $v_{oc} = 8$ V.

The solutions presented in examples 5.18 and 5.19 reduce the original problem to two separate circuit analysis problems. The next two examples show how the two parameters in the Thévenin equivalent can be obtained by a single analysis.

◆ EXAMPLE 5.20

Find the Thévenin equivalent circuit parameters for the two-terminal network of figure 5.29a by applying a variable source to the network.

SOLUTION

The given network is repeated in figure 5.31a. One can, of course, compute the Thévenin equivalent circuit parameters by the method described in example 5.18 or example 5.19. With either method, the analyses of two circuits must be carried out. It is possible, however, to obtain the answers by the analysis of a single circuit: Connect a variable voltage source v to the terminals (A,B), as shown in figure 5.31b.

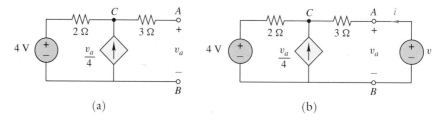

(a)   (b)

Figure 5.31 Derivation of the Thévenin's equivalent circuit by applying a variable voltage source. (a) The given circuit. (b) A variable voltage source is attached.

Step 1. *Determine the* v-i *relationship.* To find the v-i relationship, we apply KVL to the outer loop consisting of the elements (3 Ω, 2 Ω, 4 V, v). The result is

$$v = 3i + 2\left(i + \frac{v}{4}\right) + 4.$$

Solving for i yields

$$i = 0.1v - 0.8,$$

or

$$v = 10i + 8.$$

Step 2. *Compare with equation 5.10.* Comparing the latter equation with equation 5.10 implies that $R_{\text{th}} = 10$ and $v_{oc} = 8$ V.

The next example illustrates the same idea of obtaining the Thévenin equivalent circuit parameters by a single analysis in which a variable current source is applied to the network.

EXAMPLE 5.21

Find the Thévenin equivalent circuit parameters for the two-terminal network of figure 5.32a.

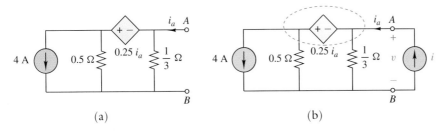

(a) (b)

Figure 5.32 Derivation of the Thévenin equivalent circuit by applying a variable current source. (a) The given circuit. (b) A variable current source is attached.

SOLUTION

We connect a variable current source i to the terminals (A,B), as shown in figure 5.32b.

Step 1. *Determine the* v-i *relationship.* To find the v-i relationship, we apply KCL to the supernode enclosed by the dashed lines. The result is

$$i = 3v + 2(v + 0.25i) + 4,$$

or

$$i = 10v + 8,$$

or again

$$v = 0.1i - 0.8.$$

Step 2. *Compare with equation 5.10.* Comparing this last equation with equation 5.10 implies that $R_{\text{th}} = 0.1\ \Omega$ and $v_{\text{oc}} = -0.8$ V.

Guide for Selecting Methods for Computing Thévenin Equivalents

The solutions of examples 5.20 and 5.21 bring out an interesting observation: Had we applied a variable current source to example 5.20 and written node equations, or applied a variable voltage source to example 5.21 and written mesh equations, the solution would be considerably more complicated, although the same answer would result. This should reinforce the statement made in chapter 4 that the choice of a good method of attack is an important part of problem solving. We shall now summarize the various methods we have illustrated and offer a guide to the selection of a method for solving a specific problem.

Examples 5.10–5.21 have illustrated a number of strategies or procedures for obtaining the Thévenin equivalent and Norton equivalent circuits of a two-terminal linear network:

Strategy 1. Use of source transformations. (See examples 5.10, 5.13.)

Strategy 2. Use of equation 5.13a: Determine R_{th} and v_{oc}, and then calculate $i_{\text{sc}} = v_{\text{oc}}/R_{\text{th}}$. (See examples 5.14, 5.15, and 5.19.)

Strategy 3. Use of equation 5.13b: Determine R_{th} and i_{sc}, and then calculate $v_{\text{oc}} = i_{\text{sc}}R_{\text{th}}$.

Strategy 4. Use of equation 5.13c: Determine v_{oc} and i_{sc}, and then calculate $R_{\text{th}} = v_{\text{oc}}/i_{\text{sc}}$. (See example 5.18.)

Strategy 5. Use of equation 5.10: Determine R_{th} and v_{oc} from two measured (i, v) points. (See example 5.12.)

Strategy 6. Use of equation 5.10: Determine R_{th} and v_{oc} by applying a variable external source. (See examples 5.20 and 5.21.)

In applying strategies 2 and 3, if the dead network is not a series-parallel combination of resistances, then R_{th} may be found by applying an external voltage source (see example 5.15) or an external current source (see the exercise following example 5.15).

To simplify the solution process, the following pointers are useful:

1. One can replace portions of the network by their simpler equivalents. (See examples 5.16 and 5.17.)
2. One can use a combination of the foregoing strategies, as illustrated in examples 5.16 and 5.17.

Since, for any given circuit, one can usually use different strategies to find the Thévenin equivalent circuit, the problem of choosing a good strategy is always perplexing to beginners. The following flowchart offers some help in this respect:

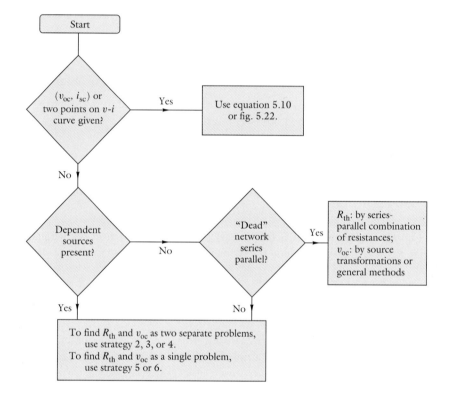

5. MAXIMUM POWER TRANSFER THEOREM

Figure 5.33 shows an approximate model of a "practical source" consisting of an ideal voltage source in series with an internal or source resistance R_s. Connected across the source is a variable load resistor R_L. The load voltage v_L, the load current i_L, and the power p_L delivered to the load are all functions of R_L.

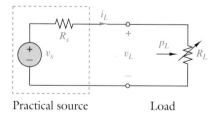

Practical source   Load

Figure 5.33 Maximum power transfer to R_L is desired in this circuit.

What value of R_L maximizes the instantaneous power delivered to the load? Such a question is of fundamental importance in communication circuits, in radio transmission circuits such as transmitters loaded by an antenna, etc. The answer to the question for the circuit configuration of figure 5.33 is given by the following theorem.

MAXIMUM POWER TRANSFER

Maximum Power Transfer Theorem. Let a practical source consisting of an independent voltage source $v_s(t)$ in series with an internal resistance R_s be attached to a variable load R_L. For fixed R_s and an arbitrary $v_s(t)$, maximum instantaneous power is transferred to the load when

$$R_L = R_s$$

and the maximum instantaneous power is given by

$$p_{L,\max}(t) = \frac{v_s^2(t)}{4R_s}.$$

In the dc case, i.e., when $v_s(t)$ is a constant, the instantaneous power is constant and has the same value as the average power (defined in chapter 1). In this case, the condition $R_L = R_s$ also assures the transfer of the maximum average power.

A derivation of the maximum power transfer theorem proceeds as follows. From figure 5.33, the power absorbed by the load in terms of $v_s(t)$, R_s, and R_L is

$$p_L(t) = i_L^2(t)R_L = \left[\frac{v_s(t)}{R_s + R_L}\right]^2 R_L. \tag{5.15}$$

Following the standard procedure of calculus for determining a maximum, we calculate dp_L/dR_L, equate it to zero, and solve for R_L:

$$\frac{dp_L}{dR_L} = \frac{v_s^2}{(R_s + R_L)^2} - 2\frac{v_s^2 R_L}{(R_s + R_L)^3} = v_s^2\frac{R_s - R_L}{(R_s + R_L)^3} = 0.$$

$R_L = R_s$ and $R_L = \infty$ are the only possible solutions of this equation. The fact that $p_L(t)$ is a maximum instead of a minimum at $R_L = R_s$ is obvious from equation 5.15, as $p_L(t) = 0$ for both $R_L = 0$ and $R_L = \infty$ and $p_L(t) > 0$ for $0 < R_L < \infty$. A rough sketch of the curve of p_L vs. R_L immediately confirms that p_L is a maximum at $R_L = R_s$.

Substituting $R_L = R_s$ into equation 5.15 yields

$$p_{L,\max}(t) = \frac{v_s^2(t)}{4R_s}. \tag{5.16}$$

If the source is a complex linear two-terminal resistive network, then its Thévenin equivalent must be found before applying the maximum power transfer theorem. If the source is represented by a Norton equivalent circuit, we can use a source transformation to obtain the Thévenin form and then apply equation 5.16.

 EXAMPLE 5.22

Consider the circuit of figure 5.34a. Find the value of R_L for maximum power transfer and the corresponding $p_{L,\max}$.

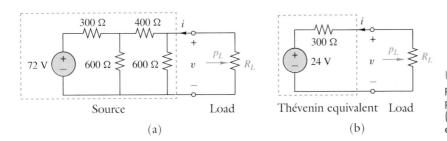

Figure 5.34 Use of the maximum power transfer theorem. (a) Maximum power transfer to R_L is desired. (b) Replacing the source by its Thévenin equivalent.

SOLUTION

In example 5.10, we computed the Thévenin equivalent circuit for the source of figure 5.34a. This Thévenin equivalent is sketched here as figure 5.34b. According to the maximum power transfer theorem, $p_{L,\max} = 24^2/1{,}200 = 0.48$ W, which occurs when $R_L = 300\ \Omega$. The tabulated results of example 5.10 verify this assertion.

Exercise. If an adjustable R_L is connected to terminal (A,B) of the circuit of figure 5.24a, find the value of R_L for maximum power transfer and the corresponding $p_{L,\max}(t)$.
ANSWERS: 6 Ω, 16.67 W.

The derivation of the maximum power transfer theorem considers R_L as the independent variable and sets dp_L/dR_L to zero according to the rules of calculus. We now present another proof with simpler mathematical operations that has a deeper consequence.

Let us consider figure 5.33 and take i_L to be the independent variable. Then

$$p_L = i_L v_L = i_L(v_s - R_s i_L) = i_L v_s - R_s i_L^2.$$

From this equation, maximum power transfer occurs at a point of the curve of p_L vs. i_L with zero slope. Hence, we set dp_L/di_L to zero and solve for i_L:

$$\frac{dp_L}{di_L} = v_s - 2R_s i_L = 0.$$

Therefore, for maximum power transfer to the load,

$$i_L = \frac{v_s}{2R_s}, \tag{5.17a}$$

from which it follows that

$$v_L = \frac{v_s}{2}. \tag{5.17b}$$

If the load is a linear resistor, then, from 5.17a and 5.17b,

$$R_L = \frac{v_L}{i_L} = \frac{0.5v_s}{\left(\dfrac{0.5v_s}{R_s}\right)} = R_s. \tag{5.17c}$$

This proves the first part of the maximum power transfer theorem. The second part follows as before.

The deeper consequence of this derivation is that the load need not be a linear resistor: As long as the source delivers a load current of $i_L = v_s/(2R_s)$, maximum power is transferred to the load, whether or not the load is a linear resistor!

◆ **EXAMPLE 5.23**

In figure 5.33, let $v_s = 4$ V and $R_s = 0.125\ \Omega$. Suppose the load resistance R_L is replaced by a nonlinear resistor R_n whose v-i relationship is

$$i_L = k v_L^3,$$

where k is an adjustable parameter. Find the value of k such that maximum power is transferred to the nonlinear resistor.

SOLUTION

From equations 5.17a and 5.17b, maximum power transfer occurs when

$$i_L = \frac{v_s}{2R_s} = \frac{4}{0.25} = 16\ \text{A}$$

and

$$v_L = \frac{v_s}{2} = \frac{4}{2} = 2\ \text{V}.$$

For (i_L, v_L) to be a point on the nonlinear resistor characteristic curve, we must have

$$i_L = 16 = kv_L^3 = k2^3 = 8k.$$

Therefore, $k = 2$. Thus, the maximum power delivered to this nonlinear resistor is

$$p_{L,\max} = \frac{4^2}{0.5} = 32 \text{ W}.$$

Exercise. Repeat example 5.23 if the load is a two-terminal device characterized by $i_L = 4v_L + kv_L^3$.

ANSWER: $k = 1$.

Exercise. Repeat example 5.23 if the load is a two-terminal network consisting of a 0.0625-Ω resistor in series with an adjustable independent voltage source v_x.

ANSWER: $v_x = 1$ V.

The following are some final remarks about maximum power transfer:

1. If R_L is fixed and R_s is adjustable, then maximum $p_L(t)$ is obtained when R_s is reduced to zero. One can show this by redoing the first proof, but differentiating with respect to R_s.
2. The curve of p_L vs. R_L is very flat in the neighborhood of $R_L = R_s$. Even at $R_L = 2R_s$ and $R_L = 0.5R_s$, the value of p_L is about 89% of its maximum value.
3. If R_s is not a physical resistor, but a resistance in the Thévenin equivalent of a two-terminal network, then the power dissipated in R_s has no physical significance, because it is not related to the actual heat loss in the physical source. This should come as no surprise, as we mentioned in section 3 that when a two-terminal network is replaced by its Thévenin or Norton equivalent, only those electrical quantities *outside* the box remain unchanged; there is no correlation between the voltages and currents inside the original circuit and those inside its equivalent.
4. A practical dc voltage source (such as a battery in an automobile) is designed to provide nearly constant output voltage for the intended load current. Accordingly, it has a rather small source resistance R_s. Any attempt to transfer the maximum power from such a source continuously will overload the source and cause damage. Thus, maximum power transfer is a factor of importance in the design of communication circuits, but not necessarily in power transmission circuits or in other electronic circuits.

6. FURTHER EXAMPLES OF APPLICATIONS OF NETWORK THEOREMS

In the preceding sections, we presented several network theorems of fundamental importance and general use. There are many more network theorems, some of which are obvious while others may be useful only for special classes of problems. We state a few obvious theorems (e.g., the E-shift and the I-shift theorems) in the problems at the end of the chapter and postpone those of limited use (e.g., the wye-delta transformation) until later chapters, as the need arises. In this section, we illustrate how the application of a *combination* of several network theorems may simplify the analysis of some useful circuits while providing better insight into their operation.

EXAMPLE 5.24

The circuit of figure 5.35a is a variation of the four-bit digital-to-analog converter shown in figure P3.12. The op amp is assumed ideal. Show that the output voltage is

$$V_{\text{out}} = -[8V_3 + 4V_2 + 2V_1 + V_0]\frac{R_f}{32R}.$$

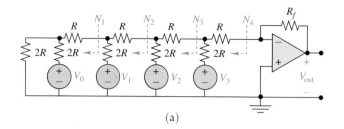

(a)

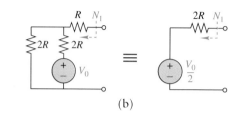

(b)

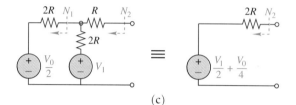

(c)

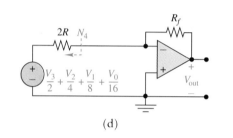

(d)

Figure 5.35 (a) A four-bit digital-to-analog converter. (b) and (c) Derivation of Thévenin equivalents of N_1 and N_2. (d) Equivalent inverting amplifier.

SOLUTION

Our strategy is to find the Thévenin equivalent of the circuit to the left of the dashed arrow labeled N_4 in figure 5.35a. To this end, we find successively the Thévenin equivalent circuits of the circuits to the left of the dashed arrows labeled N_1, N_2, N_3, and, finally, N_4. The results are shown in figures 5.35b through 5.35d. The calculation of v_{oc} for N_1 in figure 5.35b is a simple application of the voltage divider formula. The calculation of v_{oc} for N_2 in figure 5.35c is easily done with the voltage divider formula and the principle of superposition. The calculations are similar for N_3 and N_4. The Thévenin equivalent resistance is $2R$ for N_1 through N_4. This stems from the fact that in each case the Thévenin equivalent resistance is equal to

$$(2R \text{ in parallel with } 2R) \text{ in series with } R.$$

Applying the gain formula of an inverting amplifier to figure 5.35d immediately yields the desired expression for V_{out}.

Some remarks about the circuit of figure 5.35a and the method of solving it are in order:

1. If each voltage source V_k can have only one of two values—0 or E volts—then the expression for V_{out} has the same form as equation 3.8 (differing by a multiplicative constant only). This shows that the circuit may also be used as a digital-to-analog converter.

2. This circuit is superior to that of figure P3.12 in that it has a much smaller spread of resistor values: The resistors have only two values with ratio 2 to 1. This advantage is achieved at the price of having more resistors than the circuit of figure P3.12 does.

3. For a four-bit digital-to-analog converter, one can also derive the output voltage formula by writing and solving a set of simultaneous equations. However, such a method would be impractical for a 10-bit converter. Nevertheless, the present method of using voltage division and superposition can be applied to any n-bit digital-to-analog converter with the same ease as it can be applied to a four-bit converter.

Exercise. In the circuit of figure 5.35a, if the resistor R connected to the inverting input terminal is replaced by a short circuit, what is the new expression for V_{out}?

When a linear circuit exhibits some kind of symmetry, the use of the principle of superposition together with the symmetry can greatly simplify analysis. A difference amplifier is an example of such a circuit. A **difference amplifier** (or a differential amplifier) has two inputs, V_1 and V_2, both measured with respect to ground, and a single output V_o. The output terminals may have one node grounded or both nodes *floating*. Ideally, the output of a difference amplifier should depend only on the difference of the two inputs; i.e., $V_o \propto (V_1 - V_2)$. In practical difference amplifiers, however, the output is also affected to a small extent by the average level of the two inputs. Difference amplifiers are used extensively in electronic instrumentation circuits, especially for biomedical electronic measurements. Figure P3.17 depicts a difference amplifier built with an op amp. The next two examples employ a difference amplifier circuit built around two controlled sources. They will illustrate how symmetry and superposition combine to greatly simplify the response calculation of a differential amplifier circuit.

 EXAMPLE 5.25

Figure 5.36a shows the linear circuit model of a difference amplifier. Note the mirrorlike symmetry of the circuit configuration and element values. Let the desired output be the floating voltage $V_o = V_{ab}$.

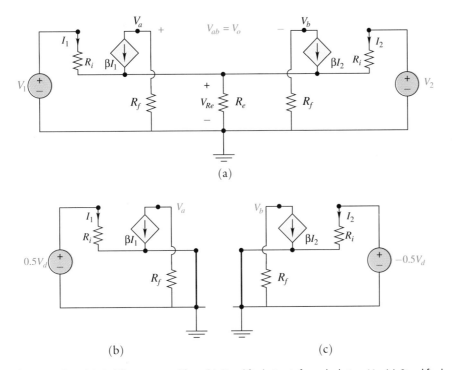

Figure 5.36 (a) A difference amplifier. (b) Simplified circuit for calculating V_a. (c) Simplified circuit for calculating V_b.

1. If $V_1 = V_2 = V_c$, find the voltage gain $A_c \equiv V_o/V_c$.
2. If $V_1 = -V_2 = 0.5V_d$, find the voltage gain $A_d \equiv V_o/V_d$.
3. Let $\beta = 100$, $R_i = 2$ kΩ, $R_f = 4$ kΩ, and $R_e = 20$ kΩ. Utilizing the results of parts 1 and 2, find V_o if $V_1 = 0.2$ V and $V_2 = 0.16$ V.

SOLUTION

Part 1. With $V_1 = V_2 = V_c$, the symmetry in figure 5.36a makes $V_a = V_b$. Consequently, $V_o = V_a - V_b = 0$ and $A_c \equiv V_o/V_c = 0$.

Part 2. Here, $V_1 = -V_2 = 0.5V_d$. From the symmetry of the circuit, V_{Re} due to the left source $0.5V_d$ and V_{Re} due to the right source $-0.5V_d$ must have the same magnitude, but opposite signs. Adding up the two component voltages leads to $V_{Re} = 0$. Since there is zero voltage across R_e, we may substitute a short circuit without disturbing the voltages and currents in all the other circuit elements. This short circuit decouples the analysis of the left and right half circuits of figure 5.36a. As a result, in calculating V_a, we need look only at the circuit shown in figure 5.36b, which readily yields

$$V_a = -\beta I_1 R_f = -\frac{\beta R_f V_d}{2R_i}.$$

In exactly the same manner, from figure 5.36c, we obtain

$$V_b = \beta I_2 R_f = \frac{\beta R_f V_d}{2R_i}.$$

Finally,

$$A_d = \frac{V_0}{V_d} = \frac{V_a - V_b}{V_d} = -\frac{\beta R_f}{R_i}.$$

Part 3. $V_1 = 0.2$ V and $V_2 = 0.16$ V. This set of inputs does not meet the conditions stated in either part 1 or part 2. However, we can decompose the set into a sum of two sets, each of which corresponds to either part 1 or part 2, as follows:

$$[V_1 \ V_2] = \underset{\text{given inputs}}{[0.2 \ 0.16]} = \underset{\text{set 1}}{[0.18 \ 0.18]} + \underset{\text{set 2}}{[0.02 \ -0.02]}.$$

The output V_o due to input set 1 is zero, according to part 1. The output V_o due to input set 2 is, according to part 2,

$$-100 \times \frac{4,000}{2,000} \times 0.04 = -8 \text{ V}.$$

By the principle of superposition, the actual output voltage is $V_o = 0 + (-8) = -8$ V. Note that, since $A_c = 0$, the circuit behaves as an ideal difference amplifier.

Exercise. In part 3 of example 5.25, if $V_1 = 20.2$ V and $V_2 = 20.16$ V, find V_o, making use of A_c and A_d calculated in parts 1 and 2.

◆ **EXAMPLE 5.26**

Repeat the three-part analysis of example 5.25 if the output is taken to be $V_o = V_a$.

SOLUTION

Part 1. To display the mirrorlike symmetry about a vertical line drawn through the middle of the circuit, replace R_e in figure 5.36a by a parallel connection of two resistors, each having resistance $2R_e$. Figure 5.37a shows the resultant circuit.

We invoke the superposition theorem to calculate the current I_x. From the symmetry of the circuit, I_x due to the left source and I_x due to the right source must have the same magnitude, but flow in opposite directions. Adding up the two component currents leads to $I_x = 0$. Since there is no current flowing through this segment of the connection, it may

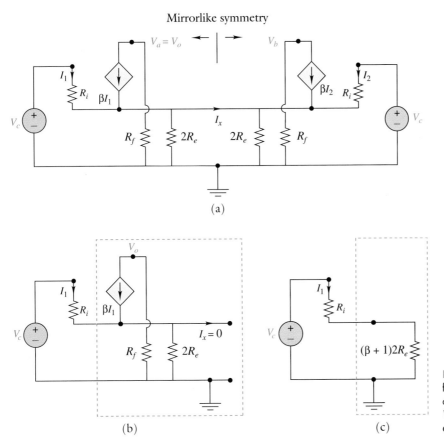

Mirrorlike symmetry

(a)

(b)   (c)

Figure 5.37 (a) A redrawing of figure 5.36a. (b) Simplified circuit for calculating V_a under the condition $V_1 = V_2 = V_c$. (c) Use of the Thévenin equivalent.

be broken without disturbing the voltages and currents in all the other circuit elements. Therefore, in calculating $V_o = V_a$, we need only look at the left half of the complete circuit. Figure 5.37b shows the pertinent circuit for calculating V_o. Next, we find the Thévenin equivalent of the circuit enclosed by the dashed lines in figure 5.37b. The result is a resistance with value $2(\beta + 1)R_e$, as shown in figure 5.37c, from which we find

$$I_1 = \frac{V_c}{R_i + (\beta + 1)2R_e}.$$

Applying this result to figure 5.37b yields

$$V_o = -\beta I_1 R_f = -\frac{\beta R_f V_c}{R_i + (\beta + 1)2R_e}$$

and

$$A_c = \frac{V_o}{V_c} = -\frac{\beta R_f}{R_i + (\beta + 1)2R_e}.$$

Part 2. By the same reasoning as in part 2 of example 5.25, we have

$$V_o = V_a = -\beta I_1 R_f = -\frac{\beta R_f V_d}{2R_i}$$

and

$$A_d = \frac{V_o}{V_d} = \frac{V_a - V_b}{V_d} = -\frac{\beta R_f}{2R_i}.$$

Part 3. Using the given circuit parameters and the results of previous parts, we find that $A_c = -0.09896$ and $A_d = -100$. The decomposition of the input signal set is the same as before, i.e.,

$$[V_1 \quad V_2] = [0.2 \quad 0.16] = [0.18 \quad 0.18] + [0.02 \quad -0.02].$$

Input set 1, $[V_1 \quad V_2] = [0.18 \quad 0.18]$, produces

$$V_o^1 = A_c \times 0.18 = -0.09896 \times 0.18 = 0.0178 \text{ V}.$$

Input set 2, $[V_1 \quad V_2] = [0.02 \quad -0.02]$, produces

$$V_o^2 = A_d \times 0.04 = -100 \times 0.04 = -4 \text{ V}.$$

By the principle of superposition, the actual output is

$$V_o = V_o^1 + V_o^2 = -0.0178 + (-4) = -4.0178 \text{ V}.$$

Exercise. In part 3 of example 5.26, if $V_1 = 20.2$ V and $V_2 = 20.16$ V, find V_o, making use of A_c and A_d calculated in parts 1 and 2.

Several remarks are in order concerning difference amplifiers:

1. The following terms are commonly used in the literature:

Difference mode input	$V_d \equiv (V_1 - V_2)$
Common mode input	$V_c \equiv 0.5(V_1 + V_2)$
Difference mode gain	$A_d \equiv V_o/V_d$, when $V_c = 0$
Common mode gain	$A_c \equiv V_o/V_c$, when $V_d = 0$
Common mode rejection ratio (CMRR)	$\rho = A_d/A_c$

   As a result of these definitions, the following relationships hold:

   $$[V_1 \quad V_2] = [V_c + 0.5V_d \quad V_c - 0.5V_d],$$
   $$V_o = A_c V_c + A_d V_d.$$

2. For an ideal difference amplifier, $A_c = 0$ and $\rho \to \infty$. The circuit of figure 5.36a, with floating output $V_o = V_a - V_b$, satisfies these properties. However, when the circuit parameters are not perfectly matched, the common mode gain A_c will be small, but not zero.

3. For the circuit of figure 5.36a, if one output node is at ground (e.g., $V_o = V_a$), then even when all circuit parameters are perfectly matched, A_c is not zero. Typically, the value of CMRR is in the range of 10^4 to 10^6. From the expression for A_c, we see that A_c can be made very small by using a large value for R_e. This, however, requires a very high dc power supply voltage to operate the circuit. Methods for circumventing the difficulty (e.g., replacing the resistor R_e by a dc current source) are discussed in most texts on electronics.

SUMMARY

In this chapter, we have discussed several basic theorems pertaining to linear resistive networks. Foremost among these is the linearity theorem, which asserts the superposition and proportionality properties for linear networks. The concept of equivalent two-terminal networks is defined as two networks having the same terminal voltage-current relationship. This concept will be extended later to equivalent *n*-terminal networks.

The superposition property allows us to pinpoint the contribution of each independent source in a linear circuit to the total response. One very interesting and significant application of the superposition property occurs in the analysis of a circuit with a single input and a single dependent source. The superposition approach leads to equation 5.6, which is the basis for the "return difference technique" used in advanced feedback amplifier theory. For beginners in circuit analysis, many difficulties are attributable to the presence of dependent sources. The use of equation 5.6 overcomes the difficulties in many cases.

One powerful strategy for analyzing a complex network is to replace portions of the network by their simpler equivalents. Thévenin's theorem assures us that no matter how many elements it has inside, any two-terminal linear resistive network is equivalent to a simple network consisting of a voltage source and a resistance. The use of Thévenin's theorem and the source transformation theorem often simplifies manual analysis. Only one application of Thévenin's theorem to circuit design is included in this chapter: The design of a load to extract maximum power from a linear resistive network containing sources. Many other practical applications of the theorem are presented in later chapters.

Although all theorems in this chapter are stated for linear resistive networks, all of them can be extended to linear dynamic networks. This extension will be presented after the introduction of two new circuit elements—the capacitor and the inductor—in chapter 7.

TERMS AND CONCEPTS

Common mode rejection ratio, CMRR: for a difference amplifier, the ratio of difference mode gain to common mode gain; i.e., CMRR $= A_d/A_c$.

Common mode voltage gain, A_c: for a difference amplifier, $A_c \equiv V_o/V_c$ when both inputs V_1 and V_2 have the same value V_c.

Deactivating (killing) an independent current source: replacing the said source by an open circuit.

Deactivating (killing) an independent voltage source: replacing the said source by a short circuit.

Difference amplifier (differential amplifier): an amplifier with two inputs V_1 and V_2 and a single output V_o that, for the ideal case, depend only on the difference of the two inputs; i.e., $V_o \propto (V_1 - V_2)$.

Difference mode voltage gain, A_d: for a difference amplifier, $A_d \equiv V_o/V_d$ when both inputs have equal magnitudes of $0.5V_d$, but opposite polarities, i.e., when $V_1 = -V_2 = 0.5V_d$.

Equivalent n-terminal networks: two n-terminal networks having the same terminal voltage-current relationship. As an alternative definition, two n-terminal networks are said to be **equivalent** when, after substituting one for the other in every possible network N, the voltages and currents in N are unaffected.

Linearity property: let the responses due to inputs u_1 and u_2, each acting alone, be y_1 and y_2. When the scaled inputs $\alpha_1 u_1$ and $\alpha_2 u_2$ are applied simultaneously, the response is $y = \alpha_1 u_1 + \alpha_2 u_2$. Linearity implies both superposition and proportionality, and vice versa.

Linear resistive element: an n-terminal circuit element whose terminal voltage and current relationships are described by a set of linear homogeneous algebraic equations.

Linear resistive network (circuit): a network consisting of only linear resistive elements and independent voltage and current sources.

Maximum power transfer theorem: let an adjustable resistor R_L be connected to a practical power source represented by an ideal voltage source V_s in series with a fixed source resource R_s. Then maximum power is absorbed by the resistor when its value is adjusted to $R_L = R_s$.

Nonlinear resistor: a two-terminal circuit element having a v-i curve that is not a straight line, but passes through the origin of the v-i plane.

Norton's equivalent circuit: circuit with an independent current source in parallel with a resistance, equivalent to a two-terminal network consisting of independent sources and linear resistive elements.

Proportionality property: property such that when an input to a linear resistive network is acting alone, increasing the input K times increases the response K times.

Source transformation: transformation in which a two-terminal network consisting of an independent voltage source in series with a resistance is equivalent to another two-terminal network consisting of an independent current source in parallel with a resistance.

Superposition property: property such that when a number of inputs are applied to a linear resistive network simultaneously, the response is the sum of the responses due to each input acting alone.

Thévenin's equivalent circuit: circuit with an independent voltage source in series with a resistance, equivalent to a two-terminal network consisting of independent sources and linear resistive elements.

Thévenin's equivalent resistance: the resistance that appears in the Thévenin equivalent circuit of a two-terminal linear network. It is also the equivalent resistance of the two-terminal network when all internal independent sources are deactivated.

Two-terminal network: an interconnection of circuit elements inside a box having only two accessible terminals for connection to other networks. The concept is extendible to n-terminal networks.

PROBLEMS

1. The v-i curve of a two-terminal black box is a displaced cubic curve as shown in figure P5.1. Construct two different circuit models for the black box: Using (a) a nonlinear resistor and an independent voltage source and (b) a nonlinear resistor and an independent current source. In each case, show the connection of the components and write the equation for the v-i relationship of the nonlinear resistor. (*Check:* (a) $|V_s| = 2$ V and $i_a = v_a^3$; (b) $|I_s| = 8$ A, $i_b = v_b^3 - 6v_b^2 + 12v_b$.)

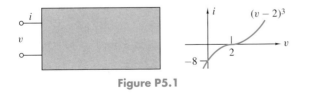

Figure P5.1

2. A nonlinear resistive circuit has two inputs, v_1 and v_2, and one output, v_o.

(a) If the input-output relationship is

$$v_o = 3v_1^2 + 4v_2^2,$$

show that the circuit satisfies the superposition property, but not the proportionality property.

(b) If the input-output relationship is

$$v_o = 3v_1 + 4v_2 + 5v_1v_2,$$

show that the circuit satisfies the proportionality property, but not the superposition property.

3. In the circuit shown in figure P5.3, $v_a(t) = 12$ V and $i_b(t) = 6$ A. Recall that if the current through a current source is zero, the source looks like an open circuit. What is the equivalent statement for a voltage source with zero V across it? Determine $v_{\text{out}}(t)$ by superposition.

ANSWER: 16 V.

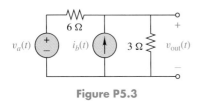

Figure P5.3

4. Consider the circuit shown in figure P5.4.

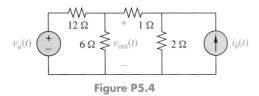

Figure P5.4

(a) If $v_a(t) = 14$ V for $t \geq 0$ and zero otherwise, and $i_b(t) = 0$, determine $v_{\text{out}}(t)$.
(b) If $v_a(t) = 0$ for all t, and $i_b(t) = 3.5$ A for $t \geq 0$ and zero otherwise, determine $v_{\text{out}}(t)$.
(c) If $v_a(t) = 14$ V for $t \geq 0$ and zero otherwise, and $i_b(t) = 3.5$ A for $t \geq 0$ and zero otherwise, determine $v_{\text{out}}(t)$.
(d) If $v_a(t) = 7$ V for $t \geq 0$ and zero otherwise, and $i_b(t) = 7$ A for $t \geq 0$ and zero otherwise, determine $v_{\text{out}}(t)$.
(e) If $v_a(t) = 14t$ V for $t \geq 0$ and zero otherwise, and $i_b(t) = 0$, determine $v_{\text{out}}(t)$.
(*Check*: In random order, 9, 6, 4, 2, 2t V for $t > 0$.)

5. Use superposition to determine the voltage V in the circuit shown in figure P5.5.

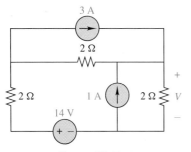

Figure P5.5

ANSWER: 8 V.

6. (a) Use superposition to determine the voltage $v_{\text{out}}(t)$ in terms of $v_1(t)$, $v_2(t)$, $i_3(t)$, and $i_4(t)$ shown in figure P5.6.
(b) If $v_1(t) = 28 \sin(t)$ V, with all other sources set to zero, determine $v_{\text{out}}(t)$.
(c) If $v_1(t) = 14 \cos(t)$ V, with all other sources set to zero, determine $v_{\text{out}}(t)$.
(d) If $v_1(t) = 14 \sin(t) + 28 \cos(t)$ V, with all other sources set to zero, determine $v_{\text{out}}(t)$.

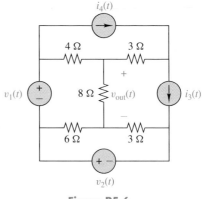

Figure P5.6

Remark: Parts (b)–(d) should be done by inspection; explain.

ANSWER to part (a): $v_{\text{out}} = 4v_1/7 + 8v_2/21 - 16i_3/7 + 16i_4/7$ V.

7. In the circuit shown in figure P5.7, suppose $R_L = 5$ Ω. Determine the value of V_L by superposition.

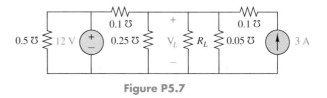

Figure P5.7

ANSWER: 7 V.

8. In the circuit shown in figure P5.8, find v_o by superposition.

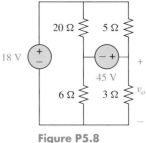

Figure P5.8

ANSWER: 19 V.

9. A linear resistive circuit has two independent sources, as shown in figure P5.9. If $v_a = 10$ V with $i_b = 0$, then $v_o = 5$ V. On the other hand, if $i_b = 10$ A with $v_a = 0$, then $v_o = 1$ V. Find v_o when $v_a = 20$ V and $i_b = 20$ A.

ANSWER: 12 V.

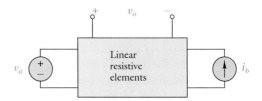

Figure P5.9 The output is determined by the linearity theorem.

10. Consider the circuit of the previous problem. If the measured data are

$$v_o = 15 \text{ V when } v_a = 10 \text{ V and } i_b = 5 \text{ A}$$

and

$$v_o = 10 \text{ V when } v_a = 5 \text{ V and } i_b = 2 \text{ A},$$

find v_o when $v_a = 5$ V and $i_b = 4$ A.

ANSWER: 0 V.

11. Consider the circuit shown in figure P5.11.
  (a) If $v_a = 52$ V and $i_b = 26$ A, determine v_{out} by superposition and loop analysis.
  (b) Observe that if $v_{\text{out}} = Av_a + Bi_b$, then

$$A = \left. \frac{v_{\text{out}}}{v_a} \right|_{i_b=0}.$$

Determine A. What is the equation for determining B? Find B.
  (c) If $v_1 = 26$ V and $i_1 = 39$ A, determine v_{out}.
  (d) Determine the power delivered by the dependent source in part (a).

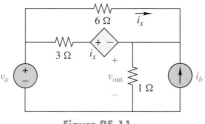

Figure P5.11

ANSWERS: (a) 34 V, (b) 4/13, 9/13, (c) 35 V, (d) −15 W.

12. Consider the linear resistive circuit shown in figure P5.12, which contains only resistors and dependent sources. If $i_{s1}(t) = 2\cos(2t)$ A and $v_{s2}(t) = 0$, then $v_{\text{out}}(t) = 14\cos(2t)$ V. If $v_{s2}(t) = 10\cos(2t)$ V and $i_{s1}(t) = 0$, then $v_{\text{out}}(t) = -5\cos(2t)$ V.
  (a) If $i_{s1}(t) = 2\cos(2t)$ A and $v_{s2}(t) = -10\cos(2t)$ V, find $v_{\text{out}}(t)$.
  (b) If $i_{s1}(t) = -4\cos(2t)$ A and $v_{s2}(t) = 20\cos(2t)$ V, find $v_{\text{out}}(t)$.

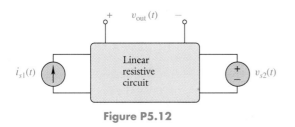

Figure P5.12

13. The shaded box in the circuit shown in figure P5.13 is a linear resistive circuit. If $V_s = 0$, the response is $I_{\text{out}} = 25$ A when $I_s = 5$ A. If $I_s = 0$ and $V_s = -10$ V, the response is $I_{\text{out}} = 10$ A. If $I_s = 1$ A and $V_s = 1$ V, determine I_{out}.

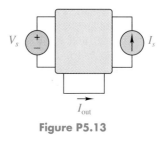

Figure P5.13

ANSWER: 4 A.

14. The linear network in the box in the circuit shown in figure P5.14 contains at most resistors and linear dependent sources. Two separate dc measurements are taken from the circuit. In the first experiment, it is found that when $V_a = 7$ V and $I_b = 3$ A, the load current is $I_{\text{load}} = 1$ A. In the second experiment, it is found that when $V_a = 9$ V and $I_b = 1$ A, the load current is $I_{\text{load}} = 3$ A.
  (a) Determine the linear relationship

$$I_{\text{load}} = AV_a + BI_b$$

between V_a, I_b, and I_{load}.
  (b) Given the equation of part (a), determine I_{load} when $V_a = 15$ V and $I_b = 9$ A.

ANSWER: (b) 0.6 A.

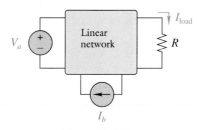

Figure P5.14

15. In the circuit shown in figure P5.15, each of the ten two-terminal elements is a 1-Ω resistor. Find V_{out} if $V_{\text{in}} = 178$ V. (*Hint*: Use the proportionality property and follow example 5.1.) (*Check*: The answer is an integer.)

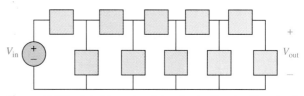

Figure 5.15 A ladder network is solved without writing simultaneous equations.

16. Find V_{out} for the circuit shown in figure P5.16. (*Hint*: Assume $V_{\text{out}} = 4$ V and then follow example 5.1) (*Check*: The answer is an integer.)

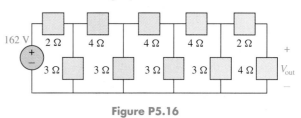

Figure P5.16

17. The circuit shown if figure P5.17 has mirror symmetry about the dashed vertical line.
   (a) If $V_a = V_b = -E$, find V_c by superposition.
   (b) If $V_a = V_b = E$, find I_d by superposition.

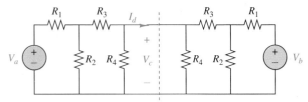

Figure 5.17 Symmetry simplifies the solution of this circuit.

Remark: In both parts, the answers can be obtained without solving simultaneous equations.

18. Consider figure P5.18. By fully utilizing symmetry and superposition, obtain the answers to the following questions without solving simultaneous equations:

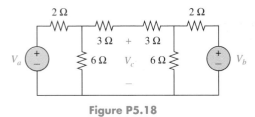

Figure P5.18

   (a) If $V_a = V_b = 16$ V, find V_c.
   (b) If $V_a = 8$ V and $V_b = -8$ V, find V_c.
   (c) If $V_a = 24$ V and $V_b = 8$ V, find V_c.
(*Hint*: For (c), decompose the two inputs as follows:

$$V_a = E_c + E_d, \qquad V_b = E_c - E_d.$$

Then $E_c = 0.5(V_a + V_b)$ and $E_d = 0.5(V_a - V_b)$. Next, find V_c' due to $(V_a = E_c, V_b = E_c)$ and V_c'' due to $(V_a = E_c, V_b = -E_c)$. Finally, by superposition, $V_c = V_c' + V_c''$.)
ANSWERS: (a) 12 V, (b) 0, (c) 12 V.

19. The circuit shown in figure P5.19 is the linear model of a *difference amplifier*. Utilize superposition and symmetry to the fullest extent to solve this problem.

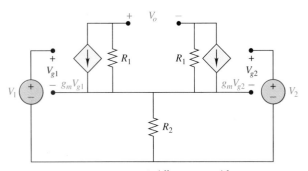

Figure P5.19 A difference amplifier.

   (a) Show that if $V_1 = V_2 = E_c$, then $V_o = 0$.
   (b) Show that if $V_1 = E_d$ and $V_2 = -E_d$, then $V_o = -2g_m R_1 E_d$.
   (c) If $R_1 = 5$ kΩ, $R_2 = 10$ kΩ, $g_m = 2$ mmho, $V_1 = 2$ V, and $V_2 = 4$ V, find V_o.

(*Hint*: Decompose the inputs as suggested in the previous problem.)
ANSWER: (c) 20 V.

20. (a) For the circuit of figure P5.20a, express V_o and V_g in terms of V_i and V_x by superposition.
   (b) For the circuit of figure P5.20b, find the voltage ratio V_o/V_i by the use of equation 5.6 and the results of part (a).

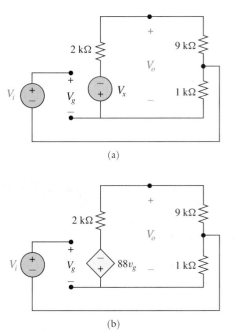

Figure P5.20 (a) A circuit solved by superposition. (b) This circuit is solved by utilizing the results in (a).

ANSWER: (b) -8.8.

21. (a) For the circuit of figure P5.21a, express V_o and V_g in terms of V_i and I_x by superposition.
   (b) For the circuit of figure P5.21b, find the voltage ratio V_o/V_i by the use of equation 5.6 and the results of part (a).

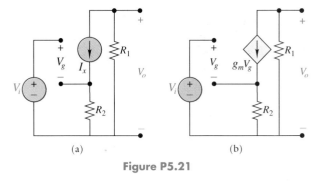

Figure P5.21

ANSWER: (b) $-g_m R_1/(1 + g_m R_2)$.

22. For the circuit shown in figure P5.22, use source transformations as an aid in finding the power in watts absorbed by the 5-Ω resistor.

ANSWER: 0 W.

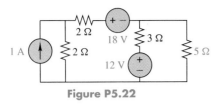

Figure P5.22

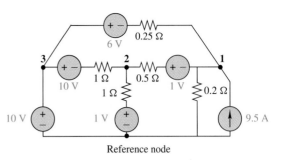

Reference node

Figure P5.26 Source transformations simplify the writing of node equations.

23. (a) Use a series of source transformations to reduce the circuit shown in figure P5.23 to one resistor and two sources. Then find $v_{\text{out}}(t)$.

(b) Determine the total power delivered by the 15-V and 10-A sources.

(c) If both sources have their values increased by a factor of two, determine the new value of $v_{\text{out}}(t)$. Can you do this by inspection? Explain.

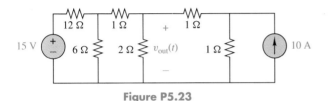

Figure P5.23

ANSWERS: 5 V, 87.5 W, 10 V.

24. For the circuit shown in figure P5.24, use source transformations to determine v_o if $i_o = 2.5$ A.

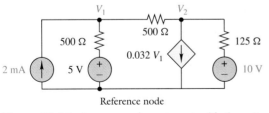

Figure P5.24

ANSWER: 28 V.

25. Apply source transformations to the circuit shown in figure P5.25. Then write two nodal equations to find V_1 and V_2.

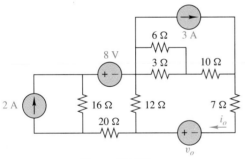

Figure P5.25 Source transformations simplify the writing of node equations.

ANSWERS: 2.8 V, −0.4 V.

26. Apply source transformations to the circuit shown in figure P5.26. Then write two nodal equations to find V_1 and V_2.

ANSWERS: 2.5 V, 2 V.

27. Use a series of source transformations to find the Thévenin equivalent of each of the circuits shown in figure P5.27. In each case, determine the current i_y. (*Hint*: In part (b), there is some extraneous information.)

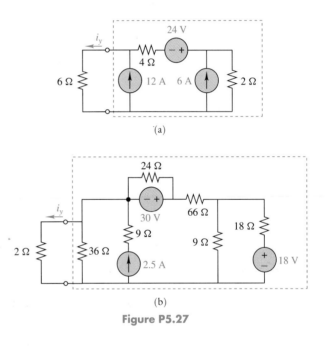

Figure P5.27

ANSWERS: (a) 60 V, 6 Ω, 5 A; (b) 52 V, 24 Ω, 2 A.

28. Find the Thévenin equivalent of the circuit shown in figure P5.28.

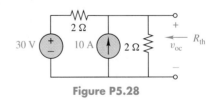

Figure P5.28

ANSWER: $R_{\text{th}} = 1$ Ω, $V_{\text{oc}} = 25$ V.

29. Find the Thévenin equivalent of the network shown in figure P5.29. (*Hint*: Identify the appropriate supernode, and use nodal analysis to find V_{oc}.)

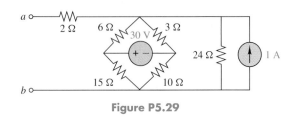

Figure P5.29

ANSWER: $R_{th} = 8 \ \Omega$, $V_{oc} = 4.5$ V.

30. Determine the Thévenin equivalent resistance of the circuit shown in figure P5.30.

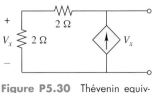

Figure P5.30 Thévenin equivalent resistance is negative for this circuit.

ANSWER: $-4 \ \Omega$.

31. Determine the Norton equivalent circuit of the circuit shown in figure P5.31.

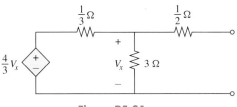

Figure P5.31

ANSWER: -1Ω, $I_{sc} = 0$.

32. (a) Assuming that $k = 2/3$, find the Thévenin and Norton equivalent circuits for the network shown in figure P5.32.
(b) For what value of k is the open-circuit voltage zero? Determine R_{th} for this value of k.

ANSWER: (a) 5/6 V, 95/6 Ω , 1/19 A; (b) $-1/3$, 20 Ω.

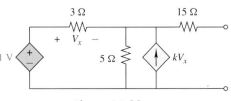

Figure P5.32

33. Use loop analysis to determine the Thévenin equivalent network of the circuit shown in figure P5.33. What is the Norton equivalent?

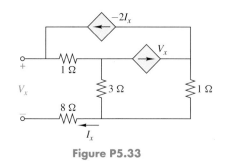

Figure P5.33

ANSWER: $V_{oc} = 0$, $R_{th} = 1 \ \Omega$.

34. Determine the Thévenin equivalent of the circuit shown in figure P5.34.

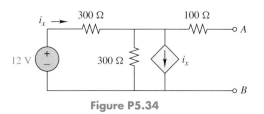

Figure P5.34

ANSWER: $R_{th} = 400 \ \Omega$, $V_{oc} = 0$.

35. Determine the Thévenin equivalent seen at the terminals A–B of the circuit shown in figure P5.35.

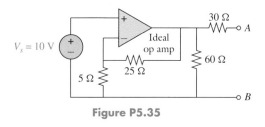

Figure P5.35

ANSWER: 60 V, 30 Ω.

36. Find the Thévenin equivalent circuit at terminals a and b of the network shown in figure P5.36.

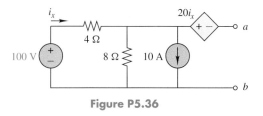

Figure P5.36

ANSWER: 16 Ω, -260 V.

37. The linear circuit shown in figure P5.37a is found experimentally to have the voltage and current relationship given in figure P5.37b. Determine its Norton equivalent.

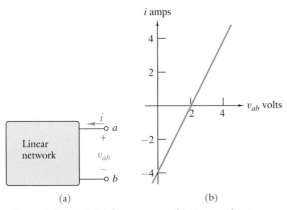

(a)   (b)

Figure P5.37 (a) A linear circuit. (b) Measured v-i curve.

ANSWER: 0.5 Ω, 4 A.

38. Find the Norton equivalent of the circuit shown in figure P5.38. (*Hint*: The equivalent resistance is negative, and $-20 \le I_{sc} \le -10$, where I_{sc} is the short-circuit current that flows across a short circuit from a to b. Recall that the Thévenin equivalent resistance can be computed as $R_{th} = V_{oc}/I_{sc}$.)

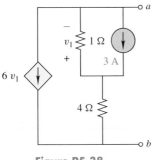

Figure P5.38

ANSWER: 15 A, downward, -1 Ω.

39. Find the Thévenin resistance of the circuit shown in figure P5.39.

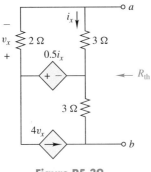

Figure P5.39

ANSWER: $R_{th} = -9$ Ω.

40. Figure P5.40a shows a circuit containing a transistor. For the purpose of determining the dc operating point, the transistor is replaced by an approximate circuit model shown in figure P5.40b. A solution to the operating point is obtained by analyzing the latter circuit.

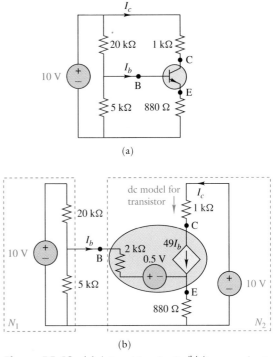

(a)

(b)

Figure P5.40 (a) A transistor circuit. (b) Its approximate linear circuit model.

(a) Find the Thévenin equivalent of the two-terminal network N_1.
(b) Find the Thévenin equivalent of the two-terminal network N_2.
(c) Using the results of (a) and (b), and referring to figure P5.40b, find I_b, I_c, V_{BE}, and V_{CE}.

ANSWERS: (c) 30 μA, 1.47 mA, 0.56 V, 7.21 V.

41. This problem describes a laboratory procedure for determining the Thévenin equivalent circuit of a linear resistive two-terminal network by measurement. The required equipment is a resistance decade box R, a dc voltmeter with meter resistance R_m, and a dc voltage source having internal resistance R_s.

Case 1. The two-terminal network being tested has no internal sources.

Step 1. Set up the circuit as shown in figure P5.41a.

Step 2. Set $R = 0$. Adjust V_s to obtain a reasonable reading of $V_m = E$.

Step 3. Increase R until the reading of V_m drops to $E/2$. Record the value of R.

Step 4. Calculate R_{th} from the relationship

$$R_{th}//R_m = R - R_s.$$

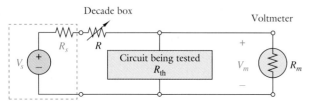

Decade box

Voltmeter

Practical dc voltage source

(a)

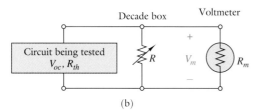

(b)

Figure P5.41   (a) Measurement of R_{th}. (b) Measurement of R_{th} and V_{oc}.

Case 2.   The two-terminal network being tested has internal sources.

Step 1.   Set up the circuit as shown in figure P5.41b.

Step 2.   Open-circuit R. Record the voltmeter reading of $V_m = E$.

Step 3.   Decrease R until the meter reading drops to $E/2$. Record R.

Step 4.   Calculate R_{th} and V_{oc} from the following relationships:

$$R_{\text{th}}//R_m = R,$$

$$V_{\text{oc}} = (1 + R_{\text{th}}/R_m)E.$$

In each case, derive the formulas given in step 4.

42. (a) The Thévenin equivalent circuit of a linear resistive network containing no sources is to be determined experimentally by the setup of case 1 of problem 41. The voltmeter has an input resistance $R_m = 20$ kΩ. The dc voltage source has an internal resistance $R_s = 2$ kΩ. The following readings are obtained:
    (1) With R set to zero, V_s is adjusted until the voltmeter reads 4 V. (Note that it is not necessary to know the actual value of V_s.)
    (2) Without adjusting V_s further, the decade box is adjusted until the meter reads 2 V. At this time, the decade box shows $R = 6$ kΩ.
    Find R_{th}.
ANSWER: 5 kΩ.
    (b) Suppose the measuring instrument has an internal resistance R_m much larger than the Thévenin resistance of the circuit being tested. Simplify the relationships given in step 4 of case 2 of problem 41.

43. (a) The Thévenin equivalent circuit of a linear resistive network containing sources is to be determined experimentally by the setup of case 2 of problem 41. The voltmeter has an input resistance $R_m = 1$ MΩ. The following readings are obtained:
    (1) When R is open-circuited, the voltmeter reads 4 V.

    (2) When R is decreased to 800 kΩ, the voltmeter reads 2 V.
    Find R_{th} and V_{oc}.
ANSWERS: 4 MΩ and 20 V.
    (b) Suppose the measuring instrument has an internal resistance R_m much larger than the Thévenin resistance of the circuit being tested. Simplify the relationships given in step 4 of case 2 of problem 41.

44. For the circuit in figure P5.44, determine the maximum power, in watts, that can be delivered to the load resistor R_L.

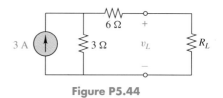

Figure P5.44

ANSWER: 2.25 W.

45. For the circuit shown in figure P5.45, determine the value of the load resistor R_L and the maximum power that can be delivered to R_L.

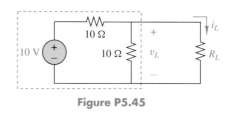

Figure P5.45

ANSWERS: 5 Ω, 1.25 W.

46. In figure P5.46, find the Norton equivalent of the circuit to the left of R_L. Determine the value of R_L required for maximum power transfer. Determine the maximum power that can be absorbed by R_L.
ANSWERS: 3 Ω, 75 W.

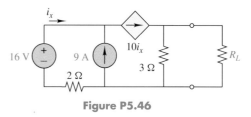

Figure P5.46

47. Consider the following definition: Two n-terminal networks N_1 and N_2 are said to be **equivalent** when, upon substituting one for the other in every possible network N, the voltages and currents in N are unaffected.
    (a) Show that the two three-terminal networks N_1 and N_2 of figure P5.47 are equivalent.

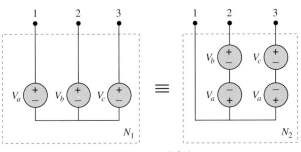

Figure P5.47 E-shift theorem.

(b) A generalization of the property of part (a) is usually called the **E-shift theorem**. Draw a circuit diagram to illustrate the E-shift theorem when both N_1 and N_2 have n accessible terminals.

48. (a) Show that the two three-terminal networks N_1 and N_2 of figure P5.48 are equivalent.

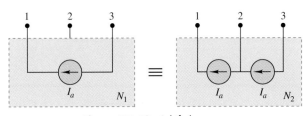

Figure P5.48 I-shift theorem.

(b) A generalization of the property described in part (a) is usually called the **I-shift theorem**. Draw a circuit diagram to illustrate the I-shift theorem when both N_1 and N_2 have n accessible terminals.

49. In the circuit of figure P5.49a, suppose the voltage $v_m(t)$ has been found by some means. For calculating volt-

ages and currents inside N_1, the network N_2 may be replaced by an independent voltage source $v_m(t)$, as shown in figure P5.49b. This property is called the **voltage source substitution theorem.** Justify the theorem. (*Hint*: Show that the equations involving only the currents and voltages of N_1 are the same for both (a) and (b).)

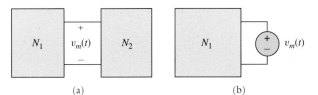

Figure P5.49 (a) Original network. (b) Voltage source substitution.

50. In the circuit of figure P5.50a, suppose the current $i_m(t)$ has been found by some means. For calculating voltages and currents inside N_1, the network N_2 may be replaced by an independent current source $i_m(t)$, as shown in figure P5.50b. This property is called the **current source substitution theorem.** Justify the theorem. (*Hint*: Same as that for problem 49.)

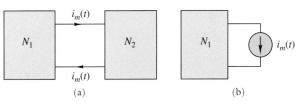

Figure P5.50 (a) Original network. (b) Current source substitution.

CHAPTER 6

CHAPTER OUTLINE

Piecewise Linear Analysis of Diode Circuits

PROTECTION AGAINST POLARITY REVERSAL AND OVERVOLTAGE

MOST of us have had the experience of replacing the batteries in some portable electronic equipment, e.g., a boom box, a tape recorder, or a calculator. Usually, the manufacturer stencils a diagram onto the plastic battery housing showing the correct positioning of the batteries. What happens if one accidentally reverses the polarities during installation? A minor consequence is that the equipment won't operate. However, in some cases, the result can be much more serious. Because many electronic circuits use voltage-sensitive transistors and integrated circuits, reversing the battery's polarities could result in much more serious damage to the circuits.

Even if the polarity of the dc power supply is correct, other electric phenomena can damage the circuits. For example, high-voltage transients due to lightning or due to the switching on and off of other electrical devices may enter the circuits causing a breakdown. Currently, the technology used for making integrated circuits cannot safeguard against such transients. Safeguards are needed to protect expensive electronic equipment from electrical damage.

This chapter discusses some circuits that protect against both voltage reversal and overvoltage (electrical surges). These circuits utilize devices called diodes and Zener diodes. An understanding of the operations of such circuits and several others is made easier through the use of the mathematical model of an ideal diode and the view of an ideal diode as an electrically controlled switch. ◆

◆ CHAPTER OBJECTIVES

- Introduce the ideal diode as a basic element in circuit analysis.
- Present approximate models for practical diodes.
- Describe how polarity reversal and overvoltage (surge) protection are achieved with diode circuits.
- Introduce the Zener diode and describe its properties.
- Describe the design of a Zener diode circuit for maintaining a dc power supply voltage at a constant level.
- Demonstrate the practical use of ideal diodes in the analysis of an op amp comparator circuit, i.e., a circuit that compares the voltage level of an input signal with a given reference voltage and outputs one voltage if the input has a higher voltage than the reference signal and another if the input has a lower voltage.

1. INTRODUCTION

Our daily experience with the lighting switches in a household gives us a feeling for the essential properties of a switch: When closed, an ideal switch behaves like a short circuit; when open, an ideal switch behaves like an open circuit. Several examples and problems in the previous chapters have included switches. There, the *presence of the switches did not alter the essential linear nature of the circuit*. By this we mean that the circuit response calculation was broken down into time intervals defined between the switching moments. During each such time interval (when there was no switching), the solution methods used coincided with those for circuits without switches. These methods included the general node and mesh analysis methods of chapter 4 and the network theorems of chapters 2 and 5.

A device similar to a switch, and a very useful element in electronic circuits, is the *diode*. A practical diode is a two-terminal element possessing a nonlinear *v-i* relationship that is approximately exponential. A deep understanding of the origin of the nonlinear *v-i* relationship requires a knowledge of device physics. An accurate analysis of circuits containing practical diodes presupposes sophisticated numerical techniques. Both backgrounds are beyond the level of this text. Fortunately, a different approach is available. First, we can treat a diode as a black box with a given terminal voltage-current relationship. With this, we are able to derive useful properties of diode circuits and postpone the study of device physics. One should not be alarmed about this approach, because we have already used it in defining a resistor in chapter 1. There, a resistor was defined as an element having the relationship $v(t) = Ri(t)$. We never investigated how the free electrons inside a conductor move about to produce the resistance.

Second, we can avoid the use of numerical solution techniques by seeking only *approximate* answers. The aim here is to achieve a qualitative understanding of circuit operation through basic linear circuit analysis. This approximation is done as follows. Roughly speaking, a diode offers a very small resistance when current flows in one direction, but a very large resistance to current flowing in the opposite direction (hence the name "electronic valve" in some texts). Since approximate answers through simple analysis is our aim, we idealize the diode into having zero resistance for the current in one direction and infinite resistance for the current in the opposite direction. Ideally, the diode behaves as a switch! However, there is one important difference between the diode switch and the mechanical switch used in earlier chapters: The mechanical switch is usually controlled by an *external agent*, whereas the on-off state of a diode switch is controlled by the voltage and current associated with the diode.

As mentioned, the solution of linear resistive circuits containing switches requires only the methods and theorems studied in chapters 1 through 5. Because the voltage across and the current through a diode determine its switching action, a strategy for computing switching instants becomes necessary. Accordingly, we will introduce such a strategy, called the *piecewise linear* (PWL) technique. In conjunction with the linear resistive techniques of chapters 1 through 5, the PWL technique will allow us to compute approximate answers to resistor-diode circuits efficiently.

By viewing a diode as a switch, we are able to understand, explore, and design some simple diode circuits using only linear circuit techniques. More advanced analysis methods and circuit applications are postponed until chapter 12. The physical structure of real diodes used in electronic circuits has evolved with technology—from the early days of vacuum tube diodes to the present-day semiconductor diodes. Although the physical principles governing these devices are entirely different, the mathematical principles of PWL circuit analysis to be discussed in this chapter and in chapter 12 remain unchanged.

2. THE IDEAL DIODE: A BASIC CIRCUIT ELEMENT

From the circuit analysis point of view, a diode is any two-terminal device that allows current to pass much more easily in one direction than the other. In early years (1900 through the 1940s), diodes were constructed most often in the form of vacuum tubes or selenium stacks. Nowadays, diodes are practically all semiconductor devices made of silicon or germanium. The symbol of a semiconductor diode and its typical v-i curve, measured using dc inputs, are shown in figures 6.1a and 6.1b, respectively.

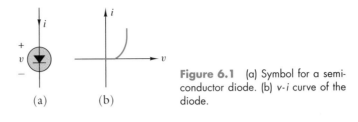

Figure 6.1 (a) Symbol for a semiconductor diode. (b) v-i curve of the diode.

(a)   (b)

If one were to connect the probes of an ohmmeter across the diode, a large reading (in the MΩ range) would appear for one way of connecting the probes, and a very small reading (less than 100 Ω) would appear when the probes are reversed. The darkened arrowhead in figure 6.1a indicates the direction of easiest current flow. In a later course in electronics, one learns that a large portion of the v-i curve can be represented by an exponential equation called *Shockley's equation*. An analysis of a diode circuit based on Shockley's equation would immediately force us to digress to a study of the numerical solution of nonlinear equations, a topic beyond the scope of the text. Rather, our aim is to apply linear analysis methods of previous chapters to the analysis of nonlinear circuits. The trade-off is less accurate answers, but a much simpler analysis. To this end, one approximates the smooth curve of figure 6.1b by a series of straight-line segments. The result constitutes the approximate PWL curve. The approximation can be made as close as possible to the original smooth curve by using more line segments. Our exposition begins with a crude two-segment approximation to figure 6.1b.

A two-terminal device with the idealized v-i curve shown in figure 6.2 is called an **ideal diode**. No standard symbol has been established in the literature for an ideal diode. We shall indicate an ideal diode by putting the word *ideal* beside the diode symbol (as done in figure 6.2a) or by declaring the ideal nature of the diode in the text.

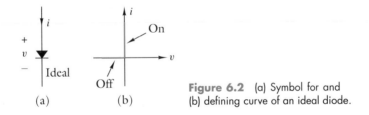

Figure 6.2 (a) Symbol for and (b) defining curve of an ideal diode.

(a)   (b)

With reference to figure 6.2b, when the current i is positive, the diode is said to be **forward biased**. When forward biased, a diode behaves as a short circuit because the diode voltage $v = 0$. When v is negative, the diode is **reverse biased** and behaves as an open circuit because $i = 0$. Thus, an ideal diode has two main states of operation, referred to by various equivalent nomenclatures: (1) forward biased, on, closed, or conducting and (2) reverse biased, off, open, or nonconducting. The on-off and closed-open terminologies emphasize an ideal diode's use as a switching element. It is important also to identify the boundary state bordering the open and closed operating states of the diode. This boundary state is called the **break point**.

It is possible to characterize an ideal diode by the following set of equations:

$$i > 0, v = 0 \text{ (forward biased)}, \tag{6.1a}$$

$$v < 0, i = 0 \text{ (reverse biased)}, \tag{6.1b}$$

$$v = 0, i = 0 \text{ (break point)}. \tag{6.1c}$$

Recall that chapter 1 describes an approximate model of a real dc source (e.g., a battery) as an independent voltage source—an ideal element that does not exist in the real world—in series with a resistance. In like manner, a real diode may be represented *approximately* as an ideal diode in series with and in parallel with other circuit elements. Figures 6.3a, b, and c represent three increasingly more accurate models of a real diode. From the accompanying

PWL v-i curves, we see that a better approximation to the true diode characteristic can be achieved with a more complicated circuit model. This is generally true in all device-modeling problems.

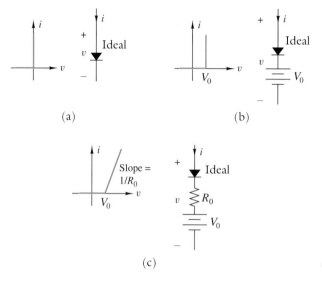

(a)

(b)

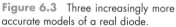

(c)

Figure 6.3 Three increasingly more accurate models of a real diode.

In figures 6.3b and c, the break point voltage V_0 is usually called the **cut-in voltage** or **threshold voltage**. It is, of course, necessary to verify that the circuit models in figure 6.3 indeed have the v-i characteristics shown to their left. We shall do this for figure 6.3c only, as the other two models are actually special cases of figure 6.3c. From the v-i curve of figure 6.3c, it is clear that no current flows through the diode when the diode terminal voltage v is less than the cut-in voltage V_0. This relationship is also evident from the circuit model of figure 6.3c: When the terminal voltage v is less than V_0, the *net* biasing voltage across the ideal diode, $v - V_0$, is negative. Therefore, the ideal diode is reverse biased, and the current i is zero. On the other hand, from the v-i characteristic of figure 6.3c, if the current i is positive, then the v-i relationship is that of a slanted line beginning at $(v, i) = (V_0, 0)$ and having a slope $1/R_0$. Now, is the circuit model consistent with this analysis? In the circuit model, if i is positive, then the ideal diode behaves as a short circuit. Hence, the terminal v-i relationship is that of a dc voltage source V_0 in series with a resistance R_0. Mathematically, $v = R_0 i + V_0$ whenever $i > 0$. (Equivalently, $i = (1/R_0)v - (1/R_0)V_0$ whenever $i > 0$.) This verifies the consistency of the v-i curve with the circuit model.

For most of the circuits analyzed in this text, the v-i curve and model of figure 6.3b yields reasonably accurate results without complicating the mathematical calculations. There are, however, situations wherein the effect of R_0 (figure 6.3c) must be included in order to obtain good agreement between calculated and measured results. Typically, R_0 is less than 100 Ω. The cut-in voltage V_0 averages 0.6 V for silicon diodes and 0.3 V for germanium diodes and depends on the operating temperature of the diode. Roughly speaking, starting at room temperature, 25°C, the value of V_0 increases by 2.3 mV for each 1°C decrease in temperature.

3. SIMPLE CIRCUITS CONTAINING IDEAL DIODES

When a diode is represented by any of the models of figure 6.3, the circuit under consideration is no longer a linear circuit. Each ideal diode, however, can be in only one of three states: on, off, or at the break point. If the ideal diode is on, it behaves as a short circuit. If the ideal diode is off, it behaves as an open circuit. For many practical circuits containing diodes, a simple inspection of the polarity and magnitude of the driving sources will identify the state of each diode. Knowledge of these states then allows the replacement of each diode by a short circuit or an open circuit, as appropriate. The resultant circuit is linear. Accordingly, the techniques studied in the previous chapters apply, making it straightforward to determine the voltages and currents in the circuit. However, because the diode states are determined by inspection, it is always necessary to check and verify that the *assumed* diode states are in fact correct.

The next few examples illustrate the solution technique and some interesting applications of diodes. For more complicated circuits, it may be difficult to determine the diode states by inspection. Chapter 12 presents a systematic method for analyzing such circuits.

 EXAMPLE 6.1

Reverse voltage protection. Suppose the operation of a piece of electronic equipment requires 11 to 12 V dc from a battery. The + and − terminals of the battery must be connected correctly to the equipment. If the polarity is reversed, the equipment will not function. For some delicate instrumentation, a reversal of the polarity may actually damage the internal circuitry of the equipment. The insertion of a diode in the path of the input current, as in figure 6.4, provides protection from incorrect connections. If the battery terminals are connected incorrectly, the diode, being reverse biased, prevents current from flowing to the equipment.

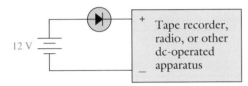

Figure 6.4 Reverse voltage protection with a diode.

Two facts should be kept in mind in this application: (1) When the battery is correctly connected, the available dc voltage to the equipment drops by about 0.7 V, a typical value of V_0 in figure 6.3b, and (2) the diode must have adequate current and power ratings to handle the full current load drawn by the equipment.

 EXAMPLE 6.2

Waveform clipper. Suppose a positive voltage $v_s(t)$, applied to a piece of electronic instrumentation such as a personal computer, has voltage bursts or surges from time to time due to some uncontrollable factors. For example, surges often occur in house current during lightning storms. To protect the internal circuitry from these surges, a circuit is used that will clip off the voltage above a certain value, i.e., the circuit will limit voltage inputs to a maximum level. In principle, the circuit of figure 6.5a may be used.

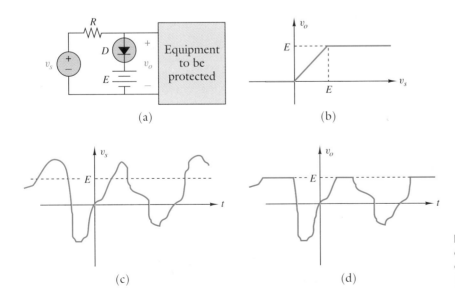

Figure 6.5 (a) A diode amplitude-clipping circuit. (b) Input-output characteristic. (c) Input waveform. (d) Clipped output waveform.

Let us now analyze the operation of the circuit of figure 6.5a, assuming an ideal diode and a very high input resistance of the equipment. When v_s is less than E, the ideal diode is reverse biased, preventing current from flowing through the diode. Thus the series combination of D-E behaves as an open circuit, leading to $v_o = v_s$. When v_s is greater than E, the ideal diode, being forward biased, behaves as a short circuit, leading to $v_o = E$. Figure 6.5b shows the transfer characteristic, v_o vs. v_s. The D-E combination is seen to limit the output to an amplitude of E, but have no effect on the circuit if v_s is less than E. For example, if v_s has the waveform shown in figure 6.5c, then the output will have the waveform shown in figure 6.5d.

◆ **EXAMPLE 6.3**

The circuit of example 6.2 clips the output voltage only at positive values of E. The circuit of figure 6.6a shows how an additional D-E combination provides clipping action in both directions. The operation of the circuit becomes clear if we first obtain the v-i curve, shown in figure 6.6b, for the dashed-line box (see problem 9) and then obtain the v_o vs. v_s transfer characteristic shown in figure 6.6c by reasoning analogous to that in example 6.2. From figure 6.6c the clipping property is easily seen.

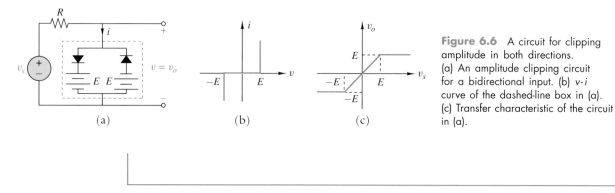

Figure 6.6 A circuit for clipping amplitude in both directions. (a) An amplitude clipping circuit for a bidirectional input. (b) v-i curve of the dashed-line box in (a). (c) Transfer characteristic of the circuit in (a).

The clipping circuit of example 6.3 requires two independent voltage sources. In practice, we use two passive components to replace the whole dashed-line box, as is discussed in the next section.

4. ZENER DIODES: CIRCUIT MODELS AND APPLICATIONS

v-i Characteristic of a Zener Diode

Figure 6.1b shows the v-i curve of a typical semiconductor diode. When the reverse bias voltage exceeds some threshold value, a phenomenon called *electron avalanche* or *breakdown* occurs inside the semiconductor material. In the breakdown region, the magnitude of the current through the diode increases very rapidly. Diodes designed with adequate power dissipation capability to operate in the breakdown region are called **breakdown** or **Zener** diodes. Figures 6.7a and b depict the symbol for a Zener diode and its typical v-i curve, respectively.

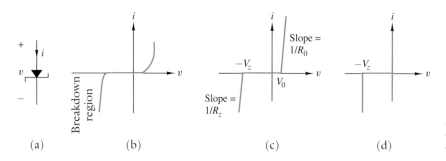

Figure 6.7 (a) Symbol for a Zener diode. (b) v-i curve. (c) PWL approximation. (d) Idealized PWL v-i curve.

CHAPTER 6   PIECEWISE LINEAR ANALYSIS OF DIODE CIRCUITS

Because of the sharp reverse-biased breakdown indicated in figure 6.7b, the diode voltage in the breakdown region is nearly constant. A Zener diode can be designed to operate in its breakdown region as a constant voltage source for control circuitry. The most common applications of Zener diodes include power supply voltage regulation and waveform clipping.

For an approximate analysis of Zener diode circuits, the three-segment piecewise linear approximation shown in figure 6.7c can replace the smooth v-i curve of figure 6.7b. Four parameters completely specify the PWL approximation:

V_z reverse breakdown voltage (or Zener voltage),

R_z reciprocal of slope in breakdown region
(or breakdown resistance),

V_0 forward cut-in voltage,

R_0 reciprocal of slope in forward-biased region.

Commercially available Zener diodes have values of V_z ranging from 2 V to about 200 V. The tolerance of the parameter specifies its variation from the nominal value due to manufacturing irregularities. Typical tolerances for V_z of Zener diodes are 10%, 5%, and 1%, with the lower tolerance unit higher priced. In addition to the tolerance, the power rating is an important parameter. Typical power ratings are 0.5, 1, 2, 5, 10, and 50 W. A unit with a higher power rating costs more. Typical values of R_z are on the order of 5 to 50 Ω.

Circuit Model for a Zener Diode

When highly accurate calculations are not necessary, we can use the ideal Zener diode v-i curve of figure 6.7d to produce a greatly simplified analysis. The idealized v-i curve requires only one characterizing parameter, the Zener voltage V_z. The analysis of Zener diode circuits in this section will proceed by using this idealized characteristic. The v-i curve can also be described by words, by equations, or by a circuit model containing ideal diodes. For example, the circuit model depicted in figure 6.8b assumes that the two diodes, D_r and D_f, are ideal.

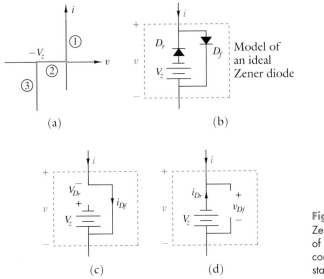

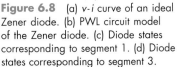

Figure 6.8 (a) v-i curve of an ideal Zener diode. (b) PWL circuit model of the Zener diode. (c) Diode states corresponding to segment 1. (d) Diode states corresponding to segment 3.

There are two different approaches for verifying that the v-i curve of figure 6.8a represents the circuit of figure 6.8b: the *graphical* method and the *algebraic* method. Example 6.5 will illustrate the graphical method. In the algebraic method, one computes all possible combinations of diode states (on or off), and then, for each combination, one reduces the piecewise linear model to a linear model and determines the corresponding v-i relationship. Such an approach is impractical when the circuit contains more than several diodes, but works quite well otherwise. For two ideal diodes, there are four possible combinations of states, each of which is analyzed as follows:

1. *Both diodes off.* Recall the ideal diode characteristics from figure 6.2. For D_f to be off, the condition is $v < 0$. For D_r to be off, the condition is $-V_z - v < 0$, or $v > -V_z$. Therefore, both diodes are off when $-V_z < v < 0$. For this case, the current i is obviously zero, resulting in segment 2 of figure 6.8a.

2. D_r *off and* D_f *on.* Replacing D_r by an open circuit and D_f by a short circuit results in figure 6.8c. For D_f to be on, it is necessary that $i = i_{D_f} > 0$. Under this condition, $v = v_{D_f} = 0$ and $v_{D_r} = -V_z$, indicating that D_r is indeed off. The fact that $v = 0$ for $i > 0$ leads to segment 1 of figure 6.8a.

3. D_r *on and* D_f *off.* Replacing D_r by a short circuit and D_f by an open circuit results in figure 6.8d. For D_r to be on, it is necessary that $i = -i_{D_r} < 0$. Under this condition, $v = v_{D_f} = -V_z$, indicating that D_f is indeed off. The fact that $v = -V_z$ for $i < 0$ leads to segment 3 of figure 6.8a.

4. *Both diodes on.* This combination is not permitted, as it violates KVL. In figure 6.8b, if both diodes are replaced by short circuits, KVL demands that $V_z = 0$, which violates the given condition that $V_z \neq 0$.

This completes the verification of the circuit model for an ideal Zener diode.

Examples and Applications

Zener diodes are very often used to provide a constant dc output voltage against the variations in load current and the dc supply voltage. They are also used to provide very precise reference voltages in measurement circuits. The next few examples illustrate the principles of operation of these circuits.

◆ EXAMPLE 6.4

Figure 6.9a shows a simple voltage regulator circuit. The dc input voltage, V_s, varies between 20 V and 24 V. The load current varies from 0 to 50 mA. Design the circuit such that the output voltage will be constant at 15 V for the stated variations. Specifically, determine V_z, the current-limiting resistance R_1, and the minimum Zener diode power rating.

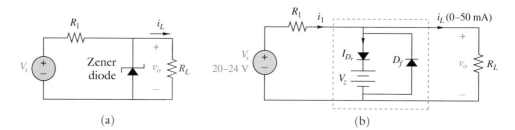

Figure 6.9 Design of a voltage regulator circuit. (a) The Zener diode holds output voltage nearly constant. (b) A PWL circuit model used in the analysis.

SOLUTION

It is obvious that to maintain a constant 15 V output voltage, the Zener diode in figure 6.9a must have $V_z = 15$ V and must operate in the breakdown region. Assuming an ideal Zener diode, we replace it with the model of figure 6.8b. This yields the circuit model of figure 6.9b as the basis for the approximate analysis.

In figure 6.9b, if D_r is on and D_f is off, then the output voltage is $v_o = V_z = 15$ V. To have D_r on, the current i_{D_r} must be positive; i.e.,

$$i_{Dr} = i_1 - i_L > 0$$

for i_L varying from 0 to 50 mA. This in turn requires that i_1 be at least 50 mA for all input source voltages; i.e.,

$$i_1 = \frac{V_s - 15}{R_1} > 0.05 \text{ A.}$$

R_1 must then satisfy

$$R_1 \leq \frac{V_s - 15}{0.05}.$$

For a satisfactory design, one always considers the worst possible case. This happens when $V_s = 20$ V, its smallest value. It follows that

$$R_1 \leq \frac{20 - 15}{0.05} = 100 \ \Omega.$$

Suppose $R_1 = 100 \ \Omega$ is chosen. Then the largest current through the Zener diode occurs when the load current is zero and the input source voltage is a maximum. In this "worst case" situation,

$$(i_{Dr})_{\max} = \frac{24 - 15}{100} = 0.09 \text{ A.}$$

The maximum power to be dissipated in the Zener diode is therefore

$$P_{\max} = 15 \times 0.09 = 1.35 \text{ W.}$$

Since 1.35 W is not a standard component value, one possible conservative design is to use a 2-W Zener diode having a 15-V breakdown voltage and a 100-Ω current-limiting resistor.

The preceding analysis of the idealized circuit shows that the load voltage is constant for all anticipated variations of supply voltage and load current. In reality, some variation in load voltage still exists, for at least two reasons: (1) The v-i curve in the breakdown region is slanted (see figure 6.7b) instead of vertical and (2) the Zener voltage V_z itself varies with temperature. Depending on the operating voltage and current of the Zener diode, the temperature coefficient may be positive or negative. Normally, it will be in the range of ± 1 percent/$^\circ$C. Near room temperature, Zener diodes with V_z in the range of 5 V can have a zero temperature coefficient. Accordingly, Zener diodes can be used to provide very precise reference voltages. (See problem 14.)

> **Exercise.** Repeat the design of example 6.4 if the load variation is from 20 to 100 mA and the input voltage variation is from 18 to 24 V. A constant output voltage of 15 V is desired.

 EXAMPLE 6.5

The bidirectional amplitude-clipping circuit of example 6.3 uses two ideal diodes and two independent voltage sources. Here, we show that these components—enclosed in the dashed-line box of figure 6.6a—can be replaced by two Zener diodes, as shown in figure 6.10a.

SOLUTION

Assuming ideal Zener diodes with $V_z = E$, we may replace the dashed-line box of figure 6.10a by its equivalent circuit, shown in figure 6.10b. It is sufficient to show that the v-i curve of figure 6.10b is given by figure 6.10e, which coincides with that of the dashed-line box of figure 6.6a.

There are four ideal diodes in figure 6.10b. The diode state combination method is impractical for manual solution, because there are $2^4 = 16$ possible cases to examine. The graphical method is advantageous in this situation. The v_1-i curve for the top Zener diode

---

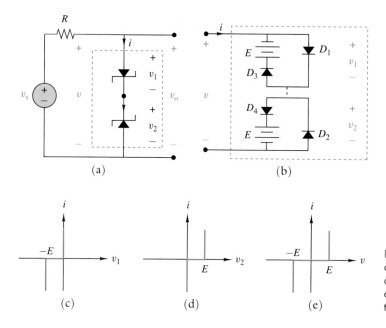

(a)   (b)

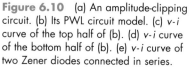

(c)   (d)   (e)

Figure 6.10 (a) An amplitude-clipping circuit. (b) Its PWL circuit model. (c) *v-i* curve of the top half of (b). (d) *v-i* curve of the bottom half of (b). (e) *v-i* curve of two Zener diodes connected in series.

is shown in figure 6.10c (which is the same as figure 6.8a). The v_2-i curve for the bottom Zener diode is shown in figure 6.10d (which is figure 6.8a turned around because of the orientation of the bottom Zener diode). From the series connection of these two Zener diodes, $v = v_1 + v_2$. One obtains the overall v-i curve by adding up v_1 and v_2 at every value of the current i. This addition is most conveniently carried out by arranging the axes of the graphs as shown in figures 6.10c, d, and e. The final result, in figure 6.10e, shows that the two circuits of figures 6.6a and 6.10a are indeed equivalent.

5. Further Application of Ideal Diodes in Circuit Models

As mentioned in chapter 3, the internal circuitry of a real op amp contains more than a dozen transistors and a dozen resistors. If only an approximate analysis of an op amp circuit is desired, we can greatly simplify the task through the use of an equivalent circuit. In fact, this is precisely what is done in chapter 3. The equivalent circuit must *mathematically* simulate the *essential* properties of the real op amp. Such an equivalent circuit is sometimes referred to as a **macro model**. Such models ignore the physics of the materials inside the device. Ideal diodes are again quite useful in developing macro models of devices such as the op amp.

The macro model of an op amp, as shown in figure 3.8a, contains a *nonlinear* controlled voltage source that accounts for the output voltage saturation effect. Using the amplitude-clipping circuit developed earlier, we can now replace the nonlinear controlled source by a linear controlled source and a clipping circuit. Figure 6.11 shows the resultant circuit model. For $|v_o| < V_{sat}$, it is easy to see that both ideal diodes are open, and therefore, $v_o = Av_d$. When the magnitude of the differential input v_d exceeds V_{sat}/A, one of the two ideal diodes will be turned on. This clamps the output voltage v_o to a maximum of V_{sat} or a minimum of $-V_{sat}$. Thus, the model of figure 6.11 has exactly the same transfer characteristic given by the curve of figure 3.6b or by equation 3.2.

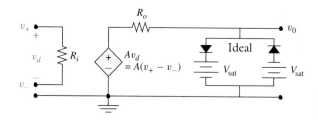

Figure 6.11 A global op amp model including saturation effect.

The next example describes an application of the foregoing op amp model. It also introduces a device known as the op amp comparator with a single power supply. A **comparator** is a device that compares the values of two voltage signals. Typically, the device compares an input signal with a reference signal, producing a constant positive voltage if the input signal exceeds the reference signal and a constant negative voltage if the input signal is smaller than the reference signal.

EXAMPLE 6.6

The op amp circuit of figure 6.12a depicts a comparator. It is powered by a single battery of $2E$ V, which, for the purpose of circuit analysis, has been drawn as a dual power supply of $\pm E$ V, as shown in figure 6.12b.

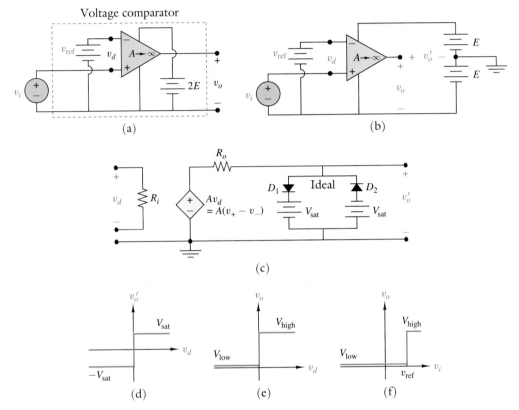

Figure 6.12 (a) A comparator circuit. (b) Dual supply representation. (c) Piecewise linear circuit model. (d) Curve of v_o' vs. v_d. (e) Curve of v_o vs. v_d. (f) Curve of v_o vs. v_i.

Assuming that the op amp has a very high gain and that $V_{\text{sat}} = E - 0.7$ V, find:

(a) the transfer characteristic for v_o' vs. v_d.
(b) the transfer characteristic for v_o vs. v_d.
(c) the transfer characteristic for v_o vs. v_i.

SOLUTION

(a) Despite the assumption of an ideal op amp, the device may not be operating in the active region. Hence, it is necessary to replace the op amp by the global model of figure 6.11. Figure 6.12c exhibits the resultant circuit. For the sake of simplicity, we let v_d be the input.

Since the open-loop gain $A \to \infty$, any positive value of v_d will produce a very large voltage Av_d of such polarity as to forward bias D_1 and reverse bias D_2. This leads to the conclusion that $v_o' = V_{\text{sat}}$ for $v_d > 0$. By a similar reasoning, we conclude that $v_o' = -V_{\text{sat}}$ for $v_d < 0$. These relationships delineate the curve of v_o' vs. v_d shown in figure 6.12d.

(b) From figure 6.12b, $v_o = v_o' + E$. A simple upward shift of the curve in figure 6.12.d yields the desired curve of v_o vs. v_d. Figure 6.12e displays this curve; note that $V_{high} = E + V_{sat}$ and $V_{low} = E - V_{sat}$.

(c) From figure 6.12b, $v_i = v_d + v_{ref}$. A simple right shift of the curve in figure 6.12e yields the desired curve of v_o vs. v_i shown in figure 6.12f.

The transfer characteristic of figure 6.12f demonstrates a very important property. The circuit compares the input voltage v_i with a reference voltage or threshold voltage v_{ref}. Whenever v_i exceeds v_{ref}, the output v_o goes to the high level, V_{high}. Whenever v_i is smaller than v_{ref}, v_o goes low to V_{low}. There are only two possible values for the output voltage: a high value, denoted by V_{high}, and a low value, denoted by V_{low} (or a positive and a negative value). Abrupt changes of the output voltage occur when the input crosses the threshold voltage. Otherwise, the output voltage stays constant, even for time-varying input signals. A circuit with this property is called a **comparator**, or more precisely, a **voltage comparator**.

Comparators have very important applications in digital circuits and instrumentation circuits. Advanced courses explore most of these applications. Our objective here is to learn the basic behavior of the circuit with a simple method of analysis. Two elementary applications of diodes, requiring minimal background, are given as problems. (See problems 14 and 15.)

SUMMARY

This chapter introduced a new basic circuit element called an ideal diode. An ideal diode is equivalent to a short circuit in its "on" state, and to an open circuit in its "off" state. Thus, an ideal diode behaves as a switch. Given a linear resistive circuit containing ideal diodes, once the states of all the diodes are ascertained, the analysis of the circuit requires no more than the methods and theorems studied in chapters 1 through 5. Treating diodes as switches enables us to sidestep studies of device physics and numerical methods and immediately investigate some simple, useful, and interesting circuits.

Through some simple examples, we illustrated how diodes may be used to provide protection against polarity reversal, overvoltage, and waveform clipping.

The chapter explored how a Zener diode may be modeled with ideal diodes and independent voltage sources. Zener diodes may be used to provide a constant voltage from a raw dc power supply. An example illustrated in detail the design of such a circuit to meet specifications on the variations of load current and fluctuation in the raw dc source voltage.

An analysis of an op amp comparator served the dual purposes of showing the use of ideal diodes in an op amp macro model and illustrating the operation of an op amp using a single power supply (as opposed to the circuits of chapter 3, which used a dual power supply).

By using the PWL analysis method, we were able to investigate, analyze, and design very practical circuits that protect our everyday electronic apparatus, help control dc power supply voltage levels, and the like.

TERMS AND CONCEPTS

Breakdown region of Zener diode: region wherein, when the reverse bias voltage applied to a Zener diode reaches a certain threshold value, there is a sharp increase in current. The voltage across the Zener diode remains essentially constant for any further increase in the reverse current. The breakdown region is what makes a Zener diode useful in electronic circuits.

Break point: the state of an ideal diode in which $i = 0$ and $v = 0$.

Comparator: a double-input, single-output circuit with the following properties: The output, V_{out}, has only two possible values, V_{high} and V_{low}; $V_{out} = V_{high}$ when $V_{in1} > V_{in2}$, and $V_{out} = V_{low}$ when $V_{in1} < V_{in2}$.

Cut-in voltage: the voltage V_o in the piecewise linear v-i curve of an ideal diode as shown in figure 6.3c.

Diode: a two-terminal element that offers a small resistance to

current in one direction, but a very large resistance to current in the other direction.

Forward biased: the state of a diode in which the externally applied source sends current through the diode in the direction of easy flow. Alternatively, the diode is said to be on or conducting.

Ideal diode: a two-terminal element that behaves as a short circuit when forward biased and as an open circuit when reverse biased.

Macro model of an op amp: a circuit model using basic circuit elements to simulate the terminal voltage-current relationships of an op amp. A macro model contains a much smaller number of components than does the actual op amp, and the components need not resemble those in the op amp.

Regulated power supply: a dc power supply that uses Zener diodes (or other types of circuitry) to maintain a constant output voltage for changes in load current or the raw dc source voltage.

Reverse biased: the state of a diode in which the externally applied source has polarities opposite to those of the forward-biased case and, hence, extremely small current results. Alternatively, the diode is said to be off or nonconducting.

Voltage-clipping circuit: a circuit employing diodes or Zener diodes to limit the amplitude of the voltage that may appear across the terminals of an electronic equipment.

Zener diode: a special type of semiconductor diode with nearly constant voltage in the reverse breakdown region.

PROBLEMS

1. The circuit of example 6.1 given in figure 6.4 offers protection for equipment against incorrect battery connections. By using three additional diodes as per figure P6.1, show that the battery connection may be reversed, but that the equipment will still receive voltage of the correct polarity.

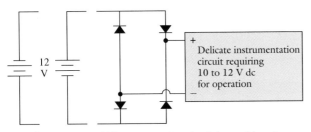

Figure P6.1 Voltage protection circuit for problem 1.

2. Plot the waveform $v(t)$ for the circuit of figure P6.2b for the following two cases:

Case 1: The diode is assumed ideal. (See figure 6.3a.)

Case 2: The diode is represented by the model of figure 6.3c with $V_o = 0.5$ V and $R_o = 50$ Ω.

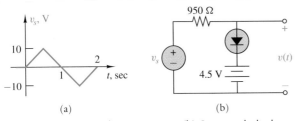

Figure P6.2 (a) Voltage excitation. (b) Circuit with diode.

3. Plot the waveform $v(t)$ for the circuit of figure P6.3 if both diodes are represented by the model of figure 6.3b with cut-in voltage $V_o = 0.5$ V.

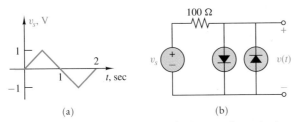

Figure P6.3 (a) Voltage excitation. (b) Circuit with two diodes.

4. The diode circuit shown in figure P6.4 is called a *full-wave rectifier*. Assume that all diodes are ideal with a v-i curve of figure 6.2. Sketch the waveform of $v_R(t)$. What is the peak reverse voltage appearing across each diode?

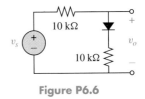

Figure P6.4 (a) Input voltage waveform. (b) A full-wave rectifier.

5. In all of the circuits shown in figure P6.5, the diodes are ideal and $v_s(t) = 10\sin(2\pi t)$ V. Sketch the waveform $v_o(t)$ for each of the four circuits.

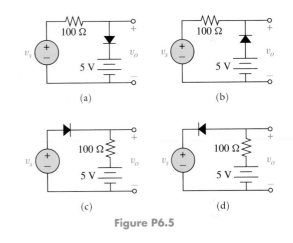

Figure P6.5

6. Assume that the the diode in figure P6.6 is represented by the model of figure 6.3b with $V_o = 0.5$ V. If $v_s(t)$ has the triangular waveform shown in problem 2, sketch the wave of $v_o(t)$.

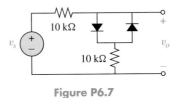

Figure P6.6

7. Repeat problem 6 for the circuit of figure P6.7.

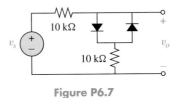

Figure P6.7

8. Repeat problem 6 for the circuit of figure P6.8.

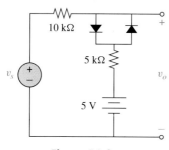

Figure P6.8

9. Show that the v-i curve for the two-terminal network enclosed in dashed lines in figure 6.6a is given by figure 6.6b. Do this by two different methods:

    (a) *Method 1* (graphical). For each of the two series connections between the diode and the source E, find the v-i curve. The two E-Diode branches are connected in parallel. Therefore, the overall v-i curve may be obtained by adding up the currents at every value of the common voltage.

    (b) *Method 2* (algebraic). Refer to the dashed-line box of figure 6.6a. Show, by considering the possible states of the two diodes, that (1) $v = E$ when $i > 0$; (2) $v = -E$ when i< 0; and (3) $i = 0$ when $-E < v < E$.

10. Generalize the design procedure in example 6.4, and show that

$$R_1 \leq \frac{(V_s)_{\min} - v_z}{(i_L)_{\max}}$$

and that the power rating of a Zener diode must be greater than or equal to

$$V_z \left[\frac{(V_s)_{\max} - V_z}{R_1} - (i_L)_{\min} \right].$$

11. A raw dc power supply is represented by a 3-Ω resistance in series with an independent voltage source V_s whose value varies between 20 and 25 V. (See figure P6.11.) The load resistance R_L varies from 100 Ω to open circuit. Zener diodes rated at 1, 2, 5, 10, and 20 W with $V_z = 15$ V are available. Design a voltage regulator circuit such that the voltage across R_L is constant at 15 V. Specifically, do the following:

    (a) Determine the range of the limiting resistor R.

    (b) Determine the largest current through the Zener diode.

    (c) Use the Zener diode with the smallest power rating for the design.

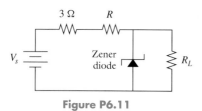

Figure P6.11

12. In example 6.4 and problem 10, the Zener diode is assumed ideal with v-i curve shown in figure 6.7d. In reality, a small amount of current must flow through the diode in order to maintain the breakdown state. This is illustrated in figure P6.12, where the small current is denoted $I_{z,\min}$. The figure is exaggerated for clarity. A typical value for $I_{z,\min}$ is a few milliamperes.

    (a) Consider $I_{z,\min}$, and modify the design formula for R_1 in problem 10.

    (b) Suppose $I_{z,\min} = 10$ mA. Redesign the voltage regulator circuit of Problem 11.

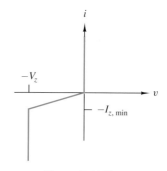

Figure P6.12

13. The speech circuit inside a telephone set uses integrated circuits and requires a minimum of 3.5 V dc (with correct polarity) for proper operation. The telephone central office uses a 48-V dc power supply. The resistance of the telephone line between the central office and a certain subscriber is 800 Ω. Figure P6.13 shows these two parts of the system.

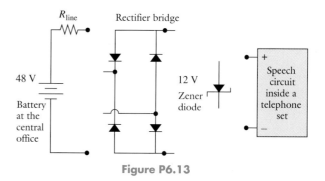

Figure P6.13

    (1) Show how to connect the rectifier bridge and the 12-V Zener diode in order to provide protection against both polarity reversal and overvoltage.

    (2) Assume that each diode has the model of figure 6.3b with $V_o = 0.5$ V and that the Thévenin equivalent of the speech circuit is a 350-Ω resistance. Find the dc voltage across the input terminals of the speech circuit.

    (3) Repeat part (2) if the telephone line between the central office and another, farther-away, subscriber has a resistance of 2,000 Ω.

14. Both of the circuits shown in figure P6.14 provide very accurate output voltage $v_o = V_z$ to be used as the voltage reference in measurement and instrumentation circuits. Compare the advantages and disadvantages of these two circuits.

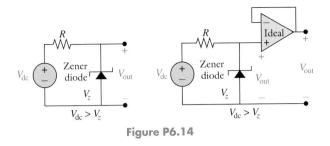

Figure P6.14

TABLE P6.15 PARAMETER INFORMATION

LED	R, Ω	V_z, V	V_{in}
1	220	2	
2	330	5	
3	470	9	
4	560	12	

15. The *light-emitting diode* (LED) is a special diode that emits light when it is forward biased. A typical forward-biased characteristic is shown in figure P6.15a. For any forward current greater than 5 mA, the diode voltage drop is about 1.6 V. LEDs are used in instruments for visual displays. The circuit of figure P6.15b may be used as a voltage level indicator, as is found in a stereo or a tape recorder. Assume that each LED becomes visibly lit at a current of about 10 mA (brighter for a higher current). The LEDs glow in sequence as the input voltage is increased from 0 to 20 V. Calculate the input voltage at which each LED is lit, and complete the last column of the table P6.15.

16. Mathematicians and engineers who do optimization have different, but equivalent, characterizations of devices. Show that the characterization of the ideal diode given in equations 6.1 is equivalent to the following set of three conditions (i.e., show that equations 6.1 imply the following set and the following set implies equations 6.1):

$$i(t) \geq 0,$$

$$v(t) \geq 0,$$

$$v(t)i(t) = 0.$$

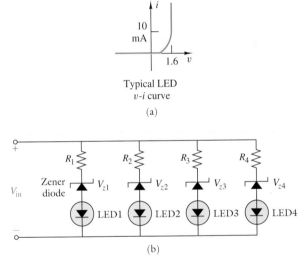

Typical LED
v-i curve

(a)

(b)

Figure P6.15 Visual voltage level indicator using LEDs.

CHAPTER 7

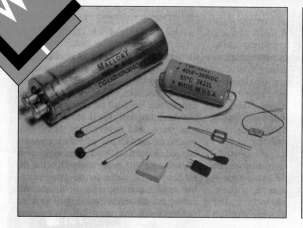

CHAPTER OUTLINE

Inductors, Capacitors, and Duality

CAPACITIVE SMOOTHING IN POWER SUPPLIES

INSIDE every nonportable personal computer is a power supply that converts the sinusoidally varying voltage of the ordinary household outlet to a regulated dc voltage. Regulated means that the dc voltage must stay within very tight limits of its nominal value, e.g., 12 ± 0.1 V. Producing a truly constant dc voltage is practically impossible, so designers create circuits that produce a voltage having a small variation between prespecified limits. The process of converting ac to dc has three main phases:

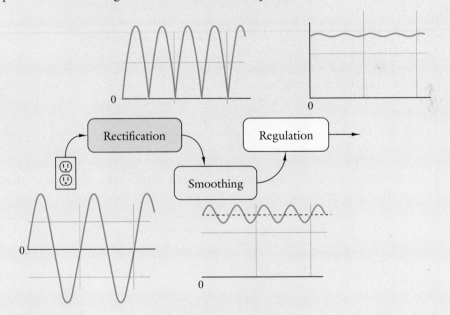

First, the ac waveform is rectified into its absolute value. Then, a smoothing operation takes place that reduces the variation in the voltage to an acceptable level. Finally, the smoothed waveform is fed into a voltage regulator, which clamps the voltage oscillation between critical levels even when the load drawn by different components of the device varies in the course of its operation. The voltage regulator is a precision subcircuit and requires a fairly constant voltage for it to perform its task. Accordingly, the rectified sine wave is smoothed before passing it along to the voltage regulator. A crude smoothing can be accomplished with a capacitor, a device studied in this chapter. Intuitively, capacitors resist voltage changes, attempting to steady their voltage at a constant level. In this chapter, we will study the capacitor and investigate a simplified smoothing operation for a power supply. ◆

◆ CHAPTER OBJECTIVES

- Define the notion of inductance and introduce a new device, called an inductor, in which the voltage developed across it is proportional to the time derivative or time rate of change of the current through it.
- Investigate properties of inductors, including the effect of an initial inductor current, the ability of an inductor to store energy, and the computation of the equivalent inductance of series-parallel connections of inductors.
- Define the notion of capacitance and introduce a new device, called a capacitor, in which the current through it is proportional to the time derivative of the voltage developed across it.
- Investigate the properties of capacitors, including the effect of an initial capacitor voltage, the ability of a capacitor to store energy, and the computation of the equivalent capacitance of series-parallel connections of capacitors.
- Define and illustrate the principle of conservation of charge.
- Explore the dual roles of current and voltage between the capacitor and the inductor, and define and explore the notion of dual circuits.

---

1. INTRODUCTION

This chapter introduces two new circuit elements, the linear inductor and the linear capacitor, hereafter referred to simply as inductor and capacitor. The inductor is a device whose voltage is proportional to the time rate of change of its current with a constant of proportionality L, called the inductance of the device. The unit of inductance is the henry, denoted by H. Macroscopically, inductance measures the magnitude of the voltage induced by a change in the current through the inductor. The capacitor is a device whose current is proportional to the time rate of change of its voltage. Here, the constant of proportionality, C, is the capacitance of the device. Capacitance measures the capacitor's ability to produce a current from the changes in the voltage across it. The unit of capacitance is the farad, F.

By adding the inductor and the capacitor to the previously studied devices (the resistor, independent and dependent sources, etc.), one discovers an entire panorama of possible circuit responses, to be explored in the next four chapters. Together, these devices allow one to design radios, transmitters, televisions, stereos, tape decks, and other electronic equipment. In this chapter, our goal is to understand the basic operation of inductors and capacitors.

2. THE INDUCTOR

Some Physics

Consider figure 7.1. From **Faraday's law**, a changing current flowing through an ideal conductor from points A to B induces a voltage v_{AB} between A and B. The induced voltage is proportional to the rate of change, i.e., the derivative, of the current.

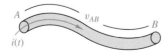

Figure 7.1 A time-varying current flowing through an ideal conductor.

One can experimentally arrive at the following observation. Suppose the conductor in figure 7.1 is 6 feet of #22 copper with resistance 16.5 $\Omega/1,000$ ft. The 6-foot length has a resistance of about 0.1 Ω. Using a current generator, we apply a pair of ramp currents (shown in figure 7.2c) to the conductor, as per figure 7.2a. The measured responses, satisfying Ohm's law, are shown in figure 7.2d.

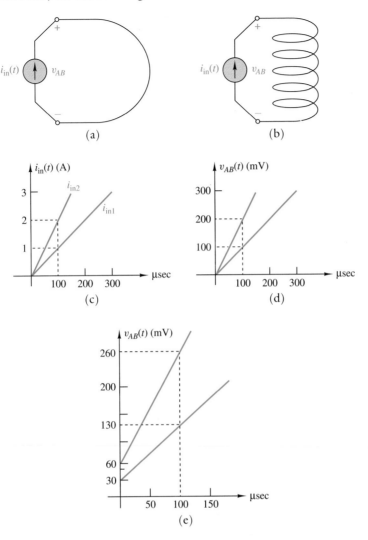

Figure 7.2 (a) Six feet of #22 wire attached to a current generator. (b) Six feet of #22 wire coiled into 45 turns 1" long and 1" in diameter. (c) Ramp current inputs to uncoiled and coiled wire. (d) Voltage responses to ramp current inputs of uncoiled wire. (e) Voltage responses to ramp current inputs of coiled wire.

Now suppose the wire is coiled (45 turns) into a cylinder 1" in diameter and 1" long, as shown in figure 7.2b. Applying the same ramp currents of figure 7.2c produces the measured responses shown in figure 7.2e. These responses are identical in shape to those of figure 7.2d, except for the offsets of 30 mV and 60 mV, respectively. These offset voltages are proportional to the derivatives of the input current ramps. The derivative of i_{in1} is 10^4 A/sec, and the derivative of i_{in2} is 2×10^4 A/sec. The proportionality constant, called the inductance L of the coil, can be computed as 3×10^{-6} henries. The **henry**, abbreviated H, is the unit of inductance. One henry equals 1 volt-sec/amp. This number is consistent with the formula given in problem 16 for the inductance of a coil in air. Thus, one can experimentally measure the voltage induced across a coil by a time-varying current and, from this observation, compute the inductance of the coil. Moreover, from this experiment, one concludes that the inductance of a cylindrical coil of wire is much greater than the inductance of a straight piece of wire.

The physics of the preceding interaction is governed by **Maxwell's equations**, which describe the interaction between electric and magnetic fields. A time-varying current flow creates a time-varying magnetic field, which in turn sets up a time-varying electric field, i.e., an electric potential or voltage. One can verify the presence of this magnetic field by bringing a compass close to a wire carrying a current. The magnetic field surrounding the wire will cause the compass needle to deflect. Physically speaking, a changing current causes a change in the storage of energy in a magnetic field surrounding the conductor.

The energy transferred to the magnetic field requires work and, hence, power. Because power is the product of voltage and current, it follows that there is an induced voltage between the ends of the conductor.

Advanced texts investigate the particulars of induced voltage. For our purposes, three facts are important: (1) Energy storage occurs, (2) the induced voltage is proportional to the derivative of the current, and (3) the constant of proportionality is called the inductance of the coil and is denoted by L.

As discussed, a straight wire has a very small inductance, whereas a cylindrical coil of the same length of wire has a much greater inductance. Moreover, if one puts an iron bar in the center of a cylindrical coil, the inductance increases many times over—possibly several thousand times. The calculation of inductance belongs to more advanced texts, on, e.g., field theory or transmission line theory. Nevertheless, problem 16 shows an empirical formula for estimating the inductance of a single-layer air-core coil. Also, using magnetic circuit concepts, we can calculate the values of various quantities pertaining to some simple iron-core inductors. (See volume II.)

Basic Definition and Examples

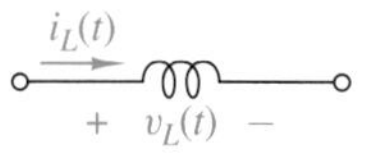

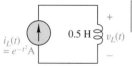

Figure 7.3 The inductor and its differential voltage-current relationship.

DEFINITION OF THE LINEAR INDUCTOR

The linear **inductor**, symbolized by a coiled wire, shown in figure 7.3, is a two-terminal energy storage device whose voltage is proportional to the derivative of the current passing through it. The constant of proportionality, denoted by L, has the unit of **henry** (H), equal to 1 volt-sec/amp. L is said to be the **inductance** of the coil. The specific voltage-current relationship of the linear inductor is given by

$$v_L(t) = L \frac{di_L(t)}{dt}. \tag{7.1}$$

In equation 7.1, we have defined the inductance of a two-terminal device strictly from its terminal voltage-current relationship. However, the value of L also depends on the physical configuration of the device. A segment of wire, such as that shown in figure 7.1, has a negligible inductance. Twisting the wire into a coil greatly increases its inductance. Because coiling a wire increases its inductance, one often calls an inductor a **coil**, which gave rise to the symbol of figure 7.3.

◆ **EXAMPLE 7.1**

Determine $v_L(t)$ for the inductor circuit of figure 7.4 when $i_L(t) = e^{-t^2}$ A.

Figure 7.4 Simple inductor driven by a current source.

SOLUTION

A direct differentiation of the inductor current $i_L(t)$, according to equation 7.1, leads to

$$v_L(t) = 0.5 \frac{d\left(e^{-t^2}\right)}{dt} = 0.5(-2t)e^{-t^2} = -te^{-t^2} \text{ V}.$$

Exercise. In example 7.1, suppose $i_L(t) = \sin(20t + \pi/3)$. Determine $v_L(t)$.
ANSWER: $10\cos(20t + \pi/3)$.

The differential equation 7.1 has a *dual* integral relationship. Safely supposing that at $t = -\infty$, the inductor had not yet been manufactured, one can take $i_L(-\infty) = 0$, in which case

$$i_L(t) = \frac{1}{L} \int_{-\infty}^{t} v_L(\tau)d\tau = \frac{1}{L} \int_{-\infty}^{t_0} v_L(\tau)d\tau + \frac{1}{L} \int_{t_0}^{t} v_L(\tau) \, d\tau$$

$$= i_L(t_0) + \frac{1}{L} \int_{t_0}^{t} v_L(\tau)d\tau. \tag{7.2}$$

The time t_0 represents an initial time that is of interest or significance, e.g., a time when a switch is thrown or a source excitation is activated. The quantity

$$i_L(t_0) = \frac{1}{L} \int_{-\infty}^{t_0} v_L(\tau)d\tau$$

specifies the initial current flowing through the inductor at t_0. This quantity sums up the entire past history of the voltage excitation across the inductor; i.e., the initial current $i_L(t_0)$ embodies the complete history of the operation of the inductor. Because of this, the inductor is said to have **memory**.

 EXAMPLE 7.2

For the circuit of figure 7.5a, determine $i_L(0)$ and $i_L(t)$ for $t \geq 0$ when

$$v_L(t) = \begin{cases} e^t \text{ V} & \text{for } t < 0 \\ e^{-t} \text{ V} & \text{for } t \geq 0 \end{cases}.$$

For reference, figure 7.5b shows a plot of $v_L(t)$.

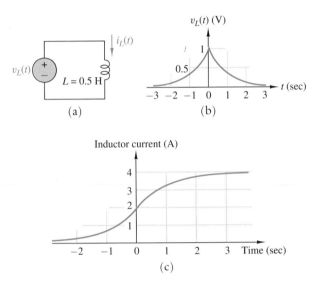

Figure 7.5 (a) Simple inductor driven by a voltage source. (b) Source waveform $v_L(t)$. (c) Resulting inductor current $i_L(t)$.

SOLUTION

A direct application of equation 7.2 leads to

$$i_L(t) = \frac{1}{L} \int_{-\infty}^{0} e^\tau d\tau + \frac{1}{L} \int_{0}^{t} e^{-\tau} d\tau = \frac{1}{L} + \frac{1}{L} \left[1 - e^{-t}\right] \text{ A}.$$

It follows that

$$i_L(0) = \frac{1}{L} = 2 \text{ A and } i_L(t) = \frac{1}{L} \left[2 - e^{-t}\right] = 4 - 2e^{-t} \text{ A}.$$

The graph of $i_L(t)$ for all t is given in figure 7.5c.

Exercise. In example 7.2, compute an expression for $i_L(t)$ for $t \leq 0$.
ANSWER: $i_L(t) = 2e^t$ A for $t \leq 0$.

Exercise. Repeat example 7.2 with $L = 1/4\pi$ H when $v_L(t) = \cos(2\pi t)$ V for $t \geq -0.25$ sec and zero otherwise.
ANSWER: $i_L(0) = 2$ A, $i_L(t) = 2 + \sin(2\pi t)$ A for $t \geq 0$.

◆ **EXAMPLE 7.3**

Consider again the circuit of figure 7.5a. Suppose figure 7.6 now depicts the source waveform $v_L(t)$ in volts. Determine the inductor current $i_L(t)$ for $t \geq 0$, assuming that $i_L(0) = 0$.

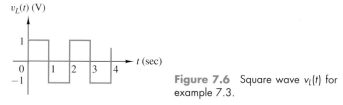

Figure 7.6 Square wave $v_L(t)$ for example 7.3.

SOLUTION

It is necessary to apply equation 7.2 to each interval, $[0,1], [1,2], \ldots, [n, n+1], \ldots$. If n is even, then the value of the inductor current over the interval $[n, n+1]$, i.e., $n \leq t \leq n+1$, is

$$i_L(t) = i_L(n) + \frac{1}{L} \int_n^t d\tau = i_L(n) + 2(t - n) = 2(t - n) \text{ A,}$$

since $i_L(n) = 0$ for all even values of n. Observe that $i_L(t) = 2t - 2n$ A is the equation of a straight line having slope 2 and y-intercept $-2n$. On the other hand, if n is odd, then for the interval $[n, n+1]$, the inductor current is

$$i_L(t) = i_L(n) - \frac{1}{L} \int_n^t d\tau = i_L(n) - 2(t - n) = 2 - 2(t - n) \text{ A.}$$

Hence, $i_L(n) = 2$ A for all odd values of n. Here, $i_L(t) = -2t + 2 + 2n$ A is again the equation of a straight line, but with slope -2 and y-intercept $2 + 2n$. Figure 7.7 sketches the response for $t \geq 0$.

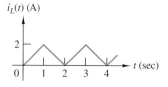

Figure 7.7 Triangular shape of inductor current for the square wave voltage excitation of figure 7.6 applied to the circuit of figure 7.5a.

It is important to recognize that the square wave voltage input of figure 7.6 is a discontinuous function, but that the current waveform of figure 7.7 is continuous. Integration is a smoothing operation: It smooths simple discontinuities. This means that the inductor current is a continuous function of t, even for discontinuous inductor voltages, provided that they are bounded. A voltage or current is **bounded** if the absolute value of the excitation remains smaller than some fixed finite constant for all time.

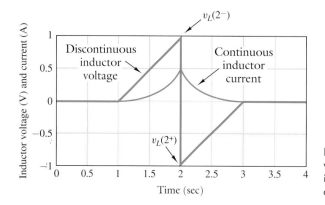

Figure 7.8 A possible discontinuous voltage $v_L(t)$ appearing across an inductor of 1 H. (b) The resulting continuous inductor current.

Equation 7.2 leads to the **continuity property of the inductor**: If the voltage $v_L(t)$ across an inductor is bounded over the time interval $t_1 \leq t \leq t_2$, then the current through the inductor is continuous for $t_1 < t < t_2$. In particular, if $t_1 < t_0 < t_2$, then $i_L(t_0^-) = i_L(t_0^+)$, even when $v_L(t_0^-) \neq v_L(t_0^+)$. The notation "$-$" and "$+$" on t_0 is used to distinguish the moment immediately before $t_0 = 2$ and immediately after $t_0 = 2$, as illustrated in figure 7.8. As in the figure, $t_0 = 2$ is a point of discontinuity of the function $v_L(t)$. The value $v_L(2^+)$ can be seen as the limiting value of $v_L(t)$ when approaching 2 from the right, whereas $v_L(2^-)$ can be seen as the limiting value of $v_L(t)$ when approaching t_0 from the left.

Power and Energy

Recall that the instantaneous power of a device is the product of the voltage across and the current through the device. For an inductor,

$$p_L(t) = v_L(t)i_L(t) = Li_L(t)\frac{di_L(t)}{dt} \text{ watts,}$$

where $v_L(t)$ is in volts, $i_L(t)$ in amps, and L in henries.

Since energy (absorbed or delivered) is the integral of the instantaneous power over a given time interval, it follows that the net energy $W(t_0, t_1)$ stored in the magnetic field surrounding the inductor over the interval $[t_0, t_1]$ is

$$W_L(t_0, t_1) = L\int_{t_0}^{t_1}\left(i_L(\tau)\frac{di_L(\tau)}{d\tau}\right)d\tau = L\int_{i_L(t_0)}^{i_L(t_1)} i_L \, di_L$$

$$= \frac{1}{2}L\left[i_L^2(t_1) - i_L^2(t_0)\right] \text{ joules,} \tag{7.3}$$

for L in henries and i_L in amps. From equation 7.3, the energy stored in the inductor over the interval $[t_0, t_1]$ depends only on the value of the inductor current at times t_1 and t_0, i.e., on $i_L(t_1)$ and $i_L(t_0)$. This means that the stored energy is independent of the particular current waveform between t_0 and t_1. If the current waveform is periodic, i.e., if $i_L(t) = i_L(t+T)$ for some $T > 0$ and all t, then over any period of length T, the value of the stored energy in the inductor is zero because $i_L(t_0+T) = i_L(t_1) = i_L(t_0)$ forces equation 7.3 to zero. One can illustrate this property by exciting a 0.1-H inductor by a periodic current $i_L(t) = \sin(2\pi t)$ A, as shown in figure 7.9a. The period of the current waveform is $T = 2\pi/2\pi = 1$. The voltage across the inductor is $v_L(t) = 0.2\pi\cos(2\pi t)$ A. The instantaneous power is $p_L(t) = 0.2\pi\cos(2\pi t)\sin(2\pi t)$ watts, which is plotted in figure 7.9b. It follows that

$$W_L(0, 1) = \int_0^1 p_L(t)dt = \frac{1}{2}Li_L^2(1) - \frac{1}{2}Li_L^2(0) = 0.$$

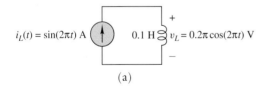

$i_L(t) = \sin(2\pi t)$ A   0.1 H $v_L = 0.2\pi\cos(2\pi t)$ V

(a)

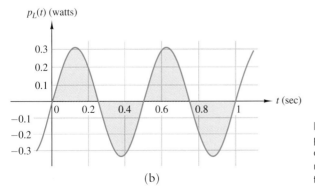

(b)

Figure 7.9 (a) Inductor excited by periodic voltage. (b) Plot of the power absorbed by the inductor. Shaded areas represent stored energy for each part of the cycle.

Since the energy over a period is zero, the area under the power curve must have equal parts of positive and negative area as illustrated by the shaded parts of figure 7.9b. This means that all the energy stored by the inductor over the part of the cycle of positive power is delivered back to the circuit over the portion of the cycle when the power is negative. This is true for all periodic signals over any period. Because no energy is dissipated, and because energy is only stored and returned to the circuit, the inductor is said to be a **lossless** device. It is convenient to define the instantaneous energy stored in an inductor as

$$W_L(t) = \frac{1}{2} L i_L^2(t) \tag{7.4}$$

for all t. Equation 7.4 can be viewed as a special case of equation 7.3 in which $t_0 = -\infty$ where one implicitly assumes that $i_L(-\infty) = 0$. Thus, equation 7.4 can be interpreted as the change in stored energy in the inductor over the interval $(-\infty, t]$.

◆ EXAMPLE 7.4

Find the instantaneous energy stored in each inductor of the circuit of figure 7.10a for the source waveform given in figure 7.10b. In figure 7.10b, observe that $i_s(t) = 0$ for $t \le 0$.

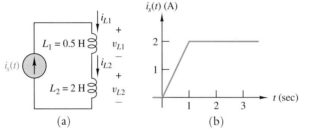

Figure 7.10 (a) Series inductors excited by a source current. (b) Graph of the source current.

SOLUTION

From KCL, $i_s(t) = i_{L1}(t) = i_{L2}(t)$ for all t. Since $i_s(t) = 2t$ A for $0 \le t \le 1$ and $i_s(t) = 2$ A for $t \ge 1$, equation 7.3 or 7.4 immediately yields the instantaneous stored energies, as plotted in figure 7.11.

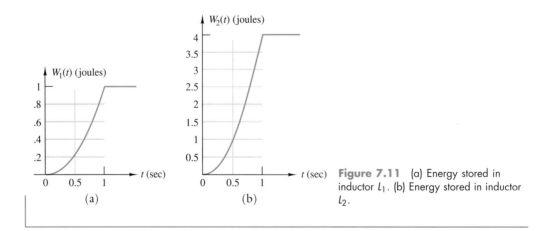

Figure 7.11 (a) Energy stored in inductor L_1. (b) Energy stored in inductor L_2.

Exercise. In example 7.4, compute analytical expressions for $W_1(t)$ and $W_2(t)$, which must be consistent with figure 7.11.

ANSWERS: $W_1(t) = t^2$ for $0 \le t \le 1$ and $W_1(t) = 1$ for $t \ge 1$; $W_2(t) = 4t^2$ for $0 \le t \le 1$ and $W_2(t) = 4$ for $t \ge 1$.

CHAPTER 7   INDUCTORS, CAPACITORS, AND DUALITY

 **EXAMPLE 7.5**

Find the input current $i(t)$ for $t \geq 0$ and the energy stored in each of the inductors for the intervals $0 \leq t \leq 1$ and $1 \leq t$ for the circuit of figure 7.12, assuming that $i_1(0) = 0$.

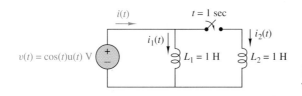

Figure 7.12 Parallel inductive circuit with switch.

SOLUTION

Step 1. Equation 7.2 implies that, for $0 \leq t < 1$,

$$i(t) = \frac{1}{L_1} \int_0^t v_1(\tau)d\tau = \int_0^t \cos(\tau)d\tau = \sin(t) \text{ A}.$$

Step 2. At $t = 1$, the switch closes. The two inductors are then in parallel, and the source voltage appears across each. Hence, the current through each inductor is again given by equation 7.2, i.e.,

$$i_1(t) = i_1(1) + \int_1^t \cos(\tau)d\tau = \sin(1) + \sin(t) - \sin(1) = \sin(t) \text{ A}$$

and

$$i_2(t) = i_2(1) + \int_1^t \cos(\tau)d\tau = \sin(t) - \sin(1) \text{ A}.$$

From KCL, the input current $i(t) = i_1(t) + i_2(t) = 2\sin(t) - \sin(1)$ A for $t \geq 1$. Note that at $t = 1$, the inductor current is continuous.

Step 3. *Compute the energy stored in the inductor over the interval $0 \leq t \leq 1$.* From equation 7.3, it follows that for $0 \leq t \leq 1$, $W_{L1}(0, t) = 0.5 \sin^2(t)$ joules, whereas $W_{L2}(0, t) = 0$.

Step 4. *Compute the energy stored in the inductor over the interval $1 \leq t$.* Again from equation 7.3, it follows that for $1 \leq t$, $W_{L1}(0, t) = 0.5 \sin^2(t)$ joules and $W_{L2}(0, t) = 0.5[\sin^2(t) - 2\sin(1)\sin(t) + \sin^2(1)]$ joules.

3. THE CAPACITOR

Definitions and Properties

DEFINITION OF THE CAPACITOR

Like the inductor, the capacitor, denoted by figure 7.13a, is an energy storage device. For the capacitor,

$$i_C(t) = C\frac{dv_C(t)}{dt}, \tag{7.5}$$

where C denotes capacitance, with unit of farads (F). One farad equals 1 amp-sec/volt. The capacitance C is a measure of the capacitor's capacity to store energy in an electric field. Physically, one can think of a capacitor as two metal plates separated by some insulating material (called a dielectric) such as air. A voltage, placed across the plates of the capacitor, will generate an electric field between the plates. This electric field stores energy. Figure 7.13b illustrates the idea. Modern-day capacitors take on all sorts of shapes and sizes. In keeping with tradition, the parallel-plate concept remains the fundamental perspective.

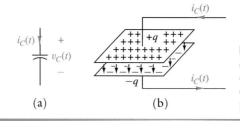

Figure 7.13 (a) The symbol for the capacitor with conventional voltage and current directions. (b) Illustration of electric field between plates of a parallel-plate capacitor.

(a) (b)

Calculating the capacitance of two arbitrarily shaped conducting surfaces separated by a dielectric is, in general, very difficult. Fortunately, the ordinary capacitor of a practical circuit is of the parallel-plate variety, with the plates separated by a thin dielectric. The two plates are often rolled into a tubular form, whence the complete structure is sealed. Problem 17 gives an approximate formula for calculating parallel-plate capacitance.

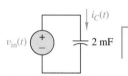

Figure 7.14
Capacitor in parallel with a voltage source.

◆ **EXAMPLE 7.6**

Consider the capacitor circuit of figure 7.14. Determine the capacitor current when $v_{in}(t) = e^{-500t} \sin(1,000t)$ V for $t > 0$.

SOLUTION

A direct application of equation 7.5 yields

$$i_C(t) = C\frac{dv_C(t)}{dt} = -e^{-500t}\sin(1,000t) + 2e^{-500t}\cos(1,000t)\text{A}.$$

Exercise. In figure 7.14, suppose $v_{in}(t) = \exp(-0.25t^4)$ V. Draw the circuit, identify the current direction of $i_C(t)$, and determine the value of $i_C(t)$. Sketch $i_C(t)$ for $-5 \le t \le 5$.

The differential relationship of equation 7.5 has the equivalent integral form

$$v_C(t) = \frac{1}{C}\int_{-\infty}^{t} i_C(\tau)d\tau = \frac{1}{C}\int_{-\infty}^{t_0} i_C(\tau)d\tau + \frac{1}{C}\int_{t_0}^{t} i_C(\tau)d\tau$$

$$= v_C(t_0) + \frac{1}{C}\int_{t_0}^{t} i_C(\tau)d\tau,$$

(7.6)

where v_C is in volts, i_C is in amps, and C is in farads and where we have taken $v_C(-\infty) = 0$ because the capacitor was not manufactured at $t = -\infty$. The time t_0 represents an initial time of interest or significance, e.g., the time when the capacitor is first used in a circuit. The quantity

$$v_C(t_0) = \frac{1}{C}\int_{-\infty}^{t_0} i_C(\tau)d\tau$$

specifies the initial voltage across the capacitor at t_0. This initial voltage sums up the entire past history of the current excitation into the capacitor; i.e., the initial voltage $v_C(t_0)$ embodies the complete history of the operation of the capacitor. Because of this, the capacitor, like the inductor, is said to have **memory**.

Exercise. Consider the capacitor circuit of figure 7.15. Suppose the current source is $i_s(t) = e^{-t}u(t)$ A and $v_C(0) = 1$ V. Compute the capacitor voltage $v_C(t)$, the resistor voltage $v_R(t)$, and the voltage $v_s(t)$ across the current source.

ANSWERS: $v_C(t) = 3 - 2e^{-t}$ V for $t \geq 0$, $v_R(t) = 2e^{-t}$ V for $t \geq 0$, and, by KVL, $v_s(t) = 3$ V for $t \geq 0$.

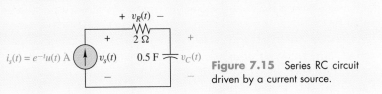

Figure 7.15 Series RC circuit driven by a current source.

EXAMPLE 7.7

Suppose a current source with (sawtooth) current given in figure 7.16b drives a relaxed 0.5F capacitor (zero initial voltage) as illustrated in figure 7.16a. Determine and plot the voltage across the capacitor.

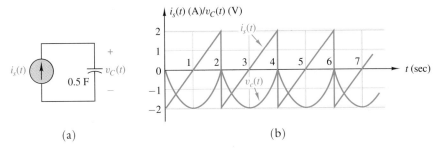

(a)   (b)

Figure 7.16 (a) Current source driving a capacitor. (b) Sawtooth current waveform and voltage response of a 0.5-F capacitor.

SOLUTION

The input waveform is periodic in that it repeats itself every 2 sec. Therefore, the solution will proceed on a segment-by-segment basis. For $0 \leq t < 2$, $i_s(t) = (2t - 2)$ A. With $v_C(0) = 0$, it follows from equation 7.6 that for $0 \leq t < 2$,

$$v_C(t) = \frac{1}{C} \int_0^t (2\tau - 2)d\tau = 2\left(t^2 - 2t\right) \text{ V.}$$

This can be written as $v_C(t) = 2(t^2 - 2t)u(t)u(2 - t)$ V. Observe that at $t = 2$, $v_C(2) = 0$; hence, the capacitor voltage over the interval $2 \leq t < 4$ is simply a right-shifted version of the voltage over the first interval. Right-shifting is achieved by replacing t by $t - 2$. In other words, for $2 \leq t < 4$, $v_C(t) = 2[(t - 2)^2 - 2(t - 2)]u(t - 2)u(4 - t)$ V. In general, for the interval $2k \leq t \leq 2(k + 1)$, $v_C(t) = 2[(t - 2k)^2 - 2(t - 2k)]u(t - 2k)u(2k + 2 - t)$, $k = 0, 1, 2, \ldots$. Lastly, observe that the voltage across the capacitor, as illustrated in figure 7.16b, is continuous despite the discontinuity of the capacitor current. Again, this follows because the capacitor voltage is the integral of the capacitor current supplied by the source.

To repeat, it is important to recognize that the sawtooth current input depicted in figure 7.16 is a discontinuous function, but that the associated voltage waveform is continuous. Again, integration is a smoothing operation. This means that the capacitor voltage is a continuous function of t even for discontinuous capacitor currents, provided they are bounded. This observation stems from equation 7.6 and leads to the **continuity property of the capacitor**: If the current $i_C(t)$ through a capacitor is bounded over the time interval $t_1 \leq t \leq t_2$, then the voltage across the capacitor is continuous for $t_1 < t < t_2$. In particular, if $t_1 < t_0 < t_2$, then $v_C(t_0^-) = v_C(t_0^+)$, even when $i_C(t_0^-) \neq i_C(t_0^+)$.

There appear to be some exceptions to the continuity property of v_C, e.g., when two charged capacitors or one charged and one uncharged capacitor are connected in parallel. In such cases, KVL takes precedence and will force an instantaneous equality in the capacitor voltages, subject to the principle of conservation of charge to be discussed shortly. Another example is when capacitors and some independent voltage sources form a loop. When any of the voltage sources has an instantaneous jump, so will the capacitor voltages. Upon closer examination, however, we see that there is really no exception to the stated continuity rule: It can be shown that in all of the cases where the capacitor voltage jumps instantaneously, an "impulse" current flows through the circuit. The current is not bounded, and consequently, the capacitor voltage may jump instantaneously. This jump does not violate the rule, which presumes that the current is bounded.

Relationship of Charge to Capacitor Voltage and Current

We have defined the capacitance of a two-terminal device strictly from its terminal voltage-current relationship—the differential equation 7.5 and the integral equation 7.6. A physical interpretation of equation 7.6 will bring out another basic property of a capacitor. The integral of $i_C(t)$ over the interval (t_0, t) represents the amount of charge passing through the top wire in figure 7.13b during that interval. Because of the insulating material (called a *dielectric*), this charge cannot pass through to the other plate. Instead, a charge of $+q$ is stored on the top plate, as shown in figure 7.13b. By KCL, if $i_C(t)$ flows into the top plate, then $-i_C(t)$ must flow into the bottom plate. This causes a charge of $-q$ to be deposited on the bottom plate. The positive and negative charges on these two plates, separated by the dielectric, produce a voltage drop v_C from the top plate to the bottom plate. For a linear capacitor, the only type studied in this text, the value of v_C is proportional to the charge q. The proportionality constant is the capacitance of the device. In other words,

$$q(t) = Cv_C(t). \tag{7.7}$$

Thus, equation 7.6 has the following physical interpretation: The first term, $v_C(t_0)$, is the capacitor voltage at t_0; the integral in the second term represents the additional charge transferred to the capacitor during the interval (t_0, t). Dividing this value by C gives the additional voltage attained by the capacitor during (t_0, t). Therefore, the sum of these two terms, i.e., equation 7.6, is the voltage of the capacitor at t. The interpretation of equation 7.5 is even simpler: Since $q(t) = Cv_C(t)$,

$$i_C(t) = \frac{dq(t)}{dt} = C\frac{dv_C(t)}{dt}. \tag{7.8}$$

The Principle of Conservation of Charge

It is important in terms of modern trends in circuit applications to investigate further the relationship of charge to capacitor voltages and currents. The principle of **conservation of charge** says that *the total charge transferred into a junction (or out of a junction) is zero*. This is a direct consequence of KCL. To see it, consider the junction indicated in figure 7.17 where four capacitors meet. By KCL,

$$i_1(t) + i_2(t) + i_3(t) + i_4(t) = 0.$$

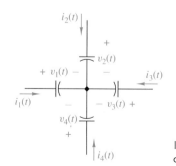

Figure 7.17 Junction of four capacitors.

Since the integral of current with respect to time represents charge, the integral of this equation over the interval $(-\infty, t]$ is

$$\int_{-\infty}^{t} \left(i_1(\tau) + i_2(\tau) + i_3(\tau) + i_4(\tau) \right) d\tau = q_1(t) + q_2(t) + q_3(t) + q_4(t) = 0, \qquad (7.9)$$

where $q_k(t)$ is the charge transferred to capacitor k. By equation 7.6, at every instant of time,

$$q_i(t) = C_i v_i(t), \qquad (7.10)$$

which defines the relationship between transported charge, capacitance, and the voltage across the capacitor. Hence, from equations 7.9 and 7.10, at every instant of time,

$$C_1 v_1(t) + C_2 v_2(t) + C_3 v_3(t) + C_4 v_4(t) = 0.$$

Thus, voltages, capacitances, and charge transport are related by this simple equation. The following example provides an application of these ideas.

◆ EXAMPLE 7.8

Consider the circuit of figure 7.18, in which $v_{C1}(0^-) = 1$ V and $v_{C2}(0^-) = 0$ V. Find $v_{C1}(t)$ and $v_{C2}(t)$ for $t \geq 0$.

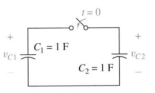

Figure 7.18 Two parallel capacitors connected by a switch.

SOLUTION

For $t > 0$, $v_{C1}(t) = v_{C2}(t)$. After the switch is closed at $t = 0$, some of the charge stored in C_1 is transferred to C_2. However, conservation of charge requires that the total charge transferred be zero, i.e.,

$$C_1[v_{C1}(0^+) - v_{C1}(0^-)] + C_2[v_{C2}(0^+) - v_{C2}(0^-)] = 0.$$

Since $v_{C1}(0^-) = 1$, $v_{C2}(0^-) = 0$, and $v_{C1}(0^+) = v_{C2}(0^+)$,

$$C_1[v_{C1}(0^+) - 1] + C_2 v_{C1}(0^+) = 0.$$

Hence, $(C_1 + C_2)v_{C1}(0^+) = 1$ implies that $v_{C1}(0^+) = 0.5$ V $= v_{C2}(0^+)$.

Example 7.8 is illustrative of charge transport that is germane to switched capacitor circuits which are of fundamental importance in the industrial world.

Energy Storage in a Capacitor

As with all devices, the energy stored or utilized in a capacitor is the integral of the power absorbed by the capacitor. The net energy entering the capacitor over the interval $[t_0, t_1]$ is

$$W_C(t_0, t_1) = \int_{t_0}^{t_1} p_C(\tau) d\tau = \int_{t_0}^{t_1} v_C(\tau)\, i_C(\tau)\, d\tau$$

$$= C \int_{t_0}^{t_1} \left(v_C(\tau) \frac{dv_C(\tau)}{d\tau} \right) d\tau = C \int_{v_C(t_0)}^{v_C(t_1)} v_C dv_C \qquad (7.11)$$

$$= \frac{1}{2} C \left[v_C^2(t_1) - v_C^2(t_0) \right] \text{ joules,}$$

for C in farads and v_C in amps. From equation 7.11, the change in energy stored in the capacitor over the interval $[t_0, t_1]$ depends only on the value of the capacitor voltage at times t_0 and t_1, i.e., on $v_C(t_0)$ and $v_C(t_1)$. This means that the change in stored energy is independent of the particular voltage waveform between t_0 and t_1. If the voltage waveform is periodic, i.e., if $v_C(t) = v_C(t+T)$ for some $T > 0$ and all t, then over any period of time T, the change in the stored energy in the capacitor is zero because $v_C(t_0+T) = v_C(t_1) = v_C(t_0)$ forces equation 7.11 to zero.

As with the inductor, it is convenient to define the instantaneous stored energy in a capacitor as

$$W_C(t) = \frac{1}{2} C v_C^2(t). \tag{7.12}$$

◆ EXAMPLE 7.9

Consider the circuit of figure 7.19, in which $v_C(0) = 0$. If it is known that the voltage across the capacitor is

$$v_C(t) = R\left(1 - e^{-\frac{t}{RC}}\right) u(t),$$

then determine the energy in joules stored in the capacitor for $t \geq 0$.

Figure 7.19 Parallel RC circuit.

SOLUTION

Since $v_C(0) = 0$, from equation 7.11 (or 7.12),

$$W_C(0, t) = \frac{1}{2} C \left[R\left(1 - e^{-\frac{t}{RC}}\right)\right]^2 = \frac{CR^2}{2}\left(1 - e^{-\frac{t}{RC}}\right)^2.$$

◆ EXAMPLE 7.10

Let us consider again example 7.7, in which a capacitor is excited by the sawtooth current waveform of figure 7.16b. The continuous voltage waveform is also plotted in that figure. Recall from the example that over the first period, $i_s(t) = (2t - 2)u(t)u(2 - t)$ A and $v_C(t) = 2(t^2 - 2t)u(t)u(2 - t)$ V. Accordingly, the power absorbed by the capacitor over $0 \leq t \leq 2$ is $p_C(t) = v_C(t)i_s(t) = (2t - 2)(2t^2 - 4t) = 4t(t - 1)(t - 2)u(t)u(2 - t)$ watts. This waveform repeats itself over subsequent intervals, leading to the absorbed power signal plotted in figure 7.20. Observe once again that the power is periodic, as is $i_s(t)$, and observe how over each period, half the waveform is positive and half negative. Integrating the power to obtain the energy entering the capacitor for each time t in the first period yields $W_C(0, t) = (t^4 - 4t^3 + 4t^2)u(t)u(2 - t) = t^2(t - 2)^2 u(t)u(2 - t)$ joules. This waveform repeats itself over subsequent periods. Again, over each period the total energy stored in the capacitor is zero, because the change in energy depends only on the capacitor voltage at the endpoints of the period; i.e., the voltage is zero at $t = 2k$, $k = 0, 1, 2, 3, \ldots$ Note that since $v_C(0) = 0$, $W_C(0) = 0$.

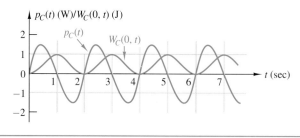

Figure 7.20 Plot of the power $p_C(t)$ absorbed by the capacitor of figure 7.16 and the energy stored in the capacitor as a function of time. Note that the curve $W_C(0,t)$ represents the accumulated area beneath the curve $p_C(t)$ from 0 to t.

Capacitance and Dielectrics

Insulating materials (dielectrics) other than air between the capacitor plates produce different capacities to store energy in the associated electric field. To compare the capacitances of different materials, one defines the notion of a dielectric constant of a material as the ratio of the capacitance of the material to that of an identical capacitor having air as its insulating material. As with all materials, if the voltage were sufficiently large, the material in question would break down; i.e., it would disintegrate and allow a spark to jump between the capacitor plates. Table 7.1 lists the dielectric constants and breakdown voltages (in volts/mil, where 1 mil = 0.001 inch) of several common substances.

TABLE 7.1 DIELECTRIC CONSTANTS AND BREAKDOWN VOLTAGES OF SOME COMMON MATERIALS		
Material	Relative Dielectric Constant	Breakdown Voltage (V/mil)
Air	1.0	
Bakelite	4.4–5.4	300
Formica	4.6–4.7	450
Glass (Window)	7.6–8.0	200–250
Glass (Pyrex)	4.8	335
Mica, Ruby	5.4	3,800–5,600
Paper	3.0	200
Plexiglass	2.8	990
Porcelain	5.1–5.9	40–100
Teflon	2.1	1,000–2,000

4. SERIES AND PARALLEL INDUCTORS AND CAPACITORS

Series Inductors

Just as resistors in series combine to form an equivalent resistance, inductors in series combine to form an equivalent inductance. As it turns out, series inductors combine in the same way as series resistors do. To see this, consider the series connection of three inductors illustrated in figure 7.21.

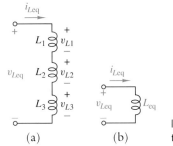

Figure 7.21 (a) Series connection of three inductors. (b) Equivalent inductance.

The voltage labeled $v_{L\mathrm{eq}}$ appears across the series connection, and, by KCL, the current $i_{L\mathrm{eq}}$ flows through each of the inductors, i.e., $i_{L1} = i_{L2} = i_{L3}$. The equivalent inductance,

L_{eq}, is defined by the relationship

$$v_{Leq}(t) = L_{eq} \frac{di_{Leq}}{dt}. \tag{7.13}$$

Our goal is to express L_{eq} in terms of L_1, L_2, and L_3. To obtain such an expression, observe that, by KVL,

$$v_{Leq} = v_{L1} + v_{L2} + v_{L3}.$$

Since each inductor satisfies the v-i relationship

$$v_{Li}(t) = L_i \frac{di_{Li}}{dt},$$

it follows that

$$v_{Leq}(t) = (L_1 + L_2 + L_3) \frac{di_{Leq}}{dt}.$$

Hence, the series inductors of figure 7.21a can be replaced by a single inductor with inductance $L_{eq} = L_1 + L_2 + L_3$.

The preceding derivation implies that inductances in series add. It is possible to prove this rigorously using mathematical induction; however, for our purposes, we need not do so. Suffice it to say that if n inductors, $L_1, L_2, \ldots, L_n$, are connected in series, then the equivalent inductance is

$$L_{eq} = L_1 + L_2 + \ldots + L_n. \tag{7.14}$$

Inductors in Parallel

The same basic question as with inductors in series arises with a parallel connection of inductors: What is the equivalent inductance? Rather than derive the general formula, let us consider the case of three inductors in parallel, as illustrated in figure 7.22a. Our goal is to show that the equivalent inductance is given by the reciprocal of the sum-of-the-reciprocals formula,

$$L_{eq} = \frac{1}{\dfrac{1}{L_1} + \dfrac{1}{L_2} + \dfrac{1}{L_3}}. \tag{7.15}$$

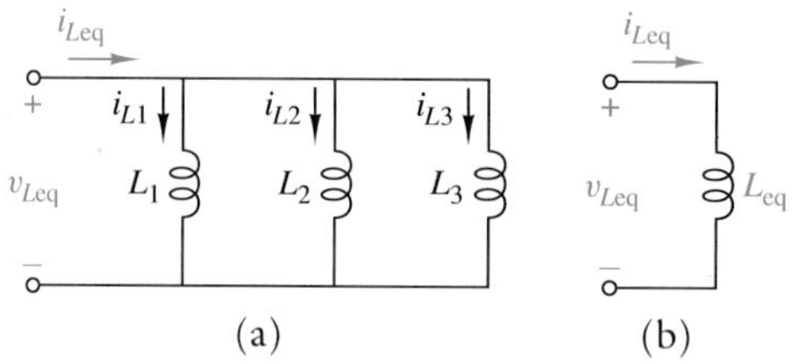

Figure 7.22 (a) Parallel connection of three inductors. (b) Equivalent inductance.

Once again, equation 7.13 defines the relationship for the equivalent inductance. The goal is achieved if we can express L_{eq} in terms of L_1, L_2, and L_3 in a way that satisfies equation 7.13. This will produce equation 7.15. The first step in this endeavor is to write KCL for the parallel connection shown in figure 7.22a. Here, by KCL

$$i_{Leq} = i_{L1} + i_{L2} + i_{L3}.$$

Differentiating both sides with respect to time yields

$$\frac{di_{Leq}}{dt} = \frac{di_{L1}}{dt} + \frac{di_{L2}}{dt} + \frac{di_{L3}}{dt}.$$

But

$$\frac{di_{Li}}{dt} = \frac{v_{Li}}{L_i};$$

hence,

$$\frac{di_{Leq}}{dt} = \frac{v_{L1}}{L_1} + \frac{v_{L2}}{L_2} + \frac{v_{L3}}{L_3} = \left(\frac{1}{L_1} + \frac{1}{L_2} + \frac{1}{L_3}\right) v_{Leq}.$$

This has the form of equation 7.13 which implies equation 7.15.

The foregoing argument is easily generalized. Suppose there are n inductors, $L_1, L_2, \ldots,$ L_n, connected in parallel. Then the equivalent inductance is given by the reciprocal of the sum-of-the-reciprocals formula,

$$L_{eq} = \frac{1}{\dfrac{1}{L_1} + \dfrac{1}{L_2} + \cdots \dfrac{1}{L_n}}. \qquad (7.16)$$

> **Exercise.** For two inductors L_1 and L_2 in parallel, show that the equivalent inductance satisfies the formula
>
> $$L_{eq} = \frac{L_1 L_2}{L_1 + L_2}. \qquad (7.17)$$

Series-Parallel Combinations

Often, inductors are connected in series-parallel combinations. In this subsection, we examine such combinations, i.e., connections of inductors that allow one to use the series and parallel formulas derived earlier.

◆ **EXAMPLE 7.11**

Find the equivalent inductance, L_{eq}, of the circuit of figure 7.23.

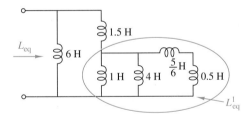

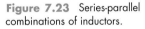

Figure 7.23 Series-parallel combinations of inductors.

SOLUTION

Step 1. In the circuit of figure 7.23 several inductors are enclosed by an ellipse. Let L_{eq}^1 denote the equivalent inductance of this combination. Observe that the series inductance of the 5/6-H and 0.5-H inductors equals 4/3 H. This inductance is in parallel with a 1-H and a 4-H inductance. Hence,

$$L_{eq}^1 = \frac{1}{\dfrac{1}{1} + \dfrac{1}{4} + \dfrac{3}{4}} = \frac{1}{2} \text{ H}.$$

Step 2. The equivalent circuit at this point is given by figure 7.24. This figure consists of a series combination of a 1.5-H and a 0.5-H inductor connected in parallel with a 6-H inductor. It follows that

$$L_{eq} = \frac{1}{\dfrac{1}{0.5 + 1.5} + \dfrac{1}{6}} = \frac{6}{4} = 1.5 \text{ H}.$$

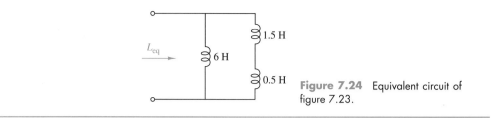

Figure 7.24 Equivalent circuit of figure 7.23.

Capacitors in Series

Capacitors in series have capacitances that combine according to the same formula for combining inductances in parallel. Similarly, capacitances in parallel combine in the same way that inductances in series combine. This means that the equivalent capacitance of a parallel combination of capacitors is the sum of the individual capacitances, and the equivalent capacitance of a series combination of capacitances satisfies the reciprocal of the sum-of-the-reciprocals rule. These ideas are made precise in the following development.

Consider the series combination of three capacitors given in figure 7.25.

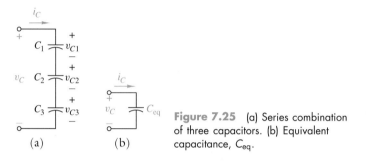

Figure 7.25 (a) Series combination of three capacitors. (b) Equivalent capacitance, C_{eq}.

The equivalent capacitance denoted in figure 7.25b must satisfy

$$i_C = C_{eq} \frac{dv_C}{dt}.$$

The goal is to generate this same v-i relationship in terms of the individual capacitances C_1, C_2, and C_3. To this end, observe that from KVL,

$$v_C = v_{C1} + v_{C2} + v_{C3}.$$

Differentiating this expression with respect to time yields

$$\frac{dv_C}{dt} = \frac{dv_{C1}}{dt} + \frac{dv_{C2}}{dt} + \frac{dv_{C3}}{dt} = \frac{i_C}{C_1} + \frac{i_C}{C_2} + \frac{i_C}{C_3}.$$

Factoring out i_C and manipulating yields

$$i_C = \left(\frac{1}{\dfrac{1}{C_1} + \dfrac{1}{C_2} + \dfrac{1}{C_3}} \right) \frac{dv_C}{dt}.$$

It follows that

$$C_{eq} = \frac{1}{\dfrac{1}{C_1} + \dfrac{1}{C_2} + \dfrac{1}{C_3}}.$$

Generalizing, we may say that capacitors in series satisfy the reciprocal of the sum-of-the-reciprocals rule. Thus, for n capacitors $C_1, C_2, \ldots, C_n$, connected in series, the equivalent capacitance is

$$C_{eq} = \frac{1}{\dfrac{1}{C_1} + \dfrac{1}{C_2} + \ldots + \dfrac{1}{C_n}}. \tag{7.18}$$

CHAPTER 7 INDUCTORS, CAPACITORS, AND DUALITY

Exercise. Show that if two capacitors C_1 and C_2 are connected in series, then

$$C_{eq} = \frac{C_1 C_2}{C_1 + C_2}.$$ (7.19)

Capacitors in Parallel

If two capacitors are connected in parallel as in figure 7.26a, there results an equivalent capacitance C_{eq}, shown in figure 7.26b.

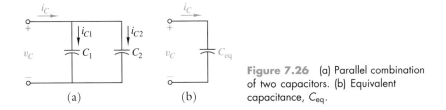

Figure 7.26 (a) Parallel combination of two capacitors. (b) Equivalent capacitance, C_{eq}.

Since the voltage v_C appears across each capacitor, and since $i_C = i_{C1} + i_{C2}$ by KCL, it follows that

$$i_C = i_{C1} + i_{C2} = C_1 \frac{dv_C}{dt} + C_2 \frac{dv_C}{dt} = (C_1 + C_2) \frac{dv_C}{dt}.$$

Hence,

$$C_{eq} = C_1 + C_2.$$

One surmises that, in general, capacitors in parallel have capacitances that add. And indeed, this is the case: If there are n capacitors $C_1, C_2, \ldots, C_n$, in parallel, the equivalent capacitance is

$$C_{eq} = C_1 + C_2 + \ldots + C_n.$$ (7.20)

Series-Parallel Combinations

We round out our discussion of capacitance by considering a simple example of a series-parallel interconnection of capacitors. Consider the circuit of figure 7.27. We seek to determine the equivalent capacitance, C_{eq}. To do so, observe that the two series capacitances of 0.5 mF and 0.5 mF combine to make a 0.25-mF capacitance. The two parallel capacitances, 0.3 mF and 0.6 mF, at the bottom of the circuit add to make a 0.9-mF capacitance. The new equivalent circuit is shown in figure 7.28.

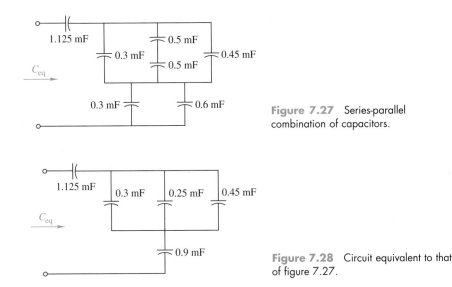

Figure 7.27 Series-parallel combination of capacitors.

Figure 7.28 Circuit equivalent to that of figure 7.27.

The three parallel capacitances of the equivalent circuit of figure 7.28 have an equivalent capacitance of 1 mF, since capacitances in parallel add. Hence, from equation 7.18,

$$C_{eq} = \frac{1}{\frac{1}{1.125} + \frac{1}{1} + \frac{1}{0.9}} = \frac{1}{3} \text{ mF.}$$

5. SMOOTHING PROPERTY OF A CAPACITOR IN A POWER SUPPLY

As mentioned in the chapter opener, a power supply converts a sinusoidal input voltage to an almost constant dc output voltage. Such devices are present in televisions, transistor radios, stereos, computers, and a whole host of other household electronic gadgets. Producing a truly constant dc voltage is virtually impossible, so engineers design special circuits called **voltage regulators** that generate a voltage with only a small variation between set limits for a given range of variation in load. The voltage regulator is a precision device whose input must be fairly smooth for proper operation. A capacitor can provide a crude, inexpensive degree of smoothing that is oftentimes sufficient for the task. This section explores the design of a capacitive smoothing circuit.

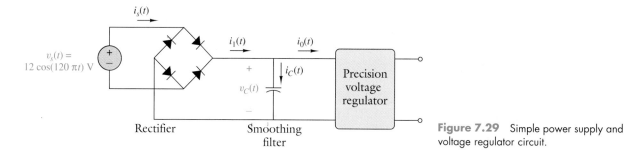

Figure 7.29 Simple power supply and voltage regulator circuit.

Consider, for example, the circuit shown in figure 7.29. The four (ideal) diodes are arranged in a configuration called a *full-wave bridge rectifier circuit*. An ideal diode allows current to pass only in the direction of the arrow. The diode configuration ensures that $i_1(t)$ remains positive, regardless of the sign of the source current. Specifically, the diodes ensure that $i_1(t) = |i_s(t)|$. Using the integral relationship (equation 7.6) of the capacitor voltage and current, it follows that

$$v_C(t) = v_C(t_0) + \frac{1}{C} \int_{t_0}^{t} i_C(\tau) d\tau = v_C(t_0) + \frac{1}{C} \int_{t_0}^{t} \left[|i_s(\tau)| - i_0(\tau) \right] d\tau. \tag{7.21}$$

Because of the difference $|i_s(\tau)| - i_0(\tau)$ inside the integrand of the integral, $i_s(t)$ tends to increase the capacitor voltage, whereas $i_0(t)$ tends to decrease the capacitor voltage. Further, because the diodes are assumed ideal, it follows that

$$v_C(t) \geq |v_s(t)|. \tag{7.22}$$

To see this, suppose the opposite were true; i.e., suppose $|v_s(t)| > v_C(t)$. One of the diodes would then have a positive voltage across it in the direction of the arrow. But this is impossible, because an ideal diode behaves like a short circuit in such a condition. The consequence is that $v_C(t)$ will be 12 V whenever $|v_s(t)|$ is 12 V. This occurs every 1/120 of a second. Thus, the rectifier output will recharge the capacitor every 1/120 of a second. Between charging times, the current, $i_0(t)$, will tend to discharge the capacitor and diminish its voltage.

The design problem for the capacitive smoothing circuit is to select a value for C which guarantees that $v_C(t)$ is sufficiently smooth to ensure proper operation of the voltage regulator. Here, "sufficiently smooth" means that the maximum and minimum voltages differ by less than a prescribed amount. To be specific, suppose that $v_C(t)$ must remain

CHAPTER 7 INDUCTORS, CAPACITORS, AND DUALITY

between 8 V and 12 V. Recall that $i_s(t)$ tends to increase the capacitor voltage, while $i_0(t)$ tends to decrease it. The design requires selecting a value for C to ensure that $i_s(t)$ can keep up with $i_0(t)$, so that the capacitor voltage remains fairly constant. The value for $i_0(t)$ is obtained from the specification sheet of the voltage regulator. Suppose this value is a constant 1 A. It remains to select C so as to ensure that $v_C(t)$ remains above 8 V between charging times. From equation 7.21, it is necessary that

$$v_C(t) = v_C(t_0) + \frac{1}{C} \int_{t_0}^{t} [|i_s(\tau)| - i_0(\tau)] \, d\tau \geq 8.$$

Now, we need consider only values for t between 0 and $1/120$, because the capacitor will recharge and the process will repeat itself every $1/120$ of a second. Thus, because $i_s(t)$ will only increase the capacitor voltage, to ensure that $v_C(t)$ remains above 8 V, it is sufficient to require that

$$v_C(t_0) - \frac{1}{C} \int_{0}^{\frac{1}{120}} i_0(\tau) d\tau \geq 8.$$

With $i_0(t) = 1$ and $v_C(0) = 12$, it follows that

$$C \geq \frac{1 \text{ A} \times \frac{1}{120} \sec}{4 \text{ V}} = 2.083 \text{ mF}.$$

A 2,100-μF capacitor satisfies this requirement. A method for computing the capacitor voltage waveform is described in chapter 22 of volume 2. However, using SPICE or one of the other available circuit simulation programs, one can generate a plot of the time-varying capacitor voltage produced by this circuit, as is shown in figure 7.30. In the figure, it is seen that the capacitor voltage varies between 12 and 9.02 V, which is smaller than the allowed variation of $(12 - 8)$V. Two factors contribute to this conservative design: (1) We used $C = 2,100$ μF instead of the calculated value, $C = 2,083$ μF, and (2) the increase in the capacitor voltage due the charging current i_s is not included in the calculation.

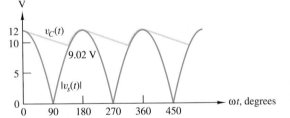

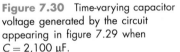

Figure 7.30 Time-varying capacitor voltage generated by the circuit appearing in figure 7.29 when $C = 2,100$ μF.

The preceding brief introduction made several simplifying assumptions to clarify the basic use of a capacitor as a smoothing or filtering device. Practical power supply design is a challenging field. A complete design would need to consider many other issues, some of which are the nonzero resistance of the source, the nonideal nature of the diodes, the current-handling ability of the components, protection of the components from high-voltage transients, and heatsinking of the components. Problem 29 asks for another application of the foregoing ideas.

6. THE DUALITY PRINCIPLE

Definition and Basic Relationship of Dual Circuits

As an introduction to the duality principle in circuit analysis, consider two simple circuits, N and N^* shown in figure 7.31. Each circuit obeys three types of constraints: KCL, KVL, and the branch v-i relationships. Table 7.2 lists the constraint equations for the two circuits side by side for easy comparison.

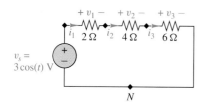
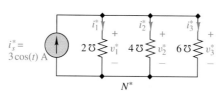

Figure 7.31 Two simple dual circuits.

TABLE 7.2 CONSTRAINT EQUATIONS FOR TWO DUAL CIRCUITS	
Constraint equations for N	Constraint equations for N^*
KVL: $$v_s = v_1 + v_2 + v_3$$	KCL: $$i_s^* = i_1^* + i_2^* + i_3^*$$
KCL: $$i_s = i_1$$ $$i_1 = i_2$$ $$i_2 = i_3$$	KVL: $$v_s^* = v_1^*$$ $$v_1^* = v_2^*$$ $$v_2^* = v_3^*$$
Branch v-i relationships: $$v_1 = 2i_1$$ $$v_2 = 4i_2$$ $$v_3 = 6i_3$$ $$v_s = 3\cos(t)$$	Branch i-v relationships: $$i_1^* = 2v_1^*$$ $$i_2^* = 4v_2^*$$ $$i_3^* = 6v_3^*$$ $$i_s^* = 3\cos(t)$$

Observe a very special relationship among the equations in table 7.2: Every constraint equation in N becomes a constraint equation in N^* (and vice versa) upon the interchange of voltage and current variables in the following manner:

$$i_k \leftrightarrow v_k^*, \qquad v_k \leftrightarrow i_k^*. \qquad (7.23)$$

Thus, since the only difference in the two systems of equations for N and N^* is a change in variable names, we expect the same relationship prevails in the solution of any variable and in any derived relationship among the variables. For example, in figure 7.31, an analysis of the circuit N yields the solution

$$v_3 = 1.5\cos(t) \text{ V}$$

and the voltage divider relationship

$$v_3 = 0.5v_s.$$

Then, using the interchange of variables of equation 7.23, we can write down (without any additional computation), for the circuit N^*, the solution

$$i_3^* = 1.5\cos(t) \text{ A}$$

and the current divider relationship

$$i_3^* = 0.5i_s^*.$$

The special relationships between the two simple circuits of figure 7.31 can be generalized to the concept of dual circuits. Formal statements of the definition and a basic relationship of dual circuits follow.

DEFINITION OF DUAL CIRCUITS

Two circuits N and N^* are called *dual* circuits if, with proper labeling of the branches, KCL, KVL, and branch v-i equations in N become, respectively, KVL, KCL, and branch i-v equations in N^* upon the interchanges $i_k \leftrightarrow v_k^*$ and $v_k \leftrightarrow i_k^*$.

The basic relationship of dual circuits is as follows: If N and N^* are dual circuits, then, with proper labeling of the branches, and upon the interchanges $i_k \leftrightarrow v_k^*$ and $v_k \leftrightarrow i_k^*$: (1) any solution in N leads to a solution in N^* and (2) any relationship among variables in N leads to a valid relationship in N^*.

The application of the duality concept is obvious: Once the results for N have been obtained, one can infer the results of its dual network *without any additional computation!* This is the most effort-effective measure one can think of in solving circuit problems.

At this point, one naturally raises the question: Given a circuit N, can we always construct a dual circuit N^*? The answer is "No": The dual circuit N^* exists if and only if the original circuit N is *planar*. (See section 6 of chapter 4 for the definition of a planar circuit.) The proof of this assertion is complex and above the level of this text. Fortunately, from the point of view of utility, what a beginner needs is not the proof of the assertion, but a simple procedure to construct the dual circuit N^* once a planar circuit N has been given. This is provided in the next subsection.

Constructing the Dual N^* of a Planar Circuit N

By the definition of dual circuits, N and N^* must have the same number of branches, but not necessarily the same number of nodes. Consider any branch in N and its counterpart in N^*. Their voltages and currents must be related by the interchange of variables given in equation 7.23. For example, if the kth branch in N is a 3-F capacitor, i.e., $i_k = 3dv_k/dt$, then its counterpart in the dual circuit N^* must have the constraint $v_k^* = 3di_k^*/dt$, implying a 3-H inductor. By the same procedure, we can establish relationships for all sorts of dual circuit elements. These are presented in table 7.3. (See chapter 3 for notations for controlled sources.)

TABLE 7.3   DUAL CIRCUIT ELEMENTS

Element in N		Element in N^*	
Resistance:	$R = a\ \Omega$	Conductance:	$G = a\ \mho$
Capacitance:	$C = a\ \mathrm{F}$	Inductance:	$L = a\ \mathrm{H}$
Inductance:	$L = a\ \mathrm{H}$	Capacitance:	$C = a\ \mathrm{F}$
VCCS:	$i_k = av_j$	CCVS:	$v_k^* = ai_j^*$
CCVS:	$v_k = ai_j$	VCCS:	$i_k^* = av_j^*$
VCVS:	$v_k = av_j$	CCCS:	$i_k^* = ai_j^*$
CCCS:	$i_k = ai_j$	VCVS:	$v_k^* = av_j^*$
Open circuit		Short circuit	
Short circuit		Open circuit	
Independent voltage source: $v_s = a\ \mathrm{V}$		Independent current source: $i_s^* = a\ \mathrm{A}$	
Independent current source: $i_s = a\ \mathrm{A}$		Independent voltage source: $v_s^* = a\ \mathrm{V}$	
Series-connected elements		Parallel-connected elements	
Parallel-connected elements		Series-connected elements	

Our next task is to connect the dual circuit elements properly in N^* to satisfy the conditions of dual circuits, namely, that KCL and KVL equations in one circuit become KVL and KCL equations respectively in the other circuit. For the simple case of series-parallel circuits without controlled sources, the problem can usually be solved by stating the condition existing in N and then generating a statement for N^* with the aid of table 7.3. For example, referring to the circuit N in figure 7.31, we can state that

The voltage source v_s sees three resistances: 2 Ω, 4 Ω, and 6 Ω, in series.

Then, for the dual network N^*, the corresponding statement would be

The current source i_s^* sees three conductances: 2 mhos, 4 mhos, and 6 mhos, in parallel.

This statement leads directly to the dual circuit N^* in figure 7.31. The next example further illustrates this method of constructing dual networks.

◆ **EXAMPLE 7.12**

Consider the circuit N shown in figure 7.32a.

(a) Construct its dual N^*.
(b) An analysis (see chapter 8) of N shows that the voltage source "sees" the dash-lined box as equivalent to a 2-Ω resistor. What can be said about N^* from duality (without additional analysis)?

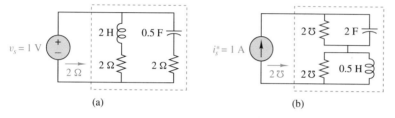

(a)   (b)

Figure 7.32 Dual of a series-parallel circuit. (a) Circuit N. (b) Circuit N^*.

SOLUTION

Part (a)   For N:

The voltage source "sees" a series-parallel network consisting of a (2-H inductance in series with a 2-Ω resistance) in parallel with a (0.5-F capacitance in series with a 2-Ω resistance).

Using the relationships of table 7.3, the above word description translates to the following statement for the dual circuit N^*:

The current source "sees" a network consisting of a (2-F capacitance in parallel with a 2-mho conductance) in series with a (0.5-H inductance in parallel with 2-mho conductance).

This description leads directly to the dual circuit N^* of figure 7.32b.

Part (b)   From the basic relationship of dual circuits, the current source in N^* "sees" the dash-lined box as equivalent to a 2-mho conductance.

The foregoing simple procedure for constructing dual networks runs into some difficulty when any of the following conditions are present:

1. The circuit configuration is not series-parallel.
2. The circuit contains controlled sources.
3. The reference directions of the circuit elements play an important role in the solution.

A geometrical method helps cope with this more general situation. Since this phase of the solution is concerned only with the KCL and KVL equations, we can represent N and N^* by graphs, as was done in section 6 of chapter 4. In the application of the graphs, it is important to remember that the arrow attached to each branch serves as the reference direction for both the current and voltage of that branch; i.e., i_k denotes the current through the kth branch *in the direction of the arrow*, and v_k denotes the voltage drop across the kth branch plus to minus *in the direction of the arrow*. Instead of formally presenting the procedure, we shall give some rationale behind it and then illustrate it with an example.

Figure 7.33a shows the graph of a non-series-parallel network N. First, consider a typical KVL equation for one of the meshes in N:

$$-v_1 + v_2 + v_3 = 0.$$

From equation 7.23, we expect the following KCL equation in N^*:

$$-i_1^* + i_2^* + i_3^* = 0.$$

This current relationship can be satisfied by placing a node of N^* inside the mesh and drawing three branches of N^* to intersect corresponding branches in N, as shown in figure 7.33b. The rule for assigning the branch arrows in N^* is as follows: If the branch arrow in N is clockwise around the node of N^*, the arrow for the corresponding branch in N^* is *away* from the node of N^*.

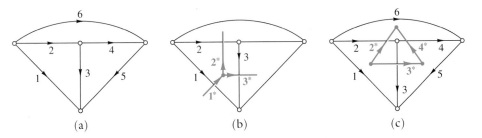

(a) (b) (c)

Figure 7.33 Determining the connection of dual circuit branches. (a) Non-series-parallel circuit. (b) Dual node-branch relationships associated with loop of branches 1-2-3 in (a). (c) Dual loop-branch relationship associated with node formed by branches 2-3-4 of (a).

Next, consider a typical KCL equation for one of the nodes in N:

$$-i_2 + i_3 + i_4 = 0.$$

Again, from equation 7.23, we expect the following KVL equation in N^*:

$$-v_2^* + v_3^* + v_4^* = 0.$$

This voltage relationship can be satisfied by the same node placements and branch connections just described and is illustrated in figure 7.33c.

Putting all these facts together, we have the procedure for determining the connection of the dual network branches:

Step 1. Draw the given planar circuit N without branch crossings, and identify the meshes (regions).

Step 2. Place a node of N^* inside each mesh and one in the outside, infinite region.

Step 3. For each branch b in N that is on the boundary of two regions, draw the dual branch b^* joining the nodes placed in each of these regions. If the branch arrow in N is clockwise around the node of N^*, the arrow for the corresponding branch in N^* is *away* from the node of N^*.

Figure 7.34a shows the graphs for N (black lines) and its dual N^* (heavy color lines), constructed according to the preceding procedure. For clarity, the graph of N^* is drawn separately in figure 7.34b.

Our final example of the chapter will illustrate the complete process of constructing a dual circuit by this geometrical procedure and the use of table 7.1.

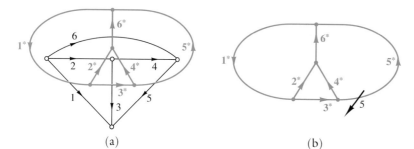

Figure 7.34 A geometrical procedure for determining the graph for N^*. (a) Basic circuit with dual circuit drawn in color. (b) Separate drawing of the dual circuit.

◆ **EXAMPLE 7.13**

Figure 7.35a shows a Colpitts oscillator, which is a linear active circuit N containing a voltage-controlled voltage source.

(a) Construct the dual circuit N^*, and list the corresponding circuit parameters in the two circuits.

(b) An analysis of N (by a method to be studied in volume II) shows that if $\mu_6 = C_2/C_1$, then the circuit has a sinusoidal current-voltage response, even though no external input is applied. Without any further analysis, use duality to state a similar result for the dual circuit N^*.

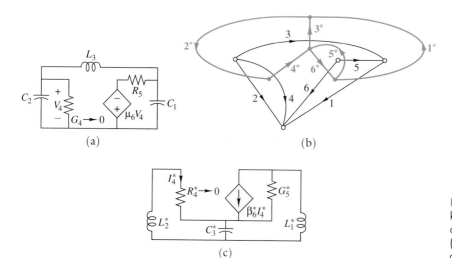

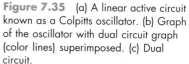

Figure 7.35 (a) A linear active circuit known as a Colpitts oscillator. (b) Graph of the oscillator with dual circuit graph (color lines) superimposed. (c) Dual circuit.

SOLUTION

(a) The geometrical method of constructing the dual circuit is detailed in figure 7.35b. The final dual circuit is shown in figure 7.35c. The corresponding quantities in the dual circuits are listed in table 7.4.

Note in particular that the voltage-controlled voltage source in N becomes a current-controlled current source in N^*. The polarity of each controlled source is of vital importance in this case. The circuit is meant to generate sinusoidal waveforms. A mistake in the connection of the controlled source will completely ruin this possibility. The geometrical procedure of constructing dual circuits assures the correct reference directions of all branches.

(b) From the basic relationship of dual circuits, we can immediately state that the circuit N^* of figure 7.35c has sinusoidal responses if $\beta_6^* = L_2^*/L_1^*$.

TABLE 7.4 LISTING OF DUAL CIRCUIT ELEMENTS FOR FIGURE 7.35

Original circuit N	Dual circuit N^*
C_1	L_1^*
C_2	L_2^*
L_3	C_3^*
$G_4 \to 0$	$R_4^* \to 0$
R_5	G_5^*
μ_6	β_6^*

Exercise. With $\mu_6 = C_2/C_1$ in figure 7.35a, any voltage or current in the circuit has the form $K \cos(\omega t + \theta)$, where

$$\omega = \frac{1}{\sqrt{L_3 \dfrac{C_1 C_2}{C_1 + C_2}}}.$$

By the use of duality, state a similar result for the circuit of figure 7.35c.

Exercise. The circuit parameters in figure 7.35a are $C_1 = 5$ mF, $C_2 = 20$ mF, and $L_3 = 4$ mH. (a) Find the values of μ_6 and ω. (b) Find the parameter values in figure 7.35c if the circuit is to produce sinusoidal oscillation at the same frequency ω.

SUMMARY

This chapter has introduced the notions of a capacitor and an inductor, each of which is a lossless energy storage device whose voltage and current satisfy a differential equation. The inductor has a voltage proportional to the derivative of the current through it; the constant of proportionality is the inductance L. The capacitor has a current proportional to the derivative of the voltage across it; the constant of proportionality is the capacitance C. It is interesting to observe that the roles of voltage and current in the capacitor are the reverse of their roles in the inductor. Because of this reversal, the capacitor and the inductor are said to be dual devices.

That the inductor and the capacitor are lossless energy storage devices means that they can store energy and deliver it back to the circuit, but they can never dissipate energy, as does a resistor. The inductor stores energy in a surrounding magnetic field, and the capacitor stores energy in an electric field between its conducting surfaces. Unlike energy in a resistor, the energy stored in an inductor over an interval $[t_0, t_1]$ is dependent only on the inductance L and the values of the inductor current at t_0 and t_1. Likewise, the energy

stored in a capacitor over an interval $[t_0, t_1]$ is dependent only on the capacitance C and the values of the capacitor voltage at t_0 and t_1.

Both the inductor and the capacitor have memory. The inductor has memory because at a particular time t_0, the inductor current depends on the past history of the voltage across the inductor. The capacitor has a voltage at, say, time t_0 that depends on the past current excitation to the capacitor. The concept of memory stems from the fact that the inductor current is proportional to the integral of the voltage across the inductor and the capacitor voltage is proportional to the integral of the current through the capacitor. This integral relationship gives rise to the important properties of the continuity of the inductor current and the continuity of the capacitor voltage under bounded excitations.

The dual capacitor and inductor relationships are capsulized in table 7.5.

Finally, we investigated the smoothing action of a capacitor in a power supply and the notion and use of the dual circuit properties of linear networks.

$i_C(t) \xrightarrow{\quad} \;\; +\, v_C(t)\, -$	$i_L(t) \xrightarrow{\quad} \;\; +\, v_L(t)\, -$
$i_C(t) = C\dfrac{dv_C(t)}{dt}$	$v_L(t) = L\dfrac{di_L(t)}{dt}$
$v_C(t) = v_C(t_0) + \dfrac{1}{C}\displaystyle\int_{t_0}^{t} i_C(\tau)\,d\tau$	$i_L(t) = i_L(t_0) + \dfrac{1}{L}\displaystyle\int_{t_0}^{t} v_L(\tau)\,d\tau$
$W_C(t_0, t_1) = \dfrac{1}{2}Cv_C^2(t_1) - \dfrac{1}{2}Cv_C^2(t_0)$	$W_L(t_0, t_1) = \dfrac{1}{2}Li_L^2(t_1) - \dfrac{1}{2}Li_L^2(t_0)$

TERMS AND CONCEPTS

Bounded voltage or current: voltage or current signal whose absolute value remains below some fixed finite constant for all time.

Capacitance of a pair of conductors: a property of conductors separated by a dielectric that permits the storage of electrically separated charge when a potential difference exists between the conductors. Capacitance is measured in stored charge per unit of potential difference between the conductors.

Capacitor (linear): a two-terminal device whose current is proportional to the time derivative of the voltage across it.

Coil: another name for an inductor.

Conservation-of-charge principle: principle that the total charge transferred into a junction (or out of a junction) is zero.

Continuity property of the capacitor: property such that if the current $i_C(t)$ through a capacitor is bounded over the time interval $t_1 \leq t \leq t_2$, then the voltage across the capacitor is continuous for $t_1 < t < t_2$. In particular, if $t_1 < t_0 < t_2$, then $v_C(t_0^-) = v_C(t_0^+)$, even when $i_C(t_0^-) \neq i_C(t_0^+)$.

Continuity property of the inductor: property such that if the voltage $v_L(t)$ across an inductor is bounded over the time interval $t_1 \leq t \leq t_2$, then the current through the inductor is continuous for $t_1 < t < t_2$. In particular, if $t_1 < t_0 < t_2$, then $i_L(t_0^-) = i_L(t_0^+)$, even when $v_L(t_0^-) \neq v_L(t_0^+)$.

Coulomb: quantity of charge that, in 1 sec, passes through any cross section of a conductor maintaining a constant one-amp current flow.

Dielectric: an insulating material often used between two conducting surfaces to form a capacitor.

Dual circuits: two planar circuits N and N^* are dual circuits if, with proper labeling of the branches, KCL, KVL, and branch v-i equations in N become, respectively, KCL, KVL, and branch i-v equations in N^* upon the interchange of voltage and current variables.

Farad: a measure of capacitance in which a charge of 1 coulomb produces a 1-V potential difference.

Faraday's law of induction: law asserting that, for a coil of wire sufficiently distant from any magnetic material, such as iron, the voltage induced across the coil by a time-varying current is proportional to the time derivative of the current; the constant of proportionality, denoted L, is the inductance of the coil. Faraday's law is usually stated in terms of flux and flux linkages, which are discussed in physics texts.

Henry: the unit of inductance; equal to 1 V-sec/amp.

Inductance: property of a conductor and its local environment (a coil with an air core or an iron core) that relates the time derivative of a current through the conductor to an induced voltage across the ends of the conductor.

Inductor (linear): a two-terminal device whose voltage is proportional to the time derivative of the current through it.

Instantaneous power: $p(t) = v(t)i(t)$; in watts when $v(t)$ is in volts and $i(t)$ in amps.

Lossless device: device in which energy can only be stored and retrieved and never dissipated.

Lossy device: a device, such as a resistor (with positive R), that dissipates energy as some form of heat or as work.

Maxwell's equations: a set of mathematical equations governing the properties of electric and magnetic fields and their interaction.

Memory: property of a device whose voltage or current at a particular time depends on the past operational history of the device; e.g., the current through an inductor at time t_0 depends on the history of the voltage excitation across the inductor for $t \leq t_0$.

Stray or parasitic capacitance: an unwanted capacitance that exists between two conducting surfaces.

Stray or parasitic inductance: an unwanted inductance affecting a circuit's performance when a time-varying current flows through a conductor in the circuit.

Unbounded voltage or current: a voltage or current whose value approaches infinity as it nears some instant of time, possibly $t = \infty$.

Voltage regulator: circuit that produces a voltage having only a small variation between set limits for a given range of load variations from a fairly smooth input signal.

PROBLEMS

1. (a) For $i_s(t) = \sin(t)u(t)$ A in figure P7.1, determine and sketch $i_{out}(t)$ for $0 \le t \le 4\pi$.
   (b) What is the instantaneous power delivered by the dependent source?
   (c) Determine and sketch the energy stored in the 2-H inductor for $0 \le t \le 4\pi$.

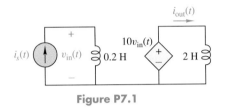

Figure P7.1

2. For $v_s(t)$ sketched in figure P7.2a, determine and sketch $v_{out}(t)$ for the circuit of figure P7.2b. What is the instantaneous power delivered by the dependent source?

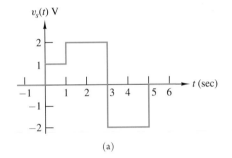

(a)

(b)

Figure P7.2 (a) Input waveform $v_s(t)$. (b) Circuit excited by $v_s(t)$.

3. In the circuit of figure P7.3, $i_{L1}(0) = 10$ mA and $i_{L2}(0) = -5$ mA.
   (a) Find L_{eq}, i_{L1}, i_{L2}, v_{L1}, and v_{L2} for $t \ge 0$ when $v_{in}(t) = 200te^{-t}u(t)$ mV.
   (b) Determine an expression for the energy stored in the 3-mH inductor.

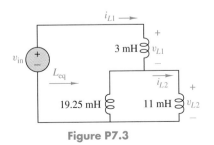

Figure P7.3

4. In this problem, we seek to develop a voltage division formula for relaxed inductor circuits. For the circuit of figure P7.4, suppose all initial currents at $t = 0$ are zero. Find v_{L1} and v_{L2} as a function of v_{in}, L_1, and L_2.

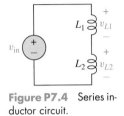

Figure P7.4 Series inductor circuit.

5. In this problem, we seek to develop a current division formula for relaxed inductor circuits. For the circuit of figure P7.5, suppose all initial currents at $t = 0$ are zero. Find i_{L1} and i_{L2} as a function of i_{in}, L_1, and L_2.

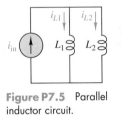

Figure P7.5 Parallel inductor circuit.

6. Consider the circuit of figure P7.6. Suppose $i_{in}(t) = 120\cos(300\pi t)$ mA.
   (a) Determine $v_{in}(t)$ and $v_{L2}(t)$.
   (b) Plot the instantaneous power delivered by the source for $0 \le t \le 14$ msec.

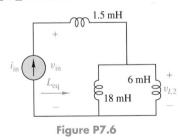

Figure P7.6

7. Repeat part (a) of problem 6 for the input current sketched in figure P7.7. In addition, compute an expression for, and plot the energy stored in, the 6-mH inductor.

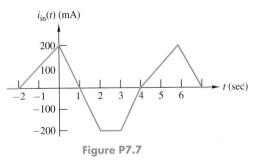

Figure P7.7

8. (a) Consider the circuit sketched in figure P7.8. Suppose $v_s(t) = \sin(\pi t)u(t)$, and suppose $v_{out}(0) = 10$ V. Determine $v_{out}(t)$. Is the output voltage independent of the initial voltage on C_1? Why?

(b) Determine the energy stored in the 2-F capacitor, assuming that the capacitor voltage is zero at $t = 0$.

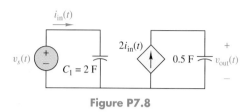

Figure P7.8

9. Suppose $v_{in}(t)$, sketched in figure P7.9a, is an input excitation to the circuit of figure P7.9b.

(a) Determine and sketch $v_{out}(t)$ in the circuit of figure P7.9b.

(b) What is the instantaneous power delivered by the dependent source?

(c) Determine and plot the energy stored in the 0.5-F capacitor.

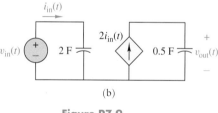

(a)

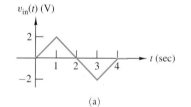

(b)

Figure P7.9

10. For the circuit of figure P7.10, suppose $i_{in}(t) = \sin(t)u(t)$ A and all capacitors are uncharged at $t = 0$. Determine and sketch $i_{out}(t)$. What is the instantaneous power delivered by the dependent source?

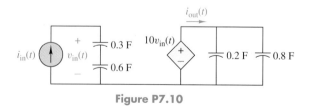

Figure P7.10

11. For the circuit of figure P7.11, compute $v_{out}(t)$ as a function of $i_s(t)$, assuming that $i_s(t)$ has been exciting the circuit for all time.

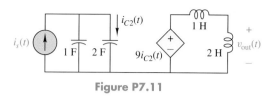

Figure P7.11

12. For the circuit of figure P7.12, compute $v_{out}(t)$ as a function of $v_{in}(t)$, assuming that $v_{in}(t)$ has been exciting the circuit for all time.

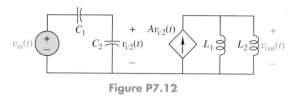

Figure P7.12

13. For the circuit of figure P7.13, determine L_{eq}.

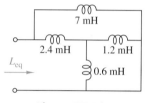

Figure P7.13

14. For the circuit of figure P7.14, determine C_{eq}.

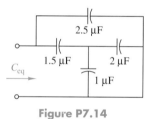

Figure P7.14

15. (a) In the circuit of figure P7.15, all capacitors and inductors are initially uncharged. If $i_s(t) = 6u(t)$ A, determine $i_{C2}(t)$.

(b) Now determine $i_{out}(t)$.

(c) Compute the energy stored in the 2-H inductor.

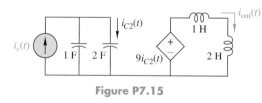

Figure P7.15

16. The approximate inductance of a single-layer coil wound on a nonmagnetic form may be calculated from the formula

$$L = \frac{4 \times 10^{-5}(\text{diameter})^2 \ (\text{\# of turns})^2}{18 \ (\text{diameter}) + 40 \ (\text{length})},$$

where L is in henries and the diameter and length of the coil are in meters. The formula provides a good estimate of the inductance if the length is equal to or greater than 0.4 times the diameter.

CHAPTER 7 INDUCTORS, CAPACITORS, AND DUALITY

A coil has 48 turns wound at 12 turns/cm. The diameter of the coil is 2 cm. Find the approximate value of the inductance. (*Check:* L is between 18 and 20 µH.)

17. The approximate capacitance of a parallel-plate capacitor may be calculated from the formula

$$C = \varepsilon_r \varepsilon_0 \frac{A}{d},$$

where C is the capacitance in farads; A is the area of one plate in m²; d is the distance between the plates in meters; $\varepsilon_0 = 8.854 \times 10^{-12}$ F/m is the dielectric constant of free space (air); and ε_r is the relative dielectric constant of the insulator.

Some typical values of ε_r are as follows: air, 1; mica, 5.4; paper, 3; glass, 4.8. A paper capacitor consists of two aluminum foils rolled into a tubular form with paper in between. Each foil is 4 cm by 80 cm. The paper is 0.1 mm thick. Find the capacitance. (*Check:* Between 8 and 10 nF.)

18. Construct the dual circuit for the circuit of problem 15. For all the results calculated in problem 15, state similar results for the dual circuit.

19. Construct the dual for the circuit of problem 1.

20. Construct the dual for the circuit of problem 14. From the result of problem 14, what is the equivalent inductance of the dual circuit?

21. In the circuit of figure P7.21, suppose $i_{in}(t) = 10 \sin(120\pi t)u(t)$ A.
   (a) Determine $v_{C1}(t)$ and $v_{C2}(t)$ for $t \geq 0$ if $v_{C1}(0) = 2$ V and $v_{C2}(0) = 4$ V.
   (b) Determine an expression for the energy stored in the 0.1-F and the 0.15-F capacitors over the interval $[0, t]$.

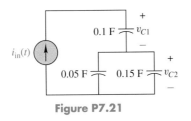

Figure P7.21

22. In the circuit of figure P7.22, suppose $v_{in}(t) = 100e^{-2t}u(t)$ V.
   (a) Determine $i_{in}(t)$, $i_{C1}(t)$, and $i_{C2}(t)$ for $t \geq 0$.
   (b) If $v_{C1}(0) = 10$ V and $v_{C2}(0) = 5$ V, determine $v_{C1}(t)$ and $v_{C2}(t)$ for $t \geq 0$. (*Note:* At $t = 0$, KVL must be satisfied.)
   (c) Determine the energy stored in the 0.03-F capacitor.

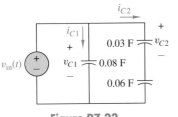

Figure P7.22

23. For the circuit of figure P7.23a, plot $i_{in}(t)$, $i_{C1}(t)$, and $i_{C2}(t)$ for $v_{in}(t)$ as specified in figure 7.23b.

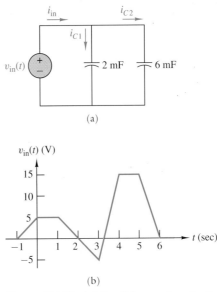

(a)

(b)

Figure P7.23 (a) Parallel capacitor circuit. (b) Voltage input waveform.

24. Suppose $i_{in}(t)$, as specified in figure P7.24a, excites the circuit of figure P7.24b.
   (a) Compute $v_{C1}(t)$ and $v_{C2}(t)$.
   (b) Plot $v_{C1}(t)$ and $v_{C2}(t)$ for $0 \leq t \leq 8$.
   (c) Determine and plot the energy stored in the 0.25-F and 0.1-F capacitors for $0 \leq t \leq 8$.

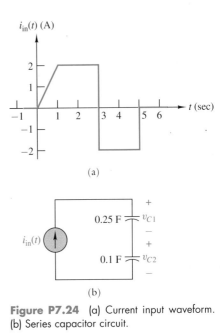

(a)

(b)

Figure P7.24 (a) Current input waveform. (b) Series capacitor circuit.

25. For the circuit of figure P7.25, determine the equivalent capacitance "seen" by the source and $v_{in}(t)$ when $i_{in}(t) = 10 \cos(2t)u(t)$ mA.

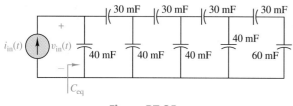

Figure P7.25

26. The purpose of this problem is to develop a voltage division formula for relaxed capacitor circuits. Assume that all capacitors are initially uncharged in figure P7.26. Determine expressions for $v_1(t)$ and $v_2(t)$ for $t \geq 0$ in terms of $v_{in}(t)$, C_1, and C_2.

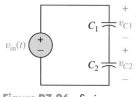

Figure P7.26 Series capacitor circuit.

27. The purpose of this problem is to develop a current division formula for relaxed capacitor circuits. Assume that all capacitors in figure P7.27 are initially uncharged. Determine expressions for $i_1(t)$ and $i_2(t)$ for $t \geq 0$ in terms of $i_{in}(t)$, C_1, and C_2.

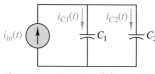

Figure P7.27 Parallel capacitor circuit.

28. (a) For the circuit of figure P7.28, find L_{eq}, v_{L1}, i_{L1}, and v_{L2} when $i_{in}(t) = e^{-t^2}$ mA.
(b) Using MATLAB or some other software program that has a numerical integration routine, construct a graph of the energy stored in the 4-mH inductor.

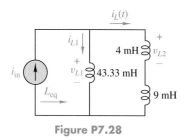

Figure P7.28

29. Using the circuit given in figure 7.29, select a capacitor value to filter the voltage for a regulator requiring $14 \text{ V} < v_C(t) < 20$. Use $v_s(t) = 20\cos(200\pi t)$ V and $i_0(t) = 2$ A.

30. Consider the circuit of figure P7.30a with voltage excitation shown in figure P7.30b.
(a) Compute and sketch the input current for $t \geq 0$.

(b) Determine the total energy stored in the battery of four inductors.
(c) Compute and sketch the current $i_L(t)$ through the 0.9-H inductor.

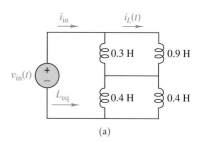

(a)

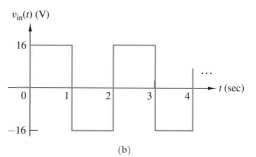

(b)

Figure P7.30 (a) Circuit for problem 30. (b) Periodic voltage excitation for inductive circuit given in (a).

31. Consider the circuit of figure P7.31, which is excited by the current waveform $i_{in}(t) = 20te^{-10t}u(t)$ A.
(a) Compute the sketch $v_L(t)$, $v_C(t)$, and $v_{in}(t)$.
(b) Compute and sketch the energy stored in the inductor for $t \geq 0$.
(c) Compute and sketch the energy stored in the capacitor for $t \geq 0$.

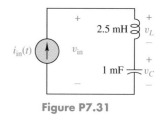

Figure P7.31

32. The circuit of figure P7.32 has two inductors in parallel. The input is $v_s(t) = 12\cos(500t)u(t)$ V. Each inductor is relaxed at $t = 0$, i.e., has no initial current. The switch between the two inductors moves down at $t = 3$ msec. Determine the currents $i_{L1}(t)$ and $i_{L2}(t)$ for $0 \leq t < 3$ msec and $3 \text{ msec} \leq t$. Also, determine the energy stored in each inductor as a function of t for the same time interval.

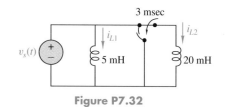

Figure P7.32

CHAPTER 7 INDUCTORS, CAPACITORS, AND DUALITY

33. The circuit of figure P7.33 has two capacitors in parallel. The input is $i_s(t) = 12e^{-500t}u(t)$ A. Each capacitor is relaxed at $t = 0$, i.e., has no initial voltage. The switch between the two capacitors opens at $t = 2$ msec. Determine the voltage $v_C(t)$ for $0 \leq t < 2$ msec and 2 msec $\leq t$. Also, determine the energy stored in each capacitor as a function of t for the same time intervals.

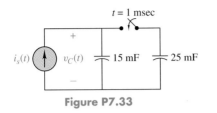

Figure P7.33

34. In the circuit of figure P7.34, both capacitors are uncharged at $t = 0^-$, and $v_s(t) = 25u(t)$ V. Determine $v_C(t)$ for $0 \leq t < 2$ sec and 2 sec $\leq t$.

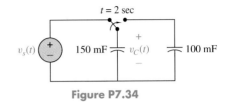

Figure P7.34

CHAPTER 8

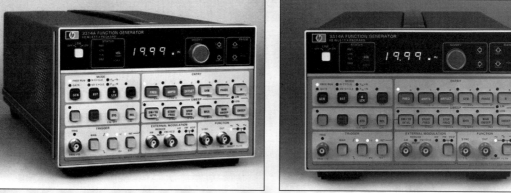

CHAPTER OUTLINE

First-Order RL and RC Circuits

SIGNAL GENERATION

W HEN watching a manufacturing process, a visitor might see a pair of robotic arms assemble an engine or machine a block of metal with perfectly timed maneuvers. Timing is a critical aspect of the manufacturing process. In television transmitters, there is a signal called the raster that is critical to the generation of the picture on the screen. In an oscilloscope, there is a timing signal called a horizontal sweep that acts as a time base allowing one to view different signals as a function of time. Each of these applications utilizes a signal called a linear voltage sweep, often called a sawtooth. This sweep waveform is pictured below together with an approximating exponential curve for comparison.

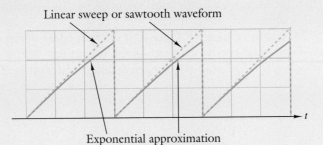

Ideally, the sawtooth voltage increases linearly with time until reaching a threshold, whereupon it immediately drops to zero, which reinitiates the process. The threshold voltage corresponds to a fixed unit of time. The linear voltage increase then acts as an electronic second hand ticking off the smaller units of time. In practice, the linear increase in voltage is approximated by the near-linear part of an exponential response of an RC circuit. When the voltage across the capacitor reaches a certain threshold, an electronic switch changes the equivalent circuit "seen" by the capacitor, allowing the capacitor to discharge very quickly. The capacitor voltage then drops to zero almost instantaneously. Once the voltage nears zero, the electronic switch reinstates the earlier circuit structure, which then causes the capacitor to charge up again, almost linearly. The process repeats itself indefinitely.

RC circuits and RL circuits have exponential responses to initial conditions and constant source excitations; i.e., their responses are proportional to $A + Be^{-Kt}$ for constants A, B, and K. The purpose of this chapter is to develop techniques for computing the exponential responses of first-order RC and RL circuits. Once a basic understanding of these ideas is reached, we will apply them to generate a crude approximation to a sawtooth in example 8.10. Problem 28 takes up the details of generating an approximation of an ideal sawtooth waveform. The sawtooth is but one of many possible complex waveforms of practical importance that can be generated by RC and RL circuits alone and in conjunction with operational amplifiers.

CHAPTER OBJECTIVES

- Introduce the exponential response of simple (first-order) RL and RC circuits without sources and with constant excitation sources present.
- Introduce the notion of a first-order linear differential equation with constant coefficients as a mathematical model for first-order RL and RC circuits.
- Derive the exponential solution form of the differential equation model that characterizes the voltage and current responses of RL and RC circuits driven by constant source excitations.
- Interpret the solution form of the differential equation model in terms of the initial and final values of the capacitor voltage and inductor current.
- Apply the preceding ideas to an improperly connected op amp circuit, to a practical op amp integrator called a leaky integrator, and as a practical way to generate sawtooth waveforms.

1. INTRODUCTION

Interconnections of resistors, capacitors, inductors, and sources lead to new and fascinating circuit behaviors. In resistive circuits, all the equations are algebraic because they are interconnections of devices satisfying Ohm's law and other algebraic constraints imposed by dependent sources. Inductors and capacitors have differential or integral voltage-current relationships. Accordingly, interconnecting resistors and capacitors or resistors and inductors leads to circuits that must satisfy both algebraic (KVL, KCL, and Ohm's law) and differential or integral relationships. Hence, nodal and loop equations will contain sums of linear-algebraic (Ohm's law) and differential terms (e.g., $L di_L/dt$). A simple example serves to explain this situation.

In the parallel RL circuit of figure 8.1, suppose there is some initial inductor current $i_L(0^-)$, where the minus sign on the zero means the instant just before zero proper. We often distinguish between 0^-, 0, and 0^+ when switching or discontinuities of excitation functions occur at $t = 0$.

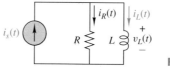

Figure 8.1 Parallel RL circuit.

A node equation at the top is

$$i_s(t) = i_R(t) + i_L(t). \tag{8.1}$$

The equation for $i_R(t)$ in terms of $i_L(t)$ is

$$i_R(t) = \frac{v_R(t)}{R} = \frac{v_L(t)}{R} = \frac{L}{R}\frac{di_L(t)}{dt}.$$

Substituting this expression into equation 8.1 and multiplying by R/L yields the differential equation

$$\frac{di_L(t)}{dt} = -\frac{R}{L}i_L(t) + \frac{R}{L}i_s(t), \tag{8.2}$$

subject to the initial condition $i_L(0^-)$. Equation 8.2 is called a **first-order linear differential equation** with constant coefficients. This equation models the behavior of the inductor current in the circuit. It is first order because only the first derivative appears. Since it comes from a linear circuit, it is linear, meaning that under certain conditions, properties such as superposition apply. In terms of details, the equation says that the derivative of the inductor current equals $-R/L$ times the inductor current plus R/L times the input current waveform. The equation, then, constrains the behavior of the inductor current. The goal of the analyst is to find (if possible) a current waveform $i_L(t)$ that satisfies the constraints imposed by equation 8.2. This type of equation is very different from the simple algebraic equations of the earlier chapters. Its added complexity leads to new and fascinating circuit behaviors that permit us to make first steps toward controlling our environment and transferring information.

Before continuing, however, it is important to point out some unstated assumptions about the specification of initial conditions. When an initial condition or set of initial conditions is specified on a circuit at $t = 0$ or, more generally, at $t = t_0$, one presumes *a priori* that the initial condition embodies the entire history of the circuit up to time t_0. One cannot presume that the functional form of an excitation actually occurred before the initial condition. Further, one cannot even presume that the given structure of the circuit existed prior to the initial condition. One can only presume that the circuit structure and excitations are legitimate for $t \geq t_0$.

> **Exercise.** For the circuit of figure 8.1, show that the inductor voltage $v_L(t)$ also satisfies the differential equation
>
> $$\frac{dv_L(t)}{dt} = -\frac{R}{L}v_L(t) + R\frac{di_s(t)}{dt}.$$

In addition to the preceding assumptions, it is often the case that a signal is nonzero only for $t \geq 0$ or, more generally, for $t \geq t_0$. For example, a voltage waveform may be defined as $v(t) = 0$ for $t < 0$ and $v(t) = 10$ V for $t \geq 0$. To represent such waveforms conveniently, one defines a signal called the **unit step function** as

$$u(t) = \begin{cases} 1 & \text{if } t \geq 0 \\ 0 & \text{if } t < 0 \end{cases}.$$

With the unit step so defined, $v(t) = 10u(t)$ V means

$$v(t) = \begin{cases} 10 \text{ V} & \text{if } t \geq 0 \\ 0 & \text{if } t < 0 \end{cases}.$$

Further, if $v(t) = 10$ V for $t \geq t_0$ and $v(t) = 0$ for $t < t_0$, then $v(t) = 10u(t - t_0)$ V would be the proper representation of $v(t)$ because the shifted step function $u(t - t_0)$ means

$$u(t - t_0) = \begin{cases} 1 & \text{if } t \geq t_0 \\ 0 & \text{if } t < t_0 \end{cases}.$$

The unit step function is a **universally used function** and will appear many times in this text.

> **Exercise.** Plot $u(t_0 - t)$. (*Hint*: For what values of t is the function 0, and for what values is it 1?)

2. SOME MATHEMATICAL PRELIMINARIES

A working model of a physical system underlies an engineer's ability to analyze, design, or modify, in a methodical fashion, the behavior of the system. Linear circuits are physical systems that have differential equation models. The RL and RC circuits investigated in this chapter will have differential equation models of the form

$$\frac{dx(t)}{dt} = \lambda x(t) + f(t), \quad x(t_0) = x_0, \tag{8.3}$$

valid for $t \geq t_0$, where $x(t_0) = x_0$ is the initial condition of the differential equation. The term $f(t)$ denotes a forcing function in the theory of differential equations. Usually, $f(t)$ will be a linear function of the input excitations to the circuit and their derivatives.

Before proceeding, it is appropriate to explore the intuitive nature of a differential equation. Equation 8.3 is a **differential equation** because of the presence of the derivative of some unknown function $x(t)$. The derivative of $x(t)$ must equal a constant, λ, times $x(t)$, plus a known forcing function $f(t)$ which incorporates the effect of all the circuit excitations. This is a pretty hefty requirement. Trying to find such unknown functions $x(t)$ might seem almost impossible. Fortunately, it is not.

The parameter λ denotes a **natural frequency** of the circuit. Natural frequencies are natural modes of oscillation. In physical objects, they are called natural modes of vibration. All physical objects have a vibrational motion, even though it may be imperceptible. Knowledge of natural modes of vibration is important for the safety and reliability of large buildings and bridges. For example, the Tacoma Narrows Bridge had natural modes of vibration that the wind excited. This caused the natural swaying motion of the bridge to increase without bound, until the bridge collapsed. In a circuit, the natural modes of oscillation are reflected in the shapes of the voltage and current waveforms the circuit produces. A more thorough and mathematical discussion of the notion of natural frequency will take place in the next chapter, when we study second-order (RLC) circuits.

Let us return now to the goal of solving the differential equation 8.3. For $t \geq 0$, the solution (to be derived shortly) has the form

$$x(t) = e^{\lambda(t-t_0)} x(t_0) + \int_{t_0}^{t} e^{\lambda(t-\tau)} f(\tau) d\tau. \tag{8.4}$$

This means that the expression on the right side of the equals sign (1) satisfies the differential equation 8.3 (its derivative equals λ times itself plus $f(t)$) and (2) satisfies the correct initial condition, $x(t_0) = x_0$. A simple example illustrates this statement.

Suppose $f(t) = u(t-1)$, a shifted unit step function, $\lambda = -1$, $t_0 = 1$, and $x(1) = 10$, in which case

$$\frac{dx(t)}{dt} = -x(t) + u(t-1), \quad x(1) = 10.$$

Then, for $t \geq 1$,

$$x(t) = e^{-(t-1)}10 + \int_{1}^{t} e^{-(t-\tau)} u(\tau-1) d\tau = 10 e^{-(t-1)} + \left(1 - e^{-(t-1)}\right)$$

$$= 9 e^{-(t-1)} + 1.$$

Observe that $9 e^{-(t-1)} + 1$ does indeed satisfy the differential equation because, for $t > 1$,

$$\frac{d}{dt}\left(9 e^{-(t-1)} + 1\right) = -9 e^{-(t-1)} = -x(t) + 1.$$

Further, at $t = 1$, $9 e^{-(t-1)} + 1 = 10$, which is the mandatory initial condition.

This example spells out the use of the solution (equation 8.4) to the differential equation 8.3. A derivation of this solution follows directly through using the **integrating factor method**. The first step of this method is to multiply both sides of equation 8.3 by the integrating factor $e^{-\lambda t}$ and then bring the term containing $x(t)$ to the left-hand side. This results in the equation

$$\left[e^{-\lambda t} \frac{dx(t)}{dt} - \lambda e^{-\lambda t} x(t) \right] = e^{-\lambda t} f(t). \tag{8.5}$$

Recognizing the sum on the left as the time derivative of $e^{-\lambda t} x(t)$, one can integrate both sides of equation 8.5 from t_0 to t:

$$\int_{t_0}^{t} \left[e^{-\lambda \tau} \frac{dx(\tau)}{d\tau} - \lambda e^{-\lambda \tau} x(\tau) \right] d\tau = \int_{t_0}^{t} \frac{d}{d\tau}\left(e^{-\lambda \tau} x(\tau) \right) d\tau = \int_{t_0}^{t} e^{-\lambda \tau} f(\tau) d\tau. \tag{8.6}$$

Evaluating the left side of equation 8.6 yields

$$e^{-\lambda \tau} x(\tau) \Big|_{t_0}^{t} = e^{-\lambda t} x(t) - e^{-\lambda t_0} x(t_0) = \int_{t_0}^{t} e^{-\lambda \tau} f(\tau) d\tau. \tag{8.7}$$

Bringing the term $e^{\lambda t_0} x(t_0)$ to the right side of equation 8.7 and multiplying through by $e^{\lambda t}$ results in the solution to the differential equation 8.3 given by the very powerful formula of equation 8.4.

There are four main points to remember about the foregoing discussion: (1) Circuits have behaviors modeled by differential equations such as equation 8.3; (2) the solution to a first-order differential equation is a waveform (also called a signal or response) that, for our needs, is given by equation 8.4; (3) the formula for equation 8.4 works for all continuous and piecewise continuous time functions $x(t)$; and (4) a solution to a differential equation is a waveform that satisfies the given differential equation with the proper initial condition.

3. THE SOURCE-FREE OR ZERO-INPUT RESPONSE

Figure 8.2 depicts the most basic RL or RC circuit, which is the parallel connection of a resistor with an inductor or a capacitor without a source. In such circuits, one assumes the presence of an initial inductor current or initial capacitor voltage. The presence of a source complicates the structure of the solution. Once the source-free or **zero-input** behavior is understood, one can understand more complicated responses that result from constant source excitations, because they incorporate the source-free solution form.

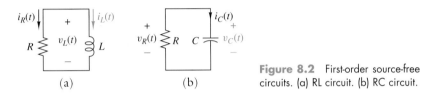

Figure 8.2 First-order source-free circuits. (a) RL circuit. (b) RC circuit.

Our first goal is to derive differential equation models for the RL and RC circuits of figures 8.2a and 8.2b, respectively. We do this in parallel:

1. At the top node of figure 8.2a, KCL implies that

$$i_R(t) = -i_L(t).$$

1. From KVL, it follows that

$$v_R(t) = v_C(t).$$

2. But

$$i_R(t) = \frac{v_L(t)}{R} = \frac{L}{R}\frac{di_L(t)}{dt}.$$

2. However,

$$v_R(t) = -Ri_C(t) = -RC\frac{dv_C(t)}{dt}.$$

3. Making the obvious substitution and multiplying by R/L yields the differential equation model

$$\frac{di_L(t)}{dt} = -\frac{R}{L}i_L(t),$$

3. Making the obvious substitution and dividing by RC yields the differential equation model

$$\frac{dv_C(t)}{dt} = -\frac{1}{RC}v_C(t), \qquad (8.8)$$

with $i_L(t_0)$ a given initial condition.

with $v_C(t_0)$ a given initial condition.

Both of these differential equation models have the same general form,

$$\frac{dx(t)}{dt} = -\frac{1}{\tau}x(t); \qquad (8.9)$$

i.e., the derivative of $x(t)$ is a constant, $(-1/\tau)$, times $x(t)$ itself. From equation 8.4, both equations in 8.8 have solutions with the same general exponential solution form, viz.,

$$x(t) = e^{-\frac{t-t_0}{\tau}}x(t_0), \quad t \geq t_0, \qquad (8.10)$$

where τ is a special constant called the **time constant** of the circuit. In particular, the responses for $t \geq t_0$ of the undriven RL and RC circuits are respectively given by

$$i_L(t) = e^{-\frac{R}{L}(t - t_0)}i_L(t_0) \quad \text{and} \quad v_C(t) = e^{-\frac{1}{RC}(t - t_0)}v_C(t_0), \qquad (8.11)$$

where the *time constant of the RL circuit is* $\tau = L/R$ and the *time constant of the RC circuit is* $\tau = RC$. The time constant of the circuit is the time it takes for the source-free circuit response to drop to $e^{-1} \cong 0.368$ of its initial value. Roughly speaking, the response value must drop to a little over a third of its initial value. This is a good rule of thumb for approximate calculations involving decaying exponentials.

The mathematics that underlies the solution given in equation 8.10 to the differential equation 8.9 is nothing more than elementary calculus. To see this, consider an exponential solution form, i.e.,

$$x(t) = Ke^{-t/\tau}, \qquad (8.12)$$

where K is an arbitrary constant. The function $e^{-t/\tau}$ has the property that its derivative is a constant, $-1/\tau$, times the function itself. This is precisely what equation 8.9 requires. Therefore, equation 8.12 satisfies the differential equation 8.9 and is said to be a solution of it. To specify $x(t)$ completely, it remains only to identify the proper value of K from the initial condition constraints. Evaluating $x(t)$ at $t = t_0$ yields

$$x(t_0) = Ke^{-t_0/\tau},$$

in which case

$$K = e^{-t_0/\tau}x(t_0).$$

Substituting this value of K into equation 8.12 produces the solution given in equation 8.10, which is adapted to specific RL and RC circuits in equation 8.11.

In sum, the circuits of figure 8.2 motivate the development of the rudimentary machinery for constructing solutions to undriven RL and RC circuits. For more general circuits—those containing multiple resistors and dependent sources—it is necessary to use the Thévenin equivalent resistance "seen" by the inductor or capacitor in place of R in equation 8.11. Figure 8.3 illustrates this idea.

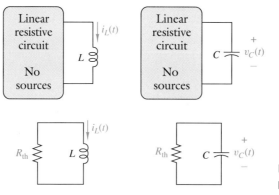

Figure 8.3 Replacement of resistive part of circuit by its Thévenin equivalent.

These facts imply that the general formulas for computing the **responses of undriven RL and RC** circuits have the form

$$i_L(t) = e^{-\frac{R_{th}}{L}(t-t_0)}i_L(t_0) \quad \text{and} \quad v_C(t) = e^{-\frac{1}{R_{th}C}(t-t_0)}v_C(t_0). \qquad (8.13)$$

The difference between equations 8.11 and 8.13 is that, in the latter, R_{th} is the Thévenin equivalent resistance "seen" by the inductor or capacitor.

◆ **EXAMPLE 8.1**

For the circuit of figure 8.4, find $i_L(t)$ and $v_L(t)$ for $t \geq 0$, given that $i_L(0^-) = 10$ A and switch S closes at $t = 0.4$ sec.

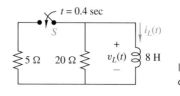

Figure 8.4 Parallel RL circuit containing a switch.

SOLUTION

Step 1. With switch S open, the relevant time interval is $0 \le t \le 0.4$ sec. Recall that $i_L(0^+) = i_L(0^-) = 10$ A. Using equation 8.13,

$$i_L(t) = e^{-\frac{R_{th}}{L}t} i_L(0^+) = 10e^{-2.5t}u(t) \text{ A.}$$

Step 2. With switch S closed, the relevant time interval is $t \ge 0.4$ sec. For this time interval, the Thévenin equivalent resistance "seen" by the inductor is $R_{th} = 20\ \Omega \| 5\ \Omega = 4\ \Omega$, i.e., the equivalent resistance of a parallel 20-Ω and 5-Ω combination. According to equation 8.13, the response for $t \ge t_0 = 0.4$ sec is

$$i_L(t) = e^{-\frac{R_{th}}{L}(t - t_0)} i_L(t_0)$$

$$= i_2(0.4)e^{-0.5(t - 0.4)}u(t - 0.4)$$

$$= 3.679e^{-0.5(t - 0.4)}u(t - 0.4) \text{ A.}$$

Figure 8.5 shows a sketch of the response. There are two different time constants present here. The 0.4-sec time constant has a much faster rate of decay than the lengthier 2-sec time constant.

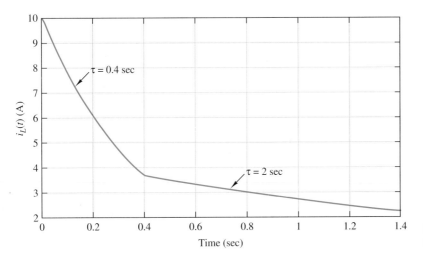

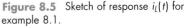

Figure 8.5 Sketch of response $i_L(t)$ for example 8.1.

It is a simple matter now to compute $v_L(t)$, since

$$v_L(t) = -R_{th}i_L(t).$$

In particular, $v_L(0^+) = -200$ V. Hence, for $0 \le t < 0.4$,

$$v_L(t) = v_L(0^+)e^{-\frac{R_{th}}{L}t} = -200e^{-2.5t} \text{ V.}$$

For $t \ge 0.4$,

$$v_L(t) = v_L(0.4^+)e^{-\frac{R_{th}}{L}(t - 0.4)} = -73.576e^{-0.5(t - 0.4)} \text{ V.}$$

◆ **EXAMPLE 8.2**

Find $v_C(t)$ for $t \ge 0$ for the circuit of figure 8.6, given that $v_C(0) = 9$ V.

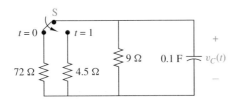

Figure 8.6 Parallel RC circuit.

SOLUTION

Step 1. The first relevant time interval is $0 \leq t < 1$. Over this interval, the equivalent circuit is a parallel RC circuit, as shown in figure 8.7a.

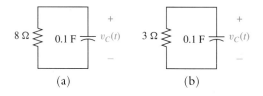

Figure 8.7 (a) Equivalent circuit for figure 8.6 for $0 \leq t < 1$. (b) Equivalent circuit for figure 8.6 for $1 \leq t$.

Since $v_C(0^+) = v_C(0^-) = 9$ V, it follows that

$$v_C(t) = v_C(0^+)e^{-\frac{t}{R_{\text{th}}C}} = 9e^{-1.25t} \text{ V}.$$

Step 2. For this part, consider $t \geq 1$. Figure 8.7b depicts the pertinent equivalent circuit. This time, the response is

$$v_C(t) = v_C(t_0^+)e^{-\frac{t-t_0^+}{R_{\text{th}}C}} = 2.58e^{-\frac{(t-1)}{0.3}} \text{ V},$$

which is valid only for $t \geq 1$.

The specification of $v_C(t)$ uses two separate time functions, each defined over a separate time interval. By using the shifted unit step function, the two expressions can be combined into a single expression:

$$v_C(t) = 9e^{-1.25t}[u(t) - u(t - 1)] + 2.58e^{-\frac{(t-1)}{0.3}}u(t - 1) \text{ V}.$$

A plot of the voltage response appears in figure 8.8. Here, the 0.3-sec time constant shows a much greater rate of decay than the longer 0.8-sec time constant.

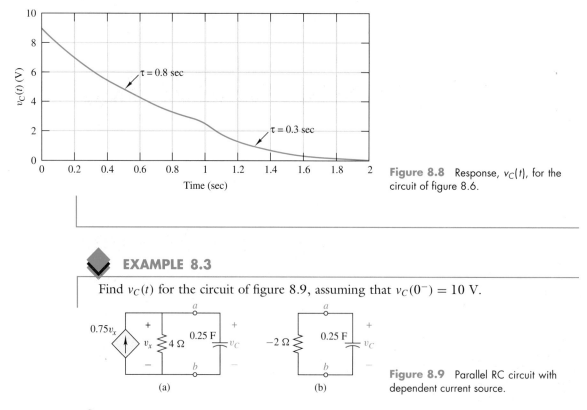

Figure 8.8 Response, $v_C(t)$, for the circuit of figure 8.6.

EXAMPLE 8.3

Find $v_C(t)$ for the circuit of figure 8.9, assuming that $v_C(0^-) = 10$ V.

Figure 8.9 Parallel RC circuit with dependent current source.

SOLUTION

It is straightforward to show that the Thévenin equivalent "seen" by the capacitor is a negative resistance, $R_{\text{th}} = -2 \ \Omega$, illustrated in figure 8.9b. Hence, the response is

$$v_C(t) = v_C(0^+)e^{-\frac{t}{R_{\text{th}}C}} = 10e^{2t} \text{ V}.$$

Because of the negative resistance, this response blows up exponentially, as the plot of figure 8.10 indicates. A response that increases without bound is said to be **unstable**.

It is important to recognize that the form of the capacitor current is similar to the form of the capacitor voltage. In particular,

$$i_C(t) = i_C(0^+)e^{-\frac{t}{R_{th}C}} = 5e^{2t} \text{ A}.$$

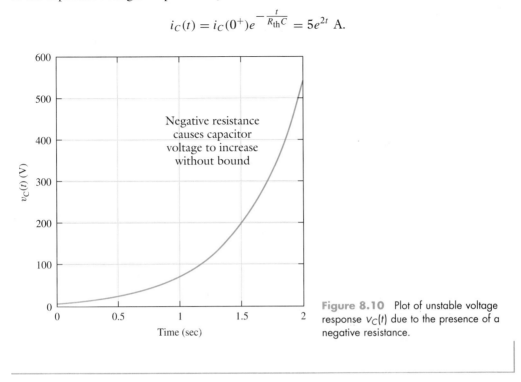

Negative resistance causes capacitor voltage to increase without bound

Figure 8.10 Plot of unstable voltage response $v_C(t)$ due to the presence of a negative resistance.

4. THE DC OR STEP RESPONSE OF FIRST-ORDER CIRCUITS

The circuits of the previous section had no source excitations. This section takes up the calculation of voltage and current responses when constant voltage and constant current sources are present. It is instructive to start with the basic series RL and RC circuits, as shown in figure 8.11.

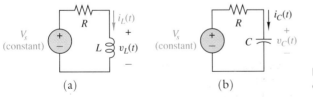

(a)   (b)

Figure 8.11 (a) Driven series RL circuit and (b) driven series RC circuit.

Given these basic circuits and initial conditions at t_0, what is the structure of a differential equation model that governs the voltage and current behavior of the circuits for $t \geq t_0$? The first objective is to derive the differential equation models characterizing the circuit's voltage and current responses. It is convenient to use $i_L(t)$ as the desired response for constructing the differential equation for the series RL circuit; for the series RC circuit, $v_C(t)$ is the more convenient variable. We have the following parallel derivations:

1. The circuit model for the inductor is

$$v_L(t) = L\frac{di_L(t)}{dt}.$$

1. The circuit model for the capacitor is

$$i_C(t) = C\frac{dv_C(t)}{dt}.$$

2. By KVL and Ohm's law, the voltage across the inductor is

$$v_L(t) = V_s - Ri_L(t).$$

2. By KCL and Ohm's law, the current through the capacitor is

$$i_C(t) = \frac{V_s - v_C(t)}{R}.$$

3. Substituting for $v_L(t)$ leads to the differential equation model,

$$\frac{di_L(t)}{dt} = -\frac{R}{L}i_L(t) + \frac{1}{L}V_s,$$

with initial condition $i_L(t_0^-) = i_L(t_0^+)$ since V_s is constant (not impulsive).

3. Substituting for $i_C(t)$ leads to the differential equation model,

$$\frac{dv_C(t)}{dt} = -\frac{1}{RC}v_C(t) + \frac{1}{RC}V_s,$$

with initial condition $v_C(t_0^-) = v_C(t_0^+)$ since V_s is constant (not impulsive).

A simple application of basic circuit principles has led to these two differential equation models. The next important question is, *What do these models tell us about the behavior of each circuit? Or equivalently, how do we find a solution to the equations?*

Exercise. Construct differential equation models for the parallel RL and RC circuits of figure 8.12. Choose $i_L(t)$ as the response for the RL circuit and $v_C(t)$ for the RC circuit.

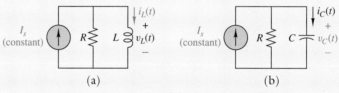

(a)   (b)

Figure 8.12 (a) Driven RL and (b) driven RC parallel circuits.

Both of the differential equations in step **3** have the same structure,

$$\frac{dx(t)}{dt} = -\frac{1}{\tau}x(t) + F, \tag{8.14}$$

which is valid for $t \geq t_0$ with initial condition $x(t_0^-) = x(t_0^+)$. The continuity of this initial condition follows because the source excitation F is a constant (nonimpulsive) forcing function. The time constant is $\tau = L/R$ (or L/R_{th}, where R_{th} is the Thévenin equivalent resistance "seen" by the inductor) for RL circuits and $\tau = RC$ (or $R_{th}C$, where R_{th} is the Thévenin equivalent resistance "seen" by the capacitor) for RC circuits. Equation 8.4, rewritten as

$$x(t) = e^{-\frac{t-t_0}{\tau}}x(t_0^+) + \int_{t_0}^{t} e^{-\frac{t-q}{\tau}}F\,dq, \tag{8.15}$$

is the general formula for solving such a differential equation. Since F is constant, it is straightforward to evaluate the integral in this formula:

$$x(t) = e^{-\frac{t-t_0}{\tau}}x(t_0^+) + Fe^{-\frac{t}{\tau}}\int_{t_0}^{t} e^{\frac{q}{\tau}}\,dq = e^{-\frac{t-t_0}{\tau}}x(t_0^+) + F\tau\left[1 - e^{-\frac{t-t_0}{\tau}}\right].$$

Some rearranging of terms yields

$$x(t) = F\tau + \left[x(t_0^+) - F\tau\right]e^{-\frac{t-t_0}{\tau}}, \tag{8.16}$$

which is valid for $t \geq t_0$. After some interpretation, this formula will serve as a basis for computing the response to RL and RC circuits driven by constant sources. Problem 23 provides a different and direct derivation of the formula.

At this point, it is helpful to interpret the quantity $F\tau$ in equation 8.16. This will provide us with a simple formula for determining the response of RL and RC circuits. If $\tau > 0$, then

$$\lim_{t \to \infty} x(t) = \lim_{t \to \infty} \left(F\tau + \left[x(t_0^+) - F\tau \right] e^{-\frac{t-t_0}{\tau}} \right) = F\tau.$$

Hence, when $\tau > 0$, $F\tau$ becomes the final value of the variable $x(t)$. This implies that the formula for the solution of equation 8.9 for constant or dc excitations is

$$x(t) = x(\infty) + \left[x(t_0^+) - x(\infty) \right] e^{-\frac{t-t_0}{\tau}}. \tag{8.17}$$

Note that $\tau > 0$ whenever all equivalent resistors, capacitors, and inductors have positive values. Thus, equation 8.17 has the rather nice physical interpretation,

$$x(t) = [\text{final value}] + \left([\text{initial value}] - [\text{final value}] \right) e^{-\frac{\text{elapsed time}}{\text{time constant}}}.$$

Several examples will illustrate the use of this formula.

◆ EXAMPLE 8.4

For the circuit of figure 8.13, suppose a 10-V excitation is applied at $t = 1$, when it is found that the inductor current is $i_L(1^-) = 1$ A. The application of the 10-V excitation has the mathematical representation $10u(t - 1)$ V in the figure. Find $i_L(t)$ for $t \geq 1$.

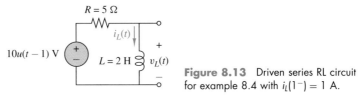

Figure 8.13 Driven series RL circuit for example 8.4 with $i_L(1^-) = 1$ A.

SOLUTION

Step 1. To link the solution with our earlier development, let us first write the differential equation and determine the time constant. Summing the voltages around the loop yields

$$Ri_L(t) + L\frac{di_L(t)}{dt} = 10,$$

which is valid for $t \geq 1$. Rearranging the terms and dividing by L produces the expected differential equation model,

$$\frac{di_L(t)}{dt} = -\frac{R}{L}i_L(t) + 10,$$

which is valid only for $t \geq 1$.

Step 2. Since $i_L(1^-) = i_L(1^+)$, equation 8.17 implies that for $t \geq 1$,

$$i_L(t) = i_L(\infty) + [i_L(1^+) - i_L(\infty)]e^{-\frac{(t-1)}{\tau}}.$$

(One cannot presume that the response is zero for $t < 1$.) Given this expression for the inductor current, it follows that the inductor voltage for $t \geq 1$ is

$$v_L(t) = L\frac{di_L(t)}{dt} = -\frac{L}{\tau}[i_L(1^+) - i_L(\infty)]e^{-\frac{(t-1)}{\tau}}u(t - 1).$$

Here, the presence of $u(t - 1)$ emphasizes that the response is valid only for $t \geq 1$. As $t \to \infty$, $v_L(t) \to 0$. Physically, zero volts means a short circuit. Hence, for dc excitations, at $t = \infty$, the inductor looks like a short circuit. Replacing the inductor in the circuit of figure 8.13 by a short permits us to compute $i_L(\infty) = 2$ A. It follows that for $t \geq 1$,

$$i_L(t) = 2 + [1 - 2]e^{-2.5(t - 1)} = 2 - e^{-2.5(t - 1)} \text{ A}.$$

A graph of $i_L(t)$, for $t \geq 1$ appears in figure 8.14.

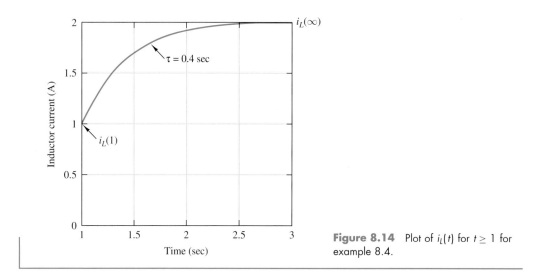

Figure 8.14 Plot of $i_L(t)$ for $t \geq 1$ for example 8.4.

An important property of RL circuits arises from an analysis of the preceding example: In a **passive RL circuit** (which can only store or dissipate energy) driven by a constant excitation, at $t = \infty$, the inductor voltage is zero and hence looks like a short circuit. The current, $i_L(\infty)$, is easily computed by the usual circuit techniques after replacing the inductor by a short. By duality, in a passive RC circuit driven by a constant excitation, the capacitor voltage approaches a constant. Since the current is proportional to the derivative of the voltage, the capacitor current is zero. Physically speaking, the capacitor looks like an open circuit. The final voltage, $v_C(\infty)$, can be computed by replacing the capacitor by an open circuit and using resistive circuit analysis techniques to compute the voltage appearing across this open circuit.

The foregoing method of determining the final values is also used to determine the *initial values* of v_C and i_L at $t = t_0$ if dc excitations have been applied to the circuit for a long time before $t = t_0$. Problems 9, 10, 12, and 13 illustrate the use of this method for establishing initial conditions.

For a rigorous justification of these ideas about the capacitor voltage and current, consider the equation

$$v_C(t) = v_C(\infty) + [v_C(t_0^+) - v_C(\infty)]e^{-\frac{t - t_0}{R_{th}C}}, \tag{8.18}$$

which is an interpretation of equation 8.17 for the case of an RC circuit and is valid only for $t \geq t_0$. Differentiating equation 8.18 for $t > t_0$ yields the capacitor current,

$$i_C(t) = C\frac{dv_C(t)}{dt} = -\frac{1}{R_{th}}[v_C(0^-) - v_C(\infty)]e^{-\frac{t - t_0}{R_{th}C}}.$$

For a passive circuit, $R_{th}C > 0$. This implies that $i_C(t) \to 0$ as $t \to \infty$. A zero current, $i_C(\infty) = 0$, translates to an open circuit. Hence, to compute $v_C(\infty)$, we replace the capacitor by an open circuit and use any of the standard circuit analysis techniques to compute the associated open-circuit voltage. The analogous result in which the inductor looks like a short circuit follows by a parallel argument.

EXAMPLE 8.5

The source in the circuit of figure 8.15 furnishes a 12-V excitation for $t < 0$ and a 24-V excitation for $t \geq 0$, denoted by $v_{in}(t) = [12u(-t) + 24u(t)]$ V. The switch in the circuit closes at $t = 10$ sec. First, determine the value of the capacitor voltage at $t = 0^-$ which, by continuity, equals $v_C(0^+)$. Next, determine $v_C(t)$ for all $t \geq 0$.

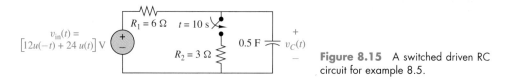

Figure 8.15 A switched driven RC circuit for example 8.5.

Step 1. *Compute the initial capacitor voltage.* For $t < 0$, the 12-V excitation has been applied for a long time. Therefore, at $t = 0^-$, the capacitor "looks like" an open circuit to the source. Hence, the entire source voltage of 12 V appears across the capacitor at $t = 0^-$; i.e., $v_C(0^-) = v_C(0^+) = 12$ V.

Step 2. *Use equation 8.17 or 8.18 to determine* $v_C(t)$ *for* $0 \le t \le 10$ *sec.* For $0 \le t \le 10$ sec, the time constant of the circuit is $R_1C = 3$ sec. Furthermore, without the switch, $v_C(\infty) = 24$ V because the capacitor looks like an open circuit for a very large t. It is important to realize here that for $0 \le t \le 10$ sec, the circuit behaves as if the switch were not present. Thus, equation 8.18 is valid over this interval. Accordingly,

$$v_C(t) = v_C(\infty) + [v_C(0^+) - v_C(\infty)]e^{-\frac{t}{R_{th}C}}u(t) = 24 + [12-24]e^{-\frac{t}{3}} = 24 - 12e^{-\frac{t}{3}}.$$

Step 3. *Determine the capacitor voltage at* $t = 10^+$ *sec.* Plugging into the preceding equation yields

$$v_C(10^+) = 24 - 12e^{-10/3} = 23.57.$$

Step 4. *Determine* $v_C(t)$ *for* $t \ge 10$. For $t \ge 10$, the resistive part of the circuit can be replaced by its Thévenin equivalent, which yields the new circuit of figure 8.16.

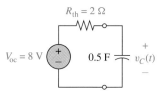

Figure 8.16 Circuit equivalent to that of figure 8.15 for $t \ge 10$. V_{oc} and R_{th} are the Thévenin open circuit voltage and equivalent resistance.

Here, equation 8.13 again applies. The value of $v_C(\infty)$, however, is now 8 V, and the new time constant is now $R_{th}C = 1$ sec. Hence, for $t \ge 10$,

$$v_C(t) = v_C(\infty) + [v_C(10^+) - v_C(\infty)]e^{-\frac{(t-10)}{\tau}} = 8 + 15.57e^{-(t-10)}.$$

With the use of step functions, the response for $t \ge 0$ is

$$v_C(t) = (24 - 12e^{-\frac{t}{3}})[u(t) - u(t-10)] + \left(8 + 15.57e^{-(t-10)}\right)u(t-10).$$

Plotting $v_C(t)$ yields the graph of figure 8.17.

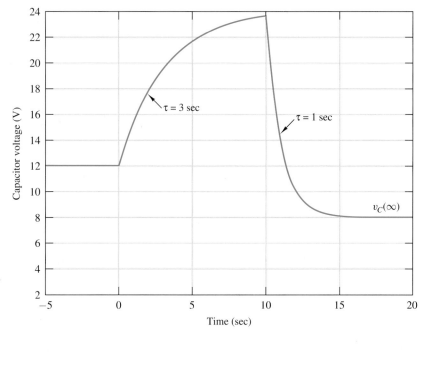

Figure 8.17 The capacitor voltage, $v_C(t)$, for $t \ge 0$.

5. Superposition and Linearity

As discussed in chapter 5, superposition, a special case of linearity, helps simplify the analysis of resistive circuits. Recall that linear resistive circuits are interconnections of resistors and sources, both dependent and independent. Does superposition still apply when capacitors and inductors are added to the circuit? The answer is yes, provided that one properly accounts for initial conditions.

To justify the use of **superposition** for RC and RL circuits, observe that resistors satisfy Ohm's law, which is a linear algebraic equation. Similarily, capacitors satisfy the linear differential relationship

$$i_C = C\frac{dv_C}{dt}.$$

To see the linearity in this equation, suppose that voltages v_{C1} and v_{C2} individually excite a relaxed capacitor, producing respective currents

$$i_{C1} = C\frac{dv_{C1}}{dt} \quad \text{and} \quad i_{C2} = C\frac{dv_{C2}}{dt}.$$

Let i_C be the current induced by the sum of the two voltages affecting the same relaxed capacitor. Then

$$i_C = C\frac{d}{dt}(v_{C1} + v_{C2}).$$

However, the linearity of the derivative implies the property of superposition:

$$i_C = C\frac{dv_{C1}}{dt} + C\frac{dv_{C2}}{dt} = i_{C1} + i_{C2}.$$

In fact, the current due to the input excitation $a_1 v_{C1} + a_2 v_{C2}$ is, by the same arguments, $a_1 i_{C1} + a_2 i_{C2}$.

On the other hand, suppose two separate currents i_{C1} and i_{C2} individually excite a relaxed capacitor C. Then each produces a voltage given by the integral relationship:

$$v_{C1} = \int_{-\infty}^{t} i_{C1}(\tau)d\tau, \quad v_{C2} = \int_{-\infty}^{t} i_{C2}(\tau)d\tau.$$

As before, the combined effect of the input, $a_1 i_{C1} + a_2 i_{C2}$, would be a voltage

$$v_C = \int_{-\infty}^{t}(a_1 i_{C1}(\tau) + a_2 i_{C2}(\tau))\,d\tau = a_1\int_{-\infty}^{t} i_{C1}(\tau)d\tau + a_2\int_{-\infty}^{t} i_{C2}(\tau)d\tau = v_{C1} + v_{C2},$$

the second equality of which follows by the distributive property of the integral. Again, linearity and hence superposition are seen to be valid.

Arguments analogous to these imply that a relaxed inductor satisfies a linear relationship, and thus superposition is valid, whether the inductor is excited by currents or voltages.

The interconnection of linear capacitors and linear inductors with linear resistors and sources through KVL and KCL produce linear circuits because KVL and KCL are linear algebraic constraints on the linear element equations. Since the interconnection is a linear circuit, the property of linearity carries over, and as a consequence, superposition holds.

To cap off this discussion, we must account for the presence of initial conditions on the capacitors and inductors of the circuit. For first-order RC and RL circuits, this need is clearly indicated by the first term of equation 8.4. For a general linear circuit, one can view each initial condition as being set up by an input that shuts off the moment the initial condition is established. Hence, the effect of the initial condition can be viewed as the effect of some input that turns off at the time the initial condition is specified. This means that, when using superposition on a circuit, one first looks at the effect of each input excitation to a circuit having no initial conditions. Then, one sets all independent sources to zero and computes the response due to each initial condition, with all other initial conditions set to zero. The sum of all the independent source responses and the individual initial condition responses yields the complete circuit response, obtained by the principle of superposition. A rigorous justification of this principle is given in volume 2, using the Laplace transform method.

The following example illustrates the application of these ideas.

EXAMPLE 8.6

The linear circuit of figure 8.18 has two source excitations applied at $t = 0$, as indicated by the presence of the step functions. The initial condition on the inductor current is $i_L(0^-) = -1$ A. Using superposition, determine the response $i_L(t)$ for $t \geq 0$.

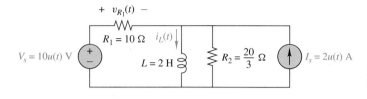

Figure 8.18 Driven RL circuit excited by two sources.

SOLUTION

Because the linearity of the circuit results in a linear differential equation, superposition is but one of several methods for obtaining the solution. A more natural approach would be to find the Thévenin equivalent "seen" by the inductor. As we will see, the superposition approach sometimes has an advantage over the Thévenin approach.

Superposition must be carefully applied, however. One superimposes or adds the response due only to the initial condition with the sources set to zero, the response due to V_s only with no initial condition, and, finally, the response due to I_s only with no initial condition to compute the complete response of the circuit.

Figure 8.19
Equivalent parallel RL circuit when sources are set to zero.

Part 1. Determine the part of the circuit response due only to the initial condition with the sources set to zero. With both sources set to zero, there results an equivalent circuit given by figure 8.19. The Thévenin equivalent resistance is $R_{th} = 4\ \Omega$, resulting from the parallel combination of R_1 and R_2. Figure 8.19 depicts an undriven RL circuit whose response is

$$i_L^1(t) = e^{-\frac{R}{L}t}i_L(0^+) = -e^{-2t}u(t) \text{ A}.$$

Part 2. Determine the response due only to $V_s = 10u(t)$ V. For this case, figure 8.20 portrays the equivalent circuit. The initial condition is set to zero, as its effect was accounted for by the calculations of part 1.

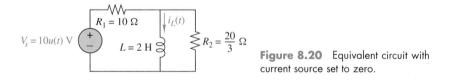

Figure 8.20 Equivalent circuit with current source set to zero.

At $t = \infty$, the inductor looks like a short circuit. This means that $i_L(\infty) = V_s/R_1 = 1$ A. Further, $R_{th} = R_1//R_2 = 4\ \Omega$, as before. Thus, the contribution to the complete response due only to the source V_s is:

$$i_L^2(t) = i_L^2(\infty) + [i_L^2(0^+) - i_L^2(\infty)]e^{-\frac{R_{th}}{L}t}u(t) = (1 - e^{-2t})u(t) \text{ A}.$$

Part 3. Compute the response due only to the current source $I_s = 2u(t)$ A. Here, the equivalent circuit is given by figure 8.21, where, again, $i_L(0^-) = 0$.

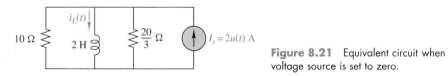

Figure 8.21 Equivalent circuit when voltage source is set to zero.

For this equivalent circuit, $i_L(\infty) = 2$ A and $R_{th} = 4\ \Omega$, as before. Hence,

$$i_L^3(t) = i_L^3(\infty) + [i_L^3 0^-) - i_L^3(\infty)]e^{-\frac{R_{th}}{L}t}u(t) = 2(1 - e^{-2t})u(t) \text{ A}.$$

Part 4. *Apply the principle of superposition.* By superposition, the complete response $i_L(t)$ of the circuit is given by the sum of the individual responses, i.e., $i_L^1(t) + i_L^2(t) + i_L^3(t)$:

$$i_L(t) = \underbrace{-e^{-2t}u(t)}_{\substack{\textit{due to} \\ \textit{initial} \\ \textit{condition}}} + \underbrace{(1 - e^{-2t})u(t)}_{\substack{\textit{due to} \\ \textit{source } V_s}} + \underbrace{2(1 - e^{-2t})u(t)}_{\substack{\textit{due to} \\ \textit{source } I_s}}.$$

This, of course, simplifies to $i_L(t) = (3 - 4e^{-2t})u(t)$ A.

The question remains as to why the superposition principle holds an advantage over the Thévenin equivalent method. The answer lies in the following questions:

Question 1. What is the new response if the initial condition is changed to $i_L(0^-) = 5$ A? Within the complete response, decomposed as above, only the part due to the initial condition changes. Since the circuit is linear, this change is linear, i.e., $i_L(0^-)_{\text{new}} = -5 \times i_L(0^-)_{\text{old}}$. Therefore, the part of the response due to the initial condition must be scaled by a factor of -5. Hence, the new response is

$$i_L(t) = 5e^{-2t}u(t) + (1 - e^{-2t})u(t) + 2(1 - e^{-2t})u(t) \text{ A}.$$

Question 2. What is the new response if the voltage source V_s is changed to $5u(t)$ V, with all other parameters held constant at their initial values?

In this case, the value of the voltage source is cut in half, i.e., scaled by a factor of 0.5. Therefore, the contribution to the response due only to V_s must be scaled by 0.5, which results in

$$i_L(t) = -e^{-2t}u(t) + 0.5(1 - e^{-2t})u(t) + 2(1 - e^{-2t})u(t) \text{ A},$$

valid for $t \geq 0$.

Question 3. What is the new response if the initial condition is changed to 5 A, the voltage source V_s is changed to $5u(t)$ V and the current source I_s is changed to $8u(t)$ A?

Again, by linearity, it is necessary to scale each of the individual responses by the appropriate factor. Thus, for $t \geq 0$, the new response is

$$i_L(t) = 5e^{-2t}u(t) + 0.5(1 - e^{-2t})u(t) + 8(1 - e^{-2t})u(t).$$

The preceding discussion argues that linearity permits one to decompose a complete circuit response into the superposition of the response due only to the initial condition and the responses due to each of the source excitations. Changes in any source or in the initial condition are easily reflected in the new response without resolving the circuit equations. This allows one to easily explore a circuit's behavior over a wide range of excitations and initial conditions.

An approach based on the Thévenin equivalent circuit "seen" by the inductor would allow one to quickly compute the complete response, but not in a way that identifies the contribution due to each of the individual sources. Answers to the foregoing three questions would then require new solutions to the circuit equations. There are times, however, when the Thévenin equivalent approach is more efficient. Indeed, the following example shows that one can solve for a variable in a circuit excited by two sources directly, completely bypassing superposition.

EXAMPLE 8.7

For the circuit of figure 8.18 of example 8.6, determine the voltage $v_{R_1}(t)$.

SOLUTION

The voltage across the resistor satisfies the same basic formula as would a capacitor voltage or inductor current. This is clear because $v_{R_1}(t) = V_s - v_L(t)$. (We shall have more to say on this in section 7.) Hence, in accordance with equation 8.17,

$$v_{R_1}(t) = v_{R_1}(\infty) + [v_{R_1}(0^+) - v_{R_1}(\infty)]e^{-\frac{R_{\text{th}}}{L}t}u(t).$$

The time constant of the circuit is the same, $\tau = 0.5$ sec. At $t = \infty$, the inductor looks like a short circuit, in which case $v_{R_1}(\infty) = 10$ V. The only quantity remaining to be found is $v_{R_1}(0^+)$. The equivalent circuit at 0^+ is given by figure 8.22.

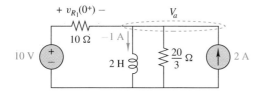

Figure 8.22 Equivalent circuit at $t = 0^+$ of that of figure 8.18.

Writing a node equation for the voltage V_a yields

$$\frac{1}{10}(V_a - 10) - 1 + \frac{3}{20}V_a = 2.$$

The solution is $V_a = 16$ V, in which case $v_{R_1}(0^+) = -6$ V. Therefore, for $t \geq 0$,

$$v_{R_1}(t) = \left(10 + [-6 - 10]e^{-2t}\right)u(t) = \left(10 - 16e^{-2t}\right)u(t) \text{ V}.$$

Again, the presence of the step function $u(t)$ emphasizes that the response is valid only for $t \geq 0$.

6. CLASSIFICATION OF RESPONSES

Having gained some understanding of the form of the responses to RL and RC circuits, we shall find that it is instructive to classify the responses into categories. The **zero-input response** of a circuit is the response to the initial conditions when the input is set to zero. The **zero-state response** is the response to a specified input signal or set of input signals, given that the initial conditions are all set to zero. The sum of the zero-input and zero-state responses is called the **complete response** of the circuit.

Frequently, texts on circuits contain two other notions of response, the natural response and the forced response. However, decomposition of the complete response into the sum of a natural and a forced response applies only when the input excitation is dc, real exponential, sinusoidal, or exponentially modulated or damped sinusoidal. Further, the exponent of the input excitation, e.g., a in $f(t) = e^{at}u(t)$, must be different from those exponents appearing in the zero-input response. Under these conditions, it is possible to define the natural and forced responses as follows: The **natural response** is that portion of the complete response which has the same exponents as the zero-input response; the **forced response** is that portion of the complete response which has the same exponent as the input excitation, provided that the input excitation has exponents different from those of the zero-input response.

Decomposition of the complete response into a natural and a forced response is important for two reasons. First, it agrees with the classical method of solving linear ordinary differential equations having constant coefficients, wherein the natural response corresponds to the complementary function and the forced response corresponds to the particular integral. Students fresh from a course in linear differential equations feel quite at home with these concepts. The second reason is that the forced response is easily calculated for dc inputs.

7. FURTHER POINTS OF ANALYSIS AND THEORY

Equation 8.17 is a physical interpretation of equation 8.16 when the time constant τ is positive. In many practical circuits that incorporate a switching phenomenon, τ is sometimes negative. For example, between switching instants, a negative time constant might be desirable to achieve a fast rise in voltage or current. When τ is negative, the constant $F\tau$ in equation 8.16 no longer has the interpretation of $x(\infty)$. Since $x(\infty)$ has infinite magnitude, as illustrated in example 8.3, a different interpretation must be given to $F\tau$. This is not difficult. By direct substitution,

$$x(t) = \text{constant} = F\tau$$

satisfies the differential equation 8.14. If $x(t)$ is a capacitor voltage, a constant voltage implies zero capacitor current. Thus, $F\tau$ is the capacitor voltage that exists after replacing the capacitor by an open circuit. By analogous reasoning, if $x(t)$ is an inductor current, then $F\tau$ is interpreted as the inductor current after replacing the inductor by a short circuit. Mathematically, any $x(t) = $ constant that satisfies a differential equation is called an **equilibrium state** of that differential equation. This means that we can denote $F\tau$ by X_e (meaning equilibrium state) and rewrite equation 8.17 as

$$x(t) = X_e + [x(t_0^+) - X_e]e^{-\frac{t-t_0}{\tau}}. \tag{8.19}$$

Equation 8.19 is valid for $t \geq t_0$ and for both positive and negative time constants. It is similar to equation 8.17 in form. The only difference is in the interpretation given to $F\tau = X_e$, i.e., the equilibrium value of $x(t)$ computed with C opened or L shorted.

In deriving equations 8.17 and 8.19, the quantity $x(t)$ has been either a capacitor voltage or an inductor current. Since all source excitations are constants, and since the derivatives of the capacitor currents and inductor voltages also have the same general form as equations 8.17 and 8.19, all remaining resistor voltages and currents must also have the same general form. However, it is necessary that the initial value be evaluated at $t = t_0^+$ instead of $t = t_0^-$. This is because only the inductor currents and the capacitor voltages are guaranteed to be continuous from one instant to the next for constant input excitations. The capacitor current and the inductor voltage, as well as other circuit voltages and currents, may not behave continuously. We emphasize this requirement by rewriting equation 8.19 as follows, valid for $t \geq t_0$:

$$y(t) = Y_e + [y(t_0^+) - Y_e]e^{-\frac{t-t_0}{\tau}}, \tag{8.20}$$

where $y(t)$ may be any voltage or current in the circuit, and Y_e is the equilibrium value of $y(t)$. The use of equation 8.20 sometimes simplifies the analysis of first-order networks, as the next two examples will illustrate.

EXAMPLE 8.8

Reconsider example 8.5. This time, the goal is to find $i_1(t)$, the current through R_1 from left to right, for $0 < t < 10$.

SOLUTION

It is possible to obtain $i_1(t) = i_C(t)$ by differentiating $v_C(t)$ obtained in step 2 of example 8.5 and multiplying by C. However, equation 8.20 offers a direct route to getting $i_1(t)$. In equation 8.20, let $x(t) = i_1(t)$. By inspection of the circuit of figure 8.15, $i_1(0^+) = 12/6 = 2$ A, the equilibrium value of the resistor current is $X_e = I_{1e} = 0$, and the time constant is $\tau = R_1C = 3$. From equation 8.20, for $0 \leq t < 10$,

$$i_1(t) = 0 + [2 - 0]e^{\frac{-t}{3}} = 2e^{\frac{-t}{3}} \text{ A}.$$

Note that $i_1(0^-) = 0$. Had the value of $i_1(0^-)$ been used in equation 8.20, the answer would be wrong.

EXAMPLE 8.9

Once again, consider example 8.5. Find $i_C(t)$, the current through the capacitor from top to bottom, for $10 < t < \infty$.

SOLUTION

Differentiating $v_C(t)$ obtained in step 2 of example 8.5 and multiplying by C leads to the answer. However, the use of equation 8.20 leads directly to the answer without differentiation. In that equation, let $x(t) = i_C(t)$. Once again, by inspection of the circuit, we find that $i_C(10^+) = (12 - 11.57)/6 - 11.57/3 = -3.785$ A, the equilibrium current is $I_{Ce} = 0$, and the time constant $\tau = (R_1//R_2)C = 1$.

From equation 8.20, for $10 < t < \infty$,

$$i_C(t) = 0 + [-3.785 - 0]e^{-(t-10)} = -3.785e^{-(t-10)} \text{ A}.$$

Note that $i_C(10^-) = (12 - 11.57)/6 = 0.07166$ A. Using $i_C(10^-)$ in equation 8.15 would have produced a wrong answer.

In several of the previous examples in section 4, the circuit contained switches by means of which a switching action occurred at *prescribed* instants of time. In some electronic circuits, the switch is a semiconductor device whose on-off state is determined by the value of a controlling voltage somewhere else in the circuit. If the controlling voltage is below a certain level, the electronic switch is off; if it is above that level, the electronic switch is on. Knowing the time it takes for a controlling voltage to rise (or to fall) from one level to another is very important because timing is as critical in electronic circuits as scheduling is for large organizations. In the case of first-order linear networks with constant excitations, the calculation is straightforward because all waveforms are exponential functions, as per equation 8.17 or equation 8.20.

Consider figure 8.23. In equation 8.17, let $x(t_1) = X_1$ and $x(t_2) = X_2$ be the two levels of interest. A simple manipulation leads to an **elapsed time formula** for first-order circuits:

$$t_2 - t_1 = \tau \ell n \left(\frac{X_1 - x(\infty)}{X_2 - x(\infty)} \right). \tag{8.21}$$

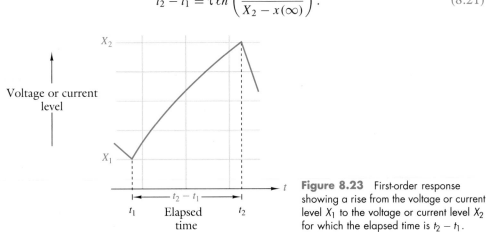

Figure 8.23 First-order response showing a rise from the voltage or current level X_1 to the voltage or current level X_2 for which the elapsed time is $t_2 - t_1$.

EXAMPLE 8.10

This example applies the elapsed time formula to example 8.5 to produce an approximate sawtooth waveform. Suppose the switch is electronically controlled so that it closes when v_C rises to 9 V and opens when v_C falls to 5 V. Determine the waveform of $v_C(t)$ for several switchings.

SOLUTION

Suppose the switch first closes at $t = t_a$, subsequently opens at $t = t_b$, and then closes again at $t = t_c$, etc. For $0 < t < t_a$, $\tau = 3$ sec, $v_C(0) = 0$, and $v_C(\infty) = 12$. Using equation 8.17 or equation 8.20, we obtain

$$v_C(t) = 12 \left(1 - e^{\frac{-t}{3}} \right) \text{ V}$$

and, from the elapsed time formula of equation 8.21,

$$t_a - 0 = 3 \ell n \left[\frac{0 - 12}{9 - 12} \right] = 3 \times 1.386 = 4.159 \text{ sec}.$$

Now, for $t_a < t < t_b$, $\tau = 1$ sec, $v_C(t_a) = 9$, and $v_C(\infty) = 4$. Using equation 8.17 or equation 8.20 and the elapsed time formula yields

$$v_C(t) = 4 + 5e^{-(t-t_a)}$$

and

$$t_b - t_a = 1 \times \ell n \left[\frac{9-4}{5-4} \right] = \ell n(5) = 1.61 \text{ sec.}$$

For the time interval $t_b < t < t_c$, $\tau = 3$ sec, $v_C(t_b) = 5$, and $v_C(\infty) = 12$. Using equation 8.17 or equation 8.20 and the elapsed time formula of equation 8.21, we get

$$v_C(t) = 12 - 7e^{\frac{-(t-t_b)}{3}} \text{ V}$$

and

$$t_c - t_b = 3 \times \ell n \left[\frac{5-12}{9-12} \right] = 3 \times \ell n \left(\frac{7}{3} \right) = 2.54 \text{ sec.}$$

The waveform of $v_C(t)$ for $0 < t < t_c$ is plotted in figure 8.24.

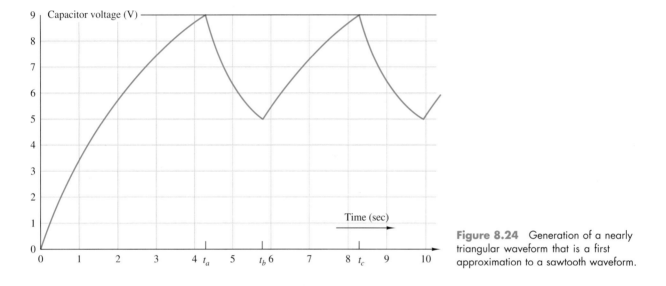

Figure 8.24 Generation of a nearly triangular waveform that is a first approximation to a sawtooth waveform.

From the preceding solution, $t_a = 4.16$ sec, $t_b = 4.16 + 1.61 = 5.77$ sec, and $t_c = 5.77 + 2.54 = 8.31$ sec. If we proceed to calculate the waveform for $t > t_c$, the waveform begins to repeat itself, as is evident from figure 8.24. Practically speaking, the first cycle of a periodic approximately triangular waveform occurs in the time interval $[t_a, t_c]$, and the period is $t_c - t_a = 8.31 - 4.16 = 4.15$ sec.

The waveform in figure 8.24 is approximately *triangular*. This is because two time constants, 1 and 3 sec, have the same order of magnitude. If we select the resistances so that the charging time constant is much larger than the discharging constant, then the capacitance voltage waveform will be closer to a **sawtooth waveform**. (See problem 28.) Sawtooth waveforms are used to drive the horizontal sweep of the electronic beam in an oscilloscope or a television picture tube.

8. FIRST-ORDER RC OP AMP CIRCUITS

RC op amp circuits have some singular characteristics that set them apart from standard passive RC and RL types of circuits. Specifically, because of the nature of the operational amplifier, the time constant of the circuit will often depend on only some of the resistances. This is usually not difficult to see. To illustrate the behavior of RC op amp circuits, we present two examples.

 **EXAMPLE 8.11**

Compute the response $v_{\text{out}}(t)$ for the ideal op amp circuit of figure 8.25, assuming that $v_C(0^-) = 0$.

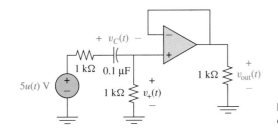

Figure 8.25 Op amp circuit for example 8.11.

SOLUTION

First, observe that the 1-kΩ resistor at the output of the op amp has no effect on $v_{\text{out}}(t)$. This is because in the ideal dependent source model of the op amp (see chapter 3), this load resistance is in parallel with a dependent voltage source whose voltage is independent of the load resistance. Second, note that $v_{\text{out}}(t)$ is equal in value to the voltage $v_+(t)$ across the 1-kΩ resistor attached to the positive terminal of the op amp. Hence, to determine $v_{\text{out}}(t)$, it is necessary to find $v_+(t)$.

To find $v_+(t)$, one first determines $v_C(t)$. This is because the circuit to the left of the op amp is a simple driven RC circuit in which $v_C(t)$ is easily found. Since (ideally) no current enters the positive terminal of the op amp, $i_C(t)$ flows through each resistor, and $v_+(t) = 10^3 i_C(t)$. If $v_C(t)$ is known, then $i_C(t)$ is simply C times the derivative of $v_C(t)$.

From our previous work,

$$v_C(t) = v_C(\infty) + [v_C(0^+) - v_C(\infty)]e^{-\frac{t}{R_{\text{th}}C}}u(t),$$

where $R_{\text{th}} = 2$ kΩ. This implies that the circuit time constant is $\tau = 0.2 \times 10^{-3}$. By assumption, $v_C(0^-) = 0$. At $t = \infty$, the capacitor acts like an open circuit, and thus, the entire source voltage appears across the capacitor at $t = \infty$; hence, $v_C(\infty) = 5$. It follows that for $t \geq 0$,

$$v_C(t) = 5(1 - e^{-5,000t})u(t) \text{ V.}$$

Differentiation yields

$$i_C(t) = C\frac{dv_C(t)}{dt} = 2.5 \times 10^{-3}e^{-5,000t} \text{ A.}$$

Finally,

$$v_{\text{out}}(t) = 10^3 i_C(t) = 2.5e^{-5,000t}u(t) \text{ V.}$$

Exercise. Obtain $v_{\text{out}}(t)$ in example 8.11 directly by the use of equation 8.20. Hint: Let $y(t) = v_+(t)$.

 EXAMPLE 8.12

This second example considers the so-called **leaky integrator circuit** of figure 8.26, which contains an ideal op amp. The input for all time is $v_s(t) = 5u(t)$ V. R_2 represents the leakage resistance of the capacitor. Given C and R_2, R_1 is chosen to achieve an overall gain, in this case, of 1. The objective is to compute the response $v_{\text{out}}(t)$, assuming that $v_C(0^-) = 0$, and compare it with a pure integrator having a gain of 1. This problem is also considered in chapter 13 of volume 2. (See example 13.28.)

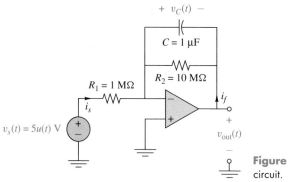

Figure 8.26 Leaky integrator op amp circuit.

SOLUTION

Since the op amp is ideal and the minus terminal is at virtual ground, $v_{out}(t) = -v_C(t)$. Because the circuit is once again a simple RC circuit, the usual formula for the response applies, with $v_{out}(t)$ substituted for $v_C(t)$:

$$v_{out}(t) = v_{out}(\infty) + [v_{out}(0^-) - v_{out}(\infty)]e^{-\frac{t}{R_{th}C}}u(t).$$

By hypothesis, $v_{out}(0^-) = 0$. At $t = \infty$, the capacitor looks like an open circuit under a dc excitation. Hence,

$$v_{out}(\infty) = -\frac{R_2}{R_1}v_s = -10v_s = -50 \text{ V}.$$

Since $R_{th}C = R_2C = 10$, it follows that the output voltage, in volts, is

$$v_{out}(t) = (-50 + 50e^{-0.1t})u(t) = 50(e^{-0.1t} - 1)u(t) \text{ V}.$$

A plot of the op amp output voltage appears in figure 8.27, along with the ideal integrator curve. One observes that the somewhat realistic leaky integrator circuit approximates an ideal integrator only for $0 \le t \le 1.5$ before the error induced by the feedback resistor R_2 becomes too large. Such integrators need to be periodically reinitialized by setting the capacitor voltage to zero.

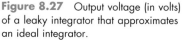

Figure 8.27 Output voltage (in volts) of a leaky integrator that approximates an ideal integrator.

![Plot showing Op amp output voltage vs Time. Vertical axis labeled "Op amp output voltage" from 0 to -20, horizontal axis labeled "Time" from 0 to 4. A solid curve labeled "Leaky integrator response" and a dashed line labeled "Ideal response."]

> ◆ **EXAMPLE 8.13**

In trying to build an inverting amplifier, a student inadvertently reverses the connection of the two input terminals, resulting in the circuit of figure 8.28a. Assume that the op

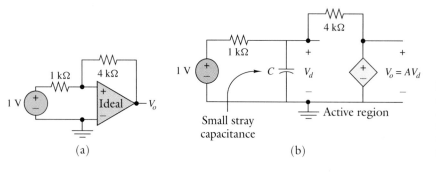

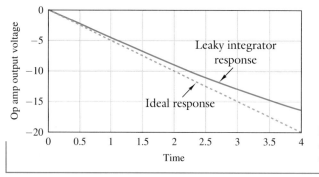

Figure 8.28 Analysis of an incorrectly connected inverting amplifier (gain $A = 10,000$) by incorporating a stray capacitance $C = 1$ pF. (a) The circuit. (b) A more accurate model is used for the op amp.

(a)   (b)

amp is ideal (see figure 3.8) with $V_{\text{sat}} = 9$ V. The student observes a 9-V output in the laboratory. Explain how this 9-V output can exist.

SOLUTION

In example 3.5, resistive models were used to represent the ideal op amp, and three *mathematical* solutions of the circuit were obtained, only one of which gave a 9-V output. The question as to why the other two solutions were not observed could not be answered satisfactorily in chapter 3.

To answer the question, we must use a more realistic model for the op amp. We do this in two ways: (1) the finite gain model of figure 3.7c with $A = 10,000$ is used to represent the op amp; (2) a small **stray capacitance** $C = 1$ pF is assumed to exist between the two input terminals. This more accurate circuit model is shown in figure 8.28b.

The equivalent resistance looking to the right of C is

$$R_{\text{in}} = \frac{4,000}{1 - 10,000} = -2.49975 \ \Omega.$$

The Thévenin equivalent resistance "seen" by C is

$$R_{\text{eq}} = 1,000 || R_{\text{in}} = 1,000 || (-2.49975) = -2.506 \ \Omega.$$

The time constant of the first-order circuit is

$$\tau = R_{\text{eq}}C = -2.506 \times 10^{-12} \ \text{sec, or} \ -2.506 \ \text{picosecond}.$$

The negative time constant spells instability. The complete response may be written directly with the use of equation 8.20 with $y = v_o$:

$$V_{\text{oc}} = V_o \ \text{with } C \text{ open circuited}$$

$$= 1 \times \frac{-2.49775}{1,000 - 2.49775} \times 1,000 = -2.504 \ \text{V}.$$

Let the initial voltage across C be $v_d = \varepsilon$ V, a very small noise voltage. Then

$$v_o(0^+) = 1,000\varepsilon.$$

Equation 8.20 gives the complete response,

$$v_o(t) = -2.504 + [1,000\varepsilon + 2.504]e^{0.399t},$$

where t is assumed to be in picoseconds.

It is clear that, for whatever small positive initial capacitance voltage ε there is, the output would increase exponentially to infinity were it not for the saturation limit of 9 V on the op amp. This is why the 9-V output voltage is observed by the student. Had the initial capacitance voltage been negative, a -9-V output would have been observed.

SUMMARY

This chapter has explored the behavior of first-order RL and RC circuits without sources and when constant excitation sources are present. In general, first-order RL and RC circuits have only one capacitor or one inductor present, although there are special conditions when more than one inductor or capacitor can be present. Our discussion here will presume that only one capacitor or one inductor is present in the circuit.

Using a first-order linear constant coefficient differential equation model of the circuit, the chapter sets forth two types of exponential responses: the source-free response and the response when constant independent sources are present. The source-free responses for the RL and RC circuits have the exponential forms

$$i_L(t) = e^{-\frac{R}{L}(t - t_0)}i_L(t_0)$$

and

$$v_C(t) = e^{-\frac{1}{RC}(t - t_0)}v_C(t_0),$$

where $i_L(t_0)$ is the initial condition for the inductor and $v_C(t_0)$ is the initial condition on the capacitor. For an RC circuit, the time constant τ equals $R_{\text{Th}}C$, where R_{Th} is the Thévenin equivalent resistance seen by the capacitor. On the other hand, for an RL circuit, the time constant τ equals L/R_{Th}, where R_{Th} is the Thévenin equivalent resistance seen by the inductor. If the time constant is positive, the solution shows that the resistor uses up all the energy initially stored in the capacitor or inductor.

When independent sources are present in the circuit, the response of a first-order RC or RL circuit has the general form

$$x(t) = x(\infty) + \left[x(t_0^+) - x(\infty)\right] e^{-\frac{t-t_0}{\tau}},$$

which can be stated equivalently in words as

$$x(t) = [\text{final value}] + ([\text{initial value}]$$
$$- [\text{final value}]) \, e^{-\frac{\text{elapsed time}}{\text{time constant}}},$$

provided that the time constant τ is positive. When τ is negative, it is necessary to modify the interpretation in the manner discussed in section 6. However, when τ is positive, as time approaches infinity, the inductor current and capacitor voltage approach constants. Physically, this means that for a very large elapsed time, the inductor looks like a short circuit and the capacitor looks like an open circuit.

The time constant properties of a circuit can be changed by switching (resistances) within the circuit. By being able to switch time constants in a circuit, one can generate different types of waveforms such as the triangular waveform of figure 8.24. As mentioned at the beginning of the chapter, wave shaping is an important application of circuit design. When inductors, resistors, and capacitors are present in the same circuit, many other wave shapes can be generated. RLC circuits are the topic of the next chapter.

As a final application of the concepts of this chapter, we looked at the leaky integrator op amp circuit. Integrators are present in a whole host of signal-processing and control applications. Unfortunately, ideal integrators do not exist in the practical world. The leaky integrator circuit of figure 8.24 provides a reasonable model of an ideal integrator and is used in practice.

TERMS AND CONCEPTS

Complete response: the sum of the zero-input and zero-state responses.

Differential equation of a circuit: an equation in which a weighted sum of derivatives of an important circuit variable (e.g., a voltage or current) is equated to a weighted sum of derivatives of the source excitations to the circuit.

Equilibrium state of a differential equation: a constant, say, $x(t) = K$, that satisfies the differential equation.

First-order differential equation of a circuit: a differential equation of a circuit in which a first-order derivative is the highest derivative present.

Forced response: that portion of the complete response that has the same exponent as the input excitation, provided that the input excitation has exponents different from those of the zero-input response under the condition that the input excitation is either dc, a real exponential, sinusoidal, or an exponentially modulated or damped sinusoidal.

Integrating factor method: a mathematical technique for finding the solution of a differential equation in which multiplication by the integrating factor, $e^{-\lambda t}$, on both sides of the differential equation leads to a new equation that can be explicitly integrated for a solution.

Leaky integrator circuit: an operational amplifier circuit having a response approximating an ideal integrator, as described in example 8.12.

Natural frequency of a circuit: a natural mode of oscillation of the circuit. For a first-order circuit having a zero-input response proportional to $e^{\lambda t}$, the natural frequency is the coefficient λ in the exponent.

Natural response: that portion of the complete response that has the same exponents as the zero-input response under the conditions outlined in section 6.

Passive RLC circuit: a circuit consisting of resistors, inductors, and capacitors that can only store and/or dissipate energy.

Sawtooth waveform: a triangular waveform resembling the teeth on a saw blade and typically used to drive the horizontal sweep of the electronic beam in an oscilloscope or a television picture tube.

Source-free response: response of a circuit in which sources are either absent or set to zero.

Step response: the response, for $t \geq 0$, of a relaxed single-input circuit to a unit step, i.e., a constant excitation of unit amplitude.

Stray capacitance: small capacitance that is always present between a conductor and ground. Stray capacitance is often modeled in a circuit by a capacitor having a very small value.

Superposition: in linear RC and RL circuits, the sum of the relaxed circuit responses due to each source with all other sources set to zero plus the responses to each initial condition when all other initial conditions are set to zero and all independent sources are set to zero.

Time constant: in a source-free first-order circuit, the time it takes for the circuit response to drop to $e^{-1} = 0.368$ of its initial value. Roughly speaking, the response value must drop to a little over a third of its initial value or rise to within a third of its final value.

Unit step function: a universally used function that is 0 for $t < 0$ and 1 for $t \geq 0$.

Unstable response: a response that increases without bound as t increases.

Zero-input response: response in which all sources are set to zero.

Zero-state response: the response to a specified input signal or set of input signals, given that the initial conditions are all set to zero.

PROBLEMS

1. (a) In the RC circuit shown in figure P8.1, $v_C(0^-)$ = 10 V. Determine $i_C(t)$ for $t \geq 0$. Plot your answer.

(b) Repeat part (a) for $v_C(0^-) = 5$ V and $v_C(0^-) = 20$ V. (*Hint*: What principle makes this part a straightforward calculation, given your answer to part (a)?)

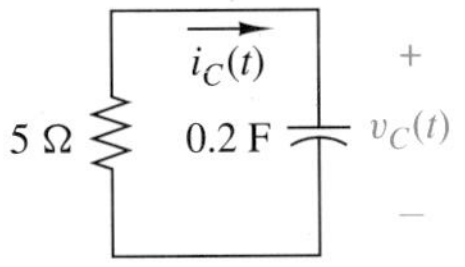

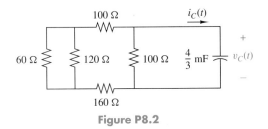

Figure P8.1

2. For the circuit in figure P8.2, determine $v_C(t)$ for $t \geq 0$ when $v_C(0^-) = 80$ V. Plot your result.

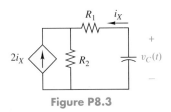

Figure P8.2

3. For the circuit in figure P8.3, find and plot $v_C(t)$ and $i_x(t)$ for $t \geq 0$ when $R_1 = 90$ Ω and $R_2 = 70$ Ω.

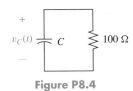

Figure P8.3

4. In figure P8.4, suppose $v_C(0^-) = 10$ V. Determine C so that at $t = 0.1$ sec, $v_C(0.1) = 10e^{-1}$ V.

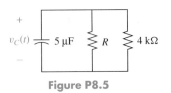

Figure P8.4

5. Consider the circuit of figure P8.5. Determine the value of R and the initial condition, $v_C(0^-)$ so that at $t = 0.012$ sec, $v_C(0.012) = 100e^{-1}$.

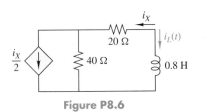

Figure P8.5

6. Consider the circuit shown in figure P8.6. If $i_L(0) = 10$ A, determine $i_L(t)$ for $t \geq 0$. (*Hint*: Find the Thévenin equivalent "seen" by the inductor for $t > 0$.)

Figure P8.6

7. In the circuit shown in figure P8.7, $i_L(0^-) = 25$ A. Determine $i_L(t)$ and $v_L(t)$ for $t \geq 0$.

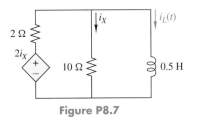

Figure P8.7

8. In figure P8.8, determine the value of L and the value of $i_L(0^-)$ for which the circuit time constant is 0.02 sec and $i_L(0.04) = 2.03$ A.

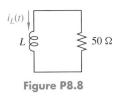

Figure P8.8

9. For figures P8.9a and b, suppose the 120-V source has been applied for a long time.
   (a) In figure P8.9a, find $i_L(t)$ for all t. Sketch $i_L(t)$ for -10 ms $\leq t \leq 10$ ms.
   (b) Find $v_C(t)$ for $t \geq 0$ for the circuit in figure P8.9b, in which the inductor of figure 8.9a has been replaced by a capacitor of 0.2 F. What has happened to the time constant?

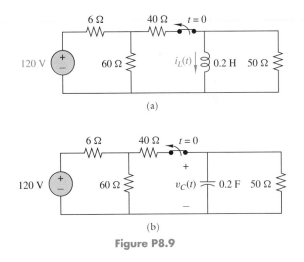

(a)

(b)

Figure P8.9

10. Consider the circuit shown in figure P8.10. The 80-V source has been applied for a long time. Determine $v_C(t)$ for all t.

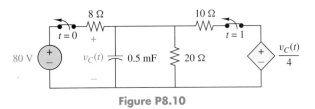

Figure P8.10

11. Consider the circuit of figure P8.11, in which $v_C(0^-) = 15$ V.
(a) Determine $v_C(t)$ for $t \geq 0$.
(b) Determine $i_R(t)$ for $t \geq 0$.

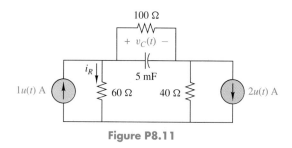

Figure P8.11

12. The circuits of figure P8.12 differ structurally only in the switch that is present in figure P8.12b, but absent in figure P8.12a. The responses, however, are quite different. Why?
(a) For the circuit of figure P8.12a, determine the response $v_C(t)$ for $t \geq 0$.
(b) For the circuit of figure P8.12b, determine the response $v_C(t)$ for $t \geq 0$.

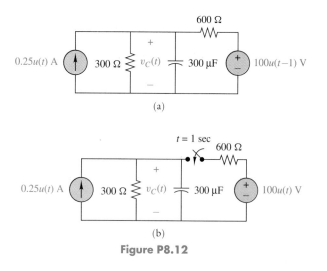

(a)

(b)

Figure P8.12

13. In the circuit shown in figure P8.13, the two voltage sources have been present for a long time. Determine $i_L(t)$ for all t. Provide a rough sketch of your results.

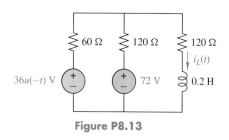

Figure P8.13

14. Consider the circuit shown in figure P8.14.
(a) Find the Thévenin equivalent "seen" by the capacitor for $t \geq 0$.
(b) Find $v_C(t)$ for $t \geq 0^+$, assuming that $v_C(0^-) = -6$ V.

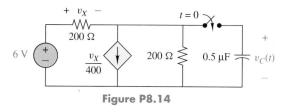

Figure P8.14

15. Figure P8.15a shows an op amp lowpass filter. (Such a filter passes the low-frequency content of signals, while attenuating the high-frequency content.) Figure P8.15b shows a highpass filter. (This filter passes the high-frequency content of signals, while attenuating the low-frequency content.) For each of the circuits, determine the response to a step, assuming that $v_C(0^-) = 0$ in each case.

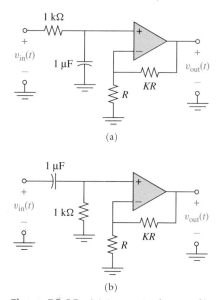

(a)

(b)

Figure P8.15 (a) An op amp lowpass filter. (b) An op amp highpass filter.

16. Consider the circuit of figure P8.16. This problem is motivated by an energy storage system using an inductor and a solar cell. During the day, the solar cell stores energy by increasing the current in the inductor. Then, during the night, the stored energy is used to power lights and so forth. Energy from the solar cell is stored in the inductor during $0 \leq t < T_1$. At $t = 0$, the beginning of the storage interval, $i_L(0^-) = 0$. At $t = T_1$, the device is switched from storing energy in the solar cell, denoted by V_{solar}, to powering a light denoted by R_1. Note that there is some overlap in the switching. This is done to ensure the continuity of the inductor current. Finally, at $t = T_2$, the television, represented by R_2, is also turned on. All answers must be in terms of V_{solar}, R_{solar}, L_{store}, R_{store}, R_1, and R_2.

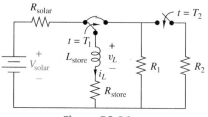

Figure P8.16

CHAPTER 8 FIRST-ORDER RL AND RC CIRCUITS

(a) Indicating all device values, draw a simplified equivalent circuit with three circuit elements for $0 \le t < T_1$.

(b) Derive an expression for $i_L(t)$, $0 \le t < T_1$.

(c) Indicating all device values, draw a simplified equivalent circuit with two circuit elements for $T_1 \le t < T_2$.

(d) Derive an expression for $i_L(t)$, $T_1 \le t < T_2$. You will need an expression for $i_L(T_1^-)$. After obtaining your expression, for simplicity, let I_{T1^-} denote $i_L(T_1^-)$.

(e) Indicating all device values, draw a simplified equivalent circuit with two circuit elements for $T_2 \le t$.

(f) Derive an expression for $i_L(t)$, $T_2 \le t$. You will need an expression for $i_L(T_2^-)$. After obtaining your expression, for simplicity, let I_{T2^-} denote $i_L(T_2^-)$.

Remark: For the remaining two parts, all answers are to be given in terms of V_{solar}, R_{solar}, L_{store}, R_{store}, R_1, R_2, and $i_L(t)$. This is done so that one will not substitute possibly incorrect answers for $i_L(t)$ from prior parts, since writing an answer to parts (g) and (h) in terms of $i_L(t)$ is fine.

(g) For each of the four devices labeled V_{solar}, R_{solar}, L_{store}, and R_{store}, write down an expression for the power absorbed at time t. Call the results $P_{V_{\text{solar}}}$, $P_{R_{\text{solar}}}$, $P_{L_{\text{store}}}$, and $P_{R_{\text{store}}}$.

(h) Give an expression for the energy $W_L(t)$ stored in the inductor at time t if $W_L(t = 0) = 0$.

17. In the circuit shown in figure P8.17, the source voltage becomes a 48-V constant at $t = 0$ and is represented by the excitation $48u(t)$ V. Determine $v_C(t)$ for $t \ge 0$. Assume that $v_C(0^-) = -12$ V. As indicated, the switch closes at $t = 9$ sec. Note that there are several ways to solve this problem.

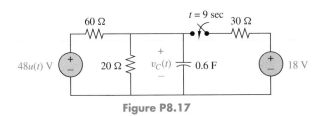

Figure P8.17

18. Consider the circuit shown in figure P8.18, in which the indicated source excitations are valid for all time.

(a) Determine the response $i_L(t)$ for $t \ge 0$.

(b) Determine the inductor voltage $v_L(t)$ for $t > 0$.

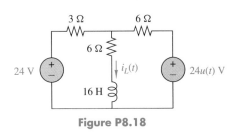

Figure P8.18

19. Consider the circuit shown in figure P8.19, for which the response $v_C(t) = 20 - 10e^{-0.4t}$ V for $t \ge 0$. Identify the natural and forced responses. Suppose further that $i_C(t) = 0.4e^{-0.4t}$ A for $t \ge 0$. Determine I_s, C, the initial capacitor voltage $v_C(0)$, and the value of R.

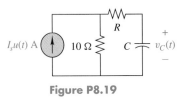

Figure P8.19

20. In the circuit shown in figure P8.20, $v_C(0^-) = 0$, and the 8-V source, denoted by $8u(t)$ V, is applied at $t = 0$.

(a) Determine $v_{\text{out}}(t)$ for $t \ge 0$.

(b) Repeat part (a) if $v_C(0^-) = 5$ V.

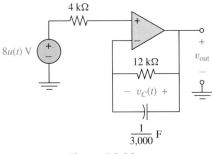

Figure P8.20

21. For the circuit shown in figure P8.21, if $i_L(0^-) = 2$ A and the two sources are turned on at $t = 0$:

(a) Determine $i_L(t)$ for $t \ge 0$.

(b) If R_1 is changed to 3 Ω, determine $i_L(t)$ for $t \ge 0$.

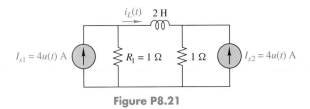

Figure P8.21

22. In the circuit of figure P8.22, suppose $v_{C1}(0^-) = 6$ V and $v_{C2}(0^-) = 24$ V. An 18-V source is applied at $t = 0$, as denoted by $18u(t)$ V. Determine $v_{\text{out}}(t)$ for $t \ge 0$, and determine the total initial and final stored energy in the circuit. (*Hint*: Obtain an equivalent capacitance with an appropriate initial voltage.)

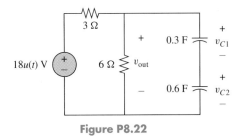

Figure P8.22

23. The derivation of equation 8.16, the solution to the basic RL or RC differential equation, builds on the integral solution of equation 8.4, which is valid for arbitrary $f(t)$. This powerful formula is studied in a course in the theory of differential equations. When $f(t) = F$, a constant, it is possible to develop an alternative derivation of equation 8.16 requiring no more than some basic knowledge of calculus. Since the solution to the source-free case is the exponential

function of equation 8.10 or equation 8.13, it is reasonable to expect (or to try) a solution for the case of constant inputs of the form

$$x(t) = K_1 e^{-\frac{t}{\tau}} + K_2, \qquad \text{(P8.1)}$$

where K_1 and K_2 are two constants to be determined. The constant K_2 arises from the constant input, suggesting that the response would contain a constant term.

(a) Substitute equation P8.1 into equation 8.14. You should obtain the result $K_2 = F\tau$.

(b) With K_2 determined, evaluate $x(t)$ at $t = t_0$ to obtain an expression for K_1. Your result should be

$$K_1 = [x(t_0) - F\tau] e^{\frac{t_0}{\tau}}. \qquad \text{(P8.2)}$$

(c) Finally, substitute $K_2 = F\tau$ and equation P8.2 into equation P8.1. This should yield the solution form given in equation 8.16.

24. An approximate sawtooth waveform can be produced by charging and discharging a capacitor with widely different time constants. The circuit of figure P8.24 illustrates the principle. The switch S is operated as follows: S has been at position B for a long time and is then moved to position A at $t = 0$ to charge the capacitor. When v_C increases to 9 V, S is moved to position B to discharge the capacitor. When v_C decreases to 1 V, S is moved to position A to recharge the capacitor. The process repeats indefinitely.

(a) Determine the waveform of $v_C(t)$ for four switchings.

(b) Plot $v_C(t)$ vs. t. Is the name "sawtooth waveform" appropriate? What is the frequency in Hz of the sawtooth waveform?

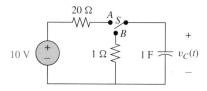

Figure P8.24 A circuit for generating a sawtooth waveform.

25. Although all of the first-order circuits considered in the text have only one capacitor or one inductor, it is possible to have a first-order circuit containing more than one energy storage element, as illustrated in figure P8.25. This problem and the next one illustrate this point. The solutions given by equation 8.10 for the source-free case and equation 8.17 for the constant input case remain valid.

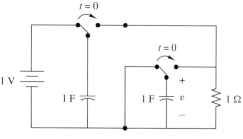

Figure P8.25

(a) Show that for $t > 0$, the voltage $v(t)$ satisfies the first-order differential equation

$$\frac{dv(t)}{dt} = -0.5v(t).$$

(b) Find $v(0^-)$ by inspection of the circuit and $v(0^+)$ by the principle of conservation of charge.

(c) Use equation 8.10 to write down directly the answer for $v(t)$ for $t > 0$. Had $v(0^-)$ been used in equation 8.10, would this answer be correct?

26. The circuit shown in figure P8.26 contains two capacitors.

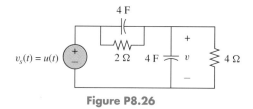

Figure P8.26

(a) Show that for $t > 0$, the voltage $v(t)$ satisfies the first-order differential equation

$$\frac{dv}{dt} = -\frac{3}{32}v + \frac{1}{16}.$$

(b) The input $v_s(t) = u(t)$ has been applied for a long time. Find $v(0^-)$ by inspection of the circuit and $v(0^+)$ by the principle of conservation of charge.

ANSWERS: 0, 0.5 V.

(c) Use equation 8.17 to write down directly the answer for $v(t)$ for $t > 0$. Had $v(0^-)$ been used in equation 8.10, would this answer be correct?

27. The circuit shown in figure P8.27 is a transistor photo timer used for timing the light in photographic enlarger and printing boxes. Briefly, the circuit operates as follows: When the relay contact closes, the lamp is lit; when the contact opens, the lamp is turned off. The relay has 4,000 Ω dc resistance and a negligible inductance. The pickup current is 2 mA, and the dropout current is 0.5 mA; i.e., the contact closes when the relay current increases from zero and reaches 2 mA, and it opens when the current drops below 0.5 mA.

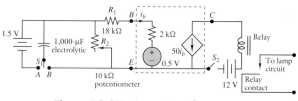

Figure P8.27 A transistor photo timer.

To use the timer, switch S_2 is closed first. Switch S_1 is normally in its B position. When it is thrown momentarily to position A, the 1,000-μF electrolytic capacitor C charges from the battery to 1.5 V. When S_1 is then thrown back to position B at $t = 0$, the capacitor discharges and produces a current i_b that, after amplification by the transistor, actuates the relay and turns on the lamp. At some later instant t_1, the amplified current drops below a point for the relay to open, and the lamp is turned off.

Determine t_1 if the 10-kΩ potentiometer is set at the middle of its full range (i.e., only 5 kΩ is used in the circuit).

28. The sawtooth waveform is used in television sets and oscillographs to control the horizontal motion of an electron beam that sweeps across the screen. One method of generating such a waveform is to charge a capacitor with a large time constant and then discharge it with a very small time constant. The circuit in the shaded box of figure P8.28a is a crude functional model of the neon bulb (figure 8.28b, type 5AB, 59 cents each) whose v-i characteristic is shown in figure 8.28c. The switch S in the model operates as follows:

    (a) S is at position A when v is less than 90 V and *increasing*, and S moves to B when v reaches 90 V (the breakdown voltage).

    (b) S is at position B when v is greater than 60 V and *decreasing*, and S moves to position A when i drops to 1 mA and v drops to 60 V.

Assume that at $t = 0$, S is at A and $v_o(0) = 60$ V. Find $v_o(t)$ for one cycle of operation, i.e., charging and discharging of the capacitor, and make a rough sketch of the waveform. What is the frequency of the sawtooth waveform?

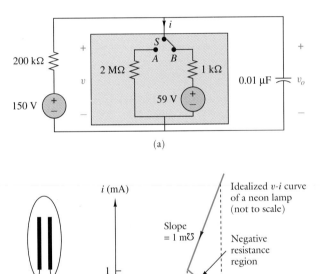

(a)

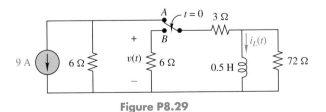

(b)   (c)

Figure P8.28   This neon bulb circuit generates a sawtooth waveform.

29. In figure P8.29, find $v(t)$ and $i_L(t)$ for $t > 0$ if the switch has only been in position A for $t < 0$ and moves to position B at $t = 0$.

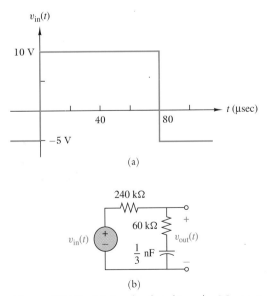

Figure P8.29

30. In the circuit of figure P8.30, the switch is thrown at time $t = 0$. The initial charges stored on the two capacitors are $v_{C1}(0^-) = 1$ V and $v_{C2}(0^-) = 0$ V.

(a) Find an expression for the current $i_R(t)$ for $t \geq 0$. On the same set of axes, plot the current for $R = 5, 1$, and 0.5 Ω.

(b) Compute the total energy dissipated in the resistor for $t \geq 0$. Does the dissipated energy depend upon the value of R? Compute the initial energy stored in the two capacitors and the final $(t \to \infty)$ energy stored in the two capacitors. Verify that conservation of energy holds for the circuit.

(c) Compute the charge stored in the system at $t = 0^-$ and at $t = \infty$, and compare the two values.

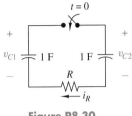

Figure P8.30

31. Explain how you could compute the steady-state values of the capacitor voltages in the circuit of figure P8.31 without knowing the value of the resistor. For $v_1(0^-) = 1$ V and $v_2(0^-) = 0.5$ V, compute $v_1(\infty)$ and $v_2(\infty)$.

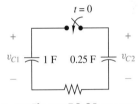

Figure P8.31

32. Determine and plot the output voltage of the circuit in figure P8.32. The circuit is driven by the 80-microsecond pulse. Notice that, for $t < 0$, the input waveform has been at -5 V for a very long time.

(a)

(b)

Figure P8.32   (a) A pulse that drives the RC circuit of (b).

33. The voltage waveform $v_{in}(t)$ shown in figure P8.33a drives the circuit of figure P8.33b. The voltage-controlled switch S_1 closes when the output voltage $v_C(t)$ goes positive and opens when $v_C(t)$ goes negative. Compute the voltage $v_C(t)$ across the capacitor. Assume that $v_{in}(t)$ has been at -10 V for $t < 0$ for a very long time.

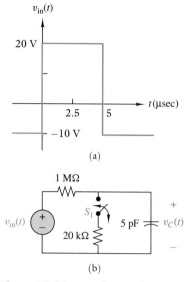

(a)

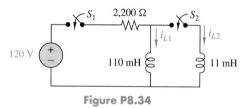

(b)

Figure P8.33 (a) Pulse waveform that excites RC circuit of (b).

34. In the circuit of figure P8.34, switch S_1 closes at $t = 0$, and switch S_2 closes 50 msec later. Determine and plot the waveforms for $i_{L1}(t)$ and $i_{L2}(t)$.

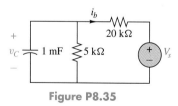

Figure P8.34

35. Consider the circuit of figure P8.35.
(a) If $v_C(0) = 1.5$ V and $V_s = 0.5$ V, find $v_C(t)$ and $i_b(t)$ for $t \geq 0$.
(b) Find the time t_1 at which the current i_b drops to 10 μA.
(c) If $v_C(0) = 1.5$ V and $V_s = 20$ V, find $v_C(t)$ for $t \geq 0$ and the time t_1 at which the voltage v_C reaches 3 V.

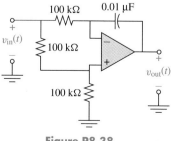

Figure P8.35

36. Consider the series circuit of figure P8.36. Suppose $v_C(0) = 0$ and $V_C(t)$ is the voltage to be found for $t \geq 0$.
(a) Find the forced response by viewing C as an open circuit.
(b) Is the forced response found in part (a) equal to $v_C(\infty)$?

(c) Find the complete response $v_C(t)$ for $t \geq 0$. What is $v_C(\infty)$?

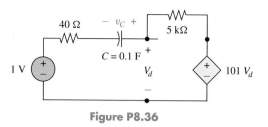

Figure P8.36

37. In design, it is often necessary to generate waveforms such as that shown in figure P8.37a, using an op amp circuit. This waveform is used in a video demodulator circuit. It is possible to generate the waveform by driving an op amp circuit composed of an integrator and a summing amplifier with the square wave shown in figure P8.37b. Develop, draw, and label a circuit that will produce the desired waveform from the square wave input, and select values of resistance and capacitance in the circuit to give the scaling shown in figure P8.37a. Choose realistic values of R and C.

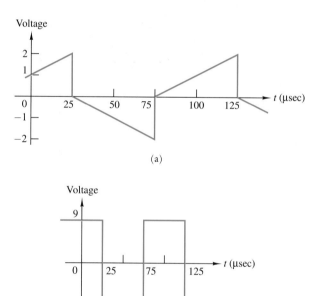

Figure P8.37 (a) Desired output waveform. (b) The square wave input.

38. For the circuit shown in figure P8.38, find $v_{out}(t)$ as a function of $v_{in}(t)$.

Figure P8.38

39. In the circuit of figure P8.39, all op amps are assumed to be ideal. All capacitors are uncharged at $t = 0$. The first two op amps are connected as differentiators, and the last is an inverting amplifier.

(a) With switch S at position A and $v_s(t) = \sin(1,000t)$ V, find $v_a(t)$, $v_b(t)$, and $v_o(t)$ for $t \geq 0$.

(b) If, at a later instant, switch S is quickly moved to positon B, what would you expect $v_o(t)$ to be?

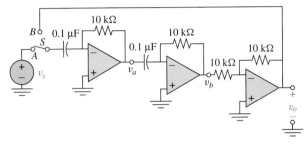

Figure P8.39 Differentiating op amp circuit.

40. Consider the circuit of figure P8.40.

(a) Figure P8.40a shows a source-free parallel RLC circuit whose past history is depicted by figure P8.40b, in which the switch S has been at position A for a long time before moving to position B at $t = 0$. Find the inital values $v_C(0^+)$ and $i_L(0^+)$.

(b) Change the values of 10 V, 1 Ω, and 4 Ω in figure P8.40b so that the initial values of the source-free circuit are $v_C(0^+) = 20$ V and $i_L(0^+) = 5$ A.

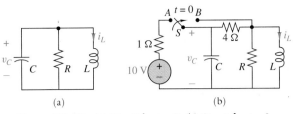

(a)　　　　　(b)

Figure P8.40 (a) Circuit for $t > 0$. (b) Circuit for $t < 0$.

41. Consider the circuit of figure P8.41.

(a) Figure P8.41a shows a source-free parallel RLC circuit whose past history is depicted by figure P8.41b, in which the switches S have been at position A for a long time before moving to position B at $t = 0$. Find the initial values $v_C(0^+)$ and $i_L(0^+)$.

(b) Change the values of 2 V and 0.1 Ω in figure P8.41b so that the initial values of the source-free circuit are $v_C(0^+) = 10$ V and $i_L(0^+) = -2$ A.

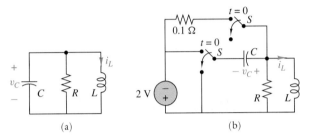

(a)　　　　　(b)

Figure P8.41 (a) Circuit for $t > 0$. (b) Circuit for $t < 0$.

CHAPTER 9

Second-Order Linear Circuits

W ARMING up snacks in a microwave oven is a common scene in student dormitories. A microwave oven works much faster than a conventional oven; heating up a sandwich takes about 30 sec. How does the microwave oven do this? While a precise explanation is beyond the scope of this text, the basic principle can be understood through the properties of a simple LC circuit.

Recall that two conducting plates separated by a dielectric (insulating material) form a capacitor. Suppose some food were placed between the plates instead of an ordinary dielectric. The food itself will act as a dielectric. Ordinary food contains a great number of water molecules. Each water molecule has a positively charged end and a negatively charged end. Their orientations are totally random for uncharged plates, as illustrated in figure (a) on this page. Applying a sufficiently high dc voltage to the plates sets up an electric field produced by the charge deposited on the plates. This causes the water molecules to align themselves with the field, as illustrated in figure (b). If the polarity of the dc voltage is reversed, the molecules will realign themselves in the opposite direction, as illustrated in figure (c). If the polarity of the applied voltage is reversed repeatedly, then the water molecules will repeatedly flip their orientations. In doing so, the molecules encounter considerable friction, resulting in a buildup of heat, which cooks the food. Microwave cooking is therefore very different from conventional cooking. Instead of heat coming from the outside, the heat is generated inside the food itself.

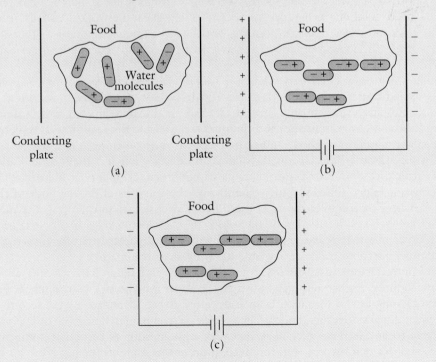

Reversals of the polarity of the applied voltage at a low frequency can be easily achieved with the circuit elements studied in earlier chapters: the resistor, the capacitor, and the inductor. However, the friction-induced heat production is inefficient at low frequencies. To have a useful amount of heat for cooking purposes, very high frequencies must be used. The typical frequency used in a microwave oven is 2.45 terahertz; i.e., the water molecules reverse their orientations $2 \times 2.45 \times 10^{12}$ times per second. At such a high frequency, capacitors and inductors are quite different from their conventional forms. For example, the LC circuit becomes a "resonant cavity," and the connecting wire becomes a "wave guide." These microwave components are studied in a course in field theory. The theory studied in this chapter will enable us to understand the low-frequency version of the phenomenon, i.e., how an inductor and a capacitor connected together can produce oscillatory voltage and current waveforms. ◆

CHAPTER OBJECTIVES

- Understand and explore the phenomenon that occurs when an ideal inductor is connected to an ideal capacitor with an initial stored charge.
- Use a second-order differential equation for modeling series RLC and parallel RLC circuits.
- Develop a systematic method for writing a second-order differential equation model of a general second-order linear circuit.
- Learn how to solve a second-order differential equation circuit model excited by a dc source by first determining the natural frequencies of the circuit, then obtaining the general form of the solution, and, finally, determining the associated arbitrary constants.
- Define and understand the significance of the underdamped, overdamped, and critically damped responses.
- Investigate and understand the underlying principles of several oscillator circuits.

1. INTRODUCTION

The previous chapter developed techniques for computing the responses of first-order linear networks, either without sources or with constant-valued sources. This chapter extends these solution techniques to second-order linear networks. Usually, but not always, a second-order network contains two energy storage elements—either (L, C), (C, C), or (L, L). Second-order linear networks have second-order differential equation models.

Unlike a first-order linear network whose source-free response has only real exponential terms, second-order linear networks have a wide variety of response waveforms: exponentials, sinusoids, exponentially damped sinusoids, and exponentially growing sinusoids, among others. Many computational techniques are available for analyzing these networks. The Laplace transform method, studied in a second course in circuits, is a general analysis tool for second-order linear networks subject to arbitrary initial conditions and arbitrary input excitations. For second-order circuits without sources or for second-order circuits with constant-valued sources, some straightforward extensions of the methods of chapter 8 are sufficient to compute the responses. These extensions—and, indeed, the methods themselves—build on a basic knowledge of calculus. Unfortunately, further extensions to higher order linear networks are impractical. Again, the Laplace transform method and several other advanced tools are available for analyzing such circuits.

Many introductory texts discuss only the parallel and series RLC circuits and state separate formulas for the responses of each. These treatments lack completeness and are plagued by long lists of formulas. It appears better first to formulate a basic second-order differential equation circuit model. Then, solution techniques based on this model are applicable to any second-order linear network without sources or with constant-valued sources and, for that matter, even to a second-order mechanical system.

An oscillator circuit (section 2) serves to motivate the study of second-order linear networks. This is one of several practical examples scattered throughout the chapter. Unfortunately, most applications require a background that students at the introductory level simply do not have. Some of these applications include lowpass, highpass, and bandpass filtering covered in volume 2 and in more advanced texts on circuits. Other applications include dc motor analysis, position control, the analysis and design of a car wheel suspension, and the motion of a boat in a river. What is significant about the material in this chapter is that the ideas are common to a host of engineering problems. Hence, the techniques and concepts in the chapter will prove useful time and time again.

2. DISCHARGING A CAPACITOR THROUGH AN INDUCTOR

As explained in chapter 8, an initially charged capacitor in parallel with a resistor has a voltage which decreases exponentially to zero: The capacitor discharges its stored energy through the resistor. What voltage waveform emerges if an inductor replaces the resistor?

The circuit of figure 9.1a aids in finding an answer to this question. The switch S, originally at position A, is moved to position B at $t = 0$. Our goal is to determine an expression for the voltage, $v_C(t)$, for $t \geq 0$. Figure 9.1b shows the circuit of interest for $t \geq 0$.

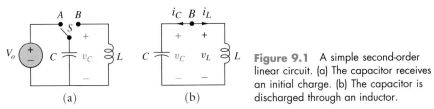

Figure 9.1 A simple second-order linear circuit. (a) The capacitor receives an initial charge. (b) The capacitor is discharged through an inductor.

From the v-i relationships for L and C (chapter 7) in conjunction with KCL and KVL, it follows that

$$\underbrace{\frac{dv_C}{dt} = \underbrace{\frac{i_C}{C}}_{\text{definition of } C} = \underbrace{-\frac{i_L}{C}}_{\text{KCL}}}$$

(9.1a)

and

$$\underbrace{\frac{di_L}{dt} = \underbrace{\frac{v_L}{L}}_{\text{definition of } L} = \underbrace{\frac{v_C}{L}}_{\text{KVL}}}.$$

(9.1b)

We may write equations 9.1a and 9.1b as a single matrix equation:

$$\begin{bmatrix} \dfrac{dv_C}{dt} \\ \dfrac{di_L}{dt} \end{bmatrix} = \begin{bmatrix} 0 & \dfrac{-1}{C} \\ \dfrac{1}{L} & 0 \end{bmatrix} \begin{bmatrix} v_C \\ i_L \end{bmatrix}.$$

(9.1c)

A set of coupled first-order differential equations such as equation 9.1 is called the **state equation** for the circuit. The functions v_C and i_L are unknowns. These unknown functions are called **state variables**. In more advanced texts, the properties of a linear system or circuit are typically investigated via state equations. At the level of circuit analysis in this text, however, in order not to rely too heavily on the background of matrix theory, one may use an alternative characterization of the network: a single second-order differential equation in one unknown. In the present case, a derivation of such a model is easy. Section 5 covers the general case.

To derive the second-order differential equation model, we first differentiate both sides of equation 9.1a to obtain

$$\frac{d^2v_C}{dt^2} = -\frac{1}{C}\frac{di_L}{dt}.$$

Substituting equation 9.1b into this equation yields

$$\frac{d^2 v_C}{dt^2} = -\frac{1}{LC} v_C. \tag{9.2}$$

This is a second-order differential equation circuit model with the function $v_C(t)$ as the unknown.

> **Exercise.** Starting from equation 9.1, derive a single second-order differential equation with the function $i_L(t)$ as the unknown. Your answer should be the same as equation 9.2, with i_L replacing v_C.

Equation 9.2 requires that the second derivative of the unknown function $v_C(t)$ equal the function itself multiplied by a *negative constant*, $-1/LC$. Even though students in a beginning course in circuits may not have a background in differential equations, the solution of equation 9.2 should present no difficulty, as it requires only a very elementary knowledge of calculus. In particular, recall the derivative properties of the sine and cosine functions:

$$\frac{d}{dt} \sin(\omega t) = \omega \cos(\omega t) \quad \text{and} \quad \frac{d}{dt} \cos(\omega t) = -\omega \sin(\omega t).$$

It follows that

$$\frac{d^2}{dt^2} \cos(\omega t + \theta) = -\omega^2 \cos(\omega t + \theta)$$

and

$$\frac{d^2}{dt^2} \sin(\omega t + \theta) = -\omega^2 \sin(\omega t + \theta).$$

Both the sine and the cosine functions have the desired property: The second derivative of the function equals the function itself multiplied by a *negative constant*. Therefore, it is reasonable to assume that a solution of equation 9.2 has the general form

$$v_C(t) = K \cos(\omega t + \theta) = A \cos(\omega t) + B \sin(\omega t). \tag{9.3}$$

The derivative of $v_C(t)$ is

$$v_C'(t) = -K \omega \sin(\omega t + \theta). \tag{9.4}$$

Differentiating equation 9.4 again yields

$$\frac{d^2 v_C}{dt^2} = -K \omega^2 \cos(\omega t + \theta) = -\omega^2 v_C. \tag{9.5}$$

Matching the coefficients of equation 9.5 with those of equation 9.2 implies that

$$\omega^2 = \frac{1}{LC}, \quad \text{or} \quad \omega = \frac{1}{\sqrt{LC}}. \tag{9.6}$$

It remains to determine the constants K and θ (or equivalently, A and B) in equation 9.3. These depend on the initial conditions as follows. When the switch is at position A, the capacitor is charged up to V_o volts and the inductor current is zero. Immediately after the switch moves to position B, i.e., at $t = 0^+$, the capacitor voltage remains at V_o and the inductor current remains at zero, since the capacitor voltage and inductor current in this circuit cannot change *instantaneously*. The initial value, $v_C'(0^+)$, is now calculated from equation 9.1a as

$$v_C'(0^+) = \frac{i_C(0^+)}{C} = \frac{-i_L(0^+)}{C} = 0.$$

Evaluating equations 9.3 and 9.4 at $t = 0^+$ yields

$$V_o = K \cos(\theta) \qquad\qquad (9.7a)$$

and

$$0 = \omega K \sin(\theta). \qquad\qquad (9.7b)$$

From these two conditions, it follows that $\theta = 0$ and $K = V_o$. Hence, the solution for $v_C(t)$ is

$$v_C(t) = V_o \cos\left(\frac{1}{\sqrt{LC}} t\right). \qquad\qquad (9.8)$$

This completes the mathematical analysis of the simple circuit of figure 9.1.

Several very interesting and significant facts about this circuit and the method of solving it must be pointed out:

1. For the source-free LC circuit of figure 9.1, the voltage and current responses are sinusoidal waveforms with an angular frequency equal to $1/\sqrt{LC}$. Since the amplitude of the sinusoidal oscillations remains constant (i.e., it does not damp out), the circuit is said to be **undamped**. The angular frequency $\omega = 1/\sqrt{LC}$ is called the **undamped resonant frequency** (or undamped oscillation frequency).

2. The frequency of the sinusoidal oscillations depends on the values of L and C only, while the amplitude K and the phase angle θ depend on L, C, and the initial values of the capacitor voltage and inductor current.

3. Although the energy stored in the capacitor, $w_C(t)$, and the energy stored in the inductor, $w_L(t)$, both vary with time, their sum is constant. (See problem 1.) There is a continuous transfer between the energy stored in the magnetic field of the inductor and that stored in the electric field of the capacitor, with no *net* energy loss. This is analogous to a frictionless pendulum swinging in a vacuum: Because of the absence of friction, the seesaw motion of the pendulum never stops. In this mechanical system, there is a continuous interchange between potential energy and kinetic energy.

Figure 9.1 shows, in theory, the simplest circuit that generates sinusoidal waveforms; it is an idealized oscillator model. An electronic circuit that generates sinusoidal waveforms as just computed is called an **oscillator circuit**. Oscillator circuits play a very important role in many communication and instrumentation systems. In practice, however, discharging a capacitor through an inductor does not produce pure sinusoidal oscillations. Rather, it produces sinusoids with decaying amplitudes. The reason is that a practical inductor has a small series resistance and a practical capacitor has a large parallel resistance in their circuit models. Figure 9.2 shows a more accurate circuit model, in which R_1, typically a small value, represents the resistance of the inductor wire and R_2, typically a large value, represents the effective value of the capacitor leakage resistance and any load

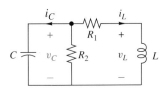

Figure 9.2 A more accurate circuit model accounting for losses.

connected to the oscillator. Since both resistors consume energy, the total energy stored in the magnetic and electric fields decreases monotonically, accompanied by a corresponding decrease in the amplitude of oscillation. Intuitively, one would expect the waveforms to be *damped sinusoids*. The next section provides a precise mathematical analysis of this valid expectation.

3. SOURCE-FREE SECOND-ORDER LINEAR NETWORKS

Development of the Differential Equation Model

The circuits of figures 9.1b and 9.2 are second-order source-free linear networks containing two energy storage elements, one capacitor and one inductor. Figure 9.3 shows two more examples of second-order driven linear circuits in which the energy storage elements are both capacitors or both inductors.

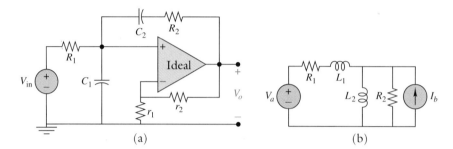

Figure 9.3 Further examples of second-order circuits. (a) A second-order circuit with two capacitors. (b) A second-order circuit with two inductors.

If all independent sources have zero values, then the second-order linear network is termed **source free**. Source-free circuits can be characterized by either of the following two forms of equations, in which the x's are typically capacitor voltages or inductor currents:

1. *State equations*

$$\frac{dx_1}{dt} = a_{11}x_1 + a_{12}x_2,$$

$$\frac{dx_2}{dt} = a_{21}x_1 + a_{22}x_2. \tag{9.9}$$

2. *A single second-order differential equation*

$$\frac{d^2x}{dt^2} + 2\sigma\frac{dx}{dt} + \omega_n^2 x = 0. \tag{9.10}$$

The constant σ is called the **damping coefficient**, and the constant ω_n is called the **undamped oscillation frequency**. The relative magnitudes of σ and ω_n determine the distinctive characteristics of the response waveforms, to be investigated in detail shortly.

For a general second-order source-free network, constructing the differential equation model of equation 9.10 often entails writing the state equations 9.9 as a first step. Section 9.5 delineates a general procedure for doing this. However, for some simple circuits, it is possible to write the single differential equation 9.10 directly without first writing the state equations. Example 9.1 illustrates this.

◆ EXAMPLE 9.1

For the series and parallel RLC circuits shown in figure 9.4, write a single second-order differential equation model, and find the constants ω_n and σ in terms of the circuit parameters.

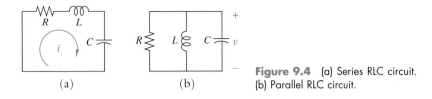

Figure 9.4 (a) Series RLC circuit. (b) Parallel RLC circuit.

SOLUTION

(1) *Determine the differential equation of the series circuit of figure 9.4a.* The series RLC circuit of figure 9.4a has a mesh equation

$$Ri + L\frac{di}{dt} + \frac{1}{C}\int_{-\infty}^{t} i(\tau)d\tau = 0.$$

Differentiating with respect to t yields

$$R\frac{di}{dt} + L\frac{d^2i}{dt^2} + \frac{i}{C} = 0.$$

Equivalently, after dividing through by L,

$$\frac{d^2i}{dt^2} + \frac{R}{L}\frac{di}{dt} + \frac{1}{LC}i = 0. \tag{9.11}$$

Equation 9.11 represents the differential equation model of the series RLC circuit.

(2) *Determine the parameters* x(t), σ, *and* ω_n *of equation 9.10.* Comparing equation 9.11 with equation 9.10, we have:

$$x(t) = i(t), \quad \omega_n = \frac{1}{\sqrt{LC}}, \quad \sigma = \frac{0.5R}{L}.$$

(3) *Determine the differential equation of the parallel RLC circuit of figure 9.4b.* Writing a node equation produces

$$\frac{v}{R} + C\frac{dv}{dt} + \frac{1}{L}\int_{-\infty}^{t} v(\tau)d\tau = 0.$$

Differentiating with respect to time yields

$$\frac{1}{R}\frac{dv}{dt} + C\frac{d^2v}{dt^2} + \frac{1}{L}v = 0,$$

or equivalently,

$$\frac{d^2v}{dt^2} + \frac{1}{RC}\frac{dv}{dt} + \frac{1}{LC}v = 0. \tag{9.12}$$

Equation 9.12 represents the differential equation model of the parallel RLC circuit.

(4) *Determine the parameters* x(t), σ, *and* ω_n. Comparing equation 9.12 with equation 9.10, we have

$$x(t) = v(t), \quad \omega_n = \frac{1}{\sqrt{LC}}, \quad \sigma = \frac{0.5}{RC}.$$

Exercise. For figure 9.4a, write a second-order differential equation with the capacitance voltage as the independent variable. (*Hint:* Substitute $i = Cdv/dt$ into the mesh equation.)

Exercise. For figure 9.4b, write a second-order differential equation with the inductance current as the independent variable. (*Hint:* Substitute $v = Ldi/dt$ into the node equation.)

Solution of the Second-Order Differential Equation Model

Having obtained the homogeneous differential equation 9.10 for a source-free second-order linear network, it is necessary to construct a solution to the differential equation. It is possible to find the solution by a method similar to that used in section 9.2 for the

---

undamped case. The method consists of the following steps:

1. Determine the differential equation model of the circuit.
2. From the differential equation model, determine the characteristic equation, and find its roots using the quadratic root formula.
3. From the nature (real or complex) of the roots of the characteristic equation, determine the general form of the solution, which contains two unknown parameters.
4. Using the initial conditions of the circuit, determine the two unknown parameters.

To illustrate this method, suppose $\sigma = 0$ in equation 9.10. The general solution has the form of a sine or cosine function, as illustrated by equation 9.3, with K and θ as two undetermined parameters. When σ is NOT zero, the form of the general solution is unclear. Following the method used in chapter 8 for first-order networks, *assume* a solution of the form

$$x(t) = Ke^{st}, \qquad (9.13)$$

and determine the allowable values for s. Substituting equation 9.13 into equation 9.10 yields

$$Ks^2e^{st} + 2K\sigma se^{st} + \omega_n^2 Ke^{st} = Ke^{st}(s^2 + 2\sigma s + \omega_n^2) = 0.$$

For nontrivial solutions, K cannot be zero. The function e^{st} is always different from zero. Hence, the quadratic factor in s must be zero, which means that s must be a root of

$$s^2 + 2\sigma s + \omega_n^2 = 0; \qquad (9.14)$$

i.e., the roots are given by

$$s_1 = -\sigma + \sqrt{\sigma^2 - \omega_n^2}$$

and

$$s_2 = -\sigma - \sqrt{\sigma^2 - \omega_n^2}. \qquad (9.15)$$

Equation 9.14 is called the **characteristic equation** of the second-order linear circuit. The roots of the characteristic equation, equation 9.15, are called the **natural frequencies** of the circuit. These are the intrinsic frequencies of the circuit response and are akin to the natural frequencies of oscillations of a pendulum or of a bouncing ball.

From elementary algebra, a quadratic equation (such as the characteristic equation above) can have two distinct real roots, two repeated real roots, or two conjugate complex roots, depending on whether the discriminant is greater than, equal to, or less than zero. This grouping separates the solution of equation 9.10 into three corresponding categories.

Case 1. *Real and distinct roots,* i.e., $\sigma^2 > \omega_n^2$.

If the roots are real and distinct, then for arbitrary constants K_1 and K_2, both

$$x(t) = x_1(t) = K_1 e^{s_1 t}$$

and

$$x(t) = x_2(t) = K_2 e^{s_2 t}$$

satisfy the second-order differential equation 9.10. Since equation 9.10 is a *linear* homogeneous differential equation, the sum $x(t) = x_1(t) + x_2(t)$ is also a solution—a fact easily verified by direct substitution. Therefore, the most general solution to equation 9.10 is

$$x(t) = K_1 e^{s_1 t} + K_2 e^{s_2 t}, \qquad (9.16)$$

where $s_1 \neq s_2$ and s_1 and s_2 are real. The arbitrary constants K_1 and K_2 depend on the initial conditions of the differential equation of the circuit. Such a response is said to be **overdamped.** The significance of the term will be explained shortly, after discussing the remaining two cases.

Case 2. *Complex and distinct roots,* i.e., $\sigma^2 < \omega_n^2$.

The complex roots are given by

$$s_{1,2} = -\sigma \pm j\sqrt{\omega_n^2 - \sigma^2} = -\sigma \pm j\omega_d, \qquad (9.17)$$

where

$$\omega_d = \sqrt{\omega_n^2 - \sigma^2} \qquad (9.18)$$

is called the **damped oscillation frequency** of the circuit. Since s_1 and s_2 are complex conjugates, so are $e^{s_1 t}$ and $e^{s_2 t}$ in equation 9.16. The constants K_1 and K_2 in that equation must be chosen to be complex conjugates in order for $x(t)$ to be real. Thus, $K_2 = K_1^*$. The two terms in equation 9.16 can be combined to yield a real-time function by the use of **Euler's formula**,

$$e^{jy} = \cos y + j \sin y.$$

Specifically,

$$
\begin{aligned}
x(t) &= K_1 e^{s_1 t} + K_2 e^{s_2 t} \\
&= K_1 e^{(-\sigma + j\omega_d)t} + K_2 e^{(-\sigma - j\omega_d)t} \\
&= e^{-\sigma t}[K_1 \cos(\omega_d t) + jK_1 \sin(\omega_d t)] + e^{-\sigma t}[K_2 \cos(\omega_d t) - jK_2 \sin(\omega_d t)] \\
&= e^{-\sigma t}[A \cos(\omega_d t) + B \sin(\omega_d t)],
\end{aligned}
$$

where $A = K_1 + K_2$ and $B = jK_1 - jK_2$ are two *real* constants. Hence, when the natural frequencies are complex, the general solution has the form

$$x(t) = e^{-\sigma t}[A \cos(\omega_d t) + B \sin(\omega_d t)] = Ke^{-\sigma t} \cos(\omega_d t + \theta), \qquad (9.19)$$

where

$$K = \sqrt{A^2 + B^2}, \qquad \theta = \tan^{-1}\left(\frac{-B}{A}\right).$$

This is a damped sinusoidal response, and the circuit is said to be **underdamped.** Note that the angular frequency of the sinusoidal function in equation 9.19 is ω_d (not ω_n) and that the oscillations are bounded by the envelope $\pm Ke^{-\sigma t}$.

Case 3.   Real, but not distinct roots, i.e., $\sigma^2 = \omega_n^2$. In this case, the two roots of the characteristic equation are equal, and equation 9.16 does not represent the general solution form. As per problem 2, the general solution for this case has the form

$$x(t) = (A + Bt)e^{-\sigma t}. \qquad (9.20)$$

Such a response is said to be **critically damped.**

Although we have obtained solution functions for a general source-free second-order network, it is illuminative to see the generic waveforms of $x(t)$ that correspond to the various cases. Figure 9.5 shows typical waveforms for equations 9.3 and 9.19 and possible

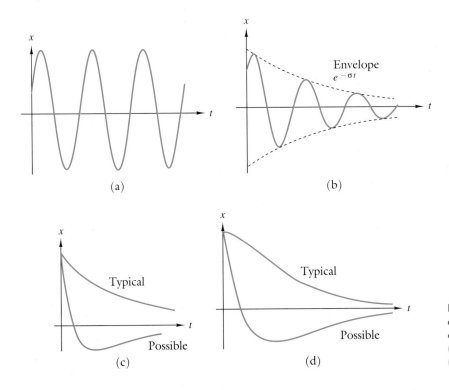

Figure 9.5 Possible waveforms corresponding to various degrees of damping. (a) Undamped. (b) Underdamped. (c) Overdamped. (d) Critically damped.

waveforms for equations 9.16 and 9.20, for the case of $\sigma \geq 0$. The waveforms are said to be *undamped*, *underdamped*, *overdamped*, and *critically damped*, respectively. These terms stem from the intuitive notion of "damping." The source-free response of an **undamped** second-order linear system, whether electrical or mechanical, has an oscillatory waveform of constant amplitude. **Damping** means a continuous decrease in the amplitude of oscillation and is caused by system elements that consume energy. In electrical circuits, the presence of resistance produces the damping effect. In mechanical systems, the presence of friction causes damping. When the amount of damping is just enough to prevent the oscillation from occurring, the system is said to be **critically damped**. Less damping corresponds to the underdamped case, in which the oscillation is present but eventually dies out. A greater amount of damping corresponds to the **overdamped** case, wherein the waveform is nonoscillatory and a very small perturbation of any circuit parameter will *not* cause oscillations to occur.

It is important to note that for the underdamped case, the frequency of oscillation is ω_d, which is smaller than ω_n. In general, it is not possible to distinguish between the overdamped response and the critically damped response by looking at the waveform plots. In many cases, the waveforms look similar: Both types of response may have no zero crossing or at most one zero crossing. (See problems 15 through 18.)

The foregoing derivations of the general solutions are summarized in table 9.1.

TABLE 9.1   GENERAL SOLUTIONS FOR SOURCE-FREE SECOND-ORDER NETWORKS

General solution of the homogeneous differential equation

$$\frac{d^2x}{dt^2} + 2\sigma\frac{dx}{dt} + \omega_n^2 x = 0$$

having characteristic equation $s^2 + 2\sigma s + \omega_n^2 = (s - s_1)(s - s_2) = 0$, where

$$s_{1,2} = -\sigma \pm \sqrt{\sigma^2 - \omega_n^2}.$$

Case 1. Real, distinct roots, i.e., $\sigma^2 > \omega_n^2$ (overdamped response for $\sigma > 0$):

$$x(t) = Ae^{s_1 t} + Be^{s_2 t}.$$

Case 2. Complex, distinct roots, i.e., $\sigma^2 < \omega_n^2$ (underdamped response for $\sigma > 0$):

$$x(t) = e^{-\sigma t}\left[A\cos(\omega_d t) + B\sin(\omega_d t)\right] = Ke^{-\sigma t}\cos(\omega_d t + \theta),$$

where, notationally,

$$s_{1,2} = -\sigma \pm j\omega_d \quad \text{and} \quad \omega_d = \sqrt{\omega_n^2 - \sigma^2}.$$

Case 3. Real, identical roots, i.e., $\sigma^2 = \omega_n^2$ (critically damped response for $\sigma > 0$):

$$x(t) = (A + Bt)e^{-\sigma t}.$$

Once the roots of the characteristic equation are known and the expression for the general solution is selected from table 9.1, the remaining task is to determine the two unknown constants A and B in the expression. This is achieved with the knowledge of $x(0)$ and $x'(0)$. Since $x(t)$ represents either a capacitor voltage or an inductor current, its initial value is usually given or can be determined from the past history of the circuit. The value of $x'(0)$, on the other hand, is usually not given. Generally, it must be calculated either from the state equations or from the given circuit. If $x(t) = v_C(t)$, then $x'(0^+) = v_C'(0^+) = i_C(0^+)/C$. If $x(t) = i_L(t)$, then $x'(0^+) = i_L'(0^+) = v_L(0^+)/L$. The problem then reduces to finding a capacitor current or an inductor voltage at $t = 0^+$. Since the initial values, $v_C(0^+)$ and $i_L(0^+)$, are known, this allows us to treat the capacitor as an independent voltage source of value $v_C(0^+)$ and the inductor as an independent current source of value $i_L(0^+)$. This treatment results in a resistive circuit. Values for $i_C(0^+)$ and $v_L(0^+)$ follow by the usual methods of resistive circuit analysis and permit us to evaluate the *expressions* $x(t)$ and $x'(t)$ at $t = 0^+$ to obtain two equations whose solution yields the parameters A and B.

EXAMPLE 9.2

Switch S in figure 9.6a has remained in position A for a long time. It moves to position B at $t = 0$. The 1-μF capacitor is assumed to be ideal, whereas the inductor is modeled by a 10-mH inductance in series with a 20-Ω resistance. Find $v_C(t)$ for $t \geq 0$ for the following three cases, and plot the corresponding waveforms: case 1: $R_2 = 0$; case 2: $R_2 = 180$ Ω; case 3: $R_2 = 405$ Ω.

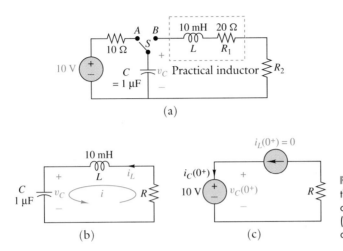

(a)

(b)   (c)

Figure 9.6 Discharge of a capacitor through a practical inductor. (a) The capacitor receives an initial charge. (b) Circuit for $t > 0$. (c) Circuit condition at $t = 0^+$.

SOLUTION

For $t > 0$, figure 9.6b shows the circuit of interest, where $R = R_1 + R_2$ represents the series combination of R_1 and R_2. The first step in the solution is to obtain a second-order differential equation for the series RLC circuit of figure 9.6b with v_C as the unknown. Applying KVL to the single loop yields

$$L\frac{di}{dt} + Ri + v_C(t) = 0.$$

Substituting the relationship $i(t) = C(dv_C/dt)$ into this equation and dividing both sides by LC yields the desired differential equation:

$$\frac{d^2v_C}{dt^2} + \frac{R}{L}\frac{dv_C}{dt} + \frac{1}{LC}v_C = 0. \tag{9.21}$$

Comparing equation 9.21 with equation 9.10 yields $\omega_n = 1/\sqrt{LC} = 10^4$ rad/sec and $\sigma = 0.5R/L = 50(R_2 + 20)$. The three cases will now be investigated separately.

Case 1. $R_2 = 0$. With $R_2 = 0$, $R = 20$ Ω, $\sigma = 10^3 < \omega_n$, and the roots of the characteristic equation are complex. Hence, the response is underdamped. From table 9.1, the general underdamped response has the form

$$v_C(t) = e^{-\sigma t}[A\cos(\omega_d t) + B\sin(\omega_d t)], \tag{9.22}$$

where

$$\omega_d = \sqrt{\omega_n^2 - \sigma^2} = 9,950 \text{ rad/sec}.$$

It remains to determine A and B in equation 9.22. From equation 9.22 and its derivative,

$$v_C(0^+) = A \tag{9.23a}$$

and

$$v_C'(0^+) = -\sigma A + \omega_d B. \tag{9.23b}$$

Since the switch S in figure 9.6a was in position A for a long time, the circuit reached a dc steady state before S moved to position B. This implies that $v_C(0^-) = 10$ V. With S at position A, there is no current in the inductor, i.e., $i_L(0^-) = 0$. Immediately after S moves to position B, the circuit becomes that of figure 9.6b. The presence of the series resistance R and the absence of any impulsive independent sources imply that the capacitor voltage

---

and inductor current cannot change instantaneously. Therefore, $v_C(0^+) = v_C(0^-) = 10$ V, and $i_C(0^+) = i_C(0^-) = 0$. Thus, $v'_C(0^+) = i_C(0^+)/C = 0$. Substituting these values into equations 9.23 yields

$$A = 10 \text{ and } B = \frac{\sigma A}{\omega_d} = 1.005 \text{ V.}$$

Finally,

$$v_C(t) = e^{-1,000t}[10\cos(9,950t) + 1.005\sin(9,950t)]$$
$$= 10.05e^{-1,000t}\cos(9,950t + 5.7°) \text{ V.} \tag{9.24}$$

Case 2. $R_2 = 180 \; \Omega$. If $R_2 = 180 \; \Omega$, then $R = 200 \; \Omega$, $\sigma = 10^4 = \omega_n$, and the characteristic equation has identical, real roots. Hence, the response is critically damped. From table 9.1, the general critically damped response has the form

$$v_C(t) = (A + Bt)e^{-\sigma t}. \tag{9.25}$$

From this expression and its derivative,

$$v_C(0^+) = A \tag{9.26}$$

and

$$v'_C(0^+) = -\sigma A + B. \tag{9.27}$$

From the circuit of figure 9.6c, as in case 1, $v_C(0^+) = v_C(0^-) = 10$ V, and $v'_C(0^+) = i_C(0^+)/C = 0$. Substituting these values into equations 9.26 and 9.27 yields

$$A = 10 \text{ V and } B = \sigma A = 10^5 \text{ V.}$$

Finally,

$$v_C(t) = 10(1 + 10^4 t)e^{-10^4 t} \text{ V.} \tag{9.28}$$

Case 3. $R_2 = 405 \; \Omega$. If $R_2 = 405 \; \Omega$, then $R = 425 \; \Omega$, $\sigma = 21{,}250 > \omega_n$, and the characteristic equation has distinct real roots,

$$s_{1,2} = -21,250 \pm \sqrt{21,250^2 - 10,000^2} = -21,250 \pm 18,750 \text{ sec}^{-1}.$$

Thus, the response is overdamped and has the general form

$$v_C(t) = Ae^{-2,500t} + Be^{-40,000t} \tag{9.29}$$

for $t > 0$. From this expression and its derivative,

$$v_C(0^+) = A + B \tag{9.30a}$$

and

$$v'_C(0^+) = -2,500A - 40,000B. \tag{9.30b}$$

From the circuit of figure 9.6c, as in case 1, $v_C(0^+) = v_C(0^-) = 10$ V, and $v'_C(0^+) = i_C(0^+)/C = 0$. Substituting these values into equations 9.30 yields

$$A = 10.667 \text{ and } B = -0.667.$$

Finally, for $t > 0$,

$$v_C(t) = 10.667e^{-2,500t} - 0.667e^{-40,000t} \text{ V.} \tag{9.31}$$

The waveforms of $v_C(t)$ for the three cases—underdamped, critically damped, and over-damped—are plotted in figure 9.7.

On a practical note, commercially available resistors come in standard values with an associated tolerance. Tolerances vary from ±1% (a precision resistor) to as much as ±20%. Further, because of heating action over a long time, original resistance values change. Given the preceding example in which the type of response depends on the resistance, one can imagine the care needed in designing such circuits.

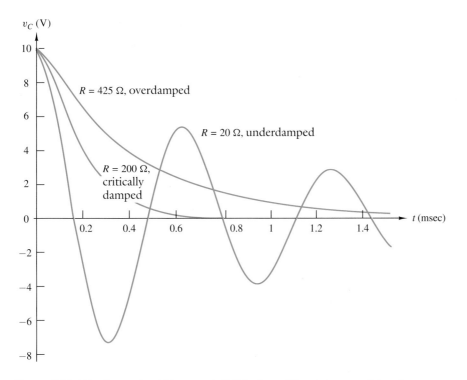

Figure 9.7 Waveforms of $v_C(t)$ in example 9.2 for three different degrees of damping.

4. SECOND-ORDER LINEAR NETWORKS WITH CONSTANT INPUTS

The previous section studied second-order linear networks without excitations. When independent sources are present, such as in figure 9.3, the network equations are the same as in the source-free case, except for an added term on the right-hand side accounting for the effect of the inputs. Either of the following two forms of equations can characterize such networks:

1. *State equations*

$$\frac{dx_1}{dt} = a_{11}x_1 + a_{12}x_2 + u_1(t),$$

$$\frac{dx_2}{dt} = a_{21}x_1 + a_{22}x_2 + u_2(t), \tag{9.32}$$

where each $u_k(t)$ is a *scaled sum* of the source excitations.

2. *A single second-order differential equation*

$$\frac{d^2x}{dt^2} + 2\sigma\frac{dx}{dt} + \omega_n^2 x = f(t), \tag{9.33}$$

where $f(t)$ is a *scaled sum of the inputs and their first-order derivatives*. Problem 24 illustrates this general form of $f(t)$. Section 9.5 delineates a general procedure for writing equations 9.32 or 9.33. The remainder of this section, however, will consider simple circuits, whose equation models can be written almost by inspection.

The solution of equation 9.33 for *arbitrary* inputs and initial conditions is best obtained by the Laplace transform method, to be studied in volume 2. However, for the very special case of *constant inputs*, the solution can be obtained almost as easily as in the source-free case. With constant inputs, the function $f(t)$ on the right-hand side of equation 9.33 is a constant (possibly zero), which we denote by F. Since the expressions in table 9.1 satisfy the **homogeneous differential equation** 9.10, the general solution to equation 9.33 follows by adding a constant X_f to the expressions in the table. Table 9.2 summarizes the resulting different cases.

TABLE 9.2 GENERAL SOLUTION FOR A SECOND-ORDER LINEAR NETWORK WITH CONSTANT INPUTS

General solution of the differential equation

$$\frac{d^2x}{dt^2} + 2\sigma\frac{dx}{dt} + \omega_n^2 x = F = \text{constant}$$

having characteristic equation $s^2 + 2\sigma s + \omega_n^2 = (s - s_1)(s - s_2) = 0$, where

$$s_{1,2} = -\sigma \pm \sqrt{\sigma^2 - \omega_n^2}.$$

Case 1. Real, distinct roots, i.e., $\sigma^2 > \omega_n^2$ (overdamped response for $\sigma > 0$):

$$x(t) = Ae^{s_1 t} + Be^{s_2 t} + X_f.$$

Case 2. Complex, distinct roots, i.e., $\sigma^2 < \omega_n^2$ (underdamped response for $\sigma > 0$):

$$x(t) = e^{-\sigma t}(A\cos\omega_d t + B\sin\omega_d t) + X_f$$

$$= Ke^{-\sigma t}\cos(\omega_d t + \theta) + X_f,$$

where, notationally,

$$s_{1,2} = -\sigma \pm j\omega_d$$

and

$$\omega_d = \sqrt{\omega_n^2 - \sigma^2}.$$

Case 3. Real, identical roots, i.e., $\sigma^2 = \omega_n^2$ (critically damped response for $\sigma > 0$):

$$x(t) = (A + Bt)e^{-\sigma t} + X_f.$$

The form of each general solution in table 9.2 is

$$x(t) = x_n(t) + X_f, \tag{9.34}$$

where $x_n(t)$ is the general solution of the homogeneous equation 9.10. It remains to determine A, B, and X_f. Two methods are available for finding the constant X_f.

Method 1. Substituting equation 9.34 into equation 9.33 with $f(t) = F$ yields

$$X_f = \frac{F}{\omega_n^2}.$$

This follows because $x_n(t)$ satisfies equation 9.10, in which case $\omega_n^2 X_f = F$.

Method 2. Since $x(t) = \text{constant} = X_f$ satisfies the nonhomogeneous differential equation, the X_f's represent constant capacitor voltages or constant inductor currents that satisfy KCL, KVL, and the element v-i characteristics. When a capacitor has a constant voltage, its current is zero; when an inductor has a constant current, its voltage is zero. Therefore, the X_f's are appropriate (capacitor) voltages and/or (inductor) currents computed when the capacitor (or capacitors) are open circuited and the inductor (or inductors) are short circuited.

Once the appropriate general solution has been selected from table 9.2 and the constant X_f determined, the parameters A and B are computed by the same methods as are used in the source-free case. The following example illustrates the procedure for a series RLC circuit.

◆ **EXAMPLE 9.3**

In the series RLC circuit of figure 9.8, the voltage across the capacitor is the desired output. Assuming a step voltage input, find $v_C(t)$, $t \geq 0$, for the following three cases: *case 1: R = 0.2 Ω; case 2: R = 2 Ω; case 3: R = 4.25 Ω.*

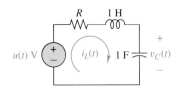

Figure 9.8 Series RLC circuit excited by a step voltage input.

SOLUTION

Writing a mesh equation and using $i_L = C(dv_C/dt) = dv_C/dt$ in this circuit, we have, for $t > 0$,

$$\frac{d^2v_C}{dt^2} + R\frac{dv_C}{dt} + v_C = 1,$$

with characteristic equation

$$s^2 + Rs + 1 = (s - s_1)(s - s_2) = 0.$$

Comparing the latter with table 9.2, one identifies $\omega_n = 1$, $\sigma = 0.5R$, and $F = 1$. The three cases will now be investigated.

Case 1. R = 0.2 Ω. If $R = 0.2$ Ω, the characteristic equation is

$$s^2 + Rs + 1 = s^2 + 0.2s + 1 = 0.$$

Using the quadratic root formula, we find that the roots are complex:

$$s_{1,2} = -0.1 \pm j0.995.$$

From table 9.2, case 2, the response has the form

$$v_C(t) = e^{-0.1t}\left[(A\cos(0.995t) + B\sin(0.995t)\right] + X_f \text{ V}. \tag{9.35}$$

To determine X_f, we use method 2 described earlier. Opening the capacitor and shorting the inductor leads to $X_f = v_C = 1$. To determine A and B in equation 9.35, we make use of the initial conditions. The voltage input is a **unit step function**, which has the value zero for $-\infty < t < 0$. Any energy stored in L and C would have been completely dissipated in R by $t = 0^-$. Therefore, $v_C(0^-) = 0$ and $i_L(0^-) = 0$. Since an impulsive voltage or current does not exist in this series RLC circuit, the capacitor voltage and inductor current cannot change instantaneously. (See chapter 7.) Therefore, $v_C(0^+) = v_C(0^-) = 0$ and $i_L(0^+) = i_L(0^-) = 0$. From figure 9.8,

$$i_L(t) = i_C(t) = C\frac{dv_C(t)}{dt} = \frac{dv_C(t)}{dt}.$$

Therefore, $v_C'(0^+) = i_L(0^+) = 0$. Evaluating equation 9.35 and its derivative at $t = 0^+$ results in the two equations

$$A + 1 = 0 \tag{9.36a}$$

and

$$-0.1A + 0.995B = 0. \tag{9.36b}$$

Solving equation 9.36 yields $A = -1$ and $B = -0.1005$. Therefore, the solution of $v_C(t)$ for $t > 0$ is

$$v_C(t) = e^{-0.1t}[-\cos(0.995t) - 0.1005\sin(0.995t)] + 1$$

$$= 1.005e^{-0.1t}\cos(0.995t + 174.3°) + 1 \text{ V}. \tag{9.37}$$

This voltage is termed the **step response** of the circuit because it is the response to a step input when all initial capacitor voltages and inductor currents are zero.

Case 2. R = 2 Ω. Then the roots of the characteristic equation are real and identical; i.e., the response is critically damped. From table 9.2, the general response has the form

$$v_C(t) = (A + Bt)e^{-\sigma t} + X_f, \tag{9.38}$$

where $X_f = 1$, as calculated in case 1, and $\sigma = 0.5R = 1$.

Evaluating equation 9.38 and its derivative at $t = 0^+$ yields

$$v_C(0^+) = A + 1, \tag{9.39a}$$

$$v_C'(0^+) = -\sigma A + B. \tag{9.39b}$$

---

4. SECOND-ORDER LINEAR NETWORKS WITH CONSTANT INPUTS

Equating these expressions to $v_C(0^+) = 0$ and $v'_C(0^+) = 0$, as determined in case 1, and solving for A and B produces $A = -1$ and $B = -1$. Hence, for $t > 0$, the solution is

$$v_C(t) = -(1 + t)e^{-t} + 1 \text{ V}. \tag{9.40}$$

Case 3. R $= 4.25$ Ω. Here, the roots of the characteristic equation are real and distinct:

$$s_{1,2} = -2.125 \pm \sqrt{2.125^2 - 1} = -2.125 \pm 1.875.$$

The response is overdamped. From table 9.2, the general response has the form

$$v_C(t) = Ae^{-0.25t} + Be^{-4t} + X_f, \tag{9.41}$$

where, again, $X_f = 1$, as calculated in case 1. Evaluating equation 9.41 and its derivative at $t = 0^+$ produces

$$v_C(0^+) = A + B + 1 \tag{9.42a}$$

and

$$v'_C(0^+) = -0.25A - 4B. \tag{9.42b}$$

Equating these expressions to $v_C(0^+) = 0$ and $v'_C(0^+) = 0$, as determined in case 1, and solving for A and B yields $A = -1.0667$ and $B = 0.667$. For $t > 0$, the solution is

$$v_C(t) = -1.0667e^{-0.25t} + 0.0667e^{-4t} + 1 \text{ V}. \tag{9.43}$$

Figure 9.9 sketches the three different step responses for the series RLC circuit of figure 9.8.

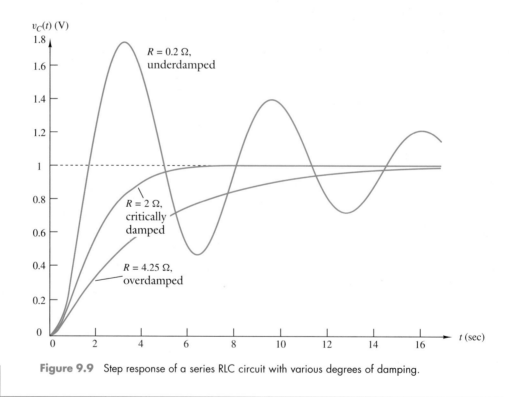

Figure 9.9 Step response of a series RLC circuit with various degrees of damping.

Exercise. Find $v_C(t)$ in example 9.3 if the resistance R is changed to 0.5 Ω.

In a linear system or circuit, the response to a step input is often an indicator of the quality of the system's performance. Consider the example of measuring the voltage of a battery using a multimeter, as discussed in chapter 2. Here, the input is the battery dc voltage and the output is the position of the pointer of the meter. Connecting the meter probes to the battery terminals amounts to applying a step input to the multimeter circuit. Naturally, one would like to read the voltage in a short time. If the system is underdamped, then the pointer oscillates for a short period of time before coming to rest at its final position. This is undesirable. On the other hand, if the system is overdamped, the pointer will not oscillate, but may take a long time to reach its final position. This is also undesirable. For such an application, it is necessary to design the system to be close to the critically damped case.

* 5. FORMULATION OF A SINGLE SECOND-ORDER DIFFERENTIAL EQUATION

The examples of second-order circuits in the previous two sections are relatively simple and the corresponding derivation of equation 9.10 or equation 9.33 straightforward. General second-order circuits necessitate a systematic method of formulating the equation model. This section describes a method for constructing a single second-order differential equation model for a general second-order circuit whose energy storage elements may be two inductors, two capacitors, or an inductor and a capacitor. The procedure has two stages:

Stage 1. Write the state equations; i.e., using the capacitor voltages and/or the inductor currents as the unknowns, write two first-order differential equations.

Stage 2. Eliminate the unwanted variable from the state equations.

The next two subsections explain the details of the procedure.

Writing the State Equations

Examine equations 9.9 and 9.32. State equations contain only capacitor voltages, inductor currents, their derivatives, and inputs as variables. The first step in writing them begins with the defining v-i relationships for C and/or L. Here, the capacitor voltage v_C and inductor current i_L are taken as the desired **state variables**. On the other hand, the capacitor current i_C and inductor voltage v_L are unwanted variables, to be eliminated. The elimination is achieved by finding expressions for i_C and v_L *in terms of* v_C, i_L, and the inputs; this is the second step. In the third step, one then substitutes the results into the L-C defining equations written in step 1.

In finding expressions for i_C and v_L in step 2, the variables v_C and i_L are treated as known quantities. This fact suggests a "trick" for finding the expression: We replace the capacitor by an independent voltage source with value v_C and replace the inductor by an independent current source with value i_L; the resultant network is a *linear resistive circuit* with multiple sources. Using either nodal or mesh analysis or some other method, one can then solve the circuit for expressions for i_C and v_L in terms of the independent source variables. Equation 9.9 or equation 9.32 follows after some additional straightforward arithmetic. This method of writing state equations reduces the problem to that of solving a resistive network. The next example illustrates the procedure.

◆ **EXAMPLE 9.4**

Find the state equations for the circuit of figure 9.10a.

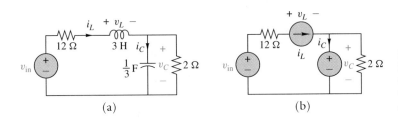

Figure 9.10 (a) RLC network for development of state equations. (b) Intermediate linear resistive network, which aids in finding expressions for i_C and v_L.

---

*Denotes a more advanced topic.

SOLUTION

From the defining equations for L and C, we have

$$\frac{dv_C}{dt} = \frac{1}{C}i_C = 3i_C \qquad (9.44a)$$

and

$$\frac{di_L}{dt} = \frac{1}{L}v_L = \frac{1}{3}v_L. \qquad (9.44b)$$

To express i_C and v_L *in terms of* v_C and i_L, we replace C by a voltage source and L by a current source, producing the resistive circuit of figure 9.10b. By the principle of superposition, i_C and v_L in figure 9.10b should each be a *scaled sum* of the inputs v_C, i_L, and v_{in}. To obtain the coefficients in the scaled sum, we solve this circuit by any method of chapter 4, to obtain

$$i_C = -0.5v_C + i_L \qquad (9.45a)$$

and

$$v_L = -v_C - 12i_L + v_{\text{in}}. \qquad (9.45b)$$

After the substitution of equations 9.45 into equations 9.44, some simple multiplication and division yield the desired state equation model of figure 9.10a:

$$\frac{dv_C}{dt} = 3i_C = -1.5v_C + 3i_L, \qquad (9.46a)$$

$$\frac{di_L}{dt} = \frac{1}{3}v_L = -\frac{1}{3}v_C - 4i_L + \frac{1}{3}v_{\text{in}}. \qquad (9.46b)$$

Reduction of State Equations to a Single Second-Order Differential Equation

Following the procedure outlined above, the state equations of any second-order linear network are of the form

$$\frac{dx_1}{dt} = a_{11}x_1 + a_{12}x_2 + u_1(t), \qquad (9.47a)$$

$$\frac{dx_2}{dt} = a_{21}x_1 + a_{22}x_2 + u_2(t), \qquad (9.47b)$$

where $u_1(t)$ and $u_2(t)$ are each a *scaled sum* of the actual circuit inputs. (This is clear from applying the principle of superposition to the associated resistive circuit, as illustrated by figure 9.10b.) Our goal now is to eliminate one of the state variables—say, x_2—and obtain a single second-order differential equation in x_1. The method to be used is an elementary one called "elimination by addition or subtraction" in algebra. The essence of the method is as follows:

> *To eliminate a variable x from a pair of linear equations, multiply the equations by numbers that will make the coefficients of x in the resulting equations equal. Then subtract or add, according to whether these coefficients have like or unlike signs.*

In applying this method to differential equations, the operations on each equation include differentiation, besides multiplication by a number. To achieve the elimination of the unwanted variable, we first rewrite equations 9.47 as

$$\left(\frac{d}{dt} - a_{11}\right)x_1(t) - a_{12}x_2(t) = u_1(t) \qquad (9.48a)$$

and

$$-a_{21}x_1(t) + \left(\frac{d}{dt} - a_{22}\right)x_2(t) = u_2(t). \qquad (9.48b)$$

To eliminate x_2, we perform the operation $(d/dt - a_{22})$ on equation 9.48a, and multiply equation 9.48b by a_{12}. After properly grouping terms, the results are

$$\frac{d^2x_1}{dt^2} - (a_{11} + a_{22})\frac{dx_1}{dt} + a_{11}a_{22}x_1 - a_{12}\frac{dx_2}{dt} + a_{12}a_{22}x_2 = \frac{du_1}{dt} - a_{22}u_1 \qquad (9.49a)$$

and

$$-a_{12}a_{21}x_1 + a_{21}\frac{dx_2}{dt} - a_{12}a_{22}x_2 = a_{12}u_2. \qquad (9.49b)$$

Adding equation 9.49b to equation 9.49a yields the desired second-order differential equation in *explicit form:*

$$\frac{d^2x_1}{dt^2} - (a_{11} + a_{22})\frac{dx_1}{dt} + (a_{11}a_{22} - a_{12}a_{21})x_1 = \frac{du_1}{dt} - a_{22}u_1 + a_{12}u_2. \qquad (9.50a)$$

If x_2 is retained, then a similar derivation yields

$$\frac{d^2x_2}{dt^2} - (a_{11} + a_{22})\frac{dx_2}{dt} + (a_{11}a_{22} - a_{12}a_{21})x_2 = \frac{du_2}{dt} - a_{11}u_2 + a_{21}u_1. \qquad (9.50b)$$

From a consideration of symmetry, one can also obtain equation 9.50b directly from equation 9.50a by interchanging the subscripts 1 and 2.

Comparing equations 9.50 with equation 9.33, we obtain *explicit formulas* for σ, ω_n^2, and $f(t)$:

$$\sigma = -0.5(a_{11} + a_{22}), \qquad (9.51a)$$

$$\omega_n^2 = a_{11}a_{22} - a_{12}a_{21}. \qquad (9.51b)$$

For the case of $x = x_1$,

$$f(t) = \frac{du_1}{dt} - a_{22}u_1(t) + a_{12}u_2(t), \qquad (9.51c)$$

and for the case of $x = x_2$,

$$f(t) = \frac{du_2}{dt} - a_{11}u_2(t) + a_{21}u_1(t). \qquad (9.51d)$$

When the state equations 9.47 of a second-order network have been obtained, we may write the single second-order differential equation by the use of the explicit formulas of equations 9.50 and 9.51. The student is urged to repeat the derivation of equation 9.50 independently. "Plugging and chugging" formulas without understanding their derivation is unhealthy. On the other hand, repeating the same derivation from the beginning every time a problem is to be solved is not effective either. The moral here is to use formulas judiciously.

 EXAMPLE 9.5

For the circuit of figure 9.10a, find the single second-order differential equation with v_C as the unknown.

SOLUTION

Example 9.4 and, in particular, equations 9.46 present the state equations for figure 9.10a:

$$a_{11} = -1.5, \qquad a_{12} = 3,$$

$$a_{21} = -\frac{1}{3}, \qquad a_{22} = -4,$$

$$u_1(t) = 0, \qquad u_2(t) = \frac{v_{in}}{3}.$$

Substituting these results into equation 9.50a yields a second-order differential equation for v_C:

$$\frac{d^2v_C}{dt^2} + 5.5\frac{dv_C}{dt} + 7v_C = v_{in}.$$

The next example illustrates the application of the procedure to a second-order circuit containing an op amp.

◆ **EXAMPLE 9.6**

For the circuit of figure 9.11a, find a single second-order differential equation model with v_{out} as the unknown.

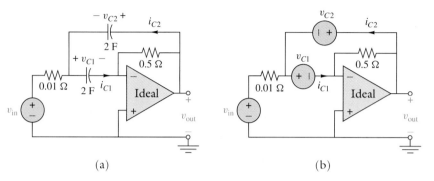

Figure 9.11 Derivation of a second-order differential equation for a circuit containing an op amp. (a) A second-order RC op amp circuit. (b) Intermediate resistive circuit for finding i_{C1} and i_{C2}.

(a)　　　　　　　(b)

SOLUTION

Step 1. Formulation of the state equation. From the defining equations for capacitors, we have

$$\frac{dv_{C1}}{dt} = 0.5i_{C1} \tag{9.52a}$$

and

$$\frac{dv_{C2}}{dt} = 0.5i_{C2}. \tag{9.52b}$$

To express i_{C1} and i_{C2} *in terms of* v_{C1}, v_{C2}, and v_{in}, we replace the capacitors by independent voltage sources. Figure 9.11b shows the resultant *resistive* circuit. Since an ideal op amp allows no current flowing into the input terminals, the current i_{C1} equals the current in the 0.5-Ω resistor, whose terminal voltage drop, from right to left, is $v_{C1} + v_{C2}$. Therefore,

$$i_{C1} = -2(v_{C1} + v_{C2}). \tag{9.53a}$$

The current i_{C2} is equal to the sum of i_{C1} and the current in the 0.01-Ω resistor (from right to left). Due to the virtual ground property of an ideal op amp, the voltage across the 0.01-Ω resistor is $v_{C1} + 0 - v_{\text{in}}$. Therefore,

$$i_{C2} = i_{C1} + 100(v_{C1} - v_{\text{in}}) = 98v_{C1} - 2v_{C2} - 100v_{\text{in}}. \tag{9.53b}$$

After the substitution of equations 9.53 into equations 9.52, some simple multiplication and division yield the desired state equations:

$$\frac{dv_{C1}}{dt} = -v_{C1} - v_{C2}, \tag{9.54a}$$

$$\frac{dv_{C2}}{dt} = 49v_{C1} - v_{C2} - 50v_{\text{in}}. \tag{9.54b}$$

Step 2. Reduction to a second-order differential equation. Comparing equation 9.54 with equation 9.47 yields

$$a_{11} = -1, \quad a_{12} = -1, \quad u_1(t) = 0,$$
$$a_{21} = 49, \quad a_{22} = -1, \quad u_2(t) = -50v_{\text{in}}.$$

Substituting these results into equation 9.50a yields a second-order differential equation for v_{C1}:

$$\frac{d^2v_{C1}}{dt^2} + 2\frac{dv_{C1}}{dt} + 50v_{C1} = 50v_{\text{in}}. \tag{9.55a}$$

Similarly, equation 9.50b yields the second-order differential equation with v_{C2} as the unknown:

$$\frac{d^2v_{C2}}{dt^2} + 2\frac{dv_{C2}}{dt} + 50v_{C2} = -50\frac{dv_{\text{in}}}{dt} - 50v_{\text{in}}. \tag{9.55b}$$

From figure 9.11a, $v_{out} = v_{C1} + v_{C2}$. To obtain the desired differential equation with v_{out} as the unknown, we add equation 9.55b to equation 9.55a and replace $v_{C1} + v_{C2}$ by v_{out}. The final result is

$$\frac{d^2v_{out}}{dt^2} + 2\frac{dv_{out}}{dt} + 50v_{out} = -50\frac{dv_{in}}{dt}. \qquad (9.56)$$

The characteristic equation for the circuit described by equation 9.56 has complex roots. Therefore, the circuit is of the underdamped variety. Example 9.6 demonstrates two important facts: (1) Any response of a second-order RLC network can also be produced with a network containing resistors, capacitors, and an op amp (but no inductors); (2) when a general second-order linear network is characterized by a single differential equation, the derivative of the input may appear on the right-hand side. (See equations 9.55 and 9.56 and also problem 24.)

> **Exercise.** If $v_{in}(t) = 0$, determine an expression for $v_{out}(t)$, $t \geq 0$, when $v_{out}(0) = 10$ V and $v'_{out}(0) = 7$ V. Now find the response to a step voltage input with the same initial conditions.

6. FURTHER EXAMPLES AND APPLICATIONS

An important difference between the behaviors of first-order and second-order linear networks is the possibility of oscillatory responses in the latter. In some applications, sinusoidal oscillations are precisely what the circuit is intended for, while in other applications oscillations are highly undesirable. In this section, we present some practical examples, utilizing the methods of the previous sections.

Suppose we wish to build an electronic circuit that generates a pure sinusoidal voltage waveform at a specified frequency. As was shown in section 2, in theory, the simplest way to achieve this goal is to discharge a capacitor through an inductor. But in practice, both the capacitor and the inductor have losses (as modeled in figure 9.2) that cause the amplitude of oscillation to decrease and eventually die out completely. To have sustained sinusoidal oscillations, energy must be replenished to the circuit by some means. One simple method is to connect a so-called negative-resistance device to the circuit. A negative resistance (see chapter 1) delivers power to its embedding network. While not an off-the-shelf component, it can be built (see chapter 3) from controlled sources or op amps, all of which require dc power supplies to operate. The following example illustrates the principle behind a negative-resistance oscillator circuit.

 EXAMPLE 9.7

In the circuit shown in figure 9.12, R_e represents the combined load resistance and capacitor leakage resistance. The resistance r represents the wire resistance of the inductor. Find the value of the negative resistance, $-R_n$, required for sustained sinusoidal oscillations. Is the frequency of oscillation equal to $1/\sqrt{LC}$ rad/sec?

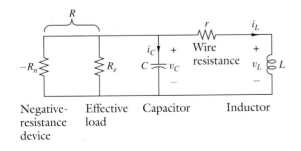

Figure 9.12 A negative-resistance oscillator circuit.

SOLUTION

For analysis purposes, R_e and $-R_n$ may be combined in parallel and denoted by R, i.e., $R = R_e//(-R_n)$. Using the same procedure as in example 9.4, the circuit state equations are

$$\frac{dv_C}{dt} = -\frac{1}{RC}v_C - \frac{1}{L}i_L, \tag{9.57a}$$

$$\frac{di_L}{dt} = \frac{1}{L}v_C - \frac{r}{L}i_L. \tag{9.57b}$$

From equation 9.50a, the associated second-order differential equation is

$$\frac{d^2 v_C}{dt^2} + \left(\frac{1}{RC} + \frac{r}{L}\right)\frac{dv_C}{dt} + \left(1 + \frac{r}{R}\right)\frac{1}{LC}v_C = 0. \tag{9.58}$$

Comparing equation 9.58 with equation 9.10, one observes that

$$2\sigma = \frac{1}{RC} + \frac{r}{L}, \tag{9.59a}$$

$$\omega_n^2 = \left(1 + \frac{r}{R}\right)\frac{1}{LC}. \tag{9.59b}$$

To have sustained sinusoidal oscillations, σ must be zero; i.e.,

$$\frac{1}{RC} + \frac{r}{L} = 0, \tag{9.60}$$

or equivalently,

$$\frac{1}{R} = -\frac{C}{L}r. \tag{9.61}$$

Recall that R is the parallel combination of $-R_n$ and R_e, i.e.,

$$\frac{1}{R} = \frac{1}{R_e} - \frac{1}{R_n}. \tag{9.62}$$

It follows from equations 9.61 and 9.62 that the condition for sustained oscillation is

$$R_n = \frac{1}{\dfrac{C}{L}r + \dfrac{1}{R_e}}. \tag{9.63}$$

Under the condition of equation 9.61, the frequency of oscillation given by equation 9.59b becomes

$$\omega_n = \sqrt{1 - \frac{C}{L}r^2}\,\frac{1}{\sqrt{LC}}. \tag{9.64}$$

It is important to note the following:

1. Even if a negative-resistance device has been connected to the circuit to replenish the lost energy, the frequency of oscillation, as given by equation 9.64, is smaller than $1/\sqrt{LC}$, the oscillation frequency for the undamped case.
2. Using a value of R_n smaller (greater) than that calculated from equation 9.63 will lead to exponentially growing (decreasing) oscillations.

Exercise. In example 9.7, $L = 1$ mH, $r = 5\ \Omega$, $C = 0.1\ \mu$F, and $R_e = 500\ \Omega$. Find the negative resistance for sustained oscillation and the corresponding frequency of oscillation.

ANSWER: $R_n = 400\ \Omega$, frequency of oscillations $= 99,875$ rad/sec.

In order for the oscillations to start, the negative resistance $-R_n$ should be designed to have a magnitude slightly less than that given by equation 9.63. Then the value of σ in equation 9.10 will be *negative*, producing an exponentially growing sinusoidal response. If all circuit parameters are truly constant, the amplitude of oscillation will grow to be infinite, at least in theory. In physical oscillators, such growth is limited to some finite amplitude by nonlinearities that appear when the voltage overloads the circuit elements.

The waveform is then no longer a pure sine wave. The study of such an oscillator is beyond the scope of this book.

There is, however, one special design that can restrict the oscillator to nearly linear operation and therefore produce a nearly sinusoidal output. Recall the discussion of the temperature coefficient of a resistor in chapter 1. For most resistors, the resistance increases with the temperature, which in turn increases with the magnitude of current through the resistor. If such a temperature-sensitive resistor R_t is inserted in series with the inductor in figure 9.12, then in all the previously derived formulas (equations 9.57 through 9.64), r should be replaced by $r + R_t$. The condition for sustained oscillations, equation 9.63, becomes

$$R_n = \frac{1}{\frac{C}{L}(r + R_t) + \frac{1}{R_e}}. \tag{9.65}$$

As the amplitude of oscillation grows, the value of the temperature-sensitive resistance R_t also increases, changing equation 9.65 into an inequality with a "greater than" ($>$) sign. This condition leads to $\sigma > 0$ in equation 9.10 and a decrease in the amplitude of oscillation. Thus, there is a tendency for the amplitude to return to a smaller value. On the other hand, should the amplitude decrease for some reason, R_t will also decrease. This changes equation 9.65 into an inequality with a "less than" ($<$) sign, which leads to $\sigma < 0$, accompanied by an increase in the amplitude of oscillation. With a proper selection of the temperature-sensitive resistor R_t, the amplitude of oscillation will stabilize at some constant value. If the output swing is within the linear range of the negative-resistance device, the oscillatory waveform will be sinusoidal. The circuit is then called a *linear oscillator*. This principle has actually been used in a very well-known circuit called the *Wien bridge oscillator*, described in the next example.

 ## EXAMPLE 9.8

Figure 9.13a shows a Wien bridge oscillator constructed with an op amp as the active (negative-resistance) element. R_2 is a lamp with a positive temperature coefficient that is used to stabilize the amplitude of oscillation. Find the condition for sustained oscillation and the frequency of oscillation of the circuit.

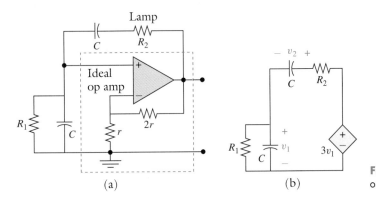

Figure 9.13 (a) Wien bridge oscillator. (b) Its equivalent circuit.

SOLUTION

In chapter 3, it was shown that the noninverting amplifier enclosed in dashed lines is equivalent to a voltage-controlled voltage source with a gain equal to $(2r + r)/r = 3$. Thus, for the purpose of circuit analysis, we may look at the simplified circuit of figure 9.13b. Following the procedure for writing state equations described in section 9.5, we have

$$\frac{dv_1}{dt} = \left(-\frac{1}{CR_1} + \frac{2}{CR_2}\right)v_1 - \frac{1}{CR_2}v_2, \tag{9.66a}$$

$$\frac{dv_2}{dt} = \frac{2}{CR_2}v_1 - \frac{1}{CR}v_2. \tag{9.66b}$$

By the use of the explicit answer given in equation 9.50a, the second-order differential equation is

$$\frac{d^2 v_1}{dt^2} + \left(\frac{1}{R_1 C} - \frac{1}{R_2 C} \right) \frac{dv_1}{dt} + \frac{1}{R_1 R_2 C^2} v_1 = 0, \tag{9.67}$$

from which it follows that

$$2\sigma = \frac{1}{R_1 C} - \frac{1}{R_2 C}, \tag{9.68a}$$

$$\omega_n^2 = \frac{1}{R_1 R_2 C^2}. \tag{9.68b}$$

To have sustained sinusoidal oscillation, σ must be zero, or

$$R_2 = R_1. \tag{9.69a}$$

Hence, the frequency of oscillation is

$$\omega_n = \frac{1}{C R_1}. \tag{9.69b}$$

SUMMARY

This chapter has explored the modeling and response computation of general second-order linear circuits having either no input or constant input excitation. The main model examined was the second-order differential equation model of a circuit. The idea of a state model was introduced, primarily as an aid in computing a differential equation model. More advanced texts explore the use of the state model, not only in representing circuits but also in representing systems in general. A salient characteristic of these models was the possibility of sinusoidal and damped sinusoidal oscillations not present in first-order circuits.

With hindsight, our investigation broke down into two phases. Phase 1 entailed the formulation of the second-order differential equation. For many simple circuits, this may be done by inspection, as illustrated by the examples in sections 2–4. However, for the general case of non-series-parallel RLC circuits or circuits containing active elements, a more systematic method is needed. The chapter outlined the use of the state equations in writing the second-order differential equation model.

The second phase of the analysis focused on the solution of the second-order differential equation. Recognizing the fact that many electrical engineering students taking a first course in circuits have not yet had a full course in differential equations, we based the solution technique on only some basic principles of calculus, namely, the derivatives of exponential and sinusoidal functions in conjunction with the location of the roots of a quadratic equation called the *characteristic equation* of the circuit or differential equation. As there are three types of roots to a quadratic—real and distinct, real and identical, and complex—the responses are divided into three categories: underdamped, which coincides with the case of complex roots; overdamped, which coincides with the case of real, distinct roots; and critically damped, which accompanies the case of real, identical roots. These results are applicable to linear circuits or any other second-order linear system, mechanical or electromechanical.

Since sinusoidal waveforms are used extensively in electrical systems, the chapter presented several circuits that generate such waveforms. For high-power applications, an oscillator circuit constructed with an RLC network and an active element proves useful. For low-power applications, an oscillator circuit constructed with an RC network and an op amp avoids the use of an inductor.

TERMS AND CONCEPTS

Characteristic equation: the algebraic equation $s^2 + a_1 s + a_0 = 0$, as related to a linear circuit described by a second-order differential equation $x''(t) + a_1 x'(t) + a_0 x(t) = f(t)$.

Characteristic roots: roots of a characteristic equation; also called the natural frequencies of the linear circuit.

Critically damped circuit: a second-order linear circuit having damping coefficient σ equal to ω_n. The source-free response of such a circuit has a nonoscillatory waveform, but is on the border of becoming oscillatory.

Damped oscillation frequency: the angular frequency ω_d in the source-free response $K e^{-\sigma t} \cos(\omega_d t + \theta)$ of an underdamped second-order linear circuit.

Damping coefficient: the parameter σ in the differential equation $x''(t) + 2\sigma x'(t) + \omega_n^2 x(t) = f(t)$ for a second-order linear circuit. In the case of an underdamped source-free circuit, the curves $\pm K e^{-\sigma t}$ form the envelope for the damped sinusoidal oscillation.

Homogeneous differential equation: a differential equation in which there are no forcing terms—for example, equation 9.33 when $f(t) = 0$.

Natural frequencies: roots of the characteristic equation of a linear circuit.

Oscillator circuit: an electronic circuit designed to produce sinusoidal voltage or current waveforms.

Overdamped circuit: a second-order linear circuit having a damping coefficient σ greater than ω_n. The source-free response of such a circuit has a nonoscillatory waveform.

Scaled sum of waveforms: an expression of the form $f(t) = a_1 f_1(t) + \ldots + a_n f_n(t)$ for real (or possibly complex) scalars $a_1, \ldots, a_n$, where $f_1(t), \ldots, f_n(t)$ are a set of waveforms.

Second-order linear circuit: a circuit whose input-output relationship may be expressed by a second-order differential equation $x''(t) + a_1 x'(t) + a_0 x(t) = f(t)$.

Source free: property of a circuit such that all independent sources have zero values.

State equations: a set of first-order coupled differential equations characterizing a circuit. In matrix form, state equations are written as $x'(t) = Ax(t) + f(t)$, where A is a square matrix and $x'(t)$, $x(t)$, and $f(t)$ are vectors.

State variables: the variables in the state equations, or entries of $x(t)$.

Step function: a function equal to 0 for $t < 0$ and equal to 1 for $t \geq 0$.

Step response: the response of a circuit to a step function input when all capacitor voltages and inductor currents are initially zero.

Undamped circuit: a second-order linear circuit whose damping coefficient σ is zero, producing a sinusoidal response.

Undamped oscillation (resonant) frequency: the quantity ω_n in the differential equation $x''(t) + 2\sigma x'(t) + \omega_n^2 x(t) = f(t)$, for a second-order linear circuit. It is the frequency of oscillation if the damping coefficient σ is made zero.

Underdamped circuit: a second-order linear circuit having damping coefficient σ smaller than ω_n. The source-free response of such a circuit has an oscillatory waveform.

PROBLEMS

1. Find the expressions of the instantaneous energy stored in C and L for the circuit of figure 9.1b. Show that the sum of $w_C(t)$ and $w_L(t)$ is constant.

2. By direct substitution, show that

$$x(t) = (A + Bt)e^{-\sigma t},$$

where A and B are arbitrary constants, satisfies the differential equation

$$x''(t) + 2\sigma x'(t) + \sigma^2 x(t) = 0.$$

3. For the circuit shown in figure P9.3, find $v_C(0^+)$, $i_L(0^+)$, $i_C(0^+)$, and $v_L(0^+)$ in two steps:

Step 1. Find $v_C(0^-)$ and $i_L(0^-)$ by open-circuiting C and short-circuiting L.

Step 2. Construct a resistive circuit valid at $t = 0^+$. From this circuit, find $i_C(0^+)$, and $v_L(0^+)$.

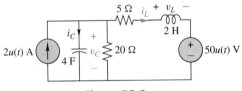

Figure P9.3

4. Repeat the previous problem with the independent voltage source changed to $50u(-t)$ V.

5. The element values in the circuit shown in figure P9.5 are $R = 0.2\ \Omega$, $C = 1$ F, and $L = 0.25$ H. If $i_L(0) = 8$ A and $v_C(0) = 20$ V, find $v_C(t)$ for $t > 0$.

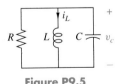

Figure P9.5

6. (a) Consider case 1 of example 9.2 again. The circuit is underdamped, and the response is given by equation 9.24. How many cycles of oscillations occur in

the voltage waveform before the peak value drops to $1/e = 0.368 = 36.8$ percent of the largest peak value of 10.05?

(b) Prove that for the underdamped case, the response will oscillate for N cycles before the amplitude decreases to $1/e$ of its initial value, where

$$N = \frac{\omega_d}{2\pi\sigma}.$$

7. Consider figure P9.7. Switch S has been at position A for a long time and is moved to position B at $t = 0$. Find $v_C(t)$ for $t > 0$.

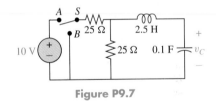

Figure P9.7

8. The switch S in the circuit of figure P9.8 has been closed for a long time and is opened at $t = 0$. Find $v_C(t)$ for $t \geq 0$.

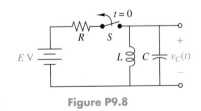

Figure P9.8

9. In the circuit of figure P9.9, if $i_L(t) = 50e^{-10t} \sin(10\sqrt{3}t)$ A for $t \geq 0$, determine the proper value of C.

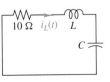

Figure P9.9

10. Consider the circuit of figure P9.10. The response of this circuit is $v_C(t) = [10 - 600t]e^{-20t}$ V for $t \geq 0$.
   (a) Determine the value of the initial condition, $i_L(0^-)$.
   (b) Determine the value of L.

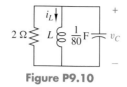

Figure P9.10

11. The switch S in figure P9.11 has been closed for a long time and is opened at $t = 0$. If $R = 10\ \Omega$, find $v_C(t)$ for $t \geq 0$.

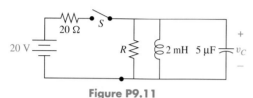

Figure P9.11

12. Repeat problem 11 with R changed to $100\ \Omega$.

13. Repeat problem 11 with R changed to $87/17\ \Omega$.

14. The capacitance voltage of a source-free parallel RLC circuit with $R = 10\ \Omega$ has the form

$$v_C(t) = Ae^{-3t}\cos(4t + \theta)\ \text{V}.$$

If the values of L and C remain unchanged, find the value of R for the circuit to be critically damped and the general form of the source-free response $v_C(t)$ under this condition.

15. The voltage or current in a second-order source-free *overdamped* circuit has the general form

$$x(t) = Ae^{\sigma_1 t} + Be^{\sigma_2 t},$$

where all coefficients are real (but not necessarily positive).
   (a) Prove that $x(t) = 0$ at some $t = T$, $0 < T < \infty$, only if A and B have opposite signs.
   (b) Prove that the curve of $x(t)$ vs. t has at most one zero crossing for $t > 0$.
   (c) State the *necessary and sufficient* conditions on the coefficients A, B, σ_2, and σ_1 for the presence of one zero crossing.

16. The voltage or current in a second-order source-free *overdamped* circuit has the general form

$$x(t) = (A + Bt)e^{\sigma t},$$

where all coefficients are real (but not necessarily positive). Prove that $x(t) = 0$ at some $t = T < \infty$ if and only if A and B have opposite signs.

17. Figure P9.17 shows an overdamped source-free circuit.

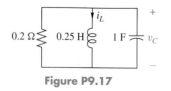

Figure P9.17

   (a) If $v_C(0) = -1$ V and $i_L(0) = 5$ A, find $v_C(t)$ for $t \geq 0$. Use SPICE or any other circuit simulation

program to plot the $v_C(t)$ waveform, and verify that there is no zero crossing.
   (b) If $v_C(0) = 1$ V and $i_L(0) = 5$ A, find $v_C(t)$ for $t \geq 0$. Plot the $v_C(t)$ waveform, and verify that there is one zero crossing.

18. Figure P9.18 shows a critically damped source-free circuit.

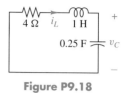

Figure P9.18

   (a) If $v_C(0) = -5$ V and $i_L(0) = 1$ A, find $i_L(t)$ for $t \geq 0$. Use SPICE or any other circuit simulation program to plot the $v_C(t)$ waveform, and verify that there is no zero crossing.
   (b) If $v_C(0) = 5$ V and $i_L(0) = 1$ A, find $v_C(t)$ for $t \geq 0$. Plot the $v_C(t)$ waveform, and verify that there is one zero crossing.

19. A second-order circuit for $t > 0$ is shown in figure P9.19. It is given that $v_C(0^+) = 20$ V and $i_L(0^+) = 8$ A. Find $v_C(t)$ for $t > 0$.

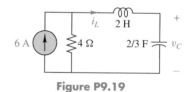

Figure P9.19

20. When a dc voltage of E V is applied to a series LC circuit with no initial stored energy, the voltage across the capacitor reaches a peak value of twice the source voltage. This problem investigates this phenomenon. The switch S in the circuit of figure P9.20 is closed at $t = 0$. Both the inductor current and the capacitor voltage are zero at $t = 0$. Show that for $t \geq 0$,

$$i(t) = \frac{E}{\sqrt{\frac{L}{C}}}\sin\left(\frac{1}{\sqrt{LC}}t\right)\ \text{A}$$

and

$$v_C(t) = E\left(1 - \cos\frac{t}{\sqrt{LC}}\right)\ \text{V}.$$

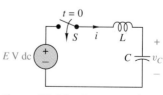

Figure P9.20 Capacitor voltage reaches a maximum of $2E$.

21. The circuit in figure P9.21 is a dual of the one in the previous problem. The switch S is opened at $t = 0$, at which

time both the inductor current and the capacitor voltage are zero. Show that for $t \geq 0$,

$$v(t) = I_{dc}\sqrt{\frac{L}{C}} \sin\left(\frac{1}{\sqrt{LC}}t\right) \text{ V}$$

and

$$i_L(t) = I_{dc}\left(1 - \cos\frac{t}{\sqrt{LC}}\right) \text{ A.}$$

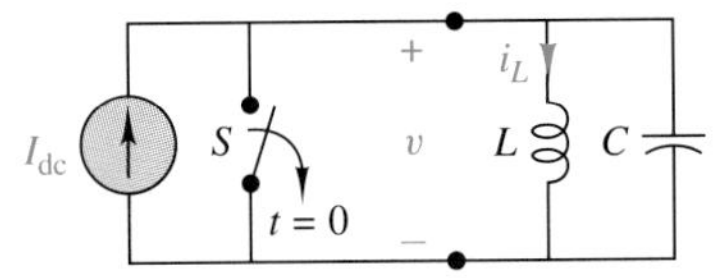

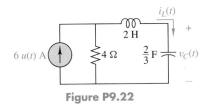

Figure P9.21 Inductor current reaches a maximum of $2I_{dc}$.

22. Consider the RLC circuit of figure P9.22. Here, $v_C(0^-) = 20$ V and $i_L(0^-) = 8$ A. Determine the response, $v_C(t)$, for $t \geq 0$.

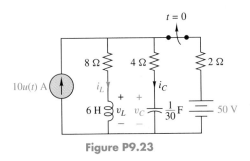

Figure P9.22

23. Consider the circuit shown in figure P9.23.
(a) Find $v_C(0^-)$ and $v_C(0^+)$.
(b) Find $i_L(0^-)$ and $i_L(0^+)$.
(c) Find $v_L(0^+)$ and $i_C(0^+)$.
(d) Find the characteristic equation and natural frequencies of the circuit.
(e) Determine the general form of the response, $v_C(t)$.
(f) Determine all coefficients in the general form of the response.

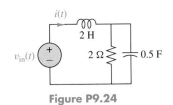

Figure P9.23

24. Consider the circuit shown in figure P9.24. Write a second-order differential equation with $i(t)$ as the unknown.
Answer: $i''(t) + i'(t) + i(t) = 0.5v'_{in}(t) + 0.5v_{in}(t)$.

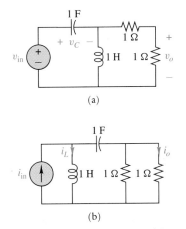

Figure P9.24

25. For the circuit of figure P9.25a, $v_C(0) = 0$, $i_L(0) = 0$, and the output for $t \geq 0$ is $v_o(t) = g(t)$ V when the input for $t \geq 0$ is $v_{in}(t) = f(t)$ V. The circuit of (b) is the dual of the circuit of (a). What is the output $i_o(t)$ for $t > 0$ of the circuit of figure P9.25b when $v_C(0) = 0$, $i_L(0) = 0$, and the input for $t \geq 0$ is $i_{in}(t) = f(t)$ amperes? Note that you do not need to know the functions $f(t)$ and $g(t)$ in order to work the problem.

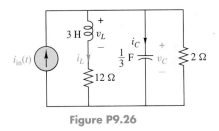

(a)

(b)

Figure P9.25 (a) A circuit with known input and output. (b) The dual circuit of (a).

26. Consider the circuit shown in figure P9.26.
(a) Write a second-order differential equation with v_C as the unknown. Find the values of σ and ω_n.
(b) Repeat part (a), with i_L as the unknown.

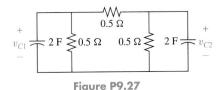

Figure P9.26

27. The second-order circuit shown in figure P9.27 contains two capacitors.
(a) Find the second-order differential equation with v_{C1} as the unknown. Give the values of σ and ω_n.
(b) If $v_{C1}(0) = 2$ V, and $v_{C2}(0) = 4$ V, find $v_{C1}(t)$ for $t > 0$.

Figure P9.27

28. In the circuit shown in figure P9.28, $i_L(0^-) = 0$ A and $v_C(0^-) = 0$ V. Determine the responses $v_C(t)$ and $i_L(t)$.

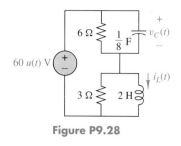

Figure P9.28

29. In the circuit of figure P9.29, the voltage-controlled voltage source has a gain $A > 0$. Find the ranges of A for the circuit to be (a) overdamped, (b) underdamped, (c) critically damped, and (d) undamped.

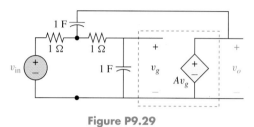

Figure P9.29

30. The second-order circuit shown in figure P9.30 is of the overdamped variety. Find the step response, i.e., the expression for $v_o(t)$ for $t > 0$, when the input is $v_i(t) = u(t)$ V and the capacitors are initially uncharged. Sketch roughly the waveform of $v_o(t)$. (*Remark*: The waveform $v_o(t)$ consists of a very fast rise toward 1 V and then a relatively slow exponential decrease toward 0 V. This can be explained using the first-order RC circuit properties studied in chapter 8. During the first few microseconds, the 0.1-μF capacitor behaves almost as a short circuit, and the 1-nF capacitor is charged up with a time constant of about 10 nsec. After the smaller capacitor is charged up to nearly 1 V in about 5 nsec, it behaves approximately as an open circuit. The 0.1-μF capacitor is then charged up with a time constant of about 1 msec. As the larger capacitor is charged up, the output across the smaller decreases toward zero.)

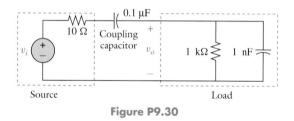

Figure P9.30

31. Find the value of the negative resistance, $-R_n$, required for the circuit shown in figure P9.31 to generate sinusoidal oscillations with constant amplitude.

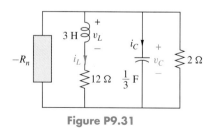

Figure P9.31

32. For the Wien bridge oscillator of example 9.8, suppose the op amp has saturations voltages ±12 V. Let $C = 1$ μF and $R_1 = 500$ Ω. R_2 is a temperature-sensitive resistor whose resistance is a function of the amplitude of the sinusoidal current passing through it. The relationship between the device's resistance and current is given the equations

$$i_2(t) = I_m \sin(\omega t + \theta)$$

and

$$R_2 = 500 + 100(I_m - 0.01).$$

(a) Find the frequency of oscillation ω_n.
(b) Find the amplitude of voltage at the op amp output terminal (with respect to the ground).

33. In the Wien bridge oscillator of figure 9.13a, let $R_1 = R_2 = 1$ kΩ, $r = 10$ kΩ, and $C = 0.1$ μF.
(a) Determine the frequency of oscillation in Hz.
(b) If $v_{C1}(0) = 5$ V, and $v_{C2}(0) = 0$ V, find $v_1(t)$ for $t \geq 0$.
(c) Use any circuit simulation (e.g., SPICE) to verify the waveform of $v_1(t)$ in part (b).

34. In the Wien bridge oscillator of figure 9.13a, let $R_1 = 1$ kΩ, $r = 10$ kΩ, and $C = 0.1$ μF. The lamp resistance R_2 is a function of the peak value of the sinusoidal current, as shown in figure P9.34.

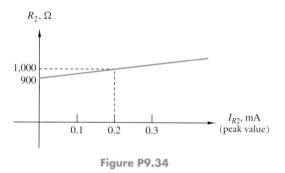

Figure P9.34

(a) Determine the frequency of oscillation.
(b) Determine the amplitude of the sinusoidal waveform $v_1(t)$.

35. A *failure* of a circuit is a situation wherein a circuit dramatically fails to perform as designed. Almost 75% of failures in circuits are due to opens and shorts of individual circuit elements. Heating, cycling a circuit on and off, etc., cause degradation in the circuit parameters, resistances, capacitances, and inductances, which often precipitates the short or open. For example, the material inside a resistor might become brittle over a period of time and finally crumble, leaving a break in the circuit. On the other hand, the material might congeal or become dense, decreasing the resistance of the resistor. In what follows, you are to determine the length of time it takes for a circuit to move from an overdamped behavior to an underdamped behavior due to changes in the resistor characteristic as a function of time.

(a) For the parallel RLC circuit in figure P9.35a, suppose $R = R_0 + \exp(t - 5)$ Ω, where $t \geq 0$ constitutes time in years. Determine the time t' during which the circuit changes its behavior from overdamped to underdamped.

CHAPTER 9   SECOND-ORDER LINEAR CIRCUITS

(b) For the series RLC circuit of figure P9.35b, the resistor satisfies $R = R_0/[1 + \exp(t - 5)]$ Ω, where again, t is time in years. Repeat part (a). Here, one presumes that the circuit is part of a larger piece of electronic apparatus, such as a television set, which is used extensively over a period of years. In that case, the time t' is not connected with the response time of the circuit.

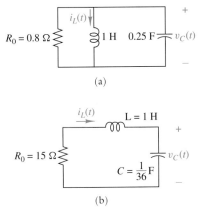

(a)

(b)

Figure P9.35 (a) Parallel and (b) series RLC circuit subject to resistor degradation over time.

36. The operational amplifier circuit of figure P9.36 is a second-order circuit.

(a) Determine the values of C_1 and C_2 that produce a characteristic equation having natural frequencies at -4 and -12 (in standard units).

(b) Adjust the value of R_1 so that for a step function input voltage, the value of the output voltage for large t is, for all practical purposes, 10 V.

(c) Determine $v_{out}(t)$ when $v_s(t) = u(t)$ and all capacitor voltages are zero at $t = 0$.

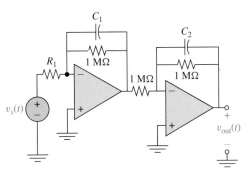

Figure P9.36 Cascade of leaky integrator circuits having a second-order response.

CHAPTER 10

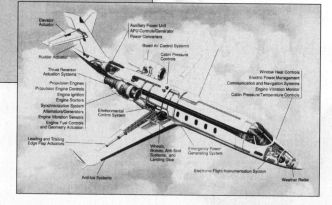

CHAPTER OUTLINE

Sinusoidal Steady-State Analysis by Phasor Methods

A HIGH-ACCURACY PRESSURE SENSOR APPLICATION

THE control of high-performance jet engines requires highly accurate pressure measurements—less than a tenth of a percent of a full-range measurement—over a wide range of temperatures, from −65°F to 200°F. The pressure range may be as low as 20 PSIA or as high as 650 PSIA. In jet (turbine) engine applications, knowing pressure and temperature allows one to compute the mass (volume) airflow, a critical aspect of the engine's performance. A pressure sensor is also a critical component in the regulation of aircraft cabin pressure. Such a sensor is pictured below, along with a functional block diagram of its operation. Here, a diaphragm consisting of two fused quartz plates separated by a vacuum has a capacitance that changes as a function of pressure and temperature. This quartz capacitive diaphragm is an element in a bridge circuit. It is the bridge circuit, in conjunction with a detailed knowledge of the characteristics of a pair of quartz capacitors over the required operating range of pressure and temperature, that delivers accurate pressure measurements.

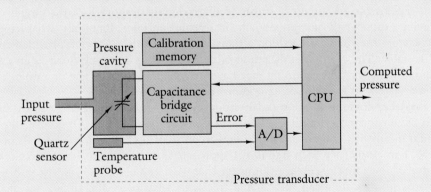

Because of the small capacitances—on the order of picofarads—associated with the quartz (diaphragm) capacitor, the bridge circuit is driven by an ac source and is called an ac bridge. Driving the bridge by an ac source moves its analysis outside the realm of dc and step response techniques studied in earlier chapters. New methods of analysis are necessary. Phasor methods, the primary focus of this chapter, will allow us to analyze capacitive

279

and inductive circuits excited by sinusoidal (ac) inputs. In particular, phasor techniques will permit us to analyze an ac bridge circuit. Although the analysis of the pressure sensor just described is beyond the scope of this text, the chapter will end with an analysis of a simplified pressure sensor circuit motivated by the one pictured above. ◆

◆ CHAPTER OBJECTIVES

- Review and elaborate on the basic arithmetic and essential properties of complex numbers pertinent to the sinusoidal steady-state analysis of circuits.
- Develop two complementary techniques for computing the response of simple RL, RC, and RLC circuits excited by sinusoidal inputs and modeled by differential equations.
- Define the notion of a (complex) phasor for representing sinusoidal currents and voltages in a circuit.
- Using the notion of a phasor, introduce the concepts of impedance (a frequency-dependent resistance), admittance (a frequency-dependent conductance), and a generalized Ohm's law for two-terminal circuit elements having phasor currents and voltages.
- Utilizing the methods of nodal and loop analysis and the network theorems of chapter 5, analyze passive and op amp circuits by the phasor method.
- Introduce and utilize the phasor diagram for determining the qualitative behavior of circuits in the sinusoidal steady state.
- Introduce the notion of frequency response for linear circuits; i.e., investigate the behavior of a circuit driven by a sinusoid as its frequency ranges over a given band.

---

1. INTRODUCTION

Mechanical oscillations exemplified in a swinging pendulum or felt in a car with broken shock absorbers are relatively familiar experiences. Oscillations are present in electric circuits also. For example, sources having sinusoidal voltages or currents will often drive a circuit. Such sources will have an oscillatory voltage or current of the form $V_s \cos(\omega t + \theta)$ or, perhaps, $V_s \sin(\omega t + \theta)$. Ordinarily, driving a circuit with sinusoidal excitations sets up sinusoidal voltages and currents in the circuit. This may not be evident at first, but if the circuit is stable, after a long period of time one can measure such voltages or currents with an oscilloscope. A circuit is **stable** if all zero-input responses of the circuit consist of decaying exponentials or exponentially decreasing sinusoids. Physically, what happens is that at startup, the circuit will exhibit a transient behavior that dies out, provided that the circuit is stable. The often-seen flickering of lights during a lightning storm is an example of transient behavior. The lightning may have struck a transmission line or a pole, causing the power system to waver from its nominal behavior.

Sinusoidal excitations and sinusoidal responses are so common that their study falls under the circuit-theoretic heading of *sinusoidal steady-state analysis*. Of course, the word *sinusoidal* means that the excitations to the circuit have the form $V_s \cos(\omega t + \theta)$ or $V_s \sin(\omega t + \theta)$. For consistency with traditional approaches, we take $V_s \cos(\omega t + \theta)$ as the general form of the input excitation, knowing that $V_s \sin(\omega t + \theta) = V_s \cos(\omega t + \theta - 90°)$. The words *steady state* mean that all the transient behavior of the circuit has died out, which presumes the circuit is stable. A detailed study of stability is taken up in chapter 15 of volume 2. To emphasize our assumption of stability, we will assume that any transient behavior of the circuit will decay to zero, so that there is a well-defined steady-state behavior.

This type of study is important for a number of reasons. First, the analysis of power systems normally occurs in the steady state, as voltages and currents are sinusoidal. Also, as we know, music is composed of different notes; mathematically, a musical signal can be decomposed into a sum of sinusoidal voltages or currents of different frequencies. Finally, the analysis of a sound system builds around the steady-state behavior of the microphone, the amplifier, and the speakers, driven by sinusoidal excitations whose frequency varies from around 40 Hz (1 Hz = 1 hertz = 1 cycle per second) to 20 kHz. Indeed, almost any form of transmission of speech or music requires an understanding of steady-state circuit behavior.

As mentioned earlier, stable circuits driven by sinusoidal excitations of the form $V_s \cos(\omega t + \theta)$ produce sinusoidal voltages and currents in the circuit. A particular steady-state voltage response will then have the form $V_m \cos(\omega t + \phi)$, as illustrated in figure 10.1. This means that a linear, time-invariant (all parameters, resistances, capacitances, inductances, gains, etc., are constant with time), stable circuit can change only the magnitude (V_s goes to V_m) and the phase angle (θ goes to ϕ) of the sinusoid. The frequency ω remains the same. For circuits that are not linear, ω can and usually does change.

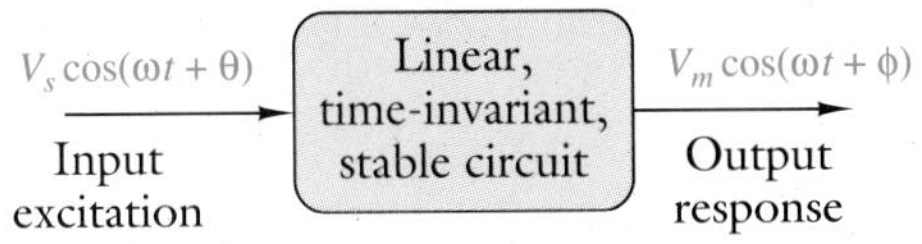

Figure 10.1 Graphical illustration of steady-state sinusoidal circuit behavior. (V_s and V_m could just as well be I_s and I_m or any combination thereof.)

In figure 10.1, the steady-state voltage response is $V_m \cos(\omega t + \phi)$. Alternatively, this could have been a current response of the form $I_m \cos(\omega t + \phi)$. This waveform has the equivalent form $A \cos(\omega t) + B \sin(\omega t)$, where $A = V_m \cos(\phi)$ and $B = -V_m \sin(\phi)$. This can be shown using trigonometric identities as follows:

$$V_m \cos(\omega t + \phi) = V_m \cos(\phi) \cos(\omega t) - V_m \sin(\phi) \sin(\omega t)$$
$$= A \cos(\omega t) + B \sin(\omega t). \tag{10.1}$$

Conversely, by summing the squares of A and B, one obtains

$$V_m = \sqrt{A^2 + B^2}. \tag{10.2a}$$

By taking the inverse tangent of the ratio of B to A, one obtains

$$\phi = -\tan^{-1}\left(\frac{B}{A}\right). \tag{10.2b}$$

In using equation 10.2b, it is important to adjust the resulting angle for the proper quadrant of the complex plane. Equation 10.1 turns out to be useful in developing a conceptually simple technique (see section 3) for computing the steady-state response from a differential equation model of the circuit.

This chapter will introduce three techniques for computing the sinusoidal steady-state response. The first two are precursors for the third, very powerful, technique of *phasor analysis*, which builds on ideas from the theory of complex variables and the basic principles of circuits studied thus far. As a preliminary to these methods of circuit analysis, the next section reviews some fundamental concepts of complex arithmetic. A background in complex variables is helpful in the discussions that follow.

2. A BRIEF REVIEW OF COMPLEX NUMBERS

Let $z_1 = a + jb$ be an arbitrary complex number. The real number a is the **real part** of z_1, denoted $a = \text{Re}[z_1]$. The real number b is the **imaginary part** of z_1 and is denoted $b = \text{Im}[z_1]$. It is simple to verify that

$$a = \text{Re}[z_1] = \frac{z_1 + \bar{z}_1}{2}$$

and

$$b = \text{Im}[z_1] = \frac{z_1 - \overline{z}_1}{2j},$$

where $\overline{z}_1 = a - jb$ is the complex conjugate of z_1. The **magnitude** or **modulus** of z_1, denoted $|z_1|$, satisfies

$$|z_1|^2 = a^2 + b^2.$$

The number $z_1 = a + jb$ is said to be represented in **rectangular coordinates.** Another representation of z_1, called polar form or **polar coordinates**, follows from the simple geometry illustrated in figure 10.2. In figure 10.2, the number z_1 can be thought of as a vector of length $\rho = \sqrt{a^2 + b^2} = |z_1|$, which makes an angle $\theta = \tan^{-1}(b/a)$ with the horizontal in the counterclockwise direction. (Again, in computing $\tan^{-1}(b/a)$, it is important to adjust the angle to be in the proper quadrant of the complex plane.) Hence, $z_1 = a + jb$ is completely specified by its magnitude ρ and angle θ, where

$$a = \rho \cos(\theta) \quad \text{and} \quad b = \rho \sin(\theta).$$

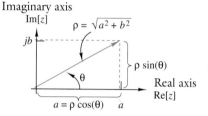

Figure 10.2 Diagram showing relationship between polar and rectangular coordinates representing a complex number.

Exercise. Determine the polar coordinates of $z_1 = -1 - j$ and $z_2 = 1 + j$.
ANSWERS: $\sqrt{2}e^{-j135°}$ and $\sqrt{2}e^{j45°}$.

From the preceding discussion, it follows that

$$z_1 = a + jb = \rho \cos(\theta) + j\rho \sin(\theta) = \rho[\cos(\theta) + j \sin(\theta)] = \rho e^{j\theta} \equiv \rho \angle \theta,$$

where $\angle \theta$ is a shorthand notation for $e^{j\theta}$ and

$$e^{j\theta} = \cos(\theta) + j \sin(\theta) \tag{10.3}$$

is the famous **Euler identity**. (This identity can be demonstrated by writing the Taylor series for $e^{j\theta}$ and recognizing it as the sum of the Taylor series for $\cos(\theta)$ added to j times the Taylor series for $\sin(\theta)$.)

Exercise. Let $z = 6e^{j\pi/6}$, where $\pi/6$ has units of rads (e.g., π rads equals $180°$). Find the real and imaginary parts of z.
ANSWER: $z = 5.1962 + j3$.

The properties of the exponential function immediately imply that

$$e^{j(\theta_1 + \theta_2)} = e^{j\theta_1} e^{j\theta_2}. \tag{10.4}$$

Exercise. Show by direct computation that if $e^{j\theta_1} = \cos(\theta_1) + j \sin(\theta_1)$ and $e^{j\theta_2} = \cos(\theta_2) + j \sin(\theta_2)$, then $e^{j(\theta_1 + \theta_2)} = \cos(\theta_1 + \theta_2) + j \sin(\theta_1 + \theta_2) = e^{j\theta_1} e^{j\theta_2}$.

With the foregoing simple definitions, the product of two complex numbers, $z_1 = a + jb = \rho_1 e^{j\theta_1}$ and $z_2 = c + jd = \rho_2 e^{j\theta_2}$, can be found using rectangular coordinates. The result is

$$z_1 z_2 = (a + jb)(c + jd) = ac - bd + j(bc + ad).$$

In polar coordinates, we have

$$z_1 z_2 = \rho_1 e^{j\theta_1} \rho_2 e^{j\theta_2} = \rho_1 \rho_2 e^{j(\theta_1 + \theta_2)}$$
$$= \rho_1 \rho_2 \cos(\theta_1 + \theta_2) + j\rho_1 \rho_2 \sin(\theta_1 + \theta_2),$$

which, in shorthand notation, is

$$z_1 z_2 = \rho_1 \rho_2 \angle(\theta_1 + \theta_2).$$

EXAMPLE 10.1

Suppose $z_1 = 3 - j4 = 5\angle - 53.13°$ and $z_2 = 8 + j6 = 10\angle 36.87°$. Then

$$z_1 z_2 = (24 + 24) + j(18 - 32) = 48 - j14.$$

Equivalently,

$$z_1 z_2 = 5\angle - 53.13° \times 10\angle 36.87° = 50\angle(-53.13° + 36.87°)$$

$$= 50e^{j(-53.13° + 36.87°)}$$

$$= 50\cos(16.26°) - j50\sin(16.26°) = 48 - j14.$$

Exercise. Let $z_1 = 2 + j2$ and $z_2 = -2 + j6$.

1. Compute the polar form of z_1 and z_2.
2. Compute $z_1 z_2$ in rectangular coordinates.
3. Compute $z_1 z_2$ in polar coordinates.

ANSWERS: $2.8284e^{j45°}$, $6.3246e^{j108.43°}$, $-16 + j8$, $17.8885e^{j153.43°}$.

In rectangular coordinates, the arithmetic for the division of two complex numbers is

$$\frac{z_1}{z_2} = \frac{a + jb}{c + jd} = \frac{(a + jb)(c - jd)}{(c + jd)(c - jd)} = \frac{(a + jb)(c - jd)}{c^2 + d^2}$$

$$= \frac{(ac + bd) + j(bc - ad)}{c^2 + d^2}.$$

In polar coordinates, the calculation is more straightforward:

$$\frac{z_1}{z_2} = \frac{\rho_1 e^{j\theta_1}}{\rho_2 e^{j\theta_2}} = \frac{\rho_1}{\rho_2} e^{j(\theta_1 - \theta_2)} = \frac{\rho_1}{\rho_2} \cos(\theta_1 - \theta_2) + j\frac{\rho_1}{\rho_2} \sin(\theta_1 - \theta_2).$$

In our shorthand notation,

$$\frac{z_1}{z_2} = \frac{\rho_1}{\rho_2} \angle(\theta_1 - \theta_2),$$

where $\angle\theta$ is a shorthand notation for $e^{j\theta}$.

Of particular concern in this chapter are equations involving mixed representations of complex numbers. For example, suppose an unknown complex number $z = Ve^{j\theta}$ satisfies the equation

$$Ve^{j\theta}(a + jb) = c + jd.$$

Then dividing through by $a + jb$ yields

$$Ve^{j\theta} = \frac{c + jd}{a + jb}.$$

Since V is the magnitude of the complex number on the right side of the equals sign, it follows that

$$V = \frac{|c + jd|}{|a + jb|} = \frac{\sqrt{c^2 + d^2}}{\sqrt{a^2 + b^2}}.$$

Here, we have used the fact that a complex number that is the ratio of two other complex numbers has a magnitude equal to the ratio of the magnitudes. To determine the angle θ, one uses the property that θ equals the angle of the number in the numerator minus the angle of the number in the denominator, i.e.,

$$\theta = \angle(c + jd) - \angle(a + jb) = tan^{-1}\left(\frac{d}{c}\right) - tan^{-1}\left(\frac{b}{a}\right).$$

Exercise. Let $z_1 = 2 - 2j$ and $z_2 = 5.5 + j2.4$. Determine V and θ when $Ve^{j\theta} = z_1/z_2$.

 **EXAMPLE 10.2**

Suppose

$$-5Ve^{j\phi} + j6Ve^{j\phi} - 3Ve^{j\phi} = 20e^{j45°}.$$

Find V and ϕ.

SOLUTION

Factoring $Ve^{j\phi}$ out to the left and dividing by $-8 + j6$ yields

$$Ve^{j\phi} = \frac{20e^{j45°}}{-8 + j6} = 2e^{-j98.13°}.$$

Hence, $V = 2$ and $\phi = -98.13°$.

Often in circuit analysis, a function $v(t)$ is a complex number for each t, e.g., $v(t) = Ve^{j(\omega t + \phi)}$. If $v(t)$ were a voltage in a circuit, it would have to satisfy some specific differential equation. Because of this, it is possible to find values for V and ϕ. The next example illustrates such a computation.

EXAMPLE 10.3

Suppose a particular function or circuit waveform

$$v(t) = Ae^{j(\omega t + \phi)}$$

satisfies the differential equation

$$\frac{d^2 v}{dt^2} + 2\frac{dv}{dt} + 2v = 10e^{j(\omega t + 60°)}.$$

Determine the values of A and ϕ if ω is known to be 2 rad/sec.

SOLUTION

Since the waveform $v(t)$ must satisfy the differential equation, the first step is to substitute into the differential equation. Substituting $Ae^{j(\omega t + \phi)}$ into the differential equation and taking appropriate derivatives yields

$$-\omega^2 Ae^{j(\omega t + \phi)} + j2\omega Ae^{j(\omega t + \phi)} + 2Ae^{j(\omega t + \phi)} = 10e^{j(\omega t + 60°)}.$$

The $e^{j\omega t}$ term, which is always nonzero, cancels on both sides, leaving

$$Ae^{j\phi}[2 - \omega^2 + j2\omega] = 10e^{j60°}.$$

Since $\omega = 2$, one can equate magnitudes and angles to obtain

$$Ae^{j\phi} = \frac{10e^{j60°}}{2 - \omega^2 + j2\omega} = \frac{10e^{j60°}}{-2 + j4} = 2.236e^{-j56.57°}.$$

Exercise. Repeat the preceding example if $\omega = 3$ and $10e^{j(\omega t + 60°)}$ is changed to $18.44e^{j(\omega t - 81.2°)}$. .

ANSWER: $A = 2$ and $\phi = 139.4°$.

Throughout much of this chapter, the techniques of circuit analysis will require complex number arithmetic. The quantities of practical interest are generally real. The complex arithmetic is a shortcut to computing the real quantities, which are obtained by taking the real part of the complex number or complex function. The various manipulations depend on some general properties related to the real parts of complex numbers.

Property 1. $\text{Re}[z_1 + z_2] = \text{Re}[z_1] + \text{Re}[z_2]$.

This property has a particularly nice application to summing trigonometric waveforms. Let $v_1(t) = \cos(\omega t + 55°)$ and $v_2(t) = 10\sin(\omega t - 30°) = 10\cos(\omega t - 120°)$. Notice that a $-90°$ shift converts the cosine to a sine. Hence,

$$v_1(t) + v_2(t) = \cos(\omega t + 55°) + 10\cos(\omega t - 120°)$$

$$= \text{Re}[e^{j(\omega t + 55°)}] + \text{Re}[10e^{j(\omega t - 120°)}]$$

(by the Euler identity, equation 10.3)

$$= \text{Re}[e^{j(\omega t + 55°)} + 10e^{j(\omega t - 120°)}]$$

(by the properties of real parts of a complex number)

$$= \text{Re}[e^{j\omega t}(e^{j55°} + 10e^{-j120°})]$$

(by factoring out $e^{j\omega t}$)

$$= \text{Re}\{e^{j\omega t}[(0.5736 + j0.8192) + (-5 - j8.66)]\}$$

(after conversion to rectangular form)

$$= \text{Re}[e^{j\omega t}(-4.426 - j7.841)] = \text{Re}[9e^{j(\omega t - 119.4°)}]$$

(after combining the rectangular forms and converting back to polar form)

$$= 9\cos(\omega t - 119.4°)$$

(after taking the real part).

This sequence of manipulations shows that the magnitudes and phases of two cosines at the same frequency ω can be represented by distinct complex numbers. One can then add the complex numbers and determine the magnitude and angle of a third cosine equal to the sum of the original two cosines. The procedure is a shortcut for adding two cosines together.

Property 2. (Proportionality Property) $\text{Re}[\alpha z_1] = \alpha \text{Re}[z_1]$, for all real scalars α.

Properties 1 and 2 taken together imply that

$$\text{Re}[\alpha_1 z_1 + \alpha_2 z_2] = \alpha_1 \text{Re}[z_1] + \alpha_2 \text{Re}[z_2],$$

which is a **linearity property** for complex numbers multiplied by real scalars. The next property, which underpins the techniques of this chapter, defines how differentiation can be interchanged with the operation $\text{Re}[\cdot]$.

Property 3. (Differentiation Property) Let $A = \rho e^{j\theta}$. Then

$$\frac{d}{dt}\text{Re}\left[Ae^{j\omega t}\right] = \text{Re}\left[\frac{d}{dt}\left(Ae^{j\omega t}\right)\right] = \text{Re}\left[j\omega Ae^{j\omega t}\right].$$

Exercise. Find

$$\text{Re}\left[\frac{d}{dt}\left((10 + j5)e^{j100t}\right)\right].$$

ANSWER: $1,118\cos(100t + 116.57°)$.

The next property tells us the conditions for the equality of two complex-valued time functions.

Property 4. For all complex numbers A and B, $\text{Re}[Ae^{j\omega t}] = \text{Re}[Be^{j\omega t}]$ for all t if and only if $A = B$.

Lastly, taken together, the foregoing four properties imply a fifth very important property.

Property 5. The sum of any number of (1) complex exponentials (also referred to as complex sinusoids), say, $A_i e^{j\omega t}$, or (2) derivatives of any order of complex exponentials of the same frequency ω, or (3) indefinite integrals of any order of a complex exponential of the same frequency ω is a complex exponential of the same frequency ω.

This property is another foundation stone on which the phasor analysis of the chapter builds.

Table 10.1 summarizes the properties of complex numbers.

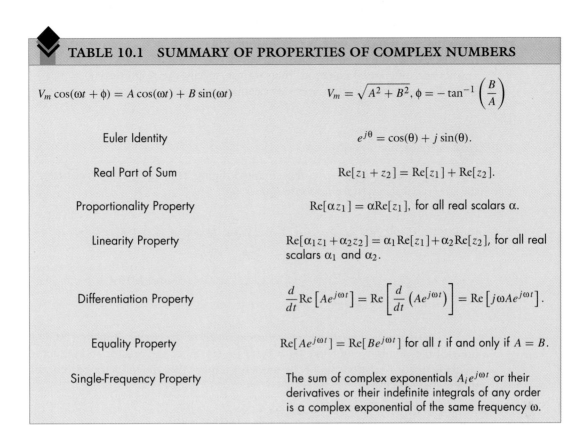

TABLE 10.1   SUMMARY OF PROPERTIES OF COMPLEX NUMBERS

$V_m \cos(\omega t + \phi) = A \cos(\omega t) + B \sin(\omega t)$	$V_m = \sqrt{A^2 + B^2}, \phi = -\tan^{-1}\left(\dfrac{B}{A}\right)$
Euler Identity	$e^{j\theta} = \cos(\theta) + j\sin(\theta).$
Real Part of Sum	$\mathrm{Re}[z_1 + z_2] = \mathrm{Re}[z_1] + \mathrm{Re}[z_2].$
Proportionality Property	$\mathrm{Re}[\alpha z_1] = \alpha\,\mathrm{Re}[z_1],$ for all real scalars α.
Linearity Property	$\mathrm{Re}[\alpha_1 z_1 + \alpha_2 z_2] = \alpha_1 \mathrm{Re}[z_1] + \alpha_2 \mathrm{Re}[z_2],$ for all real scalars α_1 and α_2.
Differentiation Property	$\dfrac{d}{dt}\mathrm{Re}\left[Ae^{j\omega t}\right] = \mathrm{Re}\left[\dfrac{d}{dt}\left(Ae^{j\omega t}\right)\right] = \mathrm{Re}\left[j\omega Ae^{j\omega t}\right].$
Equality Property	$\mathrm{Re}[Ae^{j\omega t}] = \mathrm{Re}[Be^{j\omega t}]$ for all t if and only if $A = B$.
Single-Frequency Property	The sum of complex exponentials $A_i e^{j\omega t}$ or their derivatives or their indefinite integrals of any order is a complex exponential of the same frequency ω.

3. A NAIVE TECHNIQUE FOR COMPUTING THE SINUSOIDAL STEADY-STATE RESPONSE

Property 5 of the previous section suggests a conceptually simple technique for determining the sinusoidal steady-state response of a circuit. Recall that RL, RC, and RLC circuits have differential equation models, as shown in chapters 8 and 9. In contrast to the dc sources presented in those chapters, suppose the source excitations are sinusoidal at a frequency ω rad/sec. A first-order circuit differential equation model with a sinusoidal excitation would then have the form

$$\frac{dx(t)}{dt} + ax(t) = V_m \cos(\omega t + \theta),$$

or, in the second-order case,

$$\frac{d^2 x(t)}{dt^2} + a\frac{dx(t)}{dt} + bx(t) = V_m \cos(\omega t + \theta),$$

where $x(t)$ is a desired circuit response, such as $v_L(t)$, $i_L(t)$, $v_C(t)$, etc. To make our derivation meaningful, an important assumption is needed—that the zero-input response consists of decaying exponentials or exponentially decaying sinusoids. The need for this assumption is taken up in chapter 15 of volume 2, at which time we will have a more sophisticated set of analytical tools available to us. For our present purpose, if the zero-input response consists of decaying exponentials or exponentially decaying sinusoids, there is a sinusoidal steady-state response.

Now, property 5 guarantees that the sum of any number of cosines or derivatives of any order of cosines of the same frequency ω is a cosine of the same frequency ω. Hence, the circuit response $x(t)$ has a steady-state form and must be a cosine of frequency ω. Further, the *sum* of the derivatives of $x(t)$ on the left-hand side of each differential equation model must equal $V_m \cos(\omega t + \theta)$, the input excitation. This implies that the steady-state circuit response must also be a cosine of the same frequency, but not necessarily the same magnitude or phase. Hence, we can assume that $x(t) = V_x \cos(\omega t + \phi) = A\cos(\omega t) + B\sin(\omega t)$. Accordingly, if we can find values for A and B so that $A\cos(\omega t) + B\sin(\omega t)$ satisfies the differential equation, then we can specify the response $x(t) = V_x \cos(\omega t + \phi)$. The following example illustrates this calculation.

EXAMPLE 10.4

Let the source excitation to the circuit of figure 10.3 be $i_s(t) = I_s \cos(\omega t)$. Compute the sinusoidal steady-state response $i_L(t)$.

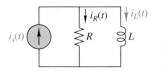

Figure 10.3 Parallel RL circuit for example 10.4.

SOLUTION

Step 1. Determine the differential equation model of the circuit. Applying KCL at the top node of the circuit, we obtain $i_s(t) = i_R(t) + i_L(t)$. Since the resistor and inductor voltages coincide, the v-i relationship of the inductor implies that the inductor current satisfies the differential equation

$$\frac{di_L(t)}{dt} + \frac{R}{L}i_L(t) = \frac{R}{L}i_s(t). \tag{10.5}$$

Step 2. Determine the form of the response. Since the input is a cosine wave, the sinusoidal steady-state response will have the sinusoidal form

$$i_L(t) = A\cos(\omega t) + B\sin(\omega t). \tag{10.6}$$

Step 3. Substitute the form of the response (equation 10.6) into the differential equation 10.5. Inserting equation 10.6 into the differential equation 10.5 and evaluating the derivatives yields

$$\frac{R}{L}I_s \cos(\omega t) = \frac{d}{dt}\left(A\cos(\omega t) + B\sin(\omega t)\right) + \frac{R}{L}\left(A\cos(\omega t) + B\sin(\omega t)\right)$$

$$= -\omega A\sin(\omega t) + \omega B\cos(\omega t) + \frac{RA}{L}\cos(\omega t) + \frac{RB}{L}\sin(\omega t).$$

Step 4. Group like terms and solve for A and B. Grouping like terms leads to

$$\left[B\omega + \frac{R}{L}A - \frac{R}{L}I_s\right]\cos(\omega t) + \left[\frac{R}{L}B - A\omega\right]\sin(\omega t) = 0. \tag{10.7}$$

To determine the coefficients A and B, we evaluate equation 10.7 at two distinct instants of time. Since equation 10.7 must be true at every instant of time, it must be true at $t = 0$, i.e., at $t = 0$,

$$\left[B\omega + \frac{R}{L}A - \frac{R}{L}I_s\right] = 0,$$

or equivalently,

$$\frac{R}{L}A + \omega B = \frac{R}{L}I_s.$$

In addition, equation 10.7 must be true for $t = \pi/(2\omega)$, in which case

$$-\omega A + \frac{R}{L}B = 0.$$

Solving these equations simultaneously for A and B yields

$$A = \frac{R^2 I_s}{R^2 + L^2 \omega^2}$$

and

$$B = \frac{\omega R L I_s}{R^2 + L^2 \omega^2}.$$

Step 5. *Determine the steady-state response.* Since A and B are known,

$$i_L(t) = \frac{R^2 I_s}{R^2 + L^2 \omega^2} \cos(\omega t) + \frac{\omega R L I_s}{R^2 + L^2 \omega^2} \sin(\omega t).$$

The alternative form of $i_L(t) = I_m \cos(\omega t + \phi)$ is more common. From equation 10.2, it follows that

$$i_L(t) = \frac{R I_s}{\sqrt{R^2 + L^2 \omega^2}} \cos(\omega t + \phi) \text{ A,} \qquad (10.8a)$$

where

$$\phi = -\tan^{-1}\left(\frac{\omega L}{R}\right) \qquad (10.8b)$$

is adjusted to reflect the proper quadrant of the complex plane.

This example has illustrated a procedure for finding the sinusoidal steady-state response of a circuit. Step 1 is to substitute an assumed sinusoidal response form, e.g., $A \cos(\omega t) + B \sin(\omega t)$, having unspecified constants A and B into the differential equation and evaluate all derivatives. Step 2 is to group like terms, and step 3 is to use the properties of complex numbers to obtain the constants A and B, which specify the form of the response. After finding A and B, it is necessary to find the magnitude and phase of the cosine, $V_m \cos(\omega t + \phi)$, via equations 10.2.

The next section offers an alternative approach. Using complex excitation signals of the form $V_s e^{j(\omega t + \theta)}$, we can find V_m and ϕ by a more direct route.

4. COMPLEX EXPONENTIAL FORCING FUNCTIONS IN SINUSOIDAL STEADY-STATE COMPUTATIONS

Complex exponential forcing functions are simply complex exponential input excitations of the form $V_s e^{j(\omega t + \theta)}$ or $I_s e^{j(\omega t + \theta)}$. From the properties of complex numbers presented in section 2, we can replace the input excitation, $V_s \cos(\omega t + \theta)$, and assumed circuit response, $V_m \cos(\omega t + \phi) = A \cos(\omega t) + B \sin(\omega t)$, by their complex counterparts, $V_s e^{j(\omega t + \theta)}$ and $V_m e^{j(\omega t + \phi)}$, respectively, without any penalty. To recover the actual real-valued responses, we simply take real parts of the complex quantities. Again, this is justified by properties 1 through 5 of section 2. This process of substitution and subsequent taking of real parts actually simplifies the calculations, because of the simple differential and multiplicative properties of the exponential function. The following example illustrates the advantages by reworking example 10.4.

EXAMPLE 10.5

The goal of this example is to show a more direct calculation of the sinusoidal steady-state response than the method of the previous section through the use of complex exponentials. Let the source current in the parallel RL circuit of figure 10.4 be the complex exponential $i_s(t) = I_s e^{j\omega t}$. Determine the steady-state inductor current $i_L(t)$.

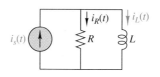

Figure 10.4 Parallel RL circuit.

SOLUTION

As before, the steady-state response is $i_L(t) = I_m e^{j(\omega t + \phi)}$. The computational objective then is to compute I_L and ϕ. Also, as in example 10.4, we substitute the response form into the circuit differential equation model

$$\frac{di_L(t)}{dt} + \frac{R}{L} i_L(t) = \frac{R}{L} i_s(t).$$

The required substitution leads to

$$I_m j\omega e^{j(\omega t + \phi)} + \frac{R}{L} \left[I_m e^{j(\omega t + \phi)} \right] = \frac{R}{L} I_s e^{j\omega t}. \tag{10.9}$$

(Differentiation of the complex exponential is more straightforward than differentiation of $A \cos(\omega t) + B \sin(\omega t)$, especially for higher order derivatives.) After canceling the $e^{j\omega t}$ term on both sides of equation 10.9, grouping like terms, and dividing by $j\omega + R/L$, we obtain the relationship

$$I_m e^{j\phi} = \frac{R I_s}{R + j\omega L} = \frac{R I_s}{R^2 + (\omega L)^2} (R - j\omega L). \tag{10.10}$$

It remains only to equate magnitudes and angles of the left and right sides of equation 10.10. Clearly, then,

$$I_m = \frac{R I_s}{\sqrt{R^2 + (\omega L)^2}} \tag{10.11a}$$

and

$$\phi = -\tan^{-1} \left(\frac{\omega L}{R} \right), \tag{10.11b}$$

with due regard to quadrant. These are precisely the same magnitude and angle obtained in example 10.4 (equations 10.8a and 10.8b) of the previous section. Further, if the input were a real-valued sinusoid of the form

$$i_s(t) = I_s \cos(\omega t) = \text{Re}[I_s e^{j\omega t}] \text{ A},$$

as in example 10.4, then the response would be given by

$$i_L(t) = \text{Re}[I_m e^{j(\omega t + \phi)}] = I_m \cos(\omega t + \phi) \text{ A},$$

with I_L and ϕ given by equations 10.11a and 10.11b, respectively.

The point of the preceding example is that the use of complex exponentials allows us to find the values of the magnitude and phase of the sinusoidal response more directly and efficiently than otherwise. A second example helps reinforce the approach.

◆ **EXAMPLE 10.6**

For the series RC circuit of figure 10.5, let $v_s(t) = V_s \cos(\omega t)$. Determine the steady-state response, $v_C(t)$.

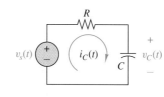

Figure 10.5 Series RC circuit for example 10.6.

CHAPTER 10 SINUSOIDAL STEADY-STATE ANALYSIS BY PHASOR METHODS

Step 1. *Compute the differential equation of the circuit.* Writing a loop equation and substituting for $i_C(t)$ yields

$$RC\frac{dv_C(t)}{dt} + v_C(t) = v_s(t). \tag{10.12}$$

Step 2. If $v_s(t)$ were to equal the complex exponential $V_s e^{j\omega t}$, then the response would be $V_m e^{j(\omega t+\phi)}$. However, from the properties of complex numbers, if $v_s(t) = \text{Re}[V_s e^{j\omega t}] = V_s \cos(\omega t)$ (as it does), then $v_C(t) = \text{Re}[V_m e^{j(\omega t+\phi)}]$. Hence, for the moment, let us set $v_s(t) = V_s e^{j\omega t}$ and assume that $v_C(t) = V_m e^{j(\omega t+\phi)}$. Later, we will take the appropriate real parts. Substituting the complex expressions into the circuit's differential equation 10.12 yields

$$j\omega RC V_m e^{j(\omega t+\phi)} + V_m e^{j(\omega t+\phi)} = V_s e^{j\omega t}.$$

After canceling the $e^{j\omega t}$ terms, factoring $V_m e^{j\phi}$ out to the left and dividing through by $j\omega RC + 1$ yields

$$V_m e^{j\phi} = \frac{V_s}{1 + j\omega RC}. \tag{10.13}$$

Step 3. *Determine the magnitude V_m and the angle ϕ.* Equating magnitudes on both sides of equation 10.13 yields

$$V_m = \frac{V_s}{\sqrt{1 + \omega^2 R^2 C^2}}, \tag{10.14a}$$

and equating angles yields

$$\phi = -\tan^{-1}(\omega RC). \tag{10.14b}$$

Step 4. *Determine the steady-state response.* From equations 10.14, the desired response is

$$v_C(t) = \text{Re}\left[\frac{V_s}{\sqrt{1 + \omega^2 R^2 C^2}} e^{j(\omega t - \tan^{-1}(\omega RC))} \right] \tag{10.15}$$

$$= \frac{V_s}{\sqrt{1 + \omega^2 R^2 C^2}} \cos\left(\omega t - \tan^{-1}(\omega RC)\right) \text{ V}.$$

In deriving equation 10.13 from the differential equation 10.12, we have used a complex exponential function as the input to the circuit. A complex exponential input is not a signal that can be generated in the laboratory. Nevertheless, it is used often in advanced circuit theory to simplify the derivation of many important results, as in the foregoing examples. If one does not mind a more lengthy derivation, then the same result (equations 10.13 through 10.15) can be obtained without the use of a fictitious complex exponential excitation. To demonstrate this, let the voltage source in figure 10.5 represent a real signal source

$$v_s(t) = V_s \cos(\omega t) = \text{Re}[V_s e^{j\omega t}].$$

Then the steady-state response has the form

$$v_C(t) = V_s \cos(\omega t + \phi) = \text{Re}[V_m e^{j(\omega t+\phi)}].$$

Substituting these expressions into the differential equation 10.12 yields

$$RC\frac{d}{dt}\left(\text{Re}\left[V_m e^{j(\omega t+\phi)}\right]\right) + \text{Re}\left[V_m e^{j(\omega t+\phi)}\right] = \text{Re}\left[V_s e^{j\omega t}\right].$$

Making use of properties 2 and 3 of section 2 (see table 10.1), we move the position of the operator Re[ ] outside the first term to obtain

$$\text{Re}\left[RCV_m \frac{d}{dt}\left(e^{j(\omega t + \phi)}\right)\right] + \text{Re}\left[V_m e^{j(\omega t + \phi)}\right] = \text{Re}\left[V_s e^{j\omega t}\right].$$

Evaluating the derivative and using the linearity property of complex numbers produces

$$\text{Re}\left[RCV_m j\omega e^{j(\omega t + \phi)} + V_m e^{j(\omega t + \phi)}\right] = \text{Re}\left[V_s e^{j\omega t}\right]. \tag{10.16}$$

By property 4 of section 2, equation 10.16 is true if and only if

$$RCV_m j\omega e^{j(\omega t + \phi)} + V_m e^{j(\omega t + \phi)} = V_s e^{j\omega t}.$$

This is precisely the equation following equation 10.12, which leads to equations 10.13, 10.14, and finally, 10.15.

As we can see, the use of complex exponentials does indeed lead to a more direct calculation of the sinusoidal steady-state response. However, this method and the method of section 3 require a differential equation model of the circuit. For multiple sources, dependent sources, and large interconnections of circuit elements, finding the differential equation model is often a nontrivial task. In the next section, we eliminate the need to do so by introducing the phasor concept.

5. PHASOR REPRESENTATIONS OF SINUSOIDAL SIGNALS

Recall that $A\angle\phi$ is shorthand for $Ae^{j\phi} = A\cos(\phi) + jA\sin(\phi)$. If the frequency ω is known, then the complex number $A\angle\phi$ completely determines the complex exponential $Ae^{j(\omega t + \phi)}$. In turn, if ω is known, then $A\angle\phi$ completely specifies $A\cos(\omega t + \phi) = \text{Re}[Ae^{j(\omega t + \phi)}]$. This means that the complex number $A\angle\phi$ can represent a sinusoidal function $A\cos(\omega t + \phi)$ whenever ω is known. Complex number representations that denote sinusoidal signals at a fixed frequency are called **phasors**. A phasor voltage or current will be denoted by a boldface capital letter. A typical voltage phasor is $\mathbf{V} = V_m\angle\phi$, and a typical current phasor is $\mathbf{I} = I_m\angle\phi$. For example, the current $i(t) = 25\cos(\omega t + 45°)$ has the phasor representation $\mathbf{I} = 25\angle45°$; the voltage $v(t) = -15\sin(\omega t + 30°) = 15\cos(\omega t + 135°)$ has the phasor representation $\mathbf{V} = 15\angle135°$.

As all voltages and currents satisfy KVL and KCL, respectively, one might expect phasor voltages and currents to do likewise. This is not patently clear. The following simple example demonstrates the truth of the statement as regards KCL.

Consider the circuit node drawn in figure 10.6. From KCL, it follows that

$$i_4(t) = i_1(t) - i_2(t) + i_3(t)$$

$$= 10\cos(\omega t) - 5.043\cos(\omega t + 7.52°) + 8\cos(\omega t - 90°).$$

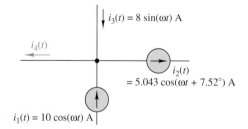

Figure 10.6 Single node having four incident branches.

Using trigonometric identities or property 1 of section 2 to combine terms on the right side leads to

$$i_4(t) = 10\cos(\omega t - 60°) \text{ A}.$$

For the corresponding phasors to satisfy KCL, it must follow that

$$10\angle{-60°} = \mathbf{I_4} = \mathbf{I_1} - \mathbf{I_2} + \mathbf{I_3}.$$

The right side of this equation requires that

$$\mathbf{I_1} - \mathbf{I_2} + \mathbf{I_3} = 10\angle{0°} - 5.043\angle{7.52°} + 8\angle{-90°}$$
$$= 10 - (5 + j0.66) + (-j8) = 5 - j8.66$$
$$= 10\angle{-60°} = \mathbf{I_4}.$$

Thus, the phasors (which have both a real and an imaginary part) appear to satisfy KCL. The reason this is true is because, in reality, $i_4(t) = i_1(t) - i_2(t) + i_3(t)$ implies that

$$\text{Re}[10e^{j(\omega t - 60°)}] = \text{Re}[10e^{j\omega t}] - \text{Re}[5.043e^{j(\omega t + 7.52°)}] + \text{Re}[8e^{j(\omega t - 90°)}]$$
$$= \text{Re}[(10 - 5.043e^{j7.52°} + 8e^{-j90°})e^{j\omega t}]$$

for all t. By property 4 of section 2, this is true if and only if

$$10\angle{-60°} = (10 - 5.043e^{j7.52°} + 8e^{-j90°}).$$

In phasor notation, this stipulates that

$$\mathbf{I_4} = \mathbf{I_1} - \mathbf{I_2} + \mathbf{I_3}.$$

It is the properties of complex numbers and the fact that an equation is true for all t which guarantees that phasors satisfy KCL. The preceding exercise can be generalized to show rigorously that phasors do indeed satisfy KCL. However, the exercise is enough for our present pedagogical purpose. A similar exercise indicates that phasor voltages satisfy KVL, as is illustrated by the following example.

 EXAMPLE 10.7

Using the phasor concept, determine the voltage across the resistor in the circuit of figure 10.7.

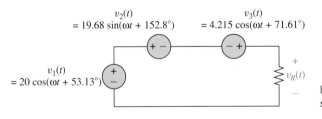

Figure 10.7 Resistive circuit with three sources.

SOLUTION
From KVL,

$$v_R(t) = v_1(t) - v_2(t) + v_3(t).$$

Since voltage phasors must satisfy KVL,

$$\mathbf{V_R} = \mathbf{V_1} - \mathbf{V_2} + \mathbf{V_3} = 20\angle{53.13°} - 19.68\angle{62.8°} + 4.215\angle{71.6°}.$$

Changing to rectangular coordinates and adding yields

$$\mathbf{V_R} = 12 + j16 - (9 + j17.5) + 1.33 + j4 = 4.33 + j2.5.$$

Equivalently, $\mathbf{V_R} = 5\angle 30°$, which implies that

$$v_R(t) = \text{Re}[5e^{j(\omega t + 30°)}] = 5\cos(\omega t + 30°) \text{ V}.$$

Given that phasor voltages and currents satisfy KVL and KCL, respectively, it is possible to develop phasor analogs of Ohm's law for resistors, capacitors, and inductors operating in the sinusoidal steady state. These would then allow us to do sinusoidal steady-state circuit analysis with techniques similar to resistive dc analysis. The next section takes up this thread by introducing the notion of impedance.

6. ELEMENTARY IMPEDANCE CONCEPTS: PHASOR RELATIONSHIPS FOR RESISTORS, INDUCTORS, AND CAPACITORS

Ohm's law-like relationships do exist for resistors, capacitors, and inductors operating in the sinusoidal steady state. The constraint that these circuit elements operate in the *sinusoidal* steady state suggests that any such analog should exhibit a dependence on the sinusoidal frequency.

The first objective of this section is to derive three analogs of Ohm's law, one each for the resistor, capacitor, and inductor. The relationships each take the form $\mathbf{V} = Z(j\omega)\mathbf{I}$, where $\mathbf{V}$ is a phasor voltage, $\mathbf{I}$ is a phasor current, and $Z(j\omega)$ is called the *impedance* of the device—$Z_R(j\omega)$ for a resistor, $Z_C(j\omega)$ for a capacitor, and $Z_L(j\omega)$ for an inductor. The fact that the phasor voltage $\mathbf{V}$ is $Z(j\omega)$ times a phasor current $\mathbf{I}$ indicates a clear kinship with Ohm's law for resistors. Indeed, the unit of impedance is the ohm, because it is the ratio of voltage to current. The impedance $Z(j\omega)$ shows explicitly that the relationship may be frequency dependent.

The derivation of these elementary impedance concepts will build on the assumption that all voltages and currents are complex sinusoids of the same frequency represented by complex phasors. This is permissible because real sinusoids can be recovered from complex sinusoids simply by taking their real parts. To this end, consider the resistive circuit of figure 10.8a.

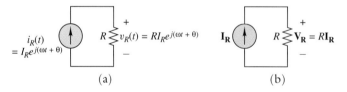

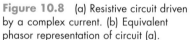

Figure 10.8 (a) Resistive circuit driven by a complex current. (b) Equivalent phasor representation of circuit (a).

From Ohm's law,

$$v_R(t) = Ri_R(t) = RI_Re^{j(\omega t + \theta)}.$$

The equivalent phasor representation of this equation is

$$\mathbf{V_R} = R\mathbf{I_R} = Z_R(j\omega)\mathbf{I_R}, \qquad (10.17)$$

where $\mathbf{I_R} = I_R\angle\theta$ and $Z_R(j\omega) = R$ is the impedance of the resistor. Ideally (however, see the remarks in chapter 1 with respect to the skin effect at high frequencies), the resistor impedance is independent of frequency. Thus, $\mathbf{V_R} = RI_R\angle\theta$. If $i_R(t) = I_R\cos(\omega t + \theta) = \text{Re}[I_Re^{j(\omega t + \theta)}]$, then $v_R(t) = RI_R\cos(\omega t + \theta) = \text{Re}[RI_Re^{j(\omega t + \theta)}]$. This phasor relationship restates Ohm's law for complex excitations. The uniqueness of phasors comes with their application to inductors and capacitors.

Now consider the inductor circuits of figure 10.9a and b. In figure 10.9a, the inductor is driven by a complex current $i_L(t) = I_Le^{j(\omega t + \theta)}$. From the *v*-*i* relationship of the inductor,

it follows that

$$v_L(t) = L\frac{di_L(t)}{dt} = (j\omega L)I_L e^{j(\omega t + \theta)}.$$

The equivalent phasor form of this equation is easily seen to be

$$\mathbf{V_L} = (j\omega L)\mathbf{I_L} = Z(j\omega)\mathbf{I_L}, \tag{10.18}$$

where $\mathbf{I_L} = I_L\angle\theta$ and $Z_L(j\omega) = j\omega L$ is the impedance of the inductor. Clearly, the inductor impedance depends on the value of the radian frequency ω. Specifically, if $\omega = 0$, then the impedance of the inductor is $0\ \Omega$; i.e., in the sinusoidal steady state, the inductor "looks like" a short circuit to dc excitations. If $\omega = \infty$, the impedance is infinite; i.e., in the steady state, the inductor "looks like" an open circuit to signals of very high frequency.

(a)   (b)

Figure 10.9 (a) Inductor driven by complex sinusoidal current. (b) Phasor relationship of the circuit of figure (a).

Equation 10.18 is a frequency-dependent analog of Ohm's law for the inductor. From the properties of the product of two complex numbers, the polar form of the voltage phasor is

$$\mathbf{V_L} = (j\omega L)\mathbf{I_L} = (\omega L I_L)\angle(\theta + 90°)\ \text{V}.$$

Hence, if $i_L(t) = \text{Re}[I_L e^{j(\omega t + \theta)}] = I_L\cos(\omega t + \theta)$ A, then $v_L(t) = \text{Re}[j\omega L I_L e^{j(\omega t + \theta)}] = \text{Re}[\omega L I_L e^{j(\omega t + \theta + 90°)}] = \omega L I_L\cos(\omega t + \theta + 90°)$ V. From this relationship, one sees that the voltage phase leads the current phase by 90°. Equivalently, one can say that the current lags the voltage by 90°. This leading and lagging takes on a more concrete meaning when viewing phasors as vectors in the complex plane, as per figure 10.10. In figure 10.10 one sees that the voltage phasor of the inductor always leads the current phasor by 90°.

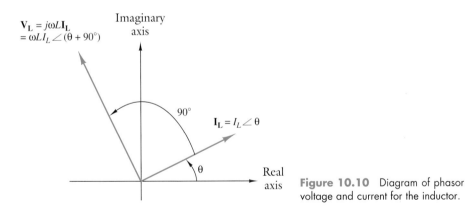

Figure 10.10 Diagram of phasor voltage and current for the inductor.

The capacitor has a similar impedance relationship, derived as follows. Assume that the circuit of figure 10.11a is in the steady state. Then the complex current and voltage associated with the capacitor are, respectively, $i_C(t) = I_C e^{j(\omega t + \theta)}$ and $v_C(t) = V_C e^{j(\omega t + \phi)}$. Substituting these expressions into the defining equation for a capacitor yields

$$I_C e^{j(\omega t + \theta)} = C\frac{d}{dt}\left[V_C e^{j(\omega t + \phi)}\right]$$

$$= j\omega C V_C e^{j(\omega t + \phi)}.$$

Canceling out $e^{j\omega t}$ on both sides yields

$$I_C e^{j\theta} = j\omega C V_C e^{j\phi}.$$

In terms of the phasors $\mathbf{I_C} = I_C e^{j\theta}$ and $\mathbf{V_C} = V_C e^{j\phi}$, this relationship is

$$\mathbf{I_C} = j\omega C \mathbf{V_C},$$

or equivalently,

$$\mathbf{V_C} = \frac{1}{j\omega C} \mathbf{I_C} = Z_C(j\omega) \mathbf{I_C}. \qquad (10.19)$$

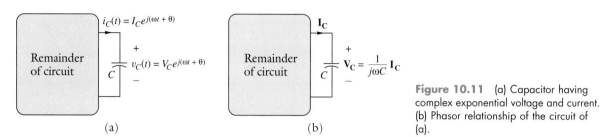

Figure 10.11 (a) Capacitor having complex exponential voltage and current. (b) Phasor relationship of the circuit of (a).

Equation 10.19 defines the impedance of the capacitor as $Z_C(j\omega) = 1/(j\omega C)$. If $\omega = 0$, the impedance of the capacitor is infinite in magnitude. This means that in the sinusoidal steady state, the capacitor "looks like" an open circuit to dc signals. On the other hand, if $\omega = \infty$, then the capacitor has zero impedance and "looks like" a short circuit to large frequencies.

Looking again at equation 10.19, we observe that

$$\mathbf{V_C} = \frac{1}{j\omega C} \mathbf{I_C} = \frac{I_C}{\omega C} \angle (\theta - 90°). \qquad (10.20)$$

Equation 10.20 has a vector interpretation in the complex plane, as shown in figure 10.12. In the figure, the capacitor voltage lags the capacitor current phasor by 90°; or, put another way, the capacitor current leads the capacitor voltage by 90°. Thus, the voltage and current of the capacitor always have a 90° phase difference that is opposite to their phase difference in the inductor.

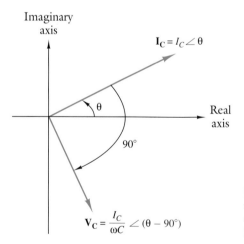

Figure 10.12 Diagram of capacitor voltage and current phasors wherein the voltage phasor lags the current phasor by 90°.

Exercise.

1. For the circuit of figure 10.13a, show that

$$\mathbf{I_L} = \frac{1}{j\omega L} \mathbf{V_L}$$

and that

$$i_L(t) = \frac{V_L}{\omega L} \cos(\omega t + \theta - 90°) \text{ A}.$$

2. For the circuit of figure 10.13b, show that

$$\mathbf{I_C} = j\omega C \mathbf{V_C}$$

and that

$$i_C(t) = \omega C V_C \cos(\omega t + \theta + 90°) \text{ A.}$$

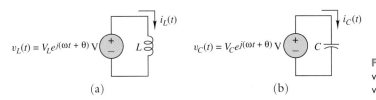

(a)

(b)

Figure 10.13 (a) Inductor driven by a voltage source. (b) Capacitor driven by a voltage source.

Recall that resistance is the reciprocal of conductance. Likewise, impedance in ohms has a reciprocal counterpart called *admittance*. Admittance has units of mhos, just like conductance. The admittance, denoted $Y(j\omega)$, associated with an impedance, $Z(j\omega)$, is defined by the inverse relationship

$$Y(j\omega) = \frac{1}{Z(j\omega)}, \tag{10.21}$$

provided that $Z(j\omega)$ is not equal to zero everywhere. What this means is that the phasor v-i relationship of a resistor, capacitor, and inductor satisfy an equation of the form $\mathbf{I} = Y(j\omega)\mathbf{V}$. Hence, the admittances of the resistor, inductor, and capacitor are respectively given by

$$Y_R(j\omega) = \frac{1}{R}, \; Y_L(j\omega) = \frac{1}{j\omega L}, \; \text{and } Y_C(j\omega) = j\omega C. \tag{10.22}$$

The impedance and admittance relationships of the resistor, capacitor, and inductor are summarized in table 10.2.

TABLE 10.2 SUMMARY OF IMPEDANCE AND ADMITTANCE RELATIONSHIPS FOR RESISTOR, CAPACITOR, AND INDUCTOR

Element	Impedance	Admittance
R	$Z_R(j\omega) = R$	$Y_R(j\omega) = \frac{1}{R}$
C	$Z_C(j\omega) = \frac{1}{j\omega C}$	$Y_C(j\omega) = j\omega C$
L	$Z_L(j\omega) = j\omega L$	$Y_L(j\omega) = \frac{1}{j\omega L}$

In the next section, the notion of impedance is applied to an arbitrary two-terminal device. This generalization will allow us to consider the impedance and admittance of interconnections of capacitors, inductors, resistors, and sources.

For the resistor, inductor, and capacitor, impedance equals the ratio of the respective phasor voltage to the phasor current. The admittance is the reciprocal of the impedance; i.e., **admittance** *is the ratio of the phasor current to the phasor voltage*. Analogously, the **impedance** *of a two-terminal device*, as illustrated in figure 10.14, *is the ratio of the phasor voltage to the phasor current*, i.e.,

$$Z_{in}(j\omega) = \frac{\mathbf{V}_{in}}{\mathbf{I}_{in}}$$

(10.23a)

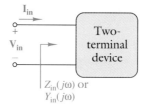

Figure 10.14 Two-terminal device with phasor voltage $\mathbf{V}_{in}$, phasor current $\mathbf{I}_{in}$, and input impedance $Z_{in}(j\omega)$.

Because *impedance* is the ratio of voltage to current, its unit is the ohm. Inverting the relationship of equation 10.23a defines the **admittance** as the ratio of the phasor current to the phasor voltage, i.e.,

$$Y_{in}(j\omega) = \frac{\mathbf{I}_{in}}{\mathbf{V}_{in}}.$$

(10.23b)

The unit of admittance is the *mho*. Provided that $Z(j\omega) \not\equiv 0$, in contrast to a short circuit,

$$Y_{in}(j\omega) = \frac{1}{Z_{in}(j\omega)}.$$

Note that the impedance of an inductor is $j\omega L$ and its admittance is $1/(j\omega L)$ which is infinite at $\omega = 0$.

In general, an impedance or an admittance of a two-terminal element is a complex-valued function of ω; i.e., at each ω, the impedance or admittance is generally a complex number. Since a complex number has a real and an imaginary part, we can further classify the real and imaginary parts of an impedance or an admittance. For an impedance $Z(j\omega)$, the terminology $\text{Im}[Z(j\omega)]$ (the imaginary part of $Z(j\omega)$) is called the **reactance** of the two-terminal element, while $\text{Re}[Z(j\omega)]$ refers to the **resistance** of the element. Further, for an admittance $Y(j\omega)$, $\text{Im}[Y(j\omega)]$ is called the **susceptance** of the two-terminal device, whereas $\text{Re}[Y(j\omega)]$ is referred to as the **conductance** of the device. These definitions are summarized in table 10.3.

Historically, impedance and admittance were first defined as per equations 10.23. However, with the widespread use of the Laplace transform (see chapter 13 of volume 2) in the past several decades, impedance and admittance have taken on a broader meaning, as set forth in chapter 14 of volume 2.

TABLE 10.3 DEFINITIONS OF VARIOUS TERMS			
Impedance: $Z(j\omega) = \dfrac{\mathbf{V}_{in}}{\mathbf{I}_{in}}$		Admittance: $Y(j\omega) = \dfrac{\mathbf{I}_{in}}{\mathbf{V}_{in}}$	
Resistance	Reactance	Conductance	Susceptance
$\text{Re}[Z(j\omega)]$	$\text{Im}[Z(j\omega)]$	$\text{Re}[Y(j\omega)]$	$\text{Im}[Y(j\omega)]$

Using definition 10.23a, one can compute the equivalent impedance, $Z_{in}(j\omega)$, of two devices in series, as shown in figure 10.15a. Here, $\mathbf{V_{in}} = \mathbf{V_1} + \mathbf{V_2}$. By Ohm's law for impedances, $\mathbf{V_1} = Z_1(j\omega)\mathbf{I_1}$ and $\mathbf{V_2} = Z_2(j\omega)\mathbf{I_2}$. But $\mathbf{I_1} = \mathbf{I_2} = \mathbf{I_{in}}$. Hence,

$$\mathbf{V_{in}} = [Z_1(j\omega) + Z_2(j\omega)]\mathbf{I_{in}} \equiv Z_{in}(j\omega)\mathbf{I_{in}};$$

i.e.,

$$Z_{in}(j\omega) = \frac{\mathbf{V_{in}}}{\mathbf{I_{in}}} = Z_1(j\omega) + Z_2(j\omega). \tag{10.24}$$

This simple derivation has a corollary: Given $Z_{in}(j\omega) = Z_1(j\omega) + Z_2(j\omega)$ and the fact that $\mathbf{V_i} = Z_i(j\omega)\mathbf{I_{in}}$, $i = 1, 2$, a simple substitution yields the **voltage division formula**,

$$\mathbf{V_i} = \frac{Z_i(j\omega)}{Z_1(j\omega) + Z_2(j\omega)}\mathbf{V_{in}}, \tag{10.25}$$

for $i = 1, 2$.

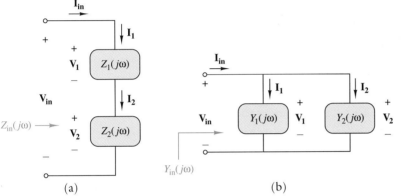
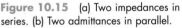

(a) (b)

Figure 10.15 (a) Two impedances in series. (b) Two admittances in parallel.

Happily, equations 10.24 and 10.25 are consistent with our early development of series resistance.

> **Exercise.** Duplicate the derivation of equation 10.24 for three impedances in series.

> **Exercise.** Derive a formula for voltage division when there are three impedances in series.

The admittance of two devices in parallel, as sketched in figure 10.15b, satisfies

$$Y_{in}(j\omega) = \frac{\mathbf{I_{in}}}{\mathbf{V_{in}}} = \frac{\mathbf{I_1} + \mathbf{I_2}}{\mathbf{V_{in}}} = \frac{\mathbf{I_1}}{\mathbf{V_1}} + \frac{\mathbf{I_2}}{\mathbf{V_2}},$$

since $\mathbf{V_{in}} = \mathbf{V_1} = \mathbf{V_2}$. Because

$$Y_1(j\omega) = \frac{\mathbf{I_1}}{\mathbf{V_1}} \quad \text{and} \quad Y_2(j\omega) = \frac{\mathbf{I_2}}{\mathbf{V_2}},$$

we conclude that

$$Y_{in}(j\omega) = Y_1(j\omega) + Y_2(j\omega). \tag{10.26}$$

> **Exercise.** Duplicate the derivation of equation 10.26 for three admittances in parallel.

Exercise. Show that the equivalent impedance of two devices, $Z_1(j\omega)$ and $Z_2(j\omega)$, in parallel is given by

$$Z_{eq}(j\omega) = \frac{Z_1(j\omega)Z_2(j\omega)}{Z_1(j\omega) + Z_2(j\omega)}. \tag{10.27}$$

Exercise. Show that the equivalent admittance of two devices, $Y_1(j\omega)$ and $Y_2(j\omega)$, in series is given by

$$Y_{eq}(j\omega) = \frac{Y_1(j\omega)Y_2(j\omega)}{Y_1(j\omega) + Y_2(j\omega)}. \tag{10.28}$$

Now, since $Y_{in}(j\omega) = Y_1(j\omega) + Y_2(j\omega)$ and since $\mathbf{I_i} = Y_i(j\omega)\mathbf{V_{in}}$, for $i = 1, 2$, one immediately obtains the **current division formula**,

$$\mathbf{I_i} = \frac{Y_i(j\omega)}{Y_2(j\omega) + Y_1(j\omega)} \mathbf{I_{in}}. \tag{10.29}$$

Because devices represented by impedances or admittances must satisfy KVL and KCL in terms of their phasor voltages and currents, and because each device so represented satisfies a generalized Ohm's law, i.e.,

$$\mathbf{V} = Z(j\omega)\mathbf{I} \qquad \text{or} \qquad \mathbf{I} = Y(j\omega)\mathbf{V},$$

it follows that *impedances can be manipulated in the same manner as resistances, and admittances can be manipulated in the same manner as conductances.* The voltage division formula of equation 10.25 and the current division formula of equation 10.29 illustrate this fact. Example 10.8 further clarifies these statements.

Exercise.

1. Derive a current division formula for three admittances in parallel.
2. Find the admittance and then the impedance of each parallel connection in figure 10.16.
   ANSWERS: Admittances are $j(C_1 + C_2 + C_3)\omega$, $1/j\omega (1/L_1 + 1/L_2 + 1/L_3)$.
3. Determine the equivalent inductance for the circuit of figure 10.16a and the equivalent capacitance for the circuit of figure 10.16b.
   SCRAMBLED ANSWERS: $C_1 + C_2 + C_3$, $1/L_1 + 1/L_2 + 1/L_3$.

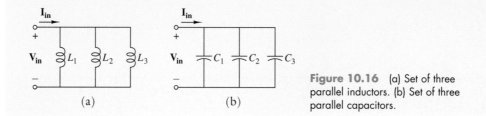

(a)　　　　　　　　　(b)

Figure 10.16 (a) Set of three parallel inductors. (b) Set of three parallel capacitors.

EXAMPLE 10.8

Compute the input impedance, $Z_{in}(j\omega)$, of the ideal op amp circuit of figure 10.17.

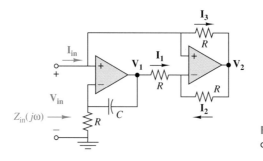

Figure 10.17 Op amp circuit called an impedance converter.

SOLUTION

The trick to solving this problem entails the full use of the properties of the ideal op amp discussed in chapter 3.

Step 1. From the properties of an ideal op amp, from KVL, and from Ohm's law,

$$\mathbf{V_2} = \mathbf{V_{in}} - R\mathbf{I_3} = \mathbf{V_{in}} - R\mathbf{I_{in}}. \tag{10.30}$$

This follows because no current enters the positive or negative terminal of each op amp. But then, $\mathbf{I_{in}} = \mathbf{I_3}$.

Step 2. From the phasor voltage division formula (equation 10.25), it follows that

$$\mathbf{V_{in}} = \frac{R}{R + \dfrac{1}{j\omega C}}\mathbf{V_1},$$

or equivalently,

$$\mathbf{V_1} = \left(1 + \frac{1}{j\omega RC}\right)\mathbf{V_{in}}. \tag{10.31}$$

Here, of course, because of the idealized properties of the op amp, i.e., the voltage across the input terminals of each ideal op amp is zero, the voltage $\mathbf{V_{in}}$ appears across the resistor R in the leftmost op amp.

Step 3. Writing a node equation at the "negative" terminal of the rightmost op amp yields

$$\frac{1}{R}(\mathbf{V_1} - \mathbf{V_{in}}) = -\frac{1}{R}(\mathbf{V_2} - \mathbf{V_{in}}),$$

again by the properties of an ideal op amp. Simplifying this equation results in

$$2\mathbf{V_{in}} = \mathbf{V_1} + \mathbf{V_2}. \tag{10.32}$$

Step 4. Substituting equations 10.30 and 10.31 into equation 10.32 yields

$$2\mathbf{V_{in}} = \mathbf{V_{in}} + \frac{\mathbf{V_{in}}}{j\omega RC} + \mathbf{V_{in}} - R\mathbf{I_{in}},$$

or equivalently,

$$Z_{in}(j\omega) = \frac{\mathbf{V_{in}}}{\mathbf{I_{in}}} = j\omega R^2 C \ \Omega. \tag{10.33}$$

Equation 10.33 suggests that the op amp circuit of figure 10.17 can replace a grounded inductor whose impedance is $j\omega L$ with a proper choice of R and C, i.e., $L = R^2 C$. In integrated circuit technology, it is not possible to build a wire-wound inductor. Instead, inductors are simulated by circuits such as that of figure 10.17.

The next section continues to develop our skill with, and deepen our understanding of, the phasor technique by computing the steady-state responses of various circuits.

8. STEADY-STATE CIRCUIT ANALYSIS USING PHASORS

This section presents a series of examples that illustrate various aspects of the phasor technique. Our purpose is not only to demonstrate how the sinusoidal steady state is computed, but also to illustrate such things as Thévenin equivalents, nodal analysis, mesh analysis, etc., in the context of phasors. Our first example considers again the parallel RL circuit of examples 10.4 and 10.5, together with the series RC circuit of example 10.6. These examples did not use the phasor technique. Example 10.9 demonstrates the superiority of the phasor technique over the methods presented in sections 3 and 4.

EXAMPLE 10.9

Determine the steady-state voltage $v_C(t)$ for the circuit of figure 10.18.

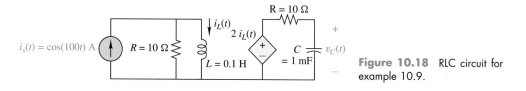

Figure 10.18 RLC circuit for example 10.9.

SOLUTION

Step 1. *Determine* $\mathbf{I_L}$. Since the phasor $\mathbf{I_s} = 1\angle 0°$, by current division it follows that

$$\mathbf{I_L} = \frac{\dfrac{1}{j\omega L}}{\dfrac{1}{R} + \dfrac{1}{j\omega L}}\,\mathbf{I_s} = \frac{1}{1 + j\dfrac{\omega L}{R}}\,\mathbf{I_s} = \frac{1}{1 + j} = \frac{1}{\sqrt{2}}\angle -45°. \qquad (10.34)$$

Step 2. *Use equation 10.34 and voltage division on the RC part of the circuit to compute* $\mathbf{V_C}$. Using voltage division and equation 10.34, we find that the capacitor voltage phasor is

$$\mathbf{V_C} = \frac{\dfrac{1}{j\omega C}}{R + \dfrac{1}{j\omega C}}\,(2\mathbf{I_L}) = \frac{1}{1 + j\omega RC}\,(2\mathbf{I_L}) \qquad (10.35)$$

$$= \frac{1}{1 + j}\,\frac{2}{\sqrt{2}}\,\angle -45° = \angle -90°.$$

Step 3. *Determine* $v_C(t)$. Converting the phasor $\mathbf{V_C}$ of equation 10.35 to its corresponding time function yields

$$v_C(t) = \cos(100t - 90°)\ \text{V} = \sin(100t)\ \text{V}.$$

The next example illustrates the computation of a Thévenin equivalent circuit with the aid of nodal analysis. Because impedances may be manipulated in the same manner as resistances and admittances may be manipulated in the same manner as conductances, Thévenin's theorem, the source transformation theorem (chapter 5) and node and mesh analysis (chapter 4) carry over directly.

EXAMPLE 10.10

(a) Find the Thévenin equivalent of the circuit of figure 10.19 if $\omega = 4$ rad/sec.

(b) Determine the voltage $v_L(t)$ when a 1.2-Ω load resistor is connected across the terminals a–b.

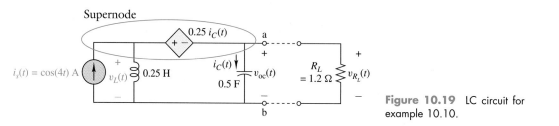

Figure 10.19 LC circuit for example 10.10.

SOLUTION

First we find the Thévenin equivalent circuit, and then we find $v_L(t)$ using the Thévenin equivalent.

Step 1. A nodal equation in terms of phasors at the supernode of figure 10.19 is

$$\mathbf{I}_s = \frac{1}{j\omega L}\mathbf{V}_L + j\omega C \mathbf{V}_{oc} = -j\mathbf{V}_L + 2j\,\mathbf{V}_{oc}. \tag{10.36}$$

Step 2. The relationship between $\mathbf{V}_L$ and $\mathbf{V}_{oc}$, as determined by the dependent source, is

$$\mathbf{V}_L - \mathbf{V}_{oc} = 0.25[\,j2\mathbf{V}_{oc}],$$

or equivalently,

$$\mathbf{V}_L = (1 + 0.5j)\mathbf{V}_{oc}. \tag{10.37}$$

Step 3. Substituting equation 10.37 into equation 10.36 yields

$$\mathbf{I}_s = (0.5 + j)\mathbf{V}_{oc}.$$

Solving for $\mathbf{V}_{oc}$ with $\mathbf{I}_s = 1\angle 0°$ yields

$$\mathbf{V}_{oc} = \frac{1}{0.5 + j}\mathbf{I}_s = (0.4 - j0.8)\mathbf{I}_s = 0.894\angle{-63.43°}\text{ V}. \tag{10.38}$$

Step 4. *Compute the Thévenin equivalent impedance, $Z_{th}(j4)$.* Consider the circuit of figure 10.20, which is the phasor version of figure 10.19 with the output terminals shorted. In figure 10.20, the short-circuit current phasor is

$$\mathbf{I}_{SC} = 1\angle 0°\text{ A}.$$

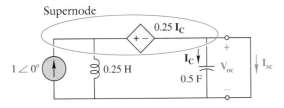

Supernode

$1\angle 0°$ 0.25 H   0.25 $\mathbf{I}_C$ $\mathbf{I}_C$ $\mathbf{V}_{oc}$ $\mathbf{I}_{sc}$ 0.5 F

Figure 10.20 Phasor version of figure 10.19 with shorted terminals.

Therefore, as per equation 10.38,

$$Z_{th}(j4) = \frac{\mathbf{V}_{oc}}{\mathbf{I}_{sc}} = (0.4 - j0.8)\ \Omega.$$

Step 5. *Interpret Z_{th} to generate the Thévenin equivalent circuit.* To interpret the Thévenin equivalent impedance physically, observe that

$$Z_{th}(j4) = (0.4 - j0.8)\ \Omega = (R_{th} + 1/j4C)\ \Omega.$$

Thus, $R_{th} = 0.4\ \Omega$ and $C = 0.3125$ F. Hence, the desired Thévenin equivalent circuit (valid at $\omega = 4$ rad/sec) has the form sketched in figure 10.21.

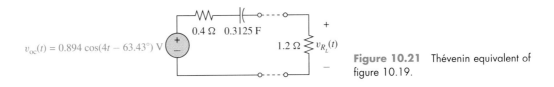

$v_{oc}(t) = 0.894\cos(4t - 63.43°)$ V   0.4 Ω   0.3125 F   1.2 Ω $v_{R_L}(t)$

Figure 10.21 Thévenin equivalent of figure 10.19.

Step 6. *Compute $v_L(t)$ by voltage division.* Using voltage division on the circuit of figure 10.21, we obtain

$$\mathbf{V}_L = \frac{1.2}{1.2 + (0.4 - j0.8)}\mathbf{V}_{in} = (0.6 + j0.3)\,(0.894\angle{-63.43°})$$

$$= 0.6\angle{-36.87°}\text{ V}.$$

Converting the load voltage phasor to its corresponding time-domain sinusoid yields

$$v_L(t) = 0.6\cos(4t - 36.87°)\ \text{V}.$$

EXAMPLE 10.11

Determine the phasor voltage $\mathbf{V_x}$ and the corresponding time function $v_x(t)$ for the circuit of figure 10.22 if $\omega = 100$ rad/sec.

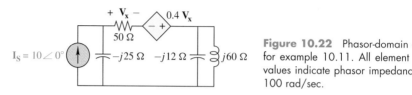

Figure 10.22 Phasor-domain circuit for example 10.11. All element values indicate phasor impedances at 100 rad/sec.

SOLUTION

To solve this problem, it is convenient to execute a source transformation on the independent current source and to combine the impedances of the parallel combination of the capacitor and inductor on the right side of the circuit. After executing these two operations, one obtains the new circuit of figure 10.23.

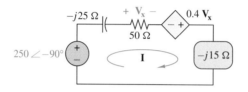

Figure 10.23 Phasor-domain equivalent circuit to that of figure 10.22. All element values indicate phasor impedances at 100 rad/sec. I denotes a phasor loop current.

For the circuit of figure 10.23, the obvious loop equation is

$$250\angle{-90°} = (50 - j25)\mathbf{I} - 0.4(50\mathbf{I}) - j15\mathbf{I} = (30 - j40)\mathbf{I}.$$

Solving for $\mathbf{I}$ yields

$$\mathbf{I} = 4 - j3 = 5\angle{-36.87°}.$$

Consequently, $\mathbf{V_x} = 50\mathbf{I} = 250\angle{-36.87°}$, and $v_x(t) = 250\cos(100t - 36.87°)$ V.

The next section takes up a geometric interpretation of the phasor technique.

9. THE PHASOR DIAGRAM

To this point, a phasor, which is a complex number, has been viewed mainly as a point in the complex plane. A phasor may also be viewed as a vector in the plane. Figure 10.2, which illustrated the polar form of a complex number, foreshadowed this vector point of view. Also, figure 10.10 depicted the orthogonal relationship of the voltage and current phasors of an inductor when viewed as vectors in the plane. Figure 10.12 did the same for the voltage and current phasors of the capacitor.

The convention used in the vector representation is that a vector (or, more precisely, a plane vector) represents a complex number that is the difference between the complex number representing the head of the arrow and the complex number representing its tail. Hence, all the vectors in figure 10.24 represent the complex number $3 + 2j$.

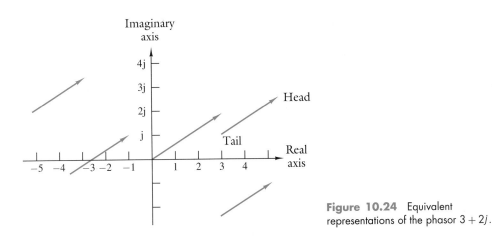

Figure 10.24 Equivalent representations of the phasor $3 + 2j$.

The vector representation of a complex number allows us to perform additions and subtractions *graphically*. Given two complex numbers v_1 and v_2, we may find the sum $v_3 = v_1 + v_2$ graphically by either of the following two different methods:

Method 1. **Parallelogram**

Draw vectors for v_1 and v_2 with their tails joined together at a point P. Draw a straight line through the arrowhead of v_1 and parallel to the vector representing v_2. Draw a straight line through the arrowhead of v_2 and parallel to the vector representing v_1. A parallelogram is formed. Denote by Q the vertex opposite to P. Then the vector PQ (directed from P to Q) represents the sum $v_1 + v_2$. As an example, figure 10.25a illustrates the parallelogram method for finding $v_3 = v_1 + v_2$.

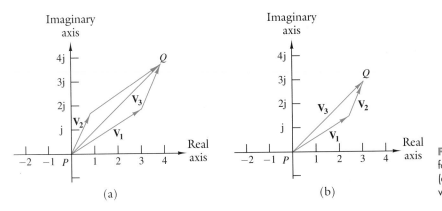
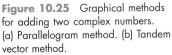

Figure 10.25 Graphical methods for adding two complex numbers. (a) Parallelogram method. (b) Tandem vector method.

Method 2. **Tandem Vectors**

This second method is an obvious variation of the parallelogram method. Instead of drawing the vectors for v_1 and v_2 with their tails joined together, we first draw the v_1 vector in an arbitrary location and then draw the v_2 vector with its tail at the head of the v_1 vector. The vector joining the tail of the v_1 vector to the head of the v_2 vector (and so directed) then represents the sum, $v_3 = v_1 + v_2$. As an example, the same problem solved in figure 10.25a by method 1 is solved again by method 2 and illustrated in figure 10.25b.

When only two complex numbers are involved, the two graphical methods are equally simple. But when more than two complex numbers are to be added together graphically, the tandem vector approach has an advantage: Fewer lines need to be drawn on the plane. We will use whichever method is more convenient for a particular problem.

Recall from chapter 2 that V_{ab}, the voltage drop from node a to node b, may be expressed as the difference of two voltage drops, i.e., $V_{ab} = V_{ac} - V_{bc}$, where c is any other node that may or may not be a network node. Clearly, there are frequent occasions for performing the subtraction of two complex numbers, i.e., $v_3 = v_1 - v_2$. And just as in the case of addition, there are two graphical methods for subtracting two complex numbers. In the first method, we let $v_3 = v_1 - v_2 = v_1 + (-v_2)$ and apply the tandem vector approach for adding two vectors $\mathbf{v}_1$ and $-\mathbf{v}_2$. The second method is to draw the $\mathbf{v}_1$ and $\mathbf{v}_2$ vectors with their tails joined together. Then the vector joining the head of the $\mathbf{v}_2$ vector to the head of the $\mathbf{v}_1$ vector (and so directed) represents the difference $\mathbf{v}_3 = \mathbf{v}_1 - \mathbf{v}_2$. Figures 10.26a and 10.26b illustrate these graphical methods, with $\mathbf{v}_1 = 3 + j2$ and $\mathbf{v}_2 = 1 + j2$. The second method is recommended for future use.

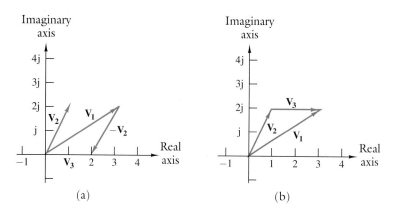

Figure 10.26 Graphical methods for subtracting two complex numbers. (a) A tandem vector approach. (b) A more direct approach.

In electrical engineering texts, when relationships among the various voltage and current phasors are described by vectors drawn on a plane, the resultant vector diagram is called a **phasor diagram**. Before the wide availability of scientific calculators, complex number arithmetic was considered a horrendous task in ac circuit analysis. The phasor diagram, which manipulates complex numbers graphically, was a valuable tool for ac circuit analysis, despite its low degree of accuracy. Nowadays, advanced scientific calculators can perform arithmetic operations on complex numbers as easily as on real numbers. From the computational point of view, the value of the phasor diagram is greatly diminished. However, it is still very useful in certain applications because of its clear portrayal of the *qualitative* properties of circuits without anyone's having to do tedious numerical calculations. The next few examples portray some of these applications. In each case, the phasor diagram approach provides a qualitative insight into the circuit's operation.

EXAMPLE 10.12

A load draws current from a practical voltage source. In the dc case, shown in figure 10.27a, voltage division implies that the load voltage V_L has a smaller magnitude than the source voltage V_s. Is the same true for the ac case, presented in figure 10.27b? Investigate the problem qualitatively with the aid of a phasor diagram.

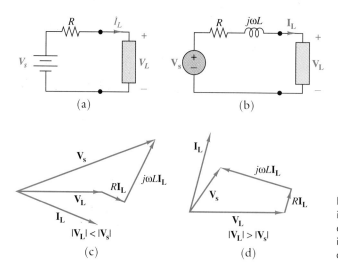

Figure 10.27 Phasor diagrams for investigating $|\mathbf{V_L}|$ relative to $|\mathbf{V_s}|$. (a) dc version of (b), which contains an inductor. (c) and (d): Phasor diagrams for different cases of (b).

SOLUTION

The phasor diagram for the ac circuit of figure 10.27b is shown in parts (c) and (d) of the same figure. It is immediately clear from the diagram, without any numerical calculation, that $|\mathbf{V_L}|$ could be smaller than, greater than, or equal to $|\mathbf{V_s}|$, depending on the phase relationship between $\mathbf{I_L}$ and $\mathbf{V_L}$.

 EXAMPLE 10.13

Two identical voltage sources are connected in series with an RC circuit, as shown in figure 10.28a. Show by the use of a phasor diagram that as R is varied from 0 to ∞, the magnitude of $\mathbf{V_o}$ remains unchanged, while the angle of $\mathbf{V_o}$ varies over a 180° range.

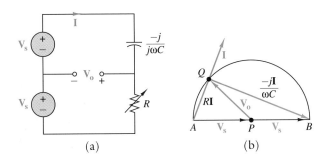

(a)　　(b)

Figure 10.28 (a) A phase shifter circuit and (b) its phasor diagram.

SOLUTION

Applying KVL to figure 10.28a yields

$$R\mathbf{I} + \frac{-j}{\omega C}\mathbf{I} = \mathbf{V_s} + \mathbf{V_s}$$

and

$$\mathbf{V_o} = R\mathbf{I} - \mathbf{V_s}.$$

The phasor diagram shown in figure 10.28b describes the two equations. To facilitate the identification of various vectors (phasors) in the diagram, we have labeled the heads and tails of some vectors with letters. Thus, AP is the vector directed from point A to point P and represents the complex number $\mathbf{V_s}$; and PQ is the vector directed from point P to point Q and is the phasor for $\mathbf{V_o}$. As the value of R varies, the point Q moves in the complex plane. However, the sum of the vectors AQ and QB is fixed and equals $2\mathbf{V_s}$. Together with the fact that the vectors AQ and QB must be perpendicular to each other, the sum of the vectors indicates (from geometry) that the locus of Q is a half-circle with center at point P and radius equal to the magnitude of $\mathbf{V_s}$. This verifies the fact that $|\mathbf{V_o}|$ remains constant and actually is equal to $|\mathbf{V_s}|$ as R varies. The range of variation in the angle of $\mathbf{V_o}$ may be determined by inspection: When R is very large, $\mathbf{I}$ is nearly in phase with $\mathbf{V_s}$, and the point Q nearly coincides with the point B. The angle of $\mathbf{V_o}$ is then the same as that of $\mathbf{V_s}$. When R is very small, $\mathbf{I}$ leads $\mathbf{V_s}$ by approximately 90 degrees. Here, the point Q nearly coincides with the point A, and the angle of $\mathbf{V_o}$ is 180 degrees plus the angle of $\mathbf{V_s}$. Therefore, the range of phase shift obtainable with the circuit is 180 degrees.

Two practical aspects of the preceding phase shift circuit must be pointed out: (1) Although two independent voltage sources are shown in figure 10.28a, in practice the two sources are derived from a single ac source with a center-tapped transformer, a device to be studied in volume 2; (2) the capacitance in the circuit may be replaced by an inductance, and the circuit will still perform as a phase shifter.

> *Exercise.* In the circuit of figure 10.28, suppose the capacitance is replaced by an inductance L. Construct a phasor diagram to show that as R is varied, the magnitude of $\mathbf{V_o}$ remains constant and the angle of $\mathbf{V_o}$ will range from 0 to -180 degrees.

EXAMPLE 10.14

Three independent voltage sources, $v_{s1}(t) = 310\cos(377t)$ V, $v_{s2}(t) = 310\cos(377t - 120°)$ V, and $v_{s3}(t) = 310\cos(377t + 120°)$ V, are connected as shown in figure 10.29a to form a three-phase source.

(a) Construct a phasor diagram that displays the phasors $\mathbf{V_{ao}}$, $\mathbf{V_{bo}}$, $\mathbf{V_{co}}$, $\mathbf{V_{ab}}$, $\mathbf{V_{bc}}$, and $\mathbf{V_{ca}}$.

(b) If an impedance $Z = (3 + j4)$ Ω is connected between nodes a and o, what are the magnitudes of the phasor voltage across and the phasor current through the impedance?

(c) Repeat part (b) if the impedance is connected between nodes a and b.

(d) Three identical impedances, $Z = 3 + j4$ Ω, are connected to the three-phase source of figure 10.29a as shown in figure 10.29b. Find the phasors $\mathbf{I_a}$, $\mathbf{I_b}$, $\mathbf{I_c}$, and $\mathbf{I_n}$ (the phasor of the current through the common wire).

(e) Removing the common wire from the previous circuit yields figure 10.29c. Find the phasors $\mathbf{I_a}$, $\mathbf{I_b}$, and $\mathbf{I_c}$.

(f) Three identical impedances, $Z = 3 + j4$ Ω are connected to the three-phase source of figure 10.29a as shown in figure 10.29d. Find the phasors $\mathbf{I_a}$, $\mathbf{I_b}$, and $\mathbf{I_c}$.

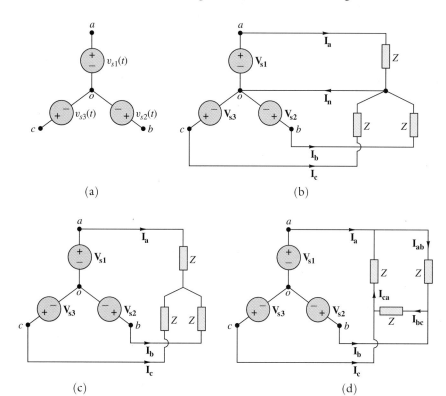

(a)　　　　　　　　　　　　(b)

(c)　　　　　　　　　　　　(d)

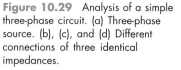

Figure 10.29 Analysis of a simple three-phase circuit. (a) Three-phase source. (b), (c), and (d) Different connections of three identical impedances.

SOLUTION

Part (a). The phasors for the three voltage sources are

$$\mathbf{V_{s1}} = 310\angle 0°, \quad \mathbf{V_{s2}} = 310\angle{-120°}, \quad \mathbf{V_{s3}} = 310\angle 120°.$$

The connection in figure 10.29a indicates that

$$\mathbf{V_{ao}} = \mathbf{V_{s1}}, \qquad \mathbf{V_{bo}} = \mathbf{V_{s2}}, \qquad \mathbf{V_{co}} = \mathbf{V_{s3}}.$$

From KVL,

$$\mathbf{V_{ab}} = \mathbf{V_{ao}} - \mathbf{V_{bo}}, \qquad \mathbf{V_{bc}} = \mathbf{V_{bo}} - \mathbf{V_{co}}, \qquad \mathbf{V_{ca}} = \mathbf{V_{co}} - \mathbf{V_{ao}}.$$

Using the rule for graphical subtraction of phasors (figure 10.26b), we obtain the desired phasor diagram, as shown in figure 10.30a.

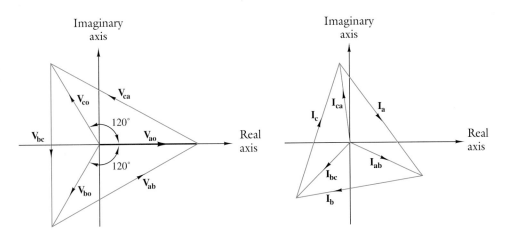

Figure 10.30 Phasor diagrams for the circuit of example 10.14. (a) Solution to part (a). (b) Solution to part (f).

Note that a slight change has been made in the drawing of vectors in the phasor diagram. Since in this case the two endpoints of each vector are clearly defined, we have moved the arrow from the head to the middle part of the vector. This way, the overcrowding of several arrows at one point is avoided.

Part (b). Obviously, the voltage phasor has a magnitude equal to $|\mathbf{V}_{ao}| = |\mathbf{V}_{s1}| = 310$ V. Since $Z = 3 + j4 = 5\angle53.1°$ Ω, the current phasor has a magnitude equal to

$$\left|\frac{\mathbf{V}_{ao}}{Z}\right| = \frac{|\mathbf{V}_{ao}|}{|Z|} = \frac{310}{5} = 62 \text{ A.}$$

Part (c). From the phasor diagram of figure 10.30a and simple geometry, $|\mathbf{V}_{ab}| = \sqrt{3}\,|\mathbf{V}_{ao}| = 534$ V. Therefore, the voltage phasor has a magnitude equal to $|\mathbf{V}_{ab}| = 537$ V, and the current phasor has a magnitude equal to

$$\left|\frac{\mathbf{V}_{ab}}{Z}\right| = \frac{|\mathbf{V}_{ab}|}{|Z|} = \frac{537}{5} = 107.4 \text{ A.}$$

Part (d). From figure 10.29b, the three impedances are each connected directly to an independent voltage source. Application of Ohm's law (generalized for the ac case) yields

$$\mathbf{I}_a = \frac{\mathbf{V}_{ao}}{Z} = \frac{310\angle0°}{5\angle53.1°} = 62\angle-53.1° \text{ A.}$$

From symmetry, the other two phasor currents can be written down directly:

$$\mathbf{I}_b = \mathbf{I}_a\angle-120° = 62\angle-173.1° \text{ A,}$$

$$\mathbf{I}_c = \mathbf{I}_b\angle-120° = 62\angle-293.1° = 62\angle66.9° \text{ A.}$$

Finally,

$$\mathbf{I}_n = \mathbf{I}_a + \mathbf{I}_b + \mathbf{I}_c = \frac{\mathbf{V}_{ao} + \mathbf{V}_{bo} + \mathbf{V}_{co}}{Z} = 0.$$

Part (e). Since $\mathbf{I}_n$ is zero, the wire carrying $\mathbf{I}_n$ can be removed (open circuited) without disturbing any current or voltage in the network. The solutions for $\mathbf{I}_a$, $\mathbf{I}_b$, and $\mathbf{I}_c$ in figure 10.29c are therefore the same as those in figure 10.29b.

Part (f). From the phasor diagram of figure 10.30a and simple geometry, we obtain the voltage phasor

$$\mathbf{V}_{ab} = \mathbf{V}_{ao} - \mathbf{V}_{bo} = 537\angle30° \text{ V.}$$

From symmetry, we can write the other two voltage phasors directly:

$$\mathbf{V_{bc}} = \mathbf{V_{ab}}\angle{-120°} = 537\angle{-90°} \text{ V},$$

$$\mathbf{V_{ca}} = \mathbf{V_{bc}}\angle{-120°} = 537\angle{-210°} = 537\angle{150°} \text{ V}.$$

The current phasor for the impedance connected between nodes a and b is then

$$\mathbf{I_{ab}} = \frac{\mathbf{V_{ab}}}{Z} = \frac{537\angle{30°}}{5\angle{53.1°}} = 107.4\angle{-23.1°} \text{ A}.$$

From symmetry, the other two current phasors are

$$\mathbf{I_{bc}} = \mathbf{I_{ab}}\angle{-120°} = 107.4\angle{-143.1°} \text{ A},$$

$$\mathbf{I_{ca}} = \mathbf{I_{bc}}\angle{-120°} = 107.4\angle{-263.1°} = 107.4\angle{96.9°} \text{ A}.$$

The phasor diagram for these currents is shown in figure 10.30b. From figure 10.29d,

$$\mathbf{I_a} = \mathbf{I_{ab}} - \mathbf{I_{ca}}, \quad \mathbf{I_b} = \mathbf{I_{bc}} - \mathbf{I_{ab}}, \quad \mathbf{I_c} = \mathbf{I_{ca}} - \mathbf{I_{bc}}.$$

The graphical subtraction of two phasors, as shown in figure 10.26b, leads to the phasors $\mathbf{I_a}$, $\mathbf{I_b}$, and $\mathbf{I_c}$ shown in figure 10.30b. Simple geometry applied to this phasor diagram indicates that $\mathbf{I_a}$ leads $\mathbf{I_{bc}}$ by 90 degrees and $|\mathbf{I_a}|$ is $\sqrt{3}$ times $|\mathbf{I_{bc}}|$, i.e.,

$$\mathbf{I_a} = \mathbf{I_{bc}}\angle{90°} = 107.4\angle{-53.1°} \text{ A}.$$

From symmetry, the other two current phasors are

$$\mathbf{I_b} = \mathbf{I_a}\angle{-120°} = 107.4\angle{-173.1°} \text{ A},$$

$$\mathbf{I_c} = \mathbf{I_b}\angle{-120°} = 107.4\angle{-293.1°} = 107.4\angle{66.9°} \text{ A}.$$

The solution of the preceding example used a phasor diagram to carry out a sinusoidal steady-state analysis by the phasor method. The circuit is actually a simplified or skeletal version of a three-phase power system. The load in figures 10.29b and 10.29c is said to be *Y connected*; the load in figure 10.29d is said to be Δ *connected*. When all three impedances are equal, the system is said to be *balanced*. Example 10.14 demonstrates that the analysis of a balanced three-phase circuit is reducible to that of a single-phase circuit. After the result pertaining to one phase is obtained, the results for the remaining two phases may be inferred from symmetry. The analysis of an unbalanced three-phase circuit is more complicated and is best handled by special techniques studied in advanced courses in power systems. Some economic aspects of a three-phase power system and additional analysis methods for balanced three-phase circuits will be discussed in the next chapter.

10. INTRODUCTION TO THE NOTION OF FREQUENCY RESPONSE

The **frequency response** of a circuit is the graph of the ratio of the phasor output to the phasor input as a function of frequency, i.e., as the frequency varies over some specified range. Since the phasor input and phasor output are complex numbers, the frequency response consists of two plots: a graph of the *magnitude* of the phasor ratio and a graph of the *angle* of the phasor ratio. Such graphs indicate the change in magnitude and the change in angle imposed on a sinusoidal input to produce a steady-state output sinusoid. In the steady state, the magnitude of the output sinusoid is the product of the magnitude of the input sinusoid and the magnitude of the frequency response at the frequency of the input. Similarly, in the steady state, the phase of the output sinusoid is the sum of the input phase and the frequency response phase at the input frequency. These properties take on a greater importance after one learns that arbitrary input signals can be decomposed into infinite sums of sinusoids of different frequencies; i.e., each signal has a frequency content. This notion is made precise in a signals and systems course wherein one studies Fourier series and Fourier transforms. The frequency response of a circuit describes the circuit behavior at each frequency component of the input signal. This permits one to isolate, enhance, or reject certain frequency components of an input signal and thereby isolate, enhance, or reject certain kinds of information.

 EXAMPLE 10.15

Plot the frequency response of the RC circuit in figure 10.31.

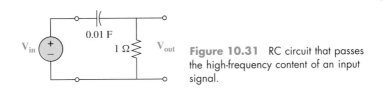

Figure 10.31 RC circuit that passes the high-frequency content of an input signal.

SOLUTION

Using voltage division, the ratio of the output phasor voltage to the input phasor voltage is

$$\frac{\mathbf{V}_{out}}{\mathbf{V}_{in}} = \frac{1}{1 + \dfrac{1}{j0.01\omega}} = \frac{j0.01\omega}{1 + j0.01\omega} = H(j\omega),$$

where we have designated the ratio as $H(j\omega)$.

The two universally important frequencies are $\omega = 0$ and $\omega = \infty$. At these frequencies, $H(j0) = 0\angle 90°$ and $H(j\infty) = 1\angle 0°$. Asymptotically, then, $|H(j\omega)| \to 1$ as $\omega \to \infty$ and $|H(j\omega)| \to 0$ as $\omega \to 0$. With regard to angle, $\angle H(j\omega) \to 0$ as $\omega \to \infty$ and $\angle H(j\omega) \to 90°$ as $\omega \to 0$. Also, a close scrutiny of $H(j\omega)$ indicates that $\omega = 100$ rad/sec is an important frequency. Here, $H(j100) = 0.707\angle 45°$. These values give us a pretty good idea of what the magnitude and phase plots look like. Using a computer program, figures 10.32a and 10.32b show the exact magnitude and phase plots, respectively. These plots are consistent with our earlier asymptotic analysis.

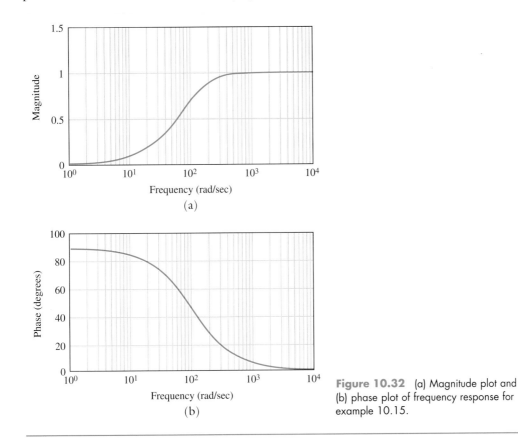

Figure 10.32 (a) Magnitude plot and (b) phase plot of frequency response for example 10.15.

Do the foregoing frequency responses make sense? They should. Going back to the circuit, observe that at $\omega = 0$, the capacitor impedance is infinite. Physically then, in steady state, the capacitor looks like an open circuit for dc, i.e., at zero frequency. The magnitude plot bears this out: For frequencies close to zero, the capacitor approximates an open circuit, and hence, the magnitude remains small. On the other hand, for large frequencies, the capacitor has a very small impedance. This means that most of the source voltage appears across the output resistor. The gain then approximates 1, as indicated by the magnitude plot. The frequency response of the circuit is such that the high-frequency content of the input signal is passed, while the low-frequency content of the input signal is attenuated. Such circuits are commonly called **highpass filters**.

EXAMPLE 10.16

Investigate the frequency response of the parallel RLC circuit of figure 10.33.

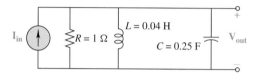

Figure 10.33 A parallel RLC circuit having a bandpass frequency response.

SOLUTION

The input admittance of the circuit is

$$Y_{in}(j\omega) = \frac{1}{R} + \frac{1}{j\omega L} + j\omega C = \frac{\dfrac{1}{LC} - \omega^2 + j\dfrac{\omega}{RC}}{j\dfrac{\omega}{C}}.$$

Inverting to obtain the input impedance yields

$$Z_{in}(j\omega) = \frac{j\dfrac{\omega}{C}}{\dfrac{1}{LC} - \omega^2 + j\dfrac{\omega}{RC}} = \frac{j4\omega}{100 - \omega^2 + j4\omega}.$$

Clearly, $\mathbf{V_{out}} = Z_{in}(j\omega)\mathbf{I_{in}}$. Hence, the ratio of the output phasor to the input phasor is simply $Z_{in}(j\omega)$. Once again, $\omega = 0$ and $\omega = \infty$ are the first two frequencies to look at. Here, $Z_{in}(0) = 0\angle 90°$ and $Z_{in}(\infty) = 0\angle -90°$. Also, at $\omega = 10$, the impedance is real, i.e., $Z_{in}(j10) = 1$. These three points provide a rough idea of the magnitude and phase response of the circuit. Two more points are necessary for a real sense of the frequency response. To obtain these points, we ask, when does the magnitude drop to 0.707 of its maximum value, or, equivalently, when does the phase angle equal $\pm 45°$? This will occur when $|100 - \omega^2| = |4\omega|$. This has the form of a quadratic equation; however, because of the absolute values, there are *two* implicit quadratics: $\omega^2 - 4\omega - 100 = 0$ and $\omega^2 + 4\omega - 100 = 0$. Solving with the quadratic formula yields $\omega = \pm 8.2, \pm 12.2$. Since the magnitude plot is symmetric with respect to the vertical axis, we consider only the positive values of ω. This information provides a good idea of the magnitude and phase plots. A computer program was used to generate the frequency response plots sketched in figures 10.34a and 10.34b. The magnitude plot shows that frequencies satisfying $8.2 \le \omega \le 12.2$ are passed with little attenuation. Frequencies outside this region are attenuated significantly. Such a characteristic is said to be of the **bandpass** variety, and the corresponding circuit is called a **bandpass circuit**. The computer-generated phase plot in figure 10.34b agrees with the behavior arrived at intuitively.

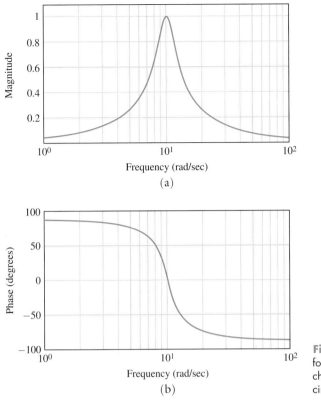

Frequency (rad/sec)

(a)

Phase (degrees)

Frequency (rad/sec)

(b)

Figure 10.34 (a) Magnitude response for example 10.16, showing a bandpass characteristic. (b) Phase portrait of the circuit of figure 10.33.

EXAMPLE 10.17

As a final example, we consider the so-called bandreject circuit. A **bandreject** circuit is the inverse of a bandpass circuit. A bandreject circuit has a central band of frequencies that are significantly attenuated while passing those frequencies outside the band with little to no attenuation. In this example, our goal is to compute the magnitude and phase of the frequency response of the bandreject circuit of figure 10.35.

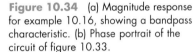

Figure 10.35 Bandreject circuit for example 10.17.

SOLUTION

Once again using voltage division, one obtains the phasor ratio

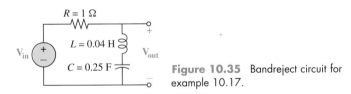

$$\frac{\mathbf{V}_{out}}{\mathbf{V}_{in}} = \frac{\dfrac{1}{LC} - \omega^2}{\dfrac{1}{LC} - \omega^2 + j\omega\dfrac{R}{L}} = \frac{100 - \omega^2}{100 - \omega^2 + j25\omega} = H(j\omega).$$

At $\omega = 0$ and $\omega = \infty$, $H(j\omega) = 1\angle 0°$. Hence, $|H(j\omega)|$ approaches 1 asymptotically as ω approaches 0 and ∞. Also, at $\omega = 10^-$, $H(j\omega) = 0\angle{-90°}$, while at $\omega = 10^+$, $H(j\omega) = 0\angle{-270°} = 0\angle 90°$. For this example, to find the frequencies at which $|H(j\omega)|$ drops to $1/\sqrt{2}$ of its maximum value of 1, it is necessary to equate the magnitude of the real and imaginary parts of the denominator. (In volume 2, we discuss this phenomenon at length.)

This produces two quadratics whose positive roots are $\omega = 3.5078$ and $\omega = 28.5078$. At these frequencies, the angles of $H(j\omega)$ are $-45°$ and $45°$, respectively. The computer-generated plots of figure 10.36 are, of course, consistent with these quickly computed values.

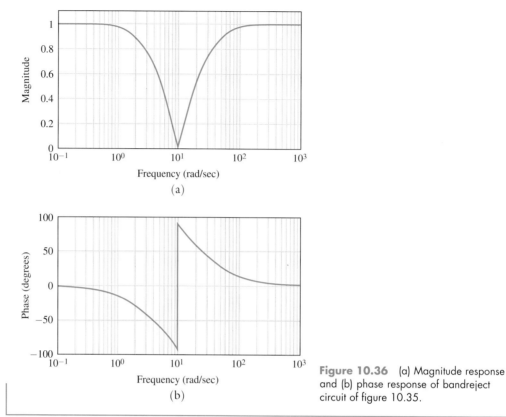

(a)

(b)

Figure 10.36 (a) Magnitude response and (b) phase response of bandreject circuit of figure 10.35.

Plainly, there is a wealth of different kinds of frequency responses one can obtain through different interconnections of resistors, inductors, and capacitors. Historically, phasor techniques were the essential tool for the analysis and design of such circuits.

11. NODAL ANALYSIS OF A PRESSURE-SENSING DEVICE

A bridge circuit, in one or another variation, has been and continues to be a widely used approach to obtaining accurate measurements. In this section, we will analyze the use of the ac bridge circuit of figure 10.37 as a pressure measurement device. The capacitance C_2 is a diaphragm capacitor consisting of a hollow cylinder capped on either side by fused quartz wafers. Between the wafers is a vacuum. The capacitance of the diaphragm changes with temperature and pressure. For our analysis, we will assume that the temperature is held constant and that the pressure being measured is constant for a time period greater than five times the shortest time constant of the circuit. This will allow the voltages and currents in the circuit to reach the steady state, thus enabling us to use phasor analysis to compute their values.

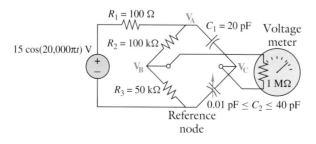

Figure 10.37 Bridge circuit diagram of pressure sensor device. The capacitance C_2 changes as a function of pressure, which causes the voltage $V_B - V_C$ to change as a function of pressure. This is registered on the attached voltage meter, which has a 1-MΩ internal impedance.

As a rule of thumb, the capacitance $C_2 \approx (0.224KA)/d$, where K is the dielectric constant of the material between the plates, A is the area of the plates, and d is the distance between them. Thus, increasing the pressure on the diaphragm decreases the distance between the wafers, thereby increasing the capacitance. Conversely, a decrease in pressure will increase the distance between the wafers and decrease the capacitance. As the capacitance changes, the magnitude of the ac voltage appearing across the voltage meter will vary accordingly. Hence, two relationships must be known: the relationship between the capacitance C_2 and the magnitude of the voltage drop $\mathbf{V_B} - \mathbf{V_C}$; and the relationship between the pressure applied to the diaphragm and the associated capacitance. Our first task will be to specify the relationship between the pressure applied to the diaphragm and the resulting capacitance. Following this, we will use nodal phasor analysis to determine the magnitude of $\mathbf{V_B} - \mathbf{V_C}$ as a function of C_2 and, finally, the relationship between pressure and the magnitude of $\mathbf{V_B} - \mathbf{V_C}$.

Pressure is measured in various units. Millimeters of mercury (mm Hg) is a common standard. One millimeter of mercury equals 1 torr, and 760 torr = 1 atmosphere (atm). One atm is the pressure of the earth's atmosphere at sea level, which supports 760 mm of mercury in a special measuring tube. Suppose it has been found experimentally that the capacitance C_2 in pF varies as a function of pressure according to the formula

$$C_2(p) \text{ (in pF)} = C_0 + K \log_{10}\left(\frac{P_0 + \Delta P}{P_0}\right)$$

$$= 26.5 + 68 \log_{10}\left(\frac{760 + \Delta P}{760}\right).$$

(10.39)

A plot of C_2 as a function of ΔP is given in figure 10.38.

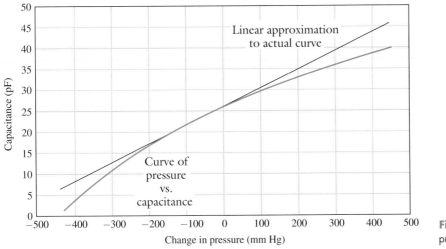

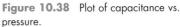

Figure 10.38 Plot of capacitance vs. pressure.

Our next task is to develop the relationship between the capacitance of the bridge circuit and the magnitude of the phasor voltage, $\mathbf{V_B} - \mathbf{V_C}$. In our analysis, $G_1 = (R_1)^{-1}$, $G_2 = (R_2)^{-1}$, $G_3 = (R_3)^{-1}$, and $G_m = 10^{-6}$ mhos is the conductance of the meter. As per figure 10.37, $C_1 = 20$ pF. We will let C_2 range as $0 \le C_2 \le 40$ pF. Finally, $\omega = 2\pi \times 10^4$. The following phasor analysis will be done symbolically so as not to obscure the methodology.

Summing the phasor currents leaving node A leads to the phasor voltage relationship

$$(G_1 + G_2 + j\omega C_1)\mathbf{V_A} - G_2\mathbf{V_B} - j\omega C_1\mathbf{V_C} = G_1 15.$$

Similarly, summing the currents leaving node B leads to the relationship

$$-G_2\mathbf{V_A} + (G_2 + G_3 + G_m)\mathbf{V_B} - G_m\mathbf{V_C} = 0.$$

Lastly, summing the currents leaving node C produces

$$-j\omega C_1 \mathbf{V_A} + j\omega(C_1 + C_2 + G_m)\mathbf{V_C} - G_m\mathbf{V_B} = 0.$$

Writing these three equations in matrix form yields

$$\begin{bmatrix} G_1 + G_2 + j\omega C_1 & -G_2 & -j\omega C_1 \\ -G_2 & G_2 + G_3 + G_m & -G_m \\ -j\omega C_1 & -G_m & G_m + j\omega(C_1 + C_2) \end{bmatrix} \begin{bmatrix} \mathbf{V_A} \\ \mathbf{V_B} \\ \mathbf{V_C} \end{bmatrix} = \begin{bmatrix} 15G_1 \\ 0 \\ 0 \end{bmatrix}. \quad (10.40)$$

The matrix on the left is said to be a *nodal admittance matrix*. Its entries can be real or complex, as indicated. It is not advisable to solve such a set of equations by hand over the range of possible values of C_2. However, using MATLAB or its equivalent, one can solve this matrix equation over the range 0 pF $\leq C_2 \leq$ 40 pF to produce the plot of figure 10.39.

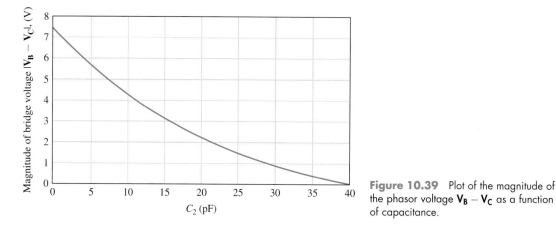

Figure 10.39 Plot of the magnitude of the phasor voltage $\mathbf{V_B} - \mathbf{V_C}$ as a function of capacitance.

Of course, one could measure the voltage appearing across the meter, determine the associated value of C_2 from figure 10.39, refer to figure 10.38 for ΔP, and then determine $P = 760 + \Delta P$. However, this is a long route. To complete our analysis, then, we need to develop the relationship between pressure and bridge voltage. As we have the relationship between C_2 and ΔP and the relationship between C_2 and $|\mathbf{V_B} - \mathbf{V_C}|$, it is a matter of using equation 10.39 to drive the value of C_2 in equation 10.40. This is best done with a simple MATLAB routine, which yields the plot given in figure 10.40.

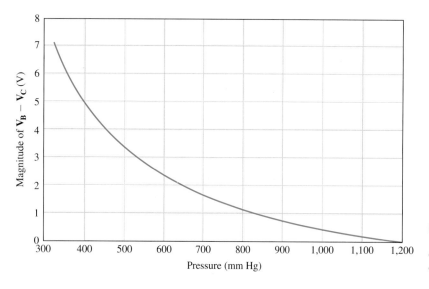

Figure 10.40 Relationship between magnitude of the bridge output voltage and pressure applied to the diaphragm capacitor C_2.

An actual pressure sensor would, of course, be more complex. For example, there would probably be a differential amplifier, such as that shown in figure 10.41, across the terminals of the bridge circuit, and this would probably drive a peak (ac) detector to determine the maximum value of the ac signal that appears at the output of the amplifier. Further, the peak value would probably be read by a digital voltmeter. Nevertheless, our analysis illustrates the basic principles involved in such a measurement. Of course, one could just as easily use loop analysis to solve the problem. This is left as an exercise in the problems at the end of the chapter.

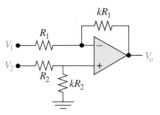

Figure 10.41 Differential amplifier having output voltage $V_o = k(V_2 - V_1)$.

Exercise. Prove that $V_0 = k(V_2 - V_1)$ for the differential amplifier of figure 10.41.

SUMMARY

The two primary goals of this chapter were to develop the phasor technique for analyzing circuits having a sinusoidal steady state and to illustrate how this technique leads to the idea of a circuit frequency response, which characterizes a circuit's behavior in response to the frequency content of an input excitation. In the development, sinusoids were first represented as the real part of a complex sinusoid. To motivate the delineation of the phasor method, we showed how complex sinusoids could be utilized to compute the sinusoidal steady-state response using differential equation circuit models. We then pointed out that a complex (voltage or current) sinusoid is specified by a complex number or phasor representing its magnitude and phase. After introducing the notions of impedance and admittance for the capacitor, the inductor, the resistor, and a general two-terminal circuit element, we showed how phasor voltage and phasor current for each such element satisfies a frequency-dependent analog of Ohm's law. This then allowed us to adapt the analysis techniques and network theorems of chapters 1 through 5 to the steady-state analysis of circuits excited by sinusoidal inputs. For example, there are voltage division formulas, current division formulas, source transformations, Thévenin and Norton theorems, etc., all valid for phasor representations. These all permit us to effectively analyze circuits that have a steady-state response.

The phasor technique opens a door to seeing how circuits respond to sinusoidal excitations. Given that input excitations are composed of sinusoids of different frequency, e.g., a music signal, phasor analysis shows why a circuit will behave differently to the different frequencies present in the input signal. This differential behavior prompts the notion of a circuit's frequency response, which is defined as the ratio of the phasor output to the phasor input excitation in the single-input, single-output case. The frequency response consists of two plots. The magnitude plot shows the magnitude of the gain of the circuit's response to sinusoids of different frequencies, and the phase plot shows the phase shift the circuit introduces to sinusoids of different frequencies. The notion of the frequency response will be generalized in volume 2, when we introduce the notion of the Laplace transform.

TERMS AND CONCEPTS

Admittance: ratio of the phasor current into a two-terminal device to the phasor voltage across the device; mathematically, $Y_{in}(j\omega) = 1/[Z_{in}(j\omega)] = \mathbf{I_{in}}/\mathbf{V_{in}}$.

Bandpass circuit: circuit that passes frequencies within a specified band, while attenuating frequencies outside the band.

Bandreject circuit: circuit in which a central band of frequencies are significantly attenuated while passing those frequencies outside the band with little to no attenuation.

Complex exponential forcing function: a function of the form $v(t) = \mathbf{V}e^{st}$, where $\mathbf{V} = V_m e^{j\phi}$ and $s = \sigma + j\omega$ are complex numbers. A special case ($\sigma = 0$) resulting in $v(t) = V_m e^{j(\omega t + \phi)}$ is used throughout this chapter as a shortcut in sinusoidal steady-state analysis.

Conductance: the real part of a possibly complex admittance.

Current division: property wherein, in a parallel connection of admittances driven by a current source, the current through a particular branch is proportional to the ratio of the admittance of the branch to the total parallel admittance.

Euler identity: $e^{j\theta} = \cos(\theta) + j\sin(\theta)$.

Frequency response: ratio of the phasor output to the phasor input of a circuit as a function of frequency. A graph of the frequency response of a circuit consists of two parts: a graph of the magnitude of the phasor ratio and a graph of the angle of the phasor ratio.

Highpass circuit: circuit with a frequency response such that the high-frequency content of the input signal is passed, while the low-frequency content of the input signal is attenuated.

Imaginary part: the variable b in a complex number $z = a + jb$ for real numbers a and b; denoted $\text{Im}[z]$.

Impedance: an ordinarily complex, frequency-dependent quantity defined as $Z(j\omega) = \mathbf{V}/\mathbf{I}$, where $\mathbf{V}$ is the phasor voltage across a two-terminal device and $\mathbf{I}$ is the phasor current through the device. For the resistor, $Z_R(j\omega) = R$ Ω; for the capacitor, $Z_C(j\omega) = 1/(j\omega C)$ Ω; and for the inductor, $Z_L(j\omega) = j\omega L$ Ω.

Magnitude (modulus): $\sqrt{a^2 + b^2}$, where a and b are real numbers making up the complex number $z = a + jb$. Also denoted $|z|$.

Phasor: complex number representation that denotes a sinusoidal signal at a fixed frequency. Boldface capital letters denote phasor voltages or currents; e.g., a typical voltage phasor is $\mathbf{V} = V_m \angle \phi$, and a typical current phasor is $\mathbf{I} = I_m \angle \phi$.

Phasor diagram: a vector diagram drawn on a plane to show the relationships among various voltage and current phasors in a circuit.

Polar coordinate representation: representation of a complex number z as $\rho e^{j\theta}$, where $\rho > 0$ is the magnitude of z and θ is the angle z makes with respect to the positive horizontal (real) axis of the complex plane.

Reactance: the imaginary part of an impedance.

Real part: the variable a in a complex number $z = a + jb$ for real numbers a and b; denoted $\mathrm{Re}[z]$.

Rectangular coordinate representation: representation of a complex number z as coordinates in the complex plane, i.e., as $a + jb$ for real numbers a and b.

Resistance: real part of a possibly complex impedance.

Sinusoidal steady-state response: the response of a circuit to a sinusoidal excitation after all transient behavior has died out. This definition presumes that the zero-input response of the circuit contains only terms that have an exponential decay.

Stable circuit: a circuit in which all zero-input responses consist of decaying exponentials or exponentially decaying sinusoids.

Susceptance: the imaginary part of an admittance.

Voltage division: property wherein, in a series connection of impedances driven by a voltage source, the voltage appearing across any one of the impedances is proportional to the ratio of the particular impedance to the total impedance of the connection.

Zero-input response: the response of a circuit when all source excitations are set to zero.

PROBLEMS

1. Determine the differential equation model of the parallel RL circuit of figure P10.1 in terms of $i_L(t)$ and $i_{in}(t)$. Then use the method of section 4 to determine the steady-state response when $i_{in}(t) = 2\cos(25t)$ A.

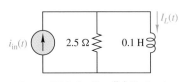

Figure P10.1 Parallel RL circuit.

2. Determine the differential equation model of the series RC circuit of figure P10.2 in terms of $v_{in}(t)$ and $v_C(t)$. Write $v_{out}(t)$ as a function of $v_{in}(t)$ and $v_C(t)$. Then use the method of section 4 to determine the steady-state response when $v_{in}(t) = 20\sin(25t)$ V.

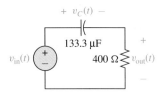

Figure P10.2 Series RC circuit with output across the resistor.

3. Determine the differential equation model of the series RL circuit of figure P10.3. Then use the method of section 4 to determine the steady-state response when $v_{in}(t) = 20\cos(400t)$ V.

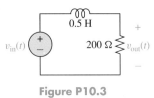

Figure P10.3

4. Determine the differential equation model of the parallel RC circuit of figure P10.4. Then use the method of section 4 to determine the steady-state response when $i_{in}(t) = 2\sin(400t)$ A.

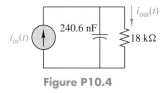

Figure P10.4

5. Determine the differential equation model of the series RLC circuit of figure P10.5. Then use the method of section 4 to determine the steady-state response when $v_{in}(t) = 120\sin(120\pi t)$ V.

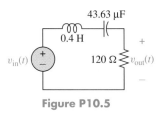

Figure P10.5

6. Determine the differential equation model of the parallel RLC circuit of figure P10.6. Then use the method of section 4 to determine the steady-state response when $v_{in}(t) = 20\cos(400\pi t)$ V.

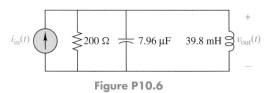

Figure P10.6

7. (a) Using the phasor method, determine the steady-state response of the circuit of problem 3. Discuss the relative advantages of the phasor method.
(b) Using the phasor method, determine the steady-state response of the circuit of problem 4. Discuss the relative advantages of the phasor method.

8. Using the phasor method, determine the response of the circuit of problem 5. Discuss the overall advantages of the phasor approach.

9. (a) Using the phasor method, determine the response of the circuit of problem 6.
(b) Repeat part (a) if L is changed to 79.58 mH. Does anything unusual happen?

10. (a) Determine the values of the phasors $\mathbf{I_L}$ and $\mathbf{V_{out}}$ in figure P10.10 when $\mathbf{I_{in}} = 2\angle 45°$ A and $\omega = 1{,}000$ rad/sec. Specify the corresponding time functions.
(b) Repeat part (a) when $\omega = 618$ rad/sec.

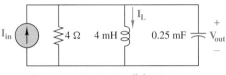

Figure P10.10 Parallel RLC circuit.

11. For the circuit of figure P10.11, use the phasor method to determine $v_{out}(t)$ when $v_{in}(t) = 50\cos(3.33 \times 10^3 t)$ V.

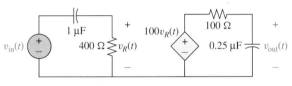

Figure P10.11 Two coupled RC circuits.

12. For the circuit of figure P10.12, use the phasor method to determine $v_C(t)$ when $i_{in}(t) = 10\cos(10^4 t)$ mA.

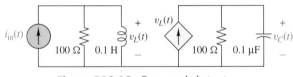

Figure P10.12 Two coupled circuits.

13. For the circuit of figure P10.13, use the phasor method to determine $v_C(t)$ when $i_{in}(t) = 10\cos(10^4 t)$ mA.

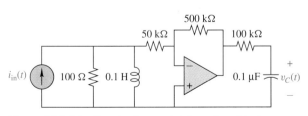

Figure P10.13 Two circuits coupled through an ideal op amp.

14. (a) For the circuit of figure P10.14a, determine $v_{out}(t)$ when $v_{in}(t) = \sin(400t)$ mV.
(b) For the circuit of figure P10.14b, find C so that when $v_{in}(t) = \cos(200t)$ mV, $v_{out}(t) = -\sin(200t)$ mV.

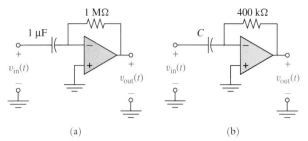

Figure P10.14 (a) and (b) are op amp differentiation circuits.

15. (a) For the circuit of figure P10.15a, determine $v_{out}(t)$ when $v_{in}(t) = \sin(800t)$ V.
(b) For the circuit of figure P10.15b, find C so that when $v_{in}(t) = -\sin(200t)$ V, $v_{out}(t) = 10\cos(200t)$ V. This represents an integration of the input with gain.

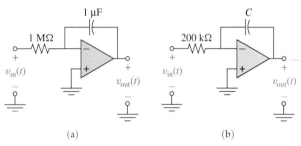

Figure P10.15 (a) and (b) are op amp integrators.

16. (a) If a 700-Hz sine wave of unit amplitude excites the leaky integrator circuit of figure P10.16a, determine the steady-state output voltage.
(b) For the circuit of figure P10.16a, plot the magnitude of the frequency response as a function of $\omega = 2\pi f$, where f is in Hz and ω is in radians per second.
(c) If the input to the circuit of figure P10.16b is $v_{in}(t) = \cos(2{,}000\pi t)$ V, determine the values of R and C so that $v_{out}(t) = 10\cos(2{,}000\pi t - 225°)$ V.

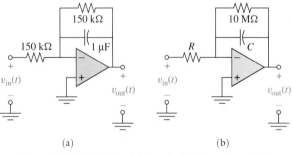

Figure P10.16 (a) and (b) are leaky integrator circuits.

17. (a) For the operational amplifier circuit of figure P10.17, write two node equations and solve them to determine a relationship between the output phasor $\mathbf{V_{out}}$ and the input phasor $\mathbf{V_{in}}$ at the frequency $f = 1{,}000$ Hz. Note that the voltage from the negative terminal of the op amp to ground is $\mathbf{V_{out}}$, which equals the voltage from the positive terminal to ground, assuming that the op amp is ideal.
(b) Repeat the calculation at $f = 1{,}500$ Hz and $f = 2{,}000$ Hz. What happens as the frequency increases?

(c) (Optional) Compute the ratio $\mathbf{V_{out}}/\mathbf{V_{in}}$ as a function of ω, and plot the frequency response of the circuit.

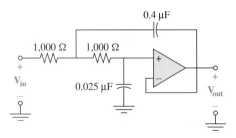

Figure P10.17 Ideal op amp circuit.

18. For the circuit of figure P10.18:
(a) Use nodal analysis to compute the ratio $\mathbf{V_{out}}/\mathbf{V_{in}}$ at $f = 2,500$ Hz.
(b) Compute the frequency response of the circuit over the range $0 \le f \le 10$ kHz.

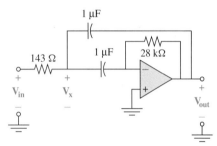

Figure P10.18 Op amp circuit having a bandpass type of response. Note that $\mathbf{V_x}$ is an intermediate variable that is useful in the nodal analysis of the circuit.

19. Compute the magnitude and phase plots of the frequency response of the circuit shown in figure P10.19. Before sketching the response, determine the asymptotic behavior for large ω and for at least one other frequency without a computer or calculator. List these behaviors in writing, along with your reasoning.

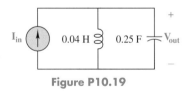

Figure P10.19

20. Compute the magnitude and phase plots of the frequency response of the circuit shown in figure P10.20. Determine the behavior of the plots intuitively, with no computer or calculator calculations. Write all this information down before sketching the response.

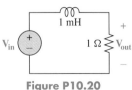

Figure P10.20

21. In the circuit of figure P10.21, at $\omega = 4$ rad/sec, it is necessary to have

$$Z_{in}(j4) = \left(\frac{1}{4} + j\frac{1}{4}\right) \ \Omega.$$

Determine the necessary value of L.

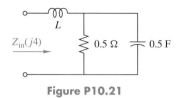

Figure P10.21

22. In the circuit of figure P10.22, $|\mathbf{V_{out}}|/|\mathbf{V_{in}}|$ is to be $1/5$ at $\omega = 4$ rad/sec. Find the necessary value(s) of C (in F).

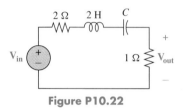

Figure P10.22

23. The *linear* circuit of figure P10.23 is such that in the *steady state*, if $i_{in}(t) = 10\cos(2\pi t)$ A with $v_{in}(t) = 0$, then $v_1(t) = 20\cos(2\pi t + 45°)$ V. On the other hand, if $i_{in}(t) = 0$ with $v_{in}(t) = 10\cos(2\pi t + 45°)$ V, then $v_1(t) = 5\cos(2\pi t + 90°)$ V. If $i_{in}(t) = 5\cos(2\pi t - 45°)$ A and $v_{in}(t) = 20\cos(2\pi t)$ V, then determine $v_1(t)$ in the steady state.

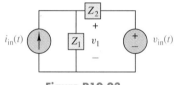

Figure P10.23

24. The circuit shown in figure P10.24 operates at a frequency of 60 Hz. Voltage and current sources are in volts and amps, respectively. Determine the current that flows through the inductor as a function of time.

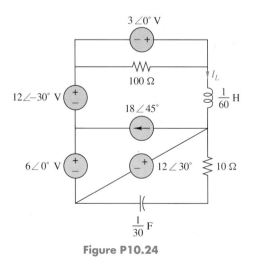

Figure P10.24

25. Find the Thévenin equivalent impedance of the network in figure P10.25 with respect to the terminals A-B.

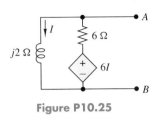

Figure P10.25

26. The circuit of figure P10.26 operates in the sinusoidal steady state at $\omega = 1$ rad/sec. Determine the Norton equivalent circuit.

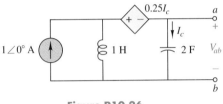

Figure P10.26

27. The circuit of figure P10.27 is operating in the sinusoidal steady state. Determine the current $i_x(t)$.

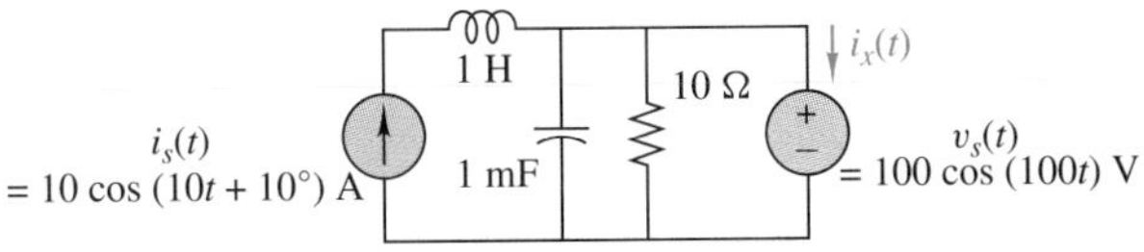

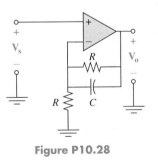

Figure P10.27

28. For the circuit of figure P10.28, determine the expression for the phasor transfer function $H(j\omega) = \mathbf{V_o}/\mathbf{V_s}$. Assume an ideal operational amplifier.

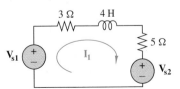

Figure P10.28

29. The ideal op amp circuit of figure P10.29 operates in the sinusoidal steady state with an input voltage $v_i(t) = V_m \cos(\omega t)$. At what angular frequency ω will the output voltage satisfy $v_o(t) = -(R_2/R_1)v_{in}(t)$? (*Hint*: Your answer should be in terms of L and C.)

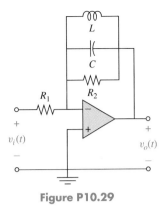

Figure P10.29

30. Consider the circuit of figure P10.30.
(a) Determine the input impedance $Z_{in}(j\omega)$ and the input admittance $Y_{in}(j\omega)$.
(b) At $\omega = 5$ rad/sec, determine the steady-state current $i_L(t)$.

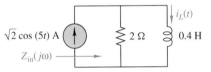

Figure P10.30 Parallel RL circuit for problem 30.

31. In the RC circuit of figure P10.31, the source input is $v_s(t) = 14.14\cos(10t)$ V, and the steady-state response is $v_C(t) = 10\cos(10t - 45°)$ V. Determine the necessary value of C.

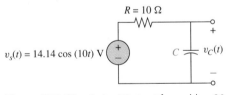

Figure P10.31 Series RC circuit for problem 31.

32. In the circuit of figure P10.32, $\mathbf{I_1} = 0.5\angle 90°$ A and $v_{s2}(t) = 4\sin(2t)$ V. Determine the source voltage $v_{s1}(t)$.

Figure P10.32 Circuit for problem 32.

33. Consider the circuit of figure P10.33. If $\omega = 10$ rad/sec, determine the Thévenin equivalent open circuit voltage $\mathbf{V}_{oc}$ and the Thévenin equivalent impedance $Z_{th}(j10)$. Represent $Z_{th}(j10)$ as a series connection of two circuit elements, and then represent it as a parallel connection of two circuit elements.

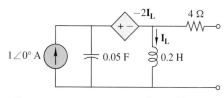

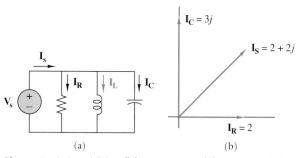

Figure P10.33 RLC circuit for problem 33.

34. The circuit of figure P10.34 is to have an input impedance $Z_{in}(j\omega) = 25j$ at $\omega = 100$ rad/sec. Determine the value of C necessary to achieve $Z_{in}(j100) = 25j$.

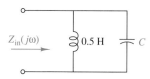

Figure P10.34 Parallel LC circuit for problem 34.

35. Consider the circuit and phasor diagram of figure P10.35. Determine the inductor current phasor.

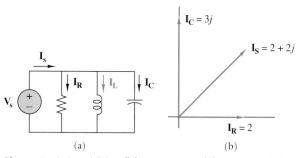

Figure P10.35 (a) Parallel RLC circuit and (b) associated phasor diagram.

36. Consider the current-voltage relationship indicated in figure P10.36a, and suppose the magnitude and phase of the admittance $Y(j\omega)$ are given by the sketches of figures P10.36b and c, respectively. If $v_s(t) = 25\cos(2,000t)$ V, determine $i(t)$.

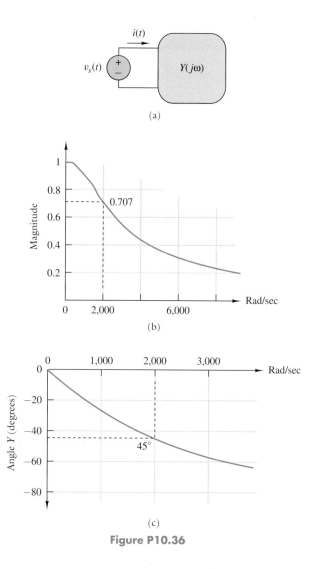

Figure P10.36

37. Consider the circuit of figure P10.37. At $\omega = 20$ rad/sec, the input admittance is $Y_{in}(j\omega) = 0.05 + j0.0866$ mhos.
 (a) Find R.
 (b) Find L.
 (c) If $i_s(t) = 20\cos(20t + 30°)$ A, determine $v_C(t)$.
 (d) If $i_s(t) = 20\cos(20t + 30°)$ A, determine $i_L(t)$.
 (e) At $\omega = 20$ rad/sec, determine the Thévenin equivalent circuit phasors, $\mathbf{V}_{oc}$, and $Z_{th}(j20)$.
 (f) At $\omega = 20$ rad/sec, determine the Thévenin equivalent circuit in which $Z_{th}(j20)$ is a series combination of two circuit elements seen at the terminals a-b.

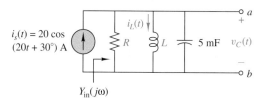

Figure P10.37 Parallel RLC circuit for problem 37.

38. This problem tests whether you can synthesize ideas from two different parts of the text. In the circuit of figure P10.38, $v_s(t) = 10\cos(t)u(t)$ (notice the step function) and $i_L(0^+) = 1$ A. If the response for $t \geq 0$ has the form

$$i_L(t) = A\cos(t + \phi) + Be^{-0.577t} \text{ A,}$$

determine the constants A, ϕ, and B.

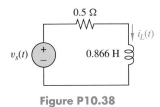

Figure P10.38

39. For the network shown in figure P10.39, determine the value of the phasor current $\mathbf{I_x}$.

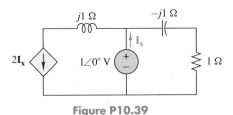

Figure P10.39

40. For the circuit of figure P10.40, determine $i_R(t)$ in steady state.

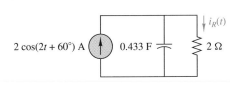

Figure P10.40

41. Consider figure P10.41. If the source voltage $v_s(t) = [20\cos(5t) + 30\cos(10t)]$ V, determine the inductor current $i_L(t)$ in steady state.

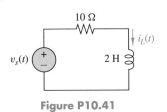

Figure P10.41

42. Consider figure P10.42. Suppose $v_s(t) = 120\cos(t)$ V. Determine $v_C(t)$ and $i_L(t)$ in steady state.

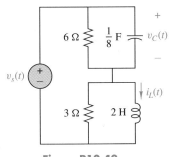

Figure P10.42

43. (a) For the circuit of figure P10.43, find the Thévenin equivalent at the terminals a-b at $f = 204$ Hz.
    (b) If the circuit is terminated with a series connection of a 9-Ω resistor and a 1.17-mH inductor, determine the voltage across the load.

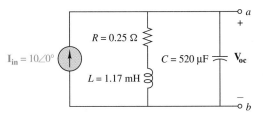

Figure P10.43 Circuit for problem 43.

ANSWER: $\mathbf{V_{oc}} = 91.24\angle{-9.3°}$.

44. (a) Determine the values of the phasors $\mathbf{I_L}$ and $\mathbf{V_{out}}$ in figure P10.44 when $\mathbf{V_{in}} = 20\angle45°$ V and $\omega = 1{,}000$ rad/sec. Specify the corresponding time functions.
    (b) Determine the value of ω for which the magnitude of the output voltage is 10% of its value at dc.

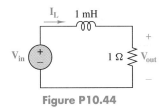

Figure P10.44

45. In the circuit of figure P10.45, assume that the operational amplifier is ideal. Compute the gain of the circuit as a function of ω. Then plot the magnitude and phase of the frequency response as the logarithm of the frequency for $1 \leq \omega \leq 10^4$.

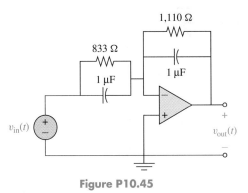

Figure P10.45

46. In the circuit of figure P10.46, find the reactance X such that the impedance "seen" by the source is real. For this case, find the steady-state current $i(t)$ corresponding to the phasor current $\mathbf{I}$ if $\omega = 10$ rad/sec.

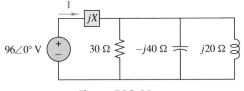

Figure P10.46

47. Audio applications ordinarily utilize multiple stages of amplification. With multiple stages, dc offsets in successive amplifier stages accumulate and cause problems. One way to circumvent these problems is to use capacitive coupling. Another approach is to use a special amplifier stage such as the one depicted in figure P10.47.

    (a) Assuming that $v_{in}(t) = \cos(2\pi f t)$ V, find the steady-state output, expressed in the form $v_{out}(t) = V_m \cos(2\pi f t + \theta)$ V. Note that the frequency f should appear as a parameter in your answer.

    (b) Plot the gain or magnitude of the output, V_m, as a function of f over the range $0 \le f \le 2$ Hz.

    (c) Explain why this amplifier circuit is useful for eliminating dc offset.

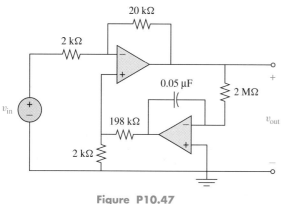

Figure P10.47

48. Figure P10.48 depicts a three-phase circuit with unbalanced loads. Write node equations at node N to find $\mathbf{V}_{N0}$ and then the phasor currents $\mathbf{I}_A$, $\mathbf{I}_B$ and $\mathbf{I}_C$. What happens to the values of the currents if each of the polarities of the voltage sources is reversed? (*Hint*: Recall the homogeneity property.) (*Check*: $\angle \mathbf{I}_A = -43°$; $|\mathbf{I}_B| = 15.5$ A; $\angle \mathbf{I}_C = 88.6°$.)

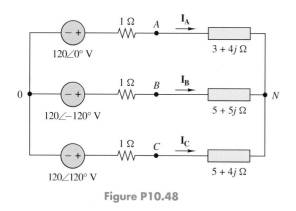

Figure P10.48

49. Consider figure P10.49. Find the voltage $v_1(t)$ in the steady state as follows:

    (a) Draw the phasor-domain circuit diagram.

    (b) Find the phasor $\mathbf{V}_1$.

    (c) Write the associated time function $v_1(t)$.

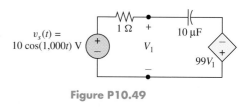

Figure P10.49

50. Find the Thévenin equivalent circuit (in the phasor domain) at terminals A and B for the circuit of figure P10.50. Now use the Thévenin equivalent to find the magnitude of the voltage V_{AB} when the 10-mH and 1-kΩ series-connected load is connected to A and B. (*Check*: $190 < |V_{oc}| < 205$ V; $|V_{AB}| \cong 0.5|V_{oc}|$.)

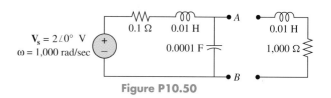

Figure P10.50

51. Consider figures P10.51a and b.

    (a) Write two mesh equations to find the currents $\mathbf{I}_A$, $\mathbf{I}_B$, and $\mathbf{I}_C$. (*Remark*: This is an example of a three-phase circuit with unbalanced loads. You may use any available software (such as MATLAB, a TI-68 calculator, or other tools) in solving the problem. But you can also do without these advanced tools for the present case of only two equations.) (*Check*: Angle of $\mathbf{I}_A \cong -43°$; magnitude of $\mathbf{I}_B \cong 15.5$ A; angle of $\mathbf{I}_C \cong 88.6°$.)

    (b) Find Z_{th}, V_2, V_3, V_4, and V_{34}. What are the new values for these quantities if a resistor R is connected across terminal 3 and 4? (*Hint*: Utilize the answer for V_{34}, and draw a conclusion without computation.)

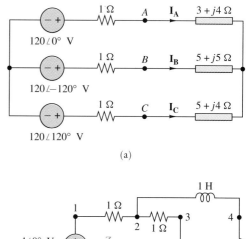

(a)

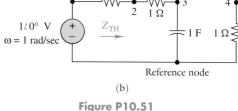

(b)

Figure P10.51

52. The impedance $\mathbf{Z} = R + jX$ in the circuit of figure P10.52a has $X > 0$ meaning that the circuit has an inductor-like behavior. The phasor diagram depicting the general relationships among various voltage and current phasors is given in figure P10.52b. Note that the parallelogram method is used for adding current phasors, and the tandem vector method is used for adding voltage vectors. The choice is arbitrary.

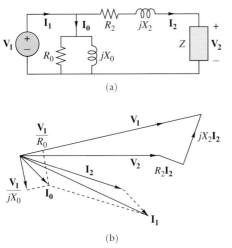

(a)

(b)

Figure P10.52 (a) Circuit for analysis. (b) Phasor diagram of circuit variables.

(a) Suppose you are given the following information:

$$\mathbf{V_2} = 500 + j0 \text{ V}, \qquad \mathbf{I_2} = 30 - j40 \text{ A},$$

$$R_2 = 2 \text{ }\Omega, \qquad\qquad X_2 = 6 \text{ }\Omega,$$

$$R_0 = 40 \text{ }\Omega, \qquad\qquad X_0 = 80 \text{ }\Omega.$$

Construct the phasor diagram accurately, using graph paper. Suggested scales are 100 V/inch for voltage phasors and 10 A/inch for current phasors. Use other suitable scales if your graph paper is in the metric system.

(b) From the phasor diagram, read the approximate magnitudes of the voltage $\mathbf{V_1}$ and the current $\mathbf{I_1}$. (*Check*: V_1 is about 800 V and I_1 is about 70 A.)

53. Consider the RLC circuit given in figure P10.53.

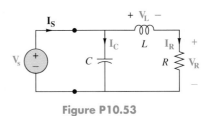

Figure P10.53

(a) Construct a phasor diagram to show the relationships among $\mathbf{I_R}$, $\mathbf{V_R}$, $\mathbf{V_L}$, $\mathbf{V_s}$, $\mathbf{I_C}$, and $\mathbf{I_s}$. (*Suggestion*: Build up the diagram with phasors in the indicated order, with the angle of $\mathbf{I_R}$ arbitrarily set to zero.)

(b) If $\omega = 6$ rad/sec, $R = 0.8 \text{ }\Omega$, $L = 0.1$ H, and $C = 0.1$ F, find the difference between the phase angles of $\mathbf{I_s}$ and $\mathbf{V_s}$, using the phasor diagram.

54. Consider again the pressure sensor example of section 11. Specify a set of mesh currents, and write a set of mesh equations that describe the circuit. Using MATLAB or some other equivalent software program, solve the equations for $1 \text{ pF} \le C_2 \le 40 \text{ pF}$. Plot the magnitude of $\mathbf{V_B} - \mathbf{V_C}$ as a function of C_2. Now construct a plot of the magnitude of $\mathbf{V_B} - \mathbf{V_C}$ as a function of pressure in mm of Hg.

CHAPTER 11

SMALL EDISON DYNAMO OR MOTOR.

Sinusoidal Steady-State Power Calculations

HISTORICAL NOTE

TODAY, all lighting and ordinary household appliances run on alternating current (ac). Direct current (dc) usage is found mainly in electronic equipment or special types of motor applications. No one questions the widespread use of ac. But the picture was actually quite different a century ago. In the late 1880s, both dc and ac systems were used for lighting and other applications. A serious debate ensued between the advocates of each type of system. The dc advocates put forth three main reasons for their views: (1) dc systems permitted the use of batteries, assuring reliability in case of a breakdown of the voltage-generating machinery; (2) a dc distribution system was safer than an ac system when humans or animals accidentally came into contact with live wires (which, as we now know, lacks supporting evidence), and (3) an efficient and reliable dc motor was available, but there was no such suitable motor to operate on ac at that time.

It was a simple, well-known fact that for either dc or ac transmission of electrical power, loss of power in the transmission lines could be reduced by using a higher voltage. In the 1880s, dc generators could produce a maximum voltage of about 2,000 V. With ac, however, the voltage could easily be raised or lowered with transformers (a device to be studied in volume 2). For example, transformers with output voltages up to 10,000 V were designed in the late 1880s.

The successful design of commercial transformers and, later, the invention of the ac induction motor made the ac power distribution system the winner of what was then known as the "battle of the systems" to that generation of electrical engineers. The battle actually did not end suddenly. Resistance to change from dc to ac systems was natural, because the low-voltage dc systems had already been installed in many cities at a very heavy outlay of capital. The dc system for transmission and distribution gradually died out by attrition. Although in recent years there has been a resurgence of interest in very high-voltage dc transmission of power, over 99 percent of bulk power transmission is still done with ac.

This chapter introduces us to the study of ac circuits, from the methods developed for their use to the calculation of power in ac systems. To round out these investigations, the economical aspects of ac power transmission (in particular, the advantage of three-phase over single-phase power transmission) are discussed in detail. ◈

1. INTRODUCTION

Chapter 1 first defined the concept of power. Although in subsequent chapters more attention was directed toward the calculations of voltages and currents, this should not be construed to mean that the consideration of power is of secondary importance. In fact, the opposite is true. For example, a consumer pays for energy used, not for voltage or current. The integral of power over a time interval, say, 30 days for a monthly bill, determines the energy consumed. Hidden in this cost is an adjustment to cover the power losses incurred by transmitting energy from the generating station to the customer's location. Reducing such losses is of prime concern to utility companies.

A second reason for understanding power usage in ac systems is safety. Each appliance and its cord that plugs into the wall outlet has a maximum safe power-handling capacity. A misunderstanding of such information or the misuse of an appliance can lead to equipment breakdown, a fire, or some other disaster.

Even for electronic equipment whose power consumption is small, power is still a very important design factor. For example, nowadays hand-held calculators are usually operated from lithium batteries. Power drainage directly determines the operating time of the calculator before needing a battery replacement. During an exam, one would want sufficient power to operate the calculator.

In this chapter, we will investigate different notions of power in ac circuits and discuss their significance and application. The term "ac circuits" has a narrow meaning here. It refers to those linear circuits having all sinusoidal sources at the same frequency and all responses in the steady state. The phasor method of chapter 10 is the basic analytic tool.

We round out our discussion in section 6, which describes various reasons for choosing a three-phase system, instead of a single-phase system, for transmitting bulk ac power over long distances. Three-phase systems break down into two categories: the balanced load variety and the unbalanced load variety. While the analysis of general unbalanced three-phase circuits belongs to a special course in power engineering, the analysis of balanced three-phase circuits is simple enough even for beginners when proper methods are used. Section 7 presents several methods that reduce the analysis of balanced three-phase circuits to that of single-phase circuits.

Throughout the chapter W will denote watts and it should not be confused with stored energy.

2. Instantaneous Power and Average Power

Figure 11.1 shows an isolated two-terminal circuit element that is part of a larger circuit. With the voltage and current reference directions as shown, the **instantaneous power** absorbed by the element, $p(t)$, is

$$p(t) = v(t)i(t), \qquad (11.1)$$

where the standard units of volt, ampere, and watt are presupposed. (This formula was presented as equation 1.12 earlier.)

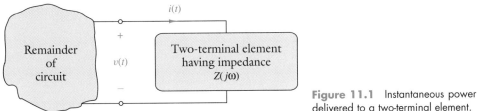

Figure 11.1 Instantaneous power delivered to a two-terminal element.

Assuming that all excitations are at the same frequency ω, if a linear circuit is in the sinusoidal steady state, then all voltages and currents are sinusoids *at the same frequency*. However, this is not true of the instantaneous power $p(t)$. To see this, let $v(t) = V_m \cos(\omega t + \theta_v)$ and $i(t) = I_m \cos(\omega t + \theta_i)$. Then the associated instantaneous power is

$$p(t) = v(t)i(t) = V_m \cos(\omega t + \theta_v) \times I_m \cos(\omega t + \theta_i)$$

$$= \frac{V_m I_m}{2} \cos(\theta_v - \theta_i) + \frac{V_m I_m}{2} \cos(2\omega t + \theta_v + \theta_i). \qquad (11.2)$$

Power Engineer

Power engineers are electrical and mechanical engineers specializing in the generation, transmission, and distribution of electric power. Some civil and chemical engineers also specialize in power engineering. Power engineers can work for electric utility companies, or for manufacturers of the hardware and software for the power industry. Still others are employed by government and research agencies. Power engineering is highly interdisciplinary, making use of engineering, computer technology, economics and a knowledge of government and government regulations very important in this field. Computers are widely used in the power industry, and many power engineers specialize in the analysis and control of power systems via computers. The field also includes international aspects in the development of energy sources and power systems in developing countries.

Power engineers design and develop methods of converting fuel and other energy sources to electrical power. They also devise ways to transmit and control the flow of electric power over long distances and at very high power levels. In many cases, computers are used for direct digital control. Power engineers design electric circuits that distribute electrical power to residences, commercial businesses, and industry. Power engineers ensure that the present facilities are adequate to handle the normal system loads, as well as unplanned conditions such as overloads and natural emergencies. System reliability is an important issue in power engineering. A system failure can present a dramatic problem for communities, especially when the failure causes a loss of electicity to thousands of residences and industry.

It is the responsibility of power engineers to determine which energy source and method of conversion is best and will reliably and economically meet the demand for energy. They can use a variety of sources to generate power including: a) water, from which power is harnessed by allowing water to pass through large turbines; b) fossil fuels such as coal, oil, and gas; c) nuclear sources (fission of uranium); d) geothermal sources that use heat found deep in the earth; and e) solar energy sources that convert either the luminous energy of sunlight directly to electricity (photovoltaics) or the heat of the sun to make steam and operate a steam turbine.

Power engineers are confronted with a variety of social, environmental, and ethical issues as they strive to meet the increasing demand for electrical power. Generating power from sources of energy such as nuclear and fossil fuels requires power engineers to discover ways of disposing of waste and reducing air pollution. Power engineers must effectively manage and use renewable energy sources such as geothermal and solar energy, as well as nonrenewable energy sources such as coal and uranium. They must design and implement safeguard systems involving strict checks and balances to avoid an uncontrolled accident. These challenges, along with the increasing demand for electricity, are expected to generate job growth for power engineers during the next decade.

Equation 11.2 follows from the trigonometric identity

$$\cos(x)\cos(y) = \frac{1}{2}\cos(x-y) + \frac{1}{2}\cos(x+y). \tag{11.3}$$

Equation 11.2 indicates that the instantaneous power $p(t)$ consists of a constant term plus another component varying with time at *twice the input frequency*. Figure 11.2 shows a typical plot of $p(t)$, $v(t)$, and $i(t)$.

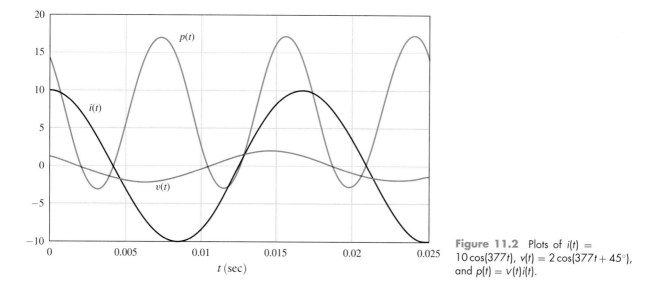

Figure 11.2 Plots of $i(t) = 10\cos(377t)$, $v(t) = 2\cos(377t + 45°)$, and $p(t) = v(t)i(t)$.

Figure 11.2 and equation 11.2 imply that $p(t)$ is a periodic waveform which may be positive for some portion of one period and negative for the remaining portion. When $p(t)$ is positive, the element absorbs power. When $p(t)$ is negative, the element delivers power to the remainder of the circuit. A quantity of fundamental importance is the **average power** P (or P_{ave} for emphasis), defined as the average value of $p(t)$ over one period T, i.e.,

$$P = P_{\text{ave}} = \frac{1}{T}\int_0^T p(t)dt. \tag{11.4}$$

For the sinusoidal steady-state case, the calculation of the average power is easy. Examine the decomposition of $p(t)$ given in equation 11.2. The average value of the second term over one period is obviously zero. The first term, being a constant, has an average value equal to itself. Therefore, for the sinusoidal steady-state case,

$$P_{\text{ave}} = 0.5V_m I_m \cos(\theta_v - \theta_i) = 0.5V_m I_m \cos(\theta_z), \tag{11.5}$$

where $\theta_z = \theta_v - \theta_i$ is the angle of the impedance Z of the two-terminal element of figure 11.1. If the two-terminal element is a resistance R, then $v(t) = Ri(t)$ and $\theta_z = \theta_v - \theta_i = 0$. It follows from equation 11.5 and Ohm's law that

$$P_{\text{ave},R} = 0.5V_m I_m = 0.5I_m^2 R = 0.5\frac{V_m^2}{R}. \tag{11.6}$$

If the two-terminal element is an inductance L, then, since $\mathbf{V_L} = (j\omega L)\mathbf{I_L}$, $\theta_z = \theta_v - \theta_i = 90°$. If the two-terminal element is a capacitance, then $\theta_z = \theta_v - \theta_i = -90°$, since $\mathbf{I_C} = (j\omega C)\mathbf{V_C}$. In either case, since $\cos(\pm90°) = 0$, equation 11.5 implies that

$$P_{\text{ave},L} = P_{\text{ave},C} = 0; \tag{11.7}$$

i.e., the average power consumed or delivered by a capacitor or an inductor is zero. Note that even though an ideal capacitor and an ideal inductor neither consume nor generate *average* power, each may absorb or deliver a large amount of instantaneous power during some particular time interval.

Equation 11.6 resembles equation 1.20b for the dc power absorbed by a resistor connected to a dc source, except for a difference in the leading coefficient. By introducing a new concept called the *effective value of a periodic waveform*, we can make the formulas for the average power absorbed by a resistor the same for a dc, a sinusoidal, or any general periodic input waveform.

The **effective value** of any periodic current $i(t)$, denoted by I_{eff}, is a positive constant such that a dc current $i_{dc} = I_{eff}$ flowing through a resistor of R ohms dissipates the same amount of energy in one period as the periodic current $i(t)$ flowing through a resistor of R ohms. In other words, both currents $i(t)$ and $i_{dc} = I_{eff}$ result in the same average power dissipation in R or the same heating effect in R. Equating the energy dissipated in R over one period for the two cases leads to

$$I_{eff}^2 = \frac{1}{T}\int_{t_0}^{t_0+T} i^2(t)\,dt,$$

from which it follows that

$$I_{eff} = \sqrt{\frac{1}{T}\int_{t_0}^{t_0+T} i^2(t)\,dt}. \tag{11.8a}$$

The effective value of a periodic voltage $v(t)$ has a similar definition:

$$V_{eff} = \sqrt{\frac{1}{T}\int_{t_0}^{t_0+T} v^2(t)\,dt}. \tag{11.8b}$$

More generally, for any periodic waveform $f(t)$, its effective value is

$$F_{eff} = \sqrt{\frac{1}{T}\int_{t_0}^{t_0+T} f^2(t)\,dt}. \tag{11.8c}$$

The last equation gives rise naturally to the alternative name of the effective value as the **root-mean-square** (abbreviated **rms**) value, as F_{eff} is the square *root* of the *mean* value of the *square* of $f(t)$ over one period.

In terms of effective values, the average power dissipated in a resistor of R ohms in the presence of a periodic current or voltage is

$$\begin{aligned} P_{ave,R} &= \frac{1}{T}\int_0^T i^2(t)R\,dt = I_{eff}^2 R \\ &= \frac{1}{T}\int_0^T \frac{v^2(t)}{R}\,dt = \frac{V_{eff}^2}{R} \\ &= V_{eff}I_{eff}. \end{aligned} \tag{11.9}$$

For a sinusoidal function $f(t) = F_m\cos(\omega t + \theta)$, the effective value is easily calculated. Using the identity

$$\cos^2(x) = \frac{1}{2} + \frac{1}{2}\cos(2x),$$

the square of $f(t)$ is

$$f^2(t) = F_m^2\cos^2(\omega t + \theta) = \frac{F_m^2}{2} + \frac{F_m^2}{2}\cos(2\omega t + 2\theta).$$

Since ω is assumed to be nonzero, the mean value of the second term on the extreme right is obviously zero. The mean value of the first (constant) term is the same constant.

Hence, by equation 11.8c,

$$F_{\text{eff}} = \sqrt{\frac{F_m^2}{2}} = \frac{F_m}{\sqrt{2}}.$$

(11.10)

Thus, for a sinusoidal waveform, the **effective or rms value** is 0.707 times the maximum value—a basic fact well worth remembering. The ac voltage and current ratings of all electrical equipment, as given on the identification plate, are rms values unless explicitly stated otherwise. For example, household ac voltage is 110 V, with a maximum voltage of $110 \times \sqrt{2} = 156$ V. A typical appliance such as a coffeemaker may have a rating of 110-V ac and 900 W. The effective values of a few other periodic waveforms are listed in figure 11.3, with their derivations assigned as exercises.

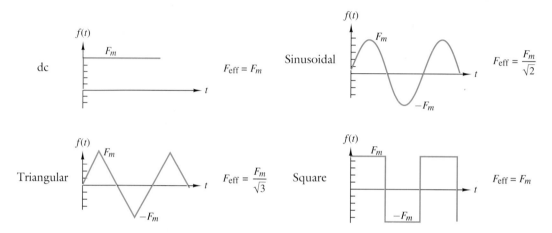

Figure 11.3 Effective values of some common periodic waveforms.

Exercise. Derive the formula for the effective value of the triangular waveform shown in figure 11.3.

Let us return now to the case of single-frequency sinusoidal steady-state analysis. The average power absorbed by a two-terminal element, given by equation 11.5, may now be rewritten in terms of effective values:

$$P_{\text{ave}} = V_{\text{eff}} I_{\text{eff}} \cos(\theta_v - \theta_i) = V_{\text{eff}} I_{\text{eff}} \cos(\theta_z).$$

(11.11a)

For the remainder of the chapter, all voltage and current phasors will be interpreted as effective values, unless a subscript m or max appears to indicate the maximum value. The subscript eff is added sometimes for purposes of emphasis. This practice is widely accepted in power engineering literature. Omitting the subscript eff in equation 11.11a yields

$$P_{\text{ave}} = V I \cos(\theta_v - \theta_i) = V I \cos(\theta_z),$$

(11.11b)

where $V = 0.707 V_m$ and $I = 0.707 I_m$.

Note that average power is defined for any periodic signal, whether sinusoidal or not. For example, if a 100-Ω resistor is connected to a voltage source having a 400-Hz, 120-V (peak-to-peak) triangular waveform, then, by equation 11.9 and figure 11.3, the average power dissipated in the resistor is

$$\frac{\left(\dfrac{60}{\sqrt{3}}\right)^2}{100} = 12 \text{ W}.$$

However, when the periodic signal is sinusoidal, the phasor diagram provides another way to calculate the average power. Let $\mathbf{V}_{\text{eff}}$ and $\mathbf{I}_{\text{eff}}$ be the phasors of $v(t)$ and $i(t)$ for the two-terminal element of figure 11.1. With these phasors, equation 11.11a for the average power has a very simple interpretation in terms of the phasor diagram shown in figure 11.4.

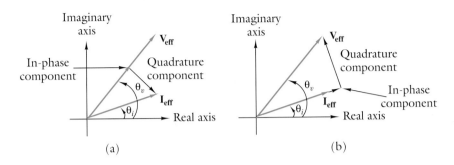

Figure 11.4 Calculation of average power from a phasor diagram. (a) The current phasor is decomposed into two components. (b) The voltage phasor is decomposed into two components.

From figure 11.4a and equation 11.11, the average power is equal to the product of the magnitude of the voltage phasor, V_{eff}, and the magnitude of the *in-phase* component of the current phasor, $I_{\text{eff}}\cos(\theta_v - \theta_i)$, where the current vector has two components, one in phase with $\mathbf{V}_{\text{eff}}$, i.e., $I_{\text{eff}}\cos(\theta_v - \theta_i)$, and the other perpendicular to $\mathbf{V}_{\text{eff}}$, i.e., $I_{\text{eff}}\sin(\theta_v - \theta_i)$. The latter component is sometimes called the **quadrature component**. Alternatively, from figure 11.14b, the average power is equal to the product of the magnitude of the current phasor, I_{eff}, and the magnitude of the *in-phase* component of the voltage phasor, $V_{\text{eff}}\cos(\theta_v - \theta_i)$, where this time, the voltage vector has been decomposed into two components, one in phase with $\mathbf{I}_{\text{eff}}$ and the other perpendicular to $\mathbf{I}_{\text{eff}}$. Such an interpretation is very helpful in visualizing the power relationship in some applications, to be described shortly. For example, the voltage and current phasors of a capacitor (or an inductor) are always 90 degrees apart. Because of this, neither the capacitor nor the inductor consumes average power.

Equation 11.11 gives the average power in terms of the voltage and current of the two-terminal element. Alternative expressions using impedance and admittance can be derived. Let the impedance and the admittance of a two-terminal element (see figure 11.5) be $Z = R + jX$ and $Y = G + jB$, respectively. Note that although $Y = 1/Z$, in general, $G \neq 1/R$. Let the phasor voltage and current (rms values) of the two-terminal element be $\mathbf{V}$ and $\mathbf{I}$, respectively. From Ohm's law,

$$\mathbf{V} = Z\mathbf{I} = (R + jX)\mathbf{I} = \underset{\substack{\text{(in phase}\\ \text{with } \mathbf{V})}}{R\mathbf{I}} \quad + \quad \underset{\substack{(90^\circ \text{ out of}\\ \text{phase with } \mathbf{V})}}{jX\mathbf{I}}.$$

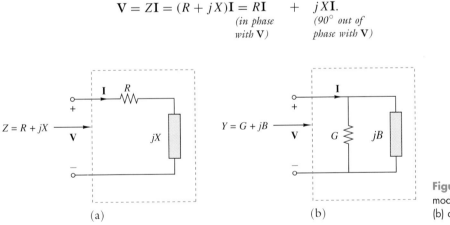

(a) (b)

Figure 11.5 Two-terminal elements modeled via (a) impedance and (b) admittance.

According to the phasor diagram of figure 11.4b, the average power delivered to the impedance equals the magnitude of the current times the magnitude of the in-phase component of the voltage. In this case, the latter is $R\mathbf{I}$. Thus,

$$P_{\text{ave}} = |\mathbf{I}| \times |R\mathbf{I}| = |\mathbf{I}|^2 R. \tag{11.12a}$$

One could also have obtained equation 11.12a by merely recalling the fact that a reactance jX consumes zero average power.

A dual development using admittance yields the equivalent relationship for average power:

$$P_{\text{ave}} = |\mathbf{V}| \times |G\mathbf{V}| = |\mathbf{V}|^2 G. \tag{11.12b}$$

Observe that the voltages and currents in equations 11.11 are rms values. If maximum values are used, then the right-hand sides of those equations must be preceded by a factor of 0.5.

> **Exercise.** Derive equation 11.12b in a manner that is dual to the derivation of equation 11.12a.

Some examples will illustrate the use of the various power relationships.

◆ **EXAMPLE 11.1**

Figure 11.6 shows three types of household loads connected to a 110-V, 60-Hz ac source. When connected to the source, each incandescent lamp consumes 100 W, the two-tube fluorescent light consumes 90 W and draws 1.2 A of line current, and the air conditioner consumes 900 W and draws 12 A of line current.

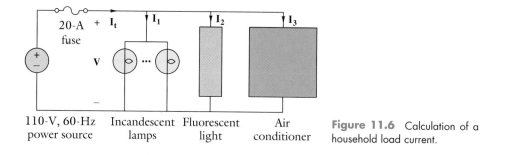

Figure 11.6 Calculation of a household load current.

110-V, 60-Hz power source — Incandescent lamps — Fluorescent light — Air conditioner

(a) If there are six lamps in use at the same time, besides the fluorescent light and the air conditioner, will the 20-A fuse blow?

(b) In order not to blow the fuse, what is the maximum number of incandescent lamps that may be connected?

SOLUTION

(a) The source voltage is $\mathbf{V} = 110\angle 0°$ V. The current drawn by the incandescent lamps is

$$\mathbf{I}_1 = 6\frac{100}{110} = 5.45\angle 0° \text{ A}.$$

From figure 11.4b for the fluorescent light, the in-phase component of $\mathbf{I}_2$ is

$$\frac{90}{110} = 0.82 \text{ A},$$

and the quadrature component of $\mathbf{I}_2$ is

$$\sqrt{1.2^2 - 0.82^2} = 0.88 \text{ A}.$$

It is known that the impedance of a fluorescent light is inductive, due to the ferromagnetic material used in the ballast. Therefore, $0 < \theta_v - \theta_i < 90°$, and

$$\mathbf{I}_2 = 0.82 - j0.88 \text{ A}.$$

By a similar reasoning, the in-phase component of I_3 is

$$\frac{900}{110} = 8.18 \text{ A},$$

and the quadrature component of I_3 is

$$\sqrt{12^2 - 8.18^2} = 8.78 \text{ A}.$$

Because of the motors used to drive the compressor and the fan of the air conditioner, its equivalent impedance is inductive (i.e., $0 < \theta_v - \theta_i < 90°$), and

$$I_3 = 8.18 - j8.78 \text{ A}.$$

The total current drawn from the power source is

$$I_t = I_1 + I_2 + I_3 = 5.45 + (0.82 - j0.88) + (8.18 - j8.78) = 14.45 - j9.66 \text{ A}.$$

Since $|I_t| = \sqrt{14.45^2 + 9.66^2} = 17.38$ A is below the 20-A rating of the fuse, the fuse will not blow.

(b) If there are m incandescent lamps,

$$I_1 = m \frac{100}{110} = m0.91\angle 0° \text{ A}.$$

The total current drawn from the power source will be

$$I_t = I_1 + I_2 + I_3 = 0.91m + (0.82 - j0.88) + (8.18 - j8.78) = (0.91m + 9) - j9.66 \text{ A}.$$

For the magnitude of I_t to be 20, we must have

$$(0.91m + 9)^2 + 9.66^2 = 20^2,$$

from which $m = 9.35$. Thus, the maximum number of lamps allowed is 9.

EXAMPLE 11.2

The previous example neglected the resistance of the wire connecting the various loads to the 110-V source. This example accounts for the wire's resistance. Suppose 100 feet of #12 standard annealed copper wire having a resistance of 1.588 ohms per 1,000 ft is used to connect the loads to the power source. Assume that the impedance of all loads remains constant under moderate variations in voltage.

(a) Find the nominal impedance of each load.
(b) Find the voltage across the lamps and the fluorescent light when the air conditioner is turned off; repeat for the case when it is turned on.
(c) Find the power dissipated in the connecting wire with all loads connected.

SOLUTION

(a) Using the results of example 11.1 and the assumption that impedances remain constant under moderate voltage variations,

$$Z_1 = \frac{V}{I_1} = \frac{110}{5.45} = 20.18 + j0 \ \Omega,$$

$$Z_2 = \frac{V}{I_2} = \frac{110}{0.82 - j0.88} = 62.34 + j66.91 \ \Omega,$$

and

$$Z_3 = \frac{V}{I_3} = \frac{110}{8.18 - j8.78} = 6.25 + j6.71 \ \Omega$$

are nominal values of impedance for each of the loads.

(b) The wire has a resistance of $2 \times 100 \times 0.001588 = 0.3176 \ \Omega$. Accordingly, if the air conditioner is turned off, the load impedance is

$$Z_L = Z_1 // Z_2 = 17.20 + j2.41 \ \Omega.$$

From the voltage division formula

$$\mathbf{V} = \frac{17.20 + j2.41}{(17.20 + j2.41) + 0.3176} \times 110 = 108.04 + j0.27$$

$$= 108.04 \angle 0.14° \ \text{V}.$$

If the air conditioner is turned on, the load impedance is

$$Z_L = Z_1 // Z_2 // Z_3 = 5.26 + j3.51 \ \Omega.$$

Again, from the voltage division formula,

$$\mathbf{V} = \frac{5.26 + j3.51}{(5.26 + j3.51) + 0.3176} \times 110 = 105.52 + j2.82$$

$$= 105.53 \angle 1.53° \ \text{V}.$$

Consequently, turning on the air conditioner reduces the lamp voltage from 108 V to 105 V.

(c) With all loads connected, the current drawn from the source is

$$\mathbf{I_t} = \frac{\mathbf{V}}{Z_L} = \frac{105.52 + j2.82}{5.26 + j3.51} = 16.69 \angle -32.2° \ \text{A}.$$

Therefore, the power dissipated in the wire is

$$P = 16.69^2 \times 0.3176 = 88.47 \ \text{W}.$$

This dissipated energy is transformed into heat.

3. APPARENT POWER AND POWER FACTOR IMPROVEMENT

The average power calculated in the previous section is also called the **real power**. Resistors transform the absorbed real power into heat. In a motor, most of the absorbed real power is transformed into mechanical power, say, to run a fan, while only a small portion is dissipated as heat. The product $V_{\text{eff}} I_{\text{eff}}$ is called the **apparent power**. The ratio of the real power to the apparent power is called the **power factor**, denoted by **pf**; i.e., for ac circuits,

$$\text{Average power} = (\text{apparent power}) \times (\text{power factor}). \tag{11.13}$$

From equation 11.11a, the power factor equals $\cos(\theta_v - \theta_i)$; i.e., the cosine of the difference of the angles of the voltage phasor $\mathbf{V}$ and the current phasor $\mathbf{I}$. Mathematically,

$$pf = \cos(\theta_v - \theta_i) = \cos(\theta_z). \tag{11.14}$$

Clearly, the value of pf lies between 0 and 1. Since $\cos(x) = \cos(-x)$, the information on the sign of $\theta_v - \theta_i$ is lost in the power factor. For example, both $\theta_z = 60°$ and $\theta_z = -60°$ yield a power factor of 0.5, or 50%. In order to carry along the information on the relative phase angle, one adds the word "lagging" if $180° > \theta_v - \theta_i = \theta_z > 0$ (or, the current lags the voltage) and the word "leading" if $180° > \theta_i - \theta_v = -\theta_z > 0$ (or, the current leads the voltage). Practically all electrical apparatus have lagging currents. Some typical power factor values are listed in table 11.1.

TABLE 11.1 TYPICAL POWER FACTORS OF ELECTRICAL APPARATUS

Type of load	Power factor (lagging)
Incandescent lighting	1.0
Fluorescent lighting	0.5 to 0.95
Single-phase induction motor, up to 1 hp	0.55 to 0.75, at rated load
Large three-phase induction motor	0.9 to 0.96, at rated load

Since, in ac circuits, the apparent power $V_{eff}I_{eff}$ does not measure the true power (except for the very special case of a purely resistive load), the unit of apparent power is not called watt, but volt-ampere, abbreviated VA. The unit VA is most often used in the rating of transformers (to be studied later), ac generators, and most large industrial ac equipment. This is because most ac machinery is built to operate at a specified voltage, from a consideration of insulation strength. The safe current of a machine depends on the heating effects of the current. Also, the cost and physical size of most ac equipment is more closely proportional to the VA rating than any other parameter. Accordingly, stating the capacity of a piece of ac machinery in terms of VA makes more sense than stating it in terms of anything else. As a specific example, consider a 60-Hz, 115-V ac generator having a capacity of 4 kVA. The maximum allowable load current is then $4,000/115 = 34.78$ A. If the load is a resistor of 3.306 Ω, then the average power delivered to the load is $115^2/3.306 = 4,000$ W. The current supplied by the generator is $115/3.306 = 34.78$ A, indicating that the generator is operating at its full rating. On the other hand, if the load is a capacitor of 802.35 μF, then the current supplied is

$$V_s \times \omega C = 115 \times (2\pi \times 60 \times 802.35 \times 10^{-6}) = 34.78 \text{ A},$$

which is also the rated current of the generator. In this case, the generator has reached its full current capacity, although the real power delivered to the load is zero! If the generator is driven by a gasoline engine, then in the case of a capacitive load, the generator requires much less fuel to run the engine than in the case of a resistive load.

A power company charges the residential consumer only for the actual electrical energy used. A watt-hour meter measures this energy, in units of kWh (kilowatt-hours). As mentioned earlier, most electrical loads have lagging currents. The 90° out-of-phase current contributes neither any useful power (heat or mechanical power) nor any revenue to the power company. Yet it is *vectorially* added to the in-phase current (see figure 11.4) and leads to a total current larger in magnitude than the in-phase current. This has two undesirable effects, from the power company's point of view: The heat loss in the connecting wires is increased due to a larger current, and a larger volt-ampere capacity of the ac machinery is used without any charge being levied. Therefore, it is desirable to reduce the magnitude of the quadrature component of current. This amounts to reducing the voltage-current phase angle difference, or raising the power factor.

Since power companies can supply more power with the same equipment if the power factor is high, they adjust their rates so that energy costs are less with a high power factor and more with a low one. For industrial and some commercial consumers, the power company charges not only for electrical energy (kWh), but for peak power demand (kW) in a given month. A multiplier on these two charges may be applied if the customer has a low power factor (e.g., below 0.86).

For industrial and commercial consumers to take advantage of the lower rate (a lower multiplier applied to energy and peak demand charges), various methods are used to improve the power factor. One of the more commonly used methods is to add capacitors in parallel with inductive loads. The leading quadrature current drawn by the capacitors will then partially or totally offset the lagging current drawn by an inductive load, improving the power factor. The following example illustrates the principle.

EXAMPLE 11.3

Figure 11.7a shows a 60-Hz, 120-V generator supplying a load that consumes 50 kW at a power factor of 0.7 lagging.

(a) Determine the value of the parallel capacitance needed to raise the power factor to 0.9 lagging.
(b) Determine the apparent power of the capacitor.
(c) If the generator has a 250-kVA rating, how many identical loads of the kind described can it safely supply when no capacitor is used to improve the pf?
(d) How many identical loads can the generator supply when each load has a parallel capacitor, as calculated in part (b)?

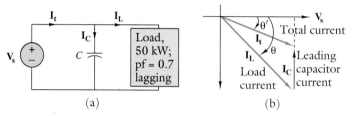

Figure 11.7 Example of power factor improvement. (a) A capacitor is used to improve the power factor. (b) A phasor diagram for the circuit.

SOLUTION

(a) Let $\mathbf{V_s} = 120\angle 0°$ be used as the reference vector. In the phasor diagram of figure 11.7b, the current $\mathbf{I_L}$ lags the voltage $\mathbf{V_s}$ by an angle

$$\theta = \cos^{-1}(0.7) = 45.57°.$$

Since only the in-phase component of the current $\mathbf{I_L}$ contributes to average power,

$$(\mathbf{I_L})_{\text{in-phase}} = \frac{50,000}{120} = 416.67 \text{ A}.$$

From the phasor diagram of figure 11.7b,

$$(\mathbf{I_L})_{\text{quadrature}} = \tan(\theta) \times (\mathbf{I_L})_{\text{in-phase}} = 1.02 \times 416.67 = 425 \text{ A}.$$

For an improved power factor of 0.9, the current $\mathbf{I_t} = \mathbf{I_L} + \mathbf{I_C}$ must lag $\mathbf{V_s}$ by

$$\theta' = \cos^{-1}(0.9) = 25.84°.$$

From figure 11.7b, this requires that

$$(\mathbf{I_t})_{\text{quadrature}} = \tan(\theta') \times (\mathbf{I_L})_{\text{in-phase}} = 0.4843 \times 416.67 = 201.8 \text{ A}.$$

Therefore, the quadrature leading current drawn by the capacitor is

$$I_C = 425 - 201.8 = 223.2 \text{ A}.$$

Using $I_C = \omega C V_s$, the required value of C is

$$C = \frac{223.2}{2\pi \times 60 \times 120} = 0.004933 \text{ F}.$$

(b) The apparent power of the capacitor is $120 \times 223.2 = 26,784$ VA.
(c) Each load alone has an apparent power of $50/0.7 = 71.43$ kVA. If m identical loads are connected in parallel, the total apparent power is $71.43m$ kVA. This value must be no greater than the generator's kVA rating. Therefore, $m \leq 250/71.43 = 3.5$. The conclusion is that at most three identical loads can be supplied by the generator.
(d) With the improved power factor, each load has an apparent power of $50/0.9 = 55.55$ kVA. From the requirement that $55.55m \leq 250$, we have $m \leq 4.5$, indicating that four identical loads can be supplied by the generator after the improvement of the power factor.

The solution of part b of example 11.3 shows that to improve the power factor to 0.9, a capacitor bank of 26.8 kVA must be installed. Whether it is desirable to install such a capacitor bank depends on cost considerations (among which is the severity of the penalty for low power factor loads). The usual way to evaluate whether further power factor correction is warranted is to calculate the annual electric energy plus peak demand charges with the power factor multiplier applied. Several cases of increasing the power factor correction are evaluated, and the annualized cost of the capacitors is considered to identify whether a lower total annual cost will be achieved. These considerations are best discussed in an engineering economics course.

4. REACTIVE POWER AND CONSERVATION OF POWER

If the two-terminal element of figure 11.1 is a pure inductance or a pure capacitance, then $\theta_v - \theta_i = 90°$ or $-90°$. From equation 11.11, the average real power delivered to an inductor or a capacitor is zero. Another way to reach this conclusion is to note that there are no in-phase components of the voltage and current phasors for a pure reactance. In this case, power engineers find it very useful to consider the inductor to be in a state of consuming a new type of power, called **reactive power** and the capacitor to be in a state of generating reactive power. More precisely, the reactive power delivered to a two-terminal element of figure 11.1, having

$$v(t) = \sqrt{2}V_{\text{eff}}\cos(\omega t + \theta_v)$$

and

$$i(t) = \sqrt{2}I_{\text{eff}}\cos(\omega t + \theta_i)$$

is defined to be

$$Q = V_{\text{eff}}I_{\text{eff}}\sin(\theta_v - \theta_i) = V_{\text{eff}}I_{\text{eff}}\sin(\theta_z). \tag{11.15}$$

The phasor diagram of figure 11.4 offers a geometrical interpretation of the quantity Q: Q is simply the product of the voltage and the quadrature current (figure 11.4a) or the product of the current and the quadrature voltage (figure 11.4b). If $\theta_v - \theta_i > 0$, as in the case of an inductive element, Q is positive, and the element consumes reactive power. On the other hand, if $\theta_v - \theta_i < 0$, as in the case of a capacitive element, Q is negative, and the element consumes a negative amount of reactive power or, equivalently, delivers reactive power.

When a two-terminal element absorbs an average power P_{ave}, there is a transformation of electrical energy into other forms of energy, for example, heat or kinetic energy. In contrast, when a two-terminal element absorbs reactive power Q, there is no real useful power derived from it. The energy transferred into the element is merely stored and later on returned to the embedding network. (See problems 13 and 14 for the relationship between Q and W_{ave}.) To distinguish it from the real power, we use **var** (volt-ampere reactive, abbreviated VAR) instead of watt as the unit for the reactive power Q.

Equations 11.11 and 11.15 and the definition of the apparent power lead to a right-triangle relationship shown in figure 11.8, which is helpful in solving ac power problems. All quantities in the diagram are associated with the same two-terminal device.

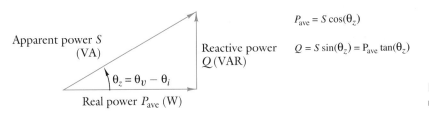

Apparent power S (VA)

Reactive power Q (VAR)

$\theta_z = \theta_v - \theta_i$

Real power P_{ave} (W)

$P_{\text{ave}} = S\cos(\theta_z)$

$Q = S\sin(\theta_z) = P_{\text{ave}}\tan(\theta_z)$

Figure 11.8 Relationships among real, reactive, and apparent powers.

It is natural to ask the following question: Since the reactive power performs no useful work (in the sense of generating heat or turning a fan), why bother to calculate it at all? The answer lies in the following basic principle:

PRINCIPLE OF CONSERVATION OF POWER IN AC CIRCUITS:

In ac circuits in the steady state, instantaneous power, average power, and reactive power are each conserved.

Two equivalent statements of this principle are: (1) The algebraic sum of the powers absorbed by all elements is zero, and (2) the power supplied by a group of elements equals the power consumed by the remaining elements.

A proof of this principle is given at the end of the section.

The following examples illustrate the terminology just introduced and the application of the principle of conservation of power.

EXAMPLE 11.4

Figure 11.9 shows five two-terminal elements connected to a 60-Hz, 120-V voltage source. The following impedances (in ohms) are given:

$$Z_1 = 5 + j0 = 5\angle 0°, \quad Z_2 = 0 + j4 = 4\angle 90°, \quad Z_3 = 0 - j5 = 5\angle -90°,$$

$$Z_4 = 3 + j4 = 5\angle 53.13°, \quad Z_5 = 8 - j6 = 10\angle -36.87°.$$

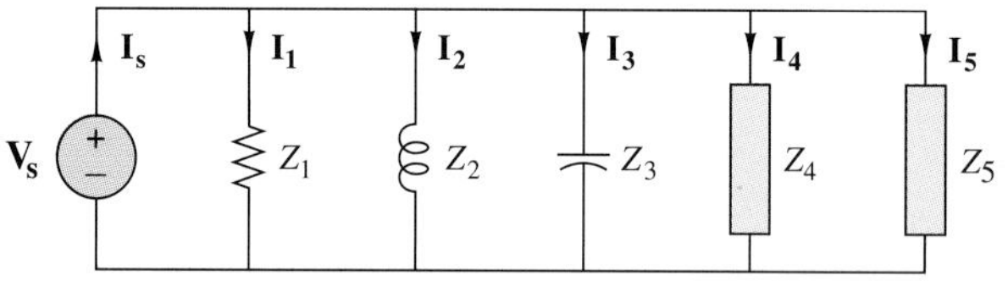

Figure 11.9 Circuit for the calculation of reactive powers.

(a) Calculate the real power and reactive power absorbed or delivered by each impedance.
(b) Verify the conservation of the real power and the reactive power.

SOLUTION

(a) Let the phase angle of the voltage source be set at zero, i.e., $\mathbf{V_s} = 120\angle 0°$ V. For each impedance Z_k, we calculate $\mathbf{I_k} = \mathbf{V_s}/Z_k$ and $\theta_{vk} - \theta_{ik} = \theta_{zk}$. The real power and reactive power are then calculated using equations 11.11a and 11.15, respectively. The results are listed in the following table:

k	I_k	θ_{zk}	$\cos(\theta_{zk})$	$\sin(\theta_{zk})$	P_k	Q_k
	(A)	(degrees)			(watts)	(vars)
1	24	0	1	0	2,880	0
2	30	90	0	1	0	3,600
3	24	−90	0	−1	0	−2,880
4	24	53.13	0.6	0.8	1,728	2,304
5	12	−36.87	0.8	−0.6	1,152	−864

(b) The current delivered by the source is

$$\mathbf{I_s} = \mathbf{I_1} + \mathbf{I_2} + \mathbf{I_3} + \mathbf{I_4} + \mathbf{I_5}$$

$$= (24 + 0 + 0 + 24 \times 0.6 + 12 \times 0.8) + j(0 - 30 + 24 - 24 \times 0.8 + 12 \times 0.6)$$

$$= 48 - j18 \text{ A}.$$

The voltage source is thus *delivering*

$$\text{Real power } = P_s = 120 \times 48 = 5{,}760 \text{ W}$$

and

$$\text{Reactive power } = Q_s = 120 \times 18 = 2{,}160 \text{ VAR}.$$

The total real power consumed by the five impedances is

$$P_z = P_1 + P_2 + P_3 + P_4 + P_5 = 5{,}760 \text{ W}.$$

The total reactive power consumed by the five impedances is

$$Q_z = Q_1 + Q_2 + Q_3 + Q_4 + Q_5 = 2{,}160 \text{ VAR}.$$

Since $P_s = P_z$ and $Q_s = Q_z$, the principle of conservation of power for ac circuits is verified.

Exercise. Repeat example 11.4 if the impedance Z_3 is disconnected.

◆ EXAMPLE 11.5

A 60-Hz, 120-V ac source supplies a group of loads, as shown in figure 11.10. The power and power factor of each load (but not the voltages or currents) are indicated in the diagram. If the voltage source is $\mathbf{V}_s = 120\angle 0°$ V, find the apparent power of the source and the current $\mathbf{I}_s$.

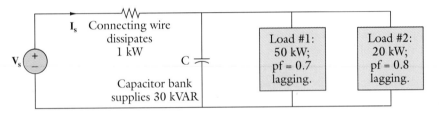

Figure 11.10 Circuit diagram for illustrating the application of the principle of conservation of power.

SOLUTION

Using the triangular relationship in figure 11.8, we calculate

$$Q_1 = 50\tan(\cos^{-1} 0.7) = 51 \text{ kVAR},$$

$$Q_2 = 20\tan(\cos^{-1} 0.8) = 15 \text{ kVAR},$$

$$Q_c = -30 \text{ kVAR},$$

$$Q_{\text{wire}} = 0.$$

The total reactive power absorbed by all of the loads is

$$Q_t = 51 + 15 - 30 = 36 \text{ kVAR}.$$

The total real power absorbed by all of the loads is

$$P_t = 50 + 20 + 1 = 71 \text{ kW}.$$

From the principle of conservation of power, the source must supply 71 kW of real power and 36 kVAR of reactive power. Figure 11.8 implies that the apparent power of the source is $\sqrt{36^2 + 71^2} = 79.6$ kVA. Since the terminal voltage is 120 V, the source current must be $79{,}600/120 = 663.33$ A.

Note that without the use of the principle of conservation of power, it is not possible to determine the source current in this case.

The real power and reactive power are both real numbers, and their definitions explicitly contain the angle difference $\theta_z = \theta_v - \theta_i$. By introducing the concept of a **complex power**, it is possible to obtain the real and reactive power directly in terms of the voltage and current phasors. This makes the phasor method of chapter 10 more attractive in ac circuit problems. The complex power absorbed by a two-terminal element, is defined as

$$\mathbf{S} = P + jQ. \tag{11.16}$$

Unlike the concept of average power which is defined for any periodic signal, *the concepts of reactive power Q and complex power S are defined only for the case of single-frequency ac circuits in the steady state.*

Let a two-terminal element (figure 11.1) have voltage phasor $\mathbf{V} = V \angle \theta_v$ and current phasor $\mathbf{I} = I \angle \theta_i$, both in rms values. We use an asterisk to indicate the conjugate of a complex number, i.e., $\mathbf{I}^* = I \angle -\theta_i$. Then we can write

$$P = VI \cos(\theta_v - \theta_i) = \mathrm{Re}\left\{ V \angle \theta_v \times I \angle -\theta_i \right\} = \mathrm{Re}\left\{ \mathbf{VI}^* \right\} \tag{11.17a}$$

and

$$Q = VI \sin(\theta_v - \theta_i) = \mathrm{Im}\left\{ V \angle \theta_v \times I \angle -\theta_i \right\} = \mathrm{Im}\left\{ \mathbf{VI}^* \right\}. \tag{11.17b}$$

It follows from equations 11.16 and 11.17 that

$$\text{Complex power } \mathbf{S} = \mathbf{VI}^*, \tag{11.18a}$$

$$\text{Average power } P = \mathrm{Re}\{\mathbf{S}\}, \tag{11.18b}$$

$$\text{Reactive power } Q = \mathrm{Im}\{\mathbf{S}\}, \tag{11.18c}$$

$$\text{Apparent power} = |\mathbf{S}| = S. \tag{11.18d}$$

Since the principle of conservation of power holds for both real and reactive powers, it is easy to see that the complex power is also conserved. However, the conservation principle does not hold for the apparent power, i.e., for the magnitude of the complex power. We give a formal statement of the principle of conservation of complex power, followed by an application thereof, and, finally, a rigorous proof of the principle based on KCL and KVL.

PRINCIPLE OF CONSERVATION OF COMPLEX POWER IN AC CIRCUITS:
For ac circuits in the steady state, complex power is conserved.

To use the complex power concept in analysis, we first solve for all voltage and current phasors by the methods described in chapter 10 and then use equations 11.18 to compute various powers. The principle of conservation of power may be used to simplify some calculations. The following example illustrates the use of this approach in the study of a simple power flow problem.

EXAMPLE 11.6

A load having impedance $8 + j6 \ \Omega$ is supplied by two sources as shown in figure 11.11, one of which is $\mathbf{V}_{s1} = 660 \angle 0° \ $V. The impedances of the connecting lines are also shown.

(a) If the second source, $\mathbf{V}_{s2}$, is disconnected, find the magnitude of the load voltage, the real power and reactive power absorbed by the load, and those supplied by the source.

(b) Repeat part (a) if $\mathbf{V}_{s2} = 662 \angle -0.5° \ $V is also connected.

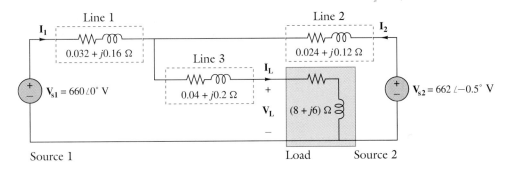

Figure 11.11 Example of complex power calculation.

SOLUTION

(a) With source 2 disconnected (open circuited), the current $\mathbf{I_L}$ is

$$\mathbf{I_L} = \frac{660}{(0.032 + j0.16) + (0.04 + j0.2) + (8 + j6)} = 64.22 e^{-j38.23°} \text{ A}.$$

By Ohm's law,

$$\mathbf{V_L} = (8 + j6)\mathbf{I_L} = 642.2\angle{-1.36°} \text{ V}.$$

Thus, load voltage has a magnitude of 642.2 V. The various powers delivered *to* the load are as follows:

$$\text{Complex power: } \mathbf{S_L} = \mathbf{V_L}\mathbf{I_L^*} = 41,247\angle36.87°,$$

$$\text{Average power: } P = \text{Re}\{\mathbf{S_L}\} = 32,998 \text{ W},$$

$$\text{Reactive power: } Q = \text{Im}\{\mathbf{S_L}\} = 24,748 \text{ VAR}.$$

The various powers delivered *by* the source are as follows:

$$\text{Complex power: } \mathbf{S_s} = \mathbf{V_{s1}}\mathbf{I_L^*} = 42,388\angle38.23°,$$

$$\text{Average power: } P_s = \text{Re}\{\mathbf{S_s}\} = 33,295 \text{ W},$$

$$\text{Reactive power: } Q_s = \text{Im}\{\mathbf{S_s}\} = 26,233 \text{ VAR}.$$

(b) The circuit analysis with source 2, $\mathbf{V_{s2}} = 662\angle{-0.5°}$ V, connected to the system requires two mesh equations:

$$\begin{bmatrix} 8.072 + j6.36 & 8.04 + j6.2 \\ 8.04 + j6.2 & 8.064 + j6.32 \end{bmatrix} \begin{bmatrix} \mathbf{I_1} \\ \mathbf{I_2} \end{bmatrix} = \begin{bmatrix} 660 \\ 662e^{-j0.5°} \end{bmatrix}.$$

Solving the equations (e.g., with MATLAB) yields

$$\mathbf{I_1} = 40.31 - j6.41 = 40.81\angle{-9.04°} \text{ A},$$

$$\mathbf{I_2} = 10.62 - j33.63 = 35.27\angle{-72.48°} \text{ A},$$

$$\mathbf{I_L} = \mathbf{I_1} + \mathbf{I_2} = 50.92 - j40.04 = 64.78\angle{-38.18°} \text{ A},$$

and

$$\mathbf{V_L} = (8 + j6)\mathbf{I_L} = 647.8\angle{-1.31°} \text{ V}.$$

The load voltage thus increases to 647.8 V. The various powers delivered *to* the load are as follows:

$$\text{Complex power: } \mathbf{S_L} = \mathbf{V_L}\mathbf{I_L^*} = 41,964\angle36.87°,$$

$$\text{Average power: } P_L = \text{Re}\{\mathbf{S_L}\} = 33,571 \text{ W},$$

$$\text{Reactive power: } Q_L = \text{Im}\{\mathbf{S_L}\} = 25,178 \text{ VAR}.$$

The various powers delivered *by* the sources are as follows:

$$\text{Complex power: } \mathbf{S}_{s1} = \mathbf{V}_{s1}\mathbf{I}_1^* = 26,934\angle 9.04°,$$

$$\mathbf{S}_{s2} = \mathbf{V}_{s2}\mathbf{I}_2^* = 23,348\angle 71.98°,$$

$$\text{Average power: } P_{s1} = \text{Re}\{\mathbf{S}_{s1}\} = 26,599 \text{ W},$$

$$P_{s2} = \text{Re}\{\mathbf{S}_{s2}\} = 7,223 \text{ W},$$

$$\text{Reactive power: } Q_{s1} = \text{Im}\{\mathbf{S}_{s1}\} = 4,232 \text{ VAR},$$

$$Q_{s2} = \text{Im}\{\mathbf{S}_{s2}\} = 22,203 \text{ VAR}.$$

Adjusting the magnitude and phase angle of the second source adjusts the amount of real and reactive powers supplied by the sources. Suppose the second source is an ac generator driven by a gasoline engine. Then we can also preset the amount of real power to be supplied by the generator by setting the opening of the throttle. The magnitude and phase angle of $\mathbf{V}_{s2}$ are then controlled by an external dc current, which establishes the magnetic field in the machine. A detailed study of ac machinery and the methods of controlling the real and reactive power flow belongs to an upper-level electrical engineering course. Nevertheless, under some simplified assumptions, the circuit analysis aspect of the problem requires no more than the phasor methods studied in chapter 10 and the various power concepts introduced in this chapter.

Exercise. In example 11.6, part (b), determine the real power and reactive power absorbed by each line, using the currents already calculated. Verify the conservation of the real power, reactive power, and complex power.

In section 6 of chapter 1, the conservation of instantaneous power was accepted as a natural outcome of the conservation of energy. This point of view, however, is of little help in establishing the conservation of reactive power and complex power. We shall now show that the conservation of instantaneous, reactive, and complex power are all consequences of KCL and KVL. We prove the conservation principle using a specific circuit configuration, general enough to bring out all the essential concepts, yet simple enough to avoid complicated notations. Consider the four-node, six-branch circuit shown in figure 11.12.

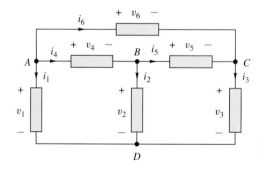

Figure 11.12 A circuit for demonstrating conservation of power.

Applying KCL to nodes A, B, and C yields

$$i_1 + i_4 + i_6 = 0,$$

$$-i_4 + i_2 + i_5 = 0,$$

$$-i_5 - i_6 + i_3 = 0. \tag{11.19}$$

Recall from chapter 4 that an alternative form of applying KVL is to express every branch voltage as the difference between two node-to-datum voltages. With node D as the reference node, this alternative form of KVL yields

$$v_1 = v_A, \qquad v_2 = v_B, \qquad v_3 = v_C,$$

$$v_4 = v_A - v_B, \qquad v_5 = v_B - v_C, \qquad v_6 = v_A - v_C. \tag{11.20}$$

The total instantaneous power absorbed by all branches is

$$p_{\text{total}} = v_1 i_1 + v_2 i_2 + v_3 i_3 + v_4 i_4 + v_5 i_5 + v_6 i_6. \tag{11.21}$$

Substituting the KVL equation 11.20 into equation 11.21 and regrouping terms leads to

$$p_{\text{total}} = v_A i_1 + v_B i_2 + v_C i_3 + (v_A - v_B)i_4 + (v_B - v_C)i_5 + (v_A - v_C)i_6$$
$$= v_A(i_1 + i_4 + i_6) + v_B(-i_4 + i_2 + i_5) + v_C(-i_5 - i_6 + i_3). \tag{11.22}$$

Note in equation 11.22 that the coefficient for each node voltage v_k is simply the sum of all currents leaving that node. According to the KCL equation 11.19, the total instantaneous power is then zero. The same reasoning applied to a general network (whether linear or not) of n_b branches leads to

$$\sum_{k=1}^{n_b} v_k(t) i_k(t) = 0. \tag{11.23}$$

This proves the conservation of instantaneous power, using KCL and KVL. Equation 11.23 is a special case of a more general theorem called **Tellegen's Theorem**, which states that equation 11.23 is valid even for the case in which the voltages and currents belong to two different networks, as long as the networks have the same node-branch relationship. That is, as long as they have the same topology. In the case of two different networks with the same topology, the summation in the equation no longer possesses the meaning of total instantaneous power; yet the relationship is mathematically correct and is useful in the calculation of network sensitivities, a topic that will be discussed in a more advanced course.

Consider next the case of a linear network in the sinusoidal steady state, assuming that all inputs are at the same frequency. The preceding derivation can be repeated with phasors $\mathbf{V_k}$ and $\mathbf{I_k}$ replacing the time functions $v_k(t)$ and $i_k(t)$. Equation 11.23 becomes

$$\sum_{k=1}^{n_b} \mathbf{V_k I_k} = 0.$$

From the properties of complex numbers discussed in chapter 10, this equation implies that

$$\sum_{k=1}^{n_b} \mathbf{V_k I_k^*} = 0. \tag{11.24}$$

According to equation 11.18a, each term $\mathbf{V_k I_k^*}$ represents the complex power $\mathbf{S_k}$ absorbed by the kth branch. Consequently, equation 11.24 can be rewritten as

$$\sum_{k=1}^{n_b} \mathbf{S_k} = 0, \tag{11.25}$$

which is a statement of the conservation of complex power.

Equation 11.25 requires that both its real and imaginary parts be zero. Taking the real part and using the properties of complex numbers yields

$$\text{Re} \sum_{k=1}^{n_b} \{\mathbf{S_k}\} = \sum_{k=1}^{n_b} \text{Re}\,\{\mathbf{S_k}\} = \sum_{k=1}^{n_b} \{P_k\} = 0. \tag{11.26}$$

This is a statement of the conservation of real power or average power. Similarly, taking the imaginary parts of both sides of equation 11.25 and making use of equation 11.18c implies that

$$\text{Im} \sum_{k=1}^{n_b} \{\mathbf{S_k}\} = \sum_{k=1}^{n_b} \text{Im}\,\{\mathbf{S_k}\} = \sum_{k=1}^{n_b} \{Q_k\} = 0. \tag{11.27}$$

This is a statement of the conservation of reactive power.

In other words, conservation of power is a direct consequence of KVL and KCL.

5. MAXIMUM POWER TRANSFER IN THE SINUSOIDAL STEADY STATE

Chapter 5 first outlined the basics of the maximum power transfer problem for the case of linear resistive networks. Having introduced the energy storage elements L and C, and having studied methods of sinusoidal steady-state analysis, we can extend the results on maximum power transfer to general linear networks in the sinusoidal steady state.

MAXIMUM POWER TRANSFER THEOREM FOR AC CIRCUITS

Let a practical ac source be represented by an independent voltage source $\mathbf{V_s}$ (a voltage phasor with rms value) in series with an impedance $Z_s = R_s + jX_s$. An adjustable load impedance $Z_L = R_L + jX_L$, with $R_L > 0$, is connected to the source (figure 11.13). For fixed Z_s, $\mathbf{V_s}$, and ω, in the steady state, the average power delivered to the load is maximum when Z_L is the complex conjugate of Z_s, i.e., when

$$R_L = R_s, \tag{11.28a}$$

$$X_L = -X_s, \tag{11.28b}$$

and the maximum average power is given by

$$P_{\max} = \frac{V_{s,\text{eff}}^2}{4R_s}. \tag{11.28c}$$

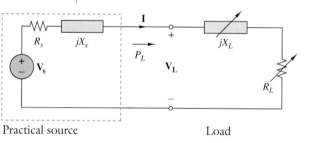

Practical source   Load

Figure 11.13 Maximum power transfer to an adjustable impedance.

The proof the maximum power transfer theorem is as follows. The current phasor is

$$\mathbf{I} = \frac{\mathbf{V_s}}{(R_s + R_L) + j(X_s + X_L)}. \tag{11.29}$$

The average power delivered to the load is

$$P = P_{\text{ave}} = |\mathbf{I}|^2 R_L = \frac{V_s^2 R_L}{(R_s + R_L)^2 + (X_s + X_L)^2}. \tag{11.30}$$

Here, P_{ave} is a function of two real variables R_L and X_L. To maximize P_{ave}, we follow the standard procedure in calculus: Set both partial derivatives, $\partial P/\partial X_L$ and $\partial P/\partial R_L$, to zero and solve for R_L and X_L. From equation 11.30,

$$\frac{\partial P}{\partial R_L} = \frac{V_s^2[(R_s + R_L)^2 + (X_s + X_L)^2 - 2R_L(R_s + R_L)]}{[(R_s + R_L)^2 + (X_s + X_L)^2]^2} = 0, \tag{11.31a}$$

$$\frac{\partial P}{\partial X_L} = \frac{V_s^2[-2R_L(X_s + X_L)]}{[(R_s + R_L)^2 + (X_s + X_L)^2]^2} = 0. \tag{11.31b}$$

From equation 11.31b, the only physically meaningful solution is

$$X_L = -X_s.$$

Substituting this result into the numerator of equation 11.31a yields

$$\frac{V_s^2(R_s - R_L)}{(R_s + R_L)^3} = 0.$$

The only physically meaningful solution is $V_s \neq 0$ and

$$R_L = R_s.$$

Substituting these results into equation 11.30 produces equation 11.28c, i.e.,

$$P_{max} = \frac{V_{s,eff}^2}{4R_s},$$

which verifies the theorem.

The maximum power transfer theorem can also be established less formally as follows. With any existing Z_L connected to the source, if the total reactance $X_L + X_s$ is not zero, we can always increase the magnitude of the current, and hence the power delivered to the load, by "tuning out" the reactance, i.e., by adjusting X_L to be $-X_s$. But this is no more than equation 11.28b. Under such a condition, the circuit becomes resistive, and the maximum power transfer theorem of chapter 5 may be applied to obtain equations 11.28a and c.

The maximum power obtainable with a passive load, given by equation 11.28c, is called the **available power** of the fixed source.

The conditions for maximum power transfer as given by equations 11.28a and b are valid when both R_L and X_L are adjustable. If X_L is fixed and only R_L is adjustable, then the condition for maximum power transfer is

$$R_L = \sqrt{R_s^2 + (X_s + X_L)^2}, \tag{11.32}$$

which is obtained by solving equation 11.31a.

If the source is a general two-terminal linear network, then its Thévenin equivalent must be found before applying the maximum power transfer theorem. If the source is represented by a Norton equivalent circuit, we can use a source transformation to obtain the Thévenin form and then apply equations 11.28.

As pointed out in chapter 5, maximum power transfer is not the objective in electric power systems, as the sources usually have very low impedances. On the other hand, it is a very important factor to be considered in the design of many communication circuits, as illustrated in the following example.

 EXAMPLE 11.7

The radio receiver shown in figure 11.14a is connected to an antenna. The antenna intercepts the electromagnetic waves from a broadcast station operating at 1 MHz. For circuit analysis purposes, the antenna is represented by a Thévenin equivalent circuit, shown in figure 11.14b.

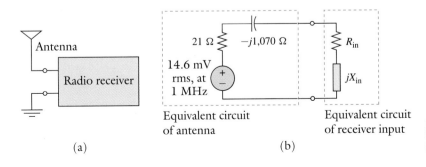

Equivalent circuit of antenna

Equivalent circuit of receiver input

Figure 11.14 Example of maximum power transfer. (a) An antenna connected to a radio receiver. (b) A circuit model for ac analysis.

(a)   (b)

(a) Find the input impedance $R_{in} + jX_{in}$ of the receiver if maximum power is to be transferred from the antenna to the receiver.

(b) Under the condition of (a), find the magnitude of the voltage across the receiver terminals and the average power delivered to the receiver.

SOLUTION

(a) From the maximum power transfer theorem, the answers are $R_{in} = 21\ \Omega$ and $X_{in} = j1,070\ \Omega$.

(b) Since the reactances in the circuit have been "tuned out," the input current to the receiver is simply $14.6/(21 + 21) = 0.348$ mA. The input impedance has a magnitude

$$|Z_{in}| = \sqrt{21^2 + 1,070^2} = 1,070.2\ \Omega.$$

Therefore, the magnitude of the voltage across the receiver terminals is $0.348 \times 1,070.2 = 372.4$ mV (rms). The power transferred from the antenna to the receiver is $0.348^2 \times 21 = 2.54\ \mu$W.

In the foregoing discussions of maximum power transfer, we have assumed that the load is adjustable. In practice, the load is often fixed, for example, with a loudspeaker having a 4-Ω voice coil. In such cases, one designs coupling networks consisting of lossless passive components. These coupling networks transform the fixed load impedance into one whose conjugate matches the fixed source impedance. This permits maximum power transfer to the load. The following example illustrates the principle; design procedure for some simple coupling networks will be discussed in volume 2.

EXAMPLE 11.8

A fixed load resistance $R_L = 100$ kΩ, representing the input resistance of an amplifier, is connected to the source of example 11.7 through a lossless coupling network, i.e., a network that does not consume or generate average power.

(a) Show that the maximum possible voltage that can be developed across R_L in figure 11.15 is 0.504 V.

(b) Show that with $L = 400.9\ \mu$H and $C = 109.8$ pF, the coupling network of figure 11.15 achieves this maximum voltage across R_L.

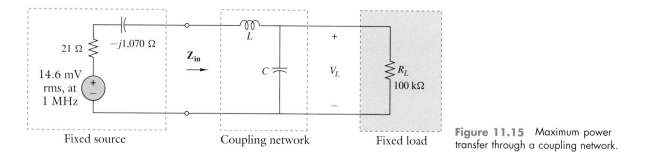

Figure 11.15 Maximum power transfer through a coupling network.

SOLUTION

(a) From equation 11.28c, as used in example 11.7, the available power from the source is 2.54 μW. Since all the power is delivered to R_L, then the voltage must be

$$V_L = \sqrt{P_{max} R_L} = \sqrt{2.54 \times 10^{-6} \times 100,000} = 0.504\ \text{V}.$$

(b) The input impedance Z_{in} of the coupling network with load must be the conjugate of the source impedance. Specifically,

$$Z_{\text{in}} = j\omega L + \cfrac{1}{j\omega C + \cfrac{1}{R_L}}.$$

Substituting the numbers $\omega = 10^6$, $L = 400.9 \times 10^{-6}$, $C = 109.8 \times 10^{-12}$, and $R = 100 \times 10^3$ into this expression yields

$$Z_{\text{in}} = 21 + j1{,}070 \ \Omega,$$

which is indeed the conjugate of the source impedance. Since Z_{in} is conjugate-matched to the source impedance, the maximum power of 2.54 μW is transferred to the coupling network. Because the coupling network consists of L and C, neither of which consumes average power, the 2.54-μW power must be transferred out of the coupling network and into the load resistance. The voltage across the load resistor is

$$V_L = \sqrt{PR_L} = \sqrt{2.54 \times 10^{-6} \times 100{,}000} = 0.504 \text{ V}.$$

This verifies that the coupling network of figure 11.15 enables the largest voltage to appear across the load resistor.

6. SOME ECONOMIC ASPECTS OF TRANSMITTING ELECTRIC POWER

One of the reasons that electricity dominates our everyday life more than any other form of energy is its ease of transmission. A pair of conducting wires connected between a power source and a load transmits electrical energy at the speed of light! In doing so, some energy dissipates in the form of heat due to the I^2R loss of the wires. In all the examples and problems of previous chapters, we have neglected this loss and assumed the connecting wires to be resistanceless. When transmitting large amounts of power, this loss can no longer be neglected, as it increases customer costs. In this section, we shall describe some basic economic factors influencing the transmission of electric power.

Since the average power absorbed by a load equals VI for dc and $V_{\text{eff}} I_{\text{eff}} \cos(\theta_z)$ for ac, it is clear that the same amount of power can be transmitted with a smaller current by making the voltage higher. A smaller current in turn means a smaller I^2R loss in the transmission lines. In volume 2, we will learn about a device called a *transformer* that can readily step up or step down an ac voltage. Changing the level of a dc voltage is a much more difficult and costly task. This fact explains why all household lighting and appliances use ac instead of dc.

Although raising the voltage can reduce the transmission power line loss, there is an upper voltage limit. This limit depends on the insulating strength of the dielectric material used to confine currents to the conducting wires. Typical insulation materials in electrical apparatus are air, rubber, cotton, paper, mica, porcelain, ethylene-propylene, cross-linked polyethylene (both plastics), and other materials. Even if no current is flowing through the wire, when the voltage between wires (or between a wire and ground) exceeds some threshold value, the insulating material breaks down and causes a sparking over and, subsequently, a short circuit. The line-to-line voltage used in long-distance transmission of electric power can be as high as 765 kV, with the wires hanging on steel towers supported by strings of porcelain insulators. The line-to-line voltage used in the distribution of electricity in residential areas is typically 2,400 to 13,800 V. The term **transmission** refers to the long-distance, bulk power, high-voltage transport of power—usually using overhead conductors. The term **distribution** refers to the transport of power over short distances, at lower power levels and lower voltages. Distribution circuits use either overhead conductors or underground insulated cables. The actual voltage supplied to a residential consumer is typically 110 or 220 volts in the United States.

For the purpose of comparing the efficiency of various transmission schemes, suppose that:

1. The line-to-line voltages at the load site have the same magnitudes V_L volts rms.
2. The *total* power absorbed by the loads is P_L watts.
3. The loads are represented by resistances, or the power factor is unity.
4. Each single line connecting the load to the power source is represented by a resistance, as the reactance of the line is not considered in the power loss calculation.

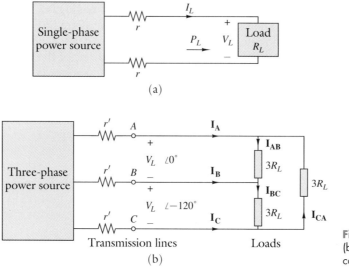

(a)

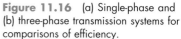

(b)

Figure 11.16 (a) Single-phase and (b) three-phase transmission systems for comparisons of efficiency.

We will first find the transmission line power loss in terms of P_L, V_L, and the resistance of the transmission line. Figure 11.16a depicts a single-phase system in which two wires connect the load to a single-phase source. In this system,

$$P_{\text{loss}} = 2 \times (I_L^2 r) = \frac{P_L^2}{V_L^2} 2r. \tag{11.33}$$

Now suppose the same amount of power is supplied by the three-phase system shown in figure 11.16b. The voltages between any two lines have the same magnitude V_L as in the single-phase system. However, their phase angles are 120 degrees apart, as indicated by the phasor diagram of figure 11.17a. Note that specification of the three line-to-line voltages satisfies KVL, as we have $\mathbf{V_{AB}} + \mathbf{V_{BC}} + \mathbf{V_{CA}} = V_L\angle 0° + V_L\angle -120° + V_L\angle 120° = 0$.

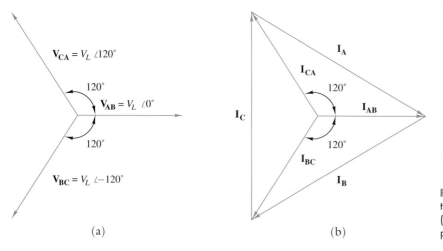

(a)

(b)

Figure 11.17 Phasor diagrams for the three-phase system of figure 11.16b. (a) Voltage phasor diagram. (b) Current phasor diagram.

By assumption 2, both systems of figure 11.16 have the same *total* load power P_L. Hence each load in figure 11.16b draws only $P_L/3$, or equivalently, each of the three identical loads in figure 11.16b has a resistance three times the load resistance in 11.16a. The magnitude of each load current is

$$|\mathbf{I_{AB}}| = |\mathbf{I_{BC}}| = |\mathbf{I_{CA}}| = \frac{P_L}{3V_L}. \tag{11.34}$$

From figure 11.16b,

$$\mathbf{I_A} = \mathbf{I_{AB}} - \mathbf{I_{CA}}.$$

According to the principle discussed in section 9 of chapter 10, the vector representing $\mathbf{I_A}$ is a vector that joins the heads of the vectors for $\mathbf{I_{AB}}$ and $\mathbf{I_{CA}}$, as drawn in figure 11.17b. The simple geometry in the figure shows that

$$|\mathbf{I_A}| = |\mathbf{I_B}| = |\mathbf{I_C}| = \sqrt{3}\,|\mathbf{I_{AB}}| = \frac{\sqrt{3}}{3}\frac{P_L}{V_L}. \tag{11.35}$$

Since three wires, each having a resistance r' ohms, are used to transmit the power, the line loss for the three-phase case is

$$P_{\text{loss}}' = 3 \times |\mathbf{I_A}|^2 r' = 3 \times \frac{3}{9}\frac{P_L^2}{V_L^2}r' = \frac{P_L^2}{V_L^2}r'. \tag{11.36}$$

Equations 11.33 and 11.36 provide a basis for comparing the merits of the two systems. Two conclusions may be drawn from the analysis:

1. If both systems of figure 11.16 have the same total delivered power, the same line-to-line voltage, and the same percentage of power loss in the transmission lines, then the amount of material (copper) used in the three-phase system is only 75% of that used in the single-phase system.
2. If both systems of figure 11.16 have the same total delivered power, have the same line-to-line voltage, and use the same amount of material (copper) in the lines, then the power loss in the transmission lines for the three-phase system is only 75% of that of the single-phase system.

To justify the first conclusion, we equate equation 11.33 to equation 11.36 and find $r' = 2r$. Since both systems have the same distance of transmission and the resistance of a wire is inversely proportional to the cross-sectional area of the wire, the condition $r' = 2r$ implies that the cross section A' of each wire in the three-phase system need be only half of the area A of the wire in the single-phase system. But there are two wires in the single-phase system and three wires in the three-phase system. Therefore, the ratio of the materials used is

$$\frac{\text{Material in three-phase system}}{\text{Material in single-phase system}} = \frac{3A'}{2A} = \frac{3}{2} \times \frac{1}{2} = \frac{3}{4} = 75\%.$$

To justify the second conclusion, we first note that if the two systems of lines have the same amount of material (copper), then $2 \times A = 3 \times A'$, or $A/A' = 1.5$ implying that $r' = 1.5r$. Dividing the respective sides of equation 11.36 by 11.33, we have

$$\frac{\text{Power loss in three-phase lines}}{\text{Power loss in single-phase lines}} = \frac{P_{\text{loss}}'}{P_{\text{loss}}} = \frac{\dfrac{P_L^2 r'}{V_L^2}}{\dfrac{P_L^2 2r}{V_L^2}} = \frac{r'}{2r} = \frac{1.5}{2} = 75\%.$$

There is another reason for using three-phase circuits for high power applications. Let the maximum current capability of a conductor be I_{max}. When $I > I_{\text{max}}$, the conductor losses will be so high that the conductor will overheat and perhaps burn until the conductor breaks. For a single-phase circuit, the maximum power-handling capability is proportional to VI_{max}. For a three-phase supply, the maximum power is proportional to $3VI_{\text{max}}$ (where

V is the line to neutral voltage). Thus, with the addition of one conductor to the single-phase circuit, one obtains a power-handling advantage of three, assuming the same power factor for all loads. Note that to carry triple power using a single-phase circuit, one would have to use a total of six conductors.

For the single-phase system, the instantaneous power $p(t)$, as given by equation 11.2, has a constant component and another component varying with time at twice the input frequency. For a three-phase system having three identical loads symmetrically connected, the *total* instantaneous power is constant. To see this, let the three identical loads be connected as shown in figure 11.18. Then let $\mathbf{Z} = Z\angle\theta$. Using equation 11.2, we can calculate the total instantaneous power delivered to the loads as follows:

$$P_{\text{total}}(t) = v_{AB}(t)i_{AB}(t) + v_{BC}(t)i_{BC}(t) + v_{CA}(t)i_{CA}(t)$$

$$= \sqrt{2}V_L \cos(\omega t)\frac{\sqrt{2}V_L}{Z_L}\cos(\omega t - \theta)$$

$$+ \sqrt{2}V_L \cos(\omega t - 120°)\frac{\sqrt{2}V_L}{Z_L}\cos(\omega t - 120° - \theta)$$

$$+ \sqrt{2}V_L \cos(\omega t + 120°)\frac{\sqrt{2}V_L}{Z_L}\cos(\omega t + 120° - \theta)$$

$$= \frac{3V_L^2}{Z_L}\cos(\theta)$$

$$+ \frac{V_L^2}{Z_L}[\cos(2\omega t - \theta) + \cos(2\omega t - 240° - \theta) + \cos(2\omega t + 240° - \theta)].$$

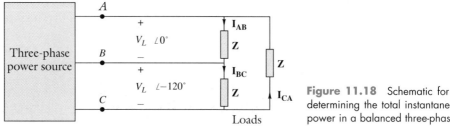

Figure 11.18 Schematic for determining the total instantaneous power in a balanced three-phase load.

From an identity in trigonometry, the three terms within the brackets add up to zero. Therefore, for a balanced three-phase load,

$$P_{\text{total}}(t) = 3\frac{V_L^2}{Z_L}\cos(\theta) = \text{ a constant.}$$

Besides a better efficiency in power transmission and constant instantaneous power, a three-phase system has some other advantages over a single-phase system. These include better starting characteristics of motors and a more flexible connection of transformers. For industrial users, motors account for the greater portion of the loads. Therefore, a three-phase power source is superior to a single-phase power source. For ordinary residential consumers, however, the power is most commonly made available at the service panel in the form of a three-wire single-phase system supplying two voltages, 110 and 220 volts. The higher voltage, 220V, is used for heavy loads such as a dryer, oven, range, and central air conditioning system. The lower voltage, 110V, is used for lighting and other small appliances.

7. ANALYSIS OF BALANCED THREE-PHASE CIRCUITS

Three-phase ac systems are used worldwide for the transmission, distribution, and utilization of large amounts of electrical energy. Section 6 explained some of the reasons for this choice. A comprehensive study of three-phase systems belongs to the special field of power engineering. Nevertheless, for the case of a balanced three-phase circuit (to be defined shortly), the analysis reduces to that of a single-phase circuit and is solvable by the phasor method of chapter 10. This section explains the basics of three-phase systems from the circuit analysis point of view.

The service panel of a customer is a point at which three-phase power is supplied through either a set of four wires or a set of three wires, as shown in figures 11.19a and b, respectively. In both cases, when all loads are removed, $V_{AB} = V_{BC} = V_{CA}$. The single value of each is called the **line-to-line voltage**, or simply, the **line voltage**, denoted by V_L. The difference in phase angle between any two of the voltages, however, is $120°$. For the four-wire system of figure 11.19a, it is further true that $V_{AN} = V_{BN} = V_{CN}$. The value here is called the **line-to-neutral voltage.** The difference in phase angle between any two of these voltages is again $120°$. When we speak of the system voltage in three-phase circuit problems, it is understood to be the rms value of the line-to-line voltage, unless specifically stated otherwise. The phasor diagram of figure 10.30a shows that

$$\text{Line-to-line voltage} = \sqrt{3}(\text{line-to-neutral voltage}).$$

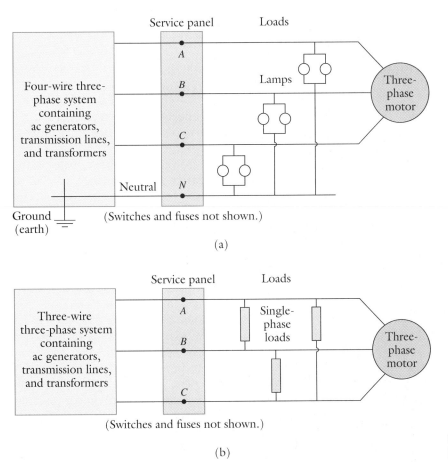

Figure 11.19 (a) Four-wire and (b) three-wire three-phase power systems.

The conductors labeled A, B, and C are called hot conductors, while the conductor labeled N is called the neutral. For safety reasons, the National Electrical Code requires the neutral conductor to be grounded, i.e., connected to a ground rod or a water pipe within the transmission system and also at the service panel.

If the loads connected to the system are indistinguishable with respect to the terminals A, B, and C, then the circuit is called a **balanced three-phase circuit.** If such symmetry does not exist, then the circuit is called an **unbalanced three-phase circuit.** In this section, our emphasis is on balanced three-phase circuits in the steady state.

Circuit Model for Practical Three-Phase Sources

The first step in the analysis of a three-phase circuit is to obtain circuit models for the power sources delineated in figures 11.19a and b. An ideal three-phase voltage source can be represented by a connection of three ideal voltage sources. Figures 11.20a and b show the connections. Ideal voltage sources connected in parallel or in a loop are generally not allowed. (See sections 2 and 3 of chapter 2.) The model shown in figure 11.20b, however, is a special case that does not violate KVL, because the three source voltages add up to zero. In fact, to calculate the voltages and currents *external* to the sources, one can remove any one of the three sources in figure 11.20b.

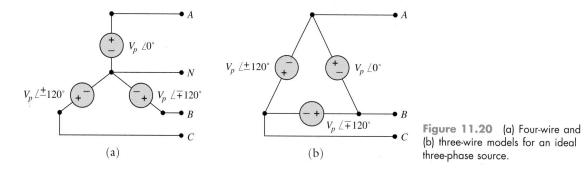

Figure 11.20 (a) Four-wire and (b) three-wire models for an ideal three-phase source.

In figure 11.20a, the angle of $\mathbf{V_{AN}}$ has been arbitrarily chosen to be zero. There are two possible phase relationships among the three line-to-neutral voltages:

1. $\mathbf{V_{BN}}$ lags $\mathbf{V_{AN}}$ by 120°, and $\mathbf{V_{CN}}$ lags $\mathbf{V_{BN}}$ by 120° (upper signs for the angles in the figure). In this case, the system is said to have a **positive phase sequence**, or an *A–B–C* phase sequence.
2. $\mathbf{V_{CN}}$ lags $\mathbf{V_{AN}}$ by 120°, and $\mathbf{V_{BN}}$ lags $\mathbf{V_{CN}}$ by 120° (lower signs for the angles in the figure). In this case, the system is said to have a **negative phase sequence**, or *A–C–B* phase sequence.

A similar definition for phase sequence applies to the three-wire system of figure 11.20b. If the upper signs of the angles in the figure prevail, the system is said to have a positive, or *A–B–C*, phase sequence; if the lower signs of the angles prevail, the system is said to have a negative, or *A–C–B*, phase sequence. When one connects a number of single-phase loads to a three-phase system, the phase sequence is of no importance. However, when one connects a three-phase motor to the system, a change in phase sequence results in a reversal of the direction of rotation of the motor! Furthermore, to put two three-phase sources in parallel to meet a larger power demand, a *necessary* condition is that the two systems have the same phase sequence, positive or negative.

In ac generators, each voltage source of magnitude V_p in figure 11.20 is actually produced by a *winding* or coil placed in the *stator* of a generator. The rotating magnetic field from the *rotor* then induces a sinusoidal voltage in each winding. Since the three windings have identical construction, the induced voltages have the same magnitude V_p, called the **phase voltage**. (The subscript p stands for phase here.) The windings are located in the stator in such a manner that the induced voltages reach their maximum values successively, with a time lag or delay of one-third of a cycle. This accounts for the 120° differences in the phase angles of the phasors of figure 11.20. Since each winding has a nonzero resistance R_g, as well as a nonzero inductance L_g, a more accurate model for the three-phase ac generator would include an impedance $\mathbf{Z}_g = R_g + jX_g$ in series with the ideal voltage source.

In figure 11.20, the conductors feeding into the service panel usually do not come directly from an ac generator. Between the ac generator and the customers' loads, there are a step-up transformer, a transmission line, and a step-down transformer. These affect the level of the voltages and impedances, but maintain the 120° phase difference. Thus, for circuit analysis purposes, without going into the details of the theory of ac machinery, we can assume that the *practical* three-phase power system of figure 11.19 can be represented by the linear models shown in figure 11.21 for the case of a positive phase sequence. The angle of one of the voltage phasors has been arbitrarily set to zero. The parameters in these models may be determined by measurements. Note that for figures 11.21a and b, the line-to-line voltage is $V_L = 1.732\,V_s$, whereas for figure 11.21c, $V_L = V_s$. For the case of a positive phase sequence, figure 10.30a is the phasor diagram that shows the relationships among all line-to-neutral and line-to-line voltages.

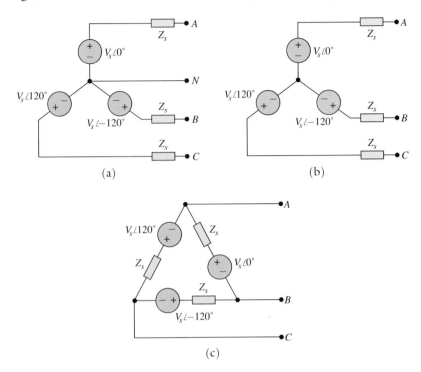

(a)   (b)

(c)

Figure 11.21 Models for practical three-phase sources with a positive phase sequence: (a) Four-wire. (b) Three-wire, Y connected. (c) Three-wire, delta connected.

Wye-Delta Transformation

Section 4 of chapter 5 defined the concepts of a two-terminal network and an equivalent network. Here, we shall extend these concepts in a straightforward manner to n-terminal networks in the sinusoidal steady state: Two n-terminal networks are **equivalent** if they have the same relationships among all terminal voltage phasors and current phasors. In sinusoidal steady-state analysis, two-terminal circuit elements are represented by impedances. We can easily establish the equivalence between two types of three-terminal networks called the **Y-network** and the **Δ-network**. The letters Y and Δ suggest vividly how the impedances are connected. For the special case of three equal impedances, shown in figure 11.22, the equivalence requires a simple relationship, $\mathbf{Z}_\Delta = 3\mathbf{Z}_Y$. (See problem 25.) The analysis of balanced three-phase circuits makes use of this simple relationship. For the general case of three unequal impedances, the more complicated formulas are given in problem 26, and the relationship is called the **wye-delta transformation**. The same transformation is also called the **T-π transformation** in circuit theory, as one can reshape the letters Y and Δ (inverted) to look like T and π. When the Y and Δ networks contain internal independent voltage sources, the equivalence can be established in a similar manner. (See problem 27.) Figure 11.23 shows the result, which is useful in the analysis of some balanced three-phase circuits.

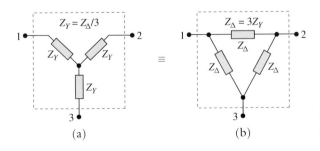

(a)   (b)

Figure 11.22 A special case of Y-Δ or T-π network transformation. (a) Y-network. (b) Δ-network.

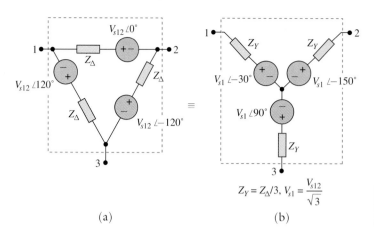

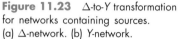

$$Z_Y = Z_\Delta/3, \quad V_{s1} = \frac{V_{s12}}{\sqrt{3}}$$

Figure 11.23 Δ-to-Y transformation for networks containing sources. (a) Δ-network. (b) Y-network.

(a)

(b)

Analysis of Balanced Three-Phase Circuits

When all devices that are connected to a three-phase system are linear, an analysis of the system can be carried out using the conventional nodal or mesh equations in phasor format. Further, if the loads are balanced, then the problem can be reduced to a single-phase circuit problem wherein one need only work out the detailed solution for one phase. The answers for the remaining two phases are inferred directly by introducing the 120° phase shift in proper sequence. Example 10.14 illustrated the procedure for the case of an *ideal* three-phase source supplying Y-connected and Δ-connected balanced loads. The next few examples illustrate the analysis when source impedances and connecting wire impedances must be considered for more accurate results. The models shown in figures 11.21 through 11.23 will be utilized.

◆ EXAMPLE 11.9

Figure 11.24a shows a three-phase system referred to as a **Y–Y circuit**, because both the sources and the loads are connected in the form of a Y. The following parameters are known:

Source phase voltage	$\mathbf{V}_{s1} = 104\angle 0°$ V (rms),
Source impedance	$Z_s = 0.05 + j0.15 = 0.158\angle 71.56°$ Ω,
Load impedance	$Z = 4 + j3 = 5\angle 36.87°$ Ω,
Feeder impedance	$Z_1 = 0.1 + j0.2 = 0.224\angle 63.43°$ Ω,
Neutral conductor	$Z_n = 0.02 + j0.08 = 0.082\angle 75.96°$ Ω.

(a) Find the magnitudes of the line currents.
(b) Find the magnitude of the voltage across each load Z.
(c) Find the total average power delivered to the loads.
(d) If $\mathbf{V}_{s1}$ is taken as the reference for phase angles, i.e., $\mathbf{V}_{s1} = 120\angle 0°$ V, find the phasors $\mathbf{I}_a$, $\mathbf{I}_b$, $\mathbf{I}_c$, $\mathbf{V}_{an}$, $\mathbf{V}_{bn}$, $\mathbf{V}_{cn}$, $\mathbf{V}_{ab}$, $\mathbf{V}_{bc}$, and $\mathbf{V}_{ca}$.

SOLUTION

Part (a). Applying KCL at node n in figure 11.24a yields

$$\frac{\mathbf{V}_{no} - V_{s1}\angle 0°}{Z + Z_1 + Z_s} + \frac{\mathbf{V}_{no} - V_{s1}\angle -120°}{Z + Z_1 + Z_s} + \frac{\mathbf{V}_{no} - V_{s1}\angle 120°}{Z + Z_1 + Z_s} + \frac{\mathbf{V}_{no}}{Z_n}$$

$$= \frac{1}{Z + Z_1 + Z_s}[3\mathbf{V}_{no} - V_{s1}(1\angle 0° + \angle -120° + \angle 120°)] + \frac{\mathbf{V}_{no}}{Z_n} = 0.$$

Since $\angle 0° + \angle -120° + \angle 120° = 0$, the preceding equation implies that $\mathbf{V}_{no} = 0$. Under this condition, the impedance Z_n may be given any value without affecting the other voltages

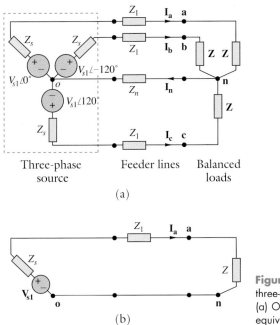

Three-phase source   Feeder lines   Balanced loads

(a)

(b)

Figure 11.24 Analysis of a balanced three-phase system with Y–Y connection. (a) Original system. (b) Single-phase equivalent circuit for phase **a**.

and currents in the circuits. In particular, if we replace $\mathbf{Z}_n$ by a short circuit, then the original circuit is seen to consist of three simpler circuits joined at one node. Accordingly, the analysis may be treated as three separate problems, one of which, associated with phase a, is shown in figure 11.24b and is called a **single-phase equivalent circuit**. It is not necessary to draw the single-phase equivalent circuits for the other two phases, because their solutions can be inferred from that of phase a by properly applying the 120° phase shift in the voltages.

From figure 11.24b,

$$\mathbf{I_a} = \frac{\mathbf{V}_{s1}}{Z + Z_1 + Z_s} = \frac{\mathbf{V}_{s1}}{(4 + j3) + (0.1 + j0.2) + (0.05 + j0.15)} = \frac{120\angle 0°}{5.333\angle 38.9°}$$

$$= 22.5\angle -38.9° \text{ A},$$

and all line currents have the same magnitude, equal to 22.5 A.

Part (b). The magnitude of the voltage across each load is

$$|\mathbf{I_a}| \times |\mathbf{Z}| = 22.5 \times 5 = 112.5 \text{ V}.$$

Part (c). From equation 11.11b,

$$P_a = P_b = P_c = 112.5 \times 22.5 \times \cos(36.87°) = 2{,}025 \text{ W},$$

and the total average power delivered to all three loads is $3 \times 2{,}025 = 6{,}075$ W.

Part (d). From part (a), $\mathbf{I_a} = 22.5\angle -38.9°$ A. Then

$$\mathbf{V}_{an} = \mathbf{Z}\mathbf{I_a} = 5\angle 36.87° \times 22.5\angle -38.9° = 112.5\angle -2.03° \text{ V}.$$

Incorporating the 120° phase shift, we have the answers for the other two phases: $\mathbf{I_b} = 22.5\angle -158.9°$ A, $\mathbf{I_c} = 22.5\angle 81.1°$ A, $\mathbf{V}_{bn} = 112.5\angle -122.03°$ V, and $\mathbf{V}_{cn} = 112.5\angle 117.97°$ V.

To compute the line-to-line voltage, we refer to figure 11.24a. Here, we have

$$\mathbf{V}_{ab} = \mathbf{V}_{an} - \mathbf{V}_{bn} = 1.732\mathbf{V}_{an}\angle 30° = 194.8\angle 27.97°\text{ V}.$$

Adding the $120°$ phase shift yields the other two line-to-line voltages: $\mathbf{V}_{bc} = 194.8 \angle-92.03°$ V and $\mathbf{V}_{ca} = 194.8\angle 147.97°$ V.

Exercise. Repeat example 11.9 if the source has a negative phase sequence and $Z = 3 + j4\ \Omega$.

◆ EXAMPLE 11.10

Suppose the Y-connected load in figure 11.24a is replaced by the Δ-connected load shown in figure 11.25a. There is no connection from the loads to the neutral conductor. The resulting circuit is called a $Y-\Delta$ **circuit** for obvious reasons. Let $\mathbf{V}_{s1} = 120\angle 0°$ V and $Z_\Delta = 12 + j9\ \Omega$.

(a) Find the line currents $\mathbf{I}_a$, $\mathbf{I}_b$, and $\mathbf{I}_c$.
(b) Find the load currents $\mathbf{I}_{ab}$, $\mathbf{I}_{bc}$, and $\mathbf{I}_{ca}$.
(c) Draw a phasor diagram to show the line and load currents.

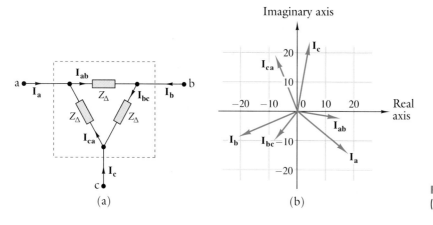

Figure 11.25 (a) Δ-connected loads. (b) Phasor diagram for currents.

SOLUTION

Part (a). Apply the $\Delta-Y$ transformation of figure 11.22 to the Δ-connected load. The result is a Y-connected load with $Z_Y = 4 + j3\ \Omega$. This is precisely the load used in example 11.9. Therefore, all of the previous calculations carry over. The line currents are $\mathbf{I}_a = 22.5\angle-38.9°$ A, $\mathbf{I}_b = 22.5\angle-158.9°$ A, and $\mathbf{I}_c = 22.5\angle 81.1°$ A.

Part (b). To find the load currents, we first note from symmetry that $\mathbf{I}_{bc} = \mathbf{I}_{ab}\angle-120°$ and $\mathbf{I}_{ca} = \mathbf{I}_{ab}\angle 120°$. Referring to figure 11.25a, we have

$$\mathbf{I}_a = \mathbf{I}_{ab} - \mathbf{I}_{ca} = (1 - \angle 120°)\mathbf{I}_{ab} = \sqrt{3}\mathbf{I}_{ab}\angle-30°,$$

or equivalently,

$$\mathbf{I}_{ab} = \frac{1}{\sqrt{3}}\mathbf{I}_a\angle 30°.$$

Using the value calculated in part (a), we have $\mathbf{I}_{ab} = 13\angle-8.9°$ A, $\mathbf{I}_{bc} = 13\angle-128.9°$ A, and $\mathbf{I}_{ca} = 13\angle 111.1°$ A.

Part (c). Figure 11.25b shows the phasor diagram for the line currents and load currents.

EXAMPLE 11.11

Figure 11.26a shows a Δ-connected three-phase source supplying power to a Δ-connected load. Such a circuit is referred to as a **Δ–Δ circuit**. The following parameters are known:

Source phase voltage	$\mathbf{V}_{s12} = 180\angle 0°$ V (rms),
Source impedance	$Z_s = 0.15 + j0.45 = 0.474\angle 71.56°$ Ω,
Load impedance	$Z = 12 + j9 = 15\angle 36.87°$ Ω,
Feeder impedance	$Z_1 = 0.1 + j0.2 = 0.224\angle 64.43°$ Ω.

(a) Find the line currents $\mathbf{I}_a$, $\mathbf{I}_b$, and $\mathbf{I}_c$.
(b) Find the magnitude of the voltage across each load Z.
(c) Find the total average power delivered to the loads.

SOLUTION

Part (a). Had we tried to solve the problem without exploiting the balanced conditions, we would have to write a set of four mesh equations or a set of eight node equations. For a balanced three-phase circuit such as the present one, the use of a Y–Δ transformation renders the solution much easier. Applying the transformation of figure 11.22 to the loads and figure 11.23 to the sources yields the equivalent circuit of figure 11.26b, with the following parameters:

Source phase voltage	$\mathbf{V}_{s1Y} = 60\angle -30°$ V (rms),
Source impedance	$Z_{sY} = 0.05 + j0.15 = 0.158\angle 71.56°$ Ω,
Load impedance	$Z_Y = 4 + j3 = 5\angle 36.87°$ Ω,
Feeder impedance	$Z_1 = 0.1 + j0.2 = 0.224\angle 63.43°$ Ω.

With these circuit parameters, figure 11.26b is the same as figure 11.24a, except for the values of the independent voltage sources. Every independent voltage source phasor in

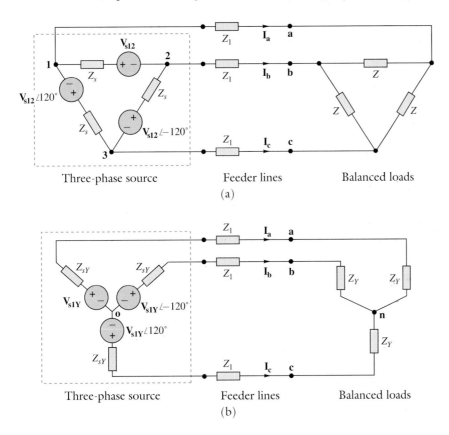

Figure 11.26 (a) A Δ–Δ circuit. (b) Its equivalent Y–Y circuit.

figure 11.26b is the corresponding voltage source phasor in figure 11.24a multiplied by the complex constant

$$\frac{\mathbf{V}_{s1Y} \text{ (figure 11.26b)}}{\mathbf{V}_{s1} \text{ (figure 11.24a)}} = \frac{60\angle -30°}{120\angle 0°} = 0.5\angle -30°.$$

According to the linearity theorem discussed in section 2 of chapter 5, every current (voltage) in figure 11.26b is the corresponding current (voltage) multiplied by the same constant. Therefore, we can use the results of example 11.9 and directly write down the answers for figure 11.26b: $\mathbf{I}_a = 11.25\angle -68.93°$ A, $\mathbf{I}_b = 11.25\angle 171.1°$ A, and $\mathbf{I}_c = 11.25\angle 51.1°$ A. Since the feeder lines are outside of the replaced equivalent circuits, these solutions of the line currents are also those for figure 11.26a.

Part (b). The magnitude of the voltage across each load Z in figure 11.26a is equal to the line-to-line voltage. For the same reasons as stated in part (a), the voltage is equal to $0.5 \times 194.8 = 97.4$ V.

Part (c). The circuits of figures 11.26b and 11.24a have the same impedances and, hence, the same power factors. According to the average power equation 11.11a, the powers in the two circuits differ by a factor of $0.5^2 = 0.25$. Therefore, the total power delivered to the loads of figure 11.26a is $0.25 \times 6,075 = 1,518.75$ W.

Exercise. Suppose the Δ-connected load in figure 11.26a is replaced by a Y-connected load with each $Z_Y = 3 + j4$ Ω. The resultant circuit is called a **Δ–Y circuit**. Find the line-to-line voltages $\mathbf{V}_{ab}$, $\mathbf{V}_{bc}$, and $\mathbf{V}_{ca}$.

All of the previous examples have loads that are linear and balanced. If the loads are linear but unbalanced, then either a node analysis or a mesh analysis is required. In this case, the analysis must be performed on the entire circuit. One can no longer obtain the answers to one phase from those of another simply by adding a 120° phase shift. Several problems at the end of the chapter illustrate this point. For unbalanced three-phase systems, the analysis may also be done by a special computational technique called "symmetrical components," a topic studied in power engineering courses.

When rotating machines are a major part of the load, the circuit no longer belongs to the class of linear circuits studied in this text. A motor is not characterized by some fixed impedance. However, if the motor is known to be operating under some specific conditions (for example, a rated voltage, a rated power, and a typical power factor), then its behavior can be modeled approximately by a linear circuit, and the analysis may be done by the methods described in this section. Several problems at the end of the chapter belong to this category. The study of more accurate models for ac machinery is beyond the scope of the text.

SUMMARY

Fundamental to the material of this chapter is the definition of the effective value (rms value) of a periodic voltage or current waveform. For a sine wave, the effective value is the maximum value divided by $\sqrt{2}$. For a general periodic voltage or current, the effective value is the value of a dc waveform that will produce the same amount of heat as the periodic waveform when both are applied to the same resistance. Using the definition of the effective value of a waveform, we set forth and derived formulas for the average power absorbed by a linear two-terminal network in the ac steady state. Recall that for a two-terminal element with a sinusoidal voltage $v(t) = \sqrt{2}V_{eff}\cos(\omega t + \theta_v)$ and current $i(t) = \sqrt{2}V_{eff}\cos(\omega t + \theta_i)$, the absorbed average power is $P = V_{eff}I_{eff}\cos(\theta_v - \theta_i)$, assuming the passive sign convention. Next, we presented the definitions of reactive power and complex power and calculated these powers using the phasors introduced in chapter 10.

Again, for a two-terminal element with a sinusoidal voltage $v(t)$ and current $i(t)$ as before, the reactive power absorbed is defined to be $Q = V_{eff}I_{eff}\sin(\theta_v - \theta_i)$vars (volt-ampere reactive). We then introduced the notion of the power factor, which is the ratio of average power to apparent power. The value of the power factor lies between 0 and 1. The need for improving a low power factor and one method for achieving this are illustrated in detail with an example.

On the theoretical side, we proved the principle of conservation of complex power from KCL and KVL. The conservation of average power and reactive power are just special cases of this general principle. It is instructive to point out that among the three laws—KCL, KVL, and conservation of average power (or instantaneous power)—any two can be used as a basis for deriving the third. In fact, some texts have derived KVL from KCL and the conservation of instantaneous power.

But when it comes to reactive power and complex power, such an approach encounters some difficulty. Unlike the principle of conservation of instantaneous power, of which nature provides abundant examples to reinforce our intuition, the definition of reactive power, although mathematically rigorous, is not amenable to an intuitive explanation. Since reactive power is not associated with real work (i.e., generating heat, running a fan, etc.), it is unclear why it obeys the conservation law. For this reason, we have used KCL and KVL to derive the principle of conservation of power in its various forms.

The maximum power transfer theorem, first studied in chapter 5 for the resistive network case, is taken up again in this chapter for the sinusoidal steady-state case. As pointed out earlier, the theorem has no application in electric power systems. However, for communication circuits, it is of extreme importance. The power that can be extracted from the antenna of a radio receiver is usually very small (in the microwatts range). It is then necessary to get as much power as possible from the antenna system. A specific example illustrates this principle, with the design of the circuit for maximum power transfer left to volume 2.

The economic aspects of three-phase vs. single-phase ac systems for the transmission of bulk power over long distances are explained in detail. The last section describes several methods for the analysis of balanced three-phase circuits containing only static loads. The analysis of three-phase circuits containing rotating machines is too advanced for inclusion here. Residential customers are typically supplied by a single-phase three-wire system. Some practical aspects of this system are discussed in the problems at the end of the chapter.

TERMS AND CONCEPTS

Apparent power: for a two-terminal element, the quantity $V_{eff}I_{eff}$, assuming that a passive sign convention is used. The unit is the VA (volt-ampere).

Average power: the average value of the instantaneous power. For a two-terminal element with sinusoidal voltage $v(t) = \sqrt{2}V_{eff}\cos(\omega t + \theta_v)$ and current $i(t) = \sqrt{2}V_{eff}\cos(\omega t + \theta_i)$, the absorbed average power is $P = V_{eff}I_{eff}\cos(\theta_v - \theta_i)$, assuming that a passive sign convention is used.

Balanced three-phase circuit: a three-phase source connected to loads that are equivalent to three identical impedances connected in a Y or Δ arrangement.

Complex power: for a two-terminal element absorbing average power P and reactive power Q, the quantity $S = P + jQ$. The unit of measurement of complex power is the VA (volt-ampere). The magnitude of S is the apparent power.

Conservation of power: for any network, the summation of the instantaneous powers absorbed by all elements is zero. For any linear network in the sinusoidal steady state, the summation of the average powers or reactive powers or complex powers absorbed by all elements is zero. This property is a consequence of KCL and KVL.

Δ-connection: three two-terminal networks connected in the form of the letter Δ. Also called a π-network in circuit theory.

Effective value (rms value): for a sine wave, the maximum value divided by $\sqrt{2}$. For a general periodic voltage or current, the value of a dc waveform that will produce the same amount of heat as the periodic waveform when both are applied to the same resistance.

Equivalent n-terminal networks: two n-terminal networks that have the same relationships among all terminal voltages and currents.

Ground wire: the conductor that has one end (or both ends) grounded for safety reasons. The ground wire does not carry a load current under normal operation.

Grounding: in power systems, connecting a certain point of the system to earth through a buried water pipe or grounding rod.

Instantaneous power: the power associated with a circuit element as a function of time. The instantaneous power absorbed by a two-terminal element is $p(t) = v(t)i(t)$, assuming that a passive sign convention is used.

Line-to-line voltage: the voltage between any two hot (ungrounded) conductors of a three-phase system or a single-phase three-wire system.

Maximum power transfer theorem: if a variable load $Z_L = R_L + jX_L$ is connected to a fixed source $\mathbf{V}_s$ having a source impedance $Z_s = R_s + jX_s$, then the largest average power is transferred to the load when Z_L is the complex conjugate of Z_s, i.e., when $R_L = R_s$ and $X_L = -X_s$.

Neutral conductor (wire): the wire connected to the neutral terminal. The neutral conductor carries a load current under normal operation.

Neutral terminal: the point common to all voltage sources in a three-phase four-wire system or a single-phase three-wire system.

Power factor: the ratio of average power to apparent power. The value of the power factor lies between 0 and 1. For a passive load, the power factor is said to be lagging when $90° > \theta_v - \theta_i > 0$ and leading when $90° > \theta_i - \theta_v > 0$.

Reactive power: for a two-terminal element with a sinusoidal voltage $v(t) = \sqrt{2}V_{eff}\cos(\omega t + \theta_v)$ and current $i(t) = \sqrt{2}I_{eff}\cos(\omega t + \theta_i)$, the quantity $Q = V_{eff}I_{eff}\sin(\theta_v - \theta_i)$, assuming that a passive sign convention is used. The unit of measurement of reactive power is the var (volt-ampere reactive).

Real power: in ac circuits, the average power. It is the real part of the complex power.

Single-phase three-wire system: two in-phase ac voltage sources of E volts connected in series with three accessible terminals such that both E and $2E$ volts are available.

Three-phase four-wire system: a three-phase source connected in a Y arrangement with the common point (called neutral) brought out to external circuits.

Three-phase source: three voltage sources $V_m \cos(\omega t)$, $V_m \cos(\omega t - 120°)$, and $V_m \cos(\omega t - 240°)$ connected in the form of a Y or Δ.

Three-phase three-wire system: a three-phase source connected in either a Δ or a Y arrangement with the common point inaccessible for external connection.

Y-connection: three two-terminal networks connected in the form of the letter Y. Also called a T-network in circuit theory.

Y–Δ transformation: the equivalence between two three-terminal networks, one with internal elements connected in a Y arrangement and the other with internal elements connected in a Δ arrangement. Conversion formulas between the two are given in problem 26.

PROBLEMS

1. In the circuit shown in figure P11.1, $\omega = 3$ rad/sec, and the source voltage is a maximum value. Determine the steady-state instantaneous power.

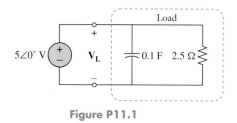

Figure P11.1

2. Determine the average power delivered by the source in the circuit shown in figure P11.2.

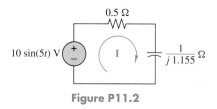

Figure P11.2

3. In the circuit shown in figure P11.3, $\omega = 64$ rad/sec. The magnitude of the source voltage is the effective (rms) value. Determine the average power delivered to the load.

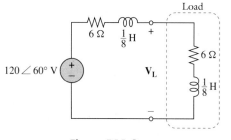

Figure P11.3

4. Determine the average power (in watts) delivered to the load in the circuit shown in figure P11.4.

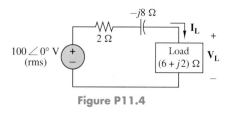

Figure P11.4

5. A coil is modeled by a series connection of L and R. When connected to a 110-V, 60-Hz source, the coil absorbs 300 W of average power. If an 8-Ω resistor is connected in series with the coil, and the combination is connected to a 220-V, 60-Hz source, the coil also absorbs 300 W of average power. Find L and R.

6. Consider figure P11.6. The inductance L is adjusted such that $V_{coil} = 150$ V and $P_{coil} = 250$ W. Find the values of R and L.

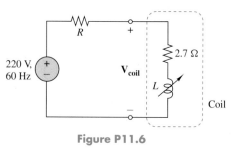

Figure P11.6

7. Determine the effective value of the periodic signal $i(t)$ sketched in figure P11.7.

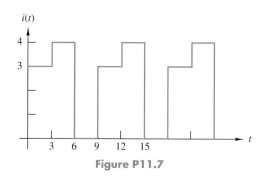

Figure P11.7

8. Find the effective value (rms) of the periodic signal $v(t)$ shown in figure P11.8. The period is 3 sec.

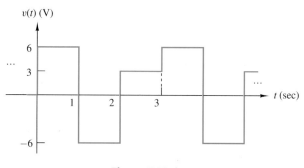

Figure P11.8

9. Prove that the effective value of the periodic function

$$f(t) = F_o + \sqrt{2}F_1 \cos(\omega_1 t + \theta_1)$$
$$+ \sqrt{2}F_2 \cos(\omega_2 t + \theta_2) + \dots$$

is

$$F_{eff} = \sqrt{F_0^2 + F_1^2 + F_2^2 + \dots},$$

provided that $0 \neq \omega_1 \neq \omega_2 \neq \dots$.

10. The circuit shown in figure P11.10 is operating in the sinusoidal steady state. Device 1 absorbs 360 W with a power factor equal to unity. Device 2 absorbs 1,440 W with a power factor of 0.8 lagging. Find the value of the capacitor C so that the magnitude of the source current equals 15 A (rms).

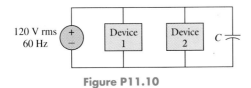

Figure P11.10

11. A group of induction motors is drawing 7 kW from a 240-V power line. The power factor is measured and found to be 0.65.

    (a) What kVA capacitor bank is needed to raise the power factor to 0.8?

    (b) Determine the annual saving resulting from installing the capacitor bank if a demand charge (besides the charge for the kilowatt-hours used) is $20.00 per kVA per month.

12. Consider example 11.1 again. The magnitude of the current is 1.2 A for the fluorescent light and 12 A for the air conditioner. Each lamp draws $100/110 = 0.909$ A. An electrician estimates the maximum number of lamps that can be used without blowing the 20 A fuse by the simple formula $(20 - 12 - 1.2)/0.909 = 7.48$ and concludes that connecting eight lamps to the line will blow the fuse. Is the electrician's conclusion correct? Why or why not?

13. Let the voltage across a capacitance C be $v(t) = V_m \sin(\omega t)$ V.

    (a) Find $p(t)$, the instantaneous power delivered to the capacitance, and show that $p(t)$ has a peak value of $0.5\omega C(V_m)^2$ W and an average value of 0.

    (b) Find $W_C(t)$, the instantaneous energy stored in C (or rather, in the electric field around C), and show that $W_C(t)$ has a peak value of $0.5C(V_m)^2$ joules and an average value of $0.25C(V_m)^2$ joules.

    (c) Let Q_C be the reactive power absorbed by C. Show that

$$W_{C,\text{ave}} = -\frac{Q_C}{2\omega}.$$

14. Let the current flowing through an inductance L be $i(t) = I_m \sin(\omega t)$ A.

    (a) Find $p(t)$, the instantaneous power delivered to the inductance, and show that $p(t)$ has a peak value of $0.5\omega L(I_m)^2$ W and an average value of 0.

    (b) Find $W_L(t)$, the instantaneous energy stored in L (or rather, in the magnetic field around L), and show that $W_L(t)$ has a peak value of $0.5L(I_m)^2$ joules and an average value of $0.25L(I_m)^2$ joules.

    (c) Let Q_L be the reactive power absorbed by L. Show that

$$W_{L,\text{ave}} = \frac{Q_L}{2\omega}.$$

15. The series RLC circuit shown in figure P11.15 has reached the steady state.

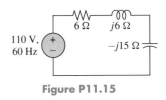

Figure P11.15

    (a) Find the stored energy W_L and W_C at the moment when the terminal voltage of the source is zero.

    (b) Find W_L at the moment when $W_C = 0$.

    (c) Find W_C at the moment when $W_L = 0$.

16. Consider figure P11.16. In the steady state, the voltages across the capacitors have been found to be (rms values)

$$\mathbf{V}_1 = (10 + j2) \text{ V}, \quad \mathbf{V}_2 = (2 + j2) \text{ V}.$$

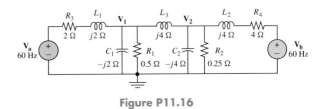

Figure P11.16

    (a) Find $\mathbf{V}_a$ and $\mathbf{V}_b$. Express the answers in polar and rectangular forms.

    (b) Find the real power absorbed by each of the 11 network elements.

    (c) Find the reactive power absorbed by each of the 11 network elements.

    (d) Use the results of parts (b) and (c) to verify the conservation of real power and reactive power.

17. In the circuit shown in figure P11.17,

$$Z_1 = (0.1 + j0.1) \ \Omega, \quad Z_2 = (0.4 + j2.2) \ \Omega,$$

$$Z_3 = (0.2 + j0.2) \ \Omega,$$

$$\mathbf{V}_a = (104 + j50) \text{ V}, \quad \mathbf{V}_b = (106 + j48) \text{ V}.$$

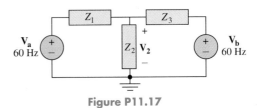

Figure P11.17

(a) Find the voltage $\mathbf{V_2}$ in polar and rectangular forms.
(b) Find the complex power absorbed by each of the five network elements.
(c) Using the results of part (b), verify the conservation of complex power.

18. Repeat example 11.5 with the capacitor bank C removed.

◆ 19. Consider figure P11.19. R_L is adjusted to achieve different goals.

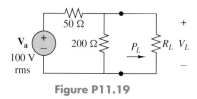

Figure P11.19

(a) Find the value of R_L that maximizes P_L. What is the value of $P_{L,\max}$?
(b) Find the value of R_L that maximizes V_L. What is the value of $V_{L,\max}$?

20. Consider figure P11.20. Determine the value of the load Z_L for maximum average power transfer to the load.

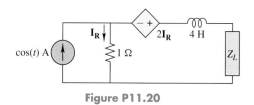

Figure P11.20

◆ 21. Choose the proper value of C to deliver maximum average power to the load in the circuit shown in figure P11.21 at $\omega = 10$ rad/sec.

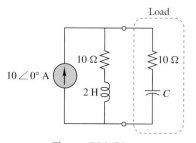

Figure P11.21

◆ 22. The circuit shown in figure P11.22 is driven at $\omega = 60$ rad/sec. What value of R should be chosen so that maximum power is delivered to the load? (*Note*: It is the source resistor here, not the load resistance, that is being varied.)

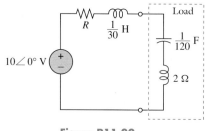

Figure P11.22

23. In the three-phase circuit shown in figure P11.23, loads are balanced and $Z = 30 + j15\ \Omega$.

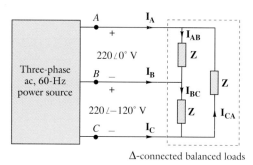

Figure P11.23 A balanced three-phase system with Δ-connected loads.

(a) Find the currents $\mathbf{I_{AB}}$, $\mathbf{I_{BC}}$, $\mathbf{I_{CA}}$, $\mathbf{I_A}$, $\mathbf{I_B}$, and $\mathbf{I_C}$. Exploit the symmetry of a balanced three-phase system fully to save computational effort. The phasor diagrams of figure 11.17 are particularly useful here.
(b) Find the total average power and reactive power delivered by the source.
(c) Repeat parts (a) and (b) if the load connected between lines A and C is changed to $20 + j20\ \Omega$. The loads are no longer balanced.

24. In the three-phase circuit shown in figure P11.24, loads are balanced and $Z = 10 + j5\ \Omega$.

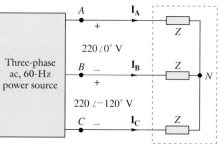

Figure P11.24 A balanced three-phase system with Y-connected loads.

(a) Exploit the symmetry of a balanced three-phase system fully to construct a phasor diagram that displays the relationships among the voltages $\mathbf{V_{AN}}$, $\mathbf{V_{BN}}$, $\mathbf{V_{CN}}$, $\mathbf{V_{AB}}$, $\mathbf{V_{BC}}$, and $\mathbf{V_{CA}}$.
(b) Find the currents $\mathbf{I_A}$, $\mathbf{I_B}$, and $\mathbf{I_C}$.
(c) Find the total average power and reactive power delivered by the source.

25. Prove the equivalence of the two three-terminal networks shown in figure 11.22. The transformation, a special case of the general Y–Δ transformation, is useful in the analysis of balanced three-phase circuits. (*Hint*: Write node equations with node 3 as the reference node. Eliminate any

node voltages other than $\mathbf{V_1}$ and $\mathbf{V_2}$. Show that the two networks have the same set of equations when the impedances satisfy the relationship $Z_\Delta = 3Z_Y$.)

26. Prove that the two three-terminal networks shown in figure P11.26 are equivalent, provided that the impedances are related by the given formulas. The relationship is called a Y–Δ or T–π **transformation.** Observe the cyclic pattern of the subscripts in these formulas: One formula may be obtained from another by the substitutions $1 \rightarrow 2$, $2 \rightarrow 3$, and $3 \rightarrow 1$. When extended to n-terminal networks, the transformation is called a *star-mesh transformation*, which is discussed in volume 2. (*Hint*: Same as that for problem 25.)

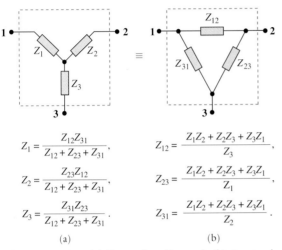

$$Z_1 = \frac{Z_{12}Z_{31}}{Z_{12} + Z_{23} + Z_{31}},$$

$$Z_2 = \frac{Z_{23}Z_{12}}{Z_{12} + Z_{23} + Z_{31}},$$

$$Z_3 = \frac{Z_{31}Z_{23}}{Z_{12} + Z_{23} + Z_{31}}.$$

(a)

$$Z_{12} = \frac{Z_1Z_2 + Z_2Z_3 + Z_3Z_1}{Z_3},$$

$$Z_{23} = \frac{Z_1Z_2 + Z_2Z_3 + Z_3Z_1}{Z_1},$$

$$Z_{31} = \frac{Z_1Z_2 + Z_2Z_3 + Z_3Z_1}{Z_2}.$$

(b)

Figure P11.26 (a) Y-network or T-network. (b) Δ-network or π-network.

27. Prove the equivalence of the two three-terminal networks shown in figure 11.23. These transformations are useful in the analysis of balanced three-phase circuits. (*Hint*: Write node equations with node 3 as the reference node. The right-hand side of the equations contains the independent voltage sources. Eliminate any node voltages on the left-hand side other than $\mathbf{V_1}$ and $\mathbf{V_2}$. Show that the two networks have the same set of equations when the impedances and voltage sources satisfy the stated relationships.)

28. Consider the three-phase circuit of example 11.9. With the feeder impedance $Z_1 = 0.1 + j0.2\ \Omega$, the voltage across each load impedance Z has been found to be 112.5 V, showing a 6.25% drop from the source voltage of 120 V. If the load is connected directly to the source, what is the percent of the voltage drop? (*Hint*: Repeat the calculation of V_{an} with $Z_1 = 0$.)

29. Consider problem 28 again. Assume that the load cannot be moved to the source location and that a feeder cable must be used. One can reduce the voltage drop by using a cable with a smaller impedance. Suppose it is desired to reduce the drop to 2% by using a feeder with a smaller impedance $Z_1 = \alpha(0.1 + j0.2)\ \Omega$. Determine the value of α.

30. A 300-hp, 2,300-V, 60-Hz three-phase motor is running at its rated load with a power factor of 0.88 lagging and an efficiency of 93.5%.
   (a) Find the total average power absorbed by the motor. Recall that 1 hp (horsepower) = 746 W.
   (b) Find the magnitude of the line current. You may assume that the three windings of the motor are connected in a Y or a Δ arrangement. The same answer for the line current should be obtained in either case.

31. In problem 30, three capacitors of C farads each may be connected across the three pairs of terminals of the motor to raise the power factor. If it is desired to raise the power factor to 0.95 lagging, what should the value of C be?

32. Consider the three-phase system of figure 11.24a, with the impedance in phase a changed from $Z = (4 + j3)\ \Omega$ to $Z = (5 + j2)\ \Omega$. The system is unbalanced, and hence, the shortcuts described in example 11.9 are not applicable. Repeat parts a through d of example 11.9. (*Hint*: First find $\mathbf{V_{no}}$ by applying KCL to node n.)

33. Consider the three-phase system of figure 11.26a, with the load connected between lines a and b changed from $Z = (12 + j9)\ \Omega$ to $Z = (15 + j6)\ \Omega$. The system is unbalanced, and hence, the shortcuts described in example 11.9 are not applicable. Repeat parts a through c of example 11.11. (*Hint*: First apply the general Δ–Y transformation of problem 26 to the unbalanced load and the special Δ–Y transformation of figure 11.23 to the source. The equivalent system, now Y–Y connected, may be solved by the same procedure used in problem 32.)

Note: Problems 34–37 explore grounding aspects of the single-phase three-wire ac power supply in residential buildings and the importance of grounding for safety, a topic rarely discussed in texts on circuits. We investigate the safety aspect by estimating the voltage between any selected point in the circuit and the ground. Some background information, necessary for this discussion, is presented next, before stating the problems.

1. In circuit analysis, the term "ground" usually means a reference node for measuring voltages, such as in nodal analysis. The ground node may or may not be connected to the earth. In automotive electric circuits, "ground" refers to the metal frame of the car, not the earth. In power systems, however, a ground always is a true ground. The point to be grounded must be connected to a ground rod driven into the earth or connected to a buried pipe of a water system, according to the electrical codes. Although the earth conducts electric charges, it is not a perfect conductor: Between any two grounded points in a system, there is some impedance to the ac current flow. This impedance plays a very important role in determining the noise that creeps into sensitive measuring instruments. However, for the purpose of investigating the safety of human operators of equipment, the effect of the earth's impedance is usually neglected. Therefore, in the problems that follow, we will assume that there is a zero voltage drop between any two grounded points.

2. A human body is also a conductor. When a person standing on ground touches a high-voltage point P, some

current flows through the body. The magnitude of the current determines the degree of shock, very roughly as follows:

Less than 1 mA	barely perceptible
1 to 8 mA	strong surprise
8 to 15 mA	unpleasant, victim able to detach
Greater than 15 mA	muscular freeze, victim cannot let go
Greater than 75 mA	usually fatal

In addition, victims may suffer injuries by being thrown against furniture or other fixtures by their own reflexes. Suppose the voltage between a point P and the ground is E volts (rms). The current flowing through the body is determined by E and the total impedance in the circuit. The latter depends on many factors, including whether the person is standing on a dry surface, a damp surface, or in water and whether the person is wearing rubber shoes or has bare feet. Anyone trying to operate a piece of 115-V electrical equipment while in a bath is certainly risking his or her life.

34. Figure P11.34 shows a typical single-phase three-wire power supply found in a residential building. We should not be misled and think that any three-wire ac power supply is a three-phase system! The 115-V branch circuits are for lighting and small appliances. The 115/230-V branch circuits are for heavy appliances, such as an electric range, a large air conditioner, etc. An electric range requires 230 V for its heating elements and 115 V for a range-top clock and lights.

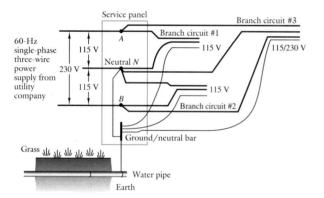

Figure P11.34 A typical single-phase three-wire system in a residential building.

(a) Branch circuit #1 employs type NM 14/3 cable (nonmetallic sheath, #14 gauge, three wires) and is protected by a 15-A fuse or circuit breaker in the hot wire. The neutral wire is never "fused" (interrupted by any device). The circuit supplies electricity to lights and receptacles in a room located 40 feet from the service panel. If the total load current on this circuit is 10 A, determine whether the cable is of adequate size so that the voltage at the light is within 2% of the supply voltage of 115 V. The resistance of copper wire at 60 Hz and 25° C is as follows: #12 gauge, 1.619 Ω per 1,000 feet, 20-A current-carrying capacity; #14 gauge, 2.575 Ω per 1,000 feet, 15-A capacity.

(b) Branch circuit #2 employs type NM 12/3 cable (nonmetallic sheath, #12 gauge, three wires) and is protected by a 20-A fuse or circuit breaker. The circuit supplies electricity to a larger room 50 feet from the service panel. If the total load current on this circuit is 14 A, determine whether the cable is of adequate size to ensure a drop of less than 2% from the supply voltage of 115 V. Use the data on the resistance of copper wire given in part (a).

(c) In parts (a) and (b), under normal conditions (i.e., there are no defective devices), does the ground wire (usually bare, and of a smaller size than the other wires) carry any current? Compare the magnitudes of the currents carried by the hot wire (insulated and colored black or red) and the neutral wire (insulated and colored white).

(d) Construct a phasor diagram to show the voltages V_{AN}, V_{BN}, and V_{AB}, arbitrarily taking the phase angle of V_{AN} as zero.

35. A portable electric drill, rated at 115 V, 3 A, is plugged into a receptacle supplied by branch circuit #1 of figure P11.34. The circuit model of figure P11.35, although very crude, is adequate for the purpose of investigating the safety problem. Suppose the plug is not fully in, and a person's finger comes in touch with some metal part. Determine the voltage with respect to ground that the person is subjected to when the metal part is:

(a) Blade A.
(b) Blade N.
(c) Blade G (round in shape).
(d) Metal case of the drill.

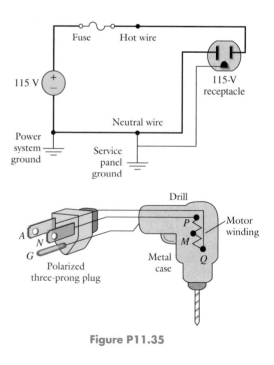

Figure P11.35

36. Refer to figure P11.35 again. The insulation of the motor winding inside the metal case may fail due to overheating or extended usage, among other reasons. The exposed copper wire then touches the metal case and becomes

"grounded." If a person operates such a defective drill, what voltage does he or she experience if the defective point is:

(a) P, one terminal of the winding, which is connected to neutral.

(b) M, the middle point of the motor.

(c) Q, the terminal of the winding that is connected to the hot wire.

Note that the fuse should blow in this case; the problem demonstrates clearly that a properly grounded power tool provides safety to the operator.

37. Refer to the circuit of figure P11.35 again. Consider what would happen if a sloppy electrician (or, more likely, a do-it-yourselfer) forgot to connect the round metal hole of the 115-V receptacle to the ground wire that comes with a three-wire cable. Find the voltage between the metal case and ground if the insulation fails at (a) point Q, (b) point M, and (c) point P. Will the fuse blow in any of these scenarios? This problem demonstrates clearly the danger of operating power equipment without proper grounding.

CHAPTER 12

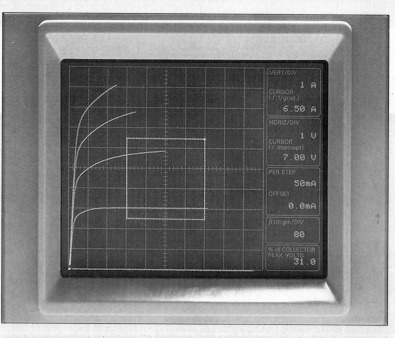

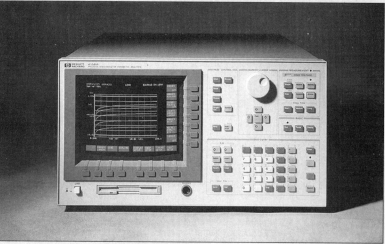

Piecewise Linear Analysis of Transistor Amplifier Circuits

HISTORICAL NOTE

WHILE linear circuit analysis is of fundamental importance, real-world devices are often nonlinear. Many useful and interesting properties of practical circuits cannot be satisfactorily explained using linear circuit theory. One remedy would be to initiate a study of nonlinear circuits early. Pedagogically, however, this is difficult to do, because nonlinear problems are inherently much more difficult than linear problems.

Fortunately, a compromise is possible. The diode and other nonlinear resistors are characterized by single v-i curves. A smooth v-i curve can be approximated by a set of straight-line segments. The process is called piecewise linear (PWL) approximation. Of course, there will always be an error between the approximation and the actual v-i curve. Nevertheless, this error can be reduced by using more line segments in the approximation. When the nonlinear resistor operates on a particular segment, the v-i relationship can be modeled with a linear resistor and independent sources. The resultant linear circuit is then amenable to all the solution methods developed for linear circuits. The process is repeated for each segment of the operation. Further, the technique can be extended to multiterminal devices, such as transistors.

The PWL method offers a means to understand a circuit's operation without resorting to advanced mathematics. The approach played an important role in the early study of electronic circuits. In fact, in the 1960s, several circuit texts appeared based entirely on the piecewise linear approach.

Today, software programs are widely available to simulate nonlinear circuits, but the PWL approach is necessary so that designers can continually test the results of such programs for reasonableness. Thus, the PWL method is a necessary tool for intelligent design. Indeed, many electronics texts implicitly use the method to discuss the transistor circuit operating-point problem. Further, in the area of control systems design, the PWL method underlies the so-called gain-scheduling method. ◆

CHAPTER OBJECTIVES

- Describe the different ways of characterizing a three-terminal resistive device by its terminal voltage-current relationships.
- Use simple two-terminal resistor-diode-source circuit models of basic two-segment v-i curves as the building blocks for circuit models of three-terminal resistive devices.
- Describe and illustrate the construction of piecewise linear circuit models for bipolar junction transistors from their input and output characteristics.
- Learn how to analyze a piecewise linear circuit by the elementary method of "assumed diode states."
- Understand the significance of an ac equivalent circuit and the origin of a controlled source through the analysis of a complete single-stage transistor amplifier.

1. INTRODUCTION

In this age of electronics and computers, it is safe to assume that the ordinary reader of this text has had some experience with audio equipment. In a turntable setup, a needle running through the groove of a record produces a weak voltage by electromagnetic induction. The weak signal is amplified many times by an electronic circuit whose output signal drives the speakers to reproduce the music. From common experience, one knows that if the volume control is set too high (the amplification is too high), the sound that comes out of the device becomes distorted and unpleasant to hear.

How is signal amplification achieved? As early as chapter 1, we showed how controlled sources might produce voltage gain and power gain. Unlike a resistor, a controlled source is not a simple component that one can pick off the shelf of an electronic parts house. Rather, it is constructed from a semiconductor three-terminal device and requires a dc power supply for proper operation.

This chapter investigates the representation of three-terminal semiconductor (resistive) devices and the techniques for analyzing circuits containing such devices. A deep understanding of these topics requires a background in device physics and the solution of nonlinear equations. Fortunately, we are able to present the basic principles at a more elementary level and produce useful results, even though our knowledge of the topic is incomplete. We achieve this by the following means:

1. Using the *measured* terminal voltage-current relationships of the device as a starting point for understanding the operation of the circuit. The study of device physics may then be postponed to a later time.
2. Representing the measured characteristic curves of the device by approximate piecewise linear (PWL) curves. Accuracy is sacrificed to some extent, but we need not rely on nonlinear circuit theory; all resulting calculations take place on linear circuits, using the methods set forth in earlier chapters.

By the end of the chapter, the reader will be able answer the question posed in the second paragraph of this section.

2. CHARACTERIZATION AND MODELS OF THREE-TERMINAL RESISTIVE DEVICES

The goals of this section are to characterize three-terminal resistive devices by their terminal voltage-current relationships and apply these characterizations to practical semiconductor devices in the next section.

Figure 12.1a shows a general three-terminal device with standard voltage and current labelings. Although there are three terminal currents, only two need to be considered explicitly, because KCL constrains the third, i.e., $i_1 + i_2 + i_3 = 0$. Similarly, among the three terminal voltage differences, only two need be explicit, as KVL, yielding $v_{12} = v_{13} - v_{23}$, constrains the third. Figure 12.1b shows a common choice of the voltage and current variables for inclusion in the characterizing circuit equations.

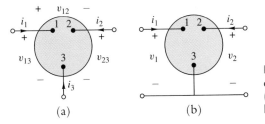

Figure 12.1 (a) Labeling of voltages and currents for a three-terminal device. (b) A common circuit configuration and labeling.

Recall that a two-terminal resistive device is completely specified by one equation or one v-i curve. A three-terminal resistive device, on the other hand, requires two equations to describe the terminal voltage-current relationships. Among the four necessary variables (i_1, i_2, v_1, v_2), two may be selected as *dependent* variables and expressed in terms of the remaining two *independent* variables. There are six possible representations, two of which have the following forms:

1. With v_1 and v_2 as independent variables, one has

$$i_1 = y_1(v_1, v_2), \tag{12.1a}$$

$$i_2 = y_2(v_1, v_2). \tag{12.1b}$$

2. With i_1 and v_2 as independent variables, one has

$$v_1 = h_1(i_1, v_2), \tag{12.2a}$$

$$i_2 = h_2(i_1, v_2). \tag{12.2b}$$

In both forms, y_i and h_i, $i = 1, 2$, are appropriate functions.

Exercise. Write down the remaining four sets of fundamental representations for a three-terminal resistive device.

Each of the preceding equations contains one dependent variable and two independent variables. In some cases, specific *mathematical* relationships are possible, e.g.,

$$i_1 = v_1 + v_2|v_2|, \tag{12.3a}$$

$$i_2 = 2v_1 + v_2^3. \tag{12.3b}$$

In other cases, *visual* or tabular displays are necessary. Strictly speaking, a visual display of a relationship such as equation 12.3a requires a three-dimensional graph having three axes, one each for i_1, v_1, and v_2, respectively. Such three-dimensional pictures are inconvenient to show on the printed page. An alternative and popular *graphical* method is to choose one of the two independent variables as a *running (or controlling) parameter* and plot a set of curves that relate the remaining two variables, one curve for each important value of the second independent variable. Each equation is then determined by a *family* of curves. As an example, equation 12.3a is plotted in figure 12.2a with v_2 as the running parameter. Figure 12.2b represents equation 12.3b with v_1 as the running variable.

As mentioned earlier, an approximate analysis of nonlinear resistive circuits proceeds by replacing each smooth curve by a set of straight-line segments. The accuracy of the approximation depends on the number of segments. For example, figure 12.2c shows a three-segment approximation to each curve in figure 12.2b. This is precisely the piecewise linear (PWL) approximation method used in chapter 6 for analyzing circuits containing diodes.

Characteristic curves, such as those of figures 12.2a and b, are very useful for *graphically* determining the responses of simple circuits. For an *analytical* solution of more complex circuits, either manually or by digital computer, one must use equations (e.g., equations 12.3) or simulate a circuit model based on the characteristic equations, as is done in the program SPICE.

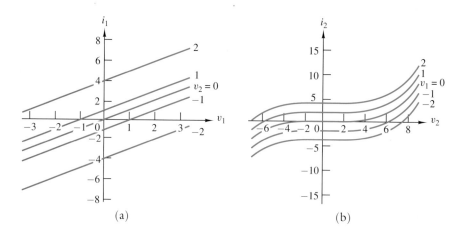

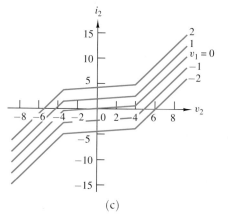

Figure 12.2 (a) v_1-i_1 curves with v_2 as the running parameter. (b) v_2-i_2 curves with v_1 as the running parameter. (c) Piecewise linear approximation to the curves of (b).

If the v-i curve of a two-terminal resistive device is piecewise linear, then a circuit model can be constructed for the device using linear resistors, dc independent sources, and ideal diodes. The construction of the model for a general PWL v-i curve is beyond the scope of this chapter; chapter 6 developed models for some simple cases. Several basic PWL curves and their circuit models appear in table 12.1 for reference purposes. We will use these as building blocks for developing families of curves for three-terminal devices. As an exercise, the reader might verify that the circuit models depicted have the associated v-i characteristic, by utilizing either the graphical method or the "assumed diode states" method, as described in chapter 6.

An examination of the two items in each row of table 12.1 reveals an interesting property: The v-i curve of one item is obtained from the v-i curve of the other item by rotating 180° about the origin of each axis, while the circuit model of one item is obtained from the circuit model of the other item by "flipping over" all the circuit elements. This property holds for any v-i relationship, piecewise linear or not.

Let us now consider the problem of creating a circuit model for a three-terminal resistive device. As shown in figure 12.2, for each such device, two *families of curves* are needed for its characterization. If each curve in the family is related to the other by a horizontal shift or a vertical shift, then the effect may be simulated by adding a properly located controlled source. The controlling variable equals the second independent variable (the running parameter) in figure 12.2. A controlled voltage source will shift the

TABLE 12.1   CIRCUIT MODELS FOR SOME SIMPLE PWL v-i CURVES

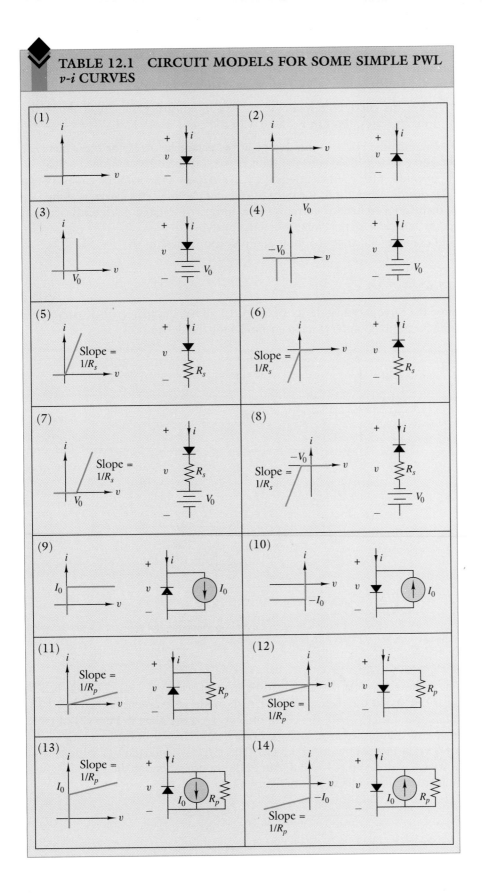

curves horizontally, while a controlled current source will shift the curves vertically. If, for every equal change in the running parameter, the corresponding v-i curves are displaced by equal distances, then only linear controlled sources are needed. For the case of nonuniform displacement between curves, a nonlinear controlled source is required.

Consider a family of PWL curves represented by $i_k = f(v_k, p)$, where p is the running parameter. Assume uniform spacing between the curves for each increment Δp. The procedure for constructing the PWL circuit model is as follows:

Step 1. Select the curve corresponding to $p = 0$; construct the circuit model for $i_k = f(v_k, 0)$. In many cases, this can be done with the aid of table 12.1.

Step 2. Examine the effect of the change in the parameter p. If, for every increment Δp in p, the curve of step 1 shifts to the right by Δv_k, then add a controlled voltage source, properly oriented, with coefficient equal to $\Delta v_k / \Delta p$, *in series with* the circuit obtained in step 1. If, on the other hand, for every increment Δp in p, the curve is shifted upward by Δi_k, then add a controlled current source, properly oriented, with coefficient equal to $\Delta i_k / \Delta p$, *in parallel with* the circuit obtained in step 1.

Several examples illustrate the ideas and procedures described.

◆ EXAMPLE 12.1

Suppose the three-terminal device of figure 12.1b has the characteristic curves shown in figures 12.3a and b. Construct a circuit model for the device.

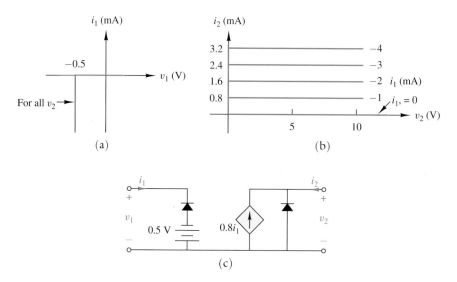

Figure 12.3 (a) v_1-i_1 curves with v_2 as the running parameter. (b) v_2-i_2 curves with i_1 as the running parameter. (c) Piecewise linear circuit model.

SOLUTION

Consider figure 12.3a first. We find it to be item (4) of table 12.1. The left half-portion of figure 12.3c shows the corresponding circuit model. Next, consider figure 12.3b. With the running parameter $i_1 = 0$, the curve of i_2 vs. v_2 is the same as item (2) of table 12.1. Note that i_1 is always negative, and for every 1-mA change in the magnitude of i_1, the curve of i_2 vs. v_2 moves upward by 0.8 mA. This effect is simulated by a current-controlled current source with $\beta = 0.8/1 = 0.8$. After due consideration for the sign of i_1, we obtain the complete circuit model shown in figure 12.3c.

◆ EXAMPLE 12.2

Suppose the three-terminal device of figure 12.1b has the characteristic curves shown in figures 12.4a and b. Construct a circuit model for the device.

SOLUTION

First consider figure 12.4a. With the running parameter $v_2 = 0$, the curve is of the form of item (7) in table 12.1, with $V_0 = 1$ V. From the given coordinates of $P(2$ V, 10 mA),

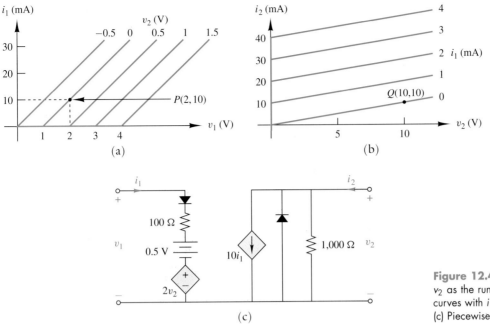

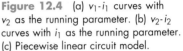

Figure 12.4 (a) v_1-i_1 curves with v_2 as the running parameter. (b) v_2-i_2 curves with i_1 as the running parameter. (c) Piecewise linear circuit model.

the slope is

$$\frac{0.01}{2-1} = 0.01 \text{ mho.}$$

Therefore, the series resistance in item (7) is $R_s = 100 \ \Omega$.

Each time the running parameter v_2 increases by 0.5 V, the characteristic curve moves to the right by 1 V. Putting a voltage-controlled voltage source with $\mu = 1/0.5 = 2$ in series with circuit model (7) accommodates this movement. The left half of the model of figure 12.4c simulates the v-i curve of figure 12.4a.

Next, consider figure 12.4b. With the running parameter $i_1 = 0$, the curve has the form of item (11) in table 12.1. From the given coordinates of $Q(10 \text{ V}, 10 \text{ mA})$, the slope is

$$\frac{0.01}{10} = 0.001 \text{ mho.}$$

Therefore, the parallel resistance needed to complete the specification of item (11) is $R_p = 1 \text{ k}\Omega$.

Each time the running parameter i_1 increases by 1 mA, the characteristic curve moves up by 10 mA. This indicates that a current-controlled current source with $\beta = 10/1 = 10$ must be inserted in parallel with model (11). The situation is depicted in the right half of figure 12.4c. As such figure 12.4c shows the complete model, which simulates the v-i curves of figures 12.4a and b.

3. PIECEWISE LINEAR MODELS OF BIPOLAR JUNCTION TRANSISTORS

The two basic types of transistors are the bipolar and the field-effect transistor. This section considers piecewise linear modeling of the former only. A **bipolar junction transistor (BJT)** is a three-terminal semiconductor device capable of amplifying voltage, current, and power. There are two types of BJTs: the *npn* **type** and the *pnp* **type**, each ordinarily made from silicon or sometimes germanium wafers. Figure 12.5 shows the basic structures and symbols for these two types of BJT. Three distinct conduction regions are created by introducing controlled amounts of certain chemical impurities into the materials of which the BJTs are composed. A region is designated as **N type** when the material is rich in free electrons and **P type** when the material is deficient in free electrons. Figure 12.5 presents an enlarged illustration of the *npn* and *pnp* configurations. The *pnp* and *npn* regions are extremely thin—on the order of several millionths of a meter. A wire is attached to each of these regions. The entire silicon wafer is hermetically sealed in a protective casing, with the three leads brought out for circuit connections. The three terminals are designated as emitter (E), base (B), and collector (C).

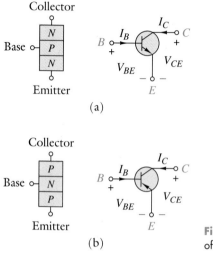

Figure 12.5 (a) *npn* and (b) *pnp* types of BJT and their symbols.

Note that in figure 12.5 the only difference in the two BJT symbols is the direction of the arrow within the circle: *entering* the emitter terminal for the *pnp* type and *leaving* the emitter terminal for the *npn* type. The basic properties of the two types of BJT are the same. The only difference is that *corresponding electrical quantities* have opposite signs. Therefore, when one type of BJT has been investigated in detail, the properties of the other can be inferred easily with appropriate changes of the signs of all electrical quantities. For this reason, in this section we shall concentrate on only the *npn* type of BJT.

At first sight, a BJT may appear to be simply two diodes connected back to back. But due to the extreme thinness of the base region, interesting interactions between the diodes occur. In a more advanced course in semiconductor devices, you will learn about the physical properties of the BJT and the hows and whys of its operation. Our present purpose emphasizes its use as a circuit element having specified terminal voltage-current relationships. With this in mind, our starting point is the device characteristic curves or equations, as described in section 2.

In section 2, *generic* three-terminal devices had terminals labeled by the integers 1, 2, and 3. When considering a specific device, such as the BJT, it is customary to replace the integer subscripts by letter labels. For the BJT, these are as follows:

B or b for base,
E or e for emitter, and
C or c for collector.

Uppercase and lowercase subscripts have distinct meanings. In a typical electronic circuit containing semiconductor devices, there are dc voltage sources (called power supplies), which establish the **quiescent** dc voltages and currents (called operating points) of the devices. There are also time-varying voltages or currents (called signals), which are applied to the circuit for processing. Therefore, the total instantaneous voltage or current is the sum of two terms: a constant and a function of time. The convention in texts on electronics is to use

1. a lowercase variable and uppercase subscript for the total instantaneous quantity, e.g., v_{BC};
2. an uppercase variable and uppercase subscript for the dc quantity, e.g., V_{BC};
3. a lowercase variable and lowercase subscript for the time-varying (usually sinusoidal) portion of the total instantaneous quantity, e.g., v_{bc}; and

4. an uppercase variable and lowercase subscript for the magnitude of the sinusoidal component of the total instantaneous quantity, e.g., V_{bc}.

As a more illustrative example, suppose the applied signal is sinusoidal. Then the base current of a BJT may be expressed, using the foregoing notations, as

$$i_B(t) = I_B + i_b(t),$$

$$i_b(t) = I_b \cos(\omega t + \theta),$$

$$\mathbf{I_b} = I_b e^{j\theta},$$

where

i_B is the total instantaneous value of the response, a function of t;
I_B is the dc operating point, a constant;
i_b is the ac time response, a function of t;
I_b is the magnitude of the sinusoidal component; and
$\mathbf{I_b}$ is the phasor of the sinusoidal component.

Such notations appear complicated, but are necessary for distinguising different quantities and for understanding the related concepts.

Most circuit diagrams follow the convention that the input is drawn on the left and the output on the right. Thus, for the configuration of figure 12.5, the curves of I_B vs. V_{BE} (with either V_{CE} or I_C as the running parameter) specify the **input characteristics**, while the curves of I_C vs. V_{CE} (with either I_B or V_{BE} as the running parameter) specify the **output characteristics**. A **common-emitter configuration** has the emitter terminal common to both input and output terminal pairs. Similarly, a **common-base configuration** has the base terminal common to both input and output terminal pairs. For the common-emitter configuration of figure 12.5, we are explicitly dealing with I_B, I_C, V_{BE}, and V_{CE}.

Figure 12.6 shows the first-quadrant input and output characteristic curves of a typical *npn* type of transistor in the common-emitter configuration with its voltages and currents marked as shown in figure 12.5a. Using a transistor curve tracer, one can experimentally obtain such curves in the laboratory. Such a determination is a stock experiment in undergraduate laboratories dealing with circuits.

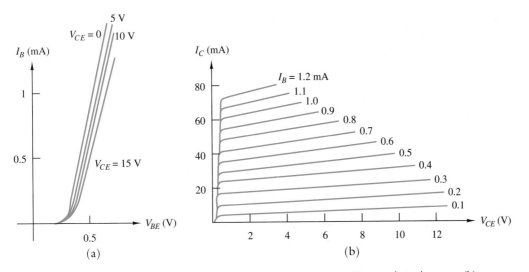

Figure 12.6 (a) Input characteristics of a common-emitter *npn* BJT; note their closeness. (b) Output characteristics.

The typical usage of a common-emitter *npn* BJT does not involve the third quadrant. For practical reasons, we develop a model for the transistor only in the quadrant of actual usage. This greatly simplifies the development.

For an approximate analysis of a transistor circuit, we replace the smooth curves in figure 12.6 by piecewise linear segments. For the input characteristics of figure 12.6a, although there is a family of curves, they are very close together. It is customary to replace this whole family of curves by a single curve through the middle. This approximation greatly simplifies the analysis at the expense of less accurate, but very useful, results. Figure 12.7a shows the two-segment piecewise linear approximation to figure 12.6a. Similarly, figure 12.7b shows the approximation to the output characteristics of figure 12.6b by a family of *uniformly spaced* PWL curves. A more accurate approximation in which all breakpoints lie on a slanted line is given in problem 6.

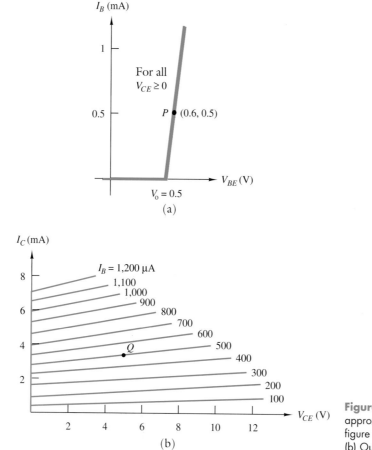

Figure 12.7 Piecewise linear approximation to the curves in figure 12.6. (a) Input characteristics. (b) Output characteristics.

Our next goal is to obtain a circuit model that produces the characteristics of figure 12.7. The PWL input characteristic is the same as item (7) of table 12.1. Therefore, the circuit model consists of a series connection of a dc voltage source V_0, a resistor R_s and an ideal diode. From $V_0 = 0.5$ V and the coordinates of P (0.6 V, 0.5 mA), we calculate

$$R_s = \frac{0.6 - 0.5}{0.5 - 0} = 0.2 \text{ k}\Omega, \text{ or } 200 \text{ }\Omega.$$

In this equation, we understand that the voltage is expressed in V and the current in mA. Consequently, the resistance is expressed in kΩ. A word about the units used in all calculations in this chapter is in order. Practical electronic circuits usually have currents in the mA range, resistances in the kΩ range, and capacitances in the μF or pF range. To avoid the frequent appearances of powers of 10 in the equations, two conventions are widely used in texts on electronic circuits:

1. *The use of an implied consistent set of units.* For example, instead of using the standard units of (V, A, Ω) for (voltage, current, resistance), one can use an alternative consistent set of units, (V, mA, kΩ). Another consistent set of units is (mV, μA, kΩ). The chosen set of units is usually not explicitly stated, but *implied* by the context.

2. *The use of standard units in conjunction with numerical prefixes, e.g., k for 10^3, m for 10^{-3}, and μ for 10^{-6}.* (See the inside cover of this text for a more complete list of prefixes.) In subsequent calculations, both conventions will be used to familiarize students with the common practice.

    CHAPTER 12   PIECEWISE LINEAR ANALYSIS OF TRANSISTOR AMPLIFIER CIRCUITS

The PWL output characteristics have the same form as the curves in figure 12.4b. Therefore, the model is a parallel connection of a resistance, a current-controlled current source, and an ideal diode. Since the curves in figure 12.7b are *uniformly* spaced nearly parallel lines, we can determine the resistance from the slope of the straight line and the controlled source parameter from the spacing in the neighborhood of an arbitrary point Q, as identified in the center of the figure. A calculation of the approximate model parameters yields

$$R_p \cong \frac{5 - 0}{3.33 - 2.75} = 8.620 \text{ k}\Omega,$$

$$\beta = \frac{70}{1.2} = 58.3.$$

Figure 12.8 shows the complete piecewise linear circuit model for approximating the characteristic curves of figure 12.7. Both diodes are assumed ideal.

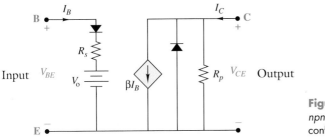

Figure 12.8 PWL model for *npn* transistor in common-emitter configuration.

An alternative BJT configuration is the common-base configuration shown in figure 12.9. By redrawing the model of figure 12.8, one obtains a PWL circuit model for the BJT in the common-base configuration. Figure 12.10 depicts this redrawing.

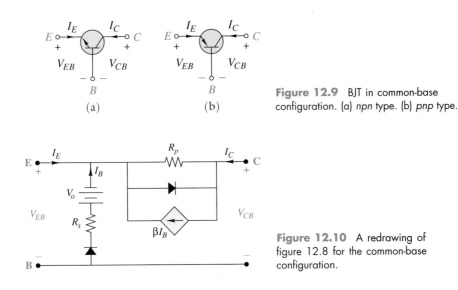

Figure 12.9 BJT in common-base configuration. (a) *npn* type. (b) *pnp* type.

Figure 12.10 A redrawing of figure 12.8 for the common-base configuration.

For a manual analysis, the use of figure 12.10 has two undesirable features:

1. The controlled source is not controlled by the input current i_E.
2. R_s appears in the input and output meshes.

To obtain a model more suitable for manual analysis in the common-base configuration, we first obtain the input and output characteristics of the transistor when used in

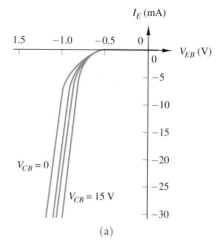

(a)

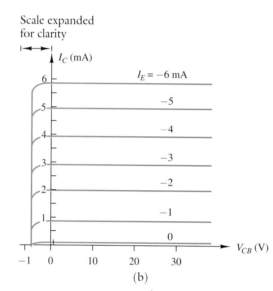

(b)

Figure 12.11 Typical characteristics of an *npn* BJT in the common-base configuration. (a) Input characteristic. (b) Output characteristic.

the common-base configuration—say, by a transistor curve tracer. Figure 12.11 illustrates typical input and output characteristics for the *npn* type of BJT.

An approximate analysis proceeds by replacing the smooth curves in figure 12.11 by piecewise linear segments. Since the input characteristics of figure 12.11a are very close together, a single curve through the middle provides a simple approximation. Figure 12.12 shows the piecewise linear approximation to figure 12.11.

The input characteristic of figure 12.12a is item (8) of table 12.1. From figure 12.12a, we read $-V_0 = -0.7$ V and $R_s = (1 - 0.7)/0.02 = 15$ Ω. From figure 12.12b, and from the readings taken near the point Q, one can estimate that

$$\text{coefficient for controlled source} = \alpha \cong 0.98,$$

$$R_p \cong \infty.$$

The complete PWL circuit model for approximating the characteristic curves of figure 12.12 is shown in figure 12.13, in which the two diodes are assumed ideal.

For a preliminary analysis of a BJT transistor circuit, the recommended models are figure 12.8 for the common-emitter configuration and figure 12.13 for the common-base configuration. For reference purposes, table 12.2 summarizes these PWL circuit models. The next two sections will illustrate the use of the models in an approximate analysis of BJT circuits. A more accurate PWL model that accounts for a nonzero output voltage in the saturation region is given in problem 7.

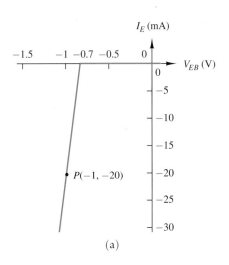

(a)

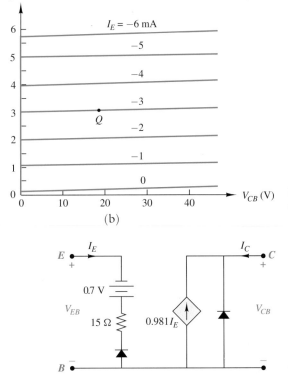

(b)

Figure 12.12 Piecewise linear approximation to figures 12.11(a) and (b).

Figure 12.13 PWL model for *npn* transistor in common-base configuration.

4. PIECEWISE LINEAR ANALYSIS OF RESISTIVE BIPOLAR TRANSISTOR CIRCUITS

In chapters 2 and 3, we illustrated the use of controlled sources to achieve the amplification of an input voltage, current, or power. The previous section showed that a controlled source is a critical part of the model for a multiterminal device. The next few examples demonstrate why a properly connected dc power supply is necessary for proper operation of a multiterminal device such as a bipolar transistor. The building blocks for our analysis are the PWL models listed in table 12.2.

The **operating point** or **quiescent point** of an electronic circuit is the set of constant voltage and current values set up by the dc power supply under the condition that no time-varying input signal is applied. The operating point of a semiconductor device is of particular importance to a circuit designer because it critically affects the performance of the circuit when a signal is applied. A real semiconductor device has an approximately linear voltage-current relationship only over some restricted range of voltages and currents. Improperly placing the operating point reduces the fidelity of amplification and also increases

---

TABLE 12.2 PIECEWISE LINEAR MODELS FOR BIPOLAR JUNCTION TRANSISTORS

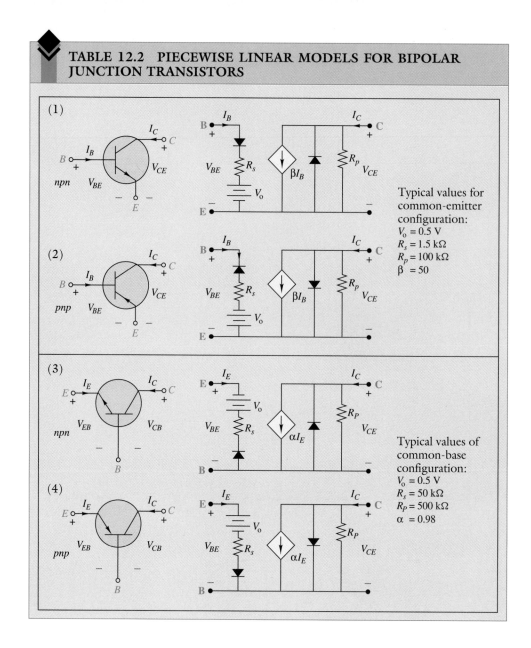

Typical values for common-emitter configuration:
$V_o = 0.5$ V
$R_s = 1.5$ kΩ
$R_p = 100$ kΩ
$\beta = 50$

Typical values of common-base configuration:
$V_o = 0.5$ V
$R_s = 50$ kΩ
$R_p = 500$ kΩ
$\alpha = 0.98$

the potential for *clipping* of the output signal, a severe distortion. Some forthcoming examples will illustrate the clipping effect. Nonlinear amplitude distortion is discussed in chapter 22 of volume 2. The models of table 12.2 are useful in calculating the dc operating point and the maximum output swing before clipping of the output waveform occurs. As before, the presence of ideal diodes makes the model nonlinear. However, with each ideal diode assumed to be either on or off, the actual calculations are performed on linear circuits.

Three Regions of BJT Operation

The next example illustrates the use of the PWL model of a BJT to determine the operating point and to define the three regions of operation of the device.

◆ EXAMPLE 12.3

The *npn* transistor in figure 12.14a has the PWL model of figure 12.14b with $V_0 = 0.5$ V, $R_s = 1$ kΩ, $R_p \cong \infty$, and $\beta = 20$, all obtained using item (1) of table 12.2. Find I_B and V_{CE} for the following three cases: (a) $V_i = 0.7$ V; (b) $V_i = 0.3$ V; (c) $V_i = 1$ V.

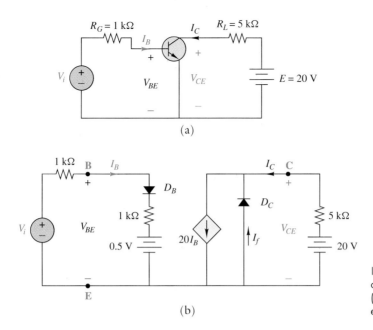

(a)

Figure 12.14 Determination of operating point by use of the PWL model. (a) Transistor amplifier circuit. (b) PWL equivalent circuit.

(b)

SOLUTION

Case (a). With $V_i = 0.7$ V, the ideal diode D_B is obviously forward biased. Replacing D_B by a short circuit produces

$$I_B = \frac{0.7 - 0.5}{1 + 1} = 0.1 \text{ mA}.$$

The positive value of I_B indicates that D_B is indeed forward biased.

The state of the ideal diode D_C is by no means obvious. Persons experienced in analyzing electronic circuits can usually guess the right state. Let us, however, not presuppose any experience and use the "assumed diode state" method. According to this method, we assume that D_C is forward biased and replace it by a short circuit. This results in the linear circuit of figure 12.15a. The current through D_C is

$$I_f = 20I_B - I_C = 20 \times 0.1 - \frac{20 - 0}{5} = -2 \text{ mA}.$$

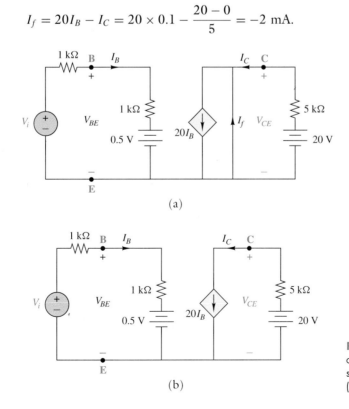

(a)

(b)

Figure 12.15 Analysis of PWL circuit of figure 12.14b by the "assumed diode state" method. (a) D_B on and D_C on. (b) D_B on and D_C off.

The negative answer for I_f indicates that D_C is *not* forward biased. Now, assuming D_C is reverse biased and replacing it by an open circuit yields the linear circuit of figure 12.15b. The resulting voltage is

$$V_{CE} = 20 - 5I_C = 20 - 5 \times 20 \times 0.1 = 10 \text{ V.}$$

The positive value for V_{CE} verifies a reverse-biased state for D_C. Thus, a valid solution for this case is $I_B = 0.1$ mA and $V_{CE} = 10$ V. A BJT with the input-side diode in the PWL model forward biased and the output-side diode reverse biased, such as in the present case, is said to be operating in the **active region**.

Case (b). With $V_i = 0.3$ V, the ideal diode D_B is reverse biased, forcing $I_B = 0$. Since the controlled current source $20I_B$ has a zero value, the ideal diode D_C becomes reverse biased by the 20-V power supply. Therefore, $I_C = 0$ and $V_{CE} = 20$ V. A BJT with both diodes in the PWL model reverse biased, such as in this case, is said to be operating in the **cutoff region**.

Case (c). With $V_i = 1$ V, the ideal diode D_B is forward biased. After replacing D_B in figure 12.14b by a short circuit, we find the current to be

$$I_B = \frac{1 - 0.5}{1 + 1} = 0.25 \text{ mA.}$$

The positive value of I_B confirms the "on" state of the diode D_B. The state of the diode D_C is not obvious. Assume, as in case 1, that D_C is reverse biased. Figure 12.15b shows the linear circuit with D_C reverse biased. Here,

$$V_{CE} = 20 - 5 \times I_C = 20 - 5 \times 20 \times 0.25 = -5 \text{ V.}$$

The negative value calculated for V_{CE} indicates that the assumption that D_C is reverse biased is false. Since one of the assumed diode states is wrong, the solution is not valid. The remaining possibility is to assume that D_C is forward biased. Figure 12.15a depicts the associated linear circuit. Here,

$$I_f = 20I_B - I_C = 20 \times 0.25 - \frac{20 - 0}{5} = 1 \text{ mA.}$$

The positive value for I_f verifies the assumption that D_C is forward biased. Therefore a valid solution for this case is $I_B = 0.25$ mA and $V_{CE} = 0$ V.

A BJT with both diodes in the PWL model forward biased, such as in this third case, is said to be operating in the **saturation region**. Table 12.3 summarizes the calculated results of this example.

TABLE 12.3   CALCULATED RESULTS FOR EXAMPLE 12.3

Case	V_i (V)	I_B (mA)	V_{CE} (V)	D_B	D_C	Region
(a)	0.7	0.1	10	On	Off	Active
(b)	0.3	0	20	Off	Off	Cutoff
(c)	1.0	0.25	0	On	On	Saturation

Graphical Method for a Simple Configuration

For the simple circuit configuration of figure 12.14, the graphical method, important during the slide rule era, offers a simple, straightforward solution method. Nowadays, practical circuits are usually simulated on digital computers using programs that employ the *algebraic method*. On the other hand, the graphical method provides additional insight into the circuit's operation. To illustrate the method, consider again case (a) of example 12.3.

From figure 12.14a, we see that, at the input terminals of the BJT, I_B and V_{BE} must satisfy two constraints: (1) The transistor requires that (V_{BE}, I_B) be a point on the input characteristic, which in this case is a single PWL curve in figure 12.16a; (2) the source and resistor on the left require that (V_{BE}, I_B) be a point on the straight line representing $V_{\mathrm{BE}} = V_i - R_G I_B$. This straight line, called the **input load line**, is constructed by joining two points, $(V_i, 0)$ and $(0, V_i/R_G)$, on the same graph as the input characteristic. The solution is the point of intersection of the two constraint curves. Figure 12.16a shows these two curves and their point of intersection $P = (V_{\mathrm{BE}}, I_B) = (0.6 \text{ V}, 0.1 \text{ mA})$.

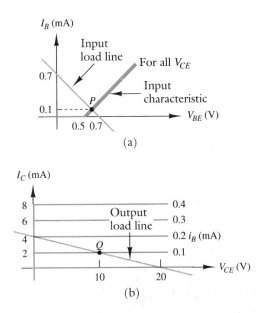

Figure 12.16 The graphical method of solving example 12.3. (a) Construction of input load line. (b) Construction of output load line.

At the output terminals of the BJT, I_C and V_{CE} must satisfy two similar constraints: (1) The transistor operation requires that (V_{CE}, I_C) lie on the output characteristic corresponding to $I_B = 0.1 \text{ mA}$; (2) the dc source and resistor on the right require that (V_{CE}, I_C) be a point on the straight line representing $V_{CE} = E - R_L I_C$. This straight line, called the **output load line**, is constructed by joining two points $(E, 0)$ and $(0, E/R_L)$ by a straight line on the same graph as the output characteristic. Figure 12.16b illustrates this procedure. The intersection point of the load line and the curve $I_B = 0.1 \text{ mA}$ is $Q = (V_{CE}, I_C) = (10 \text{ V}, 2 \text{ mA})$.

> **Exercise.** Solve cases (b) and (c) of example 12.3 by the graphical method, using the input and output characteristics plotted in figure 12.16.

The graphical method has several advantages: (1) It works even when the input and output characteristic curves are *not* piecewise linear; (2) it offers a visual aid for circuit analysis and design; and (3) when a manufacturer provides the device characteristics in the form of graphs, the graphical method is a more direct solution technique.

The most serious drawback of the graphical method is that it can handle only some very simple circuit configurations, such as the source-device-load configuration of figure 12.14a. Even in this simple configuration, the input characteristic must be a single curve instead of a family of curves for the method to be practical. The PWL method, on the other hand, does not have this limitation.

Stabilization of the Operating Point

From the solution of example 12.3, by either the PWL method or the graphical method, we see that the operating point $Q = (V_{CE}, I_C)$ is very sensitive to changes in the parameter β. In practice, even transistors of the same type number can have β-values that differ by as much as 100%. A common technique for stabilizing the operating point against β-variations is to insert a resistor, R_E, between the emitter terminal and the common ground. This will require a higher power supply voltage, a price to be paid for a more stable operating point. Example 12.4 demonstrates the stabilization effect.

EXAMPLE 12.4

The amplifier circuit of figure 12.17a contains an *npn* transistor that has a PWL model with $V_0 = 0.6$ V, $R_s = 2$ kΩ, $R_p \cong \infty$, and $\beta = 400$.

(a) Find I_C and V_{CE}.

(b) With R_E and β kept as variables, and all other parameters fixed as given, find an expression for I_C in terms of R_E and β.

(c) Discuss the effect of R_E on maintaining a constant I_C.

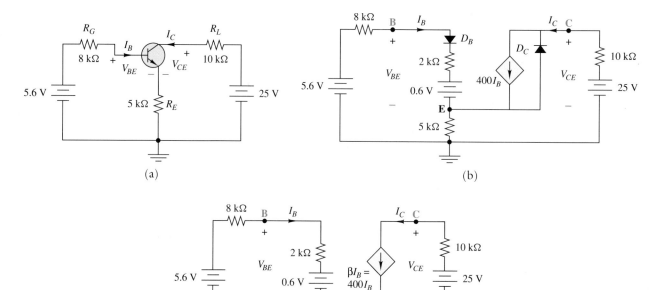

(a)　　　　　　　　　　　　　　　　　(b)

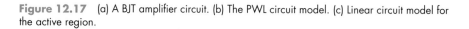

(c)

Figure 12.17　(a) A BJT amplifier circuit. (b) The PWL circuit model. (c) Linear circuit model for the active region.

SOLUTION

Part (a).　Replacing the transistor in figure 12.17a by its PWL model (item (1) of table 12.2) yields the PWL circuit of figure 12.17b. Reflecting on the experience gained in solving example 12.3, we assume the two diode states D_B "on" and D_C "off." The *linear* circuit of figure 12.17c results after replacing D_B by a short circuit and D_C by an open circuit. It is on this *linear* circuit that all calculations are to be done. Applying KVL to the input mesh yields

$$-5.6 + (8 + 2)I_B + 0.6 + 5 \times (400 + 1)I_B = 0,$$

from which it follows that $I_B = 0.002481$ mA > 0, indicating that D_B is indeed on. Consequently,

$$V_{BE} = 2 \times 0.002481 + 0.6 = 0.605 \text{ V} > 0,$$

and

$$I_C = 400 I_B = 0.9924 \text{ mA}.$$

Application of KVL to the output mesh yields

$$V_{CE} = 25 - 10 \times 0.9924 - 5 \times 401 \times 0.002481 = 10.1 \text{ V}.$$

The positive value of V_{CE} indicates that D_C is indeed off. Therefore, the answers for V_{CE} and I_C are valid.

Part (b). Repeating the derivations of part (a), but keeping R_E and β as variables, yields

$$I_C = \frac{5\beta}{(\beta+1)R_E + 10{,}000}. \qquad (12.4)$$

Part (c). If the resistor R_E is not used, i.e., if $R_E = 0$, then equation 12.4 reduces to $I_C = 0.0005\beta$ A. This indicates that the percentage change in I_C is the same as that of β, a very undesirable situation. On the other hand, if R_E is present, and the values of R_E and β are large enough so that $(\beta+1)R_E >> (R_G + R_0) = 10{,}000$, then equation 12.4 reduces to

$$I_C \cong \frac{\beta}{(\beta+1)} \frac{5}{R_E}. \qquad (12.5)$$

Since $\beta = 400 >> 1$ in the present case, the fraction $\beta/(\beta+1)$ approximates 1 for all reasonable changes in β (i.e., provided that β remains large). Therefore, equation 12.4 further reduces to the approximation

$$I_C \cong \frac{5}{R_E}. \qquad (12.6)$$

This indicates a predominant dependence of the collector current on R_E, making it essentially independent of β.

Equations 12.4 through 12.6 are for the circuit configuration of figure 12.17a, which has specific values for the voltage sources and some resistors. Similar formulas for the general case are given in problem 17. These formulas are for investigating the effect of changes in β on the operating point. Note that there are other causes of operating point drift as well, a very important one being the change in the ambient temperature. A study of this problem belongs to a later course in electronics.

In part (a) of example 12.4, suppose that we try to find I_C and V_{CE} by the load line method. Refering to figure 12.17a and examining the input side, we observe that when $R_E = 0$, the 5.6-V source and 8-kΩ resistance require that $V_{BE} = 5.6 - 8{,}000I_B$. This equation for a straight line represents the load line, as illustrated in figure 12.16a. When $R_E \neq 0$, the constraint equation becomes

$$V_{BE} = 5.6 - 8{,}000I_B - R_E(I_B + I_C).$$

Using the special relationship $I_C = 400I_B$ in the present circuit, we find that this equation becomes

$$V_{BE} = 5.6 - (8{,}000 + 401R_E)I_B,$$

which is again a straight line in the (V_{BE}, I_B) plane. However, the presence of I_C in the equation for V_{BE} generally means that the plot of I_B vs. V_{BE} is no longer a single straight line. Accordingly, the simple load-line construction of figure 12.16a is not applicable. The load-line method can be modified and used to solve the problem; however, when the circuit configuration becomes more complex (for example, by adding another resistor across the top terminals B and C in figure 12.17b), the method breaks down, while the PWL method still works well.

PWL Model for Determining Amplifier Gain and Maximum Output Swing

In the previous two examples, the only independent sources present were the dc power supplies. These power supplies established an operating point, or quiescent point, in the active region of the transistor. We shall now add an ac input signal to the circuit in order to study signal amplification. Our analysis will again utilize the piecewise linear approach.

EXAMPLE 12.5

Figure 12.18a depicts a single-stage transistor amplifier. The *npn* transistor has the piece-wise linear model of item (1) in table 12.2, with $V_0 = 0.5$ V, $R_s = 500$ Ω, $R_p \cong \infty$, and $\beta = 100$.

(a) Find the dc operating point.

(b) If $v_i(t) = V_m \sin(1,000t)$ V and $V_m = 0.1$ V, find $v_o(t)$ and the magnitude of the voltage gain.

(c) What is the largest input amplitude V_m such that neither the top nor the bottom of the output waveform is clipped? What is the largest output swing without clipping?

(d) If the dc voltage source $V_{bb} = 1.6$ V in figure 12.18a is adjustable, what should the new value be in order to maximize the output voltage swing?

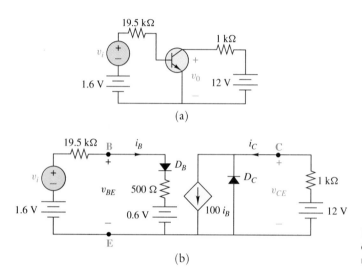

(a)

(b)

Figure 12.18 (a) A simple transistor amplifier. (b) Piecewise linear circuit model.

SOLUTION

Part (a). We represent the BJT in figure 12.18a by its piecewise linear model, shown in figure 12.18b. To determine the dc operating point, we set $v_i = 0$ in figure 12.18b. The polarities and magnitudes of the dc independent sources in figure 12.18b suggest the obvious conclusion: The diode D_B is in the "on" state. Replacing D_B by a short circuit and calculating I_B yields

$$I_B = \frac{1.6 - 0.6}{19.5 + 0.5} = 0.05 \text{ mA} > 0.$$

The calculated positive value of I_B confirms the correctness of the assumed state of D_B. It follows that

$$V_{BE} = 0.5 I_B + 0.6 = 0.625 \text{ V.}$$

Since the circuit is intended as an amplifier, we assume that D_C is reverse biased and replace it by an open circuit. Figure 12.19 sketches the resultant *linear* circuit.

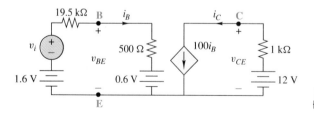

Figure 12.19 Linear circuit derived from figure 12.18.

From the figure, with $v_i = 0$, the voltage

$$V_{CE} = 12 - 1 \times I_C = 12 - 1 \times 100 \times 0.5 = 7 \text{ V.}$$

The positive value of V_{CE} shows that the ideal diode D_C is indeed in the "off" state. In sum, the operating point of the transistor is as follows:

$$V_{CE} = 7 \text{ V}, \qquad I_C = 5 \text{ mA},$$

$$V_{BE} = 0.625 \text{ V}, \qquad I_B = 50 \text{ } \mu\text{A}.$$

Part (b). With the small input voltage $v_i = 0.1 \sin(1{,}000t)$ V, the states of the two ideal diodes remain the same as in the quiescent case ($v_i = 0$). This permits using the equivalent circuit of figure 12.19 to find the *total response*, as long as the diode states agree with the circuit assumptions. Using figure 12.19, we obtain

$$i_B(t) = \frac{1.6 + 0.1 \sin(1{,}000t) - 0.6}{19.5 + 0.5} = [0.5 + 0.005 \sin(1{,}000t)] \text{ mA}, \quad (12.7)$$

$$v_0(t) = v_{CE}(t) = 12 - 1 \times 100 i_B(t) = 7 - 0.5 \sin(1{,}000t) \text{ V}. \qquad (12.8)$$

This shows that $v_o(t) = v_{CE}(t) > 0$ for all t, confirming the assumption that the ideal diode D_C remains off all the time. The input signal has a magnitude of 0.1 V, while the ac component of $v_o(t)$ has a magnitude of 0.5 V. Therefore, the magnitude of the voltage gain is $0.5/0.1 = 5$.

Part (c). For the model of figure 12.19 to be valid, D_B must remain on and D_C off. A borderline situation occurs when either D_B or D_C reaches the breakpoint. The conditions for reaching the breakpoints may be found from the linear circuit model of figure 12.19 by leaving v_i as a variable and forcing v_{CE} or i_B to be 0. In this case,

$$i_B(t) = \frac{1.6 + v_i - 0.6}{19.5 + 0.5} = (1 + v_i) \times 0.05 \text{ mA} \geq 0, \qquad (12.9)$$

producing the condition $v_i \geq -1$ V. Thus, if v_i is more negative than -1 V, the top of the output waveform will be clipped. Next, from the constraint

$$v_o(t) = v_{CE}(t) = 12 - 1 \times 100 \ i_B(t) = 7 - 5v_i \geq 0, \qquad (12.10)$$

one has the condition $v_i \leq 1.4$ V. Hence, if v_i is more positive than 1.4 V, the bottom of the output waveform will be clipped. Combining these two conditions produces

$$-1 \leq v_i = V_m \sin(100t) \leq 1.4.$$

This implies that the largest value of V_m without output waveform clipping is 1 V, and the largest output swing is $2 \times 5 \times 1 = 10$ V, peak to peak.

Part (d). The best way to investigate this problem is to plot the *transfer characteristic*, v_o vs. v_i. Equations 12.9 and 12.10 show that

$$v_o = 7 - 5v_i \quad \text{for } -1 < v_i < 1.4 \text{ V},$$

$$v_o = 0 \qquad \text{for } v_i \geq 1.4 \text{ V},$$

$$v_o = 12 \text{ V} \qquad \text{for } v_i \leq -1 \text{ V}.$$

Figure 12.20a is a plot of this relationship.

From the plot (with $V_{bb} = 1.6$ V), the operating point is Q_1. For a sinusoidal input voltage, the output may vary from 2 to 12 V, without the bottom or top of the waveform being clipped. Therefore, the maximum peak-to-peak output is 10 V. On the other hand, if the operating point is chosen to be the middle point of the sloped line, marked Q_2 in figure 12.20a, then the output may vary from 0 to 12 V without clipping. This is also the maximum output swing possible, corresponding to a sinusoidal input with amplitude $V_m = 1.2$ V. Q_2 indicates a constant 0.2-V component for $v_i(t)$. This additional 0.2 V actually is obtained by adjusting the dc power supply V_{bb} from 1.6 V to 1.8 V. With V_{bb}

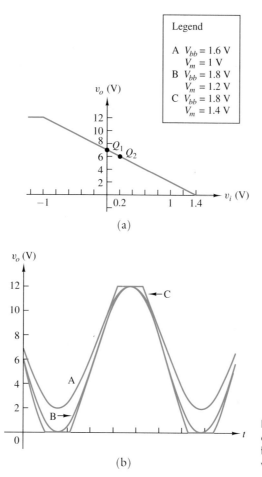

Legend

A $V_{bb} = 1.6$ V
  $V_m = 1$ V
B $V_{bb} = 1.8$ V
  $V_m = 1.2$ V
C $V_{bb} = 1.8$ V
  $V_m = 1.4$ V

(a)

(b)

Figure 12.20 (a) Transfer characteristic for the amplifier of figure 12.18a. (b) Output waveforms for various values of V_{bb} and V_m.

adjusted to 1.8 V, an input with amplitude $V_m > 1.2$ V will lead to clipping both at the top and the bottom of the output waveform. Figure 12.20b illustrates the clipping when $V_m = 1.4$ V and $V_{bb} = 1.8$ V.

EXAMPLE 12.6

Consider the amplifier circuit model of figure 12.18b. Will the controlled source, $100i_B$, provide amplification if both the dc power supplies are dead, i.e., if the dc sources are replaced by short circuits?

SOLUTION

When both power supplies are dead, the active region linear circuit model of figure 12.19 is invalid. So we return to figure 12.18b, which has a global PWL model for the BJT, and replace the dc power supplies by short circuits. Figure 12.21 depicts the result.

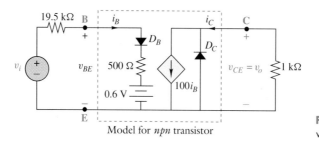

Model for *npn* transistor

Figure 12.21 PWL circuit model when the dc power supplies are dead.

Because of the ideal diode D_B, the current i_B is never negative. For any positive i_B, the controlled current source $100i_B$ always forward biases the ideal diode D_C. Therefore, no matter what value v_i has, the output voltage v_o is always zero, and no amplification is provided by the controlled source.

Although the transistor model in figure 12.21 contains a controlled source, it has no effect externally unless a dc power supply is applied to reverse bias the ideal diode D_C. Similar conditions exist in other semiconductor device models that contain controlled sources. This substantiates the statement in chapter 3 that a controlled source functions only in conjunction with a dc power supply.

Significance of ac Equivalent Circuits

After ascertaining the states of the ideal diodes in a PWL circuit and replacing them by open and short circuits accordingly, we obtain a *linear* circuit. In general, this linear circuit contains both dc independent sources and input signal sources, as illustrated in figure 12.19. Calculations proceed using the linear methods studied in earlier chapters. Each response found in the linear circuit is a *total response* that contains a constant term and a term due to the input signal, as illustrated by equations 12.7 and 12.8. The dc independent sources contribute to the constant term, which specifies the quiescent point. The **ac response** is the part due to the input signal. A better name for it is the **signal response**, because the applied signal need not be a sinusoidal waveform. When analyzing an amplifier's performance, we take note of the important quantities, such as voltage gain, current gain, power gain, etc. All such performance measures pertain to the response to an ac sinusoidal input. By the principle of superposition (chapter 5), we can kill all the dc independent sources when finding the ac or signal response. After setting all the dc independent sources to zero, we call the resultant circuit an **ac equivalent circuit.** Figure 12.22a shows the ac equivalent circuit derived from figure 12.19. The currents and voltages are the time-varying part of the total response. If a sinusoidal steady-state analysis is carried out using phasors, then the notations in figure 12.22a are further changed to those shown in figure 12.22b, in accordance with the rules described at the beginning of section 3.

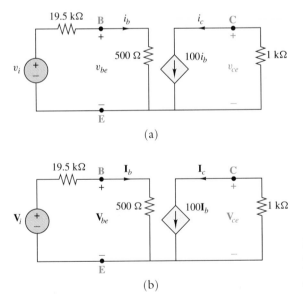

(a)

(b)

Figure 12.22 (a) ac equivalent circuit derived from figure 12.19. (b) Change in notation to indicate the phasors of voltages and currents.

The PWL circuit models (table 12.2) for a BJT predict uniformly spaced input and output characteristic curves. Measured characteristic curves, however, show that the spacing is uniform only over a small region of the v-i plane. This discrepancy is the only source of error in using the PWL technique; the analysis phase does not introduce any additional errors.

For a more accurate ac analysis, the model parameters in the ac equivalent circuit should be determined using the readings of (v, i) values in a small neighborhood (a nearly linear region) of the actual dc operating point. The resultant ac equivalent circuit, which yields more accurate results for inputs of small magnitude, is called a **small-signal equivalent circuit.** Such equivalent circuits, while more accurate for ac analysis, cannot be used to determine the maximum output swing. In contrast, a circuit model that is valid for inputs of either large or small magnitude, such as any of those shown in table 12.2, is called a **large-signal equivalent circuit.** Different notations are usually used for the parameters of these two types of equivalent circuits. For example, the parameter for the controlled current sources may be denoted as β_{ac} in the former and as β_{dc} in the latter. We need not be burdened by such ramifications here because our analysis is based on PWL models and assumes uniformly spaced input and output characteristic curves.

Generally, circuits containing controlled sources have the form shown in figure 12.22, i.e., with all dc independent sources omitted. Hence, all calculated quantities are actually the ac components of the total response. To build the circuit, then, one has to restore the dc biasing circuitry.

5. PIECEWISE LINEAR ANALYSIS OF A PRACTICAL AMPLIFIER STAGE

The transistor amplifier of figure 12.18 shows the input signal $v_i(t)$ in series with the 1.6-V dc supply. This is done mainly to simplify the calculations and to better illustrate the principles of circuit operation. Such a configuration is avoided in practice because of the following two undesirable traits:

1. The dc current will flow through the signal source.
2. The input signal source, v_i, is floating instead of grounded.

Figure 12.23a shows a circuit configuration that allows us to circumvent these adverse effects. The capacitor at the top of the ac signal input blocks dc, but allows an ac signal to pass. This capacitor should be large enough so that its impedance at the signal frequency is small enough to approximate a short circuit. With such a capacitor, it is also possible to ground the input source together with the dc power supplies. The capacitor is called a *blocking capacitor* from the dc operation point of view and a *coupling capacitor* from the ac operation point of view.

The function of the 2-kΩ resistor connected to the emitter in the figure, as explained in section 4, is to stabilize the dc operating point. An analysis (see problem 16) shows that it also causes a reduction in the ac voltage gain of the amplifier. To restore the voltage gain, a large capacitor—100 μF in the present case—is connected in parallel with the 2-kΩ resistor. At signal frequencies, this capacitor has a very low impedance. It is replaced by a short circuit in the ac equivalent circuit. Consequently, the 2-kΩ resistor has no effect as far as signal amplification is concerned. Such a capacitor is called a *bypass* capacitor because almost all of the ac current out of the emitter terminal will flow through it instead of through the 2-kΩ resistor.

The next example shows the configuration of a practical single-stage transistor amplifier and its analysis. The solution of this example represents a synthesis of the entire course material, as it utilizes many concepts and methods studied in previous chapters. The example serves as a yardstick for gauging your grasp of the material studied thus far.

◆ **EXAMPLE 12.7**

Consider again the common-emitter amplifier of figure 12.23a. The two capacitors behave as open circuits for dc and as short circuits over a desirable range of input frequencies. The *npn* transistor has a piecewise linear model (item (1), table 12.2) with $V_0 = 0.6$ V, $R_s = 2$ kΩ, $R_p \cong \infty$, and $\beta = 50$.

(a) Find the dc operating point.
(b) If $v_i(t) = 0.1 \sin(1,000t)$ V, find $v_o(t)$ and the voltage gain.

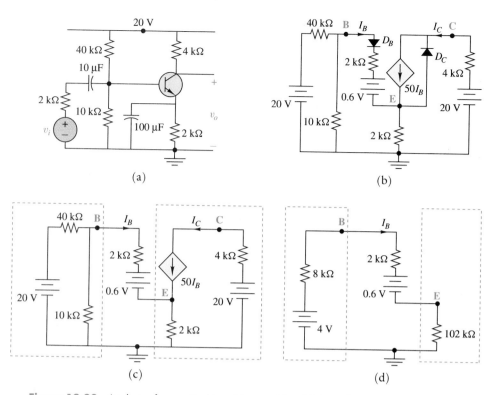

(a)

(b)

(c)

(d)

Figure 12.23 Analysis of a practical transistor amplifier. (a) A single-stage BJT amplifier. (b) The PWL dc equivalent circuit. (c) Equivalent circuit in the active region. (d) Use of Thévenin equivalents.

SOLUTION

Part (a). In figure 12.23b, the single 20-V power supply has been split into two for analysis purposes. Intuitively, such action is justified by noticing that in both of figures 12.23a and b, one of the terminals of both the 40-kΩ and 4-kΩ resistors is maintained at 20 V above the ground, while the connection of the other terminal is unchanged.

For the dc operating analysis, set $v_i = 0$ and open-circuit both capacitors, as per figure 12.23b. For amplifier operation, one can reasonably assume D_B to be forward biased and D_C to be reverse biased. In any case, if a wrong assumption is made, it will be uncovered in the solution process. Replacing D_B by a short circuit and D_C by an open circuit produces the *linear* circuit shown in figure 12.23c.

Example 12.4 illustrates one of a variety of methods for solving this circuit. Now let us utilize the equivalent circuit approach of chapter 5. The Thévenin equivalent circuits for the two dashed-lined two-terminal networks in figure 12.23c appear in figure 12.23d. After some straightforward computation,

$$I_B = \frac{4 - 0.6}{8 + 2 + 102} = 0.03036 \text{ mA} > 0.$$

The calculated positive value of I_B confirms the correctness of the assumed D_B state. It follows from figure 12.23c that

$$V_{BE} = 2I_B + 0.6 = 0.661 \text{ V}.$$

Again using figure 12.23c, we obtain

$$I_C = 50I_B = 1.518 \text{ mA}$$

and

$$V_{CE} = 20 - 4I_C - 2(I_C + I_B) = 20 - 6.07 - 3.1 = 10.83 \text{ V}.$$

The positive value of V_{CE} confirms the "off" state for the ideal diode D_C. In sum, the operating point for the transistor is as follows:

$$V_{CE} = 10.83 \text{ V}, \qquad I_C = 1,518 \text{ mA},$$

$$V_{BE} = 0.661 \text{ V}, \qquad I_B = 30.36 \text{ μA}$$

Part (b). With the small input voltage $v_i(t) = 0.1 \sin(1,000t)$ V, the states of the two ideal diodes remain the same as in the quiescent case, as will be verified soon. Replacing D_B and D_C by a short circuit and an open circuit, respectively, yields the equivalent circuit of figure 12.24a.

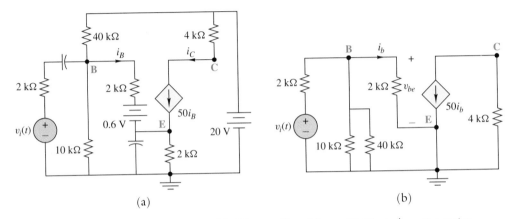

Figure 12.24 (a) Linear circuit model of the amplifier of figure 12.23a in the active region. (b) The equivalent circuit for ac calculations.

The *linear* circuit of figure 12.24a contains three independent sources: a 20-V external dc power supply, the 0.6-V dc source in BJT model (figure 12.23b), and the input signal $v_i(t)$. The principle of superposition may be used to find any response in the circuit. The responses due to the two dc independent sources, however, were found in part (a). These responses generate the dc operating point of the circuit. To find the responses due to the input $v_i(t)$, we set the two dc independent sources to zero according to the principle of superposition. Also, we replace the two capacitors by short circuits because their impedances are very small at the signal frequency of 1,000 rad/sec. Figure 12.24b shows the resultant ac equivalent circuit. This figure underlies the calculations of all ac-related performance measures. Since the circuit model contains no reactive elements, the use of phasors is unnecessary. Specifically, the calculations proceed in the time domain as follows.

$$10 \text{ k} || 40 \text{ k} || 2 \text{ k} = 1.6 \text{ k}, \qquad \text{(three resistances in parallel)}$$

$$v_{be} = v_i \frac{1.6}{1.6 + 2} = 0.444 v_i, \qquad \text{(voltage division)}$$

$$i_b = \frac{0.444 v_i}{2} \text{ mA}, \qquad \text{(Ohm's law)}$$

$$v_o = -50 i_b \times 4 = -44.44 v_i. \text{ (controlled source and Ohm's law)}$$

Therefore, the output voltage (ac component) of the amplifier is

$$v_o(t) = 4.444 \sin(1,000t + 180°) \text{ V},$$

and the voltage gain is

$$\frac{v_0}{v_i} = -44.44.$$

SUMMARY

Section 2 of this chapter discussed the characterization of a general three-terminal resistive device by its terminal voltage-current relationships. Here we used circuit models for simple v-i curves (listed in table 12.1) as building blocks for modeling the more complex three-terminal behavior. This input-output approach obviates the need for investigating device physics in an introductory text. The v-i relationships are in general nonlinear. In order to apply all of the elegant methods of linear network analysis learned in previous chapters, we approximate a smooth curve by a set of straight-line segments. The process is called piecewise linear (PWL) approximation, and the resultant circuit is called a PWL circuit. For any PWL circuit, a mathematical model can be constructed using linear circuit elements and *ideal diodes*. The solution of a general PWL circuit requires some sophisticated numerical algorithms that are beyond the level of the text. However, for the simple circuits considered in this chapter, the number of ideal diodes is small, so that the "assumed diode states" method works well.

Section 3 developed PWL models for bipolar junction transistors in the common-emitter and common-base configurations. Section 4 used these models to investigate several fundamental problems in a transistor amplifier circuit, including the determination of the operating point, the calculation of voltage gain, and the determination of the maximum output voltage swing. With the PWL method, approximate answers to these problems are obtained without resorting to numerical algorithms for solving nonlinear equations.

In previous chapters (and later in volume 2), circuits containing controlled sources generally represented only the ac equivalent circuit; i.e., the biasing circuitry was omitted from the circuit diagram. While adequate for calculating ac performance measures such as voltage gain, ac equivalent circuits do not provide any information about the maximum output swing or correct polarity of the dc power supply. Accordingly, section 5 presented the analysis of a practical BJT amplifier stage with all the circuit components present. Among these components are the dc power supply, the transistor, the resistors and the coupling and bypass capacitors. A PWL model for the transistor is used. Step by step, we show how to determine the operating point, how to use the active region model for voltage gain calculations, and how to use the global model for determining the maximum output voltage swing. In the solution process, the meanings of a large-signal equivalent circuit and a small-signal equivalent circuit become very clear. The reliance on dc power supplies for a controlled source to function is also made clear by the PWL model for transistors.

TERMS AND CONCEPTS

ac equivalent circuit: the linear circuit model suitable for the determination of the ac portion of the total response. Independent dc sources are set to zero in ac equivalent circuits.

Active region: in the piecewise linear model of a bipolar transistor, the state in which the base-emitter diode is conducting and the other diode is off.

"Assumed diode state" method: an elementary method for analyzing circuits containing ideal diodes. If an ideal diode is assumed to be in the "on" state, it is replaced by a short circuit; if the ideal diode is assumed to be in the "off" state, it is replaced by an open circuit.

Bipolar junction transistor (BJT): a three-terminal semiconductor device having the salient property that a small change in the current of one terminal causes a much larger change in the current of another terminal.

Bypass capacitor: a capacitor placed in parallel with a resistor for the purpose of diverting the ac current at the signal frequency through the capacitor, essentially shorting out the resistor at the signal frequency. At the frequency of interest, a bypass capacitor should have an impedance much smaller than its parallel resistance.

Common-emitter configuration: a way of using the BJT in which the emitter is common to both the input and output.

Coupling capacitor: a capacitor used to prevent dc current in the biasing circuitry from entering the signal source or load. A coupling capacitor should have a relatively small impedance at the signal frequency so that it may be treated approximately as a short circuit.

Cutoff region: in the piecewise linear model of a bipolar transistor, the state in which both ideal diodes are off.

Input characteristics: for a three-terminal resistive device, the family of curves showing the input voltage-current relationship, with the output voltage (or current) as the running parameter.

Input load line: in a graphical method of determining the operating point of a device, the v-i locus arising from the components external to the device on the input side.

Large-signal equivalent circuit: a piecewise linear circuit model suitable for approximate analysis of the circuit response when the amplitude of the input is large.

Operating point (quiescent point): point that represents the solution of all voltages and currents in an electronic circuit with only dc sources turned on and no time-varying signal applied.

Output characteristics: for a three-terminal resistive device, the family of curves showing the output voltage-current relationship, with the input current (or voltage) as the running parameter.

Output load line: in a graphical method of determining the operating point of a device, the v-i locus arising from the components external to the device on the output side.

Piecewise linear (PWL) approximation: the process of replacing a smooth curve by a set of straight-line segments hugging the original curve. With more segments, a PWL curve can approximate the original curve to any desired degree of accuracy.

Piecewise linear (PWL) circuit model: a circuit model containing linear circuit elements and elements having piecewise linear v-i relationships that, in turn, can be modeled by resistors, independent sources, and ideal diodes.

Saturation region: in the piecewise linear model of a bipolar transistor, the state in which both ideal diodes are on.

Small-signal equivalent circuit: same as ac equivalent circuit.

PROBLEMS

1. The input and output characteristics of a three-terminal resistive device are shown in figure P12.1. The curves are not to scale. Use the coordinates given, and model the device with the following types of components: resistors, linear controlled sources, independent sources, and ideal diodes. Show the circuit configurations and all parameter values.

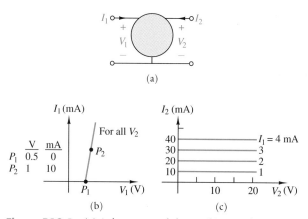

(a)

	V	mA
P_1	0.5	0
P_2	1	10

(b)   (c)

Figure P12.1 (a) A three-terminal device. (b) Input characteristic. (c) Output characteristic.

2. Repeat problem 1 for the device characteristics shown in figure p12.2.

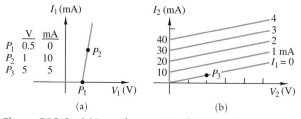

	V	mA
P_1	0.5	0
P_2	1	10
P_3	5	5

(a)   (b)

Figure P12.2 (a) Input characteristic. (b) Output characteristic.

3. Repeat problem 1 for the device characteristics shown in figure P12.3.

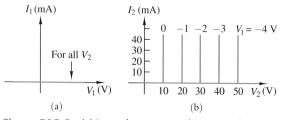

(a)   (b)

Figure P12.3 (a) Input characteristic. (b) Output characteristic.

4. Plot the v-i curves for each of the circuits shown in figure P12.4. (*Check*: For (c), the breakpoint is at (10 V, 5 A).)

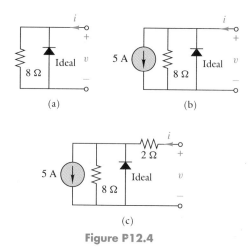

(a)   (b)

(c)

Figure P12.4

5. (a) Consider the circuit of figure P12.5a. Determine the range of i for which the ideal diode is on.
(b) Verify the v-i curve shown in figure P12.5b.

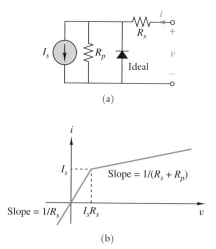

(a)

(b)

Figure P12.5 (a) A diode circuit. (b) Associated PWL v-i curve.

6. A three-terminal device has input and output characteristic curves in the first quadrant, as shown in figure P12.6. The curves are not to scale. Use the coordinates given, and construct a PWL circuit model. (*Hint*: For the output characteristics, make use of the results of problem 5, and replace the independent current source by a controlled current source.)

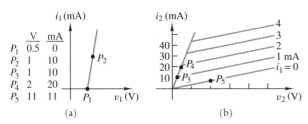

	V	mA
P_1	0.5	0
P_2	1	10
P_3	1	10
P_4	2	20
P_5	11	11

(a)   (b)

Figure P12.6 (a) Input characteristic. (b) Output characteristics.

7. Figure P12.7a is a certain three-terminal device whose input characteristic is shown in figure P12.7b. The output characteristic is a family of uniformly spaced parallel lines. The associated PWL circuit model is shown in figure P12.7d. State, in words, how to determine the parameters from the curves. (*Hint*: Make use of the experience obtained in solving problem 6.)

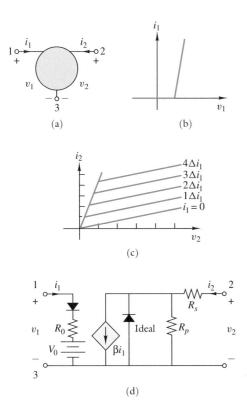

Figure P12.7 (a) Three-terminal device. (b) Input characteristic. (c) Output characteristic. (d) Piecewise linear circuit model.

8. Consider the piecewise linear circuit shown in figure P12.8. Find the operating point of the ideal diode D.

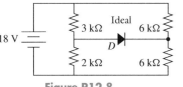

Figure P12.8

9. Consider the piecewise linear circuit of figure P12.9.
 (a) Find V_{out} if $V_{in} = 18$ V and $R = 2$ kΩ.
 (b) Find V_{out} if $V_{in} = 18$ V and $R = 12$ kΩ.
 (c) Find the V_{out} vs. V_{in} transfer curve for the case $R = 2$ kΩ.

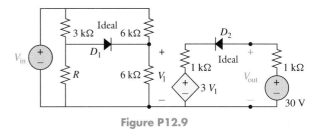

Figure P12.9

10. The three-terminal device in the amplifier circuit of figure P12.10 has the piecewise linear input and output characteristics described in problem 1.
 (a) Find V_2 for the following values of V_{in}: 0, 0.5, 2, 3.5, and 5 V.
 (b) Find the transfer characteristic V_2 vs. V_{in} for the range $-5 \le V_{in} \le 5$ V.

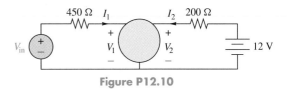

Figure P12.10

11. Repeat problem 10 if the device characteristic curves are those given in problem 2.

12. The input V_{in} in problem 10 is replaced by a dc voltage source of 1.5 V in series with a sinusoidal voltage source $v_s(t) = V_m \sin(1,000t)$ V, as shown in figure P12.12. Determine the largest value of V_m such that the output $v_2(t)$ is an undistorted sinusoid (with a dc offset). What is the maximum peak-to-peak output voltage?

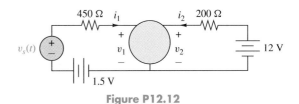

Figure P12.12

13. Figure P12.13 shows a single-stage transistor amplifier and the piecewise linear model for the transistor.

    (a) If $R_E = 0$ and $V_{th} = 0.82$ V, find the operating point of the transistor.

    (b) If $R_E = 2$ kΩ, and $V_{th} = 2.86$ V, find the operating point of the transistor. (*Check:* Both cases have $I_B = 20$ μA.)

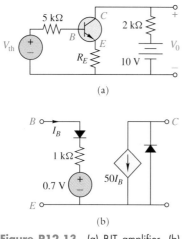

(a)

(b)

Figure P12.13 (a) BJT amplifier. (b) PWL model for BJT.

14. (a) Consider the amplifier circuit of problem 13a (with $R_E = 0$). Find the transfer characteristic V_o vs. V_{th} for the input range $-1 < V_{th} < 1$ V.

    (b) An ac input signal $v_i(t) = V_m \cos(2{,}000t)$ V is applied to the circuit of problem 13a (with $R_E = 0$), as shown in figure P12.14. What should V_{th} be in order to have maximum undistorted output? What are the maximum output voltage swing (peak-to-peak value) and the required V_m? You may solve this problem by the use of the PWL circuit model or the graphical method. (*Check:* (b) $V_{th} = 1$ V.)

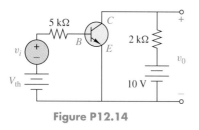

Figure P12.14

15. An ac input signal $v_i(t) = V_m \cos(2{,}000t)$ V is applied to the circuit of problem 13a (with $R_E = 2$ kΩ), as shown in figure P12.15. What should V_{th} be in order to have maximum undistorted output? What are the maximum output voltage swing (peak-to-peak value) and the required V_m? You may solve this problem by the use of the PWL circuit model.

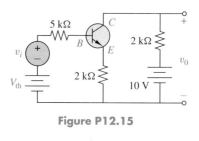

Figure P12.15

16. A small input signal $v_i(t) = 0.1 \cos(200t)$ V is applied to the amplifier circuits of figures P12.14 and P12.15. Find the voltage gain in each case. How much reduction in the voltage gain is caused by the presence of $R_E = 2$ kΩ in this case?

17. Figure P12.17 shows a single-stage BJT amplifier and its PWL circuit model. The purpose of the resistance R_E is to stabilize the operating point.

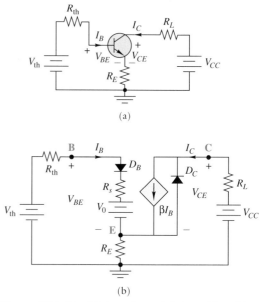

(a)

(b)

Figure P12.17 (a) A single-stage BJT amplifier. (b) Associated PWL circuit model.

(a) Show that

$$I_C = \frac{\beta(V_{th} - V_0)}{R_{th} + R_s + (\beta + 1)R_E}.$$

(b) Show that if R_E is chosen large enough so that $(\beta + 1)R_E \gg (R_{th} + R_s)$, then

$$I_C \cong \frac{V_{th} - V_0}{R_E}.$$

This result indicates that I_C is essentially immune to the variation in β.

18. Repeat the analysis of example 12.7 with the BJT model changed to that shown in figure P12.18.

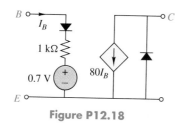

Figure P12.18

19. Consider the common-emitter amplifier circuit of example 12.7. If $v_i(t) = V_m \sin(1{,}000t)$, determine the largest value of V_m such that no clipping occurs in the output voltage waveform.

20. Consider the common-emitter amplifier circuit of example 12.7. If the signal source has an internal resistance of 1 kΩ (instead of 2 kΩ), and if $v_i(t) = 0.1 \sin(1{,}000t)$ V, find $v_o(t)$.

21. The input and output characteristics of the transistor in figure P12.21 are approximated by the piecewise linear curves of problem 2. Sketch the waveform of $v_o(t)$ if $v_i(t) = 10u(t)$ V. (*Hint*: Find the Thévenin equivalent of the input circuit external to the transistor, and make use of the PWL circuit model for the transistor derived in problem 2.)

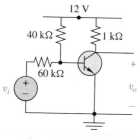

Figure P12.21

22. A BJT may be used as a switch. When it operates in the saturation region, V_{CE} is nearly zero, and the BJT approximates the "switch closed" condition. When it operates in the cutoff region, $I_C = 0$, and the BJT simulates the "switch open" condition. The advantage of using a BJT as

a switch is that a very small current can control the switching of a much larger current. A simple circuit demonstrates the principle.

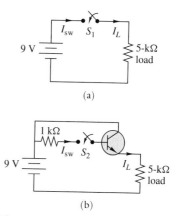

Figure P12.22 (a) A simple circuit containing a mechanical switch. (b) The transistor is used as a switch.

The 5-kΩ load in figure P12.22 represents the dc resistance of a certain relay whose on/off condition is controlled by a mechanical switch. The inductance of the relay coil is neglected in dc steady-state analysis. The relay is adjusted for a 1-mA pickup, i.e., a dc current greater than 1 mA is enough to actuate the relay and close its contacts.

(a) In the circuit of figure P12.22a, the relay is operated from a 9-V dc source through a mechanical switch S_1. Determine the magnitudes of the currents I_{sw} and I_L that are switched on and off by the operation of S_1.

(b) In the circuit of figure P12.22b, a BJT is used to operate the relay. Assume that the BJT has the PWL model shown in figure P12.18. Determine the magnitudes of the currents I_{sw} and I_L that are switched on and off by the operation of the mechanical switch S_2.

23. In the circuits of problem 22, if the 9-V dc source is changed to 6 V, will the switches control the relay satisfactorily?

CHAPTER 13

Pierre Laplace 1749-1827

Leonhard Euler 1707-1783

CHAPTER OUTLINE

Laplace Transform Analysis, 1: Basics

HISTORICAL NOTE

THE Laplace transform is an integral that transforms a time function into a new function of a complex variable. The name *Laplace transform* comes from the name of a French mathematician, Pierre Simon Laplace (1749–1827). Pierre Laplace adapted the idea from Joseph Louis Lagrange (1736–1813), who in turn had borrowed the notion from Leonhard Euler (1707–1783). These early mathematicians set the stage for converting complicated differential equation models of physical processes and of circuits into simpler algebraic equations. This allowed electrical engineers to analyze circuits and calculate responses quickly and efficiently. In turn this helped engineers design better circuits for radio communication, the telephone, and other, earlier electronic conveniences. The current chapter introduces the notion of the Laplace transform, a mathematical tool that is ubiquitous in its application to an army of engineering problems. ◆

◆ CHAPTER OBJECTIVES

- Explain and illustrate the benefits of using the Laplace transform tool for solving circuits.
- Develop a basic understanding of the Laplace transform tool and its mathematical properties.
- Develop some skill in applying the Laplace transform to differential equations and circuits modeled by differential equations.

1. INTRODUCTION

This chapter introduces a powerful mathematical tool for circuit analysis and design named the Laplace transform. Design aspects are left for future courses. The use of the Laplace transform is commonplace in engineering, specifically electrical engineering. A student might ask why such a potent tool is necessary for basic circuits, especially since many texts use an alternative technique called generalized phasor analysis. This technique, however, does not permit general transient analysis, but rather allows only for sinusoids, exponentials, damped sinusoids, and dc signals as source excitations. This class of signals is small and does not begin to encompass the broad range of excitations necessary for general circuit analysis and the related area of signal processing. The Laplace transform framework, on the other hand, permits both steady-state and transient analysis of circuits in a single setting. In addition, it affords general, rigorous definitions of *impedance, transfer function,* and various response classifications that are the bases of more advanced courses on system analysis and signal processing. There are many advantages to the Laplace method, and there will be an entire semester to practice using the tool.

The next section of this chapter introduces some of the difficulties associated with the methods of circuit analysis introduced in volume I when applied to more general circuits. Following this, we present an overview of the idea of Laplace transform analysis in section 3, define important basic signals in section 4, and introduce the formal definition of the one-sided Laplace transform in section 5. The inverse Laplace transform and important properties of the transform process are introduced in sections 6 and 7, with numerous examples to illustrate their use. Section 8 applies the technique to circuits modeled by differential equations. Such models were developed in volume I. To help the reader focus on the important points, a summary section and a section on terms and concepts are provided. Numerous problems are included at the end of the chapter for further practice.

2. REVIEW AND DEFICIENCIES OF "SECOND-ORDER" TIME DOMAIN METHODS

Circuits are interconnections of basic components such as sources, resistors, capacitors, inductors, and the like. These components shape or modify signals, voltages, and currents supplied by the independent sources. Sources model excitations to the circuit. Interconnections constrain the behavior of component voltages and currents according to the well-known Kirchhoff voltage and current laws. Voltages and currents of primary importance are typically labeled outputs. The body of knowledge called circuit theory contains numerous techniques for the computation of these output signals. This volume explores various fundamental techniques of circuit analysis through the vehicle of a mathematical tool called the Laplace transform.

The output or response of a circuit depends on the independent source excitations, on the initial capacitor voltages, and on the initial inductor currents. Calculation of the output often begins by writing an algebraic or a differential equation model of the circuit for the output variable in terms of the source excitations or inputs. For simple circuits with simple source excitations, including dc or purely sinusoidal voltages or currents, the solution of the differential equation model has a known general form containing arbitrary constants. The arbitrary constants depend on the initial conditions and the magnitude of the dc excitation or on the magnitude and phase of the sinusoidal excitation. For such a restricted class of circuits and excitations, calculating a response reduces to determining these arbitrary constants. Example 13.1 reviews this straightforward recipe. For more complicated circuits, such an approach breaks down. Example 13.2 demonstrates how the approach breaks down with a simple third-order circuit.

◆ EXAMPLE 13.1

In the series RLC circuit of figure 13.1, the capacitor voltage $v_{out}(t)$ is the important circuit response. The goal of the problem is to review the straightforward recipe for computing the responses of such simple circuits. Specifically, we will determine the differential equation model of the circuit, the form of the response, and the values of the arbitrary constants in the response form.

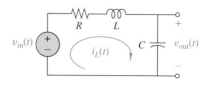

Figure 13.1 Basic series RLC circuit having capacitor voltage as output. Such simple circuits have a known response form.

SOLUTION

Step 1. Generate a differential equation model of the circuit. From Kirchhoff's voltage law (KVL), summing the voltages around the loop yields the integrodifferential equation

$$Ri_L(t) + L\frac{di_L}{dt}(t) + \frac{1}{C}\int_{-\infty}^{t} i_L(q)dq = v_{in}(t),$$

where q is simply a dummy variable of integration. Making the substitution

$$i_L(t) = i_C(t) = C\frac{dv_C}{dt} = C\frac{dv_{out}}{dt}$$

and dividing through by LC yields the following differential equation model:

$$\frac{d^2v_{out}}{dt^2}(t) + \frac{R}{L}\frac{dv_{out}}{dt}(t) + \frac{1}{LC}v_{out}(t) = \frac{1}{LC}v_{in}(t). \tag{13.1}$$

Step 2. Determine the form of the solution. If $v_{in}(t)$ is a dc voltage of magnitude V_0 volts for $t \geq 0$ and zero for $t < 0$, then there are only two possible general forms of the solution:

$$v_{out}(t) = Ae^{at} + Be^{bt} + D, \quad t \geq 0 \tag{13.2a}$$

and

$$v_{out}(t) = Ae^{at} + Bte^{at} + D, \quad t \geq 0. \tag{13.2b}$$

The solution to equation 13.1 is said to be specified whenever the constants, A, B, D, a, and b in equation 13.2 have fixed values.

Step 3. Determine the characteristic equation of the circuit and the values of a *and* b. The constants a and b are the (possibly complex) roots of the circuit's **characteristic equation**, which comes from the differential equation 13.1. To obtain the characteristic equation, replace each derivative of the output variable in equation 13.1 by s^k, where k is the order of the derivative, and set all sources [e.g., $v_{in}(t)$] to zero. The result for the differential equation 13.1 is

$$s^2 + \frac{R}{L}s + \frac{1}{LC} = 0,$$

which has roots

$$s = -0.5\frac{R}{L} \pm 0.5\sqrt{\frac{R^2}{L^2} - \frac{4}{LC}}. \tag{13.3}$$

For convenience, set

$$a = -0.5\frac{R}{L} - 0.5\sqrt{\frac{R^2}{L^2} - \frac{4}{LC}} \quad \text{and} \quad b = -0.5\frac{R}{L} + 0.5\sqrt{\frac{R^2}{L^2} - \frac{4}{LC}}.$$

If $R^2/L^2 - 4/LC \neq 0$, then $a \neq b$, and the solution is given by equation 13.2a. If, on the other hand, $R^2/L^2 - 4/LC = 0$, then $a = b$, and the solution has the form of 13.2b. *For simplicity in this example, assume that* a $\neq$ b, *i.e., there are distinct (possibly complex) roots.*

Step 4. Determine the constant D. For dc excitations to passive RLC circuits, a well-known rule of thumb states that after a long time, capacitors appear open and inductors look like shorts. For the series RLC circuit of figure 13.1, this principle forces $v_{out}(t) = V_0$ for large t. This in turn requires that the roots, a and b, in equation 13.2a have negative real parts; otherwise $v_{out}(t)$ would blow up for large t. Is the preceding arithmetic analysis consistent with this intuition? With $R > 0$, $C > 0$, and $L > 0$ (as expected for a typical RLC circuit), the roots, a and b, always have negative real parts. Hence, for infinite t, $v_{in}(t) = D = V_0$. For engineering purposes, for large t, $v_{in}(t) = V_0$.

Step 5. Determine A *and* B. It remains only to find A and B, which depend explicitly on the initial conditions. To compute these constants, notice that $v_{out}(0) = v_C(0)$, the initial capacitor voltage. Setting $t = 0$ in equation 13.2a yields

$$v_C(0) - V_0 = A + B. \tag{13.4}$$

The next step is to relate the derivative of the output to the initial conditions of the circuit. For the series circuit, the inductor and capacitor currents coincide, in which case

$$\frac{1}{C}i_C(t) = \frac{1}{C}i_L(t) = \dot{v}_C(t) = \dot{v}_{\text{out}}(t) = aAe^{at} + bBe^{bt}.$$

The dot over the functions v_{out} and v_C indicates a first-order time derivative. Two dots would indicate a second-order derivative. Evaluating the preceding equation at $t = 0$ implies that

$$\frac{1}{C}i_L(0) = aA + bB. \tag{13.5}$$

Equations 13.4 and 13.5 constitute two equations in two unknowns. This allows one to solve for A and B. In matrix form, equations 13.4 and 13.5 are

$$\begin{bmatrix} 1 & 1 \\ a & b \end{bmatrix} \begin{bmatrix} A \\ B \end{bmatrix} = \begin{bmatrix} v_C(0) - V_0 \\ \frac{1}{C}i_L(0) \end{bmatrix}.$$

Solving for A and B using the known form of the 2×2 matrix inverse (see appendix A-1) yields

$$\begin{bmatrix} A \\ B \end{bmatrix} = \begin{bmatrix} 1 & 1 \\ a & b \end{bmatrix}^{-1} \begin{bmatrix} V_C(0) - V_0 \\ \frac{1}{C}i_L(0) \end{bmatrix} = \frac{1}{b-a} \begin{bmatrix} b(v_C(0) - V_0) - \frac{1}{C}i_L(0) \\ a(v_C(0) - V_0) + \frac{1}{C}i_L(0) \end{bmatrix}.$$

As mentioned earlier, the foregoing technique, although quite useful for simple circuits, has serious drawbacks for circuits with more than two C's or L's. Drawbacks occur because higher order derivatives of circuit output variables generally have little or no physical meaning. Such derivatives are complicated linear combinations of initial capacitor voltages and initial inductor currents. The following example illuminates the difficulties.

◆ **EXAMPLE 13.2**

Figure 13.2 shows three RC circuits coupled together through dependent voltage sources. The goal of the example is to construct a differential equation model, determine the solution form in terms of arbitrary constants, and demonstrate the difficulties with the simple recipe of example 13.1 by attempting to relate the arbitrary constants to the initial conditions.

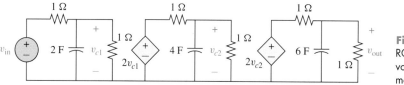

Figure 13.2 A string of three RC circuits coupled through dependent voltage sources. The differential equation model of the circuit is third order.

SOLUTION

Step 1. Construct the differential equation of the circuit. To construct the circuit's differential equation model, first write a differential equation relating v_{c1} to v_{in}. Then write one relating v_{c2} to v_{c1}, and finally, write one relating v_{out} to v_{c2}. Some straightforward algebra leads to the following three differential equations:

$$\frac{dv_{c1}}{dt} + v_{c1}(t) = 0.5v_{\text{in}} \tag{13.6}$$

$$2\frac{dv_{c2}}{dt} + v_{c2}(t) = v_{c1} \tag{13.7}$$

$$3\frac{dv_{\text{out}}}{dt} + v_{\text{out}}(t) = v_{c2}. \tag{13.8}$$

Successively substituting equation 13.7 into equation 13.6 and equation 13.8 into the result produces the input-output differential equation model

$$\frac{d^3 v_{out}}{dt^3} + \frac{11}{6}\frac{d^2 v_{out}}{dt^2} + \frac{dv_{out}}{dt} + \frac{1}{6}v_{out} = \frac{1}{12}v_{in}. \qquad (13.9)$$

Step 2. *Determine the characteristic equation.* The characteristic equation for this differential equation is

$$s^3 + \frac{11}{6}s^2 + s + \frac{1}{6} = 0,$$

which has roots $a = -0.5$, $b = -\frac{1}{3}$, and $d = -1$.

Step 3. *Determine the form of the solution.* If $v_{in}(t) = V_{in}$ volts dc for $t \geq 0$ and 0 for $t < 0$, then the complete solution has the form

$$v_{out}(t) = Ae^{at} + Be^{bt} + De^{dt} + E$$

for $t \geq 0$. As computed in step 2, $a = -0.5$, $b = -\frac{1}{3}$, and $d = -1$.

Step 4. *Compute A, B, D, and* E. Using the rule of thumb mentioned earlier, the simple calculation yields $E = 0.5V_{in}$. Calculation of A, B, and D specifies the solution. Applying the technique of example 13.1, one computes

$$v_{out}(0) - V_0 = A + B + D$$

$$\dot{v}_{out}(0) = aA + bB + dD$$

$$\ddot{v}_{out}(0) = a^2 A + b^2 B + d^2 D.$$

Again, *one dot over a variable means a first-order time derivative and two dots a second-order time derivative.* A, B, and D are computed by solving this set of equations. The difficulty is in specifying $\dot{v}_{out}(0)$ and $\ddot{v}_{out}(0)$. First, $v_{out}(0)$ is simply the initial capacitor voltage on the third capacitor, and $\dot{v}_{out}(t)$ is proportional to the current through it. But what is $\ddot{v}_{out}(t)$? And how do $\dot{v}_{out}(0)$ and $\ddot{v}_{out}(0)$ relate to the other two initial capacitor voltages? The relationship is complex and lacks any meaningful physical interpretation. Last, even for this simple example, computation of the differential equation 13.9 proves tedious.

Exercise. For example 13.2, compute expressions for $\dot{v}_{out}(0)$ and $\ddot{v}_{out}(0)$.
ANSWERS: $\dot{v}_{out}(0) = [v_{C2}(0) - v_{out}(0)]/3$; $\ddot{v}_{out}(0) = v_{out}(0)/9 - 5v_{C2}(0)/18 + v_{C1}(0)/6$.

One of the advantages of Laplace transform analysis is that it does not destroy the physical meaning of the circuit variables in the analysis process. Chapter 2 addresses how the Laplace transform approach explicitly accounts for initial capacitor voltages and initial inductor currents.

3. OVERVIEW OF LAPLACE TRANSFORM ANALYSIS

What is Laplace transform analysis? It is a technique that transfers or transforms the time domain analysis of a circuit, system, or differential equation to the **frequency domain**. In the frequency domain, solution of the equations is generally much easier. Hence, obtaining the output responses of a circuit to known inputs proceeds more smoothly.

What is the general solution methodology? To apply the technique, one takes the Laplace transform of the time-dependent input signal or signals to produce new signals dependent only on a new frequency variable s. In an intuitive sense, precisely defined later, one also takes the Laplace transform of the circuit. Assuming no initial conditions, one multiplies these two transforms together to produce the Laplace transform of the output signal. Taking the inverse Laplace transform of the output signal by known algebraic and table look-up formulas yields the desired response of the circuit. The effect of initial conditions is easily incorporated.

Figure 13.3 is a pictorial rendition of the method. As just mentioned, one transforms the input signal, transforms the circuit to obtain an equivalent circuit in the Laplace transform world, and computes the Laplace transform of the output by "multiplying" the two transforms together. Then, inverting this transform with the aid of a look-up table produces the desired output signal.

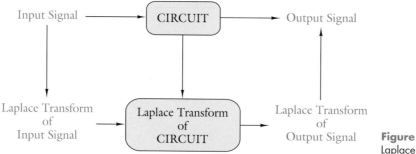

Figure 13.3 Diagram showing flow of Laplace transform circuit analysis.

In a mathematical context, one executes the same type of procedure on a differential equation model of a circuit and, indeed, differential equations in general. Figure 13.4 illustrates the idea.

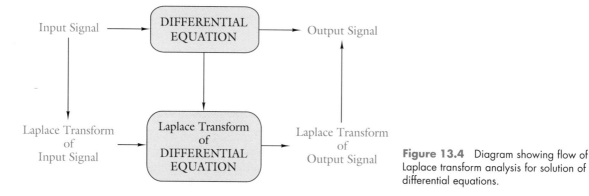

Figure 13.4 Diagram showing flow of Laplace transform analysis for solution of differential equations.

The benefit of this type of analysis lies in its numerous uses. Some of these uses include steady-state and transient analysis of circuits driven by complicated as well as the usual basic signals, a straightforward look-up table approach for computing solutions, and explicit incorporation of capacitor and inductor initial conditions in the analysis. The forthcoming sections will flesh out these applications.

4. BASIC SIGNALS

Several basic signals are fundamental to circuit analysis, as well as to future courses in system analysis. Perhaps the most common signal is the **unit step** function, denoted $u(t)$. The heavy line in figure 13.5 looks like a step on a household staircase and represents the graph of $u(t)$.

Mathematically the unit step is defined as

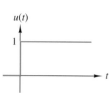

$$u(t) = \begin{cases} 1, & t \geq 0 \\ 0, & t < 0 \end{cases}. \tag{13.10}$$

Figure 13.5 Graph of the unit step function. It often represents a constant voltage or current level.

The unit step function has many practical uses, including the mathematical representation of dc voltage levels. Any type of sustained, constant physical phenomenon, such as constant pressure, constant heat, or the constant thrust of a jet engine has a steplike representation. In the case of jet engine thrust, a pilot will send a command signal through the control panel to the engine, requesting a given amount of thrust. The step function models this command signal.

The **shifted step**, shown in figure 13.6, models a delayed unit step signal.

Shifted steps, $u(t - T)$, often represent voltages that turn on after a prescribed time period T. The **flipped step function**, $u(T - t)$, of figure 13.7 depicts yet another variation on the unit step. Here the step takes on the value of unity for time $t \leq T$. Often, it provides an idealized model of signals which have excited the circuit for a long long time and turn off at time T.

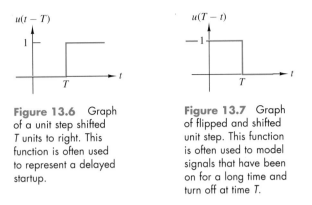

Figure 13.6 Graph of a unit step shifted T units to right. This function is often used to represent a delayed startup.

Figure 13.7 Graph of flipped and shifted unit step. This function is often used to model signals that have been on for a long time and turn off at time T.

Exercise. Represent each of the following functions as sums of step functions:

(*i*) $\quad f(t) = \begin{cases} 1, & 0 \leq t \leq 2 \\ 0, & \text{otherwise} \end{cases}$ $\qquad$ (*ii*) $\quad f(t) = \begin{cases} 1, & -3 < t < 6 \\ 0, & \text{otherwise} \end{cases}$

(*iii*) $\quad f(t) = \begin{cases} 1, & t \leq -1 \\ 0, & -1 < t < 1 \\ 1, & t \geq 1 \end{cases}$

ANSWERS, in random order: $u(-1-t) + u(t-1)$, $u(t) - u(t-2)$, $u(t+3) - u(5-6)$.

The **pulse function**, $p_T(t)$, of figure 13.8 is the product of a step and a flipped step or, equivalently, the difference of a step and a shifted step. Specifically, a pulse of height A and width T is

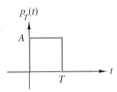

Figure 13.8 Pulse of width T and height A. This function is often used to model signals of fixed magnitude and short duration.

$$p_T(t) = Au(t)u(T-t) = A[u(t) - u(t-T)]. \qquad (13.11)$$

Another signal sharing a close kinship to the unit step is the **ramp**. The ramp function displayed in figure 13.9 is the integral of the unit step, i.e.,

$$r(t) = \int_{-\infty}^{t} u(q)dq = tu(t), \qquad (13.12)$$

where q is simply a dummy variable of integration.

Newtonian physics provides a good motivation for defining the ramp signal. Applying a constant force to an object causes a constant acceleration having the functional form $Ku(t)$. The integral of acceleration is velocity, which has the form $Kr(t)$, a ramp function.

A very common and conceptually utile signal is the (Dirac) **delta function**, or **unit impulse function**, implicitly defined by its relationship to the unit step as

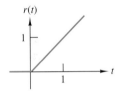

Figure 13.9 Graph of the ramp function, $r(t) = tu(t)$. Ramp functions conveniently model signals having a constant rate of increase.

$$u(t) = \int_{-\infty}^{t} \delta(q)dq. \qquad (13.13)$$

This relationship prompts a natural inclination to define

$$\delta(t) = \frac{d}{dt}u(t) = \lim_{h \to 0} \frac{u(t) - u(t-h)}{h}. \qquad (13.14)$$

Strictly speaking, the derivative does not exist at $t = 0$, due to the discontinuity of $u(t)$ at that point. Without delving into the mathematics, one typically interprets equations 13.13 and 13.14 as follows: Define a set of continuous differentiable functions $u_\Delta(t)$, as illustrated in figure 13.10a. The derivatives, $\delta_\Delta(t) = \frac{d}{dt}u_\Delta(t)$, of these functions are depicted in figure 13.10b.

Clearly, $\delta_\Delta(t)$ has a well-defined area of 1, has a height $1/\Delta$, and is zero outside the interval $0 \leq t \leq \Delta$. In addition, $u(t) = \lim u_\Delta(t)$, and $\lim \delta_\Delta = \delta(t)$ as $\Delta \to 0$. Hence, although the definition of equation 13.14 is not mathematically rigorous, one can interpret

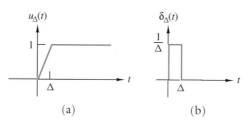

Figure 13.10 (a) Continuous differentiable approximation to the unit step. (b) Derivative of $u_\Delta(t)$: The integral of $\delta_\Delta(t)$ produces $u_\Delta(t)$.

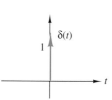

Figure 13.11
Standard graphical illustration of a unit impulse function having a well-defined area of 1. The function often represents a very large energy transfer over a very short time duration, such as might occur when a bat hits a baseball.

the delta function as a limit of a set of well-behaved functions. In fact, the delta function can be viewed as the limit of a variety of different sets of functions. A problem at the end of the chapter explores this phenomenon.

Despite the preceding mathematics, the delta function is not a function at all, but a distribution,[1] and its rigorous definition (in terms of so-called testing functions) is left to more advanced mathematics courses. Nevertheless, we shall still refer to it as the delta or impulse function. The standard graphical illustration of the delta function appears in figure 13.11, which shows a pulse of infinite height, zero width, and a well-defined area of unity, as identified by the "1" next to the spike. Visualization of the delta function by the spike in the figure will aid our understanding, explanations, and calculations that follow.

The unit area property comes about because of equation 13.13, i.e.,

$$\int_{-\infty}^{\infty} \delta(t)dt = \int_{0^-}^{0^+} \delta(t)dt = u(0^+) - u(0^-) = 1,$$

where 0^- is infinitesimally to the left, and 0^+ infinitesimally to the right, of zero. *If the area is different from unity, a number K alongside the spike will designate the area, i.e., the spike will be a signal $K\,\delta(t)$.*

One motivation for defining the delta function is its ability to "ideally" represent phenomena in nature of relative immediate energy transfer (i.e., the elapsed time over which energy transfer takes place is very small compared to the macroscopic behavior of the physical process). An exploding shell inside a gun chamber causing a bullet to change its given initial velocity from zero to some nonzero value "instantaneously" is such an example. So is a batter who hits a pitched ball, "instantaneously" transferring the energy of the swung bat to the ball. Also, the delta function provides a mathematical setting for representing the sampling of a continuous signal. Suppose, for example, that a continuous signal $v(t)$ is to be sampled at discrete time instants $t_1, t_2, t_3, \ldots$, i.e., $v(t_i)$ is to be physically measured at these time instants. The mathematical representation of this measuring process is given by the **sifting property** of the delta function, i.e.,

$$v(t_i) = \int_{-\infty}^{\infty} v(q)\delta(q - t_i)dq. \tag{13.15}$$

In other words, the value of the integral is the nonimpulsive part of the (continuous) integrand, $v(q)$, replaced by that value of q which makes the argument of the impulse zero, in this case $q = t_i$. Verifying equation 13.15 depends on an application of the definition given in equation 13.13. Specifically, if $v(t)$ is continuous at $t = t_i$ then

$$\int_{-\infty}^{\infty} v(q)\delta(q - t_i)dq = \int_{t_i^-}^{t_i^+} v(q)\delta(q - t_i)dq$$

$$= v(t_i) \int_{t_i^-}^{t_i^+} \delta(q - t_i)dq = v(t_i)$$

by equation 13.13, where $t_i^{\pm}$ are infinitesimally to the right and left of t_i.

These basic signals are the building blocks for constructing more complicated signals, as the next example illustrates.

---

[1]See A. H. Zemanian, *Distribution Theory and Transform Analysis* (New York: McGraw Hill, 1965).

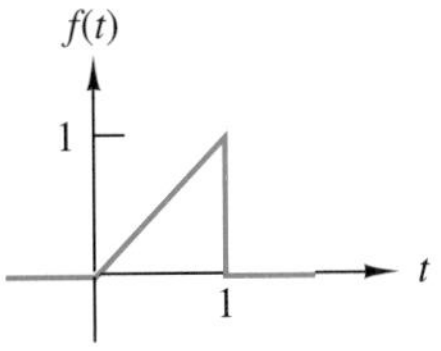

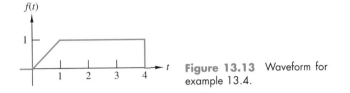

Figure 13.12 A sawtooth waveform for example 13.3. Sequences of sawtooth waveforms are used as timing signals in televisions and other electronic devices.

EXAMPLE 13.3

Figure 13.12 depicts a **sawtooth** waveform denoted by $f(t)$. It is possible to represent the sawtooth as the sum of a ramp, a shifted ramp, and a shifted step. Specifically, the sawtooth $f(t)$ of figure 13.12 has the representation

$$f(t) = r(t) - r(t-1) - u(t-1) = tu(t) - [(t-1)u(t-1) + u(t-1)].$$

EXAMPLE 13.4

Represent the signal of figure 13.13 in terms of the basic signals described earlier.

Figure 13.13 Waveform for example 13.4.

SOLUTION

By inspection for $0 \le t \le 1$, $f(t) = r(t)$. For $1 \le t \le 4$, $f(t) = 1$, which may be expressed as $r(t) - r(t-1)$. As such, $f(t) = r(t) - r(t-1)$, over $[0,4]$. It follows immediately that

$$f(t) = r(t) - r(t-1) - u(t-4).$$

Exercise. Compute the derivatives of the signals in figures 13.12 and 13.13.
ANSWERS, in random order: $u(t) - u(t-1) - \delta(t-4)$ and $u(t) - u(t-1) - \delta(t-1)$.

5. THE ONE-SIDED LAPLACE TRANSFORM

Intuitively, a transform is like a prism that breaks white light apart into its colored spectrum. The one-sided, or **unilateral, Laplace transform** is an integral mapping, somewhat like a prism, between time-dependent signals $f(t)$ and functions $F(s)$ that are dependent on a complex variable s.

LAPLACE TRANSFORM

Mathematically, the **one-sided Laplace transform** of $f(t)$ is

$$\mathcal{L}[f(t)] = F(s) = \int_{0^-}^{\infty} f(t)\, e^{-st}\, dt, \tag{13.16}$$

where $s = \sigma + j\omega (j = \sqrt{-1})$ is a complex variable (usually called the complex frequency).

As the equation makes plain, the Laplace transform integrates out time to obtain a new function, $F(s)$, displaying the frequency content of the original time function $f(t)$. In the vernacular, $F(s)$ is the frequency domain counterpart of $f(t)$. Analysis using Laplace transforms is often called frequency domain analysis.

A number of questions about the transform promptly arise:

Question 1: Why is it called one-sided or unilateral?

Answer: It is unilateral because the lower limit of integration is 0^-, as opposed to $-\infty$. If the lower limit of integration were $-\infty$, equation 13.16 would be called the **two-sided Laplace transform**, which is not covered in this text.

Question 2: Why use 0^- instead of 0^+ or 0 as the lower limit of integration?

Answer: Our future circuit analysis must account for the effect of "instantaneous energy transfer" and, hence, impulses at $t = 0$. The use of 0^+ would exclude such direct analysis, since the Laplace transform of the impulse function would be zero. Using $t = 0$ is simply ambiguous.

Question 3: What about functions that are nonzero for t < 0?

Answer: Because the lower limit of integration in equation 13.16 is 0^-, the Laplace transform does not distinguish between functions that are different for $t < 0$ but equal for $t \geq 0$ (e.g., $u(t)$ and $u(t+1)$ would have the same unilateral Laplace transform). However, since $t = 0$ designates the universal starting time of a circuit or system, the class of signals dealt with will usually be zero for $t < 0$ and thus will have a unique Laplace transform. Conversely, each Laplace transform $F(s)$ will determine a unique time function $f(t)$ with the property that $f(t) = 0$ for $t \leq 0$. Because of this dual uniqueness, the one-sided Laplace transform is said to be bi-unique for signals $f(t)$ with $f(t) = 0$ for $t < 0$.

Question 4: Does every signal f(t) *such that* f(t) $= 0$ *for* t < 0 *have a Laplace transform?*

Answer: No. For example, the function $f(t) = e^{t^2}u(t)$ does not have a Laplace transform because the integral of equation 13.16 does not exist for that function. To see why, one must study the integral closely. If one were to try to find the Laplace transform, then one would try to evaluate

$$\int_{0^-}^{\infty} e^{t^2} e^{-st} dt = \int_{0^-}^{\infty} e^{t^2 - st} dt = \int_{0^-}^{\infty} e^{t^2 - \sigma t} e^{-j\omega t} dt. \qquad (13.17)$$

Observe that $e^{-j\omega t} = \cos(\omega t) - j\sin(\omega t)$ is a complex sinusoid. As t approaches infinity, the real and imaginary parts of the integrand in equation 13.17 must blow up, due to the $e^{t^2-\sigma t}$ term. Hence, the area underneath the curve $e^{t^2-\sigma t}$ grows to infinity, and the integral does not exist for any value of σ. Whenever $f(t)$ is piecewise continuous, a sufficient condition for the existence of the Laplace transform is that

$$|f(t)| \leq k_1 e^{k_2 t} \qquad (13.18)$$

for some constants k_1 and k_2. This bound restricts the growth of a function; i.e., the function cannot rise more rapidly than an exponential. Such a function is said to be **exponentially bounded**. The condition, however, is not necessary. Specifically, the transform exists whenever the integral exists, even if the function $f(t)$ is unbounded. Without belaboring the mathematical rigor underlying the Laplace transform, we will presume throughout the book that the functions we are dealing with are Laplace transformable.

Question 5: Why does the existence of the Laplace transform integral depend on the value of σ, *mentioned in the answer to question 4?*

Answer: If the condition in equation 13.18 is satisfied, then there is a range of σ's (recall that $s = \sigma + j\omega$) over which the Laplace transform integral is convergent; i.e., the product $f(t)e^{\sigma t}$ is absolutely integrable over some range of σ's. Perhaps the best way to explain this is with an example. Let us find the Laplace transform of the unit step. By equation 13.16,

$$\mathcal{L}[u(t)] = U(s) = \int_{0^-}^{\infty} u(t)e^{-st} dt = \int_{0^-}^{\infty} e^{-st} dt = \int_{0^-}^{\infty} e^{-(\sigma + j\omega)t} dt$$

$$= \frac{e^{-\sigma t}e^{-j\omega t}}{\sigma + j\omega} \bigg]_{0^-}^{\infty} = \frac{1}{\sigma + j\omega} = \frac{1}{s}, \qquad (13.19)$$

provided that $\sigma > 0$. Notice that if $\sigma > 0$, then $e^{-\sigma t}e^{-j\omega t} \to 0$ as $t \to \infty$. This keeps the area underneath the curve finite. For $\sigma \leq 0$, the Laplace transform integral will not exist for the unit step. The smallest number σ_0 such that for all $\sigma > \sigma_0$ the Laplace transform integral exists is called the **abscissa of (absolute) convergence**. In the case of the unit step, the integral exists for all $\sigma > 0$; hence, $\sigma_0 = 0$ is the abscissa of convergence. The region σ > 0 is said to be the **region of convergence (ROC)** of the Laplace transform of the unit step. Figure 13.14 illustrates the ROC for the unit step.

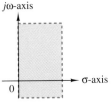

Figure 13.14
Region of convergence, σ > 0, of the Laplace transform of the unit step function (i.e., the Laplace transform integral will exist for all σ > 0).

Question 6: Is the unilateral Laplace transform valid only in its region of convergence?

Answer: Again, the answer is no. There is a method in the theory of complex variables called "analytic continuation" which, although beyond the scope of this text, permits us to uniquely and analytically extend the transform to the entire complex plane.[2] *Analytically* means smoothly and also that the extension is valid everywhere except at the poles (to be discussed later) of the transform. Thus, the region of convergence goes unmentioned in the standard mathematical tables of one-sided Laplace transform pairs.

The preceding discussion sets up the mathematical framework of the Laplace transform method. Our eventual focus rests on its application to circuit theory, which builds on two fundamental laws: Kirchhoff's voltage law (the KVL) and Kirchhoff's current law (the KCL). The KVL requires that the voltage drops around any closed loop sum to zero, and the KCL requires that the sum of all the currents entering a node be zero. For the Laplace transform technique to be useful, it must distribute over such sums of voltages and currents. Fortunately, it does.

Linearity Property The Laplace transform operation is linear, i.e.,

$$\mathcal{L}[a_1 f_1(t) + a_2 f_2(t)] = a_1 L[f_1(t)] + a_2 L[f_2(t)]. \qquad (13.20)$$

Or equivalently, if $f(t) = a_1 f_1(t) + a_2 f_2(t)$, then $F(s) = a_1 F_1(s) + a_2 F_2(s)$.
This property is easy to verify:

$$\mathcal{L}[a_1 f_1(t) + a_2 f_2(f)] = \int_{0^-}^{\infty} [a_1 f_1(t) + a_2 f_2(t)] e^{-st} dt$$

$$= a_1 \int_{0^-}^{\infty} f_1(t) e^{-st} dt + a_2 \int_{0^-}^{\infty} f_2(t) e^{-st} dt.$$

This is precisely what equation 13.20 states. Hence, our curiosity may rest peacefully in the knowledge that the Laplace transform technique conforms with the basic laws of circuit theory.

Before exploring structural properties of the Laplace transform further, several examples are in order.

◆ EXAMPLE 13.5

Find $F(s)$ when $f(t) = e^{-at} u(t)$, where a is real.

SOLUTION

Applying equation 13.16 yields

$$F(s) = \int_{0^-}^{\infty} e^{-at} e^{-st} dt = \int_{0^-}^{\infty} e^{-(s+a)t} dt = \frac{1}{s+a} \qquad (13.21)$$

The integral exists if $\text{Re}[s+a] > 0$, which implies that the ROC is σ > −a. As mentioned in the answer to question 6, by **analytic continuation**, $F(s) = 1/(s+a)$ is valid and analytic in the entire complex plane, except at the point $s = -a$. The point $s = -a$ is a **pole** of the rational function $1/(s+a)$ because as s approaches $-a$, the value of the function becomes infinitely large.

[2]John B. Conway, *Functions of One Complex Variable* (New York: Springer-Verlag, 1973).

EXAMPLE 13.6

Find $F(s)$ when $f(t) = 3u(t) + 2e^{-at}u(t)$, where a is a real number.

SOLUTION

The Laplace transform of $u(t)$ is $1/s$ by equation 13.19 and that of $e^{-at}u(t)$ is $1/(s+a)$, by equation 13.21. By the linearity property (equation 13.20),

$$F(s) = \frac{3}{s} + \frac{2}{s+a},$$

with region of convergence $\{\sigma > 0\} \cap \{\sigma > -a\}$, where $\cap$ denotes intersection. Again, by analytic continuation, the transform is valid in the entire complex plane except at the poles, $s = 0$ and $s = -a$.

Exercise. Find the Laplace transform of (i) $f(t) = e^{at}u(t) + e^{-at}u(t) + 2u(t)$ and (ii) $f(t) = -2u(t) + e^{at} - 2e^{-at}u(t)$.

ANSWERS: (i) $\dfrac{3s-a}{s^2-a^2} + \dfrac{2}{s}$; (ii) $\dfrac{3a-s}{s^2-a^2} - \dfrac{2}{s}$.

EXAMPLE 13.7

Suppose $f(t) = te^{-at}u(t)$, where a is real. Find $F(s)$.

SOLUTION

To find $F(s)$, apply integration by parts as follows:

$$F(s) = \int_{0^-}^{\infty} te^{-(s+a)}dt = \int_{0^-}^{\infty} v\ du = uv\ \Big]_{0^-}^{\infty} - \int_{0^-}^{\infty} u\ dv,$$

where $v = t$ and $du = e^{-(a+s)t}dt$. Thus,

$$F(s) = -\frac{te^{-(s+a)t}}{s+a}\Bigg]_{0^-}^{\infty} + \int_{0^-}^{\infty} \frac{e^{-(s+a)t}}{s+a}dt. \qquad (13.22)$$

The ROC is $\sigma > -a$, in which case the first term on the right-hand side of equation 13.22 is zero. Thus,

$$\mathcal{L}[te^{-at}u(t)] = \int_{0^-}^{\infty} \frac{e^{-(s+a)t}}{s+a}dt = \frac{1}{(s+a)^2}. \qquad (13.23)$$

Equation 13.23 is a special case of the more general formula

$$\mathcal{L}[t^n e^{-at}u(t)] = \frac{n!}{(s+a)^{n+1}}. \qquad (13.24)$$

Table 13.1 lists this transform pair, as well as numerous other such pairs without mention of the underlying region of convergence. In fact, we shall dispense with any mention of the ROC unless drawn by necessity.

EXAMPLE 13.8

Find $F(s)$ for the sawtooth $f(t)$ sketched in figure 13.15.

SOLUTION

Applying equation 13.16 and integration by parts produces the following string of equalities:

$$F(s) = \int_{0^-}^{1} t e^{-st} dt = -\left[\frac{t e^{-st}}{s}\right]_{0^-}^{1} + \int_{0^-}^{1} \frac{e^{-st}}{s} dt$$

$$= -\frac{e^{-s}}{s} - \frac{e^{-st}}{s^2}\Big]_{0^-}^{1} = -\frac{e^{-s}}{s} - \frac{e^{-s}}{s^2} + \frac{1}{s^2}.$$

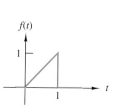

Figure 13.15
Sawtooth waveform of example 13.8.

EXAMPLE 13.9

Find $F(s)$ when $p(t) = Au(t)u(T - t)$, as sketched in figure 13.16.

SOLUTION

The integration here is not difficult:

$$P(s) = A \int_{0}^{T} e^{-st} dt = A \frac{1 - e^{-sT}}{s}. \tag{13.25}$$

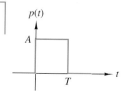

Figure 13.16 A pulse of height A and width T for example 13.9.

EXAMPLE 13.10

The circuit of figure 13.17a has two source excitations, $i_1(t)$ and $i_2(t)$, shown in figures 13.17b and c. Determine $V_{\text{out}}(s)$.

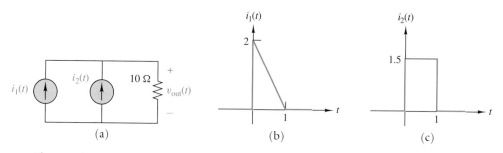

Figure 13.17 (a) Resistive circuit driven by two current sources. (b) Triangular signal, $i_1(t)$. (c) Pulse signal, $i_2(t)$.

SOLUTION

Step 1. Determine the form of $V_{out}(s)$. By superposition and Ohm's law,

$$v_{\text{out}}(t) = 10i_1(t) + 10i_2(t).$$

Using the linearity of the Laplace transform, we obtain

$$V_{\text{out}}(s) = 10I_1(s) + 10I_2(s).$$

Step 2. *Determine* $I_1(s)$. Some reflective thought yields $i_1(t) = (2 - 2t)u(t)u(1 - t)$. With this analytic expression, the transform computation proceeds as follows:

$$I_1(s) = \int_{0^-}^{1} (2 - 2t)e^{-st}dt = 2\int_{0^-}^{1} e^{-st}dt - 2\int_{0^-}^{1} te^{-st}dt.$$

The integral on the far right coincides with the one computed in example 13.8, except for the scale factor of 2. Its evaluation requires integration by parts:

$$I_1(s) = 2\frac{1 - e^{-s}}{s} + 2\frac{e^{-s}}{s} + 2\frac{e^{-st}}{s^2}\Big]_{0^-}^{1} = 2\left(\frac{1}{s} - \frac{1}{s^2} + \frac{e^{-s}}{s^2}\right).$$

Step 3. *Determine* $I_2(s)$. The signal in example 13.9 has the same shape as $i_2(t)$. From equation 13.25,

$$I_2(s) = 1.5\frac{1 - e^{-s}}{s}.$$

Step 4. *Determine* $V_{out}(s)$. Since $V_{out}(s) = 10I_1(s) + 10I_2(s)$, it follows that

$$V_{out}(s) = \frac{35}{s} - \frac{20}{s^2} + e^{-s}\left(\frac{20}{s^2} - \frac{15}{s}\right).$$

Exercise. Find the Laplace transform of (*i*) $f_1(t) = e^{at} + e^{-at}$, (*ii*) $f_2(t) = e^{-at} + e^{-bt} + te^{-bt} + te^{-at}$, and (*iii*) $f_3(t) = e^{-at}u(t) - e^{-a(t-2)}u(t - 2) + u(t - 3)$.

ANSWERS, in random order: $\frac{1}{s+a} + \frac{1}{s+b} + \frac{1}{(s+a)^2} + \frac{1}{(s+b)^2}$, $\frac{1}{(s+a)} + \frac{1}{(s-a)}$, and $\frac{1 - e^{-2s}}{s+a} + \frac{e^{-3s}}{s}$.

---

6. THE INVERSE LAPLACE TRANSFORM

For the Laplace transform tool to effectively analyze circuits, one must be able to uniquely reconstruct time functions $f(t)$ from their frequency domain partners $F(s)$. This is attained by the inverse Laplace transform integral.

INVERSE LAPLACE TRANSFORM
Intuitively, if $\mathcal{L}[f(t)] = F(s)$, then $f(t) = \mathcal{L}^{-1}[F(s)]$. Rigorously speaking, the **inverse Laplace transform** integral is a complex line integral defined as

$$f(t) = \mathcal{L}^{-1}[F(s)] = \frac{1}{2\pi j}\int_{\Gamma} F(s)e^{st}ds \qquad (13.26)$$

over a particular path Γ in the complex plane. The path Γ is typically taken to be the vertical line $\sigma_1 + j\omega$, where ω ranges from $-\infty$ to $+\infty$ and σ_1 is any real number greater than σ_0, the abscissa of absolute convergence.

This integral uniquely reconstructs the time structure of $F(s)$ to obtain $f(t)$ in which $f(t)$ is zero for $t < 0$. Conceptually, the process resembles the reverse action of a prism, to produce white light from its spectrum. An appreciation for the power of this integral requires a solid background in complex variables and would not even aid our purpose, the analysis of circuits. In fact, the evaluation of the integral is carried out using the famous residue theorem of complex variables. Further discussion is beyond the scope of this text.

Just as the Laplace transform is **linear**, so, too, is the inverse Laplace transform, as its integral structure would suggest; i.e., $\mathcal{L}^{-1}[a_1 F_1(s) + a_2 F_2(s)] = a_1 f_1(t) + a_2 f_2(t)$. Also, the unilateral transform pair $\{f(t), F(s)\}$ is **unique**, where, by unique, we mean the following: Let $F_1(s) = \mathcal{L}[f_1(t)]$ and $F_2(s) = \mathcal{L}[f_2(t)]$ coincide in any small open region of the complex plane. Then $F_1(s) = F_2(s)$ over their common regions of convergence, and $f_1(t) = f_2(t)$ for almost all $t > 0$. "Almost all" means except for a small or thin set of isolated points that are of no engineering significance. Hence, there is a one-to-one correspondence between time functions $f(t)$ for which $f(t) = 0$ for $t < 0$ and one-sided Laplace transforms. Linearity and this uniqueness cause the Laplace transform technique to be a productive tool for circuit analysis.

Virtually all the transforms of interest to us have a **rational function structure**; i.e., $F(s)$ is the ratio of two polynomials. Rational functions may be decomposed into sums of simple rational functions. These simple rational functions are called **partial fractions** and their sums are known as **partial fraction expansions**. Two of the more common "simple" terms in partial fraction expansions have the form K/s and $K/(s+a)$. Such simple rational functions correspond to the transforms of steps, shifted steps, exponentials, and the like. Table 13.1 lists these known inverse transforms. With the table, direct evaluation of the line integral in equation 13.26 becomes unnecessary. Our goal is to describe techniques to compute the simple rational functions in a partial fraction of $F(s)$. Once these are found, the transform dictionary in table 13.1, in conjunction with some well-known properties of the Laplace transform, will allow us quickly to determine the time function $f(t)$.

Partial Fraction Expansions: Distinct Poles

Our focus will center on proper[3] rational functions, i.e.,

$$F(s) = \frac{a_m s^m + a_{m-1} s^{m-1} + \cdots + a_1 s + a_0}{s^n + b_{n-1} s^{n-1} + \cdots + b_1 s + b_0} = \frac{a_m s^m + a_{m-1} s^{m-1} + \cdots + a_1 s + a_0}{(s - p_1)(s - p_2) \cdots (s - p_n)},$$

where $m \le n$ and $p_1, \ldots, p_n$ are the **zeros** of the denominator polynomial, $s^n + b_{n-1} s^{n-1} + \cdots + b_1 s + b_0$, and are called the **finite poles** of $F(s)$. For the most part, rational functions are sufficient for the study of basic circuits. There are three cases of partial fraction expansions to consider:

(1) the case of distinct poles, i.e., $p_i \ne p_j$ for all $i \ne j$;
(2) the case of repeated poles, i.e., $p_i = p_j$ for at least one $i \ne j$; and
(3) the case of complex poles.

Although case (3) is a subcategory of case (1) or (2) or both, its attributes warrant special recognition.

If $F(s)$ is a proper rational function with **distinct** (equivalently, **simple**) poles $p_1, \ldots, p_n$, then

$$F(s) = K + \frac{A_1}{(s - p_1)} + \frac{A_2}{(s - p_2)} + \cdots + \frac{A_n}{(s - p_n)}, \qquad (13.27)$$

where

$$K = \lim_{s \to \infty} F(s) \qquad (13.28\text{a})$$

and where the **residue** A_i of the pole p_i is

$$A_i = \lim_{s \to p_i} [(s - p_i) F(s)] = [(s - p_i) F(s)]_{s = p_i}, \qquad (13.28\text{b})$$

in which the rightmost equality is valid only when the numerator factor $(s - p_i)$ has been canceled with the factor $(s - p_i)$ in the denominator of $F(s)$; otherwise, one will obtain zero

---

[3]In the engineering literature *proper* refers to $m \le n$ and *strictly proper* $m < n$; in the mathematics literature, *proper* means $m < n$ and *improper* $m \ge n$.

---

divided by zero which, in general, is undefined. As intimated earlier, this partial fraction expansion should enable a straightforward computation of $f(t)$. Indeed, by inspection,

$$f(t) = K\delta(t) + A_1 e^{p_1 t} u(t) + A_2 e^{p_2 t} u(t) + \cdots + A_n e^{p_n t} u(t). \tag{13.29}$$

◆ EXAMPLE 13.11

Find $f(t)$ when

$$F(s) = \frac{1}{s(s+a)}.$$

SOLUTION

The solution proceeds by executing a partial fraction expansion (equation 13.27) on $F(s)$ to produce the Laplace transform of two elementary signals, a step and an exponential. Specifically,

$$F(s) = \frac{1}{s(s+a)} = \frac{A}{s} + \frac{B}{s+a},$$

where A/s is the Laplace transform of a weighted step, $Au(t)$, and $B/(s+a)$ is that of a weighted exponential, $Be^{-at}u(t)$. To find A, multiply both sides by s, cancel common numerator and denominator factors, and evaluate the result at $s = 0$, to produce $A = 1/a$. Similarly, to find B, multiply both sides by $s + a$, cancel common numerator and denominator factors, and evaluate the result at $s = -a$, to obtain $B = -1/a$. Recall that, by linearity, $\mathcal{L}^{-1}[aF(s)] = a\mathcal{L}^{-1}[F(s)]$. Hence,

$$f(t) = \frac{1}{a}u(t) - \frac{1}{a}e^{-at}u(t) = \frac{1}{a}(1 - e^{-at})u(t).$$

◆ EXAMPLE 13.12

Find $f(t)$ when

$$F(s) = \frac{cs^2 + b}{s(s+a)}.$$

SOLUTION

Here, the degrees of the numerator and denominator coincide. As s approaches infinity, $F(s)$ approaches a constant, meaning that a partial fraction expansion of $F(s)$ must include the constant term, K, of equation 13.27. Specifically,

$$F(s) = \frac{cs^2 + b}{s(s+a)} = K + \frac{A}{s} + \frac{B}{s+a}. \tag{13.30}$$

The value of K is determined by the behavior of $F(s)$ at infinity (equation 13.28a), i.e.,

$$K = \lim_{s \to \infty} \frac{cs^2 + b}{s(s+a)} = c.$$

A is found by multiplying equation 13.30 by s and evaluating the result at $s = 0$; i.e., as per equation 13.28b,

$$\left[\frac{cs^2 + b}{s+a}\right]_{s=0} = \left[Ks + A + \frac{Bs}{s+a}\right]_{s=0},$$

in which case $A = b/a$. B is found in a similar way: Multiply both sides by $s + a$ and evaluate at $s = -a$, to obtain

$$B = -\frac{b + a^2 c}{a}.$$

Using table 13.1 to take the inverse Laplace transform of the right-hand side of equation 13.30 with the values of K, A, and B properly inserted yields

$$f(t) = c\delta(t) + \frac{b}{a}u(t) - \frac{b + a^2c}{a}e^{-at}u(t).$$

Note: The inverse Laplace transform of a constant is the delta function.

Partial Fraction Expansions: Repeated Poles

Proper rational functions with repeated roots have a more intricate partial fraction expansion, and calculation of the residues often proves cumbersome. For example, suppose

$$F(s) = \frac{n(s)}{(s - a)^k d(s)},$$

where the denominator factor $(s - a)^k$ determines a repeated root of order k, $d(s)$ is the remaining factor in the denominator of the rational function $F(s)$, and $n(s)$ is the numerator of $F(s)$. The structure of a partial fraction expansion with repeated roots is

$$F(s) = \frac{A_1}{s - a} + \frac{A_2}{(s - a)^2} + \cdots + \frac{A_k}{(s - a)^k} + \frac{n_1(s)}{d(s)}, \tag{13.31}$$

where $A_1, \ldots, A_k$ are unknown constants associated with $s - a, \ldots, (s - a)^k$, respectively, and $n_1(s)$ and $d(s)$ are whatever remains in the partial fraction expansion of $F(s)$. The formulas for computing the A_i of equation 13.31 are

$$A_k = (s - a)^k F(s)\big]_{s=a} = \frac{n(s)}{d(s)}\bigg]_{s=a}, \tag{13.32a}$$

$$A_{k-1} = \frac{d}{ds}\left((s - a)^k F(s)\right)\bigg]_{s=a} = \frac{d}{ds}\left(\frac{n(s)}{d(s)}\right)\bigg]_{s=a}, \tag{13.32b}$$

and, in general,

$$A_{k-i} = \frac{1}{i!}\frac{d^i}{ds^i}\left((s - a)^k F(s)\right)\big]_{s=a} = \frac{1}{i!}\frac{d^i}{ds^i}\left(\frac{n(s)}{d(s)}\right)\bigg]_{s=a}. \tag{13.32c}$$

Of these expressions, only the first looks like the case with distinct roots; the others require derivatives of $(s - a)^k F(s)$. Computation of high-order derivatives borders on the tedious and is prone to numerical error. Computer implementation circumvents these difficulties. An example that illustrates the preceding formulas, as well as a useful trick, comes next.

 EXAMPLE 13.13

The goal here is to illustrate the computation of $f(t)$ when

$$F(s) = \frac{s + 2}{s^2(s + 1)^2} = \frac{A_1}{s} + \frac{A_2}{s^2} + \frac{B_1}{s + 1} + \frac{B_2}{(s + 1)^2}. \tag{13.33}$$

The two easiest constants to find are A_2 and B_2 as they require no differentiations for their solution. In particular, from equation 13.32a,

$$A_2 = s^2 F(s)\big]_{s=0} = \left[\frac{s + 2}{(s + 1)^2}\right]_{s=0} = 2,$$

and

$$B_2 = (s + 1)^2 F(s)\big]_{s=-1} = \left[\frac{s + 2}{s^2}\right]_{s=-1} = 1.$$

Finding A_1 and B_1 is more difficult, since their solution requires differentiation. According to equation 13.32b,

$$A_1 = \frac{d}{ds}[s^2 F(s)]_{s=0}.$$

To implement this formula multiply both sides of equation 13.33 by s^2, take the derivative of the resulting expression with respect to s, and evaluate at $s = 0$:

$$\frac{d}{ds}\left[\frac{s+2}{(s+1)^2}\right]_{s=0} = \frac{d}{ds}\left[A_1 s + A_2 + \frac{B_1 s^2}{s+1} + \frac{B_2 s^2}{(s+1)^2}\right]_{s=0} = A_1.$$

Observe that it is unnecessary to differentiate the terms that contain A_2, B_1, and B_2, since they disappear at $s = 0$, as the formula for A_1 requires. Consequently, in this case,

$$A_1 = \left[\frac{1}{(s+1)^2} - 2\frac{s+2}{(s+1)^3}\right]_{s=0} = -3.$$

Similarly,

$$B_1 = \frac{d}{ds}\left[(s+1)^2 F(s)\right]_{s=-1} = \frac{d}{ds}\left[\frac{s+2}{s^2}\right]_{s=-1} = \left[\frac{1}{s^2} - 2\frac{s+2}{s^3}\right]_{s=-1} = 3.$$

The reader should note that since A_1, A_2, and B_2 were known, a simple trick allows a more direct computation of B_1: Merely evaluate equation 13.33 at $s = 1$ (in fact any value of s, excluding the poles, will do), to obtain

$$0.75 = -3 + 2 + 0.25 + 0.5 B_1,$$

which again yields $B_1 = 3$.

The derivative formulas of equations 13.32 are often difficult to evaluate for complicated rational functions, such as

$$F(s) = \frac{n(s)}{(s+1)(s+2)(s+3)(s+7)^3}$$

$$= \frac{A}{s+1} + \frac{B}{s+2} + \frac{C}{s+3} + \frac{D_1}{s+7} + \frac{D_2}{(s+7)^2} + \frac{D_3}{(s+7)^3}.$$

For these functions, it is very efficient to find A, B, C, and D_3 directly. Then one evaluates $F(s)$ at two values of s, e.g., $s = 0$ and $s = 1$, to obtain two equations in the unknowns D_1 and D_2. Typically, solving the resulting two equations simultaneously is much easier than evaluating D_1 and D_2 directly by equations 13.32.

Partial Fraction Expansions: Distinct Complex Poles

Distinct complex roots present challenges different from the repeated root case. Since the roots are distinct but not real, the methods of equations 13.27 through 13.29 apply. Unfortunately, the resulting partial fraction expansion has complex residues, and the resulting inverse transform has complex exponentials multiplied by complex constants. Such imaginary time functions lack meaning in the real world unless their imaginary parts cancel to yield real time functions. When they do, our goal is to find a direct route for computing the associated real time signals. To do this, consider a rational function having a pair of distinct complex poles as in the following equation:

$$F(s) = \frac{n(s)}{\left[(s+a)^2 + b^2\right] d(s)} = \frac{n(s)}{(s+a+jb)(s+a-jb)\, d(s)}. \tag{13.34}$$

Since the poles $-a - jb$ and $-a + jb$ are distinct, the partial fraction expansion of equation 13.27 is valid. Since the poles are complex conjugates of each other, the residues of each pole are complex conjugates. Therefore, it is possible to write the partial fraction expansion of $F(s)$ as

$$F(s) = \frac{A+jB}{s+a+jb} + \frac{A-jB}{s+a-jb} + \frac{n_1(s)}{d(s)} \tag{13.35}$$

for appropriate polynomials $n_1(t)$ and $d(s)$. As per equation 13.28b, the residue is

$$A + jB = (s + a + jb)F(s)]_{s=-a-jb}. \qquad (13.36)$$

With A and B known, executing a little algebra on equation 13.35 to eliminate complex numbers results in an expression more amenable to inversion, i.e.,

$$F(s) = \frac{C_1 s + C_2}{(s+a)^2 + b^2} + \frac{n_1(s)}{d(s)}, \qquad (13.37)$$

where

$$C_1 = 2A \qquad (13.38a)$$

and

$$C_2 = 2aA + 2bB, \qquad (13.38b)$$

with A and B given in equation 13.36. With C_1 and C_2 given by equations 13.38, the rational function

$$\frac{C_1 s + C_2}{(s+a)^2 + b^2} \qquad (13.39)$$

is easily manipulated into a form that inverts to a sum of a sine and a cosine, as demonstrated in the following example.

EXAMPLE 13.14

Find $f(t)$ when

$$F(s) = \frac{3s^2 + s + 3}{(s+1)(s^2+4)} = \frac{D}{s+1} + \frac{C_1 s + C_2}{s^2 + 4}.$$

The first step is to find D by the usual techniques:

$$D = \left. \frac{3s^2 + s + 3}{s^2 + 4} \right]_{s=-1} = 1.$$

Given that $D = 1$, to find C_2, we evaluate $F(s)$ at $s = 0$, in which case $0.75 = 1 + 0.25C_2$, or $C_2 = -1$. With $D = 1$ and $C_2 = -1$, we evaluate $F(s)$ at $s = 1$ to obtain $0.7 = 0.5 + 0.2(C_1 - 1)$, or equivalently, $C_1 = 2$. Thus,

$$F(s) = \frac{1}{s+1} + 2\frac{s}{s^2+4} - 0.5\frac{2}{s^2+4}.$$

Using table 13.1 to compute the inverse transform yields

$$f(t) = [e^{-t} + 2\cos(2t) - 0.5\sin(2t)]u(t).$$

Example 13.14 illustrates not only the computation of an inverse transform having complex poles, but also the computation of C_1 and C_2 without resorting to complex arithmetic, as was needed in equation 13.36 to evaluate equations 13.38. The trick again was to evaluate $F(s)$ at two distinct s-values different from the poles of $F(s)$. This yields two equations that can be directly solved for C_1 and C_2.

A number of other methods, useful in computing partial fraction expansions of rational functions with an eye toward numerical implementation, are given in the literature.[4]

---

[4]See J. F. Mahoney and B. D. Sivazlian, "Partial Fraction Expansion: A Review of Computational Methodology and Efficiency," *Journal of Computational and Applied Mathematics*, Vol. 9, 1983, pp. 247–269; and J. J. Bongioro, Jr., "A Recursive Algorithm for Computing Partial Fraction Expansion of Rational Functions Having Multiple Poles," *IEEE Transactions on Automatic Control*, Vol. AC–29, No. 7, July 1984, pp. 650–651.

TABLE 13.1 LAPLACE TRANSFORMS OF BASIC FUNCTIONS

Item Number	$f(t)$	$\mathcal{L}[f(t)]$
1	$K\delta(t)$	K
2	$Ku(t)$ or K	$\dfrac{K}{s}$
3	$r(t)$	$\dfrac{1}{s^2}$
4	$t^n u(t)$	$\dfrac{n!}{s^{n+1}}$
5	$e^{-at}u(t)$	$\dfrac{1}{s+a}$
6	$te^{-at}u(t)$	$\dfrac{1}{(s+a)^2}$
7	$t^n e^{-at}u(t)$	$\dfrac{n!}{(s+a)^{n+1}}$
8	$\sin(\omega t)u(t)$	$\dfrac{\omega}{s^2+\omega^2}$
9	$\cos(\omega t)u(t)$	$\dfrac{s}{s^2+\omega^2}$
10	$e^{-at}\sin(\omega t)u(t)$	$\dfrac{\omega}{(s+a)^2+\omega^2}$
11	$e^{-at}\cos(\omega t)u(t)$	$\dfrac{s+a}{(s+a)^2+\omega^2}$
12	$t\sin(\omega t)u(t)$	$\dfrac{2\omega s}{(s^2+\omega^2)^2}$
13	$t\cos(\omega t)u(t)$	$\dfrac{s^2-\omega^2}{(s^2+\omega^2)^2}$
14	$\sin(\omega t+\phi)u(t)$	$\dfrac{s\sin(\phi)+\omega\cos(\phi)}{s^2+\omega^2}$
15	$\cos(\omega t+\phi)u(t)$	$\dfrac{s\cos(\phi)-\omega\sin(\phi)}{s^2+\omega^2}$
16	$e^{-at}[\sin(\omega t)-\omega t\cos(\omega t)]u(t)$	$\dfrac{2\omega^3}{\left[(s+a)^2+\omega^2\right]^2}$
17	$te^{-at}\sin(\omega t)u(t)$	$2\omega\dfrac{s+a}{\left[(s+a)^2+\omega^2\right]^2}$

7. Elementary Properties and Examples

The transform integral of equation 13.16 has various properties. These properties provide shortcuts in the transform computation of complicated, as well as simple, signals. For example, the Laplace transform of a right shift of the signal $f(t)$ always has the form $e^{-st}F(s)$. Shifts are important for two reasons:

1. Many signals can be expressed as the sum of simple signals and shifts of simple signals.
2. Excitations of circuits are often delayed from $t = 0$.

Hence, provisions for shifts must be built into analysis techniques. Another example is multiplication of $f(t)$ by t. This always results in a Laplace transform that is the derivative of $F(s)$. Also, recall linearity (equation 13.20), the first property of the Laplace transform set apart for examination. This property is fundamental to circuit analysis because of the linear structure of the KVL and KCL. Our discussion begins with the time shift property.

Time Shift Property If $\mathcal{L}[f(t)u(t)] = F(s)$, then, for $T > 0$,

$$\mathcal{L}[f(t - T)u(t - T)]) = e^{-sT}F(s). \qquad (13.40)$$

The verification of this property begins with a direct calculation of the Laplace transform for the shifted function, i.e.,

$$\mathcal{L}[f(t - T)u(t - T)] = \int_{0^-}^{\infty} f(t - T)u(t - T)e^{-st}dt = \int_{T^-}^{\infty} f(t - T)e^{-st}dt.$$

Let $q = t - T$, with $dq = dt$. Noting that the lower limit of integration becomes 0^- with respect to q produces

$$\mathcal{L}[f(t - T)u(t - T)] = \int_{0^-}^{\infty} f(q)e^{-sq}e^{-sT}dq = e^{-sT}\int_{0^-}^{\infty} f(q)e^{-sq}dq = e^{-sT}F(s).$$

Observe that if $T < 0$, the property fails to make sense, since $f(t - T)u(t - T)$ would then shift left. Since the transform ignores information to the left of 0^-, one cannot, strictly speaking, recover $f(t)$ from the resulting transform.

Exercise. Find $\mathcal{L}[f(t - T)]$ when $f(t)$ is: (*i*) $K\delta(t)$, (*ii*) $Ku(t)$, (*iii*) $e^{-at}u(t)$, and (*iv*) $te^{-at}u(t)$.

ANSWERS, in random order: $e^{-sT}/(s + a)^2, e^{-sT}/(s + a), Ke^{-sT}, Ke^{-sT}/s$.

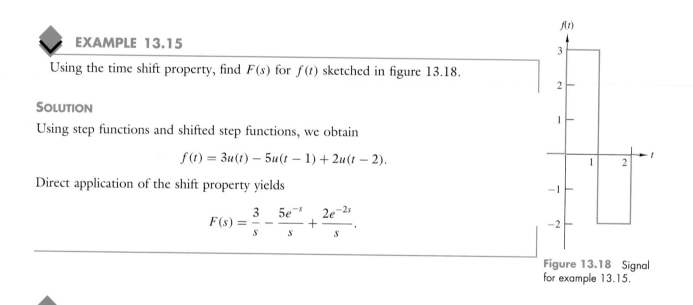

◆ EXAMPLE 13.15

Using the time shift property, find $F(s)$ for $f(t)$ sketched in figure 13.18.

SOLUTION

Using step functions and shifted step functions, we obtain

$$f(t) = 3u(t) - 5u(t - 1) + 2u(t - 2).$$

Direct application of the shift property yields

$$F(s) = \frac{3}{s} - \frac{5e^{-s}}{s} + \frac{2e^{-2s}}{s}.$$

Figure 13.18 Signal for example 13.15.

◆ EXAMPLE 13.16

Compute the inverse transform of the function

$$F(s) = \frac{1 - e^{-s}}{s(s + 1)} = \frac{1}{s(s + 1)} - \frac{e^{-s}}{s(s + 1)}.$$

SOLUTION

From example 13.10, or using a direct calculation, we obtain

$$\mathcal{L}^{-1}\left[\frac{1}{s(s + 1)}\right] = (1 - e^{-t})u(t).$$

Using the shift theorem yields

$$\mathcal{L}^{-1}\left[\frac{e^{-s}}{s(s+1)}\right] = (1 - e^{-(t-1)})u(t-1).$$

By the linearity of the inverse Laplace transform,

$$f(t) = (1 - e^{-t})u(t) - (1 - e^{-(t-1)})u(t-1).$$

A sketch of $f(t)$ appears in figure 13.19.

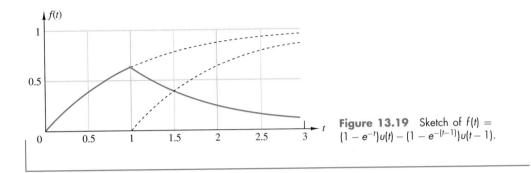

Figure 13.19 Sketch of $f(t) = (1 - e^{-t})u(t) - (1 - e^{-(t-1)})u(t-1)$.

Multiplication-by-t Property Let $F(s) = \mathcal{L}[f(t)u(t)]$. Then

$$\mathcal{L}[tf(t)u(t)] = -\frac{d}{ds}F(s). \qquad (13.41)$$

Verification of this property follows by a direct application of the Laplace transform integral to $tf(t)u(t)$ with the observation that $te^{-st} = -d/ds(e^{-st})$. In particular,

$$\int_{0^-}^{\infty} tf(t)e^{-st}dt = -\int_{0^-}^{\infty} f(t)\left[\frac{d}{ds}e^{-st}\right]dt = -\frac{d}{ds}\int_{0^-}^{\infty} f(t)e^{-st}dt = -\frac{d}{ds}F(s).$$

EXAMPLE 13.17

Recall that $\mathcal{L}[u(t)] = U(s) = 1/s$. Using the multiplication-by-t property,

$$\mathcal{L}[tu(t)] = -\frac{d}{ds}\left[\frac{1}{s}\right] = \frac{1}{s^2}.$$

EXAMPLE 13.18

Compute $F(s)$ for the pulse $f(t)$ and the triangular waveform $g(t) = tf(t)$ sketched in figure 13.20.

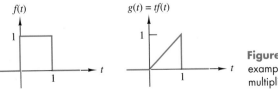

Figure 13.20 Signals for example 13.18 illustrating the multiplication-by-t property.

SOLUTION

By inspection, $f(t) = u(t) - u(t-1)$. Again by inspection or by example 13.8, $F(s) = (1 - e^{-s})/s$. Since $g(t) = tf(t)$, the multiplication-by-t property implies that

$$G(s) = -\frac{d}{ds}\left[\frac{1 - e^{-s}}{s}\right] = -\frac{e^{-s}}{s} + \frac{1 - e^{-s}}{s^2}.$$

Exercise. What are the Laplace transforms of $f(t) = tr(t)$ for the ramp function $r(t)$ and $f(t) = t^2 r(t)$?

ANSWERS: $2/s^3$, $6/s^4$.

Another handy property is the frequency shift property, which permits one to readily compute the transform of functions premultiplied by an exponential. With knowledge of the transforms of the step, $u(t)$, $\sin(\omega t)$, and other functions, computation of $e^{-at}u(t)$ and $e^{-at}\sin(\omega t)u(t)$ becomes quite easy.

Frequency Shift Property Let $F(s) = \mathcal{L}[f(t)]$. Then

$$\mathcal{L}[e^{-at} f(t)] = F(s + a). \qquad (13.42)$$

Proof: By direct calculation,

$$\mathcal{L}\left[e^{-at} f(t)\right] = \int_{0^-}^{\infty} f(t)e^{-(s+a)t}\,dt = F(s+a),$$

where we have viewed the sum, $s + a$, in the integral as a new variable p, which leads to $F(p)$ with p replaced by $s + a$.

◆ EXAMPLE 13.19

Suppose it is known that

$$\mathcal{L}[\sin(\omega t)u(t)] = \frac{\omega}{s^2 + \omega^2} \triangleq F(s).$$

If $g(t) = e^{-at}\sin(\omega t)u(t)$, compute $G(s)$. By the frequency shift property, $G(s) = F(s+a)$, or

$$\mathcal{L}[e^{-at}\sin(\omega t)u(t)] = \frac{\omega}{(s+a)^2 + \omega^2}.$$

Exercise. It is known that the Laplace transform of $\cos(\omega t)u(t)$ is $s/(s^2 + \omega^2)$. Determine $\mathcal{L}\left[e^{-at}\cos(\omega t)\right]$.

ANSWER: $(s+a)/\left[(s+a)^2 + \omega^2\right]$.

Another property of particularly widespread applicability is the time differentiation formula. Its utility resides not only in obtaining shortcuts to transforms of signals, but in the solution of differential equations. Differential equations provide a ubiquitous setting for modeling a large variety of physical systems—mechanical, electrical, chemical, etc. In terms of signal computation, recall that the velocity of a particle is the derivative of its position as a function of time. The acceleration is the derivative of the velocity. After computing the Laplace transform of the position as a function of time, one finds that a differentiation formula allows direct computation of the transforms of the velocity and acceleration. Also, as discussed at the very beginning of this chapter, circuits have differential equation models. For example, weighted sums of derivatives of the response of the circuit are equated

---

to weighted sums of derivatives of the input signal. Therefore, a differentiation formula is an essential ingredient in the analysis of circuits.

> **First-Order Time Differentiation Formula** Let $\mathcal{L}[f(t)] = F(s)$. Then
>
> $$\mathcal{L}\left[\frac{d}{dt}f(t)\right] = sF(s) - f(0^-). \qquad (13.43)$$
>
> Verification of the differentiation property proceeds via integration by parts as follows:
>
> $$\mathcal{L}\left[\frac{d}{dt}f(t)\right] = \int_{0^-}^{\infty}\left[\frac{d}{dt}f(t)\right]e^{-st}dt = f(t)e^{-st}\Big|_{0^-}^{\infty} - \int_{0^-}^{\infty}(-s)f(t)e^{-st}dt$$
>
> $$= -f(0^-) + sF(s).$$

The following examples explore some clever uses of the first-order time differentiation formula.

◆ EXAMPLE 13.20

Find $\mathcal{L}[\cos(\omega t)u(t)]$ using the time differentiation formula if it is known from table 13.1 that

$$\mathcal{L}[\sin(\omega t)u(t)] = \frac{\omega}{s^2 + \omega^2} \overset{\Delta}{=} F(s).$$

SOLUTION

Since

$$\cos(\omega t)u(t) = \frac{1}{\omega}\frac{d}{dt}\sin(\omega t)u(t),$$

the differentiation property immediately implies that

$$\mathcal{L}[\cos(\omega t)u(t)] = \frac{1}{\omega}\frac{s\omega}{s^2 + \omega^2} = \frac{s}{s^2 + \omega^2}.$$

> **Exercise.** Reverse the procedure of example 13.20; i.e., given the Laplace transform of the cosine function, compute the Laplace transform of the sine function. First relate the sine function to the derivative of the cosine function. Then apply the differentiation rule. Be careful to include the initial value of the cosine function.

◆ EXAMPLE 13.21

Recall that $\delta(t) = d/dt[u(t)]$. Using the sifting property, a direct calculation yields $\mathcal{L}[\delta(t)] = 1$. Is this consistent with the differentiation property? Interpreting the delta function as the derivative of the step function and applying the differentiation formula yields

$$\mathcal{L}[\delta(t)] = \mathcal{L}\left[\frac{d}{dt}u(t)\right] = s\mathcal{L}[u(t)] = s\left(\frac{1}{s}\right) = 1, \qquad (13.44)$$

which demonstrates the expected consistency.

◆ EXAMPLE 13.22

Let $f(t)$ and its derivative have shapes as in figure 13.21. The goal of this example is to explore the relationship between the Laplace transforms of $f(t)$ and $f'(t)$ in light of the differentiation property.

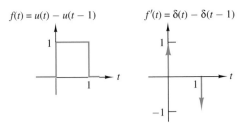

$f(t) = u(t) - u(t - 1)$ $f'(t) = \delta(t) - \delta(t - 1)$

Figure 13.21 A pulse and its derivative for example 13.22. Observe how the derivative of the pulse leads to a pair of delta functions.

Using linearity and the shift theorem on $f(t)$ yields

$$\mathcal{L}[f(t)] = \mathcal{L}[u(t) - u(t - 1)] = \mathcal{L}[u(t)] - \mathcal{L}[u(t - 1)] = \frac{1}{s}(1 - e^{-s}).$$

Applying the linearity of the Laplace transform to $f'(t)$ yields

$$\mathcal{L}[f'(t)] = \mathcal{L}[\delta(t) - \delta(t - 1)] = 1 - e^{-s}.$$

From the differentiation formula, it must follow that $\mathcal{L}[f'(t)] = s\mathcal{L}[f(t)]$. Thus,

$$\mathcal{L}[f'(t)] = s\mathcal{L}[f(t)] = s\frac{1}{s}(1 - e^{-s}) = 1 - e^{-s},$$

demonstrating consistency.

As is expected, the formula for the first derivative is a special case of the more general differentiation rule

$$\mathcal{L}\left[\frac{d^n}{dt^n}f(t)\right] = s^n F(s) - s^{n-1} f(0^-) - s^{n-2} f^{(1)}(0^-) - \cdots - f^{(n-1)}(0^-). \qquad (13.45)$$

This rule proves useful in the solution of general nth-order differential equations. Of particular use is the second-order formula

$$\mathcal{L}[f''(t)] = s^2 F(s) - sf(0^-) - f'(0^-). \qquad (13.46)$$

The inverse of differentiation is integration. The following property proves useful for quantities related by integrals.

Integration Property Let $F(s) = \mathcal{L}[f(t)]$. Then for $t \geq 0$,

$$\mathcal{L}\left[\int_{0^-}^{t} f(q)dq\right] = \frac{F(s)}{s}, \qquad (13.47a)$$

and

$$\mathcal{L}\left[\int_{-\infty}^{t} f(q)dq\right] = \frac{F(s)}{s} + \frac{\int_{-\infty}^{0^-} f(q)dq}{s}. \qquad (13.47b)$$

Proof: As with most of the proofs of the properties, integration by parts plays a key role. By direct computation (using equation 13.16),

$$\mathcal{L}\left[\int_{-\infty}^{t} f(q)dq\right] = \int_{0^-}^{\infty}\left[\int_{-\infty}^{t} f(q)dq\right]e^{-st}dt.$$

To use integration by parts, let

$$u = \int_{-\infty}^{t} f(q)dq \text{ and } dv = e^{-st}dt$$

Then

$$\mathcal{L}\left[\int_{-\infty}^{t} f(q)dq\right] = uv\Big]_{0^-}^{\infty} - \int_{0^-}^{\infty} v\,du$$

$$= \left[-\frac{e^{-st}}{s}\int_{-\infty}^{t} f(q)dq\right]_{0^-}^{\infty} + \frac{1}{s}\int_{0^-}^{t} f(t)e^{-st}dt.$$

For the appropriate region of convergence, the first term to the right of the equals sign reduces to

$$\left[-\frac{e^{-st}}{s}\int_{-\infty}^{t}f(q)dq\right]_{0-}^{\infty}=\frac{\int_{-\infty}^{0-}f(q)dq}{s}.$$

Since the second term to the right of the equals sign is $F(s)/s$, as per equation 13.47a, the property is verified.

◆ EXAMPLE 13.23

Find the Laplace transform of the signal $f(t)$ sketched in figure 13.22a using the integration property.

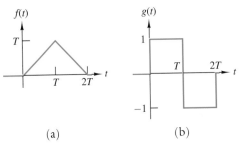

Figure 13.22 (a) A triangular signal $f(t)$ for example 13.23. (b) The derivative of $f(t)$.

SOLUTION

Observe that the triangular waveform $f(t)$ of figure 13.22a is the integral of the square wave $g(t)$. Since $g(t)$ is easily represented in terms of steps and shifted steps as

$$g(t) = u(t) - 2u(t-T) + u(t-2T),$$

its Laplace transform follows from an application of linearity and the time shift property:

$$\mathcal{L}[g(t)] = \frac{1}{s}\left[1 - 2e^{-sT} + e^{-2sT}\right].$$

The integration property implies that

$$\mathcal{L}[f(t)] = \mathcal{L}\left[\int_{-\infty}^{t}g(q)dq\right] = \frac{1}{s}\mathcal{L}[g(t)] = \frac{1}{s^2}\left[1 - 2e^{-sT} + e^{-2sT}\right].$$

◆ EXAMPLE 13.24

This example explores the voltage-current $(v\text{-}i)$ relationship of a capacitor in the frequency domain by way of the integration property. Recall the integral form of the voltage-current dynamics of a capacitor:

$$v_C(t) = \frac{1}{C}\int_{-\infty}^{t}i_C(q)dq.$$

Taking the Laplace transform of both sides and applying the integration property produces

$$\mathcal{L}[v_C(t)] = \mathcal{L}\left[\frac{1}{C}\int_{-\infty}^{t}i_C(q)dq\right] = \frac{1}{Cs}I_C(s) + \frac{1}{Cs}\int_{-\infty}^{0-}i_C(q)dq.$$

But this expression depends on the initial condition $v_C(0^-)$, because

$$v_C(0^-) = \frac{1}{C}\int_{-\infty}^{0-}i_C(q)dq.$$

Therefore,

$$V_C(s) = \frac{1}{Cs} I_C(s) + \frac{1}{s} v_C(0^-). \tag{13.48}$$

Equation 13.48 says that the voltage $V_C(s)$ is the sum of two terms: a term dependent on the frequency domain current $I_C(s)$ and a term that looks like a step voltage source and depends on the constant initial condition $v_C(0^-)$. The quantity $1/Cs$ looks like a generalized resistance: "generalized" because it depends on the frequency variable s and a "resistance" because it multiplies the Laplace current $I_C(s)$ to obtain a Laplace voltage $V_C(s)$. The ideas are analogous to Ohm's law. This prompts a series circuit interpretation of equation 13.48, as depicted in figure 13.23. An application of this equivalent frequency domain circuit to general network analysis comes about in the next chapter.

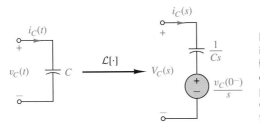

Figure 13.23 Equivalent circuit interpretation of a capacitor in the frequency domain. This equivalent is arrived at by applying the integration property of the Laplace transform to the capacitor voltage, seen as the integral of the capacitor current.

A second example interpreting the v-i characteristics of the capacitor in the frequency domain ensues from the differentiation rule. Instead of winding up with a series circuit, one obtains a parallel circuit. The interpretation is thus said to be dual to the one just described.

 EXAMPLE 13.25

This example has two goals: (*i*) Verify that equation 13.48 is consistent with the differentiation formula interpretation of the capacitor, and (*ii*) build a dual frequency domain interpretation of the v-i characteristic of a capacitor analogous to that of example 13.23.

As a first step, recall equation 13.48,

$$V_C(s) = \frac{1}{Cs} I_C(s) + \frac{1}{s} v_C(0^-),$$

which, after some algebra, becomes

$$I_C(s) = Cs V_C(s) - C v_C(0^-). \tag{13.49}$$

Notice that equation 13.49 is consistent with the application of the derivative formula to $i_C(t) = C[d(v_C/dt)]$. This consistency offers some security in the accuracy of our development. The interpretation of equation 13.49, however, is quite different from that of equation 13.48. In the latter equation, the current $I_C(s)$ equals the sum of two currents, $Cs V_C(s)$ and $-C v_C(0^-)$. This suggests a nodal interpretation, resulting in an equivalent circuit having two parallel branches. One branch contains a capacitor with voltage $V_C(s)$. The other, parallel, branch contains a current source with amperage $C v_C(0^-)$. The current through the capacitive branch is $(Cs)V_C(s)$, where "Cs" now acts like a generalized conductance because it multiplies a voltage, similar to Ohm's law. "Cs" is generalized because it depends on s. Figure 13.24 presents the equivalent circuit of the capacitor in the frequency domain and is dual to the circuit of figure 13.23. Chapter 2 covers in detail the role of these equivalent circuits in analysis.

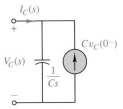

Figure 13.24 Equivalent circuit of a capacitor in the frequency domain using the differentiation formula.

The last elementary property of the Laplace transform that we consider in this chapter is the time-scaling property, also called the frequency-scaling property. Its importance is fundamental to network synthesis. Here, numerical problems, such as roundoff, prevent engineers from directly designing a circuit to meet a given set of specifications. Instead, the design engineer will normalize the specifications through a frequency-scaling technique. Once the normalized circuit is designed, frequency-scaling techniques are reapplied in an inverse fashion to obtain a circuit meeting the original specifications.

Time-/Frequency-Scaling Property Let $a > 0$ and $\mathcal{L}[f(t)] = F(s)$. Then

$$\mathcal{L}[f(at)] = \frac{1}{a}F\left(\frac{s}{a}\right), \tag{13.50}$$

or equivalently, $F(s/a) = a\mathcal{L}[f(at)]$.

Since the proof of this property is straightforward, it is left as an exercise at the end of the chapter.

EXAMPLE 13.26

Figures 13.25a and 13.25b show impulse trains that model sampling in signal-processing applications. The impulse train of figure 13.25b is the time-scaled counterpart to that of figure 13.25a.

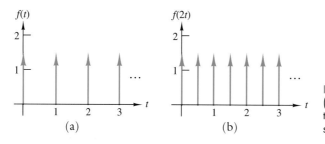

Figure 13.25 (a) Impulse train. (b) Time-scaled impulse train. Impulse trains such as these model sampling in signal-processing applications.

The time-scaled impulse train in figure 13.25b increases the frequency at which the impulses occur (twice as often as in the original signal). This is reflected in the Laplace transforms of the two signals:

$$\mathcal{L}[f(t)] = \mathcal{L}\left[\sum_{k=0}^{\infty}\delta(t-k)\right] = \sum_{k=0}^{\infty}e^{-sk} = \frac{1}{1-e^{-s}}. \tag{13.51}$$

By the time-scaling property,

$$\mathcal{L}[f(2t)] = 0.5\frac{1}{1-e^{-0.5s}}. \tag{13.52}$$

Notice that what occurs at, say, s_0 in equation 13.51 now occurs at $2s_0$ in equation 13.52. Hence, time scaling by numbers greater than 1 concentrates more of the frequency content of the signal in the higher frequency bands.

Exercise. Verify, by direct calculation that $\mathcal{L}[f(2t)]$ is given by the right side of equation 13.52.

Several more properties of the Laplace transform are germane to our purpose. However, these properties have a systems flavor and are postponed until chapter 15. We close this section by presenting table 13.2, which lists the Laplace transform properties and the associated transform pairs.

TABLE 13.2 LAPLACE TRANSFORM PROPERTIES

Property	Transform Pair
Linearity	$\mathcal{L}[a_1 f_1(t) + a_2 f_2(t)] = a_1 F_1(s) + a_2 F_2(s)$
Time Shift	$\mathcal{L}[f(t-T)u(t-T)] = e^{-sT} F(s), \; T > 0$
Multiplication by t	$\mathcal{L}[tf(t)u(t)] = -\dfrac{d}{ds} F(s)$
Multiplication by t^n	$\mathcal{L}[t^n f(t)] = (-1)^n \dfrac{d^n F(s)}{ds^n}$
Frequency Shift	$\mathcal{L}[e^{-at} f(t)] = F(s+a)$
Time Differentiation	$\mathcal{L}\left[\dfrac{d}{dt} f(t)\right] = sF(s) - f(0^-)$
Second-Order Differentiation	$\mathcal{L}\left[\dfrac{d^2 f(t)}{dt^2}\right] = s^2 F(s) - sf(0^-) - f^{(1)}(0^-)$
nth-Order Differentiation	$\mathcal{L}\left[\dfrac{d^n f(t)}{dt^n}\right] = s^n F(s) - s^{n-1} f(0^-) - s^{n-2} f^{(1)}(0^-)$ $\qquad\qquad - \cdots - f^{(n-1)}(0^-)$
Time Integration	(i) $\mathcal{L}\left[\displaystyle\int_{-\infty}^{t} f(q)dq\right] = \dfrac{F(s)}{s} + \dfrac{\int_{-\infty}^{0^-} f(q)dq}{s}$ (ii) $\mathcal{L}\left[\displaystyle\int_{0^-}^{t} f(q)dq\right] = \dfrac{F(s)}{s}$
Time/Frequency Scaling	$\mathcal{L}[f(at)] = \dfrac{1}{a} F\left(\dfrac{s}{a}\right)$

8. SOLUTION OF INTEGRODIFFERENTIAL EQUATIONS BY THE LAPLACE TRANSFORM

Differential equations provide a cross-disciplinary mathematical modeling framework. Although differential equation models may model only the dominant behavioral facets of a circuit and may in fact hide certain physical attributes, such as initial voltages or currents, their widespread utility and importance to circuits and control systems warrant special discussion.

To begin, recall the time differentiation formulas of equations 13.43 and 13.45 and the integration formulas of equations 13.47a and 13.47b. Also, recall that a differential equation relates a sum of derivatives of an output signal to a sum of derivatives of an input signal. For example, if the input and output signals are voltages, then

$$\frac{d^n v_{\text{out}}}{dt^n} + a_1 \frac{d^{n-1} v_{\text{out}}}{dt^{n-1}} + \cdots + a_n v_{\text{out}} = b_0 \frac{d^m v_{\text{in}}}{dt^m} + b_1 \frac{d^{m-1} v_{\text{in}}}{dt^{m-1}} + \cdots + b_n v_{\text{in}},$$

for constants a_i and b_i, might model the behavior of a linear circuit. We may use the following steps to solve this differential equation for $v_{\text{out}}(t)$ by the Laplace transform method:

1. Take the Laplace transform of both sides of the equation, using the appropriate derivative formulas.
2. Manipulate the resulting expression for $V_{\text{out}}(s)$.
3. Execute a partial fraction expansion.
4. Inverse transform the partial fraction expansion to obtain the answer.

If the equation is an integrodifferential equation, i.e., a mixture of both derivatives and integrals of the input and output signals, then we simply execute the same algorithm, except that we use the integral formula where appropriate. Some examples serve to illuminate the technique.

EXAMPLE 13.27

Consider the pulse current excitation (figure 13.26a) to the RC circuit of figure 13.26b. The goals of this example are: (*i*) to use and illustrate Laplace transform techniques to solve a differential equation derived from a simple RC circuit and (*ii*) to find the response voltage $v_C(t)$, $t \geq 0$, when $v_C(0^-) = 1$ V.

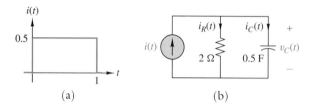

Figure 13.26 Excitation current (a) for a simple RC circuit (b) for example 13.27. What are the reference polarities for $v_C(t)$, given the indicated $i_C(t)$?

SOLUTION

Step 1. Find $\mathcal{L}[i(t)]$. Since $i(t) = 0.5u(t) - 0.5u(t-1)$,

$$\mathcal{L}[i(t)] = 0.5 \frac{1 - e^{-s}}{s}.$$

Step 2. Find the circuit's differential equation model that links the excitation current i(t) to the response voltage, $v_C(t)$. Since $i_R(t) = 0.5v_C(t)$ and $i_C(t) = 0.5dv_C/dt$, summing the currents into the top node of the circuit yields

$$i(t) = i_R(t) + i_C(t) = 0.5v_C(t) + 0.5 \frac{dv_C}{dt}(t).$$

After multiplying through by 2, the desired differential equation for the circuit becomes

$$\frac{dv_C}{dt}(t) + v_C(t) = 2i(t).$$

Step 3. Take the Laplace transform of both sides, apply the differentiation rule to the left-hand side, and solve for $V_C(s)$. Applying the Laplace transform to both sides yields

$$sV_C(s) - v_C(0^-) + V_C(s) = 2I(s).$$

Solving for $V_C(s)$ produces

$$V_C(s) = \frac{2}{s+1}I(s) + \frac{v_C(0^-)}{s+1} = \frac{1 - e^{-s}}{s(s+1)} + \frac{1}{s+1}.$$

Some straightforward calculations show that

$$\frac{1}{s(s+1)} = \frac{1}{s} - \frac{1}{(s+1)}.$$

Thus, with the aid of the shift property and the transform pairs of table 13.1, we obtain

$$v_C(t) = \mathcal{L}^{-1}[V_C(s)] = \mathcal{L}^{-1}\left[\frac{(1 - e^{-s})}{s(s+1)} + \frac{1}{s+1}\right]$$

$$= \left(1 - e^{-t}\right)u(t) - \left(1 - e^{-(t-1)}\right)u(t-1) + e^{-t}u(t)$$

$$= u(t) - \left(1 - e^{-(t-1)}\right)u(t-1).$$

Figure 13.27 presents a graph of this response. Because of the initial condition and the magnitude of the pulse input, the capacitor voltage is constant for $0 \leq t \leq 1$ second. At $t = 1$ second, the pulse magnitude drops to zero, making the circuit equivalent to a source-free RC circuit in which the capacitor voltage decays to zero as shown in the figure.

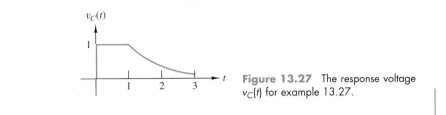

Figure 13.27 The response voltage $v_C(t)$ for example 13.27.

EXAMPLE 13.28

The goal of this example is to compute the response, denoted here by the input current $i_{in}(t)$, to the input voltage excitation $v_{in}(t) = \delta(t)$, given the series RLC circuit of figure 13.28. Suppose the initial conditions are $i_L(0^-) = 1$ A and $v_C(0^-) = -2$ V.

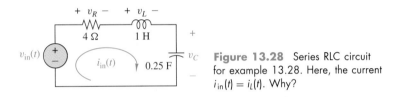

Figure 13.28 Series RLC circuit for example 13.28. Here, the current $i_{in}(t) = i_L(t)$. Why?

SOLUTION

Step 1. Compute the Laplace transform of the input. From the tables or by inspection, $\mathcal{L}[\delta(t)] = 1$.

Step 2. Compute the integrodifferential equation of the circuit of figure 13.28. The first task is to sum the voltages around the loop to obtain

$$v_R + v_L + v_C = v_{in}.$$

Substituting for each of the element voltages using the mesh current, $i_{in}(t)$, yields the desired integrodifferential equation,

$$Ri_{in}(t) + L\frac{di_{in}}{dt}(t) + \frac{1}{C}\int_{-\infty}^{t} i_{in}(q)dq = v_{in}(t). \tag{13.53}$$

Step 3. Take the Laplace transform of both sides, substitute for R, L, C, $i_L(0^-)$, $v_C(0^-)$, and $V_{in}(s)$, and solve for $I_{in}(s)$. With the aid of the differentiation and integration formulas, taking the Laplace transform of both sides of equation 13.53 produces

$$RI_{in}(s) + LsI_{in}(s) - Li_L(0^-) + \frac{1}{Cs}I_{in}(s) + \frac{v_C(0^-)}{s} = V_{in}(s).$$

This has the form

$$L\frac{s^2 + \frac{R}{L}s + \frac{1}{LC}}{s}I_{in}(s) = V_{in}(s) + Li_L(0^-) - \frac{v_C(0^-)}{s}.$$

Plugging in the required quantities and solving for $I_{in}(s)$ produces

$$I_{in}(s) = \frac{s}{s^2 + 4s + 4} + \frac{1}{s+2} = \frac{2}{s+2} - \frac{2}{(s+2)^2}.$$

Step 4. *Find* $i_{in}(t)$. Taking the inverse Laplace transform yields the desired result:

$$i_{in}(t) = (2 - 2t)e^{-2t}u(t).$$

A plot of this response appears in figure 13.29.

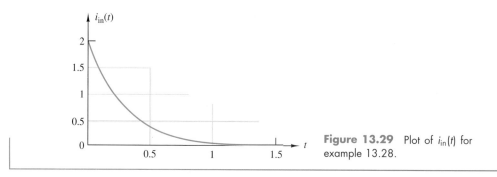

Figure 13.29 Plot of $i_{in}(t)$ for example 13.28.

⬥ **EXAMPLE 13.29**

This final example looks at the leaky integrator circuit of figure 13.30, which contains an ideal operational amplifier (op amp). Example 8.12 of volume I first considered this circuit. R_2 represents the leakage resistance of the capacitor. Given C and R_2, R_1 is chosen to achieve an overall gain constant, in this case, of 1. The objective is to compute the response $v_{out}(t)$, assuming that $v_C(0^-) = 0$, and compare it with that of a pure integrator having a gain constant of 1.

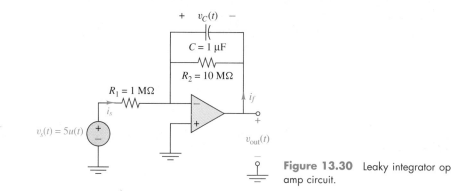

Figure 13.30 Leaky integrator op amp circuit.

SOLUTION

First, note that since the op amp is ideal, $-v_C(t) = v_{out}(t)$. The goal, then, is to write a differential equation that relates v_s to v_C and solve for v_C using the Laplace transform method.

Step 1. *Determine the differential equation.* Since the op amp is ideal, it follows that $i_f = -i_s$. From Ohm's law, $i_s = v_s/R_1$. On the other hand,

$$i_f(t) = -\frac{v_C(t)}{R_2} - C\frac{dv_C(t)}{dt}.$$

This leads to the differential equation model of the op amp circuit, viz.,

$$\frac{dv_{out}(t)}{dt} + \frac{v_{out}(t)}{R_2 C} = -\frac{v_s(t)}{R_1 C},$$

where, as indicated before, $v_{out}(t) = -v_C(t)$. Note that if $R_1 C = 1$ and R_2 is infinite, then the circuit works as a simple integrator. The circuit is called a leaky integrator because $R_2 C$ is large but finite. Since $R_1 C = 1$, one expects the gain constant to be 1 as well.

Step 2. *Substitute values, take the Laplace transform of both sides, and solve for* $V_{out}(s)$. Taking the Laplace transform of both sides, one obtains

$$s V_{out}(s) - v_{out}(0^-) + 0.1 V_{out}(s) = -5/s.$$

Since $v_{out}(0^-) = v_C(0^-) = 0$, it follows that

$$V_{out}(s) = -\frac{5}{s(s+0.1)} = \frac{-50}{s} + \frac{50}{s+0.1}.$$

Step 3. *Invert $V_{out}(s)$ to obtain $v_{out}(t)$.* Solving for $v_{out}(t)$ produces

$$v_{out}(t) = 50(e^{-0.1t} - 1)u(t).$$

A plot of the op amp output voltage appears in figure 13.31, along with the ideal integrator curve. Observe that the somewhat realistic leaky integrator circuit approximates an ideal integrator only for the approximate time interval $0 \le t \le 1.5$ before the error induced by the feedback resistor R_2 becomes too large. Such integrators need to be reinitialized by setting the capacitor voltage to zero.

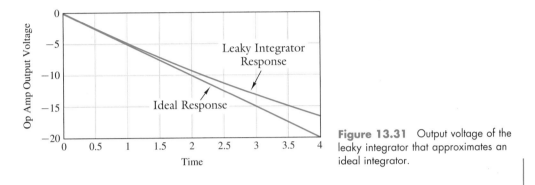

Figure 13.31 Output voltage of the leaky integrator that approximates an ideal integrator.

This concludes our discussion of the basic Laplace transform toolbox pertinent to circuit analysis. The next chapter takes up the application of the tools to the problem of basic circuit analysis.

SUMMARY

Chapter 1 has introduced the definitions of the Laplace transform of a signal and the inverse transform, which reconstructs the original signal under the assumption that the original signal was zero for $t < 0$. For the types of signals normally dealt with by electrical engineers, the Laplace transform is typically a rational function or an exponential of the form e^{-sT} times a rational function. If the transform is a rational function, then one performs a partial fraction expansion of the rational function. This produces a sum of terms that have simple catalogued inverse transforms, as listed in table 13.1. If the transform is multiplied by e^{-sT}, $T > 0$, then the resulting signal is simply a time-shifted version of the signal resulting when e^{-sT} is absent from the transform. Various properties

aiding the computation of the Laplace transform and its inverse are given in table 13.2.

Of special interest are the derivative and integral properties of the Laplace transform, because these allow one to transform differential and integrodifferential equation models of a circuit to the frequency domain. One can then solve for the Laplace transform of the output variable and invert it to obtain the time response of the circuit. This method allows one to easily incorporate the initial conditions of the differential equation into the solution. The example of a practical integrator, termed a leaky integrator, served to tie these concepts together. In the next chapter, we develop a more direct method for computing the circuit response.

TERMS AND CONCEPTS

Abscissa of convergence: the smallest value σ_0 of $s = \sigma + j\omega$ such that for all $\sigma > \sigma_0$, the Laplace transform integral converges.

Analytic continuation: the process of smoothly extending a complex-valued function, analytic in an open region of the complex plane, to the entire complex plane.

Characteristic equation of an ordinary differential equation: a polynomial, derived from an ordinary differential equation, whose zeros determine the characteristic exponents (e.g., p in e^{-pt}) in the response or output.

(Ordinary) Differential equation model: a model of the behavior of a circuit in which the input excitation and the desired response are related by an equation in which a scaled sum of time derivatives of the output signal is set equal to a scaled sum of derivatives of the input signal.

Dirac delta function: see unit impulse function.

Dot notation: one dot over a function means a first-order derivative, two dots a second-order derivative, etc.

Frequency domain: the s-domain, which refers to the analysis of circuits using Laplace transform methods.

Frequency shift: frequency translation of $F(s)$ to $F(s-a)$.

Inverse Laplace transform: a transformation whereby one takes a special complex-valued function $F(s)$ and uses it to compute a time function $f(t)$.

KCL (Kirchhoff's current law): law stating that, whether we are dealing with time signals or the transform of time signals, the sum of the currents entering (or leaving) a node is zero.

KVL (Kirchhoff's voltage law): law stating that whether we are dealing with time signals or the transform of time signals, the sum of the voltages around a loop is zero.

Laplace transform: a special integral transformation of a time function $f(t)$ to a function $F(s)$, where s is a complex variable.

Partial fraction expansion: the expansion of a rational function into a sum of terms. Each term is a constant (possibly complex), divided by a term of the form $(s+p)^k$.

Pole (simple) of rational function: zero of order 1 in a denominator polynomial.

Poles (finite) of rational function: zeros of a denominator polynomial.

Ramp function, $r(t)$: integral of unit step function; has the form $Ktu(t)$ for some constant K.

Rational function: ratio of two polynomials; also, a polynomial itself is a rational function.

Region of convergence (ROC): the region of the complex plane in which the Laplace transform integral is valid.

Shifted step function, $u(t-T)$: translation of a step function by T units. If T is positive, the translation is to the right; if T is negative, the translation is to the left.

Sifting property: the property of a delta function to simplify an integral, i.e., $f(T) = \int_{-\infty}^{+\infty} f(t)\delta(t-T)dt$, provided that $f(T)$ is continuous at T.

Time shift: time translation of a function $f(t)$ to $f(t-T)$.

Unit impulse function: a so-called distribution whose integral from $-\infty$ to t is the step function; it is nonrigorously interpreted as the derivative of the step function.

Unit step function, $u(t)$: $u(t) = 0$ for $t < 0$, and $u(t) = 1$ for $t \geq 0$.

Zeros of rational function: values that make a numerator polynomial zero.

PROBLEMS

1. *Difficulty with differential equation models of circuits.* Consider the RL one-port circuit in figure P13.1a. The differential equation relating the input voltage to the input current (the desired circuit response) is

$$i''(t) + 16i'(t) + 48i(t) = v'(t) + 8v(t).$$

Recall that a differential equation is an equilibrium relationship (similar to an energy balance equation) equating a linear combination of the input and its derivatives (in this case, $v(t)$ and $v'(t)$) with a linear combination of the output and its derivatives (in this case, $i(t)$, $i'(t)$, and $i''(t)$).

(a) If $v(t) = v'(t) = 0$ (a source-free circuit), then the response current has the form $i(t) = Ae^{at} + Be^{bt}$, $t \geq 0$, for appropriate a, b, A, and B. If $i_1(0) = 6$ A $= i_2(0)$, find $i(t)$ for $t \geq 0$ as follows: (*i*) Determine the characteristic equation, which is a quadratic. (*ii*) Determine the roots of the equation and thus a and b. (*iii*) Draw the equivalent circuit at $t = 0$, and label all initial inductor currents. (*iv*) Using the circuit constructed in part (*iii*), determine numerical values for $i(0)$ and $i'(0)$. (*Hint*: Express $i'(t)$ as a voltage.) (*v*) Determine A and B. *Check*: One term is $4.5e^{-4t}$.

(b) If $v(t) = 12$ V (a dc excitation), the solution has the general form $i(t) = Ae^{at} + Be^{bt} + C$, with a and b as in part (a). If $i_1(0) = i_2(0) = 0$, find $i(t)$ for $t \geq 0$. *Note*: First determine C by considering the circuit and the structure of $i(t)$ at $t = \infty$. *Check*: One term is $-1.5e^{-4t}$.

(c) If the input is the sawtooth sketched in figure P13.1b, can the methods just used be easily applied to find the response? Why? Spend no more than 15 minutes on this part. Doing so should illustrate the difficulty of finding the response of a circuit to inputs which are not sinusoidal or not exponential in nature. This provides a justification for Laplace transform techniques.

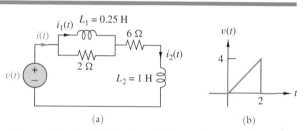

Figure P13.1 (a) Circuit and (b) waveform for problem 1.

2. *Review of transient circuit analysis by the method of differential equations.* Consider the circuit of figure P13.2.
(a) Show that the response $v_{\text{out}}(t)$ is related to the input $v_{\text{in}}(t)$ by the second-order differential equation

$$\ddot{v}_{\text{out}}(t) + 4\dot{v}_{\text{out}}(t) + 3v_{\text{out}}(t) = v_{\text{in}}(t).$$

(Recall that one dot means first derivative and two dots mean second derivative.)
(b) If $v_{c1}(0) = 7$ V, $v_{c2}(0) = 1$ V, and $v_i(t) = 6$ V dc, find $v_{\text{out}}(0)$ and $\dot{v}_{\text{out}}(0)$.
(c) Under the same conditions as in part (b), find $v_{\text{out}}(t)$ for $t \geq 0$ by the methods described in section 2. *Check*: $v(t) = Ae^{-t} + Be^{bt} + 2$ V.

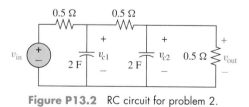

Figure P13.2 RC circuit for problem 2.

3. In the circuit of figure P13.3, the input current $i(t)$ and the output voltage $v(t)$ satisfy the differential equation

$$v^{(3)}(t) + 14v^{(2)}(t) + 52v^{(1)}(t) + 24v(t) = 4i^{(1)}(t) + 24i(t)$$

where the superscript (k) indicates the kth-order derivative (a third common notation for differentiation.)

(a) For the source-free case $i(t) = 0$, the solution $v(t)$ has the general form

$$v(t) = Ae^{at} + Be^{bt} + Ce^{ct}$$

because the characteristic equation of the differential equation is now a cubic having three roots a, b, and c. Determine the characteristic equation and the roots a, b, and c. (*Check*: Let $a = -6$; then $-0.6 \le b \le -0.5$, and $-8 \le c \le -7$.)

(b) Suppose $v_{c1}(0) = 12$ V, $v_{c2}(0) = 9$ V, and $v_{c3}(0) = 6$ V. Determine numerical values for (i) $v(0)$, (ii) $v'(0)$, and (iii) $v''(0)$. (This is straightforward, but very tedious; do not spend more than 15 minutes on this part.) With these values (if computed), determine A, B, and C. (*Hint*: Use a computer or a calculator that solves three equations in three unknowns. Do not attempt to do this part by hand.)

(c) Answer these questions: (i) Is $v'(0)$ proportional to a single voltage or current of the circuit? (ii) How does this circuit compare with the second-order circuit of problem 1? (iii) Is $v''(0)$ proportional to a single voltage or current? (iv) Would you want to solve circuit equations using such methods?

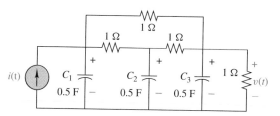

Figure P13.3 Third-order RC circuit for problem 3.

4. Determine the value of
(a) $\int_{-\infty}^{\infty} f(q + T_0)\delta(t - q)dq$.
(b) $\int_{-\infty}^{\infty} e^{-5q} \cos(0.5\pi q + \frac{\pi}{4})\delta(2t - q)dq$.

5. Sketch the indicated waveforms:
(a) $r(t - T)$ for $T > 0$.
(b) $-r(T - t)$ for $T > 0$.
(c) $r(t)r(1 - t)$.
(d) $r(t) - tu(1 - t)$.
(e) $\sum_{i=0}^{5} \delta(t - i)$.
(f) $2r(t)\delta(\tau - 5)$.
(g) $p_T(t)$ when $T = 1, 0.5, 0.25$.

6. Find simple expressions for the waveforms shown in figure P13.6, in terms of steps, shifted steps, terms of the form $Atu(t)$, and shifts of such terms. Some of the answers are

$$v(t) = u(-t + 1),$$

$$v(t) = (-0.5t + 1)u(t) + 0.5u(t - 1)$$
$$= -0.5tu(t) + 0.5tu(t - 1) + u(t) - u(t - 1),$$

and

$$v(t) = tu(t) - (t - 1)u(t - 1)$$
$$= tu(t) - tu(t - 1) + u(t - 1).$$

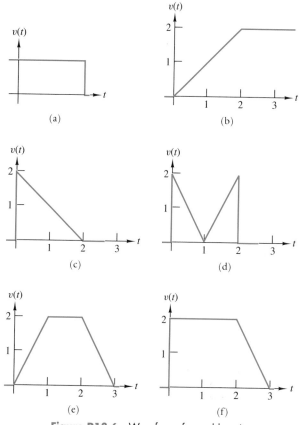

Figure P13.6 Waveforms for problem 6.

7. For each of the following time functions, first sketch the curve of $f(t)$ vs. t, and then compute the Laplace transform of $f(t)$ by evaluating the Laplace transform integral:
(a) $f(t) = 5e^{-4t}u(t)$.
(b) $f(t) = 5e^{-4t}u(t - 1)$.
(c) $f(t) = 5e^{-4(t-1)}u(t - 1)$.
(d) $f(t) = 5\delta(t)^{-4t} \cos(4t + 25\pi)$.

8. Compute the Laplace transform of the following signals by evaluating the Laplace transform integral:
(a) $f_1(t) = -2\delta(t) \cos(17\pi t - 0.25\pi)$.
(b) $f_2(t) = 2e^{-0.5t}u(t)u(1 - t)$.
(c) $f_3(t) = f_1(t) - 0.5 f_2(t)$.
(d)

(e)

Remark: Use an integral table as needed.

9. (a) For the circuit of figure P13.9a, use the Laplace transform tables to determine $\mathcal{L}[i_{out}(t)]$ for the input

$$i_{in}(t) = 4u(t) - 12tu(t) + 6e^{-3t}u(t)$$

$$+ 18e^{-t}\sin(2\pi t)u(t).$$

(b) For the circuit of figure P13.9b and the input of figure P13.9c, determine $\mathcal{L}[v_{out}(t)]$.

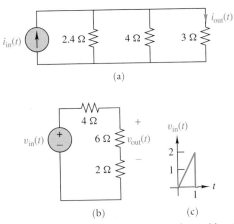

(a)

(b)   (c)

Figure P13.9 Circuits and waveform for problem 9.

10. Compute the Laplace transform of the signals (a) and (b) sketched in figure P13.10 by evaluating the Laplace transform integral.

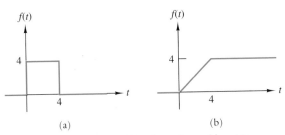

(a)   (b)

Figure P13.10 Waveforms for problem 10.

11. (a) By definition, $e^{j\omega t} = \cos(\omega t) + j\sin(\omega t)$. Find an expression for $\cos(\omega t)$ and $\sin(\omega t)$ in terms of the complex exponential $e^{j\omega t}$.
(b) Using the formulas developed in (a) and the Laplace transform of $e^{j\omega t}$, show that

(1) $\mathcal{L}[\cos(\omega t)] = \dfrac{s}{s^2 + \omega^2}$.

(2) $\mathcal{L}[\sin(\omega t)] = \dfrac{\omega}{s^2 + \omega^2}$.

(c) Show that

$$\mathcal{L}[\sin(\omega t) - \omega t\cos(\omega t)] = \frac{2\omega^2}{\left(s^2 + \omega^2\right)^2}.$$

12. Show that
(a) $\mathcal{L}[\sinh(at)u(t)] = \dfrac{a}{s^2 - a^2}$ with ROC $\text{Re}[s] > a$.

(b) $\mathcal{L}[\cosh(at)u(t)] = \dfrac{s}{s^2 - a^2}$ with ROC $\text{Re}[s] > a$.

13. Find $f(t)$ when $F(s)$ equals:
(a) $\dfrac{4}{s(s^2 + 2s + 2)}$.

(b) $\dfrac{4s + 2}{s^2(s + 1)^2}$.

(c) $\dfrac{2s}{s^4 + 1}$.

(d) $\dfrac{2}{(s + 1)^3(s + 2)}$.

14. Using partial fraction expansions and your knowledge of the Laplace transform of simple signals, find $f(t)$ when $F(s)$ equals:
(a) $\dfrac{2s + 4}{s(s + 6)}$.

(b) $\dfrac{s + 3}{(s + 2)(s + 4)}$.

(c) $\dfrac{2s^2 + 5s + 6}{(s + 2)(s + 4)}$.

(d) $\dfrac{2s^2 + 5s + 6}{(s + 2)(s + 1)^2}$.

(e) $\dfrac{1,728}{(s + 1)(s + 2)(s + 3)(s + 4)(s + 5)^3}$.

15. Find $f(t)$ if $F(s)$ equals:
(a) $\dfrac{s + 4}{s^2 + 4}$.

(b) $\dfrac{6s + 26}{4s^2 + 24s + 40}$.

(c) $\dfrac{s^2 + 44s}{(s^2 + 64)(s + 4)}$.
   (*Hint*: If a quadratic factor $q(s) = s^2 + bs + c$ of the denominator has complex roots, do not break it into linear factors. Instead, keep the quadratic factor in the partial fraction expansion, and use table 13.1 of the text.)

16. Using partial fraction expansions and the inverse transform table, determine the inverse Laplace transform of the following rational functions:
(a) $\dfrac{48}{(s + 1)(s + 2)(s + 3)(s + 4)}$.

(b) $3\dfrac{s^3 + 8s^2 + 13s + 5}{s^3 + 4s^2 + 5s + 2}$.

(c) $\dfrac{s^3 + 7s^2 + 18s + 16}{s(s^2 + 4s + 8)}$.

(d) $\dfrac{s^3 + 5s^2 + 12s + 18}{(s^2 + 4)(s^2 + 2s + 10)}$.
   (*Hint*: If a quadratic factor $q(s) = s^2 + bs + c$ in the denominator has complex roots, do not break it into linear factors. Instead, keep the quadratic factor in the partial fraction expansion, and use table 13.1 of the text.)

17. Consider the resistive circuit in figure P13.17. Use table 13.1 and the shift property to find and sketch $v_{out}(t)$ when
(a) $V_{in}(s) = 25\dfrac{e^{-s} + e^{-2s} - e^{-3s} - e^{-4s}}{s}$.

(b) $V_{in}(s) = 10\dfrac{1 - e^{-s}}{s^2} - 20\dfrac{e^{-4s}}{s}$.

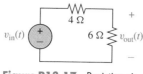

Figure P13.17 Resistive circuit used in problem 17.

18. (a) Find the Laplace transform of the function $f(t)$ sketched in figure P13.18a.

(b) Identify a relationship between $f(t)$ and the function $g(t)$ sketched in figure P13.18b. Use your answer to part (a) and the appropriate property from table 13.2 to determine the Laplace transform of $g(t)$.

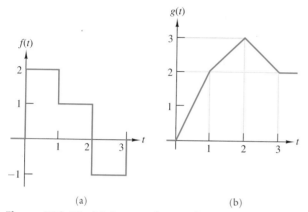

(a)   (b)

Figure P13.18 (a) Staircase function for problem 18. (b) A function related to that of (a).

19. The purpose of this problem is to develop some skill in applying the Laplace transform properties listed in table 13.2 of the text. Suppose

$$\mathcal{L}[f(t)u(t)] = F(s) = \frac{s^2}{s^2 + 4}.$$

Find the Laplace transform of

(a) $f(t-1)u(t-1)$.
(b) $tf(t)u(t)$.
(c) $(t-1)f(t-1)u(t-1)$.
(d) $tf(t-1)u(t-1)$.

List the property or properties you used to solve each part.

20. As in problem 19, the purpose of this problem is to develop some skill in applying the Laplace transform properties listed in table 13.2 of the text. As before, list the property or properties you used to solve each of the following.

(a) Suppose

$$\mathcal{L}[f(t)u(t)] = F(s) = \ell n \left[\frac{s^2 - 4}{s^2 + 4} \right].$$

Find the Laplace transform of

(1) $-tf(t)u(t)$
(2) $-te^{-2t}f(t)u(t) = -e^{-2t}tf(t)u(t)$.

(b) With $F(s)$ as in part (a), determine the function $g'(t)$ when $g(t) = -tf(t)u(t)$.

21. Use only Laplace transform properties to answer the following question. Suppose that for $t \leq 0$, $f(t) = e^t u(-t)$ and the one-sided Laplace transform of $f(t)$ is $\mathcal{L}[f(t)] =$ $F(s) = (s+1)/s^3$. Let $g(t) = f(t)u(t)$. Find the Laplace transform of $v(t)$ when

(a) $v(t) = 2g''(t) - g'(t)$.
(b) $v(t) = 2f''(t) - f'(t)$.
(c) $v(t) = g'(t) + \int_{-\infty}^{t} g(q)dq$.
(d) $v(t) = f'(t) + \int_{-\infty}^{t} f(q)dq$.

It is not necessary to have the answer be a rational function.

22. Using only the properties listed in table 13.2, derive the Laplace transform of $5e^{at}\sin(\omega t + \theta)$ from the Laplace transform of $\cos(t)$. Identify which property was used in each step in your derivation.

23. Let $f(t)$ and $g(t)$ be as sketched in figure P13.23. Determine $G(s)$ in terms of $F(s)$, e.g., $G(s) = KF(s+T)$ or some other modification of $F(s)$.

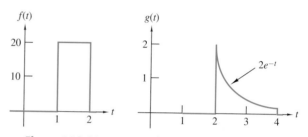

Figure P13.23 Functions $f(t)$ and $g(t)$ for problem 23.

24. The purpose of this problem is to apply the basic Laplace transform properties of table 13.2 and the transform pairs of table 13.1 to several situations. In all cases, identify which specific properties you used to obtain your answer.

(a) Find $F(s)$ when $f(t)$ is a half cycle of a sine wave, as shown in figure P13.24. (*Hint*: Can you express $f(t)$ as the difference of two sine waves?)

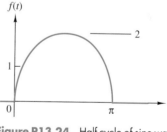

Figure P13.24 Half cycle of sine wave for problem 24.

(b) Find the Laplace transform of $f(t) = t\cos(bt)$, given the Laplace transform of $\cos(t)$ from table 13.1.
(c) Now suppose $f(t) = te^{at}\cos(bt)$. Find $F(s)$.

25. Suppose a time function $f(t)$ has zero value for negative t. If $F(s) = (s+2)/(s+1)$, find the Laplace transforms of

(a) $g_1(t) = 5f(t-2)$.
(b) $g_2(t) = 5exp(-2t)f(t)$.
(c) $g_3(t) = 5exp(-3t)f(t-2)$.
(d) $g_4(t) = 5tf(t)$.

Identify each of the properties you used to compute your answers.

26. Find the Laplace transform of each of the time functions shown in figure P13.26. Make use of the time domain derivative-integral relationships among the given time functions.

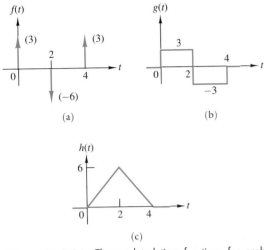

(a)

(b)

(c)

Figure P13.26 Three related time functions for problem 26.

27. This problem reviews basic operational amplifier properties typically covered in a first course on circuits. An ideal op amp has several properties among which are (*i*) a zero input current i_{in}, and (*ii*) a zero input voltage v_{in} (a virtual ground). Use these two properties to prove the relationships indicated in each circuit shown in figure P13.27.

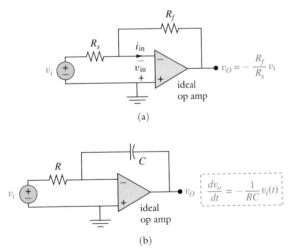

(a)

(b)

Figure P13.27 (a) Inverting amplifier. (b) Ideal integrator.

28. The op amp in the circuit in figure P13.28 is assumed to be ideal.
(a) Use nodal analysis to construct a first-order differential equation describing the input-output relationship of the voltages.
(b) If $v_i(t) = 6u(t)$ V and $v_c(0) = 4$ V, find $V_0(s)$ and then $v_0(t) = \mathcal{L}^{-1}[V_0(s)]$. Sketch the response.

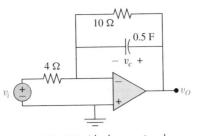

Figure P13.28 Ideal operational amplifier circuit (leaky integrator) for problem 28.

29. Regard the 10 Ω resistor in problem 28 as R and the capacitor as C. Show that if the RC product becomes large, then the output is proportional to the integral of the input, assuming no initial capacitor voltage. (*Hint*: Review example 13.28.)

30. Consider the RC active circuit shown figure 13.2 of example 13.2.
(a) Write node equations in the time domain.
(b) Laplace transform each of the three node equations, accounting for initial conditions.
(c) If no input is applied (i.e., $v_{in}(t) = 0$) and the initial capacitance voltages are $v_{c1}(0) = 12$ V, $v_{c2}(0) = 6$ V, and $v_{c3}(0) = 3$ V, find $v(t)$ for $t \geq 0$.

Remark: This problem is similar to example 13.2, which could not be solved easily by the time domain differential equation methods.

31. Prove the time-/frequency-scaling property of equation 13.51 by direct calculation via equation 13.17.

32. Consider the circuit of figure P13.32.

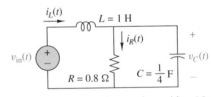

Figure P13.32 RLC circuit for problem 13.32.

(a) Use KVL and KCL to show that the differential equation relating the input voltage to the capacitor voltage is

$$\frac{d^2 v_C}{dt^2} + \frac{1}{RC}\frac{dv_C}{dt} + \frac{1}{LC}v_C = \frac{1}{LC}v_{in}.$$

(b) Take the Laplace transform of both sides of this equation to show that

$$V_C(s) = \frac{1}{LC}\frac{1}{s^2 + \frac{1}{RC}s + \frac{1}{LC}}V_{in}(s)$$

$$+ \frac{(s+5)v_C(0^-) + \dot{v}_C(0^-)}{s^2 + \frac{1}{RC}s + \frac{1}{LC}}.$$

(c) Assuming that $v_C(0^-) = \dot{v}_C(0^-) = 0$, $R = 0.8$ Ω, $L = 1$ H, $C = 0.25$ F, and $v_{in}(t) = \delta(t)$, show that

$$v_C(t) = \frac{4}{3}\left[e^{-t} - e^{-4t}\right]u(t) \text{ V}.$$

33. (a) For the circuit of figure P13.33, use mesh analysis to determine an integrodifferential equation in terms of the input current $i(t)$ and the inductor current $i_L(t)$.
(b) Take the Laplace transform of the integrodifferential equation of (a), accounting for initial conditions by using the time differentiation and time integration properties of the Laplace transform. *Hint*:

$$\frac{1}{C}\int_{-\infty}^{0^-} i_C(q)dq = v_C(0^-).$$

(c) Solve for $I_L(s)$ in terms of $I(s)$ and the initial conditions. (*Hint*: The answer should contain the sum of two terms; one depending on $I(s)$ and the other on the ICs. Also, the coefficient of s^2 in the denominators should be 1.)
(d) Find $I(s)$ as a single rational function when $i(t) = \pi^2(1 - e^{-t})u(t)$ A.
(e) Let $R = 2\ \Omega$, $L = 1$ H, and $C = (1 + \pi^2)^{-1}$. If $v_C(0^-) = \pi$ V, $i_L(0^-) = 1$ A, and $i(t)$ is as in part (d), find $i_L(t)$.

Remark: Your answer to part (e) should have two parts; one due to the forcing function and one due to the initial conditions. In linear circuits, it is always possible to separate the part due to the initial conditions from the part due to the forcing function.

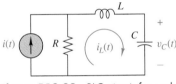

Figure P13.33 RLC circuit for problem 33.

34. In equation 13.13, the delta function, $\delta(t)$, was implicitly defined by the property

$$u(t) = \int_{-\infty}^{t} \delta(q)dq,$$

i.e., the property that the function's integral over $[-\infty, t]$ is given by the value of the unit step at t. Usually, we define a function by its value at each point. With delta functions and other related signals (so-called distributions), this is not possible, as was illustrated with equation 13.14, $\delta(t) = d/dt[u(t)]$, whose derivative fails to exist at $t = 0$. To circumvent this difficulty, we interpreted equation 13.14 to mean that $\delta(t) = \delta_\Delta(t)$ as $\Delta \to 0$. The function $\delta_\Delta(t)$ was given by figure 13.10b. This is one of many functions, say, $f_\Delta^i(t)$, that converge to $\delta(t)$ in the sense that

$$u_\Delta^i(t) \triangleq \int_{-\infty}^{t} f_\Delta^i(q)dq$$

has the property that

$$u(t) = \lim_{\Delta \to 0} u_\Delta^i(t)$$

(a) For the signals sketched in figure P13.34, show that

$$u(t) = \lim_{\Delta \to 0} u_\Delta^i(t).$$

(b) For which functions does $\delta_\Delta^i(0) = 0$ for all Δ? Many tests define $\delta(t) = \infty$ at $t = 0$. Caution should be exercised in such a definition.

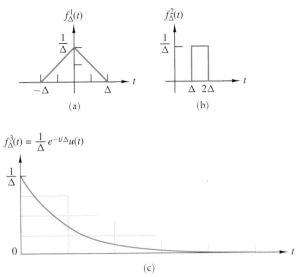

(a)   (b)

$$f_\Delta^3(t) = \frac{1}{\Delta}e^{-t/\Delta}u(t)$$

(c)

Figure P13.34 Functions for problem 34. As Δ gets small, successive functions converge to the delta function.

35. The ROC of the one-sided Laplace transform of a signal does not have a great deal of use, as the text points out and as is evidenced by its absence from various tables of transforms. However, for the so-called two-sided Laplace transform of a signal $f(t)$, i.e.,

$$\mathcal{L}_{II}[f(t)] = F_{II}(s) = \int_{-\infty}^{\infty} f(t)e^{-st}dt,$$

the ROC is critical to the correct transform of the function.
(a) Compute the two-sided Laplace transform of $f_1(t) = e^{-at}u(t)$ for $a > 0$, and determine the ROC.
(b) Compute the two-sided Laplace transform of $f_2(t) = -e^{-at}u(-t)$ for $a > 0$, and determine the ROC.
(c) Compare your answers to parts (a) and (b). Note that (a) and (b) should have the same transform, but different ROCs.

36. Complex distinct poles in a partial fraction expansion have various, but equivalent, inverse transform time representations. Show that

$$\mathcal{L}^{-1}\left[\frac{A + jB}{s + a + jb} + \frac{A - jB}{s + a - jb}\right] = Ke^{at}\cos(bt + \phi)u(t),$$

where $K = |A + jB|$ and $\phi = \tan^{-1}(B/A)$.

37. Determine the Laplace transform of the function $f(t)$ sketched in figure P13.37. If only $F(s)$ were known, determine and sketch the inverse transform of $F(s)$.

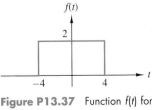

Figure P13.37 Function $f(t)$ for problem 37.

38. Given that $F(s) = \mathcal{L}\{f(t)\} = (3s + 12)/[(s + 2)^2 + 9]$, it follows that

$$f(t) = Ae^{-at}\cos(\omega t) + Be^{-at}\sin(\omega t).$$

Determine a, ω, A, and B.

39. The inductor current $i(t)$ in a second-order RLC circuit satisfies the following integrodifferential equation for $t > 0$:

$$4\frac{di}{dt} + 8i + \left[2 + 12\int_{0^-}^{t} i(\tau)d\tau\right] = 6.$$

(a) If $i(0^-) = 8$ A, determine the Laplace transform of $i(t)$.
(b) Solve your answer in part (a) for $i(t)$.

40. If $f(t) = \delta(t) + \delta(t-1)$, then $f(2t) = a[\delta(t) + \delta(t-b)]$.
(a) Determine a and b.
(b) Determine $\mathcal{L}[f(2t)]$ in two different ways, first, by computing $F(s)$ and then using the scaling property to compute $\mathcal{L}[f(2t)]$, and second, by direct calculation using your answer in part (a).

41. A pair of differential equations that represent a circuit is given as

$$\frac{dx(t)}{dt} + x(t) = y(t),$$

$$\frac{dy(t)}{dt} + y(t) = u(t),$$

with initial conditions $x(0^-) = 1$ and $y(0^-) = 2$. The solution of the pair has the form

$$x(t) = [A + Be^{-t} + Cte^{-t}]u(t).$$

Determine A, B, and C.

42. Determine the Laplace transform of the signal, part of a sine wave, sketched in figure P13.42.

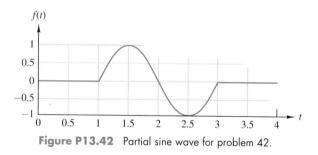

Figure P13.42 Partial sine wave for problem 42.

(*Hint*: Represent one cycle of a sine wave as the difference of a sine and a shifted sine; shift the answer.)

43. Determine the Laplace transform of the signal sketched in figure P13.43.

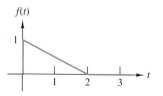

Figure P13.43 Signal for problem 43.

44. The inverse Laplace transform of a signal is

$$G(s) = \frac{12s^2 + 32s + 40}{s^3 + 2s^2 + 5s}.$$

It is known that $g(t) = [A + Be^{-t}\sin(2t) + Ce^{-t}\cos(2t)]u(t)$. Determine A, B, and C.

45. The circuit of figure P13.45 has an input of $10u(t - 1)$ V, valid for $t \geq 0$, and has an initial inductor current of $i_L(0^-) = 10$ A.

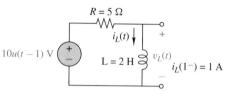

Figure P13.45 Driven series RL circuit for problem 45.

(a) Show that the differential equation for the circuit is

$$\frac{di_L(t)}{dt} + \frac{R}{L}i_L(t) = 10u(t - 1).$$

(b) Use the Laplace transform method to compute the inductor current, $i_L(t)$, for $t \geq 0$.

46. The circuit of figure P13.46 has two source excitations applied at $t = 0$, as indicated by the presence of the step functions. The initial condition on the inductor current is $i_L(0^-) = -1$ A.
(a) Determine a differential equation between each source and the response $i_L(t)$.
(b) Determine the complete response, $i_L(t)$, for $t \geq 0$ using the Laplace transform method and superposition applied to the two differential equations computed in part (a). Be careful not to account for the initial condition twice. Your answer should be

$$i_L(t) = -e^{-2t}u(t) + (1 - e^{-2})u(t)$$

$$+ 2(1 - e^{-2t})u(t) \text{ A}.$$

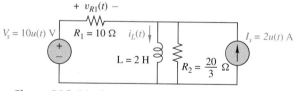

Figure P13.46 Driven RL circuit excited by two sources.

47. Consider the LC circuit of figure P13.47, in which $i_L(0^-) = 5$ A and $v_C(0^-) = 5$ V. Since there is no resistance present in the circuit, there is no damping; hence, one expects a purely sinusoidal response. Such circuits are called *lossless*.
(a) Determine a differential equation in the inductor current, $i_L(t)$.

(b) Solve the differential equation by the Laplace transform method, and show that $i_L(t) = 5\cos(t) + 5\sin(t)$ for $t \geq 0$.

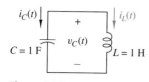

Figure P13.47 LC circuit, having a purely sinusoidal response, for problem 47.

48. The circuit of figure P13.48 is a series (lossless) LC circuit driven by a voltage source. Suppose $v_C(0^-) = 0$ and $i_L(0^-) = 0$.

(a) Determine the differential equation of the circuit in terms of the capacitor voltage, $v_C(t)$.

(b) Solve the differential equation using the Laplace transform method, and show that

$$v_C(t) = 10 - 10\cos(0.5\pi t) \text{ for } t \geq 0.$$

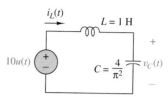

Figure P13.48 Series LC circuit for problem 48.

CHAPTER 14

CHAPTER OUTLINE

Laplace Transform Analysis, 2: Circuit Applications

A Fluorescent Light Application

THE word *fluorescence* describes a process in which one type of light is converted into another. In a fluorescent light, an electric current heats up electrodes at each end of a tube. The hot electrodes emit free electrons, which, for a sufficiently high voltage between the electrodes, initiate an arc causing mercury contained in the tube to vaporize. In turn, the energized mercury vapor emits invisible ultraviolet light, which strikes a phosphorus coating on the inside of the tube. The phosphorus absorbs this invisible short-wavelength energy and gives off light that has a longer and visible wavelength.

A starter circuit must quickly generate sufficient free electrons to initiate the arc that vaporizes the mercury inside the tube. One type of starter circuit contains a special heat-sensitive switch in series with an inductor. We will model this special switch by an ideal heat-sensitive (bimetal) switch in parallel with a capacitor. The concepts developed in this chapter will allow us to analyze the operation of such a starter circuit. Example 14.9 presents the analysis. ◆

CHAPTER OBJECTIVES

- Define a generalized (frequency-dependent) resistance called an impedance, denoted $Z(s)$, and a generalized (frequency-dependent) conductance called an admittance, denoted $Y(s)$, in terms of the Laplace transform variable s. Impedances and admittances will satisfy a type of Ohm's law. These ideas are generalizations of the phasor-based notions of impedance and admittance introduced in volume I, chapter 10, section 6.
- Learn to manipulate impedances and admittances in the Laplace transform domain in the same way that resistances and conductances are manipulated in the time domain.
- Define s-domain equivalent circuits of initialized capacitors and inductors for the purpose of the transient analysis of circuits.
- Use the new concepts of impedance and admittance to redevelop the ideas of voltage division, source transformations, linearity, and Thévenin and Norton equivalent circuits in the Laplace transform or complex-frequency domain.
- Use the impedance and admittance of circuit elements for nodal and loop analysis of circuits in the Laplace transform or complex-frequency domain.
- Utilize the Laplace transform technique, and especially the s-domain equivalent circuits of initialized capacitors and inductors, for the solution of switched RLC circuits.
- Introduce the notion of a switched capacitor circuit, which has an important place in real-world filtering applications.

1. INTRODUCTION

Chapter 1 cultivated the Laplace transform as a mathematical tool particularly useful for circuits modeled by differential equations. The current chapter adapts the Laplace transform tool to the peculiar needs and attributes of circuit analysis. With the Laplace transform methods described in this chapter, the intermediate step of constructing a circuit's differential equation, as was done in chapter 1, can be eliminated.

Available for the analysis of resistive circuits is a wide assortment of techniques: Ohm's law, voltage and current division, nodal and loop analysis, linearity, etc. For the sinusoidal steady-state analysis of RLC circuits, phasors served as a natural generalization of these techniques of resistive circuits. The Laplace transform tool permits us to extend the steady-state phasor analysis methods to a much wider setting where transient and steady-state analysis are both possible for a very large range of input excitations, not amenable to phasor analysis. Recall also that transient analysis is not possible with phasors.

The keys to this generalization are the notions of impedance and its inverse, admittance. Instead of defining impedance in terms of $j\omega$, as in phasor analysis, we will define it in terms of the Laplace transform variable s. With this definition, it will be possible to evolve a frequency- or s-dependent Ohm's law, s-dependent voltage and current division formulas, and s-dependent nodal and loop analysis; in short, all of the basic circuit analysis techniques have analogous s-dependent formulations. What is most important, however, is that with the s-dependent formulation, it will be possible to define s-dependent equivalent circuits for initialized capacitors and inductors. These equivalent circuits make transient analysis natural in the Laplace transform domain. Since all this analysis will be done with respect to s, we will use the term s-domain to characterize s-dependent analysis and circuits.

In the final section of the chapter, we will introduce the notion of a switched capacitor circuit. Switched capacitor circuits contain switches and capacitors, and possibly some op amps, but no resistors or inductors. Present-day integrated circuit technology allows us to build switches, capacitors, and op amps on chips easily and inexpensively. This has fostered an important trend in circuit design toward switched capacitor circuits. A thorough investigation of switched capacitor circuits is beyond the scope of this text. Nevertheless, it is important to introduce the basic ideas and thereby lay the foundation for more advanced courses on the topic.

2. NOTIONS OF IMPEDANCE AND ADMITTANCE

Peculiar to a resistor, a capacitor, and an inductor are quantities called *impedance* and *admittance*. Chapter 10, section 7, of volume I introduced an intermediate definition of impedance as the ratio of phasor voltage to phasor current, with admittance the inverse of this quantity. In the Laplace transform context, impedances and admittances are *s*-dependent generalizations of resistance and conductance. Such generalizations do not exist in the time domain. To crystallize this idea, we Laplace transform the standard differential *v-i* relationship of an inductor,

$$v_L(t) = L \frac{di_L}{dt}(t),$$

to obtain

$$V_L(s) = Ls I_L(s), \tag{14.1}$$

assuming $i_L(0^-) = 0$. Here, the quantity Ls multiplies a complex-frequency domain current, $I_L(s)$, to yield a complex-frequency domain voltage, $V_L(s)$, in a manner similar to Ohm's law for resistor voltages and currents. The units of Ls are ohms. The quantity Ls depends on the frequency variable s. It generalizes the concept of a fixed resistance and is universally called an impedance. This complex-frequency domain concept has no time-domain counterpart.

Although the inductor served to motivate this notion, an impedance can be defined for any two-terminal device whose input-output behavior is linear and whose parameters do not change with time. A device whose characteristics or parameters do not change with time is called *time invariant*.

IMPEDANCE

The *impedance*, denoted $Z(s)$, of a linear time-invariant two-terminal device, as illustrated in figure 14.1, relates the Laplace transform of the input current, $I(s)$, to the Laplace transform of the input voltage, $V(s)$, assuming that all independent sources inside the device are set to zero and that there is no internal stored energy at $t = 0^-$. Under these conditions,

$$V(s) = Z(s)I(s), \tag{14.2a}$$

and, where defined,

$$Z(s) = \frac{V(s)}{I(s)}, \tag{14.2b}$$

in units of ohms.

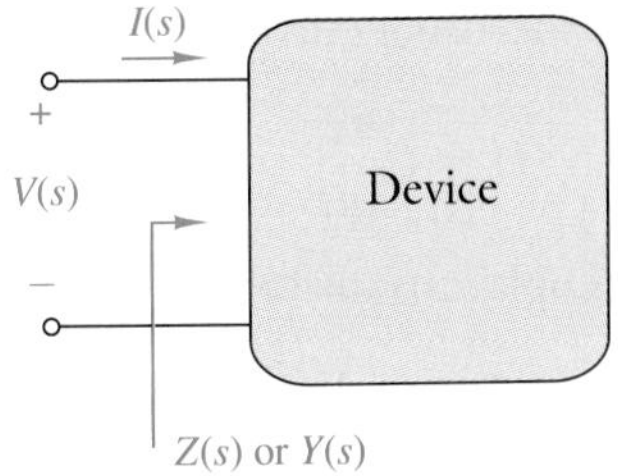

Figure 14.1 A two-terminal device having impedance $Z(s)$ or admittance $Y(s)$.

Inverse to resistance is conductance, and inverse to impedance is admittance. For example, if we divide both sides of equation 14.1 by Ls, we obtain

$$I_L(s) = \frac{1}{Ls} V_L(s). \tag{14.3}$$

This suggests that $1/Ls$ acts as a *generalized conductance* universally called an *admittance*, defined as follows.

ADMITTANCE

The *admittance*, denoted $Y(s)$, of a linear time-invariant device, as illustrated in figure 14.1, relates the Laplace transform of the input voltage, $V(s)$, of the device and the Laplace transform of the current, $I(s)$, into the device, assuming that all internal independent sources are set to zero and that there is no internal stored energy at $t = 0^-$. Under these conditions,

$$I(s) = Y(s)V(s),$$ (14.4a)

and, where defined,

$$Y(s) = \frac{I(s)}{V(s)},$$ (14.4b)

in units of mhos.

From equations 14.2 and 14.4, impedance and admittance satisfy the inverse relationship

$$Y(s) = \frac{1}{Z(s)}.$$ (14.5)

As a first step in deepening our understanding of these notions, some standard calculations of the impedance and admittance of the three basic circuit elements, as sketched in figure 14.2, prove useful.

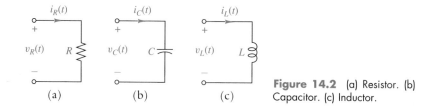

Figure 14.2 (a) Resistor. (b) Capacitor. (c) Inductor.

From Ohm's law, the simple resistor of figure 14.2a satisfies $v_R(t) = Ri_R(t)$. Laplace transforming both sides of this equation yields the obvious, $V_R(s) = RI_R(s)$. From equations 14.1 and 14.2, the **impedance of the resistor** becomes

$$Z_R(s) = R,$$

and from equation 14.5, the **admittance of the resistor** is

$$Y_R(s) = \frac{1}{R}.$$

Here, the kinship of impedance/admittance with resistance/conductance is clear.

The capacitor of figure 14.2b has the usual current-voltage relationship,

$$i_C(t) = C\frac{dv_C}{dt}(t).$$

Assuming no initial conditions, the Laplace transform relationship is

$$I_C(s) = CsV_C(s).$$

From equation 14.4, the **admittance of the capacitor** is

$$Y_C(s) = Cs,$$

and from equation 14.5, the **impedance of the capacitor** is

$$Z_C(s) = \frac{1}{Cs}.$$

As per equation 14.1, the **impedance and admittance of the inductor** (figure 14.2c) are

$$Z_L(s) = Ls \quad \text{and} \quad Y_L(s) = \frac{1}{Ls}.$$

Exercise. Given the integral form of the v-i capacitor relationship, assume no initial stored energy, and take the Laplace transform of both sides to derive the impedance of the capacitor. This provides an alternate, more basic means of deriving the impedance characterization.

Exercise. Given the integral form of the v-i inductor relationship, assume no initial stored energy, and take the Laplace transform of both sides to derive the admittance of the inductor.

The foregoing impedance representations allow us to analyze circuits by manipulating the impedances and admittances of simple circuit elements according to KVL and KCL. The techniques of analysis of resistive circuits now apply to RLC circuits in the context of impedances and admittances.

3. MANIPULATION OF IMPEDANCE AND ADMITTANCE

Recall that the Laplace transform is a linear operation with respect to sums of signals, possibly multiplied by constants. KVL and KCL are conservation laws in which, respectively, sums of voltages around a loop must add to zero and sums of all currents entering (or leaving) a node must add to zero. Since the Laplace transform is linear, it distributes over these sums, so that the sum of the Laplace transforms of the voltages around a loop must be zero and the sum of the Laplace transforms of all the currents entering a node must be zero. In other words, complex-frequency domain voltages satisfy KVL and complex-frequency domain currents satisfy KCL. Because of this, and because impedances and admittances generalize the notions of resistance and conductance, one intuitively expects their manipulation properties to be similar. In fact, this is the case.

Manipulation Rule Because impedances map complex-frequency domain currents, $I(s)$, to complex-frequency domain voltages, $V(s)$, and because all complex-frequency domain currents must satisfy KCL and all complex-frequency domain voltages must satisfy KVL:

1. Impedances, $Z(s)$, can be manipulated just like resistances, and like resistances, impedances have units of ohms.
2. Admittances, $Y(s)$, can be manipulated just like conductances, and like conductances, admittances have units of mhos.

This manipulation rule suggests, for example, that admittances in parallel add. The following example verifies this property for the case of two admittances in parallel.

 EXAMPLE 14.1

Determine the equivalent impedance, $Z_{in}(s)$, of two general admittances, $Y_1(s)$ and $Y_2(s)$, in parallel, as illustrated in figure 14.3.

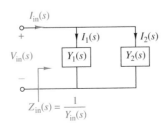

Figure 14.3 Two general admittances, $Y_i(s)$, in parallel, having an equivalent admittance $Y_{in}(s)$ or impedance $Z_{in}(s)$.

SOLUTION

From the definition of the admittance of a two-terminal device, $I_1(s) = Y_1(s)V_{in}(s)$, $I_2(s) = Y_2(s)V_{in}(s)$, and $I_{in}(s) = Y_{in}(s)V_{in}(s)$. This last relationship implicitly defines the equivalent admittance $Y_{in}(s)$. The goal is to relate $I_{in}(s)$ and $V_{in}(s)$ to $Y_1(s)$ and $Y_2(s)$, thereby finding $Y_{in}(s)$. The relationship follows from KCL and is

$$I_{in}(s) = I_1(s) + I_2(s) = [Y_1(s) + Y_2(s)]V_{in}(s) = Y_{in}(s)V_{in}(s),$$

in which one identifies $Y_{in}(s) = Y_1(s) + Y_2(s)$, i.e., admittances in parallel add.

Using the inverse relationship between impedance and admittance, one obtains

$$Z_{in}(s) = \frac{1}{Y_1(s) + Y_2(s)}.$$

Exercises.

1. Show that the impedance of two capacitors in parallel is

$$Z(s) = \frac{1}{C_1 s + C_2 s} = \frac{1}{(C_1 + C_2)s}.$$

2. Derive the following formula for the impedance of two inductors in parallel:

$$Z(s) = \frac{L_1 L_2}{L_1 + L_2} s.$$

3. A 2-F and a 0.5-F capacitor are in series. Find the equivalent admittance.
4. A 2-H inductor is connected in parallel with a 0.5-F capacitor. Find the equivalent impedance.

ANSWERS: $0.4s$, $2s/(s^2 + 1)$.

Example 14.1 suggests that a set of parallel admittances add to form an equivalent admittance. To see this, consider adding together the first two admittances in a parallel set to form a new admittance. Then add this new admittance to the third admittance in the set. Iterating the process verifies the intuition that parallel admittances add.

◆ EXAMPLE 14.2

Compute the input impedance of the parallel RLC circuit sketched in figure 14.4.

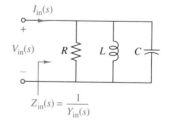

$$Z_{in}(s) = \frac{1}{Y_{in}(s)}$$

Figure 14.4 Parallel RLC circuit for example 14.2.

SOLUTION

For parallel circuits, it is convenient to work with admittances, since **parallel admittances add**. For the circuit of figure 14.4,

$$Y_{in}(s) = \frac{1}{R} + \frac{1}{Ls} + Cs = C \frac{s^2 + \frac{1}{RC}s + \frac{1}{LC}}{s}.$$

As per example 14.1, a rigorous justification follows by applying the KCL:

$$I_{in}(s) = I_R(s) + I_L(s) + I_C(s) = \left[\frac{1}{R} + \frac{1}{Ls} + Cs\right]V_{in}(s).$$

Since impedance is the inverse of admittance,

$$Z_{in}(s) = \frac{1}{Y_{in}(s)} = \frac{1}{C} \frac{s}{s^2 + \frac{1}{RC}s + \frac{1}{LC}}, \qquad (14.6)$$

which is the equivalent input impedance of a parallel RLC circuit.

Exercise. Compute the equivalent impedance of a parallel connection of six elements: two resistors of 6 Ω and 3 Ω, two inductors of 3 H and 6 H, and two capacitors of 0.2 F and 0.05 F.

ANSWER: $4s/(s^2 + 2s + 2)$.

 EXAMPLE 14.3

Compute the input impedance of a series connection of two pairs of parallel elements, as shown in figure 14.5.

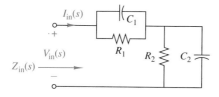

Figure 14.5 Series-parallel connection of RC elements for example 14.3.

SOLUTION

The strategy of this example is to compute each impedance, $Z_i(s)$, using the technique of example 14.1 and then use the property that impedances in series add to compute the equivalent input impedance.

Observe that figure 14.5 has the series impedance structure illustrated in figure 14.6.

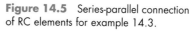

Figure 14.6 Conceptual series structure of the circuit in figure 14.5.

Since $V_{in}(s) = V_1(s) + V_2(s) = [Z_1(s) + Z_2(s)]I_{in}(s)$, it follows that $Z_{in}(s) = Z_1(s) + Z_2(s)$, i.e., impedances in series add. The first step, then, is to determine $Z_1(s)$ and $Z_2(s)$. Making the obvious identifications between figures 14.5 and 14.6, and applying the technique of example 14.1 yields

$$Z_1(s) = \frac{1}{Y_1(s)} = \frac{1}{C_1 s + \frac{1}{R_1}} = \frac{1}{C_1} \frac{1}{s + \frac{1}{R_1 C_1}},$$

and by symmetry,

$$Z_2(s) = \frac{1}{C_2} \frac{1}{s + \frac{1}{R_2 C_2}}.$$

Since impedances in series add,

$$Z_{in}(s) = Z_1(s) + Z_2(s) = \frac{1}{C_1} \frac{1}{s + \frac{1}{R_1 C_1}} + \frac{1}{C_2} \frac{1}{s + \frac{1}{R_2 C_2}}.$$

Note that this form (rather than a single expression with a common denominator) is often more convenient for computing responses because here the impedance is given as a partial fraction expansion.

Exercises.

1. Show that the equivalent admittance of two capacitors in series is

$$Y(s) = \frac{C_1 C_2}{C_1 + C_2} s.$$

2. Determine the impedance of two capacitors in series.

ANSWER: The inverse of the preceding formula.

Since one can manipulate impedances and admittances like resistances and conductances, there must be voltage and current division formulas to help simplify the analysis of basic circuits.

Voltage Division Formula For the circuit of figure 14.7, any voltage $V_j(s)$, where $j = 1$, 2, or 3, depends on the input voltage $V_{in}(s)$ via the formula

$$V_j(s) = \frac{Z_j(s)}{Z_1(s) + Z_2(s) + Z_3(s)} V_{in}(s). \tag{14.7}$$

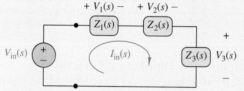

Figure 14.7 Series impedance circuit for illustrating voltage division.

To derive this formula, note that $V_{in}(s) = Z_{in}(s) I_{in}(s) = [Z_1(s) + Z_2(s) + Z_3(s)] \times I_{in}(s)$, or, equivalently,

$$I_{in}(s) = \frac{1}{Z_1(s) + Z_2(s) + Z_3(s)} V_{in}(s).$$

Equation 14.7 results by observing that for each device,

$$V_j(s) = Z_j(s) I_{in}(s).$$

The voltage division formula is easily extended to the case of n devices in series. Parallel devices have an analogous formula for current division.

Current Division Formula For the circuit of figure 14.8, any current $I_j(s)$ through the jth ($j = 1$, 2, or 3) device is given by

$$I_j(s) = \frac{Y_j(s)}{Y_1(s) + Y_2(s) + Y_3(s)} I_{in}(s). \tag{14.8}$$

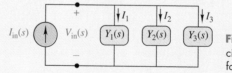

Figure 14.8 Parallel admittance circuit for developing current division formula.

The justification for equation 14.8 parallels that of equation 14.7. Equation 14.8 has the obvious generalization to any number of parallel elements.

Exercise. Derive the current division formula of equation 14.8. (*Hint:* First write $I_{in}(s)$ in terms of the $I_j(s)$. Then, using the relationship between current, voltage, and admittance, express $I_{in}(s)$ in terms of $Y_j(s)$ and $V_{in}(s)$.)

Exercise. Show that the current through the inductive branch of figure 14.4 (the parallel RLC circuit) is

$$I_L(s) = \frac{1}{LCs^2 + \dfrac{L}{R}s + 1} I_{in}(s).$$

Exercise. Suppose that in example 14.3 $R_1 = 1\ \Omega$, $C_1 = 1$ F, $R_2 = 1\ \Omega$, and $C_2 = 0.5$ F. Determine the voltage across C_2 using voltage division.

ANSWER: Depending on the reference polarity, the voltage across C_2 is $\pm 2(s+1)/(3s+4)$.

Another basic and useful circuit analysis technique is the **source transformation property**, exhibited now in terms of impedances and admittances. The first such transformation is the voltage-to-current source transformation, illustrated in figure 14.9.

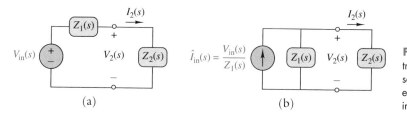

Figure 14.9 Illustration of the transformation of a voltage source in series with $Z_1(s)$, as shown in (a), to an equivalent current with a current source in parallel with $Z_1(s)$, as shown in (b).

Often, voltage-to-current source transformations provide an altered circuit topology that is more convenient for hand or calculator analysis. Mathematically, the goal is to change the voltage source series impedance structure to a current source parallel admittance structure while keeping $V_2(s)$ and $I_2(s)$ fixed. To justify this, one starts with figure 14.9a, in which voltage division implies that

$$V_2(s) = \frac{Z_2(s)}{Z_1(s) + Z_2(s)} V_{in}(s) = Z_2(s) I_2(s).$$

Hence, if $Z_1(s) \neq 0$,

$$V_2(s) = \frac{Z_2(s) Z_1(s)}{Z_1(s) + Z_2(s)} \left(\frac{V_{in}(s)}{Z_1(s)} \right) = \frac{1}{Y_1(s) + Y_2(s)} \left(\frac{V_{in}(s)}{Z_1(s)} \right), \qquad (14.9)$$

where $Y_i(s) = [Z_i(s)]^{-1}$. This equation identifies the parallel structure of figure 14.9b, i.e., figure 14.9b is a circuit equivalent of equation 14.9.

Reversing these arguments leads to the current-to-voltage source transformation, illustrated in figure 14.10.

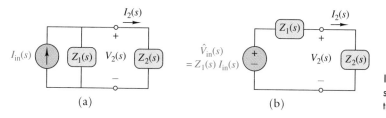

Figure 14.10 Illustration of (a) current source to (b) equivalent voltage source transformation.

Clearly, the manipulation of impedances and admittances parallels that of resistances and conductances, as suggested earlier.

4. NOTION OF TRANSFER FUNCTION

Besides impedances and admittances, other quantities such as voltage gains and current gains are critically important in, for example, amplifier circuits. The term **transfer function** is a catchall phrase for the different ratios that might be of interest in circuit analysis. Impedances and admittances are special cases of the transfer function concept.

TRANSFER FUNCTION

Mathematically, if a single-input, single-output circuit or system has all of its internal independent sources set to zero and there is no internal stored energy at $t = 0^-$, then the transfer function of the circuit or system is

$$H(s) = \frac{\mathcal{L}\,[\text{designated response signal}]}{\mathcal{L}\,[\text{designated input signal}]}. \tag{14.10}$$

In a circuit that is both relaxed (i.e., has no internal stored energy) and has no internal independent sources, if the input is $f(t)$ and the response is $y(t)$, then $Y(s) = H(s)F(s)$, which is a handy formula for computing responses. Notice that if the input is the delta function, then $F(s) = 1$ and $Y(s) = H(s)$. This means that the transfer function is the Laplace transform of the impulse response of the circuit, i.e., the response due to an impulse applied to the circuit input source when there are no initial conditions present.

◆ EXAMPLE 14.4

The circuit of figure 14.11 has zero initial conditions at $t = 0^-$. Find

$$H(s) = \frac{\mathcal{L}[i_{\text{out}}(t)]}{\mathcal{L}[v_{\text{in}}(t)]} = \frac{I_{\text{out}}(s)}{V_{\text{in}}(s)}.$$

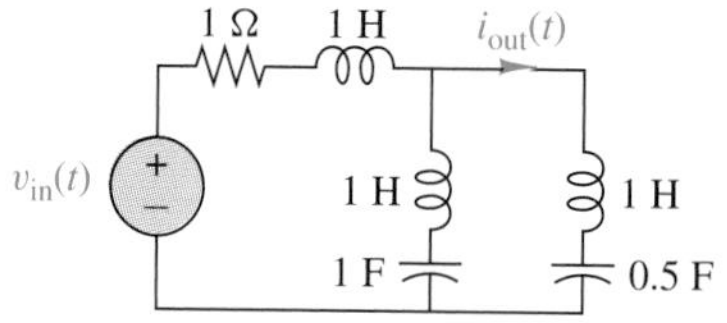

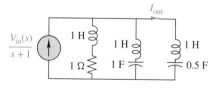

Figure 14.11   Circuit for example 14.4.

SOLUTION

Of course, there are many ways to solve this problem. One approach is to execute a source transformation on the R-L impedance in series with the voltage source. After the source transformation, one can use current division to obtain the necessary transfer function.

Step 1.   *Execute a source transformation to obtain three parallel branches, as per figure 14.12.*

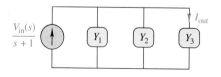

Figure 14.12   Circuit equivalent to figure 14.11 after source transformation.

This circuit has the parallel structure of figure 14.13.

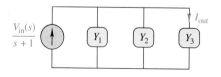

Figure 14.13   Parallel admittance form of figure 14.12.

Step 2. *Use current division.* Since the output current, $I_{out}(s)$, is a current through one of three parallel branches, the current division formula, equation 14.8, applies, producing

$$I_{out}(s) = \left(\frac{Y_3(s)}{Y_1(s) + Y_2(s) + Y_3(s)} \right) \frac{V_{in}(s)}{s+1}.$$

Hence,

$$H(s) = \frac{I_{out}(s)}{V_{in}(s)} = \left(\frac{Y_3(s)}{Y_1(s) + Y_2(s) + Y_3(s)} \right) \frac{1}{s+1}. \qquad (14.11)$$

Step 3. *Compute* $Y_1(s)$, $Y_2(s)$, *and* $Y_3(s)$. As impedances in series add, and as admittance is the inverse of impedance (equation 14.5), some straightforward algebra yields

$$Y_1(s) = \frac{1}{s+1}; \quad Y_2(s) = \frac{1}{s + \dfrac{1}{s}} = \frac{s}{s^2 + 1}; \quad Y_3(s) = \frac{1}{s + \dfrac{2}{s}} = \frac{s}{s^2 + 2}.$$

Step 4. *Substitute into equation 14.11 to obtain* $H(s)$:

$$H(s) = \frac{\dfrac{s}{s^2 + 2}}{\dfrac{1}{s+1} + \dfrac{s}{s^2 + 1} + \dfrac{s}{s^2 + 2}} \left(\frac{1}{s+1} \right) = \frac{s(s^2 + 1)}{3s^4 + 2s^3 + 6s^2 + 3s + 2}.$$

EXAMPLE 14.5

Determine the transfer function of the ideal operational amplifier circuit of figure 14.14, where $Z_f(s)$ and $I_f(s)$ denote a feedback impedance and feedback current, respectively.

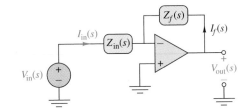

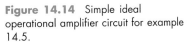

Figure 14.14 Simple ideal operational amplifier circuit for example 14.5.

SOLUTION

Since no current enters at the inputs of an ideal op amp, $I_{in}(s) = -I_f(s)$. Further, the voltage at the negative op amp terminal is driven to virtual ground; hence, $V_{in}(s) = Z_{in}(s)I_{in}(s)$, and $V_{out}(s) = Z_f(s)I_f(s)$. Combining these relationships with $I_{in}(s) = -I_f(s)$ yields

$$H(s) = \frac{V_{out}(s)}{V_{in}(s)} = -\frac{Z_f(s)}{Z_{in}(s)} = -\frac{Y_{in}(s)}{Y_f(s)}. \qquad (14.12)$$

This is a very handy formula for computing the transfer functions and responses of many op amp circuits.

> **Exercise.** Suppose $Z_f(s)$ is the impedance of a 0.1-mF capacitor. Determine R so that the transfer function is $H(s) = -1/s$.
>
> **ANSWER:** $R = 10$ KΩ.

EXAMPLE 14.6

The ideal op amp circuit of figure 14.15 is called a *leaky integrator*. If the input to the leaky integrator circuit is $v_{in}(t) = e^{-t}u(t)$, determine the values of R_1, R_2, and C leading to an output response $v_{out}(t) = -2te^{-t}u(t)$, assuming that $v_c(0^-) = 0$.

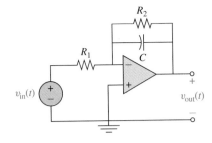

Figure 14.15 Ideal operational amplifier circuit known as the leaky integrator.

SOLUTION

Step 1. From the given data, determine the actual transfer function of the circuit. By definition of the transfer function,

$$H(s) = \frac{\mathcal{L}\,[\text{response}]}{\mathcal{L}\,[\text{input}]} = \frac{V_{\text{out}}(s)}{V_{\text{in}}(s)} = -\frac{\dfrac{2}{(s+1)^2}}{\dfrac{1}{s+1}} = -\frac{2}{s+1}. \tag{14.13}$$

Step 2. Using figure 14.15, determine the transfer function of the circuit in terms of R_1, R_2, *and* C. Here, observe that figure 14.15 has the same topology as figure 14.14, where

$$Z_f(s) = \frac{1}{Cs + \dfrac{1}{R_2}} \quad \text{and} \quad Z_{\text{in}}(s) = R_1.$$

From the result of example 14.5, i.e., equation 14.12,

$$H(s) = -\frac{Z_f(s)}{Z_{\text{in}}(s)} = -\frac{\dfrac{1}{R_1}}{Cs + \dfrac{1}{R_2}}. \tag{14.14}$$

Step 3. Match coefficients in equations 14.13 and 14.14 to obtain the desired values of R_1, R_2, *and* C. Equating the coefficients yields

$$\frac{\dfrac{1}{R_1}}{Cs + \dfrac{1}{R_2}} = \frac{2}{s+1}.$$

One possible solution is $R_1 = 0.5\ \Omega$, $R_2 = 1\ \Omega$, and $C = 1$ F. Other solutions are also possible. In fact, $R_{1\text{new}} = K_m R_1$, $R_{2\text{new}} = K_m R_2$, and $C_{\text{new}} = C/K_m$, for any positive nonzero value of K_m, is a valid (theoretical) solution. In chapter 17, we shall meet a notion called *magnitude scaling*. K_m is called a *magnitude scale factor*, which leaves the transfer function unchanged, but produces more realistic values for the circuit elements.

Energy Conservation Engineer

Energy conservation engineering is a specialized area within power systems engineering that determines the best way to use limited resources to produce power. These engineers design efficient power distribution systems to benefit society, while safeguarding the environment. Energy conservation engineers are often part of teams that develop and implement environmental monitoring systems that continually assess how advances in technology impact society and the environment.

Energy conservation engineers also explore new avenues for harnessing nature's power to meet the world's increasing demand for energy. Within the next few decades, many see a critical need to develop effective applications of solar energy to meet growing energy demands and reduce air pollution caused by the use of fossil fuels. These engineers are searching for ways to use stored-up volcanic heat to generate electricity, which would provide underdeveloped nations with an inexpensive energy source. Some energy conservation engineers are working with other experts to incorporate greater efficiency and improved safety into the design and operation of nuclear power plants. While working on these projects, energy conservation engineers must plan for an acceptable means of disposing of nuclear waste.

Exercise. In the preceding example, let $K_m = 10^4$. Verify that $R_{1\text{new}} = K_m R_1$, $R_{2\text{new}} = K_m R_2$, and $C_{\text{new}} = C/K_m$ are valid solutions. Can you verify that $R_{1\text{new}} = K_m R_1$, $R_{2\text{new}} = K_m R_2$, and $C_{\text{new}} = C/K_m$ are valid solutions for arbitrary K_m?
(*Hint*: Just plug in.)

5. Equivalent Circuits for L's and C's

The notions of impedance, admittance, and transfer function do not account for the presence of initial capacitor voltages and initial inductor currents. *How can one incorporate initial conditions into various analysis schemes?* The answer is to look at the transform of an initialized capacitor and inductor and interpret the resulting equation as an equivalent circuit in the complex-frequency domain. For the capacitor and the inductor, two equivalent circuits result for each: a series circuit containing a relaxed (no initial condition) capacitor/inductor in series with a source and a parallel circuit with a relaxed capacitor/inductor in parallel with a source. Example 13.23 of chapter 13 previewed this notion.

The capacitor, as drawn in figure 14.2b, has the usual voltage-current relationship:

$$C\frac{dv_c(t)}{dt} = i_c(t).$$

Taking the Laplace transform and allowing for a nonzero initial condition $v_C(0^-)$ yields

$$CsV_C(s) - Cv_C(0^-) = I_C(s). \tag{14.15}$$

The left-hand side of equation 14.15 is the difference of two currents, one given by the product of the capacitor admittance with the capacitor voltage and the other by $Cv_C(0^-)$. The interpretation, then, is a relaxed capacitor in parallel with a current source, as illustrated by figure 14.16.

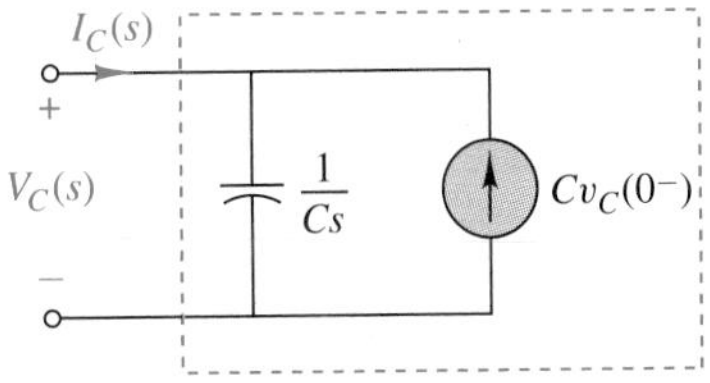

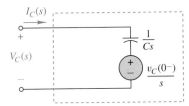

Figure 14.16 Parallel form of an equivalent circuit for an initialized capacitor. Here, the capacitor is assumed to be relaxed, with the current source $Cv_C(0^-)$ accounting for the initial condition.

Rearranging equation 14.15 yields

$$V_C(s) = \frac{1}{Cs}I_C(s) + \frac{v_C(0^-)}{s}. \tag{14.16}$$

Here, the right-hand side is the sum of two voltages, one of which is the product of the capacitor impedance with the capacitor current and the other $\frac{v_C(0^-)}{s}$. This implies a series circuit, as sketched in figure 14.17.

Figure 14.17 Series form of an equivalent circuit for an initialized capacitor. Here, the capacitor is relaxed, and the voltage source accounts for the effect of the initial condition.

Initialized inductors have equivalent complex-frequency domain circuits analogous to the capacitor equivalent circuits. With the voltage and current directions indicated in figure 14.2c, the usual inductor current-voltage relationship is

$$v_L(t) = L\frac{di_L(t)}{dt}.$$

Transforming both sides yields

$$V_L(s) = LsI_L(s) - Li_L(0^-). \tag{14.17}$$

Again, this equation manifests a sum of voltages that translates to a series circuit, as displayed in figure 14.18.

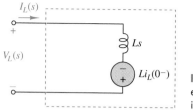

Figure 14.18 Series form of an equivalent circuit for an initialized inductor.

Dividing equation 14.17 by Ls and bringing the initial condition term to the other side leads to

$$I_L(s) = \frac{1}{Ls}V_L(s) + \frac{i_L(0^-)}{s}. \tag{14.18}$$

Equation 14.18 determines a parallel equivalent circuit, as sketched in figure 14.19.

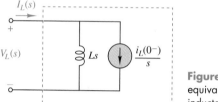

Figure 14.19 Parallel form of an equivalent circuit for an initialized inductor.

Of course, one can take the inverse transform of these circuits to obtain the time-domain counterparts, which, with superposition, clearly show the effect of the initial conditions on the time-domain response.

Several examples illustrate the use of equivalent circuits in analysis.

◆ EXAMPLE 14.7

Consider the circuit of figure 14.20, in which $v_{in}(t) = 5u(t)$, $v_{C1}(0^-) = 3$ V, $v_{C2}(0^-) = 0$ V, and $i_L(0^-) = 2$ A. Find $v_{out}(t)$.

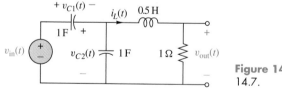

Figure 14.20 RLC circuit for example 14.7.

SOLUTION

There are, of course, many ways to solve this problem. One method is to replace each capacitor and inductor by its complex-frequency domain equivalent circuit that accounts for the initial condition. One can then combine sources and proceed to execute a series of source transformations that will simplify the circuit to a point where the transfer function is straightforward to calculate.

Step 1. *Draw the equivalent complex-frequency domain circuit, and combine series sources.* Figure 14.21 shows the equivalent complex-frequency domain circuit of figure 14.20. In this circuit, we have chosen the series equivalent replacements for both the capacitor and the inductor, as the circuit topology allows a straightforward simplification of the circuit by source transformations. Combining the two voltage sources leads to a single source of $2/s$ V.

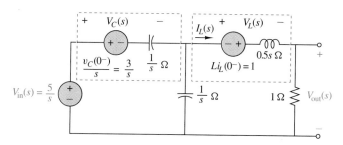

Figure 14.21 Complex-frequency domain equivalent of the circuit of figure 14.20.

Step 2. *Perform a source transformation, and combine parallel capacitors.* This is easily done, and the result is illustrated in figure 14.22.

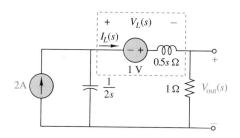

Figure 14.22 Circuit equivalent to the one in figure 14.21 after source transformation.

Step 3. *Execute a second source transformation and combine series voltage sources to produce a series circuit.* Again, this is easily accomplished and figure 14.23 displays the result.

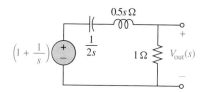

Figure 14.23 Simplified version of figure 14.21.

Step 4. *Use voltage division to determine* $V_{out}(s)$ *and take the inverse transform to obtain* $v_{out}(t)$. By voltage division, we obtain

$$V_{out}(s) = \left(\frac{1}{\frac{1}{2s} + 0.5s + 1} \right) \left(1 + \frac{1}{s} \right) = \frac{2}{s+1}.$$

Taking the inverse transform produces the desired response:

$$v_{out}(t) = \mathcal{L}^{-1} \left(\frac{2}{s+1} \right) = 2e^{-t}u(t) \text{ V}.$$

◆ **EXAMPLE 14.8**

This example illustrates how superposition and a transfer function can team up to compute a desired voltage response. In the RLC circuit of figure 14.24, suppose $v_C(0^-) = 1$ V, $i_L(0^-) = 2$ A, and $v_{in}(t) = u(t)$ V. Find $v_C(t)$ for $t \geq 0$.

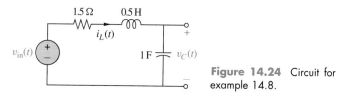

Figure 14.24 Circuit for example 14.8.

SOLUTION

In this example, it is convenient to replace the inductor by its complex-frequency domain voltage source equivalent (series) circuit, because the inductor is in series with the input voltage source. On the other hand, it is convenient to replace the capacitor by its complex-frequency domain current source equivalent circuit, because the desired output is the capacitor voltage. This results in a three-source, or multiinput, circuit. Once the equivalent circuits are in place, superposition shows that the overall response is simply the sum of the responses of each source to the relaxed network.

Step 1. *Using the voltage source model for the inductor and the current source model for the capacitor,* draw the equivalent complex-frequency domain circuit. Using the voltage source equivalent for the initialized inductor and the current source equivalent for the capacitor produces figure 14.25.

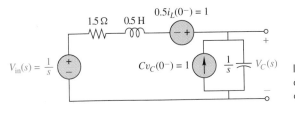

Figure 14.25 Complex-frequency domain equivalent accounting for initial conditions of the circuit of figure 14.24.

With three sources present in the figure, computing the capacitor voltage by superposition seems most reasonable. By superposition, the voltage output has the decomposition

$$V_C(s) = V_{1C}(s) + V_{2C}(s) + V_{3C}(s),$$

where $V_{1C}(s)$ is the response due to $V_{in}(s)$ given that the other sources are set to zero, $V_{2C}(s)$ is the response due to the initial condition on the inductor, i.e., due to $0.5i_L(0^-)$ acting alone, and $V_{3C}(s)$ is due to the capacitor initial condition acting alone.

Step 2. *Determine the form of* $V_{1C}(s)$, *the response due to the input excitation.* By voltage division,

$$V_{1C}(s) = H(s)V_{in}(s) = \frac{2}{s(s+1)(s+2)}.$$

Step 3. *Determine the form of* $V_{2C}(s)$, *the response due to the initial condition on the inductor.* Here, the transfer function $H(s)$ from $0.5i_L(0^-)$ to $V_{C2}(s)$ coincides with the one in step 2 because the two voltage sources are simply in series with each other and could have been combined. We obtain

$$V_{2C}(s) = H(s)[0.5i_L(0^-)] = H(s) = \frac{2}{(s+1)(s+2)}.$$

Step 4. *Determine the form of* $V_{3C}(s)$, *the response due to the initial condition on the capacitor.* At this point, it would seem best to compute the equivalent impedance, say, $Z_{eq}(s)$, across the current source, with all other sources set to zero. Once $Z_{eq}(s)$ is known, we have

$$V_{3C}(s) = Z_{eq}(s)[Cv_C(0^-)].$$

By the usual manipulation techniques,

$$Z_{eq}(s) = \frac{1}{Y_{eq}(s)} = \frac{1}{s + \dfrac{1}{1.5 + 0.5s}} = \frac{s+3}{(s+1)(s+2)}.$$

Since $Cv_C(0^-) = 1$,

$$V_{3C}(s) = \frac{s+3}{(s+1)(s+2)}.$$

Step 5. *Execute a partial fraction expansion on* $V_C(s)$, *and take the inverse transform to obtain* $v_C(t)$. Using superposition on the preceding results yields

$$V_C(s) = V_{1C}(s) + V_{2C}(s) + V_{3C}(s) = \frac{2 + 2s + s(s+3)}{s(s+1)(s+2)} = \frac{1}{s} + \frac{2}{s+1} + \frac{-2}{s+2}.$$

Inverting this transform determines the desired time response, which is the capacitor voltage

$$v_C(t) = [1 + 2e^{-t} - 2e^{-2t}]u(t) \text{ V}.$$

EXAMPLE 14.9

The chapter opened with a discussion of the operation of a fluorescent light. For a fluorescent light to begin operating, there must be sufficient free electrons in the tube to allow arcing to occur. Once arcing starts, mercury particles in the tube vaporize and give off ultraviolet light. This excites a coating of phosphorus on the inside of the tube to emit light in the visible range.

A starter initializes the arc in the tube. As shown in figure 14.26, the starter is modeled by an ideal heat-sensitive switch in parallel with a small capacitance. Initialization of the arc requires a very high voltage surge across the electrodes to generate a sufficient density of free electrons. The starter, in conjunction with the inductive ballast, generates the appropriate high voltage surge.

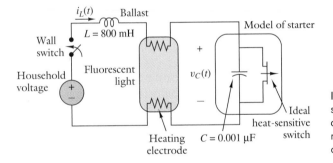

Figure 14.26 Wiring diagram of simple fluorescent light circuit, including an inductive ballast and a starter modeled by an ideal heat-sensitive switch and a capacitor.

For a simplified analysis, assume that all resistances are negligible. The source $v_{in}(t)$ is 120 V, 60 Hz, i.e., ordinary house current. When the wall switch is turned on, the ideal heat-sensitive switch in the starter is initially closed. Immediately, however, a current $i_L(t)$ (whose magnitude is limited by the inductive ballast) flows, heating the starter switch. At some point in time, which we will call $t = 0$, the inductor has an initial current $i_L(0^-)$, and the heat-sensitive switch in the starter opens because of the heating current. A very high voltage then appears across the electrodes of the lamp, resulting in ignition of the lamp. After the lamp ignites, the ac current flows between the two electrodes. The ballast again serves to limit the current.

Suppose that $L = 800$ mH, $C = 0.001$ μF, and $i_L(0^-) = 0.1$ A. For $t > 0$, we find the component of $v_C(t)$ due to the initial inductor current, i.e., the so-called zero-input response. The other component, the zero-state response, is not as important for ignition purposes. Our strategy will be to use the *s*-domain equivalent circuit for L, as illustrated in figure 14.27.

Since we are assuming that all resistances are negligible and that the internal resistance (between electrodes) of the lamp prior to arcing approximates infinity, the voltage division

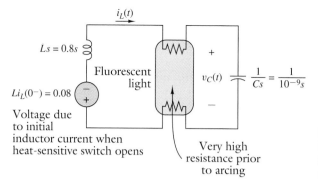

Figure 14.27 Equivalent complex-frequency domain circuit immediately prior to arcing and normal lamp operation in fluorescent lighting.

formula in terms of impedances yields

$$V_C(s) = -\frac{\dfrac{1}{Cs}}{\dfrac{1}{Cs} + Ls} Li_L(0^-) = -\frac{1}{C}\frac{i_L(0^-)}{s^2 + \dfrac{1}{LC}} = -\frac{10^8}{\sqrt{1.25 \times 10^9}}\frac{\sqrt{1.25 \times 10^9}}{s^2 + 1.25 \times 10^9}$$

$$= -2,828\,\frac{\sqrt{1.25 \times 10^9}}{s^2 + 1.25 \times 10^9}.$$

Hence, immediately prior to arcing, the capacitor voltage approximates

$$v_C(t) = -2,828\sin(35,355t),$$

which is sufficient to induce arcing and cause the lamp to operate.

Problem 70 extends the analysis of this example to the case where the ballast model includes a resistance of 100 Ω.

6. NODAL AND LOOP ANALYSIS IN THE *S*-DOMAIN

This section develops complex-frequency domain formulations of node and loop analysis. These techniques have universal applicability for computing circuit responses, input-output impedances or admittances, and transfer functions. Before continuing, however, recall that KCL requires that the sum of the currents leaving any node of a circuit be zero. Further, KVL requires that the voltages around any loop of a circuit sum to zero. Nodal analysis of circuits builds around KCL, whereas mesh and loop analysis utilize KVL.

In volume I—and indeed, in any beginning course on circuits—loop and nodal analysis are taught first in the context of resistances and conductances and later in the phasor setting. Because of the linearity of the Laplace transform, the techniques have an *s*-domain formulation where elements are characterized by impedances and/or admittances.

For loop analysis, one writes a KVL equation for each loop in terms of the transformed loop currents and element impedances. The set of all such equations then characterizes the circuit's loop currents, which determine the currents through each of the elements. Knowledge of the loop currents and the element impedances permits the computation of any of the element voltages.

In nodal analysis, one writes a KCL equation at each node in terms of the Laplace transform of the node voltages with respect to a reference, the transform of the independent excitations, and the element admittances. The set of all such equations then characterizes the node voltages of the circuit in the complex-frequency domain. Solving the set of circuit node equations yields the set of transformed node voltages. Knowledge of these permits the computation of any of the element voltages. With knowledge of the element admittances, one may compute all the element currents. Since nodal analysis has a more extensive use than loop analysis, our focus will center on nodal analysis.

The ideal operational amplifier circuit shown in figure 14.28 is a Sallen and Key normalized low-pass Butterworth filter. (See chapter 21 for a full discussion of this kind of filter.) A normalized low-pass filter passes frequencies below 1 rad/sec and attenuates larger frequencies. As we will see later in the text, the 1-rad/sec frequency can be moved to any desired value by frequency scaling the parameter values of the circuit. (See chapter 17 for a discussion of frequency scaling.) The goal here is to utilize the techniques of nodal analysis to determine the transfer function of this circuit.

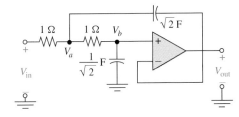
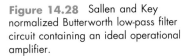

Figure 14.28 Sallen and Key normalized Butterworth low-pass filter circuit containing an ideal operational amplifier.

SOLUTION

The solution proceeds in several steps that utilize nodal analysis techniques in conjunction with the properties of an ideal op amp. Recall that for an ideal op amp, the voltage across the input terminals is zero and the current into any of the input terminals is also zero. Lastly, note that one does not write a node equation at the output, as the output appears across a dependent voltage source whose value depends on other voltages in the circuit.

Step 1. Determine V_b. Because the voltage across the input terminals of an ideal operational amplifier is zero, $V_b = V_{out}$.

Step 2. Write a node equation at the node identified by the node voltage V_a. Summing the currents leaving the node yields

$$(V_a - V_{in}) + (V_a - V_b) + \sqrt{2}s(V_a - V_{out}) = 0.$$

Substituting V_{out} for V_b and grouping like terms produces

$$(\sqrt{2}s + 2)V_a - (\sqrt{2}s + 1)V_{out} = V_{in}. \tag{14.19}$$

Step 3. Write a node equation at the node identified by the node voltage V_b. By inspection, the desired node equation is

$$-V_a + \left(\frac{1}{\sqrt{2}}s + 1\right)V_{out} = 0. \tag{14.20}$$

Step 4. Write the foregoing two node equations in matrix form. In matrix form, equations 14.19 and 14.20 combine to give

$$\begin{bmatrix} \left(\sqrt{2}s + 2\right) & -\left(\sqrt{2}s + 1\right) \\ -1 & \left(\frac{1}{\sqrt{2}}s + 1\right) \end{bmatrix} \begin{bmatrix} V_a \\ V_{out} \end{bmatrix} = \begin{bmatrix} V_{in} \\ 0 \end{bmatrix}. \tag{14.21}$$

Step 5. Solve equation 14.21 for V_{out} *in terms of* V_{in} *using Cramer's rule.* From Cramer's rule,

$$V_{out}(s) = \frac{\det\begin{bmatrix} \left(\sqrt{2}s + 2\right) & V_{in} \\ -1 & 0 \end{bmatrix}}{\det\begin{bmatrix} \left(\sqrt{2}s + 2\right) & -\left(\sqrt{2}s + 1\right) \\ -1 & \left(\frac{1}{\sqrt{2}}s + 1\right) \end{bmatrix}}.$$

The resulting transfer function is

$$H(s) = \frac{V_{out}(s)}{V_{in}} = \frac{1}{\left(\sqrt{2}s + 2\right)\left(\frac{1}{\sqrt{2}}s + 1\right) - \left(\sqrt{2}s + 1\right)} = \frac{1}{s^2 + \sqrt{2}s + 1}.$$

Notice that for small values of $s = j\omega$ (i.e., low frequencies), the magnitude of $H(s)$ approximates 1, and for large values of $s = j\omega$ (i.e., high frequencies, where $|j\omega| \gg 1$), the magnitude of $H(s)$ is small. Since $V_{out}(j\omega) = H(j\omega)V_{in}(j\omega)$, such a circuit blocks high-frequency input excitations and passes low-frequency input excitations. As mentioned at the beginning of the example, the circuit passes low frequencies and attenuates high frequencies.

The preceding example used matrix notation, which is common to much of advanced circuit analysis. In one sense, matrix notation is a shorthand way of writing n simultaneous equations: The n variables are written only once. More generally, matrix notation and the associated matrix arithmetic allow engineers to handle and solve large numbers of equations in numerically efficient ways. Further, the theory of matrices allows one to develop insights into large circuits that would otherwise remain hidden. Hence, many of the examples that follow will utilize the elementary properties of matrix arithmetic. (See appendix A-1 for a more detailed discussion.)

The next example uses nodal analysis to compute the response to an initialized circuit. The example combines the equivalent circuits for initialized capacitors and inductors with the technique of nodal analysis.

◆ **EXAMPLE 14.11**

In the circuit of figure 14.29, suppose $i_{in}(t) = \delta(t)$, $i_L(0^-) = 1$ A, and $v_C(0^-) = 1$ V. Determine the voltages $v_C(t)$ and $v_L(t)$.

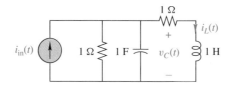

Figure 14.29 Two-node RLC circuit for example 14.11. Given the indicated current direction of $i_L(t)$, what is the implied voltage polarity for $v_L(t)$?

Step 1. *Draw the complex-frequency domain equivalent circuit with an eye toward nodal analysis.* Inserting the equivalent current source models for the initialized capacitor and inductor in figure 14.29, one obtains the complex-frequency domain equivalent circuit shown in figure 14.30.

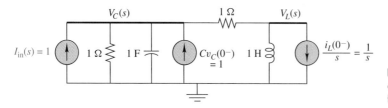

Figure 14.30 Complex-frequency domain equivalent of the circuit of figure 14.29.

Step 2. *Write the two node equations, and put them in matrix form.* At the node labeled $V_C(s)$ (identified with a bold line), KCL implies that

$$(1 + s)V_C(s) + [V_C(s) - V_L(s)] = 2.$$

Simplifying produces the first node equation:

$$(s + 2)V_C(s) - V_L(s) = 2.$$

At the node labeled $V_L(s)$, $[V_L(s) - V_C(s)] + (1/s)V_L(s) = -(1/s)$, or equivalently,

$$-V_C(s) + \frac{s+1}{s}V_L(s) = -\frac{1}{s}.$$

The matrix form of these two node equations is

$$\begin{bmatrix} s+2 & -1 \\ -1 & \dfrac{s+1}{s} \end{bmatrix} \begin{bmatrix} V_C(s) \\ V_L(s) \end{bmatrix} = \begin{bmatrix} 2 \\ -\dfrac{1}{s} \end{bmatrix}.$$

Step 3. *Solve the matrix equation of step 2 for the desired voltages.* Using Cramer's rule, computing the inverse, or simultaneously solving the equations gives

$$\begin{bmatrix} V_C(s) \\ V_L(s) \end{bmatrix} = \frac{s}{s^2 + 2s + 2} \begin{bmatrix} \dfrac{s+1}{s} & 1 \\ 1 & s+2 \end{bmatrix} \begin{bmatrix} 2 \\ -\dfrac{1}{s} \end{bmatrix} = \begin{bmatrix} \dfrac{2(s+1)-1}{(s+1)^2+1} \\ \dfrac{(s+1)-3}{(s+1)^2+1} \end{bmatrix}. \quad (14.22)$$

Step 4. *Take the inverse Laplace transform to obtain time-domain voltages.* Breaking up equation 14.22 into its components yields

$$V_C(s) = \frac{2(s+1)}{(s+1)^2+1} - \frac{1}{(s+1)^2+1},$$

in which case

$$v_C(t) = e^{-t}[2\cos(t) - \sin(t)]u(t).$$

Also,

$$V_L(s) = \frac{s+1}{(s+1)^2+1} - \frac{3}{(s+1)^2+1},$$

so that

$$v_L(t) = e^{-t}[\cos(t) - 3\sin(t)]u(t).$$

Figure 14.31 presents a plot of $v_C(t)$ and $v_L(t)$.

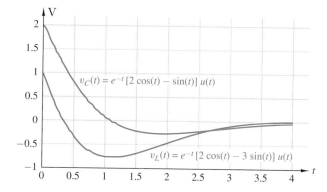

Figure 14.31 Plot of the capacitor and inductor voltages for example 14.11.

In a third and final example of nodal analysis, the parameters of a circuit with two capacitors and one inductor are kept as literals until the final computation of the transfer function. This allows the reader to observe the construction of the equations more carefully.

 EXAMPLE 14.12

Use nodal analysis to construct the transfer function

$$H(s) = \frac{V_2(s)}{V_{in}(s)}$$

for the circuit of figure 14.32.

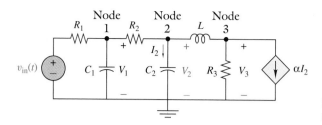

Node 1   Node 2   Node 3

Figure 14.32 Three-node circuit for example 14.12.

SOLUTION

Once again, our strategy is to write three node equations, put them in matrix form, and solve.

Step 1. Summing all currents leaving node 1 and grouping like terms yields

$$\left(\frac{1}{R_1} + \frac{1}{R_2} + C_1 s\right) V_1 - \frac{1}{R_2} V_2 = \frac{1}{R_1} V_{in}.$$

Step 2. Repeating step 1 for node 2 produces

$$-\frac{1}{R_2} V_1 + \left(\frac{1}{R_2} + \frac{1}{Ls} + C_2 s\right) V_2 - \frac{1}{Ls} V_3 = 0.$$

Step 3. Finally, for node 3,

$$0\, V_1 + \left(\alpha C_2 s - \frac{1}{Ls}\right) V_2 + \left(\frac{1}{R_3} + \frac{1}{Ls}\right) V_3 = 0.$$

Step 4. Writing the three node equations in matrix form, we obtain

$$\begin{bmatrix} \dfrac{1}{R_1} + \dfrac{1}{R_2} + C_1 s & -\dfrac{1}{R_2} & 0 \\[2mm] -\dfrac{1}{R_2} & \dfrac{1}{R_2} + \dfrac{1}{Ls} + C_2 s & -\dfrac{1}{Ls} \\[2mm] 0 & \alpha C_2 s - \dfrac{1}{Ls} & \dfrac{1}{R_3} + \dfrac{1}{Ls} \end{bmatrix} \begin{bmatrix} V_1 \\[2mm] V_2 \\[2mm] V_3 \end{bmatrix} = \begin{bmatrix} \dfrac{V_{in}}{R_1} \\[2mm] 0 \\[2mm] 0 \end{bmatrix}.$$

Step 5. For the sake of convenience, suppose all parameters are 1 unit in magnitude. The matrix equation of step 4 then reduces to

$$\begin{bmatrix} s+2 & -1 & 0 \\[2mm] -1 & \dfrac{s^2+s+1}{s} & -\dfrac{1}{s} \\[2mm] 0 & -\dfrac{1}{s}+s & \dfrac{s+1}{s} \end{bmatrix} \begin{bmatrix} V_1 \\[2mm] V_2 \\[2mm] V_3 \end{bmatrix} = \begin{bmatrix} V_{in} \\[2mm] 0 \\[2mm] 0 \end{bmatrix}.$$

Step 6. Solving the matrix equation of step 5 for $V_2(s)$ in terms of $V_{in}(s)$ by Cramer's rule leads to

$$V_2(s) = \frac{\det \begin{bmatrix} s+2 & V_{in}(s) & 0 \\[2mm] -1 & 0 & -\dfrac{1}{s} \\[2mm] 0 & 0 & \dfrac{s+1}{s} \end{bmatrix}}{\det \begin{bmatrix} s+2 & -1 & 0 \\[2mm] -1 & \dfrac{s^2+s+1}{s} & -\dfrac{1}{s} \\[2mm] 0 & -\dfrac{1}{s}+s & \dfrac{s+1}{s} \end{bmatrix}},$$

in which case

$$H(s) = \frac{V_2(s)}{V_{\text{in}}(s)} = \frac{s+1}{(s+1)(s^2+4s+3)} = \frac{1}{s^2+4s+3}.$$

The preceding examples have avoided special circuits that have dependent voltage sources between nodes. As developed in volume I or in a first course on circuits, the concept of a *supernode* aids the nodal analysis of such networks. The problems at the end of the chapter offer some skill renewal challenges in this area. Lastly, software packages such as Mathematica® and Theorist® are available that symbolically solve algebraic equations in s.

As developed in volume I or in a first course on circuits, dual to nodal analysis is loop analysis. In loop analysis, one defines loop currents and writes KVL equations in terms of these loop currents. The following example illustrates the method of loop analysis for computing the input impedance of a bridge-T network.

◆ EXAMPLE 14.13

Use loop analysis to compute the input impedance of the bridge-T network illustrated in figure 14.33.

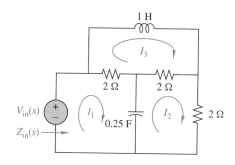

Figure 14.33 Bridge-T network of example 14.13.

SOLUTION

Define $I_{\text{in}}(s) = I_1(s)$ of figure 14.33. Since $Z_{\text{in}}(s) = V_{\text{in}}(s)/I_{\text{in}}(s)$, the goal is to use loop analysis to find $I_{\text{in}}(s) = I_1(s)$ in terms of $V_{\text{in}}(s)$.

Step 1. *Sum the voltages around loop 1.* Dropping the specific s-dependence for convenience, one obtains

$$2(I_1 - I_3) + \frac{4}{s}(I_1 - I_2) = 2\frac{s+2}{s}I_1 - \frac{4}{s}I_2 - 2I_3 = V_{\text{in}}.$$

Step 2. *Sum the voltages around loop 2.* By inspection,

$$\frac{4}{s}(I_2 - I_1) + 2(I_2 - I_3) + 2I_2 = -\frac{4}{s}I_1 + 4\frac{s+1}{s}I_2 - 2I_3 = 0.$$

Step 3. *Sum the voltages around loop 3.* Again by inspection,

$$2(I_3 - I_1) + sI_3 + 2(I_3 - I_2) = -2I_1 - 2I_2 + (s+4)I_3 = 0.$$

Step 4. *Put the three loop equations in matrix form, and solve.* In matrix form, the three loop equations are

$$\begin{bmatrix} 2\dfrac{s+2}{s} & -\dfrac{4}{s} & -2 \\[2mm] -\dfrac{4}{s} & 4\dfrac{s+1}{s} & -2 \\[2mm] -2 & -2 & s+4 \end{bmatrix} \begin{bmatrix} I_1 \\ I_2 \\ I_3 \end{bmatrix} = \begin{bmatrix} V_{\text{in}} \\ 0 \\ 0 \end{bmatrix}.$$

Using Cramer's rule to solve this equation for $V_{in}(s)$ in terms of $I_{in}(s) = I_1(s)$ yields

$$Z_{in}(s) = \frac{V_{in}(s)}{I_{in}(s)} = \frac{V_{in}(s)}{I_1(s)} = \frac{8\dfrac{s^2 + 4s + 4}{s}}{4\dfrac{(s+2)^2}{s}} = 2\ \Omega.$$

It may be a little surprising that $Z_{in}(s)$ is independent of s, despite the appearance of an inductance and a capacitance inside the network. Such a network is called a *constant-resistance network*. Problem 68 at the end of the chapter shows the conditions on the elements of a bridge-T network for it to be a constant-resistance network.

Once again, the foregoing example has avoided the presence of dependent current sources in the loop. The presence of a dependent current source in a loop or in two adjacent loops constrains the loop currents. Solution techniques are the same as those developed in volume I. Some problems at the end of the chapter contain such circuits.

7. SWITCHING IN RLC CIRCUITS

Switches control the lighting in homes, furnaces, the ignition of a car, traffic lights, and numerous other devices. Circuit elements called *diodes* are modeled by ideal switches. There are switched power supplies and switched capacitor filters. All have a functional element called a *switch* that affects and, indeed, shapes the behavior of the circuit. This section investigates the behavior of switching in simple RLC circuits, as preparation for the understanding of switching in more elaborate and sophisticated electronic circuits. Our immediate task is to apply the Laplace transform method to compute the responses of switched RLC circuits. The following example motivates a general procedure.

◆ **EXAMPLE 14.14**

In the circuit of figure 14.34, suppose $v_{in}(t) = 5\sin(t)u(t)$, $v_c(0^-) = 0$, and switch S moves from position A to position B at $t = 1$ sec and from position B to position A at $t = 2$ sec, where it remains for all subsequent time. Find $v_C(t)$ for $t \geq 0$.

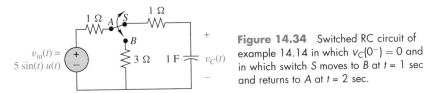

Figure 14.34 Switched RC circuit of example 14.14 in which $v_C(0^-) = 0$ and in which switch S moves to B at $t = 1$ sec and returns to A at $t = 2$ sec.

SOLUTION

Because of the switching at $t = 1$ sec, the first step is to determine $v_c(t)$ over the time interval $0 \leq t < 1$. This allows us, in turn, to determine $v_c(1^-)$, which will serve as the initial condition over the time interval $1 \leq t < 2$. This then produces $v_C(2^-)$, the initial condition for the final interval, $2 \leq t$.

Step 1. Compute the response for $0 \leq t < 1$. Over the interval $0 \leq t < 1$, the circuit of figure 14.34 is a simple RC circuit, illustrated in figure 14.35.

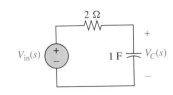

Figure 14.35 Equivalent circuit of figure 14.34 over $0 \leq t < 1$.

For the circuit of figure 14.35, voltage division implies

$$V_C(s) = \frac{\dfrac{1}{s}}{2 + \dfrac{1}{s}} V_{in}(s) = \frac{2.5}{(s+0.5)(s^2+1)} = \frac{2}{s+0.5} + \frac{-2s+1}{(s^2+1)}.$$

Hence, for $0 \le t < 1$,

$$v_C(t) = 2e^{-0.5t} - 2\cos(t) + \sin(t) = 2e^{-0.5t} + 2.2361\cos(t - 2.6779).$$

Step 2. *Compute the response over* $1 \le t < 2$. After the switch moves from position A to position B, the source is decoupled from the right half of the circuit; the response then depends only on the initial condition at $t = 1^-$, i.e., $v_C(1^-) = 2e^{-0.5} + 2.2361\cos(-1.6779) = 0.9739$. The goal is to compute $v_C(t)$ over the interval $1 \le t < 2$, or equivalently, over the interval $0 \le t' < 1$ where $t' = t - 1$. The s-domain equivalent circuit that models the behavior of the time-domain circuit of figure 14.34 over $1 \le t < 2$ has the form illustrated in figure 14.36.

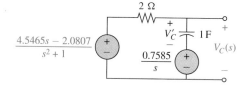

Figure 14.36 Equivalent circuit of figure 14.34 for the time interval $1 \le t < 2$.

Using current division and the voltage-current relationship of the capacitor, we obtain

$$V_C(s) = \frac{1}{s}I_C(s) = \frac{1}{s}\frac{s}{s+0.25}(0.9739) = \frac{0.9739}{s+0.25}.$$

For $0 \le t' < 1$, taking the inverse transform yields

$$v_C(t') = 0.9739e^{-0.25t'}u(t'),$$

or equivalently,

$$v_C(t) = 0.9739e^{-0.25(t-1)}u(t-1),$$

which is valid only for $1 \le t < 2$.

Step 3. *Determine* $v_C(t)$ *for* $t \ge 2$. The input for this time interval, from the Laplace transform viewpoint, is $5\sin(t)$ for $t \ge 2$, or equivalently, $5\sin(t)u(t-2)$. Also, from the result of step 2, the initial condition on the capacitor is $v_C(2^-) = 0.9739e^{-0.25} = 0.7585$. Repeating the procedure of step 2 produces a new equivalent circuit, shown in figure 14.37, which models the circuit behavior for $t \ge 2$ or for a new $t' = t - 2 \ge 0$.

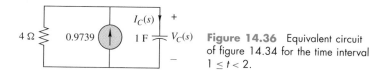

Figure 14.37 Equivalent frequency-domain circuit valid for $t \ge 2$.

From table 13.1 of chapter 13, the Laplace transform of the input excitation as a function of t' is

$$\mathcal{L}\left[5\sin(t'+2)u(t')\right] = 5\frac{\sin(2)s + \cos(2)}{s^2+1} = \frac{4.5465s - 2.0807}{s^2+1}.$$

Using superposition,

$$V_C(s) = \frac{0.5}{s+0.5}\left(\frac{4.5465s - 2.0807}{s^2+1}\right) + \frac{s}{s+10.5}\frac{0.7585}{s}$$

$$= \frac{0.5}{s+0.5}\left(\frac{4.5465s - 2.0807}{s^2+1}\right) + \left(\frac{0.7585}{s+0.5}\right).$$

Executing a partial fraction expansion yields

$$V_C(s) = \frac{-0.9831}{s + 0.5} + \frac{1.74165s + 1.4024}{s^2 + 1}.$$

Taking the inverse transform results in

$$v_C(t') = -0.9831e^{-0.5t'}u(t') + 1.7417\cos(t') + 1.4024\sin(t')$$

$$= -0.983e^{-0.5t'}u(t') + 2.236\cos(t' - 0.678),$$

which is valid for $t' \geq 0$, or equivalently, for $t \geq 2$. Thus,

$$v_C(t) = -0.983e^{-0.5(t-2)} + 2.236\cos(t - 2.678).$$

In sum,

$$v_C(t) = \begin{cases} 2e^{-0.5t} + 2.236\cos(t - 2.678), & 0 \leq t < 1 \\ 0.974e^{-0.25(t-1)}, & 1 \leq t < 2, \\ -0.983e^{-0.5(t-2)} + 2.236\cos(t - 2.678), & 2 \leq t \end{cases}$$

where all phase shifts are in rads. A plot of the response appears in figure 14.38.

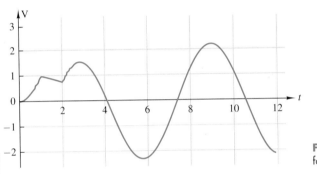

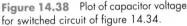

Figure 14.38 Plot of capacitor voltage for switched circuit of figure 14.34.

The method illustrated by this example can be summarized as follows:

Suppose that switching occurs at three instants of time, $t = 0, t = t_1 > 0$, and $t = t_2 > t_1$, and we wish to compute the response for $0 \leq t < \infty$. Then:

Step 1. *Divide the computation into three time intervals:* $(0, t_1), (t_1, t_2),$ *and* (t_2, ∞).

Step 2. *Construct the* s*-domain equivalent circuit, making use of the given initial conditions at* t $= 0^-$. This circuit is valid for the time interval $0 \leq t < t_1$.

Step 3. *Find the response by the Laplace transform method.*

Step 4. *Evaluate the capacitor voltages and inductor currents at* t $= t_1^-$.

Step 5. (1) For $t_1 \leq t < \infty$, let $t' = t - t_1$ (i.e., consider the time interval $0 \leq t' < \infty$), and construct the s-domain equivalent circuit, making use of the *calculated* initial conditions at $t' = 0^-$ (i.e., at $t = t_1^-$). (2) Determine the proper form of the input excitation(s) (if there are any), in terms of t'. (3) Find the response by the Laplace transform method. Note that the time variable is t'. To obtain the solution in t, substitute $(t - t_1)$ for t'.

Step 6. *Evaluate the capacitor voltages and inductor currents at* t $= t_2^-$.

Step 7. (1) For $t_2 \leq t < \infty$, let $t' = t - t_2$ (i.e., consider the time interval $0 \leq t' < \infty$), and construct the s-domain equivalent circuit, making use of the *calculated* initial conditions at $t' = 0^-$ (i.e., at $t = t_2^-$). (2) Determine the proper form of the input excitation(s) (if there are any), in terms of t'. (3) Find the response by the Laplace transform method. Again, the time variable is t'. As in step 5, to obtain the solution in t, substitute $(t - t_2)$ for t'.

CHAPTER 14   LAPLACE TRANSFORM ANALYSIS, 2: CIRCUIT APPLICATIONS

In steps 5 and 7, it is important to realize that under the time shift $t' = t - t_i$, an input excitation, say, $f(t)$, is left shifted to become $f(t' + t_i)u(t')$.

The extension of this method to more than three switching times should be clear. A restriction to one switching time is just as clear. Furthermore, although the preceding example uses a KCL circuit analysis, the same strategy or algorithm is applicable to the calculation of switching transients in any linear dynamic circuit.

In some situations, the first switching occurs at $t = 0$, but the capacitor voltages and inductor currents at $t = 0^-$ are not given. Instead, the problem specifies that dc and sinusoidal sources have excited the circuit for a long time. If the network is passive—i.e., if it consists of L's, C's, or *lossy* elements—then the circuit will have reached a steady state at $t = 0^-$. The procedure then is, first, to find the steady-state solution and then to evaluate the capacitor voltages and inductor currents at $t = 0^-$. It is instructive for the reader to review the dc and sinusoidal steady-state analysis methods studied in a first course in circuits, usually in association with phasor analysis. Recall that under certain stability conditions (to be studied in chapter 15) on the network:

1. For dc steady-state analysis, open-circuit all capacitances and short-circuit all inductors to find the steady-state voltages and currents; and
2. For sinusoidal steady-state analysis, use the phasor method to find the steady-state responses.

EXAMPLE 14.15

Consider the circuit of figure 14.39. Suppose that switch S_1 has been closed and S_2 has been open for a long time. At $t = 0$, S_1 is opened and S_2 is closed. At $t = 1$ sec, S_1 is closed and S_2 is opened. Find $v_C(t)$ for $t \geq 0$.

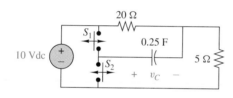

Figure 14.39 Switched circuit for example 14.15.

SOLUTION

This problem differs from the earlier ones in that it is necessary first to compute the initial conditions at $t = 0^-$ and then to proceed as in the algorithm described on page 468.

Step 1. Compute the initial condition at 0^-. At $t = 0^-$, the circuit has reached dc steady state. By inspection (with C open-circuited), $v_C(0^-) = 8$ V.

Step 2. For $0 \leq t < 1$, the s-domain equivalent circuit (with nonzero initial condition) is given in figure 14.40.

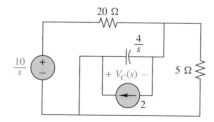

Figure 14.40 Equivalent s-domain circuit valid for $0 \leq t < 1$.

Step 3. Compute $v_C(t)$ for $0 \leq t < 1$, and evaluate at $t = 1^-$. To compute $v_C(t)$, observe that

$$V_C(s)\frac{8s - 2}{s(s + 1)} = \frac{-2}{s} + \frac{10}{s + 1}.$$

Hence, for $0 \leq t < 1$,

$$v_C(t) = (-2 + 10e^{-t}) \text{ V},$$

and at $t = 1^-$, $v_C(1^-) = 1.679$ V.

Step 4. *Compute* $v_C(t)$ *for* $1 \le t < \infty$, *or equivalently, for* $0 \le t' < \infty$, *where* $t' = t - 1$. In terms of t', the initial condition is $v_C(t' = 0^-) = 1.679$ V. The corresponding s-domain equivalent circuit, with the nonzero initial condition accounted for, is given in figure 14.41.

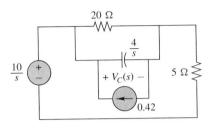

Figure 14.41 Equivalent circuit valid for $0 \le t' < \infty$.

Using the usual transform techniques,

$$V_C(s) = \frac{1.69s + 8}{s(s+1)} = \frac{8}{s} - \frac{6.321}{s+1}.$$

Inverting the Laplace transform yields

$$v_C(t') = 8 - 6.321e^{-t'},$$

in volts for $0 \le t' < \infty$, or

$$v_C(t) = 8 - 6.321e^{-(t-1)}$$

for $1 \le t < \infty$. In sum, the capacitor voltage, in volts, satisfies

$$v_C(t) = \begin{cases} -2 + 10e^{-t}, & 0 \le t < 1 \\ 8 - 6.321e^{-(t-1)}, & 1 \le t \end{cases}.$$

8. SWITCHED CAPACITOR CIRCUITS AND CONSERVATION OF CHARGE

In addition to its many uses already described, the Laplace transform method is applicable to a special class of circuits called *switched capacitor* (abbreviated SC) networks. These circuits contain only capacitors, switches, independent voltage sources, and possibly some operational amplifiers. No resistors or inductors are present. One can dispense with resistors because it is possible to approximate the effect of a resistor with two switches and a capacitor. Similarily, inductors can be approximated by circuits containing only switches, capacitors, and operational amplifiers. These facts, coupled with the easy and relatively inexpensive fabrication of switches, capacitors, and operational amplifiers in MOS (metal-oxide-semiconductor) technology, made switched capacitor filters an attractive alternative to classical filters. Given this scenario, the purpose of this section is to lay a foundation (i.e., introduce the principles) upon which switched capacitor circuit design builds. More advanced courses delve into the actual analysis and design of real-world switched capacitor circuits.

Besides the Laplace transform approach, an alternative method for analyzing switched capacitor networks builds on the principle of conservation of charge.

PRINCIPLE OF CONSERVATION OF CHARGE

The total charge transferred into a junction (or out of a junction) of a circuit at any time is zero.

This principle is a direct consequence of Kirchhoff's current law. For example, in figure 14.42, KCL implies that

$$i_1(t) + i_2(t) + i_3(t) + i_4(t) = 0. \tag{14.23}$$

Since charge is the integral of current over a time interval, integrating both sides or equation 14.23 from $-\infty$ to the present, t, yields

$$\int_{-\infty}^{t} [i_1(t) + i_2(t) + i_3(t) + i_4(t)]\,dt = 0,$$

or equivalently,

$$q_1(t) + q_2(t) + q_3(t) + q_4(t) = 0,$$

which is just another expression of the principle of conservation of charge. A simple SC circuit will now serve as a test-bed for comparing the merits of the foregoing analysis techniques.

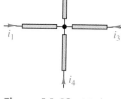

Figure 14.42 Node to which KCL applies.

 EXAMPLE 14.16

Consider the circuit shown in figure 14.43a. The switch S is closed at $t = 0$. Just before the closing of S, the initial conditions are known to be $v_{C1}(0^-) = 1$ V and $v_{C2}(0^-) = 0$. Determine the voltages $v_{C1}(t)$ and $v_{C2}(t)$ for $t > 0$.

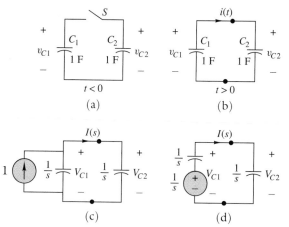

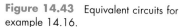

Figure 14.43 Equivalent circuits for example 14.16.

SOLUTION

Method 1. Using the model of a capacitor shown in figure 14.16, we obtain the s-domain equivalent circuit shown in figure 14.43c. By inspection,

$$V_{C1}(s) = V_{C2}(s) = \frac{1}{s}I(s) = \frac{1}{s}\frac{s}{s+s} = \frac{0.5}{s}.$$

Therefore, $v_{C1}(t) = v_{C2}(t) = 0.5u(t)$ V, and $i(t) = 0.5\delta(t)$ A.

Method 2. Using the model of a capacitor shown in figure 14.17, we have the s-domain equivalent circuit shown in figure 14.43d. Again, by inspection,

$$V_{C1}(s) = V_{C2}(s) = \frac{\dfrac{1}{s}}{\dfrac{1}{s} + \dfrac{1}{s}}\frac{1}{s} = \frac{0.5}{s}; \quad I(s) = \frac{\dfrac{1}{s}}{\dfrac{1}{s} + \dfrac{1}{s}} = 0.5.$$

This is the same answer as obtained with method 1.

Method 3. *Conservation-of-charge approach.* For $t > 0$, the network is shown in figure 14.43b. Clearly, $v_{C1}(t) = v_{C2}(t)$. After S is closed, some charge is transferred from

C_1 to C_2. However, according to the principle of conservation of charge, the *total* charge transferred out of the junction must be zero. Hence,

$$C_1[v_{C1}(0^+) - v_{C1}(0^-)] + C_2[v_{C2}(0^+) - v_{C2}(0^-)] = 0,$$

and

$$v_{C1}(0^+) = v_{C2}(0^+).$$

Solving these two equations for the two unknowns, $v_{C1}(0^+)$ and $v_{C2}(0^+)$, results in

$$v_{C1}(0^+) = v_{C2}(0^+) = \frac{C_1 v_{C1}(0^-) + C_2 v_{C2}(0^-)}{C_1 + C_2}.$$

Since there is no external input applied, the voltages remain constant once the equilibrium condition has been reached. Therefore,

$$v_{C1}(t) = v_{C2}(t) = v_{C1}(0^+) = v_{C2}(0^+)$$

for $t > 0$. For the specific capacitance values given in figure 14.43a, we obtain $v_{C1}(t) = v_{C2}(t) = 0.5$ V for $t > 0$.

Exercise. In the circuit of figure 14.43a, let $C_1 = 4$ F and $C_2 = 6$ F, $v_{C1}(0^-) = 8$ V, and $v_{C2}(0^-) = 3$ V. Find, by at least two methods, the capacitor voltages after the switch is closed. (Rework the problem if the answers do not agree.)

Computationally, the Laplace transform method is more straightforward. On the other hand, the conservation-of-charge method is more basic and provides better insight into what happens to the charges stored in various capacitors. It is particularly useful for the purpose of checking answers obtained by other methods: The answers are correct when the conservation-of-charge condition is met at every node.

EXAMPLE 14.17

The initial conditions at $t = 0^-$ of an SC network are shown in figure 14.44. Switches S_1 and S_2 are closed at $t = 0$, thereby connecting the two dc voltage sources to the network. Find the node voltages for $t > 0$.

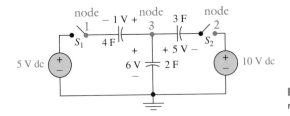

Figure 14.44 Switched capacitor network for example 14.17 at $t = 0^-$.

SOLUTION

We first construct the s-domain equivalent circuit. The result is shown in figure 14.45.

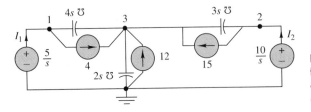

Figure 14.45 Equivalent circuit for figure 14.44, accounting for initial conditions. ℧ indicates mhos. All currents are in amps.

Applying KCL to node 3 generates the node equation

$$4s\left(V_3 - \frac{5}{s}\right) + 3s\left(V_3 - \frac{10}{s}\right) + 2sV_3 = 4 + 15 + 12,$$

from which $V_3 = 9/s$ and $v_3(t) = 9$ V for $t > 0$. Obviously, $V_1 = 5$ V and $V_2 = 10$ V. To verify that $V_3 = 9$ V is indeed the correct solution, we check for conservation of charge. From $t = 0^-$ to $t = 0^+$, the voltage of the 4-F capacitor jumps from 1 V to 4 V (note that $4 = 9 - 5$), indicating that $4 \times (4 - 1) = 12$ coulombs of charge have been transferred to this capacitor. The voltage of the 2-F capacitor jumps from 6 V to 9 V, indicating that $2 \times (9 - 6) = 6$ coulombs of charge have been transferred here. Finally, the voltage of the 3-F capacitor changes from 5 V to -1 V (note that $-1 = 9 - 10$), indicating that $3 \times (-1 - 5) = -18$ coulombs of charge have been transferred to this capacitor. As a check for conservation of charge, we have $12 + 6 + (-18) = 0$, and the solution is assured to be correct.

Exercise. Solve example 14.16 again, with all capacitors initially uncharged. Check your answers by conservation of charge.

In both of the preceding examples, there was no resistance in the circuit. These circuits are certainly *idealized* and unrealistic, because in any *practical* circuit, the wire connecting capacitors together must have some resistance. Why should we be interested in the analysis of an idealized circuit? The reason is that an idealized circuit can be analyzed much more easily than a realistic circuit and can at the same time provide answers that are often quite close to the true solution. As a case in point, consider example 14.17 again. Had we represented the circuit by a more realistic model, say, by inserting a 0.1-Ω resistance in series with every capacitor, a Laplace transform analysis would have resulted in a rational function with a cubic polynomial (a third-order network) in the denominator whose factorization for a partial fraction expansion would require the use of a root-finding program. In sharp contrast, the idealized circuit of example 14.17 was analyzed by writing just one node equation, making a partial fraction expansion unnecessary.

Idealizations of circuit models sometimes lead to phenomena that defy intuitive explanations. An interesting case is given by example 14.16. Before S is closed, the energy stored in the electric field is $0.5 \times 1^2 + 0.5 \times 0^2 = 0.5$ joule. After S is closed, the stored energy becomes $0.5 \times (0.5)^2 + 0.5 \times (0.5)^2 = 0.25$ joule. Apparently, 0.25 joule of energy has been lost. Since there is no resistance in the circuit to dissipate the energy, what accounts for the lost energy? Is energy not conserved?

An explanation of this paradox is as follows. Instead of considering a zero-resistance circuit, place a resistance R in series with all capacitances. Then analyze the circuit, and let R *approach* zero. The result shows that no matter what value R takes on, the total energy dissipated in the resistance for $0 < t < \infty$ exactly equals the difference of the total stored energies before and after the closing of the switch. This accounts for the apparent lost energy. In actuality, part of the 0.25 joule of energy would be lost in the form of radiated energy. However, the principles of field theory would be necessary to explain the radiation phenomena.

From the solutions of the previous SC circuits, as long as the independent voltage sources are *piecewise* constant, no matter how the switches are operated, all capacitor voltages are *piecewise* constant, and all currents in the circuit are impulses. These properties remain valid for more general switched capacitor circuits that allow the inclusion of voltage-controlled voltage sources, current-controlled current sources, and ideal operational amplifiers. The reason is that the parameters characterizing these components are dimensionless and hence do not result in a time constant. All voltage changes are *instantaneous*. On the other hand, if resistances, transconductances, or transresistances are included in the circuit, we have what is called a *lossy switched capacitor circuit*, whose voltages are no longer piecewise constant. The transient analysis of a lossy SC circuit requires the general transform methods we are currently studying.

One reason for our interest in SC circuits is that a switch-capacitor combination can be used to *approximate* a resistor. As a result, any RC-op amp circuit used for signal processing can be *approximated* by an SC-op amp circuit. A study of the general theory of such SC-op amp circuits is beyond the level of this book. We shall merely illustrate the approximation property with a simple integrator circuit.

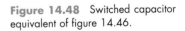 **EXAMPLE 14.18**

Consider the RC-op amp integrator circuit shown in figure 14.46.

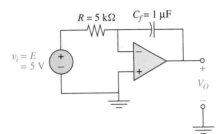

Figure 14.46 Op amp circuit of example 14.18.

Using the result of example 14.5 and the integration property of the Laplace transform, it is straightforward to show that

$$v_o(t) = -\frac{1}{RC_f} \int_0^t v_i(\tau)d\tau + v_o(0).$$

If the input is a constant voltage $v_i = E = 5$ V, and if $v_o(0) = 0$, then the output waveform is a ramp function $v_o(t) = -1,000tu(t)$ (as long as the output has not reached the saturation level), as shown in figure 14.47.

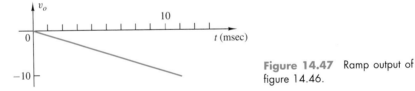

Figure 14.47 Ramp output of figure 14.46.

What we would like to do is construct an SC approximation to this circuit. The idea is to replace the resistor with a switch and a capacitor, as in figure 14.48.

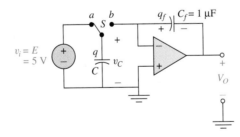

Figure 14.48 Switched capacitor equivalent of figure 14.46.

The switch S is operated in the following manner:

1. At $t = 0$, S is at position a.
2. At $t = T$, S is moved to position b.
3. At $t = 2T$, S is moved to position a.
4. At $t = 3T$, S is moved to position b, etc.

The output waveform may be determined very easily by the principle of conservation of charge as follows: For $0 \leq t < T$, $v_c = E$, $q = CE$, $q_f = 0$, and $v_o = 0$. At $t = T$, switch

S is moved to position b. Because the op amp is assumed to be ideal, the voltage across the input terminals is zero, and so is v_c. Thus, C cannot store any charge. The charge CE previously stored on C must be transferred out of C. Since the op amp is ideal, the input impedance is infinity and the input current is zero. Therefore, none of the charge can flow into the op amp. Instead, the charge must be transferred to the capacitor C_f. This leads to $q_f = CE$, and $v_o(T^+) = -CE/C_f$.

At $t = 2T$, switch S is moved back to position a, causing C to be charged to E V again. Since the charge q_f is "trapped" on C_f, the output voltage v_o remains unchanged until S is moved to position b again. At that time, another CE coulombs of charge are transferred to C_f, and v_o is *incremented* by $-CE/C_f$. Subsequent switching is similar.

In order to make the output waveforms of the circuits of figures 14.46 and 14.47 *approximately* the same, the *average* of the charges transferred to C_f must be the same in both cases. For figure 14.46, the current flowing into C_f is at a constant value of E/R. Therefore, every $2T$ sec, the charge transferred to C_f is equal to $2TE/R$. On the other hand, for figure 14.48, the charge transferred to C_f every $2T$ sec is CE. Equating these two quantities, we have $CE = 2TE/R$, or $RC = 2T$. Thus, there is no unique combination of C and T that will produce the approximate effect of a resistance. A smaller T (i.e., a higher operating frequency of the switch) in figure 14.48 produces a staircase output waveform that "hugs" closer to the ramp output of figure 14.47. For the purpose of comparison, the output waveform corresponding to $T = 1$ msec and $C = 0.4$ µF is shown in figure 14.49, together with the ramp output from the RC-op amp integrator. It is worthwhile to note that each of the circuits of figures 14.46 and 14.48 drains the same average amount of charge from the voltage source and puts the same average amount of charge on the capacitor C_f. The only difference is that in the former the process is *continuous*, whereas in the latter the process occurs in quantized steps.

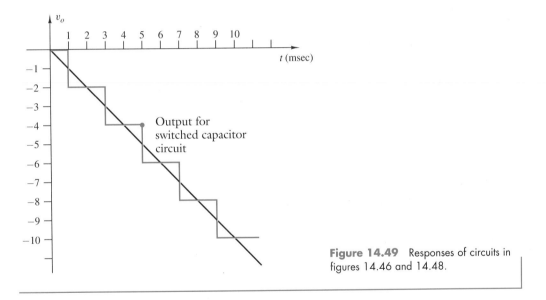

Figure 14.49 Responses of circuits in figures 14.46 and 14.48.

Exercise. Plot the output waveform of the SC circuit of figure 14.48 for $0 \leq t \leq 2$ msec if $T = 0.1$ msec and $C = 0.04$ µF. Also, plot the ramp output on the same graph for comparison.

As recently as a decade ago, switched capacitor circuits like those just discussed were considered impractical. But they are not anymore. As mentioned earlier, advances in semiconductor technology have made fabrication of op amps as cheap as resistors. Switches are cheaply constructed out of MOS devices rather than mechanical components such as relays. Furthermore, numerous switches, capacitors, and op amps can be fabricated on a single chip. Consequently, switched capacitor circuits hold an important place in future signal-processing applications. Although we cannot delve into the practical aspects of the design of such circuits, we have at least outlined the basic principles needed for their approximate or exact analysis.

SUMMARY

This chapter has presented the basic principles and techniques of circuit analysis in the complex-frequency domain. Impedance, admittance, Thévenin equivalents, superposition, linearity, voltage division, current division, source transformations, and nodal and mesh analysis are all well-defined tools of analysis in the complex-frequency domain. The simple starter circuit of a fluorescent light illustrates the importance of these techniques for the analysis and design of simple, everyday electrical conveniences. Various op amp applications were also presented. As subsequent chapters will illustrate, complex circuits and advanced analysis methods build on these principles and techniques.

The chapter also introduced the notion of switched capacitor circuits. Integrated circuit technology has made such circuits easy and inexpensive to produce. They find application in a wide range of filtering problems—for example, speech processing. Although a full-scale analysis of such circuits is beyond the scope of this text, once again the basic principles of their operation are presented and are the foundation upon which more advanced methods can build.

TERMS AND CONCEPTS

Admittance: the ratio of the Laplace transform of the input current to the Laplace transform of the input voltage with the two-terminal network initially relaxed.

Admittance of capacitor: the quantity Cs.

Admittance of inductor: the quantity $1/Ls$.

Admittance of resistor: the quantity $1/R$.

Current division: a formula for determining how currents distribute through a set of parallel admittances.

Impedance: the ratio of the Laplace transform of the input voltage to the Laplace transform of the input current with the two-terminal network initially relaxed.

Impedance of capacitor: the quantity $1/Cs$.

Impedance of inductor: the quantity Ls.

Impedance of resistor: the quantity R.

Impulse response: the response of a circuit having a single input excitation of a unit impulse; equal to the inverse Laplace transform of the transfer function.

Source transformation property: in the complex-frequency domain, voltage sources in series with an impedance are equivalent to the same impedance in parallel with a current source whose value equals the transform voltage divided by the series impedance.

Steady-state analysis: analysis of circuit behavior resulting after excitations have been on for a long time; often refers to finding the sinusoidal or constant parts of the response when the circuit is excited by sinusoids or dc.

Switched capacitor network: a circuit containing switches, capacitors, independent voltage sources, and possibly op amps, but no resistors or inductors.

Time-invariant device: a device whose characteristics do not change with time.

Transient analysis: analysis of circuit behavior for a period of time immediately after independent sources have been turned on.

Transfer function: the ratio of the Laplace transform of the output quantity to the Laplace transform of the input quantity with the network initially relaxed.

Voltage division: a formula for determining how voltages distribute around a series connection of impedances.

PROBLEMS

Impedance and Admittance

1. Determine the input impedance and admittance of each of the networks shown in figures P14.1a and b.

2. Determine the input impedance and admittance of each of the networks shown in figures P14.2a, b, and c.

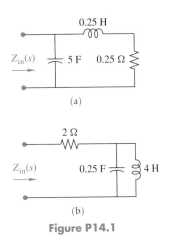

(a)

(b)

Figure P14.1

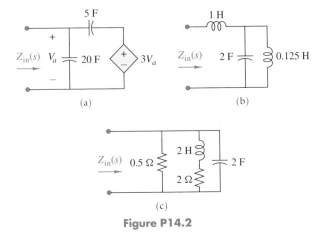

(a)

(b)

(c)

Figure P14.2

3. Recall the definitions of input impedance, input admittance, and the transfer function of a device.

   (a) For each of the circuits shown in figure P14.3, calculate the input impedance and input admittance.

   (b) For each of the circuits shown in figure P14.3, calculate the indicated transfer function.

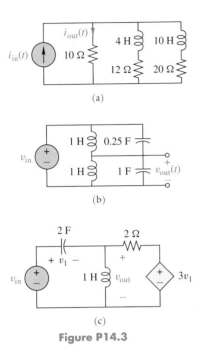

(a)

(b)

(c)

Figure P14.3

4. Determine the input impedance, $Z_{in}(s)$, and input admittance for each of the circuits shown in figure P14.4.

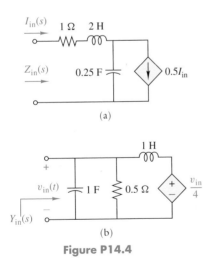

(a)

(b)

Figure P14.4

5. The two-terminal linear network in figure P14.5 has an impedance function

$$Z(s) = \frac{s+2}{s+4} \; \Omega.$$

The network is initially at rest (zero initial conditions.) A constant voltage source of 2 V is applied at $t = 0$. Determine $i(t)$ for $t \geq 0$.

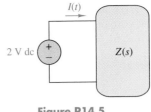

Figure P14.5

6. Find $Z_{in}(s)$ in the circuit of figure P14.6. Now suppose that the capacitor and inductor are initially ($t = 0^-$) unenergized and a dc source of 2 amperes is applied to the impedance. Find the voltage across the current source as a function of t, $t \geq 0$.

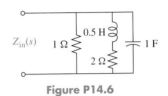

Figure P14.6

7. Compute the input impedance and input admittance of the circuit in figure P14.7.

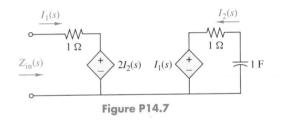

Figure P14.7

8. Find the input impedance of the circuit of figure P14.8.

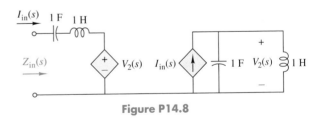

Figure P14.8

Transfer Functions

9. Determine the transfer function $H(s) = V_{out}(s)/V_{in}(s)$ for the circuit of figure P14.9.

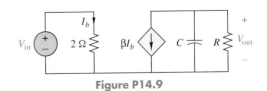

Figure P14.9

10. Determine $H(s) = V_C(s)/V_{in}(s)$ in figure P14.10. (*Check*: The real part of the pole is -0.25.) Then determine $v_C(t)$ when $v_{in}(t) = 16u(t)$ V.

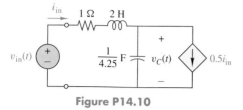

Figure P14.10

11. A particular circuit has an integrodifferential equation model relating v_{out} to v_{in} given by

$$v'_{out}(t) + 4v_{out}(t) + 1.75 \int_{0^-}^t v_{out}(q)dq$$
$$= 14v_{in}(t) - 14 \int_{0^-}^t v_{in}(q)dq.$$

Find the transfer function and the response when $v_{in}(t) = te^{-t}u(t)$ V and $v_{out}(0^-) = 0$.

12. Determine $H(s) = V_{out}(s)/I_{in}(s)$ for the circuit of figure P14.12.

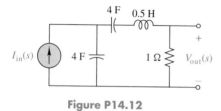

Figure P14.12

13. A certain circuit has input $v_{in}(t) = \cos(t)u(t)$ V and output $i_{out}(t) = 2\sin(t)u(t)$ A. Determine the transfer function of the circuit, assuming that the circuit had no internal stored energy at $t = 0^-$.

14. The input to a relaxed (no initial conditions) linear active circuit is $v_{in}(t) = te^{-t}u(t)$ V. The response is measured in volts as

$$v_{out}(t) = (1 + t - 0.5t^2)e^{-t}u(t) + \sin(t)u(t) - \cos(t)u(t).$$

(a) Determine the transfer function, $H(s)$, as the ratio of two polynomials $n(s)/d(s)$, where $d(s)$ is in factored form, i.e., $d(s) = (s + a)(s + b) \cdots$.
(b) Determine the response of the circuit to the new input $v_{in}(t) = (1 + t)u(t)$ V.

15. Determine the transfer function of each of the (ideal) op amp circuits of figure P14.15.

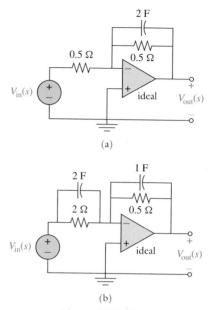

(a)

(b)

Figure P14.15

16. Determine the transfer function of the circuit of figure P14.16, assuming that all op amps are ideal. (*Hint*: It is important to label all outputs and determine the input to each operational amplifier according to those outputs as well as the designated input and output. By understanding the action of each operational amplifier circuit on its inputs, it is possible to determine, more or less, the transfer function by successive substitutions.)

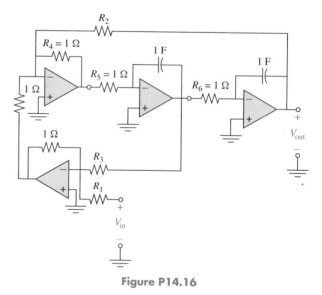

Figure P14.16

17. For the circuit of figure P14.17:
(a) Find the transfer function.
(b) If $v_{in}(t) = 4u(t)$ V, find $v_{out}(t)$.

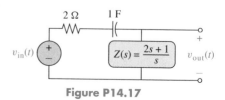

Figure P14.17

18. Compute the indicated transfer functions of each circuit of figure P14.18.

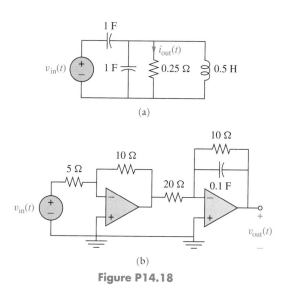

(a)

(b)

Figure P14.18

19. Consider the circuit of figure P14.19.

(a) Use a source transformation and the current divider formula to show that the transfer function between $V_{in}(s)$ (the input) and $I_C(s)$ (the output) is

$$H(s) = \frac{1}{R}\frac{s^2}{s^2 + \frac{1}{RC}s + \frac{1}{LC}}.$$

Note that the computation of transfer functions presumes no initial internal stored energy.

(b) If $L = 1$ H, $C = 0.5$ F, $R = 2/3$ Ω, and $v_{in}(t) = e^{-t}u(t)$ V, show that

$$i_C(t) = [6e^{-2t} - 4.5e^{-t} + 1.5te^{-t}]u(t)\text{A}$$

assuming that all IC's are zero.

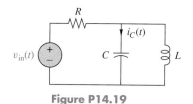

Figure P14.19

◆ **20.** Assuming that all initial conditions are zero in figure P14.20,

(a) Find the transfer function $H(s) = V_{out}(s)/V_{in}(s)$.

(b) If the ratio $R_1/R_2 = 4$, find R_1 and R_2 so that the response is $v_{out}(t) = [2+(4/3)e^{-1.6667t}]u(t)$ V. (*Check:* $R_1 = 2$ Ω.)

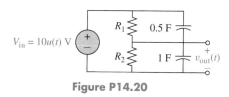

Figure P14.20

21. For each of the circuits shown in figure P14.21, compute the indicated transfer function.

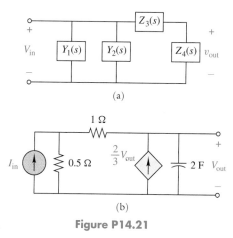

(a)

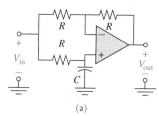

(b)

Figure P14.21

◆ **22.** The circuit of figure P14.22a is called an *all-pass* network. This means that the circuit does not alter the magnitudes of sinusoidal waveforms, but it does introduce a new phase shift depending on the frequency of the sinusoid. The circuit of figure 14.22b is the Sallen and Key low-pass filter, which can be used to eliminate unwanted high-frequency noise. For each of these circuits, compute the indicated transfer function, assuming that all op amps are ideal.

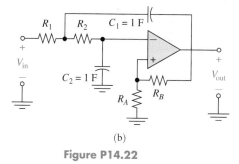

(a)

(b)

Figure P14.22

(*Hint:* For part (b), you will need to write three equations at each of the indicated nodes.) *Check:* One coefficient in the denominator is

$$\frac{1}{R_1} + \frac{1}{R_2} - \frac{K}{R_2},$$

for an appropriate constant K depending on R_A and R_B.)

Now let $R_1 = R_2 = R_A = 1$ Ω and $R_B = 1.8$ Ω. Use SPICE or some other circuit analysis program to simulate the circuit of figure 14.22b for sinusoids of frequency $\omega = 0.1$ rad/sec, 0.4 rad/sec, 0.7 rad/sec, 1.0 rad/sec, 1.5 rad/sec, 2 rad/sec, 10 rad/sec, and 100 rad/sec. Does the circuit behave like a low-pass filter?

23. Consider the circuit of figure P14.23.

(a) Compute the transfer function $H(s)$. Note: $H(s)$ should have the form $H(s) = Kn(s)/d(s)$, where $n(s)$ and $d(s)$ are polynomials whose leading coefficients are 1; i.e., the highest power of s has coefficient 1.

(b) If $2C_1 = C_2 = 0.2$ F, find R_1 and R_2 so that the transfer function has a pole at $s = -5$ and a zero at $s = -0.1$. The problem of finding element values to realize a given transfer function is called the *synthesis problem*.

(c) Let $s = j\omega$. A plot of the magnitude, $|H(j\omega)|$, and phase, $\angle H(j\omega)$, versus ω of the transfer function is called the frequency response of the circuit. Make a rough sketch of the frequency response.

(d) Check your answer to part (c), using MATLAB®, SPICE, or some other circuit analysis program.

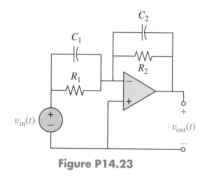

Figure P14.23

24. Determine the transfer function of the op amp circuit in figure P14.24, assuming an ideal op amp.

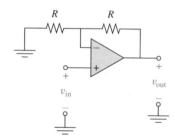

Figure P14.24 Ideal op amp circuit for problem 24.

Initial Conditions

25. Consider the circuit of figure P14.25.

(a) Suppose $i_{in}(t) = 2e^{-t}u(t)$ A and $v_C(0^-) = 1$ V. For $t \geq 0$, determine $v_C(t)$.

(b) If $v(0^-) = 8$ V and $i(t) = 4\delta(t)$ A, then, for $t > 0$, determine $v_C(t)$.

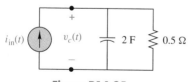

Figure P14.25

26. Consider the circuit of figure P14.26.

(a) If $i(0^-) = 1$ A and $v_{in}(t) = 2e^{-t}u(t)$ V, determine $i(t)$ for $t \geq 0$.

(b) If $i(0^-) = 1$ A and $v_{in}(t) = 10\cos(t)u(t)$ V, determine $i(t)$ for $t \geq 0$.

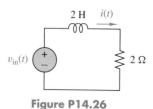

Figure P14.26

27. Consider the circuit of figure P14.27. Given that $i(0^-) = 1$ A and $v(0^-) = 2$ V, determine $V_C(s)$ and $v_C(t)$.

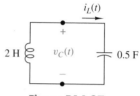

Figure P14.27

28. Consider the circuit of figure P14.28.

(a) The capacitor is initially charged, having $v_C(0^-) = 10$ V. Suppose $v_{in}(t) = 5e^{-0.25t}u(t)$ V. Determine $i(t)$ and $v_C(t)$ for $t > 0$.

(b) Using SPICE or some other circuit analysis program, verify the responses computed in (a).

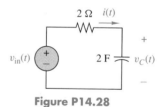

Figure P14.28

29. Consider the circuit of figure P14.29. Given the initial capacitor voltages $v_{C1}(0^-) = a$ V and $v_{C2}(0^-) = b$ V,

(a) Find $i(t)$, $v_{C1}(t)$, and $v_{C2}(t)$ for $t \geq 0$.

(b) Find $W(0^-)$ and $W(0^+)$, where $W(t)$ denotes the total energy stored in the capacitors at time t.

(c) Using the integral

$$\int_0^\infty Ri^2(t)dt,$$

calculate the energy dissipated in the resistor R, and verify that the integral is equal to $[W(0^+) - W(0^-)]$.

(d) Find $i(t)$, $v_{C1}(t)$, and $v_{C2}(t)$ as R approaches zero.

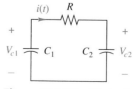

Figure P14.29 RC circuit for problem 29.

30. This problem combines the concepts of source transformations, parallel impedances/admittances, and equivalent frequency-domain circuits for L's and C's with initial conditions into a single problem. Compute the response, $v_{out}(t), t \geq 0$, for the circuit sketched in figure P14.30. Assume that $v_1(t) = 30\delta(t)$ V, $v_{out}(0^-) = 20$ V, $v_2(t) = 20e^{-4t}u(t)$ V, $v_{C1}(0^-) = 0$, $i_{L1}(0^-) = 0$, and $i_{L2}(0^-) = 10$ A.

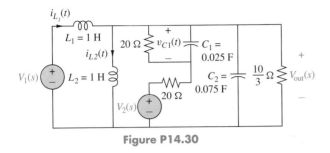

Figure P14.30

31. For each circuit of figure P14.31, determine $v_{out}(t), t \geq 0$. For the circuit of figure P14.31a, assume that $v_{C1}(0^-) = -1$ V and $v_{C2}(0^-) = 0$ V.

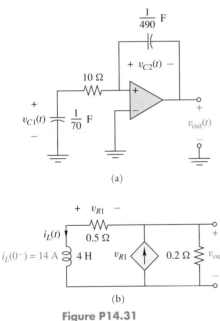

(a)

(b)

Figure P14.31

32. Compute $v_{out}(t)$ in the circuit of figure P14.32, assuming that $v_{in}(t) = u(t)$ V, $i_L(0) = 0, v_C(0^-) = 1$ V, and $v_{out}(0^-) = 1$ V. (*Hint*: Consider a source transformation, and then draw the equivalent circuit in the Laplace transform domain so that you can combine sources in the front half of the circuit.)

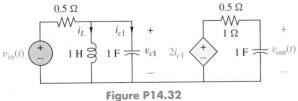

Figure P14.32

33. Consider the circuit figure P14.33.
(a) Determine the transfer function in terms of L.
(b) If the response to the input $i_{in}(t) = Atu(t)$ A is

$$i_L(t) = (15t - 3)u(t) + 3e^{-5t}u(t),$$

in amps assuming no initial inductor current, determine the values of L and A.
(c) Assuming a nonzero $i_L(0^-)$, draw the equivalent frequency-domain circuit. If the input $i_{in}(t) = \delta(t)$ A produces a response $i_L(t) = 10e^{-5t}u(t)$ A, determine the value of $i_L(0^-)$.

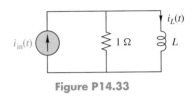

Figure P14.33

34. Consider figure P14.34. Suppose $v_{in}(t) = 2u(t)$ V. Suppose also that the initial conditions are $v_C(0^-) = 2$ V and $i_L(0^-) = 4$ A. Construct the Laplace transform domain equivalent circuit that accounts for the initial conditions. From this circuit, find $V_{out}(s)$ and then $v_{out}(t)$ for $t \geq 0$.

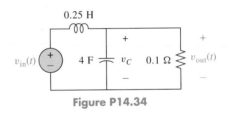

Figure P14.34

Node and Loop Analysis

35. (a) In the circuit of figure P14.35, $v_C(t)$ and $v_R(t)$ represent node voltages. Assume that $v_C(0^-) = 2$ V, $i_L(0^-) = 0$ A, and $v_{in}(t) = u(t)$ V and $i_{in}(t) = u(t)$ A. Find the response $v_R(t)$ as follows: (1) Draw the circuit in the frequency domain, accounting for initial conditions. (2) Write a set of node equations. (3) Write the node equations in matrix form. (4) Solve the equations either by using Cramer's rule or by inverting the 2×2 matrix.
(b) Repeat part (a), using a mesh analysis.

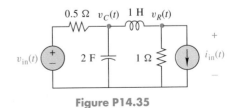

Figure P14.35

36. The circuit of figure P14.36 has initial conditions $v_C(0) = i_L(0) = 0$. The input is $v_{in}(t) = 10e^{-at}u(t)$ V.
   (a) Write node equations in the frequency domain.
   (b) Find $V_{out}(s)$.
   (c) Find $v_{out}(t)$ for $t \geq 0$ for the two cases $a = 0.5$ and $a = 0.25$.

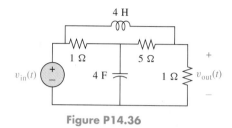

Figure P14.36

37. Figure P14.37 depicts the so-called Twin-T RC circuit. This network is often used as a band reject filter, i.e., a circuit that stops or rejects signals at certain frequencies and allows signals at all other frequencies to pass with little attenuation. This can easily be seen, since the transfer function has the form

$$H(s) = \frac{V_o(s)}{V_{in}(s)} = \frac{s^2 + \dfrac{1}{R^2C^2}}{s^2 + \dfrac{4}{RC}s + \dfrac{1}{R^2C^2}}.$$

Observe that if $s = j\omega_0 = j/RC$, then $H(j\omega_0) = 0$, i.e., $V_o(j\omega_0) = 0$ for input sinusoids of the form $V_{in}(j\omega_0) = A\sin(\omega_0 t + \phi)$.
   (a) Use nodal analysis to derive the foregoing transfer function.
   (b) Show that the poles (zeros of the denominator) are $(-2 \pm \sqrt{3})/RC$.
   (c) Let $R = 1.5915$ Ω and $C = 0.01$ F. Assuming no initial conditions, determine the response to
      (1) $v_{in}(t) = \sin(2\pi 100t)u(t)$ V.
      (2) $v_{in}(t) = \sin(2\pi 10t)u(t)$ V.
      (3) $v_{in}(t) = \sin(2\pi 5t)u(t)$ V.

   (d) Given your answers to (c), what does $v_o(t)$ approximate after several seconds in each case; this is the so-called steady-state output voltage. Can you say anything about item (2) of part (c)?

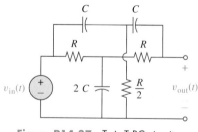

Figure P14.37 Twin-T RC circuit.

38. Use nodal analysis to compute the transfer function $H(s) = V_{out}(s)/V_{in}(s)$ for the circuit of
   (a) Figure P14.38a;
   (b) Figure P14.38b.

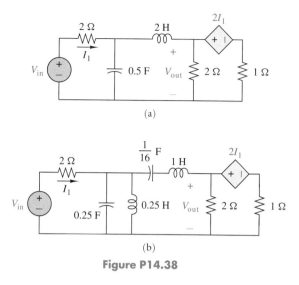

(a)

(b)

Figure P14.38

(*Check*: The transfer function of figure P14.38b is that of figure P14.38a with s replaced by $(s^2 + 16)/(2s)$.)

39. Find the transfer function of the circuit in figure P14.39 by
   (a) The method of node analysis.
   (b) The method of loop analysis.

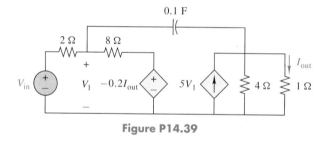

Figure P14.39

40. Consider figure P14.40. Use loop analysis to find the transfer functions
   (a) $H_1(s) = V_{out}(s)/V_1(s)$ when $V_2 = 0$;
   (b) $H_2(s) = V_{out}(s)/V_2(s)$ when $V_1 = 0$.

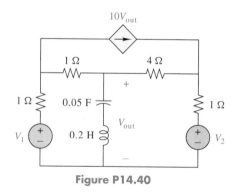

Figure P14.40

41. In the circuit shown in figure P14.41, $v_{in}(t) = 12$ V dc, $v_{C1}(0) = 6$ V, and $v_{C2}(0) = 2$ V.
   (a) Construct the equivalent circuit in the Laplace transform domain, accounting for an initialized capacitor.

(b) Write a nodal equation for the circuit constructed in part (a).
(c) Find $V_{C2}(s)$.
(d) Find $v_{C2}(t)$ for $t \geq 0$.

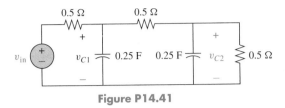

Figure P14.41

42. Consider the third-order RC circuit shown in figure P14.42. Suppose that the initial capacitance voltages are $v_{C1}(0) = 0$, $v_{C2}(0) = 6$ V, $v_{C3}(0) = 2$ V, and $v_{in}(t) = 12$ V dc.
   (a) Construct the Laplace transform domain equivalent circuit.
   (b) Find $V_{C3}(s)$.
   (c) Find $v_{C3}(t)$ for $t \geq 0$.

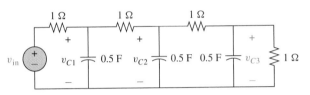

Figure P14.42 Third-order circuit for problem 42.

Remarks: Consider using CAD software as an aid in solving this problem:
   (1) A root-finding program is available on MATLAB and on the software available with this text.
   (2) A partial fraction expansion program is available on MATLAB and on the software included with this text.

43. The ideal op amp circuit shown in figure P14.43 has initial capacitor voltages of $v_{C1}(0) = 2$ V and $v_{C2}(0) = 4$ V.
   (a) Construct the Laplace transform domain equivalent circuit, accounting for these initial conditions.
   (b) Find $V_o(s)$.
   (c) Find $v_o(t)$ for $t \geq 0$ and make a rough sketch of the response.
   (d) Use SPICE or some other simulation program to verify your rough sketch found in part (c).

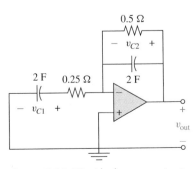

Figure P14.43 Ideal op amp circuit.

44. For the bridge-T network of figure P14.44:
   (a) Draw the equivalent frequency-domain circuit (accounting for ICs) pertinent to nodal analysis.
   (b) Write a set of nodal equations, and solve them for $V_{out}(s)$ as a function of $V_{in}(s)$ and the ICs.
   (c) Find $v_{out}(t), t \geq 0$, given that $v_c(0^-) = 1.5$ V, $i_L(0^-) = 0.5$ A, and $v_{in}(t) = 4u(t) - 3e^{-t}u(t)$ V.

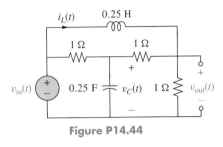

Figure P14.44

Switching

45. The switch in the circuit of figure P14.45 is in position A for a very long time and then moves to position B at time $t = 0$. Determine $v_{out}(t)$.

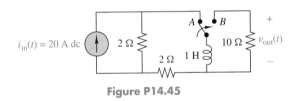

Figure P14.45

46. In the circuit in figure P14.46, the switch is in position a at $t = 0^-$ sec and moves to position b at $t = 1$ sec. The input current is $i_{in}(t) = 0.5\pi u(t)$ A. All ICs are zero at $t = 0^-$.

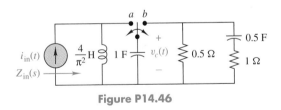

Figure P14.46

   (a) Determine $Z_{in}(s)$ for $0 \leq t < 1$.
   (b) Determine $V_C(s)$ for $0 \leq t < 1$.
   (c) Determine $v_C(t)$ for $0 \leq t < 1$.
   (d) Draw the equivalent frequency-domain circuit that is valid for $t \geq 1$.
   (e) Determine an expression for $V_C(s)$ for $t \geq 1$.
   (f) Determine a partial fraction expansion of $V_C(s)$ for $t \geq 1$.
   (g) Determine $v_C(t)$ for $t \geq 1$.

47. Repeat problem 46 for $i_{in}(t) = 0.5\pi \sin(0.5\pi t)u(t)$ A.

48. The circuit of figure P14.48 contains a switch that moves from position a to position b at $t = 1$ sec.

Assume that all ICs are zero at $t = 0^-$ and that $i_{in}(t) = 7e^{-0.5t}u(t)$ A.

    (a) Determine $v_{out}(t), 0 \leq t \leq 1$.
    (b) Determine $v_{out}(t), 1 < t$.
    (c) Make a rough sketch of $v_{out}(t)$ for $t \geq 0$.

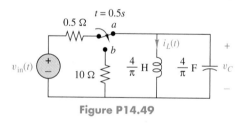

Figure P14.48

49. Consider the switched RLC circuit of figure P14.49. Suppose the switch changes position from point a to point b at $t = 0.5$ sec. Assume that all ICs are zero at $t = 0^-$ and that $v_{in}(t) = 5u(t)$ V.

    (a) Determine $i_L(t)$ and $v_C(t), 0 \leq t \leq 0.5$.
    (b) Determine $v_C(t), 0.5 < t$.
    (c) Make a rough sketch of $v_C(t)$ for $t \geq 0$.
    (d) Use MATLAB and your answers to parts (a) and (b) to check your rough sketch of part (c).

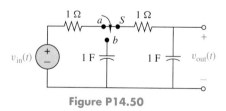

Figure P14.49

50. The switch in the circuit in figure P14.50 has been in position a for a long time, during which $v_{in}(t) = 6e^t u(t)$ V. At $t = 1$ sec, the switch moves to position b. Determine $v_{out}(t)$ for $t \geq 0$.

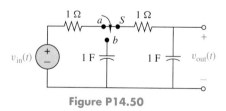

Figure P14.50

51. Refer again to the circuit of figure P14.50. Repeat problem 50, assuming that $v_{in}(t) = 10u(t)$ V for all time t.

52. The circuit in figure P14.52 contains a switch that moves from position a to position b at $t = 1$ sec. Assume that all ICs are zero at $t = 0^-$ and that $v_{in}(t) = 10u(t)$ V.

    (a) Determine $v_{out}(t), 0 \leq t \leq 1$.
    (b) Determine $v_{out}(t), 1 < t$.
    (c) Make a rough sketch of $v_{out}(t)$ for $t \geq 0$.

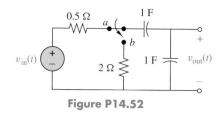

Figure P14.52

53. In each of the circuits shown in figure P14.53, the switch S has been closed for a long time and is opened at $t = 0$. It is then closed again at $t = 2$ sec. Find $v_o(t)$ for $t \geq 0$, and make a rough sketch of the waveform for each of the three circuits.

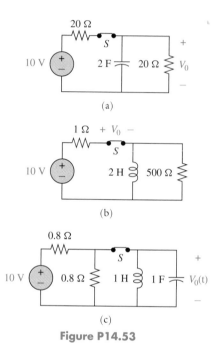

Figure P14.53

Switched Capacitor Networks

54. Consider the switched capacitor circuit of figure P14.54. The voltage labels on each capacitor indicate that capacitor's initial values at $t = 0^-$. Find the new capacitor voltages if the switch closes at $t = 0$. (Scrambled Answers: 8, 5, and 11 V.)

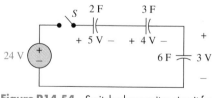

Figure P14.54 Switched capacitor circuit for problem 54.

55. All of the capacitors in the circuit of figure P14.55 have zero voltage at $t = 0^-$. The switch is closed at $t = 0$.

    (a) For figure P14.55a, determine the value of v_a at $t = 0^+$.
    (b) For figure P14.55b, determine the value of v_a at $t = 0^+$.

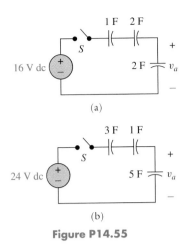

(a)

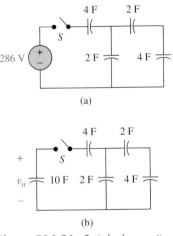

(b)

Figure P14.55

56. Consider the switched capacitor circuits of figure P14.56. Suppose the 10-F capacitor in (b) has an initial voltage $v_a(0^-) = 286$ V. Suppose all other capacitor voltages are zero at $t = 0^-$. Find all capacitor voltages for $t > 0$ in both figures (a) and (b) if the switch S is closed at $t = 0$.

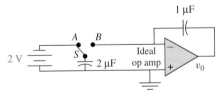

(a)

(b)

Figure P14.56 Switched capacitor circuits for problem 56.

57. In figure P14.57, $t = 0^-$, $v_o = 0$, and the switch S is at position A. At $t = 0$, S is moved to position B. Determine the output voltage at $t = 0^+$.

Figure P14.57 Ideal op amp circuit.

58. Suppose $k = 1$ in figure P14.58 and that the capacitors are initially uncharged. Suppose switch S is moved to position a at $t = 0, 2, 4, \ldots$ (even integer values) msec and to position b at $t = 1, 3, 5, \ldots$ (odd integer values) msec.
   (a) Find $v_o(t)$, and plot the waveform for $0 \le t < 20$ msec.
   (b) Repeat part (a) for $k = 0.5$.
   (c) Repeat part (a) for $k = 2$.

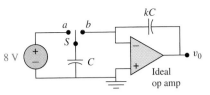

Figure P14.58 Switched capacitor op amp circuit of problem 58.

59. In the switched capacitor network of figure P14.59, the switches S are moved to position a at $t = 0$ and to position b at $t = 1$. All capacitor voltages are zero at $t = 0^-$. Determine $v_1(t)$ and $v_2(t)$ for $0 < t < 2$.

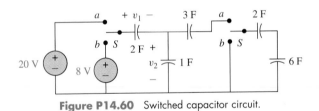

Figure P14.59

60. In the switched capacitor circuit of figure P14.60, switches S are moved to position a at $t = 0$ sec and to position b at $t = 1$ sec. All capacitor voltages are zero at $t = 0^-$. Determine $v_1(t)$ and $v_2(t)$ for $0 < t < 2$.

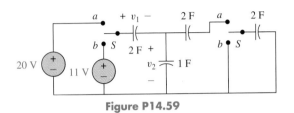

Figure P14.60 Switched capacitor circuit.

61. The opposite of analysis is synthesis. In synthesis, one has an impedance characteristic, an admittance, or a transfer function, and one seeks a circuit that has the desired characteristic. The following problems are exercises in synthesis and can be done with the techniques studied in this chapter.
   (a) Determine an RC circuit having input impedance

$$Z_{in}(s) = \frac{2s + 4.5}{(s + 0.5)(s + 4)}.$$

   (*Hint*: Do a partial fraction expansion first, which can be interpreted as a sum of impedances.)
   (b) Determine an RL circuit having input admittance

$$Y_{in}(s) = \frac{12s + 440}{(s + 120)(s + 20)}.$$

   (c) Determine an RC circuit having input admittance

$$Y_{in}(s) = \frac{0.225s^2 + 0.075s}{(s + 0.2)(s + 0.5)}.$$

   (*Hint*: First expand $Y_{in}(s)/s$ in partial fractions, and then determine $Y_{in}(s)$. Decide whether each term represents a series RC circuit or a parallel RC circuit.)
   (d) Determine an LC circuit with input admittance

$$Y_{in}(s) = \frac{0.5s}{s^2 + 1} + \frac{2s}{s^2 + 1}.$$

Miscellaneous

62. Suppose each network shown in figure P14.62 is initially at rest, i.e., $v_C(0^-) = i_L(0^-) = 0$. In each case, suppose the switch S is closed at $t = 0$ and then opened at $t = 2$ sec.

   (a) For the circuit in part (a), find $v_o(t)$ for $t \geq 0$ and make a rough sketch of the waveform.

   (b) For the circuit in part (b), find $v_o(t)$ for $t \geq 0$ and make a rough sketch of the waveform.

   (c) For the circuit in part (c), find $v_o(t)$ for $t \geq 0$ and make a rough sketch of the waveform.

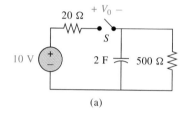

(a)

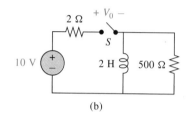

(b)

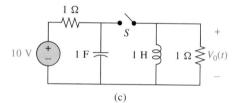

(c)

Figure P14.62 Various RLC circuits for problem 62.

63. This problem uses some simple switching techniques in an RC circuit to generate an approximate sawtooth waveform. These waveforms are common to a number of devices, such as televisions and test equipment. Consider the network of figure P14.63.

   (a) Assume the network is initially at rest. Beginning at time $t = 0$, the switch, S, is alternately closed to the left position, A, for 1 msec and then to the right position, B, for 50 μsec. Find $v_0(t)$ for $0 \leq t \leq 1.05$ msec (one cycle of operation), and sketch the waveform. (*Check*: $v_0(10^{-3}) \cong 19$ V.)

   (b) The circuit in the shaded box is a crude model of a neon lamp (59 cents apiece) and operates as follows: (1) S is at position A when v_0 is increasing but less than 80 V and is moved to position B when v_0 reaches 80 V; (2) S remains at position B when v_0 is decreasing but is greater than 5 V and is moved to position A when v_0 reaches 5 V. Find $v_0(t)$, and make a rough sketch of the waveform for one cycle of operation. (*Check*: Charging time $\cong 4.??$ msec; discharging time $\cong 3?$ μsec.)

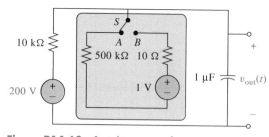

Figure P14.63 Switching circuit for generating a sawtooth waveform.

64. In the circuit of figure P14.64, all initial conditions at $t = 0^-$ are zero.

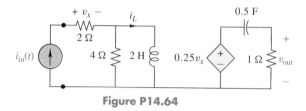

Figure P14.64

   (a) If $i_{in}(t) = 2e^{-3t}u(t)$ A, find $i_L(t)$ for $t \geq 0$.

   (b) If $i_{in}(t) = 2e^{-t}u(t)$ A, find $v_{out}(t)$ for $t \geq 0$.

   (c) If $i_{in}(t) = 2$ A dc, find $v_{out}(t)$ for $t \geq 0$.

65. In the circuit of figure P14.65, switches S_1 and S_2 have been closed for a long time. At $t = 0$, both switches are opened. At $t = 1$ sec, S_2 is closed and S_1 remains open. The value of the resistance is $R = 1/\log_e(2) = 1.4427 \ \Omega$ ($G = 1/R = 0.6931 \ \mho$).

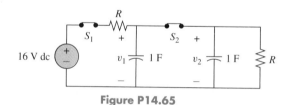

Figure P14.65

   (a) Find $v_1(0^-)$ and $v_2(0^-)$.

   (b) Find $v_1(t)$ and $v_2(t)$ for $0 \leq t < 1$ sec.

   (c) Find $v_1(1^-)$ and $v_2(1^-)$.

   (d) Find $v_1(3)$ and $v_2(3)$.

   (e) Find $v_1(1^+)$ and $v_2(1^+)$.

   (f) Do any capacitance voltages change abruptly in this circuit? If so, at what value of t?

66. For the circuit sketched in figure P14.66:

   (a) Determine the input admittance $Y_{in}(s)$ as a ratio of polynomials with leading coefficient 1 in the denominator.

   (b) Determine the transfer function $H(s) = I_{out}(s)/I_{in}(s)$ as a ratio of polynomials with leading coefficient 1 in the denominator.

   (c) If $i_{in}(t) = e^{-2t}u(t)$ A, determine $I_{out}(s)$, assuming that all ICs are zero.

   (d) Determine $i_{out}(t)$, assuming that all ICs are zero.

   (e) If $i_L(0^-) = 2$ A and $v_C(0^-) = 0$ V, determine the zero-input response.

   (f) Determine the complete response.

   (*Hint*: On parts (c) and (d), recall the frequency shift theorem.)

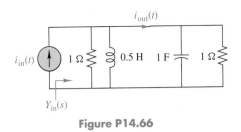

Figure P14.66

67. Consider the circuit of figure P14.67. Both capacitances are initially $(t = 0^-)$ uncharged, and $v_{in}(t) = 4u(t)$ V. Determine the voltage across the RC combination for $t \geq 0$.

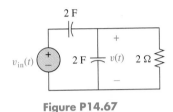

Figure P14.67

68. For the bridge-T network of figure P14.68, show that if $Z_1(s)Z_2(s) = R^2$, then $Z_{in}(s) = R$.

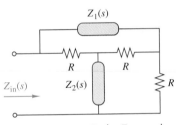

Figure P14.68 Bridge-T network.

69. For the network of figure P14.69a, the initial conditions are $v_{C1}(0^-) = 1$ V and $v_{C2}(0^-) = 0$. The switch S closes at $t = 0$.

(a) Show that for $t > 0$,

$$v_{C1}(0^+) = v_{C2}(0^+) = \frac{C_1 v_{C1}(0^-) + C_2 v_{C2}(0^-)}{C_1 + C_2} \text{ V}.$$

(b) Show that $v_{C1}(t) = v_{C2}(t) = 0.5$ V for $t > 0$.
(c) Determine the stored energy in the capacitors before and after the switch S is closed. (Answers: 0.25 and 0.5 joule.)
(d) The discrepancy between the stored energy before and after the switch closes can be explained by considering figure P14.69b, which is a more realistic model of figure P14.69a. Indeed, when the switch S is closed, figure P14.69a can be seen as a limiting case of figure P14.69b when $R \to 0$.

(1) Show that

$$i(t) = \frac{1}{R} e^{-\frac{2t}{R}} \text{ A}, \quad v_{C1}(t) = \frac{1 + e^{-\frac{2t}{R}}}{2} \text{ V},$$

$$v_{C2}(t) = \frac{1 - e^{-\frac{2t}{R}}}{2} \text{ V}.$$

(2) Show that, as long as $R > 0$, the energy dissipated in the resistance is 0.25 joule.
(3) Show that, as $R \to 0$, for $t > 0$, $i(t) \to 0.5\delta(t)$ A,

$$v_{C1}(t) \to 0.5u(t) \text{ V}, \text{ and } v_{C2}(t) \to 0.5u(t) \text{ V}.$$

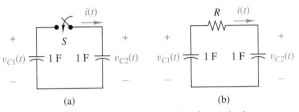

Figure P14.69 (a) Idealized model of switched capacitor network. (b) More realistic model of network when switch is closed.

Remark: As these results agree with the computations in part (a), the point here is that the idealized network of figure P14.69a yields a good approximation to the solution of the more realistic network model of figure P14.69b and that analysis of the more realistic model is extremely complicated, even for small-sized circuits.

70. Reconsider the fluorescent light starter circuit of example 14.9. In this problem, suppose that the ballast is more realistically modeled as an ideal inductor in series with a 100-Ω resistor, as shown in figure P14.70. Using SPICE or some other circuit simulation program, determine the starting voltage, $v_C(t)$, due to the initial condition on the inductor, as depicted in figure P14.70. Estimate the starting voltage from a plot of the response over one half-period. Determine the difference between the peak voltages with and without the 100-Ω resistor present. Is the lossless circuit of example 14.9 a good approximation to the starter response?

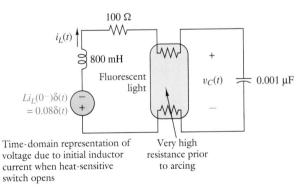

Time-domain representation of voltage due to initial inductor current when heat-sensitive switch opens

Very high resistance prior to arcing

Figure P14.70 Model of fluorescent light starter circuit that includes a ballast resistance of 100 Ω. Note that in the time domain, the effect of the initial inductor current appears as an impulse in this model.

CHAPTER 15

Laplace Transform Analysis, 3: Transfer Function Applications

APPLICATION TO ELECTRIC MOTOR ANALYSIS

ELECTRIC motors efficiently convert electrical energy into mechanical energy. An electric motor, for example, runs the fan on forced air furnaces and air conditioners. There are two broad categories of electric motors: ac and dc. As a particular application of the material of this chapter, we will develop an equivalent circuitlike model of a dc motor and analyze its operation, using the transfer function approach.

A dc motor converts direct current into mechanical energy for a variety of residential and industrial applications. For example, dc traction motors are used in electric-powered transit cars, often with two motors per axle to drive the car. A typical rating of such a motor is 140 hp, 310 V, 2,500 revolutions per minute. Another type of dc motor is a high-performance dc servomotor, found in computer disk drives and microprocessor-controlled machinery. These motors are very useful in applications where starts and stops need to be made quickly and accurately.

Of several kinds of dc motors, the type most pertinent to the analysis techniques of this chapter is a permanent magnet dc motor. Such motors are typically found in low-horsepower applications. They tend to be reliable and efficient. Further, for a permanent magnet dc motor, a plot of the torque produced on the rotating shaft of the motor versus the input current to the motor is almost a linear curve. Hence, the motor has a linear circuitlike model that most nearly describes its performance over a large range of operating characteristics. Since the output of the motor is a mechanical quantity, e.g., angular velocity of the motor shaft, the transfer function is the only viable modeling tool available to us at this stage of our development. Section 9 presents the circuit model of the motor and develops the transfer function analysis of its operation. Several problems at the end of the chapter extend the analysis. In more advanced courses, time domain analysis is developed and used. ◆

489

- Characterize the transfer function of a circuit in terms of its poles, zeros, and gain constant.
- Use knowledge of the pole locations of a transfer function to categorize generic kinds of responses (steps, ramps, sinusoids, exponentials, etc.) that are due to different kinds of terms in the partial fraction expansion of the transfer function.
- Identify, categorize, define, and illustrate various classes of circuit responses, including the zero-state (zero initial conditions) response, the zero-input response, the step response, the impulse response, the transient and steady-state responses, and the natural and forced responses.
- Define the notion of frequency response of a circuit, explore its meaning in terms of the transfer function, and introduce the concept of a Bode plot, which is an asymptotic graph of a circuit's frequency response.
- Illustrate the applicability of the transfer function concept to a circuit model of a dc motor.

1. INTRODUCTION

The experience of using the Laplace transform to calculate responses suggests that the pole-zero structure of the transfer function sets up generic kinds of circuit behaviors such as sinusoidal, exponential, constant, etc. In fact, knowledge of the location of the transfer function's poles and zeros and its gain produces knowledge of the circuit or system performance and leads to a qualitative understanding of the circuit's response. To better understand the qualitative behavior of a circuit, as exhibited by its transfer function, it is necessary first to characterize the transfer function in terms of its poles and zeros. The relative locations of the poles and zeros lead to the aforementioned generic responses induced by different terms in the partial fraction expansion of the transfer function. Among the generic responses are steps, ramps, exponentials, sinusoids, exponentially modulated sinusoids, and other kinds of responses. In addition, the location of the poles of the transfer function allows us to determine whether the response becomes large as t increases, whether it becomes periodic as t increases, whether it decays as t increases, etc. Identifying this kind of behavior allows us to define the notion of stability of a circuit or system and to categorize and compute various special types of responses, including transient, steady-state, natural, forced, step, and impulse responses. Coupling the transfer function with the presence of initial conditions in the circuit permits us to define two further types of responses fundamental to both this text and advanced courses in circuits, systems, and control: the zero-input response (due only to the initial conditions of the circuit or system) and the zero-state response (due only to the input excitation, assuming that all initial conditions are zero.)

These time domain notions are balanced by the so-called frequency response of the circuit or system. Briefly, the frequency response is the evaluation of the transfer function for $s = j\omega$. Since $H(j\omega)$ has a magnitude and phase, the frequency response breaks down into a magnitude response and a phase response. A technique for obtaining asymptotic (straight-line) approximations (called *Bode plots*) is also outlined in this chapter. Finally, the analysis of the dc motor application is presented in the last section.

As a final introductory remark, unless stated otherwise, all circuits in this chapter are linear and have constant parameter values. Such circuits are said to be *linear* and *time invariant*. When necessary in defining a particular concept, we will emphasize the need for linearity and constant parameter values. Also, for convenience in this chapter, the symbol $\angle$ will mean either the angle of a complex number, e.g., $\angle(a + jb) = \arg(a + jb) = \tan^{-1}(b/a)$, with due regard to quadrant, or $e^{j\phi}$, where, in this case, $\phi = \arg(a + jb)$. The context will determine the actual usage of the symbol.

2. POLES, ZEROS, AND THE s-PLANE

In all our circuits, impedances $Z(s)$, admittances $Y(s)$, and transfer functions $H(s)$ are rational functions of s, i.e., they are ratios of a numerator polynomial $n(s)$, divided by a denominator polynomial $d(s)$. Mathematically,

$$H(s) = \frac{n(s)}{d(s)} = K \frac{(s - z_1)(s - z_2) \cdots (s - z_m)}{(s - p_1)(s - p_2) \cdots (s - p_n)}, \tag{15.1}$$

where $s = z_i$ is a **finite zero** of $H(s)$, $s = p_i$ is a **finite pole** of $H(s)$, and K is the so-called gain constant of the transfer function. We assume that any common factors of $n(s)$ and $d(s)$ have been canceled. A finite zero satisfies $H(z_i) = 0$, i.e., the transfer function takes on the value zero at each z_i, and a finite pole satisfies $H(p_i) = \infty$, which is shorthand for $H(s) \to \infty$ as $s \to p_i$. If $z_i = z_j$, $i \neq j$, the zero is said to be a **repeated zero**. A zero repeated twice is second order, repeated three times is third order, etc. The terminology is the same for poles. Also, transfer functions sometimes have infinite poles or infinite zeros. If $m < n$, then as s approaches infinity, $H(s)$ goes to zero, suggesting the term "zero at infinity." If such is the case, $H(s)$ is said to have a zero of order $n - m$ at infinity. If, on the other hand, $n < m$, $H(s)$ is said to have a pole of order $m - n$ at infinity.

Out of all this terminology comes one striking fact: transfer functions, impedances, and admittances are characterized by their finite poles, their finite zeros, and their gain constant.

 EXAMPLE 15.1

A transfer function $H(s)$ has poles at $s = 0$, -2, and -4, with zeros at $s = -1$ and -3. At $s = 1$, $H(s) = 4/3$. Determine $H(s)$.

SOLUTION

From equation 15.1 and the given location of its poles and zeros, the transfer function must have the form

$$H(s) = K \frac{(s - z_1)(s - z_2)}{(s - p_1)(s - p_2)(s - p_3)} = K \frac{(s + 1)(s + 3)}{s(s + 2)(s + 4)}. \tag{15.2}$$

Since $H(1) = 4/3$, evaluating equation 15.2 at $s = 1$ yields

$$\frac{4}{3} = H(1) = K \frac{(1 + 1)(1 + 3)}{1(1 + 2)(1 + 4)} = K \frac{8}{15}.$$

This implies that $K = 2.5$. The transfer function is then given by equation 15.2 with $K = 2.5$.

Exercise. Suppose a transfer function has a zero at $s = 1$ and a pole at $s = -1$ and that as $s \to \infty$, $H(s) \to -3$. Determine $H(s)$.
ANSWER: $H(s) = -3(s - 1)/(s + 1)$.

Exercise. A transfer function $H(s)$ has poles at $s = -1$ and -2 and zeros at $s = -3$ and -5. It is further known that $H(0) = 15$. Find $H(\infty)$.
ANSWER: 2.

Because the essential information about transfer functions resides with the poles and zeros, a plot of these locations in the s-plane, called a **pole-zero plot**, proves informative.

EXAMPLE 15.2

Draw the pole-zero plot of the transfer function of example 15.1.

SOLUTION

The transfer function given in equation 15.2 has the pole-zero plot shown in figure 15.1.

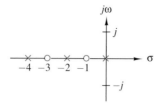

Figure 15.1 Pole-zero plot of $H(s)$ given by equation 15.2, where the poles are flagged by "x" and the zeros by "o." Since $s = \sigma + j\omega$, the real axis is labeled σ and the imaginary axis $j\omega$.

As a second illustration, consider the transfer function

$$H(s) = \frac{(s+1)(s+3)}{[(s+1)^2+1](s+2)}, \tag{15.3}$$

which has the pole-zero plot shown in figure 15.2.

Figure 15.2 Pole-zero plot of $H(s)$ given by equation 15.3. Again, the σ-axis represents the real part of the pole or zero and the $j\omega$-axis the imaginary part.

Plots such as those in figures 15.1 and 15.2 communicate much about the nature of the impedance, admittance, or transfer function of a circuit. For example, an RC input impedance, $Z_{in}(s)$, satisfies the following properties: (*i*) all of its poles and zeros are on the non-positive σ-axis of the complex plane; (*ii*) all of its poles are simple (of multiplicity 1) with real positive residues, i.e., the coefficient in a partial fraction expansion is real and positive; (*iii*) $Z_{in}(s)$ does not have a pole at $s = \infty$; (*iv*) poles and zeros alternate along the σ-axis. Proof of these assertions is beyond the scope of this text.

Exercise. Compute the input impedance of figure 15.3, and show the pole-zero plot if $R_1 = R_2 = 1\ \Omega$, $C_1 = 0.25$ F, and $C_2 = 0.5$ F. Are the poles on the non-positive σ-axis? Do the poles and zeros alternate? Is there a pole at $s = \infty$? Plot $Z_{in}(\sigma)$, where $s = \sigma + j0$. Such plots are useful in explaining network synthesis procedures.

ANSWER: $\dfrac{\dfrac{1}{C_1}}{s + \dfrac{1}{R_1C_1}} + \dfrac{\dfrac{1}{C_2}}{s + \dfrac{1}{R_2C_2}}$

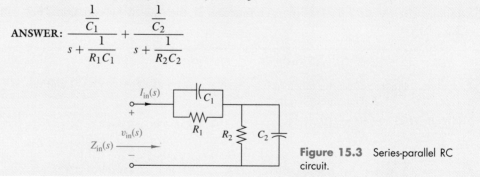

Figure 15.3 Series-parallel RC circuit.

More commonly, pole-zero locations provide important qualitative information about the response of the circuit. Pole locations determine the inherently natural behavior of the circuit, and the poles are commonly called *natural frequencies*. However, the complete set of natural frequencies of the circuit may be larger than the set of poles of the transfer function. This is because there might have been a pole-zero cancellation in constructing the transfer function. The canceled pole would amount to a natural frequency of the circuit that is not present in the poles of the transfer function. (See example 15.6.)

The terms in a partial fraction expansion of the response establish the types of behavior present in the response. Each term has only one of several possible forms. Four very common terms are K/s, K/s^j, $K/(s - p_i)$, and $K/(s - p_i)^j$, with p_i *real* in each case. Figure 15.4 sketches each of the associated responses. In figure 15.4a, the term K/s leads to a dc response and K/s^j to a polynomial response proportional to t^{j-1}. In figure 15.4b, the term $K/(s - p_i)$ leads to an exponential response that is increasing if $p_i > 0$ and decreasing if $p_i < 0$. Finally, in figure 15.4c, if $p_i < 0$, the response curve has a hump.

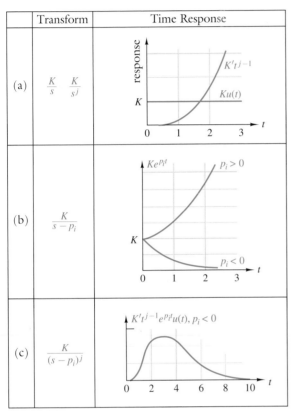

Figure 15.4 Response types common to partial fraction expansion terms. (a) The term K/s leads to a dc response and K/s^j to a polynomial response proportional to t^{j-1}. (b) The term $K/(s - p_i)$ leads to an exponential response that is increasing if $p_i > 0$ and decreasing if $p_i < 0$. (c) If $p_i < 0$, the curve has a hump.

These qualitative behaviors suggest that one of the primary applications of the transfer function is determining the "stability" of the circuit response, i.e., determining under what conditions the circuit response will remain finite for all time.

In addition to the preceding response types, there is the sinusoidal response associated with terms of the form

$$\frac{As + B}{(s + \sigma)^2 + \omega^2}.$$

Here, if $\sigma > 0$, the response is an exponentially decaying sinusoidal, as sketched in figure 15.5a, and if $\sigma = 0$, the response of figure 15.5b results. If $\sigma < 0$, an exponentially increasing sinusoidal response occurs, as shown in figure 15.5c.

Referring again to figure 15.5, the real part of a pole, i.e., σ, determines the "decay" rate of the response. Often, the word *damping* is used. If $\sigma < 0$, then the response is damped and oscillations die out. The farther σ is to the left of the imaginary axis, the larger is the damping. If $\sigma = 0$, there is no damping in the response, i.e., the response is a sustained oscillation. If $\sigma > 0$, the response is negatively damped, i.e., the response is unstable and increases without bound.

One concludes that knowledge of the locations of the poles of a transfer function identifies the behavior of the circuit with time. Within the class of all circuit behaviors is the very important behavior called *stability*.

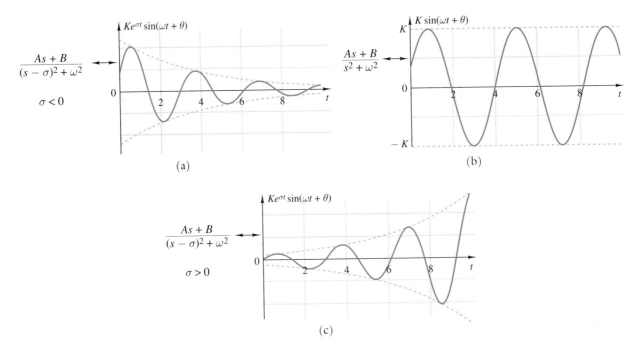

Figure 15.5   Various Laplace transforms and corresponding sinusoidal responses. (a) Exponentially decaying, $\sigma < 0$. (b) Pure sinusoidal, $\sigma = 0$. (c) Exponentially increasing, $\sigma > 0$.

STABILITY AND BOUNDEDNESS

A circuit represented by a transfer function $H(s)$, is called *stable* if every bounded input signal yields a bounded response signal. A signal, say, $f(t)$, is *bounded* if $|f(t)| < K < \infty$ for all t and for some constant K. In other words, a signal is bounded if its magnitude has a maximum finite height. Interpreting this definition in terms of the poles of the transfer function, one discovers that a circuit or system is stable if and only if all the poles of the transfer function lie in the open left half complex plane. This makes sense, because if any were in the right half-plane, the response would contain an exponentially increasing term; if any were on the imaginary axis with multiplicity two or higher, then the response would contain an unbounded term proportional to $t^{j-1}u(t)$ for $j \geq 2$; and, finally, if there were an imaginary axis pole with multiplicity one, excitation of the pole by an input of the same frequency would yield a term of multiplicity two. This would produce a response signal proportional to t, which is not bounded and which represents an unstable behavior. What this means is that, for example, a unit step current source in parallel with a 1-F capacitor would produce a voltage proportional to $tu(t)$. This voltage grows without bound and would destroy the capacitor and possibly the surrounding circuitry. Such phenomena are considered unstable.

A transfer function with first-order poles on the imaginary axis is sometimes called *metastable*. Such a classification has no practical or physically meaningful significance, since the ubiquitous presence of noise would excite the mode and cause instability of the circuit. Moreover, in power systems engineering, i.e., the study of the generation and delivery of electricity to homes and industry, transfer function poles that are in the left half-plane, but close to the imaginary axis, are highly undesirable. Such poles cause wide fluctuations in power levels. The situation is analogous to the way a car without shock absorbers would bounce. Much work has been done on how to move the poles that are close to the imaginary axis farther to the left, so as to increase the damping of the system and maintain more stable power levels. Therefore, the requirement that the transfer function have no poles on the imaginary axis is both theoretically and physically meaningful.

◆ EXAMPLE 15.3

During a laboratory experiment, a student tried to build an inverting amplifier, as shown in figure 15.6a. The student accidentally reversed the connection of the two input terminals and obtained the circuit of figure 15.6b. The student was greatly surprised that the circuit no longer behaved as expected. Explain this phenomenon in terms of the stability theory just developed.

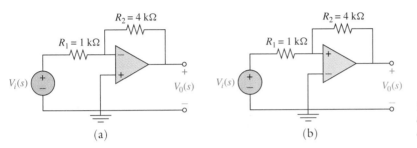

Figure 15.6 (a) Correct wiring of op amp circuit. (b) Accidental wiring of op amp circuit.

SOLUTION

Assume that the op amp is modeled as a voltage-controlled voltage source with a finite gain of 10^4 and that there is a very small stray capacitance of 1 pF across the input terminals. Figure 15.7 illustrates the equivalent circuit model for each of the circuits in figure 15.6.

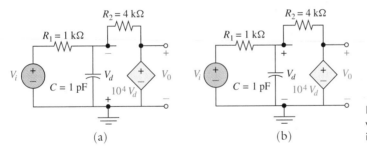

Figure 15.7 (a) Model of correctly wired op amp circuit. (b) Model of incorrectly wired op amp circuit.

Writing a node equation at V_d to compute the transfer function of figure 15.7a yields

$$-Cs V_d + \frac{1}{R_1}(-V_d - V_i) + \frac{1}{R_2}(-V_d - V_o) = 0.$$

Some algebra leads to

$$-\left(s + \frac{1}{CR_1} + \frac{1}{CR_2}\right) V_d - \frac{1}{CR_1} V_i - \frac{1}{CR_2} V_o = 0.$$

Substituting $V_o \times 10^{-4}$ for V_d reduces this expression to the transfer function

$$H(s) = \frac{V_o(s)}{V_i(s)} = -\frac{10^4}{R_1 C}\left(\frac{1}{s + \frac{1}{R_1 C} + \frac{1 + 10^4}{R_2 C}}\right) = -\frac{10^{13}}{s + 2.501 \times 10^{12}}. \qquad (15.4)$$

Observe that the transfer function is stable and that at dc the gain is $-R_2/R_1$, as expected.

Now, to compute the transfer function of the circuit of figure 15.7b, a node equation at V_d yields

$$Cs V_d + \frac{1}{R_1}(V_d - V_i) + \frac{1}{R_2}(V_d - V_o) = 0,$$

2. POLES, ZEROS, AND THE s-PLANE

which produces the transfer function

$$H(s) = \frac{V_o(s)}{V_i(s)} = \frac{10^4}{R_1 C} \left(\frac{1}{s + \frac{1}{R_1 C} + \frac{1 - 10^4}{R_2 C}} \right) = \frac{10^{13}}{s - 2.499 \times 10^{12}}. \tag{15.5}$$

The transfer function of equation 15.5 has a right half-plane pole, in contrast to that of equation 15.4. This implies that the incorrectly wired circuit is unstable, which explains the student's concern over the surprising performance of the op amp.

A brief interpretation of the zeros of a transfer function ends this section. This is best done in terms of a simple example. Suppose

$$H(s) = \frac{V_{out}}{V_{in}} = \frac{(s + 1)^2 + 1}{(s + 1)(s + 2)(s + 3)}.$$

Let the input to the circuit be $v_{in}(t) = e^{-t} \sin(t) u(t)$ V, so that

$$V_{in}(s) = \frac{1}{(s + 1)^2 + 1}.$$

Assuming that the system is initially "relaxed," i.e., that it has zero initial conditions, we obtain

$$V_{out}(s) = H(s) V_{in}(s) = \frac{1}{(s + 1)(s + 2)(s + 3)}.$$

The time response is given by

$$v_{out}(t) = [A_1 e^{-t} + A_2 e^{-2t} + A_3 e^{-3t}] u(t) \text{ V}$$

for appropriate constants A_i. Observe that the response dies out very quickly and does not have any term similar to the input signal. This follows because the input signal has transform poles, $s = -1 \pm j$, that coincide with the zeros of the transfer function. One can think of the pole locations in the transform of the input signal as identifying frequencies that are present in the input. Hence, the effect of these input signal frequencies (poles) is zeroed out by the transfer function zeros, and such frequencies are absent from the circuit response.

3. CLASSIFICATION OF RESPONSES

In addition to the various response behaviors discussed in section 2, there are other general response classifications. Three fundamentally important general response classifications germane to all of circuit and system theory are the zero-input response, the zero-state response, and the complete response.

ZERO-INPUT RESPONSE
The response of a circuit to a set of initial conditions with the input set to zero.

ZERO-STATE RESPONSE
The response of a circuit to a specified input signal, given that the initial conditions are all set to zero. Figure 15.8 illustrates this idea.

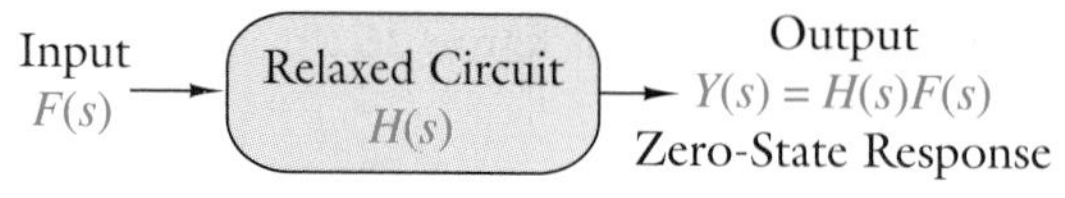

Input $F(s)$ → Relaxed Circuit $H(s)$ → Output $Y(s) = H(s)F(s)$ Zero-State Response

Figure 15.8 Relaxed circuit having transfer function H(s) and zero-state response.

COMPLETE RESPONSE

The total response of a circuit to a given set of initial conditions and a given input signal.

For linear circuits, the complete response equals the sum of the zero-input and zero-state responses. A circuit is **linear** if, for any two inputs, $f_1(t)$ and $f_2(t)$, whose zero-state responses are $y_1(t)$ and $y_2(t)$, respectively, the response to the new input $[K_1 f_1(t) + K_2 f_2(t)]$ is $[K_1 y_1(t) + K_2 y_2(t)]$, where K_1 and K_2 are arbitrary scalars. The circuits studied in this book are linear unless otherwise stated.

This decomposition of the complete response into the sum of the zero-input and zero-state responses is important for three reasons: It is defined for arbitrary input signals; the zero-state response is given by

$$\mathcal{L}^{-1}[H(s)F(s)],$$

which points up a second important application of the Laplace transform method; and it illustrates a proper application of the principle of superposition for linear dynamic networks having initial conditions. The following example illustrates this third point. Strictly speaking, the second point is not possible when the transfer function, $H(s)$, is defined by the use of a complex exponential excitation, as is done in a number of books.

EXAMPLE 15.4

Consider two linear networks: (i) a linear resistive network, as sketched in figure 15.9, and (ii) a linear dynamic network, as sketched in figure 15.10.

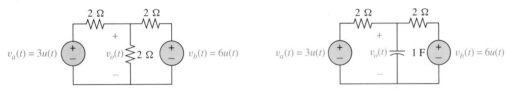

Figure 15.9 Linear resistive network. **Figure 15.10** Linear dynamic network.

For the resistive network of figure 15.9, the contribution to $v_o(t)$ due to $v_a(t)$ with v_b shorted is $v_{oa}(t) = 1u(t)$, and the contribution due to $v_b(t)$ with v_a shorted is $v_{ob}(t) = 2u(t)$. By superposition,

$$v_o(t) = v_{oa}(t) + v_{ob}(t) = 3u(t) \text{ V}.$$

Now consider the dynamic network of figure 15.10. Suppose the capacitor has an initial voltage of 2 V at $t = 0$, i.e., $v_o(0) = 2$. With $v_a(t)$ applied, assuming that v_b is shorted, the response is

$$v_{oa}(t) = (0.5e^{-t} + 1.5)u(t) \text{ V}.$$

With the input $v_b(t)$ applied under the condition that v_a is shorted, the resulting response is

$$v_{ob}(t) = (-e^{-t} + 3)u(t) \text{ V}.$$

An incorrect application of superposition implies that

$$v_o(t) = v_{oa}(t) + v_{ob}(t) = (4.5 - 0.5e^{-t})u(t) \text{ V}.$$

The reason this last answer is wrong is because the response due to the initial condition has been added in twice. A correct application of superposition would entail (i) computation of the zero-state response due to $v_a(t)$, (ii) computation of the zero-state response due to $v_b(t)$, and (iii) computation of the zero-input response due to $v_o(0)$. By superposition, the complete response is the sum of all three. In particular, the zero-state response due to $v_a(t)$ is

$$v_{oa}(t) = 1.5(1 - e^{-t})u(t) \text{ V},$$

and the zero-state response due to $v_b(t)$ is

$$v_{ob}(t) = 3(1 - e^{-t})u(t) \text{ V},$$

implying that the complete zero-state response, by superposition, is the sum of $v_{oa}(t)$ and $v_{ob}(t)$. Further, the zero-input response is $2e^{-t}u(t)$. Hence,

$$v_o(t) = 1.5(1 - e^{-t})u(t) + 3(1 - e^{-t})u(t) + 2e^{-t}u(t) = [4.5 - 2.5e^{-t}]u(t) \text{ V}.$$

It is important to note that the transfer function is defined only for circuits whose input-output behavior is linear. In terms of the zero-state response, if a circuit has a linear input-output behavior characterized by a transfer function $H(s)$, then $H(s)[K_1F_1(s) + K_2F_2(s)] = K_1H(s)F_1(s) + K_2H(s)F_2(s) = K_1Y_1(s) + K_2Y_2(s)$, where $Y_1(s) = H(s)F_i(s)$ is the zero-state response of the network to $F_i(s)$. This says that the zero-state response to $[K_1f_1(t) + K_2f_2(t)]$ is $[K_1y_1(t) + K_2y_2(t)]$. Hence, the transfer function model reflects the underlying linearity of the circuit.

The complete response has a second structural decomposition in terms of the transient and steady-state responses. The notion of a periodic signal is intrinsic to these classifications. A signal $f(t)$ is **periodic** if there exists a positive constant T such that $f(t) = f(t+T)$ for all $t \geq 0$. (The restriction to $t \geq 0$ exists because our Laplace transform analysis constrains our function class to those which are zero for $t < 0$.) If a signal is periodic, there are many positive constants for which $f(t) = f(t+T)$ for all $t \geq 0$. For example if T satisfies $f(t) = f(t+T)$ for all $t \geq 0$, then so does $2T$, $3T$, etc. We define the **fundamental period**, often called just the period, of $f(t)$ to be the smallest positive constant T for which $f(t) = f(t+T)$ for all $t \geq 0$. Sinusoids are periodic signals because, for example, $\sin(2\pi t) = \sin(2\pi t + 2\pi)$. The square wave of figure 15.11 is periodic.

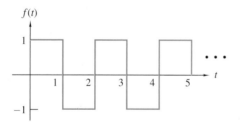

Figure 15.11   A periodic square wave with fundamental period $T = 2$.

This notion of periodicity and, by default, nonperiodicity allows us to define the transient and steady-state responses of a circuit.

TRANSIENT RESPONSE
That part of the complete response which is not periodic for $t \geq 0$, i.e., which does not satisfy the definition of a periodic function for $t \geq 0$. Note that a constant response satisfies the definition of a periodic function.

STEADY-STATE RESPONSE
That part of the complete response which satisfies the definition of periodicity for $t \geq 0$. This includes a constant response.

A circuit response may have no transient part. This is the case for the unstable circuit of figure 15.12a where the response is a sustained sinusoidal oscillation. Further, the steady-state part of the response may be zero, as per figure 15.12b, where $v_{\text{out}}(t)$ is a damped sinusoid. Transient and steady-state responses may be present in the response of any circuit. For large t, the circuit response approaches a steady-state response only if the circuit is stable.

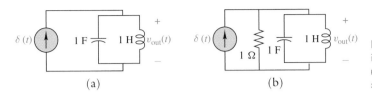

Figure 15.12 (a) Unstable circuit illustrating the possibility of no transient response. (b) Stable circuit having a zero steady-state response.

If a circuit is stable and the input is dc, sinusoidal, or periodic, then the complete response can be written as the sum of the transient response and the steady-state response. This decomposition is meaningful only when the input is constant or periodic. It is important for two reasons. First, for stable networks, the transient response dies out asymptotically; hence, for large t, only the steady-state response is crucial. Second, when the input is sinusoidal, the steady-state response is easily computed via the transfer function, $H(s)$, or by the phasor method; this is a third application of the transfer function concept. Details of the calculation are presented in section 4.

As noted, the transient response of a stable system dies out and the output approaches the steady-state response for large t. The steady state could be zero. On the other hand, if the system is unstable, the transient response has a magnitude that grows with time. The steady-state response is never reached. Hence, one cannot link the term "steady-state response" with the notion of a response for large t, unless the circuit is stable. As should be obvious, the word "transient" does not mean a response that diminishes in importance with time.

EXAMPLE 15.5

Computing the response of the circuit of figure 15.13 provides a simple illustration of the decomposition of the complete response into the sum of the zero-input and zero-state responses. Also, some rearrangement of the terms identifies the transient and steady-state responses.

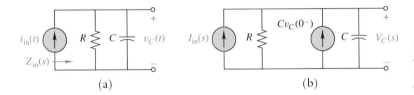

Figure 15.13 RC circuit for example 15.5. (a) Time domain circuit. (b) Frequency domain equivalent, accounting for initial condition.

SOLUTION

The input impedance of figure 15.13a is the transfer function:

$$H(s) = Z_{in}(s) = \frac{V_C(s)}{I_{in}(s)} = \frac{1}{Cs + \frac{1}{R}} = \frac{1}{C} \frac{1}{s + \frac{1}{RC}}.$$

Letting $i_{in}(t) = I_0 u(t)$ and $v_C(0^-) = 0$, from figure 15.13b, a partial fraction expansion of the zero-state response is

$$H(s)I_{in}(s) = \frac{RI_0}{s} - \frac{RI_0}{s + \frac{1}{RC}}.$$

Now, supposing that $v_c(0^-) \neq 0$ and again using figure 15.13b, we find, by superposition, that

$$v_C(t) = \underbrace{RI_0 \left(1 - e^{-\frac{t}{RC}}\right) u(t)}_{\textit{zero-state response}} + \underbrace{v_C(0^-)e^{-\frac{t}{RC}} u(t)}_{\textit{zero-input response}}.$$

3. **CLASSIFICATION OF RESPONSES**   499

As a final point, since the input was dc, a step function, the response breaks down into its transient and steady-state parts as

$$v_C(t) = \underbrace{\left(v_C(0^-) - RI_0\right) e^{-\frac{t}{RC}} u(t)}_{\text{transient response}} + \underbrace{RI_0 u(t)}_{\text{steady-state response}}.$$

Oftentimes, it is mistakenly said that the zero-input response contains only those frequencies represented by poles of the transfer function. The following example illustrates the fallacy of this statement.

EXAMPLE 15.6

Consider the RC bridge circuit of figure 15.14. Determine the zero-state and zero-input responses.

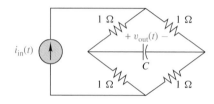

Figure 15.14 RC bridge circuit for example 15.6.

SOLUTION

The transfer function of this circuit is $H(s) = V_{\text{out}}(s)/I_{\text{in}}(s) = 0$, which has no poles. Hence, the zero-state response is always zero. On the other hand, if $v_C(0^-) \neq 0$, then

$$v_{\text{out}}(t) = v_C(0^-) \exp\left[-\frac{t}{R_{\text{eq}}C}\right] u(t) \text{ V},$$

where $R_{\text{eq}} = 1 \ \Omega$. Thus, the zero-input response is a decaying exponential whenever $v_C(0^-) \neq 0$ and $C > 0$. In this case, the zero-input response is also the transient response, with the steady-state response being zero.

On the other hand, notice that if $C < 0$, then for $v_C(0^-) \neq 0$, the zero-input response is unstable even though the transfer function is still $H(s) = 0$.

The phenomenon illustrated by example 15.6 occurs because the symmetry of the resistor values precludes excitation by the current source. Moving the current source to a different position, say, in parallel with one of the resistors, or changing the value of one of the resistors to 0.5 Ω, will result in a nonzero transfer function.

EXAMPLE 15.7

This example looks at a simple initialized series RL circuit driven by a cosine wave, as shown in figure 15.15. Let $i_L(t)$ denote the circuit response. Suppose $v_{\text{in}}(t) = 4\cos(t)u(t)$ V and $i_L(0^-) = 1$ A. The objective is to isolate the transient and steady-state responses from the zero-input and zero-state responses.

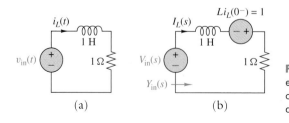

(a) (b)

Figure 15.15 RL circuit for example 15.7. (a) Time domain series RL circuit. (b) Frequency domain equivalent, accounting for initial condition.

SOLUTION

Using figure 15.15b and the principle of superposition leads to the response

$$I_L(s) = Y_{in}(s)V_{in}(s) + Y_{in}(s)[Li_L(0^-)]$$

$$= \frac{4s}{(s+1)(s^2+1)} + \frac{1}{s+1}.$$

The term $Y_{in}(s)V_{in}(s)$ is the transform of the zero-state response and the term $Y_{in}(s)Li_L(0^-)$ is the transform of the zero-input response. Specifically, the term $1/(s+1)$ is the Laplace transform of the zero-input response, while the term $4s/[(s+1)(s^2+1)]$ is the Laplace transform of the zero-state response. Thus, the zero-input response is $e^{-t}u(t)$. A partial fraction expansion of the zero-state response yields

$$\frac{4s}{(s+1)(s^2+1)} = \frac{-2}{s+1} + \frac{2s+2}{s^2+1}.$$

It follows that the zero-state response is $-2e^{-t}u(t) + 2[\cos(t) + \sin(t)]u(t)$. Notice that both the zero-input and the zero-state response contains a transient part, the part proportional to e^{-t}. A little rearranging shows that the complete response is

$$i_L(t) = -e^{-t}u(t) + 2[\cos(t) + \sin(t)]u(t) \text{ A},$$

which implies that the transient response of the circuit is $-e^{-t}u(t)$ and that the steady-state response of the circuit is $2[\cos(t) + \sin(t)]u(t)$ A.

Exercise. An RLC network has transfer function $H(s) = V_{out}/I_{in} = 1/(s+1)$. If an input $i_{in}(t) = \cos(t)$ A is applied, then for very large t, $v_{out}(t)$ approaches a cosine wave of what form?

ANSWER: $0.707\cos(t - 45^o)$.

Frequently, books on elementary circuits contain two other notions of response: the *natural response* and *forced response*.

NATURAL RESPONSE

The portion of the complete response that has the same exponents as the zero-input response.

FORCED RESPONSE

The portion of the complete response that has the same exponent as the input excitation, provided that the input excitation has exponents different from those of the zero-input response.

It would seem natural to try to decompose the complete response into the sum of the natural and forced responses. Unfortunately, such a decomposition applies only when the input excitation is (*i*) dc, (*ii*) real exponential, (*iii*) sinusoidal, or (*iv*) exponentially modulated or damped sinusoidal. Further, the exponent of the input excitation, e.g., a in $f(t) = e^{at}u(t)$, must be different from the exponents appearing in the zero-input response. The natural and forced response are properly defined only under these conditions.

The decomposition of a complete response into a natural and a forced response is important for two reasons. First, it agrees with the classical method of solving ordinary differential equations having constant coefficients, where the natural response corresponds to the complementary function and the forced response corresponds to the particular integral. Readers fresh from a course in differential equations feel quite at home with these concepts. The second reason is that the forced response is easily calculated for any of the special inputs—dc, real exponential, sinusoidal, or damped sinusoidal. For example, if the transfer function is $H(s)$, and the input is Ve^{at}, then the forced response is simply

$$H(a)Ve^{at}. \tag{15.6}$$

To justify equation 15.6, note that the Laplace transform of the input is $V/(s-a)$. Since the complete response is the sum of the zero-input and zero-state responses, we have

$$Complete\ response = [zero\text{-}input\ response] + \mathcal{L}^{-1}\left[H(s)\frac{V}{s-a}\right].$$

The zero-input response terms all have exponents different from a, the exponent of the input. The second term, $\mathcal{L}^{-1}[H(s)\ V/(s-a)]$, has only one term with exponent equal to a. Executing a partial fraction expansion of this term yields

$$H(s)\frac{V}{s-a} = \frac{K}{s-a} + [terms\ corresponding\ to\ poles\ of\ \mathrm{H(s)}].$$

Using the residue formula to calculate K produces $K = H(a)V$. Thus,

$$H(s)\frac{V}{s-a} = \frac{H(a)V}{s-a} + [terms\ corresponding\ to\ poles\ of\ \mathrm{H(s)}],$$

and $\mathcal{L}^{-1}[H(s)V/(s-a)]$ has a term $H(a)Ve^{at}u(t)$, which we identify as the forced response.

By using exactly the same arguments, it is possible to show that, if the input is a complex exponential function Ve^{s_pt}, where both V and s_p are complex numbers, then the forced response is simply

$$H(s_p)Ve^{s_pt}.$$

A complex exponential such as Ve^{st} is a mathematical entity. It cannot be generated in the laboratory. However, the real part, $\mathrm{Re}[Ve^{st}]$ (or the associated imaginary part), is simply an exponentially modulated sinusoidal signal, as shown in figure 15.5, and is readily generated in a laboratory. A derivation similar to the preceding leads to the conclusion that, if the input is $\mathrm{Re}[Ve^{st}]$, then the forced response is

$$\mathrm{Re}[H(s)Ve^{st}].$$

This relationship of the input to the forced response prompts some textbooks to define the transfer function $H(s)$ as the ratio of the forced response to the input, under the condition that the input is a complex exponential Ve^{st}. Doing so, however, is not natural and makes one wonder at the applicability of such a definition to the broad class of inputs for which the concept is most naturally defined, as per chapter 14 of the text.

Control Systems Engineer

Control systems engineers are architects of strategy for the automatic control of appliances, car and jet engines, anti-lock brakes, motor drives, manufacturing processes, and a host of other services for the modern age. These engineers produce devices that assist in manufacturing a variety of products, assembling parts, cutting out patterns, moving objects, and controlling the environment. In addition, control systems engineers must implement their strategies by developing the logic and pertinent computer software to coordinate the sensors, actuators, microprocessors, etc.

that direct equipment behavior.

Many control systems engineers specialize in artificial intelligence and expert systems, which aids in the design of robots that carry out manufacturing tasks such as assembling automobiles. By using expert systems technology, these engineers attempt to incorporate in their strategies the experience of experts in the field. The technology of artificial intelligence allows control systems engineers to implement learning capability into their strategies so controllers can automatically adapt to new situations.

4. COMPUTATION OF THE SINUSOIDAL STEADY-STATE RESPONSE FOR STABLE NETWORKS AND SYSTEMS

Suppose that a *stable* transfer function $H(s)$ models a linear circuit containing a total of n capacitors and inductors. In addition, suppose that there are no common factors in the numerator and denominator of $H(s)$ and that the *degree of the denominator of* $H(s)$ *is* n. The goal of this section is to develop the following formula: If $H(s)$ satisfies the aforementioned assumptions and the input to the circuit has the form $A\cos(\omega t + \theta)$, then the steady-state circuit output response has the form $B\cos(\omega t + \phi)$, where the **magnitude** of the response is

$$B = A \cdot |H(j\omega)| \tag{15.7a}$$

and the **phase angle** is

$$\phi = \theta + \angle H(j\omega). \tag{15.7b}$$

Here, we assume that ω is some fixed, but arbitrary, value.

From an input-output viewpoint, these formulas imply that the frequency response, in actuality, is the steady-state response of a circuit to sinusoids. To construct this formula, suppose again that a linear circuit has a stable transfer function model $H(s)$. Suppose also that the circuit input is a sinusoid $f(t)$ whose Laplace transform is $F(s)$ with zero-state response $Y(s)$, as illustrated in figure 15.16.

Since $H(s)$ is stable, all poles lie in the open left half of the complex plane. Consequently, $H(s)$ will have a partial fraction expansion containing only two types of terms: those having real poles with negative real parts and those having complex poles with negative real parts. Assuming that $H(s)$ has real, distinct poles labeled $p_1, \ldots, p_m$, and complex poles labeled $\alpha_i \pm \beta_i$, it follows that

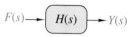

$F(s) \rightarrow \boxed{H(s)} \rightarrow Y(s)$

Transfer Function

Figure 15.16
Frequency domain representation of hypothetical circuit.

$$H(s) = \underbrace{\frac{A_1}{s - p_1} + \frac{A_2}{s - p_2} + \cdots + \frac{A_m}{s - p_m}}_{\substack{\text{real poles with} \\ \text{negative real parts}}} + \underbrace{\frac{C_1 s + D_1}{(s - \alpha_1)^2 + (\beta_1)^2}}_{\substack{\text{complex poles with} \\ \text{negative real parts}}} + \cdots \tag{15.8}$$

It is easy to account for higher order poles.

Suppose now that the circuit is excited by a sinusoidal input of the form

$$f(t) = A\cos(\omega t + \theta) = A\frac{e^{j(\omega t + \theta)} + e^{-j(\omega t + \theta)}}{2}$$

having Laplace transform

$$F(s) = \frac{0.5 A e^{j\theta}}{s - j\omega} + \frac{0.5 A e^{-j\theta}}{s + j\omega}.$$

Then partial fraction expansion of the Laplace transform of the zero-state response, $Y(s) = H(s)F(s)$, has the form

$$Y(s) = \underbrace{\frac{\hat{A}_1}{s - p_1} + \frac{\hat{A}_2}{s - p_2} + \cdots}_{\substack{\text{real poles with} \\ \text{negative real parts}}} + \underbrace{\frac{\hat{C}_1 s + \hat{D}_1}{(s - \alpha_1)^2 + (\beta_1)^2} + \cdots}_{\substack{\text{complex poles with} \\ \text{negative real parts}}} + \underbrace{\frac{R_1}{s - j\omega} + \frac{R_2}{s + j\omega}}_{\substack{\text{steady-state} \\ \text{contribution} \\ \equiv Y_{ss}(s)}}.$$

In the steady state, i.e., for large t, the only residues of interest are R_1 and R_2, since that part of the time response due to the other terms decays to zero with increasing t. By the usual methods of complex variables, we obtain

$$R_1 = \left[H(s)(s - j\omega) \left(\frac{0.5 A e^{j\theta}}{s - j\omega} + \frac{0.5 A e^{-j\theta}}{s + j\omega} \right) \right]_{s = j\omega}$$

$$= 0.5 A H(j\omega) e^{j\theta} = 0.5 A |H(j\omega)| e^{j\angle H(j\omega)} e^{j\theta}$$

$$= 0.5 A |H(j\omega)| e^{j(\angle H(j\omega) + \theta)}$$

and

$$R_2 = \left[H(s)(s+j\omega) \left(\frac{0.5Ae^{j\theta}}{s-j\omega} + \frac{0.5Ae^{-j\theta}}{s+j\omega} \right) \right]_{s=-j\omega}$$

$$= 0.5AH(-j\omega)e^{-j\theta} = 0.5A|H(-j\omega)|e^{j\angle H(-j\omega)}e^{-j\theta}$$

$$= 0.5A|H(-j\omega)|e^{j(\angle H(-j\omega)-\theta)}.$$

But $|H(-j\omega)| = |H(j\omega)|$ and $\angle H(-j\omega) = -\angle H(j\omega)$; hence,

$$R_2 = 0.5A|H(j\omega)|e^{-j(\angle H(j\omega)+\theta)}.$$

Consequently, the Laplace transform of the steady-state response when all initial conditions of the circuit are zero is

$$Y_{ss}(s) = 0.5A\,|H(j\omega)| \left[\frac{e^{j(\angle H(j\omega)+\theta)}}{s-j\omega} + \frac{e^{-j(\angle H(j\omega)+\theta)}}{s+j\omega} \right].$$

In fact, this is the Laplace transform of the actual steady-state response, provided that the zero-input (nonzero initial conditions) response makes no additional contribution. The zero-input response makes no contribution to the steady-state response when one or more of the following reasonable conditions on the circuit are met:

1. The network has only practical passive elements, meaning that there are always stray resistances present.
2. The circuit has active elements in addition to passive elements, but remains stable in the sense that every capacitor voltage and every inductor current remains bounded for any bounded circuit excitation.
3. The circuit contains a total of n capacitors and inductors, and the stable transfer function, $H(s)$, has n poles. (See problem 66.)

Under these conditions,

$$y_{ss}(t) = A\,|H(j\omega)| \cos\big(\omega t + (\angle H(j\omega) + \theta)\big) = B\cos(\omega t + \phi).$$

Equating coefficients verifies that if the input to the circuit has the form $A\cos(\omega t + \theta)$, then the steady-state circuit output response has the form $B\cos(\omega t + \phi)$, where the magnitude $B = A \cdot |H(j\omega)|$ and the phase shift $\phi = \theta + \angle H(j\omega)$.

The next question concerns the numerical calculation of B and ϕ. One method is simply to evaluate $H(s)$ at $s = j\omega$. With a calculator that easily accommodates complex numbers, this is quite straightforward. An alternative method is to use the graphical technique of the next section.

At this point, it is instructive to illustrate equation 15.7 and at the same time compare it with the phasor method studied in a first course on circuit theory.

◆ EXAMPLE 15.8

In the circuit of figure 15.17, $\mu = 0.5$ and $v_i(t) = \cos(2t)$ V. Find $v_1(t)$ for large t.

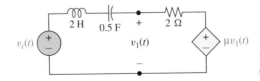

Figure 15.17 RLC circuit for steady-state computation in example 15.8.

SOLUTION

By phasor method. From the principles of phasor analysis detailed in volume I, the phasor domain circuit of figure 15.17 at $\omega = 2$ rad/sec is given by the circuit of figure 15.18.

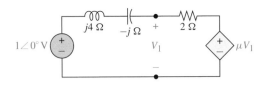

Figure 15.18 Phasor circuit equivalent of figure 15.17 at $\omega = 2$.

The single node equation for $\mathbf{V}_1$ is

$$\frac{\mathbf{V}_1 - 1}{j4 - j} + \frac{\mathbf{V}_1 - 0.5\mathbf{V}_1}{2} = 0. \tag{15.9}$$

The phasor solution to equation 15.9 is

$$\mathbf{V}_1 = \frac{1}{1 + j0.75} = 0.8\angle - 36.9°.$$

Therefore, for large t,

$$v_1(t) = 0.8\cos(2t - 36.9°) \text{ V}.$$

By Laplace transform method. The first step here is to construct the s-domain equivalent circuit, which is given in figure 15.19.

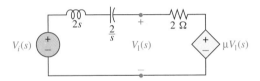

Figure 15.19 Frequency or s-domain equivalent of the circuit of figure 15.17.

The single node equation for V_1 in the s-domain is

$$\frac{V_1 - V_i}{2s + \dfrac{2}{s}} + \frac{V_1 - 0.5V_1}{2} = 0.$$

Solving for the transfer function yields

$$H(s) = \frac{V_1(s)}{V_i(s)} = \frac{2s}{(s + 1)^2}. \tag{15.10}$$

From equation 15.10, at $s = j\omega = j2$,

$$H(j\omega) = H(j2) = j4/(j2 + 1)^2 = 0.8 \angle - 36.9°.$$

According to equation 15.7, it follows that

$$v_1(t) = 0.8\cos(2t - 36.9°) \text{ V}.$$

In this example, the two methods give the same answers. Since complex numbers are easier to manipulate than rational functions, what is the motivation for such an analysis using the transfer function $H(s)$? Why not stay with the phasor method? The next example answers these questions.

◆ **EXAMPLE 15.9**

In example 15.8 with the circuit of figure 15.17, let the value of μ be increased to 1.5. Find $v_1(t)$ for large t.

SOLUTION

By phasor method. Since only the response for large t is desired, the problem appears to be one involving sinusoidal steady-state analysis. The phasor domain circuit of figure 15.18 yields the single node equation

$$\frac{\mathbf{V}_1 - 1}{j4 - j} + \frac{\mathbf{V}_1 - 1.5\mathbf{V}_1}{2} = 0.$$

Solving again for $\mathbf{V}_1$ yields

$$\mathbf{V}_1 = \frac{1}{1 - j0.75} = 0.8 \angle 36.9°.$$

Therefore, for large t,

$$v_1(t) = 0.8 \cos(2t + 36.9°) \text{ V}.$$

A beginner who has just learned sinusoidal steady-state analysis by the phasor method might accept this answer. Unfortunately, the answer is not the voltage $v_1(t)$ for large t! The reason is clear from the Laplace transform analysis, which follows.

By Laplace transform method. From the s-domain equivalent circuit of figure 15.19,

$$\frac{V_1 - V_i}{2s + \dfrac{2}{s}} + \frac{V_1 - 1.5V_1}{2} = 0.$$

Solving for the transfer function yields

$$H(s) = \frac{V_1(s)}{V_i(s)} = \frac{-2s}{(s-1)^2}.$$

Since there are poles of $H(s)$ in the right half-plane, the circuit is unstable. As t becomes very large, the magnitude of $v_1(t)$ approaches infinity, instead of $0.8 \cos(2t + 36.9°)$ V, as calculated by the phasor method.

This analysis demonstrates that the stable behavior of a circuit cannot be determined by the phasor method. It is desirable to know when to use a particular method in order to avoid unnecessary complicated calculations. The following guidelines help:

1. When the stability of the circuit has been assured by some means, and ω has a specific numerical value, the phasor method is the proper method to use for determining the response for large t, which is also the steady-state response in this case. Circuits whose stability is guaranteed include: circuits with only passive elements, such as resistors, capacitors, and inductors, and amplifier circuits of well-established configurations.
2. The circuit is known to be stable, but ω is variable. In this case, the $H(s)$ method is superior to the phasor method. To say the least, we need only write sL instead of $j\omega L$. There are other advantages to be gained from knowing the pole-zero plot of $H(s)$, which are not possible with the phasor method. The examples of frequency response calculations given in the next section clearly demonstrate this point.
3. If the stability of the circuit is not yet determined, then $H(s)$ should be calculated and its pole locations checked for stability. Then, step 1 or step 2 should be referred to, as appropriate.

Exercise. Suppose a second-order linear circuit having the transfer function

$$H(s) = \frac{V_{\text{out}}(s)}{V_{\text{in}}(s)} = \frac{s^2 - 0.5s + 5}{s^2 + 0.5s + 5.7321}$$

is driven by a sinusoidal input $v_{\text{in}}(t) = \sqrt{2} \cos(2t + 45°)u(t)$ V. Show that the steady-state response is given by $v_{\text{out}_{ss}}(t) = \cos(2t - 30°)$ V.

Exercise. Consider the LC circuit of figure 15.20, in which $i_L(0^-) = 0$, $v_C(0^-) = 0$, and $v_{\text{in}}(t) = 100u(t)$. Show that the largest voltage to appear across the capacitor for $t \geq 0$ is 200 V. (*Hint:* Show that $v_C(t) = 100u(t) - 100\cos(t)u(t)$.)

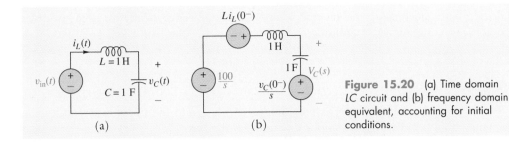

Figure 15.20 (a) Time domain LC circuit and (b) frequency domain equivalent, accounting for initial conditions.

(a)

(b)

5. FREQUENCY RESPONSE

One extremely important notion in circuit and system theory is that of the frequency response. The **frequency response** of a linear stable circuit having constant parameters measures the input-output behavior of the circuit to unit magnitude sinusoids, $\cos(\omega t)$, as ω varies from 0 to ∞. The frequency response of such stable circuits is the evaluation of the transfer function, $H(s)$, at $s = j\omega$. In terms of phasor analysis, studied in a first course, the frequency response corresponds to the ratio of the output phasor of the circuit to the input phasor of the circuit.

The motivation here is that for stable circuits, if an input has the form $\cos(\omega_0 t)$, then the steady-state response (i.e., the response for large t, after all transients have died out), has the form $B\cos(\omega_0 t + \phi)$, where $B = |H(j\omega_0)|$ and the phase shift $\phi = \angle H(j\omega_0)$. (A general formula is derived in the next section.) Here, $|H(j\omega_0)|$ is the magnitude of the complex number $H(j\omega_0)$, and $H(j\omega_0)$ is the angle of the complex number $H(j\omega_0)$. Thus, knowledge of $H(j\omega)$, for $0 \leq \omega < \infty$, tells how a linear circuit adjusts the magnitude and phase of an input sinusoidal signal to produce a steady-state output sinusoid of the same frequency, but with a different magnitude and phase.

A practical example of the importance of this information is the specification of a stereo amplifier. Here, it is important to know the frequency response and, in particular, to know that the gain, i.e., the magnitude of $H(j\omega)$, is more or less constant from 0 to 20 kHz. Why? Because musical signals are composed of sinusoids of different frequencies. Accurate amplification of the music requires that each component sinusoid be amplified with equal gain.

FREQUENCY RESPONSE

The frequency response of a stable circuit or system represented by a transfer function $H(s)$ is the complex-valued function $H(j\omega)$, for $0 \leq \omega < \infty$.

A complex-valued function $H(j\omega)$, is a function such that for each value of ω, $H(j\omega)$ is a complex number. A complex number, $a_1 + jb_1$, has a polar form, $\rho_1 e^{j\phi_1}$, in which ρ_1 is the magnitude and ϕ_1 the phase angle of the number. Moreover, if $a_2 + jb_2$ is another complex number, then

$$(a_1 + jb_1)(a_2 + jb_2) = \rho_1 \rho_2 \exp[j(\phi_1 + \phi_2)],$$

and

$$\frac{a_1 + jb_1}{a_2 + jb_2} = \frac{\rho_1}{\rho_2} \exp[\phi_1 - \phi_2].$$

In polar form, the frequency response as a function of ω is

$$H(j\omega) = \rho(\omega) \exp[j\phi(\omega)],$$

where $\rho(\omega) = |H(j\omega)|$ denotes the *magnitude response* and

$$\phi(\omega) = \angle H(j\omega) = \tan^{-1}\left(\frac{\text{Im}[H(j\omega)]}{\text{Re}[H(j\omega)]}\right)$$

is the *angle* or *phase* of the frequency response. As in other books, *magnitude response* means the magnitude of the frequency response. Typically, computation of the frequency response requires a calculator or computer. However, there is a pedagogically useful graphical technique. Mastering this technique helps concretize the meaning of the magnitude and the phase and further enables one to apply the knowledge of the locations of poles and zeros to compute the magnitude and phase.

◆ EXAMPLE 15.10

To cultivate a good grasp of the magnitude, $|H(j\omega)|$, and the phase, $\angle H(j\omega)$, of the frequency response, suppose a transfer function has the form

$$H(s) = \frac{(s - z_1)(s - z_2)}{(s + 1)(s + 1 + j)(s + 1 - j)}, \quad (15.11)$$

where $z_1 = 2j$ and $z_2 = -2j$. Figure 15.21 illustrates the pole-zero plot of $H(s)$.

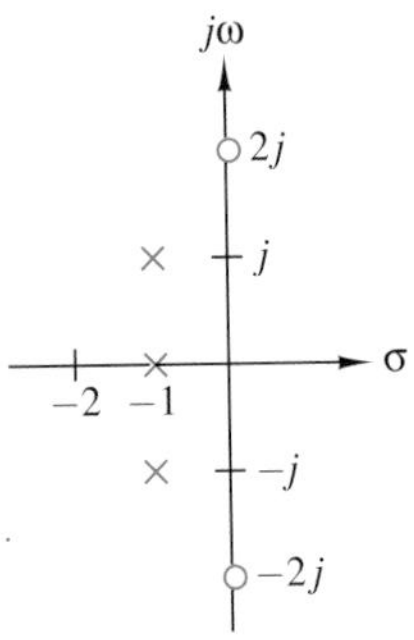

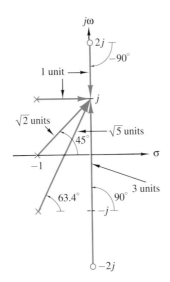

Figure 15.21
Pole-zero plot of $H(s)$ given by equation 15.11.

SOLUTION

The plan of this example is to compute the magnitude of $H(j\omega)$ graphically for $\omega = 1$, i.e., to compute $|H(j1)|$. This computation entails the following steps:

Step 1. Observe that $j1 - z_1$ defines a complex number. Think of $j1 - z_1$ as a vector in the complex plane. Similarly, $j1 - z_2$ and $j1 - p_i$, where $p_1 = -1$ and $p_{2,3} = -1 \pm j$, define vectors. Each vector has a length that can be determined either graphically, by physically measuring the distance with a ruler, or by the Pythagorean theorem. Figure 15.22 illustrates the idea.

Figure 15.22 Measuring distances graphically from zeros and poles to the point $j1$.

Step 2. As such, the magnitude of $H(j1)$ has the form

$$|H(j1)| = \frac{|j1 - j2| \cdot |j1 + j2|}{|j1 + 1| \cdot |j1 + 1 + j| \cdot |j1 + 1 - j|} = \frac{1 \times 3}{\sqrt{2} \times \sqrt{5} \times 1} = 0.95.$$

Step 3. Suppose we wish to compute the phase or angle of $H(j1)$ graphically. In figure 15.22 observe that each complex number viewed as a vector $j\omega - z_i$ or $j\omega - p_i$ can be represented in the form $\rho e^{j\psi}$, where ψ is the angle the vector makes with a horizontal line passing through its base. For example, $(j1 - j2) = -j1 = 1e^{-j(\pi/2)}$. From basic complex number theory, the angle of the product of two complex numbers is the sum of the angles, and the angle of the ratio of two complex numbers is the angle of the numerator minus the angle of the denominator. The angles associated with the problem at hand are also drawn in figure 15.22. Therefore,

$$\angle H(j1) = \angle(j1 - j2) + \angle(j1 + j2) - \angle(j1 + 1) - \angle(j1 + 1 + j) - \angle(j1 + 1 - j)$$

$$= -90° + 90° - 45° - 0° - 63.4° = -108.4°.$$

To extend the ideas of the previous example, let

$$H(s) = K \frac{(s - z_1)(s - z_2) \cdots (s - z_m)}{(s - p_1)(s - p_2) \cdots (s - p_n)}. \qquad (15.12)$$

Because the magnitude of a product of complex numbers is the product of the magnitudes of the numbers, and because the magnitude of the ratio of two complex numbers is the ratio of the magnitudes of the numbers, the general form of the magnitude response of equation 15.12 is

$$|H(j\omega)| = |K| \frac{|j\omega - z_1||j\omega - z_2| \cdots |j\omega - z_m|}{|j\omega - p_1||j\omega - p_2| \cdots |j\omega - p_n|}. \qquad (15.13a)$$

Similarly, since the angle of the product of complex numbers is the sum of the angles of the numbers, and since the angle of the ratio of two complex numbers is the difference in the angles of the numbers, the general form of the phase response of equation 15.12 is

$$\angle H(j\omega) = [\angle(j\omega - z_1) + \angle(j\omega - z_2) + \cdots + \angle(j\omega - z_m) + \angle K]$$
$$- [\angle(j\omega - p_1) + \angle(j\omega - p_2) + \cdots + \angle(j\omega - p_n)]. \qquad (15.13b)$$

Exercise. Draw an estimate of the general shape of the magnitude and phase response of the Butterworth normalized low-pass transfer function

$$H(s) = \frac{1}{s^2 + \sqrt{2}s + 1}.$$

Compute the exact magnitude and phase at $\omega = 1$. What happens to the magnitude and phase if $H(s)$ is changed to $H(s/10)$ and $H(s/100)$.

SCRAMBLED ANSWERS: -45; 0.707; the general shape is the same with $H(j10) = H(j100) = 0.707$.

Qualitatively speaking, ω's near poles force $H(j\omega)$ to have a large magnitude, and ω's near zeros force $H(j\omega)$ to have a small magnitude. This can be used to advantage in estimating the magnitude response and phase response of a transfer function.

EXAMPLE 15.11

Consider the pole-zero plot of figure 15.23. If the input is a 1-V sinusoid, what approximate value of ω leads to a maximum-amplitude steady-state response?

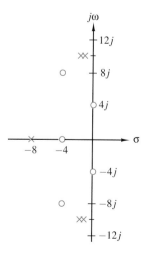

Figure 15.23 Pole-zero plot for example 15.11 in which $H(s)$ has a multiplier constant $K = 1$.

SOLUTION

Because ω's near poles lead to large magnitudes in the frequency response and ω's near zeros lead to small magnitudes in the frequency-response, $\omega = \pm 10$ approximates the frequency having a maximum steady-state response.

Exercise. For the pole-zero plot (figure 15.23) of example 15.11, suppose the poles close to the imaginary axis at $\pm j10$ have real parts of -0.1 and -0.2, respectively. Determine the approximate value of the transfer function at $s = j10$ if $H(\infty) = 1$.

ANSWER: 728.

6. IMPULSE AND STEP RESPONSES

Suppose a circuit or system has a transfer function representation given by figure 15.16, in which $Y(s) = H(s)F(s)$. Assuming that all ICs are zero, if $f(t) = \delta(t)$, then the resulting $y(t)$ is the **impulse response**. Some simple calculations indicate that the transform of the impulse response is

$$Y(s) = H(s)\mathcal{L}[\delta(t)] = H(s),$$

which is the transfer function. Hence,

$$h(t) = \mathcal{L}^{-1}[H(s)] \qquad (15.14)$$

is the impulse response of the circuit. This is another application of the transfer function concept.

Exercise. The transfer function of a certain linear network is $H(s) = (s+3)/[(s+1)(s+2)]$. Find the impulse response of the network.
ANSWER: $[2e^{-t} - e^{-2t}]u(t)$.

Exercise. Suppose $v(t) = 2\delta(t-1) - 3\delta(t-3)$ is the input to a relaxed (zero initial conditions) circuit having an impulse response $h(t) = 2u(t) - 2u(t-5)$. Determine the output $y(t)$ at time t = 7.
ANSWER: -6.

Why is the impulse response important? As we will see, it is because every linear circuit having constant parameter values for its elements can be represented in the time domain by its impulse response. This is shown in chapter 16, where we define a mathematical operation called *convolution* and show that the convolution of the input function with the impulse response function yields the circuit response, assuming that all initial conditions are zero. In addition to this significant theoretical result, the impulse response is important for identification of linear circuits or systems having unknown constant parameters. Sometimes a transfer function is unavailable or a circuit diagram is lost. In such a predicament, measuring the impulse response on an oscilloscope is quite practical.

Analogous to the impulse response is the step response. *What is the step response of a circuit?* The **step response** is merely the response of the circuit to a step function, assuming that all initial conditions are zero. Observe that, if the input $f(t)$ to the circuit is $u(t)$, then $F(s) = 1/s$ and $Y(s) = (1/s)H(s)$. By the integration property of the Laplace transform, it follows that the step response is the integral of the impulse response. Conversely, the derivative of the step response is the impulse response.

Exercises. (i) If the Laplace transform of the step response of a circuit is given by $Y(s) = 1/[s(s+1)]$, then what is the impulse response? (ii) If the step response of a circuit is $y(t) = [1 - 0.5e^{-2t} - 0.5e^{-4t}\cos(2t)]u(t)$, then what is the impulse response? (iii) If the transfer function of a circuit is $H(s) = 1/s$, then what are the impulse and step responses?
SCRAMBLED ANSWERS: $r(t)$, $e^{-t}u(t)$, $u(t)$, $[e^{-2t} + 2.24e^{-4t}\cos(2t + 26.57°)]u(t)$.

EXAMPLE 15.12

Figure 15.24a shows the impulse response of a hypothetical circuit. If a new input is $f(t) = \delta(t) + \delta(t-1)$, determine the new response, $y(t)$.

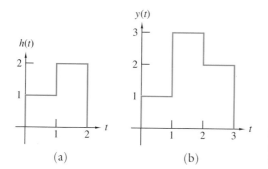

Figure 15.24 (a) Impulse response of hypothetical circuit. (b) Response to $\delta(t) + \delta(t-1)$.

SOLUTION

Since $\mathcal{L}[\delta(t) + \delta(t-1)] = 1 + e^{-s}$, the response, $y(t)$, is simply the sum of $f(t)$ and $f(t-1)u(t-1)$. Doing the addition graphically yields the waveform of figure 15.24b.

EXAMPLE 15.13

The response of a relaxed circuit to a scaled ramp, $f(t) = 8tu(t)$, is given by $y(t) = (-6 + 4t + 8e^{-t} - 2e^{-2t})u(t)$. Determine the impulse response, $h(t)$.

SOLUTION

The relationship between $f(t)$ and $\delta(t)$ identifies the strategy of the solution. If the step function is the integral of the delta function and the ramp the integral of the step, then the delta function equals the second derivative of the ramp. Hence, $\delta(t) = 0.125 f''(t)$. By the linearity of the circuit, the impulse response $h(t) = 0.125 y''(t)$, and some straightforward calculations produce

$$y'(t) = [4 - 8e^{-t} + 4e^{-2t}]u(t) + [-6 + 4t + 8e^{-t} - 2e^{-2t}]\delta(t).$$

But the right-hand-side term is zero. (Why?) Hence, $y''(t) = [8e^{-t} - 8e^{-2t}]u(t)$, and

$$h(t) = [e^{-t} - e^{-2t}]u(t).$$

To see the utility of this approach, try the alternative method of computing $F(s)$, $Y(s)$, and $H(s) = Y(s)/F(s)$. The algebra is straightforward, but tedious and prone to error.

As a final example, we compute a circuit's step response and verify that its derivative is the impulse response.

EXAMPLE 15.14

Compute the step response of the RLC circuit of figure 15.25.

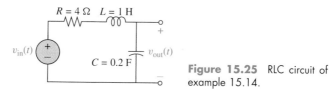

Figure 15.25 RLC circuit of example 15.14.

SOLUTION

By using voltage division, it is a simple matter to show that

$$H(s) = \frac{1}{LC} \frac{1}{s^2 + \frac{R}{L}s + \frac{1}{LC}} = \frac{5}{(s+2)^2 + 1} \tag{15.15}$$

Hence, the Laplace transform of the step response is

$$V_{out}(s) = \frac{5}{s\left[(s+2)^2 + 1\right]} = \frac{1}{s} + \frac{-s-4}{(s+2)^2 + 1}.$$

Rearranging terms yields

$$V_{out}(s) = \frac{1}{s} - \frac{s+2}{(s+2)^2 + 1} - 2\frac{1}{(s+2)^2 + 1}.$$

Taking the inverse transform produces the desired step response:

$$v_{out}(t) = [1 - e^{-2t}\cos(t) - 2e^{-2t}\sin(t)]u(t) \text{ V}. \tag{15.16}$$

As a check, observe that the derivative of equation 15.16 is

$$\frac{d}{dt}v_{out}(t) = 2e^{-2t}[\cos(t) + 2\sin(t)] - e^{-2t}[-\sin(t) + 2\cos(t)]$$

$$= 5e^{-2t}\sin(t).$$

With $H(s)$ given by equation 15.15, it is straightforward to show that the impulse response is

$$h(t) = \mathcal{L}^{-1}[H(s)] = 5e^{-2t}\sin(t)u(t),$$

as expected.

7. INITIAL- AND FINAL-VALUE THEOREMS

In system theory, and especially in control theory, engineers want the output signal of a circuit or system to track a given reference signal. The term *track* means that for large t, the reference signal and the circuit output are more or less indistinguishable. To accomplish this, the engineers generate an error signal, $e(t) = y(t) - y_{ref}(t)$, where $y(t)$ is the circuit output and $y_{ref}(t)$ is the reference signal. Since much of the analysis is done in the frequency domain, one often knows $E(s) = Y(s) - Y_{ref}(s)$ without knowing the related time functions. Moreover, many times the design engineers need to know the initial error, $e(0)$, and the final error, $e(\infty)$. Fortunately, there are two theorems, the initial-value and final-value theorems, that permit the computation of these quantities in the frequency domain.

Initial-Value Theorem Let $\mathcal{L}[f(t)] = F(s)$ be a strictly proper rational function of s, i.e., the numerator and denominator of F are both polynomials in s, with degree of the numerator less than that of the denominator. Then

$$\lim_{s\to\infty} sF(s) = f(0^+). \tag{15.17}$$

Proof: The quantity $sF(s)$ suggests a derivative operation on $f(t)$. Specifically,

$$sF(s) - f(0^-) = \mathcal{L}\left[\frac{d}{dt}f(t)\right].$$

Applying the definition of the Laplace transform to the right-hand side and taking limits as s approaches infinity yields

$$\lim_{s\to\infty}\left[sF(s) - f(0^-)\right] = \lim_{s\to\infty}\left[\int_{0^-}^{0^+} \dot{f}(t)e^{-st}\,dt + \int_{0^+}^{\infty} \dot{f}(t)e^{-st}\,dt\right] \qquad (15.18)$$

$$= \lim_{s\to\infty}\left[\int_{0^-}^{0^+} \dot{f}(t)e^{-st}\,dt\right] + \lim_{s\to\infty}\left[\int_{0^+}^{\infty} \dot{f}(t)e^{-st}\,dt\right],$$

where the dot over the function $f(t)$ means the derivative of the function. Observe that

$$\lim_{s\to\infty}\left[\int_{0^+}^{\infty} \dot{f}(t)e^{-st}\,dt\right] = \int_{0^+}^{\infty} \dot{f}(t)\,\lim_{s\to\infty}\left(e^{-st}\right)dt = 0$$

and that, because e^{-st} is continuous at $t = 0$,

$$\lim_{s\to\infty}\left[\int_{0^-}^{0^+} \dot{f}(t)e^{-st}\,dt\right] = \left[\lim_{s\to\infty} e^{-s0}\right]\int_{0^-}^{0^+} \dot{f}(t)\,dt = f(0^+) - f(0^-). \qquad (15.19)$$

Hence, the left-hand side of equation 15.18 equals the right-hand side of equation 15.19. Equation 15.17 follows by equating these two sides and canceling the $f(0^-)$ terms in both.

 EXAMPLE 15.15

Find $f(0^+)$ when

$$F(s) = \frac{1}{3s + 2}.$$

SOLUTION

By direct application of the initial value theorem,

$$f(0^+) = \lim_{s\to\infty} sF(s) = \lim_{s\to\infty} \frac{s}{3s + 2} = \frac{1}{3}.$$

 **EXAMPLE 15.16**

Find $f(0^+)$ when

$$F(s) = \frac{18s + 7}{(3s + 2)(2s - 1)}.$$

SOLUTION

Again by direct application of the initial-value theorem,

$$f(0^+) = \lim_{s\to\infty} \frac{18 + \dfrac{7}{s}}{\left(3 + \dfrac{1}{s}\right)\left(2 - \dfrac{1}{s}\right)} = 3.$$

Exercise. Suppose $F(s) = (8s + 2)/(2s^2 + 8s + 3)$. Determine $f(0^+)$.

ANSWER: 4.

The initial-value theorem and the examples that follow it illustrate the computation of initial values. However, engineers also need to know final values of, for example, error signals. Accordingly, we have the next theorem.

Final-Value Theorem Suppose $F(s)$ has poles only in the open left half complex plane, with the possible exception of a single-order pole at $s = 0$. Then

$$\lim_{s \to 0} sF(s) = \lim_{t \to \infty} f(t). \qquad (15.20)$$

Proof: The condition of the theorem, i.e., the condition that $F(s)$ has poles only in the left half complex plane, with the possible exception of a first-order pole at the origin, guarantees that the limit on the right-hand-side of equation 15.20 exists. This is because a partial fraction expansion of $F(s)$ leads to a time function $f(t)$ that is a sum of exponentially decaying signals and at most one constant signal. Since the right-hand-side limit is well defined,

$$\lim_{s \to 0} \left[sF(s) - f(0^-) \right] = \lim_{s \to 0} \left[\int_{0^-}^{\infty} \dot{f}(t) e^{-st} dt \right] = \int_{0^-}^{\infty} \dot{f}(t) dt$$

$$= \left(\lim_{s \to \infty} f(t) \right) - f(0^-).$$

This implies equation 15.20.

◆ **EXAMPLE 15.17**

Let

$$F(s) = \frac{(s+2)(s-1)}{s(s+1)(s+3)}.$$

Find the final value of $f(t)$ if possible.

SOLUTION

$F(s)$ has poles that meet the conditions of the final-value theorem. Hence,

$$\lim_{t \to \infty} f(t) = \lim_{s \to 0} sF(s) = \frac{(s+2)(s-1)}{(s+1)(s+3)} = -\frac{2}{3}.$$

◆ **EXAMPLE 15.18**

What if the conditions of the final-value theorem are not met? What would go wrong? A simple example illustrates the problem. Let

$$F(s) = \frac{1}{s^2 + 1},$$

which corresponds to $f(t) = \sin(t)u(t)$. Then

$$\lim_{t \to \infty} sF(s) \lim_{s \to 0} \frac{s}{s^2 + 1} = 0,$$

but

$$\lim_{t \to \infty} f(t) = \lim_{t \to \infty} \sin(t)u(t)$$

is undefined, i.e., it does not exist. The theorem, however, presupposes that both limits exist. Again, the condition of poles in the left half complex plane with at most one pole at the origin is necessary and sufficient for both limits to exist.

Exercise. If $F(s) = (6s + 10)/(2s^2 + 4s)$, then, by the final value theorem, $f(t)$ approaches what value for large t?

ANSWER: 2.5.

8. BODE PLOTS

Section 5 described the use of the poles and zeros of $H(s)$ to compute the frequency response of a circuit. In this regard, Heindrik Bode developed a technique for computing approximate or asymptotic frequency response curves. These so-called **Bode plots** can be quickly drawn by hand. A description of the technique requires the introduction of some terms widely used in the engineering literature.

Let $H(s)$ be a transfer function that is a dimensionless voltage ratio or current ratio. As explained in section 4, for sinusoidal steady-state analysis, one replaces s by $j\omega$ to study the circuit's magnitude response, $|H(j\omega)|$, and phase response, $\angle H(j\omega)$. For convenience, let $|H(j\omega)|$ be a voltage gain, $|V_2/V_1|$. The **gain in dB** (decibels), denoted by H_{dB}, is defined by the equation

$$H_{dB}(j\omega) \equiv 20\log_{10}|H(j\omega)|. \tag{15.21}$$

For convenience, whenever we write $\log(x)$, we will mean $\log_{10}(x)$. Solving for $|H(j\omega)|$ in equation 15.21 yields the inverse relationship

$$|H(j\omega)| \equiv 10^{0.05 H_{dB}(j\omega)}. \tag{15.22}$$

Table 15.1 presents some pairs of $|H|$ and H_{dB}.

TABLE 15.1 TRANSFER FUNCTION GAIN IN MAGNITUDE AND IN DECIBELS

$|H|$	1	$\sqrt{2}$	2	10	100	1,000
H_{dB}	0	$\cong 3$	$\cong 6$	20	40	60

Thus, instead of saying that $|V_2|$ is 100 times $|V_1|$, we may say that V_2 is 20 dB *above* V_1, or that V_1 is 20 dB *below* V_2. Similarly, to say that V_2 is 3 dB above V_1 means that $|V_2|$ is 1.414 times $|V_1|$.

One of the reasons for using the dB terminology is that it simplifies the analysis and design of multistage amplifiers. Suppose an amplifier has three stages with voltage gains equal to 20, 100, and 10, respectively. The overall voltage gain is the *product* of the gains of each individual stage, which is $20 \times 100 \times 10 = 20,000$. Using the dB specification, the overall gain in dB is the *sum* of the gains in dB of the individual stages. This is $(26+40+20) = 86$ dB. It is easy to justify this claim. First,

$$|H| = |H_1| \times |H_2| \times |H_3|.$$

Taking the logarithm of both sides and multiplying by 20 yields

$$20 \log |H| = 20 \log |H_1| + 20 \log |H_2| + 20 \log |H_3|,$$

or

$$H_{dB} = H_{1,dB} + H_{2,dB} + H_{3,dB}.$$

Summation is certainly a more simple arithmetic operation than multiplication. The advantage becomes even more pronounced for repetitive calculations at many frequency points, as when plotting a magnitude response curve such as equation 15.13a. We could, of course, convert this equation to an equation having all terms in dB. However, with an eye toward a further simplification, it is desirable to first rewrite $H(s)$ in a slightly different, but equally general, form, namely,

$$H(s) = K s^{\alpha} \frac{\left(\dfrac{s}{-z_1} + 1\right)\left(\dfrac{s}{-z_2} + 1\right) \cdots}{\left(\dfrac{s}{-p_1} + 1\right)\left(\dfrac{s}{-p_2} + 1\right) \cdots}, \tag{15.23}$$

where $\{p_1, p_2, \ldots\}$ are those poles of $H(s)$ which are not at the origin and $\{z_1, z_2, \ldots\}$ are those zeros of $H(s)$ which are not at the origin; if α is positive (negative), then $H(s)$ has α zeros (poles) at the origin. For example, a transfer function

$$H(s) = 1.2 \frac{(s + 50)(s + 200)}{(s + 8)(s + 600)}$$

has the rewritten form

$$H(s) = 2.5 \frac{\left(\dfrac{s}{50} + 1\right)\left(\dfrac{s}{200} + 1\right)}{\left(\dfrac{s}{8} + 1\right)\left(\dfrac{s}{600} + 1\right)}. \tag{15.24}$$

Observing that $H_{dB} = 10 \log_{10} |H(j\omega)|^2$, setting $s = j\omega$ in equation 15.23, and noting that the magnitude squared of a complex number is the imaginary part squared plus the real part squared yields

$$H_{dB}(\omega) = |K|_{dB} + [\omega^{\alpha}]_{dB}$$

$$+ \left[\left(\frac{\omega}{z_1}\right)^2 + 1\right]^{0.5}_{dB} + \left[\left(\frac{\omega}{z_2}\right)^2 + 1\right]^{0.5}_{dB} + \cdots$$

$$- \left[\left(\frac{\omega}{p_1}\right)^2 + 1\right]^{0.5}_{dB} - \left[\left(\frac{\omega}{p_2}\right)^2 + 1\right]^{0.5}_{dB} - \cdots \tag{15.25}$$

Equation 15.25 suggests that we may compute the dB-vs.-ω curve for each term on the right-hand side and graphically sum the curves to obtain the desired H_{dB}-vs.-ω curve. However, each individual curve is easily and fairly accurately sketched by using $\log(\omega)$ instead of ω as the independent variable. This amounts to using semilog paper to plot the dB-vs.-ω curves. Such a plot, with a linear scale for the dB values on the vertical axis and a logarithmic scale for ω on the horizontal axis, is called a *Bode magnitude plot*, in honor of its inventor. Similarly, a plot of $\angle H(j\omega)$ vs. ω, with a linear scale for $\angle H(j\omega)$ and a log scale for ω, is called a *Bode phase plot*. Note that, because of the logarithmic scale for the ω-axis, the actual distance on the paper between $\omega = 1$ and $\omega = 10$ is the same as that between $\omega = 0.1$ and $\omega = 1$. (See figure 15.26a.) Note also that the $\omega = 0$ point will not appear on the graph, because $\log(\omega)$ approaches $-\infty$ as ω approaches 0.

With $\log(\omega)$ chosen as the independent variable, the plot of each term in equation 15.25 either is exactly a straight line or is a curve having two straight line **asymptotes**. This is illustrated in figure 15.26.

In figures 15.26b and 15.26c, the rising asymptote has a slope of 20 dB/decade, which means that along this line, an increase in frequency by a factor of 10 causes an increase in gain of 20 dB. The word *decade* (abbreviated dec) simply means "10 times" here. Another

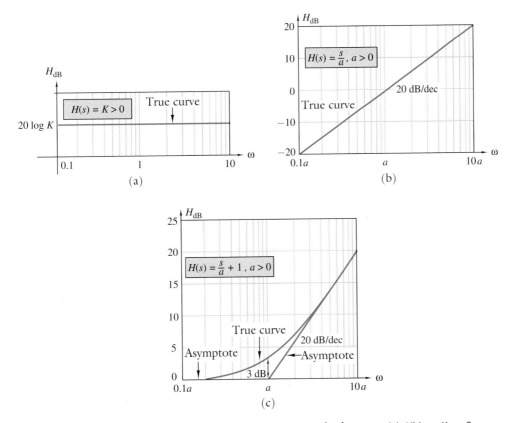

Figure 15.26 Bode magnitude plots for three basic transfer functions. (a) $H(s) = K > 0$. (b) $H(s) = s/a$, $a > 0$. (c) $H(s) = s/a + 1$, $a > 0$.

way to express the same slope is to indicate the increase of gain in dB when the frequency is doubled, or increased by an octave in music terminology. It is easy to see that 20 dB/dec is equivalent to 6 dB/octave. In general, if a frequency, ω_2, is d decades above another frequency, ω_1, then, by definition, $(\omega_2/\omega_1) = 10^d$. Conversely, if $(\omega_2/\omega_1) = r$, then we say that ω_2 is $\log(r)$ decades above ω_1.

In figure 15.26c, the left asymptote is a horizontal line and hence has a slope of 0 dB/dec. The point where the two asymptotes intersect is called a **breakpoint**, and the corresponding frequency is called a **break frequency** or a **corner frequency**.

The derivations of the true curves and asymptotes in figures 15.26a and 15.26b are very simple and are left as exercises. For figure 15.26c,

$$H(s) = \frac{s}{a} + 1,$$

in which case

$$|H(j\omega)| = \left| \frac{j\omega}{a} + 1 \right| = \sqrt{\left(\frac{\omega}{a}\right)^2 + 1},$$

and

$$|H(j\omega)|_{dB} = 20 \log \left[\sqrt{\left(\frac{\omega}{a}\right)^2 + 1} \right].$$

For $\omega \ll a$.

$$|H(j\omega)|_{dB} = 20 \log \left[\sqrt{1} \right] = 0,$$

indicating that $|H(j\omega)|_{dB}$ approaches the left asymptote in the figure. For $\omega \gg a$,

$$|H(j\omega)|_{dB} = 20 \log \left[\sqrt{\left(\frac{\omega}{a}\right)^2 + 1} \right] \cong 20 \log \left[\frac{\omega}{a} \right] = 20 \log(\omega) - 20 \log(a),$$

indicating that $|H(j\omega)|_{dB}$ approaches the right asymptote in the figure. The two asymptotes intersect at the point ($\omega = a$, $|H(j\omega)|_{dB} = 0$). At this corner frequency, the largest error,

3 dB, occurs between the true value of $|H(j\omega)|_{dB}$ and the value read from the asymptotic curve. The error at twice or half the corner frequency is about 1 dB.

The following variations of the three basic Bode magnitude plots of figure 15.26 are easily derived:

1. If $H(s) = (s/a)^n$, the Bode magnitude is similar to that shown in figure 15.26b, except that the slope is now $20n$ dB/dec. The curve still passes through the point $(\omega = a, |H(j\omega)|_{dB} = 0)$.
2. If $H(s) = (s/a + 1)^n$, the Bode magnitude is similar to that shown in figure 15.26c, except that the right asymptote has a slope of $20n$ dB/dec. The breakpoint is still at $(\omega = a, |H(j\omega)|_{dB} = 0)$, and the error at the corner frequency is $3n$ dB. If n is negative, the right asymptote points downward.

Let us now consider the Bode plot for a general transfer function $H(s)$. After expressing $H(s)$ in the form of equation 15.23, we can draw the asymptotes for each term in equation 15.25 with the aid of figure 15.26. The asymptotes for the H_{dB}-vs.-$\log(\omega)$ curve can then be constructed very easily by graphically summing the individual asymptotes. Since the asymptote for each term in equation 15.25 is a *piecewise linear curve*, the graphical sum of all the asymptotes is also a piecewise linear curve. Accordingly, it is not necessary to calculate the sum of dB values at a large number of frequencies. If there are n break frequencies, then the summation need only be carried out at these frequencies and for the slopes of the leftmost and the rightmost segments of the piecewise linear asymptote. The following example illustrates this procedure.

◆ **EXAMPLE 15.19**

Obtain the asymptotes for the Bode plot of $H(s)$ of equation 15.24.

SOLUTION

There are four break frequencies. Rewriting $H(s)$ as a product of five factors with break frequencies appearing in *ascending order* yields

$$H(s) = 2.5 \left[\frac{s}{8} + 1 \right]^{-1} \left[\frac{s}{50} + 1 \right] \left[\frac{s}{200} + 1 \right] \left[\frac{s}{600} + 1 \right]^{-1}$$

$$= H_1 \times H_2 \times H_3 \times H_4 \times H_5. \tag{15.26}$$

Figure 15.27a shows the asymptotes for the five individual terms in equation 15.24, and figure 15.27b shows the asymptotes for H_{dB}.

The calculation of the asymptote in figure 15.27b proceeds as follows:

1. The slope of the leftmost segment, i.e., the segment to the left of $\omega = 8$, is obviously zero, from figure 15.27a.
2. There is a breakpoint P_1 at $\omega = 8$. The only factor contributing to a dB value at this frequency is $H_1(s) = 2.5$, for which $H_{1.dB} = 20\log(2.5) \cong 8$.
3. There is a breakpoint P_2 at $\omega = 50$. Since $H_2(s)$ contributes -20 dB/dec to the slope for $\omega > 8$, and since 50 rad/sec is 0.796 decade above 8 rad/sec (from $\log(50/8) = 0.796$), the dB value corresponding to P_2 is

$$8 - 20 \times 0.796 = -7.92.$$

4. There is a breakpoint P_3 at $\omega = 200$. Since $H_3(s)$ contributes an *additional* 20 dB/dec to the slope for $\omega > 50$, resulting in a slope of zero, we have -7.9 dB for P_3.
5. There is a breakpoint P_4 at $\omega = 600$. Since $H_4(s)$ contributes an *additional* 20 dB/dec to the slope for $\omega > 600$, resulting in a slope of 20 dB/dec, and since 600 rad/sec is 0.477 decade above 200 rad/sec (because $\log(600/200) = 0.477$), the dB value corresponding to P_4 is

$$-7.92 + 20 \times 0.477 = 1.63.$$

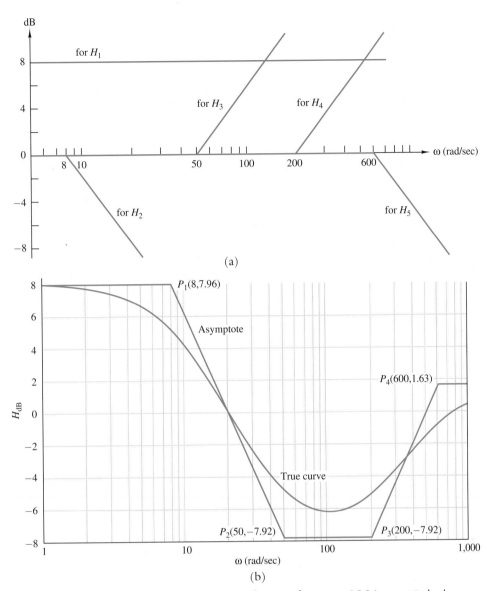

Figure 15.27 Asymptote and Bode magnitude for $H(s)$ of equation 15.24 or, equivalently, equation 15.26. (a) Bode plots for H_1 through H_5. (b) Bode plot for H_{dB} equal to sum of those for H_1 through H_5.

6. Finally, consider the slope of the rightmost segment. Since $H_5(s)$ contributes an *additional* -20 dB/dec to the slope for $\omega > 600$, the resulting slope of the rightmost segment is zero.

The complete specification of the piecewise linear asymptote is shown in figure 15.27b.

Exercise. Construct the piecewise linear asymptote for the transfer function given by $H(s) = 40s/[(s+2)(s+20)]$.

Once the piecewise linear asymptote for the Bode plot has been constructed, the true curve can be sketched approximately by noticing that the error at each corner frequency is about 3n dB (provided that the two neighboring corner frequencies are more than five times larger or smaller). Figure 15.27b shows such a rough sketch. In the precomputer era, the ability to draw such a curve by hand—even a crude approximation—was considered valuable. Nowadays, with the wide availability of personal computers and CAD software, one might just as well get the exact Bode plot without ever looking at the straight-line asymptotes. From the perspective of circuit analysis, the value of constructing the asymptotes for a Bode plot is greatly diminished, but the technique is still important for its application in the design of feedback control systems. Such an application utilizes both the Bode magnitude plot and the Bode phase plot. Some background in control systems is required to understand the use of the Bode technique in this kind of application. We will relegate the discussion of the topic to a more advanced course in feedback control. Our objective in mentioning it in this text is twofold: (1) to introduce the definitions of some commonly used terms, such as *decibels*, *decade*, and *octave*, and (2) to demonstrate a highly systematic procedure for adding up several piecewise linear curves to obtain a desired curve, as described in example 15.27.

9. TRANSFER FUNCTION ANALYSIS OF A DC MOTOR

A permanent magnet dc motor is an electromechanical device that converts direct current or voltage into mechanical energy. The motor consists of a shaft on which is mounted a coil of wire called the *armature winding* of the motor. This shaft rotates freely on bearings. Enclosing the coil are permanent magnets that interact with a magnetic field produced when a current flows through the armature winding. This interaction of magnetic fields forces the shaft to rotate causing energy conversion to take place. Here current flowing through the armature coils rotating through the magnetic field produced by the permanent magnets encasing the coils produces a torque on the motor shaft, which drives a load. Hence, power is delivered from the source to the load.

Because of the electromechanical characteristics of the permanent magnet dc motor, it has a simple circuitlike model amenable to Laplace transform analysis. The model is given in figure 15.28 and consists of a voltage source (a wall outlet, for example) in series with a resistor R_a and inductance L_a, and a device labeled "motor." The resistance R_a represents the resistance present in the armature winding, and L_a represents the equivalent inductance of the wire coil. The device labeled "motor" has the current $i_a(t)$ as an input and the angular velocity $\omega(t)$ of the rotating shaft as an output. The interaction between the electrical part of the model and the mechanical part of the model occurs at this symbol. The voltage $v_m(t)$ is an induced voltage proportional to the angular velocity $\omega(t)$. Because $\omega(t)$ is not a circuit variable, classical notions of impedance, admittance, voltage gain, etc., do not fit the problem, whereas the more general notion of a transfer function does. Although we have a circuitlike model of the motor, it is necessary to move slightly beyond the confines of circuit theory proper to analyze the system. This is precisely why we have introduced the transfer function concept as an extension of the basic ideas of impedance, admittance, voltage gain, etc.

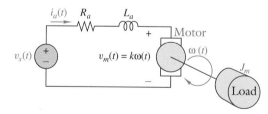

Figure 15.28 Equivalent circuitlike model of a permanent magnet dc motor.

The torque $T(t)$ produced on the rotating shaft by the current flowing through the armature coils is proportional to the armature current $i_a(t)$, i.e.,

$$T(t) = ki_a(t). \tag{15.27}$$

The mechanical rotation of the motor affects the electrical portion of the system by inducing a voltage $v_m(t)$. This voltage is proportional to the motor's rotational speed, or angular velocity, $\omega(t)$. Since the motor converts electrical energy to mechanical energy, conservation of energy dictates that the constant of proportionality be equal to the same constant that relates torque and current, i.e.,

$$v_m(t) = k\omega(t). \tag{15.28}$$

Our first goal is to find the transfer function of the motor, $H(s) = \mathcal{L}[\omega(t)]/V_s(s)$. For convenience, let $\Omega(s)$ denote $\mathcal{L}[\omega(t)]$. As a first step, sum the voltages around the loop of elements in the circuit model of figure 15.28. This results in the differential equation

$$v_s(t) = R_a i_a(t) + L_a \frac{di_a(t)}{dt} + k\omega(t). \tag{15.29}$$

Assuming zero initial conditions, the Laplace transform of equation 15.29 is

$$V_s(s) = (R_a + L_a s)\, I_a(s) + k\Omega(s). \tag{15.30}$$

From basic mechanics, the differential equation governing the mechanical portion of the system is

$$T(t) = J_m \frac{d\omega(t)}{dt} + B\omega(t), \tag{15.31}$$

where J_m is the moment of inertia of the combined armature, rotating shaft, and load, B is the coefficient of friction, and $T(t)$ is the torque produced by the motor to turn the load. Recalling equation 15.27, $T(t) = k i_a(t)$, the Laplace transform of equation 15.31, assuming zero initial conditions, is

$$k I_s(s) = (J_m s + B)\, \Omega(s). \tag{15.32}$$

Solving equation 15.32 for $I_s(s)$, substituting the result into equation 15.30, and then solving for $\Omega(s)$ leads to the expression

$$\Omega(s) = \frac{\dfrac{k}{L_a J_m}}{s^2 + \left(\dfrac{B}{J_m} + \dfrac{R_a}{L_a}\right) s + \dfrac{R_a B + k^2}{L_a J_m}} V_s(s). \tag{15.33}$$

Equation 15.33 characterizes the pertinent dynamics of the permanent magnet dc motor and allows one to find the angular velocity of the motor shaft as a function of time for given inputs. To see how the motor responds to step inputs, suppose $v_s(t) = A u(t)$. As objectives, let us find (1) the steady-state value, or final angular velocity, of the shaft and (2) the steady-state value of the armature current. The final speed of the shaft is important because, for example, one needs a fixed speed for the rotation of a compact disc or a fan. The final or steady-state current is important for determining the necessary power delivered by the source.

If $v_s(t) = A u(t)$, then $V_s(s) = A/s$. It follows from equation 15.33 that

$$\Omega(s) = \frac{\dfrac{k}{L_a J_m}}{s\left(s^2 + \left(\dfrac{B}{J_m} + \dfrac{R_a}{L_a}\right) s + \dfrac{R_a B + k^2}{L_a J_m}\right)} A. \tag{15.34}$$

Applying the final-value theorem to equation 15.34 results in

$$\omega_{ss} = \lim_{t \to \infty} \omega(t) = \frac{k}{R_a B + k^2} A.$$

To isolate the armature current $I_a(s)$, apply the final-value theorem again to determine the steady-state value of $i_a(t)$. Then combine equations 15.32 and 15.33 to obtain

$$I_a(s) = \frac{\dfrac{1}{L_a} s + \dfrac{B}{L_a J_m}}{s^2 + \left(\dfrac{B}{J_m} + \dfrac{R_a}{L_a}\right) s + \dfrac{R_a B + k^2}{L_a J_m}} V_s(s). \tag{15.35}$$

Equation 15.35 allows us to find the armature current as a function of time for a given input voltage. As before, if the input is a step, i.e., if $v_s(t) = Au(t)$, it follows that

$$I_a(s) = \frac{\dfrac{1}{L_a}s + \dfrac{B}{L_a J_m}}{s\left(s^2 + \left(\dfrac{B}{J_m} + \dfrac{R_a}{L_a}\right)s + \dfrac{R_a B + k^2}{L_a J_m}\right)} A. \tag{15.36}$$

Once again, application of the final-value theorem to this expression leads to the value of the steady-state armature current:

$$i_{a,ss} = \lim_{t \to \infty} i_a(t) = \frac{B}{R_a B + k^2} A.$$

The preceding analysis suggests the utility of the Laplace transform as a tool for analyzing the dynamic behavior of electromechanical systems. In fact, system transfer functions of the form of equations 15.33 and 15.35 are often starting points for further analysis. In this vein, problems 68 through 72 extend the analysis.

SUMMARY

This chapter has expanded the notion of a transfer function from its definition in chapter 14 into a tool for modeling not only circuits, but other practical systems, such as a dc motor. The transfer function is seen to characterize circuit or system behavior by the location of its poles and zeros. For example, if a transfer function has a pole on the imaginary axis or in the right half-plane, the associated circuit or system is said to be *unstable*, because there is an input or, possibly, simply an initial condition in the case of a second-order pole that will excite the pole and cause the response to grow without bound, eventually destroying the circuit or system. Further, the ubiquitous presence of noise will always excite poles on the imaginary axis and cause the response to be unstable.

The chapter categorized various types of responses. For example, associated with a circuit or system is the zero-state response—the response to an input, assuming zero initial conditions. This response is simply the inverse Laplace transform of the product of the transfer function and the Laplace transform of the input excitation. Associated with the zero-state response of a circuit is the zero-input response due to initial conditions on the capacitors or inductors of the circuit.

The complete response for linear circuits having constant parameter values is simply the sum of the zero-input and zero-state responses. This decomposition generalizes to the broad class of linear systems studied in advanced courses. Under reasonable conditions, other decompositions are possible, such as a decomposition into the natural and forced responses or into transient and steady-state responses. Other important responses are the impulse and step responses.

For stable circuits, the frequency response provides important information about the circuit. Recall that the frequency response is a plot of the magnitude and phase of $H(j\omega)$ as ω varies from 0 to ∞. A Bode plot is an asymptotic approximation to the magnitude response that is often straightforward to compute by hand. The information in such a plot tell us how a circuit behaves when excited by sinusoids of different frequencies. Chapters 17, 21, and 22 will apply this idea to various applications.

Finally, the chapter introduced the initial- and final-value theorems, which help one determine initial and final values of a time function from knowledge of its Laplace transform. Such theorems have wide application in control system analysis, as evidenced in our analysis of the dc servomotor.

TERMS AND CONCEPTS

Asymptote: a limiting straight-line approximation to a curve.

Bode plot: an approximate or asymptotic frequency-response curve composed of a series of straight-line segments.

Bounded: the condition wherein, a signal $f(t)$, satisfies $|f(t)| < K < \infty$ for all t; i.e., the signal has a maximum finite height.

Breakpoint: the point at which two asymptotes of a Bode plot intersect.

Complete response: the total response of a circuit to a given set of initial conditions and a given input signal.

Corner (break) frequency: frequency at which two asymptotes of a Bode plot intersect.

Decade: a frequency band whose endpoint is a factor of 10

larger than its beginning point.

Decibel (dB): a log-based measure of gain equal to $20 \log_{10} |H(j\omega)|$.

Dirac delta function: see unit impulse function.

Final-value theorem: theorem that says the following: Suppose $F(s)$ has poles only in the open left half complex plane, with the possible exception of a single-order pole at $s = 0$. Then the limit of $f(t)$ as $t \to \infty$ equals the limit of $sF(s)$ as $s \to 0$.

Forced response: the portion of the complete response that has the same exponent as the input excitation, provided the input excitation has exponents different from those of the zero-input response.

Frequency response: measure of circuit behavior to unit magnitude sinusoids, $\cos(\omega t)$, as ω varies from 0 to ∞. Equals the evaluation of the transfer function $H(s)$ at $s = j\omega$ for all ω.

Fundamental period of periodic $f(t)$: the smallest positive number T such that $f(t) = f(t + T)$.

Impulse response: assuming all ICs are zero, if the circuit or system input is $\delta(t)$, then the resulting $y(t)$ is called the *impulse response*. The inverse transform of the transfer function equals the impulse response.

Initial-value theorem: theorem that says the following: Let $\mathcal{L}[f(t)] = F(s)$ be a strictly proper rational function of s; i.e., the numerator and denominator of $F(s)$ are both polynomials in s, with degree of the numerator less than that of the denominator. Then $f(0^+)$ is the limiting value of $sF(s)$ as $s \to \infty$.

Linear circuit: circuit such that for any two inputs $f_1(t)$ and $f_2(t)$, whose zero-state responses are $y_1(t)$ and $y_2(t)$, respectively, the response to the new input $[K_1 f_1(t) + K_2 f_2(t)]$ is $[K_1 y_1(t) + K_2 y_2(t)]$, where K_1 and K_2 are arbitrary scalars.

Magnitude response: the magnitude of the frequency response as a function of ω.

Natural response: the portion of the complete response that has the same exponents as the zero-input response.

Octave: a frequency band whose endpoint is twice as large as its beginning point.

Periodic $f(t)$: function satisfying the condition that there exists a positive constant T such that $f(t) = f(t + T)$ for all $t \geq 0$.*

Phase response: the angle of the frequency response as a function of ω.

Piecewise linear curve: an unbroken curve composed of straight-line segments.

Pole (simple) of rational function: zero of order 1 in denominator polynomial.

Poles (finite) of rational function: zeros of denominator polynomial.

Ramp function, $r(t)$: integral of unit step function having the form $Ktu(t)$ for some constant K.

Rational function: ratio of two polynomials; a polynomial is a rational function.

Stable transfer function: a transfer function for which every bounded input signal yields a bounded response signal; i.e., all poles are in open left half complex plane.

Steady-state response: that part of the complete response which either is constant or satisfies the definition of periodicity for $t \geq 0$.

Step response: the response of the circuit to a step function, assuming that all initial conditions are zero.

Transient response: that part of the complete response which is neither constant nor periodic for $t \geq 0$, i.e., the transient response does not satisfy the definition of a periodic function for $t \geq 0$.

Unit impulse function: a so-called distribution whose integral from $-\infty$ to t is the step function; the unit impulse function is nonrigorously interpreted as the derivative of the step function.

Unit step function, $u(t)$: function whose value is 0 for $t < 0$ and 1 for $t \geq 0$.

Zero-input response: the response of a circuit to a set of initial conditions with the input set to zero.

Zero-state response: the response of a circuit to a specified input signal, given that the initial conditions are all set to zero.

Zeros of rational function: values that make the numerator polynomial zero.

PROBLEMS

Poles and Zeros

1. Determine a transfer function $H(s)$, that has the pole-zero plot figure P15.1 and that has gain 2 at $s = 0$.

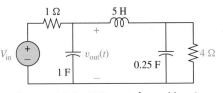

Figure P15.2 RLC circuit for problem 2.

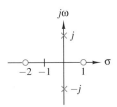

Figure P15.1 Transfer function pole-zero plot for problem 1.

2. Consider the circuit of figure P15.2.

(a) Find the poles and zeros of the transfer function $H(s) = V_{out}/V_{in}$, and sketch the pole-zero plot. *Check:* One pole is at $s = -1$.

(b) If the network has zero initial conditions and $v_{in}(t) = e^{-t}u(t)$ V, write the general form of the output, $v_{out}(t)$, as the sum of several real time functions, each correct within a multiplicative constant (e.g., $Ae^{-2t} + Bt^3 \cos(5t) + C$). *Check:* There are four terms. One term is $Ke^{-0.5t} \cos(0.866t)u(t)$.

3. An impedance function $Z(s)$ of a certain linear active network (i.e., a network containing controlled sources) has poles at $s = -1$ and $-3 \pm j4$, with zeros at $s = -2$ and $\pm j2$. It is known that $Z(0) = 8$.

---

*This (nonstandard) definition has been adapted for one-sided Laplace transform analysis.

(a) Write $Z(s)$ as the ratio of two polynomials in s.
(b) If the network has zero initial conditions and the input a current source of $u(t)$ A, find the general form of the voltage across the independent source. (The multiplicative constant in each term need not be calculated.) Does the output remain finite as $t \to \infty$?
(c) If the network has zero initial conditions and the input is a voltage source of $u(t)$ V, find the general form of the current through the independent source. (The multiplicative constant in each term need not be calculated.) Does the output remain finite as $t \to \infty$?

4. Repeat problem 3 when the zeros of $Z(s)$ are 2 and $\pm j2$.

5. The pole-zero plot of a transfer function $H(s)$ is given in figure P15.5.
(a) If the dc gain is -1, determine $H(s)$.
(b) Determine the response to a step function input.
(c) Determine the response to the impulse function. *Check*: Your answer to (c) should be the derivative of your answer to (b), since the delta function is the derivative of the step function.
(d) (Challenge!) Determine the response to the input $\sin(3t)u(t)$.
(e) If the input is $e^{-at}u(t)$, determine the positive number a such that the response does not have a term of the form $Ke^{-at}u(t)$. Find the response under this condition.

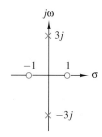

Figure P15.5 Pole-zero plot of transfer function for problem 5.

6. Determine the transfer function $H(s)$ from the pole-zero plot of figure P15.6 if $H(s)$ has gain 2 at $s = 0$.

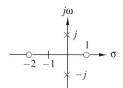

Figure P15.6 Pole-zero plot of transfer function for problem 6.

7. A transfer function $H(s)$ of a particular active circuit has poles at $-1 \pm 2j$ and -2. When this circuit is excited by the input waveforms $v_{in}(t) = \sin(t)$ and $v_{in}(t) = u(t)$, it is found that the output is zero after a long time. Also, at $s = -1$, the transfer function gain was found to be 4, and $H(\infty)$ is known to be finite.
(a) Determine the zeros of the transfer function.
(b) Determine the transfer function.

(c) If the input to the circuit is $v_{in}(t) = tu(t)$, sketch the approximate response for large t.
(d) Repeat part (c) for $v_{in}(t) = t^2u(t)$.
(e) If $G(s) = 1/H(s)$ and the input to $G(s)$ is $v_{in}(t) = \sin(t)u(t)$ V, sketch the approximate response for large t.

8. Consider the circuit of figure P15.8.
(a) Compute the transfer function $H(s)$. (*Note*: $H(s)$ should have the form $H(s) = Kn(s)/d(s)$, where $n(s)$ and $d(s)$ are polynomials whose leading coefficients are 1; i.e., the highest power of s has coefficient 1.)
(b) If $C_1 = C_2 = 1$ F, find R_1 and R_2 so that the poles of the transfer function are at $s = -0.2192$ and -2.2808. The problem of finding element values to realize a given transfer function is called the *synthesis problem*.
(c) Let $s = j\omega$. Make a rough plot of the magnitude of the frequency response; i.e., plot $|H(j\omega)|$ as a function of ω.

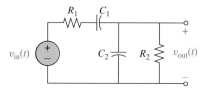

Figure P15.8 RC circuit for problem 8.

9. For the circuit of figure P15.9:
(a) Compute the transfer function $H(s)$.
(b) If $C_1 = C_2 = 1$ F, find R_1 and R_2 so that there is a pole at $s = -1$ and a zero at $s = -0.5$.
(c) Let $s = j\omega$. Make a rough plot of the magnitude of the frequency response; i.e., plot $|H(j\omega)|$ as a function of ω.

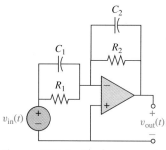

Figure P15.9 Ideal op amp circuit for problem 9.

10. The pole-zero plot of the transfer function of an active dilithium crystal amplifier circuit is shown in figure P15.10. If the gain at $s = 0$ is 1, determine the transfer function $H(s)$.

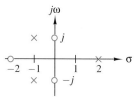

Figure P15.10 Pole-zero plot for problem 10.

Response Classifications

11. Consider the input and circuit of figure P15.11a.
    (a) Find the transfer function, and determine its pole(s) and zero(s).
    (b) Express $v_{in}(t)$ as a sum of possibly shifted simple functions such as steps, ramps, etc.
    (c) Find the zero-state response for the input of figure P15.11b.
    (d) Draw the equivalent frequency domain circuit, and find the zero-input response when $v_c(0^-) = -1$ V. Be careful! What is the time constant of the circuit when the input is zero?
    (e) Determine the complete response.
    (f) If $v_{in}(t)$ were changed to $f(t)$, as in figure P15.11c, what would be the response of the circuit? (*Hint*: What is the relationship between $v_{in}(t)$ and $f(t)$? Note that differentiation and integration are linear operations when the ICs are zero. For linear circuits, then, a linear operation on the input induces the same linear operation on the response, provided that the ICs are zero. Combine this concept with the structure of the decomposition to obtain the answer.)

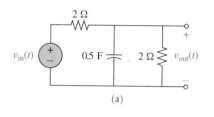

(a)

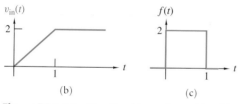

(b)   (c)

Figure P15.11 Circuit and inputs for problem 11.

12. Consider the circuit of figure P15.12. Let $v_{in}(t) = 15e^{-t}u(t)$ V, $v_C(0^-) = 6$ V, and $i_L(0^-) = 12$ A.
    (a) Find the zero-state response (i.e., the response due to $v_{in}(t)$).
    (b) Find the zero-input response (i.e., the response due to $v_C(0^-)$ and $i_L(0^-)$).
    (c) Find the complete response.
    (d) Write the general form of the natural response.
    (e) Which term in the complete response is called the forced response?

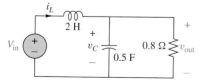

Figure P15.12 RLC circuit for problem 12.

13. (a) Repeat problem 12 when $v_{in}(t) = 15e^{-3t}u(t)$ V.
    (b) Repeat parts (a)–(d) of problem 12 if the the input is a rectangular pulse given by $v_{in}(t) = 15[u(t) - u(t-1)]$ V.

14. The initial conditions in the circuit of figure P15.14 are $v_{C1}(0^-) = 8$ V and $v_{C2}(0^-) = 0$. The input is $v_{in}(t) = 4u(t)$ V.
    (a) Find the transfer function $H(s) = V_{out}(s)/V_{in}(s)$. (*Hint*: Make use of the properties of an ideal op amp. Write two node equations, one at the output node of the circuit and one at the input node of the op amp.) *Check*: $H(s)$ has repeated poles, and $H(0) = -2.5$.
    (b) Find the zero-state response.
    (c) Find the zero-input response.
    (d) Find the complete response.
    (e) Write the general form of the natural response.

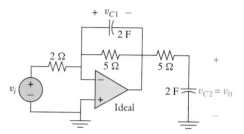

Figure P15.14 Ideal op amp circuit for problem 14.

15. Consider the circuit in figure P15.15.
    (a) $v_{in}(t) = 15u(t)$ V, and $v_C(0^-) = 6$ V.
       (1) Find the zero-state response, the zero-input response, and the complete response.
       (2) Find the forced response and the natural response.
       (3) Find the steady-state response and the transient response.
    (b) Repeat part (a) if $v_C(0^-) = 6$ V and the input is changed to $v_{in}(t) = 15\cos(2t)u(t)$ V.
    (c) Repeat part (a) if $v_C(0^-) = 6$ V and the input is changed to $v_{in}(t) = 15\exp(-2t)$ V.

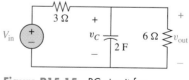

Figure P15.15 RC circuit for problem 15.

16. If you properly utilize the results of problem 15 and the principle of linearity, the answers to this problem can be written down directly. Find the complete response for $t \geq 0$ for the circuit of figure P15.15 for each of the following circuit conditions:
    (a) $v_C(0^-) = 4$, $v_{in}(t) = 5\exp(-2t)u(t)$ V.
    (b) $v_C(0^-) = 8$, $v_{in}(t) = 20\exp(-2t)u(t)$ V.

(c) $v_C(0^-) = 2$, $v_{in}(t) = 5\cos(2t)u(t)$ V.

(d) $v_C(0^-) = 3$, $v_{in}(t) = 8u(t)$ V.

17. Consider the circuit of figure P15.17.

(a) Find the transfer function $H(s) = V_{out}(s)/V_{in}(s)$.

(b) Find the zero-state response (i.e., the response of the circuit to the input when all internal stored energy is zero) to the input $v_{in}(t) = e^{-2t}u(t)$ V.

(c) Draw an equivalent s-plane circuit that accounts for ICs. There are four possibilities, but one is superior.

(d) Find the zero-input response (i.e., the response of the circuit to the initial conditions when the input is zero) for $v_C(0^-) = 0$ and $i_L(0^-) = -1$ A.

(e) Find the zero-input response when $i_L(0^-) = 0$ and $v_C(0^-) = 1.9519$ V.

(f) Find the complete response when $v_{in}(t) = e^{-2(t-1)}$ $u(t-1)$ V, $v_C(0^-) = 1.9519$ V, and $i_L(0^-) = -1$ A.

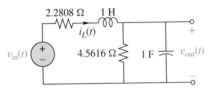

Figure P15.17 RLC circuit for problem 17.

18. An admittance function $Y(s)$, sketched in figure P15.18, has a double pole at $s = 0$ and double poles at $s = -2 \pm j$. Its zeros are at $s = -1$, 1, 3, and -3. It is known that $Y(2) = -0.22059$.

(a) Find $Y(s)$ and $Z(s)$, and draw the respective pole-zero plots.

(b) If the input is $v_{in}(t)$ and the response $i_{in}(t)$, determine the form of the natural response, assuming that there is internal stored energy.

(c) Find the actual natural response if the output is now taken to be $v_{in}(t)$ and the input $i_{in}(t)$, assuming that $v_{in}(0) = 5$ V, $v'_{in}(t) = 0$, $v''_{in}(0) = 20$ V, and $v'''_{in}(0) = 0$.

(d) If the input to the network is $v_{in}(t) = [3e^{-t} + 6te^{-t} + 3e^{-3t}]u(t)$ V, determine $V_{in}(s)$ as a rational function of the form $Kn(s)/d(s)$. What are the poles and zeros of $v_{in}(s)$? Are there any cancellations with $Y(s)$? What is the resulting zero-state response?

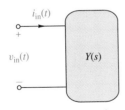

Figure P15.18 Unknown admittance for problems 18 and 19.

19. An admittance function $Y(s)$, sketched in figure P15.18, has poles at $s = 0$, -2, and -5. Its zeros are at $s = -1$ and -3. It is known that $Y(1) = 1$.

(a) Find $Y(s)$ and $Z(s)$.

(b) If the input is $v_{in}(t)$ and the response $i_{in}(t)$, determine the form of the natural response, assuming that there is internal stored energy.

(c) Determine the form of the natural response if the output is now taken to be $v_{in}(t)$, again assuming that there is initial stored energy. (*Note*: Natural responses are determined by the finite poles of the network functions considered in this class.)

(d) If the input to the network is $v_{in}(t) = 2(1 - t)e^{-t}$ $u(t)$ V, determine $i_{in}(t)$, $t \geq 0$, if it is known that $i_{in}(0^-) = 1$ A, $i'_{in}(0^-) = -1$ A, and $i''_{in}(0^-) = 47$ A.

20. Consider the circuit of figure P15.20a.

(a) Find the transfer function, and determine its pole(s) and zero(s).

(b) Express $v_{in}(t)$ as a sum of possibly shifted simple functions such as steps.

(c) Find the zero-state response for the input of figure P15.20b.

(d) Draw the equivalent frequency domain circuit, and find the zero-input response when $v_C(0^-) = 1$ V.

(e) Determine the complete response.

(f) If $v_{in}(t)$ were changed to $f(t)$ as in figure P15.20c, what would be the response of the circuit? (*Hint*: What is the relationship between $v_{in}(t)$ and $f(t)$? Note that differentiation and integration are linear operations when the ICs are zero. For linear circuits, then, a linear operation on the input induces the same linear operation on the response, provided that the ICs are zero. Combine this concept with the structure of the decomposition to obtain the answer.)

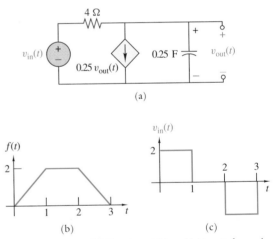

Figure P15.20 (a) Circuit and (b) and (c) inputs for problem 20.

21. Consider the circuit of figure P15.21a.

(a) Find the transfer function $H_1(s)$ if the input is a current source connected between nodes 1 and 2.

(b) Find the transfer function $H_2(s)$ if the input is a current source connected between nodes 1 and 3.

(c) Utilize the transfer functions obtained in parts (a) and (b) to find $v_{out}(t)$ in the circuit of figure P15.21b for $t \geq 0$. The initial conditions are $v_C(0^-) = 4$ V and $i_L(0^-) = 5$ A.

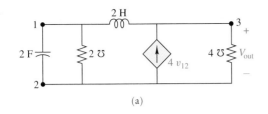

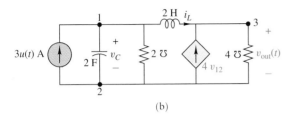

(a)

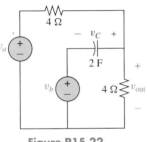

(b)

Figure P15.21 (a) Circuit for problem 21, parts (a) and (b). (b) Circuit for problem 21, part (c).

22. In the circuit shown in figure P15.22, $v_a(t) = 2u(t)$ V, $v_b(t) = 2te^{-2t}u(t)$ V, and $v_C(0^-) = 2$ V.

Figure P15.22

(a) Find $v_{out}(t)$ for $t \geq 0$ by the Laplace transform method.
(b) Write $v_{out}(t)$ as the sum of two components: the zero-state response and the zero-input response.
(c) Separate the zero-state response into one term due to v_a and another term due to v_b.
(d) Separate the complete response into the steady-state response and the transient response.
(e) Utilize the answers of previous parts to write down the complete response for $t \geq 0$ for the following two cases (all values in volts):

	$v_c(0^-)$	$v_a(t)$	$v_b(t)$
Case 1	4	$2u(t)$	$2te^{-2t}u(t)$
Case 2	8	$4u(t)$	$6te^{-2t}u(t)$

23. In figure 15.17 of example 15.8, let $\mu = 1.4$, and recall that $v_i(t) = \cos(2t)$ V.
(a) Find the transfer function $H(s) = V_1(s)/V_i(s)$.
(b) If the circuit is initially at rest, find an expression for $v_1(t)$ that is valid for all t (small or large). You might consider using the software program (included with this text) to do partial fraction expansion and inverse Laplace transformation. Alternatively, MATLAB will aid doing the partial fraction expansion of the transfer function.

Frequency Response

24. Draw to scale the pole-zero plot for $H(s) = (s+1)$ $/[s(s^2+s+10]$, and determine graphically the magnitude and phase of $H(j\omega)$ for $\omega = 0.2, 0.5, 1$, and 10. What are the limiting values of the magnitude and phase, i.e., for $\omega = \pm\infty$?

25. Draw to scale (on graph paper) the pole-zero plot of the transfer function

$$H(s) = \frac{\left((s+1)^2 + 16\right)\left(s^2 - 1\right)}{\left((s+1)^2 + 4\right)\left((s+1)^2 + 36\right)},$$

and determine graphically the magnitude and phase of $H(j\omega)$ for $\omega = 0, 2, 4, 6$, and ∞. Do your answers make sense physically? Explain.

26. Draw to scale (on graph paper) the pole-zero plot of the transfer function

$$H(s) = 2\frac{s^2 + 9}{(s+1)(s^2 + 0.4s + 4.04)},$$

and determine graphically the magnitude and phase of $H(j\omega)$ for $\omega = 0, 1, 1.5, 2, 2.5, 3$, and 5.

27. (a) Sketch the magnitude and phase shift (frequency response) for the second-order low-pass Butterworth filter transfer function

$$H(s) = \frac{1}{s^2 + 1.414s + 1}.$$

(b) Do the same for the second-order high-pass Butterworth filter transfer function

$$H(s) = \frac{s^2}{s^2 + 1.414s + 1}.$$

(c) Find the steady-state phase and magnitude of the output of each filter when the input is (1) $\sin(0.5t)$, (2) $\sin(t)$, and (3) $\sin(10t)$.
(d) Do your answers to part (c) make sense? Why? *Remark*: The *gain*, or the magnitude of $H(j\omega)$, is typically given in decibels (dB), computed as gain (dB) $= 20\log_{10}|H(j\omega)|$.

28. Frequency responses are typically plotted as $|H(j\omega)|$ vs. $\log_{10}|\omega|$ or as gain $20\log_{10}|H(j\omega)|$ dB vs. $\log_{10}\omega$.
(a) Make an accurate sketch, on log graph paper, of the magnitude $|H(j\omega)|$ and phase shift of the frequency response as a function of $\log_{10}\omega$ for $0.1 \leq \omega \leq 500$ rad for a second-order low-pass Butterworth filter transfer function

$$H(s) = \frac{25}{s^2 + 7.071s + 25}.$$

(b) Do the same for a second-order low-pass Chebyshev filter transfer function

$$H(s) = \frac{51.35}{s^2 + 8.983s + 52.85}.$$

(c) Find the steady-state phase and magnitude of the output of each filter when the input is (1) $\cos(t)$, (2) $\cos(10t)$, and (3) $\cos(100t)$.
(d) For large ω, which filter provides better attenuation, i.e., lowest gain?

29. A linear network has the transfer function

$$H(s) = 4\frac{s(s+4)}{(s+2)(s+3)(s^2+2s+101)}.$$

(a) Show the pole-zero plot in the s-plane.

(b) By calculating the lengths and angles of appropriate vectors, determine the magnitudes and angles of $H(j\omega)$ at $\omega = 9$, 10, and 11 rad/sec.

(c) Without any numerical calculations, by simply inspecting the pole-zero plot, determine an approximate value of ω at which $|H(j\omega)|$ reaches a maximum. Make a rough sketch of the $|H(j\omega)|$-vs.-ω curve.

Steady-State Calculation

30. A transfer function with gain constant $K = 6$ has the pole-zero plot given in figure P15.30. Determine the magnitude and phase of the frequency response at $\omega = 4$ rad/sec.

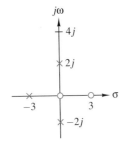

Figure P15.30 Pole-zero plot for problem 30.

31. A linear circuit with a transfer function

$$H(s) = \frac{2s+6}{s^2+5s+6}$$

has an input $x(t) = 2\cos(2t+45°)$. Determine the magnitude and phase (in degrees) of the output of the circuit in the steady state.

32. A stable active circuit has the transfer function

$$H(s) = \frac{s^2+1.5s}{s^2+3s+12}.$$

If the input is $v(t) = 4\cos(2t+45°)$ V, determine the phase and the magnitude of the steady-state response.

33. The pole-zero plot of a certain RLC network transfer function $H(s) = V_{out}/V_{in}$ is shown in figure P15.33. If the input is a sinusoidal voltage of 1-V amplitude, then, in the steady state, the output voltage has the greatest amplitude at approximately what ω?

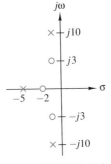

Figure P15.33 Pole-zero plot for transfer function of problem 33.

34. The circuit shown in figure P15.34 has zero initial conditions (i.e., $v_c(0^-) = 0$ and $i_L(0^-) = 0$). For $t \geq 0$, determine the largest voltage that can appear across the capacitance.

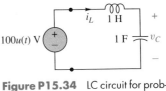

Figure P15.34 LC circuit for problem 34.

35. The voltage source in problem 34 is changed to $100\sin(t)u(t)$ V. For $t \geq 0$, determine the largest voltage that can appear across the capacitance.

36. At $t = 0$, the energy stored in the LC elements in figure P15.36 is zero. The current source is $i_{in}(t) = \cos(2t)$ A. As $t \to \infty$, what does $v_{out}(t)$ approach?

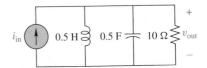

Figure P15.36 RLC circuit for problem 36.

37. In the circuit of figure P15.37, all stored energy is zero at $t = 0$. The current source is $i_{in}(t) = u(t)$ A. Determine $v_{out}(t)$ for large t.

Figure P15.37 Parallel LC circuit for problem 37.

38. Following are the transfer functions of some linear networks that contain controlled sources. The networks are initially at rest, i.e., with zero initial conditions. An input $v_{in}(t) = 2\cos(2t)u(t)$ V is applied at $t = 0$. Determine the output, $v_{out}(t)$, for very large values of t. *Remark*: The merit of your solution lies in getting the answers without performing unnecessarily complicated calculations. You may have to use the root-finding program.

(a) $H_1(s) = \dfrac{-3.75s+5}{s^2+4s+3}.$

(b) $H_2(s) = \dfrac{7s^2+s+4}{s^3+s^2+6s+6}.$

(c) $H_3(s) = \dfrac{2.5s-3}{s^2+2s+3}.$

(d) $H_4(s) = \dfrac{s^3-5.5s^2+14s-12}{s^3+5.5s^2+14s+12}.$

(e) $H_5(s) = \dfrac{2s^2+s+6}{s^3+s^2+5s+6}.$

(f) $H_6(s) = \dfrac{7s^2+s+4}{s^3+s^2+8s+7}.$

39. Following are transfer functions of some linear networks containing controlled sources. The networks are initially at rest, i.e., with zero initial conditions. An input $v_{in}(t) = 2\sin(t)u(t)$ V is applied at $t = 0$. Determine the outputs for very large values of t.

(a) $H(s) = \dfrac{3s + 4}{s^2 + 4s + 3}$.

(b) $H(s) = \dfrac{7s^2 + \sqrt{7}s + 4}{s^3 + s^2 + 7s + 6.2915}$.

(c) $H(s) = \dfrac{s^2 + 4s + 4}{s^4 + 5s^3 + 5s^2 + 9s + 7}$.

40. A stable circuit has the transfer function

$$H(s) = \frac{V_{out}(s)}{V_{in}(s)} = \frac{s^2 - 0.5s + 5}{s^2 + 0.5s + 5.7321}.$$

If the input to the circuit is $v_{in}(t) = \sqrt{2}\cos(2t + 45°)$ V, then the response has the form $A\cos(2t + \phi)$, where A and ϕ equal what values?

41. A stable active circuit has the transfer function

$$H(s) = \frac{V_{out}(s)}{V_{in}(s)} = \frac{s^2 + 1.5s}{s^2 + 3s + 12}.$$

If the input is $v_{in}(t) = 4\cos(2t + 45°)$ V, find the phase and magnitude of the steady-state response.

42. If the impulse response of the network in figure P15.42 is

$$v_{out}(t) = [2e^{-t} + 3\cos(2t) - 1.5\sin(2t)]u(t) \text{ V}:$$

(a) Compute the transfer function.

(b) Determine all poles and zeros, and draw the pole-zero plot.

(c) If there is no initial stored energy in the network, i.e., no initial conditions, determine the network response to $v_{in}(t) = \sin(t)u(t)$ V. Does your answer make sense? Why or why not?

(d) Determine the form of the response for large t when $v_{in}(t) = \cos(2t)u(t)$ V. Note that it is not necessary to compute the answer exactly. Your answer should be of the form $v_{out}(t)$ and $Af(t)$, where A represents some constant (not to be determined) and $f(t)$ is one of the functions in the Laplace transform table of chapter 13, possibly multiplied by t. Does your answer remain finite as $t \to \infty$? If it does, the network is said to have a stable response. If not, the network has an unstable response.

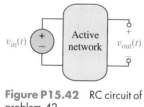

Figure P15.42 RC circuit of problem 42.

43. For the circuit shown in figure P15.43, the op amp is assumed to be ideal.

(a) Find the transfer function $H(s)$.

(b) If $v_C(0^-) = 0$ and $v_{in}(t) = \delta(t) + \delta(t - 1)$ V, find $v_{out}(t)$ for $t \geq 0$ by the Laplace transform method.

(c) If the impulses are applied to the circuit every second, i.e., if $v_{in}(t) = [\delta(t) + \delta(t - 1) + \delta(t - 2) + \ldots]$

V, find the output $v_{out}(t)$ for $n < t < (n + 1)$ when the integer n becomes very large (i.e., after the circuit reaches a steady state). Sketch the waveform.

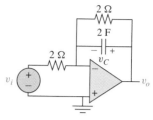

Figure P15.43 Ideal op amp circuit of problem 43.

Step and Impulse Response

44. The pole-zero plot of a transfer function of an active circuit is sketched in figure P15.44. Determine a bounded input that will make the zero-state response unbounded (increasing with time). (Answer: Any signal having a term of the form $A\cos(t - \phi)$ for $A \neq 0$.)

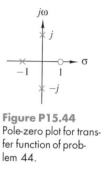

Figure P15.44
Pole-zero plot for transfer function of problem 44.

45. A time-shift differentiator circuit has the property that its zero-state output is always equal to the derivative of its input, properly shifted in time. Hence, its impulse response is $h(t) = \delta'(t - T)$, a shifted version of the derivative of the delta function. Suppose $T = 1$ and the input to the circuit is given by $f(t)$, as sketched in figure P15.45. Determine the zero-state response $y(t)$ of the circuit at $t = 2.5$ sec. (Answer: 0.)

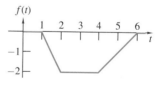

Figure P15.45

46. (a) If the zero-state step response of the circuit shown in figure P15.46 is $i_{out}(t) = e^{-2t}u(t)$ A, determine the zero-state response to $i_{in}(t) = \delta(t - 1)$ A. (Answer: $-2e^{-2t}u(t) + e^{-2t}\delta(t)$.)

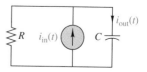

Figure P15.46

(b) Determine the zero-state response for $t \geq 0$ to a ramp input $r(t)$. Recall that the ramp function is the integral of the step function.

47. For the circuit in figure P15.47, if $v_C(0^-) = 3$ V and the response of the circuit for $t > 0$ is $v_C(t) = 0.5u(t) + 2.5e^{-2t}u(t)$ V, find the value of A (in amps).

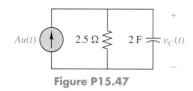

Figure P15.47

48. The Laplace transform of the response of a relaxed circuit to a ramp input $v_1(t) = r(t-1)$ V is

$$Y(s) = \frac{e^{-2s}}{s^3}.$$

Find the response of this same relaxed circuit to an input $v_2(t) = 5\delta(t-2)$ V.

49. Consider the circuit of figure P15.49.
   (a) Find the transfer function $H(s)$ and the impulse response $h(t)$.
   (b) Tabulate $h(t)$ for $t = 0, 1, 2, \ldots, 8$. Use this table to do the next three parts, assuming that $v_C(0^-) = 0$.
   (c) If $v_{in}(t) = 32\delta(t-2)$ V, find $v_{out}(5)$.
   (d) If $v_{in}(t) = 64\delta(t-2) + 16\delta(t-5) + 2\delta(t-8)$ V, find $v_{out}(8^+)$.
   (e) If the input is a rectangular pulse of duration 4 msec and amplitude 2,000 V, and if it is applied at $t = 2$ sec, find the value of $v_{out}(t)$ at $t = 5$ sec. (Note that in the RC circuit, the pulse width is much smaller than the time constant.)
   Check: All answers are close to integers.

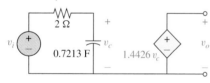

Figure P15.49   Circuit for problem 49.

50. Consider the circuit of figure P15.50.

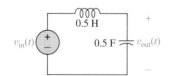

Figure P15.50   LC circuit for problem 50.

   (a) Find the transfer function $H(s)$ and the impulse response $h(t)$.
   (b) Find the response due to a unit step function input, by two different methods:
      (1) Inverse Laplace transformation of $V_{out}(s)$.
      (2) Integration of $h(t)$.
   (c) Find the response due to a unit ramp function input, by two different methods:

(1) Inverse Laplace transformation of $V_{out}(s)$.
(2) Integration of the unit step response found in part (b).
(d) Find the maximum value of $|v_{out}(t)|$ due to an input $v_{in}(t) = 200u(t)$ V. Does the answer surprise you?

51. The circuit of figure P15.51 is a scaled small-signal (i.e., signals with small amplitudes) equivalent of a common-emitter bipolar junction transistor amplifier. Assume that all initial conditions are zero.
   (a) Find the transfer function representing the forward voltage gain, i.e., $H(s) = V_{out}(s)/V_{in}(s)$. (*Hint*: Use nodal analysis and Cramer's rule. Round any poles to nearest integer.) *Check*: one zero, z_1, and two poles, p_1 and p_2, satisfying $300 \leq z_1 \leq 350$, $-5 \leq p_1 \leq -8$, and $-160 \geq p_2 \geq -100$.
   (b) Find the step response, i.e., the circuit response to an input $v_{in}(t) = u(t)$ V.
   (c) Find the impulse response.
   (d) Determine the relationship between the time domain step response and the time domain impulse response.
   (e) If the input is $v_{in}(t) = 3\delta(t-1)$ V, determine the response.
   (f) If the input is $v_{in}(t) = 3u(t-1)$ V, determine the response.
   Remark: Assuming zero initial conditions, let $v_{in}(t)$ produce a response $v_{out}(t)$. Again assuming that all ICs are zero, if for all T, the response to every $v_{in}(t-T)$ is $v_{out}(t-T)$, then the circuit is said to be *time invariant*.

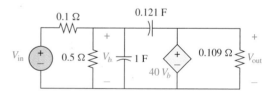

Figure P15.51   Small-signal model of a bipolar transistor.

52. Consider the circuit in figure P15.52.

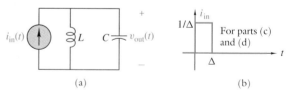

(a)   (b)

Figure P15.52   (a) Circuit and (b) pulse input for problem 52.

   (a) Find the transfer function $H(s)$ and the impulse response $h(t)$.
   (b) Find the step response.
   (c) Find $v_{out}(t)$ due to a rectangular pulse input as shown, with $\Delta = 2/(LC)$. (*Hint*: Make use of the result of part (b).)
   (d) Find $v_{out}(t)$ in part (c) for the case $\Delta \to 0$. (*Note*: The result should agree with part (a), since $i_{in}(t)$ approaches $\delta(t)$.)

53. Consider the circuit shown in figure P15.53. Suppose $i_L(0^-) = 31.5$ A, $v_C(0^-) = 7$ V, and $i_{in}(t) = 7\sin(t)u(t)$ A. The desired output is $v_C(t)$.

(a) Determine the transfer function $H(s)$ relating $I_{in}(s)$ to $V_C(s)$.

(b) Determine the poles and zeros of the transfer function. Is the circuit stable?

(c) Determine the zero-input response.

(d) Determine the zero-state (zero-IC) response.

(e) Determine the complete response.

(f) Determine the steady-state response.

(g) Determine the transient response.

(h) If $i_{in}(t)$ is changed to $i_{in}(t) = 7u(t)$ A, determine the zero-input, zero-state, complete, transient, and steady-state responses.

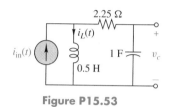

Figure P15.53

54. The point of this problem is to force you to think, rather than do arithmetic. If you must do a lot of arithmetic, sit back and reflect; there is an easier way. Consider again the circuit of problem 53. Somehow, make use of your answers to problem 53.

(a) If all initial conditions are zero, determine the response to $i_{in}(t) = u(t)$ A.

(b) If $i_{in}(t) = 0$ for all t, but $v_C(0^-) = -7$ V and $i_L(0^-) = -31.5$ A, determine $v_C(t)$.

(c) If $i_{in}(t) = 0$ for all t, but $v_C(0^-) = 1$ V and $i_L(0^-) = 4.5$ A, determine $v_C(t)$.

(d) If all ICs are zero, determine the response to $i_{in}(t) = 7\delta(t)$ A and then $i_{in}(t) = \delta(t)$ A. (*Hint:* What is the relationship between the Laplace transforms of the step and delta functions? What does multiplication by s mean? How can you reflect this in the time domain?)

(e) Repeat part (d) for $i_{in}(t) = \cos(t)u(t)$ A.

(f) If $i_{in}(t) = \cos(t)u(t)$ A, $v_C(0^-) = -7$ V, and $i_L(0^-) = -31.5$ A, determine $v_C(t)$.

(g) Find the response to $i_{in}(t) = 28r(t)$ A, a ramp function, assuming that all ICs are zero.

Additional Laplace Transform Properties

55. The output, $v_{out}(t)$, of a particular circuit is engineered to track different reference signals, $v_{ref}(t)$. After the circuit overheated, it was found that

$$V_{ref}(s) - V_{out}(s) = \frac{1}{s} - \frac{1}{s}\ln\left(\frac{s+1}{s+0.2865}\right).$$

Find the difference, $v_{ref}(t) - v_{out}(t)$, for large t.

56. The Laplace transform of a signal $f(t-1)u(t-1)$ is

$$e^{-s}\frac{(5s-1)(4s-5)(6s-2)}{s(2s+1)(3s+2)(5s+4)}.$$

Find the value of $f(0^+)$.

57. (a) If $f(t)$ and $g(t)$ are as sketched in figure P15.57a, determine $F(s)$ and $G(s)$. (b) Repeat part (a) for figure P15.57b.

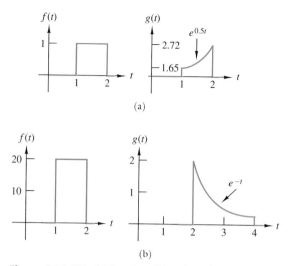

(a)

(b)

Figure P15.57 (a) Functions $f(t)$ and $g(t)$ for problem 57, part (a). (b) Functions $f(t)$ and $g(t)$ for problem 57, part (b).

58. (a) If $F(s) = [e^{-s}/s]\ln[(2.7183s+1)/(s+3)]$, determine $f(1^+)$.

(b) If $F(s) = \{(e^{-2s}/s)\ln[(s+3)/(0.135335s+1)]\}$, determine $f(2^+)$.

59. For this problem you are to use the properties of Laplace transforms as listed in chapter 13. The time function $f(t)$ under consideration has zero value for negative t. If $F(s) = (s+2)/[(s+1)(s+3)]$, find the Laplace transforms of:

(a) $g_1(t) = f(4t)$,

(b) $g_2(t) = \exp(-4t)f(4t-8)$.

60. Given the following functions, $F_i(s)$ find $f_i(0^+)$ and $f(\infty)$ for those cases where the initial-value and final-value theorems are applicable (a root-finding program is useful in this endeavor):

$$F_1(s) = \frac{3s+5}{s^2+4s+3}.$$

$$F_2(s) = \frac{2s^3+7s^2+s+4}{s(s^3+s^2+7s+6)}.$$

$$F_3(s) = \frac{s^2+4s+3}{s(s^4+5s^3+5s^2+4s+4)}.$$

61. The capacitor voltages, $V_C(s)$, of various networks have the following Laplace transforms:

(a) $\dfrac{(7s-1)^2}{s^2(7s+2)}.$

(b) $\dfrac{(2s-7)(1-e^{-6s})}{s(s+1)(1-e^{-3s})}.$

(c) $\dfrac{(s-1)(s+1)}{s(s^2+1)}.$

(d) $\dfrac{(8s-1)}{(2s+1)^2}$

$$\left[1+\frac{2s-1}{s+1}+\left(\frac{2s-1}{s+1}\right)^2+\left(\frac{2s-1}{s+1}\right)^3\right].$$

Use only the initial- and final-value theorems to determine $v_C(0^+)$ and $v_C(\infty)$ for each of the preceding. If the final-value theorem is not applicable, explain why.

62. The capacitor voltages, $V_C(s)$, of various networks have the following Laplace transforms:

(a) $\dfrac{s-1}{s(2s+9)}$.

(b) $\dfrac{(3s+1)(2-e^{-7s})}{s^2(1-e^{-2s})}$.

(c) $4s^{-1} - \dfrac{\exp\left(-\sqrt{s}\right)}{s(s+0.5)}$.

(d) $2s^{-1}e^{-\frac{2}{s}}$.

Use only the initial- and final-value theorems to determine $v_C(0^+)$ and $v_C(\infty)$ for each of the preceding. If the final-value theorem is not applicable, explain why.

Additional Problems

63. Consider figure P15.63. Suppose $i_L(0^-) = 0$. Let $i_{in}(t)$ be as given.
(a) Determine $V_L(s)$ and $v_L(t)$. Sketch $v_L(t)$.
(b) Use the shift property to determine

$$\mathcal{L}[v_L(t-1)u(t-1)].$$

(c) Use the time scale property to determine

$$\mathcal{L}[v_L(2t)u(2t)].$$

(d) In a careful manner, use appropriate Laplace transform properties to determine $\mathcal{L}[v_L(2t-1)u(2t-1)]$.
(e) Verify your answer to (d) by drawing

$$v_L(2t-1)u(2t-1)$$

and taking the Laplace transform.

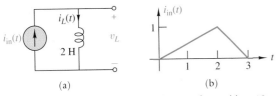

Figure P15.63 (a) Circuit and (b) input for problem 63.

64. Consider figure P15.64. Suppose $v_{out}(0^-) = 0$. Let $i_{in}(t)$ be as given.
(a) Determine $V_{out}(s)$ and $v_{out}(t)$. Sketch $v_{out}(t)$.
(b) Use the shift property to determine

$$\mathcal{L}[v_{out}(t-1)u(t-1)].$$

(c) Use the time scale property to determine

$$\mathcal{L}[v_{out}(2t)u(2t)].$$

(d) In a careful manner, use the Laplace transform properties to determine $\mathcal{L}[v_{out}(2t-1)u(2t-1)]$.
(e) Verify your answer to (d) by drawing $v_{out}(2t - 1)u(2t - 1)$ and taking the Laplace transform.

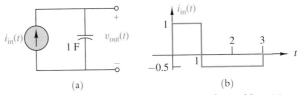

Figure P15.64 (a) Circuit and (b) input for problem 64.

65. The circuit in figure P15.65 contains a switch that moves from position a to position b at $t = 2$ sec. Suppose $i_{in}(t) = 5u(t)$ A.
(a) Determine the zero-state response of $v_{out}(t)$ for $t \geq 0$.
(b) Determine the zero-input response if each capacitor has an initial voltage of 10 V at $t = 0^-$.
(c) Make a rough sketch of the zero-state response.

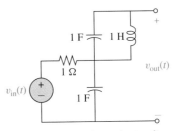

Figure P15.65 Switched circuit for problem 65.

66. Consider the circuit of figure P15.66. Compute the transfer function. (It should be first order.) Determine an expression for the natural response of the circuit. Suppose that at $t = 0^-$ each capacitor voltage is 1 V and the inductor current is zero. Determine the zero-input response. Observe that this response has a steady-state part. This phenomenon illustrates why equations 3.7 require more than just a stable transfer function as an underlying assumption. It is necessary that the transfer function model all the dynamics of the circuit, which is not the case in the circuit of figure P15.66.

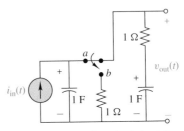

Figure P15.66 Circuit having first-order transfer function.

67. (a) Show the pole-zero plot of each of the following transfer functions, and determine whether the system (circuit) is stable:

$$H_1(s) = \dfrac{s-3}{s^2 + 4s + 3},$$

$$H_2(s) = \dfrac{s^2 + s + 1}{s^3 + s^2 + 6s + 5},$$

$$H_3(s) = \dfrac{s^2 + s + 1}{s^3 + s^2 + 5s + 6}.$$

(b) An impedance function $Z(s)$ of a certain linear two-terminal network has poles at $s = -2$ and $-6 \pm j8$, with zeros at $s = -1$, -2, and -4. It is known that $Z(0) = 16\ \Omega$.

(1) Write $Z(s)$ as the ratio of two polynomials of s.
(2) If the network has zero initial conditions, and a voltage source of $5u(t)$ V is applied to the two terminals, find the current $i(t)$ flowing into the network for $t \geq 0$.

68. Consider the network of figure P15.68. With $v_a(t) = 4u(t)$ V and $v_C(0^-) = 4$ V, a decomposition of the complete response $v_{out}(t)$ has been found to be:

$$\text{zero-state response} = (2 - e^{-t})u(t) \text{ V.}$$
$$\text{zero-input response} = 2e^{-t}u(t) \text{ V.}$$

If $v_a(t)$ is changed to $2u(t)$ V and $v_C(0^-)$ changed to 1 V, determine the complete response $v_{out}(t)$ for $t \geq 0$.

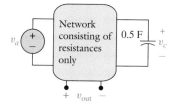

Figure P15.68

69. Suppose a dc motor is modeled as per figure 15.28 of the text. The motor parameters are $R_a = 25\ \Omega$, $J_m = 0.005$ kg-m^2, $k = 0.02$ N-m/A, $B = e^{-5}$ N-m-s, and $L_a = 100$ mH. Calculate the steady-state angular velocity of the motor and the steady-state armature current for the input $10u(t)$ V.

70. Consider the dc motor modeled in figure 15.28.
(a) Using the same parameters as in problem 69 with $v_s(t) = 5u(t)$ V, determine the armature current and the angular velocity as a function of time.
(b) Change R_a to $50\ \Omega$, and recompute the armature current as a function of time.
(c) What effect does R_a have on the steady-state speed for a step input of a fixed amplitude?
(d) What effect does R_a have on the rate at which the armature approaches a steady-state for a step input of fixed amplitude?

71. (a) For the dc motor of figure 15.28, suppose $k = 0.05$ N-m/A and $R_a = 50\ \Omega$. Plot the steady-state current as a function of B as B ranges from 0 to ∞.
(b) Using $R_a = 25\ \Omega$ and $B = e^{-4}$ N-m-s, plot ω_{ss} as k ranges from 0 to ∞. Recall that increasing k increases the torque per ampere of armature current. Explain why increasing k reduces ω_{ss}.

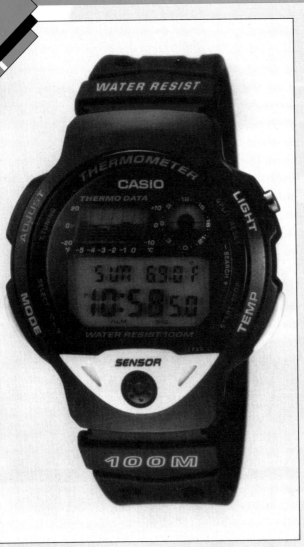

CHAPTER 16

"This Casio watch measures temperature as averaged over a short time interval."

CHAPTER OUTLINE

1. Introduction
2. Definition, Basic Properties, and Simple Examples
3. Convolution and Laplace Transforms
4. Time Domain Derivation of the Convolution Integral for Linear Time-Invariant Circuits
5. Circuit Response Computations Using Convolution
6. Convolution Properties Revisited
7. Graphical Convolution and Circuit Response Computation
8. Convolution Algebra
   Summary
   Terms and Concepts
   Problems

Time Domain Circuit Response Computations: The Convolution Method

AVERAGING BY A FINITE TIME INTEGRATOR CIRCUIT

OFTEN, a task calls for the computation of the average of some quantity. For example, one may need to know the average value of some light intensity over the last five minutes, the average value of the temperature in a room over the last hour, or the average value of a voltage over the last 50-millisecond time interval. If such averages are updated continuously, the average is said to be a *running average*. The idea is that the readout of the device which averages these quantities always produces an updated value valid over a specified prior time interval. If one uses a voltage to represent the value of the quantity to be averaged, then it is possible to build a circuit that produces the required average. This is done by observing that an average value of a continuous time variable is simply the integral of the variable over the proper time interval, divided by the length of the time interval. A device that integrates a variable over the last, say, T seconds is called a *finite time integrator*. As an application of the ideas of this chapter, we will look at a finite time integrator circuit and how it can be used to compute the average value of a quantity. ◆

CHAPTER OBJECTIVES

- Introduce the notion of the convolution of two signals.
- Using the notion of convolution, develop a technique of time domain circuit response computation that is the counterpart of the transfer function approach in the frequency domain, presented in chapter 15. In particular, we seek to show that the convolution of an input excitation with the impulse response of a circuit or system produces the zero-state circuit or system response.
- Develop objective 2 from two angles: first, from a strict time domain viewpoint, and second, as a formal theorem relating convolution to the transfer function approach.
- Develop graphical and analytical methods—in particular, an algebra—for evaluating the convolution of two signals.
- Illustrate various properties of convolution that are pertinent to block diagrams of series, parallel, and cascade interconnections of circuits or systems modeled by transfer functions.

1. INTRODUCTION

At the beginning of chapter 13, we claimed that computation of circuit responses could take place in either the time domain or the frequency domain. Yet, except for the solution of some very elementary differential equations, the circuit response computations presented thus far have relied almost exclusively on the Laplace transform technique, with some usage of the phasor method. The role of this chapter is to develop and explore the time domain counterpart of the Laplace transform method. This is achieved by introducing the notion of the convolution of two signals to produce a third signal and then showing that the time domain convolution of the input excitation with the impulse response of a circuit will produce the zero-state circuit response. Recalling that the impulse response is the inverse transform of the transfer function enables us to justify the diagram first given in figure 13.3 and, for convenience, presented in figure 16.1 again.

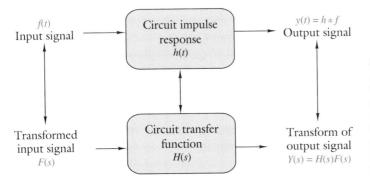

Figure 16.1 Diagram showing the symmetric relationship of time and frequency domain analysis. The upper part of the diagram asserts that the convolution, $h * f$, of the input signal $f(t)$ with the circuit impulse response $h(t)$ produces a third signal, $y(t)$. This third signal is the zero-state response, which equals $\mathcal{L}^{-1}[H(s)F(s)]$.

Justifying the diagram of figure 16.1 begins by introducing the notion of convolution. **Convolution** is an integration process between two functions to procure a third new function. Mathematically, the convolution of two functions $h(t)$ and $f(t)$, denoted by $h * f$ or $h(t) * f(t)$, is a third function,

$$y(t) = \int_{-\infty}^{\infty} h(t - \tau) f(\tau) d\tau. \tag{16.1}$$

The second step in verifying the diagram is proving that the convolution of the input signal with the impulse response produces the zero-state circuit response. This may be proved in two distinct ways. One approach is to work strictly in the time domain and construct the actual zero-state response from the impulse response and an arbitrary input excitation. The second approach is to prove that $\mathcal{L}^{-1}[H(s)F(s)]$ is the convolution of the signal $f(t)$ with the function $h(t)$. Because of the Laplace transform development of the last three chapters, this direction seems the most painless and will be taken up in section 3, after introducing the basic ideas of convolution in section 2. Section 4 will sketch the rudiments of the first avenue, wherein one constructs the zero-state response working strictly in the time domain. Section 5 will consider various properties of convolution. The properties lend themselves to different structures for designing interconnected circuits. For example, there are parallel and cascade structures that are important for designing filters or large interconnected systems. Section 6 will look at the graphical method of convolution, and section 7 will describe a convolution algebra, which yields a harvest of shortcuts for evaluating the convolution of certain types of signals.

Before closing the introduction, we would like to consider the question; *Why is convolution important?* One of the reasons is that it allows engineers to directly model the input-output behavior of circuits and general systems in the time domain, just as transfer functions model circuit behavior in the frequency domain. As mentioned, a circuit's zero-state response equals the convolution of the input signal with the circuit impulse response.

Often, circuit diagrams are unavailable or even lost. How would one generate a circuit model for analysis? One way is to display the impulse response on a CRT and approximate $h(t)$ by some interpolation function. Figure 16.2 illustrates a rectangular approximation. This process of constructing the impulse response of an unknown circuit or system is called *system identification* and is a vibrant area of current research. By storing the measured impulse response data as a table in a computer, one can numerically compute the zero-state responses of the circuit or system to any arbitrary input signal. This simulation process lets an engineer investigate a circuit's behavior off-line. For example, simulating a circuit destined for use in a hazardous environment provides a cost-effective means for evaluating the performance of the circuit in that environment. It also helps determine design improvements prior to constructing and testing prototypes.

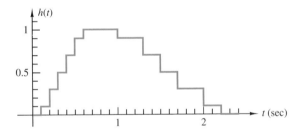

Figure 16.2 A rectangular approximation to a hypothetical impulse response obtained from measured data on an unknown circuit or system.

A third important reason for studying convolution in the context of circuits lies with a deficiency of the one-sided Laplace transform: Function segments that are nonzero for $t < 0$ are ignored by the transform technique. Hence, time domain convolution offers a direct means of computing the circuit response when signals are nonzero over the entire time interval, $-\infty < t < \infty$.

2. DEFINITION, BASIC PROPERTIES, AND SIMPLE EXAMPLES

As mentioned, convolution is an integral operation between two functions to produce a third; i.e., the convolution $y = h*f$ of two functions h and f produces a third function, y. One might expect that the convolution $h * f$ equals the convolution $f * h$, i.e., that the operation of convolution **commutes**. In fact, this is the case. To emphasize this property, we restate the mathematical definition of convolution in its more general form:

CONVOLUTION

The convolution of two signals $f(t)$ and $h(t)$ produces a third signal, $y(t)$, defined according to the formula

$$y(t) = \int_{-\infty}^{\infty} h(t - \tau) f(\tau) d\tau = \int_{-\infty}^{\infty} h(\tau) f(t - \tau) d\tau, \qquad (16.2)$$

which is well defined, provided that the integral exists. This formula emphasizes the property that convolution is a commutative operation, i.e., $h * f = f * h$.

The equivalence expressed in equation 16.2 comes about in a straightforward manner by a change of variable, i.e., $\tau' = t - \tau$. In addition to being commutative, convolution is **associative**, i.e.,

$$h * (f * g) = (h * f) * g. \qquad (16.3)$$

Exercise. Verify equations 16.2 and 16.3.

Exercise. Verify the **distributive property** of convolution, i.e., $h * (f + g) = h * f + h * g$. (*Hint*: Integration is distributive.)

In section 5, we will revisit the commutative, associative, and distributive properties and interpret them physically for circuits and systems. Some simple examples serve to demonstrate the actual calculation process. But first recall the **sifting property** (equation 13.15) of the delta function: If $h(t)$ is continuous at $t = T$, then

$$h(T) = \int_{T^-}^{T^+} h(\tau)\delta(T - \tau)d\tau.$$

◆ EXAMPLE 16.1

Determine the convolution of $h(t) = \delta(t - T)$ with $f(t) = \sin(2\pi t + \theta)$.

SOLUTION

By the definition of equation 16.1 and the sifting property of the delta function,

$$y(t) = h * f = \int_{-\infty}^{\infty} \delta(t - T - \tau)\sin(2\pi\tau + \theta)d\tau = \sin(2\pi(t - T) + \theta).$$

In the preceding example, the convolution of the sine function with the shifted delta produces a shifted sine function; the sifting property of the delta function causes the sine function to shift in time. The next example reinforces this observation.

◆ EXAMPLE 16.2

Determine the convolution of $h(t) = -2\delta(t - 3)$ with the function $f(t)$ sketched in figure 16.3.

SOLUTION

From the sifting property of the delta function (equation 13.15),

$$y(t) = h(t) * f(t) = \int_{-\infty}^{\infty} -2\delta(\tau - 3)f(t - \tau)d\tau = -2f(t - 3).$$

In other words, $f(t)$ is shifted to the right by three units and multiplied by -2, as illustrated in figure 16.4.

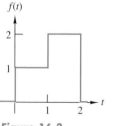

Figure 16.3 Function $f(t)$ for example 16.2.

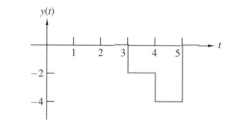

Figure 16.4 The resulting convolution, $y = h * f$, of example 16.2.

Exercise. Show that $\delta(t - T) * f(t) = f(t - T)$ for all T. In other words, the convolution of a function $f(t)$ with a delta function shifted in time by T produces an exact replica of $f(t)$ shifted in time by T. This property is particularly useful for the convolution of geometric shapes with impulses.

EXAMPLE 16.3

Find $y(t) = h * f$ when $h(t) = u(t)$ and $f(t) = u(t + 1) - u(t - 1)$. For convenience, $h(t)$, $f(t)$, and the solution, $y(t) = h * f$, are plotted in figure 16.5.

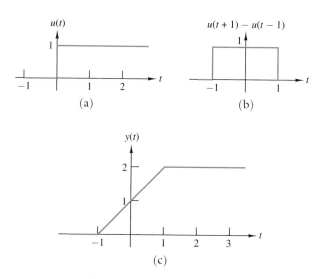

Figure 16.5 (a) $h(t) = u(t)$, the step function. (b) $f(t) = u(t + 1) - u(t - 1)$. (c) Resulting $y(t)$ for the convolution $h * f$.

SOLUTION

From the definition of the convolution integral, equation 16.2,

$$y(t) = h * f = \int_{-\infty}^{\infty} h(t - \tau)f(\tau)d\tau = \int_{-\infty}^{\infty} u(t - \tau)[u(\tau + 1) - u(\tau - 1)]d\tau.$$

Observing that $u(\tau + 1) - u(\tau - 1)$ is nonzero only for $-1 \le \tau \le 1$ yields

$$y(t) = \int_{-1}^{1} u(t - \tau)d\tau. \qquad (16.4)$$

In addition, the integrand in equation 16.4 is nonzero only when $\tau \le t$. This suggests three regions of consideration: $t < -1$, $-1 \le t \le 1$, and $t > 1$.

Case 1: $t < -1$. Here, $u(t - \tau) = 0$, since τ is restricted to the region $[-1, 1]$. Therefore, $y(t) = 0$ for $t < -1$.

Case 2: $-1 \le t \le 1$. Here, $u(t - \tau) = 1$, provided that $\tau \le t$. Therefore, for $-1 \le t \le 1$,

$$y(t) = \int_{-1}^{1} u(t - \tau)d\tau = \int_{-1}^{t} d\tau = t + 1.$$

Case 3: $1 \le t$. Here, $u(t - \tau) = 1$ for all τ in $[-1, 1]$, and evaluation of the integral over $[-1, 1]$ yields $y(t) = 2$ for $t \ge 1$.

Combining these three cases leads to the plot shown in figure 16.5c.

Exercise. Show by direct calculation that the convolution $u(t) * u(t)$ is the ramp function, $r(t)$.

◆ EXAMPLE 16.4

Compute the convolution $y(t)$ of the signals given by $h(t) = u(-t)$ and $f(t) = e^{-t}[u(t) - u(t - T)]$, which are sketched in figure 16.6, and plot $y(t)$ for $T = 1$.

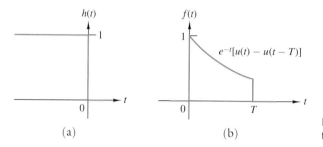

(a)   (b)

Figure 16.6 Functions $h(t)$ and $f(t)$ for the convolution of example 16.4.

SOLUTION

Step 1. *Apply definition and adjust limits of integration.* Applying the definition of equation 16.2 yields

$$y(t) = h * f = \int_{-\infty}^{\infty} h(t - \tau)f(\tau)d\tau = \int_{-\infty}^{\infty} u(\tau - t)e^{-\tau}[u(\tau) - u(\tau - T)]d\tau.$$

Since $u(\tau) - u(\tau - T)$ is nonzero only for $0 \le \tau \le T$, the lower and upper limits of integration become 0 and T, respectively:

$$y(t) = h * f = \int_0^T e^{-\tau}u(\tau - t)d\tau. \tag{16.5}$$

Step 2. *Determine the regions of t over which the integral is to be evaluated.* Based on the experience gained from example 16.3, the three regions are $t < 0$, $0 \le t \le \tau < T$, and $T \le t$.

Step 3. *Evaluate the convolution integral, equation 16.5, over the given regions.*

Case 1: $t < 0$. Here, $t < 0$ implies that $\tau - t \ge 0$ over $0 \le \tau \le T$. Hence, $u(\tau - t) = 1$ over $0 \le \tau \le T$, and

$$y(t) = \int_0^T e^{-\tau}d\tau = 1 - e^{-T}.$$

Case 2: $0 \le t < T$. For this case, $u(\tau - t)$ in equation 16.5 is nonzero only when $\tau \ge t$. Hence, in the region $0 \le t < T$, it must also be true that $0 \le t \le \tau < T$ for the integral of equation 16.5 to be nonzero. Thus, the lower and upper limits of integration with respect to the variable τ become t and T, respectively:

$$y(t) = \int_t^T e^{-\tau}u(\tau - t)d\tau = \left[-e^{-\tau}\right]_t^T = e^{-t} - e^{-T}.$$

Case 3: $t \ge T$. A simple calculation shows that $y(t) = 0$ in this region.

Step 4. Plot $y(t)$ for $T = 1$. Combining the results of step 3 with $T = 1$ implies that $y(t)$ has the graph sketched in figure 16.7.

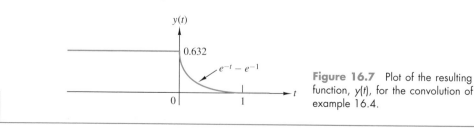

Figure 16.7 Plot of the resulting function, $y(t)$, for the convolution of example 16.4.

Exercise. Find the convolution, say, $y(t)$, of the signal $f(t) = e^{-at}u(t)$ with $h(t) = u(-t)$.
ANSWER: $a^{-1}u(-t) + a^{-1}e^{-at}u(t)$.

Exercise. Check the associativity property of convolution by computing $f_1 * f_2 * f_3$ when
(i) $f_1 = \delta(t - 1)$, $f_2 = u(t) - u(t - 2)$, and $f_3(t) = \delta(t - 3)$; and (ii) $f_1 = \delta(t - 1)$,
$f_2 = u(t) - u(t - 2)$, and $f_3(t) = \delta(t + 1)$.

3. CONVOLUTION AND LAPLACE TRANSFORMS

As claimed in the introduction, circuit analysis in the time domain by convolution and circuit analysis in the frequency domain by Laplace transformation are equivalent in terms of zero-state response calculations. The purpose of this section is to rigorize the equivalence between the time and frequency domain analysis methods by formally showing that $\mathcal{L}[h * f] = H(s)F(s)$. Section 4 presents a time domain development.

> **The Convolution Theorem.** Suppose a signal $h(t) = 0$ for $t < 0$. Suppose further that $f(t) = 0$ for $t < 0$. It follows that
>
> $$\mathcal{L}[h(t) * f(t)] = H(s)F(s), \tag{16.6}$$
>
> i.e., convolution in the time domain is equivalent to multiplication of transforms in the frequency domain.

Proof:

Step 1. Given equation 16.6, the first step is to write down the transform of $h * f$. Specifically,

$$\mathcal{L}[h(t) * f(t)] = \int_{0^-}^{\infty} \left(\int_{0^-}^{\infty} h(t - \tau)u(t - \tau)f(\tau)d\tau \right) e^{-st}dt. \tag{16.7}$$

A couple of points are in order: (i) The inner integral, surrounded by parentheses, represents the convolution $h(t) * f(t)$. (ii) The presence of the step function $u(t - \tau)$ is added as an aid to emphasize the fact that $h(t) = 0$ for $t < 0$.

Step 2. The goal at this point is to manipulate the integral of equation 16.7 into a form that can be identified as the product of the Laplace transforms of two functions, i.e., as $\mathrm{H(s)F(s)}$. *The only possible strategy is to interchange the order of integration and group appropriate terms.* Note that the $\mathrm{Re}[s]$ must be chosen sufficiently large to ensure the existence of the Laplace transforms of both $h(t)$ and $f(t)$. Under certain conditions that are typically met, it is possible to interchange the order of integration within a common domain of convergence of $H(s)$ and $F(s)$. Interchanging the order and regrouping the t-dependent terms inside a single parenthetical expression produces

$$\mathcal{L}[h(t) * f(t)] = \int_{0^-}^{\infty} f(\tau) \left(\int_{0^-}^{\infty} h(t - \tau)u(t - \tau)e^{-st}dt \right) d\tau. \tag{16.8}$$

Step 3. Observe that the interior integral, surrounded by parentheses, of equation 16.8 is simply the Laplace transform of a time-shifted $h(t)$, i.e., $\mathcal{L}[h(t - \tau)u(t - \tau)] = e^{-\tau s}H(s)$. Substituting $e^{-\tau s}H(s)$ for the interior integral leads to the desired equivalence:

$$\mathcal{L}[h(t) * f(t)] = \int_{0^-}^{\infty} f(\tau)H(s)e^{-s\tau}d\tau = H(s)\int_{0^-}^{\infty} f(\tau)e^{-s\tau}d\tau = H(s)F(s).$$

This theorem asserts the equivalence of convolution of one-sided signals with multiplication of their transforms in the s-domain. For our purposes, $h(t)$ assumes the role of the impulse response of our circuit and $f(t)$ the role of the input excitation. Accordingly, the convolution of the impulse response of a circuit or system with an input signal, a time domain computation, equals the inverse transform of the product of the respective Laplace transforms. In other words, the diagram of figure 16.1 is correct, as claimed under the conditions of the theorem.

The conditions of the theorem, however, are somewhat restrictive in terms of computing circuit responses strictly in the time domain. Specifically, it is the one-sided Laplace transform that does not recognize function segments over the negative real axis; hence, the condition that the input excitation $f(t) = 0$ for $t < 0$. This restriction does not lend itself to the computation of initial conditions and circuit responses due to input signals extending back in time to $t = -\infty$. In general, the convolution of an input excitation with a circuit's impulse response presupposes no such restriction, as some of the examples in section 2 illustrated. However, justification of the computation of zero-state responses due to input excitations extending back in time to $t = -\infty$ cannot be based on the convolution theorem of the one-sided Laplace transform. Instead, justification will depend on a time domain construction, as outlined in the next section.

4. Time Domain Derivation of the Convolution Integral for Linear Time-Invariant Circuits

As mentioned, a deficiency in the one-sided Laplace transform technique is its inability to deal with signals whose time dependence may extend back to $-\infty$. This section develops the zero-state system response as the convolution of a not necessarily one-sided input excitation with the impulse response of the circuit or system. Throughout the development, we will assume that the circuit or system under consideration is **linear**, i.e., is composed of linear circuit elements. This implies that the zero-state response of the circuit satisfies the conditions of linearity; i.e., if the zero-state response to the excitation $f_i(t)$ is $y_i(t)$ for $i = 1, 2$, then the zero-state response to the input excitation $a_1 f_1(t) + a_2 f_2(t)$ is $a_1 y_1(t) + a_2 y_2(t)$. In addition, we assume that the circuit or system is **time invariant**, i.e., that each circuit element is characterized by constant parameter values. Mathematically, this means that if $f(t)$ is the input to a circuit element and $y(t)$ the zero-state response to the circuit element, then $y(t - T)$ is the response to $f(t - T)$ for all T and all possible input signals $f(t)$. Intuitively speaking, time invariance means that, if we shift the input in time, then the associated zero-state response is shifted in time by a like amount. These properties underlie the development that follows.

Rectangular Approximations to Signals

Let us define a pulse of width Δ and height $1/\Delta$ as $\delta_\Delta(t)$. In particular,

$$\delta_\Delta(t) = \begin{cases} \frac{1}{\Delta}, & 0 < t < \Delta \\ 0 & \text{otherwise} \end{cases} \tag{16.9}$$

Figure 16.8 sketches $\delta_\Delta(t)$ for several Δ's.

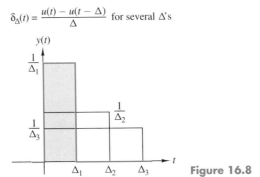

$\delta_\Delta(t) = \dfrac{u(t) - u(t - \Delta)}{\Delta}$ for several Δ's

Figure 16.8 $\delta_\Delta(t)$ for several Δ's.

Shifting the pulse of equation 16.9 yields $\delta_\Delta(t - k\Delta)$, as pictured in figure 16.9.

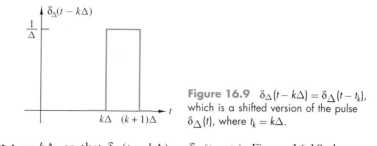

Figure 16.9 $\delta_\Delta(t - k\Delta) = \delta_\Delta(t - t_k)$, which is a shifted version of the pulse $\delta_\Delta(t)$, where $t_k = k\Delta$.

For convenience, let $t_k = k\Delta$, so that $\delta_\Delta(t - k\Delta) = \delta_\Delta(t - t_k)$. Figure 16.10 shows a rectangular approximation, $\hat{v}(t)$, to a continuous waveform $v(t)$.

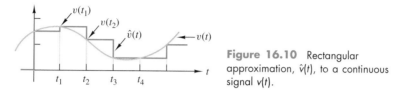

Figure 16.10 Rectangular approximation, $\hat{v}(t)$, to a continuous signal $v(t)$.

Expressing the rectangular approximation indicated in the figure analytically, using shifted versions of the pulse functions defined in equation 16.9, leads to the infinite summation

$$\hat{v}(t) = \sum_{k=-\infty}^{\infty} v(k\Delta)\delta_\Delta(t - k\Delta)\Delta = \sum_{k=-\infty}^{\infty} v(t_k)\delta_\Delta(t - t_k)\Delta. \qquad (16.10)$$

Hence, for sufficiently small Δ, it follows from equation 16.10 that

$$v(t) \approx \hat{v}(t) = \sum_{k=-\infty}^{\infty} v(t_k)\delta_\Delta(t - t_k)\Delta. \qquad (16.11)$$

One concludes that if $v(t)$ is continuous, then

$$v(t) = \lim_{\Delta \to 0} \sum_{k=-\infty}^{\infty} v(t_k)\delta_\Delta(t - t_k)\Delta = \int_{-\infty}^{\infty} v(\tau)\delta(t - \tau)d\tau, \qquad (16.12)$$

where we have interpreted the delta function as per equation 13.14, i.e.,

$$\delta(t) = \lim_{\Delta \to 0} \delta_\Delta(t).$$

Observe that the right-hand integral of equation 16.12 is precisely the convolution $v(t) * \delta(t)$.

Computation of Response for Linear Time-Invariant Systems

Suppose now that $h_\Delta(t)$ is the zero-state response of a linear time-invariant circuit to the pulse $\delta_\Delta(t)$. If the circuit's impulse response satisfies sufficient smoothness conditions, i.e., if it has sufficient continuous derivatives, then the circuit's impulse response is the limit of the $h_\Delta(t)$'s as Δ goes to zero. In particular,

$$h(t) = \lim_{\Delta \to 0} h_\Delta(t) = \lim_{\Delta \to 0} \int_{0-}^{\infty} h(t - \tau)\delta_\Delta(\tau)d\tau = \int_{0-}^{\infty} h(t - \tau)\delta(\tau)d\tau, \qquad (16.13)$$

where the fact that

$$h_\Delta(t) = \int_{0-}^{\infty} h(t - \tau)\delta_\Delta(\tau)d\tau$$

follows from the convolution theorem. Equation 16.13 is simply a prelude to the following time domain development of the fact that the zero-state response of an input to a linear time-invariant circuit is the convolution of the input with the impulse response.

Now, by the assumption of time invariance, $h_\Delta(t - k\Delta)$ is the zero-state response of a well-behaved linear time-invariant circuit to $\delta_\Delta(t - k\Delta)$. Suppose further that $y(t)$ is the zero-state response of the same circuit to $v(t)$. (See figure 16.11.)

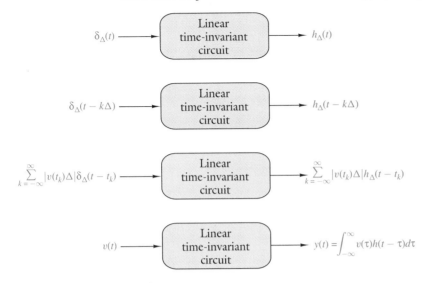

Figure 16.11 Zero-state responses to a particular linear time-invariant system, showing the framework of the derivation. Note that the bottom picture follows because as $\Delta \to 0$, $\Delta \to dt$, $t_k \to \tau$, and $\Sigma \to \int$.

It is now possible to use the approximation for $v(t)$ given in equation 16.12 to generate an approximation to $y(t)$ in terms of a summation of terms of the form $v(k\Delta)h_\Delta(t - k\Delta)\Delta$. Taking the limit as $\Delta \to 0$ will yield $y(t)$ as the convolution of $v(t)$ and $h(t)$.

To derive this, note that for each k, $v(k\Delta) = v(t_k)$ is a scalar. Hence, the zero-state response to $v(t_k)\delta_\Delta(t - t_k)\Delta$ is $v(t_k)h_\Delta(t - t_k)\Delta$. By the linearity assumption, which implies superposition, the zero-state circuit response to $\hat{v}(t)$, equation 16.11, is

$$\hat{y}(t) = \sum_{k=-\infty}^{\infty} h_\Delta(t - t_k)v(t_k)\Delta. \tag{16.14}$$

In the limit, as Δ approaches zero, t_k approaches a continuous variable τ and $\Delta \to d\tau$. Hence, if the impulse response is sufficiently smooth—i.e., if it has sufficient continuous derivatives—then

$$y(t) = \lim_{\Delta \to 0} \sum_{k=-\infty}^{\infty} v(k\Delta)h_\Delta(t - k\Delta)\Delta = \int_{-\infty}^{\infty} v(\tau)h(t - \tau)d\tau \tag{16.15}$$

Thus, we conclude that for a linear time-invariant circuit, the zero-state response $y(t)$ to an input excitation $v(t)$ is the convolution of the input $v(t)$ with the impulse response $h(t)$. We will refer to equation 16.15 as the **impulse response theorem**,[1] which says that *the zero-state response of a linear time-invariant circuit or system to a (possibly two-sided) input signal is the convolution of the input with the impulse response of the circuit.*

5. CIRCUIT RESPONSE COMPUTATIONS USING CONVOLUTION

This section contains a series of examples that illustrate the uses of the convolution approach in the computation of zero-state circuit responses.

EXAMPLE 16.5

Consider the RC circuit of figure 16.12, whose impulse response is

$$h(t) = e^{-t}u(t).$$

---

[1] The derivation of this result is, of course, not rigorous. A rigorous justification is given as theorem 4 of Sandberg's "Linear Maps and Impulse Responses," *IEEE Transactions on Circuits and Systems*, vol. 35, no. 2, February 1988, pp. 201–206.

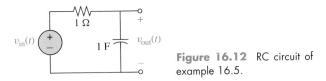

Figure 16.12 RC circuit of example 16.5.

If the input, $v_{in}(t) = e^{-at}$, has been applied for a long time (theoretically, from $t = -\infty$), determine $v_{out}(0)$ when $a = 0, 0.5$, and 2.

SOLUTION

Since the capacitor voltage, $v_{out}(t)$, is the convolution of the input with the impulse response,

$$v_{out}(0) = \int_{-\infty}^{\infty} e^{-(0-\tau)} u(0 - \tau) e^{-a\tau} d\tau.$$

Because $u(-t)$ is in the integrand, the upper limit of integration becomes 0. Hence,

$$v_{out}(0) = \int_{-\infty}^{0} e^{-(0-\tau)} e^{-a\tau} d\tau = \int_{-\infty}^{0} e^{(1-a)\tau} d\tau.$$

Case 1: If $a = 0$, then

$$v_{out}(0) = e^{\tau} \big]_{-\infty}^{0} = 1.$$

Case 2: If $a = 0.5$,

$$v_{out}(0) = \frac{e^{0.5\tau}}{0.5} \Big]_{-\infty}^{0} = 2.$$

Case 3: If $a = 2$,

$$v_{out}(0) = -e^{-\tau} \big]_{-\infty}^{0} = \infty.$$

It is interesting to observe that the input waveform is infinite at $t = -\infty$ when $a = 0.5$ or 2. Yet in one case the initial voltage is finite, and in the other it is infinite.

Some books define $H(s)$ in terms of the ratio of the forced response to an exponential input. This approach would give correct answers to the cases in which $a = 0$ and $a = 0.5$, but would give a wrong answer to the case in which $a = 2$. The erroneous solution proceeds as follows: $H(s) = 1/(s + 1)$, in which case $H(-2) = 1/(1 - 2) = -1$. This would then imply that $v_{out}(t) = H(-2)v_{in}(t) = -e^{-2t}$. Therefore, one would conclude erroneously that $v_{out}(0) = -1$.

 EXAMPLE 16.6

Reconsider the series RC circuit of figure 16.12, whose impulse response is $h(t) = e^{-t}u(t)$. Suppose here that the input excitation is $v_{in}(t) = e^{-a|t|}$, where $a \neq 1$ and $a > 0$. Determine the response $v_{out}(t)$ for all t.

SOLUTION

The solution proceeds by a direct computation of the convolution of the input with $h(t)$, where the input excitation may be written as

$$v_{in}(t) = \begin{cases} e^{at} u(-t) & \text{for } t < 0 \\ e^{-at} u(t) & \text{for } t \geq 0 \end{cases}.$$

Accordingly, for $t < 0$,

$$v_{out}(t) = h * v_{in} = \int_{-\infty}^{0} e^{-(t-\tau)} u(t - \tau) e^{a\tau} u(-\tau) d\tau = e^{-t} \int_{-\infty}^{t} e^{(1+a)\tau} d\tau. \tag{16.16}$$

Evaluating equation 16.16 leads to

$$v_{out}(t) = e^{-t} \left. \frac{e^{(1+a)\tau}}{1+a} \right]_{-\infty}^{t} = \frac{e^{at}}{1+a}, \tag{16.17}$$

which is valid for $t < 0$.

For the case when $t \geq 0$,

$$v_{out}(t) = h * v_{in} = \int_{-\infty}^{0} e^{-(t-\tau)} e^{a\tau} d\tau + \int_{0}^{t} e^{-(t-\tau)} e^{-a\tau} d\tau. \tag{16.18}$$

The first integral is simply equation 16.17 evaluated at $t = 0$ and multiplied by e^{-t}, i.e., $e^{-t}/(1+a)$. Hence, equation 16.18 reduces to

$$v_{out}(t) = \frac{e^{-t}}{a+1} + e^{-t} \left. \frac{e^{(1-a)\tau}}{1-a} \right]_{0}^{t} = \frac{e^{-t}}{a+1} + \frac{e^{-at}}{1-a} - \frac{e^{-t}}{1-a}. \tag{16.19}$$

A plot of the input waveform, $v_{in}(t)$, with $a = 2$, appears in figure 16.13a and of the impulse response in figure 16.13b. A plot of the response, equations 16.17 and 16.19, with $a = 2$, appears in figure 16.13c.

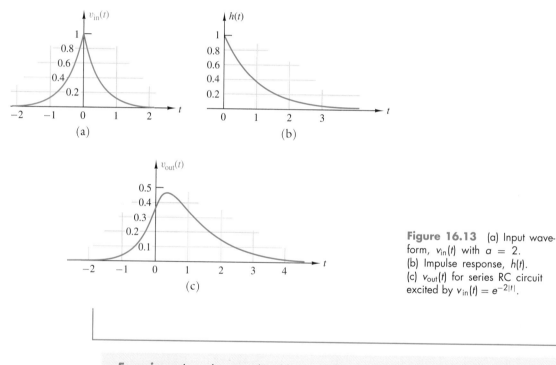

(a)

(b)

(c)

Figure 16.13 (a) Input waveform, $v_{in}(t)$ with $a = 2$. (b) Impulse response, $h(t)$. (c) $v_{out}(t)$ for series RC circuit excited by $v_{in}(t) = e^{-2|t|}$.

Exercise. An unknown relaxed linear system for which the input $v(t) = e^{-2t} u(t)$ produces a zero-state response $y(t) = 2e^{-t} u(t) + 3e^{-2t} u(t) - u(t)$. Determine the impulse response.

ANSWER: $h(t) = 4\delta(t) - 2u(t) + 2e^{-t} u(t)$.

◆ **EXAMPLE 16.7**

The goal of this example is to design a circuit that computes the running average of a voltage $v_{in}(t)$ over the interval $[t - T, t]$.

SOLUTION

To accomplish a running average, it is necessary to have a linear circuit with the property that

$$v_{out}(t) = \frac{1}{T} \int_{t-T}^{t} v_{in}(\tau) d\tau.$$

From our development of convolution, such a circuit must have an impulse response $h(t)$ satisfying the relationship

$$v_{\text{out}}(t) = \frac{1}{T}\int_{-\infty}^{\infty} h(t-\tau)v_{\text{in}}(\tau)d\tau = \frac{1}{T}\int_{t-T}^{t} v_{\text{in}}(\tau)d\tau.$$

Now, $h(t-\tau)$ must be a window function which captures that part of $v_{\text{in}}(t)$ over the interval $t - T \le \tau \le t$. Figure 16.14 depicts the proper form of $h(t-\tau)$ and $h(t)$.

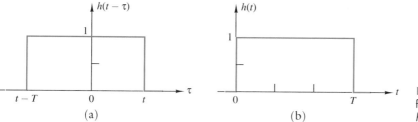

(a) (b)

Figure 16.14 (a) The windowing function $h(t-\tau)$. (b) The impulse response, $h(t)$.

The circuit design problem reduces to developing a circuit that integrates the function segment $v_{\text{in}}(\tau)$ over $t - T \le \tau \le t$. To achieve this integration, note that

$$\int_{-\infty}^{t} v_{\text{in}}(\tau)d\tau = \int_{-\infty}^{t-T} v_{\text{in}}(\tau)d\tau + \int_{t-T}^{t} v_{\text{in}}(\tau)d\tau.$$

This leads to

$$\int_{t-T}^{t} v_{\text{in}}(\tau)d\tau = \int_{-\infty}^{t} v_{\text{in}}(\tau)d\tau - \int_{-\infty}^{t-T} v_{\text{in}}(\tau)d\tau.$$

For the second integral on the right-hand side of this equation, let $\lambda = \tau - T$. Then $d\lambda = d\tau$, $\tau = \lambda - T$, and

$$\int_{-\infty}^{t-T} v_{\text{in}}(\tau)d\tau = \int_{-\infty}^{t} v_{\text{in}}(\lambda - T)d\lambda = \int_{-\infty}^{t} v_{\text{in}}(\tau - T)d\tau,$$

where the last equality follows because both λ and τ are dummy variables of integration. Hence,

$$\int_{t-T}^{t} v_{\text{in}}(\tau)d\tau = \int_{-\infty}^{t} [v_{\text{in}}(\tau) - v_{\text{in}}(\tau - T)]d\tau,$$

where $v_{\text{in}}(\tau - T)$ is simply a delayed replica of $v_{\text{in}}(\tau)$ and where, for practical reasons, we can replace the lower limit of $-\infty$ by t_0, the time the actual circuit turns on.

As a convenience, we will define a device called an **ideal delay** of T seconds, whose input is $v_{\text{in}}(t)$ and whose output is $v_{\text{in}}(t - T)$. Figure 16.15 shows the ideal delay as a device having infinite input impedance combined with a dependent voltage source whose output is a delayed version of the input. Such a device can be achieved numerically by storing the values of $v_{\text{in}}(t)$ in a digital computer or, for small T, by the use of an analog delay line.

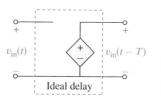

Figure 16.15 Depiction of a device called an ideal delay of T sec whose output is a delayed replica of the input.

All the pieces are now available; it is merely a matter of putting them together. Integration can occur using an inverting ideal op amp circuit having a capacitive feedback and resistive input. The input to this ideal op amp integrator can then scale and sum the voltages $-v_{\text{in}}(t)$ and $v_{\text{in}}(t - T)$ to produce the desired running average, $v_{\text{out}}(t)$, by setting $RC = T$. This will guarantee the correct scaling to achieve the desired average, since the transfer function of the integrator will be $1/(RCs)$. Figure 16.16 shows a circuit that will realize the desired running average.

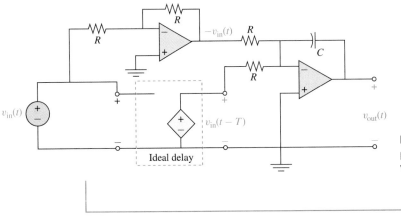

Figure 16.16 Op amp circuit that produces a running average of the input waveform, $v_{in}(t)$, provided that $RC = T$.

6. CONVOLUTION PROPERTIES REVISITED

From the perspective of the impulse response theorem, the properties of commutativity, associativity, and distributivity of convolution have important implications in terms of circuit and system configurations. For example, if $h_1(t)$ and $h_2(t)$ are the impulse responses of two systems, then **commutativity** says that $h_1 * h_2 = h_2 * h_1$. This means that the order of a cascade of circuits or systems is mathematically irrelevant, provided that there is no loading between the circuits. The idea is illustrated theoretically in figure 16.17.

Figure 16.17 Interchange of order of impulse responses in which (a) is equivalent to (b) follows from the commutativity of convolution.

A cascade operational amplifier realization of a transfer function

$$H(s) = \frac{10}{(s+1)(s+2)} \tag{16.20}$$

illustrates commutativity nicely. A designer may use either figure 16.18a or figure 16.18b to realize $H(s)$. Magnitude scaling, say, by $K_m = 10^4$, will yield more realistic resistor and capacitor values. In magnitude scaling, $R_{new} = K_m R_{old}$, $C_{new} = C_{old}/K_m$, and $L_{new} = K_m L_{old}$. The details of scaling are postponed until chapter 17.

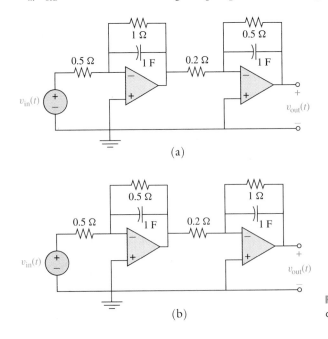

Figure 16.18 Interchange of order of cascaded circuits.

Theoretically speaking, the associative property, $h_1 * (h_2 * h_3) = (h_1 * h_2) * h_3$, means that multiple cascades of circuits or systems can be combined or realized in whatever order the designer chooses. This assumes that there is no loading between each of the circuits or systems. Operational amplifier circuits having gains of 1 between each circuit or system will provide isolation to eliminate loading effects. On the other hand, practical constraints may impose an order on the realization of a circuit that the mathematics of the associative property does not account for.

Finally, we consider the distributive property of convolution: $h_1 * (h_2 + h_3) = h_1 * h_2 + h_1 * h_3$. One interpretation of this property is that the superposition of the input signals h_2 and h_3 is valid. However, figure 16.19 presents two block diagrams, each with a different interpretation. Here, one sees two possible topologies for realizing a system.

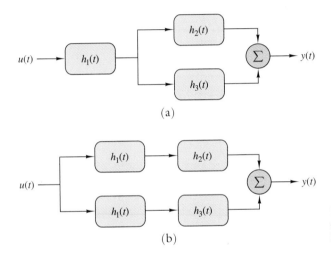

Figure 16.19 Two possible block diagram interpretations of the distributive law for convolution.

7. GRAPHICAL CONVOLUTION AND CIRCUIT RESPONSE COMPUTATION

The convolution integration formula, although explicit, has many layers of interpretation. **Graphical convolution** is a pen-and-pencil technique for figuring the convolution integral of simple, squarish waveforms. The technique often leads to a more penetrating insight into the convolution integral.

In words, computing the integral of equation 16.1 or 16.2 entails:

1. Taking $h(t)$ and forming $h(t - \tau)$.
2. Forming the product $h(t - \tau) f(\tau)$ for different regions of t, with each region determined by the specific problem.
3. Calculating the area beneath the curve $h(t - \tau) f(\tau)$ as a function of t over each t-region.

With regard to step 1, for each t, $h(t - \tau)$ is a shifted horizontal flip of $h(\tau)$: As t moves from $-\infty$ to ∞, the picture, $h(t - \tau)$, moves along the τ-axis from $\tau = -\infty$ to $\tau = \infty$; a simple illustration is $h(t - \tau) = u(t - \tau)$, which is sketched in figure 16.20. Part (a) of the figure pictures $u(\tau)$, part (b) pictures $u(-\tau)$, and part (c) pictures $u(t - \tau)$, which slides to the right, along the τ-axis, as t increases.

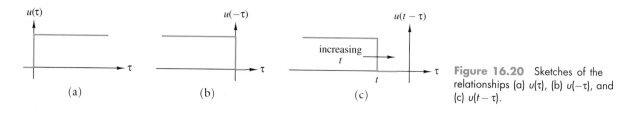

EXAMPLE 16.8

Graphically determine the convolution $y = h*f$ of the two waveforms $h(t)$ and $f(t)$ sketched in figure 16.21.

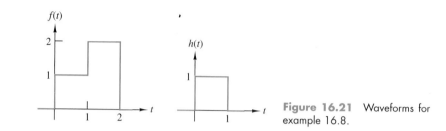

Figure 16.21 Waveforms for example 16.8.

SOLUTION

The graphical solution to this convolution depends on a partitioning of the time line into special segments over which the graphical convolution is easily done. There are five regions to consider: (1) $-\infty \leq t < 0$, (2) $0 \leq t < 1$, (3) $1 \leq t < 2$, (4) $2 \leq t < 3$, and (5) $3 \leq t$. The impetus for this partitioning becomes clear as we walk through the steps to the final answer.

Step 1. *Consider the region $-\infty \leq t < 0$.* Figure 16.22 shows $h(t-\tau)$ and $f(\tau)$ on the same τ-axis. Their product, $h(t-\tau)f(\tau)$, is clearly zero; hence,

$$y(t) = \int_{-\infty}^{\infty} h(t-\tau)f(\tau)d\tau = 0$$

in the first t-region.

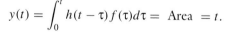

Figure 16.22 The functions $h(t-\tau)$ and $f(\tau)$ in the region $-\infty \leq t < 0$.

Step 2. *Consider the region $0 \leq t < 1$.* In figure 16.23a, the functions $h(t-\tau)$ and $f(\tau)$ are superimposed on the τ-axis. Figure 16.23b shows their product. The goal is to compute the area of the shaded region as a function of t, i.e., to compute $y(t)$. Clearly, for $0 \leq t < 1$,

$$y(t) = \int_{0}^{t} h(t-\tau)f(\tau)d\tau = \text{ Area } = t.$$

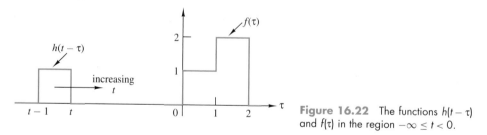

Figure 16.23 (a) The functions $h(t-\tau)$ and $f(\tau)$ in the region $0 \leq t < 1$. (b) Product of $h(t-\tau)$ and $f(\tau)$.

Step 3. *Now consider the region* $1 \le t < 2$. As in step 2, one looks at the functions $h(t - \tau)$ and $f(\tau)$ superimposed on the τ-axis, as per figure 16.24a. The product, given in figure 16.24b, suggests that

$$y(t) = \int_0^2 h(t - \tau) f(\tau) d\tau = \text{Area } 1 + \text{Area } 2 = t,$$

where Area 1 and Area 2 are as shown in the figure.

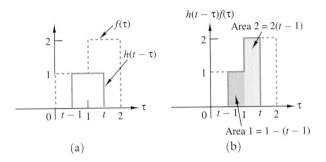

Figure 16.24 (a) The functions $h(t - \tau)$ and $f(\tau)$ in the region $1 \le t < 2$. (b) Product of $h(t - \tau)$ and $f(\tau)$.

Step 4. *Now consider the region* $2 \le t < 3$. Repeating the procedure of step 3 yields figure 16.25. From the figure, computation of the area implies that

$$y(t) = \text{Area} = 2[2 - (t - 1)] = 2(3 - t) = 6 - 2t.$$

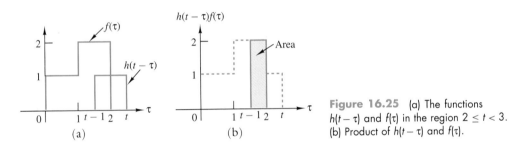

Figure 16.25 (a) The functions $h(t - \tau)$ and $f(\tau)$ in the region $2 \le t < 3$. (b) Product of $h(t - \tau)$ and $f(\tau)$.

Step 5. *Finally, consider the region* $3 \le t$. Clearly, $h(t - \tau) f(t) = 0$ in this region, in which case $y(t) = 0$ for $t \ge 3$.

Step 6. *Combine the foregoing calculations into a plotted waveform.* Figure 16.26 illustrates the resulting convolution.

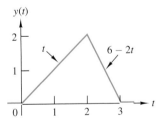

Figure 16.26 Plot of the resulting function, $y(t)$.

Another, more complex example will end our illustration of the graphical convolution technique.

◆ **EXAMPLE 16.9**

Compute the convolution $y(t)$ of the triangular pulse $h(t)$ with the square pulse $f(t)$ as sketched in figure 16.27.

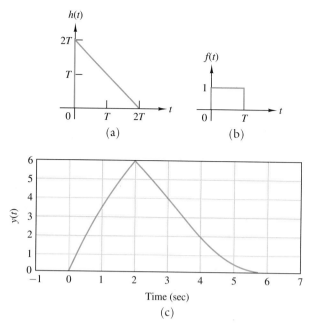

(a)

(b)

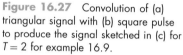

Figure 16.27 Convolution of (a) triangular signal with (b) square pulse to produce the signal sketched in (c) for $T = 2$ for example 16.9.

(c)

The goal, of course, is to graphically evaluate the convolution integral of equation 16.2 using the following steps:

1. Since τ is the variable of integration, draw $h(t - \tau)$ and $f(\tau)$ on the τ-axis.
2. Evaluate the product $h(t - \tau)f(\tau)$ for various regions of t.
3. Compute the area under each of the products for each region determined in step 2.

Step 1. Draw h($t-\tau$) *and* f(τ) *on the τ-axis for* t < 0 *and compute the area of their product.* Figure 16.28 shows $h(t - \tau)$ and $f(\tau)$ on the τ-axis. From the figure, it is clear that $h(t - \tau)f(\tau) = 0$ for $t < 0$; hence, $y(t) = 0$ in the first region.

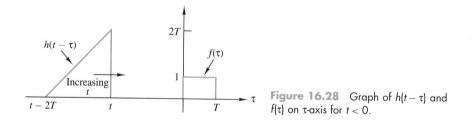

Figure 16.28 Graph of $h(t-\tau)$ and $f(\tau)$ on τ-axis for $t < 0$.

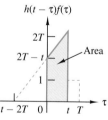

Figure 16.29 Graph of $h(t-\tau)f(\tau)$ on τ-axis for $0 \le t < T$.

Step 2. Consider the region $0 \le t < T$, *as illustrated in figure 16.29.* The shaded area of the figure is the difference between the area of the large triangle, defined as

$$\text{Area } A = 0.5(2T)^2,$$

and the area of the smaller triangle to the left of the vertical axis, defined as

$$\text{Area } B = 0.5(2T - t)^2.$$

Hence,

$$y(t) = \text{ Area } = \text{ Area } A - \text{ Area } B = 2Tt - 0.5t^2$$

for $0 \le t < T$. Alternatively, one may use the formula for the area of a trapezoid, i.e., the average height times the base, in which case one immediately obtains $0.5(2T + 2T - t)t = \text{Area} = 2Tt - 0.5t^2$.

Step 3. Now consider the region $T \le t < 2T$, *as depicted in figure 16.30.* In this figure, the shaded area, which determines $y(t)$, is again the difference of two triangular areas; specifically,

$$y(t) = 0.5[T - (t - 2T)]^2 - 0.5[2T - t]^2 = 2.5T^2 - Tt$$

for $T \le t < 2T$.

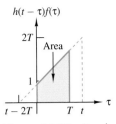

Figure 16.30 Graph of $h(t - \tau)f(\tau)$ on τ-axis for $T \le t < 2T$.

Step 4. Figure 16.31 shows the next region, $2T \le t < 3T$. Another straightforward calculation yields

$$y(t) = 0.5(3T - t)^2 = 4.5T^2 - 3Tt + 0.5t^2$$

for $2T \le t < 3T$.

Step 5. Consider the region $3T \le t$. Here, the product $h(t - \tau)f(\tau) = 0$, in which case $y(t) = 0$ for $t > 3T$.

In sum,

$$y(t) = \begin{cases} 0, & t < 0 \\[2mm] 2Tt - \dfrac{t^2}{2}, & 0 \le t < T \\[2mm] 2.5T^2 - Tt, & T \le t < 2T \\[2mm] 4.5T^2 - 3Tt + \dfrac{t^2}{2}, & 2T \le t < 3T \\[2mm] 0, & t \ge 3T \end{cases}$$

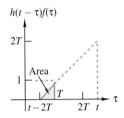

Figure 16.31 Graph of $h(t - \tau)f(\tau)$ on τ-axis for $2T \le t < 3T$.

A plot of $y(t)$ appears in figure 16.27c for $T = 2$.

Exercise. Repeat the calculations of the preceding example, except flip and shift $f(t)$ instead of $h(t)$. Here, one would look at $f(t - \tau)$ sliding through $h(\tau)$. The calculations should be easier and the result the same.

Sometimes the foregoing graphical techniques prove difficult to execute. Nevertheless, their understanding offers fundamental insight into the meaning of the convolution integral. A useful set of techniques for quickly evaluating convolution integrals arises from the properties of a convolution algebra, discussed in the next section.

8. CONVOLUTION ALGEBRA

A set of functions, together with operations called addition and multiplication, is called an *algebra*. The set of all functions that can be convolved with each other also constitutes an algebra with respect to the operations of addition ($+$) and convolution, denoted by $*$. This set, together with the two operations, will be called a **convolution algebra**. In this context, operations such as differentiation and integration are inverses of each other. For example, integrating a function and then differentiating the result returns the original function. Within the convolution algebra, the convolution $f * g$ is equivalent to the convolution of the integral of f with the derivative of g. The advantage here is that, by successive integrations and differentiations, it is often possible to reduce an apparently difficult convolution down to a more simple one.

For a set of functions to be an **algebra** with respect to $+$ and $*$, several arithmetic operations must hold. In particular, $+$ and $*$ must be both commutative and associative. The commutativity and associativity of $+$ is clear: $f + g = g + f$ and $f + (g + h) = (f + g) + h$. Similarly, the commutativity and associativity of $*$ is equally clear: $f * g = g * f$ and $f * (g * h) = (f * g) * h$. To be an algebra, the set of all functions that are mutually convolvable must also satisfy the distributive law, $f*(g+h) = f*g + f*h$. Besides these laws, algebras of functions must contain identity elements. For +, the zero function serves as the identity. For convolution, the delta function plays this role. The delta function is an identity element because of the sifting property,

$$f(t) * \delta(t) = \delta(t) * f(t) = f(t) \tag{16.21}$$

and

$$f(t) * \delta(t-a) = f(t-a). \tag{16.22}$$

For our purposes, the interesting aspects of a convolution algebra of functions rests with the interrelationship of convolution, differentiation, and integration. To map out this kinship, let

$$f^{(-1)}(t) \equiv \int_{-\infty}^{t} f(\tau)d\tau \tag{16.23}$$

and

$$h^{(1)}(t) = \frac{dh(t)}{dt}. \tag{16.24}$$

Then it can be shown that

$$f * h = f^{(1)} * h^{(-1)} \tag{16.25}$$

if $f(-\infty) = 0$ and $h^{(-1)}$ exists. The constraint that $f(-\infty) = 0$ simply means that the derivative of $f(\cdot)$ is zero at $t = -\infty$, and the constraint that $h^{(-1)}$ exists means that the integral of $h(\cdot)$ has finite area over the semiinfinite interval $(-\infty, t]$ for all finite t. Similarly,

$$f * h = f^{(-1)} * h^{(1)} \tag{16.26}$$

if $h(-\infty) = 0$ and $f^{(-1)}$ exists.

◆ **EXAMPLE 16.10**

Find the convolution $y = f * h$ for $f(t) = u(t) + u(t-1) - 2u(t-2)$ and $h(t) = u(t) - u(t-1)$, as sketched in figure 16.21. This example presents an easier solution than the graphical method of example 16.8.

SOLUTION

The goal is to use equation 16.25 to evaluate the convolution, i.e., $f * h = f^{(1)} * h^{(-1)}$, where $f^{(1)}$ and $h^{(-1)}$ are sketched in figure 16.32.

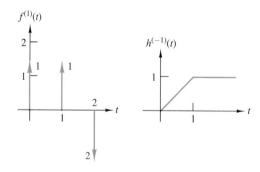

Figure 16.32 The derivative of $f(t) = u(t) + u(t-1) - 2u(t-2)$ and the integral of $h(t) = u(t) - u(t-1)$, as given example 16.10.

Since $f'(t) = \delta(t) + \delta(t-1) - 2\delta(t-2)$, a sum of impulse functions, the sifting property of the impulse function implies that

$$y(t) = f'(t) * h^{(-1)}(t) = h^{(-1)}(t) + h^{(-1)}(t-1) - 2h^{(-1)}(t-2). \tag{16.27}$$

With the picture of $h^{(-1)}(t)$ given in figure 16.32, the right side of equation 16.27 interprets as a graphical sum of (shifted) pictures of $h^{(-1)}(t)$, as illustrated in figure 16.33.

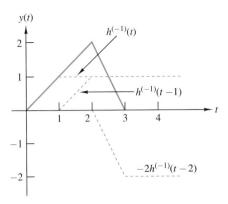

Figure 16.33 $y(t)$ equal to superposition of shifted replicas of $h^{(-1)}(t)$.

Another example reiterates the techniques of equation 16.25. Besides the reiteration, the salient pedagogical aim is to perform the integration of $h(t)$ given in figure 16.34b.

◆ **EXAMPLE 16.11**

Find the convolution $y(t) = f(t) * h(t)$, where $f(t)$ and $h(t)$ are sketched in figure 16.34.

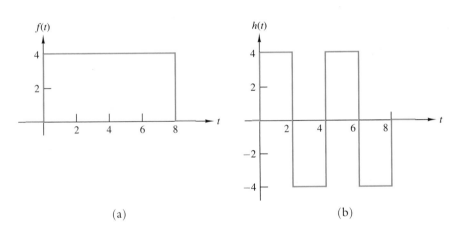

Figure 16.34 The functions (a) $f(t)$ and (b) $h(t)$ for the convolution of example 16.11.

SOLUTION

Again, the key to the solution is differentiation of $f(t)$ and integration of $h(t)$. From equation 16.25, as $f(-\infty) = 0$, $f * h = f^{(1)} * h^{(-1)}$. The derivative of $f(t)$ and the integral of $h(t)$ are obtained easily by inspection and are sketched in figure 16.35.

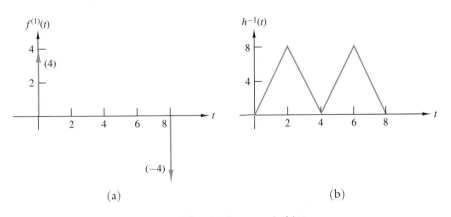

Figure 16.35 (a) The derivative of $f(t)$. (b) The integral of $h(t)$.

Since the delta function is the identity of convolution algebra (see equations 16.21 and 16.22), the convolution $y(t) = f * h$ is simply $4h^{(-1)}(t) - 4h^{(-1)}(t-8)$. A picture of $y(t)$ is sketched in figure 16.36.

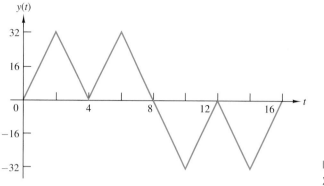

Figure 16.36 The waveform $y(t) = 4h^{(-1)}(t) - 4h^{(-1)}(t-8)$.

Equations 16.25 and 16.26, as illustrated in the preceding two examples, are special cases of more general formulas. Specifically, let $y = f * h$. Then

$$y^{(j+k)} = f^{(j)} * h^{(k)}, \tag{16.28}$$

where j and k are integers and the notation $f^{(j)}$ means the jth integral of f over $[-\infty, t]$ if $j < 0$ and the jth derivative if $j > 0$. Of course, $f^{(0)} = f$. An application of this formula to the special case where $j = -k$ with $j = 2$ is given in the following example.

◆ **EXAMPLE 16.12**

Find the convolution of $g(t) = \pi^2 \cos(\pi t)u(t)$ with $f(t) = r(t) - r(t-2)$, sketched in figure 16.37.

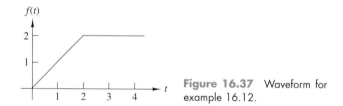

Figure 16.37 Waveform for example 16.12.

SOLUTION

Some preliminary thought suggests that evaluation of the convolution integral might proceed more smoothly via equation 16.28; i.e.,

$$f * g = f^{(2)} * g^{(-2)},$$

where

$$g^{(-1)}(t) = \pi^2 \int_{0^-}^{t} \cos(\pi q)dq = \pi \sin(\pi t)u(t)$$

and

$$g^{(-2)}(t) = \pi \int_{0^-}^{t} \sin(\pi q)dq = [1 - \cos(\pi t)]u(t).$$

Differentiating $f(t)$ twice leads to

$$f^{(2)}(t) = \delta(t) - \delta(t-2).$$

Hence,

$$f * g = f^{(2)} * g^{(-2)} = [1 - \cos(\pi t)]u(t) - [1 - \cos(\pi(t-2))]u(t-2).$$

But $[1 - \cos(\pi(t - 2))]u(t - 2)$ is just a right shift by one period of $[1 - \cos(\pi t)]u(t)$. Therefore,

$$f * g = f^{(2)} * g^{(-2)} = [1 - \cos(\pi t)]u(t)u(2 - t).$$

This function is sketched in figure 16.38.

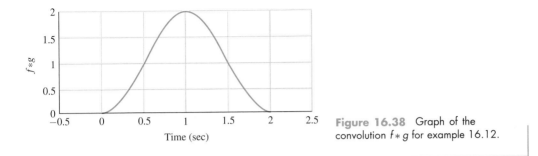

Figure 16.38 Graph of the convolution $f * g$ for example 16.12.

Convolution algebra and graphical convolution lend themselves to a second application of the convolution technique: the computation of circuit responses from a step approximation to a circuit impulse response. If a circuit schematic is lost, such an approximation could result from a CRT readout of the circuit impulse response measured in a laboratory. The following example illustrates this application.

EXAMPLE 16.13

Suppose the schematic diagram of a very old linear circuit is lost. However, the circuit impulse response is measured in the laboratory and approximated by the staircase waveform of figure 16.39. If the input to the circuit is $v_{in}(t) = 100u(t)$, compute the output voltage, $v_{out}(t)$, at $t = 0$ and at $t = 0.5$ sec.

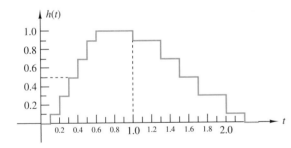

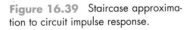

Figure 16.39 Staircase approximation to circuit impulse response.

SOLUTION

The objective is to convolve the input with the impulse response. The technique of convolution algebra wherein one differentiates the impulse response to obtain a sum of shifted impulse functions and integrates the input to obtain a ramp function seems to be the most straightforward for this calculation.

First, observe that

$$h(t) = 0.1u(t - 0.1) + 0.2u(t - 0.2) + 0.2u(t - 0.3) + 0.2u(t - 0.4)$$
$$+ 0.2u(t - 0.5) + 0.1u(t - 0.6) - 0.1u(t - 1) - 0.2u(t - 1.3)$$
$$- 0.2u(t - 1.5) - 0.2u(t - 1.7) - 0.2u(t - 2) - 0.1u(t - 2.2).$$

Hence,

$$h'(t) = 0.1\delta(t - 0.1) + 0.2\delta(t - 0.2) + 0.2\delta(t - 0.3) + 0.2\delta(t - 0.4)$$
$$+ 0.2\delta(t - 0.5) + 0.1\delta(t - 0.6) - 0.1\delta(t - 1) - 0.2\delta(t - 1.3)$$
$$- 0.2\delta(t - 1.5) - 0.2\delta(t - 1.7) - 0.2\delta(t - 2) - 0.1\delta(t - 2.2).$$

Now, since the integral of the input is $100r(t)$, we can compute the output voltage as

$$v_{\text{out}}(t) = 10r(t - 0.1) + 20r(t - 0.2) + 20r(t - 0.3) + 20r(t - 0.4)$$
$$+ 20r(t - 0.5) + 10r(t - 0.6) - 10r(t - 1) - 20r(t - 1.3)$$
$$- 20r(t - 1.5) - 20r(t - 1.7) - 20r(t - 2) - 10r(t - 2.2)$$

At $t = 0$, $v_{\text{out}}(0) = 0$ and at $t = 0.5$, $v_{\text{out}}(0.5) = 4 + 6 + 4 + 2 = 16$.

Of course, it is also possible to obtain the solution as easily in this case by the graphical method. Simply flip the $v_{\text{in}}(t)$ curve, and slide it through the $h(t)$ curve. The area under the product curve is simply the sum of the rectangular areas, which are easy to compute.

Exercise. Compute $v_{\text{out}}(t)$ of example 16.13 at $t = 1$ and $t = 1.5$ sec.

SUMMARY

This chapter has introduced the notion of the convolution of two signals. The convolution can be evaluated analytically (by direct computation of the convolution integral) or graphically. Often, by applying the properties of convolution algebra, it is possible to implement shortcuts for calculating the convolution of two signals, resulting in a simplification of the analytical calculation or a simplification of the graphical calculation.

Using the notion of convolution, the chapter developed a technique of computing circuit responses in the time domain. This technique is the direct counterpart of computing the transfer function in the frequency domain, the approach, presented in chapter 15. Using the convolution approach, one can compute the zero-state response of circuits excited by sig-

nals that extend back in time to $-\infty$, something not directly possible with the one-sided Laplace transform. However, for one-sided signals—which constitute the great majority of signals that are relevant to circuit analysis—the convolution and Laplace transform approaches are completely equivalent, as demonstrated by the convolution theorem. The chapter gave an example of the design of a circuit to compute a running average. In addition, it presented an application of the convolution technique to the computation of circuit responses for circuits whose impulse response was approximated on a CRT. Future courses will expand the seeds planted in this chapter. For example, convolution is pertinent to an understanding of radar techniques, commonly used to identify speeding motorists.

TERMS AND CONCEPTS

Algebra: a set of functions with respect to two operations, $+$ and $*$, satisfying the commutative, associative, and distributive laws. In addition, there must be an identity with respect to each operation. For $+$, the zero function serves as the identity. For convolution, the delta function plays this role. The delta function is an identity element because of its sifting property.

Associativity: for convolution, the property that $h * (f * g) = (h * f) * g$.

Commutativity: for convolution, the property that $h * f = f * h$.

Convolution: an integration process between two functions to produce a third, new function in accordance with equation 16.1.

Convolution algebra: The algebra of functions with respect to the operations of addition ($+$) and convolution.

Distributivity: for convolution, the property that $h*(f+g) = h * f + h * g$.

Graphical convolution (flip-and-shift method): a pen-and-

pencil technique for figuring the convolution integral of simple, squarish waveforms.

Ideal delay of T seconds: waveform with input $f(t)$ and output $f(t - T)$, a delayed replica of $f(t)$.

Impulse response theorem: theorem which states that the zero-state response of a linear time-invariant circuit or system to a (possibly two-sided) input signal is the convolution of the input with the impulse response of the circuit.

Linearity: property whereby, if the zero-state response to the excitation $f_i(t)$ is $y_i(t)$ for $i = 1, 2$, then the zero-state response to the input excitation $a_1 f_1(t) + a_2 f_2(t)$ is $a_1 y_1(t) + a_2 y_2(t)$.

Sifting property: the property of a delta function to simplify an integral, i.e., $f(T) = \int_{-\infty}^{+\infty} f(t)\delta(t - T)dt$, provided that $f(T)$ is continuous at T.

Time invariance: property such that, if $f(t)$ is the input to a circuit element and $y(t)$ is the zero-state response to the circuit element, then $y(t - T)$ is the response to $f(t - T)$ for all T and all possible input signals $f(t)$.

PROBLEMS

1. (a) A particular active circuit has the transfer function

$$H(s) = \frac{-0.2}{s + 0.2}.$$

Suppose the input to the circuit is $v(t)$, shown in figure P16.1. Using convolution techniques, determine the response $y(t)$.

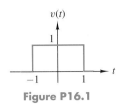

Figure P16.1

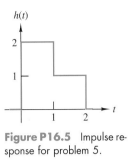

Figure P16.5 Impulse response for problem 5.

(b) Realize the transfer function $H(s)$ as a leaky integrator circuit in which the smallest resistor is $10\ k\Omega$.

2. The impulse response of a particular circuit is measured approximately on an oscilloscope, as illustrated in figure P16.2. If the input to the circuit is $v(t) = u(t) - u(t-1)$, determine the response $y(t)$.

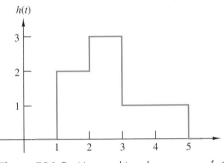

Figure P16.2 Measured impulse response of circuit.

3. (a) Recall that the convolution integral is evaluated from $-\infty$ to ∞. If $f(t) = e^{at}u(-t)$, $a > 0$, and $h(t) = Ku(t)$, then determine $y(t) = h(t) * f(t)$.
(b) Repeat part (a) if $f(t) = Ku(-t)$ and $h(t) = e^{-at}u(t)$.

4. (a) The input $f(t)$ and the impulse response $h(t)$ of a particular circuit are sketched in figure P16.4. Using one of the convolution methods, determine $y(t)$.
(b) Let $x(t)$ be an arbitrary continuous function. Determine an integral expression for the response of the circuit having the preceding impulse response $h(t)$ to the input $x(t)$. (*Hint*: Your answer should represent the running area under $x(t)$ over the interval $[t-4, t]$.)

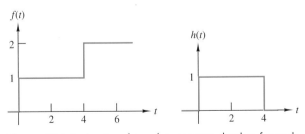

Figure P16.4 Input and impulse response sketches for problem 4.

5. A circuit has input $f(t) = \delta(t) - \delta(t-1)$ and impulse response $h(t)$, as sketched in figure P16.5. Determine the response $y(t)$ by graphical convolution.

6. (a) Suppose $v_{in}(t) = u(-t) + 2e^{-2t}u(t)$. Determine the output voltage of the circuit in figure P16.6.
(b) Repeat part (a) for $v_{in}(t) = te^{-a|t|}$ for $a \neq 0$ with $a > 0$. Note that this input is defined for all time.

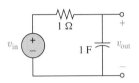

Figure P16.6 Series RC circuit for problem 6.

7. Suppose $f_1(t) = 2u(t)$, $f_2(t) = e^{-2t}u(t)$, and $f_3(t) = \delta(t-2)$. Determine each of the indicated convolutions by evaluating the convolution of the two given time functions, where, as usual, $*$ stands for convolution:
(a) $f_4(t) = f_2(t) * f_3(t)$.
(b) $f_5(t) = f_1(t) * f_2(t)$.
(c) $f_6(t) = f_2(t) * f_2(t)$.
(d) $f_7(t) = f_1(t) * f_1(t)$.

8. Suppose $f_1(t) = Ku(t)$, $f_2(t) = e^{-at}u(t)$, $f_3(t) = e^{at}u(-t)$, $a > 0$, and $f_4(t) = \delta(t-4)$. Determine each of the indicated convolutions by evaluating the convolution of the two given time functions.
(a) $f_5(t) = f_2(t) * f_4(t)$, and sketch $f_5(t)$ for $-\infty < t < \infty$.
(b) $f_6(t) = f_1(t) * f_1(t)$, and sketch $f_6(t)$ for $-\infty < t < \infty$.
(c) $f_7(t) = f_1(t) * f_2(t)$, and sketch $f_7(t)$ for $-\infty < t < \infty$.
(d) $f_8(t) = f_1(t) * f_3(t)$, and sketch $f_8(t)$ for $-\infty < t < \infty$.

9. (a) If $f_1(t) = 4u(t)$ and $f_2(t)$ is as sketched in figure P16.9a, determine $f_3(t) = f_1(t) * f_2(t)$ by evaluating the defining integral for the convolution of two time functions. Sketch $f_3(t)$.
(b) Repeat part (a) for $f_2(t)$ given in figure P16.9b.

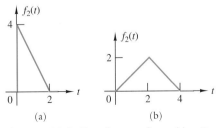

Figure P16.9 Time functions for problem 9.

10. Solve this problem with the aid of the graphical interpretation of convolution, i.e., the flip–shift–multiply–find area method. Let $f_1(t) = [\delta(t) + \delta(t-4)]$, and let $f_2(t)$ and $f_3(t)$ be as given in figure P16.10. Determine
  (a) $f_3(t) = f_1(t) * f_2(t)$.
  (b) $f_4(t) = f_2(t) * f_2(t)$.
  (c) $f_5(t) = f_2(t) * f_3(t)$.

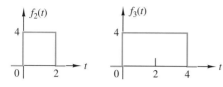

Figure P16.10 Functions $f_2(t)$ and $f_3(t)$ for problem 10.

11. Solve each of the following problems with the aid of the graphical interpretation of convolution, i.e., the flip-shift-multiply-find area method.
  (a) With $f_1(t)$ and $f_2(t)$ as given in figure P16.11a, find $f_3(t) = f_1(t) * f_2(t)$, and plot $f_3(t)$ vs. t.
  (b) With $f_1(t)$ and $f_2(t)$ as given in figure P16.11b, find $f_3(t) = f_1(t) * f_2(t)$, and plot $f_3(t)$ vs. t.
  (c) With $f_1(t)$ and $f_2(t)$ as given in figure P16.11c, find $f_3(t) = f_1(t) * f_2(t)$, and plot $f_3(t)$ vs. t.
  (d) Check your answers to (b) and (c) by using Laplace transform techniques, i.e., $f_3(t) = \mathcal{L}^{-1}[F_1(s)F_2(s)]$.

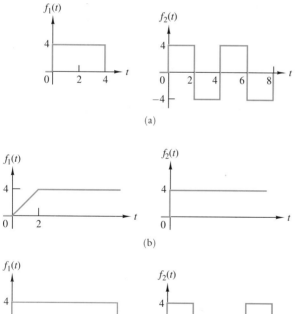

(a)

(b)

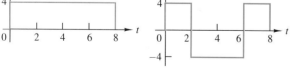

(c)

Figure P16.11 (a) Signals for part (a) of problem 11. (b) Signals for part (b) of problem 11. (c) Signals for part (c) of problem 11.

12. Redo problem 16.11 with $f_2(t)$ in parts (a), (b), and (c) changed to the signal shown in figure P16.12.

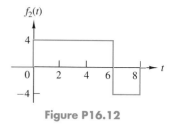

Figure P16.12

13. Determine the convolution of the signals $p(t)$ and $q(t)$ shown in figure P16.13.

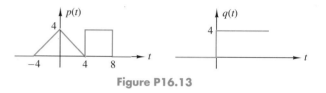

Figure P16.13

14. Given that $h(t) = 2e^{-t}u(t)$, $e(t) = 5u(t)$, and $y(t) = h(t) * e(t)$, find and plot $y(t)$.

15. Given that $h(t) = 2e^{-t}u(t)$, $e(t) = 5u(t)$, and $y(t) = h(t) * e(t)$, find $y(t)$.

16. The schematic diagram of a very old linear circuit is lost. The impulse response, $h(t)$, is measured in the laboratory and approximated by the staircase waveform of figure P16.16. Based on the available information, solve each of the following problems:
  (a) If the input is $v_{in}(t) = 100u(t)$, find the output $v_{out}(t)$ at $t = 0, 0.5, 1$, and 1.5 sec by the convolution method.
  (b) If the input is $100tu(t)$, find the output at $t = 1$ by the method of your choice.

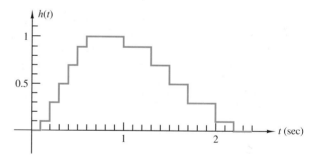

Figure P16.16 Measured impulse response for problem 16.

17. Redo problem 16.16 with the inputs changed as follows:

For part (a), let $v_{in}(t) = 100[u(t) - u(t - 0.2)]$.
For part (b), let $v_{in}(t)$ be the signal sketched in figure P16.17.

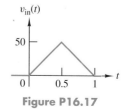

Figure P16.17

18. Use convolution algebra techniques to determine the convolution $f(t) * g(t) = y(t)$ for $f(t)$ and $g(t)$ as sketched in figure P16.18. In each case, plot the resulting $y(t)$. *Remark:* The key to determining the convolution is in choosing which function to take derivatives of and which functions to integrate.

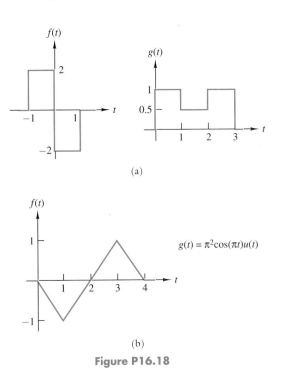

(a)

(b)

$g(t) = \pi^2 \cos(\pi t) u(t)$

Figure P16.18

19. A crude approximation to the impulse response of a relaxed active RLC circuit is given by $h(t)$, as sketched in figure P16.19a. The response of the relaxed circuit to an input signal is the convolution of the impulse response with the input signal. If the input signal is $v(t)$, as sketched in figure P16.19b, determine the value of the response $y(t)$ at $t = 0.5, 1.5, 2.5, 3.5,$ and 4.5 sec, using the techniques of the convolution algebra. Plot the response for $0 \leq t \leq 2$ sec.

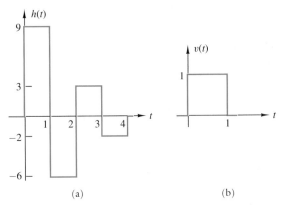

(a) (b)

Figure P16.19 (a) Crude approximation to impulse response for problem 19. (b) Input signal $v(t)$ to circuit.

20. Figure P16.20 gives two configurations for representing an interconnection of active circuits. The function inside each box is the impulse response of the circuit. Using any of the convolution techniques you have learned, determine and sketch the overall impulse response of each configuration.

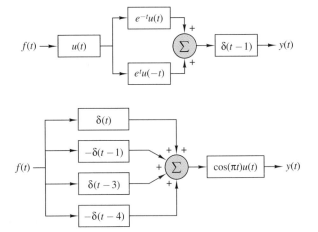

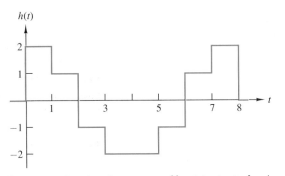

Figure P16.20 Two interconnections of active networks.

21. The impulse response of a certain relaxed futuristic circuit is given by the approximation in figure P16.21 taken from a CRT readout.

(a) Without doing a convolution or frequency domain analysis, determine the step response of the circuit.

(b) If the input to the circuit is $v(t) = e^t u(-t)$, use convolution algebra techniques to determine the response at $t = 7.5, 6.5, 5.5,$ and 0.5 sec.

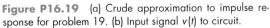

Figure P16.21 Impulse response of futuristic circuit of problem 21.

22. Use the convolution integral, and only the convolution integral, to determine the convolution of the input signal $v(t)$ with the impulse response signal $h(t)$ when:

(a) $v(t) = e^{-0.5t} u(t)$ and $h(t) = e^{-0.5t} u(t)$.

(b) $v(t) = e^{-0.5t} u(t)$ and $h(t) = e^{-t} u(t)$.

(c) $v(t) = u(t-1) - u(t-3)$ and $h(t) = 2e^{-2t} u(t) u(2-t)$. (Sketch your resulting waveform.)

(d) $v(t) = 2\delta(t) - 2\delta(t-1) + \delta(t-2)$ and $h(t)$ is as in figure P16.22. (Sketch your resulting waveform.)

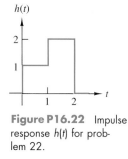

Figure P16.22 Impulse response $h(t)$ for problem 22.

Remark: Take particular note of what happens to figure P16.22 after convolution. (*Hint*: The convolution of a figure with a shifted delta function produces a shifted figure.)

23. Recall that the convolution of an input, denoted $v(t)$, with the impulse response, denoted $h(t)$, of a circuit yields the output response. Determine the step response and the zero-state response for the indicated waveforms in figure P16.23a and b.

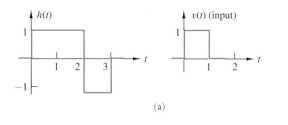

(a)

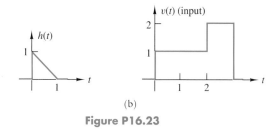

(b)

Figure P16.23

24. Consider the op amp circuit in figure P16.24.

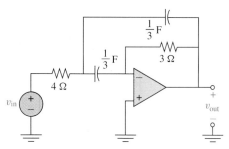

Figure P16.24 Ideal op amp circuit for problem 24

(a) Determine the transfer function $H(s)$. *Check*: Poles are at $s = -0.5, -1.5$.
(b) If the input is $v_{in}(t) = [u(t) - u(t - T)]$ for $T > 0$, determine $v_{out}(t)$, using Laplace transform methods. (*Hint*: Use the shift theorem.)
(c) Determine the impulse response $h(t)$.
(d) Analytically (not graphically) evaluate the convolution integral to determine the response $v_{out}(t)$ for the input $v_{in}(t) = u(t) - u(t - T)$.

(e) Do your answers to (b) and (d) coincide? Why or why not? Explain.

25. Consider the circuit sketched in figure P16.25a.
(a) Determine the transfer function $H(s)$.
(b) Determine the impulse response $h(t)$.
(c) Use only convolution algebra techniques to determine the response $v_{out}(t)$ for the input $i_{in}(t)$ sketched in figure 16.25b.
(d) Plot your answer.

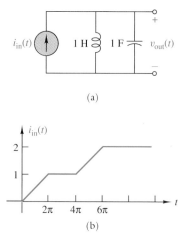

Figure P16.25 (a) LC circuit and (b) input waveform for problem 25.

26. Consider the diagram of figure P16.26a, in which $v_{in}(t) = (1 - e^{-t} - e^{-2t})u(t)$ and the zero-state response is $v_{out}(t) = (1 - 2e^{-t} - te^{-t} + e^{-2t})u(t)$.
(a) Determine the transfer function $H(s)$ and then a simple RC circuit that is represented by this transfer function. *Check*: There is only one pole.
(b) Determine the impulse response.
(c) If the input is changed to the one sketched in figure P16.26b, use only convolution algebra techniques to determine the zero-state response $v_{out}(t)$. (*Hint*: What is the double integral of the impulse response?)

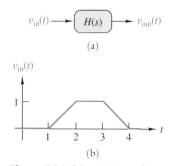

Figure P16.26 (a) General circuit or system with (b) input waveform $v_{in}(t)$.

27. Solve this problem by the use of convolution algebra. Given $f_1(t) = u(t) - u(t - 1)$, $f_2(t) = 10\cos(2\pi t)u(t)$, and $f_3(t)$ and $f_4(t)$ in figure P16.27, determine:
(a) $f_1(t) * f_2(t)$.
(b) $f_3(t) * f_4(t)$.

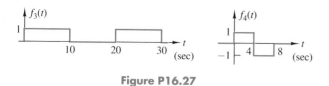

Figure P16.27

28. Consider the circuit of figure P16.28.
   (a) Find the transfer function $H(s) = V_o(s)/V_i(s)$ and the impulse response $h(t)$.
   (b) If the network is initially at rest, and $v_{in}(t) = u(t) - u(t - \pi)$, find $v_{out}(t)$ by the convolution method and sketch the waveform. You may use any method to do the convolution.
   (c) Repeat part (a) with the 2-Ω resistance replaced by a 0.5-H inductance.

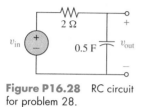

Figure P16.28 RC circuit for problem 28.

29. Consider figure P16.29, which contains an LC circuit and a rectangular input waveform.
   (a) Find the transfer function $H(s)$ and the impulse response $h(t)$.
   (b) Find the step response.
   (c) Find $v_{out}(t)$ due to the rectangular pulse input in figure P16.29b. (*Hint*: Make use of the result of part (b). Sketch the output waveform if $T = 2\pi\sqrt{(LC)}$.)
   (d) Find $v_{out}(t)$ in part (c) for the case $T \approx 0$. The result should agree with part (a), since $v_{in}(t)$ approaches $\delta(t)$.

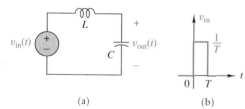

(a)   (b)

Figure P16.29 (a) LC circuit and (b) input waveform for problem 29.

30. The circuit of figure P16.30 is initially at rest. The input $v_{in}(t) = \delta(t) + \delta(t-1) + \delta(t-2) + \dots$, is a periodic impulse train.
   (a) Find the impulse response $h(t)$.
   (b) Find the exact solution of $v_o(t)$ for $0 < t < 1$.
   (c) Find the exact solution of $v_o(t)$ for $1 < t < 2$.
   (d) Find the exact solution of $v_o(t)$ for $4 < t < 5$. (*Note*: Terms having the same exponent may be summed.)
   (e) Find the solution of $v_o(t)$ for $n < t < (n + 1)$, as n becomes very large, and sketch the waveform. (This

is the so-called steady-state solution.) (*Hint*: Make use of the geometric series.)

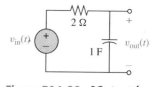

Figure P16.30 RC circuit for computing periodic solution in problem 30.

31. An op amp with finite gain has the equivalent circuit shown in figure P16.31a. Both input terminals may be "floating," or one of them may be grounded. For each circuit shown in figure P16.31b:
   (a) Find $H(s)$.
   (b) Find $v_{out}(t)$ for $t > 0$ if $v_{in}(t) = 0.1e^{-t}$ V and $v_C(0^-) = 0$. Which circuit behaves approximately as an integrator?

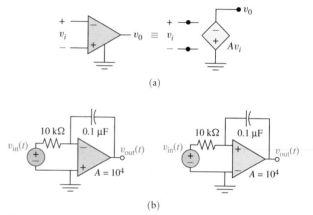

Figure P16.31 (a) Simplified equivalent circuit of nonideal op amp. (b) Operational amplifier circuits for problem 31.

32. (a) Prove that the operation of convolution is commutative; i.e., prove that $h * f = f * h$.
   (b) Prove that convolution is associative; i.e., prove that $(h * f) * g = h * (f * g)$.

33. Repeat example 16.4 with $h(t) = u(t)$.

34. Solve the following problem with the aid of either the graphical interpretation of convolution, i.e., the flip–shift–multiply–find area method, or the techniques of convolution algebra: Find $f_3(t) = f_1(t) * f_2(t)$, and plot $f_3(t)$ vs. t where f_1 and f_2 are shown in figure P16.34.

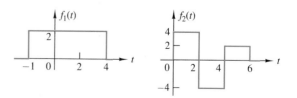

Figure 16.34 Signals for problem 34.

35. The transfer function of a particular time-invariant linear circuit is

$$H(s) = \frac{1}{s+1} + \frac{2}{s+2} + \frac{4}{s+4}$$

(a) Determine the impulse response of the circuit.

(b) Determine the step response of the circuit by the convolution method. (*Hint*: Determine the convolution of the step function with $\exp(at)$ for arbitrary $a < 0$; use this result and linearity to determine the final answer.)

(c) Compute the zero-state response of the circuit to the input $f(t) = 8u(-t) - 8u(-T-t)$ for $T > 0$. (*Hint*: Compute the response to $u(-t)$ and use linearity and time-invariance to compute the result.)

(d) Compute the zero-state response of the circuit to the input $f(t) = 8\exp(at)u(-1-t)$.

36. A clever electrical engineering student designed a time-warp current source that simulated the production of excitations from $t = -\infty$ to the present. In the circuit shown in figure P16.36, if $i_{in}(t) = 3e^{-t}$, determine $v_c(t)$.

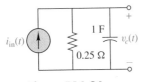

Figure P16.36

37. Solve this problem by evaluating the defining integral for the convolution of two time functions. Suppose a circuit is an integrator having a gain of 4, i.e., $H(s) = 4/s$. Suppose further that $f_1(t) = e^{-4t}u(t)$, $f_2(t) = e^{4t}u(-t)$, and $f_3(t) = \delta(t-4)$.

(a) Find the impulse response $h(t)$, and sketch it for $-\infty < t < \infty$.

(b) Find the zero-state response $y_1(t)$ to $f_1(t)$, and sketch it for $-\infty < t < \infty$.

(c) Find the zero-state response $y_2(t)$ to $f_2(t)$, and sketch it for $-\infty < t < \infty$.

(d) Find the zero-state response $y_3(t)$ to $f_3(t)$, and sketch it for $-\infty < t < \infty$.

38. The RLC circuit in figure P16.38 has input voltage $v_{in}(t) = 10te^{-|t|}$. Find $v_{out}(t)$ by the convolution method.

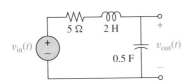

Figure P16.38 RLC circuit excited by two-sided input voltage.

39. Repeat problem 38 when the input excitation is $v_{in}(t) = \cos(2t)u(-t) + u(t)$. How can you check your answer by adapting the phasor and Laplace transform methods to this problem?

40. In courses on signal processing, there is a phenomenon called ideal sampling that closely approximates a process commonly used in the transmission of information by digital means. The ideal sampling of a signal $f(t)$ is represented by the product of the signal with an impulse train

$$h(t) = \sum_{k=0}^{\infty} \delta(t - kT).$$

The goal of this problem is to find a representation of $\mathcal{L}[f(t)h(t)]$; i.e., the transform of the product of the functions.

(a) Show that $\mathcal{L}[f(t)h(t)] = F(s) * H(s)$; i.e., the Laplace transform of the product of two functions is the convolution of their transforms.

(b) Use the result of (a) to determine an expression for $Y(s) = F(s)H(s)$. Observe that one obtains an infinite sum of replications of $F(s)$ shifted in frequency. This has a number of interesting properties pertinent to the digitization and reproduction of voice and music.

41. This problem deals with the derivation of a formula for the convolution of two *piecewise constant* time functions $h(t)$ and $v(t)$ by polynomial multiplication. Assume that both $h(t)$ and $v(t)$ are zero for negative t. In figure P16.41:

(a) Let $y(t) = h(t) * v(t)$. Using the graphical method of section 16.7, find $y(0)$, $y(T)$, $y(2T)$, and $y(3T)$ in terms of $v_0, v_1, \ldots, h_0, h_1, \ldots$. Also, find $y(t)$ for $0 \le t \le 2T$.

(b) Define two polynomials in x as follows:

$$p(x) = h_0 + h_1 x + h_2 x^2.$$

$$q(x) = v_0 + v_1 x + v_2 x^2 + v_3 x^3.$$

Find the product $r(x) = p(x)q(x)$ as a single polynomial in x.

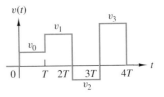

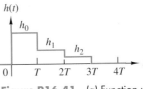

Figure P16.41 (a) Function $v(t)$ and (b) function $h(t)$ for problem 41.

(c) Compare the results of parts (a) and (b), and verify that in general, $y(0) = 0$ and $y(kT) = T \times$ {coefficient of x^{k-1} term in $r(x) = p(x)q(x)$}.

(d) From the experience gained in parts (a) through (c), state in general how the curve of $y(t)$ vs. t may be drawn once the values of $y(0)$, $y(T)$, $y(2T)$, $y(3T)$, ... have been calculated.

42. Use the results of problem 41 to solve each of the following:

(a) Consider the piecewise constant waveforms $f(t)$ and $h(t)$ of figure 16.34 in example 16.11, with $T = 2$. Find $y(t) = f(t) * h(t)$ by the polynomial multiplication method outlined in parts (c) and (d) of problem 41. The result should, of course, agree with that obtained by the method of convolution algebra (with $f(t)$ differentiated and $h(t)$ integrated).

(b) Consider the piecewise constant waveform $h(t)$ shown in figure 16.39 of example 16.13, with $T = 0.1$. If $v(t) = h(t)$, find $y(t) = f(t) * h(t)$ by the polynomial multiplication method outlined in parts (c) and (d) of problem 41. Use MATLAB to perform the tedious poly-

nomial multiplication and to plot the waveform of $y(t)$ for $0 \leq t < 4$. The MATLAB commands required are:

$p = [0 \ 0.1 \ 0.3 \ \ldots$ complete the entries];

$q = p; T = 0.1;$

$r = \text{conv} \, (p, q); y = [0 \ r \ 0] * T;$

$t = (0 : T : (\text{length} \, (y) - 1) * T); \text{ plot} \, (t, y)$

CHAPTER 17

Resonant and Bandpass Circuits

How a Touch-Tone Phone Signals the Numbers Dialed

MAKING telephone calls is a daily experience. When dialing a number, the information is sent to the central office by one of two methods: tone dialing or pulse dialing. Tone dialing is much faster than pulse dialing. For example, the electronic processing of the pulse-dialed long-distance number 555-555-5555 requires about 11 seconds, while the electronic processing for tone dialing takes only about 1 second.

The keypad of an ordinary touch-tone phone has 12 buttons arranged in four rows and three columns, as shown in the following diagram:

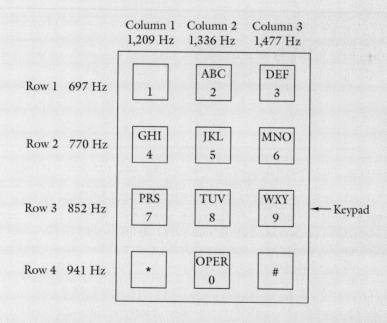

Pressing any one button generates two tones with the frequencies selected by an electronic circuit inside the telephone set. For example, pressing the number 5 generates tones at 770 and 1,336 Hz. The row and column arrangements and the dual tone method permit the representation of 10 digits (0, . . . , 9) and two symbols (*, #) using only seven tones. These seven tones are divided into two groups: the low-frequency group, from 697 to 941 Hz, and the high-frequency group, from 1,209 to 1,477 Hz.

567

An LC resonant circuit can easily produce each tone. The four tones in the low-frequency group are produced by connecting a capacitor to four different taps of a single coil. A similar connection generates the three tones in the high-frequency group. When a button is pressed to the halfway point, a dc current from the central office is sent through the coil in the tank circuit. When the button is fully pressed, the dc current is interrupted. This action initiates sinusoidal oscillations in the LC resonant tank circuit at a frequency inversely proportional to $\sqrt{LC}$. The presence of small resistances causes the oscillations of the tank circuit to die out. However, pressing the button fully also connects the tank circuit to a transistor circuit that replenishes the lost energy and sustains the oscillations.

At the central office, the equipment used to detect the presence of the tones and to decide their frequencies is much more sophisticated. Two filters are required, one for each of the frequency groups. Each filter must pass the frequencies within ±2% of their nominal values (697 to 941 Hz for the one filter and 1,209 to 1,477 Hz for the other), and reject the signal if the frequencies are outside of ±3% limits. The output tone from each filter is then processed digitally to determine its frequency.

The concepts and methods developed in this chapter will allow us to understand the properties of resonant circuits and the design of various basic types of bandpass circuits, as used in the touch-tone telephone system. ◆

CHAPTER OBJECTIVES

- Define and investigate basic properties of resonant circuits consisting of one inductor, one capacitor, and a few resistors.
- Use resonant circuits for designing simple bandpass circuits and for matching source and load impedances at a particular frequency.
- Learn various approximations made when using practical inductors and capacitors.
- Understand bandpass circuits from the viewpoint of transfer functions and pole-zero plots.
- Apply the transfer function viewpoint to the design of inductorless bandpass filters using operational amplifiers.
- Understand the principles of magnitude scaling and frequency scaling and apply these techniques to the analysis and design of linear circuits.

1. INTRODUCTION

By merely turning the knob of a radio receiver, we are able to select our favorite stations for listening. Have you ever wondered what makes this possible? Some expensive sets have a very clear reception, while with some cheaper models, other stations chatter in the background. What circuitry inside the radio makes this difference? The answer is the quality of the bandpass circuits used in the radio. The ability to select a particular broadcast station depends on the design of a bandpass circuit that will pass signals within a narrow band of frequencies, while rejecting or significantly attenuating others outside of that band. To see why this is important, note that audio signals have frequency components up to about 3 kHz for voice and up to about 15 kHz for high-fidelity music. These frequencies are far too low for wireless transmission. However, in an AM radio transmitter, the audio signal modulates the amplitude of a carrier signal, which is a sinusoidal waveform at a much higher frequency. The resultant modulated waveform contains many frequency components centered about the carrier signal frequency, but extending over a range of frequencies equal to twice the highest audio frequency. For example, the radio station WBAA, at Purdue University, has a carrier frequency of 920 kHz and occupies a band from approximately 915 to 925 kHz. These high-frequency signals can be transmitted efficiently as electromagnetic waves. A good bandpass filter will effectively pass this band, while significantly rejecting others, i.e., the bands of neighboring radio stations. It is the properties of this very important notion of a bandpass filter that are studied in this chapter.

In its simplest form, a bandpass circuit consists of only one capacitor and one inductor, connected either in series or in parallel. In a first course on circuits, you learned how to analyze simple RLC circuits. There, the emphasis was on (1) transient behavior under dc excitation and (2) sinusoidal steady-state behavior at a single frequency. This chapter will investigate the behavior of circuits over bands of sinusoidal frequencies. Many useful results may be obtained with the phasor and impedance concepts studied earlier. We will continue to use these techniques in sections 2 through 5. However, rapid advances in technology have made it possible to have a bandpass circuit without any of the usual RLC circuit components. Therefore, a study of the bandpass property of a transfer function $H(s)$ is presented in sections 6 and 8 to provide a systems point of view. The results presented in those sections are readily applicable to general linear systems, whether they are electrical, mechanical, or any other kind of linear systems.

2. RESONANT FREQUENCY OF SIMPLE CIRCUITS WITH APPLICATIONS

Suppose a circuit contains one inductance L, one capacitance C, and an arbitrary number of linear resistive elements. Regardless of the placement of L and C, we denote by ω_o the frequency at which the two **reactances** $X_c = -1/\omega_o C$ and $X_L = \omega_o L$ have equal magnitudes, i.e.,

$$\frac{1}{\omega_o C} = \omega_o L, \tag{17.1}$$

in which case

$$\omega_o = \frac{1}{\sqrt{LC}}. \tag{17.2}$$

Often in a first course on circuits, $\omega_o = 1/\sqrt{LC}$ is called the **undamped natural frequency**. The name stems from the fact that if all resistive elements are absent (i.e., the circuit is undamped), then a parallel or series connection of L and C produces a natural response of the form $K\cos(\omega_o t + \theta)$. In the jargon, the LC circuit of figure 17.1b is called a **tank circuit**, and ω_o is called the **tank frequency**.

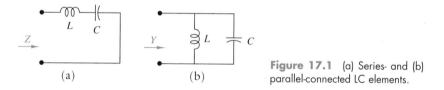

Figure 17.1 (a) Series- and (b) parallel-connected LC elements.

In sinusoidal steady-state analysis, ω_o takes on an additional interpretation: At the frequency $\omega = \omega_o$, the series-connected LC and the parallel-connected LC behave as a short circuit and an open circuit, respectively. The verification of these behaviors is easy. From the definition of ω_o, the input impedance of figure 17.1a is

$$Z(j\omega_o) = j\omega_o L - \frac{j}{\omega_o C} = 0,$$

indicating a short circuit. Similarly, for figure 17.1b, the input admittance is

$$Y(j\omega_o) = j\omega_o C - \frac{j}{\omega_o L} = 0,$$

which specifies an open circuit.

Cancellation of reactances or **susceptances** is usually achieved by adjusting the value of C or L. The net reactances (or susceptances) are then *tuned out*. Tuning out reactances (or susceptances) finds application in radio frequency amplifiers (i.e., amplifiers intended for frequencies above, say, 20 kHz), as illustrated by the following example.

EXAMPLE 17.1

Figure 17.2 represents an amplifier modeled by a controlled current source with $g_m = 2$ millimhos. The load is represented by the parallel combination of a 20-kΩ resistance with a 40-pF capacitance. This capacitance accounts for or models the wiring capacitance, the device input capacitance, and other embedded capacitances. Consequently, it cannot be removed from the circuit. Suppose the applied sinusoidal voltage, $V_{in}(j\omega)$, has a magnitude of 0.1 V at 10 MHz.

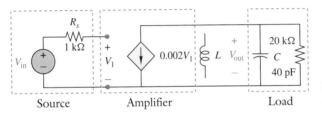

Figure 17.2 Illustration of the need for a tuned circuit as load.

The objectives of the example are as follows:

(a) With the load connected directly as shown (without L), find the magnitude of the output voltage.
(b) If an inductance L is connected across the load to tune out the effective capacitance, determine the value of L and the resulting $|V_{out}|$.

SOLUTION

(a) At 10 MHz, the load has an impedance

$$Z = \frac{1}{0.00005 + j2\pi \times 10^7 \times 40 \times 10^{-12}}$$

$$= 397.8 \angle -88.9° \ \Omega$$

Therefore, the magnitude of the output voltage is

$$|V_{out}| = 0.1 \times 0.002 \times 397.8 = 0.0796 \text{ V}.$$

Here, the voltage gain is less than 1 due to the low impedance of C at the high operating frequency.

(b) The inductance needed to tune out the capacitance is calculated from equation 17.2:

$$L = \frac{1}{\omega_0^2 C} = \frac{1}{4\pi^2 \times 10^{14} \times 40 \times 10^{-12}} = 6.33 \times 10^{-6} \text{H}.$$

With this inductance connected across the load, the parallel LC behaves as an open circuit at 10 MHz. Therefore, the output voltage becomes

$$|V_{out}| = 0.1 \times 0.002 \times 20,000 = 4 \text{ V},$$

and a voltage gain of 4/0.1 = 40 results.

CHAPTER 17   RESONANT AND BANDPASS CIRCUITS

The concept of reactance cancellation underlies the notion of resonant frequency germane to general RLC circuits. To define the concept of resonant frequency precisely, suppose a sinusoidal source excites a linear circuit containing one L, one C, and other linear resistive elements. The frequency at which the steady-state source voltage and source current are in phase is called the **resonant frequency** of the circuit; it is denoted by ω_r in this book.

For a given circuit, the determination of ω_r requires the specification of a pair of input nodes, which may be explicitly marked on the circuit diagram or may just be understood from the context of the problem statement.

 EXAMPLE 17.2

Determine the resonant frequency for the parallel RLC circuit of figure 17.3a, with the input nodes as shown.

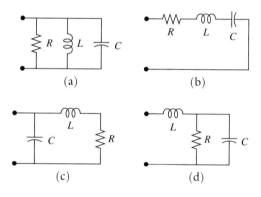

(a) (b)

(c) (d)

Figure 17.3 Parallel and series resonant circuits. (a) Parallel RLC circuit. (b) Series RLC circuit. (c) A variation of parallel resonant circuit. (d) A variation of series resonant circuit.

SOLUTION

Let the voltage across and current through the input nodes be V and I, respectively. For V and I to be in phase, the admittance Y must be a real number, or equivalently, $\text{Im}[Y] = 0$. Mathematically, for the circuit of figure 17.3a,

$$\text{Im}[Y] = \text{Im}\left[\frac{1}{R} + j\left(\omega_r C - \frac{1}{\omega_r L}\right)\right] = \omega_r C - \frac{1}{\omega_r L} = 0.$$

Solving for ω_r yields

$$\omega_r = \frac{1}{\sqrt{LC}}.$$

Observe that for the parallel RLC circuit, $\omega_r = \omega_o$. Similarly, for the series RLC circuit of figure 17.3b, $\omega_r = \omega_o = 1/\sqrt{LC}$. For this reason, ω_o is also called the **LC resonant frequency**. This means that ω_o is the resonant frequency of a circuit whenever the single L and single C are connected either in parallel or in series, as shown in figures 17.3a and 17.3b, respectively.

Although for parallel and series RLC circuits, ω_r and ω_o have the same value, such is not the case for circuits in general. The next example illustrates this point.

◆ **EXAMPLE 17.3**

Find the resonant frequency for the circuit shown in figure 17.3c and the impedance of the two-terminal network at the resonant frequency.

SOLUTION

Step 1. *Calculate the admittance "looking into" the input node pair.* By the usual techniques,

$$Y = j\omega_r C + \frac{1}{R + j\omega_r L} = j\omega_r C + \frac{R - j\omega_r L}{R^2 + (\omega_r L)^2}. \tag{17.3}$$

Step 2. Resonance occurs when Y is real, i.e.,

$$\text{Im}\{Y\} = \omega_r C - \frac{\omega_r L}{R^2 + (\omega_r L)^2} = 0.$$

Solving for ω_r and then expressing ω_r as a function of ω_o yields

$$\omega_r = \sqrt{\frac{1}{LC} - \frac{R^2}{L^2}} = \omega_o \sqrt{1 - \frac{CR^2}{L}}. \tag{17.4}$$

Step 3. To obtain the values of the admittance and impedance at resonance, substitute this value of ω_r into equation 17.3 to obtain

$$Y(j\omega_r) = \frac{RC}{L} \quad \text{and} \quad Z(j\omega_r) = \frac{L}{RC}. \tag{17.5}$$

Three conclusions can be drawn from equations 17.4 and 17.5:

1. If $(CR^2)/L > 1$, then there is no real solution for ω_r. This implies that the source voltage and current cannot be in phase at any frequency.
2. If $(CR^2)/L < 1$, then there is a unique nonzero resonant frequency ω_r that is strictly smaller than ω_o.
3. At $\omega = \omega_r$, the source "sees" a pure resistance, the value of which equals (L/CR) and is greater than R.

Results similar to equations 17.4 and 17.5 can be derived for the circuit of figure 17.3d. (See problem 2.)

If ω_r exists, then the circuit is a *resonant circuit.* Again, the word *resonance* reflects the condition that the source voltage and current are in phase (both rise and fall at the same time), or resonate with each other. The four circuits shown in figure 17.3 are examples of resonant circuits. Because of the ways their LC elements are connected, figures 17.3a and 17.3c are called *parallel-resonant circuits,* and figures 17.3b and 17.3d are called *series-resonant circuits.* Note that in figures 17.3c and 17.3d, due to the presence of the resistance R, the LC elements are not, strictly speaking, connected in parallel or in series. The LC elements become parallel or series connected when there is no energy dissipated, i.e., when $R = 0$ in figure 17.3c or $R \to \infty$ in figure 17.3d. In practice, resonance is often achieved by varying C or L until the input admittance or impedance is real. This process is called *tuning.* Accordingly, resonant circuits with the provision for adjusting L, C, or both are also called **tuned circuits.**

In most practical circuits, ω_r and ω_o are nearly equal. For some circuits in which ω_r and ω_o are widely different, useful properties of the circuit usually occur at $\omega = \omega_r$, rather than at $\omega = \omega_o$. The next example shows a significant application of resonant circuits in the context of maximum power transfer from source to load at a given fixed frequency.

 ## EXAMPLE 17.4

The output stage of a certain radio transmitter is represented by a 1-MHz sinusoidal voltage source having a fixed magnitude of 50 V rms and an internal resistance of 100 Ω. The load resistance R_L models an antenna connected to the transmitter. The purpose of this example is to show how a matching network based on the principle of resonance can be designed to maximize the power delivered to the load, i.e., to the antenna.

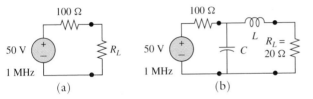

Figure 17.4 Matching load to source by a resonant circuit. (a) Load connected directly to source. (b) A coupling network designed to maximize the load power.

(a) Refer to figure 17.4a. If R_L is adjustable, find the value of R_L yielding the maximum average power P_L absorbed by the load. What is the value of $(P_L)_{max}$?

(b) If $R_L = 20$ Ω in figure 17.4a, find the value of P_L.

(c) Suppose again that R_L is fixed at 20 Ω, but a coupling network consisting of LC elements is inserted between the source and load to increase the power P_L. Design the coupling network (choose L and C) so that $(P_L)_{max}$ of part (a) is again obtained.

SOLUTION

(a) From the maximum power transfer theorem,

$$R_L = R_s = 100 \ \Omega,$$

and

$$(P_L)_{max} = \frac{V_s^2}{4R_s} = \frac{2,500}{400} = 6.25 \text{ W}.$$

(b) With $R_L = 20$ Ω,

$$V_L = \frac{50 \times 20}{100 + 20} = 8.33 \text{ V},$$

in which case

$$P_L = \frac{8.33^2}{20} = 3.472 \text{ W}.$$

(c) If we can make the impedance at the input terminals of the LC coupling network equal to $(100 + j0)$ Ω, then maximum power will be drawn from the source. Since LC elements consume zero average power, the same maximum power will be delivered to the load resistance. The resonant circuit shown in figure 17.4b provides a possible design. To calculate the element values, use equations 17.4 and 17.5 as follows:

$$(2\pi \times 10^6)^2 = \frac{1}{LC} - \frac{20^2}{L^2} \quad \text{(from equation 17.4, squared)};$$

$$100 = \frac{L}{20C} \quad \text{(from equation 17.5).}$$

Solving these equations simultaneously to obtain L and C produces

$$L = 6.37 \text{ μH} \quad \text{and} \quad C = 3.18 \text{ nF}.$$

Some remarks about this design are in order:

1. $f_o = 1/(2\pi\sqrt{LC}) = 1.12$ MHz, and $f_r = 1$ MHz $\neq f_o$. The 20-Ω resistance is transformed into a 100-Ω resistance at $f = f_r$ (not at $f = f_o$).
2. Should the source resistance be smaller than the fixed load resistance (see problem 7), C is moved to be in parallel with R_L. In that case, the formulas derived in problem 2 are used in place of equations 17.4 and 17.5.

> **Exercise.** Redesign the coupling network in example 17.3 if the resistors are $R_s = 300$ Ω and $R_L = 50$ Ω.
>
> **ANSWERS:** 17.79 μH and 1.18 nF.

3. FREQUENCY RESPONSE OF A PARALLEL RLC CIRCUIT

The preceding section looked at the properties of RLC circuits at the resonant frequency, ω_r. It is equally important to investigate the behavior of these circuits over a range of frequencies near ω_r. As stated in section 1, a primary reason for using resonant circuits is to design band-selective, or *bandpass*, filter characteristics that extract information from signals.

Let $H(s)$ be a voltage gain or some other type of network function. The curves of the magnitude $|H(j\omega)|$ and the angle $H(j\omega)$ vs. ω are called the *frequency response*, the former being the magnitude response and the latter the phase response, as described in chapter 15. An ideal bandpass circuit has a rectangular magnitude response curve, as shown in figure 17.5a. Here "ideal" means that all frequency components of the input signal within the range $\omega_1 < \omega < \omega_2$ are amplified with equal gain (in magnitude), and all frequency components outside of the range are totally eliminated from the output. Actually, for a bandpass circuit to transmit a signal with frequency components in the range $\omega_1 < \omega < \omega_2$ without distortion, there is also a requirement on the phase response. You will learn about this requirement later, in a course on signal analysis. For the present, we shall focus on simple circuits that approximate the ideal bandpass magnitude characteristic.

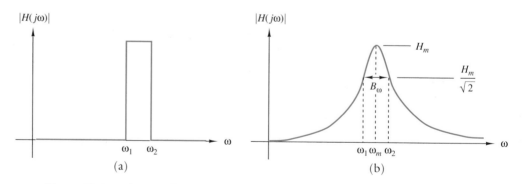

Figure 17.5 Definitions of peak frequency, ω_m, and bandwidth, B_ω. (a) Ideal bandpass characteristic. (b) Approximate bandpass characteristic.

Unfortunately, an ideal rectangular bandpass characteristic is not realizable. One can, however, approximate the ideal characteristic reasonably well. For example, with a simple tuned circuit, one can produce a bell-shaped magnitude response curve, illustrated in figure 17.5b, that is a crude approximation to the rectangle. With more complex circuits, one can improve the approximation. (How to do so is a topic to be studied in advanced courses.) The bell-shaped curve has several important features. The frequency at which $|H(j\omega)|$ reaches its maximum value, H_m, is called the **peak frequency** and is denoted by ω_m. The two side frequencies at which $|H(j\omega)|$ is $1/\sqrt{2}$ of its maximum value are called the **half-power frequencies,** denoted ω_1 and ω_2, respectively. The term "half power" comes from the fact that if the output is a voltage across a fixed resistance, then a drop in voltage by the factor $1/\sqrt{2}$ means a drop in power equal to $(1/\sqrt{2})^2 = 1/2$. As explained in section 15.8, half-power frequencies are also called -3dB frequencies. For obvious reasons, ω_1 is called the **lower half-power frequency** and ω_2 the **upper half-power frequency**. The difference $\omega_2 - \omega_1$ is the so-called **half-power bandwidth** (or simply, **bandwidth**) of the bandpass circuit and is denoted by B_ω. An electrical engineer will design the circuit so that all frequencies of interest fall within the pass band $[\omega_1, \omega_2]$.

One way of categorizing and comparing different bandpass circuits is by their **selectivity**, i.e., their relative capability to discriminate between frequencies inside the pass band and signals outside the pass band. Generally, the selectivity of a bandpass circuit is taken to be the ratio of the peak frequency to the bandwidth, i.e., ω_m/B_ω. This ratio is called the **quality factor,** Q, of the circuit. Q is sometimes denoted by Q_{cir} or $Q_{circuit}$ when it is necessary to distinguish it from other quality factors. A high-Q circuit passes only a very narrow band of frequencies relative to ω_m, whereas a low-Q circuit has a broad band and a less selective characteristic.

It is important to note that the concepts of ω_m, B_ω, and Q of a circuit are all based on the magnitude function $|H(j\omega)|$ and, therefore, on how the transfer function $H(s)$ is defined. Even for the same circuit, these values are different when the output is associated with different branches or when the input is changed from a voltage source to a current source. (See problem 12f.) Further, for the investigation of the frequency-selective characteristic of the circuit, the foregoing definition of Q is most appropriate because it directly assesses the sharpness of the magnitude response curve. Therefore, it can be determined experimentally in the laboratory. In certain other applications, where only one fixed frequency is of interest, there is another definition of a circuit's Q based on the following energy relationship. A sinusoidal voltage or current source at frequency ω_o is applied to the circuit, and a steady-state solution is obtained. Then the quality factor is defined as

$$Q = 2\pi\frac{\text{maximum energy stored}}{\text{total energy lost per period}}.$$

For parallel RLC and series RLC circuits, the two definitions of Q agree with each other. (See problem 13.) However, for a general resonant circuit, such as those of figures 17.3c and 17.3d, the two definitions of Q are conceptually and numerically different. The Q defined on the basis of energy is a quantity that is not amenable to calculation or measurement and not applicable to bandpass circuits constructed with RC and active elements. (See example 17.13.) Consequently, in this text, $Q \equiv \omega_m/B_\omega$.

Let us now determine ω_m, ω_1, ω_2, B_ω, and Q for the simple parallel RLC circuit shown in figure 17.6a. The current source provides the input, and the output is the voltage across the input nodes. Then $H(j\omega) = V_{\text{out}}/I_{\text{in}} = Z(j\omega)$.

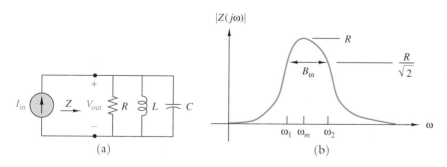

Figure 17.6 (a) A parallel RLC circuit. (b) Its magnitude response curve.

Without any calculations, we can predict that the magnitude response curve starts at the origin because the inductance becomes a short circuit at dc; moreover, the curve approaches the ω-axis as ω approaches infinity, because the capacitor becomes a short at infinite frequency. Since the parallel LC tank behaves as an open circuit at the resonance frequency $\omega = \omega_o = \omega_r = 1/\sqrt{LC}$, it follows that the maximum value of $|H(j\omega)|$ occurs at $\omega = \omega_o$, with $H(j\omega_o) = H_m = |Z(j\omega)|_{\max} = R$. Therefore, we have

$$\omega_m = \omega_o = \frac{1}{\sqrt{LC}}. \tag{17.6}$$

Figure 17.6b shows a rough sketch of the magnitude response. The bandwidth, in contrast, cannot be obtained by inspection. To compute B_ω, observe that

$$Z(j\omega) = \frac{1}{\dfrac{1}{R} + j\omega C + \dfrac{1}{j\omega L}}. \tag{17.7a}$$

Computing the magnitude yields

$$|Z(j\omega)| = \frac{1}{\sqrt{\left(\dfrac{1}{R}\right)^2 + \left(\omega C - \dfrac{1}{\omega L}\right)^2}}. \tag{17.7b}$$

This equation shows very clearly that for fixed R, the maximum value of $|Z(j\omega)|$ is achieved when the second term under the square root sign in the denominator is zero, i.e., when $\omega = \omega_m = 1/\sqrt{LC}$, in which case $Z(j\omega_m) = |Z(j\omega)|_{\max} = R$. At the half-power frequencies, the second term must have a magnitude equal to that of the first term under the square root sign, so that the magnitude is decreased by a factor of $\sqrt{2}$. Mathematically, this means that

$$\frac{1}{R} = \left(\omega_2 C - \frac{1}{\omega_2 L}\right) \tag{17.8a}$$

and

$$\frac{1}{R} = -\left(\omega_1 C - \frac{1}{\omega_1 L}\right). \tag{17.8b}$$

The negative sign in front of the parenthesis in equation 17.8b arises from the fact that $\omega_1 < \omega_m = \omega_o = 1/\sqrt{LC}$. Therefore, $\omega_1 C - 1/(\omega_1 L)$ is a negative quantity. Hence, a minus sign must be attached before equating $\omega_1 C - 1/(\omega_1 L)$ to the positive quantity $1/R$. Rewriting equations 17.8a and 17.8b as quadratics produces

$$\omega_2^2 - \frac{\omega_2}{RC} - \frac{1}{LC} = 0$$

and

$$\omega_1^2 + \frac{\omega_1}{RC} - \frac{1}{LC} = 0.$$

Applying the quadratic root formula yields

$$\omega_2 = \frac{1}{2RC} + \sqrt{\left(\frac{1}{2RC}\right)^2 + \frac{1}{LC}} \tag{17.9a}$$

and

$$\omega_1 = -\frac{1}{2RC} + \sqrt{\left(\frac{1}{2RC}\right)^2 + \frac{1}{LC}}. \tag{17.9b}$$

Applying equations 17.9a and 17.9b to the definition of bandwidth, we obtain

$$B_\omega = \omega_2 - \omega_1 = \frac{1}{RC} \tag{17.10}$$

and

$$Q = \frac{\omega_m}{B_\omega} = \omega_o RC = R\sqrt{\frac{C}{L}}. \tag{17.11}$$

The three equations 17.6, 17.10, and 17.11 are critical for the analysis and design of parallel RLC circuits. Typically, three of the six variables R, L, C, ω_m, B_ω, and Q are specified, and the remaining three solved for from these three equations. These are *exact* relationships for parallel RLC circuits. Oftentimes, an approximate analysis suffices. For example, once the exact value of B_ω is known, *approximate* values of ω_1 and ω_2 for high-Q circuits are given by

$$\omega_1 \cong \omega_o - \frac{B_\omega}{2}, \quad \omega_2 \cong \omega_o + \frac{B_\omega}{2}. \qquad (17.12)$$

To see this, we rewrite the radical sign term in equations 17.9 as follows:

$$\sqrt{\left(\frac{1}{2RC}\right)^2 + \frac{1}{LC}} = \frac{1}{\sqrt{LC}}\sqrt{\frac{L}{4R^2C} + 1} = \omega_o\sqrt{\frac{1}{4Q^2} + 1}$$

$$\cong \omega_o \text{ for high } Q.$$

This approximation, together with equation 17.10, leads to equation 17.12. One uses equations 17.9 only when exact values of ω_1 and ω_2 are necessary. For typical high-Q circuits ($Q > 10$) encountered in radio frequency amplifiers, the error due to the use of equation 17.12 instead of 17.9 is less than 1%. (See example 17.5 and problem 12c.) Lastly, note that the exact relationship of equations 17.9 shows ω_o to be the geometric mean of ω_1 and ω_2 (i.e., $\omega_o = \sqrt{\omega_1\omega_2}$. In contrast, the approximate relationships of equation 17.12 represent ω_o as the arithmetic mean of ω_1 and ω_2 (i.e., $\omega_o = 0.5(\omega_1 + \omega_2)$).

Although the phase response has an insignificant role in this chapter, it should be pointed out that at the half-power frequencies ω_1 and ω_2, the angle of $Z(j\omega)$ is $+45$ and -45 degrees, respectively. This phase relationship is sometimes used as a basis for determining the half-power frequencies by experimental measurements.

◆ **EXAMPLE 17.5**

Let the circuit parameters in figure 17.6 be $L = 20$ mH, $C = 0.05$ μF, and $R = 10$ kΩ.

(a) Find ω_m, B_ω, and Q.
(b) Estimate ω_1 and ω_2 from the results of part (a).
(c) Find the exact values of ω_1 and ω_2. What are the percentage errors in the estimates of part (b)?
(d) If $i_{in}(t) = 0.001\cos(\omega t)$, find $v_{out}(t)$ in the steady state at $\omega = \omega_r, \omega_1, \omega_2$, and 40,000 rad/sec.

SOLUTION

(a) Using the formulas just derived, we obtain

$$\omega_m = \omega_o = \frac{1}{\sqrt{LC}} = \frac{1}{\sqrt{0.02 \times 0.05 \times 10^{-6}}} = 31,622.77 \text{ rad/sec},$$

$$B_\omega = \frac{1}{RC} = \frac{1}{10,000 \times 0.05 \times 10^{-6}} = 2,000 \text{ rad/sec},$$

and

$$Q = \frac{\omega_m}{B_\omega} = \frac{31,622.77}{2,000} = 15.8.$$

(b) Using equation 17.12 results in

$$\omega_1 \cong \omega_m - \frac{B_\omega}{2} = 31,622.77 - 1,000 = 30,622.77 \text{ rad/sec}$$

and

$$\omega_2 \cong \omega_m + \frac{B_\omega}{2} = 31,622.77 + 1,000 = 32,622.77 \text{ rad/sec}.$$

(c) Substituting RLC values into equations 17.9 yields the exact values,

$$\omega_1 = 30,638.58 \text{ rad/sec}$$

and

$$\omega_2 = 32,638.58 \text{ rad/sec}.$$

The percentage errors in the estimates of part (b) are

$$100 \left| \frac{30,622.77 - 30,638.58}{30,638.58} \right| \cong 0.05\%$$

for ω_1 and

$$100 \left| \frac{32,622.77 - 32,638.58}{32,638.58} \right| \cong 0.05\%$$

for ω_2.

(d) The complex impedance $Z(j\omega)$ is calculated from equations 17.7. The output voltage phasor is $V_{out} = (0.001\angle 0) \times Z(j\omega)$, and thus, at $\omega = \omega_r$,

$$Z(j\omega_r) = 10,000 \ \Omega, \ v_{out}(t) = 10\cos(\omega_r t) \ V;$$

at $\omega = \omega_1$,

$$Z(j\omega_1) = 7,071\angle 45° \ \Omega, \ v_{out}(t) = 7.07\cos(\omega_1 t + 45°) \ V;$$

at $\omega = \omega_2$,

$$Z(j\omega_2) = 7,071\angle -45° \ \Omega, \ v_{out}(t) = 7.07\cos(\omega_2 t - 45°)V;$$

and at $\omega = 40,000$ rad/sec,

$$Z(j40,000) = \frac{1}{10^{-4} + j40,000 \times 0.05 \times 10^{-6} - \dfrac{j}{40,000 \times 0.02}}$$

$$= \frac{10,000}{1 + j7.5} = 1,321\angle -82.5°,$$

in which case

$$v_{out}(t) = 1.321\cos(40,000t - 82.5°) \ V.$$

Exercise. Repeat parts (a) through (c) of example 17.5 with the new parameter values $L = 40$ mH, $C = 0.25$ µF, and $R = 4$ kΩ.

ANSWERS: 10,000 rad/sec, 1,000 rad/sec, 10, 9,500 rad/sec, 10,500 rad/sec, 9,512.49 rad/sec, 10,512.49 rad/sec, 0.13%, 0.12%.)

The preceding example and exercise demonstrate that for high-Q ($Q \geq 10$) circuits, there is really no need to use the exact equations 17.9 to determine ω_1 and ω_2, as the much simpler estimates given by equation 17.12 are sufficiently close to the true answers.

In many practical circuits, the independent source could be a voltage source in series with a resistor. Before applying any of the foregoing formulas, it is necessary to transform the circuit into the form of figure 17.6a by the use of Norton equivalent circuit studied in a first course on circuits. The resistance R in figure 17.6a then is not a physical resistor, but rather the equivalent resistance of several resistances in parallel. The following example illustrates this point.

EXAMPLE 17.6

In the circuit of figure 17.7a, an independent voltage source V_{in} in series with an internal resistance $R_s = 40$ kΩ models a real-world sinusoidal excitation. Suppose $L = 20$ mH, $C = 0.05$ µF, and $R_L = 10$ kΩ.

(a) Find the exact values of ω_m, B_ω, and Q.
(b) Estimate the values of ω_1 and ω_2.

SOLUTION

The given circuit is not in the parallel form of figure 17.6a, so we replace the series $R_s - V_{in}$ by its Norton equivalent, as depicted in figure 17.7b. The parallel combination of R_s and R_L is identified as R in figure 17.6a. By the usual formula for two parallel resistors,

$$R = \frac{R_s R_L}{R_s + R_L} = \frac{40,000 \times 10,000}{40,000 + 10,000} = 8,000 \ \Omega.$$

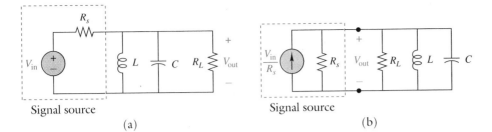

Figure 17.7 Tuned circuit driven by a practical voltage source. (a) Thévenin equivalent for the source. (b) Norton equivalent for the source.

(a) From equations 17.6, 17.10, and 17.11,

$$\omega_m = \omega_o = \frac{1}{\sqrt{LC}} = \frac{1}{\sqrt{0.02 \times 0.05 \times 10^{-6}}} = 31,622.77 \text{ rad/sec,}$$

$$B_\omega = \frac{1}{RC} = \frac{1}{8,000 \times 0.05 \times 10^{-6}} = 2,500 \text{ rad/sec,}$$

$$Q = \frac{\omega_m}{B_\omega} = \frac{31,622.77}{2,500} = 12.6.$$

(b)

$$\omega_1 \cong \omega_m - \frac{B_\omega}{2} = 31,622.77 - 1,250 = 30,372.77 \text{ rad/sec,}$$

$$\omega_2 \cong \omega_m + \frac{B_\omega}{2} = 31,622.77 + 1,250 = 32,872.77 \text{ rad/sec.}$$

From this example, it is seen that putting an external resistance in parallel with the LC tank circuit reduces the value of R, which in turn causes a lower circuit Q and a larger bandwidth, while keeping the peak frequency ω_m unaffected.

> **Exercise.** Repeat example 17.6 with the element values changed to $R_s = 36$ kΩ, $L = 40$ mH, $C = 0.25$ μF, and $R_L = 4$ kΩ.
>
> **ANSWERS:** 10,000 rad/sec, 1,111.11 rad/sec, 9, 9,444.44 rad/sec, 10,555.55 rad/sec.

So far, we have illustrated only the analysis aspect of parallel RLC circuits. In the design of a parallel-tuned circuit, we must also give attention to many other factors, such as available component sizes, desired voltage gain, cost, etc. In practice, the design specifications often impose fewer number of constraints than the number of circuit parameters to be determined. Consequently, realistic design problems usually do not have a unique answer, as illustrated in the next example.

 EXAMPLE 17.7

Design a parallel RLC circuit, as shown in figure 17.7a, to have a magnitude response with $f_m = 200$ kHz and a bandwidth of 20 kHz. Only inductors in the range 1 to 5 mH are available. The source has a resistance $R_s = 50$ kΩ.

SOLUTION

In the circuit of figure 17.7a, since there is a restriction on the inductors available, we keep L as a variable, subject to the condition that $0.001 \leq L \leq 0.005$ H. Using the specified peak frequency and equation 17.6,

$$C = \frac{1}{\omega_o^2 L} = \frac{1}{4\pi^2 \times 4 \times 10^{10} \times L} = 6.33 \times 10^{-13} L.$$

From the specified bandwidth and equation 17.10,

$$R = \frac{1}{B_\omega C} = \frac{4\pi^2 \times 4 \times 10^{10} \times L}{2\pi \times 2 \times 10^4} = 1.257 \times 10^7 L.$$

As explained in example 17.6, R is the parallel combination R_s and R_L, i.e.,

$$\frac{1}{R} = \frac{1}{R_L} + \frac{1}{R_s},$$

or

$$\frac{1}{R_L} = \frac{1}{R} - \frac{1}{R_s}.$$

Now, once a specific value of L is chosen, we can calculate successively the values of C, R, and R_L. Since R, which is the parallel combination of R_L and R_s, must be no greater than $R_s = 5 \times 10^4$ Ω, the upper limit for L is

$$L_{\max} = \frac{R_{\max}}{1.257 \times 10^7} = \frac{50,000}{1.257 \times 10^7} = 0.00398 \text{ H}.$$

Numerical answers corresponding to the extremal values of L are given in table 17.1.

TABLE 17.1

L (mH)	C (pF)	R (kΩ)	R_L (kΩ)
1	633	12.57	16.77
3.98	159	50	infinite

This shows clearly that there is no unique answer to the design problem. The freedom in choosing a value for L in the range 1 to 3.98 mH can be utilized to accommodate another design specification, e.g., a value for H_m.

Exercise. In example 17.7, if the bandwidth requirement is changed to 10 kHz, determine the maximum possible value of L.
ANSWER: 1.99 mH.

The preceding discussion is a rigorous mathematical analysis of the parallel RLC circuit of figure 17.6. One important conclusion is that if R is absent, then, as the frequency of the sinusoidal input current approaches the resonant frequency ω_r, the magnitude of the output voltage grows very large. The same phenomenon occurs even if the input is not sinusoidal. To see this, consider, for example, the parallel LC circuit shown in figure 17.8a.

The natural response of the circuit has a sinusoidal waveform with period $T = 2\pi$ sec. Suppose the circuit is initially relaxed and the input current is a periodic impulse train having the same period as the natural response of the circuit, i.e.,

$$i_s(t) = \delta(t) + \delta(t - 2\pi) + \delta(t - 4\pi) + \ldots.$$

Let the output of interest be the current through the inductance. Using the Laplace transform method of chapter 14, we find that the impulse response of the circuit is

$$h(t) = \mathcal{L}^{-1}\left[\frac{1}{s^2 + 1}\right] = \sin(t)u(t).$$

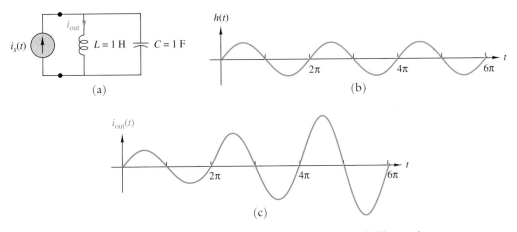

Figure 17.8 (a) LC circuit driven by a periodic impulse current source. (b) The impulse response. (c) Building up of the inductor current.

Figure 17.8b shows the waveform of the impulse response. By the principle of super-position, then, the output current due to $i_s(t)$ is

$$i_{\text{out}}(t) = \sin(t)u(t) + \sin(t - 2\pi)u(t - 2\pi) + \sin(t - 4\pi)u(t - 4\pi) + \dots,$$

where the output due to each impulse $\delta(t - 2k\pi)$, $k = 0, 1, 2, \dots$, in the input is a delayed or shifted version of the waveform of figure 17.8b. Because the input impulses are applied in synchrony with the natural response, the sinusoidal components in the output are in phase. This causes a fast buildup of the amplitude, as shown in figure 17.8c.

Such a resonance phenomenon occurs so often in real life that one utilizes the property without any mathematical knowledge. Just watch a five-year-old child on a swing. By instinct, the child knows when to flex his or her knees in synchrony with the swing's pendulum motion to swing higher and higher. That is precisely analogous to the waveform buildup shown in figure 17.8c.

Communications Engineer

Communications engineers design systems that receive, transmit, and deliver information, especially audio and video information. One can safely argue that developments in the accuracy and quality of radio, television, telephone, and video have resulted from the efforts of communications engineers making the fruit of this engineering field very familiar to the general consumer. Communications engineers often work as team members on projects involving manufacturing equipment, power transmission, switching networks, process control systems, traffic control, and acoustical engineering. In addition, they are employed by the military to develop intelligence and surveillance communications systems that protect our air, land, sea, and space personnel and facilities.

Communications engineers have been instrumental in the development of the high-speed, global computer networking system known as Internet. Through Internet, scientists, business people, educators, and others can enter in the "information superhighway" that enables them to send and receive communications worldwide, access elec-

tronic bulletin boards, log onto distant databases, and use computers and other equipment in remote locations. Since Internet is rapidly changing, it is virtually impossible to predict what the future may hold—but it is guaranteed to be exciting!

Some communications engineers believe the world is on the verge of a remarkable transition in communications. Incredible advances in the field will be largely due to the merging of communications engineering with computer principles, microchip technology, and optical communications. Already available or soon to be available within the next decade are: wristwatch radios to transmit calls, cellular phones that obey commands using voice recognition technology, phone numbers belonging to one person for life that will allow individuals to receive calls dialed to their number wherever they are, and phones that call out the name of the person calling rather than ringing. Communications engineers hope to design a solar-powered radio transmitter/receiver that can be used for emergencies in remote villages with limited access to electricity.

4. FREQUENCY RESPONSE OF A SERIES RLC CIRCUIT: APPLICATION OF DUALITY

Practical signal sources have both a Thévenin equivalent and a Norton equivalent circuit, as shown in figures 17.7a and b, respectively. Since $Q = \omega_o RC$, where R is the parallel combination of R_s and R_L, good selectivity is not achievable with the bandpass circuits of these figures when the source has a very low internal resistance. To overcome this difficulty, we rearrange the LC components into a series RLC circuit, as per figure 17.9b, whose properties will now be investigated.

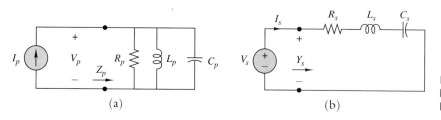

Figure 17.9 Dual circuits: (a) Parallel RLC driven by a current source. (b) Series RLC driven by a voltage source.

Three approaches are open to us for studying the series RLC circuit:

1. We may adapt and then repeat the entire analysis of the parallel RLC case to this series circuit.
2. We may apply the principle of duality to directly write down formulas for the series RLC case that are dual to those derived in section 3 for the parallel RLC case.
3. We may investigate a general second-order network function having both the series and the parallel RLC circuit, as well as several other arrangements, as special cases.

Approach 1 is inelegant and time consuming. Approach 3 will be taken up in section 6. Approach 2 is most rewarding and satisfying at this point, as it enables us to write down results for a new circuit based on those for an old circuit without additional derivations. What can be more "effort effective" than that?

At this time, the reader should review the principle of duality and the construction of dual networks, as presented in a first course on circuits. Our goal here is merely to apply the ideas to the study of resonant circuits. In particular, the parallel RLC and series RLC circuits shown in figure 5.9 are dual circuits. The dual quantities are listed in table 17.2, where the subscript p means parallel and the subscript s means series:

TABLE 17.2   DUAL QUANTITIES IN FIGURE 17.9

Parallel RLC, Figure 17.9a	Series RLC, Figure 17.9b
L_p	C_s
C_p	L_s
R_p	$1/R_s$
Z_p	Y_s
V_p	I_s
I_p	V_s

As indicated in the table, all formulas derived for parallel RLC circuits become formulas for series RLC circuits after an interchange of the dual variables. Thus, from equations 17.6, 17.10, and 17.11, for the series RLC circuit of figure 17.9b, with V_s as the input and I_s as the output, the dual formulas become:

$$\text{Peak frequency} \quad \omega_m = \omega_o = \frac{1}{\sqrt{L_s C_s}}; \tag{17.13}$$

$$\text{Half-power bandwidth} \qquad B_\omega = \frac{R_s}{L_s}; \tag{17.14}$$

$$\text{Circuit} \qquad Q = \frac{\omega_m}{B_\omega} = \frac{\omega_o L_s}{R_s} = \frac{1}{R_s}\sqrt{\frac{L_s}{C_s}}. \tag{17.15}$$

Exercise. Making use of equations 17.9 and 17.12, as well as table 17.2, write down the exact and approximate formulas for the half-power frequencies of the series RLC circuit shown in figure 17.9b.

◆ EXAMPLE 17.8

In the series resonant circuit of figure 17.9b, $R_s = 5\ \Omega$, $L_s = 1$ mH, $C_s = 0.1\ \mu$F, and V_s has a fixed magnitude.

(a) If I_s is the desired output, find the peak frequency ω_m, the bandwidth, and the circuit's Q.

(b) Prove that at $\omega = \omega_o$, the magnitude of the capacitance voltage is Q times larger than that of the source voltage V_s.

SOLUTION

(a) From equations 17.13 through 17.15, we immediately have

$$\omega_m = \omega_o = \frac{1}{\sqrt{L_s C_s}} = \frac{1}{\sqrt{10^{-3} \times 10^{-7}}} = 10^5 \text{ rad/sec},$$

$$B_\omega = \frac{R_s}{L_s} = \frac{5}{10^{-3}} = 5,000 \text{ rad/sec},$$

$$Q = \frac{\omega_m}{B_\omega} = \frac{10,000}{5,000} = 20.$$

(b) At $\omega = \omega_o$, the series $L_s - C_s$ is equivalent to a short circuit. Therefore,

$$|I_s| = \frac{|V_s|}{R_s}$$

and

$$|V_{Cs}| = \frac{|I_s|}{\omega_o C_s} = \frac{|V_s|}{R_s \omega_o C_s} = \frac{|V_s|\,\omega_o L_s}{R_s} = Q\,|V_s|.$$

The result of part (b) in the preceding example shows that it is possible to obtain an output voltage that is Q times the source voltage without the use of any active elements (transistors, for example). This is, in fact, another important reason (besides the selectivity consideration discussed earlier) for using resonant circuits in radio receivers. Needless to say, properties that hold for a series RLC circuit have analogous dual properties for the parallel RLC case.

Exercise. Using the result of example 17.8, part (b), state a dual result for the parallel RLC circuit.

ANSWER: The inductor current is Q times larger than the source current.

Up to now, determination of the peak amplitude response has centered on the case of variable frequency, with all other circuit parameters fixed. Because of the special manner in which ω appears in the impedance expression of equation 17.7b, the maximization problem is solved by inspection. In general, the solution of a maximization problem requires some knowledge of calculus. The usual way of finding the extremal (maximum or minimum) values of $f(x)$ is to set $f'(x)$ to zero and then solve for x. The analysis that follows utilizes this method.

Consider the series RLC of figure 17.10. Here, the voltage source has a fixed magnitude $|V_s|$ and a *fixed frequency* ω. With R and L fixed, we seek the value of the variable capacitance C that maximizes the magnitude of the voltage across the capacitor.

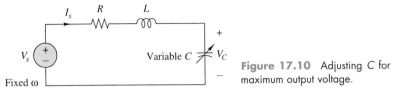

Figure 17.10 Adjusting C for maximum output voltage.

The first step is to compute the magnitude of the voltage across the capacitor:

$$|V_C| = \frac{|V_s|}{\sqrt{R^2 + \left(\omega L - \dfrac{1}{\omega C}\right)^2}} \cdot \frac{1}{\omega C}. \tag{17.16}$$

This expression follows using voltage division and then taking magnitudes. Maximizing $|V_C|$ is equivalent to maximizing $|V_C|^2$ or to minimizing $|V_s|^2/|V_C|^2$. To obtain this last expression, we square equation 17.16 and rewrite as

$$f(C) = \frac{|V_s|^2}{|V_C|^2} = (\omega C R)^2 + (\omega^2 L C - 1)^2.$$

To minimize $f(C)$, set $f'(C)$ to zero, i.e.,

$$f'(C) = 2C(\omega R)^2 + 2(\omega^2 L C - 1)\omega^2 L = 0.$$

Solving for C produces

$$C = \frac{L}{R^2 + (\omega L)^2} = \frac{1}{L\omega^2}\left[\frac{1}{\dfrac{R^2}{L^2\omega^2} + 1}\right]. \tag{17.17}$$

Although this value of C produces a maximum capacitor voltage, the circuit is *not in resonance*, as the value of $1/\sqrt{LC}$ is not equal to the signal source frequency ω. However, for a high-Q circuit ($\omega L/R > 10$), the condition given by equation 17.17 is practically the same as $1/\sqrt{LC} = \omega$.

◆ EXAMPLE 17.9

In figure 17.10, let $|V_s| = 1$ V, $\omega = 10^5$ rad/sec, and $R = 5\ \Omega$. Let C be variable. Find $|V_C|_{\max}$ and the corresponding value of C for each of the following cases: (a) $L = 1$ mH; (b) $L = 100$ μH.

SOLUTION

(a) For $L = 1$ mH, $Q = \omega L/R = 10^5 \times 10^{-3}/5 = 20$. This is a high-$Q$ circuit. Therefore, C is given *approximately* by

$$C = \frac{1}{\omega^2 L} = \frac{1}{10^{10} \times 10^{-3}} = 0.1 \times 10^{-6}\ \text{F},$$

and $|V_C|_{\max}$ is given *approximately* by

$$|V_C|_{\max} = Q|V_s| = 20\ \text{V}.$$

Exact solutions follow from equations 17.17 and 17.16:

$$C = \frac{0.001}{5^2 + (10^5 \times 0.001)^2} = 0.09975 \times 10^{-6}\ \text{F},$$

and

$$|V_C|_{max} = \cfrac{1}{\sqrt{5^2 + \left(10^5 \times 0.001 - \cfrac{1}{10^5 \times 0.09975 \times 10^{-6}}\right)^2}}$$

$$\times \frac{1}{10^5 \times 0.09975 \times 10^{-6}} = 20.025 \text{ V}.$$

Plainly, the approximate solutions are very close to the exact solutions.

(b) For $L = 100$ μH, $Q = \omega L/R = 10^5 \times 10^{-4}/5 = 2$. This is a low-$Q$ circuit, requiring the use of equations 17.17 and 17.16. Here,

$$C = \frac{0.0001}{5^2 + (10^5 \times 0.0001)^2} = 0.8 \times 10^{-6} \text{ F},$$

and

$$|V_C|_{max} = \cfrac{1}{\sqrt{5^2 + \left(10^5 \times 0.0001 - \cfrac{1}{10^5 \times 0.8 \times 10^{-6}}\right)^2}} \cfrac{1}{10^5 \times 0.8 \times 10^{-6}}$$

$$= 2.236 \text{ V}.$$

5. QUALITY FACTOR OF COMPONENTS AND APPROXIMATE ANALYSIS OF HIGH-Q CIRCUITS

The preceding sections examined either parallel or series interconnections of inductance and capacitance. These models do not accurately account for the behavior of circuits with practical (i.e., non-ideal) components. Recall that an ideal voltage source in series with an internal resistance models a practical voltage source. In the same fashion, a practical inductor (or a practical capacitor) is modeled by an inductor (or a capacitor), together with some other "parasitic" elements to account for losses and coupling effects. Figure 17.11a illustrates a fairly accurate model of a practical inductor.

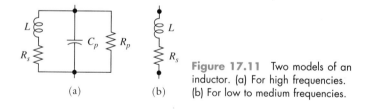

Figure 17.11 Two models of an inductor. (a) For high frequencies. (b) For low to medium frequencies.

The primary parameter is, of course, the inductance L. The remaining elements, R_s, R_p, and C_p, are viewed as "parasitic." Since an inductor usually consists of a coil of wire, the model uses R_s to represent the wire's resistance. Also, a capacitance always exists between any adjacent turns of wire. For convenience, this is represented by C_p across the two end terminals of L. The resistance R_p accounts for the energy loss in the magnetic core material (if present). Such complex models, although important, if used for every inductor in a circuit, would make the analysis of even a simple tuned circuit a horrendous task. Fortunately, for low to medium frequencies of up to a few megahertz, the simpler model of figure 17.11b is sufficiently accurate. This simpler model will underlie the analysis that follows.

Figure 17.12a shows a fairly accurate model of a practical capacitor. Again, the primary parameter here is the capacitance C, whereas R_p, R_s, and L_s are "parasitic." The resistance R_p, often called a **leakage resistance,** accounts for the energy loss in the dielectric; the inductance L_s and resistance R_s are due mainly to the connecting wires of the capacitor. At frequencies above $1/\sqrt{L_sC}$, the capacitor actually behaves as an inductor! For low to medium frequencies of up to a few megahertz, the simpler model shown in figure 17.12b is sufficiently accurate. As in the case of an inductor, this simpler model of a capacitor will underlie the analysis that follows. Very often, we will even omit R_p.

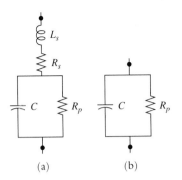

Figure 17.12 Two models of a capacitor. (a) For high frequencies. (b) For low to medium frequencies.

(a)   (b)

To investigate the effect of "parasitics," note that the impedance function $Z(j\omega)$ of a two-terminal network such as a practical inductor has the form

$$Z(j\omega) = \text{Re}\{Z(j\omega)\} + j\text{Im}\{Z(j\omega)\} = R + jX.$$

If the two-terminal network is a practical inductor or capacitor, then the reactance X is the primary parameter of concern, and R will represent the "parasitic effect" that we would rather not have, but cannot avoid. In such a case, the magnitude of X is usually much greater than the magnitude of R. The ratio $|X|/|R|$ provides a measure of how close the network element is to an ideal inductor or capacitor. This suggests defining the **quality factor** of $Z(j\omega)$ as

$$Q_Z(\omega) \equiv \frac{|X|}{R}. \tag{17.18}$$

The inclusion of ω in equation 17.18 is to emphasize the fact that Q_Z depends on the frequency of operation. The subscript Z indicates a generic impedance and may be replaced by more specific descriptors such as "coil" or "capacitor."

For an inductor (or a coil) modeled as shown in figure 17.11b, the quality factor, denoted more descriptively now by Q_L or Q_{coil}, may be calculated in accordance with equation 17.18:

$$Q_L(\omega) = \frac{\omega L}{R_s}. \tag{17.19}$$

Unlike the Q of a circuit, which depends on the values of the elements of the circuit and on the circuit's configuration, the quality factor Q_L of a coil varies with the operating frequency ω and remains unchanged irrespective of the location of the coil in the circuit. Higher Q_L implies a better quality coil in the sense that the energy loss in the component is smaller. Infinite Q_L represents an idealized lossless inductor modeled by a pure inductance L. In the audio frequency range, Q_L may vary from 5 to 20, whereas in the radio frequency range, it may reach above 100 for practical circuits. Although R_s also varies with frequency (due to the so-called skin effect), we shall, at the level of this text, treat R_s as a constant independent of ω.

Similarly, the capacitor, modeled as per figure 17.12b, has a quality factor Q_C or Q_{cap}, calculated in accordance with equation 17.18 as follows:

$$Y(j\omega) = \frac{1}{R_p} + j\omega C \equiv G_p + jB,$$

in which case

$$Z(j\omega) = \frac{1}{Y(j\omega)} = \frac{1}{G_p + jB} = \frac{G_p - jB}{G_p^2 + B^2} = \frac{G_p}{G_p^2 + B^2} - j\frac{B}{G_p^2 + B^2}.$$

On the other hand,

$$Z(j\omega) = R + jX.$$

Equating these two expressions yields

$$R = \frac{G_p}{G_p^2 + B^2}$$

and

$$X = \frac{-B}{G_p^2 + B^2}.$$

Therefore, by equation 17.18,

$$Q_C(\omega) = \frac{|X|}{R} = \frac{|-B|}{G_p} = \omega C R_p. \tag{17.20}$$

Just like Q_L, the quality factor Q_C varies with the operating frequency ω and remains unchanged, no matter where the capacitor is located in the circuit. A higher Q_C, of course, implies a better quality capacitor, again in the sense that the energy loss in the device is smaller. Infinite Q_C represents an idealized lossless capacitor modeled by a pure capacitance. In practice, Q_C is usually much greater than Q_L, and its effect on the performance of a circuit is often neglected (i.e., Q_C is assumed to be infinite). The reciprocal of Q_C is called the **dissipation factor** of the capacitor and is denoted by d_C. A lower dissipation factor means a better quality capacitor.

Inductors and capacitors with finite Q_L and Q_C are termed **lossy components**. Of course, all real-world components are lossy. The determination of Q_L and Q_C requires the specification of the operating frequency ω. However, if the value of ω is unspecified, the analysis of a tuned circuit proceeds under the assumption that $Q_L = Q_L(\omega_o)$ and $Q_C = Q_C(\omega_o)$, where, as usual $\omega_o = 1/\sqrt{LC}$. We will use these ideas shortly in more complicated resonant circuits.

When one uses the inductor and capacitor models of figures 17.11b and 17.12b, the circuit models for the parallel- and series-tuned circuits have the configurations of figures 17.13a and 17.13b, respectively. These models have neither the parallel RLC circuit structure of figure 17.6 nor the series RLC circuit structure of figure 17.9b. Therefore, the formulas derived earlier for ω_m and B_ω are not applicable.

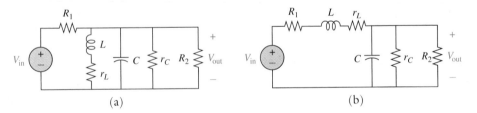

(a)   (b)

Figure 17.13 (a) Parallel- and (b) series-tuned circuits using practical inductor and capacitor models.

From a topological point of view, the circuits of figure 17.13 are called **series-parallel circuits**, meaning that the input impedance "seen" by the source consists of a sequence of series connections and parallel connections of simpler networks. Exact analysis of such series-parallel resonant circuits to obtain the values of ω_m and B_ω is extremely complicated and is usually not worth the effort. There is, however, a simpler, more efficient method widely used by engineers to obtain approximate solutions. (A third technique of approximate analysis, based on the concept of the pole-zero plot, will be presented in section 6.) This method employs two different, but equivalent, representations of an impedance $Z(j\omega)$ or admittance $Y(j\omega)$, as explained next.

Let us follow the steps used to define the Q of a practical inductor and capacitor and write a complex impedance of the two-terminal network as

$$Z(j\omega) = R_s + jX_s. \tag{17.21}$$

This suggests the representation of $Z(j\omega)$ as a reactance X_s in series with a resistance R_s, as shown in figure 17.14a. The admittance of the same network has a similar expression,

$$Y(j\omega) = G_p + jB_p, \tag{17.22}$$

which leads naturally to the representation of $Y(j\omega)$ as a susceptance B_p in parallel with a conductance G_p, as per figure 17.14b.

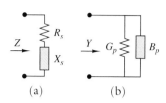

Figure 17.14 Conversion of (a) a series R_s-X_s combination to (b) a parallel G_p-B_p combination. No specific functional form is imposed upon X_s or B_p here.

Since $Y(j\omega) = [Z(j\omega)]^{-1}$, manipulating equation 17.21 into the form of equation 17.22 produces

$$G_p + jB_p = \frac{1}{R_s + jX_s} = \frac{R_s - jX_s}{R_s^2 + X_s^2}. \qquad (17.23)$$

Equating the real and imaginary parts of equation 17.23 identifies the following relationships among the four parameters R_s, X_s, R_p, and B_p:

$$B_p = -\frac{X_s}{R_s^2 + X_s^2}, \qquad (17.24a)$$

$$G_p = \frac{R_s}{R_s^2 + X_s^2}, \qquad (17.24b)$$

$$X_s = -\frac{B_p}{G_p^2 + B_p^2}, \qquad (17.24c)$$

$$R_s = -\frac{G_p}{G_p^2 + B_p^2}. \qquad (17.24d)$$

These formulas can be used to obtain approximate parallel circuits for given series circuits or approximate series circuits for given parallel ones. By a sequence of approximations, a complicated circuit can be reduced to either an approximate series or an approximate parallel RLC circuit, amenable to our earlier formulas for calculating ω_m, B_ω, and Q. To see this, consider the model of the practical inductor given in figure 17.11b. The preceding formulas make it possible to convert the series connection to an approximate parallel connection, as illustrated in figure 17.15.

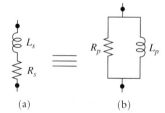

Figure 17.15 Conversion of an inductor model from (a) a series connection to (b) a parallel connection exact at a fixed frequency.

To find L_p and R_p in the parallel representation, we use equations 17.24a and b and express the results in terms of Q_L, as defined by equation 17.19. Specifically, for fixed nonzero ω, we may apply equation 17.24a to obtain

$$-\frac{1}{\omega L_p} = -\frac{\omega L_s}{R_s^2 + \omega L_s^2}$$

and apply equation 17.24b to obtain

$$\frac{1}{R_p} = \frac{R_s}{R_s^2 + X_s^2}.$$

From these two equations, at fixed ω one readily obtains

$$L_p = L_s \left(1 + \frac{1}{Q_L^2}\right) \text{(exact)}$$

$$\cong L_s \quad \text{(for } Q_L > 10)$$ (17.25a)

and

$$R_p = \left(1 + Q_L^2\right) R_s \text{ (exact)}$$

$$\cong Q_L^2 R_s = Q_L \omega L_s \text{ (for } Q_L > 10).$$ (17.25b)

Exercise. A 2-mH coil purchased from an electronic parts store has a Q_{coil} of 50 at 100 kHz. Find the element values in the series representation and the parallel representation at 100 kHz.

ANSWERS: $L_s = 2$ mH, $R_s = 25.13\ \Omega$, $L_p \simeq 2$ mH, $R_p = 62.83$ kΩ.

Similarly, the model for a capacitor, given in figure 17.12b, can be converted from a parallel connection to a series connection at a given frequency as illustrated in figure 17.16. To find C_s and R_s in the series representation, we use equations 17.24c and d and express the results in terms of Q_C, as defined by equation 17.20. The results are

$$C_s = C_p \left(1 + \frac{1}{Q_C^2}\right) \quad \text{(exact)}$$

$$\cong C_p \quad \text{(for } Q_C > 10)$$ (17.26a)

and

$$R_s = \frac{R_p}{1 + Q_C^2} \quad \text{(exact)}$$

$$\simeq \frac{R_p}{Q_C^2} = \frac{1}{Q_C \omega C} \text{ (for } Q_C > 10).$$ (17.26b)

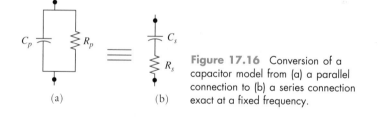

Figure 17.16 Conversion of a capacitor model from (a) a parallel connection to (b) a series connection exact at a fixed frequency.

With the foregoing formulas, we can now easily convert the parallel-tuned circuit of figure 17.13 to an approximate parallel RLC circuit and convert a series-tuned circuit to an approximate series RLC circuit valid in a neighborhood of a given frequency. Consequently, it becomes possible to analyze the circuit with the techniques described in the previous two sections. Of course, all of the results based on these converted circuits will be approximate primarily because we have assumed that $X_s = \omega L_s$ and $B_p = 1/\omega L_p$ or $X_s = 1/\omega C_s$ and $B_p = \omega C_p$ in equations 17.24a-d which is legitimate at a particular nonzero frequency. The approximation arises in two ways: (1) from neglecting the $1/Q_z^2$ term in equations 17.25a and 17.26a and (2) from ignoring the dependence of Q_z on ω, i.e., using the same value of Q_z calculated at $\omega = \omega_o$ for all frequencies near ω_o. For high-Q circuits (say, $Q_{cir} > 10$), an approximate analysis using this conversion technique ordinarily yields quite satisfactory results (an error of less than 1%). The next example will confirm this assertion. If very accurate results are desired, we will of course have to use one of the many circuit simulation programs (see, for example, SPICE in the appendix) now available to obtain the desired results.

EXAMPLE 17.10

In the circuit shown in figure 17.17a, the sinusoidal source is represented by an independent voltage source V_{in} in series with an internal resistance $R_s = 40 \text{ k}\Omega$. The capacitor has $C = 0.05 \text{ µF}$ and a dissipation factor of 0.01 at ω_o. The coil has an inductance of 20 mH, and $Q_L(\omega_o) = 40$. The external load resistance is $R_L = 10 \text{ k}\Omega$.

Find approximate values of ω_m, B_ω, Q_{cir}, ω_1, and ω_2.

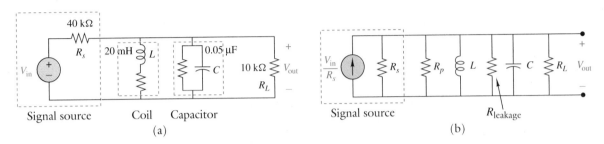

Figure 17.17 Approximate analysis of a high-Q circuit. (a) Original circuit. (b) Approximate parallel RLC circuit.

SOLUTION

The first step is to use equations 17.25 to obtain a parallel representation of the coil. This requires computing

$$\omega_o = \frac{1}{\sqrt{LC}} = \frac{1}{\sqrt{0.02 \times 0.05 \times 10^{-6}}} = 31,623 \text{ rad/sec}$$

and then

$$R_p = Q_L \omega_o L = 40 \times 31,623 \times 0.02 = 25,298 \ \Omega.$$

The capacitor has a quality factor $Q_c = 1/0.01 = 100$. The leakage resistance of the capacitor, calculated from equation 17.20, is

$$R_{\text{leakage}} = \frac{Q_C}{\omega_o C} = \frac{100}{31,623 \times 0.05 \times 10^{-6}} = 63,247 \ \Omega.$$

Replacing the series $R_s - V_{in}$ by its Norton equivalent circuit produces the network of figure 17.17b. The parallel combination of R_s, R_p, R_{leakage}, and R_L is identified with R in figure 17.6. Using the notation "//" to indicate the parallel combination of two impedances, we can write

$$R = 40,000//25,298//63,247//10,000 = 5,545 \ \Omega,$$

$$\omega_m \cong \omega_o = 31,623 \text{ rad/sec},$$

$$B_\omega = \frac{1}{RC} = \frac{1}{5,545 \times 0.05 \times 10^{-6}} = 3,607 \text{ rad/sec},$$

$$Q_{cir} = \frac{\omega_m}{B_\omega} = \frac{31,623}{3,607} = 8.77,$$

$$\omega_1 \cong \omega_m - \frac{B_\omega}{2} = 31,623 - 1,803.5 = 29,819 \text{ rad/sec},$$

$$\omega_2 \cong \omega_m + \frac{B_\omega}{2} = 31,623 + 1,803.5 = 33,426 \text{ rad/sec}.$$

Although the circuit of example 17.10 has a only a medium $Q(Q_{cir} = 8.76)$, the results should still be fairly accurate. To get some assurance that they are, let us calculate the exact value from the expression given later in problem 38. The result is $\omega_m(\text{exact}) = 31,658$ rad/sec. There are no exact formulas available for B_ω, ω_1, and ω_2 for the series-parallel-tuned circuits of figure 17.13. We therefore have to use a CAD program (for example, SPICE) to obtain

$$B_\omega = 3,607, \quad \omega_1 = 29,906, \quad \text{and} \quad \omega_2 = 33,513 \text{ (all in rad/sec)}.$$

Comparing the approximate solutions with the exact solutions, we see that for Q_{cir} around 10, the errors in ω_m, ω_1, and ω_2 are less than 0.5%. The error in B_ω is practically nil, because the approximation yields lower values for ω_1 and ω_2 at nearly the same percentage, so the effect hardly shows up in B_ω, which is the difference between ω_2 and ω_1. For circuits with higher Q, as is typically the case in radio frequency circuits, the errors are even smaller.

In the parallel RLC and series RLC circuits studied in the previous sections, $\omega_m = \omega_r = \omega_o$. For the series-parallel circuits of figure 17.13, these frequencies are different (see problems 30 and 31), even though for high-Q circuits, the differences may be very small. This was clearly seen in example 17.10.

6. BANDPASS TRANSFER FUNCTIONS WITH ONE PAIR OF COMPLEX POLES

The Structure of a Bandpass Transfer Function

The previous sections utilized the techniques of sinusoidal steady-state analysis to describe the analysis and design of some simple tuned circuits. There the network functions of interest are impedances and admittances, taken as frequency-dependent complex numbers. The powerful transfer function concept $H(s)$ based on the Laplace transform methods of chapters 13 through 15 allows us to broaden our viewpoint. The advantages of this $H(s)$ approach are that it offers

1. A unified treatment of various circuit configurations.
2. A better way of performing *approximate* analysis of high-Q circuits.
3. A starting point for realizing the desired bandpass characteristic with the so-called active elements (i.e., op amps, transistors, etc.).

At this point, it is helpful to review section 3 of chapter 15 thoroughly before proceeding.

Since all of the circuits considered in this section contain only one inductance and one capacitance, any associated network function will obviously have at most a *second-degree* polynomial in s as the denominator. Thus, the general transfer function $H(s)$ (which includes the impedance function $Z(s)$ and the admittance function $Y(s)$ as special cases) of a tuned circuit has the biquadratic form

$$H(s) = \frac{n(s)}{d(s)} = \frac{a_2 s^2 + a_1 s + a_o}{s^2 + 2\sigma_p s + \omega_p^2}.$$

Only the case of complex poles is of interest, because a reasonably sharp bandpass characteristic requires that $H(s)$ have complex poles, i.e., $\sigma_p < \omega_p$. The finite zeros of $H(s)$, which are roots of $n(s) = 0$, may be real or complex. The case of complex zeros, corresponding to more advanced filter types, such as the inverse Chebyshev or elliptic types, is beyond the scope of this text. We consider only the case when $H(s)$ has real zeros. Before proceeding, it is necessary to introduce some new terms related to the complex poles of $H(s)$. Suppose $H(s)$ has one finite zero. Then the transfer function has the form

$$H(s) = \frac{a_1 s + a_o}{s^2 + 2\sigma_p s + \omega_p^2} = \frac{a_1 s + a_o}{s^2 + \frac{\omega_p}{Q_p} s + \omega_p^2}, \tag{17.27}$$

with the pole-zero plot shown in figure 17.18.

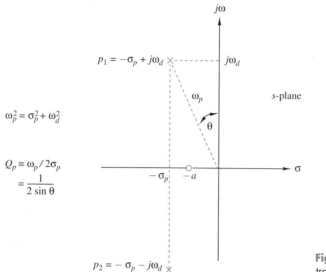

$$\omega_p^2 = \sigma_p^2 + \omega_d^2$$

$$Q_p \equiv \omega_p / 2\sigma_p$$
$$= \frac{1}{2 \sin \theta}$$

Figure 17.18 Pole-zero plot of the transfer function $H(s)$ of equation 17.27.

Equation 17.27 contains a new quality factor, Q_p, mathematically called the **pole Q**, defined as

$$Q_p \equiv \frac{\omega_p}{2\sigma_p} = \frac{1}{2 \sin(\theta)}, \tag{17.28}$$

where the angle θ is as shown in figure 17.18. From chapter 15, the impulse response of a system characterized by equation 17.27 has the form

$$h(t) = \mathcal{L}^{-1}\{H(s)\} = K e^{-\sigma_p t} \cos(\omega_d t + \theta).$$

The waveform is a damped sinusoid, and the quantity ω_d (not ω_p) determines the frequency of oscillations. For this reason, ω_d is referred to as the **damped natural frequency.**

In a pole-zero plot, Q_p measures how close the pole is to the $j\omega$-axis: A higher Q_p means a smaller θ, implying a pole closer to the $j\omega$-axis. In the magnitude response curve, as will be shown shortly, Q_p is related to the circuit Q and serves as a quick estimate of the sharpness of the response curve. For some special cases, Q_p equals the circuit Q. As we have now talked about four different kinds of Q's (Q_{cir}, Q_{coil}, Q_{cap}, and Q_p), it is important to keep in mind their conceptual differences, even though in practical circuits their numerical values may be very close. The need for distinguishing between Q_p and Q_{cir} will become apparent in section 8, when more than one pair of complex poles appear in the transfer function.

For a transfer function having a pair of complex poles, our goal is to determine several key quantities that are indicative of the circuit's behavior: ω_m, $|H(j\omega)|_{\max}$, B_ω, and the frequency at which the angle of $H(j\omega)$ is zero. There are four cases to consider.

Case 1: $H(s)$ Has a Single Zero at the Origin

For this case, $a_2 = a_o = 0$ and $a_1 \neq 0$. Equation 17.29 specifies the form of $H(s)$, whose pole-zero plot is shown in figure 17.19:

$$H(s) = K \frac{s}{s^2 + \frac{\omega_p}{Q_p} s + \omega_p^2} = K \frac{s}{s^2 + 2\sigma_p s + \omega_p^2}. \tag{17.29}$$

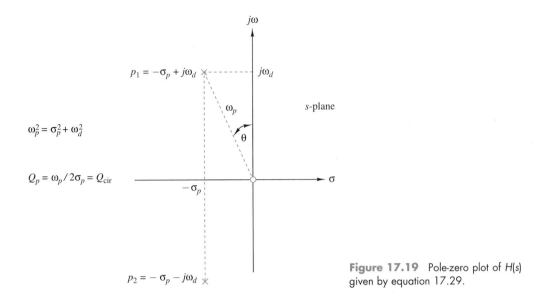

$$\omega_p^2 = \sigma_p^2 + \omega_d^2$$

$$Q_p = \omega_p/2\sigma_p = Q_{\text{cir}}$$

Figure 17.19 Pole-zero plot of $H(s)$ given by equation 17.29.

The parallel RLC and series RLC circuits shown in figure 17.9 have this structure. The reason is that the parallel RLC circuit of figure 17.9a has the transfer function

$$H(s) = \frac{V_p}{I_p} = Z_p(s) = \cfrac{1}{\cfrac{1}{R_p} + \cfrac{1}{sL_p} + sC_p} = \frac{1}{C_p} \cdot \frac{s}{s^2 + \cfrac{1}{R_pC_p}s + \cfrac{1}{L_pC_p}}$$

and the series RLC circuit of figure 17.9b has the analogous transfer function

$$H(s) = \frac{I_s}{V_s} = Y_s(s) = \cfrac{1}{R_s + sL_s + \cfrac{1}{sC_s}} = \frac{1}{L_s} \cdot \frac{s}{s^2 + \cfrac{R_s}{L_s}s + \cfrac{1}{L_sC_s}}.$$

Clearly, both transfer functions are of the form of equation 17.29.

Using the same methods as in sections 2 and 3, one can prove the following *exact* results for $H(s)$ given by equation 17.29:

(a) $\quad \omega_m = \omega_p$ $\hfill$ (17.30a)

(b) $\quad |H(j\omega)|_{max} = |H(j\omega_m)| = \dfrac{KQ_p}{\omega_m}.$ $\hfill$ (17.30b)

(c) $\quad B_\omega = 2\sigma_p = \dfrac{\omega_p}{Q_p}.$ $\hfill$ (17.30c)

(d) $\quad Q_{\text{cir}} = Q_p.$ $\hfill$ (17.30d)

(e) $\quad \omega_{1,2} = \pm\sigma_p + \sqrt{\sigma_p^2 + \omega_p^2}.$ $\hfill$ (17.30e)

(f) $\quad$ zero phase-shift frequency $= \omega_p.$ $\hfill$ (17.30f)

The derivation of these formulas is left as an exercise. (See problem 25.)

Figure 17.20 shows a plot of normalized $|H(j\omega)|$ vs. ω for different values of $Q_{\text{cir}} = Q_p$. The ordinate is the ratio $|H(j\omega)|/|H(j\omega)|_{max}$, while the abscissa shows the ratio ω/ω_m. These curves are called **universal resonance curves**, as they are applicable to parallel RLC circuits, to series RLC circuits, or to any system having a transfer function of the form of equation 17.29.

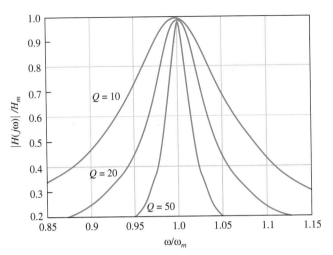

Figure 17.20 Normalized magnitude response for equation 17.29

Case 2: $H(s)$ Has a Single Zero off the Origin

In this case, $a_2 = 0$, $a_0 \neq 0$, and $a_1 \neq 0$, which causes the zero at the origin of case 1 to move into the left half-plane. The form of the bandpass transfer function is

$$H(s) = K \frac{s+a}{s^2 + \dfrac{\omega_p}{Q_p}s + \omega_p^2} = K \frac{s+a}{s^2 + 2\sigma_p s + \omega_p^2}. \qquad (17.31)$$

The series-parallel RLC circuits shown in figure 17.13 belong to this case. Figure 17.18 sketches the pole-zero plot for the transfer function of those circuits. This plot differs from figure 17.19 only in the displacement of the zero from the origin to the left. Intuitively speaking, the closer the zero is to the origin, the more the magnitude response resembles that of case 1. Derivations of *exact* values of ω_m and the zero-phase-shift frequency are possible. (See problems 26 and 27.) The results are:

$$\omega_m = \sqrt{-a^2 + \sqrt{(\omega_p^2 + a^2)^2 - (2\sigma_p a)^2}}, \qquad (17.32)$$

$$\omega \text{ (for zero phase shift)} = \sqrt{\omega_p^2 - 2\sigma_p a}. \qquad (17.33)$$

No exact expression is available for the half-power bandwidth B_ω. For the high-Q_p case *approximate* answers are

$$\omega_m \cong \omega_p, \quad Q_{\text{cir}} \cong Q_p, \quad \text{and} \quad B_\omega \cong 2\sigma_p \quad \text{for} \quad Q_p > 10. \qquad (17.34)$$

◆ **EXAMPLE 17.11**

Consider the series-parallel RLC circuit given in example 17.10. Find the *exact* value of ω_m by the use of equation 17.32.

SOLUTION

The first step in this approach is to derive $H(s)$. A straightforward analysis of the circuit yields

$$H(s) = \frac{V_{\text{out}}}{V_{\text{in}}} = \frac{1}{R_s C} \frac{s + \dfrac{R_{\text{coil}}}{L}}{s^2 + \left(\dfrac{R_{\text{coil}}}{L} + \dfrac{1}{R_g C}\right)s + \left(1 + \dfrac{R_{\text{coil}}}{R_g}\right)\dfrac{1}{LC}}, \qquad (17.35)$$

where $R_g = R_s // R_L // R_{leakage}$. Matching coefficients with equation 17.31, we have

$$\omega_p^2 = \frac{1 + \dfrac{R_{coil}}{R_g}}{LC},$$

$$2\sigma_p = \frac{R_{coil}}{L} + \frac{1}{R_g C},$$

$$a = \frac{R_{coil}}{L}.$$

Substituting numerical values into the preceding expressions yields

$$R_{leakage} = \frac{Q_c}{\omega_o C} = \frac{100}{31,622 \times 0.05 \times 10^{-6}} = 63,247 \ \Omega,$$

$$R_{coil} = \frac{\omega_o L}{Q_L} = \frac{31,622 \times 0.02}{40} = 15.81 \ \Omega,$$

$$R_g = 40,000 // 10,000 // 63,247 = 7,101.7 \ \Omega,$$

$$\omega_p^2 = \frac{1 + \dfrac{15.81}{7,101.7}}{0.02 \times 0.05 \times 10^{-6}} = 1.00222 \times 10^9 (\text{rad/sec})^2,$$

$$2\sigma_p = \frac{15.81}{0.02} + \frac{1}{7,101.7 \times 0.05 \times 10^{-6}} = 3,606.73 \ \text{rad/sec},$$

$$a = \frac{15.81}{0.02} = 790.5 \ \text{rad/sec}.$$

Finally, we substitute these values into equation 17.32 and obtain

$$\omega_m = \sqrt{-790.5^2 + \sqrt{\left(1.00222 \times 10^9 + 790.5^2\right)^2 - \left(3,606.73 \times 790.5\right)^2}}$$

$$= 31,657.79 \ \text{rad/sec}.$$

Case 3: $H(s)$ Has No Finite Zero

In this case, $a_2 = a_1 = 0$ and $a_o \neq 0$, which implies that

$$H(s) = \frac{K}{s^2 + \dfrac{\omega_p}{Q_p} s + \omega_p^2} = \frac{K}{s^2 + 2\sigma_p s + \omega_p^2}. \tag{17.36}$$

The pole-zero plot of this $H(s)$ differs from that of figure 17.19 only in that the single zero is now absent. The series RLC circuit (see problem 17a) with capacitance voltage as the output, falls into this category. So does the parallel RLC circuit, with the inductor current as the output.

It is possible for the transfer function 17.36 to display a lowpass or a bandpass magnitude response, depending the value of Q_p. To see this, let us proceed to find ω_m, the frequency at which $|H(j\omega)|$ is maximum. Setting the derivative of $|H(j\omega)|$ to zero and solving for ω produces the exact formula

$$\omega_m = \omega_p \sqrt{1 - \frac{1}{2Q_p^2}}.$$

If $Q_p < 1/\sqrt{2} = 0.707$, then there is no real solution for ω_m. In this case, although the poles are complex, the magnitude response does not display a peak at any nonzero frequency. Rather, the magnitude function $|H(j\omega)|$ equals H_m at $\omega = 0$ and decreases *monotonically* toward zero as ω increases. Hence, it has a lowpass characteristic. The borderline case of $Q_p = 1/\sqrt{2}$ results in a "maximally flat" lowpass magnitude response that will be discussed in greater detail in chapter 21.

When Q_p is greater than $1/\sqrt{2}$, the magnitude response of equation 17.36 starts from a nonzero value at $\omega = 0$, rises to the peak value at $\omega = \omega_m$, and finally decreases to zero as the frequency approaches infinity. If Q_p is only slightly greater than 0.707, then the magnitude response is essentially that of the lowpass type, with a slight hump in the passband.

For the high-Q_p case ($Q_p > 10$), the magnitude response near ω_m is essentially that of a bandpass circuit. The preceding exact expression for ω_m reduces to $\omega_m \cong \omega_p$ for the high-Q_p case. Unfortunately, no exact expression for the bandwidth B_ω is available for equation 17.36. In practical applications of bandpass circuits, Q_p in equation 17.36 is typically greater than 10, permitting the use of the high-Q approximation, to be discussed in the next subsection. The result shows that $B_\omega \cong 2\sigma_p$.

Case 4. *H(s)* Has Two Real Zeros

In this case, the transfer function has the form

$$H(s) = \frac{n(s)}{d(s)} = \frac{(s + a)(s + b)}{s^2 + \dfrac{\omega_p}{Q_p}s + \omega_p^2}.$$

Resonant circuits having more accurate models for the inductor and the capacitor give rise to a transfer function in this category. (See problem 41.)

Depending on the values of Q_p and the coefficients a and b, the magnitude response of this transfer function may have a lowpass, highpass, or bandpass characteristic. For the high-Q_p case ($Q_p > 10$), the transfer function has a bandpass characteristic, and the magnitude response near ω_m is again essentially that of case 1. No exact expression for ω_m or B_ω is available. Accurate values of ω_m and B_ω can be found only by the use of circuit simulation programs, such as SPICE. However, for the high-Q_p case, the method of the next subsection provides an easy *estimate* of ω_m and B_ω.

Computation of Approximate Solutions

In all of the preceding four cases of bandpass transfer functions, if the pole Q is high, say, $Q_p > 10$, then *approximate* answers to all the questions raised before can be obtained quite easily from the pole-zero plot without complicated formulas. In particular, we will show that the cluster of poles and zeros near the point $s = j\omega_o$ in the upper half s-plane dominates the shape of the magnitude response curve. The clusters of poles and zeros near the origin and those in the lower half s-plane, being very far from the point $s = j\omega_o$, contribute very little to the variation of $|H(j\omega)|$ for the frequency $\omega \cong \omega_m$. It is important at this point to review and thoroughly understand the material of section 15.5—in particular, example 15.10.

First, reconsider equation 17.31, in the form

$$H(s) = K\frac{s - z_1}{(s - p_1)(s - p_2)}. \tag{17.37}$$

Taking appropriate magnitudes with $s = j\omega$ yields

$$|H(j\omega)| = |K|\frac{|j\omega - z_1|}{|j\omega - p_1| \times |j\omega - p_2|}, \tag{17.38}$$

which has the geometric interpretation shown in figure 17.21.

The three vector lengths in equation 17.38 vary with the frequency ω. However, the variation of $|j\omega - p_1|$ dominates the shape of the magnitude response curve. This is because, for the high-Q_p case, the pole p_2 is very close to the $j\omega$-axis (more precisely, the angle of the vector $j\omega - p_2$ is nearly 90 degrees), resulting in $j\omega - p_2 \cong j(\omega + \omega_d)$. Suppose that the zero z_1 is close to the origin, as is the case in most practical bandpass circuits. Then

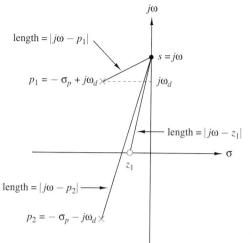

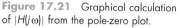

Figure 17.21 Graphical calculation of $|H(j\omega)|$ from the pole-zero plot.

$j\omega - z_1$ approximately equals $j\omega$. As long as the frequency ω of interest is near ω_d (recall that for a high-Q circuit, $\omega_d \cong \omega_p \cong \omega_m$), it follows that

$$\frac{j\omega - z_1}{j\omega - p_2} \cong \frac{j\omega}{j(\omega + \omega_d)} \cong \frac{1}{2}.$$

Substituting this approximate relationship into equations 17.37 and 17.38 yields the following simplified expressions for $\omega \cong \omega_p$:

$$H(j\omega) \cong \frac{0.5K}{j\omega - p_1}, \tag{17.39}$$

$$|H(j\omega)| \cong \frac{0.5\,|K|}{|j\omega - p_1|}. \tag{17.40}$$

From equation 17.40, the variation of the vector length $|j\omega - p_1|$ alone *approximately* determines the variation of $|H(j\omega)|$ for ω near $\omega_d \cong \omega_p$. To visualize this variation, refer to the pole-zero plot of the simplified equation 17.40 near the point p_1, as shown in figure 17.22.

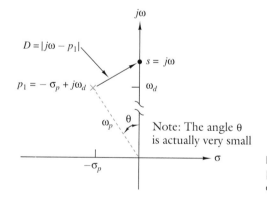

Figure 17.22 Approximate value of $|H(j\omega)|$ from the pole-zero plot (high-Q case).

Let us denote the distance from the point p_1 to the point $s = j\omega$ by D. From basic geometry, it follows that:

1. D is smallest when $\omega \doteq \omega_d$, and $D_{\min} = \sigma_p$.
2. $D = \sigma_p\sqrt{2}$ when $\omega = \omega_1 = \omega_d - \sigma_p$ and $\omega = \omega_2 = \omega_d + \sigma_p$.

Substituting these relationships into equation 17.40 leads to the conclusions that

$$\omega_m \cong \omega_d \cong \omega_p, \tag{17.41a}$$

(with ω_p being closer to the true solution),

$$H_m = |H(j\omega)|_{max} \cong \frac{|K|}{2\sigma_p},$$ (17.41b)

and

$$B_\omega = \omega_2 - \omega_1 \cong 2\sigma_p$$ (17.41c)

whenever $H(s)$ of equation 17.31 has $Q_p > 10$. The next example will illustrate the use of pole-zero plots for the approximate manual analysis of tuned circuits.

◆ EXAMPLE 17.12

Consider again the circuit of example 17.11. Use the pole-zero plot method described above to find *approximate* solutions of ω_m, H_m, B_ω, and the ratio $|H(j\omega)|/H_m$ at $\omega = 34,000$ rad/sec.

SOLUTION

From example 17.11—specifically, equation 17.35—the transfer function is

$$H(s) = \frac{V_{out}}{V_{in}} = 500\frac{s + 790.5}{s^2 + 3,606.8s + 1.00222 \times 10^9}$$

$$= 500\frac{s + 790.5}{(s - 1,803.4 - j31,606.5)(s - 1,803.4 + j31,606.5)},$$

after substitution of numerical values. Accordingly, $K = 500$, $\sigma_p = 1,803.4$, $\omega_d = 31,606.5$, and $\omega_p = 31,657.8$. Substituting these values into equation 17.41 yields the desired answers:

$$\omega_m \cong \omega_p = 31,657.8 \text{ rad/sec},$$

$$H_m = |H(j\omega)|_{max} \cong \frac{|K|}{2\sigma_p} = \frac{500}{3,606.8} = 0.1386,$$

$$B_\omega = \omega_2 - \omega_1 \cong 2\sigma_p = 3,606.73 \text{ rad/sec}.$$

Finally, to calculate the ratio $|H(j\omega)|/H_m$ at $\omega = 34,000$ rad/sec, we rewrite the approximate expression (equation 17.40) as

$$|H(j\omega)| \cong \frac{0.5|K|}{|j\omega - p_1|} = \frac{\sigma_p}{|j\omega - p_1|}\frac{|K|}{2\sigma_p} = \frac{\sigma_p}{|j\omega - p_1|}H_m.$$

This leads to

$$\frac{|H(j\omega)|}{|H(j\omega)|_{max}} \cong \frac{\sigma_p}{|j\omega - p_1|}.$$ (17.42)

Substituting numerical values into equation 17.42 produces

$$\frac{|H(j34,000)|}{H_m} = \frac{1,803.4}{|j34,000 - (-1,803.4 + j31,606.5)|} = 0.6018.$$

We have studied two methods for the approximate analysis of high-Q tuned circuits. The first makes use of circuit transformations, as shown in figures 17.15 and 17.16, and the second makes use of the pole-zero plot of $H(s)$. Both give numerical answers of comparable accuracy. This is seen by comparing the answers for $\omega_m, |H(j\omega)|_{max}$, and B_ω in example 17.12 with those obtained for these same quantities in example 17.10. They agree quite well. Some remarks about the advantages of each method are in order. The circuit transformation method of section 5 avoids the calculation of $H(s)$ as a function of s, which is required in the pole-zero plot method. The major advantages of the pole-zero plot method are that

1. It is applicable to non-series-parallel RLC circuits. (See problem 50.)
2. It is valid for general second-order linear circuits containing controlled sources. (See example 17.13.)

An Active Bandpass Circuit

To conclude this section, we present an example of a bandpass circuit that utilizes active elements. Because inductance is absent in such a circuit, none of the formulas developed in sections 2 through 5 are applicable. The only alternative is the transfer function approach described in this section. The bandpass circuit illustrated here is only one of more than a dozen configurations in use. You can learn a lot more about these active filters in a more advanced course. The example here is meant only to arouse your curiosity about the possibility of eliminating inductances and yet producing the same kind of frequency response as that of RLC resonant circuits.

 EXAMPLE 17.13

The operational amplifier in figure 17.23 is assumed to be ideal.

1. Find the transfer function $H(s) = V_o(s)/V_{in}(s)$.
2. Find a set of element values if the bandpass circuit is to have peak frequency $\omega_m = 1,000$ rad/sec and bandwidth $B_\omega = 100$ rad/sec.

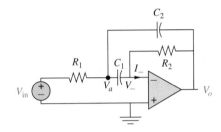

Figure 17.23 An active bandpass circuit without inductance.

SOLUTION

1. From the assumption that the operational amplifier is ideal, we have $V_- = 0$ (virtual ground) and $I_- = 0$ (infinite input impedance). Applying KCL to the inverting input node V_-, we get

$$sC_1 V_a + \frac{V_o}{R_2} = 0,$$

which yields

$$V_a = -\frac{1}{sC_1 R_2} V_o.$$

Next, we apply KCL to node V_a to obtain

$$\frac{V_a - V_{in}}{R_1} + sC_1 V_a + sC_2(V_a - V_o) = 0.$$

Substituting the previous expression for V_a into this equation, regrouping terms, and solving for V_o results in

$$H(s) = \frac{V_o(s)}{V_{in}(s)} = \frac{\dfrac{-s}{R_1 C_2}}{s^2 + \left(\dfrac{1}{R_2 C_1} + \dfrac{1}{R_2 C_2}\right)s + \dfrac{1}{R_1 R_2 C_1 C_2}}.$$

2. Since the transfer function is of the form of equation 17.29, we have the exact peak frequency and bandwidth given by equations 17.30a and c. Equating these expressions to the specified values yields

$$\omega_m = \omega_p = \frac{1}{\sqrt{R_1 R_2 C_1 C_2}} = 1,000 \text{ rad/sec} \qquad (17.43)$$

and

$$B_\omega = 2\sigma_p = \frac{\omega_p}{Q_p} = \frac{1}{R_2 C_1} + \frac{1}{R_2 C_2} = 100 \text{ rad/sec.} \qquad (17.44)$$

---

6. **BANDPASS TRANSFER FUNCTIONS WITH ONE PAIR OF COMPLEX POLES**

Since there are four unknowns—R_1, R_2, C_1, and C_2—but only two constraint equations, the solution is not unique. For a solution, choose any two unknowns and then solve for the remaining two. However, some choices may lead to negative element values, which are not passive elements. One simple solution is to let

$$C_1 = C_2 = 10^{-6} \text{ F.}$$

Then, from equation 17.44, $R_2 = 20{,}000 \ \Omega$, and from equation 17.43, $R_1 = 50 \ \Omega$.

Exercise. In example 17.13, if $C_1 = 0.5 \ \mu$F and $C_2 = 1 \ \mu$F, find the values of R_1 and R_2 required to meet the specifications on ω_m and B_ω.

ANSWER: $R_1 = 66.67 \ \Omega$, $R_2 = 30 \ k\Omega$.

---

7. MAGNITUDE SCALING AND FREQUENCY SCALING

Example 17.6 analyzed the bandpass circuit shown in figure 17.24. The results give a center frequency at $\omega_m = 31{,}622$ rad/sec and a bandwidth $B_\omega = 2{,}500$ rad/sec. At least two engineering scenarios may occur in practical applications: (1) An application might require a modified circuit for use with a signal source having a 20-kΩ internal resistance, while maintaining the same center frequency and bandwidth. (2) Another application might require an increase in the center frequency to 50,000 rad/sec, while maintaining the same circuit Q. How can the existing circuit be modified to achieve these objectives?

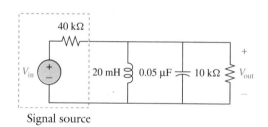

Signal source

Figure 17.24 The bandpass circuit analyzed in example 17.6.

The desired new circuits can be obtained in a very simple manner by the methods of *magnitude scaling* and *frequency scaling*, to be described next. We begin with some simple circuits that demonstrate the principles underlying these methods in a very intuitive manner. For general linear circuits, we shall only *state* the procedures for scaling and their effects, leaving a rigorous mathematical justification to a more advanced text on circuit theory.

Figure 17.25a shows a voltage divider circuit consisting of two 10-kΩ resistances. The voltage ratio is $V_o/V_i = 0.5$. Suppose both resistances are made 10 times larger, as shown in figure 17.25b. Clearly, the voltage ratio remains unchanged.

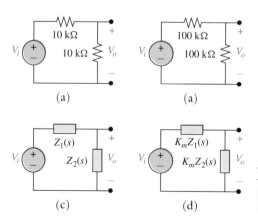

Figure 17.25 Magnitude scaling does not affect the voltage ratio. (a) A voltage divider. (b) Magnitude scaling of (a) by 10. (c) A general voltage divider. (d) Magnitude scaling of (c) by K_m.

As a more general example, figures 17.25c and 17.25d differ only in that every impedance in the latter is K_m times the corresponding impedance in the former. It is easy to see that the voltage ratios are the same for both circuits:

$$\left[\frac{V_o}{V_i}\right]_{(d)} = \frac{K_m Z_2}{K_m Z_1 + K_m Z_2} = \frac{Z_2}{Z_1 + Z_2} = \left[\frac{V_o}{V_i}\right]_{(c)}.$$

The network of figure 17.25(d) is said to be obtained from that of figure 17.25(c) by *magnitude scaling*, with a scale factor of K_m. If Z_1 is an inductance, then $Z_1(s) = sL$, and $K_m Z_1(s) = K_m sL = s(K_m L)$, i.e., the inductance is *multiplied* by K_m. On the other hand, if Z_1 is a capacitance, then $Z_1(s) = 1/sC$, and $K_m Z_1(s) = K_m/sC = s/(C/K_m)$, i.e., the capacitance is *divided* by K_m.

It is possible to generalize the preceding simple observations. Our goal is to state formally a procedure for magnitude scaling a network and then determine the effect on the circuit transfer functions.

Procedure for magnitude scaling To magnitude scale a linear network by a scale factor K_m:

1. Multiply all resistances and inductances by K_m.
2. Divide all capacitances by K_m.
3. For r_m type controlled sources (i.e., current-controlled voltage sources), multiply the parameter r_m by K_m.
4. For g_m type controlled sources (i.e., voltage-controlled current sources), divide the parameter g_m by K_m.
5. Parameters for voltage-controlled voltage sources and current-controlled current sources remain unchanged.
6. Ideal operational amplifiers remain unchanged.

Effect of magnitude scaling If $H(s)$ is a voltage ratio or current ratio, magnitude scaling has no effect on $H(s)$. If $H(s)$ has units of ohms, then the magnitude-scaled network has $H_{\text{new}}(s) = K_m H(s)$. If $H(s)$ has units of reciprocal ohms (i.e., mhos or siemens), then the magnitude-scaled network has $H_{\text{new}}(s) = H(s)/K_m$.

These stated effects are particularly obvious when rules 3 through 5 are applied to a circuit containing a single controlled source. For slightly more complex circuits, we can verify the effects by actually calculating the new transfer functions for the scaled network.

EXAMPLE 17.14

The series resonant circuit of figure 17.26a has

$$Z(s) = \frac{1}{s} + s + 0.1 = \frac{s^2 + 0.1s + 1}{s}$$

and

$$H(s) = \frac{V_{\text{out}}}{V_{\text{in}}} = \frac{0.1}{Z(s)} = \frac{0.1s}{s^2 + 0.1s + 1}.$$

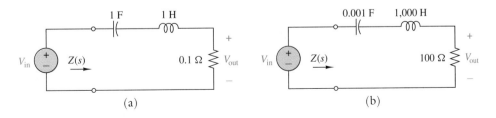

Figure 17.26 Magnitude scaling. (a) Original network. (b) Scaled network.

1. Magnitude scale the network by a factor of $K_m = 1,000$.
2. Calculate the new $Z(s)$ and $H(s)$ to verify the effects just stated.

SOLUTION

1. Following the procedure for scaling R's, L's, and C's, we obtain the scaled network shown in figure 17.26b.

2. For the simple circuit of figure 17.26b, the new input impedance is

$$Z_{new}(s) = \frac{1}{0.001s} + 1,000s + 100 = 1,000\frac{s^2 + 0.1s + 1}{s} = 1,000Z(s) = K_m Z(s).$$

By the voltage divider formula, the new transfer function is

$$H_{new}(s) = \frac{100}{1,000\dfrac{s^2 + 0.1s + 1}{s}} = \frac{0.1s}{s^2 + 0.1s + 1} = H(s).$$

There is no change in the transfer function, as it is a voltage ratio. These results clearly verify the stated effects of magnitude scaling.

Exercise. Magnitude scale the circuit of figure 17.24 by $K_m = 5$.
ANSWER: 200 kΩ, 0.1 H, 0.01 μF, 50 kΩ.

Magnitude scaling affects all RLC elements. The next type of scaling affects L and C elements only. As a simple example, consider again the series resonant circuit of figure 17.26a. Using equations 17.13 and 17.15, we find that the center frequency and Q of the circuit are $\omega_m = 1$ rad/sec and $Q = 10$. For reference, the pole-zero plot of $H(s)$ and the curve of $|H(j\omega)|$ vs. ω are shown in figures 17.27a and 17.27b, respectively.

Consider the effect of a frequency-scaling process in which we divide all capacitances and inductances of figure 17.26a by a factor $K_f = 10^6$, while keeping resistance values unchanged. This process yields the new circuit shown in figure 17.28.

Using the voltage divider formula, we find that the new transfer function is

$$H_{new}(s) = \frac{0.1}{0.1 + 10^{-6}s + \dfrac{1}{10^{-6}s}} = \frac{0.1\left(\dfrac{s}{10^6}\right)}{\left(\dfrac{s}{10^6}\right)^2 + 0.1\left(\dfrac{s}{10^6}\right) + 1} = H\left(\frac{s}{10^6}\right).$$

This new transfer function is related to the original $H(s)$ through the replacement of s by s/K_f. The implication of this relationship is that whatever happened at $s = j\omega_{old}$ before scaling must now happen at $s = j\omega_{new}$, where $\omega_{new} = K_f\omega_{old}$; hence the term *frequency scaling*. This is borne out by comparing the pole-zero plot and the magnitude response curve for $H_{new}(s)$, shown in figure 17.27c and d, with their counterparts for $H(s)$, shown in figures 17.27a and b. For example, in figure 17.27b for $H(s)$, the peak response occurs at $\omega_{old} = 1$ rad/sec, whereas in figure 17.27d for $H_{new}(s)$, the peak response occurs at $\omega_{new} = K_f\omega_{old} = 10^6$ rad/sec.

For general linear networks, the results for frequency scaling are as follows.

Procedure for frequency scaling To frequency scale a linear network by a scale factor K_f:
1. Divide all inductances and capacitances by K_f.
2. All resistance values and controlled source parameters remain unchanged.

Effect of frequency scaling $H_{new}(s) = H(s/K_f)$. The locations of the poles and zeros of $H(s)$ are expanded by a factor K_f. The scale on the frequency axis of the magnitude (phase) response curve is expanded by a factor of K_f; i.e., new critical frequencies equal K_f times old critical frequencies.

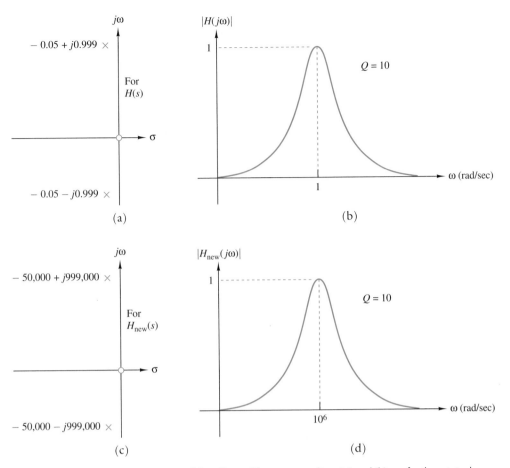

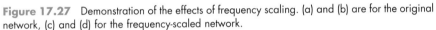

Figure 17.27 Demonstration of the effects of frequency scaling. (a) and (b) are for the original network, (c) and (d) for the frequency-scaled network.

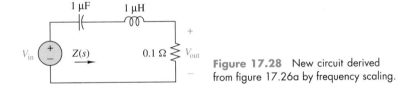

Figure 17.28 New circuit derived from figure 17.26a by frequency scaling.

Exercise. Frequency scale the network of figure 17.24 by a factor $K_f = 100$.

If an RLC network undergoes both magnitude and frequency scaling, then the order of scaling is immaterial. This is easily shown. For example, if the network of figure 17.26a is magnitude scaled by a factor of $K_m = 100$ and frequency scaled by a factor of 1,000, then, regardless of the order of scaling, the network of figure 17.29 results.

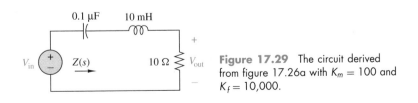

Figure 17.29 The circuit derived from figure 17.26a with $K_m = 100$ and $K_f = 10,000$.

The foregoing frequency- and magnitude-scaling techniques now allow us to readily solve the problem posed at the beginning of this section.

EXAMPLE 17.15

Using the circuit of figure 17.24 as a basis, derive new parallel RLC circuits for the following cases.

1. Source resistance = 20 kΩ, $\omega_m = 31,622$ rad/sec, and $Q = 12.6$.
2. Source resistance = 20 kΩ, $\omega_m = 50,000$ rad/sec, and $Q = 12.6$.

SOLUTION

Case 1. Magnitude scaling the original network with $K_m = 0.5$ will change the source resistance from 40 kΩ to 20 kΩ. The new network is shown in figure 17.30.

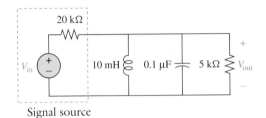

Signal source

Figure 17.30 The network for case 1 of example 17.15.

Case 2. Here, we frequency scale the network of figure 17.30 by $K_f = 50,000/31,622 = 5.1812$. Figure 17.31 shows the new network.

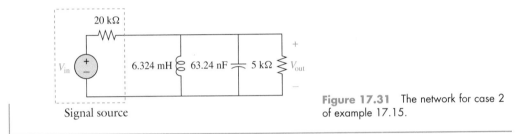

Signal source

Figure 17.31 The network for case 2 of example 17.15.

So far, we have only shown some examples of the scaling of RLC networks. The next example illustrates the procedure for a network containing active elements.

EXAMPLE 17.16

The amplifier circuit shown in figure 17.32a consists of two stages. The gain of the first stage is calculated by the method described in example 14.10, with the result

$$\frac{V_a(s)}{V_i(s)} = \frac{1}{s^2 + s + 1}.$$

The gain of the second stage is easily seen to be

$$\frac{V_o}{V_a} = \frac{-2}{s + 1}.$$

The transfer function of the amplifier is therefore

$$\frac{V_o}{V_i} = \frac{V_a}{V_i}\frac{V_o}{V_a} = \frac{1}{(s^2 + s + 1)}\frac{-2}{(s + 1)} = \frac{-2}{s^3 + 2s^2 + 2s + 1}.$$

In chapter 21, it will be shown that the amplifier has a third-order maximally flat lowpass magnitude response with the 3-dB frequency at $\omega = 1$ rad/sec. This extremely low 3-dB frequency ($f_{3dB} = 0.159$ Hz) and the 1-Ω resistors make the circuit unsuitable for practical applications. However, this "reference" amplifier circuit is just several multiplications away from a practical amplifier circuit. In particular, suppose that we wish to have the 3-dB frequency at $f_{3dB} = 1,000$ Hz and all resistors in the kΩ range (in order not to draw too

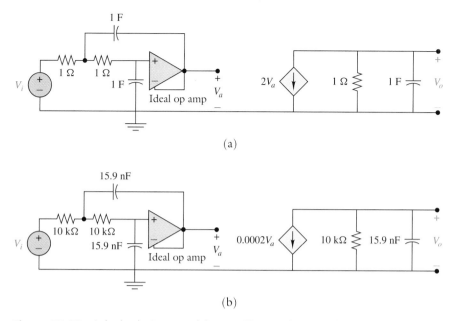

(a)

(b)

Figure 17.32 A third-order Butterworth lowpass filter. (a) The original network. (b) The scaled network with $K_f = 2,000\pi$ and $K_m = 10,000$.

much current from the signal source). To achieve this, all we need do is frequency scale the circuit by $K_f = 2,000\pi$ and magnitude scale by, say, $K_m = 10,000$. The scaled circuit meeting the requirements is shown in figure 17.32b.

Examples 17.15 and 17.16 provide one of several reasons for scaling a linear network. Further applications of this kind will be illustrated in chapter 21, where we study some elementary filter design methods. Another particularly important reason for scaling a network pertains to designing or analyzing a circuit by a digital computer. Scaling improves the accuracy of the calculated results. Generally speaking, very large and very small parameter values should be avoided. We can achieve this by suitable magnitude scaling and frequency scaling. The analysis is performed on the scaled network, and the results are transformed back to the original network. Even in the case when accuracy is not a serious problem, the proper use of scaling will greatly simplify numerical calculations. The following example illustrates this last point.

 EXAMPLE 17.17

Find the poles and zeros of the impedance Z(s) in figure 17.33a.

SOLUTION

We can, of course, find the impedance $Z(s)$ using the given element values. But then, cumbersome powers of 10 will appear almost everywhere. To simplify the calculations, let us magnitude scale the circuit by $K_m = 1/1,000$ and frequency scale the circuit by $K_f = 10^{-6}$. The scaled circuit is shown in figure 17.33b.

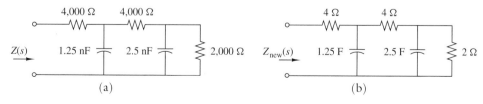

Figure 17.33 Proper use of scaling simplifies numerical calculations. (a) Original circuit. (b) Scaled circuit with $K_m = 10^{-3}$ and $K_f = 10^{-6}$.

A routine analysis of the new circuit yields

$$Z_{new}(s) = \frac{4(s^2 + 0.7s + 0.1)}{s^2 + 0.5s + 0.04} = \frac{4(s + 0.2)(s + 0.5)}{(s + 0.1)(s + 0.4)}.$$

Therefore $Z_{new}(s)$ has:

$$\text{poles at } s = -0.1, -0.4;$$

$$\text{zeros at } s = -0.2, -0.5.$$

Since $K_f = 10^{-6}$ and $K_m = 10^{-3}$ have been used to obtain figure 17.33b from figure 17.33a, frequency scaling figure 17.33b with $K_f = 10^6$ and $K_m = 10^3$ should result in figure 17.33a. From the effects of frequency scaling on poles and zeros, we conclude that $Z(s)$ of figure 17.33a has

$$\text{poles at } s = -100,000, -400,000$$

and

$$\text{zeros at } s = -200,000, -500,000,$$

with all answers in standard units.

Exercise. In figure 17.33a, if the capacitances are changed from 1.25 nF and 2.5 nF to 5 μF and 10 μF, respectively, find the poles and zeros of $Z(s)$.
  Hint: Utilize the results obtained for figure 17.33b in example 17.17.

ANSWER: Poles are at $s = -25, -100$; zeros are at $s = -50, -125$.

---

8. IMPROVING BANDPASS CHARACTERISTICS WITH STAGGER-TUNED CIRCUITS

We have studied the analysis of some simple bandpass circuits. The magnitude response curves of these circuits, illustrated in figure 17.20, appear to have different degrees of *sharpness*. The circuit Q measures this sharpness, or frequency selectivity. A higher Q yields a more selective circuit in passing or amplifying the frequency components of a signal. However, in trying to approximate the ideal bandpass magnitude characteristic, shown in figure 17.5a, all of the simple circuits fail to the same extent. To see this, we replot the curves in figure 17.20 with a new independent variable x that is the ratio of the deviation from the center frequency to half the bandwidth, i.e.,

$$x \equiv \frac{\omega - \omega_m}{0.5 B_\omega}.$$

As long as the circuit Q is high, say, $Q > 10$, the magnitude response for $H(s)$ of equation 17.29 is given approximately by equation 17.40. Using this equation, we see that the normalized gain in the neighborhood of the center frequency is

$$H_{normalized} = \frac{|H|}{H_m} \cong \frac{\sqrt{\left(\frac{B_\omega}{2}\right)^2}}{\sqrt{(\omega - \omega_m)^2 + \left(\frac{B_\omega}{2}\right)^2}} = \frac{1}{\sqrt{1 + x^2}}.$$

A plot of $|H|/H_m$ vs. x is shown as curve (a) in figure 17.34.

Curve (c) of figure 17.34 is better than curve (a) as an approximation to the ideal rectangular shape (b). Here, "better" means that within the passband the magnitude of the gain is more constant, and outside the passband unwanted frequency components are attenuated more sharply. The objective of this section is to show how the addition of a pair of complex poles in the transfer function $H(s)$ markedly improves the magnitude response curve.

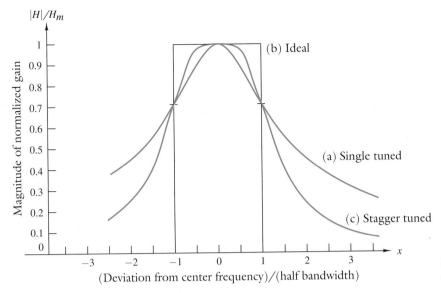

$|H|/H_m$

Magnitude of normalized gain

(b) Ideal

(a) Single tuned

(c) Stagger tuned

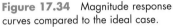

(Deviation from center frequency)/(half bandwidth)

Figure 17.34 Magnitude response curves compared to the ideal case.

A simple scheme for obtaining the better bandpass characteristic of curve (b) in figure 17.34 is to use a two-stage amplifier in which the stages are not tuned to the same frequency, but are staggered at frequencies above and below the desired center frequency of the complete amplifier. The overall gain of the amplifier is the product of the gains at each stage. The magnitude of the resulting response curve can take on a variety of shapes, depending on the bandwidth of each stage and the difference between the two center frequencies. In order to avoid getting bogged down in complicated equations, we illustrate the effects of stagger tuning with specific examples of response curves. Suppose that each stage has a bandwidth $B_\omega = 0.05$ rad/sec. The center frequencies, ω_{m1} and ω_{m2}, are in the neighborhood of 1 rad/sec. Then Q_{p1} and Q_{p2} will be in the neighborhood of 20, justifying the use of high-Q approximations. Figure 17.35 shows three possible overall magnitude response curves resulting from different values of $\Delta\omega_m = \omega_{m2} - \omega_{m1}$, i.e., the separation of the center frequencies. The response curves of each stage have been normalized to 1 at the overall amplifier center frequency of 1 rad/sec, for the purpose of comparing the bandwidth and flatness within the passband.

In figure 17.35a, with $\omega_{m2} = \omega_{m1}$, the overall response appears worse than that of each individual stage in approximating the ideal rectangular shape. In part (b) of the figure, with $\Delta\omega_m = 2B_\omega > B_\omega$, double humps appear. In part (c), with $\Delta\omega_m = B_\omega$ (i.e., the separation of the center frequencies equals the bandwidth of each individual stage), the overall response is seen to be quite flat.

We shall now investigate the pole-zero plots corresponding to the response curve of figure 17.35c. Figure 17.36a shows the poles of the transfer function for this case, with all pertinent information indicated in figure 17.36b for an enlarged portion of the upper half-plane. The two poles are symmetrically located on a circle centered at $(0, j\omega_m)$. Chapter 21 will describe how such a pole distribution leads to a magnitude response of the second-order maximally flat variety. For now, the only results needed from figure 17.36 for designing an improved bandpass circuit are the following:

1. Each stage should have a bandwidth equal to the overall bandwidth, B_ω, divided by $\sqrt{2}$.
2. The center frequencies of the two stages should be $\omega_m \pm \frac{1}{2}B_\omega/\sqrt{2}$.

It should be emphasized that the foregoing approximate results are derived under the assumption of a high circuit Q. Circuits with such a Q are said to be *narrow band*. The design of a *wide-band* circuit requires more advanced theory that is beyond the scope of this text.

The preceding discussion shows clearly that, for a multistage bandpass circuit, it is necessary to distinguish among the concepts of $Q_{circuit}$, Q_p, Q_{coil}, and Q_{cap}. The next example illustrates the design procedure.

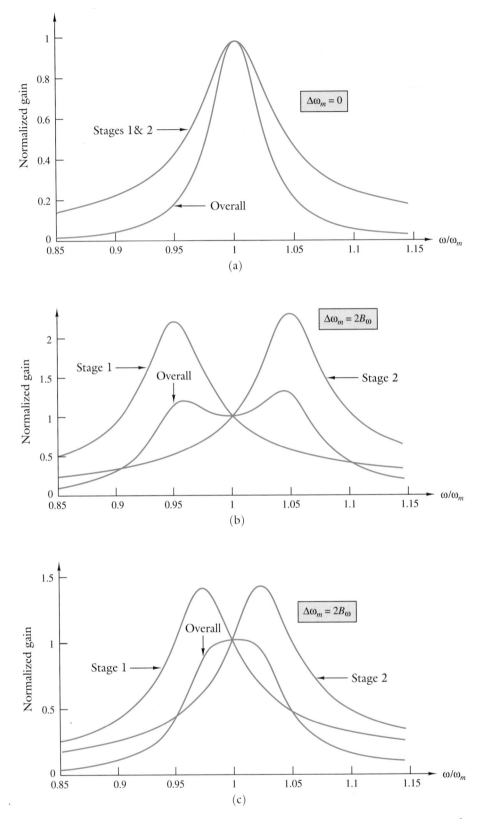

Figure 17.35 Possible response curve shapes from staggered tuning. (a) Zero separation of center frequencies. (b) Wide separation. (c) Critical separation for flattop passband.

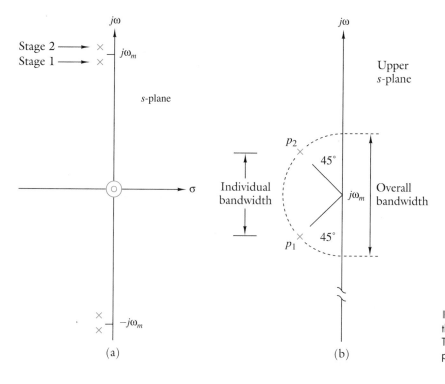

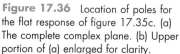

Figure 17.36 Location of poles for the flat response of figure 17.35c. (a) The complete complex plane. (b) Upper portion of (a) enlarged for clarity.

◆ EXAMPLE 17.18

Using one voltage follower (i.e., a unity-gain voltage-controlled voltage source), two 1-mH inductors (assumed ideal), and some R and C components, design a stagger-tuned amplifier to have transfer function poles placed as shown in figure 17.36b, with $f_m = 200$ kHz and overall bandwidth $B_f = 20$ kHz. The input signal is from a voltage source with internal resistance $R_s = 50$ kΩ.

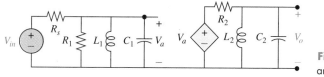

Figure 17.37 A stagger-tuned amplifier.

SOLUTION

Figure 17.37 shows the suggested circuit. Since the overall bandwidth is 20 kHz, each individual stage should have a bandwidth of $20/\sqrt{2} = 14.14$ kHz. From figure 17.36, the two poles should be at

$$p_1 = -\frac{B_{\omega 1}}{2} + j\left(\omega_m - \frac{B_{\omega 1}}{2}\right) = 2\pi \times 10^3 \left[-\frac{14.14}{2} + j\left(200 - \frac{14.14}{2}\right)\right]$$

$$= 2\pi \times 10^3(-7.07 + j192.93) \text{ rad/sec}$$

and

$$p_2 = -\frac{B_{\omega 1}}{2} + j\left(\omega_m + \frac{B_{\omega 1}}{2}\right) = 2\pi \times 10^3 \left[-\frac{14.14}{2} + j\left(200 + \frac{14.14}{2}\right)\right]$$

$$= 2\pi \times 10^3(-7.07 + j207.07) \text{ rad/sec}.$$

Now, referring to figure 17.19 for notation and geometrical relationships, we find that the two ω_{pk} values should be

$$\omega_{p1} = 2\pi\sqrt{192.93^2 + 7.07^2} = 193.06 \text{ krad/sec}$$

and

$$\omega_{p2} = 2\pi\sqrt{207.07^2 + 7.07^2} = 207.2 \text{ krad/sec}.$$

The two stages have slightly different Q's, with

$$Q_{p1} = \frac{\omega_{p1}}{B_{\omega 1}} = \frac{193.06}{14.14} = 13.65$$

and

$$Q_{p2} = \frac{\omega_{p2}}{B_{\omega 2}} = \frac{207.07}{14.14} = 14.65.$$

The complete amplifier has a circuit Q equal to $200/20 = 10$. Once ω_{pk} and Q_{pk} ($k = 1, 2$) are known, each stage is designed using the equations described in section 6. The complete results are given in table 17.3. Note in particular that $Q \neq Q_{p1}$, $Q \neq Q_{p2}$, $f_{p1} \neq f_m$, and $f_{p2} \neq f_m$.

TABLE 17.3   RESULTS FOR EXAMPLE 17.18

Stage k:	f_{pk} (kHz)	Q_{pk}	C_k (pF)	R_k (kΩ)	L_k (mH)
1	193.06	13.65	679.6	23.82	1
2	207.2	14.65	590.0	19.06	1

To verify that the circuit meets the design requirements, one can use any circuit simulation program (e.g., SPICE) to plot the magnitude response curve in figure 17.38. For the purpose of comparison, the response curve obtained with a single tuned circuit is also plotted in the same figure.

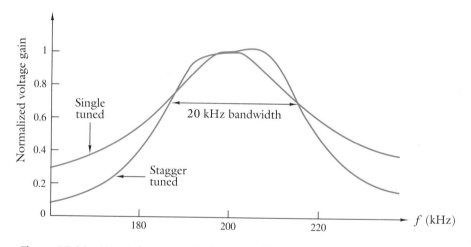

Figure 17.38   Magnitude response for the circuit of figure 17.37.

A couple of remarks about the design are in order:

1. The inductors can be eliminated so that the resultant circuit contains only resistors, capacitors, and some active elements. (See example 17.13.) Nowadays, the most popular active elements for such applications are operational amplifiers. In chapter 21, we describe an elementary method, called the *coefficient-matching method*, for designing low-order active filters.

2. The desired pole locations can also be achieved without the use of an active element as the buffer. Chapter 18 describes a method of utilizing the mutual inductance between two inductors to vary the pole locations.

SUMMARY

This chapter began with a study of the resonance phenomena of a second-order RLC circuit from the frequency domain point of view. (The time domain investigation was done in chapter 9 of volume 1.) Two important applications of resonant circuits were investigated in detail. The first application was the design of a matching network that produces maximum power transfer, at a single frequency, from a source with fixed internal resistance to a fixed resistance load. The second application was the design of bandpass circuits for a specified center frequency and bandwidth. The concept of quality factor for various components and for the entire circuit was carefully defined.

A discussion of parallel and series RLC circuits for providing bandpass magnitude response curves was followed by the more general transfer function approach to the problem, where proper location of the poles and zeros of the transfer function determines the desired magnitude response. This approach is necessary in the design of inductorless bandpass filters. We also demonstrated how the use of two pairs of complex poles can improve the shape of the bandpass magnitude response curve. Students are often confused and perplexed by the many different ω's and Q's encountered in the study of

bandpass circuits. The significance of and the need for these concepts become clear only after transfer functions of bandpass circuits having more than one pair of complex poles are studied.

A number of design formulas for parallel and series RLC bandpass circuit were derived in this chapter. In doing so, we assumed ideal components. When practical inductors and capacitors are used, their models contain resistances. Exact manual analysis of such circuits then becomes impractical. To overcome this difficulty, engineers use reasonable approximations. In particular, one can replace a series $R_s - X_s$ by a parallel $R_p - X_p$, the two being approximately equivalent circuits near the center frequency of the bandpass circuit. Alternatively, one can find the transfer function of the circuit. For high-Q circuits, the pole locations indicate the approximate center frequency and bandwidth.

When exact results of a bandpass circuit with nonideal components are required, the only practical method is to use a circuit simulation program. Presently, there are more than a dozen software packages available on the market for the simulation of linear circuits. Several problems at the end of the chapter require the use of such software.

TERMS AND CONCEPTS

Bandpass circuit: a circuit that passes signals within a band of frequencies, while rejecting other frequency components outside of the band.

Bandwidth (3-dB bandwidth), B_ω: $\omega_2 - \omega_1$, the difference between the two half-power frequencies.

Damped natural frequency, ω_d: frequency given by the condition that if the transfer function of a second-order linear circuit has complex poles at $s = -\sigma_p \pm j\omega_d$, then the impulse response has the form $K \cos(\omega_d t + \theta)$. The constant ω_d is called the *damped natural frequency.*

Frequency scaling: for a linear passive network dividing all inductances and capacitances by a factor, K_f, while keeping all resistance values unchanged.

Half-power frequencies: see lower and upper half-power frequencies.

LC resonance frequency: frequency at which the reactances of L and C have the same magnitude; equals $1/\sqrt{LC}$ rad/sec. (See also tank frequency.)

Lower half-power frequency, ω_1: the radian frequency below the center frequency at which the magnitude response is 0.707 times the maximum value.

Magnitude scaling: for a linear passive network, multiplying all resistances and inductances by a factor, K_m, and dividing all conductances and capacitances by K_m.

Matching network: an LC network that transforms a resistance R_L into a resistance of a different, specified value at one frequency or a band of frequencies.

Peak frequency (center frequency), ω_m: the radian frequency at which the magnitude response curve reaches its peak.

Q_p (pole Q): for a pair of complex poles $s_{1,2} = \omega_p e^{\pm j\theta}$, $Q_p \equiv 1/(2 \sin\theta)$.

Quality factor $Q(Q_{cir})$ of a bandpass circuit: the ratio of the center frequency to the bandwidth, i.e., $Q = \omega_m/B_\omega$.

Quality factor Q_L (Q_{coil}) of a coil: for a coil modeled by an inductance L in series with a resistance R_s, $Q_L = \omega L/R_s$ and is frequency dependent.

Quality factor Q_C (Q_{cap}) of a capacitor: for a capacitor modeled by a capacitance C in parallel with a resistance R_p, $Q_C = \omega C R_p$ and is frequency dependent.

Quality factor Q_z of a reactive component: for an impedance expressed as $Z = R + jX$, $Q_z = |X|/R$.

Reactance: in sinusoidal steady-state analysis, the imaginary part of an impedance. For L, the reactance is ωL; for C the reactance is $-1/(\omega C)$.

Resonance frequency, ω_r: the unique radian frequency at which the input impedance of a two-terminal linear circuit becomes purely resistive.

Selectivity of a bandpass circuit: the circuit Q, defined as the ratio of the center frequency to the bandwidth. A higher Q corresponds to better selectivity.

Stagger-tuned circuit: a multistage amplifier whose stages are not tuned to the same frequency, but are staggered appropriately to achieve a bandpass magnitude response curve better than that obtained with a single tuned circuit.

Susceptance: in sinusoidal steady-state analysis, the imaginary part of an admittance. For C, the susceptance is ωC. For L, the susceptance is $-1/(\omega L)$.

Tank circuit: the parallel connection of an inductor and a capacitor. In the idealized case (no resistance), the total energy stored in a tank circuit remains constant, although there is a continuous interchange of the energy stored in each component.

Tank frequency, ω_o: defined as $1/\sqrt{LC}$ in this text, regardless of the connection of the single L and single C with other components in the circuit.

Tuned circuit: a second-order circuit containing one inductance and one capacitance, at least one of which is adjustable to reach a condition of near resonance.

Undamped natural frequency: the natural frequency of a circuit consisting of lossless inductors and capacitors. For the case of one inductor and one capacitor connected together, this frequency is the same as the LC resonance frequency or the tank frequency and is equal to $1/\sqrt{LC}$).

Universal resonance curve: a normalized magnitude response curve of a bandpass transfer function having one pair of complex poles and a single zero at the origin. The magnitude is normalized with respect to the maximum gain, and the frequency is normalized with respect to the center frequency.

Upper half-power frequency, ω_2: the radian frequency above the center frequency at which the magnitude response is 0.707 times the maximum value.

PROBLEMS

1. Consider the idealized circuit shown in figure P17.1. The moment the inductance current reaches zero is chosen to be the reference point $t = 0$, at which time the capacitance voltage is E V.

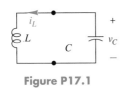

Figure P17.1

(a) Find $v_c(t)$ and $i_L(t)$ for $t \geq 0$ by the Laplace transform method.
(b) Find the energy stored in C as a function of t.
(c) Find the energy stored in L as a function of t.
(d) Show that the total energy stored in the LC tank is constant and is equal to $0.5CE^2$.

2. Prove that, for the circuit shown in figure 17.3d,

$$\omega_r = \sqrt{\frac{1}{LC} - \frac{1}{R^2 C^2}} = \omega_o\sqrt{1 - \frac{L}{CR^2}}$$

and

$$Z(j\omega_r) = \frac{L}{CR}.$$

3. Find the resonant frequency ω_r, in rad/sec, of the circuit sketched in figure P17.3.
(ANSWER: 0.48 rad/sec.)

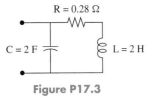

Figure P17.3

4. Consider the circuit shown in figure P17.4. At resonance, find $v_c(t)$, the voltage across the capacitor in the steady state due to the input $i_{in}(t) = 3\cos(\omega_r t)$ A.
(ANSWER: $100\cos(96,000t)$ V.)

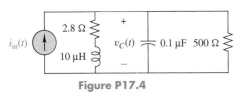

Figure P17.4

5. For the circuit shown in figure P17.5, find the value of C that makes the circuit resonant at $\omega = 3$ rad/sec.

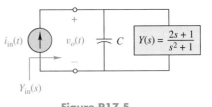

Figure P17.5

6. For each circuit shown in figure P17.6, find the resonant frequency ω_r and the impedance Z at $\omega = \omega_r$.

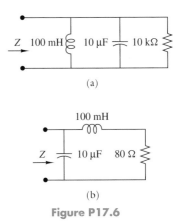

Figure P17.6

7. The voltage source for the circuits shown in figure P17.7 is $v_{in}(t) = 100\sqrt{2}\cos(2\pi \times 10^6 t)$ V. For each circuit, determine the values of L and C such that the average power delivered to the load resistance R_L is maximized. What is P_{max} in each case?
(ANSWERS: 17.9 μH, 1,186 pF, (a) 25/3 W, (b) 50 W.)

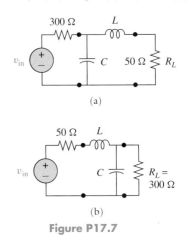

Figure P17.7

*8. Equations 17.4, 17.5, and those derived in problem 2 can be combined to produce a set of design formulas for a lossless network that matches two unequal resistances at a single frequency. The matching network consists of only one capacitance C and one inductance L, as shown in figure P17.8. At a specified frequency ω, it is desired to have matching at both ends, i.e.,

$$Z_1(j\omega) = R_1 + j0, \quad \text{and} \quad Z_2(j\omega) = R_2 + j0.$$

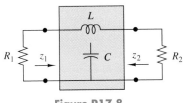

Figure P17.8

Let R_{small} denote the smaller one of (R_1, R_2) and R_{large} denote the larger one of (R_1, R_2). Prove that:

(a) C should be connected in parallel with R_{large}, and L should be connected between the two top terminals and thus in series with R_{small}.

(b) The element values are given by

$$L = \frac{1}{\omega}\sqrt{R_{small}(R_{large} - R_{small})},$$

$$C = \frac{1}{\omega R_{large}}\sqrt{\frac{R_{large} - R_{small}}{R_{small}}}.$$

9. Both of the circuits shown in figure P17.9 have $\omega_o = 0.25$ rad/sec. The purpose of this problem is to show that the resonant frequency ω_r depends on the choice of the input terminals and that $\omega_r \neq \omega_o$ in general. Find the resonant frequency ω_r if the input is connected across: (1) A and B; (2) B and C; (3) A and C; (4) E and F; (5) D and E.
(ANSWERS: 0.25, $\sqrt{3}/8 = 0.2165$, $\sqrt{3}/6 = 0.2886$, $\sqrt{3}/8 = 0.2165$, $\sqrt{21}/20 = 0.2291$ rad/sec.)

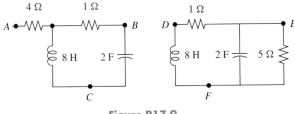

Figure P17.9

10. The equivalent circuit of a radio frequency amplifier in an AM receiver is shown in figure P17.10.

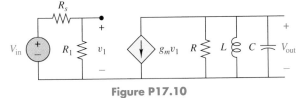

Figure P17.10

---

*Asterisk signifies a more difficult problem.

(a) If $L = 100$ µH, $f_m = 1{,}040$ kHz, and $B_f = 10$ kHz, find C and R.
(b) If $f_m = 920$ kHz, $B_f = 10$ kHz, and $C = 250$ pF, find L and R.
(ANSWERS: (a) 234.2 pF, 67.96 kΩ; (b) 119.7 µH, 63.66 kΩ.)

11. The following ideal components are available:

Inductors: 200 µH, 300 µH, 400 µH.
Variable capacitors (in pF): 20–200, 30–200, 30–300, 40–400.
Resistors: all values.

Using these components, design a parallel resonant circuit such that:
(a) The circuit can be tuned from 550 to 1,650 kHz (standard AM broadcast band).
(b) When the circuit is tuned to 920 kHz (WBAA, Purdue), the bandwidth is 20 kHz.

Figure P17.11

(c) With the components as selected, what is the bandwidth when the circuit shown in figure 17.11 is tuned to the low end and the high end of the AM broadcast band?
(ANSWERS: 300 µH, 30–300 pF, 79.78 kΩ, 7.16 kHz, 64.2 kHz.)

12. For the circuit shown in figure P17.12:
(a) Find the *exact* value of the maximum voltage gain and the corresponding frequency (in Hz).
(b) Find the *exact* 3-dB bandwidth (in Hz).
(c) Find the *exact* values of the upper and lower half-power frequencies (in Hz). What are the percent errors of the estimates given by equation 17.12?
(d) Find Q of the circuit;
(e) Make a rough sketch of the curve of $|H(j\omega)|$ vs. ω.
(f) What is the new bandwidth if the input is changed from V_{in} to an independent current source I_{in}?

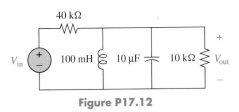

Figure P17.12

(ANSWERS: (a) 0.20 at 159.1 Hz; (b) 1.99 Hz; (c) 160.152 Hz, 158.16 Hz; (d) 80; (f) 1.591 Hz.)

13. The independent current source in figure P17.13 is $i_{in}(t) = I_m(\cos \omega_o t)$ A, and the circuit is in a steady state.

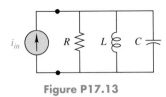

Figure P17.13

(a) Find $w_C(t)$, the instantaneous energy stored in the capacitance.
(b) Find $w_L(t)$, the instantaneous energy stored in the inductance.
(c) Show that $w_C(t) + w_L(t) = \text{constant} = 0.5C(I_m R)^2$.
(d) Find the energy dissipated in the resistance per period ($T = 2\pi/\omega_o$).
(e) Show that, for a parallel RLC circuit,

$$Q = \text{quality factor} = 2\pi \frac{\text{maximum energy stored}}{\text{total energy lost per period}}.$$

*14. The alternative definition of Q given in the preceding problem for a parallel RLC circuit is also applicable to a series RLC circuit. However, it is not suitable for the study of other tuned circuits. There are two major difficulties with the definition:

1. The total energy stored in L and C is not constant, and the calculation of the maximum stored energy is extremely difficult.
2. The definition of Q makes no mention of the location of the output, whereas the selectivity of a bandpass circuit depends on how the input and the output are defined.

This problem illustrates these difficulties with a simple circuit. The input voltage source in figure P17.14 is $v_{in}(t) = A \cos(\omega t)$, and the circuit is in a steady state.
(a) Let $R_s = 0$ and $R = 1\ \Omega$. Find $w_C(t)$ and $w_L(t)$, and show that $w_C(t) + w_L(t)$ is not a constant for any nonzero ω.
(b) Let $R_s = 10\ \Omega$ and $R = 0$. Make a rough sketch of the curves of $|V_{24}/V_{in}|$ vs. ω and $|V_{14}/V_{in}|$ vs. ω, and note the marked difference between them.

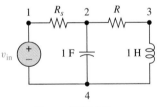

Figure P17.14

15. The parallel RLC circuit of figure 17.6 has $\omega_m = 1$ Mrad/sec, $Z(\omega_m) = 20$ kΩ, and $|Z(0.9\omega_m)| = 10$ kΩ. Find R, L, C, B_ω, and Q of the circuit.
(ANSWER: 20 kΩ, 2.438 mH, 410.2 pF, 121.9 krad/sec, 8.2.)

16. For the series resonant circuit shown in figure P17.16:
(a) Find the *exact* value of the maximum voltage gain and corresponding frequency (in Hz).
(b) Find the *exact* 3-dB bandwidth (in Hz).
(c) Find the *exact* values of the upper and lower half-power frequencies.

(d) Find Q of the circuit.
(e) Make a rough sketch of the curve of $|H(j\omega)|$ vs. ω.

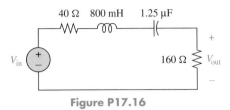

Figure P17.16

(ANSWERS: (a) 0.8 at 159.15 Hz; (b) 39.79 Hz; (c)180.29, 140.51 Hz; (d) 4.)

17. Consider the circuit in figure P17.17.

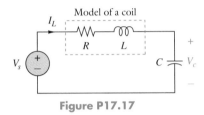

Figure P17.17

(a) Find $H_1(s) = I_L/V_s$ and $H_2(s) = V_c/V_s$.
(b) For $H_1(s)$, determine the exact values of ω_m, $H_1(\omega_m)$, B_ω, and Q.
(c) For $H_2(s)$, consider the case where the ac voltage source has fixed V_s and ω, but the capacitance C is adjustable. Find the exact value of C (in terms of R, L, and ω) such that $|V_c|$ is maximized. Show that if the coil has a high Q, then

$$|V_c|_{max} \cong Q_{coil}|V_s|.$$

This result provides a practical way of measuring Q_{coil}.

18. Determine the approximate Q of the circuit in figure P17.18.
(ANSWER: $Q_{cir} \cong 8$.)

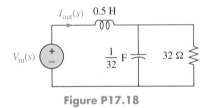

Figure P17.18

19. In the RLC circuit shown in figure P17.19, the source current is the output of interest. Recall that the Q of a lossy capacitor near resonance is $Q_{cap} \cong R_L C\omega_o$. Determine the approximate value of the Q of the RLC circuit.
(ANSWER: $Q_{cir} \cong 8$.)

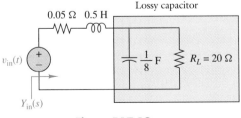

Figure P17.19

20. For the RLC circuits in figure P17.20:
    (a) Determine the approximate value of the half-power frequency ω_1 (in rad/sec).
    (b) Suppose that at ω_m, $|I_{in}(j\omega_m)| = 1$ amp. Determine, approximately, the value of $|V_{out}(j\omega_m)|$ (in volts).

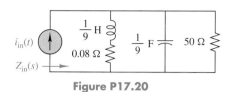

Figure P17.20

(ANSWER: (a) 8.55 rad/sec; (b) 10 V.)

21. For the amplifier circuit shown in figure P17.21:
    (a) Find Q of the coil at $\omega = 10^5$ rad/sec.
    (b) Represent the coil by a parallel RL circuit that is valid for frequencies near 10^5 rad/sec.
    (c) Find *approximate* values of the 3-dB bandwidth, $|V_{out}/V_{in}|_{max}$, and the corresponding frequency (in Hz).

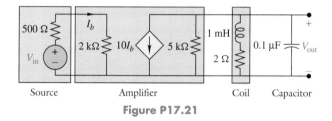

| Source | Amplifier | Coil | Capacitor |

Figure P17.21

(ANSWERS: (a) 50; (b) 1 mH, 50 kΩ; (c) 636.6 Hz, 10, 15.91 kHz.)

22. Repeat problem 21 for the circuit shown in figure P17.22.

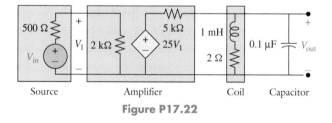

| Source | Amplifier | Coil | Capacitor |

Figure P17.22

23. A 1-mH inductor has a Q of 40 at 100 kHz. This inductor is used in the parallel resonant circuit shown in figure P17.23. The magnitude response is to have a peak at $f_m \cong 100$ kHz.

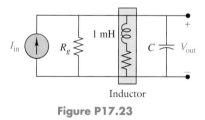

Inductor

Figure P17.23

    (a) Specify the value of the capacitor C.
    (b) Specify the value of R_g so that the bandwidth is approximately 10 kHz.
    (c) What is Q of the circuit?
(ANSWERS: 2,533 pF, 8,378 Ω, 10.)

24. Consider the circuit in figure P17.24, which contains a nonideal capacitor, a nonideal inductor, and a meter to measure the current response, $i_{out}(t)$. The 1 Ω resistor representing the meter is a precision resistor. The voltage across the resistor, $v_{out}(t)$, equals the current through the resistor. Thus, a practical way of measuring current is by measuring the voltage across a small resistance in the circuit. If the resistance of the meter is sufficiently small, this should have little effect on the behavior of the circuit. Nevertheless, in analyzing the circuit, account must be taken of the resistance of the meter.
    (a) Use various approximation techniques to develop an approximating series RLC circuit for the given circuit. Justify the use of all approximations.
    (b) Determine approximate values for B_ω, Q, ω_r, ω_m, ω_1, and ω_2.
    (c) At resonance, determine the approximate steady-state current response, $i_{out}(t)$, when the input voltage is $10 \sin(\omega_r t)u(t)$.

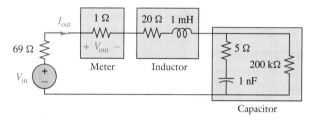

Capacitor

Figure P17.24

(ANSWERS: (b) $B_\omega \cong 10,000$ rad/sec, $Q \cong 10$, $\omega_r \cong \omega_m \cong 10^6$ rad/sec, $\omega_1 \cong 950,000$ rad/sec, $\omega_2 \cong 1,050,000$ rad/sec; (c) $v_{out}(t) \cong 0.1 \sin(\omega_r t)$ V.)

25. Derive equations 17.30a through 17.30f.

*26. Derive equation 17.32. (*Hint*: Instead of finding the maximum of $|H(j\omega)|$, try to maximize $|H(j\omega)|^2$, and consider ω^2 as the independent variable.)

*27. Derive equation 17.33. (*Hint*: Write $1/H(s)$ as $[(1/K + F(s)]$, set Im$\{F(j\omega)\} = 0$, and solve for ω.)

28. For the series resonant circuit in figure P17.28:
    (a) Find the transfer function $H(s) = V_{out}/V_{in}$.
    (b) Find the poles and zeros of $H(s)$.
    (c) Find the *exact* value of the maximum voltage gain and corresponding frequency (in Hz).
    (d) Find the *exact* 3-dB bandwidth.
    (e) Find the *exact* values of the upper and lower half-power frequencies (in Hz).
    (f) Find the quality factor Q of the circuit;
    (g) Make a rough sketch of the curve $|H(j\omega)|$ vs. ω.

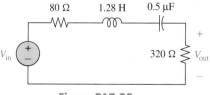

Figure P17.28

29. Consider the circuit in figure P17.29.

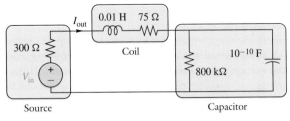

Figure P17.29

(a) Find the input admittance of the circuit, $Y_{in}(s)$, and determine its poles and zeros. (You may use any symbolic program, if available, to check your answer for $H(s)$.)

(b) By inspection of the pole-zero plot, determine the 3-dB bandwidth (approximate value) of the circuit.

(c) Determine Q_{coil} and $Q_{capacitor}$ at $\omega = 10^6 \text{rad/sec}$. Then use the approximate method to find the approximate bandwidth. Compare the result with that obtained in part (b).

30. This problem will illustrate the conceptual differences among the various frequencies encountered in the study of resonant circuits. For practical high-Q circuits, the numerical values of these frequencies are all very close. To see the differences, we choose a low-Q circuit. For the circuit of figure P17.30, show that

(a)
$$H(s) = \frac{0.6(s + 0.8)}{s^2 + 1.4s + 1.48}.$$

(b) The LC tank frequency $\omega_o = 1$ rad/sec.
The resonant frequency $\omega_r = 0.6$ rad/sec.
The peak frequency $\omega_m = \sqrt{29/25} = 1.0770$ rad/sec.
The pole frequency $\omega_p = \sqrt{1.48} = 1.2165$ rad/sec.

Hint:
1. To find ω_r, let $Z_{in}(j\omega) = K + j0$ and solve for ω.
2. To find ω_m, use your knowledge of calculus. Get $|H(j\omega)|^2$ first. Then set the derivative to zero and solve for ω_m.

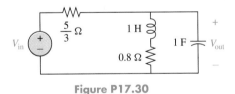

Figure P17.30

31. Consider the network N shown in figure P17.31.

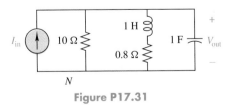

Figure P17.31

(a) Find the transfer function $H(s)$, and show the pole-zero plot.

(b) Find the *exact* values of ω_o, ω_p, ω_d, ω_r, and ω_m. Since this is a low-Q circuit, do not use the high-Q approximations.
(ANSWERS: (a) $(s+0.8)/(s^2+0.9s+1.08)$; (b) $\omega_o = 1$, $\omega_p = 1.0392$, $\omega_d = 0.9367$, $\omega_r = 0.6$, and $\omega_m = 0.9602$, all in rad/sec.)

32. Suppose that the current source in the circuit of the previous problem is $i_{in}(t) = 2\cos(t)$ A and that C is variable. Find the value of C such that $|V_{out}|$ is maximum.

33. For the circuit shown in figure P17.33:
(a) Find the transfer function $H(s)$.
(b) Sketch the pole-zero plot of $H(s)$.
(c) By inspection of the pole-zero plot, obtain approximate values of the peak frequency and 3-dB bandwidth (in Hz).
(d) Making use of the pole-zero plot, find an *approximate* answer for $H(j1.05\omega_m)$.

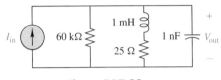

Figure P17.33

34. The circuit sketched in figure P17.34 contains an ideal op amp.

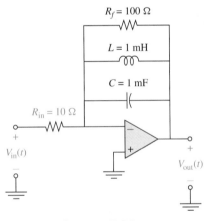

Figure P17.34

(a) Determine the transfer function $H(s) = V_{out}(s)/V_{in}(s) = -Z_f(s)/Z_{in}(s)$ in terms of the circuit elements R_{in}, R_f, L, and C, and place it in the general form

$$H(s) = K \frac{s}{s^2 + 2\sigma_p s + \omega_p^2}.$$

(b) Determine the value of K, the frequency ω_p, and the quality factor Q.
(c) Determine the value of the half-power bandwidth B_ω and the half-power frequencies ω_1 and ω_2. (Approximate values are acceptable.)
(d) Sketch the pole-zero diagram that represents the circuit, and note the *exact* locations of all the poles and zeros.
(e) Determine the value of H_m = the magnitude of $H(j\omega_m)$.

(f) Using the pole-zero diagram, determine the approximate value of the magnitude of $H(j\omega)$ at $\omega = 980$ rad/sec.

(g) If $v_{in}(t) = 100\sin(t)$ V, determine the magnitude of $v_{out}(t)$ to the nearest volt.

35. The switch S in figure P17.35a has been closed for a long time and is opened at $t = 0$.

(a) Use the Laplace transform method of chapter 14 to show that if $Q = \omega_o RC > 0.5$, then for $t > 0$, the capacitance voltage is

$$v_c(t) = V_m e^{-at}\sin(\omega_d t + \theta),$$

where

$$a = \frac{\omega_o}{2Q}$$

and

$$\omega_d = \omega_o\sqrt{1 - \frac{1}{4Q^2}}.$$

(b) Show that the peak amplitude of the damped sinusoidal waveform of part (a) decreases to $1/e = 0.368$ of the highest peak approximately after Q/π cycles if Q is large (and, hence, $\omega_o \cong \omega_d$).

(c) The opening section of this chapter discussed the generation of dial tones by resonant circuits. The circuit for generating the three tones in the high-frequency group is shown in figure P17.35b. Using the results of part (a), explain how the tone generation circuit works.

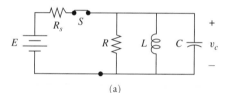

(a)

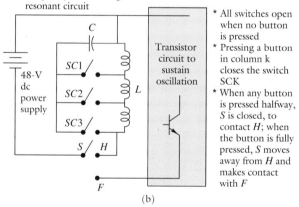

Frequency determining resonant circuit

* All switches open when no button is pressed
* Pressing a button in column k closes the switch SCK
* When any button is pressed halfway, S is closed, to contact H; when the button is fully pressed, S moves away from H and makes contact with F

(b)

Figure P17.35 (a) Circuit producing damped oscillations. (b) Circuit producing dialing tones.

36. Consider the circuit in figure P17.36.

(a) Find the poles and zeros of the transfer function $H(s) = V_{out}(s)/I_{in}(s)$. From these, determine ω_m, B_ω, and Q (approximate values). Is this a high Q circuit?

(b) The impulse response for the circuit is of the form

$$h(t) = Ae^{-at}\cos(\omega t) + Be^{-at}\sin(\omega t).$$

Show that $|A| >> |B|$ for the high-Q case and, therefore, $h(t) \cong Ae^{-at}\cos(\omega t)$.

(c) Make a rough sketch of $h(t) \cong Ae^{-at}\cos(\omega t)$. Show that the peak will decrease to $1/e$ of the first peak in approximately $Q/\pi = 20$ cycles of oscillations.

(d) Find approximate values of ω_m, H_m, B_ω, and Q by changing the series L-R to an approximate parallel L-R connection. Do the results agree with those obtained in part (a) by the transfer function approach?

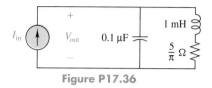

Figure P17.36

37. Consider the circuit in figure P17.37, which contains a nonideal inductor and a variable capacitor.

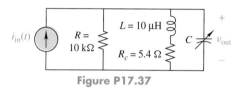

Figure P17.37

(a) Determine the approximate range of the capacitance C, i.e., $C_0 \le C \le C_1$ such that the circuit can be tuned to resonance over the AM radio band (from 550 kHz to 1,650 kHz).

(b) Given the correct answer to (a), determine the approximate minimum and maximum circuit Q's, the circuit bandwidths, and the lower and upper half-power frequencies.

(c) Suppose

$$Z_{in}(s) = \frac{s + \alpha}{s^2 + 2\sigma s + \omega_p^2}.$$

(1) Determine the imaginary part of $Y_{in}(j\omega)$.

(2) Under what conditions is the imaginary part zero.

(3) Find an expression for ω_r in terms of ω_p, σ, and α.

(d) Repeat part (a), but determine the exact range of C, making use of your answer to part (c). Compare your answer to that of part (a). They should be close.

Remark: Two answers to this problem are 8.47 nF and 8.687 nF.

38. For the circuit shown in figure P17.38:

(a) Find the transfer function $H(s)$.

(b) By the use of equation 17.32, show that

$$\omega_m = \sqrt{\sqrt{\frac{1}{(LC)^2}\left(1 + \frac{2R}{R_g}\right) + \frac{2R^2}{L^3 C} - \frac{R^2}{L^2}}}$$

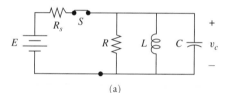

Figure P17.38

39. For the circuit shown in figure P17.39:
  (a) Find the transfer function $H(s)$.
  (b) By the use of equation 17.32, show that

$$\omega_m = \sqrt{\frac{1}{LC}\left[1 - \frac{1}{2LC}\left(R^2C^2 + \frac{L^2}{R_L^2}\right)\right]}.$$

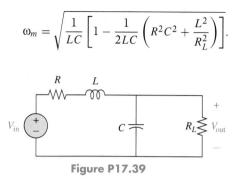

Figure P17.39

40. For the circuit shown in figure P17.40:
  (a) Find the transfer function $H(s)$.
  (b) By the use of equation 17.30, find ω_m, H_m, and B_ω.

Answer for part (b):

$$\omega_m = \sqrt{\frac{\beta+1}{LC}}, \quad H_m = \frac{1}{R_s}, \quad B_\omega = \frac{R_s}{L}.$$

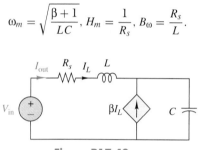

Figure P17.40

41. In the circuit in figure P17.41, let $L = 1$ H, $C = 1$ F, $r_L = 0.08$ Ω, $r_C = 0.02$ Ω, and $R_s = 40$ Ω.
  (a) Find approximate answers for ω_m and B_ω by the use of series-parallel circuit transformations, described in section 5.
  (b) Find approximate answers for ω_m and B_ω by the pole-zero plot method, described in section 6.

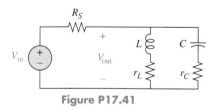

Figure P17.41

42. Consider the circuit of problem 12.
  (a) Magnitude scale by $K_m = 0.1$, and frequency scale by $K_f = 10$. Show the connection and all element values of the scaled circuit.
  (b) For the scaled circuit, repeat parts (a) through (e) of problem 12. You can make use of the results obtained earlier and write the answers directly.

43. Consider the circuit of problem 31.
  (a) Magnitude scale by $K_m = 1.25$, and frequency scale by $K_f = 10^6$. Show the connection and all element values of the scaled circuit.

  (b) For the scaled circuit, repeat parts (a) and (b) of problem 31. You can make use of the results obtained earlier and write the answers directly.

44. Consider the op amp circuit in figure P17.44.

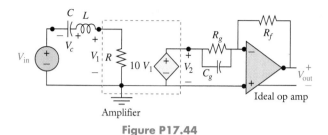

Figure P17.44

  (a) Given that $L = 1$ H, $C = 0.5$ F, $R = 3$ Ω, $R_g = R_f = 1$ Ω, and $C_g = 1$ F,
    (1) Find $H_1(s) = V_1/V_{in}$ and its poles and zeros.
    (2) Find $H_2(s) = V_2/V_1$ and its poles and zeros.
    (3) Find $H_3(s) = V_{out}/V_2$ and its poles and zeros.
    (4) Find the transfer function $H(s) = V_{out}/V_{in}$ and its poles and zeros.
  (b) The circuit of part (a) is magnitude scaled by a factor $K_m = 1,000$ and frequency scaled by a factor $K_f = 10^6$. Repeat parts 1 through 4, making use of the results obtained in part (a).

45. Design the active bandpass circuit in figure P17.45 to have $\omega_m = 1,000$ rad/sec and $B_\omega = 12.5$ rad/sec. (*Hint:* Follow the procedure described in example 17.13.)

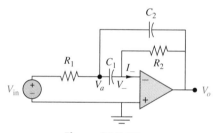

Figure P17.45

46. Redesign the stagger-tuned amplifier of figure 17.37 to meet the following specifications: $f_m = 100$ kHz, $B_f = 12.5$ kHz. The source resistance is $R_s = 50$ kΩ, and 1 mH inductors are to be used. (*Hint:* Follow the procedure described in example 17.18.)

Note: The next three problems are to be solved with aid of a circuit simulation program. Many such programs are now available on the market for personal computers. If you do not have access to such programs for a PC, you can always use SPICE, which is available at almost all electrical engineering departments. A brief user's manual for SPICE is included in the appendix of this text.

47. Consider the amplifier circuit of problem 21. Use any circuit simulation program (e.g., SPICE) to plot the curve of $|V_{out}/V_{in}|$ vs. f, and determine the 3-dB bandwidth within ± 10 Hz. Compare the result with your answer in part (c) of problem 21.

48. Consider the circuit of problem 24. Use any circuit simulation program to obtain accurate values for ω_m, B_ω, and V_{out}. Compare the result with the answers in problem 24.

49. Consider the circuit of problem 28. Verify the answers to parts (c), (d), and (e) of that problem with any circuit simulation program.

50. The analysis of the non-series-parallel circuit shown in figure P17.50 requires writing node or loop equations.

    (a) Find the transfer function $H(s)$ either by a manual analysis or by the use of a symbolic circuit analysis program if available.

    (b) Obtain the pole-zero plot of $H(s)$ in the s-plane.

    (c) By inspection of the pole-zero plot, determine the 3-dB bandwidth. (An approximate answer is acceptable.)

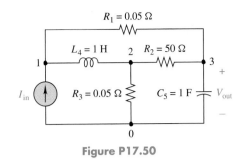

Figure P17.50

CHAPTER 18

CHAPTER OUTLINE

Magnetically Coupled Circuits and Transformers

WHAT IS INSIDE THE AC ADAPTOR?

THE majority of electronic equipment operates with dc power sources. For portable equipment, such as a cordless phone, cordless electric drill, etc., batteries supply the dc power. Using nonrechargeable batteries becomes expensive. Furthermore, replacing batteries in special equipment is a task not easily handled by ordinary consumers. These two factors have prompted manufacturers to install rechargeable nickel-cadmium batteries in portable equipment. By connecting several batteries in series, the available dc voltage may range from 1.5 to 12 V. Whenever the battery runs low, it must be recharged.

Recharging a battery requires a low dc voltage source (1.5–12 V). An adaptor houses a device called a *transformer* that changes the 110-V ac voltage at the household outlet to a much lower ac voltage. The lower ac voltage is then rectified to become a dc voltage that charges the battery. Some adaptors contain the transformer only, while others may also contain the rectifier circuit.

Typical specifications appearing on the case of an adaptor may be as follows:

model: AC9131
input:
   ac 120 V, 60 Hz, 6 W
output:
   ac 3.3 V, 500 mA

model: KX-A10
input:
   ac 120 V, 60 Hz, 5 W
output:
   dc 12 V, 100 mA

The concepts and methods developed in this chapter will allow us to understand how a transformer works to change the ac voltage level and also perform some other important functions in electronic equipment.

- Understand how the mutual inductance M accounts for the induced voltage in an inductor due to the change of current in another inductor.
- Develop a systematic method for writing time domain and frequency domain equations for circuits containing mutual inductances.
- Understand why $M \leq \sqrt{L_1 L_2}$ from energy calculation.
- Expand the repertoire of basic circuit elements to include ideal transformers, and learn how to analyze circuits containing ideal transformers.
- Learn how to model a pair of coupled inductors by an ideal transformer and at most two self-inductances.
- Understand how a practical transformer can be modeled by an ideal transformer and some additional RLC elements.
- Understand some important applications of transformers and coupled inductors in power engineering and communication engineering.

1. INTRODUCTION

When someone presses the "on" switch of a flashlight, he or she does not expect another's flashlight to turn on. Similarly, Kirchhoff's voltage and current laws imply that two separate RLC circuits, or two RLC circuits joined only at one node, will not affect each other; i.e., an excitation in one circuit should not be expected to produce a response in the other.

Now consider figure 18.1. Figure 18.1a shows two coils of wire in close proximity, and figure 18.1b shows two wire coils wound around a single ferromagnetic core. In both cases, a voltage source excites coil 1 while coil 2 is left open-circuited. In contrast to the speculation of the previous paragraph, experimental evidence shows that a change in the current i_1 will generate a voltage v_2, called the **induced voltage,** across the open circuit. Although physically isolated, each pair of coils in figures 18.1a and 18.1b are said to be *magnetically coupled.*

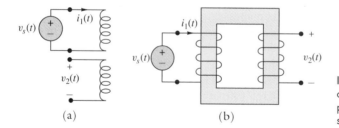

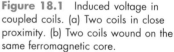

Figure 18.1 Induced voltage in coupled coils. (a) Two coils in close proximity. (b) Two coils wound on the same ferromagnetic core.

(a)   (b)

How does one quantitatively account for magnetic coupling? The answer is to introduce a new circuit quantity called *mutual inductance* for coupled coils. The next few sections will describe basic analysis methods for circuits containing mutual inductances. Before we proceed, however, some words as to the relevance of our study are in order.

Magnetically coupled circuits have many important engineering applications. An extremely important magnetically coupled device is the *transformer*, which is used to transform voltages and currents from one level to another. For example, in electric power systems, transformers are used to step up ac voltages from 10 kV at a generating station to over 240 kV for the purpose of transmitting electric power efficiently over long distances. At a customer's site, such as a home, transformers step these high voltage levels down to 220 V or 110 V for safe, everyday uses. In addition, transformers have numerous uses in electronic systems, including (1) stepping ac voltages up or down, (2) isolating parts of a circuit from dc voltages, and (3) providing impedance level changes to achieve maximum power transfer between devices. After setting forth the basic analysis methods, some examples will illustrate these uses.

2. MUTUAL INDUCTANCE AND THE DOT CONVENTION

Experimental evidence demonstrates that if the two coils in figure 18.1a are *stationary*, the induced voltage, $v_2(t)$, is proportional to the *rate of change* of $i_1(t)$. Confining our attention initially to the magnitudes of the voltage and current only, we can write the linear relationship as

$$|v_2(t)| = M_{21} \left| \frac{di_1}{dt} \right|, \tag{18.1}$$

where M_{21} is a constant of proportionality linking v_2 to the change in i_1. If, in figure 18.1a, the voltage source is moved to coil 2 while coil 1 is open-circuited, one observes a similar relationship, viz.,

$$|v_1(t)| = M_{12} \left| \frac{di_2}{dt} \right|. \tag{18.2}$$

As verified in a later section, M_{21} and M_{12} are equal and therefore can be designated by a single positive proportionality constant

$$M \equiv M_{21} = M_{12} \tag{18.3}$$

called the **mutual inductance** of the coupled inductors. The unit for mutual inductance is the henry (H), the same as that for self-inductance.

Equation 18.1 also suggests a possible experimental procedure for determining the value of M: Apply a ramp current $i_1(t) = Kr(t)$ ($K > 0$) to coil 1, and measure the constant open-circuit voltage v_2 induced in coil 2. Then $M = |v_2|/K$.

Network equations for circuits containing mutual inductances require $v_1(t)$ and $v_2(t)$, not their magnitudes $|v_1(t)|$ and $|v_2(t)|$. Removing the absolute value signs from equations 18.1 and 18.2, and using equation 18.3, implies that

$$v_1(t) = \pm M \frac{di_2}{dt} \qquad \text{with } i_1 \equiv 0, \tag{18.4}$$

$$v_2(t) = \pm M \frac{di_1}{dt} \qquad \text{with } i_2 \equiv 0. \tag{18.5}$$

The immediate question is: Which $\pm$ sign is to be used? For our present development, suppose the coupled inductors are sealed in a case with four accessible terminals. The problem must be solved by measurements, a typical scenario when using components bought from a supply store.

To determine the sign, consider figure 18.2 and apply an *increasing* current $i_1(t)$ to coil 1. Since $di_1/dt > 0$ for all t, one of the terminals of coil 2 must be at a higher potential relative to the other at all times. This terminal is easily identified by connecting a voltmeter to coil 2. As a usual laboratory procedure, we record what was observed for future reference. However, for convenience, we record the same information on the coupled coils as illustrated in figure 18.2 by marking the terminals with black dots according to the following convention: *Place a dot on the terminal at which* $i_1(t)$ *enters. Let* $i_1(t)$ *be increasing, i.e.,* $di_1/dt > 0$. *Place another dot on the terminal with higher potential that is on the other open-circuited coil.*

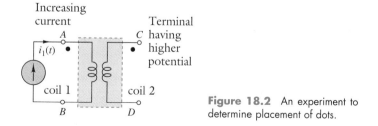

Figure 18.2 An experiment to determine placement of dots.

The dot convention allows us to easily decide the choice of $\pm$ sign in equation 18.5. For example, consider the specific dot placements in figure 18.2. If v_2 is the voltage drop from the dotted terminal to the undotted terminal (i.e., from C to D), then $v_2(t) > 0$, indicating that the plus sign must be used in equation 18.5, since $di_1/dt > 0$ and $M > 0$. On the other hand, if v_2 is the voltage drop from the undotted terminal to the dotted terminal, (i.e., from D to C), then the minus sign must be used. Interchanging the source and the voltmeter and repeating the experiment will result in the same dot placements.

The dots can be simultaneously moved to opposite ends of the coils with no change in the information they convey about relative polarity. (Problem 18.1 at the end of the chapter confirms this statement.)

Given a pair of coupled inductors with dots already marked, one still has the freedom to assign the voltage and current reference directions arbitrarily. If both currents are entering at the dotted terminals and both voltage drops are from the dotted to the undotted terminals, then, as just explained, the plus sign is used for the induced voltage due to mutual inductance. On the other hand, if you or a design engineer decide to change the polarities of the voltage labels or the directions of the current labels, then the sign of the induced voltages must be consistent with the new labeling. For example, a current may be leaving the dotted terminal, or a voltage drop may be from the undotted terminal to the dotted terminal. The following rule, which is simply a combination of the dot convention and equation 18.1, is very helpful in such cases. (See problem 18.2.)

RULE FOR THE INDUCED VOLTAGE DROP DUE TO MUTUAL INDUCTANCE

> The voltage drop across one coil, from the dotted terminal to the undotted terminal, equals M times the derivative of the current through the other coil, from the dotted terminal to the undotted terminal. (18.6a)

Or, equivalently,

> The voltage drop across one coil, from the undotted terminal to the dotted terminal, equals M times the derivative of the current through the other coil, from the undotted terminal to the dotted terminal. (18.6b)

This rule gives the voltage drop due to the mutual inductance. To obtain the total voltage drop across an inductor, one must add in the voltage drop induced by the self-inductance of the individual coil.

EXAMPLE 18.1

In figure 18.2, if $i_1(t) = 2tu(t)$ A, a voltage $v_{CD} = -10u(t)$ mV is observed. Determine the placement of the dots and the value of M.

SOLUTION

Since $i_1(t)$ is increasing and D is at a higher potential than C, the dots must be placed at (A, D) or (B, C). From equation 18.1, the value of the mutual inductance is

$$M = M_{21} = \frac{|v_2(t)|}{\left|\dfrac{di_1}{dt}\right|} = \frac{0.01}{2} = 0.005 \text{ H}.$$

The ramp input used in example 18.1 illustrates the principle of determining M by measurement. In practice, ones applies a triangular waveform input, which is available from many signal generators. Then the repetitive output waveform can be displayed on an oscilloscope for accurate measurement. Example 18.2 illustrates the procedure.

EXAMPLE 18.2

Figure 18.3 shows two waveforms, $i_1(t)$ and $v_2(t)$, as displayed on an oscilloscope. Determine the placements of the dots and the value of the mutual inductance.

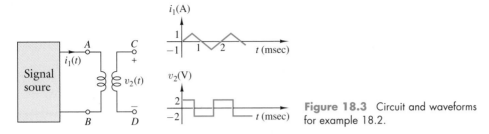

Figure 18.3 Circuit and waveforms for example 18.2.

SOLUTION

For the time interval $0 < t < 0.5$ sec, $i_1(t)$ is a ramp function and $v_2(t)$ is constant. The information is similar to that given in example 18.1. Therefore, we can solve the problem in the same way. The current i_1 is increasing, and v_2 is positive. According to figure 18.2, the dots must be placed at (A, C) or (B, D). Equation 18.5 becomes

$$v_2(t) = M\frac{di_1}{dt}.$$

The measured values during $0 < t < 0.5$ sec give $v_2 = 2$ V and $di_1/dt = 1/0.0005 = 2{,}000$ A/sec. Thus, $M = 2/2{,}000 = 0.001$ H.

> **Exercise.** Repeat example 18.2 if the observed quantities are $i_1(t) = 2t$ A and $v_2(t) = 4$ V.
> **ANSWER:** Dots are on (A, C) or (B, D), $M = 2$ H.

EXAMPLE 18.3

In the circuit of figure 18.3, if $i_1(t) = 2(1 - e^{-100t})u(t)$ A, find $v_2(t)$.

SOLUTION

In this case, $di_1/dt = 200e^{-100t}u(t)$. From example 18.2, $M = 0.001$ H. From equation 18.5 (with $+$ sign as determined in example 18.2),

$$v_2(t) = M\frac{di_1}{dt} = 0.2e^{-100t}u(t) \text{ V}.$$

> **Exercise.** In the circuit of figure 18.3, if $i_1(t) = 2e^{-100t}u(t)$ A, find $v_2(t)$ for $t > 0$.
> **ANSWER:** $-0.2e^{-100t}$ V.

The preceding treatment of the mutual inductance M has not referred to the physical construction of the circuit element at all: Equations 18.4 and 18.5 define M, and its value is determined by measurements. The coupled inductors are assumed to be enclosed in a sealed box. From the circuit analysis point of view, the treatment is adequate. However, for designing a pair of coupled inductors, or for a better understanding of mutual inductance, one must know how the physical construction determines the values of L_1, L_2, and M. A study of this problem requires a background in magnetic circuits. A brief review of magnetic circuit principles is given in the appendix. Appendix B3 and section 18.8 show several examples of the calculation of L_1, L_2, and M. Without getting involved with details, it is worthwhile to remember the following basic facts:

1. Let the numbers of turns of the two inductors in figure 18.1 be N_1 and N_2. Then the self- and mutual inductances have *approximately* the ratio

$$L_1 : L_2 : M = N_1^2 : N_2^2 : N_1 N_2.$$

2. If two inductors are placed in a nonmagnetic medium (e.g., air), bringing the inductors closer together increases the value of M.

3. If one inductor of a pair is rotated, then a larger value of M results when the axes of the inductors are parallel to each other. The smallest value of M occurs when the axes are perpendicular to each other.

4. Changing the core on which two inductors are wound from a nonmagnetic material (e.g., air, plastic, etc.) to a ferromagnetic material may increase the values of L_1, L_2, and M by several thousand times.

3. DIFFERENTIAL EQUATION, LAPLACE TRANSFORM, AND PHASOR MODELS OF COUPLED INDUCTORS

Figure 18.4 shows a pair of inductors with mutual inductance M. Besides the mutual inductance, each inductor in the pair also has a self-inductance, denoted by L_1 and L_2, respectively. The reference directions for voltages and currents may be arbitrarily chosen (see problem 18.2), with those shown in figure 18.4 being typical. Our objective here is to develop equations relating the currents and voltages for the dot placements of figures 18.4a and 18.4b.

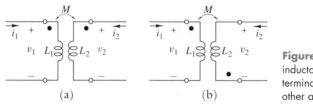

Figure 18.4 A pair of coupled inductors. (a) Both dots at the top terminals. (b) One dot at top and the other at bottom terminal.

The voltage developed across each inductor is the sum of the voltage due to the self-inductance, $L_k(di_k/dt)$, and the voltage due to the mutual inductance, $M(di_j/dt)$, with the $\pm$ signs duly considered. In figure 18.4a, currents are entering at the dotted terminals, and voltage drops are from dotted terminals to undotted terminals; hence, the mutual terms have plus signs, according to equation 18.6a. Applying superposition leads to the set of equations governing the circuit of figure 18.4a:

$$v_1(t) = L_1 \frac{di_1}{dt} + M \frac{di_2}{dt}, \tag{18.7a}$$

$$v_2(t) = M \frac{di_1}{dt} + L_2 \frac{di_2}{dt}. \tag{18.7b}$$

In figure 18.4b, $v_1(t)$ is the voltage drop from the dotted terminal to the undotted terminal of coil 1. The current through coil 2, from the dotted terminal to the undotted terminal, is $-i_2$. Therefore, according to equation 18.7a,

$$v_1(t) = L_1 \frac{di_1}{dt} + M \frac{d(-i_2)}{dt} = L_1 \frac{di_1}{dt} - M \frac{di_2}{dt}. \tag{18.8a}$$

On the other hand, the voltage $v_2(t)$ in figure 18.4b is from the undotted terminal to the dotted terminal of coil 2. The current through coil 1, from the undotted terminal to the dotted terminal, is $-i_1$. Therefore, according to equation 18.7b,

$$v_2(t) = M \frac{d(-i_1)}{dt} + L_2 \frac{di_2}{dt} = -M \frac{di_1}{dt} + L_2 \frac{di_2}{dt}. \tag{18.8b}$$

Again, the development based on figure 18.1 presupposes linearity. If the two inductors are placed in a nonmagnetic medium, this is true. If the inductors are coupled through a ferromagnetic medium (e.g., an iron core), then the linear relationships hold only if both currents are sufficiently small so that the magnetic medium has not reached **saturation,** a phenomenon to be discussed briefly in appendix B3. Our investigation in this chapter considers only the linear case.

EXAMPLE 18.4

A pair of coupled inductors is connected in two different ways, as shown in figure 18.5. In each case, find the differential equation relating v and i, and then find the equivalent inductance "seen" at the two terminals of each box.

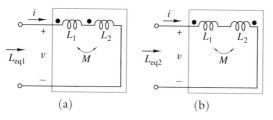

(a) (b)

Figure 18.5 Equivalent inductance of two series-connected inductors. (a) $L_{eq1} = L_1 + L_2 + 2M$. (b) $L_{eq2} = L_1 + L_2 - 2M$.

SOLUTION

For figure 18.5a, we can directly apply equations 18.7a and b to obtain

$$v(t) = \left(L_1\frac{di}{dt} + M\frac{di}{dt}\right) + \left(M\frac{di}{dt} + L_2\frac{di}{dt}\right) = (L_1 + L_2 + 2M)\frac{di}{dt}.$$

Comparing this relationship with

$$v(t) = L_{eq1}\frac{di}{dt},$$

we obtain

$$L_{eq1} = (L_1 + L_2 + 2M).$$

Similarly, for figure 18.5b, we apply equations 18.8a and b to obtain

$$v(t) = \left(L_1\frac{di}{dt} - M\frac{di}{dt}\right) + \left(-M\frac{di}{dt} + L_2\frac{di}{dt}\right) = (L_1 + L_2 - 2M)\frac{di}{dt}.$$

Therefore,

$$L_{eq2} = (L_1 + L_2 - 2M).$$

The preceding example implies that

$$L_{eq1} - L_{eq2} = 4M. \tag{18.9}$$

This relationship suggests another way of determining M and dot markings from measurements. If an instrument for measuring self-inductance is available, we can use the instrument to measure L_{eq1} and L_{eq2}. From equation 18.9, the difference is then $4M$. The connection that corresponds to L_{eq1} shows that dots may be marked on the terminals of both inductors where the current i enters, as per figure 18.5a.

When placing dots is the only objective, a much simpler experiment than that of examples 18.2 and 18.4 is possible.

EXAMPLE 18.5

In the circuit of figure 18.6, the terminal of L_1 that is connected to the + terminal of the dc voltage source (through switch S and resistor R) is given the dot marking. When the switch is closed, the meter pointer moves, either to the right or to the left. Interchange the meter leads, if necessary, so that the pointer moves to the right when S is closed at $t = 0$. Then the terminal connected to the + terminal of the voltmeter is given the dot marking.

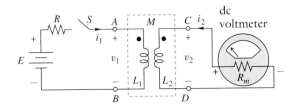

Figure 18.6 A simple setup for determining dot placements.

The justification of this procedure is very easy if we assume the meter resistance to be infinite. In that case, L_2 is open-circuited, i.e., $i_2 = 0$, and the differential equation for i_1 is

$$E = v_1 + Ri_1 = L_1 \frac{di_1}{dt} + M \frac{di_2}{dt} + Ri_1 = L_1 \frac{di_1}{dt} + Ri_1.$$

As this has the form of an RL circuit driven by a dc source, the solution of the equation for $t > 0$ is

$$i_1(t) = \frac{E}{R} \left(1 - e^{-\frac{R}{L_1}t} \right).$$

This current is monotonically increasing. Since $i_1(t)$ is increasing, and $v_2(t)$ is positive, the dot markings are in agreement with the convention illustrated in figure 18.2. It can be shown that even if the meter resistance is finite, the foregoing procedure still yields correct dot markings. (See example 18.6 and problem 18.15.)

The preceding two examples illustrate, among other things, how to write differential equations for circuits containing mutual inductances. Since the Laplace transform method is emphasized in this text for solving linear circuits, it is highly desirable to write the Laplace transform equations directly, without the intermediate step of writing differential equations. This is easy for the case of zero initial currents in all the inductors. Applying the Laplace transform to equations 18.7 and 18.8, while invoking the assumption that $i_1(0^-) = i_2(0^-) = 0$, yields

$$V_1 = sL_1I_1 \pm sMI_2, \tag{18.10a}$$

$$V_2 = \pm sMI_1 + sL_2I_2, \tag{18.10b}$$

where the $+$ sign is for figure 18.4a and the $-$ sign for figure 18.4b. (Refer to equations 18.6a and b for other cases.)

Comparing equations 18.10 with equations 18.7 and 18.8 yields only two differences:

1. $V(s)$ and $I(s)$ replace $v(t)$ and $i(t)$.
2. The Laplace transform variable s replaces the operator d/dt.

The impedance concept developed in chapter 14 is also applicable here in writing loop equations, except that it is necessary to make one extension. The voltage drop across any inductor now consists of two terms, as indicated by equations 18.10: one term due to self-inductance, sL_kI_k, and the other term due to mutual inductance, $\pm sMI_j$. The choice of the sign in "$\pm sMI_j$" is determined by the same rule, equations 18.6, given for the differential equation case.

◆ EXAMPLE 18.6

In the circuit of figure 18.7, $R_1 = 4\ \Omega$, $R_2 = 2\ \Omega$, $L_1 = 8$ H, $L_2 = 6$ H, and $M = 4$ H. If $E = 36$ V, $i_2(0^-) = 0$, and S is closed at $t = 0$:

1. Find $i_1(t)$ and $i_2(t)$.
2. Show that $v_2(t)$ and di_1/dt are positive for all $t > 0$.

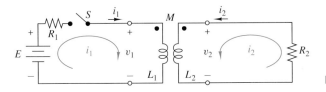

Figure 18.7 Circuit for example 18.6.

SOLUTION

Part 1. Writing two loop equations in the s-domain produces

$$4I_1 + (8sI_1 + 4sI_2) = (8s + 4)I_1 + 4sI_2 = \frac{36}{s},$$

$$(4sI_1 + 6sI_2) + 2I_2 = 4sI_1 + (6s + 2)I_2 = 0.$$

In matrix notation,

$$\begin{bmatrix} 8s + 4 & 4s \\ 4s & 6s + 2 \end{bmatrix} \begin{bmatrix} I_1 \\ I_2 \end{bmatrix} = \begin{bmatrix} \dfrac{36}{s} \\ 0 \end{bmatrix}.$$

Solving for I_1 and I_2 by Cramer's rule, we obtain

$$I_1(s) = \frac{6.75\left(s + \frac{1}{3}\right)}{s(s + 0.25)(s + 1)} = \frac{9}{s} - \frac{3}{s + 0.25} - \frac{6}{s + 1},$$

$$I_2(s) = \frac{-4.5}{(s + 0.25)(s + 1)} = \frac{-6}{s + 0.25} + \frac{6}{s + 1}.$$

Inverse Laplace transforming $I_1(s)$ and $I_2(s)$ yields

$$i_1(t) = (9 - 3e^{-0.25t} - 6e^{-t})u(t),$$

$$i_2(t) = 6(e^{-t} - e^{-0.25t})u(t).$$

Part 2. Using the expressions found in part 1 implies

$$v_2(t) = -R_2 i_2(t) = 12(e^{-0.25t} - e^{-t})u(t) \tag{18.11}$$

and

$$\frac{di_1}{dt} = 0.75e^{-0.25t} + 6e^{-t} \text{ for } t > 0. \tag{18.12}$$

From equations 18.11 and 18.12, $v_2(t) > 0$ and $di_1/dt > 0$ for all $t > 0$.

The result of example 18.6 implies that the simple experimental procedure of example 18.4 will establish the correct dot marking even if coil 2 is connected to a finite resistance.

The preceding basic formulation of Laplace transform equations presupposes zero initial conditions. If the initial currents in some inductors are not zero, two approaches are available. The first approach is to write differential equations and then apply the Laplace transform to those equations. This process will automatically bring in all of the initial conditions. The second approach is to derive the s-domain equivalent circuits for coupled inductors with initial currents. These circuits are generalized versions of those in figures 14.18 and 14.19. Since these equivalent circuits are less basic and are not frequently needed in practice, we have included their development in problem 18.19.

The Laplace transform method is very useful for finding the complete solution. In practice, however, there are many occasions where the excitation to a *stable* circuit is sinusoidal and only the *steady-state* solution is of interest. The phasor method of analysis proves easier than the Laplace transform method in such cases. Phasor equations resemble Laplace transform equations under the assumption of zero initial conditions. The only difference is that V and I are now phasors (complex numbers), with s replaced by $j\omega$. (See section 3 of chapter 15). In phasor form, equations 18.10 become

$$\mathbf{V}_1 = j\omega L_1 \mathbf{I}_1 \pm j\omega M \mathbf{I}_2 \tag{18.13a}$$

and

$$\mathbf{V}_2 = \pm j\omega M \mathbf{I}_1 + j\omega L_2 \mathbf{I}_2, \tag{18.13b}$$

where the + sign is for figure 18.4a and the − sign for figure 18.4b. (Refer to equations 18.6 for other cases.)

Since $M = M_{12} = M_{21}$ (equation 18.3), one can easily show that the coupled inductors neither consume nor generate average power in sinusoidal steady state; they are said to be *lossless*. (See problem 27.)

The following example illustrates the use of the phasor method.

◆ EXAMPLE 18.7

Consider the circuit of figure 18.8. Find the steady-state components of $v_1(t)$ and $v_2(t)$ at the frequency 1 rad/sec for the following two cases: (1) the 2-F capacitor is disconnected. (2) The 2-F capacitor is connected.

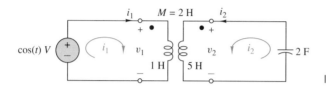

Figure 18.8 Circuit for example 18.7.

SOLUTION

First we note that $\omega = 1$ rad/sec, and that the phasor for the voltage source is $1\angle 0$ V.

Part 1. With the capacitor disconnected, $I_2 = 0$. Hence, I_2 induces no component in V_1, and the equation for V_1 is

$$V_1 = 1\angle 0 = j1 \times I_1,$$

from which it follows that

$$I_1 = 1\angle -90°$$

and

$$V_2 = j\omega M I_1 = 2\angle 0.$$

Observe that $v_1(t) = \cos(t)$ and $v_2(t) = 2\cos(t)$ are *in phase*. This observation might suggest that the voltage drops of coupled inductors measured from the dotted terminal to the undotted terminal are always in phase (i.e., they rise and fall at the same time). While this is true for a great majority of circuits, it may fail for some cases, as follows.

Part 2. With the 2-F capacitor connected, the two loop equations are

$$V_1 = 1 = (j\omega L_1 I_1 + j\omega M I_2) = jI_1 + j2I_2$$

and, since $V_2 = \dfrac{-I_2}{j\omega C}$,

$$(j\omega M I_1 + j\omega L_2 I_2) + \frac{I_2}{j\omega C} = j2I_1 + j4.5I_2 = 0.$$

The solutions of these equations are

$$I_1 = -j9 \quad \text{and} \quad I_2 = j4.$$

Using equation 18.13b,

$$V_2 = j\omega M I_1 + j\omega L_2 I_2 = j2(-j9) + j5(j4) = -2,$$

in which case $v_2(t) = 2\cos(t + 180°)$ V. Clearly, $v_1(t) = \cos(t)$ V and $v_2(t) = 2\cos(t + 180°)$ V are 180° out of phase!

This example is a drastic one contrived to bring up an underlying property: The dot markings for coupled inductors (L_1, L_2, M) determine the ± sign in the equation relating v_2 to di_1/dt and the equation relating v_1 to di_2/dt. No intrinsic relations between the polarities of v_1 and v_2 are conveyed by the dot convention. For most practical circuits, however, it is true that the voltage drops of coupled inductors from the dotted terminal to the undotted terminal are in phase or nearly in phase.

4. APPLICATIONS: AUTOMOBILE IGNITION AND RF AMPLIFIER

The fundamental principles and techniques of the previous section allow us to investigate some interesting applications. The automobile ignition circuit is one such application, and a radio frequency amplifier is another. The following two examples present a simplified analysis of these circuits.

◆ EXAMPLE 18.8

Figure 18.9 shows an automobile ignition system and a simplified circuit model thereof. This ignition system is typical of older model cars. The ignition coil is a pair of inductors wound on the same iron core. The coil connected to a power source is called the **primary**, while the coil connected to a load is called the **secondary**. The primary has a few hundred turns of heavy wire, the secondary about 20,000 turns of very fine wire. When the ignition point (or contact) opens by cam action, a voltage exceeding 20,000 V is induced across the secondary, causing the spark plug to fire. Today's ignition systems do not typically have points and a condenser; switching is done electronically. The box in dashed lines in figure 18.9a is replaced by an electronic ignition module. Nevertheless, the basic idea of generating a high voltage to cause a spark to occur at the plug is accomplished by a basic RLC circuit containing a switch that represents the point of the ignition system.

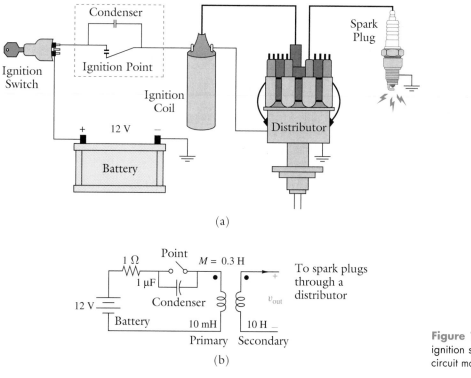

Figure 18.9 (a) An automobile ignition system and (b) its simplified circuit model.

Again, figure 18.9b is a simplified circuit model for the ignition system. Suppose that the switch has been closed for a long time. At $t = 0$, the switch opens, and a high voltage (exceeding 20,000 V) will momentarily occur at the output terminals, causing a spark plug to fire.

Since the secondary is open-circuited, it has no effect on the solution for the primary current. Accordingly, at $t = 0^-$, we have $i_1 = E/R = 12$ A. Using the model for an initialized inductor given in figure 14.18 results in the s-domain equivalent circuit of figure 18.10.

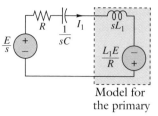

Model for
the primary

Figure 18.10 s-domain circuit for
example 18.8.

The primary current is simply the net driving voltage divided by the total impedance in the series circuit, i.e.,

$$I_1(s) = \frac{\dfrac{E}{s} + L_1 \dfrac{E}{R}}{R + \dfrac{1}{sC} + sL_1} = \frac{E}{R} \cdot \frac{s + \dfrac{R}{L_1}}{s^2 + s\dfrac{R}{L_1} + \dfrac{1}{L_1C}}. \tag{18.14}$$

Substituting the given component values into equation 18.14 yields

$$I_1(s) = \frac{12(s + 100)}{s^2 + 100s + 10^8} = \frac{12[(s + 50) + 0.005 \times 9,999.875]}{(s + 50)^2 + (9,999.875)^2}. \tag{18.15}$$

Taking the inverse Laplace transform of $I_1(s)$ yields

$$i_1(t) = 12e^{-50t}\cos(9,999.875t) + 0.06e^{-50t}\sin(9,999.875t) \tag{18.16a}$$

$$\cong 12e^{-50t}\cos(10,000t). \tag{18.16b}$$

Having obtained $i_1(t)$, we calculate $v_2(t)$ from the basic relationship of equation 18.5, using the plus sign in the present case of dot markings:

$$v_2(t) = \frac{M di_1}{dt}$$

$$= 0.3 \times 12[-50e^{-50t}\cos(10,000t) - 10^4 e^{-50t}\sin(10,000t)]$$

$$\cong -36,000e^{-50t}\sin(10,000t) \text{ V}, \quad \text{for } t > 0.$$

From this expression, the voltage $v_2(t)$ reaches a magnitude of 36,000 V in about 157 μs (one-fourth of a cycle of the oscillations). This voltage is high enough to cause the spark plug to fire.

In going from equation 18.16a to 18.16b, we have neglected the second term in 18.16a in comparison with the first. With practical component values used in ignition circuits, this approximation is usually valid. In terms of equation 18.14, the approximation is as follows:

$$I_1(s) = \frac{E}{R} \cdot \frac{\left(s + \dfrac{R}{2L_1}\right) + \dfrac{R}{2L_1}}{s^2 + \dfrac{R}{L_1}s + \dfrac{1}{L_1C}} \cong \frac{E}{R} \cdot \frac{\left(s + \dfrac{R}{2L_1}\right)}{s^2 + \dfrac{R}{L_1}s + \dfrac{1}{L_1C}}. \tag{18.17}$$

Applying the inverse Laplace transform to equation 18.17 yields

$$I_1(t) \cong \frac{E}{R}e^{-\sigma_p t}\cos(\omega_d t),$$

where

$$\sigma_p = \frac{R}{2L_1} \quad \text{and} \quad \omega_d = \sqrt{\frac{1}{L_1C} - \left(\frac{R}{2L_1}\right)^2} \cong \frac{1}{\sqrt{L_1C}}.$$

For the first few cycles of oscillations, the value of $-\sigma t$ is nearly zero and the value of the exponential factor is nearly 1. Therefore, $i_1(t)$ may be further simplified to

$$i_1(t) \cong \frac{E}{R}\cos\left(\frac{t}{\sqrt{L_1C}}\right) \quad \text{for small } t.$$

Finally, we compute $v_2(t)$ from equation 18.1:

$$v_2(t) = M\frac{di_1}{dt} = M\frac{E}{R}\left[\frac{-1}{\sqrt{L_1C}}\sin\left(\frac{t}{\sqrt{L_1C}}\right)\right]$$

for small t. Thus,

$$|v_2(t)|_{\max} \cong M\frac{E}{R}\frac{1}{\sqrt{L_1C}} = \frac{M}{L_1}QE \qquad (18.18)$$

approximates the maximum value of v_2, where

$$Q = \sqrt{\frac{L_1}{C}}\frac{1}{R}$$

is the quality factor of the series RL_1C circuit. (See equation 17.15.) Section 18.6 will show that M/L_1 is approximately equal to the ratio of the number of turns of the secondary to that of the primary.

Equation 18.18 is a simple formula for estimating the maximum voltage that occurs at the secondary. It shows that although the battery voltage E is only 12 V, what appears at the secondary for a brief moment is very much higher due to the switching action. The voltage is stepped up by two factors: (1) the Q of the series RL_1C circuit and (2) the turns ratio of the ignition coil. From equation 18.18, a smaller R produces a higher voltage across the secondary. But a small R causes a larger current to flow in the primary circuit and therefore shortens the life of the breaker point. In practice, when the engine is running, a resistance wire or an actual resistor is placed in the primary circuit to limit the amount of current flow through the breaker point. The capacitor (condenser) serves a similar purpose—that of protecting the breaking point by suppressing the arc that results when the point opens.

Exercise. The ignition circuit of figure 18.9b has $E = 12$ V, $L_1 = 1$ mH, and $C = 0.01$ μF. If the total resistance in the primary circuit is $R = 8$ Ω and $M/L_1 = 50$, estimate the maximum voltage appearing at the open-circuited secondary when the switch opens.

ANSWER: About 23.7 kV.

◆ EXAMPLE 18.9

The front end of a radio receiver typically has circuitry such as that shown in figure 18.11a. The RLCM circuit of figure 18.11b is a simplified model of this circuitry. V_s and the resistance $R_1 = 300$ Ω together represent the antenna. Suppose the other

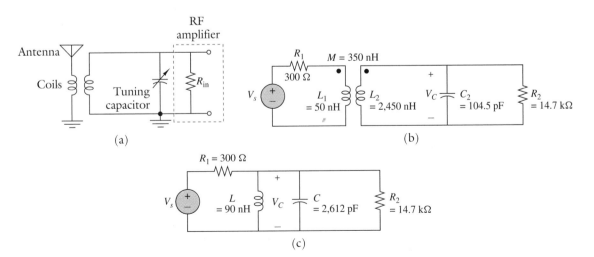

(a)

(b)

(c)

Figure 18.11 Circuits for example 18.9. (a) The original circuit. (b) A simplified circuit model in which $L_1L_2 = M^2$. (c) A design without coupled inductors.

element values are $R_2 = R_{in} = 14.7 \text{ k}\Omega$, $L_1 = 50 \text{ nH}$, $L_2 = 2{,}450 \text{ nH}$, $M = 350 \text{ nH}$, and $C_2 = 104.5 \text{ pF}$. Find the maximum value and the bandwidth of the magnitude response (i.e., the curve of $|V_C/V_s|$ vs. ω).

SOLUTION

The first step in the solution is to find the transfer function $H(s) = V_C(s)/V_s(s)$ from figure 18.11b. The self- and mutual inductances in this circuit satisfy the condition $L_1 L_2 = M^2$. The physical significance of this condition will be discussed in the next section. Applying the analysis method described in section 18.3, and utilizing the condition that $L_1 L_2 = M^2$ yields

$$H(s) = \sqrt{\frac{L_1}{L_2}} \frac{1}{R_1 C} \frac{s}{s^2 + \frac{1}{RC}s + \frac{1}{L_2 C}}, \qquad (18.19)$$

where $R = R_2 // [R_1(L_2/L_1)]$, i.e., R is the parallel combination of R_2 and $R_1(L_2/L_1)$. (The details of the algebraic manipulations are left as an exercise.) Section 18.7 discusses a shortcut to this derivation. For the present case,

$$R = R_2 // \left[\frac{L_2}{L_1}\right] R_1 = (14{,}700) // \left[\frac{2{,}450}{50}\right] 300 = 7{,}350 \ \Omega.$$

This transfer function has the bandpass characteristic of equation 17.29 studied in section 6 of chapter 17. Utilizing the results given in equation 17.30 leads to

$$\omega_m = \frac{1}{\sqrt{L_2 C}} = \frac{1}{\sqrt{(2{,}450 \times 10^{-9})(104.5 \times 10^{-12})}} = 62.5 \times 10^6 \text{ rad/sec,}$$

$$|H(j\omega_m)| = |H(j\omega)|_{max} = \sqrt{\frac{L_1}{L_2} \frac{R}{R_1}} = \sqrt{\frac{50}{2{,}450} \frac{7{,}350}{300}} = 3.5.$$

The bandwidth is

$$B_\omega = \frac{1}{RC} = \frac{1}{7{,}350 \times 104.5 \times 10^{-12}} = 1.302 \times 10^6 \text{ rad/sec.}$$

Instead of the coupled inductors of figure 18.11a, the same ω_m and B_ω can be obtained with the parallel resonant circuit of figure 18.11c. Following the design method described in example 17.7, we find the required element values to be $L = 98 \text{ nH}$ and $C = 2{,}612 \text{ pF}$. The maximum voltage gain, however, has a much lower value of $|H(j\omega_m)| = 14{,}700/(14{,}700 + 300) = 0.98$, compared with 3.5 for the coupled circuit of figure 18.11a.

The higher voltage gain achieved in figure 18.11b can be explained by the concept of maximum power transfer. A routine analysis will reveal that at $\omega = \omega_m$, the input impedance "seen" by the source is a pure resistance equal to the source resistance R_1. Thus, maximum power has been extracted from the source. Since the inductors and the capacitor together form a lossless coupling network (see problem 18.27), the same maximum power is transferred to the load resistor R_2. The average power absorbed by R_2 is $|V_{R2}|^2/R_2$. Since R_2 is fixed, maximum power to it implies the highest possible voltage across it.

Besides having a higher voltage gain, the coupled inductor circuit of figure 18.11b has another important advantage over the simple resonant circuit of figure 18.11c: There is no direct connection between its two coils. (This is sometimes referred to as the *dc isolation property.*) Consequently, dc and ac components of currents are separated, eliminating the need for dc blocking capacitors and also simplifying the design of the amplifier biasing circuitry.

Figure 18.11b contains two inductors and one capacitor. Accordingly, one would normally consider it a third-order circuit. Yet the denominator of equation 18.19 is only a second-degree polynomial in s. This simplification is due to the condition $M^2 = L_1 L_2$. The significance of this condition will be discussed in the next section.

5. COEFFICIENT OF COUPLING AND ENERGY CALCULATION

We begin this section with a justification of our assumption that $M_{12} = M_{21} = M$. The justification stems from a physical property that a pair of stationary coupled coils cannot generate average power. In appendix B1, we will justify $M_{12} = M_{21} = M$ by the principles of magnetic circuits. As a consequence, we will also show that the mutual inductance M has upper bound $\sqrt{L_1 L_2}$; i.e., the mutual inductance can never exceed the geometric mean of the self-inductances.

Justification that $M_{12} = M_{21} = M$

This property is a consequence of the principles of electromagnetic field theory and cannot be proven by a pure mathematical argument. The principles of field theory, however, are not commonly known by students studying introductory circuit analysis. To make our approach more accessible, we will build our justification on the passivity principle for inductors.

THE PASSIVITY PRINCIPLE FOR INDUCTORS

A pair of stationary coupled inductors is a passive system; i.e., they cannot generate energy and, hence, cannot deliver average power to any external network.

Our justification that $M_{12} = M_{21} = M$ will use this passivity principle and the method of proof by contradiction. Showing that $M_{12} = M_{21} = M$ is equivalent to showing that if $M_{12} \neq M_{21}$, then the passivity principle is violated, which contradicts the physical assumption about the coupled inductors. Hence, it must be true that $M_{12} = M_{21}$.

Suppose that $M_{12} \neq M_{21}$. Then, instead of having equations 18.7, the differential equations for the coupled inductors of figure 18.4a must account for this difference and take the form

$$v_1 = L_1 \frac{di_1}{dt} + M_{12} \frac{di_2}{dt}, \tag{18.20a}$$

$$v_2 = M_{21} \frac{di_1}{dt} + L_2 \frac{di_2}{dt}. \tag{18.20b}$$

Let us apply $i_1 = \sin(t)$ and $i_2 = \cos(t)$ to the inductors. From equations 18.20, the terminal voltages are

$$v_1 = L_1 \frac{d}{dt} \sin(t) + M_{12} \frac{d}{dt} \cos(t) = L_1 \cos(t) - M_{12} \sin(t)$$

and

$$v_2 = M_{21} \frac{d}{dt} \sin(t) + L_2 \frac{d}{dt} \cos(t) = M_{21} \cos(t) - L_2 \sin(t).$$

The total instantaneous power delivered to the inductors is, therefore,

$$
\begin{aligned}
p(t) &= v_1 i_1 + v_2 i_2 \\
&= L_1 \cos(t) \sin(t) - M_{12} \sin^2(t) + M_{21} \cos^2(t) - L_2 \sin(t) \cos(t). \tag{18.21}
\end{aligned}
$$

To calculate P_{ave}, the average power delivered to the coupled inductors, we use the identities $\sin(t) \cos(t) \equiv 0.5 \sin(2t)$, $\sin^2(t) \equiv 0.5[1 - \cos(2t)]$, and $\cos^2(t) \equiv 0.5[1 + \cos(2t)]$. It follows immediately that the first and the last terms in equation 18.21 make no contribution to P_{ave}, whereas the terms involving M_{12} and M_{21} lead to

$$P_{\text{ave}} = 0.5(M_{21} - M_{12}). \tag{18.22}$$

This result shows very clearly that if $M_{12} > M_{21}$, then the excitations $i_1 = \sin(t)$ and $i_2 = \cos(t)$ will lead to a negative P_{ave}, violating the passivity principle. Similarly, if $M_{12} < M_{21}$, then the new (transposed) excitations $i_1 = \cos(t)$ and $i_2 = \sin(t)$ will again lead to a negative P_{ave}, violating the passivity principle. Therefore, we conclude that $M_{12} = M_{21}$.

With $M_{12} = M_{21} = M$, the average power, P_{ave}, is always zero for arbitrary sinusoidal excitations. (See problem 18.27.)

Calculation of Stored Energy

Having proved that $M_{12} = M_{21} = M$, we shall now show that there is a limit to the value of M that is attainable once L_1 and L_2 are specified. This is again done by the use of the passivity principle.

Consider the coupled inductors shown in figure 18.4. The voltage-current relationships at the terminals are given by

$$v_1(t) = L_1 \frac{di_1}{dt} \pm M \frac{di_2}{dt} \qquad (18.23a)$$

and

$$v_2(t) = \pm M \frac{di_1}{dt} + L_2 \frac{di_2}{dt}, \qquad (18.23b)$$

with the upper signs for figure 18.4a and the lower signs for figure 18.4b. Let us assume that the inductor currents are initially zero (at $t = 0$). In this state, there is no energy stored in the system. To see this, we connect a pair of resistors to the inductors. A routine analysis of the circuit (say, by the Laplace transform method) will show that no current flows in either resistor. Thus no energy is *delivered* to the resistors, indicating that the inductors have no stored energy at $t = 0$.

Next, we apply driving sources to the inductors to bring the currents up to $i_1 = I_1$ and $i_2 = I_2$ at $t = T$. The energy *delivered* to the inductors during the time interval $(0,T)$, is

$$W(T) = \int_0^T (v_1 i_1 + v_2 i_2) dt = \int_0^T \left[\left(L_1 \frac{di_1}{dt} \pm M \frac{di_2}{dt} \right) i_1 + \left(\pm M \frac{di_1}{dt} + L_2 \frac{di_1}{dt} \right) i_2 \right] dt$$

$$= \frac{1}{2} L_1 I_1^2 + \frac{1}{2} L_2 I_2^2 \pm \int_0^T M \left(i_1 \frac{di_2}{dt} + i_2 \frac{di_1}{dt} \right) dt$$

$$= \frac{1}{2} L_1 I_1^2 + \frac{1}{2} L_2 I_2^2 \pm \int_0^{I_1, I_2} M d(i_1 i_2) = \frac{1}{2} L_1 I_1^2 + \frac{1}{2} L_2 I_2^2 \pm M I_1 I_2, \qquad (18.24)$$

where $d(i_1 i_2)$ is the total derivative of the product $i_1 i_2$ and is equal to $i_1 di_2 + i_2 di_1$.

A couple of things about this result are worth noticing:

1. The final integral in equation 18.24 depends only on the final values I_1 and I_2. The exact waveforms of $i_1(t)$ and $i_2(t)$ during $0 \le t \le T$ are immaterial.
2. The energy $W(T)$ delivered by the sources during $0 \le t \le T$ is not lost, but merely *stored* in the system.

To see property 2, we may adjust the sources so that the currents are brought back from I_1 and I_2 at $t = T$ to zero at some $t = T' > T$. Then, the energy *delivered* by the sources to the inductors during $T < t < T'$ may be calculated in a manner similar to that shown in equation 18.24, giving the result (see problem 28)

$$W(T') - W(T) = -\frac{1}{2} L_1 I_1^2 - \frac{1}{2} L_2 I_2^2 \mp M I_1 I_2, \qquad (18.25)$$

with the upper ($-$) sign for figure 18.4a and the lower ($+$) sign for figure 18.4b. Equation 18.25 is precisely the negative of equation 18.24. Thus, all of the energy delivered by the sources during $0 \le t \le T$ has been returned to the sources during $T \le t \le T'$. For this reason, the energy given by equation 18.24 is called the *stored energy*. Another way of recovering the stored energy is described in problem 29. The physics of the situation shows that the energy is stored in the magnetic field produced by the currents in the inductors.

Upper Bound for M and the Coefficient of Coupling

The energy $W(T)$ must be nonnegative for arbitrary values of I_1 and I_2. Otherwise the inductors will be *generating* energy during the time interval $0 \le t \le T$ and thus violate the passivity principle. To ensure a nonnegative $W(T)$ for all I_1 and I_2, the values of L_1, L_2, and M must satisfy the inequality

$$M^2 \leq L_1 L_2,$$ (18.26a)

or

$$M \leq \sqrt{L_1 L_2}.$$ (18.26b)

To see this, we rewrite equation 18.24 in the form

$$W(T) = \frac{1}{2} I_2^2 \left[L_1 \left(\frac{I_1}{I_2} \right)^2 \pm 2M \left(\frac{I_1}{I_2} \right) + L_2 \right]$$

$$= \frac{1}{2} I_2^2 \left[L_1 x^2 \pm 2Mx + L_2 \right] \equiv \frac{1}{2} I_2^2 f(x),$$ (18.27)

where $x \equiv I_1/I_2$ is a current ratio. Equation 18.27 shows that $W(T)$ is negative whenever $f(x)$ is negative. Now, $f(x)$ is a second-degree polynomial in x with a positive coefficient for the x^2 term. Consequently, the curve of $f(x)$ vs. x will be a parabola opening upwards. From analytic geometry, depending on the sign of the discriminant $D \equiv (M^2 - L_1 L_2)$, the curve may or may not intersect the $f(x) = 0$ axis, as illustrated in figure 18.12. From the figure, it is obvious that if $D > 0$, there will be some current ratio that yields a negative $f(x)$ and hence a negative $W(T)$, again violating the passivity principle. Therefore, $D = (M^2 - L_1 L_2) \leq 0$, which yields equation 18.26.

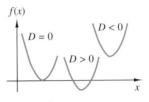

Figure 18.12 A plot of the stored energy vs. the current ratio x.

The degree to which M approaches its upper bound $\sqrt{L_1 L_2}$ is expressed by a positive number called the **coefficient of coupling**, defined as

$$k \equiv \frac{M}{\sqrt{L_1 L_2}}.$$ (18.28)

From equations 18.27 and 18.28,

$$0 \leq k \leq 1.$$ (18.29)

When $k = 0$, M is also zero, and the inductors are uncoupled. When $k = 1$, the inductors have *unity coupling*, an idealized situation impossible to realize in practice.

◆ EXAMPLE 18.10

In both circuits of figure 18.4, suppose that $L_1 = 5$ H, $L_2 = 20$ H, $M = 8$ H, $I_1 = 2$ A, and $I_2 = 4$ A. Find:

1. The coupling coefficient k.
2. The stored energy.

SOLUTION

1. For both circuits, $k = M/\sqrt{L_1 L_2} = 8/\sqrt{5 \times 20} = 0.8$.
2. For figure 18.4a,

$$\text{stored energy} = 0.5 L_1 I_1^2 + 0.5 L_2 I_2^2 + M I_1 I_2$$

$$= 0.5 \times 5 \times 2^2 + 0.5 \times 20 \times 4^2 + 8 \times 2 \times 4 = 234 \text{ joules.}$$

For figure 18.4b,

$$\text{stored energy} = 0.5L_1 I_1^2 + 0.5L_2 I_2^2 - M I_1 I_2$$

$$= 0.5 \times 5 \times 2^2 + 0.5 \times 20 \times 4^2 - 8 \times 2 \times 4 = 106 \text{ joules.}$$

Exercise. Repeat example 18.10 with $L_1 = 4$ H, $L_2 = 16$ H, $M = 6$ H, $I_1 = 4$ A, and $I_2 = -2$ A.
SCRAMBLED ANSWERS: 112, 0.75, 16.

EXAMPLE 18.11

In the circuit of figure 18.13, $I_1 = 6$ A. Find the minimum value of the stored energy and the corresponding value of I_2.

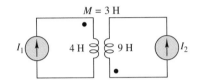

Figure 18.13 Coupled inductors for calculating stored energy in example 18.11.

SOLUTION

From equation 18.24,

$$W = 0.5L_1 I_1^2 + 0.5L_2 I_2^2 - M I_1 I_2$$

$$= 0.5 \times 4 \times 36 + 0.5 \times 9 \times I_2^2 - 3 \times 6 \times I_2$$

$$= 4.5 I_2^2 - 18 I_2 + 72 \triangleq f(I_2).$$

Following the standard method in calculus for finding the maximum and minimum, we set df/dI_2 to zero and solve for I_2:

$$\frac{dW}{dI_2} = \frac{df}{dI_2} = 9I_2 - 18 = 0.$$

This yields $I_2 = 2$ A, and the corresponding minimum stored energy $W_{\min} = 18 - 36 + 72 = 54$ joules.

6. IDEAL TRANSFORMER AS A CIRCUIT ELEMENT AND APPLICATIONS

Two coupled *differential equations* containing three parameters L_1, L_2, and M characterize the coupled coils of figure 18.4. By imposing two idealized conditions on these parameters, some very useful circuit properties result.

Idealization 1: The coupled inductors have unity coupling; i.e., $M^2 = L_1 L_2$, or the coupling coefficient $k = 1$.

Under this idealized condition of unity coupling, the pair of coils has the voltage transformation property

$$\frac{v_1(t)}{v_2(t)} = \frac{L_1}{M} = \frac{M}{L_2} = a, \tag{18.30}$$

where a is a constant and both v_1 and v_2 are the *voltage drops from dotted to undotted terminals* of the coils, as per figure 18.4a.

To derive the condition of equation 18.30, note that the constraint $M^2 = L_1 L_2$ implies that $L_1/M = M/L_2$. Denoting the ratio L_1/M by a leads to $L_1 = aM$ and $M = aL_2$.

Substituting these relationships into equation 18.7, and dividing 18.7a by equation 18.7b yields

$$\frac{v_1(t)}{v_2(t)} = \frac{aM\dfrac{di_1}{dt} + aL_2\dfrac{di_2}{dt}}{M\dfrac{di_1}{dt} + L_2\dfrac{di_2}{dt}} = a.$$

A unity coupling coefficient is an idealization that is not achievable in practice. However, coupling coefficients near unity are achievable by winding the turns of two inductors very closely together. The constant $a = L_1/M = M/L_2 = \sqrt{L_1/L_2}$ in equation 18.30 is then simply the ratio of the numbers of turns, denoted by N_1 and N_2, of the two coils and is usually referred to as the **turns ratio**, i.e., $a = N_1/N_2$. Appendix B3 offers a more complete explanation of this relationship and examines the parameters (L_1, L_2, M) from the point of view of magnetic circuits. There, we show that

$$L_1 : L_2 : M = N_1^2 : N_2^2 : N_1 N_2$$

when $k = 1$. Problem 34 outlines a novel derivation of this relationship without the use of magnetic circuit concepts.

With unity coupled coils, equation 18.30 indicates that the voltages $v_1(t)$ and $v_2(t)$, both *from dotted to undotted terminals*, always have the same polarity. With coupling less than unity, it is possible for v_1 and v_2 to have opposite polarities at some time instants, as shown in example 18.7.

Idealization 2: In addition to unity coupling, the coupled coils have infinite mutual and self-inductances.

Under these two idealized conditions, the pair of coils has the current transformation property

$$\frac{i_1(t)}{i_2(t)} = -\frac{1}{a}, \tag{18.31}$$

where both i_1 and i_2 are the *currents entering the dotted terminals* of the coils, as per figure 18.4a.

To derive equation 18.31, we make use of the unity coupling condition $L_1/M = M/L_2 = a$ and rewrite equations 18.7 as

$$\frac{v_1}{M} = \frac{L_1}{M}\frac{di_1}{dt} + \frac{di_2}{dt} = \frac{d}{dt}(ai_1 + i_2) \tag{18.32a}$$

and

$$\frac{v_2}{L_2} = \frac{M}{L_2}\frac{di_1}{dt} + \frac{di_2}{dt} = \frac{d}{dt}(ai_1 + i_2). \tag{18.32b}$$

Letting $L_1 \to \infty$, $L_2 \to \infty$, and $M \to \infty$, equations 18.32a and 18.32b each lead to

$$\frac{d}{dt}(ai_1 + i_2) = 0,$$

whose solution is obviously

$$ai_1(t) + i_2(t) = C \tag{18.33}$$

for some constant C. Assume that the coils are unenergized prior to applying excitations; i.e., at some time $t = t_o$ in the past, $i_1(t_o) = i_2(t_o) = 0$, which must be true for any real circuit. It follows that the constant C in equation 18.33 is zero, and consequently, $i_1/i_2 = -1/a$.

The negative sign in equation 18.31 implies that at any time, if a current *enters* one coil at the dotted terminal, then the current in the other coil must *leave* the dotted terminal.

The condition of infinite inductances (L_1, L_2, M) is another idealization that is not realizable in practice. However, appendix B3 shows that this condition can be *approximated* by using a magnetic material with very high permeability as the common core for the two coils.

THE IDEAL TRANSFORMER

Two coupled coils satisfying the relationships

$$\frac{v_1(t)}{v_2(t)} = \frac{N_1}{N_2} = a \quad \text{and} \quad \frac{i_1(t)}{i_2(t)} = -\frac{N_2}{N_1} = -\frac{1}{a}$$

are said to be an *ideal transformer*, shown in figure 18.14a. In the figure, the two vertical bars serve as a reminder of the presence of a ferromagnetic core in the physical device. The word "ideal" may or may not appear in the schematic diagram. Again, the mathematical model of an ideal transformer depends only on the turns ratio $a{:}1$ and the relative dot positions. To avoid the negative sign in the current relationship, an alternative labeling of voltages and currents shown in figure 18.14b may be used. The subscript p stands for the "primary" coil, connected to a power source, and s for the "secondary" coil, connected to a load. Note that i_p is entering at the dotted terminal and i_s is leaving the dotted terminal. The notation of figure 18.14b is more commonly used in the study of electric power flow.

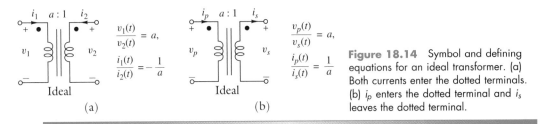

Figure 18.14 Symbol and defining equations for an ideal transformer. (a) Both currents enter the dotted terminals. (b) i_p enters the dotted terminal and i_s leaves the dotted terminal.

For our present purposes in circuit analysis, we have defined an ideal transformer strictly from its terminal voltage-current relationships. In section 8, we show how a practical transformer can be constructed to achieve *approximately* the idealized conditions $k = 1$ and $(L_1, L_2, M) \rightarrow \infty$.

One important simplification resulting from the idealizations is that an ideal transformer is characterized by two *algebraic* equations containing a single parameter a, the turns ratio. This is to be contrasted with a pair of coupled coils, characterized by two differential equations containing three parameters L_1, L_2, and M.

The input-output power properties of an ideal transformer are very interesting. From equations 18.30 and 18.31 applied to figure 18.14a, the instantaneous power delivered to an ideal transformer is

$$p(t) = v_1 i_1 + v_2 i_2 = (a v_2)\left(-\frac{i_2}{a}\right) + v_2 i_2 = 0.$$

Considered as a single unit, an ideal transformer neither generates nor consumes instantaneous power: Whatever instantaneous power is received at one side must transfer to the other side. Furthermore, since $p(t)$ is identically zero, so is its integral with respect to t. Thus, an ideal transformer cannot store any energy. Also, unlike L or C, the presence of ideal transformers in a circuit does not increase the number of poles of any network function. This follows because equations 18.30 and 18.31 suggest representing an ideal transformer by two controlled sources, as shown in figure 18.15. Such a representation is useful when a circuit simulation program does not include the ideal transformer as a permissible element, but does admit all four types of controlled sources. (See problem 56.)

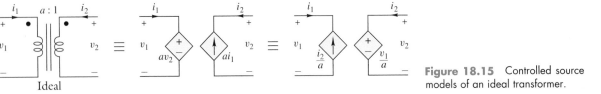

Figure 18.15 Controlled source models of an ideal transformer.

Besides having the voltage and current transformation properties, an ideal transformer also has the following property, which is extremely useful in designing electronic circuits for achieving maximum power transfer.

IMPEDANCE TRANSFORMATION PROPERTY

If the secondary of an ideal transformer is terminated in a load impedance $Z(s)$, then the impedance looking into the primary is $Z_{in}(s) = a^2 Z(s)$, where a is the turns ratio taken in the direction from source to load. (See figure 18.16; the dots are not marked on the figure because their positions are immaterial for this application.)

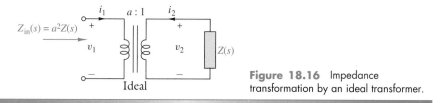

Figure 18.16 Impedance transformation by an ideal transformer.

Deriving this property is straightforward. We merely apply equations 18.30 and 18.31 to figure 18.16, to obtain the desired result:

$$Z_{in}(s) = \frac{V_1(s)}{I_1(s)} = \frac{aV_2(s)}{\dfrac{-I_2(s)}{a}} = a^2 \frac{V_2(s)}{-I_2(s)} = a^2 Z(s). \qquad (18.34)$$

Coupled inductors are often used for one of the three transformation properties: voltage transformation, current transformation, and impedance transformation. When used in such manner, they are called **transformers**. In analyzing circuits containing transformers, we may initially assume these devices to be ideal and use the models of figure 18.14. This makes it much easier to understand the basic functions of the transformers in the circuits, while avoiding complicated mathematics. When accurate results are desired, one uses the more realistic models described in section 8.

We now examine some applications of the three properties of ideal transformers in circuit analysis.

 **EXAMPLE 18.12**

The transformer in the system of figure 18.17 is assumed to be ideal. The loads represented by resistors consist of 10 incandescent lamps in parallel, each drawing 0.5 A. (1) Find the voltage (magnitude) across each lamp, and (2) find the current (magnitude) delivered by the source.

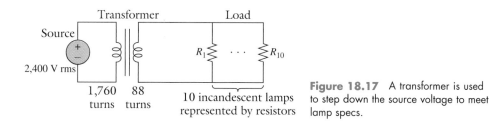

Figure 18.17 A transformer is used to step down the source voltage to meet lamp specs.

SOLUTION

Since only magnitudes are involved in this problem, dot positions on the transformer and reference directions for voltages and currents are immaterial. The turns ratio is $a = 1,760/88$. From equations 18.30 and 18.31, the voltage across each lamp is

$$\frac{88}{1,760} \, 2,400 = 120 \text{ V},$$

and the current magnitude through the power source is

$$10 \times 0.5 \times \frac{88}{1,760} = 0.25 \text{ A}.$$

EXAMPLE 18.13

Figure 18.18 shows a simplified model of an audio amplifier containing an ideal transformer. The input voltage is at 2 kHz with a magnitude of 1 V rms. The load is a loudspeaker, represented by a 4-Ω resistance.

1. Find the average power delivered to the 4-Ω load if it is connected directly to the amplifier (i.e., with the transformer removed).
2. With the transformer connected, and with a turns ratio $a = 5$, find the average power delivered to the load.
3. If the turns ratio a is adjustable, what should it be in order to have maximum power delivered to the load? What is the value of the maximum power?

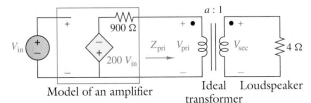

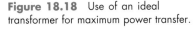

Model of an amplifier Ideal Loudspeaker **Figure 18.18** Use of an ideal
transformer transformer for maximum power transfer.

SOLUTION

Part 1. The (magnitude of the) current through the load is

$$I_{4\Omega} = 1\frac{200}{900 + 4} = 0.2212 \text{ A}.$$

Therefore, the average power is

$$P_{4\Omega} = (0.2212)^2 \times 4 = 0.1958 \text{ W}.$$

Part 2. Looking into the primary of the transformer, we find the impedance is

$$Z_{\text{pri}} = 5^2 \times 4 = 100 \ \Omega.$$

From the voltage divider formula, the voltage across the primary is

$$\left|V_{\text{pri}}\right| = 1 \times 200 \times \frac{100}{900 + 100} = 20 \text{ V}.$$

The voltage across the secondary is

$$|V_{\text{sec}}| = \frac{20}{5} = 4 \text{ V}.$$

Therefore,

$$P_{4\Omega} = \frac{4^2}{4} = 4 \text{ W}.$$

Part 3. From the maximum power transfer theorem (see volume 1, chapter 11, section 5), the turns ratio a should be such that the 4-Ω resistance reflects back to the primary as 900 Ω to match the internal resistance of the amplifier; i.e., $a^2 \times 4 = 900$, or $a = 15$. With $a = 15$, a repeat of the calculations of part 2 yields $P_{\text{max}} = 11.11$ W.

This example clearly demonstrates how a transformer can raise the output power when the load resistance and the internal resistance of the amplifier are widely different. In the present case, the transformer is connected at the output end of the amplifier, so it is called an **output transformer.** A transformer connected at the input end of the amplifier for the same purpose of matching two different resistances is called an **input transformer.**

Exercise. Repeat all parts of example 18.13 if the loudspeaker has a resistance of 16 ohms.
ANSWERS: 0.7628 W, 9.467 W, 11.11 W.

Although the positions of the dots for the ideal transformer are immaterial in examples 18.12 and 18.13, they play an important role in the next example.

EXAMPLE 18.14

The circuit of figure 18.19a has the following values for its passive elements:

$$R_1 = 100 \ \Omega, R_2 = 100 \ \Omega, L = 1 \ \text{mH}, C = 0.1 \ \mu\text{F}, a = 3.$$

If $i_L(0^-) = 0$ and $v(0^-) = 1$ V, find the voltage $v(t)$ for $t \geq 0$ by the use of the Laplace transform method for each of the following three cases: (1) $g_m = 1/33$ mho. (2) $g_m = 1/40$ mho. (3) $g_m = 1/30$ mho.

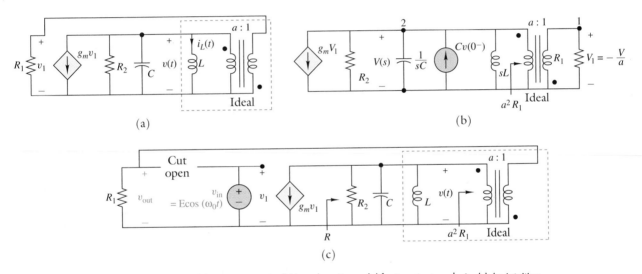

(a)   (b)

(c)

Figure 18.19 (a) An active circuit. (b) Its s-domain model for transient analysis. (c) An intuitive explanation of sustained oscillation.

SOLUTION

Step 1. *Derive the expression for* $V(s)$. Using the model of figure 14.16 for an initialized capacitor, we produce the frequency domain equivalent circuit shown in figure 18.19b. The impedance and voltage transformation properties of an ideal transformer are clearly indicated in the figure. Applying KCL to node 2 yields

$$-g_m \frac{V}{a} + \frac{V}{R_2} + sCV + \frac{V}{sL} + \frac{V}{a^2 R_1} = Cv(0).$$

Let R be the parallel combination of R_2 and $a^2 R_1$; i.e., $R = R_2//a^2 R_1 = a^2 R_1 R_2/(a^2 R_1 + R_2)$. For the present case, $R = 100//9,900 = 99 \ \Omega$. Solving for $V(s)$ from the preceding equation yields

$$V(s) = \frac{sv(0^-)}{s^2 + \left(-\dfrac{g_m}{a} + \dfrac{1}{R}\right)\dfrac{s}{C} + \dfrac{1}{LC}}.$$

Step 2. *Obtain the inverse Laplace transform of $V(s)$.* The method of chapter 14 is now used to obtain $v(t)$ by the inverse Laplace transform. The results are summarized as follows:

Case 1. $g_m = 1/33$ mho. Here,

$$V(s) = \frac{s}{s^2 + 10^{10}}$$

and

$$v(t) = \cos(10^5 t).$$

Case 2. $g_m = 1/40$ mho. For this case,

$$V(s) = \frac{s}{s^2 + 17,677s + 10^{10}} = \frac{(s + 8,838) - 0.089 \times 99,609}{(s + 8,838) + 99,609^2},$$

in which case

$$v(t) = e^{-8,838t}[\cos(99,609t) - 0.089 \sin(99,609t)] \cong e^{-8,838t} \cos(99,609t).$$

Case 3. $g_m = 1/30$ mho. In this last case,

$$V(s) = \frac{s}{s^2 - 10,101s + 10^{10}} = \frac{(s - 5,050) - 0.05 \times 99,872}{(s - 5,050)^2 + 99,872^2}$$

and

$$v(t) = e^{5,050t}[\cos(99,872t) - 0.05 \sin(99,872t)] \cong e^{5,050t} \cos(99,872t).$$

The previous example appears to be just another routine instance of linear circuit analysis using the Laplace transform method. But the results convey some very important information. The example actually demonstrates the underlying principle of one kind of oscillator circuit, i.e., a circuit that generates sinusoidal waveforms. Observe that when g_m has the critical value of $1/33$ mho (more generally, $g_m = a/R$), sinusoidal oscillations are sustained even though no external excitation is applied. When g_m is greater than the critical value, $V_o(s)$ has right half-plane poles and $v_o(t)$ has oscillations that grow exponentially in amplitude, as illustrated in case 3. The expression for $V(s)$ was derived under the assumption that the initial capacitor voltage $v(0^-) = 1$ V. In fact, *any* nonzero initial capacitor voltage, no matter how small, will start the oscillations. Suppose $v(0^-) = \varepsilon$ V, a very small voltage due to noise. Then, by the linearity principle, the voltage $v(t)$ is simply the previous answer multiplied by ε. Since $\varepsilon \neq 0$, the voltage $v(t)$ in case 3 will grow exponentially in magnitude. The point to be made here is that the ubiquitous noise will cause the oscillations to start whenever a transfer function has poles in the right half-plane. Some questions naturally arise in any inquisitive mind: Will the amplitude becomes infinitely large, as indicated in case 3? If not, what determines the amplitude? A rigorous study of the phenomenon requires the theory of nonlinear circuits. At present, an intuitive, plausible explanation is the best we can offer. If g_m is greater than the critical value needed to sustain oscillations, the amplitude grows. As the amplitude grows, the operating point of the active device swings into its nonlinear region, with an accompanying decrease in the value of g_m. The amplitude is stabilized when the "average g_m" equals the critical value.

It is important to note that if the polarities of the ideal transformer in figure 18.19 are reversed, then the circuit is incapable of producing sustained oscillations. This is because the sign in front of g_m in the general expression derived for $V(s)$ will be changed from negative to positive. The poles of $V(s)$ are in the left half-plane for all values of g_m, implying that $v(t)$ is a damped sinusoid.

There is also an intuitive explanation for the sustained sinusoidal oscillations in case 1, where $g_m = a/R$, or $g_m R = a$. Assume, in figure 18.19a, that the circuit is in sustained oscillation and a voltage $v_1(t) = E \cos(\omega_o t)$ appears. Suppose we cut open the connection to the input of the controlled source and apply a voltage $v_{in}(t) = E \cos(\omega_o t)$, as shown in figure 18.19c. Let us compute the steady-state voltage $v_{out}(t)$ across R_1. At the frequency $\omega_o = 1/\sqrt{LC}$, the parallel LC tank circuit behaves as an open circuit. The load "seen" by the controlled current source, $g_m V_1$, is simply the parallel combination of R_2 and $a^2 R_1$, denoted by R. The steady-state voltage across the capacitor is

$v(t) = -g_m v_1 R = -g_m RE \cos(\omega_o t)$. The ideal transformer steps down this voltage by a factor of a and introduces another $180°$ phase shift due to the dot positions. Thus, the voltage v_{out} across R_1 is $v_{\text{out}}(t) = (g_m R/a)E \cos(\omega_o t)$. In case 1, g_m has the critical value of a/R, which leads to $v_{\text{out}}(t) = (g_m R/a)E \cos(\omega_o t) = E \cos(\omega_o t) = v_{\text{in}}(t)$. Since, in figure 18.19c, the output voltage $v_{\text{out}}(t)$ is exactly the same as the input voltage $v_{\text{in}}(t)$, we may remove the applied voltage source, restore the connection, and use $v_{\text{out}}(t)$ to provide the needed excitation. The circuit becomes that of figure 18.19a, in which the sinusoidal oscillations are sustained.

As a final remark on this example, the dashed-line box in figure 18.19a is a mathematical model for a pair of coils with unity coupling. The derivation of such models is the topic of the next section.

7. COUPLED INDUCTORS MODELED WITH AN IDEAL TRANSFORMER

Given a circuit containing coupled inductors, a natural way to analyze the circuit is to write loop or node equations and then solve the *simultaneous equations* by any of the techniques studied before. Such a method has the advantage of being very general and systematic. However, it has the drawback of having grossly involved mathematical operations. This obscures the essential properties of the circuit. In the current section, we shall present some useful models for a pair of coupled inductors. These models make use of an ideal transformer. Since the three basic properties of an ideal transformer are extremely easy to comprehend, substituting one of the models for a pair of coupled inductors helps us to more easily understand the circuit operations without complicated mathematics.

As a first case, consider a pair of inductors with unity coupling (i.e., $k = 1$, or $M^2 = L_1 L_2$), as shown in figure 18.20a. Figures 18.20b and 18.20c show two equivalent circuits, each consisting of one inductor and one ideal transformer.

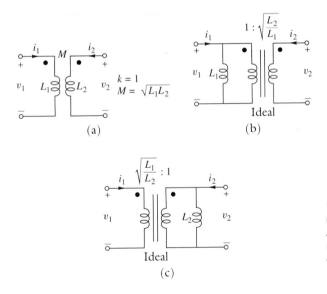

(a)

(b)

(c)

Figure 18.20 Ideal transformer models for unity coupled inductors. (a) A pair of unity coupled inductors. (b) A model consisting of one inductor and one ideal transformer. (c) An alternative model.

The proof of their equivalence is easy. The v-i relationships for the circuit of figure 18.20a are given by equations 18.7. We need only show that the circuits of figure 18.20b and 18.20c have the same v-i relationships. Consider figure 18.20b first. In this case, $M = \sqrt{L_1 L_2}$. Using the current transformation property, we obtain

$$v_1 = L_1 \frac{d}{dt}\left[i_1 + \sqrt{\frac{L_2}{L_1}}\ i_2\right] = L_1 \frac{di_1}{dt} + \sqrt{L_1 L_2}\ \frac{di_2}{dt} = L_1 \frac{di_2}{dt} + M \frac{di_2}{dt}.$$

Next, using the voltage transformation property,

$$v_2 = \sqrt{\frac{L_2}{L_1}}\, v_1 = \sqrt{L_1 L_2}\, \frac{di_1}{dt} + L_2 \frac{di_2}{dt} = M \frac{di_1}{dt} + L_2 \frac{di_2}{dt}.$$

These two equations are exactly the same as equations 18.7. Hence, the circuits of figure 18.20a and figure 18.20b are equivalent. The proof of their equivalence with the circuit shown in figure 18.20c has a similar derivation.

◆ EXAMPLE 18.15

The circuit of figure 18.21a has a unity coupling coefficient. Find the bandwidth, the center frequency, and the maximum voltage gain.

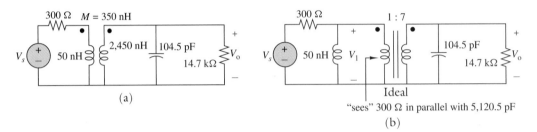

(a)

(b)

Figure 18.21 Analysis of a unity-coupled circuit. (a) A circuit containing unity coupled inductors. (b) An equivalent circuit utilizing an ideal transformer.

SOLUTION

The origin of the problem was explained in example 18.9, which used the method of section 3 to obtain the solution. It was a fairly complicated process, with all details left as an exercise. The solution that follows sheds a great deal of light on the circuit operation and is computationally much simpler.

The coefficient of coupling is $k = M/\sqrt{L_1 L_2} = 350/\sqrt{50 \times 2,450} = 1$. Replacing the coupled coils by the equivalent circuit of figure 18.20b yields the circuit of figure 18.21b. From the impedance transformation property of an ideal transformer, looking into the primary we see an admittance 49 times the load admittance. Therefore, looking into the primary, we see a resistance of $14,700/49 = 300$ Ω in parallel with a capacitance of 49×104.5 pF $= 5,120.5$ pF.

The calculation of the response curve, $|V_1/V_s|$, is of the standard type discussed in section 3 of chapter 17. Following the steps of example 17.6 one obtains the following solution:

$$R = 300//300 = 150 \ \Omega, \quad L = 50 \text{ nH}, \quad C = 5,120.5 \text{ pF},$$

$$\omega_m = \omega_o = \frac{1}{\sqrt{50 \times 10^{-9} \times 5,120.5 \times 10^{-12}}} = 62.5 \times 10^6 \text{ rad/s},$$

$$B_\omega = \frac{1}{150 \times 5,120.5 \times 10^{-12}} = 1.302 \times 10^6 \text{ rad/s},$$

$$\left| \frac{V_1}{V_s} \right|_{\max} = 0.5.$$

From the voltage transformation property of an ideal transformer, V_o is simply 7 times V_1. Thus, the desired answers, with V_o as the output are as follows: $B_\omega = 1.302 \times 10^6$ rad/s, $\omega_m = 62.5 \times 10^6$ rad/s, and $|V_o/V_s|_{\max} = 3.5$.

The answers, of course, agree with those obtained in example 18.9. But the numerical calculations are so much simpler. Furthermore, the circuit model of figure 18.21b makes it very clear that (1) although there are two inductors and one capacitor, because of unity coupling, the frequency response of the circuit is the same as that of a single-tuned circuit; and (2) the 1:7 ideal transformer provides an impedance matching with an accompanying higher voltage gain at the center frequency.

Exercise. Solve example 18.15 again, this time using the equivalent circuit of figure 18.20c.

ANSWERS: The same as example 18.15, of course.

A unity coupling coefficient is an ideal condition impossible to achieve in practice. It is desirable, therefore, to modify the models of figure 18.20 to account for coupled inductors with $k < 1$. This turns out to be a very easy manipulation. The results are shown in figure 18.22.

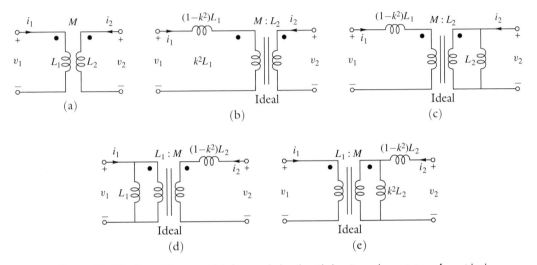

Figure 18.22 Four different models for coupled coils with $k \leq 1$, each consisting of one ideal transformer and two self-inductances. (a) Coupled inductors with $k < 1$. (b) One model using an ideal transformer and two inductances. (c), (d), and (e) Three alternative models.

The equivalent circuits are derived as follows. Since $L_1 L_2 > M^2$, it is possible to subtract a small inductance from L_1 such that the remaining inductance, L_1', satisfies the condition $L_1' L_2 = M^2$. In other words, the new inductance, L_1', together with L_2 and M, forms a unity coupling system. Since $L_1' = M^2/L_2 = k^2 L_1$, the inductance to be subtracted is equal to $L_1 - L_1' = (1-k^2)L_1$. This inductance must be added back in series with L_1' to obtain a model for the original coupled inductors. The models of figures 18.22b and 18.22c result. Repeating the process on L_2 yields the models of figures 18.22d and 18.22e. A total of four equivalent circuits is possible. Each consists of two *uncoupled* inductors and one ideal transformer. Clearly, when $k = 1$, the models in figure 18.22 reduce to those in figure 18.20.

As an application of the models of figure 18.22, let us consider the impedance transformation property of a pair of closely coupled inductors ($k \cong 1$). Unlike an ideal transformer, the impedance transformation property in this case holds only for a range of frequencies.

⬥ **EXAMPLE 18.16**

The coupled inductors in figure 18.23 have values $L_1 = 1$ H, $L_2 = 100$ H, and $M = 9.8$ H. A fixed load resistance $R = 100$ kΩ is connected to coil 2.

1. Find $|Z(j\omega)|$ as a function of ω.
2. Plot the curve of $|Z(j\omega)|$ vs. ω.
3. From the plot of part 2, determine the range of frequencies over which a *constant* impedance transformation ratio is achieved with the coils. What is this impedance ratio?

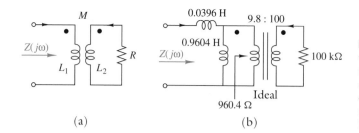

Figure 18.23 Circuits illustrating that with $k = 0.98 < 1$, the impedance transformation property holds only for a range of frequencies. (a) Closely coupled inductors terminated in a resistor. (b) Circuit model utilizing an ideal transformer.

SOLUTION

1. The coupling coefficient is $k = 9.8/\sqrt{1 \times 100} = 0.98 < 1$. Replacing the coupled coils in figure 18.23a by the equivalent circuit of figure 18.22b yields the circuit of figure 18.23b.

   From the impedance transformation property of an ideal transformer, looking into the primary, we see an impedance (resistance) equal to $(9.8/100)^2 \times 10^5 = 960.4 \ \Omega$. The input impedance, then, is a series-parallel combination of two inductances and the 960.4-Ω resistance. Hence,

$$Z(s) = 0.0396s + (0.9604s//960.4)$$

$$= 0.0396\frac{s(s + 2,523)}{s + 1,000}.$$

Taking the magnitude with $s = j\omega$ produces

$$|Z(j\omega)| = 0.0396\omega\sqrt{\frac{\omega^2 + 637.69 \times 10^6}{\omega^2 + 10^6}}.$$

2. Figure 18.24 shows a plot of $|Z|$ as a function of f.

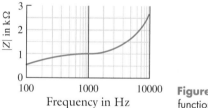

Figure 18.24 Input impedance as a function of frequency.

3. From the tabulated values of $|Z(j\omega)|$, for $457 < f < 1,380$ Hz, the magnitude, $|Z(j\omega)|$, varies from 950 to 1,050 Ω ($\pm 5\%$ from 1,000 Ω). The impedance ratio over the frequency range 457 to 1,380 Hz is then approximately $1,000/100,000 = 1/100$.

The shape of the frequency response curve in figure 18.24 is readily explained from the circuit model of figure 18.23b. The impedance transformation ratio of approximately 1:100 (for $457 < f < 1,380$ Hz) is due to the ideal transformer in the model. $|Z(j\omega)|$ drops at low frequencies because both inductances approach short circuits, whereas $|Z(j\omega)|$ increases rapidly at high frequency because the 0.0396-H inductance approaches an open circuit.

This example shows the frequency dependence of coupled coils when used for impedance transformation purposes. In fact, the useful range of frequency also depends on the load resistance. For example, a similar analysis shows that if the load resistance in example 18.16 is changed from 100 kΩ to 50 kΩ, an impedance transformation ratio of nearly 100 will be valid for a narrower frequency range, $223 < f < 707$ Hz, and $|Z(j\omega)|$ varies from 475 to 525 Ω ($\pm 5\%$ from 500 Ω). For this reason, transformers intended for impedance transformation usually have the nominal load resistance value specified, besides the applicable frequency range.

In many practical applications of coupled coils, the two coils have one terminal in common. The circuit of figure 18.21a is an example. The two undotted terminals of the coils are joined together. In using the model of figure 18.20b to represent the coupled coils, we likewise join together the two undotted terminals of the ideal transformers, as shown in figure 18.21b. The device actually has only three accessible terminals and is referred to as a *three-terminal device*. For coupled coils with a common terminal, there are two other models, called *T*- and *π-equivalent circuits*, that do not utilize ideal transformers. The derivations of these models and their application will be studied in chapter 19. However, for completeness, we indicate the equivalent circuits in figure 18.25. Note that one of the three inductances in the *T*- or *π*-equivalent circuit may have a negative value. This negative inductance appears in a mathematical model and is not the inductance of a physical component.

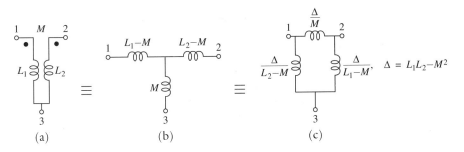

Figure 18.25 (a) Coupled coils with a common terminal. (b) *T*-equivalent circuit. (c) *π*-equivalent circuit.

The equivalent circuits shown in figure 18.25b and 18.25c are for the specific dot locations indicated in figure 18.25a. A change of one dot location in figure 18.25a will result in a change of the sign in front of M in figure 18.25b and 18.25c.

The *T*- and *π*-equivalent circuits are often helpful in manual analysis and also in using circuit simulation programs that permit inductances, but not mutual inductances. Problems 50 and 54 illustrate such applications.

* 8. MODELS FOR PRACTICAL TRANSFORMERS

One of the first basic circuit elements studied in this text is the ideal dc voltage source, which maintains a constant voltage for arbitrary loads connected between its terminals. An ideal voltage source does not exist in the real world. Nevertheless, it is very useful in circuit analysis because a practical voltage source (such as a battery) may be represented *approximately* by an ideal voltage source in series with a resistance. In the same spirit, although an ideal transformer does not exist in the real world, we can represent a practical transformer approximately by an ideal transformer and some additional circuit elements. This requires some fundamentals of magnetic circuits ordinarily studied in college physics. A brief discussion of magnetic circuits—in particular, the use of electric circuit analogies—is given in appendix B3 for those who wish to have a quick review. The terminology and basic concepts in magnetic circuits will be used in this section. In the process of developing a model for a practical transformer, we will also explore how the two idealizations of an ideal transformer mentioned in section 6 are achieved approximately in the real device.

Coupled Coils with a Magnetic Core

Consider two coils wound on a ferromagnetic core, as shown in figure 18.26. Because of the high permeability of the core, it is reasonable to assume that the flux linking both coils is confined entirely to the core (at least approximately). This common flux is denoted by ϕ_0. A flux that links with one coil, but not the other, is called a **leakage flux**. In figure 18.26, ϕ_{11} is the leakage flux that is produced by coil 1 and that links with coil 1 only. Some of

*indicates a more advanced topic.

the leakage flux may link with only a portion of a coil. To simplify our analysis, we shall interpret ϕ_{11} as the *effective* leakage flux that links with the entire N_1 turns. The leakage flux ϕ_{22} has a similar interpretation.

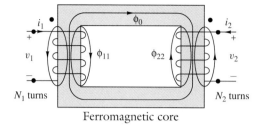

Figure 18.26 Coils coupled through a ferromagnetic core.

Assuming that the core is operating in the below-saturation region, we may determine an *approximate* value for the common flux ϕ_0 by the method described in appendix B3. For figure 18.26, we see that both of the mmfs due to coils 1 and 2 act on the core and aid each other in producing the common flux ϕ_0. Therefore, we have

$$\phi_0 = \mathbf{P}_0(N_1 i_1 + N_2 i_2), \tag{18.35}$$

where $\mathbf{P}_0$ is the permeance of the core, as calculated from equation B3.11b. Similarly, we can express the leakage fluxes as

$$\phi_{11} = \mathbf{P}_{11}(N_1 i_1) \tag{18.36}$$

and

$$\phi_{22} = \mathbf{P}_{22}(N_2 i_2). \tag{18.37}$$

Because in this case the leakage flux completes its path via the air, the permeances $\mathbf{P}_{11}$ and $\mathbf{P}_{22}$ are nearly constant and are much smaller than $\mathbf{P}_0$. The calculation of the leakage flux in the air is a very complicated mathematical problem that we cannot discuss in this book. Empirical formulas applicable to specific coil structures are available in some design handbooks. Experimental determination of the permeances $\mathbf{P}_{11}$ and $\mathbf{P}_{22}$ are discussed in advanced texts on power system devices.

The total fluxes linking each coil satisfy the formulas

$$\phi_1 = \phi_0 + \phi_{11} = (\mathbf{P}_0 + \mathbf{P}_{11})(N_1 i_1) + \mathbf{P}_0(N_2 i_2) \tag{18.38a}$$

and

$$\phi_2 = \phi_0 + \phi_{22} = (\mathbf{P}_0 + \mathbf{P}_{22})(N_2 i_2) + \mathbf{P}_0(N_1 i_1). \tag{18.38b}$$

Using Faraday's law, equation 18.35, and equations 18.38, we have

$$v_1 = N_1 \frac{d\phi_1}{dt} = N_1^2(\mathbf{P}_0 + \mathbf{P}_{11})\frac{di_1}{dt} + N_1 N_2 \mathbf{P}_0 \frac{di_2}{dt}, \tag{18.39a}$$

$$v_2 = N_2 \frac{d\phi_2}{dt} = N_2^2(\mathbf{P}_0 + \mathbf{P}_{11})\frac{di_2}{dt} + N_1 N_2 \mathbf{P}_0 \frac{di_1}{dt}. \tag{18.39b}$$

Comparing equations 18.39 with equations 18.7, we obtain the following formulas:

$$L_1 = N_1^2(\mathbf{P}_0 + \mathbf{P}_{11}), \tag{18.40a}$$

$$L_2 = N_2^2(\mathbf{P}_0 + \mathbf{P}_{22}), \tag{18.40b}$$

$$M_{12} = N_1 N_2 \mathbf{P}_0, \tag{18.40c}$$

$$M_{21} = N_1 N_2 \mathbf{P}_0. \tag{18.40d}$$

In table B3.2, the unit for permeance is given as weber/ampere. Equation 18.40 makes it evident that the unit for inductance, namely, the henry, may also be used as the unit for permeance.

The voltage and current references used in figure 18.26 lead to equations 18.39, in which the mutual terms have positive signs. Should the reference directions of v_2 and i_2 (or v_1 and i_1) be reversed, then the sign of the last term in equations 18.38 and 18.39 will be changed from plus to minus. However, the formulas given by equation 18.40 remain valid, regardless of the reference directions, as all inductances and permeances are nonnegative quantities. The signs are taken care of in the process of writing equations 18.35, 18.38, and 18.39. Comparing this result with the dot convention established in section 2, we can state an alternative rule for determining the positions of the dots when the physical construction of the coils is known:

Rule for placing dots: When the winding directions and ferromagnetic core structure are known, the two dots for the coupled coils should be placed such that currents flowing from dotted terminals to undotted terminals produce fluxes in the core that aid each other.

Exercise. Determine the placement of dots for the coupled coils shown in figure 18.27, if the core is made of (a) plastic and (b) iron.

ANSWERS: (a) $A - D$ or $B - C$; (b) $A - C$ or $B - D$.

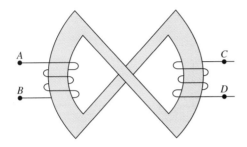

Figure 18.27 Determination of dot positions from the physical construction of the coupled coils.

◆ EXAMPLE 18.17

The core in figure 18.27 has relative permeability $\mu_r = 5,000$. The cross-sectional area is 1 cm by 1 cm, and the mean length is 125 cm. $N_1 = 50$ and $N_2 = 100$. If the permeances $\mathbf{P}_{11}$ and $\mathbf{P}_{22}$ for the leakage path are each *estimated* to be around 5% of the core permeance $\mathbf{P}_0$, find approximate values of L_1, L_2, and M.

SOLUTION

The permeance of the core is calculated from equation B3.11b:

$$\mathbf{P}_0 = \frac{\mu A}{L} = \frac{5,000 \times 4\pi \times 10^{-7} \times 0.0001}{0.125} = 5.026 \times 10^{-6} \text{ H}.$$

The estimated values of $\mathbf{P}_{11}$ and $\mathbf{P}_{22}$ are

$$\mathbf{P}_{11} = \mathbf{P}_{22} = 0.05\mathbf{P}_0 = 0.26 \times 10^{-6} \text{ H}.$$

The inductance values are now calculated from equation 18.40:

$$L_1 = 50^2 \times (0.26 + 5.026) \times 10^{-6} = 13.22 \times 10^{-3} \text{ H},$$

$$L_2 = 100^2 \times (0.26 + 5.026) \times 10^{-6} = 52.86 \times 10^{-3} \text{ H},$$

$$M = 50 \times 100 \times 5.026 \times 10^{-6} = 25.1 \times 10^{-3} \text{ H}.$$

Exercise. In example 18.17, if $N_1 = 80$ and $N_2 = 200$, find approximate values of L_1, L_2, and M.

ANSWERS: 33.84, 211.4, 80.3 mH.

The next example shows a three-leg core in which leakage flux exists mainly in the core. In this case, we use the method of appendix B3 to obtain equations 18.38, from which $\mathbf{P}_0$, $\mathbf{P}_{11}$, and $\mathbf{P}_{22}$ are determined.

EXAMPLE 18.18

In the magnetic circuit shown in figure 18.28a, coil 1 has 200 turns and coil 2 has 300 turns. Assume that $\mu_r = 5{,}000$ and the leakage flux in the air is negligible. The dimensions of the various legs are given in table 18.1. Find L_1, L_2, and M.

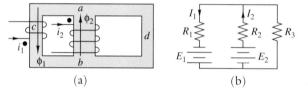

TABLE 18.1   DIMENSIONS FOR THE CORE OF FIGURE 18.28a

Part	Area (m^2)	Mean length (m)
acb	0.01	0.2
ab	0.01	0.1
adb	0.01	0.4

Figure 18.28   Calculation of self- and mutual inductances. (a) A core with three legs and two windings. (b) The electric circuit analog.

(a)   (b)

SOLUTION

The electric circuit analog of the magnetic circuit, as described in appendix B3, is shown in figure 18.28b, with various parameters calculated as follows (recall that $\mu_0 = 4\pi \times 10^{-7}$):

$$G_1 = \frac{5{,}000\mu_0 \times 0.001}{0.2} = 250\mu_0 = 0.3146 \times 10^{-3} \text{ mho}, \ R_1 = 3{,}180 \ \Omega,$$

$$G_2 = \frac{5{,}000\mu_0 \times 0.001}{0.1} = 500\mu_0 = 0.6238 \times 10^{-3} \text{ mho}, \ R_2 = 1{,}590 \ \Omega,$$

$$G_3 = \frac{5{,}000\mu_0 \times 0.001}{0.4} = 125\mu_0 = 0.1571 \times 10^{-3} \text{ mho}, \ R_3 = 6{,}360 \ \Omega,$$

$$E_1 = N_1 i_1 = 200 i_1,$$

$$E_2 = N_2 i_2 = 300 i_2.$$

Applying KVL to the loops $(E_1\text{-}R_1\text{-}R_3)$ and $(E_2\text{-}R_2\text{-}R_3)$ yields

$$3{,}180 \times I_1 + 6{,}360 \times (I_1 - I_2) = 9{,}540 \times I_1 - 6{,}360 \times I_2 = E_1,$$

$$1{,}590 \times I_2 + 6{,}360 \times (I_2 - I_1) = -6{,}360 \times I_1 + 7{,}950 \times I_2 = E_2.$$

Solving for I_1 and I_2 produces

$$I_1 = 224.6 \times 10^{-6} E_1 + 179.7 \times 10^{-6} E_2,$$

$$I_2 = 179.7 \times 10^{-6} E_1 + 269.5 \times 10^{-6} E_2.$$

Comparing these two equations with equations 18.38 (recall the analogy shown in table B3.2), we obtain the following parameters for the magnetic circuit:

$$\mathbf{P}_0 = 179.7 \times 10^{-6} \text{ H},$$

$$\mathbf{P}_{11} + \mathbf{P}_0 = 224.6 \times 10^{-6} \text{ H},$$

$$\mathbf{P}_{22} + \mathbf{P}_0 = 269.5 \times 10^{-6} \text{ H}.$$

Finally, substituting these values into equation 18.40, we have

$$L_1 = 200^2 \times 224.6 \times 10^{-6} = 8.984 \text{ H},$$

$$L_2 = 300^2 \times 269.5 \times 10^{-6} = 24.25 \text{ H},$$

$$M = 200 \times 300 \times 179.7 \times 10^{-6} = 10.78 \text{ H}.$$

The magnetic circuit approach of analyzing coupled coils provides an alternative way of proving the following two results of section 4:

1. $M_{12} = M_{21} = M$. This is obvious from equations 18.40c and 18.40d.
2. $M^2 \leq L_1 L_2$. This follows from equations 18.40, as we have

$$L_1 L_2 = N_1^2 (\mathbf{P}_0 + \mathbf{P}_{11}) N_2^2 (\mathbf{P}_0 + \mathbf{P}_{22})$$

$$\geq N_1^2 \mathbf{P}_0 N_2^2 \mathbf{P}_0 = (N_1 N_2 \mathbf{P}_0)^2 = M^2.$$

Electric Circuit Model for a Practical Transformer

In section 6, two idealized conditions on coupled coils led to the definition of an ideal transformer. These conditions will now be investigated from the magnetic circuit point of view.

> **Idealization 1:** The coupled coils have unity coupling, i.e., $M^2 = L_1 L_2$, or $k = 1$.

An examination of equation 18.40 shows that this idealization may be *approximated* by making $\mathbf{P}_{11}$ and $\mathbf{P}_{22}$ very small. It then follows from equations 18.36 and 18.37 that both ϕ_{11} and ϕ_{22} are very small, i.e., there is very little leakage flux. In practice, we try to achieve this condition by winding the turns of the two coils very closely together.

> **Idealization 2:** The coupled coils have infinite mutual and self-inductances and a finite ratio $L_1/M = M/L_2 = a$.

From equation 18.40, we see that the second idealization may be *approximated* by making $\mathbf{P}_0$ very large while keeping $\mathbf{P}_{11}$ and $\mathbf{P}_{22}$ very small. Equation B3.11b then dictates the use of magnetic core material with very high permeability, such as silicon steel.

Under these two idealized conditions, the device is called an **ideal transformer**, with its circuit symbol and defining v-i relationships given in figure 18.29a. To avoid redundancy, we will show only one of the two possible dot placements. Minor modifications in all figures for the other possible dot placement should be clear.

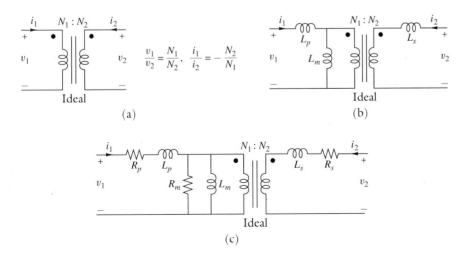

Figure 18.29 Deriving the circuit model for a practical transformer.
(a) Ideal case. (b) Leakage fluxes and finite permeability considered. (c) Wire resistance and core loss considered.

We shall now show how the basic model of figure 18.29a is modified step by step to account for the phenomena observed in a practical transformer. First of all, the permeance $\mathbf{P}_0$, although large, is not infinite. Also, the permeances $\mathbf{P}_{11}$ and $\mathbf{P}_{22}$, although small, are not zero. Consequently, for a practical transformer, the self- and mutual inductances, although large, are not infinite, and the coupling coefficient is not exactly 1. A model that accounts for these nonideal properties is shown in figure 18.29b, where L_p and L_s are called the **primary** and **secondary leakage inductances**, respectively, and L_m is called the **magnetizing inductance**. To find the parameter values, we write the differential equations for the circuit and compare them with equation 18.40. The results are:

$$L_p = N_1^2 \mathbf{P}_{11}, \tag{18.41a}$$

$$L_s = N_2^2 \mathbf{P}_{22}, \tag{18.41b}$$

$$L_m = N_1^2 \mathbf{P}_0. \tag{18.41c}$$

If $\mathbf{P}_{11} = \mathbf{P}_{22} = 0$ and $\mathbf{P}_0 \to \infty$, then figure 18.29b reduces to figure 18.29a.

Next, we need to consider the effect of the resistance of the coil wires. The resistance varies with the operating frequency. As the frequency is increased, the resistance also goes up, due to the "skin effect," a phenomenon to be studied in field theory. At one fixed frequency (e.g., 60 Hz for electric power in the United States), these resistances will be constant and are represented by R_p and R_s in figure 18.29c.

Finally, there is the power loss in the iron core due to hysteresis and eddy currents. A study of these phenomena is clearly beyond the level of this book. Suffice it to say that the power loss can be accounted for by placing a resistance R_m in parallel with the magnetizing inductance L_m. The result is the model shown in figure 18.29c, often referred to as the T-*equivalent circuit* of a practical transformer.

◆ EXAMPLE 18.19

The transformer in figure 18.30a is designed for operation at 60 Hz and 1,100/220 V. It has the following parameters (see figure 18.29c): $R_p = 0.050\ \Omega$, $R_s = 0.002\ \Omega$, $X_p = \omega L_p = 0.4\ \Omega$, $X_s = \omega L_s = 0.016\ \Omega$, $X_m = \omega L_m = 250\ \Omega$. R_m is very large, and its effect is neglected. If the load draws 100 A at a power factor of 0.6 lagging and the load voltage is 220 V, what is the magnitude of the source voltage?

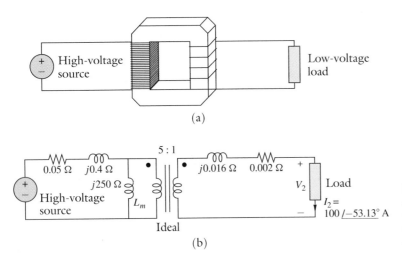

Figure 18.30 Analysis of a circuit containing a practical transformer. (a) A practical transformer for stepping down ac voltage. (b) A circuit model utilizing an ideal transformer.

SOLUTION

Substituting the model of figure 18.29c for the transformer yields the circuit shown in figure 18.30b. Using phasors for all voltages and currents, and referring to figure 18.29c for notation, we have:

Load voltage $= V_2 = 220 \angle 0°$ V.

Load current $= I_2 = 100 \angle -53.13° = 60 - j80$ A

(from $\cos(53.13°) = 0.6$, the given power factor).

Voltage drop across the impedance $(0.002 + j0.016)$ Ω

$$= (60 - j80)(0.002 + j0.016) = 1.4 + j0.8 = 1.612 \angle 29.7°$$ V.

Voltage across the secondary of the ideal transformer

$$= (1.4 + j0.8) + 220 = 221.4 + j0.8 = 221.4 \angle 0.207°$$ V.

$$\frac{N_1}{N_2} = \frac{1,100}{220} = 5 \text{ (given turns ratio)}.$$

Voltage across the primary of the ideal transformer

$$= 5 \times 221.4 \angle 0.207° = 1,107 \angle 0.207°$$ V.

Current through the primary of the ideal transformer

$$= 0.2 \times (60 - j80) = 12 - j16$$ A.

Current through the magnetizing inductance

$$= \frac{1,107 \angle 0.207°}{j250} = 4.428 \angle -89.8° = 0.015 - j4.428$$ A.

Current through the primary winding

$$= (12 - j16) + (0.015 - j4.428) = 23.7 \angle -59.54°$$ A.

Voltage drop across the impedance $(0.05 + j0.4)$ Ω

$$= 23.7 \angle -59.54° \times (0.05 + j0.4) = 9.55 \angle 23.34°$$ V

The source voltage

$$= 9.55 \angle 23.34° + 1,107 \angle 0.207° = 1,115.8 \angle 0.4°$$ V.

Thus, the magnitude of the source voltage is 1,115.8 V.

This example shows that once the transformer model and its parameters are known, the analysis of the circuit is straightforward, at least conceptually. A very important problem that must be solved prior to analyzing the circuit is determining the parameters of the transformer model. We have given the formulas for some parameters in figure 18.29 in terms of various permeances. The calculation of these permeances for any practical transformer turns out to be extremely difficult, even with modern-day digital computers. In practice, the parameters in the transformer model are usually determined by measurements. A discussion of various measurement techniques belongs to a more advanced course in power engineering.

SUMMARY

This chapter has examined the phenomenon of induced voltage in one circuit due to a change of current in another circuit. A new circuit parameter called the *mutual inductance,* denoted by M, was introduced to quantify the relationship. M was defined as a constant appearing in terminal voltage-current differential equations, often determined from measurements. This treatment avoids digressing into the study of magnetic circuits. From the circuit analysis point of view, the mathematical treatment is adequate. However, for a deeper understanding of the physical phenomena, a brief review of the principles of magnetic circuits is given in the appendix, and their application to the modeling of a practical transformer was described in section 8.

Of foremost importance in analyzing a circuit containing mutual inductance is the formulation of correct time domain or frequency domain equations—in particular, the correct signs for the mutual terms. For this reason, section 2 focused entirely on the dot convention. Once the equations are correctly written, we may use any of the solution techniques studied in chapters 13 through 15 to solve the problem.

A proof of $M_{12} = M_{21} = M$ was given that made use of the passivity principle and some trigonometric identities. After establishing this equality, computing the energy stored in the coupled inductors follows from simple integration. The expression for the stored energy and the passivity principle then lead to an upper bound for the value of the mutual inductance, namely, $M \leq \sqrt{L_1 L_2}$. The coefficient of coupling is then defined as $k = M/\sqrt{L_1 L_2}$.

An ideal transformer was defined as a device satisfying both the voltage transformation and the current transformation properties. For practical transformers, these two properties hold only approximately. Transformers are used extensively in electrical engineering. For example, in power engineering, transformers are used to step up or step down the voltages. In communication engineering, transformers are used to change a load impedance for the purpose of maximum power transfer.

An ideal transformer does not exist in the real world. Nevertheless, it is an important basic circuit element because a practical transformer or a pair of coupled inductors can be modeled by an ideal transformer and some additional RLC elements. The use of such models simplifies many analysis problems and gives better insight into the operation of the circuit. Sections 6 through 8 contain many examples illustrating the applications of ideal transformers in circuit analysis.

TERMS AND CONCEPTS

Coefficient of coupling: usually denoted by k, is equal to $M/\sqrt{L_1 L_2}$.

Controlled source model for an ideal transformer: model with a voltage-controlled voltage source and a current-controlled current source.

Coupled inductors (coils): two inductors having a mutual inductance $M \neq 0$.

Current transformation property: for unity coupled inductors having infinite inductances, the ratio $|i_1(t) : i_2(t)|$ is a constant equal to the turns ratio N_2/N_1.

Dot convention: a commonly used marking scheme for determining the polarity of induced voltages.

Energy stored in a pair of coupled inductors: $0.5L_1 I_1^2 + 0.5L_2 I_2^2 \pm M I_1 I_2$ joules, with $+$ sign for the case where both currents enter (or leave) dotted terminals and $-$ sign for the case where one current enters a dotted terminal and the other leaves the dotted terminal.

Ideal transformer: two network branches satisfying both the voltage transformation and the current transformation properties exactly.

Impedance transformation property: when the secondary of an ideal transformer is terminated in an impedance $Z(s)$, the input impedance across the primary is equal to $(N_1/N_2)^2 Z(s) = a^2 Z(s)$.

Model for coupled inductors: for a pair of unity coupled inductors, one ideal transformer and one inductance. For coupling coefficient k less than 1, one ideal transformer and two self-inductances.

Mutual inductance: a real number, usually denoted by M, that determines the induced voltage in one coil due to the change of current in another coil.

π-equivalent for coupled inductors: if two inductors have one terminal in common, then the three-terminal coupled inductors are equivalent to three uncoupled inductors (one of which may have a negative inductance) connected in the π-form.

Primary: The winding (coil) in a transformer that is connected to a power source.

Saturation: an operating state of a ferromagnetic material in which the flux does not increase linearly with the applied current. (See figure B3.4.)

Secondary: The winding (coil) in a transformer that is connected to a load.

T-equivalent for coupled inductors: if two inductors have one terminal in common, then the three-terminal coupled inductors are equivalent to three uncoupled inductors (one of which may have a negative inductance) connected in the T-form.

Transformer: a practical device that satisfies approximately the voltage transformation and current transformation properties.

Unity coupling: coefficient of coupling $k = 1$.

Voltage transformation property: for unity coupled inductors, the ratio $|v_1(t) : v_2(t)|$ is a constant equal to the turns ratio $N_1/N_2 = a$.

PROBLEMS

1. The mutual inductance of the coupled inductors shown in figure P18.1 is M. One inductor is connected to a network containing sources, and the other inductor is open-circuited.

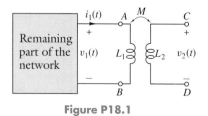

Figure P18.1

(a) Write $v_2(t)$ in terms of $i_1(t)$ for each of the following placements of dots: (a) $\{A, C\}$; (b) $\{A, D\}$; (c) $\{B, C\}$; (d) $\{B, D\}$.
(b) From the results of part (a), verify that both dots may be moved simultaneously to the opposite end of each coil with no change in the voltage-current relationships.

2. For each circuit shown in figure P18.2:
(a) Express $v_x(t)$ and $v_y(t)$ in terms of $i_a(t)$ and $i_b(t)$.
(b) Obtain the s-domain equations that contain the initial currents $i_a(0^-)$ and $i_b(0^-)$ by applying the Laplace transform to the equations of part (a).

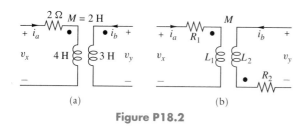

Figure P18.2

3. Write three mesh equations in the time domain for the circuit shown in figure P18.3. Be particularly careful about the signs of induced voltages due to the mutual inductance. Apply the rule given by equations 18.6 if you have any doubt about the signs.

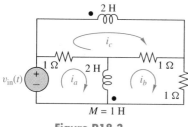

Figure P18.3

ANSWER: Coefficients for the derivative terms are as follows: mesh a, $(2, -2, -1)$; mesh b, $(-2, 2, 1)$; mesh c, $(-1, 1, 2)$.

4. Consider the circuit shown in figure P18.4. The 2-H inductor is short-circuited. The currents in the coupled inductors for $t \geq 0$ are $i_1(t) = 2e^{-t}$ A and $i_2(t) = e^{-t}$ A.
   (a) Determine the mutual inductance M (in henries).
   (b) Find $v_2(t)$ for $t > 0$.

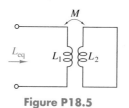

Figure P18.4

ANSWER: (a) 4 H; (b) $-e^{-t}$ V.

5. One inductor of a pair of coupled inductors is short-circuited, as shown in figure P18.5. Find L_{eq}, the equivalent inductance measured at the other coil terminals. Does the answer depend on the positions of the dots?

Figure P18.5

ANSWER: $L_1 - M^2/L_2$, independent of positions of dots.

6. Two ways to connect a pair of coupled inductors in parallel are shown in figure P18.6. For each case, find the equivalent inductance.

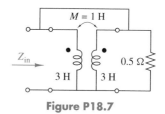

Figure P18.6

ANSWER: (a) $(L_1 L_2 - M^2)/(L_1 + L_2 - 2M)$; (b) $(L_1 L_2 - M^2)/(L_1 + L_2 + 2M)$.

7. Determine the input impedance of the circuit in figure P18.7. (*Hint*: Make use of the results of the previous problem.)

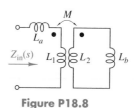

Figure P18.7

ANSWER: $0.5s/(s + 0.125)$.

8. Find the input impedance $Z_{in}(s)$ of the circuit shown in figure P18.8. Does the answer depend on the positions of the dots?

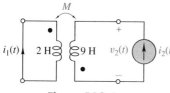

Figure P18.8

ANSWER: $s[L_a + L_1 - M^2/(L_2 + L_b)]$, independent of the dot positions.

9. Find the equivalent inductance, L_{eq}, for each of the magnetically coupled circuits shown in figure P18.9.

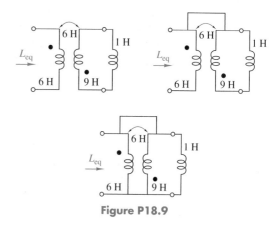

Figure P18.9

ANSWERS: 2.4 H, 2.4 H, 0.4 H.

10. A handy dandy henry counter is used to measure the inductance of a pair of coupled inductors in various configurations. After three experiments, the results are as

follows: Equivalent inductances are: (1) 37 H after a series-aiding connection, (2) 25 H after a series-opposing connection (dotted-to-dotted connection), and (3) 121/25 H after a parallel connection with dotted terminals connected together. If it is known that the primary has the larger inductance, determine L_1, L_2, and M.

ANSWERS: 26 H, 5 H, 3 H.

11. If the circuit shown in figure P18.11 is relaxed at $t = 0^-$, determine $v_2(t)$ for $t > 0$.

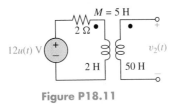

Figure P18.11

ANSWER: $30e^{-t}$.

12. If $v_{in}(t) = u(t)$ V and $i_{in}(t) = u(t)$ A, determine the indicated zero-state response for each of the circuits shown in figure P18.12. (*Hint*: The open-circuited inductor has no effect on the circuit containing the other inductor. Solve for the latter inductor current first.)

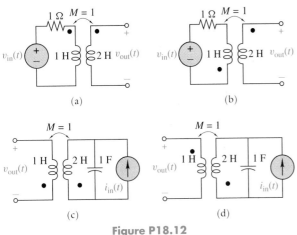

(a)   (b)

(c)   (d)

Figure P18.12

13. In the circuit shown in figure P18.13, all initial conditions are zero. If $i_{in}(t) = u(t - 1)$ A, then determine the response, $v_{out}(t)$.

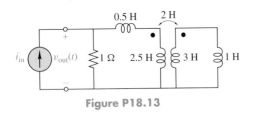

Figure P18.13

ANSWER: $e^{-0.5(t-1)}u(t - 1)$ V.

14. The switch S is closed at $t = 0$ in the circuit of figure P18.14. Find $v_2(t)$.

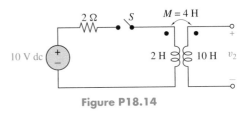

Figure P18.14

ANSWER: $20e^{-t}u(t)$.

15. Consider the circuit of figure P18.15. Before the closing of the switch at $t = 0$, the output voltage is zero.

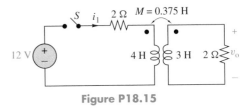

Figure P18.15

(a) Write the differential equations (two mesh equations) for $t \geq 0$.
(b) Find $v_o(t)$ by applying the Laplace transform to the preceding equations.
(c) Verify that di_1/dt and $v_o(t)$ are both positive for $t > 0$.

ANSWER: (b) $4\sqrt{2}(e^{-0.4t} - e^{-t})u(t)$.

16. Consider the circuit of figure P18.16. The switch S has been closed for a long time (i.e., the circuit has reached the dc steady state) and is then opened at $t = 0$.

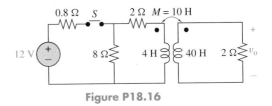

Figure P18.16

(a) Determine the currents in the two inductors at $t = 0^-$.
(b) Find $v_o(t)$ for $t \geq 0$.

ANSWER: (a) 4 A, 0 A, (b) $5.372e^{-6.75t} - 0.0393e^{-0.0494t}$.

17. The circuit shown in figure P18.17 has no initial stored energy. Determine the voltage $v_2(t)$ for $t \geq 0$.

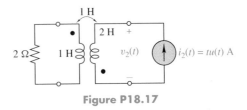

Figure P18.17

ANSWER: $2 - e^{-2t}$.

18. Consider the circuit shown in figure P18.18.
(a) Determine the input admittance. (*Check*: Poles are at -4 and -0.5.)
(b) Determine the transfer function $H(s) = V_{out}(s)/V_{in}(s)$.
(c) If $v_{in}(t) = (1 - e^{-0.5t})u(t)$ V, determine $v_{out}(t)$.

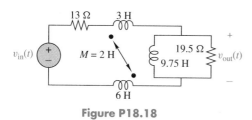

Figure P18.18

19. Consider a pair of coupled inductors with initial currents, as shown in figure P18.19a.

   (a) Derive the s-domain equivalent shown in figure P18.19b, which is a generalization of figure 14.19 for an initialized inductor.

   (b) Derive the s-domain equivalent circuit shown in figure P18.19c, which is a generalization of figure 2.18.

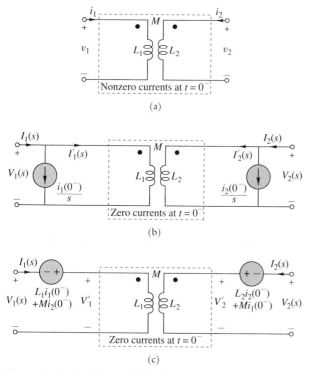

Figure P18.19 (a) Coupled inductors with initial currents. (b) An s-domain circuit model. (c) An alternative s-domain circuit model.

(*Hint*: For each circuit, write differential equations and apply the Laplace transform. Then show that the three circuits all have the same relationships among $V_1(s)$, $V_2(s)$, $I_1(s)$, and $I_2(s)$.)

20. The point of this problem is to develop an organized procedure for writing node equations for magnetically coupled circuits having initial conditions. Consider the circuit of figure P18.20. Suppose it is known that at $t = 0^-$, $i_1(0^-) = i_2(0^-) = -2$ A.

   (a) Redraw the circuit in the frequency domain accounting for the given ICs. (*Hint*: Use the equivalent circuit given in figure 18.19b.)

   (b) Determine a matrix equation of the form $V = Z(s)I'$ where $Z(s)$ is a 2 × 2 matrix with entries possibly depending on s, $V = [V_1(s) V_2(s)]^t$, and $I' = [I_1'(s) I_2'(s)]^t$.

   (c) Find a matrix solution to your equations of part (b); i.e., find I' in terms of V as $Z(s)^{-1}V = I'$.

   (d) Write nodal equations in terms of the sources, V_1, I_1', V_2, I_2', and the initial conditions as they appear in your frequency domain circuit.

   (e) Substitute your answers of part (c) into those of part (d) to obtain a pair of equations in V_1, V_2, V_s and the initial conditions. Put the equations in matrix form.

   (f) Solve the equations found in part (e) for V_1 and V_2 in terms of V_s and the initial conditions. (*Check*: Poles should be at $s = -1, -5$.)

   (g) Determine the complete response of $v_2(t)$ if $v_s(t) = 10u(t)$.

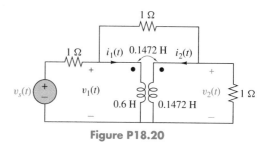

Figure P18.20

21. In the ignition circuit of figure 18.9, if the resistance and capacitance values are changed to 2 Ω and 0.25 F, respectively, find an approximate answer for the maximum voltage appearing across the spark plug.

ANSWER: 36,000 V.

22. Suppose that $i_{in}(t) = 3\cos(2t)u(t)$ A and that the initial stored energy is zero in the circuit shown in figure P18.22.

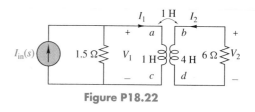

Figure P18.22

   (a) Determine the transfer function $H(s) = I_2(s)/I_{in}(s)$ if:
   (1) The dots are in positions a and b.
   (2) The dots are in positions c and d.
   (3) The dots are in positions a and d.
   (4) The dots are in positions c and b.

   (*Hint*: Apply source transformation, and then write two mesh equations. Solve the resulting equations for I_2 using Cramer's rule.) (*Check*: All transfer functions differ only by a sign.)

   (b) If the dots are in positions a and b, use phasors to determine the output in sinusoidal steady state, which must have the form $A\cos(2t + \theta)$.

   (c) If the dots are in positions a and b, determine the impulse response.

23. For the circuit shown in figure P18.23, determine (1) the equivalent inductance, L_{eq}, of the coupled coils; (2) the step response, assuming that all initial conditions are zero; (3) the resonant frequency of the circuit; and (4) the bandwidth and Q of the circuit.

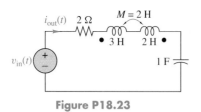

Figure P18.23

24. If $v_s(t) = 4\cos(2t)$ V, find the maximum instantaneous steady-state power delivered to the 8-Ω resistor in the circuit shown in figure P18.24.

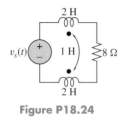

Figure P18.24

◆ 25. In the circuit shown in figure P18.25, determine the value of C (in F) that causes resonance to occur at $\omega = 2$ rad/sec.

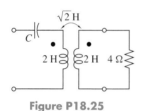

Figure P18.25

26. Shown in figure P18.26 is an RLC circuit that contains a pair of coupled inductors.

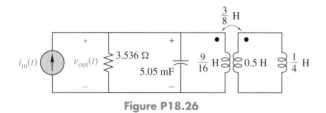

Figure P18.26

(a) Find $Y_{in}(s)$ and $Z_{in}(s)$.
(b) Find the poles and zeros, and then the natural response, of the circuit.
(c) Find the zero-input response if $v_C(0^-) = 2.667$ V and all other initial conditions are zero.
(d) Find the zero-state response if $i_{in}(t) = e^{-2t}\cos(2\pi t)$ A.

27. Consider the coupled inductors of figure 18.4. If $i_1(t) = A\cos(\omega t + \theta)$ A and $i_2(t) = B\cos(\omega t + \phi)$ A, show that the average power delivered to the coupled inductors is zero. (*Hint*: Evaluate the integral

$$P_{ave} = \frac{1}{T}\int_0^T (v_1 i_1 + v_2 i_2)dt$$

to determine the average power.)

28. Consider the coupled inductors shown in figure P18.28. Let the initial currents be $i_1 = I_1$ and $i_2 = I_2$ at $t = 0$. Both

independent sources have zero current at $t = T$ and arbitrary waveforms during $0 \le t \le T$. Show that the energy delivered by the inductors to the current sources during the interval $0 \le t \le T$ is equal to $0.5L_1 I_1^2 + 0.5L_2 I_2^2 + MI_1 I_2$.

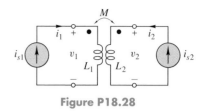

Figure P18.28

29. Consider figure P18.29. Instead of applying current sources to lead the currents to zero within a finite time, as in the previous problem, we can connect two resistors R_1 and R_2 to the inductors and let the current decrease exponentially to zero. It can then be shown that the energy delivered by the inductors to the resistors during $0 \le t \le \infty$ is equal to $0.5L_1 I_1^2 + 0.5L_2 I_2^2 \pm MI_1 I_2$. The proof for the general case is very complicated algebraically. Therefore, you are asked to demonstrate the validity of the assertion for the following specific case:

$$i_1(0) = 1 \text{ A}, \qquad i_2(0) = -3 \text{ A},$$
$$L_1 = 10 \text{ H}, \qquad L_2 = 2 \text{ H}, \qquad M = 2 \text{ H},$$
$$R_1 = R_2 = 1 \text{ } \Omega.$$

(a) Calculate the stored energy at $t = 0$.
(b) Calculate $i_1(t)$ and $i_2(t)$ for $t > 0$.
(c) Evaluate the integral $\int(v_1 i_1 + v_2 i_2)dt$ from $t = 0$ to $t = \infty$, and compare the answer with the result of part (a).
(*Check*: (a) 5 J, (b) $i_1(t) = e^{-t}u(t)$, $i_2(t) = -3e^{-t}u(t)$.)

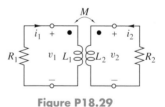

Figure P18.29

30. Consider the coupled inductors shown in figure P18.30.
(a) Determine the coefficient of coupling k.
(b) If $I_1 = 2$ A and $I_2 = -3$ A, find the stored energy W.
(c) If $I_1 = 2$ A, find the value of I_2 that will minimize W. What is the value of W_{min}?

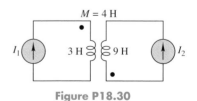

Figure P18.30

31. An electrical engineer claimed to have constructed a pair of coupled inductors with $L_1 = 10$ H, $L_2 = 8$ H, and $M = 9$ H. Rebut the claim by showing a specific set of (I_1, I_2) for which the stored energy is negative.

32. Let $i(t) = 2\cos(10t + 35°)$ A in the circuit of figure P18.32. Find the maximum instantaneous stored steady-state energy.

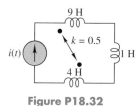

Figure P18.32

33. An independent voltage source $v_{in}(t)$ drives a network N through a pair of coupled inductors, as shown in figure P18.33.
  (a) Find k, the coefficient of coupling.
  (b) No information concerning the network N is given, other than that it is a linear network. If $v_{in}(t) = \cos(2t)$ V, can $v_o(t)$ be uniquely determined? If so, what is $v_o(t)$?
  (c) In part (b), if N is an arbitrary nonlinear network, how, if at all, would the answers be affected?

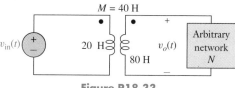

Figure P18.33

34. (a) Two identical coils having N turns each and self-inductance L are connected in series with dots as shown in figure 18.5(a) (series aiding). If the coupling coefficient is $k = 1$, show that the equivalent inductance is $4L$. The significance of this result is that it shows that when the number of turns is doubled, the inductance value is quadrupled.
  (b) In figure 18.5a, let the coupling coefficient be $k = 1$ and the numbers of turns be $N_1 = N$ and $N_2 = 2N$, respectively. If coil 1 has self-inductance $L_1 = L$, find L_2 and the equivalent inductance of the coil with $3N$ turns. Does the result indicate that, for unity coupling, the total inductance L_{eq} is proportional to the square of the number of turns?
  (c) Utilize the result of part (b) to show that for a pair of unity coupled coils with N_1 and N_2 turns each, the following ratios hold:

$$L_1 : L_2 : M = N_1^2 : N_2^2 : (N_1 N_2).$$

35. In the circuit shown in figure P18.35, the coupling coefficient k has been adjusted for the circuit to be in resonance (i.e., V_{in} and I_{in} are in phase).

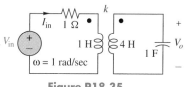

Figure P18.35

(a) Determine value of k.
(b) Find $H(s) = V_o(s)/V_{in}(s)$ and the poles and zeros of $H(s)$.
(c) Determine the approximate values of B_ω and Q of the circuit.
(d) Use any circuit simulation program (such as SPICE) to verify the result of part (c).

ANSWERS: (a) $\sqrt{3}$; (b) $1.732s/(s^3 + 4s^2 + s + 1)$, poles at -3.81, $-0.0969 \pm j0.5033$, zero at the origin; (c) $B_\omega \cong 0.1936$ rad/sec, $Q \cong 2.6$.

36. Shown in figure P18.36 is a resonant circuit that contains a pair of coupled inductors.

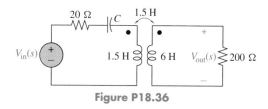

Figure P18.36

(a) Find the coupling coefficient k.
(b) Find the input impedance $Z(s)$ and then $Z(j\omega)$.
(c) Using the formula for the input impedance derived in part (b), determine the value of C that leads to a resonant frequency of $\omega_r = 1,333$ rad/sec.
(d) Determine the voltage gain at resonance; i.e., determine $V_{out}(j\omega_r)/V_{in}(j\omega_r)$.

37. Consider the circuit in figure P18.37. Determine the turns ratio n required to have maximum average power delivered to the load. What is the value of P_{max}?

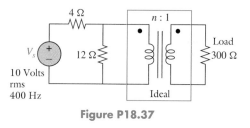

Figure P18.37

ANSWERS: 0.1, 4.6875 W.

38. Consider the circuit shown in figure P18.38.
  (a) Determine the impulse response.
  (b) Determine the step response.

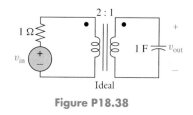

Figure P18.38

39. In the circuit shown in figure P18.39, find the phasor output V_{out}.

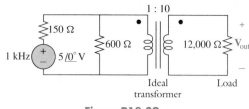

Figure P18.39

40. In the circuit shown in figure P18.40, $v_{in}(t) = 5\cos(2t)$ V. Find $i_a(t)$, $i_b(t)$, and $i_c(t)$.

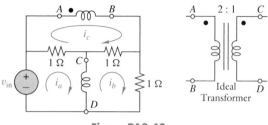

Figure P18.40

41. In the circuit shown in figure P18.41, determine each of the voltages $v_1(t)$, $v_2(t)$, and $v_3(t)$ in terms of $v_s(t)$ and other circuit parameters. In addition, determine the impedances Z_1, Z_2, and Z_3.

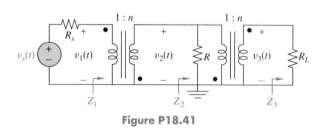

Figure P18.41

42. In the audio system shown in figure P18.42, the transformers are assumed to be ideal. The loudspeaker is represented approximately by a 4-Ω resistance.

(a) Find the turns ratios a and b such that the average power delivered to the 4-Ω loudspeaker is as large as possible. What is the maximum power?

(b) The turns ratios are those determined in part (a) for maximum power transfer. Now suppose that a loudspeaker with 16-Ω resistance is used in place of the 4-Ω speaker. What is the new power?

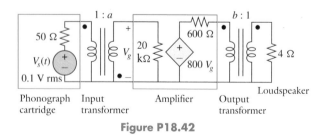

Figure P18.42

ANSWERS: (a) 20, $\sqrt{150}$, 266.67 W; (b) 170.67 W.

43. The circuit shown in figure P18.43 crudely represents an audio amplification circuit. Each 8-Ω resistor models a tweeter, and each 16-Ω resistor models a woofer. Suppose

the left and right speakers each consume 80 watts power on average. Determine:

(a) The turns ratio b and the resistance R for maximum power transfer from source to load.

(b) The voltage across and current through each woofer and tweeter (rms values).

(c) The rms source current i_{in}.

(d) The power delivered by the dependent source.

(e) The power delivered by the independent source.

(f) The power gain of the circuit, i.e., the ratio of the power delivered to the speaker load to the power delivered by the source.

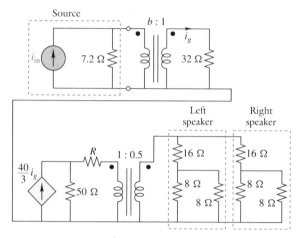

Figure P18.43

44. Consider the circuit shown in figure P18.44.

(a) Determine the output impedance $Z_{out}(s)$ as a function of the turns ratio a and the resistance R.

(b) If $v_{in}(t) = \cos(6t + 56.3°)$, determine the Thévenin equivalent circuit with the Thévenin source represented as a phasor.

(c) If $R = 2$ Ω, $a = 2$, and the output is terminated in a series LC with $L = 1$ H and $C = 1/24$ F, determine the steady-state response.

(d) Determine the complete zero-state response with the foregoing input and element values.

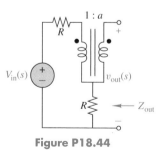

Figure P18.44

45. For the circuit shown in figure P18.45, determine the Thévenin equivalent circuit. (*Hint*: Add a voltage source V_s to the right side, and write loop equations.)

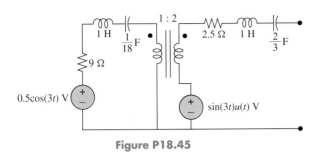

Figure P18.45

46. Determine the output voltage $v_{out}(t)$ in the circuit shown in figure P18.46 for $t \geq 0$.

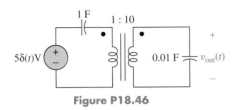

Figure P18.46

47. A certain calculator comes with an ac adaptor that is represented approximately by the circuit of figure P18.47(a). The primary is connected to a 110-V, 60-Hz household ac source. The secondary is connected to the charging circuit of the calculator. The equivalent circuit of figure P18.47(b) is recommended for use in this problem. The advantage of the approach is that the solution of simultaneous equations is avoided.

(a) Determine the parameters n, L_a, and L_b by referring to figure 18.22b.

(b) If the calculator is not connected, but the adaptor is plugged into a household ac outlet, what is the average power consumed by the adaptor?

(c) If the secondary of the adaptor is accidentally short-circuited, what are the magnitudes (rms) of the ac currents in the two windings?

(d) If a typical load, represented by a 100-Ω resistance, is connected to the secondary, what are the approximate magnitudes (rms values) of the voltage across and the current through the load?

ANSWERS: (a) 20, 0.3 H, 3.2 H; (b) 3.04 W; (c) 214 mA, 4.292 A; (d) 4.66 V, 46.6 mA.

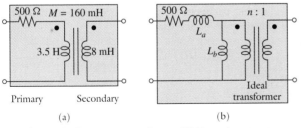

Primary Secondary

(a) (b)

Figure P18.47 (a) ac adaptor. (b) Equivalent circuit.

48. The bandpass circuit shown in figure P18.48a uses coupled coils and a capacitor.

(a) Determine the value of the coupling coefficient k.
(b) Find the transfer function $H(s) = V_o(s)/V_{in}(s)$. (*Suggestion:* Use the model of figure 18.20b or 18.20c for the coupled coils.)

(c) Determine ω_m, $|H(j\omega_m)|$, B_ω, and Q of the circuit.
Remarks:

1. Due to the mutual inductance, this passive circuit provides not only the frequency-selective property, but also a voltage gain larger than that obtainable with a parallel resonant circuit.

2. In practice, the two coils may be *approximated* by a single coil with a tap. In the present case, the tap is at $1/10$ of the total number of turns. The circuit is shown in figure P18.48b.

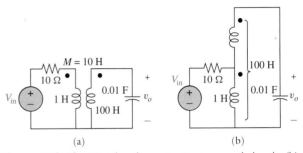

(a) (b)

Figure P18.48 (a) A bandpass circuit using coupled coils. (b) Replacing the coupled coil by a single coil with a tap.

49. The circuit shown in figure P18.49 has bandpass magnitude response.

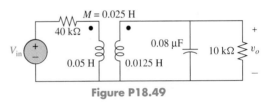

Figure P18.49

(a) Find the transfer function $H(s) = V_o(s)/V_{in}(s)$. (*Suggestion:* Use the model of figure 18.20b or 18.20c for the coupled inductors.)

(b) Find ω_m, B_ω, and H_m, and compare their values with those obtained in example 17.6 by the use of a parallel resonant circuit.

ANSWERS: (a) $625s/(s^2+2,500s+10^9)$; (b) 31,622 rad/sec, 2,500 rad/sec, 0.25. Example 17.6 has the same ω_m and B_ω, but a smaller $H_m = 0.2$.

50. Consider figure P18.50. Find the value of C such that the voltage gain is zero at $\omega = 1$ rad/sec. (*Hint:* Use the T-equivalent circuit of figure 18.25b.)

ANSWER: 1/3 F.

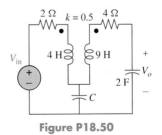

Figure P18.50

51. Consider figure P18.51. Find the value of C such that the voltage gain is zero at $\omega = 1$ rad/sec. (*Hint:* Use the π-equivalent circuit of figure 18.25c.)

ANSWER: 1/9 F.

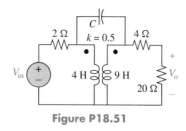

Figure P18.51

52. The two circuits shown in figure P18.52 have the same transfer function $H(s) = V_o/V_{in}$. Find the values of the three uncoupled inductances.

(*Hint*: Use the T-equivalent circuit properly.)

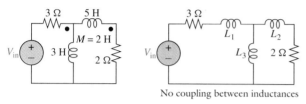

No coupling between inductances

Figure P18.52

53. For the circuit shown in figure P18.53, determine the coupling coefficient k such that at $\omega = 10$ rad/sec, the voltage gain V_o/V_{in} is zero. (*Hint*: Use the T-equivalent circuit for coupled coils, and recall that a series LC circuit behaves as does a short circuit at $\omega = 1/\sqrt{LC}$.

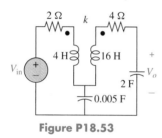

Figure P18.53

54. For the circuit shown in figure P18.54, find the value of C such that the voltage gain is zero at $\omega = 1$ rad/sec. (*Hint*: Use the π-equivalent circuit for coupled coils, and recall that a parallel LC circuit behaves as does a short circuit at $\omega = 1/\sqrt{LC}$.

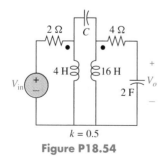

Figure P18.54

Note: The next four problems are to be solved with the aid of a linear circuit simulation program (e.g., SPICE).

55. Coupled tuned circuits can be adjusted to give a more desirable frequency response curve (e.g., a sharper cutoff and a flatter passband) than a single tuned circuit. This problem investigates this property. Consider the circuit of figure

P18.55. Roughly, this circuit works as follows. When $k = 0$, each series resonant circuit has a circuit Q and resonant frequency

$$Q = \sqrt{\frac{L}{C}}\frac{1}{R} \quad \text{and} \quad \omega_r = \omega_o = \frac{1}{\sqrt{LC}},$$

respectively. For $k \neq 0$, the transfer function $H(s) = V_o/I_{in}$ has two pairs of complex poles located in the vicinity of $\pm j\omega_o$. Increasing the value of k from 0 has the effect of separating the two poles further and thus broadening the frequency response curve. The value $k = 1/Q$ results in a curve that is the most flat. Using $k > 1/Q$ results in a curve that has two humps and a dip. The following calculations will verify these statements:

(a) If $L = 1$ H, $C = 1$ F, and $R = \beta$ Ω, prove that the transfer function is

$$H(s) = \frac{V_o}{I_{in}}$$

$$= \frac{ks}{[(1+k)s^2 + \beta s + 1][(1-k)s^2 + \beta s + 1]}.$$

(*Hint*: First apply source transformation, and then write two mesh equations.)

(b) If $L = 1$ H, $C = 1$ F, and $R = 0.02$ Ω (i.e., $Q = 50$), show the pole-zero plots for three different values of coupling coefficient $k = 0.5/Q$, $1/Q$, and $2/Q$.

(c) Use any circuit simulation program (such as SPICE) to plot the accurate frequency response curve for the three cases of part (b). The frequencies of interest are around 0.159 Hz.

(d) Let $L = 1$ H, $C = 1$ F, $R = 0.02$ Ω, and $k = 1/Q = 0.02$. Magnitude scale and frequency scale this circuit such that the frequency response curve peaks at 455 kHz (the standard intermediate frequency of an AM superheterodyne receiver) and the self-inductance of each coil is 2.35 mH. What is the 3-dB bandwidth of the scaled circuit?

(e) Consider $H(s)$ of part (a). Prove that the magnitude of $H(j\omega_o)$ as a function k and Q is

$$|H(j\omega_o)| = \frac{1}{\omega_o C}\frac{k}{\left(k^2 + \frac{1}{Q^2}\right)}.$$

(f) Consider $|H(j\omega_o)|$ of part (e). If k is the only adjustable parameter, show that $|H(j\omega_o)|$ is maximum when $k = 1/Q$ and that

$$|H(j\omega_o)|_{max} = \frac{Q}{2}\frac{1}{\omega_o C}.$$

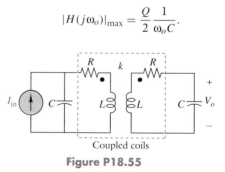

Coupled coils

Figure P18.55

56. The 1-kHz voltage source in figure P18.56a has a phasor $V_s = 10\angle 0$ V. The transformer is assumed to be ideal.

Find all node voltages by the use of a circuit simulation program (e.g., SPICE). Unfortunately, very few simulation packages include ideal transformers as a circuit element, although all of them allow controlled sources. To overcome this difficulty, we first model the ideal transformer by controlled sources, as shown in figure 18.15. Still, there is another difficulty. Most programs (e.g., SPICE) require that the controlling current be associated with an independent voltage source. When the original circuit fails to meet this condition, a dummy voltage source $V_{dummy} = 0$ has to be inserted in series with the branch in which the controlling current flows. If we now follow the suggested procedure, the circuit in figure P18.56b results. Perform a simulation of that circuit, and obtain the node voltages.

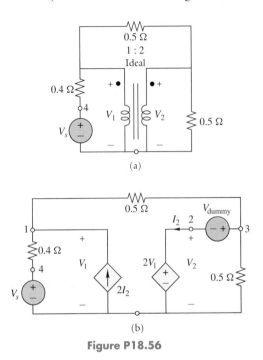

(a)

(b)

Figure P18.56

ANSWER: $V_1 = 2$ V, $V_2 = V_3 = 4$ V, $V_4 = 10$ V.

57. The basic operation of the ignition circuit shown in figure P18.57 has been explained in example 18.8. An approximate value for the voltage $v_2(t)$ is given in example 18.9. Use a circuit simulation to plot accurately the first cycle of $v_2(t)$. From this plot, determine the maximum value $|V_2|_{max}$. Is the estimate of $|v_2|_{max} \cong 36,000$ V indicated in example 18.8 following equation 18.16 reasonably accurate?

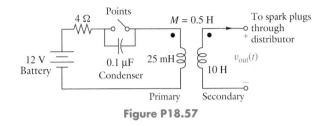

Figure P18.57

58. Consider the circuit shown in figure P18.58.

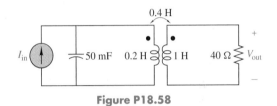

Figure P18.58

(a) Determine the coupling coefficient k.
(b) Replace the coupled inductors by the model shown in figure 18.22e. Obtain approximate answers for ω_m, B_ω, Q, and H_m by neglecting the $(1-k^2)L_2$ inductance in series with the 40-Ω resistance.

ANSWERS: $\omega_m \cong 10$ rad/sec, $B_\omega \cong 2$ rad/sec, $Q \cong 5$, $H_m \cong 20\Omega$.

(c) Use a circuit simulation program to find the accurate values of ω_m, B_ω, Q, and H_m. Comment on the accuracy of the approximate answers obtained in part (b).

CHAPTER 19

CHAPTER OUTLINE

Two-Ports

THE AMPLIFIER: A PRACTICAL TWO-PORT

A N actress speaks into a microphone. Speakers instantly replicate her voice, which resounds throughout an auditorium. What happens between the speaking and the hearing? A microphone produces a voltage signal that changes in proportion to the tenor and loudness of the voice of the actress. Amplifiers magnify this changing voltage signal perhaps a hundred or a thousand times to drive speakers whose cones reverberate in proportion to the changing voltage signal. The cones then cause the air to vibrate intensely, also in proportion to the tenor and loudness of the actress's voice. Her words become heard by thousands because of the amplifier.

Amplifier circuits are found in a huge number of appliances and in instrumentation. In radios, radio frequency amplifiers first magnify signals from an antenna. Special circuitry then extracts the audio portion from these antenna signals, and other circuitry amplifies it to drive speakers. Video signals from a video cassette recorder are amplified by special circuits to drive the picture tube inside a monitor. Amplifier circuitry enhances signals coming from sensors in various manufacturing processes. Repeater circuits, among other things, amplify phone signals whose magnitudes have attenuated during microwave transmission. There are a large number of other applications of amplifiers.

From the preceding discussion, one can surmise that an amplifier circuit has an input signal and an output signal. This configuration is represented by a device called a *two-port*, which has an input port for the input signal and an output port for the output signal. The following figure represents the idea (the reader is asked to analyze this amplifier in one of the homework problems):

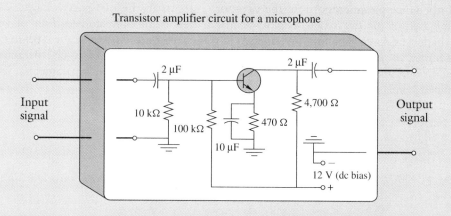

Transistor amplifier circuit for a microphone

Oftentimes, the circuit between the ports is highly complex. This chapter looks at shorthand methods for analyzing the input-output properties of two-ports without having to deal directly with a possibly highly complex circuit internal to the two port. The chapter will provide a variety of methods for analyzing the foregoing amplifier circuit. ◆

1. INTRODUCTION

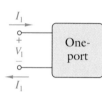

Figure 19.1
Illustration of a general one-port.

Figure 19.1 shows a general *one-port* whose two terminals satisfy the property that for any voltage V_1 across the terminals, the current entering one terminal, say, I_1, equals the current leaving the second terminal. A resistor is a one-port: The current entering one terminal equals the current leaving the other terminal. A capacitor and an inductor are also one-ports. A *general* one-port contains any number of interconnected resistors, capacitors, inductors, and other devices. In a one-port, only the relationship between the port voltage and current is of interest. For example, the port voltage and current in a resistor, capacitor, and inductor satisfy the relationships $v_R = Ri_R$, $Cdv_c/dt = i_C$, and $Ldi_L/dt = v_L$, respectively. Practically speaking, one-ports are macroscopic device models emphasizing input-output properties rather than detailed internal models. Thévenin and Norton equivalent circuits determine one-port models when only a pair of terminals of a network is of interest.

A two-port is a linear network having two pairs of terminals. Each pair behaves as a port; i.e., the current entering one terminal of a port equals the current leaving the second terminal of the same port for all voltages across the port. Figure 19.2 illustrates the idea. Coupling networks such as transformers have two pairs of terminals, each of which behaves as a one-port. Hence, transformers are two-port devices. In modeling a two-port, one must define a relationship among four variables. Different groupings of current and voltage variables lead to different kinds of characteristic parameters. For example, **admittance parameters** relate the two voltages, V_1 and V_2, to the port currents, I_1 and I_2. **Impedance parameters** relate the two-port currents to the two-port voltages. These are but two of the five types of two-port parameters commonly studied, four of which this chapter explores. The four investigated in this chapter are *impedance*, or z-parameters; *admittance*, or y-parameters; *hybrid*, or h-parameters; and *transmission*, or t-parameters. An additional set of parameters called g-parameters, the inverse of h-parameters, is not examined here.

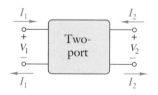

Figure 19.2 Illustration of a general two-port.

In practice, two-ports often represent coupling devices in which a source delivers energy to a load through the two-port network. For example, stereo amplifiers take a small low-power audio signal and increase its power so that it will drive a speaker system. Determining and knowing ratios such as the voltage gain, current gain, and power gain of a two-port is very important when dealing with a source that delivers power through a two-port to a load. This chapter develops various formulas for computing these gains for each type of two-port parameter.

2. ONE-PORT NETWORKS

Basic Impedance Calculations

As mentioned in the introduction, the current entering one terminal of a one-port, illustrated in figure 19.1, must equal the current leaving the second terminal for any voltage V_1 across the terminals. We begin our study of one-ports by exploring three impedance calculations that are essential to basic electronic analysis. The impedance or admittance seen at a port is fundamental to the behavior of a network to which a one-port or two-port is connected. The first two examples look at impedances of a common-collector stage of a transistor amplifier circuit.

 EXAMPLE 19.1

Consider the circuit of figure 19.3. The one-port shown models the input impedance of a common-collector stage of a transistor amplifier circuit. The goal of this example is to compute Z_{in} and interpret Z_{in} in terms of a transistor current gain parameter denoted by β.

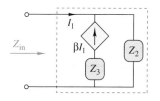

Figure 19.3 One-port representing the input characteristic of a common-collector transistor amplifier circuit.

SOLUTION

We attach a hypothetical voltage source, say, V_{in}, across the port terminals, top to bottom being plus to minus, to induce a hypothetical current I_1. From KCL,

$$I_1 + \beta I_1 - \frac{V_{in}}{Z_2} = 0.$$

Since the input impedance is the ratio of the input voltage to the input current,

$$Z_{in} = \frac{V_{in}}{I_1} = (\beta + 1)Z_2.$$

Thus, for a large β, say, 150, the input impedance can reach very high levels for reasonably sized impedances Z_2. When amplifying very low-power signals of small voltage, on the order of millivolts, high input impedance amplifiers are a necessity.

 EXAMPLE 19.2

Consider the circuit of figure 19.4. The one-port shown here models the output impedance of a common-collector stage of a transistor amplifier circuit. The goal of the example is to compute Z_{out} and interpret it in terms of a transistor gain parameter β.

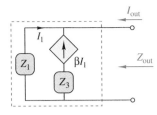

Figure 19.4 One-port representing the output characteristic of a common-collector transistor amplifier circuit.

SOLUTION

We connect a hypothetical voltage source V_{out} to the port terminals, with plus to minus being top to bottom. From the KCL, $I_1 + \beta I_1 + I_{\text{out}} = 0$. But $I_1 = -V_{\text{out}}/Z_1$. Hence,

$$Z_{\text{out}} = \frac{V_{\text{out}}}{I_{\text{out}}} = \frac{Z_1}{(\beta + 1)}.$$

Here, for a moderately sized Z_1, the output impedance Z_{out} can be made significantly smaller for large β. This allows one to match the output impedance of an amplifier to low impedance loads such as the 8-Ω or 16-Ω impedance loads of a speaker system.

The third example that is common in basic electronic analysis depicts a circuit often used in the analysis of field-effect transistor (FET) circuits.

◆ **EXAMPLE 19.3**

Consider the circuit of figure 19.5. Determine the input impedance Z_{in} in terms of Z_1, Z_2, Z_3, and g_m.

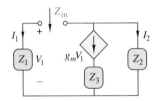

Figure 19.5 A common field-effect transistor (FET) circuit.

SOLUTION

As illustrated in figure 19.5, assume that a hypothetical voltage source V_{in} (plus to minus is right to left) has been attached across the port terminals. At the bottom node, $I_1 + g_m V_1 + I_2 = 0$. Since $V_1 = Z_1 I_1$ and $I_2 = (V_{\text{in}} + V_1)/Z_2$, it follows that

$$I_1 + g_m Z_1 I_1 + \frac{V_{\text{in}}}{Z_2} + \frac{Z_1 I_1}{Z_2} = 0.$$

Now, since $Z_{\text{in}} = V_{\text{in}}/I_{\text{in}}$,

$$Z_{\text{in}} = Z_1 + (1 + g_m Z_1)Z_2 = Z_2 + (1 + g_m Z_2)Z_1.$$

Thévenin and Norton Equivalent Circuits

Recall from chapter 5 of volume I and chapter 14 of this volume that the **Thévenin equivalent** of a network, as seen from a pair of terminals (i.e., from a one-port), is a voltage source in series with the Thévenin equivalent impedance Z_{th}. Z_{th} is simply the equivalent impedance of the one-port when all internal independent sources are set equal to zero. The value of the voltage V_{OC} of the source equals that voltage appearing at the open-circuited port terminals. A source transformation on the Thévenin equivalent produces the **Norton equivalent**, which is simply a current source in parallel with the Thévenin impedance. The value of the current source is $I_{\text{SC}} = V_{\text{OC}}/Z_{\text{th}}$. This is the current that would pass through a short circuit on the port terminals. A few examples illustrate the details best.

EXAMPLE 19.4

Consider the circuit of figure 19.6. Determine the Thévenin and Norton equivalent circuits.

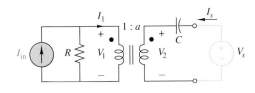

Figure 19.6 Transformer circuit for example 19.4.

SOLUTION

The voltage across the hypothetical source is

$$V_s = \frac{1}{Cs}I_s + V_2 = \frac{1}{Cs}I_s + aV_1.$$

To determine the Thévenin equivalent, we express V_1 in terms of I_{in} and I_s. From Ohm's law and the current relationship $I_1 = -aI_s$,

$$V_1 = R(I_{in} - I_1) = RI_{in} + aRI_s.$$

Therefore,

$$V_s = \left(\frac{1}{Cs} + Ra^2\right)I_s + aRI_{in}.$$

Interpreting the quantity RaI_{in} as a voltage source results in the Thévenin and Norton equivalent circuits of figure 19.7.

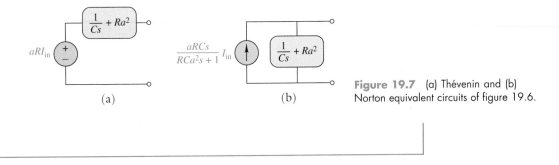

Figure 19.7 (a) Thévenin and (b) Norton equivalent circuits of figure 19.6.

A technique known as **matrix partitioning** often simplifies the calculation of Thévenin and Norton equivalent circuits. Example 19.5 illustrates the technique.

EXAMPLE 19.5

Determine the Thévenin equivalent circuit for the network of figure 19.8.

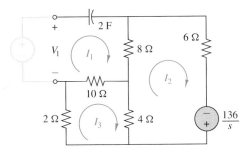

Figure 19.8 A three-loop RC circuit for example 19.5.

SOLUTION

Step 1. *Construct the loop equations.* By inspection, the loop equations for the circuit satisfy

$$\begin{bmatrix} V_1 \\ \cdots \\ \dfrac{136}{s} \\ 0 \end{bmatrix} = \begin{bmatrix} \dfrac{1}{2s} + 18 & \vdots & -8 & -10 \\ \cdots & & \cdots & \cdots \\ -8 & \vdots & 18 & -4 \\ -10 & \vdots & -4 & 16 \end{bmatrix} \begin{bmatrix} I_1 \\ \cdots \\ I_2 \\ I_3 \end{bmatrix}. \tag{19.1}$$

The goal is to find V_1 in terms of I_1; the matrix equation 19.1 is partitioned accordingly. To avoid dealing with numbers, let us rewrite this **partitioned matrix** equation as

$$\begin{bmatrix} V_1 \\ \cdots \\ V_2 \\ V_3 \end{bmatrix} = \begin{bmatrix} W_{11} & \vdots & W_{12} \\ \cdots & \vdots & \cdots \\ W_{21} & \vdots & W_{22} \end{bmatrix} \begin{bmatrix} I_1 \\ \cdots \\ I_2 \\ I_3 \end{bmatrix}, \tag{19.2}$$

where $V_2 = 136/s$, $V_3 = 0$, and the W_{ij}'s represent their obvious counterparts in equation 19.1.

Step 2. *Solve the partitioned matrix equations for* V_1 *in terms of* I_1 *and the vector* $[V_2, V_3]^t$. The matrix equation 19.2 may be rewritten as a set of two equations, namely,

$$V_1 = W_{11} I_1 + W_{12} \begin{bmatrix} I_2 \\ I_3 \end{bmatrix} \tag{19.3}$$

and

$$\begin{bmatrix} V_2 \\ V_3 \end{bmatrix} = W_{21} I_1 + W_{22} \begin{bmatrix} I_2 \\ I_3 \end{bmatrix}. \tag{19.4}$$

Since W_{22} is invertible, solving equation 19.4 for the vector $[I_2 I_3]^t$ yields

$$\begin{bmatrix} I_2 \\ I_3 \end{bmatrix} = W_{22}^{-1} \begin{bmatrix} V_2 \\ V_3 \end{bmatrix} - W_{22}^{-1} W_{21} I_1. \tag{19.5}$$

Substituting equation 19.5 into equation 19.3 produces

$$V_1 = \left[W_{11} - W_{12} W_{22}^{-1} W_{21} \right] I_1 + W_{12} W_{22}^{-1} \begin{bmatrix} V_2 \\ V_3 \end{bmatrix}. \tag{19.6}$$

MATLAB or its equivalent, or possibly a symbolic manipulation software package, allows the matrices in this equation to be conveniently computed. Observe that equation 19.6 has the form

$$V_1 = Z_{\text{th}} I_1 + V_{\text{OC}}.$$

Step 3. *Compute* Z_{th} *and* V_{OC}. After substitution for the W_{ij} matrices, the Thévenin equivalent impedance becomes

$$Z_{\text{th}} = \left[W_{11} - W_{12} W_{22}^{-1} W_{21} \right]$$

$$= \frac{36s + 1}{2s} - [-8 \quad -10] \begin{bmatrix} 18 & -4 \\ -4 & 16 \end{bmatrix}^{-1} \begin{bmatrix} -8 \\ -10 \end{bmatrix}$$

$$= 5.265 \frac{s + 0.095}{s} \ \Omega.$$

Similarly, the open-circuit voltage becomes

$$V_{\text{OC}} = W_{12} W_{22}^{-1} \begin{bmatrix} V_2 \\ V_3 \end{bmatrix} = [-8 \quad -10] \begin{bmatrix} 18 & -4 \\ -4 & 16 \end{bmatrix}^{-1} \begin{bmatrix} \dfrac{136}{s} \\ 0 \end{bmatrix} = -\frac{84}{s}.$$

Figure 19.9 shows the resulting Thévenin equivalent circuit.

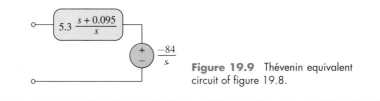

Figure 19.9 Thévenin equivalent circuit of figure 19.8.

General One-Port Analysis

Efficiently computing the Thévenin impedance Z_{th} or the open-circuit voltage V_{OC} for general one-ports necessitates an organized and general procedure such as nodal analysis or loop analysis. The generalized analysis described in this subsection builds on loop analysis. Consider figure 19.10, which contains resistors, capacitors, inductors, general impedances, and current-controlled voltage sources. Note that the figure implicitly presumes that only one loop current flows through the hypothetical voltage source, V_1.

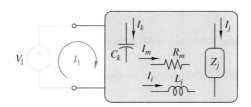

Figure 19.10 General one-port of resistors, capacitors, inductors, and current-controlled voltage sources; the source V_1 is hypothetical.

Supposing that there are N independent loops that characterize the one-port, it is possible to write N loop equations as follows:

$$
\begin{aligned}
Z_{11} I_1 + Z_{12} I_2 + \cdots + Z_{1N} I_N &= V_1, \\
Z_{21} I_1 + Z_{22} I_2 + \cdots + Z_{2N} I_N &= V_2, \\
&\vdots \\
Z_{N1} I_1 + Z_{N2} I_2 + \cdots + Z_{NN} I_N &= V_N
\end{aligned}
\tag{19.7}
$$

If there are no independent voltage sources inside the one-port, then $V_2 = V_3 = \cdots = V_N = 0$. Hence, writing equation 19.7 in partitioned matrix form yields

$$
ZI \equiv
\begin{bmatrix}
Z_{11} & \vdots & Z_{12} & \cdots & Z_{1N} \\
\cdots & & \cdots & & \cdots \\
Z_{21} & \vdots & Z_{22} & \cdots & Z_{2N} \\
& \vdots & & \vdots & \\
Z_{N1} & \vdots & Z_{N2} & \cdots & Z_{NN}
\end{bmatrix}
\begin{bmatrix}
I_1 \\
\cdots \\
I_2 \\
\vdots \\
I_N
\end{bmatrix}
=
\begin{bmatrix}
V_1 \\
\cdots \\
0 \\
\vdots \\
0
\end{bmatrix}.
\tag{19.8}
$$

Using the method of matrix partitioning, one can obtain the Thévenin impedance

$$
Z_{th} = Z_{11} - \begin{bmatrix} Z_{12} & \cdots & Z_{1N} \end{bmatrix}
\begin{bmatrix}
Z_{22} & \cdots & Z_{2N} \\
& \vdots & \\
Z_{N2} & \cdots & Z_{NN}
\end{bmatrix}^{-1}
\begin{bmatrix}
Z_{21} \\
Z_{N1}
\end{bmatrix}.
$$

On the other hand, if one were to use Cramer's rule, then one could solve equation 19.8 for the Thévenin admittance

$$
Y_{th} = \frac{\det \begin{bmatrix}
Z_{22} & \cdots & Z_{2N} \\
& \vdots & \\
Z_{N2} & \cdots & Z_{NN}
\end{bmatrix}}{\det \begin{bmatrix}
Z_{11} & Z_{12} & \cdots & Z_{1N} \\
Z_{21} & Z_{22} & \cdots & Z_{2N} \\
& & \vdots & \\
Z_{N1} & Z_{N2} & \cdots & Z_{NN}
\end{bmatrix}}.
$$

If there are internal sources present, then V_2 through V_N in equation 19.7 represent the net source voltage in each loop or mesh. It is then possible to solve for $V_{OC} = V_1$, given that $I_1 = 0$, in a manner shown in step 3 of example 19.5. Symbolic manipulation software packages are quite helpful in these calculations.

In an analogous fashion, a general analysis of one-ports can proceed using nodal analysis, with the proviso that one has voltage-controlled current sources instead of current-controlled voltage sources.

3. TWO-PORT ADMITTANCE PARAMETERS

Basic Definitions and Examples

As mentioned earlier, rather than deal with all the internal variables of a circuit, it is convenient to deal only with the terminal voltages and currents of a two-port, depicted by figure 19.2 and repeated in figure 19.11.

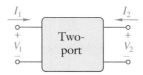

Figure 19.11 Standard two-port configuration having four external variables I_1, I_2, V_1, and V_2.

ADMITTANCE PARAMETERS

Throughout this and later discussions, assume that the two-port of figure 19.11 has no internal independent sources and that all dynamic elements are initially relaxed, i.e., have no initial conditions. Under these assumptions, the **admittance parameters** of a two-port are expressions for the terminal currents, I_1 and I_2, in terms of the port voltages, V_1 and V_2, i.e.,

$$\begin{bmatrix} I_1 \\ I_2 \end{bmatrix} = \begin{bmatrix} y_{11} & y_{12} \\ y_{21} & y_{22} \end{bmatrix} \begin{bmatrix} V_1 \\ V_2 \end{bmatrix},$$
(19.9a)

or, in scalar notation,

$$I_1 = y_{11} V_1 + y_{12} V_2,$$
$$I_2 = y_{21} V_1 + y_{22} V_2,$$
(19.9b)

Using either of these sets of equations, one can define each admittance parameter, y_{ij}, as follows:

$$y_{11} = \left.\frac{I_1}{V_1}\right|_{V_2=0}, \qquad y_{12} = \left.\frac{I_1}{V_2}\right|_{V_1=0},$$

(19.10)

$$y_{21} = \left.\frac{I_2}{V_1}\right|_{V_2=0}, \qquad y_{22} = \left.\frac{I_2}{V_2}\right|_{V_1=0}.$$

Since each admittance is defined with regard to a shorted terminal voltage $V_i = 0$, the y_{ij} are called **short-circuit admittance parameters**. The unit of an admittance parameter is the mho.

Some examples illustrate convenient methods for computing the aforementioned input-output parameters.

 EXAMPLE 19.6

Determine the short-circuit admittance parameters of the circuit in figure 19.12.

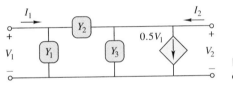

Figure 19.12 Circuit for example 19.6.

SOLUTION

The overall strategy is to write equations for I_1 and I_2 in terms of V_1 and V_2 using nodal analysis. Accordingly, by inspection,

$$I_1 = Y_1 V_1 + Y_2(V_1 - V_2) = (Y_1 + Y_2)V_1 - Y_2 V_2$$

and

$$I_2 = 0.5V_1 + Y_3 V_2 + Y_2(V_2 - V_1) = (0.5 - Y_2) + (Y_2 + Y_3)V_2.$$

In matrix form, these equations are

$$\begin{bmatrix} I_1 \\ I_2 \end{bmatrix} = \begin{bmatrix} Y_1 + Y_2 & -Y_2 \\ 0.5 - Y_2 & Y_2 + Y_3 \end{bmatrix} \begin{bmatrix} V_1 \\ V_2 \end{bmatrix},$$

in which case $y_{11} = Y_1 + Y_2$, $y_{12} = -Y_2$, $y_{21} = 0.5 - Y_2$, and $y_{22} = Y_2 + Y_3$.

In the next example, we combine the method of matrix partitioning with the use of nodal equations to compute the y-parameters.

 EXAMPLE 19.7

Compute the y-parameters for the circuit of figure 19.13.

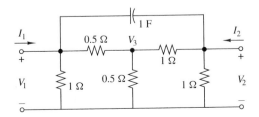

Figure 19.13 Three-node circuit for example 19.7.

SOLUTION

Computation of the y-parameters will again proceed by the method of nodal analysis. Writing the node equations of an RLC network by inspection is easy. Consider, for example, node 1, in which the current I_1 must be determined. I_1 is of the form

$$I_1 = Y_{11}V_1 + Y_{12}V_2 + Y_{13}V_3,$$

where each variable is understood to be a function of s. The coefficient Y_{11} is simply the sum of the admittances impinging on node 1. The coefficient Y_{12} is simply the negative of the sum of the admittances between nodes 1 and 2, whereas Y_{13} is the negative of the sum of the admittances between nodes 1 and 3. Similarly, $I_2 = Y_{21}V_1 + Y_{22}V_2 + Y_{23}V_3$, where Y_{22} is the sum of the admittances impinging on node 2, etc. Hence, in matrix form, the node equations of the circuit of figure 19.13 are

$$
\begin{bmatrix} I_1 \\ I_2 \\ \cdots \\ 0 \end{bmatrix} =
\left[\begin{array}{cc:c}
s+3 & -s & -2 \\
-s & s+2 & -1 \\
\hdashline
-2 & -1 & 5
\end{array} \right]
\begin{bmatrix} V_1 \\ V_2 \\ \cdots \\ V_3 \end{bmatrix}
\triangleq
\begin{bmatrix} W_{11} & W_{12} \\ W_{21} & W_{22} \end{bmatrix}
\begin{bmatrix} V_1 \\ V_2 \\ \cdots \\ V_3 \end{bmatrix}.
\tag{19.11}
$$

This nodal equation matrix is symmetric, because the RLC network contains no dependent sources. Using the method of matrix partitioning induced by equation 19.11 yields

$$
\begin{bmatrix} I_1 \\ I_2 \end{bmatrix} =
\left[W_{11} - W_{12} W_{22}^{-1} W_{21} \right]
\begin{bmatrix} V_1 \\ V_2 \end{bmatrix} =
\left(\begin{bmatrix} s+3 & -s \\ -s & s+2 \end{bmatrix} - \frac{1}{5} \begin{bmatrix} 2 \\ 1 \end{bmatrix} \begin{bmatrix} 2 & 1 \end{bmatrix} \right)
\begin{bmatrix} V_1 \\ V_2 \end{bmatrix}
$$

$$
= \begin{bmatrix} s+2.2 & -(s+0.4) \\ -(s+0.4) & s+1.8 \end{bmatrix}
\begin{bmatrix} V_1 \\ V_2 \end{bmatrix}.
$$

A last example couples a transformer with a resistive π-network.

◆ EXAMPLE 19.8

Compute the y-parameters of the circuit in figure 19.14.

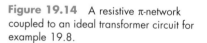

Figure 19.14 A resistive π-network coupled to an ideal transformer circuit for example 19.8.

SOLUTION

Find the y-parameters by using nodal analysis in conjunction with the ideal transformer equations. First, at port 1,

$$I_1 = V_1 + (V_1 - \hat{V}_1) = 2V_1 - \hat{V}_1 = 2V_1 - \frac{1}{a}V_2. \tag{19.12}$$

Now consider that $aI_2 = -\hat{I}_1$, a node equation at the primary of the transformer yields

$$I_2 = -\frac{1}{a}\hat{I}_1 = -\frac{1}{a}\left[-\hat{V}_1 + (V_1 - \hat{V}_1) \right] = -\frac{1}{a}V_1 + \frac{2}{a^2}V_2, \tag{19.13}$$

where the last equality uses the relationship $a\hat{V}_1 = V_2$. Putting equations 19.12 and 19.13 in matrix form yields the y-parameter relationship

$$
\begin{bmatrix} I_1 \\ I_2 \end{bmatrix} =
\begin{bmatrix} 2 & -\dfrac{1}{a} \\ -\dfrac{1}{a} & \dfrac{2}{a^2} \end{bmatrix}
\begin{bmatrix} V_1 \\ V_2 \end{bmatrix}.
$$

Two-Dependent Source Equivalent Circuit

The key to engineering analysis rests with the interpretation of appropriate mathematical equations. The key to two-port analysis is the interpretation of the two-port equations. Take, for example, the first admittance equation,

$$I_1 = y_{11}V_1 + y_{12}V_2.$$

One circuit-theoretic interpretation of this equation has the port current I_1 equal to the port voltage V_1 times an admittance y_{11} in parallel with a voltage controlled current source $y_{12}V_2$. A similar interpretation of the equation

$$I_2 = y_{21}V_1 + y_{22}V_2$$

yields an admittance branch y_{22} in parallel with a voltage-controlled current source $y_{21}V_1$.

These interpretations lead to the **two-dependent source equivalent circuit** of a two-port represented by the short-circuit admittance parameters. (See figure 19.15b.) This equivalent circuit aids the computation of input and output impedances and voltage gain formulas.

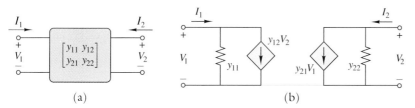

(a)　　　　　　　　　　　(b)

Figure 19.15 (a) Short-circuit admittance parameters and (b) their two-dependent source equivalent circuit. The dotted line at the bottom of (b) indicates that the two halves may not necessarily be connected.

4. Y-PARAMETER ANALYSIS OF TERMINATED TWO-PORTS

This section takes up the task of analyzing **terminated two-ports** modeled by y-parameters. A two-port is terminated when it has source load admittances. Any circuit or system in which a source provides an excitation signal to an interconnection network that modifies the signal and drives a load impedance can be represented by a teminated two-port. Such a scenario is common to numerous real-world systems. For example, the utility industry delivers power to a home from a generating facility through a transmission network, and a telephone system delivers voice information from the phone through an interconnect and transmission network to a receiver.

Input and Output Admittance Calculations

The input and output admittances of a terminated two-port are important for determining power transfer and various gain computations. In what follows next, we will show two different methods for computing the input admittance of the terminated two-port illustrated in figure 19.16. Computation of the output admittance is left as an exercise.

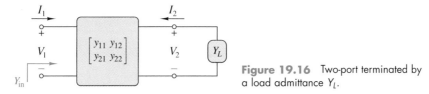

Figure 19.16 Two-port terminated by a load admittance Y_L.

The first method for computing $Y_{in} = I_1/V_1$ is a matrix method. Recall equation 19.9a, i.e.,

$$\begin{bmatrix} I_1 \\ I_2 \end{bmatrix} = \begin{bmatrix} y_{11} & y_{12} \\ y_{21} & y_{22} \end{bmatrix} \begin{bmatrix} V_1 \\ V_2 \end{bmatrix}.$$

Using the terminal conditions imposed by the load Y_L in figure 19.16, we obtain

$$I_2 = -Y_L V_2.$$

Incorporating this terminal condition into the two-port y-parameter equation yields

$$\begin{bmatrix} I_1 \\ 0 \end{bmatrix} = \begin{bmatrix} y_{11} & y_{12} \\ y_{21} & y_{22} + Y_L \end{bmatrix} \begin{bmatrix} V_1 \\ V_2 \end{bmatrix}.$$

Using Cramer's rule to solve for V_1 in terms of I_1 results in

$$V_1 = \frac{\det \begin{bmatrix} I_1 & y_{12} \\ 0 & y_{22} + Y_L \end{bmatrix}}{\det \begin{bmatrix} y_{11} & y_{12} \\ y_{21} & y_{22} + Y_L \end{bmatrix}} = \frac{(y_{22} + Y_L)I_1}{y_{11}(y_{22} + Y_L) - y_{12}y_{21}}.$$

Because Y_{in} is the ratio of I_1 to V_1, the **input admittance** of the two-port of figure 19.16 is

$$Y_{in} = y_{11} - \frac{y_{12}y_{21}}{y_{22} + Y_L}. \tag{19.14}$$

A second method for computing Y_{in} utilizes the two-dependent source equivalent circuit with load as illustrated in figure 19.17. With regard to the right side of figure 19.17, the current $y_{21}V_1$ must equal the current through the parallel admittances y_{22} and Y_L, i.e.,

$$y_{21}V_1 = -(y_{22} + Y_L)V_2.$$

Hence,

$$V_2 = -\frac{y_{21}}{y_{22} + Y_L}V_1.$$

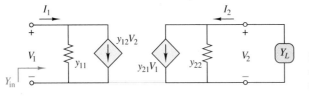

Figure 19.17 Input admittance calculation using two-dependent source equivalent circuit.

Now consider the left side of the figure. Here,

$$I_1 = y_{11}V_1 + y_{12}V_2 = \left(y_{11} - \frac{y_{12}y_{21}}{y_{22} + Y_L} \right) V_1,$$

which specifies the formula for the input admittance.

Exercise. Let Y_S denote a source admittance. Show that the **output admittance** of figure 19.18 is

$$Y_{out} = y_{22} - \frac{y_{12}y_{21}}{y_{11} + Y_S}. \tag{19.15}$$

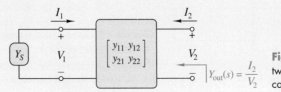

Figure 19.18 Input-terminated two-port for output admittance calculation.

$\left. Y_{out}(s) = \dfrac{I_2}{V_2} \right.$

Gain Calculations

Our objective now is to derive a formula for the voltage gain of a doubly terminated two-port, as illustrated in figure 19.19. Again, the resistance symbols denote general admittances rather than the traditional conductances.

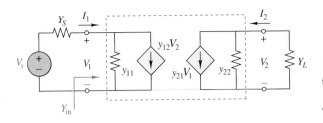

Figure 19.19 Doubly terminated two-port driven by voltage source, V_s; resistance symbols denote general admittances.

The specific goal is to derive the **voltage gain formula**

$$G_V = \frac{V_2}{V_s} = \left(\frac{Y_S}{Y_S + Y_{in}} \right) \left(\frac{-y_{21}}{y_{22} + Y_L} \right). \tag{19.16}$$

The overall gain calculation breaks down into two cascaded gain calculations according to the relationship

$$G_V = \frac{V_2}{V_s} = \frac{V_2}{V_1} \frac{V_1}{V_s}.$$

Computation of the gain, V_1/V_s, follows directly from voltage division:

$$\frac{V_1}{V_s} = \frac{Y_s}{Y_s + Y_{in}}.$$

To determine the ratio V_2/V_1, simply consider the right half of the load-terminated two-dependent source equivalent circuit model of figure 19.19. As in the derivation of the input admittance, we have

$$y_{21} V_1 = -(y_{22} + Y_L) V_2.$$

The ratio V_2/V_1 follows directly as

$$\frac{V_2}{V_1} = \frac{-y_{21}}{y_{22} + Y_L},$$

which completes the derivation of the gain formula, equation 19.16. The problems at the end of the chapter ask for extensions and various applications of this derivation and the resulting gain formula.

Exercise. Compute the current gain, $(-I_2)/I_1$, of the circuit of figure 19.19. (*Hint:* What is the relationship of I_1 to V_s and the relationship of $(-I_2)$ to V_2?)

5. IMPEDANCE PARAMETER ANALYSIS OF TWO-PORTS

Definition and Examples

The **z-parameters**, or **impedance parameters**, relate currents to voltages, as one would expect. As we will see, they are the inverse of the y-parameters in most cases.

IMPEDANCE PARAMETERS

For the two-port of figure 19.11, the z-parameters z_{ij} relate the port currents to the port voltages according to the matrix equation

$$\begin{bmatrix} V_1 \\ V_2 \end{bmatrix} = \begin{bmatrix} z_{11} & z_{12} \\ z_{21} & z_{22} \end{bmatrix} \begin{bmatrix} I_1 \\ I_2 \end{bmatrix}, \tag{19.17}$$

under the assumption of zero initial conditions and no internal independent sources. Therefore, from equation 19.17, each individual z-parameter is defined according to the formulas

$$z_{11} = \left. \frac{V_1}{I_1} \right|_{I_2=0}, \qquad z_{12} = \left. \frac{V_1}{I_2} \right|_{I_1=0},$$

$$z_{21} = \left. \frac{V_2}{I_1} \right|_{I_2=0}, \qquad z_{22} = \left. \frac{V_2}{I_2} \right|_{I_1=0}. \tag{19.18}$$

Since each z_{ij} is defined with one of the ports open-circuited, i.e., $I_1 = 0$ or $I_2 = 0$, the z_{ij}'s are called **open-circuit impedance parameters**. Their unit is the ohm.

◆ **EXAMPLE 19.9**

Compute the z-parameters for the circuit of figure 19.20.

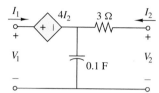

Figure 19.20 A simple *T*-circuit for computation of *z*-parameters in example 19.9.

SOLUTION

Rather than apply the z-parameter definitions of equation 19.18, we will use mesh analysis to obtain equation 19.17 directly.

Step 1. A loop equation at port 1 yields

$$V_1 = 4I_2 + \frac{10}{s}(I_1 + I_2) = \frac{10}{s}I_1 + \left(4 + \frac{10}{s}\right)I_2. \tag{19.19}$$

Step 2. Similarly, a loop equation at port 2 produces

$$V_2 = 3I_2 + \frac{10}{s}(I_1 + I_2) = \frac{10}{s}I_1 + \left(3 + \frac{10}{s}\right)I_2. \tag{19.20}$$

Step 3. By inspection of the right-hand sides of equations 19.19 and 19.20,

$$Z = \begin{bmatrix} z_{11} & z_{12} \\ z_{21} & z_{22} \end{bmatrix} = \begin{bmatrix} \dfrac{10}{s} & \dfrac{4s + 10}{s} \\ \dfrac{10}{s} & \dfrac{3s + 10}{s} \end{bmatrix}.$$

In the next example, we utilize the technique of matrix partitioning to compute the z-parameters of a π-network.

◆ **EXAMPLE 19.10**

Compute the z-parameters of the π-network of figure 19.21.

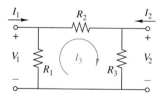

Figure 19.21 *π*-network for example 19.10.

SOLUTION

It is straightforward to write the following three loop equations:

$$V_1 = R_1 I_1 - R_1 I_3,$$

$$V_2 = R_3 I_2 + R_3 I_3,$$

$$0 = -R_1 I_1 + R_3 I_2 + (R_1 + R_2 + R_3)I_3.$$

Putting these equations in matrix form and partitioning the matrix appropriately yields

$$
\begin{bmatrix} V_1 \\ V_2 \\ \cdots \\ 0 \end{bmatrix} = \left[\begin{array}{cc:c} R_1 & 0 & -R_1 \\ 0 & R_3 & R_3 \\ \hdashline -R_1 & R_3 & R_1 + R_2 + R_3 \end{array} \right] \begin{bmatrix} I_1 \\ I_2 \\ I_3 \end{bmatrix}.
$$

Hence, using the matrix partitioning formula derived in example 19.5, we obtain

$$
\begin{bmatrix} z_{11} & z_{12} \\ z_{21} & z_{22} \end{bmatrix} = \begin{bmatrix} R_1 & 0 \\ 0 & R_3 \end{bmatrix} - \frac{1}{R_1 + R_2 + R_3} \begin{bmatrix} -R_1 \\ R_3 \end{bmatrix} \begin{bmatrix} -R_1 & R_3 \end{bmatrix}
$$

$$
= \begin{bmatrix} R_1 & 0 \\ 0 & R_3 \end{bmatrix} - \frac{1}{R_1 + R_2 + R_3} \begin{bmatrix} R_1^2 & -R_1 R_3 \\ -R_1 R_3 & R_3^2 \end{bmatrix}
$$

$$
= \frac{1}{R_1 + R_2 + R_3} \begin{bmatrix} R_1(R_2 + R_3) & R_1 R_3 \\ R_1 R_3 & R_3(R_1 + R_2) \end{bmatrix}.
$$

Relationship to y-Parameters

Since the z-parameters relate port currents to port voltages and the y-parameters relate port voltages to port currents, one might expect that the z-parameter matrix and the y-parameter matrix are inverses of each other; i.e., since

$$
\begin{bmatrix} V_1 \\ V_2 \end{bmatrix} = \begin{bmatrix} z_{11} & z_{12} \\ z_{21} & z_{22} \end{bmatrix} \begin{bmatrix} I_1 \\ I_2 \end{bmatrix},
$$

perhaps it follows that

$$
\begin{bmatrix} I_1 \\ I_2 \end{bmatrix} = \begin{bmatrix} z_{11} & z_{12} \\ z_{21} & z_{22} \end{bmatrix}^{-1} \begin{bmatrix} V_1 \\ V_2 \end{bmatrix} \equiv \begin{bmatrix} y_{11} & y_{12} \\ y_{21} & y_{22} \end{bmatrix} \begin{bmatrix} V_1 \\ V_2 \end{bmatrix}.
$$

Indeed, when both sets of parameters exist (i.e., if $z_{11}z_{22} - z_{21}z_{12} \neq 0$ and $y_{11}y_{22} - y_{21}y_{12} \neq 0$), then

$$
\begin{bmatrix} z_{11} & z_{12} \\ z_{21} & z_{22} \end{bmatrix}^{-1} = \begin{bmatrix} y_{11} & y_{12} \\ y_{21} & y_{22} \end{bmatrix} \tag{19.21}
$$

and conversely. However, some circuits have z-parameters but not y-parameters, and conversely, as illustrated by the following example.

EXAMPLE 19.11

Compute the z-parameters of the circuit of figure 19.22, and determine whether there are y-parameters.

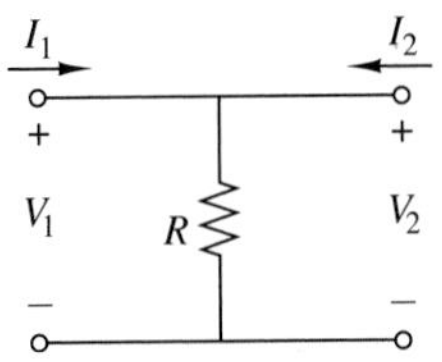

SOLUTION

By inspection, the z-parameter two-port equations are easily computed as

$$
\begin{bmatrix} V_1 \\ V_2 \end{bmatrix} = \begin{bmatrix} R & R \\ R & R \end{bmatrix} \begin{bmatrix} I_1 \\ I_2 \end{bmatrix}.
$$

Figure 19.22
Resistive two-port having z-parameters, but not y-parameters.

The z-parameter matrix, Z, is singular, since $\det[Z] = R^2 - R^2 = 0$. Because the Z-matrix does not have an inverse, the circuit fails to have y-parameters. One can check y_{11} directly

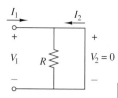

Figure 19.23
Equivalent circuit for computing y_{11} in which port 2 is shorted, so that $V_2 = 0$.

to verify this claim. Consider figure 19.23. Because $V_2 = 0$, there is also a short circuit across V_1, making the ratio

$$y_{11} = \frac{I_1}{V_1}\bigg|_{V_2=0}$$

undefined.

The Two-Dependent Source Equivalent Circuit

As with the y-parameters, the z-parameters have a two-dependent source equivalent circuit interpretation, illustrated in figure 19.24b. Consider first the equation

$$V_1 = z_{11}I_1 + z_{12}I_2.$$

Here, V_1 equals the sum of two voltages: $z_{11}I_1$ plus the voltage due to a current-controlled voltage source given by $z_{12}I_2$. This is precisely the left-hand portion of figure 19.24b. A similar interpretation follows for $V_2 = z_{21}I_1 + z_{22}I_2$, yielding the right-hand side of figure 19.24b.

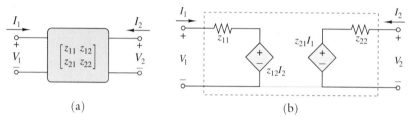

(a)   (b)

Figure 19.24 Two-dependent source equivalent circuit shown in (b) for network modeled by z-parameters shown in (a).

This equivalent circuit proves useful for computing voltage gains and input and output impedances of terminated two-ports. The reader should also note that there are other equivalent circuits that interpret the z-parameters. (See problems 43 and 44.)

6. IMPEDANCE AND GAIN CALCULATIONS OF TERMINATED TWO-PORTS MODELED BY Z-PARAMETERS

Input and Output Impedance Computations

Earlier, we derived the formula for the input admittance of a terminated two-port in terms of the y-parameters, leaving the output admittance calculation as an exercise. Here, using z-parameters we derive the output impedance of the terminated two-port of figure 19.25 and leave the input impedance calculation as an exercise. Specifically our first step is to derive the **output impedance** formula

$$Z_{\text{out}} = z_{22} - \frac{z_{12}z_{21}}{z_{11} + Z_S} \tag{19.22}$$

as a function of the network z-parameters and the source impedance. Beginning on the right-hand side of figure 19.25,

$$V_2 = z_{22}I_2 + z_{21}I_1. \tag{19.23}$$

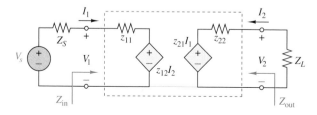

Figure 19.25 Terminated two-port modeled by z-parameters.

In calculating Z_{out}, which is the Thévenin impedance seen by the load, the independent voltage source, V_s, is set to zero. Hence,

$$-(Z_S + z_{11})I_1 = z_{12}I_2.$$

Solving for I_1 and substituting into equation 19.23 yields the output impedance formula of equation 19.22.

Exercise. Repeat the preceding derivation using a matrix method and Cramer's rule.

Exercise. Derive the following formula for the input impedance of a terminated two-port:

$$Z_{\text{in}} = z_{11} - \frac{z_{12}z_{21}}{z_{22} + Z_L}. \tag{19.24}$$

Gain Calculations

The next phase of our two-port analysis is to repeat the y-parameter derivation of the voltage gain of the two-port in the context of z-parameters. Specifically, our aim is to compute

$$G_V = \frac{V_2}{V_s} = \frac{V_2}{V_1}\frac{V_1}{V_s}.$$

The ratio V_1/V_s follows from voltage division:

$$\frac{V_1}{V_s} = \frac{Z_{\text{in}}}{Z_{\text{in}} + Z_s}. \tag{19.25}$$

To compute the ratio V_2/V_1, first apply voltage division to obtain

$$V_2 = \frac{Z_L}{Z_L + z_{22}}z_{21}I_1.$$

Then, from the definition of input impedance, $I_1 = V_1/Z_{\text{in}}$; hence,

$$\frac{V_2}{V_1} = \frac{Z_L}{Z_L + z_{22}}\frac{z_{21}}{Z_{\text{in}}}. \tag{19.26}$$

To complete the derivation, multiply equations 19.25 and 19.26 together to obtain the desired **voltage gain**:

$$G_V = \frac{V_1}{V_s}\frac{V_2}{V_1} = \left(\frac{Z_L}{z_{22} + Z_L}\right)\left(\frac{z_{21}}{Z_{\text{in}} + Z_S}\right). \tag{19.27}$$

An application of this formula and its derivation to a cascaded network of two-ports (two transistor amplifier circuits) follows in the next example.

◆ **EXAMPLE 19.12**

Consider the network of figure 19.26, which represents a two-stage (transistor) amplifier configuration. Each stage utilizes the same transistor in a different circuit configuration. The first stage is an amplification stage that will amplify a small source voltage to a much

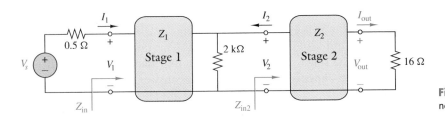

Figure 19.26 Two-stage amplifier network.

larger one. The second stage is an impedance-matching stage used to match the load to the output impedance of the amplifier circuitry to achieve maximum power transfer, at least approximately. The z-parameters, in ohms, for each stage in figure 19.26 are given by equation 19.28:

$$Z_1 = \begin{bmatrix} 1.0262 \times 10^6 & 6,790.8 \\ 1.0258 \times 10^6 & 6,793.5 \end{bmatrix}; \quad Z_2 = \begin{bmatrix} 350 & 2.667 \\ -10^6 & 6,667 \end{bmatrix}. \tag{19.28}$$

    a. Determine the input impedances, Z_{in2} and Z_{in}.
    b. Determine the voltage gain, V_{out}/V_s.
    c. Check the matching of the load and output impedance of the amplifier circuit.

SOLUTION

Step 1. Determine Z_{in2}. A straightforward application of equation 19.24 using the z-parameters of stage 2 produces the impedance Z_{in2}:

$$Z_{in2} = z_{11} - \frac{z_{12}z_{21}}{z_{22} + Z_L}$$

$$= 1.0262 \ 10^6 - \frac{6,790.8 \times 1.0258 \ 10^6}{6,793.5 + 16} = 3,159 \ \Omega. \tag{19.29}$$

Step 2. Compute the voltage gain, V_{out}/V_2, for stage 2. Equation 19.26 applies here. In particular,

$$\frac{V_{out}}{V_2} = \frac{Z_L}{Z_L + z_{22}} \frac{z_{21}}{Z_{in2}} = \frac{16(1.0258 \ 10^6)}{(16 + 6,793.5)3,195.4} = 0.7629. \tag{19.30}$$

The gain here is small, but remember that the real purpose of this stage is impedance matching rather than amplification. By proper choice of the z-parameters, the output impedance will approximately match that of the load. This allows us to dispense with an expensive and bulky impedance-matching transformer.

Step 3. Determine Z_{in} for stage 1. Here, observe that Z_{in2} (equation 19.29) in parallel with the 2-kΩ resistor between the two stages acts as a load to stage 1. Z_{in2} in parallel with 2 kΩ is $Z_{L1} = 1,224.7 \ \Omega$, where Z_{L1} denotes the load to stage 1. To compute the input impedance seen at the front end of stage 1, we use equation 19.24 and obtain

$$Z_{in} = z_{11} - \frac{z_{12}z_{21}}{z_{22} + Z_{L1}} = 350 + \frac{2.667 \times 10^{-6}}{6,667 + 1,224.7} = 687.9 \ \Omega. \tag{19.31}$$

Step 4. Compute the voltage gain, V_2/V_s, for stage 1. Using the result of equation 19.31 and applying equation 19.27 yields

$$\frac{V_2}{V_s} = \left(\frac{Z_{L1}}{Z_S + Z_{in}} \right) \left(\frac{z_{21}}{Z_{L1} + z_{22}} \right)$$

$$= \left(\frac{1,224.7}{75 + 687.9} \right) \left(\frac{-10^6}{1,224.7 + 6,667} \right) = -203.4. \tag{19.32}$$

Here, we see that the gain of stage 1 is large, leading to significant amplification of the input signal. For example, a 40-mV sine wave would be amplified to approximately 8 V, which will drive a small speaker.

Step 5. Compute the overall voltage gain, V_{out}/V_s. The desired gain is simply the product of equations 19.30 and 19.32, i.e., $G_V = -0.7629 \times 203.4 = -155.2$, which remains fairly large. Clearly, other amplification stages could be added to increase the overall gain.

Step 6. Verify that the load matches the amplifier circuitry to a reasonable degree. In this task, one first computes the output impedance of stage 1. This impedance, in parallel with the 2-kΩ resistor between the stages, looks like a source impedance to stage 2. It is then easy to compute the output impedance of stage 2. The answer is $Z_{out} = 16.4 \ \Omega$. The details are left as an exercise.

Exercise. For the circuit of figure 19.26, verify that the output impedance equals 16.4 Ω.

7. HYBRID PARAMETERS

Basic Definitions and Equivalences

As we have seen, some circuits have y-parameters, but not z-parameters, and conversely. A circuit element that has neither is the ideal transformer.

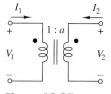

EXAMPLE 19.13

Circuit elements having neither z- nor y-parameters. Consider the ideal transformer circuit in figure 19.27. From the analysis of chapter 18, $V_2 = aV_1$ and $I_1 = -aI_2$. Clearly, V_1 and V_2 cannot be expressed as functions of I_1 and I_2; nor can I_1 and I_2 be expressed as functions of V_1 and V_2. Hence, an ideal transformer has neither y- nor z-parameters.

Figure 19.27
Ideal transformer, having neither y- nor z-parameters.

The preceding type of example necessitates an alternative modeling technique for two-port analysis. The hybrid parameters provide one of several alternatives.

HYBRID PARAMETERS

Hybrid parameters, h_{ij}, are a cross between y- and z-parameters: a voltage V_1 and a current I_2 are outputs, with V_2 and I_1 as inputs. Specifically, if the two-port of figure 19.11 contains no internal independent sources and has no initial stored energy, then the **hybrid parameters** are defined by the matrix equation

$$\begin{bmatrix} V_1 \\ I_2 \end{bmatrix} = \begin{bmatrix} h_{11} & h_{12} \\ h_{21} & h_{22} \end{bmatrix} \begin{bmatrix} I_1 \\ V_2 \end{bmatrix}. \tag{19.33}$$

This mixture of variables actually arises as a simplification of a midfrequency model of a common emitter configuration of a bipolar transistor amplifier circuit. Indeed, hybrid parameters are a commonly used description of transistor amplifier circuits. From the context of equation 19.33, h_{11} has units of ohms, h_{12} and h_{21} are dimensionless, and h_{22} has units of mhos.

As with both y- and z-parameters, we can interpret the foregoing equation as a two-dependent source equivalent circuit, as illustrated in figure 19.28.

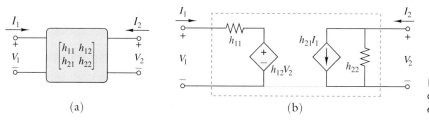

(a)

(b)

Figure 19.28 (a) Hybrid parameters and (b) the two-dependent source equivalent circuit that models them.

Exercise. Justify the two-dependent source equivalent circuit interpretation of figure 19.28, i.e., use the circuit to derive equation 19.33.

From either the preceding equivalent circuit, or from equation 19.33, the definition of each of the parameters falls out. The first h-parameter is

$$h_{11} = \left. \frac{V_1}{I_1} \right|_{V_2=0}.$$

Because h_{11} is the ratio of an input voltage to an input current, it is an input impedance. Since $V_2 = 0$, h_{11} is termed the **short-circuit input impedance**. Notice, however, that h_{11} is simply related to both the y- and the z-parameters as follows:

$$h_{11} = \left.\frac{V_1}{I_1}\right|_{V_2=0} = \frac{1}{y_{11}} = z_{11} - \frac{z_{12}z_{21}}{z_{22}}. \tag{19.34}$$

The second h-parameter, h_{21}, is called the **short-circuit forward current gain**, since it is the ratio of I_2 to I_1 under the condition that $V_2 = 0$, i.e.,

$$h_{21} = \left.\frac{I_2}{I_1}\right|_{V_2=0}.$$

From the z-parameter equation $V_2 = z_{21}I_1 + z_{22}I_2$, with $V_2 = 0$, h_{21} has a simple z-parameter interpretation, viz.,

$$h_{21} = \left.\frac{I_2}{I_1}\right|_{V_2=0} = -\frac{z_{21}}{z_{22}}. \tag{19.35}$$

The third h-parameter is

$$h_{12} = \left.\frac{V_1}{V_2}\right|_{I_1=0}. \tag{19.36}$$

Since it is the ratio of V_1 to V_2 under the condition that port 1 is open-circuited, i.e., $I_1 = 0$, b_{12} is called the **reverse open-circuit voltage gain**.

> **Exercise.** Interpret h_{12} in terms of the y-parameters.
> Lastly,
>
> $$h_{22} = \left.\frac{I_2}{V_2}\right|_{I_1=0}$$
>
> is the **open-circuit output admittance** and has units of mhos. The word **open-circuit** suggests a z-parameter interpretation. Considering that $V_2 = z_{21}I_1 + z_{22}I_2$ if $I_1 = 0$, then
>
> $$h_{22} = \frac{1}{z_{22}} = y_{22} - \frac{y_{12}y_{21}}{y_{11}}. \tag{19.37}$$
>
> This relationship is similar to equation 19.34, which determines the short-circuit input impedance.
> We will return to these equivalences later, after having gained some experience in computing them.

Computation of *h*-Parameters

◆ EXAMPLE 19.14

Pathological circuit having neither z- nor y-parameters. Consider the two-port of figure 19.29, whose front end is a short circuit and whose secondary is an open circuit. This two-port has neither z- nor y-parameters. However, by inspection, the h-parameters are

$$\begin{bmatrix} V_1 \\ I_2 \end{bmatrix} = \begin{bmatrix} 0 & 0 \\ 0 & 0 \end{bmatrix} \begin{bmatrix} I_1 \\ V_2 \end{bmatrix}.$$

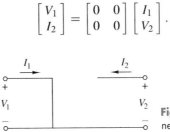

Figure 19.29 Simple two-port having neither z- nor y-parameters.

EXAMPLE 19.15

Consider the two-port of figure 19.30, and determine the *h*-parameters.

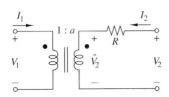

Figure 19.30 Ideal transformer circuit for example 19.15.

SOLUTION

Step 1.   Determine an equation for V_1.
From the primary and secondary voltage relationship of an ideal transformer,

$$V_1 = \frac{1}{a}\hat{V}_2.$$

However, $\hat{V}_2 = V_2 - RI_2$, in which case

$$V_1 + \frac{R}{a}I_2 = \frac{1}{a}V_2. \tag{19.38}$$

Step 2.   Determine an equation for I_2 *in terms of the other variables.* From the primary and secondary current relationship of an ideal transformer,

$$I_2 = -\frac{I_1}{a}. \tag{19.39}$$

Step 3.   Write equations 19.38 and 19.39 in matrix form, and solve for V_1 *and* I_2 *in terms of* I_1 *and* V_2. In matrix form, equations 19.38 and 19.39 are

$$\begin{bmatrix} 1 & \dfrac{R}{a} \\ 0 & 1 \end{bmatrix} \begin{bmatrix} V_1 \\ I_2 \end{bmatrix} = \begin{bmatrix} 0 & \dfrac{1}{a} \\ -\dfrac{1}{a} & 0 \end{bmatrix} \begin{bmatrix} I_1 \\ V_2 \end{bmatrix}. \tag{19.40}$$

Solving for the vector $[V_1 \quad I_2]^t$ produces the *h*-parameter equation

$$\begin{bmatrix} V_1 \\ I_2 \end{bmatrix} = \begin{bmatrix} 1 & \dfrac{R}{a} \\ 0 & 1 \end{bmatrix}^{-1} \begin{bmatrix} 0 & \dfrac{1}{a} \\ -\dfrac{1}{a} & 0 \end{bmatrix} \begin{bmatrix} I_1 \\ V_2 \end{bmatrix}$$

$$= \begin{bmatrix} \dfrac{R}{a^2} & \dfrac{1}{a} \\ -\dfrac{1}{a} & 0 \end{bmatrix} \begin{bmatrix} I_1 \\ V_2 \end{bmatrix}.$$

EXAMPLE 19.16

Find the *h*-parameters of the linear transformer circuit of figure 19.31. Assume $L_1 = L_2 = M = 1$ H.

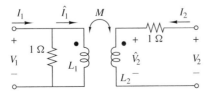

Figure 19.31 Linear transformer circuit for example 19.16.

Step 1. Write a node equation at the primary, apply the transformer relationship, and manipulate the result. Writing a node equation at the primary leads to the relationship $\hat{I}_1 = I_1 - V_1$. The voltage V_1 is given by the linear transformer equation of chapter 18:

$$V_1 = L_1 s \hat{I}_1 + M s I_2.$$

Hence, since $\hat{I}_1 = I_1 - V_1$,

$$V_1 = L_1 s I_1 - L_1 s V_1 + M s I_2.$$

Grouping like terms leads to

$$(1 + L_1 s)V_1 - M s I_2 = L_1 s I_1. \tag{19.41}$$

Step 2. Write a loop equation at the secondary, apply the transformer relationship, and simplify the result. A loop equation at the secondary implies that $V_2 = I_2 + \hat{V}_2$. But $\hat{V}_2$ is given by the linear transfer relationship, i.e., $\hat{V}_2 = M s \hat{I}_1 + L_2 s I_2$ and $I_1 = I_1 - V_1$. Hence,

$$V_2 - I_2 = M s(I_1 - V_1) + L_2 s I_2.$$

Grouping like terms with V_1 and I_2 on the left of the equal sign and I_1 and V_2 on the right leads to

$$-M s V_1 + (L_2 s + 1)I_2 = -M s I_1 + V_2. \tag{19.42}$$

Step 3. Write equations 19.41 and 19.42 in matrix form, and solve for $[V_1, I_2]^t$. The matrix form of equations 19.41 and 19.42 is

$$\begin{bmatrix} 1 + L_1 s & -M s \\ -M s & 1 + L_2 s \end{bmatrix} \begin{bmatrix} V_1 \\ I_2 \end{bmatrix} = \begin{bmatrix} L_1 s & 0 \\ -M s & 1 \end{bmatrix} \begin{bmatrix} I_1 \\ V_2 \end{bmatrix}. \tag{19.43}$$

Solving equation 19.43 with $M = L_1 = L_2 = 1$ H yields the desired h-parameter equation

$$\begin{bmatrix} V_1 \\ I_2 \end{bmatrix} = \begin{bmatrix} 1 + s & -s \\ -s & 1 + s \end{bmatrix}^{-1} \begin{bmatrix} s & 0 \\ -s & 1 \end{bmatrix} \begin{bmatrix} I_1 \\ V_2 \end{bmatrix}$$

$$= \frac{1}{2s + 1} \begin{bmatrix} s & s \\ -s & s + 1 \end{bmatrix} \begin{bmatrix} I_1 \\ V_2 \end{bmatrix}.$$

General Relations to z- and y-Parameters

The h-, y-, and z-parameters are interrelated. This subsection derives expressions that specify the h-parameters in terms of the z- and y-parameters. The formulas are, of course, important, but the methods used to obtain the formulas have a much broader application.

To express the h-parameters in terms of the z-parameters, first note that the goal is to obtain V_1 and I_2 in terms of I_1 and V_2. In this task, consider the z-parameter equations,

$$V_1 = z_{11} I_1 + z_{12} I_2$$

and

$$V_2 = z_{21} I_1 + z_{22} I_2.$$

Placing the V_1 and I_2 terms on the left and the I_1 and V_2 terms on the right leads to the reordered pair of equations

$$V_1 - z_{12} I_2 = z_{11} I_1$$

and

$$z_{22} I_2 = -z_{21} I_1 + V_2.$$

The equivalent matrix form is

$$\begin{bmatrix} 1 & -z_{12} \\ 0 & z_{22} \end{bmatrix} \begin{bmatrix} V_1 \\ I_2 \end{bmatrix} = \begin{bmatrix} z_{11} & 0 \\ -z_{21} & 1 \end{bmatrix} \begin{bmatrix} I_1 \\ V_2 \end{bmatrix}. \tag{19.44}$$

Solving equation 19.44 for the vector $[V_1 \quad I_2]^t$ under the proviso that $z_{22} \neq 0$ yields

$$\begin{bmatrix} V_1 \\ I_2 \end{bmatrix} = \begin{bmatrix} 1 & -z_{12} \\ 0 & z_{22} \end{bmatrix}^{-1} \begin{bmatrix} z_{11} & 0 \\ -z_{21} & 1 \end{bmatrix} \begin{bmatrix} I_1 \\ V_2 \end{bmatrix}$$

$$= \frac{1}{z_{22}} \begin{bmatrix} z_{11}z_{22} & z_{12} \\ -z_{21} & 1 \end{bmatrix} \begin{bmatrix} I_1 \\ V_2 \end{bmatrix}. \tag{19.45}$$

Thus, we have used matrix methods to determine, in a straightforward manner, the h-parameters in terms of the z-parameters under the condition that z_{22} is not identically zero.

Similarly, to express the h-parameters in terms of the y-parameters, consider the y-parameter equations,

$$I_1 = y_{11}V_1 + y_{12}V_2$$

and

$$I_2 = y_{21}V_1 + y_{22}V_2.$$

Again, moving the V_1 and I_2 terms to the left and the I_1 and V_2 terms to the right of the equals sign yields

$$-y_{11}V_1 = -I_1 + y_{12}V_2$$

and

$$-y_{21}V_1 + I_2 = y_{22}V_2.$$

Placing these relationships in matrix form results in

$$\begin{bmatrix} -y_{11} & 0 \\ -y_{21} & 1 \end{bmatrix} \begin{bmatrix} V_1 \\ I_2 \end{bmatrix} = \begin{bmatrix} -1 & y_{12} \\ 0 & y_{22} \end{bmatrix} \begin{bmatrix} I_1 \\ V_2 \end{bmatrix}. \tag{19.46}$$

Solving for $[V_1 \quad I_2]^t$ and the h-parameter relationship under the proviso that $y_{11} \neq 0$ yields

$$\begin{bmatrix} V_1 \\ I_2 \end{bmatrix} = \frac{1}{y_{11}} \begin{bmatrix} -1 & 0 \\ -y_{21} & y_{11} \end{bmatrix} \begin{bmatrix} -1 & y_{12} \\ 0 & y_{22} \end{bmatrix} \begin{bmatrix} I_1 \\ V_2 \end{bmatrix}$$

$$= \begin{bmatrix} \dfrac{1}{y_{11}} & -\dfrac{y_{12}}{y_{11}} \\ \dfrac{y_{21}}{y_{11}} & y_{22} - \dfrac{y_{12}y_{21}}{y_{11}} \end{bmatrix} \begin{bmatrix} I_1 \\ V_2 \end{bmatrix}. \tag{19.47}$$

Equation 19.47 holds if and only if $y_{11} \neq 0$.

Expressions for y and z in terms of the h-parameters have analogous matrix derivations. Equations 19.47 and 19.45 are listed in table 19.1, which shows the relationships among all the two-port parameters.

Impedance and Gain Calculations of Terminated Two-Ports

As in the case of both the z- and the y-parameters, computation of the input and output admittance is relatively straightforward. We focus on figure 19.32 in these calculations and in the forthcoming calculations of gain.

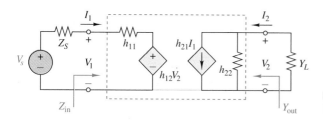

Figure 19.32 Doubly terminated hybrid equivalent circuit of two-port.

Our first calculation is that of the input impedance, $Z_{\text{in}} = V_1/I_1$. From the right half of figure 19.32, in which $h_{21}I_1 = -(h_{22} + Y_L)V_2$, it follows immediately that

$$V_2 = -\frac{h_{21}}{h_{22} + Y_L}I_1. \tag{19.48}$$

Similarly, from the left-hand side of figure 19.32, $V_1 = h_{11}I_1 + h_{12}V_2$ implies that the hybrid parameter form of the **input impedance** is

$$Z_{\text{in}} = \frac{V_1}{I_1} = h_{11} - \frac{h_{12}h_{21}}{h_{22} + Y_L}. \tag{19.49}$$

Exercise. Show that the **output admittance** is

$$Y_{\text{out}} = \frac{V_2}{I_2} = h_{22} - \frac{h_{12}h_{21}}{h_{11} + Z_s}. \tag{19.50}$$

Knowledge of the input and output impedances and admittances permits us to determine various gains of a two-port modeled by hybrid parameters. For example, consider again the right half of figure 19.32, which implies equation 19.48. Since the input impedance is known from equation 19.49, it follows that

$$I_1 = \frac{1}{Z_{\text{in}}}V_1.$$

Substituting into equation 19.48 and dividing by V_1 produces the **voltage gain** formula

$$\frac{V_2}{V_1} = -\frac{h_{21}}{Z_{\text{in}}(Y_L + h_{22})}. \tag{19.51}$$

Since voltage division implies that

$$\frac{V_1}{V_s} = \frac{Z_{\text{in}}}{Z_{\text{in}} + Z_S}, \tag{19.52}$$

the overall voltage gain is given by the product of equations 19.51 and 19.52, i.e.,

$$G_V = \frac{V_2}{V_s} = \frac{V_2}{V_1}\frac{V_1}{V_s} = -\frac{1}{(Z_{\text{in}} + Z_S)}\frac{h_{21}}{(Y_L + h_{22})}. \tag{19.53}$$

Exercise. Compute the current gain I_{out}/I_1 and the power gain P_L/P_S, where P_L is the power absorbed by the load and P_S is the power delivered by the source.

Exercise. Compute the h-parameters of each of the two stages of the amplifier in example 19.12, as illustrated in figure 19.26. Compute the amplifier gain and output impedance in terms of the h-parameters. Your answers should be the same as calculated in the example.

The inverse of the h-parameters are the g-**parameters**. These parameters are developed in problem 37.

8. Generalized Two-Port Parameters

The inputs and outputs of the two-port parameters studied thus far have similar patterns. The input variables are voltages, currents, or some hybrid, i.e., they are electrical quantities, one each from port 1 and port 2. In addition, the patterns of subscripts in all of the input and output impedance calculations are identical. This suggests defining generalized two-port parameters for a two-port, as depicted in figure 19.33. The word *immittance* is sometimes used in this context; it is a combination of the words *im*pedance and ad*mittance*, but is just a word for defining a voltage-current relationship without specifying whether it is an impedance, an admittance, or some hybrid thereof. The physical dimensions dictated by the context of the problem determine the units of each generalized two-port/immittance parameter, denoted γ_{ij}.

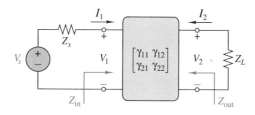

Figure 19.33 Two-port characterized by generalized two-port parameters.

In terms of **input immittance** calculations, it is straightforward to show that

$$\text{Input immittance} = \gamma_{11} - \frac{\gamma_{12}\gamma_{21}}{\gamma_{22} + [\text{load immittance}]},$$

where the load immittance units and those of γ_{22} must coincide. Also, the units of the input immittance must be consistent with those of γ_{11}. Likewise, the **output immittance** is

$$\text{Output immittance} = \gamma_{22} - \frac{\gamma_{12}\gamma_{21}}{\gamma_{11} + [\text{input immittance}]},$$

where, again, the input immittance and γ_{11} must have the same units. As mentioned earlier, the pattern of subscripts is identical to that appearing in the impedance and admittance calculation of the other two-port parameters. More advanced texts on circuits describe other properties of generalized two-port parameters.

9. TRANSMISSION PARAMETERS

Historically, **transmission** or **t-parameters** were first used by power system engineers for transmission line analysis and are still so used today. They are sometimes called **ABCD parameters**.

TRANSMISSION PARAMETERS

For the t-parameter representation, V_1 and I_1 are outputs while V_2 and $-I_2$ are inputs. The matrix relationship is

$$\begin{bmatrix} V_1 \\ I_1 \end{bmatrix} = \begin{bmatrix} t_{11} & t_{12} \\ t_{21} & t_{22} \end{bmatrix} \begin{bmatrix} V_2 \\ -I_2 \end{bmatrix}, \tag{19.54}$$

with the matrix $T = [t_{ij}]$ known as the **t-parameter matrix**. As with the y-, z-, and h-parameters the entries t_{ij} are defined in the standard way:

$$t_{11} = \frac{V_1}{V_2}\bigg|_{I_2=0}, \qquad t_{12} = \frac{V_1}{-I_2}\bigg|_{V_2=0},$$

$$t_{21} = \frac{I_1}{V_2}\bigg|_{I_2=0}, \qquad t_{22} = \frac{I_1}{-I_2}\bigg|_{V_2=0}. \tag{19.55}$$

The matrix equation 19.54 leads directly to these relationships by setting the appropriate quantity, I_2 or V_2, to zero. The units for t_{12} and t_{21} are Ω and $\mho$, respectively.

In computing a single t_{ij} with equation 19.55, some care must be exercised in exciting the circuit. By definition,

$$t_{11} = \frac{V_1}{V_2}\bigg|_{I_2=0}.$$

The ordinary interpretation of this equation is to apply an input V_2 and find an output V_1 under the condition that $I_2 = 0$. Then, t_{11} is the ratio of the Laplace transform of the response V_1 to that of the input V_2. The predicament is that an independent voltage source for V_2 causes a current I_2 to flow. Hence, it is better to consider the slightly modified formula

$$t_{11} = \frac{1}{\dfrac{V_2}{V_1}\bigg|_{I_2=0}}.$$

The ordinary interpretation of this formula is to excite port 1 by V_1 with port 2 open-circuited which forces $I_2 = 0$. It is then straightforward to calculate t_{11}. Similar interpretations must be made of the other defining formulas in equation 19.55.

A simple example now illustrates t-parameter calculations.

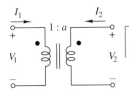

Figure 19.34
Simple transformer circuit for example 19.17.

◆ **EXAMPLE 19.17**

Consider again the ideal transformer circuit of figure 19.34. Here, $V_2 = aV_1$ and $I_1 = a(-I_2)$. This leads to the t-parameter matrix

$$\begin{bmatrix} V_1 \\ I_1 \end{bmatrix} = \begin{bmatrix} \dfrac{1}{a} & 0 \\ 0 & a \end{bmatrix} \begin{bmatrix} V_2 \\ -I_2 \end{bmatrix}.$$

Input and output impedance calculations for terminated two-ports modeled by t-parameters do not follow the pattern of the generalized two-port parameters discussed in the previous section. Nevertheless, for the two-port figure 19.35, the **input impedance** is

$$Z_{\text{in}} = \frac{t_{11} Z_L + t_{12}}{t_{21} Z_L + t_{22}}, \tag{19.56}$$

and the **output impedance** is

$$Z_{\text{out}} = \frac{t_{22} Z_S + t_{12}}{t_{21} Z_S + t_{11}}. \tag{19.57}$$

A derivation of these results is left as an exercise in the problem section.

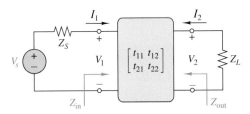

Figure 19.35 Terminated two-port modeled by t-parameters.

One of the most important characteristics of two-port parameters is the ease with which one can use them to determine the overall t-parameters of cascaded two-ports. To see this, consider figure 19.36, which is a cascade of two two-ports.

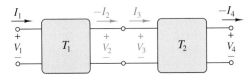

Figure 19.36 Cascade of two two-ports modeled by t-parameters.

From the definition of the t-parameters for each two-port,

$$\begin{bmatrix} V_1 \\ I_1 \end{bmatrix} = T_1 \begin{bmatrix} V_2 \\ -I_2 \end{bmatrix} \quad \text{and} \quad \begin{bmatrix} V_3 \\ I_3 \end{bmatrix} = T_2 \begin{bmatrix} V_4 \\ -I_4 \end{bmatrix}.$$

But since $V_2 = V_3$ and $-I_2 = I_3$, it follows that

$$\begin{bmatrix} V_1 \\ I_1 \end{bmatrix} = T_1 \begin{bmatrix} V_2 \\ -I_2 \end{bmatrix} \quad \text{and} \quad \begin{bmatrix} V_2 \\ -I_2 \end{bmatrix} = T_2 \begin{bmatrix} V_4 \\ -I_4 \end{bmatrix},$$

implying that the t-parameter matrix of the cascade of figure 19.36 is simply $T_{\text{new}} = T_1 T_2$.

We end this section with table 19.1 which specifies the interrelationships among all the parameters studied thus far.

	z-Parameters	y-Parameters	h-Parameters	t-Parameters
z-Parameters	$\begin{bmatrix} z_{11} & z_{12} \\ z_{21} & z_{22} \end{bmatrix}$	$\begin{bmatrix} \dfrac{y_{22}}{\Delta y} & \dfrac{-y_{12}}{\Delta y} \\ \dfrac{-y_{21}}{\Delta y} & \dfrac{y_{11}}{\Delta y} \end{bmatrix}$	$\begin{bmatrix} \dfrac{\Delta h}{h_{22}} & \dfrac{h_{12}}{h_{22}} \\ \dfrac{-h_{21}}{h_{22}} & \dfrac{1}{h_{22}} \end{bmatrix}$	$\begin{bmatrix} \dfrac{t_{11}}{t_{21}} & \dfrac{\Delta t}{t_{21}} \\ \dfrac{1}{t_{21}} & \dfrac{t_{22}}{t_{21}} \end{bmatrix}$
y-Parameters	$\begin{bmatrix} \dfrac{z_{22}}{\Delta z} & \dfrac{-z_{12}}{\Delta z} \\ \dfrac{-z_{21}}{\Delta z} & \dfrac{z_{11}}{\Delta z} \end{bmatrix}$	$\begin{bmatrix} y_{11} & y_{12} \\ y_{21} & y_{22} \end{bmatrix}$	$\begin{bmatrix} \dfrac{1}{h_{11}} & \dfrac{-h_{12}}{h_{11}} \\ \dfrac{h_{21}}{h_{11}} & \dfrac{\Delta h}{h_{11}} \end{bmatrix}$	$\begin{bmatrix} \dfrac{t_{22}}{t_{12}} & \dfrac{-\Delta t}{t_{12}} \\ \dfrac{-1}{t_{12}} & \dfrac{t_{11}}{t_{12}} \end{bmatrix}$
h-Parameters	$\begin{bmatrix} \dfrac{\Delta z}{z_{22}} & \dfrac{z_{12}}{z_{22}} \\ \dfrac{-z_{21}}{z_{22}} & \dfrac{1}{z_{22}} \end{bmatrix}$	$\begin{bmatrix} \dfrac{1}{y_{11}} & \dfrac{-y_{12}}{y_{11}} \\ \dfrac{y_{21}}{y_{11}} & \dfrac{\Delta y}{y_{11}} \end{bmatrix}$	$\begin{bmatrix} h_{11} & h_{12} \\ h_{21} & h_{22} \end{bmatrix}$	$\begin{bmatrix} \dfrac{t_{12}}{t_{22}} & \dfrac{\Delta t}{t_{22}} \\ \dfrac{-1}{t_{22}} & \dfrac{t_{21}}{t_{22}} \end{bmatrix}$
t-Parameters	$\begin{bmatrix} \dfrac{z_{11}}{z_{21}} & \dfrac{\Delta z}{z_{21}} \\ \dfrac{1}{z_{21}} & \dfrac{z_{22}}{z_{21}} \end{bmatrix}$	$\begin{bmatrix} \dfrac{-y_{22}}{y_{21}} & \dfrac{-1}{y_{21}} \\ \dfrac{-\Delta y}{y_{21}} & \dfrac{-y_{11}}{y_{21}} \end{bmatrix}$	$\begin{bmatrix} \dfrac{-\Delta h}{h_{21}} & \dfrac{-h_{11}}{h_{21}} \\ \dfrac{-h_{22}}{h_{21}} & \dfrac{-1}{h_{21}} \end{bmatrix}$	$\begin{bmatrix} t_{11} & t_{12} \\ t_{21} & t_{22} \end{bmatrix}$

10. RECIPROCITY

Writing node equations for an ordinary RLC circuit leads to a matrix equation having the form

$$\begin{bmatrix} I_1 \\ \vdots \\ I_n \end{bmatrix} = \begin{bmatrix} y_{11} & \cdots & y_{1n} \\ \vdots & \ddots & \vdots \\ y_{n1} & \cdots & y_{nn} \end{bmatrix} \begin{bmatrix} V_1 \\ \vdots \\ V_n \end{bmatrix} = Y \begin{bmatrix} V_1 \\ \vdots \\ V_n \end{bmatrix}.$$

Oftentimes, the **node admittance matrix** $Y = [y_{ij}]$ is symmetric, i.e., $y_{ij} = y_{ji}$ for $i \neq j$. How extensive is this symmetry? Does it imply anything special about the network? *Circuits that have such a symmetry in their nodal equation or loop equation representation are defined as reciprocal.* A simple example will motivate the notion of reciprocity and some very general associated circuit behaviors.

EXAMPLE 19.18

The two-port configurations of figures 19.37 and 19.38 have z-parameter equations given by

$$\begin{bmatrix} V_1 \\ V_2 \end{bmatrix} = \begin{bmatrix} R_1 + Ls & Ls \\ Ls & R_2 + Ls \end{bmatrix} \begin{bmatrix} I_1 \\ I_2 \end{bmatrix} = \begin{bmatrix} z_{11} & z_{12} \\ z_{21} & z_{22} \end{bmatrix} \begin{bmatrix} I_1 \\ I_2 \end{bmatrix}.$$

This example will show that the open-circuit zero-state response V_{1a} of figure 19.37a is the same as the open-circuit zero-state response V_{2b} of figure 19.37b. (In order to link

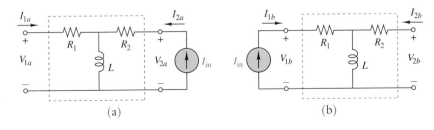

Figure 19.37 Simple *T*-network excited by same current source at secondary in (a) and at primary in (b).

(a)　　　　　(b)

each response to a particular circuit, we have labeled the circuit voltages with subscripts a and b.) In addition, the example will show that the short-circuit zero-state response I_{1a} of figure 19.38a coincides with the short-circuit zero-state response I_{2b} of the circuit in figure 19.38b.

Part 1. *Show that* $V_{1a} = V_{2b}$ *in figure 19.37.* In regard to figure 19.37a, the open-circuited primary forces $I_{1a} = 0$, in which case

$$V_{1a} = Ls\,I_{2a} = Ls\,I_{in}. \tag{19.58}$$

Switching the current source to the primary, as in figure 19.37b, leads to a symmetric relationship: With the secondary open-circuited, $I_{2b} = 0$ and

$$V_{2b} = Ls\,I_{1b} = Ls\,I_{in}. \tag{19.59}$$

Equations 19.58 and 19.59 imply that $V_{1a} = V_{2b}$. This equality occurs because of the symmetry of the z-parameters.

Part 2. *Show that* $I_{1a} = I_{2b}$ *in figure 19.38.* The symmetry of the voltage relationships of figure 19.37 has a duality with respect to voltage sources inducing identical short-circuit zero-state current responses. Figure 19.38 illustrates the idea. Here, a voltage source V_{in} attached to the secondary as in part (a) of the figure, or to the primary as in part (b), induces zero-state short-circuit current responses I_{1a} of figure 19.38a and I_{2b} of figure 19.38b, respectively. It turns out that $I_{1a} = I_{2b}$. To see that this is so, first consider figure 19.38a. Here, $V_{1a} = 0$, and from the impedance parameters

$$I_{2a} = -\frac{R_1 + Ls}{Ls}I_{1a}.$$

Again using the impedance parameter equations with $V_{2a} = V_{in}$, some algebraic simplification leads to

$$I_{1a} = -\frac{Ls}{(R_1 + R_2)Ls + R_1 R_2}V_{in}. \tag{19.60}$$

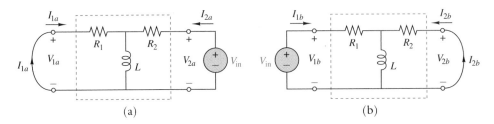

Figure 19.38 Simple *T*-network with symmetric terminations, (a) and (b), for showing equality of short-circuit current responses I_{1a} and I_{2b}.

Switching the positions of the voltage source and short circuit from their positions in figure 19.38a to those in figure 19.38b leads to $V_{1b} = V_{in}$, $V_{2b} = 0$. By symmetry

$$I_{2b} = -\frac{Ls}{(R_1 + R_2)Ls + R_1 R_2}V_{in}. \tag{19.61}$$

The right-hand sides of Equations 19.60 and 19.61 coincide, i.e., $I_{1a} = I_{2b}$. This symmetry between the zero-state responses of the circuits of figures 19.38a and 19.38b also makes sense because the symmetry of the z-parameter matrix induces a symmetry in the y-parameter matrix. Specifically,

$$\begin{bmatrix} I_1 \\ I_2 \end{bmatrix} = \begin{bmatrix} R_1 + Ls & Ls \\ Ls & R_2 + Ls \end{bmatrix}^{-1} \begin{bmatrix} V_1 \\ V_2 \end{bmatrix}$$

$$= \frac{1}{(R_1 + R_2)Ls + R_1 R_2} \begin{bmatrix} R_2 + Ls & Ls \\ Ls & R_1 + Ls \end{bmatrix} \begin{bmatrix} V_1 \\ V_2 \end{bmatrix}.$$

Equations 19.60 and 19.61 follow directly from this relationship.

In the preceding example, two open-circuit voltage responses induced by a current source coincided. Also, two short-circuit current responses induced by a voltage source were equal. In general, if the two-ports were terminated with load or source resistances, the responses would not have coincided. Suppose, for example, that $R_1 = 1 \, \Omega$, $R_2 = 2 \, \Omega$, and $L = 1$ H in figure 19.37. Suppose also that the two-port in figure 19.37a is terminated at the primary by a 1-Ω resistor. It then follows that

$$V_{1a} = (s + 1)I_{1a} + sI_{in}$$

and $I_{1a} = -V_{1a}$, in which case

$$V_{1a} = \frac{s}{s + 2} I_{in}. \tag{19.62}$$

With respect to figure 19.37b, one concludes that if the two-port is terminated with a 1-Ω load resistance,

$$V_{2b} = sI_{in} - (s + 2)V_{2b},$$

or equivalently,

$$V_{2b} = \frac{s}{s + 3} I_{in}. \tag{19.63}$$

Equations 19.62 and 19.63 differ. Therefore, adding a load or source impedance to the two-port causes the transfer functions to differ.

Would the presence of internal transformers and capacitors in the foregoing examples change the relationships suggested by the examples? No, because the z-parameters would remain symmetric. On the other hand, if dependent sources were added, the relationships in general would change. The symmetry of the z-parameters and y-parameters is typically lost when dependent sources are present.

The preceding illustrations motivate some general statements about circuits. First, *a circuit containing R's, L's, C's and transformers, but no dependent sources or independent sources, is a **reciprocal network**.* It is possible to show, using either node or loop analysis, that a reciprocal two-port network has the property that

$$z_{12} = z_{21},$$

$$y_{12} = y_{21},$$

$$h_{12} = -h_{21},$$

and

$$t_{11}t_{22} - t_{12}t_{21} = 1.$$

Conversely, any two-port that has parameters satisfying these conditions is said to be reciprocal.

Exercise. Recall that

$$\begin{bmatrix} t_{11} & t_{12} \\ t_{21} & t_{22} \end{bmatrix} = \begin{bmatrix} \dfrac{z_{11}}{z_{21}} & \dfrac{\Delta z}{z_{21}} \\[2mm] \dfrac{1}{z_{21}} & \dfrac{z_{22}}{z_{21}} \end{bmatrix}.$$

Show that if $z_{12} = z_{21}$, then $\Delta t = 1$. Then show the converse, i.e., if $\Delta t = 1$, then $z_{12} = z_{21}$.

Proving that a reciprocal two-port has, for example, symmetric z-parameters is straightforward. If the two-port is reciprocal, then the circuit that determines it has a symmetric loop equation matrix

$$\begin{bmatrix} V_1 \\ V_2 \\ \cdots \\ V_3 \\ \vdots \\ V_n \end{bmatrix} = \begin{bmatrix} z_{11} & z_{12} & \vdots & z_{13} & \cdots & z_{1n} \\ z_{12} & z_{11} & \vdots & z_{23} & \cdots & z_{2n} \\ \cdots & \cdots & \vdots & \cdots & \cdots & \cdots \\ z_{13} & z_{23} & \vdots & z_{33} & \cdots & z_{3n} \\ \vdots & \vdots & \vdots & \vdots & \ddots & \vdots \\ z_{1n} & z_{2n} & \vdots & z_{3n} & \cdots & z_{nn} \end{bmatrix} \begin{bmatrix} I_1 \\ I_2 \\ \cdots \\ I_3 \\ \vdots \\ I_n \end{bmatrix}.$$

Solving for $[V_1 \quad V_2]^t$ by the method of matrix partitioning yields

$$\begin{bmatrix} V_1 \\ V_2 \end{bmatrix} = \begin{bmatrix} W_{11} - W_{12} W_{22}^{-1} W_{21} \end{bmatrix} \begin{bmatrix} I_1 \\ I_2 \end{bmatrix}$$

$$= \begin{bmatrix} W_{11} - W_{12} W_{22}^{-1} W_{12}^t \end{bmatrix} \begin{bmatrix} I_1 \\ I_2 \end{bmatrix},$$

where the W_{ij} matrices are defined in the obvious way. W_{11} is symmetric. Since the inverse of the symmetric matrix W_{22} is symmetric, and since the sum of two symmetric matrices is symmetric, the resulting z-parameters are also symmetric. The symmetry of the y-parameters follows by the symmetry of the inverse of a symmetric matrix. The h-parameter relationship $h_{12} = -h_{21}$ follows by a similar argument.

Also, as suggested by example 19.18, any reciprocal two-port satisfies each of the following behaviors.

Reciprocity, Interpretation 1. Consider a reciprocal two-port N. If the voltage inputs V_{in} are the same as in figures 19.39a and 19.39b, then, by reciprocity, the zero-state short-circuit responses I_{2a} and I_{1b} coincide.

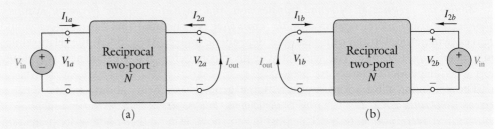

Figure 19.39 Equivalences of short-circuit zero-state responses for reciprocal networks having the symmetric terminations shown in (a) and (b).

Observe that the configuration of figure 19.39a implies that

$$\left. \frac{I_{out}}{V_{in}} \right|_{V_{2a}=0} = \left. \frac{I_{2a}}{V_{1a}} \right|_{V_{2a}=0} = y_{21}$$

and that the configuration of figure 19.39b implies that

$$\left. \frac{I_{out}}{V_{in}} \right|_{V_{1b}=0} = \left. \frac{I_{1b}}{V_{2b}} \right|_{V_{1b}=0} = y_{12}.$$

Thus, it must be the case that $y_{12} = y_{21}$, i.e., we must have symmetric y-parameters.

Reciprocity, Interpretation 2. Consider again a reciprocal two-port N. As illustrated in figure 19.40, if I_{in} is the same in both of figures 19.40a and 19.40b, then the open-circuit zero-state responses V_{2a} and V_{1b} coincide.

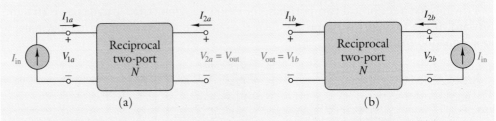

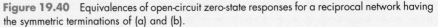

Figure 19.40 Equivalences of open-circuit zero-state responses for a reciprocal network having the symmetric terminations of (a) and (b).

Exercise. Assuming that the second interpretation of reciprocity is true, show that it must follow that $z_{12} = z_{21}$, i.e., the z-parameters are symmetric.

Reciprocity, Interpretation 3. Consider yet again a reciprocal two-port N. As illustrated in figure 19.41, if I_{in} leads to a zero-state response denoted "$-I_{2a}$," as per figure 19.41a, then, under the condition that the Laplace transforms of V_{in} and I_{in} coincide (i.e., $V_{in}(s) = I_{in}(s)$), the zero-state response V_{1b} in figure 19.41b also coincides with "$-I_{2a}$," i.e., V_{1b} and $-I_{2a}$ have the same numerical value, although different units.

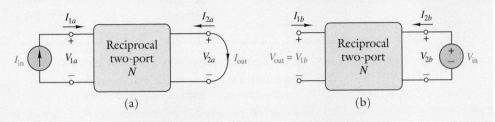

(a)   (b)

Figure 19.41 Reciprocal two-ports for illustrating h-parameter antisymmetry.

Exercise. Show that if the third interpretation of reciprocity is assumed true, then it must follow that $h_{12} = -h_{21}$.

These statements of reciprocity have rigorous proofs that are beyond the scope of the text.

Lastly, a two-port that is reciprocal has an equivalent circuit representation with no dependent sources. For example, suppose a reciprocal two-port has a z-parameter representation

$$V_1 = z_{11}I_1 + z_{12}I_2$$

and

$$V_2 = z_{12}I_1 + z_{22}I_2,$$

in which case

$$V_1 = (z_{11} - z_{12})I_1 + z_{12}(I_1 + I_2) \tag{19.64a}$$

and

$$V_2 = z_{12}(I_1 + I_2) + (z_{22} - z_{12})I_2. \tag{19.64b}$$

Equations 19.64 have the so-called **T-equivalent circuit** interpretation given by figure 19.42.

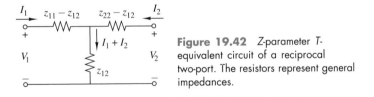

Figure 19.42 *Z*-parameter *T*-equivalent circuit of a reciprocal two-port. The resistors represent general impedances.

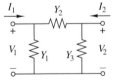

Figure 19.43 *Y*-parameter π-equivalent circuit of a reciprocal two-port. The resistors represent general impedances.

Exercise. Suppose the two-port of figure 19.43 is reciprocal and modeled by y-parameter equations. Determine Y_1, Y_2, and Y_3 in terms of the y-parameters, y_{11}, y_{12}, y_{22}.

ANSWERS: $-y_{12}$, $y_{11} + y_{12}$, $y_{22} + y_{12}$.

SUMMARY

This chapter has presented a unified setting for one-port analysis and has extended that analysis to two-ports. Two-ports are common to numerous real-world systems, including the utility grid that delivers power to a home from a generating facility through a transmission network. Another representative two-port is a telephone system, which delivers voice information from the phone through an interconnect and transmission network to a receiver. The characterization of the two-port for such systems was through input-output properties. Four sets of characterizing parameters were developed: impedance, or z-parameters; admittance, or y-parameters; hybrid, or h-parameters; and transmission, or t-parameters. To analyze various aspects of a system characterized by a two-port, formulas were derived for computing the input impedance/admittance, the output impedance/admittance, the voltage gain, etc. Quantities such as voltage and power gain are very important aspects of amplifier analysis and design, as illustrated in example 19.12, which depicts a two-stage transistor amplifier configuration. Although that example utilized the context of z-parameters, the more usual context for transistor amplifier design is h-parameters. In the case of transmission parameters, formulas for determining the t-parameters of cascades of two-ports were also developed. Finally, the chapter introduced and interpreted the notion of reciprocity in terms of the different two-port parameters. Reciprocal circuits generally contain only R's, L's, C's, and transformers. Under certain restricted conditions, a reciprocal network may contain a dependent source. (See, for example, problem 42.)

TERMS AND CONCEPTS

Admittance, or y-parameters: descriptive two-port parameters in which the port currents are functions of the port voltages.

g-parameters: the inverse of h-parameters.

Hybrid parameters or h-parameters: descriptive two-port parameters in which V_1 and I_2 are expressed as functions of I_1 and V_2.

Impedance, or z-parameters: descriptive two-port parameters in which the port voltages, V_1 and V_2, are functions of the port currents, I_1 and I_2.

Input admittance: the admittance seen at port 1 of a (possibly terminated) two-port.

Input immittance: general term for either the input impedance or input admittance of a (possibly terminated) two-port.

Input impedance: the impedance seen at port 1 of a (possibly terminated) two-port.

Matrix partitioning: the partitioning of a matrix equation into pairs of matrix equations to obtain a simplified solution in terms of the partitioned submatrices.

Norton equivalent of one-port: a current source in parallel with the Thévenin impedance.

Open-circuit impedance parameters: the impedance or z-parameters.

Open-circuit output admittance: the hybrid parameter, h_{22}.

Output admittance: the admittance seen at port 2 of a two-port possibly terminated by a source impedance.

Output immittance: a general term for either the output admittance or output impedance of a two-port.

Output impedance: the impedance seen at port 2 of a two-port possibly terminated by a source impedance.

Partitioned matrix: a matrix that is partitioned into submatrices for easier solution of original equation.

π-equivalent circuit: equivalent circuit of a reciprocal two-port containing three general impedances in the form of π, as in figure 19.43.

Primary: in this chapter, shorthand for port 1 of a two-port.

Reciprocal network: a network whose node equations or loop equations have a symmetric coefficient matrix.

Reverse open-circuit voltage gain: the hybrid parameter h_{12}.

Secondary: in this chapter, shorthand for port 2 of a two-port.

Short-circuit admittance parameters: the admittance, or y-parameters.

Short-circuit forward current gain: the hybrid parameter h_{21}.

Short-circuit input impedance: the hybrid parameter h_{11}.

T-equivalent circuit: equivalent circuit of a reciprocal two-port having three general impedances in a T-shape, as per figure 19.42.

Terminated 2-port: a two-port attached to a load impedance and to a source with, in general, a nonzero source impedance.

Thévenin equivalent of a one-port: voltage source in series with the Thévenin impedance.

Transmission, or t-parameters: parameters in which V_1 and I_1 are expressed as functions of V_2 and $-I_2$.

Two-dependent source equivalent circuit: equivalent circuit for a two-port containing impedances/admittances and two dependent sources.

PROBLEMS

One-Ports

1. The loop equations describing the one-port of figure P19.1 are given by

$$\begin{bmatrix} V_1 \\ 0 \\ 0 \end{bmatrix} = \begin{bmatrix} 10a & 1 & -a \\ a & 0.5 & 0 \\ -1 & 0 & 0.5 \end{bmatrix} \begin{bmatrix} I_1 \\ I_2 \\ I_3 \end{bmatrix}$$

Determine the input impedance of the one-port as a function of a. (*Hint*: Use Cramer's rule.)

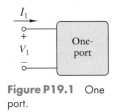

Figure P19.1 One port.

2. Consider the circuit of figure P19.2.
   (a) Determine the output impedance $Z_{\text{out}}(s)$ as a function of the turns ratio b and the resistance R.

(b) If $i_{in}(t) = \cos(10t + 45°)$ A, determine the Norton equivalent circuit with the Norton source represented as a phasor. What is the Thévenin equivalent?

(c) If $R = 25$ Ω, $b = 2$, and the output is terminated in a parallel LC, determine L and C so that the bandwidth is 10 and the resonant frequency $\omega_r = 5$ rad/sec.

(d) Determine the zero-state response with the preceding input and element values. Identify the steady-state part. If the input cosine frequency were changed to 100 rad/sec, what would happen to the steady-state magnitude?

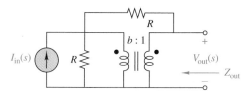

Figure P19.2   Circuit for problem 2.

3. (a) For the circuit of figure P19.3a, determine the Thévenin equivalent circuit, using the method of matrix partitioning.

(b) For the circuit of figure P19.3b, determine the Norton equivalent, using the method of matrix partitioning and node analysis.

(c) Determine the impulse and step responses of the circuit of figure P19.3b.

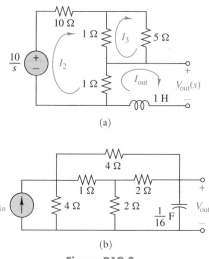

(a)

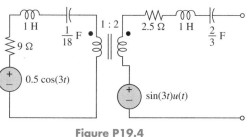

(b)

Figure P19.3

4. Determine the Thévenin equivalent circuit for the circuit of figure P19.4. (*Hint*: Add a voltage source V_s to the right side, and write loop equations.)

Figure P19.4

y-Parameters

5. (a) Determine the input admittance $Y_{in}(s)$ of the circuit in figure P19.5.

(b) If $i_{out}(t) = 5u(t)$ A, determine $v_1(t)$.

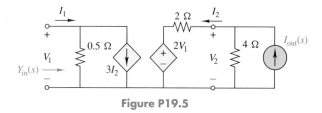

Figure P19.5

6. For the two-port configuration of figure P19.6, with *y*-parameters as indicated, the voltage gain V_1/V_s is 0.5. Determine y_{11}.

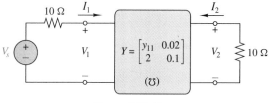

Figure P19.6

7. In a laboratory, you are asked to determine the admittance parameters of a circuit. You decide to short-circuit port 2, place a unit step current source at port 1, and measure the port voltage $v_1(t) = [1 - e^{-4t}]u(t)$ V, and the port 2 current $i_2 = -e^{-3t}u(t)$ A. Knowing that this is sufficient to determine at most two of the parameters, you then break the short circuit and terminate port 2 by a 1-Ω resistor and measure the new step responses as $v_1(t) = [1 - e^{-4t} + te^{-4t}]u(t)$ V and $i_2 = -e^{-7t}u(t)$ A.

(a) Determine the *y*-parameters of the two-port.

(b) If the two-port is terminated in the 1-Ω resistor, determine the input impedance "seen" by the current source.

(c) If the two-port is again terminated in a 1-Ω resistor, determine the steady-state magnitude of the gain, $|V_2/I_1|$, when the input current source is $i_1(t) = \cos(t)u(t)$.

8. (a) Determine the *y*-parameters for the circuit of figure P19.8.

(b) Determine the voltage gain $V_2(s)/V_s(s)$.

(c) Determine the impulse and step responses of the terminated network, assuming that $v_2(t)$ is the output.

(d) If $v_s(t) = 10 \sin(2t)u(t)$ V, determine the steady-state and transient responses, assuming that $v_2(t)$ is the output.

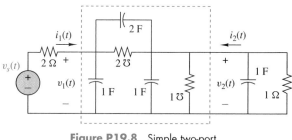

Figure P19.8   Simple two-port.

9. For the circuit of figure P19.9, determine $Z_{in}(s)$ and $Y_{out}(s)$.

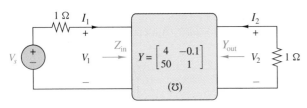

Figure P19.9 Doubly terminated two-port.

10. Figure P19.10 represents a two-stage amplifier with the indicated y-parameters. Find the voltage gain $G_v = V_{out}/V_{in}$. Although the solution may be obtained by solving a set of six simultaneous equations, a much better method that gives more insight into the performance of the amplifier and that works for any number of stages is to proceed as follows:

(a) Draw the equivalent circuit for the two-stage amplifier.

(b) Using the formula for Y_{in}, find the input admittances of the stages successively, starting from the load end. Be sure to account for the intermediate 2-kΩ resistor.

(c) Find the voltage gains of the stages successively, starting from the source end. Use this information to find the overall voltage gain.

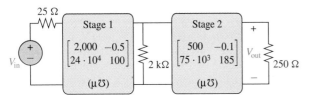

Figure P19.10 Cascaded two-port configuration for problem 10.

z-Parameters

11. Determine the z-parameters of the circuit in figure P19.11.

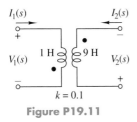

Figure P19.11

12. (a) Determine the impedance parameters for the circuit of figure P19.12.

(b) Determine the admittance parameters of the circuit if they exist.

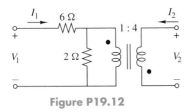

Figure P19.12

13. In the circuit of figure P19.13, determine V_2.

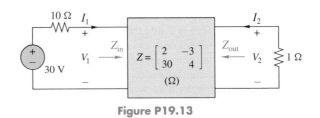

Figure P19.13

14. Consider the two-port configuration of figure P19.14, having z-parameters as indicated. Determine the value of b that yields maximum power transfer from the output of the two-port to the load.

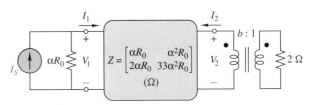

Figure P19.14 Cascaded two-port configuration for problem 14.

15. An active two-port built with resistors and ideal operational amplifiers has the following z-parameters: $z_{11} = 0$, $z_{12} = 1,000 \ \Omega$, $z_{21} = -1,000 \ \Omega$, $z_{22} = 0$. The two-port is connected as shown in figure P19.15.

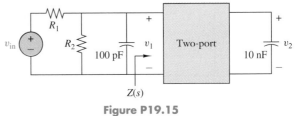

Figure P19.15

(a) Find $Z(s)$, the impedance looking into port 1. Is the result a surprise to you?

(b) If $R_1 = R_2 = 100 \ k\Omega$ and the output is V_1, find ω_m and the bandwidth of the circuit.

(c) If $R_1 = R_2 = 10 \ k\Omega$, and the output is $v_1(t)$, find the impulse response $h(t)$ of the circuit.

(d) Using the property that the z-matrix is the inverse of the y-matrix (and vice versa), find the y-parameters of the two-port.

(e) If $R_1 = R_2 = 10 \ k\Omega$ and v_2 is considered as the output, what is the impulse response? (*Hint*: Make use of the results of parts (c) and (d) and the two-dependent source equivalent circuit in terms of the y-parameters.)

Remark: The two-port shown in figure P19.15 is called a *gyrator*. A gyrator terminated in a capacitor behaves like an inductor at its input terminals. This is one of many ways to eliminate inductors in the design of an "active filter."

16. The purpose of this problem is to show that z-parameters may be used to establish several equivalent circuits of a pair of coupled inductors. Each equivalent circuit consists of one ideal transformer and two inductances. The analysis of a coupled circuit with the use of such equivalent circuits is very often more illuminating than writing and solving simultaneous equations.

(a) Find the z-parameters of the two-port N_1 of figure P19.16a.

(b) Show that two-port N_2 of figure P19.16b has the same z-parameters as N_1.

(c) Show that two-port N_3 of figure P19.16c has the same z-parameters as N_1.

(d) Use equivalent circuit N_2 of part (b) and the properties of an ideal transformer to find ω_m, the bandwidth, and $|V_o/V_{in}|_{max}$ for the coupled tuned circuit of figure P19.16d.

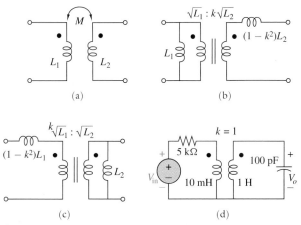

(a) (b)

(c) (d)

Figure P19.16 (a) Network N_1. (b) Network N_2 containing ideal transformer. (c) Network N_3 containing ideal transformer. (d) Coupled tuned circuit.

(*Note*: This network can also be solved by writing simultaneous equations.)

h-Parameters

17. Determine the h-parameters for the two-port of figure P19.17.

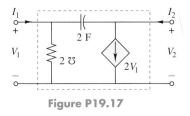

Figure P19.17

18. The h-parameters of the two-port of figure P19.18 are $h_{11} = 1$ kΩ, $h_{21} = 20$, $h_{12} = 0$, and $h_{22} = 50$ μ℧.

(a) Determine $|V_o/V_s|$.

(b) Suppose a capacitance of 0.1 μF is connected across port 1. Determine the impulse response of the network.

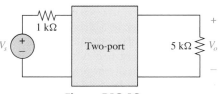

Figure P19.18

19. Consider the amplifier network of figure P19.19. Stage 1 is a common-emitter stage that drives stage 2, the common-collector stage. Such an amplifier combination might be used to drive a low-impedance load. The h-parameters of the two stages, with h_{11} in kΩ and h_{22} in m℧, are

$$H_1 = \begin{bmatrix} 2 & 0 \\ 50 & 0.05 \end{bmatrix} \quad \text{and} \quad H_2 = \begin{bmatrix} 1 & 0.966 \\ -51 & 0.8 \end{bmatrix}.$$

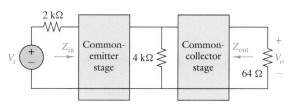

Figure P19.19 Coupled amplifier configuration.

(a) Find the input impedances of the stages successively, starting from the load end.

(b) Find the output impedance of the amplifier of the stages successively, starting from the source end.

(c) Find the voltage gains of the two stages.

(d) Find the overall voltage gain V_o/V_s.

(e) A 1-μF capacitor is inserted in series with the 2-kΩ resistance to prevent dc voltage in the power supply (not shown in the figure) from entering the signal source. Because of this capacitance, low-frequency signals will be amplified less. Determine the frequency (in Hz) at which the magnitude of V_o/V_s is 0.707 of its maximum value calculated in part (d). (*Hint*: Analyze the simple circuit consisting of V_s, R_s, C, and Z_{in}.)

20. (a) Repeat parts (a) through (d) of problem 19 for the two-stage amplifier shown in figure P19.20, assuming that the h-parameters of the two stages, with h_{11} in kΩ and h_{22} in millimhos, are

$$H_1 = \begin{bmatrix} 1 & 0 \\ 50 & 0.1 \end{bmatrix} \quad \text{and} \quad H_2 = \begin{bmatrix} 0.5 & 0.966 \\ -51 & 1.6 \end{bmatrix}.$$

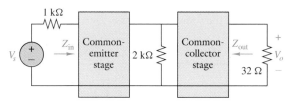

Figure P19.20 Coupled amplifier configuration.

(b) A 2-μF capacitor is inserted in series with the 1-kΩ resistance to prevent dc voltage in the power supply (not shown in the figure) from entering the signal source. Because of this capacitance, low-frequency signals will be amplified less. Determine the frequency (in Hz) at which the magnitude of V_o/V_s is 0.707 of its maximum value calculated in part (d).

21. (a) Assuming the standard port-voltage and port-current labeling, find the h-parameters for the two-ports shown in figure P19.21a. Can N_2 be described precisely by z- or y-parameters? Explain.

(b) Using the standard port-voltage and port-current labeling, find the h-parameters for the two-ports shown in figure P19.21b Display the h-parameters in matrix form. Can N_2 be described precisely by z- or y-parameters? Explain.

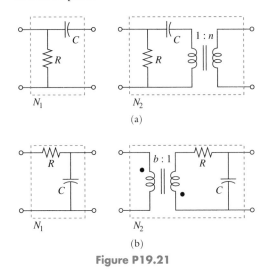

(a)

(b)

Figure P19.21

22. (a) Shown in figure P19.22 is a high-frequency model of a bipolar transistor in common-emitter configuration. The model may be viewed as a two-port. Find $h_{21}(s)$ and its poles and zeros in terms of the parameters R_x, R_π, C_π, C_μ, and g_m.
(b) Find the remaining h-parameters of the circuit.

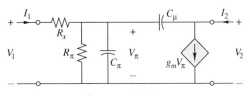

Figure P19.22

23. Consider the terminated two-port configuration in figure P19.23, with h_{ij} indicating the two-port h-parameters. Suppose that (1) the current through the admittance Y_L equals the current through h_{22}, (2) the current gain $I_2/I_1 = 150$, and (3) the source resistance is $Z_S = 9$ kΩ.

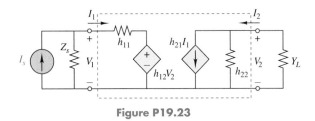

Figure P19.23

(a) Determine h_{22} in terms of Y_L and possibly other h-parameters. Explain your reasoning.
(b) Derive a formula for the current gain I_2/I_1 in terms of the h-parameters and Y_L.
(c) Determine h_{21}.
(d) Suppose now that the source at the front end of the two-port is briefly disconnected and a voltage $v_2(t) = 2u(t)$ is applied to the secondary. The measurement $v_1(t) = -u(t)$ V is made. Determine h_{12}.
(e) Suppose the source is reconnected to the two-port. If the current gain $I_1/I_s = 0.9$, and if $Y_L = 0.125$ mho, determine Z_{in} and h_{11}.

24. In this problem, you are to design an amplifier circuit represented by the doubly terminated equivalent circuit shown in figure P19.24. This means that you will be given certain amplifier specifications which will allow you to determine the parameters of the amplifier circuit. The amplifier specifications are as follows:
    (1) When I_1 is zero, the ratio $V_1/V_2 = 0$ when a source is applied to port 2.
    (2) $Z_{out} = 800$ Ω, and there must be maximum power transfer from the amplifier output to the load.
    (3) $V_1/V_s = 24/25$.
    (4) The voltage gain $V_2/V_1 = -100$.
Given these specifications,
    (a) Determine h_{12}.
    (b) Determine h_{22} and b.
    (c) Determine h_{11}.
    (d) Determine the input impedance Z_{in}.
    (e) Determine h_{21}.
    (f) Determine the overall amplifier gain V_L/V_s.
    (g) Determine the power gain, i.e., the ratio of the power delivered to the amplifier to the power delivered to the load.

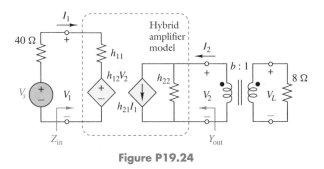

Figure P19.24

t-Parameters

25. Determine the t-parameters for each of the circuits sketched in figure P19.25.

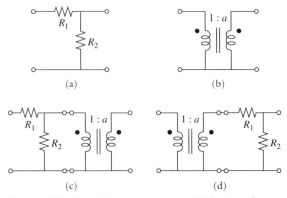

(a)

(b)

(c)

(d)

Figure P19.25 (a) Resistive two-port. (b) Ideal transformer. (c) (a) coupled to (b). (d) (b) coupled to (a).

26. (a) Determine the t-parameters of each of the circuit configurations shown in figures P19.26a–c.
(b) In the branch of circuit theory called network synthesis, one often must find circuit realizations of two-ports having specified voltage gains. The two-port of figure P19.26d is the realization of a specified voltage gain.

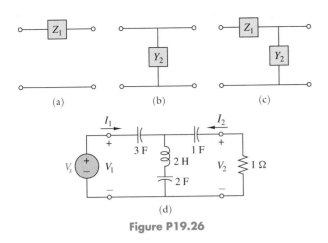

(a)   (b)   (c)

(d)

Figure P19.26

(1) Determine the t-parameters of the two-port. (*Hint*: Make use of the results of part (a).)

(2) What is the formula for $G_V(s) = V_2(s)/V_1(s)$ in terms of y-parameters?

(3) Using table 19.1, determine the y-parameters y_{12} and y_{22}.

(4) Verify that the voltage gain is

$$G_V(s) = \frac{V_2(s)}{V_1(s)} = \frac{3s(4s^2 + 1)}{12s^3 + 16s^2 + 5s + 6}.$$

27. Consider the cascaded pair of two-ports in figure P19.27. Determine the t-parameters (in standard units) of the overall cascaded network.

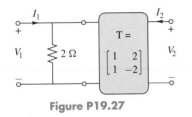

Figure P19.27

28. Determine the t-parameters (in standard units) of the cascaded two-port in figure P19.28.

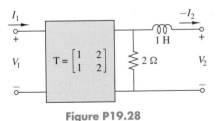

Figure P19.28

29. The two-port in figure P19.29 is a filtering section of a power supply. The filter is used to reduce ripple when converting ac to dc.

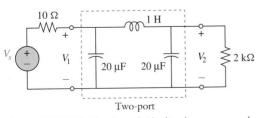

Figure P19.29 Terminated idealized power supply circuit.

(a) Determine the t-parameters of the two-port. (*Hint*: Use the results of problem 26a.)

(b) Derive formulas for $G_{V_1}(s) = V_1/V_s$, $G_{V_2}(s) = V_2/V_1$, and $G_V(s) = V_2/V_s$ in terms of the t-parameters of the circuit.

(c) Suppose $v_s(t) = 10\sin(120\pi t)$ V. Determine the magnitude of the steady-state output voltage, $v_{2ss}(t)$. In this problem, $v_s(t)$ represents a 10-V ac ripple superimposed on a dc signal. This is typical of rectified sine waves. The idea of the problem is to see how the ripple is reduced by an LC filter. Note that an inductor looks like a short to dc and a capacitor looks like an open to dc.

Reciprocity

30. Determine the condition on a and b for reciprocity in the two-port network of figure P19.30. (*Hint*: Set up a matrix equation of the form $M_1 x_1 = M_2 x_2$, where $x_1 = (V_1, V_2, 0)^t$ and $x_2 = (I_1, I_2, I_3)^t$, and eliminate I_3 by any method you want.)

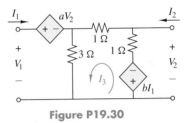

Figure P19.30

31. The two-port sketched in figure P19.31a has port 1 attached to an ideal current source. You have been informed that the two-port consists only of R's, L's, C's, and transformers. If $i_1(t) = \delta(t)$, your oscilloscope will show a waveform having the following analytic expression:

$v_2(t) = 3e^{-t} + 5e^{-t}\cos(500t - 30°)$ V.

Figure P19.31b is the same two-port with different terminations.

(a) If $i_2(t) = u(t)$ A, find the zero-state response $v_1(t)$.

(b) If the input is changed so that $i_2(t) = \cos(500t)u(t)$ A, find the steady-state response, $v_1(t)$.

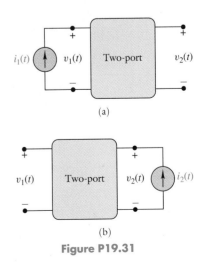

Figure P19.31

Miscellaneous

32. Let $H(s)$ be the transfer function of a linear amplifier circuit that contains only one energy storage element, C or L.

(a) Prove that if $H(\infty) = 0$ (i.e., the gain is zero at infinite frequency) then

$$H(s) = \frac{K}{s + \omega_C}$$

and that ω_c is the 3-dB bandwidth of the low pass amplifier.

(b) Prove that, for the low-pass amplifier, $\omega_c = 1/(RC)$, or $\omega_c = R/L$, where R is the Thévenin resistance of the resistive one-port "seen" by the energy storage element, C or L.

(c) A low-pass amplifier has the circuit model shown in figure P19.32. Find the 3-dB bandwidth by the use of the result of part (b).

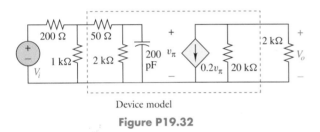

Device model

Figure P19.32

33. Let $H(s)$ be the transfer function of a linear amplifier circuit that contains only one energy storage element, C or L.

(a) Prove that if $H(0) = 0$ (i.e., the gain is zero at zero frequency) then

$$H(s) = \frac{Ks}{s + \omega_C}$$

and that ω_c is the 3-dB bandwidth of the high-pass amplifier.

(b) Prove that, for the high-pass amplifier $\omega_c = 1/(RC)$, or $\omega_c = R/L$, where R is the Thévenin resistance of the one-port "seen" by the energy storage element, C or L.

(c) A high-pass amplifier has the circuit model shown in figure P19.33. Find the 3-dB bandwidth using the results of part (b) and example 19.1.

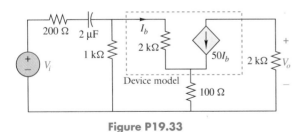

Figure P19.33

34. A high-pass amplifier has the circuit model shown in figure P19.34. Find the 3-dB bandwidth by the use of the results of part (b) of problem 33 and example 19.2.

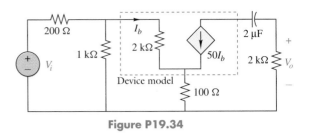

Figure P19.34

35. Let $H(s)$ be the transfer function of a linear circuit that contains only one energy storage element, C or L.

(a) Prove that

$$H(s) = \frac{H(\infty)s + H(0)\omega_c}{s + \omega_c},$$

where $\omega_c = 1/(RC)$, or $\omega_c = R/L$ and where R is the Thévenin resistance of the resistive one-port "seen" by the energy storage element, C or L.

(b) Justify the statement that $H(\infty)$ is the transfer function calculated with C short-circuited or L open-circuited and $H(0)$ is the transfer function calculated with C open-circuited or L short-circuited.

(c) The results of parts (a) and (b) show that the calculation of $H(s)$ may be reduced to an analysis of three resistive circuits. Apply this method to find $H(s)$ for each of the two circuits shown in figure P19.35.

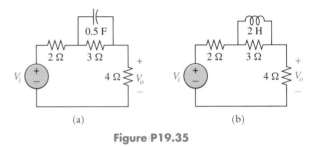

(a)   (b)

Figure P19.35

36. Determine the input impedance $Z_{in}(j\omega)$ of the circuit in figure P19.36 at $\omega = 10$ rad/sec.

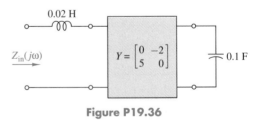

Figure P19.36

37. Consider the equivalent circuit for a terminated two-port, as sketched in figure P19.37. The numbers g_{ij} are called the g-parameters of the two-port, and the circuit, with the source, source impedance Z_S and the Z_L deleted, is called the g-parameter equivalent circuit of the two-port. Answer each of the following questions:

(a) Derive a formula for the input admittance.
(b) Derive a formula for the output impedance.
(c) Determine the voltage gain $G_1 = V_1/V_s$.
(d) Determine the voltage gain $G_2 = V_2/V_1$.
(e) Determine the voltage gain $G_v = V_2/V_s$.

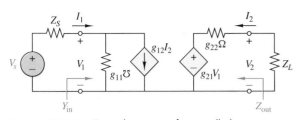

Figure P19.37 Equivalent circuit for so-called *g*-parameter model of a terminated two-port.

38. Determine the input admittance for the interconnected two-port in figure P19.38 where *Y* and *H* have standard units.

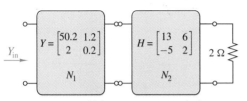

Figure P19.38 Multiparameter cascaded two-ports.

39. If the *t*-parameters of the two-port N_1 are as given in figure P19.39, determine the *t*-parameters of the cascaded two-port assuming *t* has standard units.

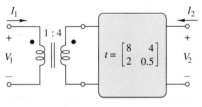

Figure P19.39

40. Determine the *y*- and *z*-parameters of the two-port of figure P19.40.

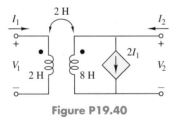

Figure P19.40

41. Determine the *g*-parameters of the circuit in figure P19.41. (See problem 37.)

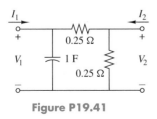

Figure P19.41

42. Consider the circuit of figure P19.42.
   (a) For what value of *a* is the circuit reciprocal?

(b) For the value of *a* determined in part (a), determine the hybrid parameters of the circuit.

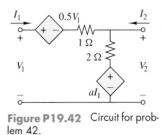

Figure P19.42 Circuit for problem 42.

43. The circuit of figure P19.43 has *z*-parameters

$$\begin{bmatrix} z_{11} & z_{12} \\ z_{21} & z_{22} \end{bmatrix}.$$

   (a) Determine R_1, R_2, R_3, and R_4 in terms of the *z*-parameters.
   (b) Flip the circuit horizontally and repeat part (a).

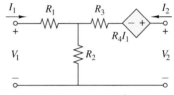

Figure P19.43 *T*-equivalent circuit for two-port modeled by *z*-parameters.

ANSWERS, in scrambled order: z_{12}, $z_{11} - z_{12}$, $z_{21} - z_{12}$, $z_{22} - z_{12}$.

44. The circuit of figure P19.44 has *y*-parameters

$$\begin{bmatrix} y_{11} & y_{12} \\ y_{21} & y_{22} \end{bmatrix}.$$

   (a) Determine G_1, G_2, G_3, and G_4 in terms of the *y*-parameters.
   (b) Flip the circuit horizontally and repeat part (a).

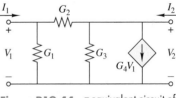

Figure P19.44 *π*-equivalent circuit of two-port modeled by *y*-parameters.

ANSWERS, in scrambled order: $(y_{21} - y_{12})$, $(y_{11} + y_{12})$, $-y_{12}$, $(y_{22} + y_{12})$.

45. A two-port is described by *z*-parameters, and it is known that the *t*-parameters exist. Determine the *t*-parameters in terms of the *z*-parameters. (*Hint*: Rewrite the *z*-parameter equations in the form

$$M_1 \begin{bmatrix} V_1 \\ I_1 \end{bmatrix} = M_2 \begin{bmatrix} -I_2 \\ V_2 \end{bmatrix}.$$

Then invert the appropriate matrix to obtain

$$\begin{bmatrix} \dfrac{z_{11}}{z_{21}} & \dfrac{\Delta z}{z_{21}} \\ \dfrac{1}{z_{21}} & \dfrac{z_{22}}{z_{21}} \end{bmatrix},$$

where $\Delta z = z_{11}z_{22} - z_{12}z_{21}$.)

46. A two-port is described by t-parameters.

    (a) Derive the input impedance relationship

$$Z_{in} = \frac{t_{11}Z_L + t_{12}}{t_{21}Z_L + t_{22}}.$$

    (b) Derive the output impedance relationship

$$Z_{out} = \frac{t_{22}Z_S + t_{12}}{t_{21}Z_S + t_{11}}.$$

47. Find the t-parameters of the circuit in figure P19.47. (*Hint*: Apply the results of problem 26.)

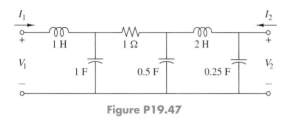

Figure P19.47

48. Determine the impedance, admittance, and h-parameters of each of the circuits of figure P19.48.

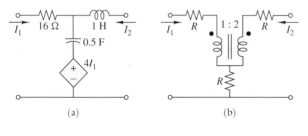

(a)   (b)

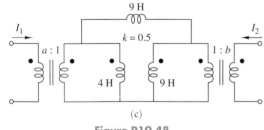

(c)

Figure P19.48

49. In the laboratory, you are given a box having two sets of leads, which you label as port 1 and port 2. After short-circuiting port 2, you measure the responses

$$v_1(t) = 0.5(1+t)e^{-t}u(t) \text{ V}$$

and

$$i_2(t) = 0.5e^{-t}u(t) \text{ A}$$

to a unit step current input. Replacing the short circuit on port 2 with a 1-Ω resistor, you repeat the test, again exciting port 1 with a unit step current input. This time, your measurements are

$$v_1(t) = -\left[\frac{8}{3}e^{2t} + \frac{1}{3}e^{-t} + \delta(t)\right] \text{ V}$$

and

$$i_2(t) = -\delta(t) - 4e^{t}u(t) \text{ A}.$$

Determine the y-parameters of the box.

50. (a) The circuit of figure P19.50, with Z representing z-parameters, in ohms, is resonant at what value of ω?

    (b) Determine the Q of the circuit.

    (c) Plot the magnitude and phase response of $Y_{in}(s)$, $s = j\omega$, using a program such as MATLAB.

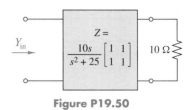

Figure P19.50

51. Determine the Y-parameters of the mutually coupled inductors in figure P19.51.

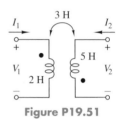

Figure P19.51

52. Consider the switched circuit of figure P19.52. For $t < 1$, the switch is in position A, and for $t \geq 1$, the switch is in position B.

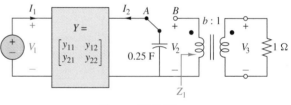

Figure P19.52

    (a) With the switch in position A, derive the voltage gain V_2/V_1 in terms of the y-parameters and the admittance of the capacitor; i.e., show that

$$\frac{V_2(s)}{V_1(s)} = \frac{-y_{21}}{y_{22} + 0.25s}.$$

    (b) If $v_1(t) = -2u(t)$ V, and the y-parameter matrix is

$$Y(s) = \begin{bmatrix} \dfrac{-1}{s} & \dfrac{1}{s} \\ \dfrac{1}{s} & \dfrac{-1}{s} \end{bmatrix},$$

determine $v_2(t)$, $0 \leq t < 1$.

    (c) Is the circuit stable? Explain your reasoning.

    (d) Determine $v_2(1^-)$.

    (e) For $t \geq 1$, after the switch has moved to position B, draw and label the frequency domain equivalent circuit that will allow one to compute $V_3(s)$.

    (f) Determine $Z_1(s)$ if $b = 2$.

    (g) Determine $v_3(t)$ for $t \geq 1$.

53. Determine the t-parameters (in standard units) of the cascaded network of figure P19.53.

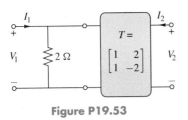

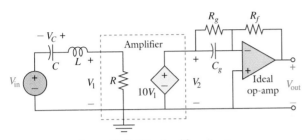

Figure P19.53

54. In the circuit of figure P19.54, suppose that $L = 1$ H, $C = 0.5$ F, $R = 3\ \Omega$, $R_g = R_f = 1\ \Omega$, and $C_g = 1$ F.

Figure P19.54 Amplifier circuit.

(a) Find $H_1(s) = V_1/V_{in}$ and its poles and zeros.
(b) Find $H_2(s) = V_2/V_1$ and its poles and zeros.
(c) Find $H_3(s) = V_{out}/V_2$ and its poles and zeros.
(d) Find the transfer function $H(s) = V_{out}/V_{in}$ and its poles and zeros.
(e) Find the impulse response $h(t)$.
(f) Find the step response, assuming that v_{out} is the output.

(g) If L and C_g have no stored energy at $t = 0^-$, and if $v_C(0^-) = 2\ V$ and $v_{in}(t) = 0$, find $v_{out}(t)$ for $t \geq 0$.

55. Consider again the circuit of problem 54. Now suppose that $L = 1$ H, $C = 0.25$ F, and $R = 0.1\ \Omega$, with all other parameter values the same. Find the bandwidth, find ω_m, and find the maximum value of the voltage gain $|V_1/V_{in}|$. (Note that the answers depend only on $H_1(s)$ with the new element values.)

56. Reconsider yet again the circuit of problem 54. If $R = 100\ \Omega$, find the z-parameters of the two-port enclosed in dashed lines.

57. Consider the two-port in figure P19.57, which depicts a transistor amplifier stage for a microphone. Suppose that the transistor has h-parameters given as $h_{11} = 4,200\ \Omega$, $h_{12} = 1$, $h_{21} = 150$, and $h_{22} = 0.1\ m\mho$.

(a) Assume that the input and output capacitors are short circuits and that the capacitor across the $470\ \Omega$ resistor is an open circuit. Determine the h-parameters of the overall two-port.
(b) Given your answer to (a), determine the value of a for maximum power transfer.
(c) Given your answer to (a), determine the voltage and power gain of the overall two-port.
(d) Using the value of a computed in (b), repeat the calculations of (a) and (c), assuming that the capacitors are present in the circuit.
(e) Plot the frequency response of the amplifier as a function of $f = 2\pi\omega$ for $0 \leq f \leq 20,000$ kHz.

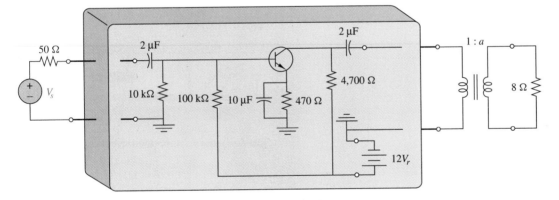

Figure P19.57 Transistor amplifier circuit for a microphone.

CHAPTER 20

CHAPTER OUTLINE

Analysis of Interconnected Two-Ports

HISTORICAL NOTE

G ENERAL methods for analyzing linear circuits require the formation and solution of a system of simultaneous equations. Manual solution of simultaneous equations is practical only for very small circuits, the type that often appears in textbooks for illustration purposes. For most real-world circuits constructed with discrete components, digital simulation becomes a necessary mode of analysis.

In the past two decades, technology has advanced so rapidly that many important analog subcircuits (e.g., intermediate-frequency amplifiers, operational amplifiers, timers, and comparators) are now manufactured as integrated circuits. Each such circuit may contain thousands of elements (resistors, capacitors, diodes, transistors, etc.). As a result, the number of simultaneous equations representing their linear models is also on the order of several thousand. If such a system of equations was solved using the usual "small system" technique, then even the most powerful digital computer currently available could not reasonably cope with the task. Large-scale circuit analysis demands special techniques to handle the large number of modeling equations, a topic of intense research in recent years. A basic approach is to (1) decompose the large system into an interconnection of many small subsystems; (2) analyze the subsystems (perhaps using parallel-processing methods); and, finally, (3) put the solutions together properly to obtain the solution to the original large-system problem. This chapter describes such a strategy for a special class of large linear circuits: those which are an interconnection of many two-ports. ◈

CHAPTER OBJECTIVES

1. Utilizing the two-port results of the previous chapter, develop a strategy for analyzing a large two-port network that is an interconnection of many component two-ports.
2. Define the notion of an indefinite admittance matrix and use it to characterize an n-terminal linear network.
3. Understand the basic mathematical properties of the indefinite admittance matrix and interpret them for interconnected networks.
4. Show how to easily derive the y-parameters of a two-port derived from a three-terminal component, characterized by its indefinite admittance matrix.

1. INTRODUCTION

Chapter 19 explored various characterizations of linear two-ports, as well as the calculation of voltages, currents, and gains when the two-port is terminated in source and load impedances. This chapter will extend these methods to the analysis of networks far more complex than the standard source–two-port–load situation. Although the analysis of a large network can be achieved by writing a system of equations based on KCL, KVL, and each component's v-i characteristic, such an approach yields little insight into the internal workings of the parts of the network. In contrast, the methods described in this chapter have the advantage of providing useful information at each intermediate step of the calculation, which contributes to the understanding of the overall circuit behavior.

The methods of this chapter view a complex circuit as an interconnection of components characterized by a set of parameters such as the two-port parameters studied in chapter 19. The special role played by each component two-port of the interconnected network can usually be identified (e.g., radio-frequency stage, intermediate-frequency stage, output amplification stage, etc.). The original large problem breaks down into several smaller problems with simpler solutions. Proper utilization of these solutions yields a solution to the original analysis or design problem. In essence, the technique is a "divide-and-conquer" strategy for handling large networks.

2. PARALLEL, SERIES, AND CASCADED CONNECTION OF TWO-PORTS

A general linear two-port has four external terminals for connection to other networks. For the special case when the input and output ports share a common terminal, there are only three external terminals. Such a two-port is often referred to as a **common-ground two-port**, although the common terminal is not necessarily grounded in the sense of being connected to earth. Figure 20.1 illustrates the terminology.

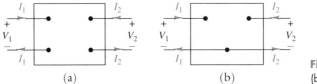

Figure 20.1 (a) A general two-port. (b) A common-ground two-port.

Interconnecting two common-ground two-ports, N_a and N_b, forms a new two-port N. This new two-port has a new set of two-port parameters (z-, y-, h-, g-, or t-parameters) that are obtained very simply from the individual two-port parameters. Figure 20.2 shows three typical interconnection structures: the parallel, the series, and the cascade interconnection. (Problem 12 illustrates two more interconnection topologies.) To avoid overcrowding of symbols, figure 20.2 omits all voltage and current reference labels. These labels are understood to conform with those shown in figure 20.1a. In particular, note that at each port, the current entering one terminal must equal the current leaving the other terminal for the two-port parameters to be valid or meaningful.

The new interconnected two-ports of figure 20.2 have new two-port parameters computed from the two-ports N_a and N_b as follows:

1. For the parallel connection of figure 20.2a,

$$Y = Y_a + Y_b;$$ (20.1)

2. for the series connection of figure 20.2b,

$$Z = Z_a + Z_b;$$ (20.2)

3. and for the cascade connection of figure 20.2c,

$$T = T_a T_b,$$ (20.3)

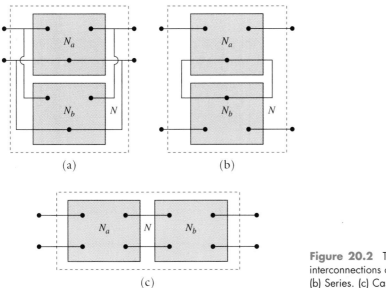

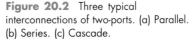

(a)

(b)

(c)

Figure 20.2 Three typical interconnections of two-ports. (a) Parallel. (b) Series. (c) Cascade.

where Y, Z, and T denote the admittance, impedance, and transmission parameter matrices, respectively, and the subscripts a and b refer to the networks N_a and N_b, respectively.

A derivation of these formulas is straightforward. For example, from the definitions of N_a and N_b,

$$I_a = Y_a V_a \qquad \text{and} \qquad I_b = Y_b V_b,$$

where

$$V_a = \begin{bmatrix} V_{1a} \\ V_{2a} \end{bmatrix}, \quad I_a = \begin{bmatrix} I_{1a} \\ I_{2a} \end{bmatrix},$$

and similarly for the voltage and current vectors of N_b. From figure 20.2a, $V_a = V_b = V$ and $I = I_a + I_b$. A substitution yields $I = (Y_a + Y_b)V$, which verifies equation 20.1.

To verify equation 20.2, consider figure 20.2b. Here, $V_a = Z_a I_a$ and $V_b = Z_b I_b$. But $V = V_a + V_b$ and $I_a = I_b = I$ imply by direct substitution that $V = (Z_a + Z_b)I$, verifying equation 20.2.

Equation 20.3 was derived in chapter 19.

The derivation of equations 20.1 and 20.3 is easily extended to the case of more than two two-ports. The results are as follows:

1. If two or more common-ground 2-ports are connected in parallel, then

$$Y = Y_a + Y_b + Y_c + \cdots \tag{20.4}$$

2. If two or more two-ports, common-ground or not, are connected in cascade, then

$$T = T_a T_b T_c \cdots \tag{20.5}$$

Equations 20.3 and 20.5 for the cascade connection hold whether or not the component two-ports are of the common-ground type. However, equations 20.1 and 20.4 for parallel connections *in general* hold only for common-ground two-port connections, as shown in figure 20.2a. Similarly, the series connection equation 20.2 holds only for the case illustrated in figure 20.2b. Series connection of two general two-ports (figure 20.1a) or series connection of more than two common-ground two-ports (figure 20.1b) requires an ideal transformer for coupling, as is demonstrated in problem 8. Examples 20.1 and 20.2 explain why equations 20.2 and 20.4 fail when two non-common-ground two-ports are connected together.

EXAMPLE 20.1

This example illustrates the difficulty with a non-common-ground series connection. Consider the two-port shown in figure 20.3, which is a series connection of two component two-ports. The z-parameters of the individual two-ports are given by

$$Z_a = \begin{bmatrix} 2 & 1 \\ 1 & 2 \end{bmatrix} \Omega \quad \text{and} \quad Z_b = \begin{bmatrix} 8 & 1 \\ 1 & 5 \end{bmatrix} \Omega.$$

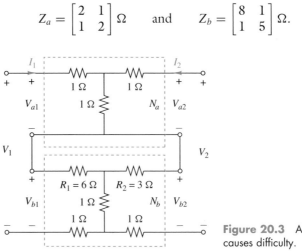

Figure 20.3 A series connection that causes difficulty.

Observing that the parallel connection of the 6-Ω and 3-Ω resistors is 2 Ω, the z-parameter matrix of the interconnected two-port is

$$Z = \begin{bmatrix} 6 & 4 \\ 4 & 6 \end{bmatrix} \neq Z_a + Z_b = \begin{bmatrix} 10 & 2 \\ 2 & 7 \end{bmatrix}.$$

Here, $Z \neq Z_a + Z_b$ because, after the interconnection, neither N_a nor N_b acts as a two-port, as defined in figure 20.1. This can be understood by inspecting figure 20.4, with the indicated loop currents.

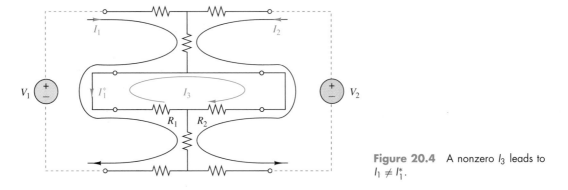

Figure 20.4 A nonzero I_3 leads to $I_1 \neq I_1^*$.

With R_1 and R_2 nonzero, the mesh current I_3 is in general nonzero. (See problem 7.) Observe that $I_1^* = I_1 - I_3 \neq I_1$. Hence, for the left terminal pair of N_a, the current entering the top terminal is not equal to the current leaving the bottom terminal. With unequal terminal currents, N_a no longer has a z-parameter characterization, because z-parameter definitions require equal currents entering and leaving the terminal pair, as shown in figure 20.1.

On the other hand, if $R_1 = R_2 = 0$ in figure 20.4, then the third mesh equation is satisfied for arbitrary values of the mesh currents I_1, I_2, and I_3, as there is no resistance at all in the third mesh. In particular, let $I_3 = 0$, in which case $I_1^* = I_1$, and the z-parameter characterization of N_a is valid. Similar arguments hold for N_b. If $R_1 = R_2 = 0$, equation 20.2 holds because figure 20.4 now has the same interconnection as depicted in figure 20.2b.

EXAMPLE 20.2

  This example illustrates the problem of a non-common-ground parallel connection. Figure 20.5 shows two two-ports connected in parallel. Before being connected in parallel, each two-port has y-parameters

$$Y_a = Y_b = \begin{bmatrix} 0.7 & -0.2 \\ -0.2 & 0.7 \end{bmatrix} \text{℧}.$$

After the connection, the new two-port has y-parameter matrix

$$Y = \begin{bmatrix} 1.625 & -0.625 \\ -0.625 & 1.625 \end{bmatrix} \text{℧}.$$

Clearly, $Y \neq Y_a + Y_b = 2Y_a$ in this case.

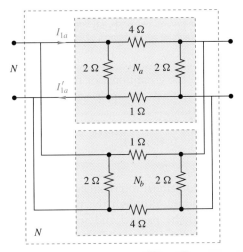

Figure 20.5 New two-port N having invalid application of equation 20.1 for non-common-ground two-ports in parallel.

  The reason for the failure of equation 20.1 under these circumstances is the same as that for the circuit of example 20.1. If a voltage source is applied to port 1 of N, we find that the currents I_{1a} and I'_{1a} are not equal. Thus, N_a cannot continue to be characterized by a set of two-port y-parameters in forming N.

  There are, however, some special cases of non-common-ground two-ports for which equation 20.1 holds for a parallel connection. The following is one such example.

EXAMPLE 20.3

  This example illustrates how to achieve a parallel interconnection of two general two-ports of figure 20.1a so that equation 20.4 remains valid. A 1:1 ideal transformer is inserted at the front end of N_a in figure 20.5. The result is shown in figure 20.6a. The Y_a and Y_b matrices are the same as in example 20.2. The way the ideal transformer forces equal currents at port terminals is shown more clearly in figure 20.6b, where components nonessential to the derivation of the result have been omitted. Four dashed-line boxes labeled #1, #2, #3, and #4, are drawn in the figure. Kirchhoff's current law is applied to each of the boxes, which represent Gaussian surfaces. From box #1, we have $I_{1a} - I'_{1a} = 0$, or $I'_{1a} = I_{1a}$. From box #3,

$$I_{1a} - I'_{1a} + I_{1b} - I'_{1b} = 0,$$

which implies that $I_{1b} = I'_{1b}$, since we have shown that $I_{1a} - I'_{1a} = 0$. In a like manner, we can show that $I'_{2a} = I_{2a}$ and $I_{2b} = I'_{2b}$ by applying KCL to boxes #2 and #4.

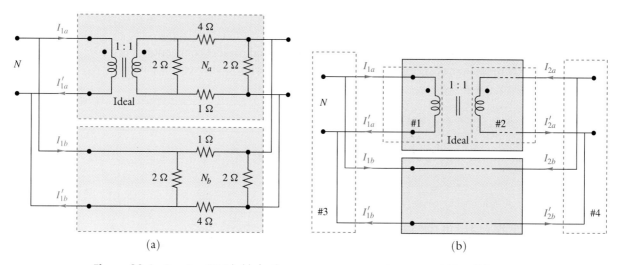

(a)   (b)

Figure 20.6 Equation 20.1 holds for these non-common-ground two-ports. (a) Parallel connection of two general two-ports. (b) Justification of equal currents at two terminals of each port.

Since both N_a and N_b continue to satisfy the conditions for two-port characterization, and their interconnection is of the parallel type, we can apply equation 20.1 to obtain

$$Y = Y_a + Y_b = \begin{bmatrix} 1.4 & -0.4 \\ -0.4 & 1.4 \end{bmatrix} \mho.$$

This example extends directly to multiple parallel interconnections and as is shown in problem 9, to multiple series interconnections. Although the precise conditions for the applicability of equations 20.1 and 20.2 to the parallel and series connections of non-common-ground two-ports are known, they are not practical enough to be included here. Our emphasis is on the interconnections of common-ground two-ports, which occur most often in practice. The next few examples illustrate some interesting applications to these interconnections.

◆ **EXAMPLE 20.4**

The two-port N shown in figure 20.7 consists of two active components, Q_a and Q_b, and three resistors. The active components have the following y-parameters (in millimhos):

$$Y_a = \begin{bmatrix} 1 & 0 \\ 51 & 0 \end{bmatrix}, \qquad Y_b = \begin{bmatrix} 1 & -1 \\ -52 & 52 \end{bmatrix}.$$

1. Find the t- and y-parameters of N.
2. Find the open-circuit voltage ratio V_2/V_1 of the two-port N.
3. Find V_2/V_1 if port 2 of N is terminated in a 0.2-kΩ resistance.

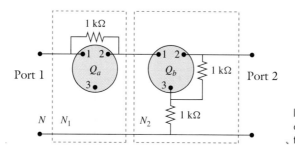

Figure 20.7 An interconnection of common-ground two-ports, N_1 and N_2, to form a new two-port N.

SOLUTION

Part 1: Five types of common-ground component two-ports (with standard voltage and current labels, as shown in figure 20.1b) can be identified in figure 20.7. Figure 20.8 shows these five component two-ports.

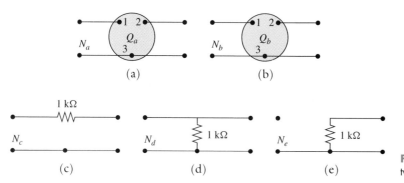

Figure 20.8 Five types of component two-ports in figure 20.7.

The overall two-port N is obtained from the component two-ports by a sequence of interconnections as follows:

$$N = N_1 \text{ cascaded with } N_2,$$

in which case the t-parameters of N are the matrix product of the t-parameters of N_1 and N_2; in addition,

$$N_1 = N_a \text{ in parallel with } N_c$$

(what does this mean in terms of y-parameters?); and, finally,

$$N_2 = (N_b \text{ in parallel with } N_e) \text{ in series with } N_d.$$

Using equations 20.1 through 20.3, and table 19.1 of chapter 19, it is straightforward to evaluate the parameters of the foregoing interconnected two-ports. To avoid the frequent appearance of powers of 10, we use kΩ as the unit for all z-parameters and t_{12} and millimhos as the unit for all y-parameters and t_{21}. First, as specified,

$$Y_a = \begin{bmatrix} 1 & 0 \\ 51 & 0 \end{bmatrix},$$

and from figure 20.8c,

$$Y_c = \begin{bmatrix} 1 & -1 \\ -1 & 1 \end{bmatrix}.$$

Applying equation 20.1 to the parallel connection of N_a and N_c yields

$$Y_1 = Y_a + Y_c = \begin{bmatrix} 2 & -1 \\ 50 & 1 \end{bmatrix}.$$

Also,

$$Y_b = \begin{bmatrix} 1 & -1 \\ -52 & 52 \end{bmatrix},$$

and from figure 20.8e,

$$Y_e = \begin{bmatrix} 0 & 0 \\ 0 & 1 \end{bmatrix}.$$

The parallel connection of N_b and N_e implies that

$$Y_{be} = Y_b + Y_e = \begin{bmatrix} 1 & -1 \\ -52 & 53 \end{bmatrix},$$

as per equation 20.1. From the conversion table 19.1,

$$Z_{be} = \begin{bmatrix} 53 & 1 \\ 52 & 1 \end{bmatrix}.$$

Next, from figure 20.8d,

$$Z_d = \begin{bmatrix} 1 & 1 \\ 1 & 1 \end{bmatrix}.$$

The z-parameters for N_2 are calculated using equation 20.2:

$$Z_2 = Z_{be} + Z_d = \begin{bmatrix} 54 & 2 \\ 53 & 2 \end{bmatrix}.$$

To execute the cascade connection of N_1 and N_2, we need the t-parameters for these networks. From table 19.1,

$$T_1 = \begin{bmatrix} -0.02 & -0.02 \\ -1.04 & -0.04 \end{bmatrix}$$

and

$$T_2 = \begin{bmatrix} \dfrac{54}{53} & \dfrac{2}{53} \\ \dfrac{1}{53} & \dfrac{2}{53} \end{bmatrix}$$

For the two-port N, as per equation 20.3,

$$T = T_1 T_2 = \begin{bmatrix} -0.0208 & -1.51 \text{ k}\Omega \\ -1.0604 \text{ m}\mho & -0.0408 \end{bmatrix}$$

Using the conversion table, the y-parameters for N are

$$Y = \begin{bmatrix} 27 & -0.5 \\ 662.5 & 13.75 \end{bmatrix} \text{ m}\mho.$$

Part 2:   From equation 19.55, the open-circuit voltage ratio is

$$\frac{V_2}{V_1} = \frac{1}{t_{11}} = -48.07.$$

Part 3:   To compute the voltage gain with load, we use the formula derived at the end of section 19.4. The result is

$$\frac{V_2}{V_1} = -\frac{y_{21}}{y_{22} + y_L} = -\frac{662.5}{13.75 + 5} = -35.33.$$

Exercise.   An alternative description of N_2 in figure 20.7 is

$$N_2 = (N_b \text{ in cascade with } N_a) \text{ in series with } N_d.$$

Use this relationship to find the t-parameters of the two-port N_2.
ANSWER: The same, of course, as that given in example 20.5.

Example 20.4 could have been solved by writing a set of simultaneous equations. This approach would bypass the calculation of the important intermediate results, such as the Y_1 matrix for N_1 and the Z_2 matrix for N_2. If figure 20.7 were a model for a two-stage amplifier, these intermediate results would provide critical information about the behavior of each of the stages.

One noteworthy point about example 20.4 is that it demonstrates a strategy for analyzing a class of large-scale networks. As long as a network can be described in terms of some parallel, series, and cascade connections of component two-ports, then, regardless of the size of the network, an analysis can be carried out using no more than the following mathematical operations:

1. Addition, multiplication, and inversion of 2×2 matrices.
2. Table 19.1 for two-port parameter conversion.

The operations may require complex-number arithmetic.

The method is conceptually simple. A computer program can be written to do all the matrix operations. The user inputs only a description of the interconnections and the parameters of the component two-ports. The computer can then carry out the tedious calculations.

Example 20.4 emphasizes the computational aspect of interconnected two-ports. The next example will illustrate an application of interconnected two-ports in the design of an oscillator. An oscillator is a device that produces a sine wave of fixed amplitude at a particular frequency. Oscillators are used extensively in communication and instrumentation systems.

EXAMPLE 20.5

One type of sinusoidal oscillator circuit, called a *Colpitts oscillator*, is a parallel connection of two common-ground two-ports, as shown in figure 20.9, where the parameters of the active two-port have the mho as their unit. Suppose that the g_i of the active two-port is negligible and all circuit parameters except g_m are fixed. Find (1) the value of g_m such that the circuit generates a sustained sinusoidal oscillation and (2) the frequency of oscillation.

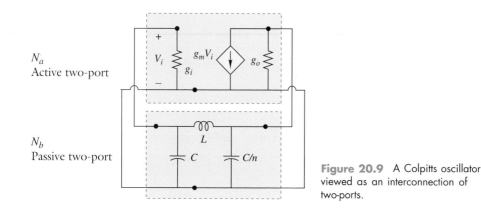

Figure 20.9 A Colpitts oscillator viewed as an interconnection of two-ports.

SOLUTION

The parallel connection of N_a and N_b suggests that we investigate the problem in terms of y-parameters, because $Y = Y_a + Y_b$. After finding the y-parameters for the overall two-port, we will invoke a result from chapter 15 to determine the condition for sustained sinusoidal oscillation.

Part 1: The y-matrices of the component two-ports are

$$Y_a = \begin{bmatrix} 0 & 0 \\ g_m & g_o \end{bmatrix}$$

and

$$Y_b = \begin{bmatrix} sC + \dfrac{1}{sL} & -\dfrac{1}{sL} \\ -\dfrac{1}{sL} & s\dfrac{C}{n} + \dfrac{1}{sL} \end{bmatrix}.$$

From equation 20.1, the Y-matrix of the interconnected two-port is

$$Y = Y_a + Y_b = \begin{bmatrix} sC + \dfrac{1}{sL} & -\dfrac{1}{sL} \\[2ex] g_m - \dfrac{1}{sL} & g_o + s\dfrac{C}{n} + \dfrac{1}{sL} \end{bmatrix}. \tag{20.6}$$

All circuit responses depend on the initial states of the LC elements, since no external input excites the circuit. Recall from chapter 15 that any zero-input response is a rational function of s. The denominator of this rational function is simply the determinant of the nodal or loop equations characterizing the circuit. Furthermore, from figure 15.5b, for the zero-input response to be sinusoidal in the steady state, the determinant must have a non-repeated conjugate pair of zeros on the $j\omega$-axis and all other zeros in the open left half-plane. For the present case, the determinant is that of the y matrix given by equation 20.6, i.e.,

$$\Delta_Y = \det[Y]$$

$$= \frac{1}{nLs}\left[LC^2s^3 + LCng_os^2 + C(n+1)s + n(g_m + g_o)\right]. \tag{20.7}$$

Equation 20.7 has a cubic numerator of the form

$$a_3s^3 + a_2s^2 + a_1s + a_0 = 0, \tag{20.8}$$

which has exactly three roots. The conditions for the roots to be $s_{1,2} = \pm j\omega_0$ and $s_3 = -b < 0$ are:

1. All a's have the same sign.
2.

$$a_3a_o = a_1a_2. \tag{20.9}$$

(Problem 10 develops these conditions.) In order to have

$$a_3[(s + b)(s^2 + \omega_0^2)] = a_3[s^3 + bs^2 + \omega_0^2 s + b\omega_0^2] = a_3s^3 + a_2s^2 + a_1s + a_0,$$

these conditions require that

$$\omega_o = \sqrt{\frac{a_1}{a_3}} = \sqrt{\frac{a_o}{a_2}} \tag{20.10}$$

and

$$b = \frac{a_2}{a_3} = \frac{a_o}{a_1}. \tag{20.11}$$

Applying equation 20.9 to equation 20.7 yields

$$LC^2n(g_m + g_o) = LCng_oC(n + 1).$$

Solving for g_m yields the required value for sustained oscillations, i.e.,

$$g_m = ng_o. \tag{20.12}$$

Part 2: The frequency of oscillation, as calculated from equation 20.10, is

$$\omega_0 = \sqrt{(n + 1)\frac{1}{LC}}. \tag{20.13}$$

At this point, several questions come to mind: (1) How can the value of g_m be maintained exactly at ng_o? (2) If g_m exceeds ng_o, will the amplitude of oscillation grow to infinity, as predicted in figure 15.5c? (3) What starts the oscillation?

A rigorous treatment of these topics requires advanced nonlinear circuit theory, which is certainly beyond the scope of this text. Nevertheless, it is possible to give an intuitive, nonrigorous explanation as follows. The active element is designed to operate at a quiescent point that has a g_m value exceeding ng_o. The inevitable noise and other electrical disturbance will provide the initial "kick" to start the oscillation, whose amplitude increases exponentially. As the amplitude increases, the operation of the active device swings into its nonlinear region. This has the effect of lowering the "average value of g_m." Thus, the amplitude of oscillation will keep increasing, and the waveform will become more distorted (in terms of its degree of departure from a pure sine wave), until the "average value of g_m" is reduced to ng_o. The circuit then reaches a steady state and continues to produce an approximate sinusoid.

Exercise. A Colpitts oscillator has the following circuit parameters: $L = 25$ μH, $C = 400$ pF, $n = 1$, $g_o = 40$ μ℧. Assuming that $g_i = 0$, find the value of g_m for sustained oscillations and the corresponding frequency.

ANSWERS: 40 μ℧, 2.25 MHz.

As a final example of the application of interconnected two-ports, consider the design of a passive circuit that will stop the transmission of a sinusoidal signal at a fixed frequency. For example, in a receiver, it might be necessary to eliminate the 10-kHz whistle tone produced by two neighboring AM radio signals reaching the receiver. This can be achieved quite easily with a parallel resonant circuit inserted between two of the receiver's amplifier stages. However, the same objective can also be achieved with a passive RC network, as illustrated in example 20.6.

EXAMPLE 20.6

Two T-networks, N_a and N_b, as shown in figures 20.10a and 20.10b, are connected in parallel to form the two-port N of figure 20.10c, called a **twin-T network.**

1. Find the parameters y_{21a}, y_{21b}, and y_{21}.
2. If a sinusoidal voltage source of $E \cos(\omega_0 t)$ V, with $\omega_0 = 1/(RC)$, is applied to port 1 of N, what is the steady-state open-circuit voltage appearing at port 2 of N?

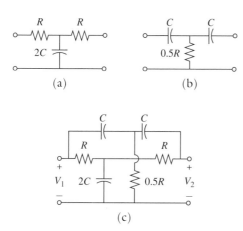

(c)

Figure 20.10 (a) A lowpass T-network, N_a. (b) A highpass T-network, N_b. (c) Parallel connection of (a) and (b), denoted N.

SOLUTION

Part 1: Let us connect a voltage source to port 1 of N_a and short-circuit port 2. Applying a source transformation and current division, we obtain

$$I_{2a} = -\frac{G}{G + G + 2sC}(GV_{1a}),$$

and then, from equation 19.10,

$$y_{21a} = \left.\frac{I_{2a}}{V_{1a}}\right|_{V_{2a}=0} = -\frac{G^2}{2(sC+G)}. \tag{20.14}$$

Applying the same steps to N_b,

$$y_{21b} = \left.\frac{I_{2b}}{V_{1b}}\right|_{V_{2b}=0} = -\frac{s^2 G^2}{2(sC+G)}. \tag{20.15}$$

It follows from equation 20.1 that

$$y_{21} = y_{21a} + y_{21b} = -\frac{C\left[s^2 + \dfrac{1}{(RC)^2}\right]}{2\left(s + \dfrac{1}{RC}\right)} = \frac{C(s^2 + \omega_0^2)}{2(s + \omega_0)} \tag{20.16}$$

Part 2: Since the voltage is sinusoidal at angular frequency $\omega_0 = 1/(RC)$, we replace s by $j\omega_0$ for steady-state analysis. From equation 20.16, $y_{21}(j\omega_0) = 0$, implying that no current is flowing through the short circuit across port 2. Under this zero-current condition, the short circuit can be replaced by any impedance without affecting the voltages and currents throughout the network. In particular, we can open-circuit port 2. The open-circuit voltage of port 2 is zero in the steady state. Thus, the twin-T network has completely stopped the transmission of a sinusoidal signal at the angular frequency $\omega_0 = 1/(RC)$. Such a network is sometimes called a **notch filter**. A calculation of the magnitude response curve of this circuit is included in problem 11.

> **Exercise.** For the two-port N of figure 20.10c, find $y_{22}(s)$.
> *Check:* $y_{22}(s) = [(sCR)^2 + ?(sCR) + ?]/[2R(sCR + ?)]$.

3. INDEFINITE ADMITTANCE MATRIX OF A THREE-TERMINAL NETWORK

The great majority of active elements used in linear circuit analysis have only three external, or accessible, terminals. When used as a two-port, one terminal must be common to the input and the output ports. Depending on which of the three terminals is chosen as the common terminal, the resultant two-port may display quite different properties. For example, figure 20.11 shows three configurations of an *npn*-type bipolar junction transistor. Some distinctive features of these configurations, to be demonstrated shortly in example 20.9, are that the CE stage provides a reasonable voltage gain, the CC stage provides a high input impedance and a low output impedance, and the CB stage provides a low input impedance and a high output impedance.

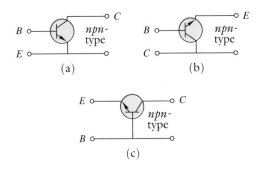

Figure 20.11 Three configurations of a transistor stage. (a) Common emitter (CE). (b) Common collector (CC). (c) Common base (CB).

A question that immediately arises is, shouldn't the two-port parameters of each configuration be related to each other? In particular, suppose we have computed the y-parameters for one of the configurations by the methods studied in chapter 19. Must we repeat the calculations in order to obtain the y-parameters of the remaining configurations? The answer is no. The simpler solution requires learning a new characterization of a three-terminal network, called the **indefinite admittance matrix** representation. This section explains the basic concepts and applications of this representation.

The essence of the new characterization is not to select *any* of the three-terminal network nodes as a reference node for writing the nodal equations. Instead, a node *outside* of the network is chosen as the reference node, and the three-terminal network is said to be "floating." Figure 20.12 illustrates how sources are applied. Note that we have labeled the nodes with the letters *a*, *b*, and *c*, instead of the usual integer numbering. We reserve integers for labeling the two-port created from the three-terminal network.

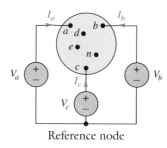

Reference node

Figure 20.12 Three-terminal network with external nodes *a,b,c* and internal nodes *d* through *n*.

Since the network is linear and there are no independent sources inside the network, by the principles of proportionality and superposition, each terminal current should be a linear combination of the three applied voltages. This linear relationship has the matrix representation

$$\begin{bmatrix} I_a \\ I_b \\ I_c \end{bmatrix} = \begin{bmatrix} y_{aa} & y_{ab} & y_{ac} \\ y_{ba} & y_{bb} & y_{bc} \\ y_{ca} & y_{cb} & y_{cc} \end{bmatrix} \begin{bmatrix} V_a \\ V_b \\ V_c \end{bmatrix}, \tag{20.17a}$$

or, in matrix notation,

$$I = Y_{\text{ind}} V \tag{20.17b}$$

Y_{ind} is called the **indefinite admittance matrix** (abbreviated IAM) of the three-terminal network. The adjective "indefinite" emphasizes the fact that the Y_{ind} characterization of the network does not favor any particular one of the external network nodes.

◆ EXAMPLE 20.7

Consider the three-terminal network *N* of figure 20.13, where node *d* is an internal node. Find the indefinite admittance matrix Y_{ind}.

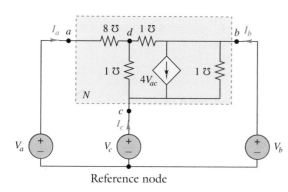

Reference node

Figure 20.13 A three-terminal network with one internal node labeled d.

SOLUTION

With the voltage sources connected to the three external nodes, and with the outside node as reference, the partitioned nodal equations are:

$$
\begin{bmatrix}
8 & 0 & 0 & \vdots & -8 \\
4 & 2 & -5 & \vdots & -1 \\
-4 & -1 & 6 & \vdots & -1 \\
\cdots & \cdots & \cdots & \vdots & \cdots \\
-8 & -1 & -1 & \vdots & 10
\end{bmatrix}
\begin{bmatrix}
V_a \\
V_b \\
V_c \\
\cdots \\
V_d
\end{bmatrix}
=
\begin{bmatrix}
I_a \\
I_b \\
I_c \\
\cdots \\
0
\end{bmatrix}.
\tag{20.18}
$$

Using the method of matrix partitioning illustrated in equation 19.11 for eliminating the internal node variables yields

$$
\begin{bmatrix}
I_a \\
I_b \\
I_c
\end{bmatrix}
=
\begin{bmatrix}
W_{11} - W_{12} W_{22}^{-1} W_{21}
\end{bmatrix}
\begin{bmatrix}
V_a \\
V_b \\
V_c
\end{bmatrix}
$$

$$
=
\left(
\begin{bmatrix}
8 & 0 & 0 \\
4 & 2 & -5 \\
-4 & -1 & 6
\end{bmatrix}
- \frac{1}{10}
\begin{bmatrix}
-8 \\
-1 \\
-1
\end{bmatrix}
\begin{bmatrix}
-8 & -1 & -1
\end{bmatrix}
\right)
$$

$$
=
\begin{bmatrix}
1.6 & -0.8 & -0.8 \\
3.2 & 1.9 & -5.1 \\
-4.8 & -1.1 & 5.9
\end{bmatrix}
\begin{bmatrix}
V_a \\
V_b \\
V_c
\end{bmatrix}.
$$

Comparing this result with equation 20.17 gives

$$
Y_{\text{ind}} =
\begin{bmatrix}
1.6 & -0.8 & -0.8 \\
3.2 & 1.9 & -5.1 \\
-4.8 & -1.1 & 5.9
\end{bmatrix}.
\tag{20.19}
$$

An examination of equation 20.19 reveals that the sum of the three elements in any row is zero, and the same is true for any column. This is not a coincidence. The two zero-sum properties are true for arbitrary three-terminal networks and, in fact, for n-terminal networks. They are formally stated next, accompanied by an informal proof.

Property 1   The sum of all elements in any column of Y_{ind} is zero.

Proof: Refer to figure 20.12 and equation 20.17. Applying KCL to the reference node yields

$$
I_a + I_b + I_c = (y_{aa} V_a + y_{ab} V_b + y_{ac} V_c)
$$

$$
+ (y_{ba} V_a + y_{bb} V_b + y_{bc} V_c) + (y_{ca} V_a + y_{cb} V_b + y_{cc} V_c)
\tag{20.20}
$$

$$
= (y_{aa} + y_{ba} + y_{ca}) V_a + (y_{ab} + y_{bb} + y_{cb}) V_b + (y_{ac} + y_{bc} + y_{cc}) V_c = 0.
$$

Since equation 20.20 is valid for arbitrary values of (V_a, V_b, V_c), the coefficient of each voltage V_a, V_b, and V_c, must be identically zero; i.e., the sum of the admittances that make up the coefficient must sum to zero. This establishes property 1.

Property 2   The sum of all elements in any row of Y_{ind} is zero.

Proof: From equation 20.17,

$$
I_a = y_{aa} V_a + y_{ab} V_b + y_{ac} V_c.
$$

If we set all voltages equal to some arbitrary constant E, i.e.,

$$V_a = V_b = V_c = E,$$

then

$$I_a = (y_{aa} + y_{ab} + y_{ac})E. \qquad (20.21)$$

Since the three applied voltages have the same value, there is no voltage difference between any two terminals of the three-terminal network. Consequently, there is no current flow into the network. Thus, I_a in equation 20.21 must be identically zero for any value of E. But this requires that $(y_{aa} + y_{ab} + y_{ac}) = 0$. The same is true for the second and third rows. This establishes property 2.

Two additional properties follow from the zero-sum property and basic matrix determinant theory.

Property 3   The determinant of Y_{ind} is zero.
   This follows because the third column is the sum of the first two and a square matrix with dependent columns has a zero determinant.

Property 4   The first-order cofactors of all elements are equal.

We only illustrate this property, without any proof. Consider the indefinite admittance matrix of equation 20.19. Using the usual determinant notation, straightforward calculations show that

$$\Delta = 0,$$

$$\Delta_{11} = 1.9 \times 5.9 - (-5.1) \times (-1.1) = 5.6,$$

$$\Delta_{23} = -[1.6 \times (-1.1) - (-0.8) \times (-4.8)] = 5.6, \text{ etc.}$$

Once the IAM of a three-terminal network N has been found, it is a very simple matter to obtain the y-parameters of a two-port constructed from N by making one terminal the common terminal. The following property specifies how to achieve this.

Property 5   If a two-port is constructed from a three-terminal network (see figure 20.1b), then the y matrix of the common ground two-port may be obtained from the indefinite admittance matrix of N by deleting the row and the column corresponding to the common ground and possibly interchanging the entries.

To justify property 5, refer to figure 20.12. Suppose that a two-port is defined with node c as common to the input terminal pair (a,c) and the output terminal pair (b,c). Choosing c as the common node shared by V_a and V_b amounts to setting the voltage V_c to zero. Then, in equation 20.17a, the third column of the coefficient matrix may be deleted, since $V_c = 0$. In the two-port characterization, we do not explicitly consider the current I_c through the common terminal. Therefore, the third row, which corresponds to the equation for I_c, may be deleted. The resulting two equations, expressing port currents in terms of port voltages, contain the entries of the y-parameters, as defined by equation 19.9. However, if node b is to be port 1 and node a to be port 2, it is necessary to interchange entries $(1, 1)$ with $(2, 2)$, and $(1, 2)$ with $(2, 1)$. A similar reasoning applies when node b or c is chosen as the common node.

 **EXAMPLE 20.8**

   Consider the three-terminal network N of figure 20.13. A two-port is defined with (c,b) as port 1 and (a,b) as port 2. Find the y-matrix for the common-ground two-port.

SOLUTION

The IAM for N was found in example 20.7 and is given by equation 20.19. Deleting the row and column corresponding to node b yields the following 2×2 matrix in which node labels are indicated on the side and top.

$$
\begin{array}{c}
 \\
a \\
c
\end{array}
\begin{array}{c}
\quad a \qquad c \\
\left[\begin{array}{cc}
1.6 & -0.8 \\
-4.8 & 5.9
\end{array}\right].
\end{array}
$$

This corresponds to the equation

$$
\begin{bmatrix} I_a \\ I_c \end{bmatrix} = \begin{bmatrix} 1.6 & -0.8 \\ -4.8 & 5.9 \end{bmatrix} \begin{bmatrix} V_a \\ V_c \end{bmatrix}.
$$

To have ports 1 and 2 as specified, we must change the order of the elements in the voltage and current vectors. In terms of the new vectors, the preceding equation becomes

$$
\begin{bmatrix} I_c \\ I_a \end{bmatrix} = \begin{bmatrix} 5.9 & -4.8 \\ -0.8 & 1.6 \end{bmatrix} \begin{bmatrix} V_c \\ V_a \end{bmatrix},
$$

and the desired admittance matrix is

$$
[y] = \begin{array}{c} c \\ a \end{array} \begin{array}{c} \quad c \qquad a \\ \left[\begin{array}{cc} 5.9 & -4.8 \\ -0.8 & 1.6 \end{array}\right]. \end{array}
$$

The new y matrix is obtained from the old one by switching the diagonal elements $(1, 1)$ with $(2, 2)$ and the off-diagonal elements $(1, 2)$ with $(2, 1)$. This demonstrates the interchange of entries mentioned in property 5.

The next example shows how to make use of properties 1, 2, and 5 to solve the problem posed at the beginning of this section on page 720.

◆ EXAMPLE 20.9

As shown in figure 20.11, a bipolar junction transistor may be used in three different configurations. Suppose typical y-parameters for the CE configuration at low frequency are

$$
y_{11} = 0.5, \quad y_{12} \cong 0,
$$

$$
y_{21} = 25, \quad y_{22} = 0.02,
$$

each in millimhos.

1. Find the y-parameters for the CC and CB configurations.
2. Suppose the transistor is inserted in the amplifier stage shown in figure 20.14. Calculate Z_{in}, Z_{out}, and V_{out}/V_s, and tabulate the results.

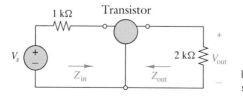

Figure 20.14 A typical amplifier stage.

SOLUTION

Part 1: From the given y-parameters for the CE stage,

$$[y]_{CE} = \begin{bmatrix} 0.5 & 0 \\ 25 & .02 \end{bmatrix} \text{m}\mho.$$

Augmenting this matrix to obtain the IAM of the transistor by utilizing properties 1 and 2 yields

$$Y_{ind} = \begin{matrix} & \begin{matrix} B & C & E \end{matrix} \\ \begin{matrix} B \\ C \\ E \end{matrix} & \begin{bmatrix} 0.5 & 0 & -0.5 \\ 25 & 0.02 & -25.02 \\ -25.5 & -0.02 & 25.52 \end{bmatrix} \end{matrix} \text{m}\mho. \qquad (20.22)$$

According to property 5, deleting the row and column corresponding to the chosen common node yields, for the common-collector (CC) configuration,

$$[y]_{CC} = \begin{bmatrix} 0.5 & -0.5 \\ -25.5 & 25.52 \end{bmatrix} \text{m}\mho.$$

and, for the common-base (CB) configuration,

$$[y]_{CB} = \begin{bmatrix} 25.52 & -0.02 \\ -25.02 & 0.02 \end{bmatrix} \text{m}\mho.$$

Part 2: Z_{in}, Z_{out}, and V_{out}/V_s are computed using the methods described in chapter 19, section 4. Table 20.1 lists the results.

TABLE 20.1 COMPARISON OF Z_{in}, Z_{out} AND VOLTAGE GAIN OF A BIPOLAR JUNCTION TRANSISTOR AMPLIFIER STAGE

Configuration	Z_{in} ($k\Omega$)	Z_{out} ($k\Omega$)	V_{out}/V_s
CE	2	50	−32
CC	100	0.058	0.97
CB	0.041	884	1.88

Since the source resistance, the load resistance, and the transistor parameters in figure 20.14 are typical, the results shown in table 20.1 are also typical of an amplifier stage using a bipolar junction transistor. Note in particular the following properties:

1. A common-collector stage has a high input impedance, a low output impedance, and a voltage gain less than unity.
2. A common-base stage has a low input impedance and a high output impedance.
3. A common-emitter stage is most suitable for obtaining a large voltage gain.

Our discussion of the application of the indefinite admittance matrix has been confined to three-terminal networks. The definition and properties of the IAM extend easily to multiterminal networks. Much more significant, but specialized applications of the IAM are found in (1) the characterization of multiterminal networks and (2) the handling of any number of arbitrarily interconnected multiterminal networks. (See problems 27 and 28 for the basic ideas.) We leave the detailed discussion of these more specialized applications to texts on advanced circuits.

SUMMARY

This chapter has investigated five different types of two-port interconnections: cascade, series, parallel, series-parallel, and parallel-series (the last two in problems at the end of the chapter). For each type of interconnection, the overall two-port parameter matrix relates directly to the component two-port matrices (e.g., $Y = Y_a + Y_b$ for the parallel connection and $Z = Z_a + Z_b$ for the series connection). Conditions under which these relationships hold were explained. An oscillator circuit and a bridge-T notch filter illustrated the practical applicability of the theory.

The simple relationship among the two-port matrices leads to a strategy for analyzing a class of large-scale networks. As long as the network decomposes into parallel, series, and cascade connections of component two-ports, then, regardless of its size, an analysis can be carried out using only (1) addition, multiplication, and inversion of 2×2 matrices and (2) a conversion table for two-port parameters. This approach avoids the formulation and solution of large systems of simultaneous equations.

If a two-port is created from a four-terminal network N, then any two-port parameter matrix is incapable of giving complete information about N. A complete characterization of an n-terminal component network requires a new type of matrix called the *indefinite admittance* matrix, Y_{ind}. This matrix has the zero-column-sum and zero-row-sum properties. An n-terminal network has an indefinite admittance matrix of order $n \times n$.

Most active elements used in circuit design are three-terminal elements. When such elements are used as a two-port, there are different ways of choosing the common terminal, as well as the input and output terminals. Suppose that the y-parameters for some two-port configuration have been measured or calculated. Then the indefinite admittance matrix method can easily be used to derive the y-parameters for any other two-port configuration. This was demonstrated through various examples.

TERMS AND CONCEPTS

Cascade connection of two-ports N_a and N_b: an interconnection, illustrated in figure 20.2c, whose essential properties are that the current leaving port 2 of N_a flows into port 1 of N_b and $T = T_a T_b$. Also called a **tandem** connection.

Colpitts oscillator: an oscillator circuit whose frequency-determining subcircuit consists of an inductor and two capacitors, as shown in figure 20.9.

Equicofactor property: the property that the first-order cofactors Δ_{ij} of an indefinite admittance matrix are equal for all i and j.

Gaussian surface: a closed curve or closed surface surrounding two or more nodes.

Indefinite admittance matrix: Y_{ind}, of an n-terminal network: a square matrix of order n relating the n terminal currents to the n terminal voltages. The node voltages are with respect to a reference node that is not one of the network nodes.

Notch filter: a frequency-selective circuit that stops the trans-

mission of signals over a very narrow band of frequencies and passes signals outside of that frequency band.

Parallel connection of two-ports N_a and N_b: an interconnection, illustrated in figure 20.2a, whose essential properties are that the port k voltage of N_a equals the port k voltage of N_b and $Y = Y_a + Y_b$.

Series connection of two-ports N_a and N_b: an interconnection, illustrated in figure 20.2b, whose essential properties are that the port k current of N_a equals the port k current of N_b, and $Z = Z_a + Z_b$.

Twin-T notch filter: a two-port made up of two T-networks connected in parallel. The components are selected such that, at one specified frequency, there is no transmission of the input signal to the output. (See figure 20.10.)

Zero-column-sum-property: the property that the sum of all elements in any column of the matrix Y_{ind} is zero.

Zero-row-sum property: the property that the sum of all elements in any row of the matrix Y_{ind} is zero.

PROBLEMS

1. Find the z-parameters of the interconnected circuit shown in figure P20.1.

2. Find the y-parameters of the interconnected network shown in figure P20.2.

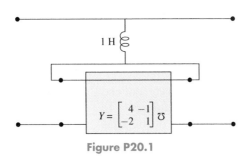

Figure P20.1

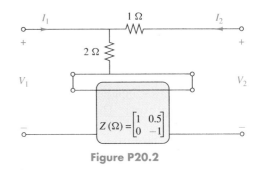

Figure P20.2

ANSWERS, in scrambled form: $s + 1, s + 2, s + 0.5, s + 0.5$.

ANSWERS, in scrambled form: 2, 2, 2.5, 3.

3. An active three-terminal component A has the y-parameters shown in figure P20.3. Find the y-parameters of the two-port.

ANSWERS, scrambled form: $1/8$, $1/8$, $5/16$, $1/16$.

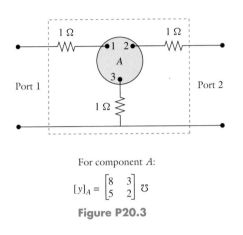

For component A:

$$[y]_A = \begin{bmatrix} 8 & 3 \\ 5 & 2 \end{bmatrix} \text{ ℧}$$

Figure P20.3

4. In the two-port N in figure P20.4, the z-parameters are as follows: $z_{11} = 7$ Ω, $z_{12} = 2$ Ω, $z_{21} = 10$ Ω, $z_{22} = 3$ Ω. Find the y-parameters of the new two-port N^*.

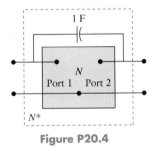

Figure P20.4

ANSWERS, in scrambled form: $(s + 7)$, $-(s + 2)$, $(s + 3)$, $-(s + 10)$.

5. Two components, $2Nyyy$, have z-parameters

$$Z = \begin{bmatrix} 3 & 1 \\ 5 & 2 \end{bmatrix} \text{k}\Omega$$

and are interconnected as shown in figure P20.5. Find the t- and y-parameters of the new two-port. (*Note:* Parallel, series, and cascade connections are present in this problem. Be careful about the sequence of these connections. Furthermore, you may use kΩ and m℧ as units, to avoid dragging along the factors 10^3 and 10^{-3}.)

ANSWERS: t-parameters: 0.6548, 0.1865 kΩ, 0.1875 m℧, 0.3125; y-parameters, all in m℧: 0.1667, −0.8889, −5.3333, 3.4444.

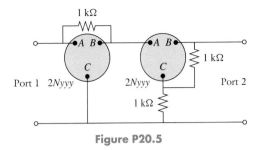

Figure P20.5

6. Determine the overall t-parameters of the cascaded pair of two-ports shown in figure P20.6.

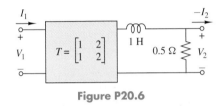

Figure P20.6

ANSWERS: $2s + 5$, $s + 2$, $2s + 5$, $s + 2$.

7. Consider the circuit of figure 20.4, with element values given in figure 20.3. Show that the mesh current I_3 is zero if and only if $V_1/V_2 = 7/8$.

8. Two two-ports N_a and N_b are connected in series to form a new two-port N, as illustrated in figure P20.8. Show that equation 20.2 holds in this case, despite the fact that neither N_a nor N_b is a three-terminal network.

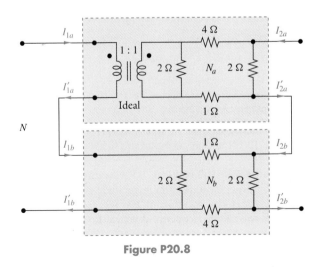

Figure P20.8

9. Three two-ports N_a, N_b, and N_c are connected in series to form a new two-port N, as illustrated in figure P20.9.

Show that equation 20.2 can be extended to apply to this case because of the presence of the ideal transformers.

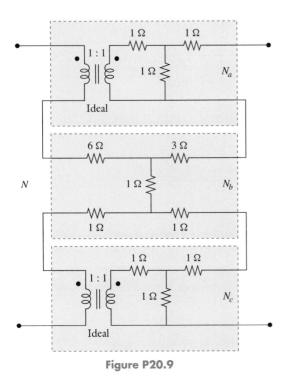

Figure P20.9

10. Consider the following cubic equation, with all coefficients nonnegative:

$$P(s) = a_3 s^3 + a_2 s^2 + a_1 s + a_0 = 0.$$

Prove that $P(s) = 0$ has two imaginary roots $s = \pm j\omega_0$ if and only if

$$a_3 a_0 = a_1 a_2$$

and that under this condition, the three roots are

$$s_{1,2} = \pm j \sqrt{\frac{a_1}{a_3}} \quad \text{and} \quad s_3 = -\frac{a_2}{a_3}.$$

(*Hint*: Let $s = j\omega_0$, and equate $\text{Re}\{P(j\omega_0)\}$ and $\text{Im}\{P(j\omega_0)\}$ to zero.)

11. For the notch filter of figure 20.10c, y_{21} is given by equation 20.16, and y_{22} is found in the exercise following example 20.7.

    (a) Let $R = 1\ \Omega$ and $C = 1$ F. Use any circuit simulation program to plot the magnitude response of the filter for $0 < f < 1$ Hz, in 10-mHz steps. From the plot, what is the frequency at which the voltage gain is zero?

    (b) Frequency scale and impedance scale the circuit so that R is 1 kΩ and the frequency for zero voltage gain is 10 kHz.

12. Two common-ground two-ports are connected to form a new two-port, as shown in figure P20.12. The names for their connections are obvious from the figure.

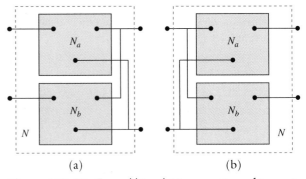

(a)   (b)

Figure P20.12 Two additional interconnections of two-ports. (a) Series-parallel connection. (b) Parallel-series connection.

    (a) Prove that for the series-parallel connection,

$$h_{11} = h_{11a} + h_{11b}, \qquad h_{12} = h_{12a} - h_{12b},$$

$$h_{21} = h_{21a} - h_{21b}, \qquad h_{22} = h_{22a} + h_{22b}.$$

    (b) Prove that for the parallel-series connection,

$$g_{11} = g_{11a} + g_{11b}, \qquad g_{12} = g_{12a} - g_{12b},$$

$$g_{21} = g_{21a} - g_{21b}, \qquad g_{22} = g_{22a} + g_{22b}.$$

(For the definitions of the g-parameters, see chapter 19, problem 37.)

13. The amplifier circuit shown in figure P20.13a is called a *feedback amplifier* because a fraction of the output voltage, obtained through the 100-kΩ potentiometer (a voltage divider) is fed back to the input. The figure depicts the amplifier as a two-port that is a series-parallel connection of two two-ports N_a and N_b shown in figure P20.13b.

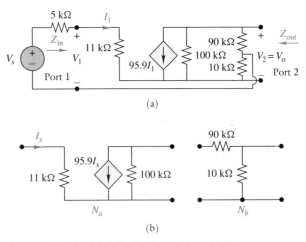

Figure P20.13 (a) A feedback amplifier. (b) Component two-ports.

    (a) Find the *h*-parameters of the two-ports N_a and N_b.

    (b) Using the formulas given in problem 12, find the *h*-parameters of the two-port representing the amplifier.

    (c) Using the methods of chapter 19, find Z_{in}, Z_{out}, and V_2/V_s.

ANSWERS: (a) 11 kΩ, 0, 95.9, 0.1 $\mu\mho$; (b) 9 kΩ, 0.1, −0.1, 0.1 $\mu\mho$; (c) 500 kΩ, 2,475.25 Ω, −9.505.

14. Find the indefinite admittance matrix $Y_{ind}(s)$ for each of the three-terminal networks shown in figure P20.14. (*Check*: Entries on the diagonal: (a) 6, $5s + 4$, $5s + 2$; (b) 0.45, 0.8, 1.05; (c) 1.875, 6.25, 7.5; (d) $s(C_{GD} + C_{GS})$, sC_{GD}, $(sC_{GS} + g_m)$.)

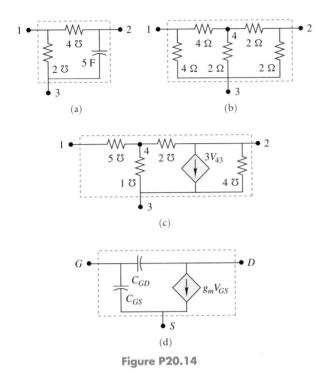

(a)

(b)

(c)

(d)

Figure P20.14

15. (1) A three-terminal network has y-parameters

$$[y] = \begin{bmatrix} 0 & 0 \\ g_m & 0 \end{bmatrix}$$

when used as a two-port in configuration (a) of figure P20.15. Find the y matrices when the same component is used as a two-port in the remaining five possible configurations (b) through (f). (Port voltages and currents are marked in the standard way.)

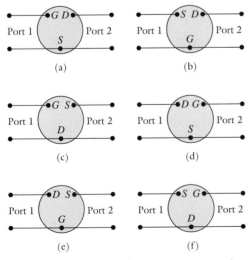

(a)   (b)

(c)   (d)

(e)   (f)

Figure P20.15 Six configurations investigated in problem 15.

(2) Identify those two-port configurations in part (1) which provide signal transmission from port 1 to port 2 (i.e., when a voltage is applied to port 1, some voltage is observed at port 2).

ANSWER: (b) configurations a, b, and c.

16. The three-terminal device $2Nyyy$ has the y-parameters given in figure P20.16 and is used in the amplifier circuit shown in the same figure.

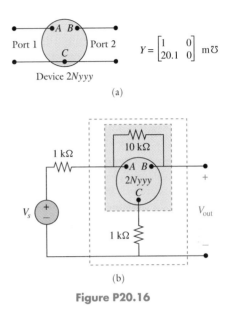

(a)

(b)

Figure P20.16

(a) Find the y-parameters of the two-port of the shaded box.
(b) Find the z-parameters of the two-port enclosed in the dashed lines.
(c) Find the voltage ratio V_{out}/V_s.

ANSWERS: (a) 1.1, −0.1, 20, 0.1 m℧; (b) 1.047, 1.047, −8.479, 1.521 kΩ; (c) −4.141.

17. The indefinite admittance matrix of the three-terminal component N is partially given in figure P20.17. Fill in the blanks in the matrix.

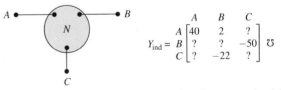

$$Y_{ind} = \begin{array}{c} \\ A \\ B \\ C \end{array} \begin{array}{c} \begin{array}{ccc} A & B & C \end{array} \\ \begin{bmatrix} 40 & 2 & ? \\ ? & ? & -50 \\ ? & -22 & ? \end{bmatrix} \end{array} ℧$$

Figure P20.17 Three-terminal network with associated indefinite admittance matrix having missing entries to be determined.

ANSWERS, scrambled: −7, −42, 20, 30, 92.

18. In figure P20.18, two-port N_a is constructed from a three-terminal network and N_b from a four-terminal network. Both two-ports have the same z-parameters

$$Z_a = Z_b = \begin{bmatrix} 7 & 4 \\ 4 & 7 \end{bmatrix} \Omega.$$

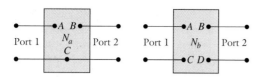

Figure P20.18 Two-port z-parameters do not give complete information about a four-terminal component.

(a) If an ohmmeter is connected across terminals A and B of N_a, what reading is expected? Is the answer unique?

(b) If an ohmmeter is connected across terminals A and B of N_b, what reading is expected? Is the answer unique?

(c) Using resistors, construct two possible two-ports N_a and two possible two-ports N_b that have the specified z-parameters.

ANSWERS: (a) 6 Ω, a unique answer; (b) not enough information given on the four-terminal network to reach a unique answer.

Remark: This problem illustrates the fact that two-port parameters are sufficient for characterizing a three-terminal network, but inadequate for characterizing a four-terminal network. In general, an n-terminal linear network requires $(n-1)^2$ parameters for its complete characterization.

19. The three-terminal network shown in figure P20.19 contains only resistors inside. Therefore, any two-port constructed from the network must be reciprocal. Some entries of the indefinite admittance matrix are as given in the figure.

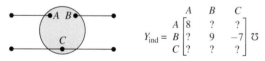

$$Y_{\text{ind}} = \begin{array}{c} \\ A \\ B \\ C \end{array} \begin{array}{c} A \quad B \quad C \\ \begin{bmatrix} 8 & ? & ? \\ ? & 9 & -7 \\ ? & ? & ? \end{bmatrix} \end{array} \mho$$

Figure P20.19 Missing entries Y_{ind} are to be determined.

(a) Find the remaining entries in Y_{ind}.

(b) If terminals A and B are short-circuited, what is the equivalent resistance seen at terminals A and C?

(c) If a voltage source of 1 V is connected across terminals A and B, what is the magnitude of the voltage appearing across the terminals B and C (without any load connected)?

ANSWERS: (b) 1/13 Ω, (c) 6/13 V

20. The admittance matrix for a transistor in its common-collector configuration is

$$\begin{bmatrix} I_{BC} \\ I_{EC} \end{bmatrix} = \begin{bmatrix} 1 & -1 \\ -100 & 100.1 \end{bmatrix} \begin{bmatrix} V_{BC} \\ V_{EC} \end{bmatrix},$$

where matrix entries are in m$\mho$. B = base, E = emitter, and C = collector designate terminals 1, 2, and 3 of the device, respectively.

(a) Find the associated indefinite admittance matrix of the transistor.

(b) Find the z-parameter that relates the collector voltage to the collector and base currents for the common-emitter configuration.

21. In the network of figure P20.21, N_a has z-parameters and N_b has y-parameters as shown.

(a) Find the input impedance Z_{in}.

(b) Find the voltage gain.

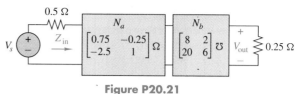

Figure P20.21

22. Figure P20.22a shows a Motorola-type 2N4223 n-channel silicon junction field-effect transistor (JFET) used in the common-source configuration. At 100 MHz, the measured two-port y-parameters are:

$$[y]_{CS} = \begin{bmatrix} 0.2 + j2.5 & -0.01 - j0.65 \\ 3.1 - j0.65 & 0.05 + j0.8 \end{bmatrix} \text{m}\mho.$$

Find the y-parameters when the same transistor is used in the common-gate configuration, as shown in figure P20.22b.

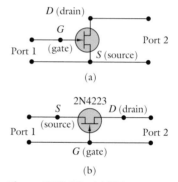

Figure P20.22 (a) FET in common-source configuration. (b) FET in common-gate configuration.

ANSWERS, in scrambled form: $3.34 + j2$, $0.05 + j0.8$, $-3.15 - j0.15$, $-0.04 - j0.15$, all in m$\mho$.

23. Part of the indefinite admittance matrix of a three-terminal component is shown in figure P20.23a. This device is utilized in forming the two-port of figure P20.23b. Find the y-parameters of the two-port in figure P20.23b.

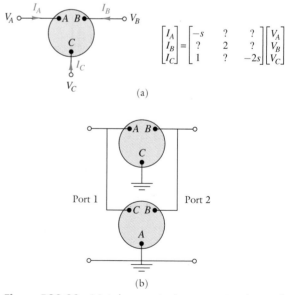

Figure P20.23 (a) A three-terminal component and part of its Y_{ind}. (b) A two-port constructed with the components of (a).

24. If the JFET, type 2N4223, of problem 22 is used in the common-drain configuration, with $G - D$ as port 1 and $S - D$ as port 2, find the two-port y-parameters.

ANSWERS, in scrambled form: $3.34 + j2$, $-0.19 - j1.85$, $0.2 + j2.5$, $-3.3 - j1.85$, all in m℧.

25. (a) Determine the indefinite admittance matrix for each of the circuits shown in figure P20.25.
  (b) For the circuit in part (a) of the figure, determine the admittance matrix Y when node C is grounded.

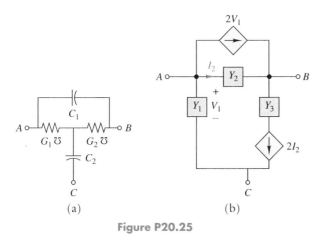

Figure P20.25

(*Check*: (a) The first entry on the diagonal is $sC_1 + G_1$ $(sC_2+G_2)/(sC_2+G_1+G_2)$, (b) The entries on the diagonal are $(Y_1 + Y_2 + 2)$, $-Y_2$, Y_1.)

26. (a) Determine the indefinite admittance matrix for each of the circuits in figure P20.26.
  (b) For the circuit in part (b) of the figure, determine the impedance matrix when point C is grounded.

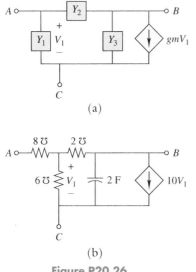

Figure P20.26

27. The indefinite admittance matrix Y_{ind} completely characterizes a linear n-terminal network. In fact, any one row and any one column may be deleted from Y_{ind} without losing information about the network. This redundancy, however, greatly simplifies the analysis of interconnected multiterminal networks. To achieve this goal, some rules are first established. Using simple examples, verify the following rules for manipulating Y_{ind}:

Rule 1, for adding an isolated node: Suppose an n-terminal network N has indefinite admittance matrix Y_{ind}. When an isolated node is added to N to form an $(n + 1)$-node network, the new $(n+1) \times (n+1)$ indefinite admittance may be obtained from Y_{ind} by adding one more row and one more column with all zeros and attaching proper row and column labels.

Rule 2, for joining two n-terminal networks: Let N_a and N_b be two n-terminal networks having the same number of nodes, labeled 1, 2, ..., n, and having the indefinite admittances $Y_{a,ind}$ and $Y_{b,ind}$, respectively. Then, when node k of N_a is joined with node k of N_b, for all k, to form a new n-terminal network N, the indefinite admittance matrix for N is determined by

$$Y_{ind} = Y_{a,ind} + Y_{b,ind}.$$

Rule 3, for suppressing a node: Suppose an n-terminal network N has indefinite admittance matrix Y_{ind}. When an accessible node m is moved inside N and made an internal node, the new $(n-1) \times (n-1)$ indefinite admittance matrix is given by

$$Y_{new,ind} = Y^* - \frac{1}{Y_{mm}} C^* R^*,$$

where

Y^* is the matrix obtained from Y_{ind} by deleting the mth row and mth column;
y_{mm} is the mth row, mth column element of Y_{ind};
C^* is the mth column of Y_{ind}, with the element y_{mm} deleted; and
R^* is the mth row of Y_{ind}, with the element y_{mm} deleted.

Note: The next few problems involve extensive calculations of complex numbers, manipulations of matrices, and manipulations of two-port parameters. We suggest that you use the program tfc (transfer function calculator), accompanying this text, or the popular MATLAB as a tool to solve the problems.

28. A four-terminal network N_a and a three-terminal network N_b are connected together, as shown in figure P20.28. To facilitate the application of the rules stated in problem 27, we add isolated nodes so that both networks are now five-terminal networks, with all five nodes accessible.
  (a) By inspection, write the indefinite admittance matrices for N_a and N_b.
  (*Note*: The presence of some isolated nodes requires the use of rule 1 of problem 27.)

(b) Apply rule 2 of problem 27 to find Y_{ind} for N, with all five nodes accessible.

(c) If nodes 4 and 5 in N are made internal, find the indefinite admittance matrix for the resulting three-terminal network by the use of rule 3 of problem 27.

(d) A two-port is constructed from the three-terminal network of part (c), with nodes (1,3) as port 1 and nodes (2,3) as port 2. Find the y- and z-parameters of the two-port.

(*Checks:* (d) y-parameters: 0.4352, −0.2778, −0.2778, 0.4167; z-parameters: 4, 2.6667, 2.6667, 4.1776.)

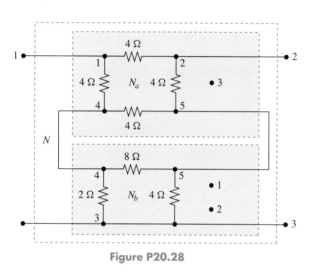

Figure P20.28

29. The network shown in figure P20.29 is a small-signal model of a transistor amplifier. It couples a source having internal resistance R_S to a resistive load R_L. A voltage feedback from the output port to the input port in the form of a parallel combination of resistor R_f and capacitor C_2 is also applied. For the element values $R_S = 50\ \Omega$, $R_1 = 500\ \Omega$, $C_1 = 2.5$ pF, $C_2 = 0.25$ pF, $R_f = 1{,}200\ \Omega$, $g_M = 95$ m℧, $R_0 = 14$ kΩ, and $R_L = 50\ \Omega$, and for an input frequency of 650 MHz, determine:

(a) The admittance matrix for network N_a, network N_b, and their interconnection.

(b) The input admittance Y_{in} seen by the source.

(c) The output admittance Y_{out} seen by the load.

(d) The steady-state voltage gain V_2/V_s.

ANSWERS: (b) $Y_{in} = 6.85 + j15.6$ m℧; (c) $Y_{out} = 5.362 + j3$ m℧; (d) $V_2/V_s = -2.41 + j1.6$ or $2.894\angle146.38°$.

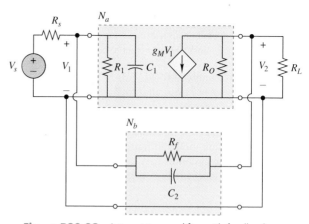

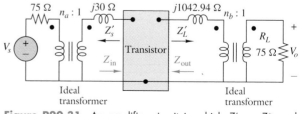

Figure P20.29 A transistor amplifier with feedback.

30. The amplifier shown in figure P20.30 is composed of a common-emitter stage driving a common-collector stage. Such an amplifier combination might be used to drive a low-impedance load. The h-parameters of the two stages are

$$H_1 = \begin{bmatrix} 1 & .001 \\ 50 & .06 \end{bmatrix}, \quad H_2 = \begin{bmatrix} 1 & .99 \\ -5 & .8 \end{bmatrix},$$

where h_{11} is in kΩ and h_{22} is in m℧.

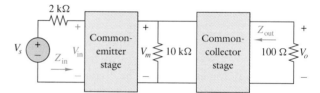

Figure P20.30 A two-stage transistor amplifier.

(a) Find the input impedances of the stages successively, starting from the load end.

(b) Find the voltage gains V_m/V_{in} and V_o/V_m.

(c) Find the overall voltage gain V_o/V_s.

(d) Find the output impedance Z_{out} of the amplifier.

31. The transistor amplifier shown in figure P20.31 operates at 50 MHz. The two-port h-parameters of the transistor, measured at 50 MHz, are:

$$h_{11} = 60 - j50\ \Omega, \qquad h_{12} = 0.01,$$

$$h_{21} = -j2, \qquad h_{22} = 0.0005 + j0.0004\ ℧.$$

The ideal transformer turns ratios are:

$$n_a = 1.1514, \qquad n_b = 3.4012.$$

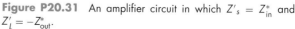

Figure P20.31 An amplifier circuit in which $Z'_s = Z^*_{in}$ and $Z'_L = -Z^*_{out}$.

(a) Find the load impedance Z_L' and the source impedance Z_s' as indicated on the circuit diagram.

(b) Find Z_{in} and Z_{out} of the transistor stage.

(c) Verify that the lossless coupling networks satisfy the conditions of conjugate matching at both ends of the transistor and therefore achieve the maximum power transfer to the 75-Ω load.

(d) Find the voltage gain V_o/V_s with the aid of tfc. (*Check*: The magnitude is between 2.5 and 3.5, the angle between 45° and 55°.)

CHAPTER 21

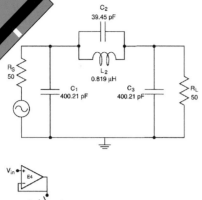

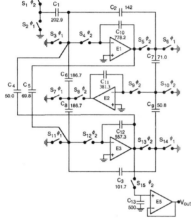

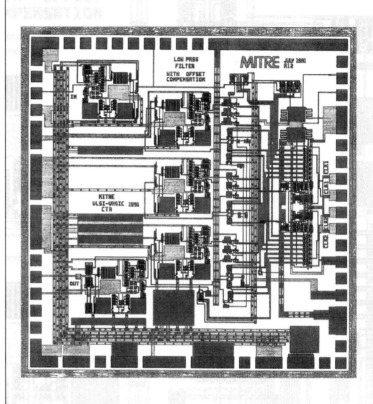

Principles of Basic Filtering

LOUDSPEAKERS AND CROSSOVER NETWORK

I N a stereo system, a power amplifier is connected to a pair of loudspeakers. The most common type of speaker system consists of one or more drivers enclosed in a wooden box. The amplifier feeds a signal to each driver's voice coil, which is placed within the field produced by a permanent magnet. The voice coil is attached to a heavy paper or plastic cone. The power amplifier generates a varying current through the coil that interacts with the magnetic field of the permanent magnet and produces motion. The coil pushes and pulls the speaker's cone, which, in turn, proportionately moves the air, making sound waves.

A single driver cannot effectively handle all the frequency components of music. High and low frequencies place opposite requirements on a loudspeaker. A good low-frequency speaker should be large, in order to push a lot of air. A good high-frequency speaker should be light, in order to move back and forth rapidly.

A two-way speaker system consists of a small, light tweeter to handle the treble signals and a large woofer to handle the bass. A better system is a three-way system, with a third, midrange speaker to handle the frequencies in the middle. The magnitude response of a typical two-way system is shown in the following figure:

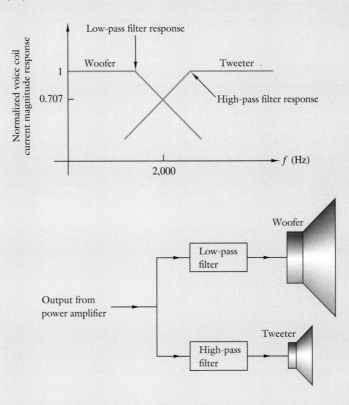

735

A circuit called a *crossover network* is used to separate the frequencies so that each driver receives only the frequencies it is supposed to reproduce. Basically, a crossover network consists of a low-pass filter and a high-pass filter with responses as shown in the figure. The low-pass filter directs low-frequency signals to the woofer, while the high-pass filter directs high-frequency signals to the tweeter. In the magnitude response plot illustrated in the figure, both curves have the same 3-dB frequency at 2,000 Hz. This frequency is called the *crossover frequency*. The current chapter explores the basic principles used in the design of low-pass and high-pass filters. Some simple crossover circuits are considered in the problem section. ◆

◆ **CHAPTER OBJECTIVES**

- Introduce the meaning and (brickwall) specification of low-pass filters in terms of dB loss.
- Introduce the notion of a digital filter.
- Develop the maximally flat Butterworth magnitude response and associated Butterworth transfer function.
- Present a step-by-step design algorithm for finding a Butterworth transfer function.
- Present basic passive and active circuits that realize a Butterworth transfer function.
- Introduce high-pass filter specifications and design.

---

1. INTRODUCTION AND BASIC TERMINOLOGY

Types of Filtering

Filtering plays an important role in circuit theory, as well as in communication theory, image processing, and control. This chapter focuses on some basic filtering concepts and techniques within the domain of circuit theory. There are three fundamental types of filters: analog, digital, and switched capacitor. Switched capacitor filters are something of a hybrid between analog and digital filters. The idea was introduced in chapter 14. The focus in this chapter, however, is on analog filtering, with a very brief introduction to the notion of digital filtering by way of an example. The different types of analog and digital filters are illustrated in figure 21.1.

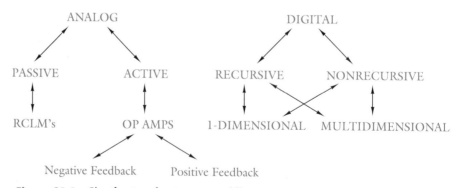

Figure 21.1 Classification of various types of filters.

Analog filters process the actual input waveform with circuits composed of discrete components such as resistors, capacitors, inductors, and op amps. Passive analog filters are composed only of resistors, capacitors, and inductors. Active analog filters are circuits whose elements are resistors, capacitors, and op amps or other types of active elements.

Filters allow one to extract different types of information from a signal. Filter design engineers work closely with communications engineers to develop filters that play a key role in the proper operation of communication and information systems. Communications technology, i.e., the storage and transmission of information (video, radio, telephone, fax, etc.) is changing rapidly. This swiftly evolving frontier provides exciting opportunities for filter design careers within the research, development, production, and installation of communication systems.

Filter design engineers develop switched capacitor filters, analog filters, and digital filters to perform signal discrimination, wherein unwanted signals and excess noise are filtered from signal transmissions. This property allows the information contained in a signal transmitted and received in one form to be reproduced in another, more usable form such as audio. For instance, recent advances in technology have made it possible for most telephone communications to be wireless, as exemplified by cellular phones and satellite communications.

Communications technology, in both the private and public sectors, is moving away from analog systems and toward digital systems under the rubric of digital signal processing (DSP). This has made it necessary for filter design engineers to broaden their area of expertise to include the principles and information coding techniques of DSP. One example of the effect of DSP technology on today's consumer market is the use of digital filters in CD-ROM players and hi-tech stereo speakers.

Conceptually and practically, digital filters are very different from analog filters. In a digital filter, the input waveform is sampled by a device called an analog-to-digital converter. This converter outputs a number that corresponds to the value of the input waveform at the particular sampling instant. The numbers are then processed by some filtering algorithm in a computer. New numbers are formed and are converted to an analog waveform by a digital-to-analog converter. Again, our focus in this chapter is on analog filtering.

Basic Terminology

A **filter** *is a device (often an electrical circuit) that shapes or modifies the frequency content (spectrum) of a signal or waveform*. Throughout our studies, a filter will have a transfer function model $H(s)$ whose frequency response is $H(j\omega)$. The *gain*, **gain magnitude**, *or* **frequency response magnitude** is $|H(j\omega)|$. If $H(j\omega)$ is normalized so that its maximum gain is 1, then the **gain in dB** (decibels) equals $20 \cdot \log_{10} |H(j\omega)|$.

Often, design specifications for a filter are expressed in terms of attenuation or loss rather than gain. In that case, several definitions result that are dual to the gain-related definitions in the previous paragraph. The loss function, denoted $\hat{H}(j\omega)$, is the reciprocal of the transfer function. Thus, the **attenuation**, or **loss magnitude** is

$$\left| \hat{H}(j\omega) \right| = \frac{1}{|H(j\omega)|}.$$

If $H(j\omega)$ is normalized so that its maximum gain is 1, then the filter loss or attenuation in dB is

$$A(\omega) = -20 \log_{10} |H(j\omega)| = 20 \log_{10} \left| \hat{H}(j\omega) \right| \qquad (21.1)$$

These definitions permit us to specify four basic types of filters: low pass, high pass, band pass, and band reject. This chapter examines only low- and high-pass filters. As a first step in exploring the design of such filters, we will derive the structure of the low-pass Butterworth transfer function. The word *Butterworth* refers to an English engineer who first developed this special class of transfer functions in his paper, "On the Theory of Filter Amplifiers."[1] The next objective is to present an algorithm for finding a Butterworth transfer function that meets a given set of low-pass filter design specifications. Once the

---

[1] *Wireless Engineer*, vol. 7, 1930, pp. 536–41.

transfer function is known, an engineer must implement this transfer function as a passive or active circuit. This chapter will outline some elementary methods for realizing such filters as passive and active circuits. The chapter will conclude with similar investigations of the Butterworth high-pass filter, which inverts the operation of the low-pass filter.

2. LOW-PASS FILTER BASICS

Question 1: What is a **low-pass filter**?

Answer: A low-pass filter is a device (typically a circuit) offering very little attenuation to the low-frequency content (low-frequency spectrum) of signals, while significantly attenuating (blocking) the high-frequency content of those signals. Figure 21.2 illustrates the loss magnitude in dB of a low-pass filter.

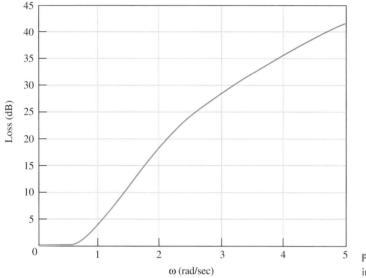

Figure 21.2 Low-pass filter loss $A(\omega)$, in dB, as a function of (normalized) ω.

Question 2: Why use a low-pass filter?

Answer: The noise in a noisy signal often has most of its energy in the high-frequency range. For example, so-called white noise has a constant frequency spectrum. Hence, low-pass filtering a sinusoidal signal corrupted by noise will generally result in a "cleaned-up" information signal, as illustrated in figure 21.3. In that figure, the thickness of the curves represents the infiltration of noise.

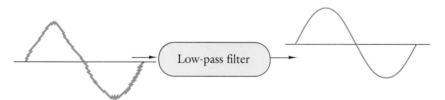

Figure 21.3 The effect of low-pass filtering on a noisy signal.

Question 3: How does an engineer specify a low-pass filter?

Answer: Figure 21.4 depicts a typical **low-pass filter (brickwall) specification**. The two pairs (ω_p, A_{max}) and (ω_s, A_{min}) characterize this brickwall specification. The related specification terminology is:

1. ω_p equals the passband edge frequency.
2. $0 \leq \omega \leq \omega_p$ is called the pass band.
3. A_{max} is maximum attenuation permitted in the pass band.

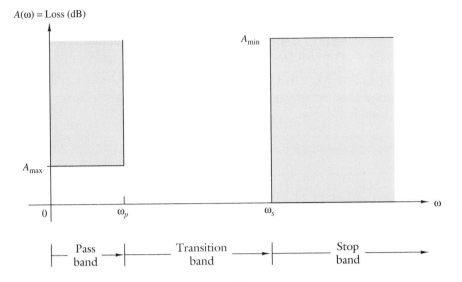

Figure 21.4 Brickwall specification of low-pass filter.

4. ω_s is the stop band edge frequency.
5. $\omega_s \leq \omega$ defines the stop band.
6. A_{min} is the minimum allowable attenuation in the stop band.
7. $\omega_p \leq \omega \leq \omega_s$ defines the transition band.

Observe that figure 21.2 has a general shape that would meet some form of the brickwall specification given in figure 21.4. Finding a low-pass filter which meets that specification involves two critical steps. The first step is to determine a loss function $\hat{H}(s)$ whose magnitude spectrum, equation 21.1, remains outside the shaded regions of figure 21.4. This is called the **approximation problem.** The second step is to construct a circuit that realizes the associated transfer function. Such a circuit is called a **realization**.

Before tackling the low-pass approximation problem, a simple example serves to illustrate the behavior of a first-order transfer function with respect to gain magnitude and loss magnitude.

 EXAMPLE 21.1

Suppose it is necessary to have a filter for which $(\omega_p, A_{max}) = (10^4, 3 \text{ dB})$ and $(\omega_s, A_{min}) = (10^6, 40 \text{ dB})$. The RC circuit in figure 21.5, with $R = 10 \text{ k}\Omega$ and $C = 0.01 \text{ µF}$, meets these specifications. Intuitively, the capacitor has little impedance at high frequencies and a large impedance at low frequencies. By voltage division, most of the input voltage will appear across the capacitor at low frequencies, and very little voltage will exist across the capacitor at high frequencies.

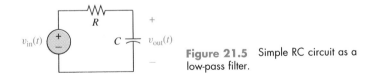

Figure 21.5 Simple RC circuit as a low-pass filter.

For a mathematical analysis, consider the transfer function or gain function,

$$H(s) = \frac{\dfrac{1}{RC}}{s + \dfrac{1}{RC}} = \frac{10^4}{s + 10^4}. \tag{21.2}$$

A plot of the loss magnitude appears in figure 21.6, where, from equation 21.1, the loss in dB is $A(\omega) = -20 \log_{10} |H(j\omega)|$. Observe that the loss magnitude response meets the desired specifications, in which case,

1. the pass band is $0 \leq \omega \leq 10^4$,
2. the transition band is $10^4 \leq \omega \leq 10^6$, and
3. the stop band is $10^6 \leq \omega$.

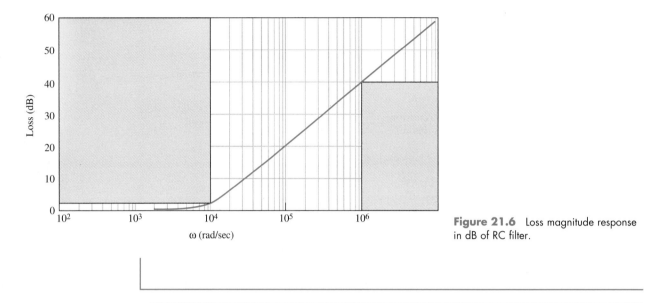

Figure 21.6 Loss magnitude response in dB of RC filter.

Exercise. Plot the phase in degrees of the gain function for $-10^8 \leq \omega \leq 10^8$.

Many filter design procedures (in particular, active realizations) build around the cascade of second-order sections, i.e., second-order transfer functions. A basic second-order gain function (section) for a low-pass filter is

$$H(s) = \frac{V_{\text{out}}(s)}{V_{\text{in}}(s)} = \frac{K}{s^2 + 2\sigma s + \omega_0^2} = \frac{K}{s^2 + \dfrac{\omega_0}{Q}s + \omega_0^2},$$

where Q is the quality factor and 2σ is the bandwidth, as described in chapter 17. Assuming that $\sigma^2 < \omega_0^2$, the quadratic formula specifies the poles as

$$s = -\sigma \pm j\omega_d,$$

where $\omega_0^2 = \sigma^2 + \omega_d^2$. Here, ω_0 denotes what is called the *pole frequency* ω_p of chapter 17. Figure 21.7 shows the resulting pole-zero pattern.

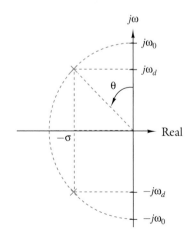

Figure 21.7 Pole-zero plot of basic second-order low-pass filter.

CHAPTER 21 PRINCIPLES OF BASIC FILTERING

When the topic of active filter design comes up, more will be said of the second-order transfer function.

Finally, note that the suspension system of a car is a low-pass filter: Slow, rhythmic (low-frequency) road variations are passed, while the effects of chuckholes and bumps (high frequencies) are filtered out.

3. THE BUTTERWORTH LOW-PASS TRANSFER CHARACTERISTIC

Introduction

In this section, the goal is to derive a transfer function $H(s)$ whose magnitude for $s = j\omega$ is flat over the pass band and has high attenuation (maximum roll-off) over the stop band. This is accomplished in two phases. The first phase constructs a magnitude response that the desired transfer function must have. The second phase actually determines the zeros of the loss function or, equivalently, the poles of the transfer function having the desired magnitude characteristics.

Before starting the derivation, it is necessary to identify two properties of the squared magnitude of a complex-valued function $H(z)$, which is real whenever z is real.

Property 1: $|H(j\omega)|^2$ *is an even function in* ω.

To see this, observe that for each ω, $H(j\omega)$ is a complex number, having the usual form

$$H(j\omega) = \text{Re}[H(j\omega)] + j\text{Im}[H(j\omega)].$$

By the properties of the complex conjugate of a complex number, the squared magnitude of $H(j\omega)$ satisfies

$$|H(j\omega)|^2 = H(j\omega)H^*(j\omega) = H^*(-j\omega)H(-j\omega) = |H(-j\omega)|^2,$$

because $H^*(j\omega) = H(-j\omega)$. Since $|H(j\omega)|^2 = |H(-j\omega)|^2$, it is an even function.

Property 2: $|H(j\omega)|^2$ *is the ratio of two polynomials of the form* $n(\omega^2)$ *and* $d(\omega^2)$.

To verify this property, observe that since $H(s)$ is a rational function in s, and since $|H(j\omega)|^2$ is an even function, $|H(j\omega)|^2$ is expressible as the ratio of two even polynomials in ω. Since each polynomial is even, only even powers of ω occur. This means that the polynomials can be written as functions of ω^2 rather than ω, i.e.,

$$|H(j\omega)|^2 = \frac{n(\omega^2)}{d(\omega^2)},$$

where $n(\omega^2)$ and $d(\omega^2)$ denote even polynomials in ω.

Phase 1: Development of the Butterworth Magnitude Response

Determine the form or structure of the squared magnitude of a rational transfer function $H(s)$ so that the magnitude for $s = j\omega$ is flat over the pass band and has maximum continuous increasing attenuation over the stop band.

Step 1. *What general structure of* $|H(j\omega)|^2$ *would permit maximum continuous increasing roll-off in the stop band?* To achieve such a roll off, $n(\omega^2)$ can have no zeros, i.e., $n(\omega^2) = n_0$, a constant. Making $n(\omega^2)$ a constant maximizes the difference between the degrees of the denominator and numerator. Maximizing this difference results in a structure that has maximum continuous increasing attenuation in the stop band. Hence, a candidate squared magnitude is

$$|H(j\omega)|^2 = \frac{n_0}{d_0 + d_2\omega^2 + d_4\omega^4 + \cdots + d_{2n}\omega^{2n}},$$

where we have chosen even-numbered subscripts on the d_i coefficients to coincide with the even powers of ω.

Step 2. *Achieve unity gain (0-dB loss) at dc.* To achieve unity gain at dc, simply set $n_0 = d_0$. Without loss of generality, these constants are normalized to unity, i.e., $n_0 = d_0 = 1$.

Step 3. To achieve maximal flatness in the pass band, set

$$d_2 = d_4 = \dots = d_{2n-2} = 0.$$

If any of these coefficients were nonzero, the denominator polynomial would not be maximally flat. (A mathematical explanation of this property appears at the end of the section.)

At this point, our candidate magnitude response is

$$|H(j\omega)|^2 = \frac{1}{1 + d_{2n}\omega^{2n}}. \tag{21.3}$$

Step 4. For a sufficiently large n, the coefficient d_{2n} is chosen so that the filter specifications at ω_p and ω_s are not violated; i.e.,

$$-10\log_{10}|H(j\omega_p)|^2 \le A_{\max}$$

and

$$-10\log_{10}|H(j\omega_s)|^2 \ge A_{\min}.$$

To clarify how this is done, rewrite equation 21.3 as

$$|H(j\omega)|^2 = \frac{1}{1 + \varepsilon^2\left(\dfrac{\omega}{\omega_p}\right)^{2n}}, \tag{21.4}$$

where $d_{2n} = (\varepsilon^{(1/n)}/\omega_p)^{2n}$. It is now the new constant, ε, that is chosen so that for an appropriate n, the specifications at ω_p and ω_s are not violated. Usually, ε is not unique. However, for any proper choice, equation 21.4 can be rewritten again as

$$|H(j\omega)|^2 = \frac{1}{1 + \varepsilon^2\left(\dfrac{\omega}{\omega_p}\right)^{2n}} = \frac{1}{1 + \left(\dfrac{\omega}{\omega_c}\right)^{2n}}, \tag{21.5}$$

where $\omega_c = \omega_p/\varepsilon^{(1/n)}$ is the 3-dB down point, or **cutoff frequency**, of the filter. The terminology arises here because, for all n,

$$10\log_{10}\left[1 + \left(\frac{\omega}{\omega_c}\right)^{2n}\right]_{\omega=\omega_c} = 10\log_{10}[2] = 3 \text{ dB};$$

i.e., there are 3 dB of loss at $\omega = \omega_c$. When $\varepsilon = 1$, the passband edge frequency ω_p and cutoff frequency ω_c coincide. If we now define the normalized frequency as $\Omega = \omega/\omega_c$, the magnitude response of equation 21.5 becomes

$$|H(j\Omega)|^2 = \frac{1}{1 + \Omega^{2n}}. \tag{21.6}$$

This denotes the nth-order *normalized Butterworth magnitude response*. The word *normalized* refers to the fact that at $\Omega = 1$, the loss is 3 dB (the gain is -3 dB). Remember that the actual filter transfer function depends on a proper choice of ω_c or ε.

Before proceeding further, a critical question comes to mind: *Does this kind of polynomial representation make sense?* To answer this question, note that the dB loss, as per equation 21.6, is

$$A(\Omega) = \text{loss (dB)} = 20\log_{10}|\hat{H}(j\Omega)| = 10\log_{10}[1 + \Omega^{2n}].$$

Plotting this function for various n's as a function of Ω for a hypothetical Ω_s yields the sketch in figure 21.8. Clearly, these polynomials have a suitable magnitude response and hence can validly be used to meet the filter specifications.

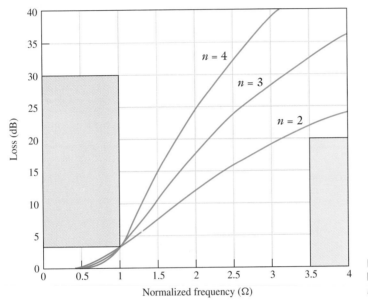

Figure 21.8 Plot of the normalized Butterworth magnitude responses for $n = 2, 3, 4$.

The next task is to find an $H(s)$ whose magnitude response at $s = j\Omega$ satisfies the normalized response of equation 21.6. Once the normalized transfer function is found, the desired transfer function that satisfies the original brickwall specification at ω_p is computed by the frequency-scaling procedure $s \to s/\omega_c = \varepsilon^{(1/n)}s/\omega_p$.

Phase 2: Development of the Butterworth Transfer Function

Derive the form of the Butterworth transfer function $H(s)$ whose magnitude satisfies the normalized Butterworth magnitude response of equation 21.6.

Step 1. Recall that the loss function $\hat{H}(s) = 1/H(s)$. From the preceding discussion, $\hat{H}(s)$ is a normalized Butterworth loss function (polynomial) if and only if it has a Butterworth loss magnitude; i.e.,

$$|\hat{H}(j\Omega)|^2 = 1 + \Omega^{2n}. \tag{21.7}$$

Step 2. Find $\hat{H}(s)$ *such that at* $s = j\Omega$, $\hat{H}(j\Omega) = 1 + \Omega^{2n}$. Observe that if $s = j\Omega$, then $\Omega = -js$, and

$$|\hat{H}(j\Omega)|^2 = \hat{H}(j\Omega)\hat{H}(-j\Omega) = \hat{H}(s)\hat{H}(-s) = 1 + (-js)^{2n} = 1 + \left(-s^2\right)^n. \tag{21.8}$$

Step 3. *Determine the zeros of* $\hat{H}(s)$. From equation 21.8, $\hat{H}(s)\hat{H}(-s)$ must satisfy

$$\hat{H}(s)\hat{H}(-s) = 1 + \left(-s^2\right)^n = 0.$$

Because a filter must be stable, the left half complex plane zeros of this function will determine $\hat{H}(s)$. Since the zeros of $\hat{H}(s)$ and $\hat{H}(-s)$ are symmetric about the vertical axis, the zeros of $\hat{H}(-s)$ lie in the right half-plane; i.e., the zeros of $\hat{H}(s)$ are the left-half plane zeros of $1 + [-s^2]^n = 0$. These zeros are given by DeMoivre's theorem (or by using the roots-command in MATLAB) and have the form

$$s_k = \exp\left[j\frac{\pi}{2}\left(\frac{2k+n-1}{n}\right)\right] = \cos\left[\frac{\pi}{2}\left(\frac{2k+n-1}{n}\right)\right] + j\sin\left[\frac{\pi}{2}\left(\frac{2k+n-1}{n}\right)\right] \quad (21.9)$$

for $k = 1, 2, \ldots, 2n$. If $s_1, \ldots, s_n$ denote the left half-plane zeros, then

$$\hat{H}(s) = (s - s_1)(s - s_2)\cdots(s - s_n). \quad (21.10)$$

This set of roots, parameterized by n and given by equations 21.9 and 21.10, are the zeros of the **normalized Butterworth loss function**. Calculation of this function is more or less straightforward. The following example illustrates how the normalized sixth-order Butterworth function is computed.

◆ **EXAMPLE 21.2**

Find the sixth-order normalized Butterworth loss function.

SOLUTION

To solve the problem, we find the 12 roots of

$$1 + [-s^2]^6 = 0.$$

From DeMoivre's theorem or equation 21.9, each root is

$$s_k = \exp\left[j\frac{\pi}{2}\left(\frac{2k+5}{6}\right)\right] \equiv \exp[j\alpha]$$

for appropriate α's. Observe that $|s_k| = 1$ for all k, implying that all roots lie on the unit circle. Table 21.1 lists the roots in the left half-plane.

TABLE 21.1 ROOTS OF SIXTH-ORDER NORMALIZED BUTTERWORTH LOSS FUNCTION

k	$j\alpha$	$\exp(j\alpha)$
1	$j7\pi/12$	$-0.2588 + j0.9659$
2	$j9\pi/12$	$-0.7071 + j0.7071$
3	$j11\pi/12$	$-0.9659 + j0.2588$
4	$j13\pi/12$	$-0.9659 - j0.2588$
5	$j15\pi/12$	$-0.7071 - j0.7071$
6	$j17\pi/12$	$-0.2588 - j0.9659$

The remaining roots are in the right half-plane and are not part of the loss function. Consequently, the sixth-order normalized Butterworth loss function is

$$\hat{H}(s) = \prod_{k=1}^{6}(s - s_k) = (s^2 + 0.5176s + 1)(s^2 + 1.414s + 1)(s^2 + 1.9319s + 1).$$

A plot of these root locations is given in figure 21.9.

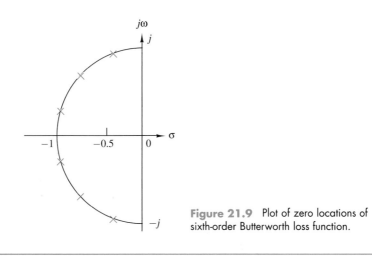

Figure 21.9 Plot of zero locations of sixth-order Butterworth loss function.

Calculation of the roots for various orders of n leads to table 21.2.

n	$\widehat{H}(s)$
	TABLE 21.2 NORMALIZED BUTTERWORTH LOSS FUNCTIONS, $N = 1, \ldots, 5$
1	$(s + 1)$
2	$(s^2 + \sqrt{2}s + 1)$
3	$(s + 1)(s^2 + s + 1)$
4	$(s^2 + 0.76537s + 1)(s^2 + 1.84776s + 1)$
5	$(s + 1)(s^2 + 0.61803s + 1)(s^2 + 1.61803s + 1)$

Properties of the Butterworth Loss Function

As mentioned earlier, the Butterworth loss function has a maximally flat passband response. **Maximally flat** means that as many derivatives as possible are zero at the origin, i.e., at $\Omega = 0$. In the case of the nth-order Butterworth loss function, it is possible to show that

$$\frac{d^k}{d\Omega^k} \left| \hat{H}(j\Omega) \right| = 0, \text{ for } k = 1, 2, \ldots, 2n - 1, \tag{21.11}$$

at $\Omega = 0$ and

$$\frac{d^{2n}}{d\Omega^{2n}} \left| \hat{H}(j\Omega) \right| = -0.5(2n!) \tag{21.12}$$

at $\Omega = 0$. This is consistent with the notion of being maximally flat. To verify equations 21.11 and 21.12, observe that if $\Omega^{2n} << 1$, then

$$\left| \hat{H}(j\Omega) \right| = \sqrt{1 + \Omega^{2n}} = 1 + \frac{1}{2}\Omega^{2n} - \frac{1}{8}\Omega^{4n} + \frac{1}{16}\Omega^{6n} - \cdots . \tag{21.13}$$

Equations 21.11 and 21.12 follow after differentiating equation 21.13 $2n$ times and evaluating each derivative at $\Omega = 0$.

Several closing remarks are now in order. First, the cutoff frequency is the **half-power point**, or the 3-dB down point. The terminology follows because $|H(j0)| = 1$ (the gain at dc is unity) and $|H(j1)| = 1/\sqrt{2} = 0.707$, yielding half power for normalized Butterworth gain functions $H(s)$. Second, for large Ω, $|H(j\Omega)| \approx 1/\Omega^n$. Hence, as Ω increases, the gain decreases according to

$$-n[20 \log_{10}(\Omega)],$$

indicating that the gain rolls off with a slope proportional to the number of poles of the gain function n. Specifically, the slope equals $-20n$ dB per decade. These statements are also valid if ω replaces Ω.

4. COMPUTATION OF BUTTERWORTH LOSS FUNCTIONS FROM BRICKWALL SPECIFICATIONS

How does the development in the previous section help us unearth an $H(s)$ meeting a given brickwall specification? Finding an n allows one to choose a normalized $H(s)$. In general, a normalized $H(s)$ will not meet a given brickwall specification. Frequency scaling, $s \to s/\omega_c = \varepsilon^{(1/n)} s/\omega_p$, adjusts the normalization to comply with the specification. This section forges an algorithm for computing n, $H(s)$, and ε or ω_c and illustrates the procedure with an example. The algorithm for computing a Butterworth loss function meeting a set of brickwall specifications has five steps.

Step 1. Identify the filter specs $(\omega_p, A_{\max})$ and $(\omega_s, A_{\min})$.

Step 2. Set $\Omega_s = \omega_s/\omega_p$, and compute the filter order via the formula

$$n \geq \frac{\log_{10}\left(\sqrt{\dfrac{10^{0.1 A_{\min}} - 1}{10^{0.1 A_{\max}} - 1}}\right)}{\log_{10}(\Omega_s)}. \tag{21.14}$$

Typically, the number given by the formula on the right side of this inequality is not an integer. The filter order is taken as the smallest integer greater than the number produced by the formula. A derivation of this formula follows at the completion of the algorithm.

Step 3. Look up the normalized Butterworth polynomial in table 21.2. Alternatively, you can use a computer program or your calculator to solve for the proper zeros of $1 + (-s^2)^n = 0$ and generate the polynomial directly.

If the filter is to be realized as a cascade of second-order sections, then express the loss function as a product of quadratics and at most one first-order polynomial. On the other hand, if the filter is to be an LC ladder network, then express the loss function as a single polynomial by multiplying out all factors in the aforesaid product. (See example 21.4.)

Step 4. Choose ω_c or, equivalently, ε so that the loss magnitude response, $\hat{H}(j\omega)$, falls within the range permitted by the brickwall specifications.

There is freedom in choosing ω_c or ε because the filter order is taken as the smallest integer greater than the numerical value produced by equation 21.14. The role of ε is to slide the loss magnitude response curve up or down within the range permitted by the brickwall specifications.

Step 5. Frequency scale the loss function by $\omega_c = \omega_p/[\varepsilon^{(1/n)}]$. This scaling process is usually called the LP-to-LP (low-pass to low-pass) type transformation.

Some discussion of the algorithm is now in order. First, to derive the filter order, equation 21.14 of step 2, observe that n must satisfy the passband and stop band edge frequency constraints. From equations 21.1 and 21.4, at the passband edge frequency,

$$A_{\min} \leq 10 \log_{10}\left|\hat{H}\left(j\frac{\omega_s}{\omega_p}\right)\right|^2 = 10 \log_{10}\left[1 + \varepsilon^2\left(\frac{\omega_s}{\omega_p}\right)^{2n}\right].$$

Hence,

$$\varepsilon^2 \left(\frac{\omega_s}{\omega_p} \right)^{2n} \geq 10^{0.1 A_{min}} - 1, \tag{21.15}$$

or equivalently,

$$\left(\frac{\omega_s}{\omega_c} \right)^{2n} \geq 10^{0.1 A_{min}} - 1. \tag{21.16}$$

Similarly, at the stop band edge frequency,

$$\left(\frac{\omega_p}{\omega_c} \right)^{2n} \leq 10^{0.1 A_{max}} - 1. \tag{21.17}$$

Dividing the left and right sides of equation 21.16 by the left and right sides of equation 21.17 (this maintains the inequality of equation 9.16) yields

$$\left(\frac{\omega_s}{\omega_p} \right)^{2n} \geq \frac{10^{0.1 A_{min}} - 1}{10^{0.1 A_{max}} - 1}. \tag{21.18}$$

Solving for n produces

$$n \geq \frac{\log_{10} \left(\sqrt{\frac{10^{0.1 A_{min}} - 1}{10^{0.1 A_{max}} - 1}} \right)}{\log_{10} \left(\frac{\omega_s}{\omega_p} \right)} = \frac{\log_{10} \left(\sqrt{\frac{10^{0.1 A_{min}} - 1}{10^{0.1 A_{max}} - 1}} \right)}{\log_{10} (\Omega_s)},$$

which is precisely equation 21.14.

As already mentioned, the right side of equation 21.14 does not generally yield an integer value for a given set of specifications. Logically, one picks the smallest integer containing the result of equation 21.14. This induces some leeway in the choice of ε. Usually, one computes ε to achieve a desired A_{max} at the passband edge frequency or a desired A_{min} at the stop band edge frequency. In the first case, $A_{max} = 10 \log_{10}[1 + \varepsilon^2]$, where

$$\varepsilon = \varepsilon_{max} = \sqrt{10^{0.1 A_{max}} - 1}. \tag{21.19}$$

In the second case, observe that $A_{min} = 10 \log_{10}[1 + \varepsilon^2 (\Omega_s)^{2n}]$. Solving for ε yields

$$\varepsilon = \varepsilon_{min} = \frac{\sqrt{10^{0.1 A_{min}} - 1}}{\Omega_s^n}. \tag{21.20}$$

Either choice, and any value in between, i.e., $\varepsilon_{min} \leq \varepsilon \leq \varepsilon_{max}$, will yield a magnitude response meeting the brickwall specifications. *The convention in this text will be to choose ε to satisfy the A_{max} specification as per equation 21.19.* In this case, the magnitude response curve will cut through the passband corner of the given brickwall specification and, in general, exceed the A_{min} specification at ω_s.

Choosing different ε's is equivalent to choosing different ω_c's for the filter. Again, the cutoff frequency, $\omega_c = \omega_p / [\varepsilon^{(1/n)}]$, is the 3-dB loss, or half-power point, of the filter. Based on equations 21.19 and 21.20, the leeway in ω_c is

$$\frac{\omega_p}{\sqrt[2n]{10^{0.1 A_{max}} - 1}} \leq \omega_c \leq \frac{\omega_s}{\sqrt[2n]{10^{0.1 A_{min}} - 1}}. \tag{21.21}$$

All such choices yield a filter meeting the brickwall specifications. Choosing the smallest value for ω_C from equation 21.21 forces the magnitude response curve to cut through the passband corner, the effect being the same as choosing $\varepsilon = \varepsilon_{max}$. Available component values might dictate different choices of ω_c.

After computing ω_c, one can scale the normalized Butterworth transfer function to obtain a desired gain function. Again, however, practical circuit design is usually not based on a frequency-scaled transfer function. Rather, the ordinary design procedure has two steps: (1) Find a circuit that realizes the normalized gain/loss function, and (2) magnitude and frequency scale the circuit parameters—R's, L's, C's, etc.—to obtain realistic parameter values and the desired cutoff frequency ω_c.

An example will help clarify these ideas.

◆ **EXAMPLE 21.3**

Find the Butterworth transfer function that meets the low-pass brickwall specification of figure 21.10.

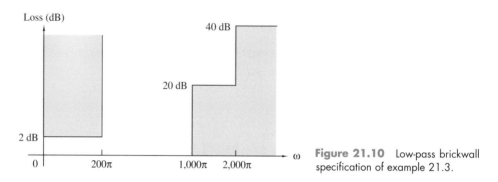

Figure 21.10 Low-pass brickwall specification of example 21.3.

SOLUTION

Step 1. From the brickwall specification, there are two cases to consider: (1) $(\omega_p = 200\pi,$ $A_{max} = 2$ dB$)$, $(\omega_{s1} = 1,000\pi, A_{1min} = 20$ dB$)$ and (2) $(\omega_p = 200\pi, A_{max} = 2$ dB$)$, $(\omega_{s2} = 2,000\pi, A_{2min} = 40$ dB$)$.

Step 2. *Compute the filter orders for each case.* For case (1),

$$n \geq \frac{\log_{10}\left(\sqrt{\dfrac{10^{0.1A_{min}} - 1}{10^{0.1A_{max}} - 1}}\right)}{\log_{10}(\Omega_{s1})} = \frac{2.2286}{1.3979} = 1.594,$$

where $\Omega_{s1} = 5$. This implies that the filter order must be at least 2. For case (2),

$$n \geq \frac{\log_{10}\left(\sqrt{\dfrac{10^{0.1A_{min}} - 1}{10^{0.1A_{max}} - 1}}\right)}{\log_{10}(\Omega_{s2})} = \frac{4.2329}{2} = 2.116,$$

where $\Omega_{s2} = 10$. This implies that the minimum filter order is $n = 3$.

Step 3. Looking up the third-order normalized Butterworth loss function and inverting to obtain the transfer function yields

$$H(s) = \frac{1}{(s + 1)(s^2 + s + 1)}. \tag{21.22}$$

As this gain function does not yet meet the specifications, it is necessary to frequency scale to obtain the necessary cutoff.

Step 4. The cutoff or 3-dB loss frequency is $\omega_c = \omega_p/[\varepsilon^{(1/n)}] = 200\pi/[\varepsilon^{(1/3)}]$. Choosing ε to meet the specification at A_{max} yields $\varepsilon = 0.7648$, in which case $\omega_c = 218.7\pi$. Frequency scaling equation 21.22 produces the Butterworth transfer function

$$H(s) = \frac{324.3 \times 10^6}{(s + 687.1)(s^2 + 687.1s + 472.1 \times 10^3)},$$ (21.23)

which meets the given magnitude specification.

It is interesting that the attenuation (equation 21.1) at $\omega = 2,000\pi$ is

$$A(\Omega_{s2}) = A(10) = 10\log_{10}\left[1 + \varepsilon^2 \Omega_{s2}^6\right] = 10\log_{10}\left[1 + \left(\frac{\omega_s}{\omega_c}\right)^6\right] = 57.7 \text{ dB},$$

which exceeds the 40-dB specification required in figure 21.10. This follows because of the choice of ε.

Exercise. In terms of ω_c, show that the magnitude function of equation 21.6 is

$$|H(j\omega)| = \frac{K}{\sqrt{1 + \left(\frac{\omega}{\omega_c}\right)^{2n}}}$$ (21.24)

for appropriate K, usually 1.

Two final questions, before closing this section, are as follows:

Question 1: Does the design engineer now find a filter that realizes the gain function of equation 21.23?

Answer: In passive and active circuit design practice, one does not usually compute this transfer function. Rather, one finds a circuit that realizes the normalized Butterworth gain or loss function and then frequency and magnitude scales the circuit element values to obtain the desired cutoff and realistic parameter values.

Question 2: How do the normalized filter pole locations move after frequency scaling by ω_c?

Answer: In frequency scaling, $s \to s/\omega_c = \varepsilon^{1/n} s/\omega_p$. A zero of the loss function satisfies $s - s_k = 0$. Thus, under the given substitution, $s/\omega_c - s_k = 0$ implies that s_k moves to $\omega_c s_k$.

This pole movement formula is useful in the construction of a loss function by a computer program and in developing digital filter implementations of a transfer function.

5. BASIC PASSIVE REALIZATION OF BUTTERWORTH TRANSFER FUNCTIONS

The second phase of analog filter design is circuit implementation, i.e., finding a circuit that realizes the desired Butterworth transfer function. In an advanced course, very systematic synthesis methods are discussed. These are beyond the scope of this text. For low-order filters, we can use a very elementary **coefficient-matching technique**. To accomplish this, one picks a candidate circuit and computes its transfer function in terms of the circuit parameters, i.e., in terms of, say, R_1, R_2, C_1, C_2, L_1, etc. The next step is to match coefficients with the appropriate normalized transfer function. This generates a set of equations whose solution yields the needed circuit parameter values. With these values, one then frequency scales to obtain the necessary ω_c and magnitude scales to obtain proper impedance levels. Example 21.4 illustrates these design steps.

EXAMPLE 21.4

In example 21.3, a third-order normalized Butterworth transfer function was found to meet the brickwall specification of figure 21.10. The normalized transfer function, equation 21.22, then has the form

$$H(s) = \frac{K}{s^3 + 2s^2 + 2s + 1}. \tag{21.25}$$

The transfer function, which actually meets the given specifications, requires frequency scaling of equation 21.25 by $\omega_c = 218.7\pi$.

The desired filter is to operate between a source having an internal resistance $R_s = 100 \ \Omega$ and a load $R_L = 100 \ \Omega$. As a first step in realizing this filter, consider the circuit of figure 21.11. A proper choice of the element values will yield the transfer function of equation 21.25. Steps 1 and 2, which follow, verify this claim.

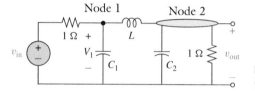

Figure 21.11 Third-order filter circuit for example 21.4.

Step 1. *Write and solve node equations for the circuit of figure 21.11.* By inspection, at node 1,

$$\left(C_1 s + \frac{1}{Ls} + 1\right) V_1 - \frac{1}{Ls} V_{out} = V_{in}. \tag{21.26}$$

At node 2,

$$-\frac{1}{Ls} V_1 + \left(C_2 s + \frac{1}{Ls} + 1\right) V_{out} = 0. \tag{21.27}$$

Solving equation 21.27 for V_1 and eliminating V_1 from equation 21.26 yields

$$H(s) = \frac{V_{out}(s)}{V_{in}(s)} = \frac{\dfrac{1}{LC_1C_2}}{s^3 + \dfrac{C_1 + C_2}{C_1C_2}s^2 + \dfrac{C_1 + C_2 + L}{LC_1C_2}s + \dfrac{2}{LC_1C_2}}. \tag{21.28}$$

Step 2. *Equate the coefficients of equation 21.28 with those of equation 21.25.* After the exercise of some algebra, the only solution possible is $C_1 = C_2 = 1$ F and $L = 2$ H. (See problem 7.) This produces a gain constant of $K = 0.5$. Remember that the important part of the design is the correct pole locations, not the overall gain. Hence, the foregoing circuit will realize the proper magnitude response, although the gain will be lower than is desired.

Step 3. *Frequency scale the circuit to obtain the desired cutoff frequency, and magnitude scale the circuit to obtain realistic impedances.* Suppose the resistance values are to be 100 Ω. Recall that $\omega_c = \omega_p/\varepsilon^{(1/3)} = (200\pi)/(0.7648)^{(1/3)} = 218.7\pi$. Thus, after frequency and magnitude scaling the circuit parameters, we find that the final circuit, shown in figure 21.12, meets the desired specifications.

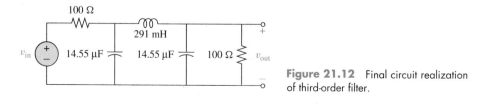

Figure 21.12 Final circuit realization of third-order filter.

Of course, there may be other passive RLC-type circuits that satisfy this particular Butterworth transfer function. Problem 11 suggests an alternative solution with two inductances and one capacitor. From a practical point of view, such a circuit is less desirable than one with two capacitors and one inductor, since capacitors are usually closer to ideal than are inductors.

Also, in step 2, one must solve three nonlinear equations in three unknowns to obtain the solution $C_1 = C_2 = 1$ F and $L = 2$ H. If you have gone through the details of working out problem 7, you soon will see that solving such a set of equations is nontrivial. For this reason, the coefficient-matching approach is usually applied only to low-order transfer functions, say, of no more than third order.

As another example of filter realization, consider the circuit of figure 21.13. This circuit meets the pole requirements of a second-order Butterworth filter. To see this, observe that a straightforward nodal analysis shows that the circuit has the transfer function

$$H(s) = \frac{V_{out}(s)}{V_{in}(s)} = \frac{\dfrac{1}{LC}}{s^2 + \left(\dfrac{1}{L} + \dfrac{1}{C}\right)s + \dfrac{2}{LC}}. \qquad (21.29)$$

By equating the denominator of equation 21.29 to $s^2 + \sqrt{2}s + 1$, one computes the only solution, $L = \sqrt{2}$ and $C = \sqrt{2}$. There then results the normalized second-order Butterworth filter with equal source and load resistors. Had the source resistance exceeded the load resistance, then the configuration of figure 21.13 could still be used to realize the second-order filter. However, in that case, there would be two possible solutions. (See problem 12 to work through these details.)

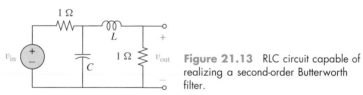

Figure 21.13 RLC circuit capable of realizing a second-order Butterworth filter.

6. BASIC ACTIVE REALIZATION OF BUTTERWORTH TRANSFER FUNCTIONS

Active circuits provide an alternative to passive circuits. Here, however, the realization process builds around a cascade of second-order sections, with at most one first-order section.

Sallen and Key Active Low-Pass Filter

Figure 21.14 shows a common positive-feedback, second-order, active, low-pass filter section known as the *Sallen and Key circuit*. It is easy to verify that this circuit actually realizes a second-order Butterworth low-pass filter. The first step is to derive the circuit transfer

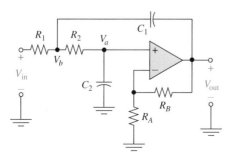

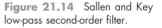

Figure 21.14 Sallen and Key low-pass second-order filter.

function $H(s)$. In deriving $H(s)$, note that the properties of an ideal op amp force the voltage across R_A to be V_a. Using voltage division, one then observes that

$$V_a = \frac{R_A}{R_A + R_B} V_{\text{out}},$$

or equivalently,

$$V_{\text{out}} = \frac{R_A + R_B}{R_A} V_a = \left(1 + \frac{R_B}{R_A} \right) V_a \equiv K V_a.$$

This observation permits us to model the Sallen and Key circuit with the equivalent circuit of figure 21.15.

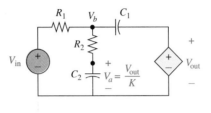

Figure 21.15 Equivalent circuit of the Sallen and Key low-pass filter of figure 21.14; V_b identifies node b and V_a identifies node a.

To construct $H(s)$, observe that at node a—the node between C_2 and R_2—

$$-\frac{1}{R_2} V_b + \left(\frac{1}{R_2} + C_2 s \right) \frac{V_{\text{out}}}{K} = 0. \tag{21.30}$$

Similarly, at node b,

$$\left(\frac{1}{R_1} + \frac{1}{R_2} + C_1 s \right) V_b - \left(\frac{1}{R_2} \right) \frac{V_{\text{out}}}{K} = \frac{V_{\text{in}}}{R_1} + C_1 s V_{\text{out}}. \tag{21.31}$$

Eliminating V_b and solving for the ratio $V_{\text{out}}/V_{\text{in}}$ yields the transfer function of the Sallen and Key low-pass filter, viz.,

$$H(s) = \frac{V_{\text{out}}(s)}{V_{\text{in}}(s)} = \frac{\dfrac{K}{R_1 R_2 C_1 C_2}}{s^2 + \left(\dfrac{1}{R_1 C_1} + \dfrac{1}{R_2 C_1} + \dfrac{1-K}{R_2 C_2} \right) s + \dfrac{1}{R_1 R_2 C_1 C_2}}, \tag{21.32}$$

where $K = V_{\text{out}}/V_a = 1 + (R_B/R_A)$.

This transfer function has the general form

$$H(s) = K \frac{\omega_0^2}{s^2 + \dfrac{\omega_0}{Q} s + \omega_0^2}, \tag{21.33}$$

where Q is the quality factor, $\omega_0^2 = \sigma^2 + \omega_d^2$, and 2σ is the bandwidth, as described in chapter 17. Assuming that $\sigma^2 < \omega_0^2$, the quadratic formula specifies the poles of equation 21.33 as

$$s_p = -\sigma \pm j\omega_d.$$

Here, ω_0 denotes what is called the *pole frequency*, ω_p, of chapter 17. In design, ω_0 is normalized to 1, yielding a normalized transfer function

$$H(s) = \frac{K}{s^2 + \dfrac{1}{Q} s + 1}. \tag{21.34}$$

Designs are constructed by equating coefficients in equations 21.34 and 21.32 and solving the resulting equations. The solutions, however, are not unique. Different solutions have the proper filtering action, but have different behaviors with respect to deviations from nominal resistance and capacitor values. Also, different designs have different ratios of element values.

Design A: $R_A = \infty$, $R_B = 0$, $K = 1$, $R_1 = R_2 = 1\ \Omega$, $C_1 = 2Q$, and $C_2 = 1/2\ Q$

One possible solution (labeled design A) having the transfer function of equation 21.34, normalizes $K = 1$ and $R_1 = R_2 = 1\ \Omega$. Equating the coefficients of equations 21.32 and 21.34 then requires that

$$H(s) = \frac{K}{s^2 + \dfrac{1}{Q}s + 1} = \frac{\dfrac{1}{C_1 C_2}}{s^2 + \dfrac{2}{C_1}s + \dfrac{1}{C_1 C_2}}. \tag{21.35}$$

Hence, $C_1 = 2Q$ and $C_2 = 1/2Q$. Notice that the ratio $C_1/C_2 = 4Q^2$, which is 400 for a circuit with $Q = 10$. Such a large variation may be undesirable. At any rate, figure 21.16 illustrates the resulting circuit.

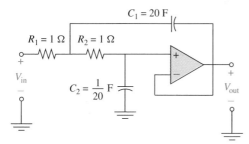

Figure 21.16 Design A for realizing the transfer function of equation 21.35.

Design B: $R_A = R_B = 1\ \Omega$, $K = 2$, $C_1 = 1\ F$, $C_2 = Q^{-1}$, $R_1 = 1$, and $R_2 = Q$

A second design, labeled design B, sets $R_A = R_B = 1\ \Omega$ (making $K = 2$) $C_1 = 1\ F$, and requires a time constant equality, $R_1 C_1 = R_2 C_2$. With these choices, $1/(R_1 R_2 C_1 C_2) = 1$ implies that $R_1 = 1$ and $R_2 = Q$. Thus, $R_1 C_1 = 1 = R_2 C_2$ implies that $C_2 = Q^{-1}$. Here, the maximum parameter ratio is $R_2/R_1 = C_1/C_2 = Q$, which is better than the $4Q^2$ ratio of design A. Figure 21.17 exemplifies the resulting circuit.

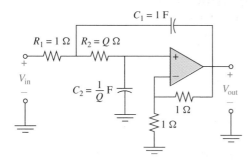

Figure 21.17 Design B for meeting a second-order low-pass characteristic.

Saraga Design: $C_1 = \sqrt{3}Q$, $C_2 = 1$, $R_1 = Q^{-1}$, $R_2 = 1/\sqrt{3}$, $R_B = R_A/3$, $K = 4/3$

A third design, known as the *Saraga design*, sets $C_2 = 1, C_1 = \sqrt{3}Q$ and maintains the ratio

$$\frac{R_2}{R_1} = \frac{Q}{\sqrt{3}}. \tag{21.36}$$

Matching coefficients with equation 21.34 yields

$$\omega_0^2 = 1 = \frac{1}{R_1 R_2 \sqrt{3} Q} = \frac{1}{R_1^2 Q^2},$$

in which case $R_1 = Q^{-1}$. From equation 21.36, $R_2 = 1/\sqrt{3}$. Solving

$$\frac{1}{R_1 C_1} + \frac{1}{R_2 C_1} + \frac{1 - K}{R_2 C_2} = \frac{1}{Q}$$

produces $K = 4/3$. Figure 21.18 pictures the design, which has certain sensitivity properties that make it attractive.

There are, of course, many other active circuits that will meet a low-pass filter specification. The interested reader should consult the numerous available texts on filter design for a discussion of them.

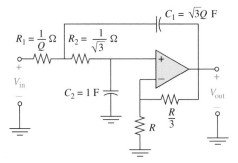

Figure 21.18 Saraga design of Sallen and Key circuit.

7. INPUT ATTENUATION AND GAIN ENHANCEMENT FOR ACTIVE CIRCUIT DESIGN

Input Attenuation

In design A in the previous section, the gain of the filter is $K = 1$. In design B, the filter gain is $K = 2$. The choice of K affects the other design parameters and, indirectly, the pole locations of the filter. Often, a designer wants a reduced gain for a given design. Perhaps the designer will tolerate a change in a resistor. How can one reduce the filter gain, but at the same time keep as many parameter values as possible at their original design values? A technique known as *input attenuation* provides an answer. In this technique, the front-end resistor is replaced with a voltage divider circuit, as illustrated in figure 21.19.

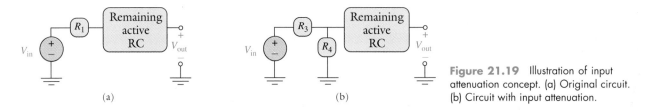

(a)   (b)

Figure 21.19 Illustration of input attenuation concept. (a) Original circuit. (b) Circuit with input attenuation.

Figure 21.19a represents the original active network, while figure 21.19b represents the modified network. For the filtering characteristics to remain invariant, the impedance seen at the input to the remaining part of the active RC network must be the same for both circuits. Thus, the parallel combination of R_3 and R_4 must equal R_1, i.e.,

$$R_1 = \frac{R_3 R_4}{R_3 + R_4}. \tag{21.37}$$

In addition, if the new gain is to be αK, $\alpha < 1$, instead of K, then

$$\alpha = \frac{R_4}{R_3 + R_4}. \tag{21.38}$$

Solving equations 21.37 and 21.38 for R_3 and R_4 yields

$$R_3 = \frac{R_1}{\alpha} \quad \text{and} \quad R_4 = \frac{R_1}{1 - \alpha}. \tag{21.39}$$

Thus, one can reduce the gain of the Sallen and Key low-pass circuit via the simple technique of input attenuation.

Gain Enhancement

Now suppose a designer wants more gain than a particular Sallen and Key design can deliver. Is it possible to modify the circuit to increase the gain? The Sallen and Key is a feedback circuit: Current feeds back through the C_1 capacitor to the input. Decreasing the amount of current feedback increases the overall gain. Decreasing the feedback current through C_1 is accomplished, in one way, by decreasing the feedback voltage via a voltage divider circuit on the filter output. This is called *resistive gain enhancement*. Figure 21.20 shows the Sallen and Key circuit with resistive gain enhancement.

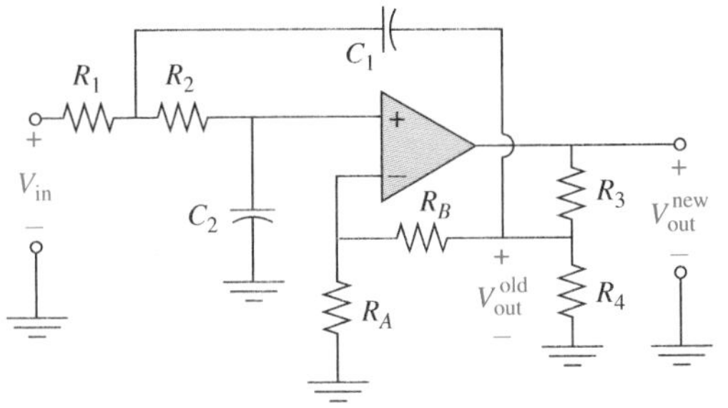

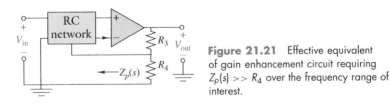

Figure 21.20 Sallen and Key circuit with resistive gain enhancement.

Under certain conditions, to be stated later, if

$$k = \frac{R_4}{R_3 + R_4}, \tag{21.40}$$

then

$$k V_{\text{out}}^{\text{new}} \cong V_{\text{out}}^{\text{old}}$$

and

$$\frac{1}{k} H(s) = \frac{1}{k} \frac{V_{\text{out}}^{\text{old}}}{V_{\text{in}}} \cong \frac{V_{\text{out}}^{\text{new}}}{V_{\text{in}}} = H_{\text{new}}(s). \tag{21.41}$$

Since $k < 1$, it follows that $1/k > 1$, and there is a gain increase proportional to $1/k$. Note that $V_{\text{out}}^{\text{old}}$ is also an approximation, which is reasonable for k not too far from 1. Moreover, the approximation in equation 21.41 requires that R_4 be "small" relative to the parallel impedance $Z_p(s)$ of figure 21.21, which depicts the remainder of the circuit over the frequency range of interest. Hence, the factor k, computed via equation 21.40, is reasonably accurate. Otherwise, the impedance $Z_p(s)$ must be incorporated into the formula. Also, the lower bound on the value of R_4 is the maximum current that the operational amplifier can provide. Too small an $R_3 + R_4$, and hence an R_4 will draw too much current, thereby changing the op amp gain and the filtering characteristics.

Figure 21.21 Effective equivalent of gain enhancement circuit requiring $Z_p(s) \gg R_4$ over the frequency range of interest.

An alternative means of gain enhancement is via the method of element splitting in the feedback loop. Here, one treats the capacitor C_1 in the same way as the resistor R_1 was treated in input attenuation. (See problem 14 for more details.)

8. Basic High-Pass Filter Design with Passive Realization

High-pass filters invert the frequency characteristics of low-pass filters. A high-pass filter is a device—usually a circuit—that significantly attenuates the low-frequency content of a signal, while passing the high-frequency content with minimal attenuation. Figure 21.22 illustrates typical brickwall specifications for a high-pass filter. Here, frequencies above the passband edge frequency ω_p have little attenuation, while frequencies below ω_s are significantly attenuated—precisely the inverse of a low-pass filter. In fact, low-pass and high-pass specifications are related by a simple inversion of ω. In particular, we define a high-pass to low-pass frequency transformation

$$\Omega = \frac{\omega_p}{\omega}. \tag{21.42}$$

This frequency transformation, applied to the brickwall specifications of figure 21.22, yields a set of normalized low-pass specifications given by figure 21.23.

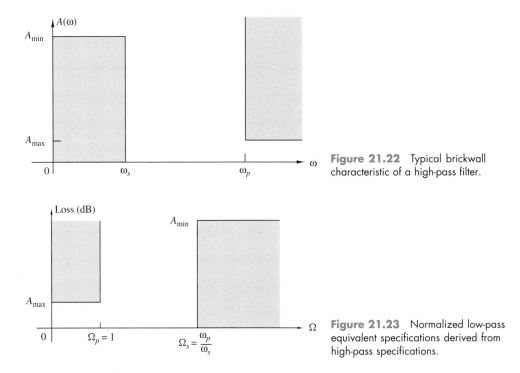

Figure 21.22 Typical brickwall characteristic of a high-pass filter.

Figure 21.23 Normalized low-pass equivalent specifications derived from high-pass specifications.

The specifications of figure 21.23 represent the equivalent normalized low-pass specifications of the given high-pass filter. If one were to find the low-pass transfer function, say, $H_{\mathrm{LP}}(s)$, meeting these specifications, then the needed high-pass transfer function would be given by

$$H_{\mathrm{HP}}(s) = H_{\mathrm{LP}}\left(\frac{\omega_p}{s}\right). \tag{21.43}$$

The operation

$$s \to \frac{\omega_p}{s} \tag{21.44}$$

is called the LP (low-pass) to HP (high-pass) transformation and has a natural interpretation in terms of the inductors and capacitors of a passive network. In particular, inductors become capacitors and capacitors become inductors, as will be shown shortly.

The Butterworth design process produces a low-pass filter circuit having a 3-dB frequency of 1 rad/sec. This circuit is frequency scaled to shift the 3-dB frequency to Ω_c, as calculated from equation 21.21. Finally, an LP to HP network transformation is carried out to obtain the desired high-pass filter.

The preceding discussion suggests that the passive realization of a high-pass filter would entail the following steps:

1. Compute the equivalent low-pass specifications, using the frequency transformation of equation 21.42, i.e.,

$$\Omega = \frac{\omega_p}{\omega}.$$

2. Using equation 21.14, determine the order of the normalized low-pass Butterworth filter, and using, for example, table 21.2, find the transfer function of the normalized low-pass filter in which the 3-dB frequency equals 1 rad/sec.
3. Realize the normalized low-pass filter as an LC network terminated in resistances by the method described earlier for low-order cases, or refer to a low-pass filter table for high-order cases.
4. Using equation 21.21, determine the 3-dB frequency Ω_c for the low-pass filter. Use the left (right) equality sign if the loss curve is to pass through the passband (stop band) corner.
5. Frequency scale the filter of step 3 so that the 3-dB frequency becomes Ω_c calculated in step 4.
6. Using ω_p, execute the LP to HP transformation on individual circuit elements. The element transformations are shown in figure 21.24.
7. Magnitude scale the network to obtain the desired load resistances.

In executing the LP to HP transformation of step 6, resistors stay as resistors with the same value, while capacitors and inductors are interchanged. A capacitor C becomes an inductor of value $(1/C\omega_p)$, and an inductor L becomes a capacitor of value $(1/L\omega_p)$. Figure 21.24 illustrates the process.

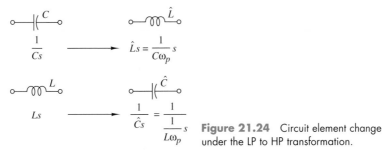

Figure 21.24   Circuit element change under the LP to HP transformation.

 EXAMPLE 21.5

Design a minimum-order Butterworth filter meeting the high-pass brickwall specifications in figure 21.25, in which the loss magnitude curve is to pass through the passband corner.

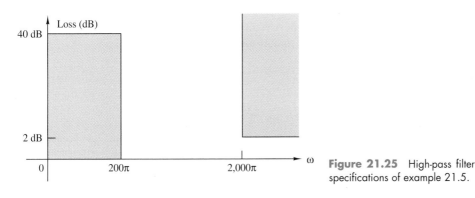

Figure 21.25   High-pass filter specifications of example 21.5.

SOLUTION

Step 1. Compute the equivalent low-pass specifications, with $\Omega = 2{,}000\pi/\omega$. Following the earlier discussion, the equivalent low-pass specifications are easily computed as

$$(\Omega_p = 1, A_{max} = 2 \text{ dB}) \quad \text{and} \quad (\Omega_s = 10, A_{min} = 40 \text{ dB}).$$

Step 2. As these specifications correspond to those of example 21.3, one computes $\varepsilon = 0.7648$ and $n = 3$. From table 20.2, the transfer function for the normalized low-pass filter is

$$H(s) = \frac{1}{(s+1)(s^2+s+1)} = \frac{1}{s^3 + 2s^2 + 2s + 1}.$$

Step 3. The normalized third-order low-pass Butterworth filter (3-dB frequency $= 1$ rad/sec), as designed in example 21.4 (see figure 21.11), with $C_1 = C_2 = 1$ and $L = 2$, is shown in figure 21.26.

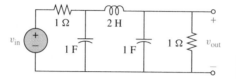

Figure 21.26 Third-order normalized Butterworth passive realization.

Step 4. The 3-dB frequency Ω_c of the low-pass filter is calculated from equation 21.21, using the left inequality:

$$\Omega_c = 1 \times (10^{0.2} - 1)^{-1/6} = 1.0935.$$

Step 5. The normalized low-pass filter is frequency scaled by a factor of 1.0935, to produce the network of figure 21.27.

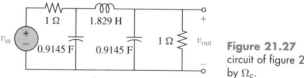

Figure 21.27 Normalized low-pass circuit of figure 21.27 frequency scaled by Ω_c.

Step 6. Executing the LP to HP circuit element transformation yields the network of figure 21.28.

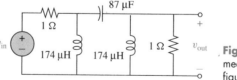

Figure 21.28 High-pass circuit meeting the specifications of figure 21.26.

Step 7. Finally, we impedance scale the circuit by a factor of 10 to obtain the desired high-pass filter with 10-Ω resistance terminations, as shown in figure 21.29.

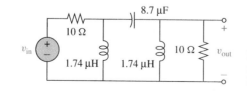

Figure 21.29 Passive circuit realizing the high-pass specifications of figure 21.25 with 10-Ω terminations.

This completes our basic discussion of passive high-pass filtering. Next, we consider pole-zero movement in going from low-pass to high-pass gain functions.

9. Pole-Zero Movement under the LP to HP Transformation

It is important and instructive to understand how the poles and zeros of the equivalent low-pass gain function move under the LP to HP transformation. This movement is intrinsic to the numerical computation of the transfer function and to other applications, such as digital filtering. To understand the pole-zero movement, consider a typical zero of the gain function, $s - z_i$. Here, under the LP to HP type transformation,

$$s - z_i \rightarrow \frac{\omega_p}{s} - z_i = -z_i \frac{\left(s - \frac{\omega_p}{z_i}\right)}{s}. \tag{21.45}$$

The pole movement is similar:

$$s - p_i \rightarrow \frac{\omega_p}{s} - p_i = -p_i \frac{\left(s - \frac{\omega_p}{p_i}\right)}{s}. \tag{21.46}$$

Combining the preceding two expressions implies that a typical zero-over-pole factor moves according to

$$\frac{s - z_i}{s - p_i} \rightarrow \frac{z_i}{p_i} \frac{\left(s - \frac{\omega_p}{z_i}\right)}{\left(s - \frac{\omega_p}{p_i}\right)}. \tag{21.47}$$

If

$$H_{\text{LP}}(s) = K_{\text{LP}} \frac{(s - z_1)(s - z_2) \cdots (s - z_m)}{(s - p_1)(s - p_2) \cdots (s - p_n)}$$

meets the normalized low-pass specifications, $(\Omega_p = 1, A_{\max})$ and $(\Omega_s, A_{\min})$, for $m \leq n$, then

$$H_{\text{HP}}(s) = K_{\text{HP}} \frac{\left(s - \frac{\omega_p}{z_1}\right)\left(s - \frac{\omega_p}{z_2}\right) \cdots \left(s - \frac{\omega_p}{z_m}\right) s^{(n-m)}}{\left(s - \frac{\omega_p}{p_1}\right)\left(s - \frac{\omega_p}{p_2}\right) \cdots \left(s - \frac{\omega_p}{p_n}\right)}, \tag{21.48}$$

where, because $H_{\text{HP}}(\infty) = H_{\text{LP}}(0)$,

$$K_{\text{HP}} = (-1)^{n-m} K_{\text{LP}} \left(\prod_{i=1}^{m} z_i\right) \left(\prod_{i=1}^{n} \frac{1}{p_i}\right). \tag{21.49}$$

If $H_{\text{LP}}(s)$ meets the specifications $(1, 3\text{dB})$ and $(\Omega_s, A_{\min})$, as in the Butterworth case, then it is necessary to replace ω_p in equations 21.45–21.49 with $\omega_c = \omega_p \varepsilon^{1/n}$.

10. Active Realization of High-Pass Filters

Of course, active realizations of high-pass filters parallel active realizations of low-pass filters, i.e., cascading of second-order sections. A typical high-pass second-order section is the Sallen and Key high-pass circuit obtained by an interchange of resistors and capacitors in the low-pass configuration. Note that inductors are absent from active high-pass circuits.

The interchange of resistors and capacitors in a low-pass active circuit with cutoff frequency ω_c to achieve a high-pass circuit with cutoff frequency ω_c is called the RC-to-CR transformation. The specific formulas are

$$\frac{1}{Cs} \rightarrow \frac{1}{\left(\frac{C\omega_c}{s}\right)s} = \frac{1}{C\omega_c} \qquad \text{(a resistor)} \tag{21.50}$$

and

$$R \rightarrow \left(\frac{R\omega_c}{s}\right) = \frac{1}{\left(\frac{1}{R\omega_c}\right)s} \qquad \text{(a capacitor).} \qquad (21.51)$$

The validity of the RC-to-CR transformation may be inferred from the following informally stated network theorem: Suppose two networks, N' and N, are composed of interconnections of only R's, C's, and voltage-controlled voltage sources (VCVS's) (op amps). Let the transfer function of N' be $H'(s') = V'_{out}(s')/V'_{in}(s')$ and the transfer function of N be $H(s) = V_{out}(s)/V_{in}(s)$. Suppose further that the network N is derived from the network N' by replacing each resistor R of N' by a capacitor of value $1/(\alpha R)$ and each capacitor C of N' by a resistor of value $1/(\alpha C)$. The op amps (VCVS's) are left intact. It follows that

$$H'(s') = H\left(\frac{\alpha^2}{s'}\right) \text{ and } H(s) = H'\left(\frac{\alpha^2}{s}\right). \qquad (21.52)$$

Hence, if $H'(s')$ is a low-pass transfer function, $H(s) = H'(\alpha 2/s)$ is a high-pass transfer function and conversely. These ideas are encapsulated in figure 21.30.

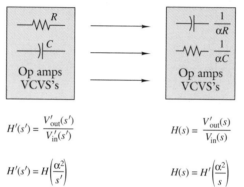

Figure 21.30 Two networks that demonstrate the RC-to-CR transformation for generating high-pass active filters from low-pass filters.

To verify the foregoing theorem, consider a typical composite branch in N'. Such a branch is a parallel connection of an R and a C with admittance

$$Y'(s) = \frac{1}{R} + s'C.$$

The corresponding composite branch in N will have an admittance

$$Y(s) = \frac{s}{\alpha R} + \alpha C = \frac{s}{\alpha}\left[\frac{1}{R} + \frac{\alpha^2}{s}C\right].$$

Except for the factor s/α, the difference between $Y'(s)$ and $Y(s)$ in the bracketed quantities is that s' is replaced by α^2/s. This observation allows us to write the node admittance matrix of N in a special way.

In order to verify equation 21.52, let $Y'(s)$ be the node admittance matrix of N'. From the relationship between a parallel RC branch in N' and one in N, the node admittance matrix for N can be written as

$$Y(s) = \frac{s}{\alpha}\left[Y'(s)\right] = \frac{s}{\alpha}\left[Y'\left(\frac{\alpha}{s^2}\right)\right], \qquad (21.53)$$

with $s' = \alpha^2/s$. By Cramer's rule, the ratio of the input and output voltages may be expressed as the ratio of two first-order cofactors of the node admittance matrix. If the specific matrix (from which the cofactor is calculated) is of order m, then $(s/a)^{m-1}$ appears in both cofactors and cancels out. Equation 21.52 then follows directly from equation 21.53. Note that the inclusion of op amps and VCVS's reduces the order of the admittance matrix, but does not change the relationship of equation 21.52.

To interpret equation 21.52, consider

$$H'(j\omega') = H\left(\frac{\alpha^2}{j\omega'}\right) = H\left(-j\frac{\alpha^2}{\omega'}\right)$$

and

$$\left|H'(j\omega')\right| = \left|H\left(-j\frac{\alpha^2}{\omega'}\right)\right| = \left|H\left(j\frac{\alpha^2}{\omega'}\right)\right|;$$

i.e., the magnitude of $H'(j\omega')$ is the same as the magnitude of $H(j\omega)$ at that ω for which $\omega' = \alpha^2/\omega$, or equivalently, for which $\omega = \alpha^2/\omega'$. Hence, if $H'(s')$ is low pass, then $H(s)$ is high pass and conversely.

This suggests the following generalized design procedure for high-pass active filters:

1. Arbitrarily select a constant α^2, e.g., $\alpha^2 = \omega_p\omega_s$, or $\alpha^2 = \omega_p^2$, or $\alpha^2 = 1$, etc.
2. Compute the frequency transformation, $\omega' = \alpha^2/\omega$.
3. Execute the frequency transformation; i.e., redescribe the high-pass specification in terms of ω', which must yield a set of low-pass specifications.
4. Design a low-pass filter, say N'.
5. Execute a network transformation; i.e., obtain a high-pass filter, say, N, by replacing each resistance R by a capacitor of value $1/(\alpha R)$ and each capacitor C by a resistance of value $1/(\alpha C)$.

As a final remark, the network N' does depend on α, but the final network N is the same for any chosen $\alpha \neq 0$.

In our Sallen and Key circuit low-pass designs, we have normalized $\omega_c = 1$. Using the RC-to-CR transformation with $\alpha = 1$, we obtain the Sallen and Key high-pass design illustrated in figure 21.31.

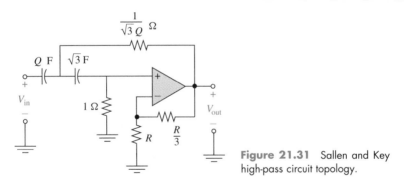

Figure 21.31 Sallen and Key high-pass circuit topology.

The realization of a high-pass filter with cutoff ω_c requires only frequency scaling by ω_c and magnitude scaling to achieve desirable impedance levels. Again, recall that for Butterworth high-pass design, $\omega_c = \omega_p\varepsilon^{1/n}$.

This completes our discussion of the active realization of high-pass circuits. There are, of course, many other circuit designs that realize a high-pass gain function. The reader should consult the many texts on active filter design for a sampling.

SUMMARY

This chapter has covered the basics of Butterworth low-pass and high-pass filter design. Such design techniques build on a set of brickwall filter specifications which require that the desired filter magnitude response lie outside certain regions. Finding Butterworth transfer functions that meet these brickwall constraints is called the *approximation problem*. This chapter has developed algorithms for finding such transfer functions for low-pass and high-pass filter specifications. In addition, basic passive and active circuit realizations were presented. In the active case, the focus was the Sallen and Key low-pass and high-pass topologies. Both the passive and active type of realization build around the coefficient-matching technique, which associates the coefficients in the circuit transfer function with the coefficients of the desired transfer function; one then solves the resulting equations for appropriate circuit parameter values.

There are, of course, many other types of filter transfer functions—Chebyshev, inverse Chebyshev, and elliptic, to name the three other well-known types. Also, in addition to low- and high-pass filters, there are band-pass, band-reject, and magnitude and phase equalizers. To add to the richness of the area of filtering, there are analog passive, analog active, recursive digital, nonrecursive digital, and switched capacitor implementations of all of these filter types. The preceding exposition is merely a drop in a very large and fascinating bucket of filter design challenges.

TERMS AND CONCEPTS

Approximation problem: the problem of finding a transfer function having a magnitude response that meets a given set of brickwall filter specifications.

Attenuation (dB): the loss magnitude expressed in dB, i.e., $A(\omega) = -20\log_{10}|H(j\omega)|$.

Coefficient-matching technique: a method of determining circuit parameter values by matching the coefficients of the transfer function of the circuit to those of the desired transfer function and solving the resulting equations for the circuit parameters.

Cutoff frequency: the frequency at which the magnitude response of the filter is 3 dB down from its maximum value.

Digital filter: a discrete-time or digital approximation to an analog filter.

Filter: a circuit or device that significantly attenuates the frequency content of signals in certain frequency bands and passes the frequency content in certain other, user-specified, frequency bands.

Frequency response magnitude: magnitude of the transfer function as a function of $j\omega$, i.e., $|H(j\omega)|$.

Gain in dB: $-20\log_{10}|H(j\omega)|$.

Gain magnitude: frequency response magnitude.

Half-power point: the point at which the magnitude response curve is 3 dB down.

High-pass filter: a filter that significantly attenuates the low-frequency content of signals and passes the high-frequency content.

HP to LP frequency transformation: a transformation that converts a given set of high-pass brickwall specifications to an equivalent set of low-pass specifications.

Loss magnitude: the inverse of the gain magnitude, i.e., $1/|H(j\omega)|$.

Low-pass (brickwall) filter specification: a filter specification which requires that the desired filter magnitude response lie outside certain "brickwall" regions.

Low-pass filter: a filter that passes the low-frequency content of signals and significantly attenuates the high-frequency content.

Maximally flat: property of a filter at a point ω wherein the magnitude response has a maximum number of zero derivatives.

Normalized Butterworth loss functions: a set of Butterworth transfer functions, ordered by degree, having 3-dB loss at the normalized frequency $s = j\Omega = j1$.

Passive analog filter: a filter composed only of resistors, capacitors, inductors, and transformers.

RC-to-CR transformation: a technique for translating a low-pass active filter to a high-pass active filter in which resistors become capacitors and capacitors become resistors.

PROBLEMS

1. Suppose $H(s)$ is the transfer function of a second-order Butterworth low-pass filter, $H(0) = 1$, and the 3-dB frequency of the filter is 5 kHz. Determine the *magnitude* of the transfer function at 10 kHz.

2. The low-pass filter shown in figure P21.2 has a maximally flat amplitude response with 3-dB frequency $\omega_c = 1,000$ rad/sec. Determine values of L and C

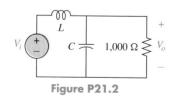

Figure P21.2

3. Suppose $H_1(s)$ and $H_2(s)$ are first- and second-order Butterworth low-pass transfer functions, respectively, with $\omega_p = \omega_c = 1$ rad/sec and $H(0) = 1$.

(a) Find $H_1(s)$ and $H_2(s)$.
(b) Find $|H_1(j\omega)|$ and $|H_2(j\omega)|$.
(c) Tabulate $|H_1(j\omega)|$ and $|H_2(j\omega)|$ at $\omega/\omega_c = 0.2$, 0.5, 0.8, 1.0, 1.25, 2.0, and 5.0, and make an accurate plot of the curves for comparison.
(d) Find the step responses for both systems, and make a *rough* sketch of the response curves.

4. (a) Consider the two-port configurations shown in figure P21.4. Prove that for figure P21.4a,

$$\frac{V_o}{V_i} = \frac{z_{21}}{z_{11} + R_s}$$

and for figure P21.4b

$$\frac{V_o}{V_i} = \frac{-y_{21}}{y_{22} + G_L}.$$

where $G_L = 1/R_L$.

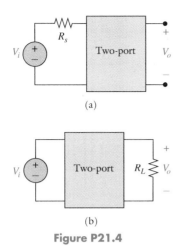

(a)

(b)

Figure P21.4

(b) Making use of the results of part (a), design a second-order Butterworth low-pass filter having $\omega_p = \omega_c = 1$ rad/sec, $R_s = 1\ \Omega$, and $R_L = \infty$. (*Hint*: Divide the numerator and denominator of $H(s)$ by s, and then equate the result with the gain expression to obtain two z- or y-parameters. Next, review the z- and y-parameters of a two-port consisting of one series element and one shunt element (a special case of a T- or π-network). Finally, put all results together to design the two-port.)

(c) Repeat part (b) for $\omega_c = 1$ rad/sec, $R_s = 0$, and $R_L = 1\ \Omega$.

(d) Frequency and magnitude scale the circuits of parts (b) and (c) so that the new filters have $\omega_c = 5{,}000$ rad/sec and the single resistance in the circuit is $1{,}000\ \Omega$.

5. Derive equation 21.18.

6. Derive equation 21.21.

7. Show that, by equating the coefficients of equations 21.25 and 21.28, the only possible solution is the one given in the text, i.e., $C_1 = C_2 = 1$ F and $L = 2$ H.

8. A low-pass filter is to have a passband edge frequency of 700 Hz and a stop band edge frequency of 5,000 Hz, with $A_{max} = 2$ dB and $A_{min} = 20$ dB.

(a) Determine the minimum order of a Butterworth filter that can meet these specifications.

(b) Determine the range of ω_c that the filter could have and still meet the specifications.

(c) Plot the filter magnitude response against the brick-wall specifications for minimum and maximum allowable values of ω_c.

9. In this problem, you will again see the design of a second-order all-pole low-pass filter when both a source and load resistance are present by the coefficient-matching method. Consider the RLC circuit of figure P21.9.

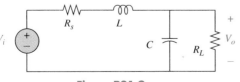

Figure P21.9

(a) Derive the transfer function $H(s) = V_o/V_i$.

(b) If $R_s = 2\ \Omega$ and $R_L = 8\ \Omega$, find the values of L and C so that $H(s)$ has a second-order Butterworth (maximally flat) response with 3-dB frequency equal to 1 rad/sec. There should be two solutions.

(c) Magnitude scale and frequency scale the networks obtained in part (b) so that in these new networks, the smaller resistance is 2 kΩ, and the 3-dB frequency is 10 kHz.

10. Consider a second-order low-pass transfer function

$$H(s) = \frac{V_o}{V_i} = \frac{K}{\left(\dfrac{s}{\omega_p} + 1\right)^2},$$

where $\omega_p = 10^5$ rad/sec.

(a) Find $|H(j\omega)|$ and the exact cutoff frequency (3-dB frequency) ω_c. Is $\omega_c = \omega_p$ in this case?

(b) Find the impulse response and step response.

(c) Using the RLC topologies shown in figure P21.10a, design a network with the voltage ratio given by H(s). The value of the resistance is to be 10 kΩ. What is the value of K attained in your solution?

(d) Using 0.01-μF capacitors, resistors of any value, and *one* ideal operational amplifier (no inductors), design the network of figure P21.10b so that it has the specified transfer function with $K = -10$. Make use of the coefficient-matching method.

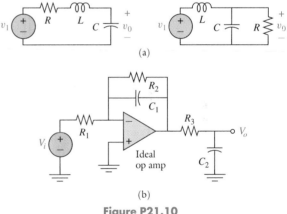

(a)

(b)

Figure P21.10

(e) If the multiplicity of the pole of the transfer function is increased by 1, i.e., if

$$H(s) = \frac{V_o}{V_i} = \frac{K}{\left(\dfrac{s}{\omega_p} + 1\right)^3},$$

find a realization of $H(s)$ using RC elements and *two* ideal op amps. Let $K = 10$ and $\omega_p = 10^5$ rad/sec. (The solution is not unique, which is often the case with practical design problems.)

11. Consider figure P21.11.

(a) Compute the transfer function of the circuit in terms of L_1, L_2, and C.

(b) Determine values so that the circuit realizes a third-order Butterworth gain function.

(c) Find the parameter values of a third-order low-pass Butterworth filter having a cutoff of 20 kHz and resistor values of $1,000\ \Omega$.

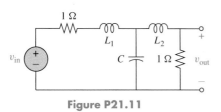

Figure P21.11

12. Find two sets of parameter values that meet the second-order normalized Butterworth gain function in the context of the circuit shown in figure P21.12.

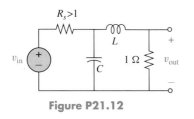

Figure P21.12

13. The circuit shown in figure P21.13 has the transfer function

$$H(s) = \frac{\dfrac{1}{LC}}{s^2 + \left(\dfrac{1}{C} + \dfrac{1}{L}\right)s + \dfrac{2}{LC}},$$

which can be used to realize the second-order Butterworth transfer function

$$H(s) = \frac{K}{s^2 + \sqrt{2}s + 1}$$

for appropriate K. A second-order high-pass Butterworth loss function has stop band edge frequency $f_s = 1,000$ Hz, passband edge frequency $f_p = 7,000$ Hz, and cutoff frequency $f_c = 5,500$ Hz.

(a) Determine the attenuation in dB at f_p and f_s.

(b) Using the circuit of figure P21.13, realize the normalized second-order low-pass Butterworth filter; i.e., determine the values of L and C.

(c) Given your answer to (b), find an appropriate high-pass filter circuit whose resistor values are $1,000\ \Omega$.

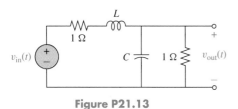

Figure P21.13

14. Consider the circuits of figure P21.14. Determine k so that the transfer function gain is enhanced by a factor of $1/k$. In terms of cost, would there be any advantage or disadvantage to such an enhancement?

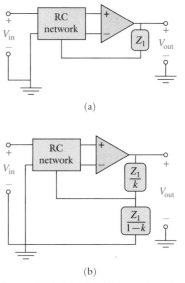

(a)

(b)

Figure P21.14 (a) Given active circuit topology. (b) Gain-enhanced active circuit topology.

15. The circuit of figure P21.15 realizes a third-order normalized low-pass Butterworth filter. Use the following procedure to determine a third-order high-pass filter circuit with $A_{\max} = 2$ dB at $f_p = 5$ kHz so that the largest capacitor equals 1 μF.

(a) Determine ε.

(b) Determine Ω_c for the normalized low-pass filter.

(c) Determine ω_c for the desired high-pass filter. Does your answer make sense? Think about this carefully.

(d) Determine the appropriate high-pass circuit whose largest capacitor is equal to 1 μF.

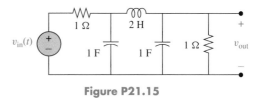

Figure P21.15

16. Suppose you have a set of tweeters that have great sound reproduction for frequencies above 7 kHz. As a lark, you decide to build an active high-pass Butterworth filter that will isolate highs from a particular audio signal. The specifications you decide on are ($f_p = 7$ kHz, $A_{\max} = 3$ dB) and ($f_s = 3$ kHz, $A_{\min} = 40$ dB). Determine the filter as a product of second-order active Sallen and Key circuits. Minimum capacitor values should be 1 μF in a Saraga design.

17. A second-order transfer function

$$H(s) = \frac{K}{s^2 + \dfrac{1}{Q}s + 1}$$

is realized by the Saraga design of a Sallen and Key circuit, shown in figure 21.18.

(a) Determine the new values of R_1, R_2, C_1, and C_2 that will realize the poles of a normalized second-order

Chebyshev gain function given by

$$H(s) = \frac{0.65378}{s^2 + 0.80382s + 0.82306}.$$

(*Hint*: Frequency scale first.)
(b) What is the dc gain, K, of the circuit. Modify the circuit to achieve the desired gain of the transfer function.
(c) If the actual passband edge frequency is to be 7,000 Hz, and if C_2 is to be 1 μF, determine the new element values.

18. A certain audio amplifier has a very low internal resistance. It is therefore represented approximately by an ideal voltage source. The woofer and tweeter each may be represented approximately by a resistance of 8 Ω. Design the simple crossover network shown in figure P21.18 such that the crossover frequency is 200 Hz.

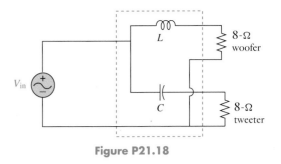

Figure P21.18

ANSWER: $L = 636$ μH, $C = 9.94$ μF.

19. The crossover network of the previous problem provides first-order Butterworth low-pass and high-pass response curves. A better quality of sound reproduction is achieved by upgrading the responses to the second-order Butterworth type. Design such a crossover network, with the same crossover frequency and loads as in the previous problem.

CHAPTER 22

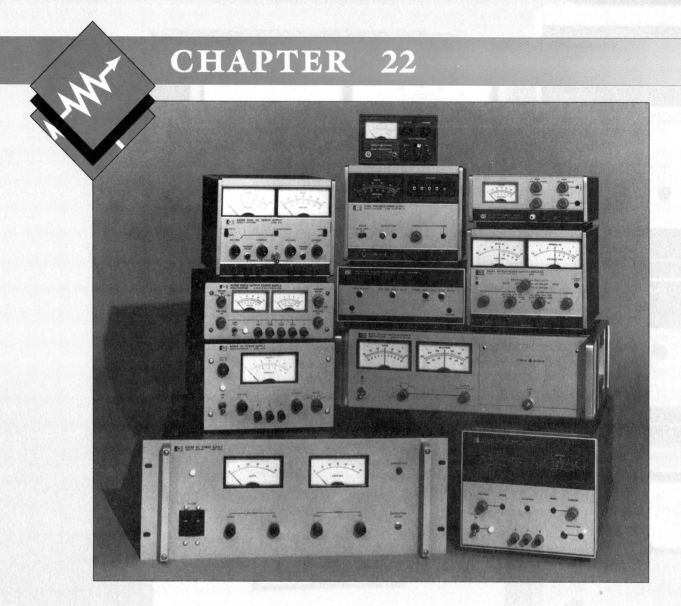

CHAPTER OUTLINE

1. Introduction
2. The Fourier Series: Trigonometric and Exponential Forms
3. Harmonic Distortion in an Amplifier
4. Ripple Factor in dc Power Supplies
5. Additional Properties of and Computational Shortcuts to Fourier Series
   Summary
   Terms and Concepts
   Problems

Fourier Series with Applications to Electronic Circuits

DC POWER SUPPLIES

RADIOS, televisions, stereos, computers, and virtually all electronic gadgets and instrumentation require a dc voltage source for their operation. Because typical household current is ac, a power supply that converts ac to dc is necessary. AC to dc power supplies are devices which do this conversion. Typically, these power supplies contain rectifier circuits and filters. The rectifier circuit rearranges the ac waveform so that only positive sinusoidal pulses appear. The pulses have an average dc value over which is superimposed a set of sinusoidal ripples, as is shown in the following diagram:

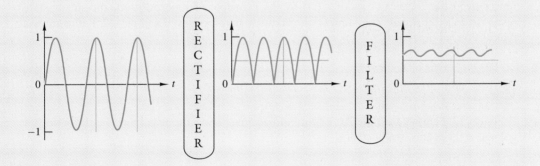

The sinusoidal ripples are unwanted. In audio amplifiers having crude or poorly designed power supplies, the ripples are not adequately removed and sometimes produce a hum in the audio output. Better quality power supplies contain filters that minimize the presence of sinusoidal ripples.

The mathematical tool of Fourier series underlies the important analysis of ripple in dc power supplies. Section 4 of this chapter applies this tool to power supply analysis, as an example of the usefulness of the Fourier series method. This is but one of the many applications of the technique. ◆

CHAPTER OBJECTIVES

- Introduce the notion and calculation of the Fourier series of a periodic signal.
- Describe the relationship between the complex and the real Fourier series representations.
- Set forth and discuss basic properties of the Fourier series.
- Show how the basic Fourier series properties can be used to compute the Fourier series of a wide range of signals from a few basic ones.
- Demonstrate the utility of the Fourier series as a tool for quantifying harmonic distortion in an audio amplifier and the ripple factor in ac-to-dc power supplies.

1. INTRODUCTION

Nonsinusoidal periodic waveforms are an important class of signals in electrical systems. Some prominent examples are the square waveform used to clock a digital computer and the sawtooth waveform used to control the horizontal motion of the electron beam of a TV picture tube. Nonsinusoidal periodic functions are also very important in nonelectrical systems. In fact, it was the study of heat flow in a metal rod that led the French mathematician J. B. J. Fourier to invent the trigonometric series representation of a periodic function. Today, the series bears his name and has wide application in the analysis of circuits and systems excited by periodic signals.

When a periodic input excites a linear circuit, there are many ways to determine the steady-state output. Using a Fourier series method of analysis, the input is first resolved into the sum of a dc component and infinitely many ac components at harmonically related frequencies. For example, a 1-kHz square wave voltage with zero mean and a 0.5π-V peak-to-peak value has the Fourier series representation

$$v_s(t) = \cos(\omega_0 t) - \frac{1}{3}\cos(3\omega_0 t) + \frac{1}{5}\cos(5\omega_0 t) - \ldots$$

where $\omega_0 = 2\pi \times 1,000$ rad/sec.

The next step in the analysis is to compute the transfer function $H(s)$ that yields $H(j\omega)$, assuming, of course, that the poles of $H(s)$ are confined to the left half complex plane. The steady-state circuit response to each ac component of $v_s(t)$ is computed from the relationship $\mathbf{V}_o = H(j\omega)\mathbf{V}_s$, where $\mathbf{V}_o$ and $\mathbf{V}_s$ are phasors representing the sinusoidal functions in the Fourier series of $v_s(t)$. By the principle of superposition, the steady-state response to $v_s(t)$ equals the sum of the individual responses. Example 22.1 illustrates a typical Fourier analysis.

EXAMPLE 22.1

Figure 22.1 shows a simple RC circuit and a square wave input voltage. Find the output voltage $v_o(t)$ in the steady state.

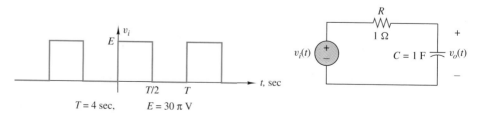

Figure 22.1 Series RC circuit excited by square wave, used to demonstrate the calculation of the steady-state response.

SOLUTION

Step 1. Determine the Fourier series representation of $v_i(t)$. The fundamental frequency of $v_i(t)$ is $f_0 = 1/T = 0.25$ Hz, or $\omega_0 = 2\pi f_0 = 0.5\pi$ rad/sec. As shown in the next section, $v_i(t)$ has the infinite (Fourier) series representation

$$v_i(t) = 15\pi + 60\sin(\omega_0 t) + 20\sin(3\omega_0 t) + 12\sin(5\omega_0 t) + \ldots.$$

Step 2. Find the circuit transfer function. The stable transfer function of the circuit is

$$H(s) = \frac{V_o}{V_i} = \frac{1}{s+1}.$$

For sinusoidal steady-state analysis, set $s = j\omega$ to obtain

$$H(j\omega) = \frac{1}{j\omega+1} = H_m e^{i\theta}, \quad H_m = \frac{1}{\sqrt{\omega^2+1}}, \quad \theta = -\tan^{-1}(\omega).$$

Step 3. Find steady-state responses to each component of $v_i(t)$. Table 22.1 lists the steady-state responses to various components of $v_i(t)$.

TABLE 22.1   SEVERAL FOURIER SERIES COMPONENTS OF $v_i(t)$ AND THE ASSOCIATED MAGNITUDE AND PHASE OF THE RESPONSES

Frequency	0	ω_0	$3\omega_0$	$5\omega_0 \ldots$
Input magnitude	15π	60	20	12
Input angle*	0	0	0	0
H_m	1	0.5370	0.2075	0.1263
θ (degrees)	0	−57.52	−78.02	−82.74
Output magnitude	15π	32.22	4.150	1.516
Output angle	0	−57.52	−78.02	−82.74

*Angles are in reference to sine functions in this example.

Step 4. Apply superposition to obtain the steady-state response. Neglecting harmonics of seventh order and higher, the approximate steady-state solution is

$$v_o(t) \cong 15\pi + 32.22\sin(\omega_0 t - 57.52°) + 4.15\sin(3\omega_0 t - 78.02°) + 1.516\sin(5\omega_0 t - 82.74°).$$

From example 22.1, the very first step in the solution is to represent a periodic waveform as a sum of sinusoidal components, called the *Fourier series*. Section 2 covers the definition and basic properties of Fourier series. Section 4 describes several shortcuts for computing the Fourier coefficients. Since, in practice, only a finite number of terms can be considered, the Fourier series method yields only an *approximate* solution.

The development of Fourier series for circuit analysis in this text adopts a practical flavor: We concentrate on applications to electronic circuits. Such applications include the calculation of the harmonic distortion in an amplifier (section 3) and the calculation of the ripple factor of a rectifier circuit containing inductors and capacitors for filtering (section 4). Many other important applications can be found in a signal analysis course.

Because many mathematical and engineering handbooks have extensive tables of the Fourier series of different waveforms, it is convenient to use these tables in much the same way as a table of integrals or a table of Laplace transforms. Table 22.3 presents the Fourier series of some basic signals. The use of this table, together with some properties and shortcuts discussed in section 5, make the study of Fourier series much more palatable to beginning students of circuit analysis. In this text, much less emphasis is placed on the calculation of the Fourier coefficients by integration.

---

2. The Fourier Series: Trigonometric and Exponential Forms

Basics

A signal $f(t)$ is **periodic** if, for some $T > 0$ and all t,

$$f(t + T) = f(t). \tag{22.1}$$

T is the period of the signal. The **fundamental period** is the smallest positive real number T_0 for which equation 22.1 holds. $f_0 = 1/T_0$ is called the **fundamental frequency** (in hertz) of the signal; $\omega_0 = 2\pi f_0 = 2\pi/T_0$ is the fundamental angular frequency (in rad/sec). The sinusoidal waveform of an ac power source and the square waveform used to clock a digital computer are some familiar periodic signals. Figure 22.2 shows a portion of a general periodic signal.

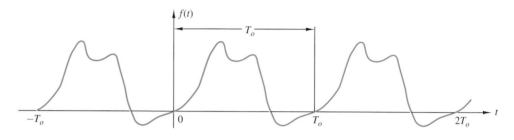

Figure 22.2 A general periodic signal.

THE TRIGOMETRIC FOURIER SERIES

Under conditions that are ordinarily satisfied by signals encountered in engineering practice, a periodic function $f(t)$ has a decomposition as the sum of sinusoidal functions:

$$f(t) = \frac{a_0}{2} + \sum_{n=1}^{\infty} (a_n \cos(n\omega_0 t) + b_n \sin(n\omega_0 t)). \tag{22.2}$$

Using the well-known trigonometric identity

$$A \cos(x) + B \sin(x) = D \cos(x + \theta),$$

where

$$D = \sqrt{A^2 + B^2} \quad \text{and} \quad \theta = \tan^{-1}\left(-\frac{B}{A}\right)$$

(with due regard to quadrant), the sine and cosine terms in equation 22.2 can be combined to yield an equivalent decomposition,

$$f(t) = d_0 + \sum_{n=1}^{\infty} d_n \cos(n\omega_0 t + \theta_n), \tag{22.3}$$

where

$$d_0 = \frac{a_0}{2} \tag{22.4a}$$

and

$$d_n = \sqrt{a_n^2 + b_n^2}, \quad \theta_n = \tan^{-1}\left(\frac{-b_n}{a_n}\right). \tag{22.4b}$$

Both infinite series, equations 22.2 and 22.3, are called the **trigonometric Fourier series** representations of $f(t)$.

From equation 22.3, observe that d_0 is the average value of $f(t)$. In electric circuit analysis, d_0 refers to the dc component of $f(t)$. The first term under the summation sign, $d_1 \cos(\omega_0 t + \theta_1)$, is called the **fundamental component** of $f(t)$, with amplitude d_1 and phase angle θ_1. The term $d_2 \cos(2\omega_0 t + \theta_2)$ is called the **second harmonic** of $f(t)$, with amplitude d_2 and phase angle θ_2, and similarly for the other terms $d_n \cos(n\omega_0 t + \theta_n)$.

As illustrated in example 22.1, given any periodic function $f(t)$, it is important to determine the coefficients in equation 22.2 or equation 22.3. For the purpose of easier calculation, it is advantageous to introduce another representation of the periodic function $f(t)$.

EXPONENTIAL COMPLEX FOURIER SERIES

Making use of the trigonometric identities $\cos(x) = (e^{jx} + e^{-jx})/2$ and $\sin(x) = (e^{jx} - e^{-jx})/2j$, we can rewrite equation 22.2 as

$$f(t) \equiv \sum_{n=-\infty}^{\infty} c_n e^{jn\omega_0 t} = \frac{a_0}{2} + \sum_{n=1}^{\infty}(a_n \cos(n\omega_0 t) + b_n \sin(n\omega_0 t))$$

$$= \frac{a_0}{2} + \sum_{n=1}^{\infty}\left(a_n \frac{e^{jn\omega_0 t} + e^{-jn\omega_0 t}}{2} + b_n \frac{e^{jn\omega_0 t} + e^{-jn\omega_0 t}}{2j}\right),$$

(22.5)

where

$$c_n = 0.5(a_n - jb_n) \text{ and } c_{-n} = 0.5(a_n + jb_n) = c_n^*.$$

(Recall that c_n^* is the complex conjugate of c_n.) The infinite series of equation 22.5 represents the **exponential (complex) Fourier series** for $f(t)$. The coefficients a_n, b_n, d_n, and θ_n are related to c_n according to the formulas

$$a_n = 2\mathrm{Re}(c_n),$$ (22.6a)

$$b_n = -2\mathrm{Im}(c_n),$$ (22.6b)

$$d_0 = c_0,$$ (22.6c)

$$d_n = 2|c_n|, n = 1, 2, \ldots,$$ (22.6d)

$$\theta_n = \text{ angle of } c_n.$$ (22.6e)

An equation for the direct calculation of c_n is given in equation 22.9.

In the trigonometric Fourier series equations 22.2 and 22.3, the summation is over positive integer values of n, whereas in the exponential Fourier series, the summation extends over integers n such that $-\infty < n < \infty$. While each term in equation 22.2 or equation 22.3 has a waveform displayable on an oscillograph, each individual term in the exponential Fourier series lacks such a clear physical picture. However, two conjugate terms in the exponential Fourier series always combine to yield a real time signal $d_n \cos(n\omega_0 t + \theta_n)$.

To find the coefficient c_n, we multiply both sides of equation 22.5, i.e.,

$$f(t) = \sum_{k=-\infty}^{\infty} c_k e^{jk\omega_0 t},$$

by $\exp(-jn\omega_0 t)$. Integrating over one whole period, starting at any $t = t_0$, produces

$$\int_{t_0}^{t_0+T_0} f(t)e^{-jn\omega_0 t}dt = \int_{t_0}^{t_0+T_0} \sum_{k=-\infty}^{\infty} (c_k e^{jk\omega_0 t})e^{-jn\omega_0 t}dt$$

$$= \sum_{k=-\infty}^{\infty}\left\{\int_{t_0}^{t_0+T_0} (c_k e^{jk\omega_0 t}e^{-jn\omega_0 t})dt\right\}$$

(22.7)

$$= \sum_{k=-\infty}^{\infty}\left\{\int_{t_0}^{t_0+T_0} c_k e^{j(k-n)\omega_0 t}dt\right\}.$$

Because $\omega_0 T_0 = 2\pi$ and $e^{j2\pi} = 1$, the integral in equation 22.7 becomes

$$\int_{t_0}^{t_0+T_0} c_k e^{j(k-n)\omega_0 t} dt = \begin{cases} 0 & \text{for } k \neq n \\ c_n T_0 & \text{for } k = n \end{cases}. \tag{22.8}$$

Substituting equation 22.8 into equation 22.7 and dividing by T_0 yields

$$c_n = \frac{1}{T_0} \int_{t_0}^{t_0+T_0} f(t) e^{-jn\omega_0 t} dt. \tag{22.9}$$

The lower limit of integration, t_0, can be any real number, but is usually chosen to be 0 or $-T_0/2$, whichever is more convenient. In addition, T_0 will sometimes be written simply as T.

A hand computation of coefficients would proceed by first computing c_n for $n = 0, 1, \ldots$, using equation 22.9. One would then obtain a_n and b_n using equations 22.6 and d_n and θ_n by equations 22.6 or 22.4. Other formulas are available for obtaining the Fourier coefficients by integrals involving sine and cosine functions. However, equation 22.9 is preferred because an integration involving exponential functions is simpler than an integration involving sinusoidal functions.

◆ **EXAMPLE 22.2**

Find the trigonometric Fourier series for the square wave signal of figure 22.3.

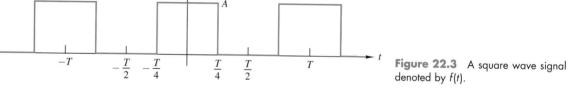

Figure 22.3 A square wave signal denoted by $f(t)$.

SOLUTION

The fundamental period of $f(t)$ is $T_0 = T$. By inspection, the average (dc) value of $f(t)$ is

$$d_0 = \frac{a_0}{2} = \text{average value of } f(t) = \frac{A}{2}.$$

To calculate other Fourier coefficients, choose $t_0 = -T/2$. Equation 22.9 yields

$$c_n = \frac{1}{T} \int_{-\frac{T}{2}}^{\frac{T}{2}} f(t) e^{-jn\omega_0 t} dt = \frac{1}{T} \int_{-\frac{T}{4}}^{\frac{T}{4}} A e^{-jn\omega_0 t} dt = \frac{A}{-jn\omega_0 T} \left[e^{-jn\omega_0 t} \right]_{-\frac{T}{4}}^{\frac{T}{4}}$$

$$= \frac{A}{\pi n} \sin\left(n\frac{\pi}{2}\right), n = 1, 2, \ldots. \tag{22.10}$$

From equations 22.6,

$$a_n = 2\text{Re}(c_n) = \frac{2A}{\pi n} \sin\left(n\frac{\pi}{2}\right), \quad b_n = 0.$$

Substituting these coefficients into equation 22.2 yields the following trigonometric Fourier series for the square wave of figure 22.3:

$$f(t) = \frac{A}{2} + \frac{2A}{\pi} \left(\cos(\omega_0 t) - \frac{1}{3} \cos(3\omega_0 t) + \frac{1}{5} \cos(5\omega_0 t) - \ldots \right). \tag{22.11}$$

CHAPTER 22 FOURIER SERIES WITH APPLICATIONS TO ELECTRONIC CIRCUITS

Properties of the Fourier Series

After computing the Fourier series of a periodic signal $f(t)$, it is straightforward to find the Fourier series of a related periodic signal $g(t)$ whose plot is a translation of the plot of $f(t)$. A **translation** of a plot is a horizontal and/or vertical movement of the figure without any rotation. A translation of the waveform in the vertical direction causes a change in the dc level and hence affects only the coefficients a_0, d_0, and c_0. A translation of the waveform in the horizontal direction causes a time shift that changes only the angles θ_n and has no effect on the amplitude d_n. We state the relationship formally as follows.

> **The Translation Property of Fourier Series:** Let c_n be the coefficients of the exponential Fourier series of a periodic function $f(t)$ and $\hat{c}_n$ be those for another periodic function $g(t)$. If $g(t)$ is a translation of $f(t)$ consisting of a dc-level increase K and a time shift (delay) to the right by t_d, then
>
> $$g(t) = f(t - t_d) + K, \tag{22.12a}$$
>
> $$\hat{c}_0 = c_0 + K, \tag{22.12b}$$
>
> $$\hat{c}_n = c_n e^{-jn\omega_0 t_d}, n = \pm 1, \pm 2, \ldots . \tag{22.12c}$$

The proof of this property is straightforward and is left as an exercise. Note that equation 22.12c indicates that a time shift of the signal affects only the phase angles of the harmonics; the amplitudes of all harmonics remain unchanged.

We will now use the translation property to obtain the Fourier series of a square wave that is a translation of figure 22.3.

◆ EXAMPLE 22.3

The square wave $g(t)$ shown in figure 22.4 is symmetrical with respect to the origin. Find the trigonometric Fourier series for $g(t)$ from the Fourier series (equation 22.11) of $f(t)$ depicted in figure 22.3.

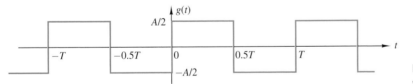

Figure 22.4 A square wave $g(t)$ symmetrical with respect to the origin.

SOLUTION

$g(t)$ is simply a translation of the waveform $f(t)$ of figure 22.3. Specifically,

$$g(t) = f\left(t - \frac{T}{4}\right) - \frac{A}{2}.$$

By equation 22.12a and equation 22.11, the desired Fourier series is

$$g(t) = \frac{2A}{\pi}\left\{ \cos\left(\omega_0 t - \frac{\pi}{2}\right) - \frac{1}{3}\cos\left(3\omega_0 t - \frac{3\pi}{2}\right) + \frac{1}{5}\cos\left(5\omega_0 t - \frac{5\pi}{2}\right) - \ldots \right\}$$

$$= \frac{2A}{\pi}\left(\sin(\omega_0 t) + \frac{1}{3}\sin(3\omega_0 t) + \frac{1}{5}\sin(5\omega_0 t) + \ldots \right). \tag{22.13}$$

For the case of a square wave signal, by choosing the time origin and dc level properly, the resultant plot displays a special kind of symmetry that results in the disappearance of all sine terms or all cosine terms. The square waves of figures 22.3 and 22.4 are special cases of the periodic functions amenable to such simplifications. The general case is given by the following statement.

The plot of an even function is symmetrical about the vertical axis. Examples of even functions include $\cos(\omega t)$ and the square wave of figure 22.3. The plot of an odd function is symmetrical about the origin. Examples of odd functions include $\sin(\omega t)$ and the square wave of figure 22.4. The proofs of the symmetry properties are left as exercises; section 5 investigates some other symmetry properties.

To simplify the calculation of the Fourier coefficients, we should *attempt* to relocate the time origin or change the dc level so that the new function $g(t)$ displays even or odd function symmetry. This may not be possible for an arbitrary periodic signal. When it is possible, we will calculate the Fourier coefficients of the new function $g(t)$, which has only cosine terms or sine terms, and then use the translation property to obtain the Fourier coefficients for the original function $f(t)$.

A waveform of particular importance in signal analysis is the periodic rectangular signal shown in figure 22.5. The fundamental period is T, and the pulse width is αT. The constant α is called the **duty cycle**, usually expressed as a percent of T. The square wave of figure 22.3 is a special case of the rectangular wave of figure 22.5 with a 50% duty cycle.

◆ **EXAMPLE 22.4**

Find the trigonometric Fourier series for the rectangular waveform of figure 22.5.

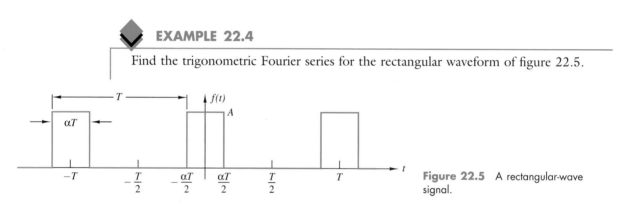

Figure 22.5 A rectangular-wave signal.

SOLUTION

The procedure is almost identical to that used in example 22.2 for a square wave. By inspection, the average value of $f(t)$, F_{av}, is

$$d_0 = \frac{a_0}{2} = F_{av} = \alpha A. \tag{22.14a}$$

To calculate the other Fourier coefficients, choose $t_0 = -T/2$. Equation 22.9 then yields

$$c_n = \frac{1}{T} \int_{-\frac{T}{2}}^{\frac{T}{2}} f(t) e^{-jn\omega_0 t} dt = \frac{1}{T} \int_{-\frac{\alpha T}{2}}^{\frac{\alpha T}{2}} A e^{-jn\omega_0 t} dt = \frac{A}{-jn\omega_0 T} \left[e^{-jn\omega_0 t} \right]_{-\frac{\alpha T}{2}}^{\frac{\alpha T}{2}}$$

$$= \frac{A}{\pi n} \sin(n\alpha\pi), n = 1, 2, \ldots. \tag{22.14b}$$

From equation 22.6, the coefficients of the trigonometric series are

$$a_n = 2\text{Re}(c_n) = \frac{2A}{\pi n} \sin(n\alpha\pi), \tag{22.14c}$$

$$b_n = 0, \tag{22.14d}$$

$$d_n = |a_n|. \tag{22.14e}$$

A very important conclusion about the rectangular wave can be drawn by examining equation 22.14: As the ratio of the pulse width to the period becomes very small, the magnitudes of the fundamental and all harmonic components converge to twice the average (dc) value. To see this, recall that $\sin(x)/x$ approaches 1 as x approaches 0. From equation 22.14c, we may rewrite a_n as

$$a_n = \frac{2A}{\pi n} \sin(n\alpha\pi) = 2\alpha A \frac{\sin(n\alpha\pi)}{n\alpha\pi}.$$

It follows that $d_n = |a_n| \to 2\alpha A = 2F_{av}$ as $\alpha \to 0$.

To give some concrete feel to this property, table 22.2 tabulates the ratios of d_n/F_{av} and d_n/d_1 for $n = 1, \ldots, 9$, for the case of $\alpha = 0.01$. Answers are rounded off to three digits after the decimal point. Note that when the periodic rectangular signal is shifted vertically, the ratio d_n/F_{av} is affected; however, the ratio d_n/d_1 remains unchanged.

TABLE 22.2   AMPLITUDE OF THE FIRST NINE HARMONICS FOR THE CASE $\alpha = 0.01$

n	1	2	3	4	5	6	7	8	9
d_n/F_{av}	2.000	1.999	1.998	1.995	1.992	1.989	1.984	1.979	1.974
d_n/d_1	1.000	1.000	0.999	0.998	0.996	0.994	0.992	0.990	0.987

The constant d_n property holds approximately when the pulse width is a very small fraction of the period T_0. *Even if a waveform is not rectangular*, if the pulse width is very small compared to the period, then the nearly constant property of d_n/F_{av} and d_n/d_1 continues to hold, as long as the pulse is of one polarity. For example, consider the periodic short pulse shown in figure 22.6. In calculating the Fourier coefficients c_n of this pulse by equation 22.9, the limits of the integral, originally $(t_0, t_0 + T)$, are changed to $(-\alpha T/2, \alpha T/2)$. As α approaches zero, the factor $\exp(-jn\omega_0 t)$ in the integrand has a value very close to 1 in the new time interval, as long as n, the harmonic order being considered, is not very high. Therefore, we have

$$c_n = \frac{1}{T} \int_{-\frac{\alpha T}{2}}^{\frac{\alpha T}{2}} f(t)e^{-jn\omega_0 t}dt \cong \frac{1}{T} \int_{-\frac{\alpha T}{2}}^{\frac{\alpha T}{2}} f(t)dt = F_{av}. \qquad (22.15a)$$

From equation 22.6d, it then follows that

$$d_n = 2\,|c_n| \cong 2F_{av}, \qquad n = 1, 2, \ldots. \qquad (22.15b)$$

This result is pertinent to the approximate analysis of a rectifier circuit, described in section 4.

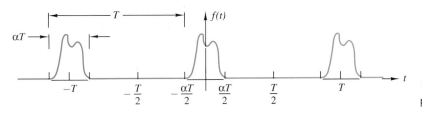

Figure 22.6 Periodic positive short pulses of an arbitrary periodic waveform.

To this point, we have calculated the Fourier coefficients only for some very simple periodic signals. The evaluation of the integral in equation 22.9 becomes much more involved when the signal $f(t)$ is not rectangular. Fortunately, many mathematical and engineering handbooks now include comprehensive tables of Fourier series. From a utility point of view, one may use these tables much the same as one uses a table of integrals or a table of Laplace transforms. In effect, the need to carry out the integration in equation 22.9 is not compelling in practice.

For reference purposes, table 22.3 specifies the Fourier series of some common periodic signals. In some simple cases, the series is given explicitly. In more complicated cases, the coefficients a_n and b_n are given. In some very complicated cases (items 3 and 7), only the coefficients d_n are given, the expressions for the angle θ_n being too complicated. This poses no serious problem, because in many applications, one is concerned only with the average power of the signal and the magnitude of its various harmonic components. In particular, from equation 22.3, the effective value of the dc component is $|d_0|$, and those for the fundamental and various harmonics are $d_n/\sqrt{2}, n = 1, 2, \ldots$. It is easy to show (see problem 9 in chapter 11 of volume 1) that the **effective value**, or the **rms value**, of $f(t)$ is

$$F_{\text{eff}} = \sqrt{d_0^2 + \frac{1}{2}d_1^2 + \frac{1}{2}d_2^2 + \ldots}, \tag{22.16}$$

where the d_i coefficients are from the Fourier series of equation 22.3. The information on the phase angle is important when one wishes to construct the time domain response in the steady state.

Convergence of the Fourier Series

Convergence of the Fourier series is an intricate mathematical problem, the details of which are beyond the scope of this text. On the other hand, it is important to be aware of the ways in which the Fourier series may or may not converge to a given $f(t)$. Our discussion will not be complete, but is adequate for our present purposes.

To begin, we define a partial sum of terms of the complex Fourier series of a function $f(t)$ as

$$S_N(t) = \sum_{k=-N}^{N} c_k e^{j\omega_0 t}.$$

From our experience thus far, $S_N(t)$ must in some way approximate $f(t)$. The difference between $f(t)$ and its approximation, $S_N(t)$, is defined as the error

$$e_N(t) = f(t) - S_N(t).$$

One way to get a handle on how well $S_N(t)$ approximates $f(t)$ is to use the so-called *total squared-error magnitude* over one period, $[t_0, t_0 + T]$, defined as

$$E_N = \int_{t_0}^{t_0+T} |e_N(t)|^2 \, dt.$$

This is often called the energy in the error signal, as energy is proportional to the integral of the squared magnitude of a function. It turns out that for functions which have a Fourier series, the choice of the Fourier coefficients minimizes E_N for each N. Further, for such functions, $E_N \to 0$ as $N \to \infty$, i.e., the energy in the error goes to zero as N gets large. This does not mean that at each t, $f(t)$ and its Fourier series are equal; it merely means that the energy in the error goes to zero.

Continuous and piecewise continuous periodic functions have Fourier series representations. A piecewise continuous function, such as a square wave, is a function that (1) has a finite number of discontinuities over each period, but is otherwise continuous, and (2) has well-defined right- and left-hand limits as the function approaches a point of discontinuity. For piecewise continuous functions, it turns out that the Fourier series converges to a value halfway between the values of the left- and right-hand limits of the function around the point of discontinuity. Even so, $E_N \to 0$ as $N \to \infty$ for piecewise continuous functions.

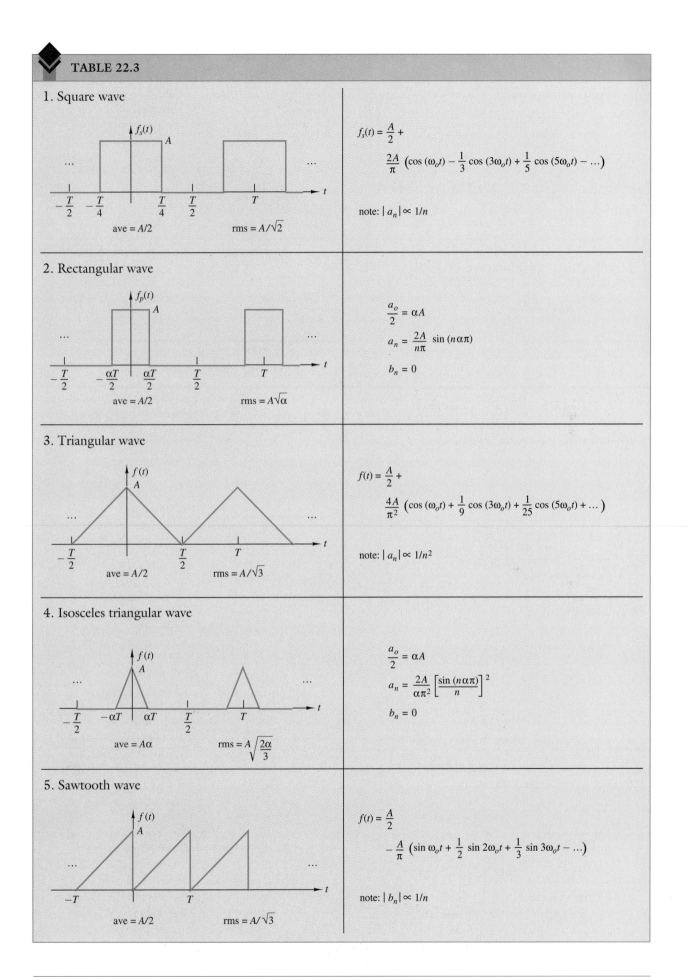

TABLE 22.3

1. Square wave

$$f_s(t) = \frac{A}{2} +$$

$$\frac{2A}{\pi}\left(\cos(\omega_o t) - \frac{1}{3}\cos(3\omega_o t) + \frac{1}{5}\cos(5\omega_o t) - \ldots\right)$$

note: $|a_n| \propto 1/n$

ave = $A/2$ rms = $A/\sqrt{2}$

2. Rectangular wave

$$\frac{a_o}{2} = \alpha A$$

$$a_n = \frac{2A}{n\pi}\sin(n\alpha\pi)$$

$$b_n = 0$$

ave = $A/2$ rms = $A\sqrt{\alpha}$

3. Triangular wave

$$f(t) = \frac{A}{2} +$$

$$\frac{4A}{\pi^2}\left(\cos(\omega_o t) + \frac{1}{9}\cos(3\omega_o t) + \frac{1}{25}\cos(5\omega_o t) + \ldots\right)$$

note: $|a_n| \propto 1/n^2$

ave = $A/2$ rms = $A/\sqrt{3}$

4. Isosceles triangular wave

$$\frac{a_o}{2} = \alpha A$$

$$a_n = \frac{2A}{\alpha\pi^2}\left[\frac{\sin(n\alpha\pi)}{n}\right]^2$$

$$b_n = 0$$

ave = $A\alpha$ rms = $A\sqrt{\dfrac{2\alpha}{3}}$

5. Sawtooth wave

$$f(t) = \frac{A}{2}$$

$$-\frac{A}{\pi}\left(\sin\omega_o t + \frac{1}{2}\sin 2\omega_o t + \frac{1}{3}\sin 3\omega_o t - \ldots\right)$$

note: $|b_n| \propto 1/n$

ave = $A/2$ rms = $A/\sqrt{3}$

TABLE 22.3 (continued)

6. Clipped sawtooth wave

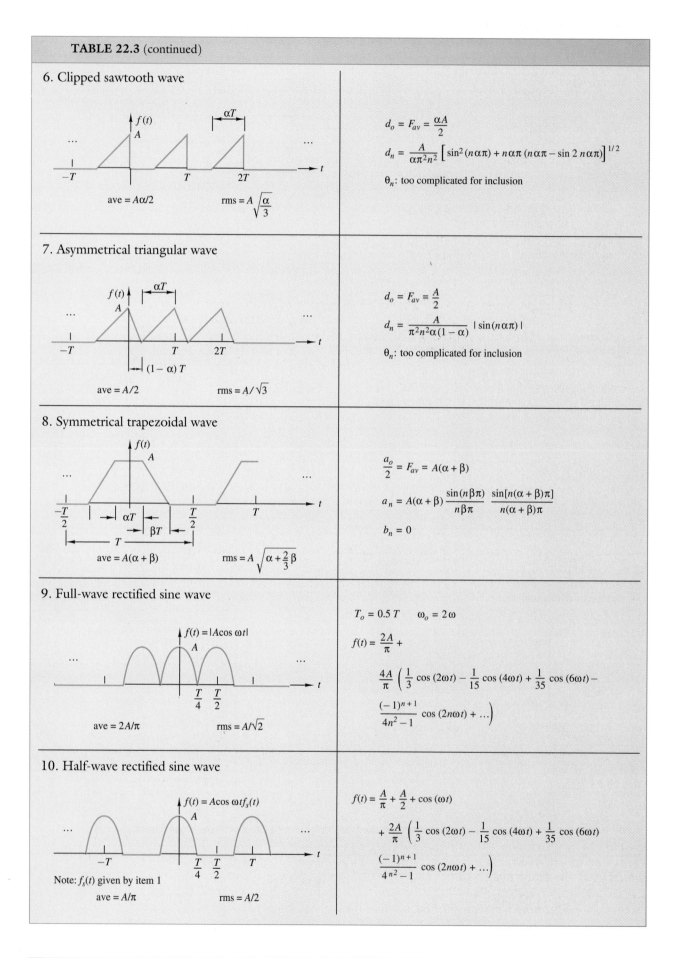

ave = $A\alpha/2$

rms = $A\sqrt{\dfrac{\alpha}{3}}$

$$d_o = F_{av} = \frac{\alpha A}{2}$$

$$d_n = \frac{A}{\alpha\pi^2 n^2}\left[\sin^2(n\alpha\pi) + n\alpha\pi\,(n\alpha\pi - \sin 2\,n\alpha\pi)\right]^{1/2}$$

θ_n: too complicated for inclusion

7. Asymmetrical triangular wave

ave = $A/2$

rms = $A/\sqrt{3}$

$$d_o = F_{av} = \frac{A}{2}$$

$$d_n = \frac{A}{\pi^2 n^2 \alpha(1-\alpha)}\,|\sin(n\alpha\pi)|$$

θ_n: too complicated for inclusion

8. Symmetrical trapezoidal wave

ave = $A(\alpha + \beta)$

rms = $A\sqrt{\alpha + \dfrac{2}{3}\beta}$

$$\frac{a_o}{2} = F_{av} = A(\alpha + \beta)$$

$$a_n = A(\alpha + \beta)\,\frac{\sin(n\beta\pi)}{n\beta\pi}\,\frac{\sin[n(\alpha+\beta)\pi]}{n(\alpha+\beta)\pi}$$

$$b_n = 0$$

9. Full-wave rectified sine wave

ave = $2A/\pi$

rms = $A/\sqrt{2}$

$$T_o = 0.5\,T \qquad \omega_o = 2\omega$$

$$f(t) = \frac{2A}{\pi} +$$

$$\frac{4A}{\pi}\left(\frac{1}{3}\cos(2\omega t) - \frac{1}{15}\cos(4\omega t) + \frac{1}{35}\cos(6\omega t) - \right.$$

$$\left. \frac{(-1)^{n+1}}{4n^2 - 1}\cos(2n\omega t) + \ldots\right)$$

10. Half-wave rectified sine wave

Note: $f_s(t)$ given by item 1

ave = A/π

rms = $A/2$

$$f(t) = \frac{A}{\pi} + \frac{A}{2} + \cos(\omega t)$$

$$+ \frac{2A}{\pi}\left(\frac{1}{3}\cos(2\omega t) - \frac{1}{15}\cos(4\omega t) + \frac{1}{35}\cos(6\omega t)\right.$$

$$\left. \frac{(-1)^{n+1}}{4\,n^2 - 1}\cos(2n\omega t) + \ldots\right)$$

There are many functions that are not piecewise continuous and yet have a Fourier series. A set of conditions that is sufficient, but not necessary, for a function to have a Fourier series representation is the so-called *Dirichlet conditions*.

Condition 1. Over any period $[t_0, t_0 + T]$, $f(t)$ must have the property that

$$\int_{t_0}^{t_0+T} |f(t)|\, dt < \infty.$$

In the language of mathematics, this means that $f(t)$ is *absolutely integrable*. The consequence of this property is that each of the Fourier coefficients c_n is finite, i.e., the c_n exist.

Condition 2. Over any period of the signal, there must be only a finite number of minima and maxima. In other words, functions like $\sin(1/t)$ are excluded. In the language of mathematics, a function that has only a finite number of maxima and minima over any finite interval is said to be of *bounded variation*.

Condition 3. Over any period, $f(t)$ can have only a finite number of discontinuities.

As mentioned, at points of discontinuity, the Fourier series will converge to a value midway between the left- and right-hand values of the function next to the discontinuity. There may be other differences as well. Despite these differences, the energy between the function $f(t)$ and its Fourier series representation is zero; i.e., E_N, with N approaching ∞, goes to zero. Thus, for all practical purposes, the functions are identical. This practical equivalence allows us to analyze how a circuit responds to a signal $f(t)$ by analyzing how the circuit responds to each of its Fourier series components.

3. HARMONIC DISTORTION IN AN AMPLIFIER

For an ideal amplifier, any input waveform should emerge magnified, but otherwise identical, at the output a very short time later. Any difference between the input and output waveforms other than a scale change indicates the presence of **distortion**. One source of distortion is the nonlinear transfer characteristic of the (nonideal) amplifier. Figure 22.7a exhibits a nonideal curve of v_o vs. v_i. Because of the curvature, a pure sinusoidal input signal (v_i in figure 22.7b) appears at the output with flattened positive and negative peaks. A Fourier analysis of the output signal will disclose the presence of many (unwanted) harmonics.

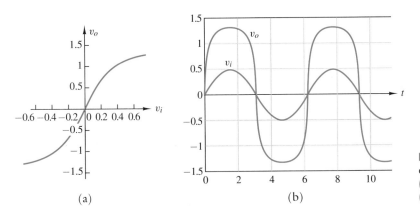

(a)   (b)

Figure 22.7 Distortion due to a nonlinear transfer characteristic. (a) A nonlinear transfer characteristic. (b) A sinusoidal $v_i(t)$ and a distorted $v_o(t)$.

The ratio of the nth ($n = 2, 3, \ldots$) harmonic amplitude to the fundamental amplitude is called the **nth-harmonic distortion**, usually expressed as a percent of the fundamental. The **total harmonic distortion** is the ratio of the effective value of the sum of all the harmonics to the effective value of the fundamental. The total amount of distortion allowable depends on the application. For audio amplifiers whose output is destined for listening, 5% or less total harmonic distortion is usually considered acceptable. For video amplifiers, the requirement is much more stringent.

Two examples will illustrate the methods for calculating harmonic distortion. The first method makes use of a power series representation of the transfer characteristic and is often convenient for moderate nonlinearities. The second method builds on an approximate piecewise transfer characteristic for which a power representation does not exist.

EXAMPLE 22.5

The amplifier of figure 22.8 has a 100 Ω load connected to it. Suppose the transfer characteristic satisfies the algebraic relationship

$$v_o = 10v_i + 0.5v_i^2 + 0.1v_i^3.$$

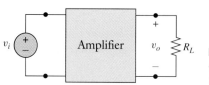

Figure 22.8 Example diagram of amplifier with distortion due to a nonlinear transfer characteristic.

If $v_i = \cos(6{,}000t)$ V, find (1) the total harmonic distortion and (2) the average output power at the fundamental frequency.

SOLUTION

The strategy for solving this problem is to use some standard trigonometric identities to obtain an expression for $v_o(t)$ as a sum of harmonics of the input. The following formulas prove useful:

$$\cos^2(x) = 0.5 + 0.5\cos(2x),$$

$$\cos^3(x) = 0.75\cos(x) + 0.25\cos(3x).$$

With $\omega = 6{,}000$ rad/sec, applications of these formulas and some straightforward grouping of terms lead to

$$v_o(t) = 0.25 + 10.075\cos(\omega t) + 0.25\cos(2\omega t) + 0.025\cos(3\omega t).$$

The effective values of various components of $v_o(t)$ are listed in table 22.4.

TABLE 22.4 HARMONIC CONTENTS OF $v_o(t)$ IN EXAMPLE 22.5				
dc	Fundamental	Second harmonic	Third harmonic	Total harmonic
0.25	$10.075/\sqrt{2}$	$0.25/\sqrt{2}$	$0.025/\sqrt{2}$	$0.2512/\sqrt{2}$

The total harmonic distortion is

$$\frac{0.2512}{10.075} \times 100\% = 2.493\%,$$

and the average power at the fundamental frequency is $0.5 \times 10.075^2/100 = 0.507$ W.

In an actual amplifier circuit, the coefficients in a power series representing $v_o = f(v_i)$ depend on the load resistance R_L. If the highest degree term in the series is v_i^m, then there are $m + 1$ coefficients in the series. In theory, it should be possible to determine these coefficients from $m + 1$ pairs of (v_i, v_o) values, either calculated from characteristic curves or from measurements. Harmonic analysis then proceeds as in example 22.5. However, an alternative and perhaps simpler approach exists. In that approach, if the highest order harmonic in v_o is m, we may write the Fourier series directly with unknown coefficients. Then, at judiciously chosen instants of time, the $m + 1$ pairs of (v_i, v_o) values are used to determine the coefficients in the Fourier series, bypassing the algebraic calculations of example 22.5.

As an example of this alternative approach, suppose the output has significant harmonics only up to third order. Then for an input $v_i(t) = V_m \cos(\omega t)$, the output $v_o(t)$ is an even function of t, since

$$v_o(-t) = f[v_i(-t)] = f[v_i(t)] = v_o(t).$$

This means that the Fourier series contains only cosine terms, i.e.,

$$v_o(t) = V_0 + V_1 \cos(\omega t) + V_2 \cos(2\omega t) + V_3 \cos(3\omega t). \tag{22.17}$$

There are four unknown coefficients to be determined in equation 22.17: V_0, V_1, V_2, and V_3. In theory, knowledge of v_o at any four instants of time should be adequate for finding these unknowns. For convenience, we choose these four instants to be $\omega t = 0$, $\pi/3$, $2\pi/3$, and π. From equation 22.17,

$$v_o(0) = V_0 + V_1 + V_2 + V_3, \tag{22.18a}$$

$$v_o\left(\frac{T}{6}\right) = V_0 + 0.5V_1 - 0.5V_2 - V_3, \tag{22.18b}$$

$$v_o\left(\frac{T}{3}\right) = V_0 - 0.5V_1 - 0.5V_2 + V_3, \tag{22.18c}$$

$$v_o\left(\frac{T}{2}\right) = V_0 - V_1 + V_2 - V_3. \tag{22.18d}$$

Putting these equations in matrix form and solving produces

$$V_0 = \frac{v_o(0) + 2v_o\left(\frac{T}{6}\right) + 2v_o\left(\frac{T}{3}\right) + v_o\left(\frac{T}{2}\right)}{6}, \tag{22.19a}$$

$$V_1 = \frac{v_o(0) + v_o\left(\frac{T}{6}\right) - v_o\left(\frac{T}{3}\right) - v_o\left(\frac{T}{2}\right)}{3}, \tag{22.19b}$$

$$V_2 = \frac{v_o(0) - v_o\left(\frac{T}{6}\right) - v_o\left(\frac{T}{3}\right) + v_o\left(\frac{T}{2}\right)}{3}, \tag{22.19c}$$

and

$$V_3 = \frac{v_o(0) - 2v_o\left(\frac{T}{6}\right) + 2v_o\left(\frac{T}{3}\right) - v_o\left(\frac{T}{2}\right)}{6}. \tag{22.19d}$$

If specific values are available for $v_o(nT/6)$, $n = 0, 1, 2, 3$, then one can compute V_0, V_1, V_2, and V_3 from equation 22.19. Many circuit simulation programs (e.g., SPICE2) implement advanced numerical techniques for computing Fourier coefficients accurately from sampled points of the signal.

The next example illustrates the use of equation 22.19.

 EXAMPLE 22.6

The transfer characteristic of the amplifier shown in figure 22.8 depends on the load resistance R_L. With the input $v_i = \cos(6,000t)$ V applied, it may be assumed that the only significant harmonics in the output are the second and the third. For two different load resistance values, suppose the measurement data of table 22.5 are obtained. Find (1) the total harmonic distortion and (2) the average power at the fundamental frequency for each load resistance used.

---

3. HARMONIC DISTORTION IN AN AMPLIFIER **781**

TABLE 22.5 OUTPUT VOLTAGES FOR TWO DIFFERENT VALUES OF R_L

	ωt	0	$\pi/3$	$2\pi/3$	π
	v_i	1	0.5	−0.5	−1
Case 1. $R_L = 16\ \Omega$	v_o	10.6	5.138	−4.888	−9.6
Case 2. $R_L = 12\ \Omega$	v_o	9.64	4.46	−4.46	−8.84

SOLUTION

Using equation 22.19 and the data from table 22.5, one can compute the data listed in table 22.6. From these data, the total harmonic distortion for the case of $R_L = 16\ \Omega$ is

$$\frac{\sqrt{0.25^2 + 0.025^2}}{10.075} \times 100\% = 2.493\%.$$

The average power at fundamental frequency is $0.5 \times (10.075)^2/16 = 3.172$ W.

TABLE 22.6 AMPLITUDES OF THE FIRST FOUR TERMS OF THE FOURIER SERIES OF $v_o(t)$

R_L	V_0	V_1	V_2	V_3
16 Ω	0.25	10.075	0.25	0.225
12 Ω	0.20	9.00	0.40	0.04

For the case of $R_L = 12\ \Omega$, the total harmonic distortion is

$$\frac{\sqrt{0.4^2 + 0.04^2}}{9} \times 100\% = 4.467\%.$$

The average power at the fundamental frequency is $0.5 \times 9^2/12 = 3.375$ W.

In the preceding example, a slight reduction of the load resistance is accompanied by an increase in output power and also a higher harmonic distortion. This correlation is not a fixed trend. Finding the optimum load resistance producing the largest output power within the allowable harmonic distortion range is a problem that requires much trial and error.

As mentioned earlier, when the transfer characteristic is piecewise linear, the power series method encounters some difficulty, as no polynomial of finite degree can create a breakpoint with an abrupt change of slope. The next example shows how to tackle such a problem with the aid of the Fourier series list given in table 22.3.

EXAMPLE 22.7

The op amp in the inverting amplifier circuit of figure 22.9a is assumed to be ideal and has a saturation voltage of ± 15 V. If the input is $v_i = \cos(6{,}000t)$ V, find the harmonic distortion in $v_o(t)$ up to the fifth order.

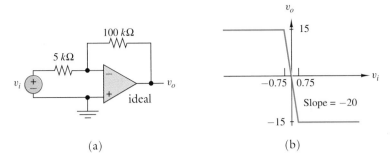

Figure 22.9 Distortion in an overdriven operational amplifier. (a) An inverting amplifier. (b) Its transfer characteristic.

SOLUTION

When the op amp is in the active region, the amplifier has a gain of −20. Together with the fact that v_o reaches saturation at ±15 V, this leads to the piecewise linear transfer characteristic shown in figure 22.9b. Figure 22.10a depicts the output waveform with $v_i = \cos(6{,}000t)$ V applied.

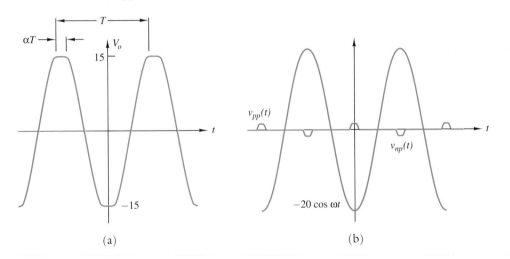

Figure 22.10 (a) Clipped sine wave and (b) its decomposition.

The positive peak of the sine wave is clipped for the duration αT and is determined as follows:

$$20\cos(0.5\omega\alpha T) = 15 \Rightarrow 0.5\omega\alpha T = \pi\alpha = \cos^{-1}(0.75) = 0.7227 \Rightarrow \alpha = 0.23.$$

Here, we view the waveform as the sum of three component waveforms: a cosine wave, $-20\cos(\omega t)$; a periodic positive pulse train, $v_{pp}(t)$; and a negative periodic pulse train, $v_{np}(t)$, as illustrated in figure 22.10b; i.e.,

$$v_o(t) = -20\cos(\omega t) + v_{pp}(t) + v_{np}(t). \tag{22.20}$$

Figure 22.10b also shows very clearly that

$$v_{np}(t) = -v_{pp}\left(t - \frac{T}{2}\right). \tag{22.21}$$

The crux of finding a Fourier series for $v_o(t)$ is to find Fourier coefficients for $v_{pp}(t)$. Since $v_{pp}(t)$ is an even function, the Fourier series has only cosine terms; i.e.,

$$v_{pp}(t) = \frac{a_0}{2} + a_1\cos(\omega t) + a_2\cos(2\omega t) + a_3\cos(3\omega t)$$
$$+ a_4\cos(4\omega t) + a_5\cos(5\omega t) + \dots. \tag{22.22}$$

Recalling the relationship $\omega T = 2\pi$, equation 22.20 implies that

$$v_{np}(t) = -\frac{a_0}{2} + a_1\cos(\omega t) - a_2\cos(2\omega t) + a_3\cos(3\omega t)$$
$$- a_4\cos(4\omega t) + a_5\cos(5\omega t) + \dots. \tag{22.23}$$

Substituting equations 22.22 and 22.23 into equation 22.20 yields

$$v_o(t) = -20\cos(\omega t) + 2a_1\cos(\omega t) + 2a_3\cos(3\omega t) + 2a_5\cos(5\omega t) + \dots \quad (22.24)$$

Here, only half of the Fourier coefficients of $v_{pp}(t)$, namely a_{odd}, need to be calculated. Call $f_p(t)$ the rectangular wave of item 2 of table 22.3, with $\alpha = 0.23$ and $A = 1$. Then we see that $v_{pp}(t)$ results by downshifting the $20\cos(\omega t)$ plot by 15 and then multiplying the resultant function by $f_p(t)$, i.e.,

$$v_{pp}(t) = [20\cos(\omega t) - 15]f_p(t). \quad (22.25)$$

The Fourier series for $f_p(t)$, as calculated from table 22.3, is

$$f_p(t) = 0.23 + 0.421\cos(\omega t) + 0.3158\cos(2\omega t) + 0.1755\cos(3\omega t)$$
$$+ 0.0396\cos(4\omega t) - 0.0578\cos(5\omega t) - 0.0987\cos(6\omega t) + \dots \quad (22.26)$$

Substituting equation 22.26 into equation 22.25 and using the trigonometric identity $\cos(x)\cos(y) = 0.5\cos(x + y) + 0.5\cos(x - y)$, we obtain

$$v_{pp}(t) = 1.443\cos(\omega t) + 0.9211\cos(3\omega t) + 0.2763\cos(5\omega t)$$
$$+ \text{higher order odd harmonics} + \text{even harmonics.} \quad (22.27)$$

Substituting the values $a_1 = 1.443$, $a_3 = 0.9211$, and $a_5 = 0.2763$ into equation 22.24 yields

$$v_o(t) = -17.94\cos(\omega t) + 1.743\cos(3\omega t) + 1.065\cos(5\omega t)$$
$$+ \text{higher order odd harmonics.} \quad (22.28)$$

From equation 22.28, the harmonic distortions are as follows:

Even order: none.

Odd order:

$$\text{Third order: } \frac{1.842}{17.11} = 0.1076 = 10.76\%.$$

$$\text{Fifth order: } \frac{0.5526}{17.11} = 0.0323 = 3.23\%$$

Note that if the total harmonic distortion is to be determined, then many more terms must be included in equation 22.26 for $f_p(t)$. This is because the Fourier coefficients for a rectangular wave (item 2 of table 22.3) are given by $2\alpha A\sin(n\alpha\pi)/(n\alpha\pi)$, and the function $(\sin(x))/x$ is not monotonically decreasing in magnitude. The fact $|a_n|$ is not monotonically decreasing is easily verified by actual calculations of the coefficients.

4. *RIPPLE FACTOR IN DC POWER SUPPLIES

All amplifiers and electronic instruments require dc voltage sources for their operation. On the other hand, almost all electric power plants generate ac voltages. This difference often makes it necessary to convert an ac voltage source into a dc voltage source. A **rectifier** circuit achieves the goal. Figures 22.11a through 22.11c show several of the most well-known rectifier circuits: the **half-wave rectifier**, the **full-wave rectifier**, and the **full-wave bridge rectifier**. As these circuits all use diodes, the reader might review chapter 6 of volume 1 for the definition of an ideal diode and the analysis of simple resistive diode circuits. Assuming ideal diodes, if $v_s(t) = V_m\cos(\omega t)$, the output waveforms for the case of a resistive load are easily obtained. These waveforms are shown in figures 22.11d and e, in which the peak output voltage is $A = V_m R_L/(R_s + R_L)$. In practice, the ac voltage source in these circuits is taken from the secondary winding of a transformer. The two voltage sources in figure 22.11b are obtained from the secondary of a transformer with a center

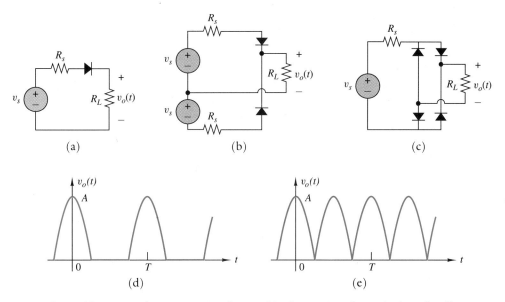

Figure 22.11 Rectifier circuits. (a) Half wave. (b) Full wave. (c) Full-wave bridge. (d) Half-wave rectified output. (e) Full-wave rectified output.

tap. The value of V_m ranges from a few volts for a low-voltage power supply to several thousand volts for the power supply of a TV picture tube.

A Fourier series analysis of the output voltage waveforms of the half-wave rectifier (see item 10 of table 22.3) gives

$$
\begin{aligned}
v_o(t) = & \frac{A}{\pi} + \frac{A}{2}\cos(\omega t) \\
& + \frac{2A}{\pi}\left(\frac{1}{3}\cos(2\omega t) - \frac{1}{15}\cos(4\omega t) + \frac{1}{35}\cos(6\omega t) - \ldots\right).
\end{aligned}
\tag{22.29}
$$

The first term on the right is the dc voltage—the desired output of the rectifier. The remaining terms represent undesirable ac harmonics. The lowest ac frequency present is that of the supply, and the rest are all at even multiples of the supply frequency.

Similarly, a Fourier series analysis of the output voltage waveforms of the full-wave rectifier (see item 9 of table 22.3) gives

$$
v_o(t) = \frac{2A}{\pi} + \frac{4A}{\pi}\left(\frac{1}{3}\cos(2\omega t) - \frac{1}{15}\cos(4\omega t) + \frac{1}{35}\cos(6\omega t) - \ldots\right).
\tag{22.30}
$$

The full-wave rectified dc output is twice the value of that of the half-wave rectifier, as expected. There is no ac component at the supply frequency in equation 22.30. The lowest ac frequency present is double the supply frequency, a fact regarded as an advantage over the half-wave rectifier, because the higher frequency component is easier to filter out than a lower frequency component. Other than the dc component, all terms in equation 22.30 are at even multiples of the supply frequency, and their magnitudes are twice those of the half-wave rectifier.

The pulsations, or **ripple,** caused by the ac components of the rectifier output often appear as an audible hum in an audio amplifier of a radio. This hum should be kept as low as possible. The term **ripple factor** denotes the purity of the rectifier output; that is,

$$
\text{Ripple factor} = \frac{\text{effective value of all ac components}}{\text{dc component}}.
\tag{22.31a}
$$

From equation 22.16,

$$
V_{\text{eff}} = \sqrt{V_{\text{dc}}^2 + V_{\text{ac,eff}}^2},
$$

or

$$
V_{\text{ac,eff}} = \sqrt{V_{\text{eff}}^2 - V_{\text{dc}}^2}.
$$

---

4. *RIPPLE FACTOR IN DC POWER SUPPLIES

Substituting these results into equation 22.31a yields the mathematical formula for the ripple factor:

$$\text{Ripple factor} = \frac{\sqrt{V_{\text{eff}}^2 - V_{\text{dc}}^2}}{V_{\text{dc}}} = \sqrt{\left(\frac{V_{\text{eff}}}{V_{\text{dc}}}\right)^2 - 1}. \tag{22.31b}$$

Let us now apply equation 22.31b to the half-wave and full-wave rectifier outputs with resistive load. Using the data in table 22.3, items 9 and 10, we have

$$\text{Half-wave rectifier ripple factor} = \sqrt{\left(\frac{\frac{A}{\pi}}{\frac{A}{2}}\right)^2 - 1} = 1.21$$

and

$$\text{Full-wave ripple factor} = \sqrt{\left(\frac{\frac{A}{\sqrt{2}}}{\frac{2A}{\pi}}\right)^2 - 1} = 0.48.$$

Both ripple factors are unsatisfactorily high. Some filtering is needed to reduce the ac components in the output waveform. Several simple and commonly used filtering circuits achieve this goal. We present and analyze these circuits, with a threefold objective in mind: (1) to explain the basic operations of these circuits; (2) to show how the Fourier series method is utilized in rectifier circuit analysis; and (3) to point up the use of some engineering approximations backed by experience.

The simplest method for filtering out the ac components is to connect a capacitor of suitable size across the output terminals. The presence of this capacitor changes the shape of $v_o(t)$ from that of figure 22.11d or e. In fact, obtaining the $v_o(t)$ waveform by manual analysis in this case is nightmarish without some simplifying assumptions. Accordingly, for our analysis, we assume that (1) the source resistance is negligible, i.e., $R_s = 0$, and (2) all diodes are ideal. Thus, each diode behaves as an ideal switch—a short circuit when forward biased and an open circuit when reversed biased. The half-wave rectifier, together with a shunt capacitor and associated waveforms, is shown in figure 22.12. For convenience, the plots use $\theta = \omega t$ as the independent variable.

For this simple circuit, it is possible to obtain exact expressions for the output voltage waveforms in figure 22.12.b. Let us assume that the source voltage $v_s(t) = V_m \cos(\omega t)$ is applied at $\theta = \omega t = -2.25\pi$ and that the initial capacitor voltage is zero at that instant. Since the source voltage is positive, the ideal diode is forward biased and behaves as a short circuit. Therefore, v_o starts to follow v_s exactly. After the capacitor is charged up to the peak positive voltage V_m, the source voltage begins to decrease. If the load resistance were not there, the diode would have been immediately switched off, and the trapped charge on the capacitor would have kept the output voltage constant at $v_o = V_m$. With the load resistance present, however, the diode remains conducting for a little longer. To see this, let us assume an "on" state for the diode and calculate the current through the diode:

$$i_d(t) = C\frac{dv_s}{dt} + \frac{v_s(t)}{R} = -\omega C V_m \sin(\omega t) + \frac{V_m \cos(\omega t)}{R}. \tag{22.32}$$

The diode is switched off at the moment the current decreases to zero. Let us denote the diode turning-off time by t_1 and the corresponding angle ωt_1 by θ_1. Setting equation 22.32 to zero and solving for θ_1 yields

$$\theta_1 = \tan^{-1}\left(\frac{1}{\omega RC}\right). \tag{22.33}$$

Two such turning-off points are marked in figure 22.12b as θ_1 and $(\theta_1 - 2\pi)$. After the diode is switched off, the capacitor voltage decreases exponentially with a time constant RC. The moment this exponential decay curve intersects the rising voltage source waveform, the diode switches on again. The angles corresponding to these switched-on points are marked as $-\theta_2 \pm 2k\pi$ in figure 22.12b. The waveform from $\theta = \theta_1 - 2\pi$ to $\theta = \theta_1$

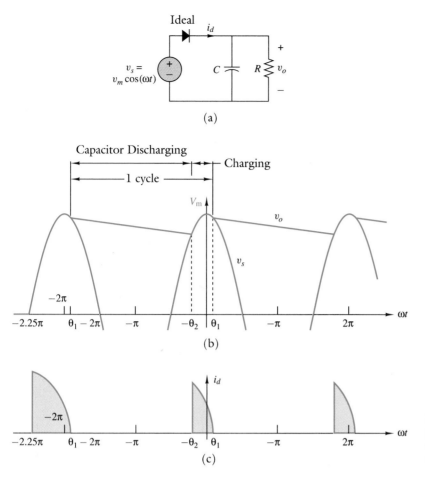

(a)

(b)

(c)

Figure 22.12 A half-wave rectifier. (a) The circuit. (b) Input and output voltage waveforms. (c) Diode current waveform.

represents one complete cycle of the steady-state waveform and repeats itself. The analysis of the rectifier circuit shows that the ideal diode serves merely as a *switch* and connects the source to the capacitor only during the charging interval $\theta = (-\theta_2, \theta_1)$. In fact, this is the reason that the idealized rectifier problem can be solved as a switching transient problem in linear circuits.

We have yet to find the value of θ_2. Again from figure 22.12b, the exponential decay of $v_o(t)$ starts with the value $V_m \cos(\theta_1)$ at $\theta = \theta_1 - 2\pi$, or $t = \theta_1/\omega - T$. Therefore, the capacitor voltage is

$$v_o(t) = (V_m \cos \theta_1)e^{-\frac{t-\left(\frac{\theta_1}{\omega}-T\right)}{RC}} = (V_m \cos \theta_1)e^{-\frac{\omega t-(\theta_1-2\pi)}{\omega RC}}. \qquad (22.34)$$

The source voltage is $v_s(t) = V_m \cos \omega t$. At $\omega t = \theta = -\theta_2$, the two curves intersect; i.e., $v_s(-\theta_2) = v_o(-\theta_2)$. This leads to

$$\cos(\theta_1)e^{-\frac{\theta_2-\theta_1+2\pi}{\omega RC}} = \cos(\theta_2). \qquad (22.35)$$

Since θ_1 has been calculated from equation 22.33, equation 22.35 is a single transcendental equation in the unknown θ_2, and the answer, like θ_1, depends on the product ωRC only. In precomputer days, the solution of equation 22.35 was considered a very difficult task. Therefore, electrical engineers compiled many charts as design aids for rectifier circuits. One such chart plots θ_1 and θ_2 vs. the parameter ωRC. Such plots still appear in many handbooks. They are valuable in providing instant information on the effect of changing ωRC. Since our main purpose here is to understand the principle and analysis of a rectifier circuit, we shall be content with a numerical solution of equation 22.35. Equipped with a scientific calculator, one can solve the equation in a few iterations.

If ωRC is very large—say, $\omega RC > 500$—then equation 22.33 indicates that θ_1 is very close to zero. Neglecting $(\theta_2 - \theta_1)$ in the exponent in equation 22.35 leads to the approximate solution

$$\theta_2 \cong \cos^{-1}\left(e^{-\frac{2\pi}{\omega RC}}\right). \tag{22.36}$$

For most estimation purposes, this answer is good enough. To obtain a more accurate answer for θ_2, we use this value as the initial guess in an iterative method for solving equation 22.35. Typically, the process converges in two or three iterations because the initial guess is already very close to the true solution.

◆ EXAMPLE 22.8

A half-wave rectifier with ac source voltage 120 V rms at 60 Hz operates into a shunt capacitor filter having $C = 4\mu F$ and $R = 20,000\ \Omega$. For what fraction of a period is the diode conducting?

SOLUTION

The first parameter to calculate is $\omega RC = 120\pi \times 20,000 \times 0.000004 = 30.16$, so that $1/\omega RC = 0.03316$. From equation 22.33,

$$\theta_1 = \tan^{-1}(0.03316) = 1.9 \text{ degrees}.$$

From equation 22.36,

$$\theta_2 \cong \cos^{-1}(e^{-2\pi/30.16}) = 0.6233 \text{ rad, or } 35.71°.$$

Using this value as an initial guess, a numerical solution of equation 22.35 yields

$$\theta_2 = 0.5951 \text{ rad, or } 34.1°.$$

The diode conducts for a total of $1.9° + 34.1° = 36°$, which is 10% of each period.

As the preceding example shows, the diode conducts only for a small fraction of each period. In other words, the current flows through the diode in the form of a train of narrow pulses, as illustrated in figure 22.12c. For a full-wave rectifier, θ_1 is still given by equation 22.33, and the formula for θ_2 differs from equation 22.35 only in a very minor way. The formulas for a full-wave rectifier are stated in problem 15.

Once the cutout angle θ_1 and the cut-in angle θ_2 are known, all other performance measures of the rectifier circuit may be calculated. Some of the quantities of interest are the dc output voltage, the ripple factor, the peak diode current, the average diode current, and the peak reverse voltage across the diode. Conceptually, there is no difficulty in determining these quantities, but computationally, the task may still involve a difficult evaluation of a definite integral. It is therefore important to know what kind of simplifying approximations will maintain an acceptable degree of accuracy. We shall use the half-wave rectifier circuit as an example of such calculations.

The $v_o(t)$ waveform in figure 22.12b shows that each cycle consists of a sinusoidal-shaped rise of short duration followed by an exponential decay of long duration. For an approximate analysis, both curves are replaced by straight-line segments joining their respective endpoints. The resultant asymmetrical triangular waveform, shown in figure 22.13, will be the basis for the approximate analysis of the half-wave rectifier circuit.

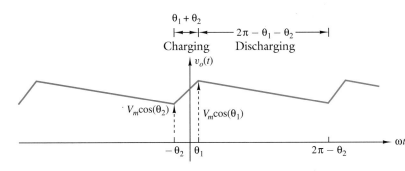

Figure 22.13 Approximate $v_o(t)$ waveform for a half-wave rectifier.

The calculations of various quantities of interest proceed as follows:

Part 1. *Computation of dc output voltage.* From figure 22.13,

$$V_{dc} \cong \frac{V_m \cos(\theta_1) + V_m \cos(\theta_2)}{2}. \tag{22.37a}$$

Further simplification is possible. For the case of $\omega RC \gg 1$, $\theta_1 \cong 0$ and $\theta_2 \ll 2\pi$. From equations 22.33 and 22.35

$$\cos(\theta_1) \cong 1,$$

$$\cos(\theta_2) \cong e^{\frac{-2\pi}{\omega RC}} \cong 1 - \frac{2\pi}{\omega RC} = 1 - \frac{2\pi T}{\omega T RC} = 1 - \frac{T}{RC}.$$

Substituting these results into equation 22.37a yields

$$V_{dc} \cong \frac{V_m \left(1 + 1 - \dfrac{T}{RC}\right)}{2} = \left(1 - \frac{\pi}{\omega RC}\right) V_m. \tag{22.37b}$$

Problem 16 suggests a direct derivation of equation 22.37b, without first deriving expressions for θ_1 and θ_2.

Part 2. *Calculation of average diode current.* The diode conducts only during the interval $-\theta_2 < \omega t < \theta_1$, over which equation 22.32 specifies the current $i_d(t)$. Finding the average value of $i_d(t)$ is a routine exercise in integration. Since we have already computed the dc output voltage, the following alternative approach is actually superior. From figure 22.12a,

$$i_d(t) = i_R(t) + i_C(t).$$

Hence,

$$[i_d]_{ave} = [i_R]_{ave} + [i_C]_{ave}.$$

Now, V_{dc}, the dc component of $v_o(t)$, pushes a dc current

$$[i_R]_{ave} = \frac{V_{dc}}{R}$$

through the resistor, with no dc current component through the capacitor. Therefore,

$$[i_d]_{ave} = \frac{V_{dc}}{R}. \tag{22.38a}$$

For the case of $\omega RC \gg 1$,

$$I_{dc} = [i_d]_{ave} \cong \left(1 - \frac{0.5T}{RC}\right)\frac{V_m}{R} = \left(1 - \frac{\pi}{\omega RC}\right)\frac{V_m}{R}. \tag{22.38b}$$

Part 3. *Calculation of ripple factor.* The waveform of figure 22.13 differs from the asymmetrical triangular waveform of item 7 of table 22.3 only by a translation. This does not alter the amplitudes of the ac components. The rms value given in the table as $A/\sqrt{3}$ is for the complete waveform. Excluding the dc component, the rms value of the ac components becomes

$$\sqrt{\left(\frac{A}{\sqrt{3}}\right)^2 - \left(\frac{A}{2}\right)^2} = \frac{A}{\sqrt{12}}.$$

To apply this result to figure 22.13, we note that

$$A = V_m(\cos\theta_1 - \cos\theta_2).$$

Therefore,

$$\text{rms value of ac components of } v_o(t) = \frac{V_m}{\sqrt{12}}(\cos(\theta_1) - \cos(\theta_2)).$$

Next, from figure 22.13 and equation 22.37a,

$$V_{dc} \cong \frac{V_m\cos(\theta_1) - V_m\cos(\theta_2)}{2}.$$

Finally, the definition of the ripple factor implies that

$$\text{ripple factor} = \frac{V_{ac,rms}}{V_{dc}} = \frac{\cos(\theta_1) - \cos(\theta_2)}{\sqrt{3}\big(\cos(\theta_1) + \cos(\theta_2)\big)}$$

$$\cong \frac{1 - \cos(\theta_2)}{\sqrt{3}\big(1 + \cos(\theta_2)\big)}. \qquad (22.39)$$

Part 4. *Calculation of peak diode current.* During the interval $-\theta_2 < \omega t < \theta_1$, the diode conducts, and equation 22.32 specifies the current. A routine phasor calculation shows that

$$i_d(t) = \frac{V_m}{R}\sqrt{(\omega RC)^2 + 1}\,\cos\left[\omega t + \tan^{-1}(\omega RC)\right]. \qquad (22.40)$$

For typical values of $\omega RC \gg 1$, the peak value of i_d during the conducting interval occurs at $\omega t = -\theta_2$. From equation 22.40, this peak value is

$$i_{d,\text{peak}} \cong \left(\frac{V_m}{R}\right)(\omega RC)\sin\theta_2 = V_m\omega C\sin\theta_2. \qquad (22.41)$$

◆ **EXAMPLE 22.9**

The half-wave rectifier circuit of figure 22.12a has $C = 30\ \mu\text{F}$, $R = 6{,}800\ \Omega$, and $V_m = 30$ V. The ac source is 60 Hz. Find the dc output voltage, the ripple factor, and the diode average and peak currents.

SOLUTION

The solution is a basic "plug and chug." First, we calculate the parameter $\omega RC = 2\pi \times 60 \times 6{,}800 \times 0.00003 = 76.91$. From equation 22.33,

$$\theta_1 = \tan^{-1}\left(\frac{1}{76.91}\right) = 0.013 \text{ rad, or } 0.745°.$$

From equation 22.36 (we make no attempt to get a more accurate value of θ_2 from equation 22.35),

$$\theta_2 \cong \cos^{-1}\left(e^{\frac{-2\pi}{76.91}}\right) = 0.399 \text{ rad, or } 22.84°.$$

From equation 22.37b,

$$V_{dc} = \left(\frac{1 - \pi}{76.91}\right) \times 30 = 0.96 \times 30 = 28.77 \text{ V}.$$

From equation 22.39,

$$\text{ripple factor} = \frac{1 - \cos(0.399)}{\sqrt{3}\,(1 + \cos(0.399))} = 0.0234, \text{ or } 2.34\%.$$

From equation 22.38a,

$$\text{diode average current} = \frac{28.77}{6,800} = 0.00423 \text{ A, or } 4.23 \text{ mA.}$$

From equation 22.41,

$$\text{diode peak current} = \frac{30}{6,800}(76.91)\sin(0.399) = 0.1318 \text{ A.}$$

In order to reduce the ripple factor to a low value, the parameter ωRC for the circuit of figure 22.12a must be large. This causes a smaller conducting angle and a large peak current for the diode. In example 22.9, the conduction angle is $23.6°$ and the diode peak current is 31.2 times the delivered dc current. This undesirable large peak current may damage the diode. For this reason, practical rectifiers usually add more filtering sections.

Two common full-wave rectifiers are shown in figure 22.14. Even with all the simplifying assumptions (ideal diode and zero source resistance), any attempt to analyze these circuits following the steps previously used for the case of a single-capacitor filter will prove extremely difficult. This is when circuit designers make use of specially prepared charts for accurate analysis. We shall content ourselves here with a manual analysis giving *approximate* answers.

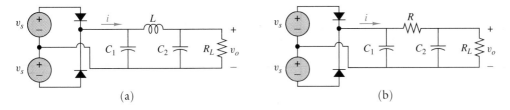

Figure 22.14   Full-wave rectifier with a three-element filter. (a) LC filter. (b) RC filter.

The key to the approximate solution is the short pulse property discussed in section 2, viz.:

As the ratio of the unipolar pulse width to the period becomes very small, the magnitudes of the fundamental and all harmonic components approach twice the average value (dc value).

We may apply this property to the input current $i(t)$ of any well-designed rectifier to determine approximately the ac components, since the dc load current is usually specified. The output voltage is calculated by the use of superposition and the transfer function of the linear filter. Example 22.10 illustrates the procedure.

EXAMPLE 22.10

In the rectifier circuit of figure 22.14b, $C_1 = C_2 = 15$ µF, $R = 2$ kΩ, $R_L = 5$ kΩ, and $v_s(t)$ is a 60-Hz, 30-V (peak value) source. Find the dc output voltage, the rms ripple voltage, and the ripple factor.

SOLUTION

As before, assume that the diodes are ideal. The capacitor C_1 is charged up to a peak value of 30 V. The dc value of v_{C1} is very close to 30 V. At dc, the capacitors behave as open circuits. Therefore, the dc current flowing through the resistors is

$$I_{dc} \cong \frac{30}{5,000 + 2,000} = 0.00428 \text{ A or } 4.28 \text{ mA.}$$

The dc output voltage is then

$$[v_0]_{dc} \cong 0.00428 \times 5,000 = 21.42 \text{ V.}$$

The input current $i(t)$ consists of very short pulses at 120 Hz. From the short pulse property, all ac components of $i(t)$ have magnitudes approximately equal to $2 \times I_{dc} = 8.56$ mA. For this full-wave rectifier, the lowest ac frequency is twice the source frequency. At this point, we can find the transfer function $H(j\omega)$ in the standard way—say, by writing the node equations. However, it is easier and more informative to recognize that the $1/\omega C$ is much smaller than the other resistances in the circuit. Accordingly, the input current "sees" an impedance

$$Z_{in}(j\omega) \cong Z_{C1} = \frac{-j}{\omega C_1}.$$

The parallel combination of C_2 and R_L has an impedance

$$Z_{load}(j\omega) \cong Z_{C2} = \frac{-j}{\omega C_2}.$$

The transfer function is

$$H(j\omega) = \frac{\mathbf{V}_o}{\mathbf{I}} = \frac{Z_{in} \times Z_{load}}{R + Z_{load}} \cong \frac{Z_{C1} Z_{C2}}{R}.$$

Therefore,

$$H(j\omega) = \left| \frac{\mathbf{V}_o}{\mathbf{I}} \right| \cong \frac{1}{\omega^2 C_1 C_2 R}. \tag{22.42}$$

From equation 22.42, the magnitude of the transfer function at 120 Hz approximates

$$\frac{1}{(2\pi \times 120 \times 0.000015)^2 \times 5,000} = 3.91 \ \Omega.$$

The magnitude of the 120-Hz component of the input current is $8.56/\sqrt{2}$ mA (rms). Therefore, the rms value of the output voltage at 120 Hz is

$$\frac{0.00856 \times 3.91}{\sqrt{2}} = 0.0335 \text{ V.}$$

Because $|H(j\omega)|$ is inversely proportional to ω^2, higher harmonics are insignificant at the output. The ripple factor is approximately

$$\frac{0.00335}{21.42} = 0.00156, \text{ or } 0.156\%.$$

This is much smaller than the ripple factor of the rectifier of example 22.9, with a comparable dc output voltage and current. The price to be paid, however, is more components and more complex circuitry.

5. ADDITIONAL PROPERTIES OF AND COMPUTATIONAL SHORTCUTS TO FOURIER SERIES

If a periodic function $f(t)$ is known only at some sampled points, e.g., by measurements, then its Fourier coefficients must be calculated by the use of numerical methods, as illustrated in example 22.6. On the other hand, if $f(t)$ has an analytic expression, then its Fourier coefficients can be calculated from equation 22.9. The Fourier coefficients for some commonly encountered periodic waveforms are given in table 22.3. Engineering and mathematical handbooks contain much more comprehensive tables. On some occasions, when a waveform does not appear in the table, the Fourier coefficients must be computed manually. The properties to be discussed next are of great value in simplifying the solution in such cases. Their proofs are fairly straightforward and are left as exercises.

The Linearity Property: Let $f_1(t)$ and $f_2(t)$ be periodic with fundamental period T. If $f(t) = K_1 f_1(t) + K_2 f_2(t)$, then the Fourier coefficients of $f(t)$ may be expressed in terms of those of $f_1(t)$ and $f_2(t)$ according to the following formulas:

$$c_n = K_1 c_{1n} + K_2 c_{2n},$$
$$a_n = K_1 a_{1n} + K_2 a_{2n}, \tag{22.43}$$
$$b_n = K_1 b_{1n} + K_2 b_{2n}.$$

In general,

$$d_n \neq K_1 d_{1n} + K_2 d_{2n},$$

unless the angle θ_n is the same for all n.

The solution of example 22.6 utilizes the linearity property.

DEFINITION

A periodic function $f(t)$ is said to be **half-wave symmetric** if

$$f(t - 0.5T) = -f(t) \text{ for all } t.$$

In words, $f(t)$ is half-wave symmetric if it equals a half-period shift of itself combined with a flip about the horizontal axis. Some simple examples of half-wave symmetric functions include $\sin(\omega t)$, $\cos(\omega t)$, and the square wave of figure 22.4.

The half-wave symmetry property: A half-wave symmetric periodic function $f(t)$ contains only odd harmonics.

The waveforms of figures 22.4, 22.10a, and 22.10b demonstrate this property.

The derivative/integral property: Let $f^{(k)}(t)$ denote the kth derivative of a periodic function $f(t)$. Then the Fourier coefficients $c_n^{(k)}$ of $f^{(k)}(t)$ satisfy

$$c_n^{(k)} = (jn\omega_0)^k c_n \text{ for all } n, \tag{22.44a}$$

or

$$c_n = \frac{c_n^{(k)}}{(jn\omega_0)^k} \text{ for all } n \text{ except } n = 0. \tag{22.44b}$$

Except for the constant term, all other terms of $f^{(k)}(t)$ derive from those of $f(t)$ by differentiating k times; conversely, all terms of $f(t)$ derive from those of $f^{(k)}(t)$ by a k-fold (indefinite) integration. The exclusion of the constant term in the relationships poses no difficulty, since the constant is simply the average value of the periodic function.

The foregoing properties help to simplify the calculation of Fourier coefficients. In fact, it is possible to find Fourier series for all of the waveforms listed in table 22.3 without carrying out the integration of equation 22.9. Achieving this, however, depends on first finding the Fourier series of a waveform for which computing c_n by equation 22.9 is extremely easy. In examples 22.11 through 22.15, the following trigonometric identities prove useful:

$$\cos(x) - \cos(y) = -2 \sin\left(\frac{x+y}{2}\right) \sin\left(\frac{x-y}{2}\right),$$ (22.45a)

$$\cos(x) \cos(y) = .5[\cos(x+y) + \cos(x-y)].$$ (22.45b)

◆ **EXAMPLE 22.11**

Find the Fourier series for the periodic impulse train $f_\delta(t)$ shown in figure 22.15.

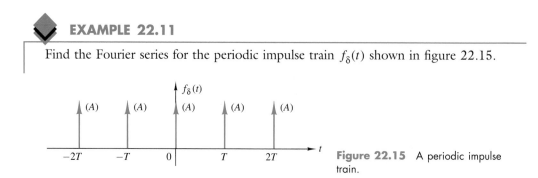

Figure 22.15 A periodic impulse train.

SOLUTION

Using the sifting property of an impulse function together with equation 22.9,

$$c_n = \frac{1}{T} \int_{-\frac{T}{2}}^{\frac{T}{2}} A\delta(t) e^{-jn\omega_0 t} dt = \frac{A}{T}$$ (22.46a)

for all n, and

$$f_\delta(t) = \frac{A}{T} + \frac{2A}{T} \{\cos(\omega_0 t) + \cos(2\omega_0 t) + \cos(3\omega_0 t) + \ldots\}.$$ (22.46b)

Equation 22.46b states that, for a periodic impulse train, all ac components have magnitude equal to twice the average value. This is the limiting case of the short pulse property stated in section 2. The next example shows an alternative derivation of the Fourier series for a periodic rectangular pulse train, derived earlier by the use of equation 22.9.

◆ **EXAMPLE 22.12**

Find the Fourier series for the periodic rectangular pulses $f_p(t)$ shown in figure 22.16.

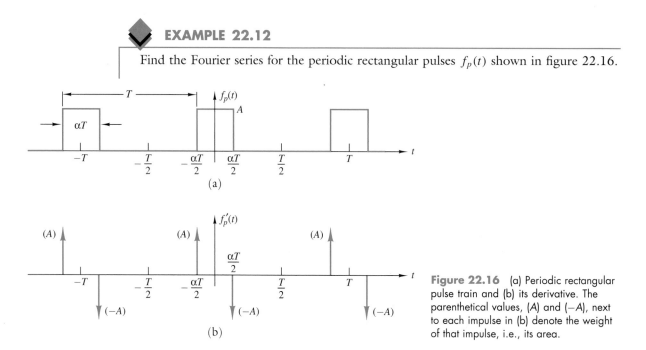

Figure 22.16 (a) Periodic rectangular pulse train and (b) its derivative. The parenthetical values, (A) and (−A), next to each impulse in (b) denote the weight of that impulse, i.e., its area.

Figures 22.16(a) and (b) show $f_p(t)$ and its derivative $f'_p(t)$. The latter may be written as the sum of two shifted impulse trains:

$$f'_p(t) = f_\delta\left(t + \frac{\alpha T}{2}\right) - f_\delta\left(t - \left(\frac{\alpha T}{2}\right)\right). \tag{22.47}$$

Using the time-shift property (translation in the horizontal direction), together with equations 22.46b and 22.45a, one obtains

$$f'_p(t) = \sum_{n=1}^{\infty} \frac{2A}{T}\left(\cos(n\omega_0 t + n\alpha\pi) - \cos(n\omega_0 t - n\alpha\pi)\right)$$

$$= \sum_{n=1}^{\infty} \frac{-4A}{T}\left(\sin(n\omega_0)t\,\sin(n\alpha\pi)\right). \tag{22.48}$$

Applying the derivative/integral property to equation 22.48 yields

$$f_p(t) = [f_p(t)]_{\text{ave}} + \sum_{n=1}^{\infty} \frac{4A}{n\omega_0 T}\left(\cos(n\omega_0 t)\,\sin(n\alpha\pi)\right)$$

$$= \alpha A + \sum_{n=1}^{\infty} \frac{2A\sin(n\alpha\pi)}{n\pi}\cos(n\omega_0 t). \tag{22.49}$$

The result agrees, of course, with equation 22.14.

EXAMPLE 22.13

Find the Fourier series for the half-wave rectified sine wave $f_{hs}(t)$ shown in figure 22.17a.

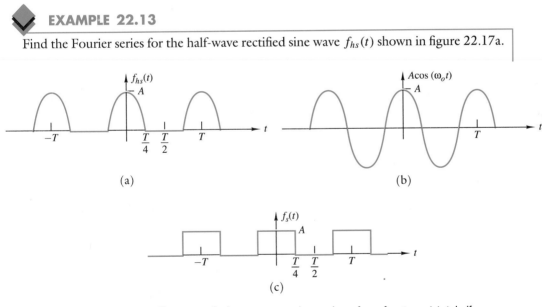

Figure 22.17 Half-wave rectified sine wave as the product of two functions. (a) A half-wave rectified cosine wave. (b) A cosine wave. (c) A square wave.

SOLUTION

The periodic function $f_{hs}(t)$ of figure 22.17a may be viewed as the product of the sinusoidal wave $A\cos(\omega_0 t)$ and the square wave $f_s(t)$, shown in figures 22.17b and c, respectively. Using the Fourier series for $f_s(t)$ given by equation 22.11, we have

$$f_{hs}(t) = [A\cos(\omega_0 t)]\,f_s(t)$$

$$= A\cos(\omega_0 t)\left\{\frac{1}{2} + \frac{2}{\pi}\left(\cos(\omega_0 t) - \frac{1}{3}\cos(3\omega_0 t) + \frac{1}{5}\cos(5\omega_0 t) - \dots\right)\right\}. \tag{22.50}$$

Applying equation 22.45b to each product in equation 22.50 yields

$$f_{hs}(t) = \frac{A}{2}\cos(\omega_0 t) + \frac{A}{\pi}\left\{(\cos(2\omega_0 t) + \cos(0\omega_0 t)) - \frac{1}{3}(\cos(4\omega_0 t) + \cos(2\omega_0 t)) + \ldots\right\}$$

$$= \frac{A}{\pi} + \frac{A}{2}\cos(\omega_0 t) + \frac{2A}{\pi}\left(\frac{1}{3}\cos(2\omega_0 t) - \frac{1}{15}\cos(4\omega_0 t) + \right.$$

$$\left. \frac{1}{35}\cos(6\omega_0 t) - \frac{(-1)^{n+1}}{4n^2 - 1}\cos(2n\omega_0 t) + \ldots\right).$$

(22.51)

Note that the fundamental component is present in $f_{hs}(t)$, and the remaining terms are all even harmonics.

◆ **EXAMPLE 22.14**

Find the Fourier series for the full-wave rectified sine wave $f_{fs}(t)$ shown in figure 22.18a.

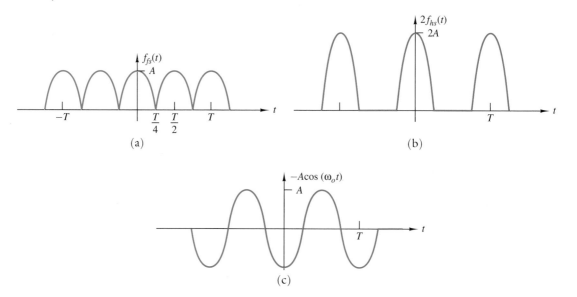

Figure 22.18 (a) Full-wave rectified sine wave as the sum of two signals, (b) and (c).

SOLUTION

One approach is to apply the same technique as in example 22.13: Specifically,

$$f_{fs}(t) = [A\cos(\omega_0 t)][2f_s(t) - 1],$$

where $f_s(t)$ is the square wave of figure 22.17c. Alternatively, to avoid repeating all the arithmetic, here note that $f_{fs}(t)$ is the sum of the two waveforms shown in figures 22.18b and c; i.e.,

$$f_{fs}(t) = 2f_{hs}(t) - A\cos(\omega_0 t). \qquad (22.52)$$

Substituting equation 22.51, the Fourier series $f_{hs}(t)$, into equation 22.52 yields the desired Fourier series:

$$f_{fs}(t) = \frac{2A}{\pi} + \frac{4A}{\pi}\left(\frac{1}{3}\cos(2\omega_0 t) - \frac{1}{15}\cos(4\omega_0 t) + \frac{1}{35}\cos(6\omega_0 t) \right.$$

$$\left. - \frac{(-1)^{n+1}}{4n^2 - 1}\cos(2n\omega_0 t) + \ldots\right).$$

(22.53)

The derivative property is particularly useful for tackling periodic piecewise linear wave-forms. Piecewise linear waveforms consist of straight-line segments. By differentiating once, or at most twice, impulses appear. The integration given by equation 22.9 is trivial if the integrand contains a shifted impulse function. This fact, together with the derivative property, reduces the complexity of calculating the Fourier coefficients for piecewise linear waveforms to some complex number arithmetic. The last example of this section illustrates the procedure.

EXAMPLE 22.15

Find the magnitudes of the dc, fundamental, second, and third harmonics of the piece-wise linear periodic waveform of figure 22.19a.

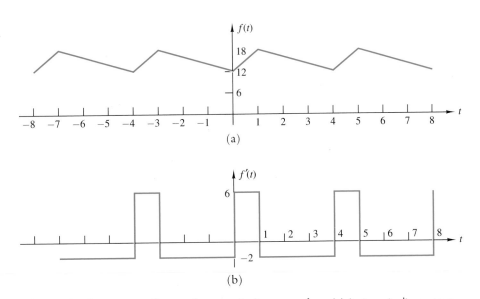

Figure 22.19 Fourier coefficients of a piecewise linear waveform. (a) A piecewise linear waveform. (b) Its derivative.

SOLUTION

The dc component of $f(t)$ is easily seen to be $F_{av} = 15$. For ac components, we differentiate $f(t)$ to obtain $f'(t)$, shown in figure 22.19b. Further differentiation of this waveform generates impulses. However, such differentiation is unnecessary, because $f'(t)$ is a translation of $f_p(t)$ of figure 22.16a with $A = 8$, $\alpha = 0.25$, $T = 4$, and $\omega_0 = 2\pi/T = 1.571$. Equation 22.49 gives the Fourier coefficients of $f_p(t)$. The translation of a waveform does not alter the magnitudes of the ac components. Equation 22.44b is then used to compute the magnitude of the ac components of $f(t)$. For example, the magnitude of the third harmonic of $f_p(t)$ and $f'(t)$, from equation 22.49, is

$$\left| \frac{2 \times 8 \times \sin(3 \times 0.25\pi)}{3\pi} \right| = 1.2.$$

Hence, the magnitude of the third harmonic of $f(t)$, from equation 22.44b, is

$$\frac{1.2}{\left(\frac{3 \times 2\pi}{4} \right)} = 0.2546$$

Other components are computed similarly. Table 22.7 lists the results.

The angle θ_n, if needed, can be calculated from the information contained in equations 22.49, 22.12c (for translation), and 22.44b (for differentiation).

TABLE 22.7 FOURIER COMPONENTS OF THE WAVEFORM OF FIGURE 22.19				
dc	Fundamental	Second harmonic	Third harmonic	Fourth harmonic
d_0	d_1	d_2	d_3	d_4
15	2.722	0.8104	0.2546	0

SUMMARY

Given that many mathematical and engineering handbooks have extensive lists of the Fourier series of common signals, this chapter has taken a practical approach to the application of Fourier series to circuit analysis, de-emphasizing the theoretical development of the mathematical technique. The idea is to use tables such as table 22.3, in the same way that engineers have come to use integral tables. The keys to using tables for the computation of the Fourier series of a waveform are the various properties of the series that allow one to convert a known series into one which fits a new waveform. The idea is to express the new waveform as a translation of a tabularized signal, a linear combination of tabularized signals, a k-fold derivative or k-fold integral of a tabularized signal, or any mixture of these operations. The Fourier coefficients of the new signal can then be expressed in terms of the Fourier coefficients of the tabularized signals.

Knowledge of the Fourier coefficients of a signal—for example, the output of an audio amplifier—allows one to investigate such things as the distortion introduced by the amplifier. Or, in the case of a dc power supply, such knowledge allows us to characterize the degree of unwanted ripple in the output of a rectifier circuit. Both problems are of immense practical significance. Both illustrate the power of the Fourier series method in the analysis of practical circuits.

In addition, the Fourier series method plays an important role in the computation of steady-state circuit responses to periodic input signals. An example illustrating its role was investigated in this chapter.

In sum, then, we have reviewed the basic structure and computation of the Fourier series of a periodic signal without delving into the theoretical subtleties of the process. Instead, we have emphasized the use of properties of the Fourier series to convert a set of basic Fourier series representations into Fourier series representations of new, modified signals. These techniques and the information they supply were applied to the problem of harmonic distortion in amplifiers and the calculation of ripple in dc power supplies.

TERMS AND CONCEPTS

Average value of a periodic function $f(t)$: d_0 in equation 22.3; also referred to as the dc component of $f(t)$.

Derivative/integral property: let $f^{(k)}(t)$ denote the kth derivative of a periodic function $f(t)$. Then the Fourier coefficients of the kth derivative, $c_n^{(k)}$, satisfy $c_n^{(k)} = (jn\omega_0)^k c_n$ for all n, and conversely, if $f(t)$ is the k-th integral of $f^{(k)}(t)$, then $c_n = c_n^{(k)}/(jn\omega_0)^k$ for all n except zero.

Distortion: for an amplifier, any difference between the input and output waveforms other than a scale change and a time delay.

Duty cycle: the constant α that determines the pulse width αT of a rectangular signal having fundamental period T.

Effective, or rms value: $F_{\text{eff}} = \sqrt{d_0^2 + \frac{1}{2}d_1^2 + \frac{1}{2}d_2^2 + \dots}$, where the d_i coefficients are from the Fourier series of equation 22.3.

Even function: function $f(t)$ such that $f(t) = f(-t)$. The plot of an even function is symmetrical about the vertical axis.

Exponential (complex) Fourier series of f(t): decomposition of $f(t)$ into a sum of complex exponentials, as given in equation 22.5.

Fundamental component (first harmonic) of $f(t)$: the first term under the summation sign in equation 22.3, i.e., $d_1 \cos(\omega_0 t + \theta_1)$, having amplitude d_1 and phase angle θ_1.

Fundamental frequency (in Hertz): $f_0 = 1/T_0$, where T_0 is the fundamental period. (*Note*: $\omega_0 = 2\pi f_0 = 2\pi/T_0$ is the fundamental angular frequency in rad/sec.)

Fundamental period: the smallest positive real number T_0 for which $f(t + T_0) = f(t)$.

Half-wave symmetric: a periodic function $f(t)$ that satisfies $f(t - 0.5T) = -f(t)$.

Half-wave symmetry property: property such that a half-wave symmetric periodic function $f(t)$ contains only odd harmonics.

Linearity property: let $f_1(t)$ and $f_2(t)$ be periodic with fundamental period T. If $f(t) = K_1 f_1(t) + K_2 f_2(t)$, then the Fourier coefficients of $f(t)$ may be expressed in terms of those of $f_1(t)$ and $f_2(t)$ according to the formulas $c_n = K_1 c_{1n} + K_2 c_{2n}$, $a_n = K_1 a_{1n} + K_2 a_{2n}$, and $b_n = K_1 b_{1n} + K_2 b_{2n}$. In general, $d_n \neq K_1 d_{1n} + K_2 d_{2n}$, unless the angle θ_n is the same for all n.

nth-harmonic distortion: the ratio of the nth ($n = 2, 3, \dots$) harmonic amplitude to the fundamental amplitude, usually expressed as a percent.

Odd function: function $f(t)$ such that $f(t) = -f(-t)$.

Periodic signal $f(t)$: A signal whose waveform repeats every T seconds. Mathematically, for some $T > 0$ and all t, $f(t + T) = f(t)$, where T is the period of the signal.

Rectifier circuit: a circuit that transforms an ac sinusoid to an approximate dc waveform.

Ripple: the presence of ac components in the output of a rectifier circuit.

Ripple factor: a measurement of the purity of a rectifier output; mathematically given by the ratio of the effective value of all ac components to the dc component.

Second harmonic of $f(t)$: The term $d_2 \cos(2\omega_0 t + \theta_2)$ in equation 22.3 having amplitude d_2 and phase angle θ_2.

Symmetry properties of the Fourier series: (1) If a periodic function $f(t)$ is an even function, then its Fourier series has only cosine terms and possibly a constant term. (2) If a periodic function $f(t)$ is an odd function, then its Fourier series has only sine terms.

Total harmonic distortion: the ratio of the effective value of the sum of all the harmonics to the effective value of the fundamental.

Translation of a plot: a horizontal and/or vertical movement of a curve without any rotation.

Translation property of the Fourier series: if $g(t)$ is a translation of $f(t)$ consisting of a dc-level increase K and a time shift (delay) to the right by t_d, then $g(t) = f(t - t_0) + K$. (See equations 22.12.)

Trigonometric Fourier series: representation of a periodic signal $f(t)$ in terms of sines and cosines, as given in equation 22.2 or 22.3.

PROBLEMS

1. The input to the low-pass filter circuit shown in figure P22.1 is

$$v_{in}(t) = 200 + \sqrt{2}[200 \cos 377t +$$

$$60 \cos(3 \times 377t + 30°) + 80 \cos(5 \times 377t + 50°)] \text{ V}.$$

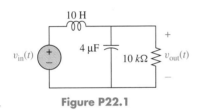

Figure P22.1

(a) Find $v_{out}(t)$ in the steady state.
(b) Find the rms value of $v_{out}(t)$ and the average power absorbed by the 10-kΩ resistor.

ANSWER: (a) $v_{out}(t) = 200 + \sqrt{2}[42.6 \cos(\omega t - 175.4°) + 1.192 \cos(3\omega t - 148.7°) + 0.56 \cos(5\omega t - 129.2°)]$ V, where $\omega = 377$; (b) 204.5 V, 4.18 W.

2. Find the Fourier series for the periodic impulse trains shown in figure P22.2. Carry out the integration in equation 22.9, and write the Fourier series in the form of equation 22.2.

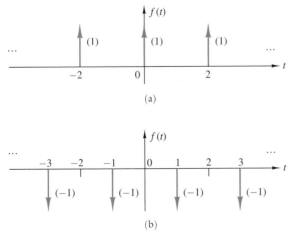

Figure P22.2 Periodic impulse trains.

ANSWER: (a) $0.5 + \cos(\pi t) + \cos(2\pi t) + \cos(3\pi t) + \cos(4\pi t) + \cos(5\pi t) + \dots$; (b) $-0.5 + \cos(\pi t) - \cos(2\pi t) + \cos(3\pi t) - \cos(4\pi t) + \cos(5\pi t) + \dots$

3. For the periodic function $f(t)$ shown in figure P22.3, $T = 1$ sec and $a = \log_e(2) = 0.693$.

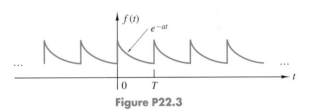

Figure P22.3

(a) Find the coefficient c_n by carrying out the integration in equation 22.9.
(b) Write the first three terms of the Fourier series, in the form of equation 22.3.

ANSWER: (a) $c_n = 0.5/(\log_e(2) + j2n\pi)$; (b) $f(t) = 0.7213 + 0.158 \cos(2\pi t - 83.7°) + 0.0795 \cos(4\pi t - 86.84°) + \dots$. Note: For problems 4–11, evaluating the integral in equation 22.9 is unnecessary. You can obtain the Fourier coefficients more easily by the use the Fourier series properties discussed in sections 2 and 5 of this chapter.

4. Find the Fourier series for the sawtooth waveform shown in figure P22.4. (Hint: Note that $f'(t) = A - Af_\delta(t)$. Use equation 22.46b for $f_\delta(t)$.)

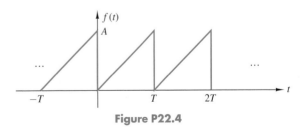

Figure P22.4

ANSWER: See item 5 of table 22.3.

5. Find the Fourier coefficients d_n and θ_n of the periodic function shown in figure P22.5. (Hint: Make use of the result of problem 4.)

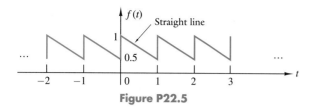

Figure P22.5

ANSWER: $f(t) = 0.75 + 0.159 \sin 2\pi t + 0.0795 \sin 4\pi t + \ldots$.

6. Find the Fourier series for the triangular wave shown in figure P22.6. (*Hint*: $f'(t)$ is a square wave whose Fourier series was calculated before.)

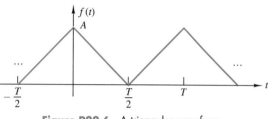

Figure P22.6   A triangular waveform.

ANSWER: See item 3 of table 22.3.

7. Consider the isoceles triangular wave shown in figure P22.7. Derive its Fourier series.

ANSWER: See item 4 of table 22.3.

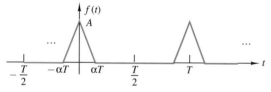

Figure P22.7   An isoceles triangular waveform.

8. Consider the clipped sawtooth wave shown in figure P22.8. Derive its Fourier series.

ANSWER: See item 6 of table 22.3.

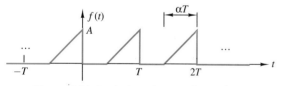

Figure P22.8   A clipped sawtooth waveform.

9. Find the Fourier series for the waveform shown in figure P22.9. (*Hint*: Make use of the result of problem 8.)

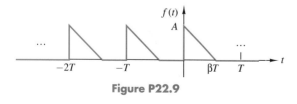

Figure P22.9

10. Consider the asymmetrical triangular wave shown in figure P22.10. Let $T = 1$ and $\alpha = 0.25$. Find the Fourier

coefficients d_0, d_1, and d_2. (*Hint*: Differentiating $f(t)$ twice results in periodic impulse trains.)

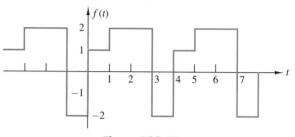

Figure P22.10   An asymmetrical triangular waveform.

11. Find the Fourier series coefficients c_0, c_1, and c_2 for the periodic waveform shown in figure P22.11, with period $T = 4$. (*Hint*: Express $f(t)$ as the sum of several shifted square waves.)

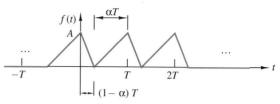

Figure P22.11

12. Repeat the harmonic distortion analysis of example 22.5 if the nonlinear input-output relationship is changed to

$$v_o = 10v_i + v_i^2 + v_i^3.$$

13. When a 1-V (rms), 1-kHz sine wave is applied to a certain amplifier, the output voltage is known to have negligible harmonics of order four and higher. From the output waveform displayed on an oscilloscope, the following readings are taken:

t	0	$T/6$	$T/3$	$T/2$
v_o	10	5.2	−4.6	−9.6

Find the total harmonic distortion.

14. Consider the operational amplifier circuit of example 22.7. If the input voltage amplitude is reduced to 0.9 V (peak), find the third-order harmonic distortion.

15. Equation 22.35 shows the relationship between the angles θ_1 and θ_2 of a half-wave rectifier. Prove that, for a full-wave rectifier, the equation becomes

$$(\cos \theta_1) e^{\frac{-\theta_2 - \theta_1 + \pi}{\omega RC}} = \cos \theta_2.$$

16. Equation 22.37b gives an approximate value of the dc output voltage for the half-wave rectifier of figure 22.12a. Derive this formula in a very simple manner by assuming that $\theta_1 = 0$, $\theta_2 = 0$, and $RC \gg T$. Under these conditions, $v_o(t)$ is approximately a straight line within the interval $0 < \omega t < 2\pi$, and the average value of $v_o(t)$ equals approximately $v_o(T/2)$. (*Hint*: Expand $v_o(t)$ as a series in t, and keep only the first two terms.)

17. (1) Suppose the half-wave rectifier circuit of figure 22.12a has $C = 20$ μF, $R = 100$ kΩ, and $V_m = 20$ V. Let the ac source be 60 Hz. Find the dc output voltage, the ripple factor, and the diode average and peak currents. (2) Determine the value of C if the ripple factor is to be reduced by one-half.

18. Repeat problem 17 for a full-wave rectifier, using the same values for C and R.

19. In the full-wave rectifier circuit shown in figure P22.19, the output dc voltage is 30 V and the load R_L draws 10 mA dc. The input voltage is $v_s(t) = 49.5\cos(377t)$ V. Other component values are $R = 1,950$ Ω and $C = 100$ μF. Assume that each diode conducts for a very small fraction of the period, and hence, $i(t)$ may be treated as a periodic impulse train. Estimate the rms value of the 120-Hz component in the output voltage.

20. In the full-wave rectifier circuit shown in figure P22.20, the output dc voltage is 30 V and the load R_L draws 10 mA dc. The input voltage is $v_s(t) = 49.5\cos(377t)$ V. Other component values are $R = 975$ Ω and $C = 16$ μF. Assume that each diode conducts for a very small fraction of the period, and hence, $i(t)$ may be treated as a periodic impulse train. Estimate the rms value of the 120-Hz component in the output voltage.

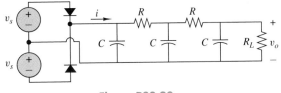

Figure P22.20

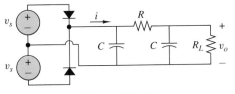

Figure P22.19

APPENDIX A1

Matrices

C HAPTER 4 introduced the node and loop analysis of circuits. Since most circuits have multiple nodes and multiple loops, computing node voltages or loop currents requires solving sets of equations simultaneously. Having to solve sets of equations simultaneously is a widespread engineering phenomenon because the modeling of real-world processes requires many variables and many constraining equations for even the simplest physical systems. A convenient organized methodology for handling sets of linear simultaneous equations is matrices and vectors. This appendix introduces some basic concepts pertaining to matrices, vectors, and the solution of vector-matrix equations. The material here is relevant to chapters 4 and 10 and to solving other types of problems throughout volumes 1 and 2 of the text.

Definitions and Basic Properties

An $m \times n$ (read "m by n") **matrix** is a rectangular array of numbers arranged in m rows and n columns (usually enclosed by brackets). A 1×1 matrix is called a **scalar**. A $1 \times n$ matrix is called a **row vector**. An $m \times 1$ matrix is called a **column vector**. If $m = n$, the matrix is called a **square matrix** of order n. A matrix with all elements equal to zero is called a **null matrix** and is denoted by $[0]$ or 0.

EXAMPLE 1

The following table shows various matrices:

2×3 matrix	Row vector	Square matrix of order 2	Column vector	Null matrix of dimension 1×3
$A_1 = \begin{bmatrix} 2 & 3 & -1 \\ 5 & 0 & 2 \end{bmatrix}$	$A_2 = [\,2 \; -3 \; 4\,]$	$A_3 = \begin{bmatrix} 2 & 3 \\ 5 & 8 \end{bmatrix}$	$b = \begin{bmatrix} 4 \\ 1 \end{bmatrix}$	$0_{1 \times 3} = [\,0 \; 0 \; 0\,]$

A standard notation for a general $m \times n$ matrix uses double subscripts as

$$A = \begin{bmatrix} a_{11} & a_{12} & \ldots & a_{1n} \\ a_{21} & a_{22} & \ldots & a_{2n} \\ \vdots & \vdots & \vdots & \vdots \\ a_{m1} & a_{m2} & \ldots & a_{mn} \end{bmatrix}$$

More compactly, $A = [a_{ik}]$, where a_{ik}, or $a(i, k)$, is the value of the entry in the ith row and kth column of A. The rows are numbered from the top down and the columns from left to right.

Two matrices A and B are said to be **equal** if and only if $a_{ik} = b_{ik}$ for all i and k, which also demands that A and B have the same dimensions m by n.

The **addition** or **subtraction** of two $m \times n$ matrices A and B, by definition, results in another $m \times n$ matrix $C = A \pm B$ whose elements are given by

$$c_{ik} = a_{ik} \pm b_{ik} \quad \text{for all } i \text{ and } k;$$

i.e., corresponding entries are added or subtracted.

The **multiplication** of two matrices is defined as follows: Let A be an $n \times s$ matrix and B be an $s \times m$ matrix. Then the product of A and B is an $n \times m$ matrix $C = AB$ whose elements are given by

$$c_{ik} = \sum_{j=1}^{s} a_{ij} b_{jk} \quad \text{for } i = 1, \ldots, n \text{ and } k = 1, \ldots, m.$$

In words, to obtain the element of C in the ith row and the kth column, take the ith row of A and the kth column of B, perform element-by-element multiplications, and sum up the resulting products. Note that AB is defined if and only if the number of columns of A is the same as the number of rows of B.

The product of a scalar α and a matrix A is obtained by multiplying every element of A by α.

◆ **EXAMPLE 2**

Consider the matrices

$$A_1 = \begin{bmatrix} 2 & 3 & -1 \\ 5 & 0 & 2 \end{bmatrix}, A_4 = \begin{bmatrix} -1 & 0 & 5 \\ 1 & -1 & 2 \end{bmatrix}, A_5 = \begin{bmatrix} 2 & 0 \\ -1 & 3 \\ 3 & 4 \end{bmatrix}.$$

The addition of A_1 and A_4 is

$$A_1 + A_4 = \begin{bmatrix} (2) + (-1) & (3) + (0) & (-1) + (5) \\ (5) + (1) & (0) + (-1) & (2) + (2) \end{bmatrix} = \begin{bmatrix} 1 & 3 & 4 \\ 6 & -1 & 4 \end{bmatrix}.$$

The product of A_4 and A_5 is

$$A_4 A_5 = \begin{bmatrix} (-1)(2) + (0)(-1) + (5)(3) & (-1)(0) + (0)(3) + (5)(4) \\ (1)(2) + (-1)(-1) + (2)(3) & (1)(0) + (-1)(3) + (2)(4) \end{bmatrix} = \begin{bmatrix} 13 & 20 \\ 9 & 5 \end{bmatrix}.$$

Finally, the scalar 2 times A_4 is

$$2A_4 = 2 \begin{bmatrix} -1 & 0 & 5 \\ 1 & -1 & 2 \end{bmatrix} = \begin{bmatrix} -2 & 0 & 10 \\ 2 & -2 & 4 \end{bmatrix}.$$

A **diagonal matrix** is a square matrix whose off-diagonal elements are all zero. For convenience, diagonal matrices are often specified by listing only the elements on the diagonal, using the notation $\text{diag}(a, b, c, \ldots)$. Further, if all elements on the diagonal are equal to 1, the special diagonal matrix $\text{diag}(1, 1, \ldots, 1)$ is called an **identity matrix**, typically denoted by I. When necessary, the order of the identity matrix may be indicated with a subscript. For example, $I_{n \times n}$ means an identity matrix having n rows and n columns.

◆ **EXAMPLE 3**

The matrix

$$A_6 = \text{diag}(3, 4, -1) = \begin{bmatrix} 3 & 0 & 0 \\ 0 & 4 & 0 \\ 0 & 0 & -1 \end{bmatrix}$$

is a diagonal matrix, and the matrices

$$I_2 = \begin{bmatrix} 1 & 0 \\ 0 & 1 \end{bmatrix} \quad \text{and} \quad I_3 = \begin{bmatrix} 1 & 0 & 0 \\ 0 & 1 & 0 \\ 0 & 0 & 1 \end{bmatrix}$$

are identity matrices.

Almost all of the familiar laws governing the addition and multiplication of scalars also hold for matrix addition and multiplication. The counterparts of 0 and 1 in scalar operations are $[0]$ and I, respectively, for matrix operations. Provided that the dimensions of all of the following matrices are compatible for additions and multiplications, we have:

$A + B = B + A$	Commutative property of addition
$(A + B) + C = A + (B + C)$	Associative property of addition
$ABC = (AB)C = A(BC)$	Associative property of multiplication
$A(B + C) = AB + AC$	Distributive property
$[0] + A = A$	Additive identity
$IA = AI = A$	Multiplicative identity

A few relationships of scalars do not hold for matrices. Some important ones are the following:

1. $ab = ba$, but in general, $AB \neq BA$. When $AB = BA$, the matrices A and B are said to *commute*. For example, suppose

$$A = \begin{bmatrix} 1 & 2 \\ 3 & 4 \end{bmatrix}, B = \begin{bmatrix} -1 & -2 \\ 3 & 4 \end{bmatrix}.$$

It follows that

$$AB = \begin{bmatrix} 5 & 6 \\ 9 & 10 \end{bmatrix} \neq \begin{bmatrix} -7 & -10 \\ 15 & 22 \end{bmatrix} = BA.$$

2. $ab = 0$ implies that at least one of a and b is zero; on the other hand, $AB = [0]$ may occur even if both A and B are not 0. For example,

$$\begin{bmatrix} 1 & 1 \\ 0 & 0 \end{bmatrix} \begin{bmatrix} 1 & 1 \\ -1 & -1 \end{bmatrix} = \begin{bmatrix} 0 & 0 \\ 0 & 0 \end{bmatrix}.$$

3. $a \neq 0$ and $ab = ac$ imply $b = c$; on the other hand, $AB = AC$ does not imply $B = C$, even if $A \neq [0]$. For example,

$$\begin{bmatrix} 1 & 1 \\ 0 & 0 \end{bmatrix} \begin{bmatrix} 1 & 1 \\ 1 & -1 \end{bmatrix} = \begin{bmatrix} 1 & 1 \\ 0 & 0 \end{bmatrix} \begin{bmatrix} 1 & 0 \\ 1 & 0 \end{bmatrix} = \begin{bmatrix} 2 & 0 \\ 0 & 0 \end{bmatrix}$$

The following two matrix operations are quite useful in linear circuit analysis:

1. **Transpose** of a matrix: If A is an $m \times n$ matrix, then the transpose of A, denoted by A' (or A^t in some texts) is a matrix obtained from A by interchanging its rows and columns.
2. **Inverse** of a matrix: Let A be a square matrix of order n; if another square matrix B of order n exists such that

$$AB = BA = I_n,$$

then B is called the **inverse** of A and is denoted by A^{-1} or inv(A).

EXAMPLE 4

From the matrices given in example 1, we have

$$A_1' = \begin{bmatrix} 2 & 5 \\ 3 & 0 \\ -1 & 2 \end{bmatrix}, \quad A_2' = \begin{bmatrix} 2 \\ -3 \\ 4 \end{bmatrix}, \quad b' = \begin{bmatrix} 4 & 1 \end{bmatrix}.$$

Further,

$$A_3^{-1} = \begin{bmatrix} 8 & -3 \\ -5 & 2 \end{bmatrix}$$

because

$$A_3 A_3^{-1} = \begin{bmatrix} 2 & 3 \\ 5 & 8 \end{bmatrix} \begin{bmatrix} 8 & -3 \\ -5 & 2 \end{bmatrix} = \begin{bmatrix} 8 & -3 \\ -5 & 2 \end{bmatrix} \begin{bmatrix} 2 & 3 \\ 5 & 8 \end{bmatrix} = \begin{bmatrix} 1 & 0 \\ 0 & 1 \end{bmatrix} = I_2.$$

A square matrix may not have an inverse, in which case it is said to be **singular**. If it does have an inverse, it is said to be **nonsingular**. In linear algebra, it is proved that a matrix has an inverse if and only if the determinant of its square array is nonzero. The inverse of a nonsingular 2×2 matrix is given by the special formula

$$\text{inv} \begin{bmatrix} a_{11} & a_{12} \\ a_{21} & a_{22} \end{bmatrix} = A^{-1} = \frac{1}{a_{11}a_{22} - a_{12}a_{21}} \begin{bmatrix} a_{22} & -a_{12} \\ -a_{21} & a_{11} \end{bmatrix},$$

where the quantity $a_{11}a_{22} - a_{12}a_{21}$ is the determinant of the 2×2 matrix. In words, this formula says: Interchange the diagonal entries, change the sign on the off-diagonal entries, and divide by the determinant. For higher order nonsingular square matrices, various methods for computing the inverse are discussed in linear algebra texts. For our present purposes, we will use commercially available software to compute inverses.

Solution of Simultaneous Equations

At the core of linear circuit analysis is the solution of a set of linear simultaneous equations. Matrix algebra offers very definite advantages in tackling this problem. First of all, a set of n linear equations in n unknowns can be expressed very compactly using matrix notation. As an example, let

$$A = \begin{bmatrix} a_{11} & a_{12} & a_{13} \\ a_{21} & a_{22} & a_{23} \\ a_{31} & a_{32} & a_{33} \end{bmatrix}, \quad x = \begin{bmatrix} x_1 \\ x_2 \\ x_3 \end{bmatrix}, \quad b = \begin{bmatrix} b_1 \\ b_2 \\ b_3 \end{bmatrix}.$$

Then the single matrix equation

$$Ax = b$$

represents the following set of three linear equations:

$$a_{11}x_1 + a_{12}x_2 + a_{13}x_3 = b_1,$$

$$a_{21}x_1 + a_{22}x_2 + a_{23}x_3 = b_2,$$

$$a_{31}x_1 + a_{32}x_2 + a_{33}x_3 = b_3.$$

To see this, one need only compute the product Ax and then equate the resultant 3×1 matrix to the matrix b. The generalization to a set of n equations is fairly straightforward.

The advantage of using matrices is more than just a simplification in notation. Recall that in algebra, two common methods for solving simultaneous equations are the Gaussian elimination method and Cramer's rule. Using matrices, we can readily derive a third method. Assume that the coefficient matrix A has an inverse. Multiplying both sides of the equation $Ax = b$ by the inverse of A produces

$$A^{-1}(Ax) = (A^{-1}A)x = Ix = x = A^{-1}b.$$

Therefore, the solution of $Ax = b$ is

$$x = A^{-1}b.$$

The inverse of a general 2×2 matrix was given earlier. To determine the inverse of a higher order square matrix (it if exists), we can use any of the mathematical software programs that are available in mainframe or personal computers or even in a calculator. The next subsection describes a popular software program for such purposes.

Solution of Linear Simultaneous Equations Using MATLAB

In MATLAB, a matrix is entered as a list. In the list, two neighboring elements are separated by blanks or a single comma, and a semicolon is used to indicate the end of a row. The list is surrounded by square brackets. For example, the matrices A_1, A_2, and A_3, given earlier are entered as

```
A1 = [2 3 -1; 5 0 2]; A2 = [2 -3 4]; A3 = [2 3; 5 8].
```

There are two ways to enter a column matrix: directly, as an $m \times 1$ matrix, or indirectly, as the transpose of a $1 \times m$ row matrix. The latter method is more convenient when the matrix elements are entered manually. For example, we might have

```
C1 = [5; 6; -3; 0; 7; 9; -4]; C2 = [5 6 -3 0 7 9 -4]'.
```

In C2, the prime to the right of the last square bracket indicates the transpose when the elements are real numbers and the conjugate transpose when the elements are complex numbers. Both C1 and C2 represent the same 7×1 column vector, or column matrix. In C2, one saves the typing of all semicolons inside the square bracket.

Special matrices may be entered in simpler ways:

1. An $m \times n$ null matrix is entered as zeros(m,n). A square null matrix of order n is entered as zeros(n).
2. eye(n) represents an identity matrix of order n.
3. A diagonal matrix with diagonal elements $[a\ b\ c\ d\ldots]$, starting from the $(1,1)$ element down to the (n, n) element, may be entered as diag([a b c d]).

EXAMPLE 5

The following are some special matrices:

$$A_7 = \begin{bmatrix} 0 \\ 0 \\ 0 \end{bmatrix}, \quad A_8 = \begin{bmatrix} 1 & 0 & 0 \\ 0 & 1 & 0 \\ 0 & 0 & 1 \end{bmatrix}, \quad A_9 = \begin{bmatrix} 5 & 0 & 0 \\ 0 & -4 & 0 \\ 0 & 0 & 2 \end{bmatrix}.$$

MATLAB commands for entering these matrices include the following:

```
A7 = [ 0 0 0 ]'; or A7 = zeros(3,1);
A8 = eye(3);
A9 = diag([ 5 -4 2]);.
```

Addition, subtraction, and **multiplication** of scalars and matrices are denoted by $+$, $-$, and $*$, respectively. To divide a scalar p by another scalar q, the notation p/q is universally used. For the case of matrices, the command inv(A) yields the inverse (if it exists) of a square matrix A. Therefore, to find the solution x in the matrix equation $Ax = b$, one simply enters the command x = inv(A) * b.

There is another computationally more efficient and numerically more accurate way of finding the solution of $Ax = b$: Rather than finding the inverse of A and then executing the multiplication inv(A)*b, one can use a modified Gaussian elimination method. Entering the MATLAB command x = A\b directs the program to find x using this elimination method.

◆ **EXAMPLE 6**

Consider the simultaneous equations

$$\begin{bmatrix} 2 & 3 \\ -4 & 5 \end{bmatrix} \begin{bmatrix} x_1 \\ x_2 \end{bmatrix} = \begin{bmatrix} 12 \\ 42 \end{bmatrix}.$$

MATLAB commands for finding the solution of these equations are:

```
A = [2 3; -4 5]; b = [12 42]'; or b = [12; 42];
x = inv(A)*b or x = A\b
```

◆ **EXAMPLE 7**

Consider the simultaneous equations with complex coefficients

$$\begin{bmatrix} 2+j3 & 4+j5 \\ -1-j2 & -2+j5 \end{bmatrix} \begin{bmatrix} x_1 \\ x_2 \end{bmatrix} = \begin{bmatrix} -11+j29 \\ -16 \end{bmatrix}.$$

MATLAB commands for finding the solution of these equations are:

```
A = [2+j*3 4+j*5; -1-j*2 -2+j*5]; b = [-11+j*29; -16];
or b = [-11+j*29 -16].';
x = inv(A)*b or x = A\b
```

MATLAB interprets both i and j as the square root of -1.

Two points deserve special attention in the preceding solution:

1. In entering a complex number $\alpha \pm j\beta$, there should be no space before and after the $+$ or $-$ sign separating the real part and the imaginary part of the number. Otherwise, one single complex number will be misinterpreted by the program as two numbers: a real number α and a pure imaginary number $j\beta$.
2. When some elements of a matrix are complex, the prime denotes the conjugate transpose. To get an unconjugated transpose, use . ' (period followed by prime).

This brief appendix should enable you to use matrix notation and computer software such as MATLAB to solve simultaneous equations, whether the coefficients are real or complex numbers.

Solving Problems With Software

O NE of the activities that sets apart engineering and science students from liberal arts students is *computation*. Engineers in particular must be able to write equations that describe a physical process and solve the equations to produce *quantitative* information with whatever tools are considered the best at the time. The engineering trademark a generation ago was the slide rule. As electronic and computer technologies advanced, slide rules became obsolete. The hand-held calculator has replaced the slide rule. Since the early 1970s, tremendous progress has occurred in the capabilities of calculators: From basic four-function calculators, to scientific calculators, to programmable calculators, to calculators having built-in software packages and matrix-vector capabilities. During the same period, software for mathematical computation and for simulating electronic circuits, developed primarily for mainframe computers, has become available on personal computers. Indeed, very powerful mathematical programs and circuit simulation programs are now available on personal computers that students can use in their dormitories or at home. These advances in technology have had a dramatic impact on engineering education.

In this book, our philosophy is to encourage the use of the best tools available, but in a manner that does not diminish the learning of basic principles. Thus, in volume 1, students are asked to use only general mathematical programs to solve simultaneous equations, to perform matrix algebra, to manipulate complex numbers, and to plot curves. Pedagogically, it appears to us that computer problems should be assigned only after the student has learned to solve similar, but smaller, problems manually. Such learned skills are often useful in other courses. However, in volume 2, we introduce some general-purpose circuit simulation programs and some special programs developed for linear circuit analysis.

This appendix contains a set of 18 problems and 10 worked-out examples. The problems are coordinated with the chapters of the text, and each category has more than one problem. Therefore, a subset of problems may be assigned successively as the course moves on. The worked-out examples are provided so that the reader can learn to solve the problem quickly by parroting the examples and reading only the pertinent parts of the user manual. Although MATLAB (1992 student version) is the program mentioned here, other similar programs (e.g., Mathematica, etc.) may be used just as well.

EXAMPLE 1

Let y and z be functions of x given by

$$y = \frac{10x}{x+5}, \quad z = \frac{100x}{(x+5)^2}.$$

If $x = 1, 2, 3, \ldots, 10,$

(a) tabulate the corresponding values of y and z, and
(b) plot the curve of z vs. x.

To solve this example, you will need the following commands:

```
+    -    *    /    ^      .*    ./    .^      plot(x,y)
```

Find out the use of these commands by reading appropriate sections of the student MATLAB manual.

SOLUTION

The required MATLAB commands are:

```
x = (1:1:10);
y = 10*x./(x + 5);
z = 100*x./(x + 5).^2;
[x' y' z']
plot (x,z)
```

(*Note*: Alternatively, the third line may be `z = 100*x./((x+5).*(x+5)).`)

1. Consider the circuit of figure PA2.1. Solve parts (a) and (b) manually. Use the student MATLAB software to solve parts (c) and (d).

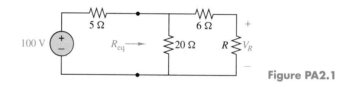

Figure PA2.1

(a) Find R_{eq} (in terms of R).
(b) Use voltage and/or current division to find an expression for V_R.
(c) The resistance R varies from 1 to 20 Ω in 1-Ω steps. Tabulate V_R and P_R (the power absorbed by R) vs. R.
(d) Plot P_R vs. R. At what value of R does P_R reach its maximum value?

EXAMPLE 2

Given

$$\begin{bmatrix} 3 & -3 & 4 \\ 1 & 6 & 5 \\ 1 & -2 & 3 \end{bmatrix} \begin{bmatrix} x_1 \\ x_2 \\ x_3 \end{bmatrix} = \begin{bmatrix} 30 \\ 7 \\ 17 \end{bmatrix},$$

find the unknowns by the use of MATLAB.

SOLUTION

From the appropriate sections of the student MATLAB manual, dealing with basic matrix operations, we have the following MATLAB commands:

```
A = [ 3 -3 4; 1 6 5; 1 -2 3]
```

```
b = [ 30 7 17]'
x = A\b
```

(*Note*: Alternatively, the third line may be x = inv(A)*b.)

2. For the circuit shown in figure PA2.2, write the node equations, and then use MATLAB to find the node voltages.

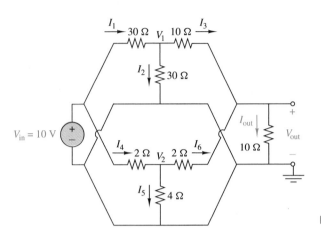

Figure PA2.2

◆ **EXAMPLE 3**

Given five resistors with values (in ohms) of

$$10, 20, 40, 80, 100,$$

and the corresponding voltages across the resistors (in volts) of

$$20, 22, 24, 26, 30,$$

find:
  (a) the currents through the resistors,
  (b) the power consumed by each resistor, and
  (c) the total power consumed by the resistors.

SOLUTION

The MATLAB commands are:

```
r = [10 20 40 80 100]
v = [ 20 22 24 26 30]
i = v./r
p = v.*i
ptotal = v*i'
```

Note that:

1. The last line may also be written as ptotal=i*v'.
2. Instead of entering the resistances as a vector, one can enter them as a diagonal matrix and compute *i* as follows:

```
r = diag([10 20 40 80 100])
g = inv(r)
i = v*g
```

3. Refer again to the circuit of problem 2. Solve the following with the aid of MATLAB. (*Suggestion:* Run example 3 to get familiar with the commands needed.)

   (a) Label the circuit resistor voltages based on the given current directions. Write each of these voltages as the difference of two node voltages. Put your equations in matrix form, and solve for the resistor voltages using MATLAB, i.e., by multiplying V_{node} computed in problem 2 by the matrix you have just written. Call the resulting vector V_r.

   (b) Define a vector $R = [R_1, R_2, \ldots, R_{out}]'$, each entry being the resistance of the appropriate resistor. Compute $I_r = [I_1, I_2, \ldots, I_{out}]'$.

   (c) Compute the power consumed by each resistor.

   (d) Compute the power delivered by the source.

4. Repeat problems 2 and 3 for the circuit of figure PA2.4. Before beginning, you should carefully label the node voltages and each of the resistor voltages and currents.

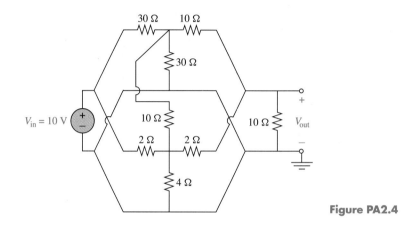

Figure PA2.4

5. Repeat problem 2 for the circuit of figure PA2.5.

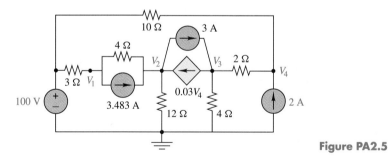

Figure PA2.5

6. For the two circuits of figure PA2.6, $V_{s1} = 240$ V, $I_{s2} = 12$ A, and $V_{s2} = 240$ V, use the same matrix techniques developed in previous problems to:

   (a) Write the matrix form of the nodal equations for each of the circuits.

   (b) For each set of nodal equations, solve the equations for V_1, V_2, and V_3 using MATLAB.

   (c) Compute the currents I_a, I_b, and I_c.

   (d) Compute the power delivered by each source and the total power delivered to the network.

   (*Remark*: Figure PA2.6a represents three interconnected balanced transmission lines to which are attached various sources and loads. Figure PA2.6b is the single-line equivalent in that the voltages V_1, V_2, and V_3 and the currents through R_a, R_b, and R_c coincide with those of figure PA2.6a.)

APPENDIX A2   SOLVING PROBLEMS WITH SOFTWARE

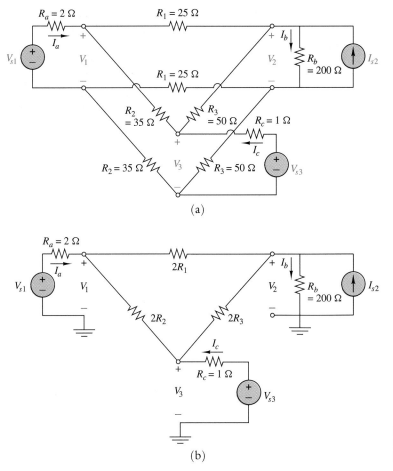

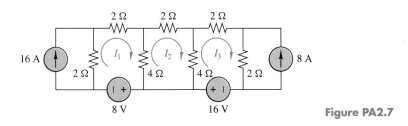

Figure PA2.6 (a) A system with balanced interconnecting lines. (b) Single-line equivalent of (a).

7. Solve this problem with the aid of MATLAB. Consider the circuit of figure PA2.7.
   (a) Write a set of mesh equations in matrix form.
   (b) Solve the mesh equations for each of the mesh currents.
   (c) Determine the power consumed by each of the resistors.
   (d) Determine the total power delivered to the resistors.

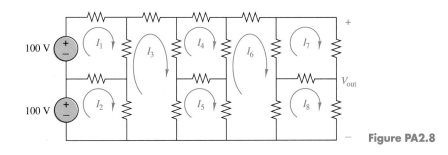

Figure PA2.7

8. Repeat problem 4 using the circuit of figure PA2.8, in which all resistances are 1 Ω. Also, determine V_{out}. (*Check*: $10 < V_{\text{out}} < 11$.)

Figure PA2.8

9. *Using manual techniques only* (e.g., do not use MATLAB),

(a) derive the following voltage gain formula for the circuit of figure PA2.9a:

$$\frac{V_o}{V_s} = \frac{A(G_b + G_c)}{(A+1)G_b + G_a + G_c}.$$

(b) Consider the difference amplifier circuit shown in figure PA2.9b with $R_{o1} = R_{o2} = 500$ kΩ, $R_1 = R_2 = 50$ kΩ, $A_1 = A_2 = 20$ kΩ, and $R_3 = 10$ kΩ.

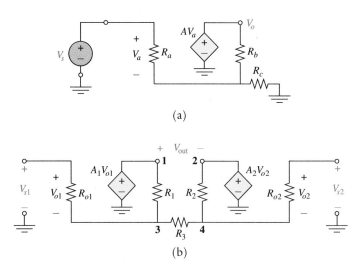

(a)

(b)

Figure PA2.9 (a) Amplifier with one input. (b) A difference amplifier.

Find the output voltage V_{out} for the following cases:

Case 1: $V_{s1} = V_{s2} = 48$ V.

Case 2: $V_{s1} = -V_{s2} = 2$ V.

Case 3: $V_{s1} = 50$ V and $V_{s2} = 46$ V.

Case 4: $V_{s1} = 40$ V and $V_{s2} = 36$ V.

The calculations should be very simple if you utilize symmetry, superposition, and the result of part (a). The theory is explained in section 6 of chapter 5.

(c) Let $A_1 = 20$ and $A_2 = 16$. (The other parameters remain the same.) The shortcuts used are now no longer applicable, because the circuit is not symmetric. Find V_{out} with the aid of MATLAB, given that $V_{s1} = 48$ V and $V_{s2} = 48$ V. (*Hint*: Write five simple equations in five unknowns V_1, V_2, V_3, V_4, and V_{out} as follows:

V_{out} in terms of V_1 and V_2;
V_1 in terms of V_3, $A_1(= 20)$, and $V_{s1}(= 48)$;
V_2 in terms of V_4, $A_2(= 16)$, and $V_{s2}(= 48)$;
KCL applied to node 3,
KCL applied to node 4.)

10. The circuit of figure PA2.10 is the equivalent circuit of an op-amp difference amplifier. The output is an amplified version of the difference of the two input voltages, V_{s1} and V_{s2}. $R_1 = 45$ kΩ, $R_2 = 10$ kΩ, $R_3 = 45$ kΩ, $R_4 = 10$ kΩ, $R_5 = 10$ kΩ, $R_6 = 100$ kΩ, and $R_7 = 100$ kΩ. Suppose $V_{s1} = 50$ V and $V_{s2} = 45$ V.

(a) For $R_{o1} = R_{o2} = R_{o3} = 500$ kΩ, determine the node voltages (in particular, V_{out}) as $A_1 = A_2 = A_3 = 10, 25, 50, 100, 500, 1000, 10^4$, and 10^5.

(b) Repeat part (a) for $R_{o1} = R_{o2} = R_{o3} = 100$ kΩ.

(c) With $R_{o1} = R_{o2} = R_{o3} = 100$ kΩ, investigate the behavior of V_{out} when A_1 is 20% higher than A_2 over the same range as before.

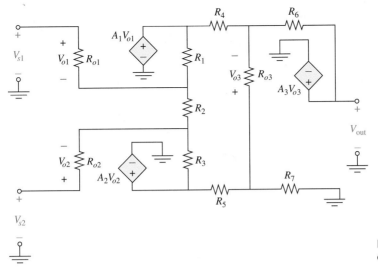

Figure PA2.10 Equivalent circuit of an op amp difference amplifier.

EXAMPLE 4

Given that $x(t) = e^{-t}$, $y(t) = t$, and $z(t) = x(t)y(t)$, use MATLAB to plot the curves of x, y, and z vs. t on the same graph, for $0 \leq t < 2$ sec in steps of 0.1 sec. Add grid lines, a title, axis labels, and curve labels to the graph. Also, tabulate t, x, y, and z for $0 \leq t \leq 2$ sec in 0.5-sec steps.

To solve this example, read appropriate pages of the student MATLAB manual about the basic graphing commands, and about subscripting. In particular, learn how to use the following commands: plot, grid, title, xlabel, ylabel, text, gtext.

SOLUTION

The MATLAB commands are:

```
t = (0: 0.1: 2);
x = exp(-t);
y = t;
z = x.*y;
plot(t,x,t,y,t,z);
grid;
title('plots of x, y and z vs. t');
xlabel('t');
ylabel('x, y, and z');
table = [t' x' y' z'];
table(1: 5: 21, :)
```

After the curves are displayed, type the command

```
gtext('x(t)')    (press return),
```

and use the mouse (if available) to place the label at the desired location. Repeat for $y(t)$ and $z(t)$. The following graph and table (for t, x, y, z) should be obtained.

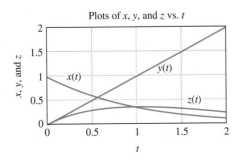

TABLE A2.1 TABULATED RESULTS OF EXAMPLE 4			
t	x	y	z
0	1.0000	0	0
0.5000	0.6065	0.5000	0.3033
1.0000	0.3679	1.0000	0.3679
1.5000	0.2231	1.5000	0.3347
2.0000	0.1353	2.0000	0.2707

11. In the circuit of figure PA2.11, the current through the inductor is $i(t) = 1.5 \sin(200\pi t)$ A.

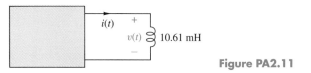

Figure PA2.11

(a) Find the expressions for $v(t)$, the voltage across the inductor, and $p(t)$, the instantaneous power absorbed by the inductor.
(b) Use MATLAB to plot the waveforms of $i(t)$, $v(t)$, and $p(t)$ on the same graph, for $0 \le t < 0.02$ sec, in time steps of 0.0001 sec. Add grid lines, a title, an X-axis label, and a Y-axis label to the graph. Also, identify the three curves.
(c) Tabulate t, $i(t)$, $v(t)$, and $p(t)$ for $0 \le t \le 0.01$ sec, in 0.0005-sec steps.

EXAMPLE 5

Let $x(t) = e^{-t}$, $y(t) = e^{-(t-1)}u(t-1)$, and $z(t) = x(t) + y(t)$. Use MATLAB to plot the curves of x, y, and z vs. t, first individually and then all three curves on the same graph, for $0 \le t < 5$ sec, in 0.02-sec steps.

To solve this example, read appropriate pages of the student MATLAB manual, about function files. Also, read about the use the function abs(x), and the use of the *signum* function, sign(x).

Student MATLAB does not have the unit step function $u(t)$. Accordingly, create a unit step function for use in this and future problems. This is done as follows:

When the screen shows that MATLAB is in command mode, click **File** *and drag to* **New.** *Type the following three lines:*

```
function f = u(t) %, defining the unit step function
t = t + 1e-12;
f = 0.5*(sign(t) + sign(abs(t)));
```

Click **File** *and drag to* **Close.** *Answer the dialog and save the document as* u. *Click* **Save.**

(*Remark*: The unit step function $u(t)$ has been created for use in MATLAB. It is now used in the following solution to example 5.)

SOLUTION

The MATLAB commands are:

```
t = (0: 0.02: 5);
x = exp(-t);
y = exp(-(t-1)).*u(t-1);
z = x+y;
```

```
plot(t,x)
plot(t,y)
plot(t,z)
plot(t,x,t,y,t,z)
grid
```

The plot of z vs. t is as follows:

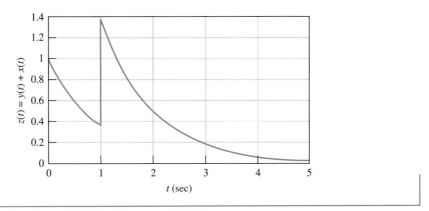

12. (a) Find the step response of the circuit of figure PA2.12 manually.

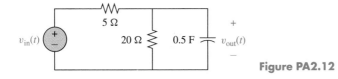

Figure PA2.12

(b) If the capacitor voltage is zero at $t = 0$ and

$$v_{in}(t) = 10u(t) + 20u(t-1) - 30u(t-2) \text{ V},$$

find a single expression for $v_{out}(t)$. Use MATLAB to plot the waveforms of $v_{in}(t)$ and $v_{out}(t)$ on the same graph, for $0 \leq t \leq 5$ sec, in 0.02-sec steps.

 **EXAMPLE 6**

Given that

$$x(t) = 2t^2 \qquad \text{for } 0 \leq t \leq 1 \text{ sec,}$$

$$x(t) = (t-1)^2 \qquad \text{for } 1 \leq t \leq 3 \text{ sec,}$$

$$x(t) = 4e^{-(t-3)} \qquad \text{for } t \geq 3 \text{ sec,}$$

use MATLAB to plot the waveform of $x(t)$ for $0 \leq t \leq 4$ sec in 0.02-sec steps.

To solve this example, review example 4 about MATLAB graphing commands and example 5 about the use of the unit step function $u(t)$.

SOLUTION

Method 1. (This method uses more computer time, but is easier for the user. It is recommended for beginning users of MATLAB.)

```
t = (0: 0.02: 4);
x = 2*(t.^2).*(u(t)-u(t-1))+((t-1).^2).*(u(t-1)-u(t-3))+ ..
4*exp(-(t-3)).*u(t-3);
plot(t,x)
grid
```

Note: If one line is not enough for the command, then finish the first line with two periods (. .) and continue on the next line, as illustrated in the second line above.

Method 2. (This method uses less computer time, but the user has to write more commands. It is recommended for experienced users of MATLAB.)

```
t = (0:0.02:1); t1 = t;
x1 = 2*t.^2;
t = (1:0.02:3); t2 = t;
x2 = (t-1).^2;
t = (3:0.02:4); t3 = t;
x3 = 4*exp(-(t-3));
t = [t1 t2 t3];
x = [x1 x2 x3];
plot(t,x)
grid
```

Both methods should produce the following graph:

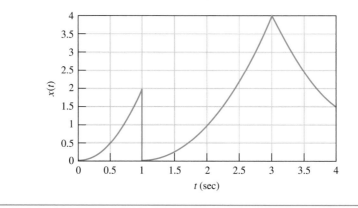

13. It is suggested that you run example 6 and understand the commands used therein before working problem 13. Analyze the RC circuit of figure PA2.13, with multiple switching operations. The switch S operates as follows:
Open for a long time;
closes at $t = 0$;
opens at $t = 2$ sec;
closes at $t = 4$ sec.

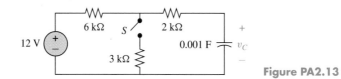

Figure PA2.13

(a) Find $v_C(t)$ for $0 \leq t \leq 2$ sec.
(b) Find $v_C(t)$ for $2 \leq t \leq 4$ sec.
(c) Find $v_C(t)$ for $t \geq 4$ sec.
(d) Utilizing shifted step functions, write a single expression for $v_C(t)$ that is valid for all $t \geq 0$.
(e) Use MATLAB to plot the waveform of $v_C(t)$ for $0 \leq t \leq 5$ sec, in 0.02-sec steps.

◆ EXAMPLE 7

Given that $v(t) = A \sin(\omega t)$ V, use MATLAB to plot the waveform of $v(t)$ for $0 \leq t \leq 1$ sec, in 0.01-sec steps, for the following three cases:

Case 1: $A = 30$ and $\omega = 10$ rad/sec.

Case 2: $A = 10$ and $\omega = 30$ rad/sec.

Case 3: $A = 6$ and $\omega = 50$ rad/sec.

Plot the three curves on the same graph, and label them properly.

To solve this example, review example 4 about MATLAB graphing commands. Then read the student MATLAB manual about the command **hold.** A method of plotting several curves on the same graph was described in example 4. An alternative method, using the command **hold,** is described next.

SOLUTION

Method 1. (This method uses more computer storage, but is easier for the user.)

```
t = (0:.01:1);
v1 = 30*sin(10*t);
v2 = 10*sin(30*t);
v3 = 6*sin(50*t);
plot(t,v1,t,v2,t,v3)
grid; gtext('Case 1');gtext('Case 2'); gtext('Case 3');
xlabel('t'); ylabel('v(t)')
```

Method 2. (This method uses less computer storage.)

```
t = (0:.01:1);
v = 30*sin(10*t);
plot(t,v);
hold
v = 10*sin(30*t);
plot(t,v);
v = 6*sin(50*t);
plot(t,v)
grid; gtext('Case 1');gtext('Case 2'); gtext('Case 3');
xlabel('t(sec)'); ylabel('v(t)')
```

Both methods should produce the following graph:

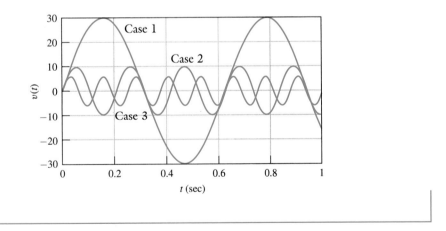

14. In this problem, we seek to discharge a capacitor through a series RL circuit. In figure PA2.14, the switch S, originally at position A, is moved to position B at $t = 0$.

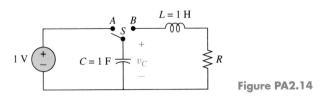

Figure PA2.14

(a) If $R = 0$, find $v_C(t)$ for $t \geq 0$ (undamped case).
(b) If $R = 2\ \Omega$, find $v_C(t)$ for $t \geq 0$ (critically damped case).
(c) If $R = 1.2\ \Omega$, find $v_C(t)$ for $t \geq 0$ (underdamped case).
(d) Use MATLAB to plot the $v_C(t)$ waveforms for the foregoing three cases, for $0 \leq t < 50$ sec, in 1-sec steps. Label the curves properly.

◆ EXAMPLE 8

Convert the following four complex numbers, either from the rectangular form $x + jy$ to the polar form $re^{j\theta}$ or from the polar form to the rectangular form. Use MATLAB or similar software.

$$z_1 = 4 + j5,\ z_2 = 3 - j8,\ z_3 = 5e^{j0.5},\ z_4 = 4\exp(-j60°).$$

To solve this example, read appropriate pages of the student MATLAB manual, about the entry of complex numbers. Also, learn the use of the following functions:

$$\text{real}(x),\ \text{imag}(x),\ \text{conj}(x),\ \text{abs}(x),\ \text{angle}(x)$$

SOLUTION

The MATLAB commands are:

```
z1 = 4+j*5, r1 = abs(z1), theta1 = angle(z1)*180/pi% angle in degrees
z2 = 3-j*8, r2 = abs(z2), theta2 = angle(z2)*180/pi% angle in degrees
z3 = 5*exp(j*0.5)
z4 = 4*exp(-j*60*pi/180)
```

◆ EXAMPLE 9

Find A and θ, given that

$$f(t) = 5\cos(\omega t) + 8\sin(\omega t) + 10\cos(\omega t + 20°) = A\cos(\omega t + \theta°).$$

SOLUTION

$$
\begin{aligned}
f(t) &= 5\cos(\omega t) + 8\sin(\omega t) + 10\cos(\omega t + 20°) \\
&= 5\text{Re}\{\exp(j\omega t)\} + 8\text{Re}\{-j\exp(j\omega t)\} + 10\text{Re}\{\exp(j(\omega t + 20\pi/180))\} \\
&= \text{Re}\{[5 - j8 + 10\exp(j20\pi/180)]\exp(j\omega t)\} \\
&= \text{Re}\{z\exp(j\omega t)\} = \text{Re}\{r\exp(j\theta)\exp(j\omega t)\} = \text{Re}\{r\exp(j\omega t + j\theta)\} \\
&= r\text{Re}\{\exp(j\omega t + j\theta)\} = r\cos(\omega t + \theta).
\end{aligned}
$$

The crux of the problem is to compute the complex number

$$z = 5 - j8 + 10\exp(j20\pi/180)$$

and express it in the polar form $re^{j\theta}$. The solution, via MATLAB, is as follows:

```
z = 5 - j*8 + 10*exp(20*pi/180)
r = abs(z)
thetad = angle(z)*180/pi
```

The answers from MATLAB are $z = 14.40 - j4.58$, $r = 15.11$, thetad $= -17.65°$. Therefore, $f(t) = 15.11\cos(\omega t - 17.65°)$, $A = 15.11$, $\theta = -17.65°$. This could also be done using a good scientific calculator.

15. (a) Write a complete expression of the form

$$v(t) = V_m \sin(\omega t + \theta^\circ),$$

given the following information:
    (1) The frequency of $v(t)$ is $f = 10$ Hz.
    (2) $v(t)$ leads $\sin(\omega t)$ by 30°.
    (3) $v(0.3) = 1.0$.
  (b) For each of the following cases, express $v(t)$ in the form

$$v(t) = A \cos(\omega t + \theta^\circ),$$

using a scientific calculator or MATLAB:
    (1) $v(t) = 3 \cos(\omega t) + 4 \sin(\omega t)$.
    (2) $v(t) = 5 \cos(\omega t) - 6 \sin(\omega t)$.
    (3) $v(t) = 5 \cos(\omega t + \pi/4) + 8 \sin(\omega t) + 10 \cos(\omega t + 120^\circ)$.
As a suggestion, run examples 8 and 9, and understand the principles and commands used, before working on the present problem. Remember that in MATLAB, the angle for trigonometric functions must be expressed in radians. MATLAB recognizes pi as the constant $\pi = 3.1416$.

16. The LC components in the dash-lined box of figure PA2.16a form a delay line. For any input signal having frequency components below some upper limit, the output signal is approximately a replica of the input signal except for being delayed by some time t_d. This problem finds the delay time t_d by solving the node equations with the aid of MATLAB. Let $v_{in}(t) = \cos(5 \times 10^4 t)$ V. The frequency domain circuit model is shown in figure PA2.16b.

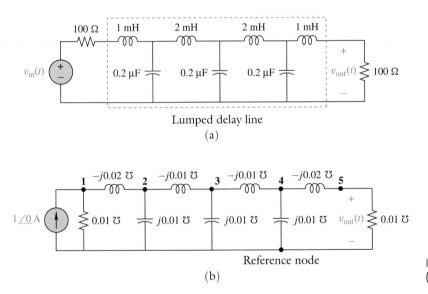

Lumped delay line
(a)

Reference node
(b)

Figure PA2.16 (a) Given circuit.
(b) Frequency domain equivalent of (a).

  (a) Write the nodal equations for the circuit. Recall that there is a very simple rule for writing nodal equations for RLC circuits.
  (b) With the aid of MATLAB, solve for the phasor $\mathbf{V}_{out}$. What is $v_{out}(t)$?
  (c) Repeat parts (a) and (b) if the input is $v_{in}(t) = \cos(2.5 \times 10^4 t)$ V.
  (d) If the output voltage of parts (b) and (c) is written in the form

$$v_{out}(t) \cong v_{in}(t - t_d),$$

what is the delay time t_d? (*Check*: The delay time is between 50 and 100 μsec.)

 **EXAMPLE 10**

For the circuit shown in the following figure, derive the gain function $H(j\omega) = V_{out}/V_{in}$ with $j\omega$ as the variable. Then use MATLAB to plot the curve of $|H|$ vs. ω for $0.8 \leq \omega \leq 1.2$ rad/sec, in 0.02-rad/sec steps, for two cases: $R = 2\ \Omega$ and $R = 50\ \Omega$. Show both curves in the same figure for the purpose of comparison. What is the effect of the resistance?

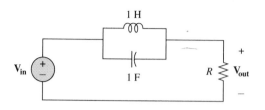

To solve this example, read in the student MATLAB manual about the command **freqs**.

SOLUTION

Use of the voltage division formula yields the following result, with $j\omega$ as the variable:

$$H(j\omega) = \frac{V_{out}}{V_{in}} = \frac{(j\omega)^2 + 1}{(j\omega)^2 + \dfrac{1}{R}j\omega + 1}.$$

The MATLAB commands to plot the frequency response curve are:

```
w = (0.8:0.02:1.2);
num = [ 1 0 1];
den = [ 1 .5 1];  % case 1
h1 = freqs(num,den,w);
magh1 = abs(h1);
den = [ 1 .02 1]; %case 2
h2 = freqs(num,den,w);
magh2 = abs(h2);
plot(w,magh1,w,magh2)
grid
%use commands title, xlabel, ylabel,
% and gtext to annotate the figure.
```

The plot is as follows:

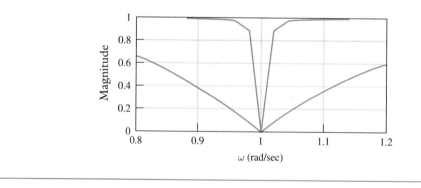

17. It is suggested that you run example 10 and understand the commands used before working on the problem. A bandpass circuit has the gain function

$$H(j\omega) = \frac{V_{out}}{V_{in}} = \frac{\dfrac{1}{Q}j\omega}{(j\omega)^2 + \dfrac{1}{Q}j\omega + 1}.$$

Plot the curve of $|H|$ vs. ω for $0.5 \le \omega \le 1.5$ rad/sec, in 0.01-rad/sec steps for two cases: $Q = 5$, and $Q = 10$. Show both curves in the same figure for the purpose of comparison.

18. For the circuit of figure PA2.18, derive the gain function $H(j\omega) = V_{out}/V_{in}$, with $j\omega$ as the variable. Then use MATLAB to plot:

(a) the curve of $|H|$ vs. ω for $60 \le \omega \le 140$ krad/sec, in steps of 1 krad/sec.

(b) the angle of H, in degrees, vs. ω for $60 \le \omega \le 140$ krad/sec, in steps of 1 krad/sec.

(c) $|H|$ vs. f—a replot of the curve in part (a)—with f as the independent variable.

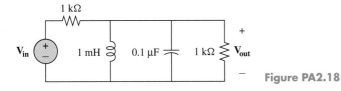

Figure PA2.18

Use of SPICE in Linear Circuits

1. INTRODUCTION

SPICE (Simulation Program with Integrated Circuit Emphasis) is a general-purpose circuit simulation program, originally developed at the University of California (Berkeley) in 1975, for nonlinear dc, nonlinear transient, and linear ac analyses. In this book, we use SPICE for linear network analysis alone. Hence, it is necessary to consider only the following types of elements: resistors, inductors, capacitors, independent voltage sources, independent current sources, all four types of dependent sources, and mutual inductors.

The input to the SPICE program is a data file describing all pertinent circuit information. The structure of the data file is as follows:

- Title (one line for identification of the problem).
- Element statements. These identify element types, element values, and node-to-node connections.
- Control statements. These specify:
  the type of analysis (dc, ac, or transient);
  the output nodes;
  the frequency range of ac analysis;
  the printing time step and final time of transient analysis;
  the type of output desired (tabular or plot).
- .END (one line, indicating the end of the data file).
- *. Any statement beginning with an asterisk is a comment. A comment statement appears in the output, but has no effect on the calculations.

Later in this appendix, we will explain how to write the element and control statements for linear circuits. Once the data file has been constructed, one can often run the problem on a UNIX Computer Network by simply typing (after the prompt sign on the screen)

```
spice < datafile > answer,
```

where **datafile** should be the name of the file describing the circuit and **answer** should be the name of the file for storing the computer output. One can examine the output on a monitor or obtain a hard copy of it in the usual way.

2. ELEMENT STATEMENTS

Each element in the circuit must have a specific element statement that contains, successively, the element name, the circuit nodes to which the element is connected, and the values of the parameter or parameters characterizing the element. The entries in each line are separated by one or more spaces.

Element Names

The first letter of the element name specifies the element type. (See table A3/B1.1.) It is followed by an alphanumeric string of one to seven characters that label the particular element. In SPICE, data file characters may be typed in either lowercase or uppercase. Some examples are R1, C2, Cbypass, VIN, VCC, gm1, L1, L2, and k12.

TABLE A3/B1.1 LIST OF ELEMENT TYPES AND IDENTIFYING LETTER IN SPICE STATEMENT

First letter	Element type
R	Resistor
L	Inductor
C	Capacitor
V	Independent voltage source
I	Independent current source
G	Voltage-controlled current source
E	Voltage-controlled voltage source
F	Current-controlled current source
H	Current-controlled voltage source
K	Coupled inductors

Node Numbering

Circuit nodes must be nonnegative integers, but need not be numbered sequentially. Zero will always designate the datum (ground) node. *Each node in the circuit must have at least two elements connected to it and also must have a dc path to ground.* For example, a node with only two capacitors connected to it is not allowed, because there is no dc path from that node to ground.

Parameter Values

Standard units (ohm, henry, farad, volt, ampere, second, hertz) are used. Parameter values may be specified in different formats: integer, floating-point number, either of these followed by an integer exponent, or followed (without a space) by a scale factor. (See table A3/B1.2.) For example, 1000, 1000.0, 1K, 1k, 1e3, 1E3, and 1.0e3 are all equivalent as an input value, and so are 1234.5, 1.2345e3, 1.2345K, and 1.2345k.

TABLE A3/B1.2 SCALE FACTORS FOR PARAMETER VALUE SPECIFICATION

Scale factor	K	MEG	G	T	M	U	N	P
Power of 10	3	6	9	12	−3	−6	−9	−12

Description of Elements

In the description of elements, N+ and N− are the nodes marked + and −, respectively, in the circuit diagram for voltage reference. Suppose an element is connected between the nodes (N+, N−). Then the voltage of the element refers to the voltage drop from N+ to N−. The current of the element refers to the current flowing from N+, through the element, to N−. In short, N+ stands for the "from-node," and N− the "to-node."

(a) Each **RLC element** is described by one statement of the general form

```
RXXXXXXX      N+      N-      VALUE
LYYYYYYY      N+      N-      VALUE      <IC=INCON>
CZZZZZZZ      N+      N-      VALUE      <IC=INCON>.
```

Examples

```
R12   2   3 10K
CBYP 13 0 50u
L3    7   5 7m  ic = 25m
```

Notes:

(1) <IC=INCON> for C or L is needed only if the user wants to specify a nonzero initial value for transient analysis. Leaving this field blank has the same effect as typing IC=0.
(2) The value of R must be nonzero, but can be negative.
(3) The values of L and C must be positive.

(b) Each **dc independent source** is described by one statement of the form

```
VXXXXXXX N+ N- VALUE
IYYYYYYY N+ N- VALUE
```

Examples

```
VCC   10 2 12
ISRC   5 7  8m
VMEAS 12 9
```

Interpretation

The dc voltage source VCC maintains a voltage drop of 12 V from node 10 to node 2. The dc current source ISRC is connected to nodes (5, 7) and injects 8 mA into node 7. The zero-valued dc voltage source VMEAS is equivalent to a short-circuit element and is connected between nodes 12 and 9.

(c) Each **ac independent source** is described by one statement of the form

```
VXXXXXX N+ N- AC ACMAG ACPHASE
IYYYYYY N+ N- AC ACMAG ACPHASE
```

ACMAG is the ac magnitude, and ACPHASE is the ac phase (in degrees). The keyword AC must be present. If only one value is specified after the keyword AC, then it is taken as ACMAG, with ACPHASE = 0. If no values are specified after AC, then the program assumes that ACMAG = 1 and ACPHASE = 0.

Examples

```
VIN 2 5 AC
VS1 2 5 AC 115
VS2 2 5 AC 115 45
IIN 3 4 AC 2    30
```

Interpretation

The ac voltage source VS2 maintains a voltage drop of $115\angle 45°$ V (a phasor) from node 2 to node 5. The ac current source IIN is connected between nodes 3 and 4 and injects an ac current $2\angle 30°$ A into node 4.

(d) Each piecewise linear independent source is described by one statement of the form

```
VXXXXXX N+ N- pwl(0 v0, t1 v1, t2 v2,...)
IYYYYYY N+ N- pwl(0 v0, t1 v1, t2 v2,...)
```

Each pair (tk vk) specifies a breakpoint. Commas are optional. They are inserted to improve readability. The time points 0, t1, t2, ... must be monotonically increasing. v0 may or may not be zero.

Refer to a complete SPICE manual for the descriptions of independent sources that are exponential, sinusoidal, or periodic pulses.

Examples

```
IS1 2 5 pwl( 0 2, 2 2, 4 0)
VS2 2 5 pwl( 0 0, 1 0, 3 2, 5 2, 5.0001 1, 7 1, 7.0001 0)
```

Interpretation

The independent sources IS1(t) and VS2(t) have waveforms as shown in figure A3/B1.1.

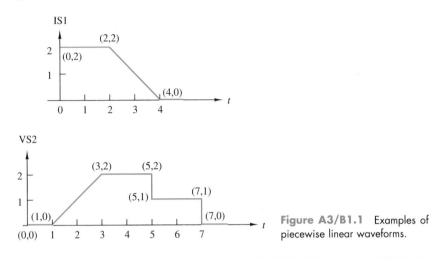

Figure A3/B1.1 Examples of piecewise linear waveforms.

Note: In VS2, the breakpoint $(5, 1)$ is entered as $(5.0001, 1)$, because SPICE does not allow repeated tk's. The situation is similar for the breakpoint $(7, 0)$.

(e) Each **voltage-controlled source** has a description of the form

```
GXXXXXXX N+ N- NC+ NC- VALUE
EYYYYYYY N+ N- NC+ NC- VALUE
```

The controlled source is connected between nodes (N+, N−). The controlling variable is the voltage drop from node NC+ to node NC−. VALUE is the parameter characterizing the element.

Examples

```
GM1 2 3 5 0 0.1
E2  2 3 5 7 2.5
```

Interpretation

Figure A3/B1.2 gives a graphical interpretation of these two types of controlled sources.

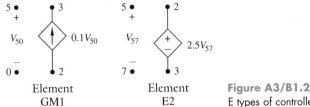

Figure A3/B1.2 Examples of G and E types of controlled sources.

(f) Each **current-controlled source** has a description of the form

```
FXXXXXXX N+ N- VNAME VALUE
HYYYYYYY N+ N- VNAME VALUE
```

The entry VNAME denotes the name of the independent voltage source through which the controlling current flows. In SPICE, the controlling current must be associated with some independent voltage source. It is often necessary to insert an independent voltage source of 0 volts to satisfy this condition.

Examples

```
F1     13 5 VSENS 5.5
VSENS 4  6
H2     5 7 VREF  4.5
VREF  3  0 .5
```

Interpretation

Figure A3/B1.3 gives a graphical interpretation of the F and H types of controlled sources.

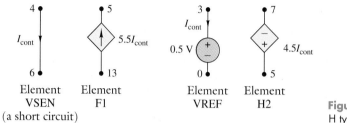

| Element VSEN (a short circuit) | Element F1 | Element VREF | Element H2 |

Figure A3/B1.3 Examples of F and H types of controlled sources.

(g) Each pair of **coupled inductors** is described by three statements: one for L1, one for L2, and one for K. The statement for the coupling coefficient K has the general form

```
KXXXXXXX LYYYYYYY LZZZZZZZ VALUE
```

Examples

```
L1 3  2  8
L2 5  4  2
K  L1 L2 0.95
```

Interpretation

Figure A3/B1.4 gives a graphical interpretation of coupled inductors.

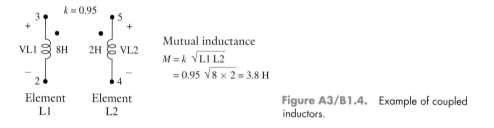

Mutual inductance
$$M = k \sqrt{L1\ L2}$$
$$= 0.95 \sqrt{8 \times 2} = 3.8 \text{ H}$$

| Element L1 | Element L2 |

Figure A3/B1.4. Example of coupled inductors.

3. CONTROL STATEMENTS

For the linear circuit problems in this book, the only analysis type of control statements needed are .AC and .TRAN.

If neither .AC nor .TRAN is present, then SPICE will automatically perform a dc analysis in which:

1. All ac independent sources are set to zero (i.e., we open all ac current sources and replace all ac voltage sources by short circuits).
2. All inductors are short-circuited.
3. All capacitors are open-circuited.
4. All resulting dc node voltages are calculated and printed.
5. All currents through dc independent voltage sources are calculated and printed.

If an ac (sinusoidal steady-state) analysis is desired, then the control statements must contain one .AC statement to specify the frequencies and one or more .PRINT (or .PLOT) statements to specify the outputs. In ac analysis, SPICE will always do a dc analysis first (without a request from the user) and then:

1. Set all dc independent sources to zero.
2. Analyze the circuit at the frequencies specified.
3. Print and/or plot the specified ac outputs.

.AC Statement

The general form of the .AC statement is

```
.AC DEC ND FSTART FSTOP
.AC LIN NP FSTART FSTOP
```

FSTART is the starting frequency and FSTOP is the final frequency (both in Hz). DEC stands for logarithmic frequency variation, and ND is the number of points per decade. With the DEC option, the ratio between two adjacent frequencies is equal to $\sqrt[ND]{10}$. LIN stands for the linear frequency variation, and NP is the number of uniformly spaced frequencies. With the LIN option, the difference between adjacent frequencies is equal to $(FSTOP - FSTART)/(NP - 1)$.

Examples

```
.AC DEC 10 1K   100MEG
.AC LIN 20 10   200
.AC LIN 1  500 500
```

The first statement requests ac analysis from 1,000 Hz to 100 MHz, at 10 frequencies per decade, with a constant ratio between two adjacent frequencies equal to $\sqrt[ND]{10} = 1.2589$. By similar reasoning, the second frequency over which the circuit is to be analyzed is 1,258.9 Hz and the third 1,584.9 Hz. The second statement requests ac analysis from 10 Hz to 200 Hz, at 20 frequencies, with a constant difference between two adjacent frequencies equal to $(200 - 10)/(20 - 1) = 10$ Hz. The third statement requests ac analysis at a single frequency equal to 500 Hz.

Generally speaking, the DEC statement is used to plot the frequency response over a wide frequency range (several decades), whereas the LIN statement is used for a narrow range (within a decade).

.TRAN Statement

The general form of the .TRAN statement is

```
.TRAN  TSTEP TSTOP UIC
```

where TSTEP is the printing time step and TSTOP is the final time of transient analysis. The initial time is assumed to be zero. The presence of the letters UIC requires the use of initial capacitor voltages and initial inductor currents, as specified in the element statements by IC=Value. If UIC is absent in the .TRAN statement, then the initial capacitor voltages and initial inductor currents are those calculated from the dc analysis. (See note 3, to follow.) For problems in linear circuit analysis, the use of UIC is recommended.

Examples

```
.TRAN  0.1  2  UIC
```

Interpretation

Transient analysis is performed for $0 \leq t \leq 2$ sec, using the initial conditions specified in the element statements. The results are printed out in steps of 0.1 sec.

Notes

1. Internally, SPICE uses variable time steps, determined by a complicated algorithm, to carry out the numerical integration. The integration time step is, in general, different from the printing time step TSTEP.

2. If piecewise linear independent sources are present, the final time TSTOP must be extended to one of the breakpoint times.
3. If UIC does not appear in the .TRAN statement, then the initial conditions given in the LC element statements are ignored. In this case, the initial conditions are determined by the dc solution. The dc solution is determined with all independent sources interpreted as follows: If a source has only dc and/or ac specification, use it as a dc source; if a source has a piecewise linear description with first breakpoint $(0,v0)$, use it as a dc source of value $v0$ in the dc solution.

.PRINT Statement

Without a request from the user, SPICE always performs a dc analysis and prints out the values of all dc node voltages and all dc currents through dc independent voltage sources. To print out the results of ac analysis, a .PRINT statement is needed.

The general form of the .PRINT statement is

```
.PRINT AC    VX(J)    VX(J,K)    IX(VSRC)
.PRINT TRAN   V(J)     V(J,K)     I(VSRC)
```

where

1. V(J) means the voltage drop from node J to ground (reference node),
2. V(J,K) means the voltage drop from node J to node K, and
3. I(VSRC) means the current through the independent voltage source named VSRC.

The letter X in the .PRINT AC statement is replaced by one of the letters (R, I, M, P, DB) to indicate the desired form of output, as shown in table A3/B1.3. Up to eight outputs may be requested, and at least one must be requested.

**TABLE A3/B1.3
TYPES OF OUTPUT
SPECIFICATION**

Indicator	Output selected
R	Real part
I	Imaginary part
M	Magnitude
P	Phase (in degrees)
DB	$20 \log_{10}$ (magnitude)

Examples

```
.PRINT AC    VM(3) VM(2,5) VDB(3) IM(VMEAS)
.PRINT TRAN V(2)  V(3)     V(2,3)
```

.PLOT Statement

The .PLOT statement is similar to the .PRINT statement, except that the outputs will be plotted instead of tabulated. We simply replace .PRINT by .PLOT.

Examples

```
.PLOT AC    VM(3) VM(2,5) VDB(3) IM(VMEAS)
.PLOT TRAN V(2)  V(3)     V(2,3)
```

4. SIMULATION EXAMPLES

Circuit 1

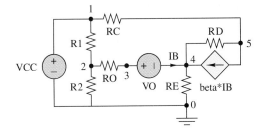

Parameter values: R1 = 6 kΩ, R2 = 1 kΩ, RE = 500 Ω, RC = 2 kΩ,
RD = 10 kΩ, RO = 100 Ω, beta = 50,
VCC = 9 V, VO = 0.7 V.

Figure A3/B1.5 dc equivalent circuit
of an amplifier.

The following data file determines the dc operating point of the circuit shown in figure
A3/B1.5 and prints out all node voltages:

```
dc equivalent circuit of an amplifier
vcc   1   0   9
r1    1   2   6k
r2    2   0   1k
ro    2   3   100
vo    3   4   .7
re    4   0   500
rc    1   5   2k
rd    4   5   10k
fbeta 5   4   vo 50
.end
```

Circuit 2

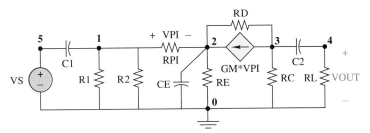

Parameter values: R1 = 1 kΩ, R2 = 6 kΩ, RE = 500 Ω, RC = 2 kΩ,
RL = 10 kΩ, RPI = 308 Ω, RD = 10 kΩ,
GM = 0.1626 mho, C1 = 5 μF, CE = 60 μF, C2 = 5 μF

Figure A3/B1.6 Low-frequency
equivalent circuit of an amplifier.

The following data file determines the magnitude and phase of the voltage gain for the
circuit of figure A3/B1.6 at the frequencies f = 1, 10, 100, 1k, 10k, 100k, 1Meg Hz:

```
frequency response of an amplifier
vs    5   0   ac
r1    1   0   1k
r2    1   0   6k
re    2   0   500
rc    3   0   2k
rL    4   0   10k
rpi   1   2   308
rd    2   3   10k
gm    3   2   1   2   162.6m
c1    5   1   5u
```

```
c2   3   4   5u
ce   2   0   60u
.ac dec   1   1   1meg
.print   ac   vm(4)   vp(4)
.end
```

Circuit 3

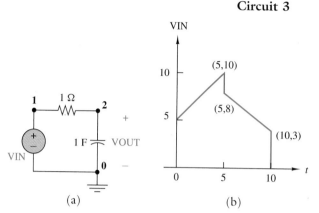

(a)

(b)

Figure A3/B1.7 Transient analysis of a circuit with piecewise linear input.

(a) Find the step response of the RC circuit shown in figure A3/B1.7a.
(b) If the input has a piecewise linear waveform like that shown in figure A3/B1.7b, and the initial capacitor voltage is $v_{out}(0) = 4$ V, find $v_{out}(0)$ for $0 \leq t \leq 10$ in 1-sec steps.

Solution

The SPICE input file is as follows:

```
Circuit 3 of appendix A3/B1.
*Part (a). Step response in which vin = 1 for t ≥0.
vin    1   0   1
r      1   2   1
c      2   0   1   ic=0
.tran 1   0   uic
.print tran v(2)
.end
*Part (b).  Piecewise linear input (figure A3/B1.7b) and nonzero IC.
vin    1   0   pwl (0 5, 5 10, 5.0001 8, 10 3)
r      1   2   1
c      2   0   ic=4
.tran      1   10 uic
.print   tran v(2)
.end
```

PROBLEMS

1. Analyze the dc equivalent of the amplifier circuit in figure PA3/B1.1 with the aid of SPICE.

(a) Number the nodes, and prepare the data file for SPICE.

(b) Run the SPICE program. Hand in your output with the following quantities marked on the circuit diagram:

(1) All node voltages with respect to the reference node.

(2) The voltage across the independent current source.

(3) The voltage across the dependent current source.

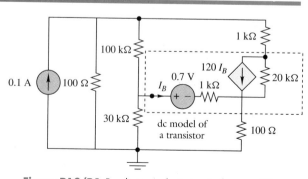

Figure PA3/B1.1 dc equivalent circuit of an amplifier.

2. An op amp with open-loop voltage gain $A = 200{,}000$ and input resistance $R_i = 2.5$ MΩ has an equivalent circuit as shown in figure PA3/B1.2(a). The op amp is a component of the active filter circuit of figure PA3/B1.2(b). Use SPICE to find the magnitude and phase of the voltage gain. Your SPICE output should contain a plot of the frequency response, with frequencies ranging from 1 Hz to 1 kHz, at 10 frequencies per decade.

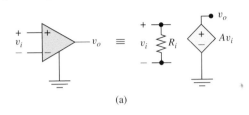

(a)

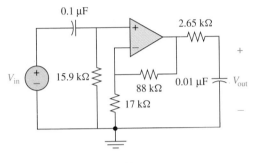

(b)

Figure PA3/B1.2 (a) Equivalent circuit of an op amp. (b) An active filter.

3. An op amp with open-loop voltage gain $A = 200{,}000$

and input resistance $R_i = 2.5$ MΩ has an equivalent circuit as shown in figure PA3/B1.2a. The op amp is a component of the active filter circuit shown in figure PA3/B1.3 (a Boctor highpass notch filter). Use SPICE to find the magnitude and phase of the voltage gain. Your SPICE output should contain a plot of the frequency response, with frequencies ranging from 1 Hz to 1 kHz, at 10 frequencies per decade.

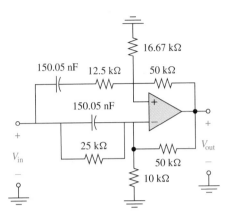

Figure PA3/B1.3 Boctor highpass notch filter.

4. Consider the circuit of figure PA3/B1.2(b).
(a) Use SPICE to plot the step response for $0 < t < 10$ msec.
(b) If the 0.1-μF capacitor is initially uncharged, $v_{out}(0) = 2$ V, and the input voltage is $v_{in}(t) = u(t) + u(t - 0.001) + u(t - 0.002) - 3u(t - 0.003)$ V, use SPICE to plot $v_{out}(t)$ for $0 < t < 5$ msec.

Software-Assisted Problems

Circuit analysis and circuit design build on a small set of basic principles: The definitions of some basic circuit elements, and KVL and KCL which govern how the elements interact. Using the mathematical tools of algebra and calculus, we generated circuit equations and solved them for circuit responses. Repeatedly solving equations leads to algorithms for computing responses and determining circuit behaviors more efficiently. Observing certain general types of behavior leads us to understand various circuit properties that derive from the mathematical models of the circuits. Over the years, circuits have become more complex, and the need for software programs to perform calculations has risen dramatically. Quality software programs automate important aspects of circuit analysis and design, relieving engineers of tedious and often impossible hand calculations.

However, numerical algorithms implemented as canned programs are no substitute for an understanding of the basic principles and properties that govern circuit behavior. In addition, they are no substitute for a firm understanding of the steps necessary to solve a problem. This becomes more apparent when one understands the fourfold task of a course in circuits: (1) to impart knowledge about circuit analysis and design, as gleaned from the accumulated experience of engineers and physicists; (2) to develop a deep understanding of circuit behavior; (3) to develop a sufficient capability for rational thinking so that the student can handle a broad variety of engineering problems with a well-stocked arsenal of perspectives, principles, and mathematical tools, and (4) to develop a familiarity with reliable numerical algorithms for analyzing and designing circuits.

From the long-term perspective, in order to maintain adaptability and avoid obsolescence, objective (3) is most important. In the near term, in order to be part of the engineering work force, it is necessary to be able to compute meaningful numbers accurately and efficiently. This appendix focuses the student's attention on the latter point through the medium of currently available software programs. Using such programs not only allows the student to calculate numbers more easily, but reinforces the delineated properties of circuit behavior covered in the text. On the other hand, the principles and properties developed throughout the text allow the students to assess the reasonableness and accuracy of answers computed using a software program. Software programs are not infallible, and it is always possible to construct an example that will cause a program to fail to produce meaningful numbers.

Many computer-aided circuit simulation programs have been developed over the past 30 years. Some are general-purpose programs (dc, ac, and transient analyses of nonlinear networks), and many are special-purpose programs (active filter design, symbolic network analysis, etc.). Some are available on mainframe computers and some on personal computers. In order for the student to gain some familiarity with such programs, we have included a set of software-assisted problems in this appendix. It is suggested that for about every six problems assigned, one should be software related. We believe that it is best to spread the use of software over the entire course. Each type of computer problem should be assigned only after the student has grasped the relevant basic principles and techniques through solving smaller size problems manually.

Since a multitude of circuit analysis programs, many possessing roughly the same capabilities, are readily available on the market, it is neither possible to list all of them nor desirable to endorse any particular one. Four such programs are mentioned next only because they are available at Purdue University. Other institutions using this text may substitute for these programs with whatever is locally available.

SPICE is a program much more powerful than is needed for the analysis of linear circuits. A simplified user's guide to linear networks is included as an appendix in volumes 1 and 2.

MATLAB is a general mathematical analysis (particularly matrix algebra) program available in an inexpensive student version for both the IBM PC and the Macintosh.

tfc is a special program developed to accompany this book. It does a better job in partial fraction expansion and inverse Laplace transformation than MATLAB does. Furthermore, the two-port manipulation capability of **tfc** is not available in any other programs. **tfc** is available at Purdue on a mainframe computer. A 20-page user's guide is issued to students. For IBM personal computers, the user's guide accompanies the disk.

NAPPE is a symbolic network analysis program available at Purdue on a mainframe computer. Instructions for using the program appear interactively at the terminal. A few other institutions have similar symbolic network analysis programs.

The set of software-assisted problems for volume 2 is divided into 10 categories. For each category, certain computer-aided design (CAD) programs are suggested. Depending on the availability of software, each instructor may wish to expand or delete a particular category and particular problems in each category. Since learning how to use these programs is basically a matter of reading the manuals and following the illustrative examples, no extra lecture time is devoted to this segment of learning. Some examples with their solutions are included in this appendix and in the SPICE appendix.

Category 1. Partial Fraction Expansion and Inverse Laplace Transforms

 EXAMPLE B2.1

Given that

$$F(s) = 50\frac{2s^4 + 28s^3 + 142s^2 + 308s + 240}{s^4 + 8s^3 + 38s^2 + 56s + 25},$$

find the partial fraction expansion of $F(s)$ and $f(t) = \mathcal{L}^{-1}[F(s)]$.

SOLUTION

Solution 1. In MATLAB, the commands are:

```
num = 50*[ 2 28 142 308 240];
den = [1 8 38 56 25];
[r, p, k] = residue(num, den)
```

MATLAB gives the answer

```
r =                    p=                      k=
    1.0e+02 *              -3.0000 + 4.0000i       100
                           -3.0000 - 4.0000i
    1.8700 + 0.3400i       -1.0000
    1.8700 - 0.3400i       -1.0000
    2.2600
    1.2000
```

from which it follows that

$$F(s) = 100 + \frac{226}{s+1} + \frac{120}{(s+1)^2} + \frac{187 + j34}{s+3-j4} + \frac{187 - j34}{s+3+j4}.$$

With the aid of table 13.1, we find that the inverse Laplace transform is

$$f(t) = 100\delta(t) + 226e^{-t} + 120te^{-t} + 374e^{-3t}\cos(4t) - 68e^{-3t}\sin(4t).$$

Solution 2. Run tfc, and respond to the questions on the screen. The same answers for $F(s)$ and $f(t)$ as above are printed out without any additional calculations by the user.

1. With the aid of MATLAB and table 13.1, find the inverse Laplace transform of the following rational functions:

$$\text{(a)}\ \frac{200(s+1)(s+4)}{(s+2)(s^2+4)}. \qquad \text{(b)}\ \frac{5s^2 + 38s + 98}{(s+4)(s^2+4s+13)}.$$

2. (a) With the aid of MATLAB and table 13.1, find the inverse Laplace transform of the following rational functions:

$$\text{(1)}\ F_1(s) = \frac{2s^4 + 14s^3 + 50s^2 + 42s + 9}{(s+2)^3(s^2+4s+13)}.$$

$$\text{(2)}\ F_2(s) = \frac{s^5 + 32s^4 + 223s^3 + 674s^2 + 976s + 614}{s^6 + 14s^5 + 88s^4 + 320s^3 + 679s^2 + 746s + 312}.$$

(b) Repeat problem 2a, using the program tfc. (Table 13.1 is not needed.)

3. Using both MATLAB and tfc, find the partial fraction expansion of

$$F(s) = \frac{s^2 + 2s - 1}{s^4 + 2s^2 + 1} = \frac{s^2 + 2s - 1}{(s^2 + 1)^2}.$$

Note that $F(s)$ has repeated poles at $s = \pm j$. The answers obtained from the two programs are different. Verify that tfc gives the correct answer and that MATLAB (version 4.1) cannot do partial fraction expansion for the case of repeated complex poles.

Category 2. Plotting and Tabulating the Transient Response from a Given Circuit (not a Closed-Form Solution)

EXAMPLE B2.2

Find the step response of the series RLC circuit shown in figure B2.1 for three cases: (a) $R = 0.2\ \Omega$ (underdamped), (b) $R = 2\ \Omega$ (critically damped), and (c) $R = 4.25\ \Omega$ (overdamped). Plot the three curves on the same graph.

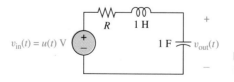

Figure B2.1 Step response of a series RLC circuit.

SOLUTION

Solution using SPICE. The input file is as follows:

```
Example of SPICE transient analysis.
*case 1.  R = 0.2 (underdamped)
r1 1 2 0.2
l1 2 3 1 ic = 0
c2 3 0 1 ic = 0
```

```
*case 2. R = 2 (critically damped)
r2 1 4 2
l2 4 5 1 ic = 0
c2 5 0 1 ic = 0
*case 3. R = 4.25 (overdamped)
r3 1 6 4.25
l3 6 7 1 ic = 0
c3 7 0 1 ic = 0
vin 1 0 1
.tran 0.1 15 uic
.print tran v(3) v(5) v(7)
.plot tran v(3) v(5) v(7)
.end
```

4. In the circuit shown in figure PB2.4, $v_{in}(t) = u(t)$ V, $v_C(0^-) = 4$ V, and $i_L(0^-) = 8$ A. Use a suitable program (SPICE, for example) to plot the waveform of $v_{out}(t)$ for $t > 0$.

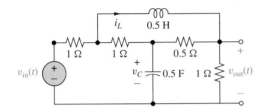

Figure PB2.4 Transient response with nonzero initial conditions.

5. The rise time of a circuit is defined from the step response as $t_2 - t_1$, where $v_{out}(t_1) = 0.1v_{out}(\infty)$ and $v_{out}(t_2) = 0.9v_{out}(\infty)$. In figure PB2.5, the insertion of the inductor in circuit 2 has the effect of decreasing the rise time.

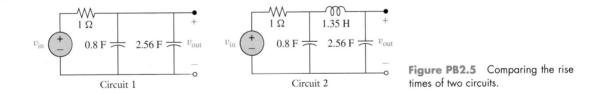

Circuit 1   Circuit 2

Figure PB2.5 Comparing the rise times of two circuits.

(a) Use an appropriate program (for example, SPICE) to plot the step responses of both circuits on the same graph.

(b) From the graph of part (a), find the ratio of the two rise times.

6. *Waveform of an RC circuit with multiple switching operations.* In the circuit of figure PB2.6, the switch S operates as follows:

    open for a long time;
    closes at $t = 0$;
    opens at $t = 2$ sec;
    closes at $t = 4$ sec.

Use SPICE or some other program to plot the waveform of $v_C(t)$ for $0 \le t \le 5$ sec, in 0.02-sec steps.

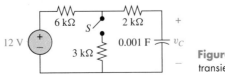

Figure PB2.6 Calculation of switching transients.

7. In the circuit of figure PB2.7, the switch S, originally at position A, is moved to position B at $t = 0$. For $R = 1.2\ \Omega$ (underdamped), use SPICE or some other program to plot $v_C(t)$, for $0 \le t < 15$ sec in 0.1-sec steps.

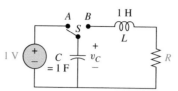

Figure PB2.7 Discharging a capacitor through a series RL circuit.

Category 3. *Closed-form* Transient Response (Use of MATLAB or tfc)

8. For the circuit of problem 4, find $V_{out}(s)$ manually. Then use the program tfc to find $v_{out}(t)$ in *closed-form*.
9. For the two circuits of problem 5, find $V_{out}(s)$ manually. Then use the program tfc to find $v_{out}(t)$ in *closed-form*.

Category 4. Magnitude Response and Phase Response When the Transfer Function H(s) is Known (Use of MATLAB or tfc)

◆ EXAMPLE B2.3

A band elimination circuit has the transfer function

$$H(s) = \frac{V_{out}}{V_{in}} = \frac{s^2 + 1}{s^2 + \dfrac{1}{Q}s + 1}.$$

Plot the curve of $|H(j\omega)|$ vs. ω for $0.8 \le \omega \le 1.2$ rad/sec in 0.02-rad/sec steps for two cases: $Q = 2$ and $Q = 50$. Show both curves in the same figure for the purpose of comparison.

SOLUTION

Solution 1. With MATLAB, the commands are shown below and the output is shown in figure B2.2.

```
w= (0.8:0.02:1.2);
num=[ 1 0 1];
den= [ 1 .5 1];   % case 1
h1= freqs(num,den,w);
magh1= abs(h1);
den=[ 1 .02 1]; %case 2
h2 = freqs(num,den,w);
magh2= abs(h2);
plot(w,magh1,w,magh2)
grid
```

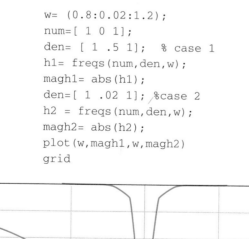

Figure B2.2 Magnitude response of a band elimination circuit obtained from MATLAB.

Solution 2. With tfc, follow the instructions on the screen to enter the rational function $H(s)$ and to tabulate the magnitude response. The present version of tfc does not have a plotting routine.

10. A bandpass circuit has the gain function

$$H(s) = \frac{V_{\text{out}}}{V_{\text{in}}} = \frac{\frac{1}{Q}s}{s^2 + \frac{1}{Q}s + 1}.$$

Plot the curve of $|H(j\omega)|$ vs. ω for $0.5 \le \omega \le 1.5$ rad/sec in 0.01-rad/sec steps for two cases: $Q = 5$ and $Q = 10$. Show both curves in the same figure for the purpose of comparison.

11. A lowpass fourth-order elliptic filter has the transfer function

$$H(s) = \frac{V_{\text{out}}}{V_{\text{in}}} = \frac{0.3015(s^2 + 1.3333)}{s^3 + 1.4054s^2 + 0.9420s + 0.4120}.$$

Plot the curve of $|H(j\omega)|$ vs. ω for $0 \le \omega \le 2$ rad/sec in 0.01-rad/sec steps.

12. A lowpass filter has the transfer function

$$H(s) = \frac{V_{\text{out}}}{V_{\text{in}}} = \frac{0.2718(s^2 + 4.0796)}{s^3 + 1.4479s^2 + 1.7950s + 1.1088}.$$

(a) Plot the curve of $|H(j2\pi f)|$ vs. f for $0 \le f \le 5,000$ Hz in 100-Hz steps.
(b) Plot the curve of the angle of $H(j2\pi f)$ vs. f over the frequency range, with the angle expressed in degrees. Note that the MATLAB command for obtaining θ, the angle (in degrees) of a complex number z, is $\theta = \text{angle}(z) * 180/\text{pi}$.

Category 5. Plotting Magnitude and Phase Responses for a Linear Circuit When the Transfer Function *H(s)* is not Known

13. Use a suitable program (e.g., SPICE) to plot the magnitude response of the circuit shown in figure PB2.13. See SPICE appendix.

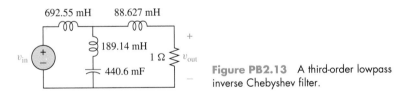

Figure PB2.13 A third-order lowpass inverse Chebyshev filter.

14. Use a suitable program (e.g., SPICE) to plot the magnitude response of the circuit shown in figure PB2.14 for the frequency range $0 < f < 3,000$ Hz.

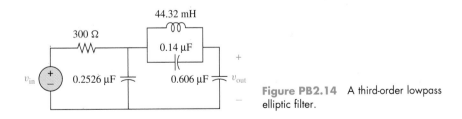

Figure PB2.14 A third-order lowpass elliptic filter.

15. The op amp circuit shown in figure PB2.15 is a normalized notch filter. All op amps are assumed ideal. Use a suitable program (e.g., SPICE) to plot the magnitude response for the frequency $0 < f < 1$ Hz.

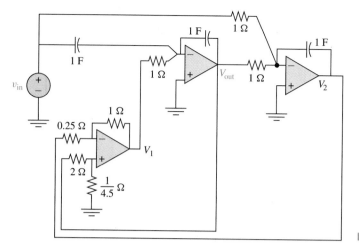

Figure PB2.15 An active notch filter.

Category 6. Convolution of Two Piecewise Constant Causal Time Functions

◆ EXAMPLE B2.4

Consider the two piecewise constant waveforms $v(t)$ and $h(t)$ given in figure B2.3. Find the convolution $y(t) = v(t) * h(t)$ by the use of MATLAB.

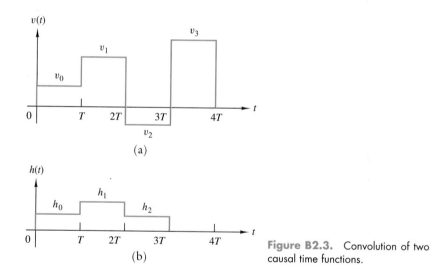

(a)

(b)

Figure B2.3. Convolution of two causal time functions.

SOLUTION

From the graphical convolution method described in section 7 of chapter 16, the waveform of $y(t)$ is piecewise linear and is completely specified by the breakpoints $\{T, y(kT)\}$. We define the following three polynomials in x:

$$p(x) = h_0 + h_1 x + h_2 x^2,$$

$$q(x) = v_0 + v_1 x + v_2 x^2 + v_3 x^3,$$

$$r(x) = p(x)q(x).$$

It can be shown that

$$y(0) = 0, \, y(7T) = 0,$$

and, for $k = 1, \ldots, 6$,

$$y(kT) = T \times \{\text{coefficient of the } x^{k-1}\text{-term in } r(x) = p(x)q(x)\}.$$

The MATLAB commands for calculating $y(kT)$ and plotting $y(t)$ are:

```
p = [v0   v1   v2   v3 ];
q = [h0   h1   h2 ];
r = conv(p,q);
y = T*[ 0 r 0];
t = (0: T : T*(length(y)-1));
plot(t,y)
```

 EXAMPLE B2.5

Find the convolution of the two piecewise constant waveforms shown in figure 16.31, and plot the resultant waveform.

SOLUTION

The MATLAB commands are:

```
T= 2;
p = [4   4   4   4  ];
q = [4  -4   4  -4 ];
r = conv(p,q);
y = T*[ 0   r   0];
t = (0: T : T*(length(y)-1));
plot(t,y)
```

The output from MATLAB is shown in figure B2.4.

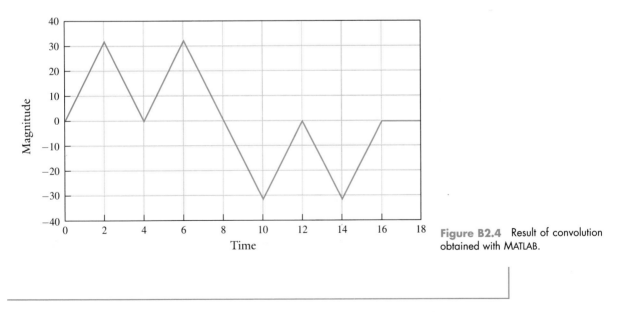

Figure B2.4 Result of convolution obtained with MATLAB.

16. Consider the piecewise constant waveform $h(t)$ shown in figure 16.39 of example 16.13, with $T = 0.1$. If $v(t) = h(t)$, find $y(t) = f(t) * h(t)$ by the use of MATLAB.

17. Find the convolution of the two piecewise constant waveforms shown in figure P16.11a. Use MATLAB to solve this problem.

18. Solve the convolution problem 16.2 by the use of MATLAB.

Category 7. Calculation of Two-Port Parameters

◆ **EXAMPLE B2.6**

Find the y-parameters for the circuit of figure 19.13 at $s = j2$.

SOLUTION

The MATLAB commands are:

```
w11 = [ 3+j*2 -j*2; -j*2 2+j*2];
w12 = [-2; -1];
w21 = [ -2 1];
w22 = [5];
y = w11 - w12*inv(w22)*w21
```

19. For the circuit shown in figure PB2.19, write the nodal equation and partition the matrix as illustrated in equation 19.11. Then use MATLAB to calculate the y-parameters.

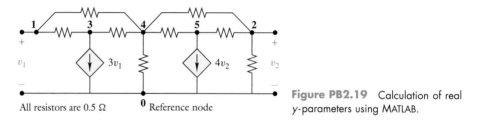

All resistors are 0.5 Ω   **0** Reference node

Figure PB2.19 Calculation of real y-parameters using MATLAB.

20. Repeat the previous problem for the circuit shown in figure PB2.20.

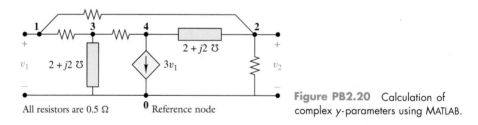

All resistors are 0.5 Ω   **0** Reference node

Figure PB2.20 Calculation of complex y-parameters using MATLAB.

Category 8. Analysis of Interconnected Two-Ports

21. Determine the overall *admittance* matrix for the two-port of figure PB2.21 by the use of the program tfc. (See examples in tfc manual.)

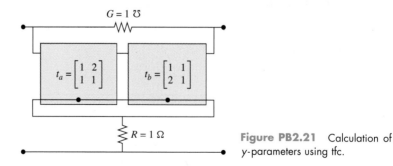

Figure PB2.21 Calculation of y-parameters using tfc.

22. The amplifier shown in figure PB2.22 is composed of two stages. The first stage provides a high input impedance, while the second stage provides the desired voltage gain. The two-port y-parameters are in millimhos.

$$Y_1 = \begin{bmatrix} 0.955 & -1 \\ -19 & -19.5 \end{bmatrix}, \quad Y_2 = \begin{bmatrix} 0.4 & -0.000625 \\ 40 & 0.2 \end{bmatrix}$$

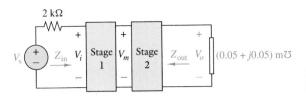

Figure PB2.22 Analysis of cascaded two-ports using tfc.

Use the program tfc to find:
(a) the input impedance Z_{in}.
(b) the overall voltage gain V_o/V_s.

Category 9. Indefinite Admittance Matrix

◆ EXAMPLE B2.7

The y-parameters of a transistor in its common-base configuration are given in figure B2.5. Determine the appropriate y-parameters when the device is used in the common-emitter configuration.

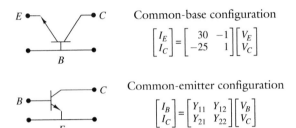

Common-base configuration

$$\begin{bmatrix} I_E \\ I_C \end{bmatrix} = \begin{bmatrix} 30 & -1 \\ -25 & 1 \end{bmatrix} \begin{bmatrix} V_E \\ V_C \end{bmatrix}$$

Common-emitter configuration

$$\begin{bmatrix} I_B \\ I_C \end{bmatrix} = \begin{bmatrix} Y_{11} & Y_{12} \\ Y_{21} & Y_{22} \end{bmatrix} \begin{bmatrix} V_B \\ V_C \end{bmatrix}$$

Figure B2.5 Two configurations using a bipolar junction transistor.

SOLUTION

First, we define some matrices:

$$\text{augment} = \begin{bmatrix} 1 & 0 & -1 \\ 0 & 1 & -1 \end{bmatrix}, \quad \text{switch} = \begin{bmatrix} 0 & 1 \\ 1 & 0 \end{bmatrix}$$

$$\text{delete1} = \begin{bmatrix} 0 & 0 \\ 1 & 0 \\ 0 & 1 \end{bmatrix} \quad \text{delete2} = \begin{bmatrix} 1 & 0 \\ 0 & 0 \\ 0 & 1 \end{bmatrix}, \quad \text{delete3} = \begin{bmatrix} 1 & 0 \\ 0 & 1 \\ 0 & 0 \end{bmatrix}.$$

We obtain the desired y-matrix by the procedure described in section 3 of chapter 20 (in particular, example 10 of chapter 20).
To augment a 2×2 matrix y_n into an indefinite admittance matrix y_{ind}, we write

$$y_{ind} = \text{augment}' * y_n * \text{augment}.$$

To delete row k and column k from y_{ind}, we set

$$y_n = \text{delete}k' * y_{ind} * \text{delete}k.$$

To switch rows and columns of a 2×2 matrix y_{old}, we put

$$y_{new} = \text{switch} * y_{old} * \text{switch}.$$

The MATLAB commands for solving the problem are:

```
augment = [ 1 0 -1; 0 1 -1];
switch  = [ 0 1 ; 1 0];
delete1 = [ 0 0; 1 0; 0 1];
delete2 = [ 1 0; 0 0; 0 1];
delete3 = [ 1 0 ; 0 1; 0 0];
yn = [ 30 -1; -25 1];
yind= augment'*yn*augment
yn = deletek'*yind*deletek
ynew = switch*yold*switch
```

23. Solve example 10 of chapter 20 by the use of MATLAB.
24. Solve problem 22 of chapter 20 by the use of MATLAB.

Category 10. *Closed-form* Solutions Involving the Use of a Symbolic Program

25. Use a symbolic program (NAPPE, for example) to find $Z(s)$ for the 10-branch network shown in figure PB2.25.

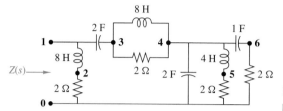

Figure PB2.25 Use of a symbolic program to find $Z(s)$.

26. In the circuit shown in figure PB2.26, all capacitor voltages are zero at $t = 0^-$.
    (a) Use a symbolic program (NAPPE, for example) to find the transfer function $V_{C3}(s)/I_{in}(s)$.
    (b) With the aid of tfc or MATLAB, find $v_{C3}(t)$ for $t \geq 0$ in *closed form* if $i_{in}(t) = 65e^{-t}\cos(2t)$ A.

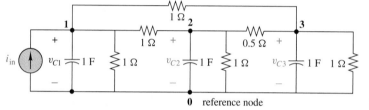

Figure PB2.26 Use of a symbolic program to find $V_{C3}(s)/I_{in}(s)$.

27. In the circuit of figure PB2.27, $v_{in}(t) = \cos(2t)u(t)$ V, $v_C(0^-) = 0$, and $i_L(0^-) = 0$.

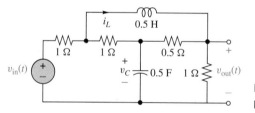

Figure PB2.27 Use of a symbolic program to find $V_{out}(s)/V_{in}(s)$.

    (a) Use a symbolic program (NAPPE, for example) to find the transfer function $H(s) = V_{in}(s)/V_{out}(s)$.
    (b) With the aid of tfc or MATLAB, find the poles and zeros of $H(s)$.
    (c) With the aid of tfc or MATLAB, find the zero-state response in closed form.
28. The step response of the circuit shown in figure PB2.28 has been plotted earlier using SPICE. Here, we shall obtain the answer in closed form.

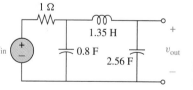

Figure PB2.28 Finding the step response in closed form.

    (a) Use a symbolic program to find the transfer function $H(s)$.
    (b) Use tfc or MATLAB to find the poles and zeros of $H(s)$.
    (c) Use the program tfc to find the step response (in closed form).
    (d) Use MATLAB to plot the step response.
    (e) Use MATLAB to plot the frequency response. Give plots of both magnitude vs. frequency and phase vs. frequency.

Introduction to Magnetic Circuit Analysis

I N sections 1–7 of chapter 18, the mutual inductance M between two coils is defined as a proportionality constant in the experimentally observed relationship between the induced voltage in one coil and the *rate of change of current* in the other coil. (See equations 18.1 through 18.3.) Chapter 18 outlined the basic methods of analysis and applications of circuits containing mutual inductances. The question of how to construct coupled coils with specified self- and mutual inductances was not discussed. Nor were the underlying physical principles of induced voltages. A rigorous treatment of these topics requires advanced electromagnetic field theory, a topic beyond the scope of the present book. Nevertheless, for the very important case of several circuits coupled through a ferromagnetic core, it is possible to closely predict the electrical properties of the coupled circuits using an *electric circuit analogy*.

In general, the magnetic circuit properties of coupled coils are analogous to the behavior of a nonlinear resistive circuit. Under certain conditions, the problem reduces to one requiring only basic linear circuit analysis. To proceed with this approach, we must first review some basic principles and terminology in physics. Readers with a reasonable background in college physics may skip the following subsection and proceed directly to the subsection entitled "Analysis of Ferromagnetic Circuits by Electric Circuit Analogy," on page 829.

Review of Some Basic Magnetic Field Concepts

In physics, we learn that a conductor carrying electric current (or any moving charge) sets up a *magnetic field* in the space around it. At any point in space, the magnetic field has two attributes: a **direction** and a **magnitude**. A widely used method to depict the *direction* of the field is to draw *closed* lines called **lines of induction** that follow the pattern assumed by iron filings sprinkled around the conductor. Figure B3.1 illustrates the idea.

If a free-rotating compass needle is placed in a magnetic field, it will align its long axis tangentially to the lines of induction, as shown in figure B3.1b. The relationship between the direction of the magnetic field and the electric current is described by the following rule:

RIGHT-HAND RULE

Using the right hand, if the thumb points in the direction of the current flow, then the fingers will curve in the direction of the magnetic field. (See figure B3.1a.) On the other hand, for a coil of wire, if the fingers point in the direction of current flow, then the thumb will point in the direction of the magnetic field, as illustrated in figure B3.1b.

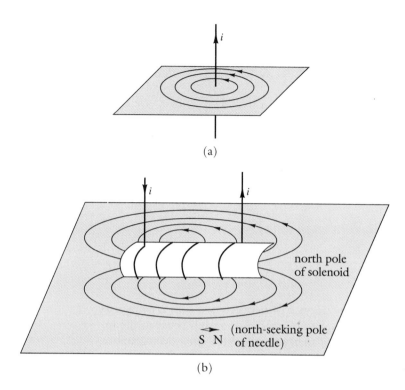

(a)

(b)

Figure B3.1 Representing a magnetic field by lines of induction. (a) Current through a straight wire. (b) Current through a coil of wire.

The total number of lines of induction threading through a surface is called the **magnetic flux** ϕ, in units of webers (abbreviated Wb). At any point in space, the amount of magnetic flux threading through one unit area perpendicular to the line of induction is called the **flux density** B, in units of teslas (abbreviated T). One tesla equals 1 weber per square meter. The flux density is considered to be the magnitude of the magnetic field. Since at any point in space the magnetic field has both direction and magnitude, it is a vector quantity, denoted by a boldface **B**. The light face letter B indicates the magnitude of **B**.

All lines of induction, or magnetic flux lines, are *closed* lines. Although there is nothing in the nature of a flow along these lines, it is useful to draw an analogy between the closed path of flux lines and a closed conducting circuit path supporting a current. This analogy proves to be a great aid in the visualization and prediction of magnetic effects in electric power apparatus. The flux density B attains a very clear physical meaning after considering the physical law governing a force acting on a moving charge. On the other hand, Faraday's law for the induced electromotive force (emf) has a very clear physical meaning for the flux ϕ. Since induced voltage is the topic under consideration, it is appropriate to restate Faraday's law in a form that is more suitable to electric circuit analysis. Consider a single-turn coil, shown in figure B3.2a, with ϕ webers of magnetic flux (produced by whatever source) passing through the surface defined by the wire. **Faraday's law** states that *the induced emf in the loop has a magnitude proportional to the rate of change of the flux through the loop.*

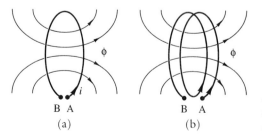

B   A
(a)

B   A
(b)

Figure B3.2 Faraday's law in terms of a voltage drop. (a) Flux through a single-turn coil. (b) Flux through a two-turn coil.

Next, consider an N-turn coil closely wound such that the same flux ϕ links with every turn, as shown in figure B3.2b. The N turns are in fact connected electrically in series. Therefore, the total induced emf is simply N times that of a single turn. The quantity $\lambda = N\phi$ is the **flux linkage** of the N-turn coil (measured in webers, since N is dimensionless). In equation form, Faraday's law is |induced emf| $= |d\lambda/dt|$. **Lenz's law** specifies the direction of the induced emf: *The direction of the induced emf is such that the current due to this emf will generate a magnetic flux that opposes the change in the flux linkage of the circuit.*

Since Kirchhoff's voltage law requires knowledge of the voltage drop across a two-terminal element, it is convenient to combine both Faraday's law and Lenz's law into a single alternative statement. In figures B3.2a and b, the coils are open-circuited. We mark the two terminals A and B and obtain the following:

FARADAY'S LAW (alternative form, for writing circuit equations):
The voltage drop from terminal A to terminal B, due to the change in flux linkage λ, is

$$v_{AB}(t) = \frac{d\lambda}{dt} = \frac{d(N\phi)}{dt} = \frac{N d\phi}{dt}, \tag{B3.1}$$

where ϕ is the flux in the same direction as the flux produced by a current if it were to flow through the coil from A to B.

EXAMPLE B3.1

In the circuit of figure B3.2b, $N = 50$, and the flux produced by some external source is $\phi(t) = 0.02\sin(120\pi t)$ Wb. Find the voltage drop $v_{AB}(t)$ of the open-circuited coil.

SOLUTION

The flux linkage of the coil is $\lambda = N\phi = 50 \times 0.02\sin(120\pi t)$ Wb. According to equation B3.1,

$$v_{AB}(t) = \frac{d\lambda}{dt} = 120\pi\cos(120\pi t) \cong 377\cos(120\pi t) \text{ V}.$$

Equation B3.1 provides a way of interpreting the unit weber for the flux ϕ: *A **weber** is the amount of flux such that a change of flux linkage at the rate of 1 weber per second at* $t = t_1$ *will induce a voltage drop of 1 volt across the open circuited coil at* $t = t_1$.

So far, we have only discussed the effect of a changing flux linkage. How is magnetic flux produced and varied? From a theoretical point of view, a magnetic field is produced by electric charges in *motion*. From a more practical point of view, a magnetic field is produced either by a permanent magnet or by a current. Only the latter will be discussed in this book.

By means of the **Biot-Savart law** in physics, one can compute the flux density B at any point of space around a circuit in which there is a current. The mathematics involved is usually very elaborate, requiring integration of a vector quantity along a curve in three-dimensional space. Only some very simple cases lend themselves to analytic solutions by the Biot-Savart law. Another law in physics, called **Ampère's circuital law**, is more often used to determine B. Ampère's circuital law is derived from the Biot-Savart law and is computationally easier to apply when the flux lines are confined in a region and possess a definite contour. We shall state Ampère's circuital law and illustrate it with several applications.

AMPÈRE'S CIRCUITAL LAW:
Around any closed curve C in free space, the integral $\int (B_t/\mu_0)ds$ is equal to the algebraic sum of currents through the area enclosed by C; i.e.,

$$\frac{1}{\mu_0}\int_C B_t ds = \sum i_k, \tag{B3.2}$$

where

ds is an infinitesimal element on the closed curve C;

B_t is the tangential component of the flux density at the point ds;

$\mu_0 = 4\pi \times 10^{-7}$ weber/ampere-m is a physical constant called the **permeability** of the free space; and

i_k is a current enclosed by C. The current i_k is considered positive (negative) if its direction and the direction of C for the line integral agree (disagree) with the right-hand rule.

If the closed curve is taken to be one of the lines of induction, then $B = B_t$, and Ampère's circuital law becomes

$$\frac{1}{\mu_0} \int_C B\,ds = \sum i_k. \tag{B3.3}$$

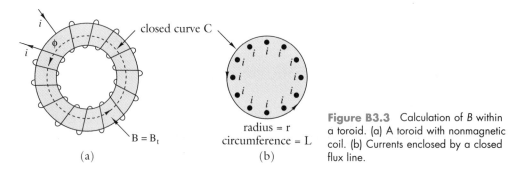

Figure B3.3 Calculation of B within a toroid. (a) A toroid with nonmagnetic coil. (b) Currents enclosed by a closed flux line.

As an application of Ampère's circuital law, consider the toroid shown in figure B3.3a, with the N turns of wire uniformly wound on a nonmagnetic core. It can be shown that in this case, provided that the turns are adjacent to each other, practically all of the magnetic flux is confined to the interior of the toroid. Furthermore, from a symmetry consideration, the flux lines will be concentric circles within the core, where B is the same at every point on a circle. If we apply equation B3.3 to the closed flux line through the center of the core (shown as a dashed circle in figure B3.3a and as a circle in figure B3.3b), we have

$$\frac{1}{\mu_0} \int_C B\,ds = \frac{B}{\mu_0} 2\pi r = \frac{B}{\mu_0} L = Ni,$$

where the right side, Ni, occurs because there are N currents, each equal to i, enclosed by C. These currents are shown as solid dots in figure B3.3b, to indicate that they are directed away from the paper and toward the reader. Note that the direction of the currents and the direction of the line integral in the figure conform to the right-hand rule.

The preceding discussion implies that, for a toroid in a nonmagnetic medium,

$$B = \frac{\mu_0(Ni)}{L}, \tag{B3.4}$$

where L is the mean *length* of the toroid. [$L = 2\pi \times$ (mean radius).] Since $\phi = BA$, it follows from equation B3.4 that

$$\phi = (Ni)\frac{\mu_0 A}{L}. \tag{B3.5}$$

An examination of equation B3.4 shows that A, the cross-sectional area of the toroid, does not affect the flux density B. The area does affect the *flux* in the toroid, however, as shown in equation B3.5. It is also evident from this equation that the driving force for the magnetic flux is the product Ni. For example, 2,000 turns at 6 amperes has the same effect as 4,000 turns at 3 amperes. For this reason, the product Ni is called the **magnetomotive force**, abbreviated mmf. The quantity Ni has the units of amperes. Since the number of turns is dimensionless, the unit of mmf is sometimes expressed as ampere-turns (AT).

EXAMPLE B3.2

A toroid has 500 turns closely wound on a plastic form. The cross section has an area of 2 cm by 2.5 cm, and the mean length is 25 cm. If a current of 4 A flows through the coil, find the flux density and the flux linkage of the toroid.

SOLUTION

From equation B3.4,

$$B = \frac{\mu_0(Ni)}{L} = 4\pi \times 10^{-7} \times 500 \times \frac{4}{0.25} \cong 0.01 \frac{Wb}{m^2}.$$

The flux within the toroid is

$$\phi = BA = 0.01 \times 0.02 \times 0.025 = 5 \times 10^{-6} \text{ Wb},$$

and the flux linkage is

$$\lambda = N\phi = 500 \times 10^{-6} = 5 \times 10^{-4} \text{ Wb}.$$

Now, suppose that we change the core of the toroid of figure B3.3a to some *ferromagnetic material,* such as iron. Then, for the same mmf, the observed flux density will be many times (several hundreds to several thousands) greater than that of the original toroid with the nonmagnetic core. A physical explanation of what happens within the iron that leads to the tremendously increased flux density is not only complicated, but also unnecessary for our present purpose of elementary magnetic circuit analysis. Suffice it to say that the presence of a ferromagnetic core greatly increases the flux lines *in a nonlinear fashion* and confines them, practically speaking, to within the core. The nonlinear relationship can be depicted as a curve of ϕ vs. Ni. The shape of the curve will depend on the ferromagnetic material used, as well as the dimensions of the core. On the other hand, if the nonlinear relationship is expressed as a curve of ϕ/A vs. Ni/L, then each type of ferromagnetic material can be described by one curve, called the **dc magnetization curve**. Such a curve for annealed iron is shown in figure B3.4.

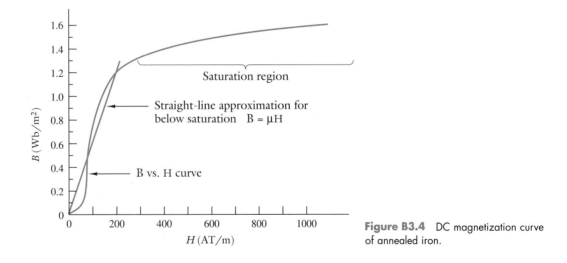

Figure B3.4 DC magnetization curve of annealed iron.

The quantity

$$H = \frac{Ni}{L}, \tag{B3.6}$$

where L is the mean length of the flux lines, is called the magnitude of the **magnetic field intensity** within the toroid and is measured in ampere-turns per meter (AT/m). The quantity B/H, denoted by μ, is called the **permeability** of the medium. Thus,

$$B = \mu H, \tag{B3.7}$$

where the unit of μ is $[Wb/m^2]/[AT/m] = Wb/[ampere\text{-}m]$. More often, it is common to speak of the ratio $\mu_r = \mu/\mu_0$, called the **relative permeability**. Thus,

$$\mu = \mu_r \mu_0. \tag{B3.8}$$

For free space (and approximately for air), $\mu_r = 1$. For a ferromagnetic medium, the value of μ_r depends on the operating point on the B-H curve.

The dc magnetization curve describes the relationship between ϕ and Ni only when the material is initially unmagnetized and Ni is steadily increased from zero. It is found that after reaching a certain flux density, if we decrease the value of H gradually, the B-H curve for decreasing H does not coincide with that for increasing H. This phenomenon is called **hysteresis**. In ac applications, hysteresis leads to power loss in the magnetic material in the form of heat, an undesirable effect. It is convenient to ignore hysteresis and consider the B-H curve as typically shown in figure B3.4 to maintain an introductory level of study.

◆ **EXAMPLE B3.3**

The toroid in example B3.2 has a core made of annealed iron, with the B-H curve shown in figure B3.4.

(a) Find ϕ, μ, and μ_r if the current is 0.05 A.
(b) Find the current needed to produce a flux of 0.0007 Wb.

SOLUTION

Part a. $H = Ni/L = (500 \times 0.05)/0.25 = 100$ AT/m. From figure B3.4, the corresponding flux density is $B \cong 0.7$ Wb/m^2. The cross-sectional area is $A = 0.02 \times 0.025 = 0.0005$m^2. Therefore, $\phi = BA = 0.7 \times 0.0005 = 0.00035$ Wb. Next,

$$\mu = \frac{B}{H} = \frac{0.7}{100} = 0.007 \text{ Wb/amp-m},$$

and

$$\mu_r = \frac{0.007}{4\pi \times 10^{-7}} \cong 5{,}570.$$

Part b. $B = \phi/A = 0.0007/0.0005 = 1.4$ Wb/m^2. From figure B3.4, the corresponding magnetic intensity is $H \cong 450$ AT/m. Therefore,

$$Ni = 450 \times 0.25 = 112.5 \text{ AT},$$

and

$$i = \frac{112.5}{500} = 0.225 \text{ ampere}.$$

For the case of a toroid with a ferromagnetic core, the integral of equation B3.3 generalizes to

$$\int_C \frac{B}{\mu} ds = \int_C H ds = \sum i_k = Ni. \tag{B3.9}$$

It must be emphasized that, to apply equation B3.9, the closed curve for the line integral must coincide with a flux line. As will be shown shortly, this equation provides the basis for a simplified analysis of magnetic circuits by electric circuit analogy.

Example B3.3 shows that to double the flux, the mmf has to be 4.5 times greater. In general, after H has reached a certain threshold, any further increase in it produces very little increase in the flux density, a phenomenon called **saturation**. In figure B3.4, the saturation region starts at around $H = 500$ AT/m.

Electrical apparatuses using ferromagnetic materials are usually designed to operate within the region below saturation. In these cases, for an *approximate* analysis, the B-H

curve within the region below saturation may be taken as a straight line through the origin, as shown in figure B3.4. Here, the nominal value for μ becomes the slope of the straight line. In the analysis that follows, we will *assume a below-saturation operation and treat* μ *and* μ$_r$ *as constants.* Table B3.1 presents typical values of μ$_r$ for some commonly used ferromagnetic materials.

TABLE B3.1 TYPICAL VALUES OF μ_r	
Material	Typical μ_r
Free space	1
Air	1
Cast iron	380
Cast steel	1,350
Silicon steel	5,630
Wrought iron	41,000

Analysis of Ferromagnetic Circuits by Electric Circuit Analogy

The toroid of figure B3.3a is the basis for the derivation of several important relationships and concepts (equations B3.4 through B3.7). When the core material is nonmagnetic, it is necessary to require that the turns of wire be uniformly distributed, to ensure that the flux lines are confined to nearly within the toroid. When a ferromagnetic material such as iron is used to construct the core, because of its high permeability, the flux lines within the core will be essentially the same whether the turns are uniformly distributed or whether they are concentrated in only a portion of the core. Furthermore, the core need not be of the toroid shape, as shown in figure B3.5a. The core can be rectangular (figure B3.5b) or may have more than one loop (figure B3.6). The magnetic flux will still be essentially confined to the core. Such a region occupied by the flux lines is called a **magnetic circuit** or, more precisely, a ferromagnetic circuit. Figures B3.5a and b show some elementary examples of ferromagnetic circuits.

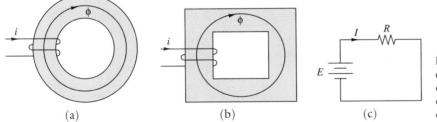

Figure B3.5 Simple magnetic circuits and electric circuit analogy. (a) Magnetic circuit on a round core. (b) Magnetic circuit on a rectangular core. (c) An electric circuit analogy.

Suppose that the cross-sectional area of each core in figure B3.5a is A. Then the flux density is constant and is equal to ϕ/A. An application of equation B3.9 leads to the following equation, which is a generalization of equation B3.5:

$$\phi = (Ni)\frac{\mu A}{L} = \frac{(Ni)}{\dfrac{L}{\mu A}} = \frac{(Ni)}{\mathbf{R}} = \mathbf{P}(Ni). \qquad \text{(B3.10)}$$

Here,

$$\mathbf{R} = \frac{L}{\mu A} \qquad \text{(B3.11a)}$$

and

$$\mathbf{P} = \frac{\mu A}{L}. \qquad \text{(B3.11b)}$$

R, with units of ampere-turns per weber, is called the **reluctance** of the simple core. **P**, the reciprocal of **R**, is called the **permeance** of the core. Note that bold face **R** and bold face **P** will denote reluctance and permeance respectively.

By comparing equations B3.10 and B3.11a with Ohm's law, $I = E/R = GE$, and the resistance for a conductor, $R = L/(\sigma A)$ (see volume 1, chapter 1, section 7), we find analogous quantities and relationships between electric and magnetic circuits, as delineated in table B3.2. The electric circuit analogy for the magnetic circuits of figures B3.5a and b is shown in figure B3.5c.

TABLE B3.2   ANALOGIES BETWEEN MAGNETIC AND ELECTRIC CIRCUITS

Magnetic circuit quantity	Electric circuit quantity
Magnetic flux ϕ (weber)	Current I (ampere)
mmf, Ni (ampere-turn)	emf, E (volt)
Reluctance **R** (ampere-turn/weber)	Resistance R (ohm)
Permeance **P** (weber/ampere-turn)	Conductance G (mho)
Permeability μ (weber/ampere-m)	Conductivity σ (mho/m)
$\phi = (Ni)/\mathbf{R} = \mathbf{P}(Ni)$	$I = E/R = GE$
$\mathbf{R} = L/(\mu A)$	$R = L/(\sigma A)$
$\mathbf{P} = (\mu A)/L$	$G = (\sigma A)/L$

To complete the analogy between magnetic and electric circuits, we need the magnetic counterparts of Kirchhoff's current and voltage laws. Consider, for example, a ferromagnetic core of the shape shown in figure B3.6. The magnetic fluxes in various legs of the core are denoted by ϕ_k. Because fluxes are assumed to be totally confined within the core, the following relationship must hold at each junction:

$$\sum (\phi)_{\text{in}} = \sum (\phi)_{\text{out}}. \tag{B3.12}$$

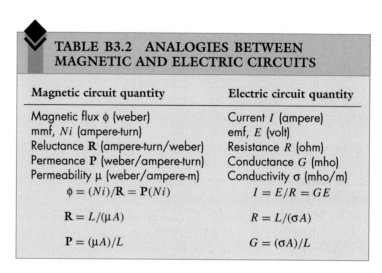

Figure B3.6 A two-loop magnetic circuit.

Equation B3.12, then, is the counterpart of Kirchhoff's current law for a magnetic circuit. For example, applying equation B3.12 to figure B3.6, we have

$$\phi_1 = \phi_2 + \phi_3.$$

The counterpart of Kirchhoff's voltage law is a consequence of equation B3.9. Suppose that in equation B3.9 the closed contour C is divided into segments $S_1, S_2, \ldots$, and the integral is separated accordingly. Recall that $H\,ds$ actually gives the mmf (in ampere-turns) needed to drive the flux over the infinitesimally short length ds. Then the integral

$$\int_{S_k} H\,ds = (Ni)_k$$

represents the mmf in ampere-turns that must act on C in order to drive ϕ over the length S_k. Equation B3.9 may be written as a summation

$$(Ni)_1 + (Ni)_2 + \ldots = (Ni)_{\text{total}}, \tag{B3.13}$$

where each $(Ni)_k$ is the mmf needed to establish the flux over the segment S_k, and $(Ni)_{total}$ is the total mmf (with directions considered) acting on the loop C. Equation B3.13 for magnetic circuits is the counterpart of one form of Kirchhoff's voltage law for electric circuits.

Equations B3.10 through B3.13 may be used to write a set of simultaneous equations for any ferromagnetic circuit. Alternatively, we may use these relationships to construct an analogous electric circuit for any given ferromagnetic circuit. All of the techniques we have learned earlier for linear resistive circuit analysis may then be applied to obtain a solution of the electric circuit. By the analogies shown in Table B3.2, the solution for the original magnetic circuit is immediately obtained. Two facts must be kept in mind, however, in applying this method:

1. It is assumed that all ferromagnetic materials are operating in the below-saturation region, with the B-H curve *approximated* by $B = \mu H$. If this is not the case, then the analogous electric circuit becomes a *nonlinear* resistive circuit. The solution of a nonlinear resistive circuit requires a graphical or an iterative method not discussed in this book.
2. The calculation of the reluctance $\mathbf{R}$ from equation B3.11a assumes constant values of A and μ. Whenever the cross-sectional area changes, or when a different medium is encountered along the loop, a separate resistor must be constructed in the electric circuit analogy. Furthermore, if an mmf of $(Ni)_k$ ampere-turns is seated in a segment with reluctance $\mathbf{R}_k$, then in the electric circuit model, an independent voltage source E_k is placed in series with the corresponding resistance R_k, with the relative reference directions for various quantities shown in figure B3.7.

We shall now illustrate the method with several examples. Since the solution of a linear resistive circuit is thoroughly discussed in a more basic course in electric circuit analysis, we will omit the details and concentrate on the construction of the electric circuit analogy from a given ferromagnetic circuit.

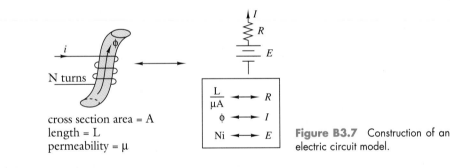

Figure B3.7 Construction of an electric circuit model.

◆ EXAMPLE B3.4

Suppose the mean length of the toroid shown in figure B3.8a is 50 cm, and the cross-sectional area is 4 cm². The air gap is one millimeter in length, and the coil has 200 turns. The core is made of a grade of silicon steel that has nominal relative permeability of 5,000 and a saturation flux density of 1.2 Wb/m². What is the current i needed to established a flux of 0.0004 Wb? Assume that there is no "fringing effect" at the air gap.

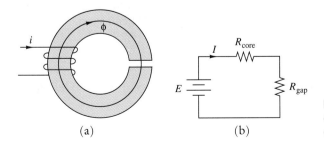

(a)   (b)

Figure B3.8 Analysis of a toroid with an air gap. (a) A magnetic core with an air gap. (b) The electric circuit analogy.

SOLUTION

The electric circuit analogy, constructed according to the rule given by figure B3.7, is shown in figure B3.8b. The various parameters in the circuit are calculated as follows:

$$\mathbf{R}_{\text{core}} \cong \frac{0.5}{5{,}000 \times 4\pi \times 10^{-7} \times 4 \times 0.0001} = 795{,}770 \frac{\text{AT}}{\text{Wb}}; \quad R_{\text{core}} = 795{,}770 \; \Omega;$$

$$\mathbf{R}_{\text{gap}} = \frac{0.001}{4\pi \times 10^{-7} \times 4 \times 0.0001} = 1.9894 \times 10^{6} \frac{\text{AT}}{\text{Wb}}; \quad R_{\text{gap}} = 1.9894 \times 10^{6} \; \Omega.$$

To have $I = 0.0004$ ampere, the required voltage source is

$$E = I \times (R_{\text{core}} + R_{\text{gap}}) = 0.0004 \times (795{,}770 + 1.9894 \times 10^{6}) = 1{,}114 \text{ V}.$$

Thus, the mmf needed in the magnetic circuit is

$$(Ni) = 1{,}114 \text{ AT},$$

and the current is $i = 1{,}114/200 = 5.57$ A. The flux density in the core is $B = \phi/A = 0.0004/(4 \times 0.0001) = 1 \text{ Wb/m}^2$. This value is below the given saturation value of 1.2 Wb/m^2. Therefore, the linear circuit analogy is valid.

◆ EXAMPLE B3.5

Suppose the magnetic circuit shown in figure B3.9a is made of silicon steel with $\mu_r \cong 5{,}000$ and has the following dimensions:

Part	Area (m^2)	Mean length (m)
acb	0.009	0.56
ab	0.0032	0.26
adb	0.0045	0.51

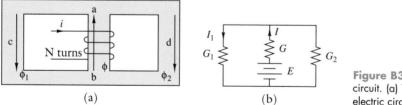

(a)   (b)

Figure B3.9 A two-loop magnetic circuit. (a) The magnetic circuit. (b) Its electric circuit analogy.

(1) If the flux ϕ in the center leg is known to be 0.01 Wb, find the flux ϕ_1 in the left leg.
(2) If $N = 300$ and $i = 4$ A, find the flux in the center leg.

SOLUTION

The electric circuit model is shown in figure B3.9b.

Part 1. Since $I = 0.01$ A, we use the current divider formula to find I_1. The circuit parameters G_1 and G_2 are calculated as follows:

$$\mathbf{P}_1 = 5{,}000 \times 4 \times \pi \times 10^{-7} \times \frac{0.009}{0.56} = 1.01 \times 10^{-5}; \quad G_1 = 10.1 \; \mu\text{mho}.$$

$$\mathbf{P}_2 = 5{,}000 \times 4 \times \pi \times 10^{-7} \times \frac{0.0045}{0.51} = 5.541 \times 10^{-5}; \quad G_2 = 55.4 \; \mu\text{mho}.$$

Therefore, we have

$$I_1 = \frac{G_1}{G_1 + G_2} I = \frac{10.1}{10.1 + 55.4} \times 4 = 0.617 \text{ A}.$$

By analogy, $\phi_1 = 0.617$ Wb. (Note that we have used the permeances instead of the reluctances in the calculation.)

Part 2. We also need R and E in this case:

$$\mathbf{R} = \frac{0.26}{5,000 \times 4 \times \pi \times 10^{-7} \times 0.0032} = 12.93 \times 10^3; \quad R = 12.93 \text{ k}\Omega;$$

$$Ni = 300 \times 4 = 1,200; \quad E = 1,200 \text{ V}.$$

Figure B3.9b is a simple series-parallel circuit. The total resistance "seen" by the voltage source is therefore

$$R_{\text{total}} = R + \frac{1}{G_1 + G_2} = (12.93 + 15.27) \times 10^3 = 12,950 \ \Omega,$$

in which case $I = 1,200/12,950 = 0.0927$ A. This implies that $\phi = 0.0927$ Wb.

Although we have solved the preceding two examples using electric circuit models, the circuits are too simple to show the power of the method. We shall conclude with a slightly more complicated magnetic circuit, to demonstrate the usefulness of the electric circuit model.

 EXAMPLE B3.6

Suppose the magnetic circuit shown in figure B3.10a is made of a certain grade of silicon steel that has a typical relative permeability of $\mu_r = 5,000$. The cross-sectional area is uniform in all segments of the core and is equal to 0.001 m^2. The air gap is 1.8 mm long. Assume no "fringing effect" at the gap. The mean lengths of various segments are as follows:

$$bae = 45 \text{ cm}, \quad bfe = 40 \text{ cm}, \quad bc = de = 8 \text{ cm}.$$

Find ϕ, the flux through the air gap, if $i_1 = 2$ A and $i_2 = 2.7$ A.

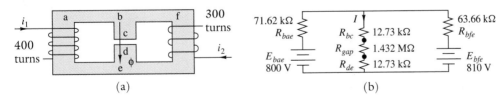

Figure B3.10 Analysis of a two-loop magnetic circuit. (a) A magnetic circuit with an air gap. (b) Its electric circuit analogy.

SOLUTION

The electric circuit model constructed according to figure B3.7 is shown in figure B3.10b. Note in particular the directions of the voltage sources. The parameters in the circuit are calculated as follows:

$$\mu = 5{,}000 \times 4 \times \pi \times 10^{-7} = 0.006283 \frac{\text{Wb}}{\text{AT}};$$

$$\mathbf{R}_{\text{gap}} = \frac{0.0018}{4 \times \pi \times 10^{-7} \times 0.001} = 1.432 \times 10^{6} \frac{\text{AT}}{\text{Wb}}; \quad R_{\text{gap}} = 1.432 \text{ M}\Omega;$$

$$\mathbf{R}_{bc} = R_{de} = \frac{0.08}{0.006283 \times 0.001} = 1.2733 \times 10^{4} \frac{\text{AT}}{\text{Wb}};$$

$$\mathbf{R}_{bc} = R_{de} = 12.73 \text{ } k\Omega;$$

$$\mathbf{R}_{bae} = \frac{0.45}{0.006283 \times 0.001} = 7.1623 \times 10^{4} \frac{\text{AT}}{\text{Wb}}, \quad R_{bae} = 71.62 \text{ k}\Omega;$$

$$\mathbf{R}_{bfe} = \frac{0.40}{0.006283 \times 0.001} = 6.3665 \times 10^{4} \frac{\text{AT}}{\text{Wb}}, \quad R_{bfe} = 63.66 \text{ k}\Omega;$$

$$(Ni)_{bae} = 400 \times 2 = 800 \text{ AT}, \quad E_{bae} = 800 \text{ V};$$

$$(Ni)_{bfe} = 300 \times 2.7 = 810 \text{ AT}, \quad E_{bfe} = 810 \text{ V}.$$

The circuit of figure B3.10b may be solved by a variety of methods. The answer for I is 5.4×10^{-4} A. Therefore, by analogy, $\phi = 5.4 \times 10^{-4}$ Wb.

Photo Credits

The photos in Chapter 1 were provided as a courtesy of DALE ELECTRONICS, VISHAY INTERTECHNOLOGY, INC., and JENN-AIR.

The photos in Chapter 2 were provided as a courtesy of DALE ELECTRONICS, VISHAY INTERTECHNOLOGY, INC., and the photo of Kirchhoff is from the Meggers Collection and was provided as a courtesy of AIP EMILIO SEGRE VISUAL ARCHIVES.

The photos in Chapter 3 were provided as a courtesy of HARRIS SEMICONDUCTOR.

The photos in Chapter 4 were provided as a courtesy of FUNDAMENTAL PHOTOGRAPHS and the author.

The photo in Chapter 6 was provided as a courtesy of MOTOROLA.

The photos in Chapter 7 were provided as a courtesy of SPRAGUE/VISHAY INTERTECHNOLOGY, INC.

The photos in Chapter 8 were provided as a courtesy of WAVETEK and HEWLETT PACKARD.

The photos in Chapter 9 were provided as a courtesy of AMERICAN RADIO RELAY LEAGUE, KINETIC CORP., and RADIO SHACK, TANDY CORP.

The photos in Chapter 10 were provided as a courtesy of ALLIEDSIGNAL.

The photos in Chapter 11 were provided as a courtesy of CULVER PICTURES and PHOTO RESEARCHERS.

The photo of the semiconductor in Chapter 12 was provided as a courtesy of HEWLETT PACKARD.

The photo of the computer screen showing the graph was photographed by Jason Kinch and was provided as a courtesy of TEKTRONIX PHOTOGRAPHIC.

The photos in Chapter 13 were provided as a courtesy of Culver Pictures and Archive Photos.

The photos in Chapter 14 were provided as a courtesy of North American Phillips and Advance Transformer Co.

The photos in Chapter 15 were provided as a courtesy of Boeing.

The photo in Chapter 16 was provided as a courtesy of Casio.

The photo in Chapter 17 was provided as a courtesy of Motorola.

The photo in Chapter 18 of the zener diode was provided as a courtesy of ABB.

The photos in Chapter 19 were provided as a courtesy of John W. Reo.

The photos in Chapter 20 were provided as a courtesy of Melinda Reo.

The diagrams in Chapter 21 were provided as a courtesy of the MITRE Corporation; work sponsored by the U.S. Air Force.

The photo in Chapter 22 was provided as a courtesy of the Hewlett-Packard Company.

Index

We hope **Linear Circuit Analysis: Time Domain, Phasor, and Laplace Transform Approaches** by DeCarlo and Lin will meet your classroom needs. Please take a moment to complete the following information, then mail the card to us so we may learn more about your department. Thank you!

Name _____ Phone _____

School Address _____

Department _____

Office Hours _____

Course Title and Number _____

Current Text _____

Length of Course _____

Enrollment: _____ Fall _____ Spring _____ Other _____

Are you likely to change books? _____ Decision Date _____

Other decision makers? _____

Are you currently using software? _____ If yes, what kind? _____

Do you want to see supplements? (circle) _____ Instructor's Manual with Problem Solutions

User's Guide for Transfer Function Calculator with Software Disk _____ MATLAB® Supplement

Comments _____

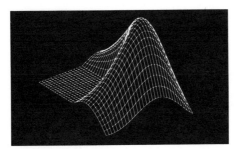

Marketing Manager
Engineering/Computer Science
PRENTICE HALL
Simon & Schuster Education Group
113 Sylvan Avenue, Route 9W
Englewood Cliffs, NJ 07632

BUSINESS REPLY MAIL
FIRST CLASS PERMIT NO. 82 NATICK, MA

POSTAGE WILL BE PAID BY ADDRESSEE

The MathWorks, Inc.
24 Prime Park Way
Natick, MA 01760-9889

LAPLACE TRANSFORMS OF BASIC FUNCTIONS

Item Number	$f(t)$	$\mathcal{L}[f(t)]$
1	$K\delta(t)$	K
2	$Ku(t)$ or K	$\dfrac{K}{s}$
3	$r(t)$	$\dfrac{1}{s^2}$
4	$t^n u(t)$	$\dfrac{n!}{s^{n+1}}$
5	$e^{-at}u(t)$	$\dfrac{1}{s+a}$
6	$te^{-at}u(t)$	$\dfrac{1}{(s+a)^2}$
7	$t^n e^{-at}u(t)$	$\dfrac{n!}{(s+a)^{n+1}}$
8	$\sin(\omega t)u(t)$	$\dfrac{\omega}{s^2+\omega^2}$
9	$\cos(\omega t)u(t)$	$\dfrac{s}{s^2+\omega^2}$
10	$e^{-at}\sin(\omega t)u(t)$	$\dfrac{\omega}{(s+a)^2+\omega^2}$
11	$e^{-at}\cos(\omega t)u(t)$	$\dfrac{(s+a)}{(s+a)^2+\omega^2}$
12	$t\sin(\omega t)u(t)$	$\dfrac{2\omega s}{(s^2+\omega^2)^2}$
13	$t\cos(\omega t)u(t)$	$\dfrac{s^2-\omega^2}{(s^2+\omega^2)^2}$
14	$\sin(\omega t+\phi)u(t)$	$\dfrac{s\sin(\phi)+\omega\cos(\phi)}{s^2+\omega^2}$
15	$\cos(\omega t+\phi)u(t)$	$\dfrac{s\cos(\phi)-\omega\sin(\phi)}{s^2+\omega^2}$
16	$e^{-at}[\sin(\omega t)-\omega t\cos(\omega t)]u(t)$	$\dfrac{2\omega^3}{\left[(s+a)^2+\omega^2\right]^2}$
17	$te^{-at}\sin(\omega t)u(t)$	$2\omega\dfrac{s+a}{\left[(s+a)^2+\omega^2\right]^2}$